Plant Polyphenols 2

Chemistry, Biology, Pharmacology, Ecology

BASIC LIFE SCIENCES

Ernest H. Y. Chu, Series Editor
The University of Michigan Medical School
Ann Arbor, Michigan

Alexander Hollaender, Founding Editor

A Continuation Order Plan is available for this series. A continuation order will bring delivery of each new volume immediately upon publication. Volumes are billed only upon actual shipment. For further information please contact the publisher.

Plant Polyphenols 2

Chemistry, Biology, Pharmacology, Ecology

Edited by

Georg G. Gross
University of Ulm
Ulm, Germany

Richard W. Hemingway
United States Department of Agriculture, Forest Service
Pineville, Louisiana

and

Takashi Yoshida
University of Okayama
Okayama, Japan

Associate Editor

Susan J. Branham
Kenner, Louisiana

Kluwer Academic / Plenum Publishers
New York, Boston, Dordrecht, London, Moscow

Library of Congress Cataloging-in-Publication Data

Plant polyphenols 2: chemistry, biology, pharmacology, ecology/edited by Georg G.
Gross, Richard W. Hemingway, and Takashi Yoshida; associate editor, Susan J. Branham.
     p. cm. — (Basic life sciences; v. 66)
    "Proceedings of the 3rd Tannin Conference, held July 14–19, 1998, in Bend,
Oregon"—T.p. verso.
   Includes bibliographical references.
  ISBN 0-306-46218-4
   1. Plant polyphenols—Congresses. I. Title: Plant polyphenols two. II. Gross, Georg G., 1941–
III. Hemingway, Richard W., 1939– IV. Yoshida, Takashi, 1939– V. North American
Tannin Conference (3rd: 1998: Bend, OR) VI. Series.

  QK898.P764 P582 1999
  572'.2—dc21

                            99-046706

Proceedings of the 3rd Tannin Conference, held July 14–19, 1998, in Bend, Oregon

ISBN 0-306-46218-4

©1999 Kluwer Academic/Plenum Publishers, New York
233 Spring Street, New York, N.Y. 10013

http://www.wkap.nl/

10 9 8 7 6 5 4 3 2 1

A C.I.P. record for this book is available from the Library of Congress

PREFACE

This book has been developed from the proceedings of the 3rd Tannin Conference honoring Professor Edwin Haslam, Department of Chemistry, University of Sheffield, for his efforts and outstanding contributions to the field of plant polyphenols. The meeting was held in Bend, Oregon, July 20–25, 1998, with the objective of promoting collaboration between chemists and biologists to improve our understanding of the biological significance and to expand possibilities for use of plant polyphenols. Special efforts were made to summarize current research on the influence of these compounds on human health. As in our 2nd Tannin Conference held in Houghton, Michigan, June 1991, we strove to build the "international" character of this meeting and were rewarded by an attendance of over 150 guests from 23 countries who enjoyed more than 50 lectures and 60 presentations. Our thanks go to all those who contributed to our meeting.

A special thanks is extended to Professor G. Wayne McGraw, Department of Chemistry, Louisiana College, Pineville, Louisiana, who served as the financial manager of our meeting. He, his wife Bobbie McGraw, and Helen Hemingway worked hard to make this meeting run as smoothly as possible. We also would like to thank the Bend Chamber of Commerce, Central Oregon Community College, The Riverhouse, Interstate Tours, Wanderlust Tours, and the people of Bend for welcoming our group to Central Oregon.

In an attempt to keep this volume a manageable size, we were forced to select about 50 of the contributions for this publication under the following topics: Hydrolyzable Tannins; Condensed Tannins and Related Compounds; Biotechnology; Antioxidant Properties and Heart Disease; Conformation, Complexation, and Antimicrobial Properties; Polyphenols and Cancer; Polyphenols in Commerce; Polyphenols and Ecology. We thank all the authors who have patiently worked with us to develop this book. Janie Gurgainers worked especially hard organizing files and correspondence. Dr. Hemingway thanks Dr. Bénédicte Berké and Dr. Timothy G. Rials for their help and patience. We also would like to thank Kluwer Academic/Plenum Publishers for their willingness to help build a continuing series of books that bring together the work of chemists and biologists from a wide array of disciplines, but with a common interest in plant polyphenols.

A comparison of the contributions to the proceedings of our 1st Tannin Conference, published by Plenum Press as "Chemistry and Significance of Condensed Tannins" in 1989, the 2nd Tannin Conference published by Plenum Press as

"Plant Polyphenols; Synthesis, Properties, Significance" in 1992, and the contents of this book show recognition of the significance of these compounds on the part of biologists and biochemists and increasing relevance in medically oriented disciplines.

To be a difference, a difference must make a difference. Over the past twelve years, our science has grown to clearly demonstrate that plant polyphenols make real differences to people's lives.

Georg G. Gross
Richard W. Hemingway
Takashi Yoshida
Susan J. Branham

CONTENTS

INTRODUCTION

HYDROLYZABLE TANNINS

CONDENSED TANNINS AND RELATED COMPOUNDS

BIOTECHNOLOGY

ANTIOXIDANT PROPERTIES AND HEART DISEASE

CONFORMATION, COMPLEXATION, AND ANTIMICROBIAL PROPERTIES

POLYPHENOLS AND CANCER

POLYPHENOLS IN COMMERCE

POLYPHENOLS AND ECOLOGY

CONCLUDING REMARKS

Plant Polyphenols 2

Chemistry, Biology, Pharmacology, Ecology

INTRODUCTION

PROFESSOR EDWIN HASLAM, RECIPIENT OF THE 3rd TANNIN CONFERENCE AWARD

Herbert L. Hergert

901 Burdan Drive
Pottstown, Pennsylvania 19464
USA

1. SYNOPSIS OF PROFESSOR HASLAM'S CAREER

It is with sincere pleasure that we dedicate this volume, *"Plant Polyphenols 2: Chemistry, Biology, Medicine, Ecology,"* to Professor Haslam who has been a beacon in the study of the chemistry and application of plant polyphenols. It is unimaginable that anyone who is active in this field or even has a mild interest would not be familiar with Professor Haslam's seminal overviews, *"Chemistry of Vegetable Tannins"* and *"Plant Polyphenols; Vegetable Tannins Revisited"*. To this distinguished duo, a third volume has appeared this past year titled *"Practical Polyphenolics—From Structure to Molecular Recognition, and Physiological Action"*. Those of us present at the 2nd Tannin Conference held June 17–21, 1991, in Houghton, Michigan, will recall his overview of the chemistry of gallic acid derivatives at that meeting, so we were looking forward to his 1998 update in this new book.

Professor Haslam received B.Sc. and Ph.D. degrees from the University of Sheffield and a subsequent Ph.D. degree from Cambridge University. Between the years of 1957 and 1980, he served as Assistant Lecturer, Lecturer, Senior Lecturer, and Reader in the Department of Chemistry at the University of Sheffield. In 1980, he was appointed to a Personal Chair (*ad hominem*) in Chemistry, and in 1985, he became the Head of the Chemistry Department at the University of Sheffield. He has received a number of honors and awards during his career. Among these are the following:

1955	Turner Research Prize in Chemistry—University of Sheffield
1955–1957	Sir William Ramsay Memorial Research Fellowship, Emmanuel College, Cambridge

Plant Polyphenols 2: Chemistry, Biology, Pharmacology, Ecology, Edited by
Gross et al. Kluwer Academic / Plenum Publishers, New York, 1999

1975	Hugh Kelly Senior Research Fellowship (First Election), Rhodes University, Grahamstown, South Africa
1977	Tate and Lyle Prize and Award, Phytochemical Society of Europe
1984	Visiting Professor, University of the South Pacific, Suva, Fiji
1987	Proctor Memorial Lecturer, Society of Leather Trades Chemists
1996	Wolstenholme Lecturer, Society of Leather Trades Chemists

During his career, he has published more than 150 research papers (see bibliography) principally in the *Journals of the Royal Society of Chemistry, Tetrahedron*, and *Phytochemistry*. He has published six books, three of which have already been mentioned, and served as Editor of Volume 5, *"Biological Compounds"* in the six-volume series, *Comprehensive Organic Chemistry*. Although he has been "retired" for some time, a glance at his list of publications does not show any letup in his research activities. At a time when some of us have reached retirement age (including the writer) or are facing "downsizing," the current American fad for supposedly improving the profitability of industry or academe, Professor Haslam serves as an excellent role model for not giving in to the social pressures to step aside on account of calendar age.

It is interesting to note that Haslam's first published paper (coauthored with R.D. Haworth) was on the constitution of sesamolin, a bifuran lignan with methylenedioxy substituted aromatic rings. This compound is one of the constituents of sesame oil and can serve as a synergist for pyrethrum insecticides. Although not a tannin, the discipline involved in isolation and structural determination of this compound was a good prelude to Haslam's subsequent work on the constituents of the gallotannins. From there, his work broadened to include the biosynthesis and interactions of plant polyphenols with proteins, carbohydrates, etc. Overall, Professor Haslam's research has fallen into four broad categories.

- Phenolic metabolism in plants. This has involved studies of the structure, chemical properties, and biosynthesis of simple phenols such as hydroxycinnamoyl esters and glycosides, gallic acid and galloyl esters, piceatannol (a stilbene derivative), larixinol, catechins, and hydroquinone. This work included the taxonomic distribution, structure, chemical properties, conformation or shape and biosynthesis of the two principal classes of vegetable tannins, i.e., hydrolyzable tannins (mainly esters of gallic and hexahydroxydiphenic acid), and condensed tannins (polymers based on proanthocyanidins and, to a lesser extent, polyphenolic norlignans, stilbenes, etc).

- Molecular recognition and interactions: phenols and polyphenols. For the attendees of the 3rd Tannin Conference, this area of Professor Haslam's work was probably of the greatest interest and importance. The means whereby simple and complex polyphenols engage in non-covalent

Figure 1. Professor Edwin Haslam who was presented with the 3ᴿᴰ Tannin Conference award.

interactions (complexation) with other molecules, such as other phenols (for example, anthocyanidins), peptides and proteins, cyclodextrins, saccharides and polysaccharides, simple heterocycles (such as caffeine, theorbromine, and theophylline) have been studied. This work has included the influence of environmental factors, i.e., metal ions and solvent composition, on these interactions. The importance of these interactions has been shown in various areas of applied science. Included are: (a) food and nutrition (taste, astringency, maturation, and appearance of beverages such as teas, wines, cocoa, etc.); (b) the interaction of phenols and polyphenols with proline-rich salivary proteins; (c) the development of astringency; (d) the mechanism of vegetable-tannages of leather and the deterioration of leather; (e) the role of polyphenols in health, food, and herbal medicines; (f) the mechanism of anthocyanin co-pigmentation in fruits and flowers; and (g) quinone tanning.

- Aromatic amino acid metabolism and the shikimate pathway. Biochemists are indebted to Professor Haslam's studies on the chemistry of intermediates in the shikimate pathway, i.e., shikimic acid, quinic acid, dehydro-derivatives, and chorismic acid. Research was carried out on the mechanism of action and stereochemistry of enzymes in the pathway: 3-dehydroquinate synthase, 3-dehydroquinate dehydratase, 5-enolpyruvylshikimate-3-phosphate synthase, chorismate mutase, L-phenylalanine ammonia lyase, and L-tryptophan synthase (alpha reaction).

- Secondary metabolism in plants and microorganisms. Studies have been made of the role and function of secondary metabolism and of the secondary metabolites themselves (alkaloids, terpenes, phenols, mycotoxins, etc.) in the life of the organisms that produce them.

Each reader in the field of plant polyphenols will find something in Professor Haslam's work that had or will have a bearing on their work. In the case of this writer, Haslam's first book on vegetable tannins was extremely helpful in unraveling some of the problems associated with the isolation and utilization of tannins from conifer bark. For other investigators, I could imagine that they would be especially interested in the role of polyphenol complexation in foods. Still others will be vitally interested in his studies of biosynthetic pathways and so on. In any event, it was this writer's wish that all attendees at the 3rd Tannin Conference availed themselves of the opportunity to become personally acquainted with Professor Haslam. There is a temptation for scientists to get so caught up in their own research that they fail to appreciate the personality of the other workers in their field. In the case of Eddie Haslam, this would be a great shame for he is one of the truly great scientists in our corner of the broad spectrum of chemistry and biology.

2. PUBLISHED WORKS OF PROFESSOR HASLAM

1. Haworth, R.D.; Haslam, E. Constituents of natural phenolic resins. XXIII. The constitution of sesamolin. *J. Chem. Soc.* :827–833 (1955).

2. Haslam, E. Perylene-3,9-quinone. *Tetrahedron* 5:99–101 (1959).

3. Haslam, E.; Haworth, R.D.; Jones, K.; Rogers, H.J. Gallotannins I. Introduction and fractionation of tannase. *J. Chem. Soc.* :1829–1835 (1961).

4. Haslam, E.; Haworth, R.D.; Mills, S.D.; Rogers, H.J.; Armitage, R.; Searle, T. Gallotannins II. Some esters and depsides of gallic acid. *J. Chem. Soc.* :1836–1842 (1961).

5. Haslam, E.; Haworth, R.D.; Jones, K.; Rogers, H.J.; Armitage, R.; Bayliss, G.S.; Gramshaw, J.W.; Searle, T. Gallotannins III. The constitution of Chinese, Turkish, sumach and tara tannins. *J. Chem. Soc.* :1842–1853 (1961).

6. Haslam, E.; Haworth, R.D.; Knowles, P.F. Gallotannins IV. The biosynthesis of gallic acid. *J. Chem. Soc.* :1864–1859 (1961).

7. Haslam, E.; Haworth, R.D.; Mackinson, G.K. Synthesis of 3-O-p-coumaroylquinic acid. *J. Chem. Soc.* :5153–5156 (1961).

8. Haslam, E. Biosynthesis of gallic acid. *In*: Ollis, W.D. (ed.) Recent developments in the chemistry of natural phenolic compounds. Pergamon Press. Oxford, New York and Paris. pp. 137–138 (1961).

9. Haslam, E.; Haworth, R.D.; Gramshaw, J.W.; Searle, T. Gallotannins V. The structure of penta and tetragalloylglucoses and some observations on the molecular weight of gallotannins. *J. Chem. Soc.* :2944–2947 (1962).

10. Haslam, E.; Haworth, R.D.; Searle T. Gallotannins VI. Turkish gallotannin. *J. Chem. Soc.* :3808–3814 (1962).

11. Haslam, E.; Haworth, R.D.; Keen, P.C. Gallotannins VII. Tara gallotannin. *J. Chem. Soc.* :3814–3818 (1962).

12. Haslam, E.; Haworth, R.D.; Lawton, D.A. Gallotannins VIII. Preparation and properties of some galloyl esters of quinic acid. *J. Chem. Soc.* :2173–22181 (1963).

13. Haslam, E.; Haworth, R.D.; Biggins, R.; Cairns, T.; Eglington, G. Hydrogen bonding in pyrocatechol monoesters and related compounds. *J. Chem. Soc.* :1750–1756 (1963).

14. Haslam, E.; Haworth, R.D.; Cunningham, J. Constitution of piceatannol. *J. Chem. Soc.* :2875–2883 (1963).

15. Haslam, E.; Haworth, R.D.; Knowles, P.F. The preparation and identification of 5-dehydroquinic and 5-dehydroshikimic acids. *Methods in Enzymology*. VI:498–501 (1963).

16. Calderbank, A.; Cameron, D.W.; Cromartie R.I.T.; Hamied, Y.K.; Haslam E.; Kingston, D.G.I.; Todd, L.; Watkins, J.C. Colouring matters of the Aphididae XX. The structure of the chrysoaphins and xanthoaphins. *J. Chem. Soc.* :80–89 (1964).

17. Haslam, E.; Makinson, G.K.; Naumann, M.O.; Cunningham, J. Synthesis and properties of some hydroxycinnamoyl esters of quinic acid. *J. Chem. Soc.* :2137–2146 (1964).

18. Haslam, E.; Naumann, M.O.; Haslam, E. Phenolic constituents of *Vaccinium vitis idaea*. *J. Chem. Soc.* :5649–5654 (1964).

19. Haslam, E.; Haworth, R.D. Vegetable tannins *In*: Cook, J.W.; Carruthers W. (eds.) Progress in organic chemistry. Butterworths, London, pp. 1–37 (1964).

20. Haslam, E.; Cornthwaite, D. Gallotannins IX. The biosynthesis of gallic acid. *J. Chem. Soc.* :3008–3011 (1965).

21. Haslam, E. Galloylesters in *Aceraceae*. *Phytochemistry* 4:495–498 (1965).

22. Haslam, E.; Marriot, J.E. 4-Dehydroquinic acid. *J. Chem. Soc.* :5755–5756 (1965).

23. Haslam, E.; Radford, T. Synthesis of 1,3-anhydro-*D*-glucitol. *J. Chem. Soc., Chem. Commun.* :632–633 (1965).

24. Biggins, R.; Haslam, E. Gallotannins X. The methanolysis reaction of pyrocatechol monoesters. *J. Chem. Soc.* :6883–6888 (1965).

25. Haslam, E.; Haworth, R.D.; Crabtree, P.W.; Mills, S.E.; Strangroom, J.E. Synthesis of m-digallic and m-trigallic acids and their derivatives. *J. Chem. Soc.* :6888–6893 (1965).

26. Cameron, D.W.; Cromartie, R.I.T.; Manied, Y.K.; Haslam, E.; Kingston, D.G.I.; Lord Todd; Watkins, J.C. Colouring matters of the Aphididae XXVI. The chrysoaphins and their reaction with periodate. *J. Chem. Soc.* :6923–6930 (1965).

27. Britton, G.; Haslam, E. Phenolic constituents of *Arctostaphylos uva-ursi*. (L.Spreng.). *J. Chem. Soc.* :7312–7319 (1965).

28. Britton, G.; Haslam E.; Naumann, M.O. Some new derivatives of arbutin. *Bull. Nat. Inst. Sci. India*:179–186 (1965).

29. Britton, G.; Crabtree, P.W.; Haslam, E.; Strangroom, J.E. The structure of Chinese gallotannin—evidence for a polygalloyl chain. *J. Chem. Soc.* :783–790 (1966).

30. Haslam, E.; Strangroom, J.E. The esterase and depsidase activities of tannase. *Biochem. J.* 99:28–31 (1966).

31. Haslam, E. The chemistry of vegetable tannins. Academic Press, London and New York, 180 p. (1966).

32. Brown, A.G.; Falshaw, C.P.; Haslam, E.; Holmes, A.; Ollis, W.D. The constitution of theaflavin. *Tetrahedron Letts* :1193–1204 (1966).

33. Haslam, E.; Radford, T. Synthesis of 1,3-anhydro-D-glucitol and some derivatives of 1,5-anhydroglucitol. *Carbohydrate Research* 2:301–314 (1966).

34. Critchlow, A.; Haslam, E.; Haworth, R.D.; Tinker, P.B.; Waldron, N.M. The oxidation of some pyrogallol and purpurogallin derivatives. *Tetrahedron* 23:2829–2847 (1967).

35. Haslam, E. Structure of the gallotannins. *J. Chem. Soc.* :1734–1738 (1967).

36. Haslam, E.; Uddin, M. Some observations on the structure of chebulinic acid. *J. Chem. Soc.* :2381–2384 (1967).

37. Haslam, E. Mass spectra of some aryl glycosides. *Carbohydrate Research* 5:161–165 (1967).

38. Dewick, P.M.; Haslam, E.; Sargent, D.; Uddin, M. The biosynthesis of gallic acid and related compounds. *Ges. Biol. Chem. (Biochemie der Phenolkorper)* :10–14 (1967).

39. Haslam, E. Microbiological chemistry, *Chemistry in Britain* 4:16–17 (1967).

40. Haslam, E.; Uddin, M. The synthesis of 2- and 3-galloyl esters of arbutin. *Tetrahedron* 24:4015–4020 (1968).

41. Dewick, P.M.; Haslam, E. Observations on the biosynthesis of gallic acid and caffeic acid. *J. Chem. Soc., Chem. Commun.* :673–675 (1968).

42. Haslam, E.; Jaggi, J. Phenols of *Salix* species. *Phytochemistry* 8:635–636 (1969).

43. Dewick, P.M.; Haslam, E. Phenol biosynthesis in higher plants: gallic acid. *Biochem. J.* 113:537–542 (1969).

44. Haslam, E. (+)-Catechin-3-O-gallate and a polymeric proanthocyanidin from *Bergenia* species. *J. Chem. Soc.* :1824–1828 (1969).

45. Haslam, E.; Tanner, R.J.N. Spectrophotometric assay of tannase. *Phytochemistry* 9:2305–2309 (1970).

46. Haslam, E.; Smith, B.W.; Turner, M.J. The stereochemical course of the 3-dehydroquinate dehydratase reaction and a novel preparation of shikimic acid labeled with isotopic hydrogen at C-2. *J. Chem. Soc., Chem. Commun.* :842–843 (1970).

47. Haslam, E. The stereochemistry of sesamolin. *J. Chem. Soc.* :2332–2334 (1970).

48. Haslam, E.; Sargent, D.; Thompson, R.S.; Turner, M.J. The shikimate pathway. Part 1, Introduction and the preparation of derivatives of 3-dehydroquinic acid stereospecifically labeled at C-2 with isotopic hydrogen. *J. Chem. Soc.* :1489–1495 (1971).

49. Haslam, E.; Turner, M.J. The shikimate pathway. Part 2, Conformational analysis of (−)-quinic acid and its derivatives by proton magnetic resonance spectroscopy. *J. Chem. Soc.* :1496–1500 (1971).

50. Haslam, E.; Ife, R.J. The shikimate pathway. Part 3. The stereochemical course of the L-phenylalanine ammonia lyase reaction. *J. Chem. Soc.* :2818–2821 (1971).

51. Haslam, E.; Jacques, D.; Tanner, R.J.N.; Thompson, R.S. Plant proanthocyanidins. Part 1. Introduction: the isolation, structure, and distribution in nature of plant proanthocyanidins. *J. Chem. Soc., Perkin Trans.* 1:1387–1399 (1972).

52. Haslam, E. Protection of phenols and catechols. *In*: McOmie, J.F.W. (ed.) Protective groups in organic chemistry. Plenum Press, London and New York, pp. 145–182 (1973).

53. Haslam, E. Protection of carboxyl groups. *In*: McOmie, J.F.W. (ed.) Protective groups in organic chemistry, Plenum Press, London and New York, pp. 183–216 (1973).

54. Haslam, E. Natural products—structure determination. *In*: Ollis, W.D. (ed.) M.T.P. International review of science, Series One, Organic chemistry-Volume 1. MTP—Butterworths, London, pp. 215–248 (1973).

55. Bedford, G.R.; Greatbanks, D.; Haslam, E.; Jacques, D. Proanthocyanidin A-2. *J. Chem. Soc. Chem. Communications* :518–520 (1973).

56. Haslam, E. Polyphenol—protein interactions. *Biochem. J.* 139:285–288 (1974).

57. Haslam, E. The shikimate pathway. Butterworths, London, 316 p. (1974).

58. Haslam, E. Biogenetically patterned synthesis of procyanidins. *J. Chem. Soc., Chem. Commun.* :594–595 (1974).

59. Haslam, E., Jacques, D. Biosynthesis of proanthocyanidins. *J. Chem. Soc., Chem. Commun.* :231–232 (1974).

60. Bedford, G.R.; Greatbanks, D.; Haslam, E.; Jacques, D. Plant proanthocyanidins. Part 2. Proanthocyanidin A-2 and its derivatives. *J. Chem. Soc., Perkin Trans.* 1:2663–2671 (1974).

61. Haslam, E.; Smith, B.W.; Turner, M.J. The shikimate pathway. Part 4. The stereochemistry of the 3-dehydroquinate dehydratase reaction and observations on 3-dehydroquinate synthetase. *J. Chem. Soc., Perkin Trans. 1* :52–55 (1975).

62. Haslam, E. Chemistry and biochemistry of plant proanthocyanidins. *In*: Farkas, l.; Gabor, M.; Kallay, F. (eds.) Topics in flavonoid chemistry and biochemistry. Akademiai Kiado, Budapest, pp. 77–97 (1975).

63. Haslam, E. Natural proanthocyanidins. *In*: Harborne, J.B.; Mabry, T.; Mabry, H. (eds.) The flavonoids. Chapman and Hall, London, pp. 505–559 (1975).

64. Haslam, E. Natural products—structure determination. *In*: Jackman, L.M. (ed.) M.T.P. International review of science, Series two, Organic chemistry—Volume 1. Butterworths, London, pp. 215–248 (1976).

65. Ball, L.F.; Haslam, E.; Ife, R.J.; Lowe, P. The shikimate pathway. Part 5. Chorismic acid and chorismate mutase. *J. Chem. Soc., Perkin Trans. 1* :1776–1783 (1976).

66. Fletcher, A.C.; Haslam, E.; Porter, L.J. Hindered rotation and helical structures in natural proanthocyanidins. *J. Chem. Soc., Chem. Commun.* :627–629 (1976).

67. Haslam, E.; Opie, C.T.; Porter, L.J. Procyanidin metabolism—a hypothesis. *Phytochemistry* 16:99–102 (1977).

68. Fletcher, A.C.; Gupta, R.K.; Haslam, E.; Porter, L.J. Plant proanthocyanidins. Part 3. Conformational and configurational studies of natural procyanidins. *J. Chem. Soc., Perkin Trans.* 1:1628–1637 (1977).

69. Haslam, E.; Jacques, D.; Opie, C.T.; Porter, L.J. Plant proanthocyanidins. Part 4. The biosynthesis of plant procyanidins and observations on the biosynthesis of cyanidin in plants. *J. Chem. Soc., Perkin Trans. 1* :1637–1643 (1977).

70. Haslam, E. Symmetry and promiscuity in procyanidin biochemistry. *Phytochemistry* 16:1625–1643 (1977).

71. Haslam, E. Structure, conformation and biosynthesis of plant procyanidins. *In*: Farkas, L.; Gabor, M.; Kallay, F. (eds.) Flavonoids and bioflavonoids, Akademiai Kiado, Budapest, pp. 97–111 (1975).

72. Gupta, R.K.; Haslam, E. Plant proanthocyanidins. Part 5. Sorghum polyphenols. *J. Chem. Soc., Perkin Trans. 1* :892–896 (1978).

73. Haslam, E. Proteins—Introduction: nature and classification *In*: Haslam, E. (ed.) Comprehensive organic chemistry, Vol. 5. Pergamon Press, London and New York, pp. 179–185 (1979).

74. Haslam, E. Photosynthesis, nitrogen fixation and intermediary metabolism. *In*: Haslam, E. (ed.) Comprehensive organic chemistry. Vol. 5. Pergamon Press, London and New York, pp. 915–925 (1979).

75. Haslam, E. Metabolites of the shikimate pathway. *In*: Haslam, E. (ed.) Comprehensive organic chemistry. Vol. 5. Pergamon Press, London and New York, pp. 1167–1205 (1979).

76. Haslam, E. Vegetable tannins. *In*: Swain, T.; Harborage, J.B.; van Summer, C. (eds.), Biochemistry of plant phenolics, Plenum Press, London and New York, pp. 475–525, (1979).

77. Haslam, E. Activation and protection of the carboxyl group. *Chemistry and Industry* :610–616, (1979).

78. Burnett, M.W.; Klyne, W.; Scopes, P.M.; Fletcher, A.C.; Haslam, E.; Porter, L.J. Plant proanthocyanidins. Part 6. Circular dichroism of procyanidins. *J. Chem. Soc., Perkin Trans.* 1:2375–2377 (1979).

79. Haslam, E. Recent developments in methods for the esterification of the carboxyl group. *Tetrahedron* 36:2409–2433 (1980).

80. Gupta, R.K.; Haslam, E. Vegetable tannins—structure and biosynthesis *In*: Hulse, J.H. (ed.) Polyphenols in cereals and legumes. International Development Research Centre, Ottawa, pp. 15–24 (1980).

81. Gupta, R.K.; Haslam, E. Plant Polyphenols. *In*: Hulse, J.H. (ed.) Sorghum and millets: their composition and nutritive value. Academic Press, London and New York, pp. 302–324 (1980).

82. Haslam, E. In vino veritas: oligomeric procyanidins and the ageing of red wines. *Phytochemistry*, 19:2577–2582 (1980).

83. Gupta, R.K.; Haslam, E. Plant proanthocyanidins, Part 7. Prodelphinidins from *Pinus sylvestris, J. Chem. Soc., Perkin Trans. 1* :1637–1643 (1981).

84. Haslam, E. Vegetable tannins *In*: Conn, E.E. (ed.) The biochemistry of plants. Vol. 7. Academic Press, London and New York, pp. 527–556 (1981).

85. Davis, K.G.; Haslam, E.; Lilley, T.H.; McManus, J.P. The association of proteins with polyphenols. *J. Chem. Soc., Chem. Commun.* :309–311 (1981).

86. Haslam, E. Plant polyphenols and their association with proteins. *Chimia* (Suisse) 36:304 (1982).

87. Anderson, W.R.; Boettger, H.G.; Doves, G.D.; Griffin, C.E.; Haslam, E.; Lee, T.D. Electron ionisation—flash desorption mass spectroscopic analysis of selected thermally unstable plant metabolites. *Spectroscopy International Journal* 1:110–119 (1982).

88. Al-Shafi, S.M.K.; Gupta, R.K.; Haddock, E.A.; Haslam, E.; Layden, K.; Magnolato, D. The metabolism of gallic acid and hexahydroxydiphenic acids in plants. Biogenetic and molecular taxonomic considerations. *Phytochemistry* 21:1049–1062 (1982).

89. Al-Shafi, S.M.K.; Gupta, R.K.; Haddock, E.A.; Haslam, E.; Layden, K.; Manolato, D. The metabolism of gallic acid and hexahydroxydiphenic acids in plants. Part 1. Introduction, naturally occurring galloyl esters. *J. Chem. Soc., Perkin Trans. 1* :2525–2524 (1982).

90. Al-Shafi, S.M.K.; Gupta, R.K.; Haslam, E.; Layden, K. The metabolism of gallic acid and hexahydroxydiphenic acids in plants. Part 2. Esters of (S)-hexahydroxydiphenic acid with D-glucopyranose ($^4C_1$). *J. Chem. Soc., Perkin Trans. 1* :2525–2534 (1982).

91. Gupta, R.K.; Haddock, E.A.; Haslam, E. The metabolism of gallic acid and hexahydroxydiphenic acid in plants. Part 3. Esters of (R) and (S)-hexahydroxydiphenic acid and dehydrohexahydroxydiphenic acid with D-glucopyranose ($^1C_4$ and related conformations). *J. Chem. Soc., Perkin Trans. 1* :2535–2545 (1982).

92. Haslam, E. The metabolism of gallic acid and hexahydroxydiphenic acid in higher plants. *In*: Herz, W.; Grisebach, H.; Kirby, G.W. (eds.) Progress in the chemistry of organic natural products. Vol. 41. Springer-Verlag, Vienna, pp. 1–46 (1982).

93. Haslam, E.; Jalal, M.A.F.; Read, D. Phenolic composition and its seasonal variation in *Calluna vugaris. Phytochemistry* 21:1397–1401 (1982).

94. Haslam, E.; Lilley, T.H.; McMannus, J.P. The enthalpy of interaction of resorcinol with glycine in water at 298.15K *J. Chem. Thermodynamics* 15:397–402 (1983).

95. Haslam, E.; Lilley, T.H. Plant polyphenols and their association with proteins. *In*: Hedin, P.A. (ed.) Plant resistance to insects. American Chemical Society, Washington, DC. pp. 123–138 (1983).

96. Haslam, E.; Opie, C.T.; Platt, R.V. Biosynthesis of flavan-3-ols and other secondary plant products from 2-(S)-phenylalanine. *Phytochemistry*, 23:2211–2217 (1984).

97. Beart, J.E.; Haslam, E.; Lilley, T.H. Plant polyphenols—secondary metabolism and chemical defence: some observations. *Phytochemistry* 24:33–38 (1985).

98. Gaffney, S.H.; Haslam, E.; Lilley, T.H. The enthalpy of dilution of aqueous solutions of cellobiose at 298.15K. *Thermochimica Acta* 86:175–182 (1985).

99. Haslam, E.; Lilley, T.H.; McManus, J.P. The enthalpy of dilution of aqueous solutions containing catechol, resorcinol, or pyrogallol at 298.15K. *Thermochimica Acta* 86:183–188 (1985).

100. Haslam, E.; Shen, Z. Proanthocyanidins from *Larix gmelini* bark. *Chemistry and Industry of Forest Products* 5:1–8 (1985).

101. Haslam, E.; Lilley, T.H. New polyphenols for old tannins. *Ann. Proc. Phytochemical Society of Europe* 25:237–256 (1985).

102. Beart, J.E.; Davis, K.G.; Gaffney, S.H.; Haslam, E.; Lilley, T.H.; McManus, J.P. Polyphenol interactions. Part 1. Introduction: some observations on the reversible complexation of polyphenols with proteins and polysaccharides. *J. Chem. Soc., Perkin Trans.* 2:1429–1438 (1985).

103. Beart, J.E.; Haslam, E.; Lilley, T.H. Polyphenol interactions. Part 2. Covalent binding of procyanidins to proteins during acid-catalyzed decomposition. Observations on some polymeric proanthocyanidins. *J. Chem. Soc., Perkin Trans.* 2:1439–1443 (1985).

104. Begley, M.J.; Falshaw, C.P.; Haslam, E.; Shen, Z. A novel spirobiflavonoid from *Larix gmelini. J. Chem. Soc., Chemical Communications* :1135–1137 (1985).

105. Haslam, E. Metabolites and metabolism. Clarendon Press, Oxford University Press. 161 p. (1985).

106. Begley, M.J.; Falshaw, C.P.; Haslam, E.; Shen, Z. Procyanidins and polyphenols of *Larix gmelini. Phytochemistry* 25:2629–2635 (1986).

107. Haslam, E. Hydroxybenzoic acids and the enigma of gallic acid. *In*: Conn, E.E. (ed.) Recent advances in phytochemistry. Vol. 20. Plenum Press, New York and London. pp. 163–200 (1986).

108. Haslam, E.; Lilley, T.H. Interactions of natural phenols with macromolecules. *In*: Codey, V.; Middleton, E.; Harborne, J.B. (eds.) Progress in clinical and biological research, Vol. 213, Plant flavonoids in biology and medicine. Alan R. Liss, New York, pp. 53–65 (1986).

109. Haslam, E.; Lilley, T.H. Polyphenol complexation. *In*: Farkas, L.; Gabor, M.; Kallay, F. Favonoids and bioflavonoids, 1985. Akademiai Kiado, Budapest, pp. 113–138 (1986).

110. Bailey, N.A.; Begley, M.J.; Falshaw, C.P.; Haslam, E.; Lilley, T.H.; Magnolato, D.; Martin, R. Polyphenol—caffeine complexation. *J. Chem. Soc., Chem. Commun.* :105–106 (1986).

111. Gaffney, S.H.; Haslam, E.; Lilley, T.H.; Magnolato, D.; Martin, R. The association of polyphenols with caffeine and α- and β-cyclodextrins. *J. Chem. Soc., Chem. Commun.* :107–109 (1986).

112. Eagles, J.; Gallentti, G.C.; Gujer, R.; Hartley, R.D.; Haslam, E.; Lea, A.G.H.; Magnolato, D.; Mueller-Harvey, I.; Richli, U.; Self, R. Fast atom bombardment mass spectrometry of polyphenols (syn. vegetable tannins). *Biomedical Mass Spectrometry.* 13:449–468 (1986).

113. Haslam, E.; Lilley, T.H. Polyphenol complexation: Astringency in fruits and beverages. *Journe'es Internationales Groupe Polyphenols.* 13:352–357 (1986).

114. Haslam, E. Secondary metabolism—fact and fiction. *J. Chem. Soc., Natural Products Reports.* 3:217–249 (1986).

115. Begley, M.J.; Falshaw, C.P.; Haslam, E.; Lilley, T.H.; Magnolato, D.; Martin, R. The caffeine—potassium chlorogenate complex. *Phytochemistry* 26:273–279 (1987).

116. Haslam, E. Plant polyphenols. *Chemistry and Industry of Forest Products* 7:1–35 (1987).

117. Haslam, E.; Layden, K.; Hashim, S. The shikimate pathway. Part 6. Development of a model for chorismate mutase—an unusual Claisen rearrangement. *S. African J. Chem.* 40:65–71 (1987).

118. Haslam, E.; Layden, K.; Milburn, P.J.; Richards, T.I.; Warnimski, E.E. The shikimate pathway. Part 7. Chorismate mutase—towards an enzyme model. *J. Chem. Soc., Perkin Trans. 1* :2765–2773 (1987).

119. Haslam, E.; Lilley, T.H.; Ozawa, T. Polyphenol interactions—astringency and the loss of astringency in ripening fruit. *Phytochemistry*, 26:2937–2942 (1987).

120. Haslam, E.; Scalbert, A. Polyphenols and chemical defence in the leaves of *Quercus robur. Phytochemistry* 26:3191–3195 (1987).

121. Haslam, E. Vegetable tannins—renaissance and reappraisal. *J. Soc. Leather Trades Chemists* 72:45–64 (1988).

122. Haslam, E. Plant polyphenols (syn. vegetable tannins) and chemical defence—a reappraisal. *J. Chem. Ecology* 14:1787–1803 (1988).

123. Haslam, E.; Lilley, T.H. Natural astringency in foodstuffs—a molecular interpretation. *CRC Critical Reviews in Food Science and Nutrition.* 27:1–40 (1988).

124. Cai, Y.; Gaffney, S.H.; Goulding, P.N.; Haslam, E.; Lilley, T.H.; Magnolato, D.; Martin, R.; Spencer, C.M. Polyphenol complexation—some thoughts and observations. *Phytochemistry* 27:2397–2409 (1988).

125. Cai, Y.; Haslam, E.; Lilley, T.H.; Magnolato, D.; Martin, R. Traditional herbal medicines—the role of polyphenols. *Planta Medica* 55:1–8 (1989).

126. Cai, Y.; Gaffney, S.H.; Goulding, P.N.; Haslam, E.; Lilley, T.H.; Magnolato, D.; Martin, R.; Spencer, C.M. Some observations on the role of polyphenol complexation in traditional herbal medicines. *Farmaceutishch Tijdscrift voor Belgie* 66:21–33 (1989).

127. Haslam, E. Plant polyphenols *In*: Society of Chemistry and Chemical Processing of Forest Products. VII:373–395 (1989).

128. Haslam E.; Shen, Z. Larixinol—In: Society of Chemistry and Chemical Processing of Forest Products. VII:460–469 (1989).

129. Haslam, E. Plant polyphenols—vegetable tannins re-visited. Chemistry and Pharmacology of Natural Products Series, Cambridge University Press, Cambridge, 230 p. (1989).

130. Haslam, E. Gallic acid derivatives and hydrolyzable tannins. *In*: Rowe, J.W. (ed.) Natural products of woody plants. Springer-Verlag, Berlin, pp. 399–438 (1989).

131. Cai, Y.; Gaffney, S.H.; Haslam, E.; Lilley, T.H. Carbohydrate—polyphenol complexation. *In*: Hemingway, R.W.; Karchesy, J.J. (eds.) Chemistry and significance of condensed tannins. Plenum Press, New York, pp. 399–438 (1989).

132. Cai, Y.; Haslam, E.; Lilley, T.H.; Martin, R.; Spencer, C.M. The metabolism of gallic acid and hexahydroxydiphenic acid in plants. Part 4. Polyphenol interactions. Part 3. Spectroscopic and physical properties of esters of (*S*)-hexahydroxydiphenic acid with D-glycopyranose ($^4C_1$). *J. Chem. Soc., Perkin Trans.* 2:651–660 (1990).

133. Cai, Y.; Haslam, E.; Lilley, T.H. Polyphenol—anthocyanin co-pigmentation. *J. Chem. Soc., Chem. Commun.* :380–383 (1990).

134. Adams, H.; Bailey, N.A.; Frederickson, M.; Haslam, E. 5,21-Dimethyl-8,24-pentamethylenedioxy-1,10,17,26-tetra-aza[2,2](mo)2-cyclophane. *J. Chem. Soc., Perkin Trans. 1* :2353–2355 (1990).

135. Cai, Y.; Haslam, E.; Lilley, T.H.; Martin, R.; Spencer, C.M. Polyphenol complexation. *Compte Rendu Groupe Polyphenols* XV:304–318 (1990).

136. Cai, Y.; Gaffney, S.H.; Haslam, E.; Lilley, T.H.; Magnolato, D.; Martin, R.; Spencer, C.M. Polyphenol interactions. Part 4. Model studies with caffeine and cyclodextrins. *J. Chem. Soc., Perkin Trans.* 2:2197–2209 (1990).

137. Cai, Y.; Haslam, E.; Lilley, T.H.; Mistry, T.V. Polyphenol interactions. Part 5. Anthocyanin co-pigmentation. *J. Chem. Soc., Perkin Trans.* 2:1287–1296 (1991).

138. Haslam, E. Gallic acid and its metabolites. *In*: Hemingway, R.W.; Laks, P.E. Plant polyphenols: synthesis, properties, significance, Basic Life Sciences Vol. 59, Plenum Press, New York, pp. 169–195 (1992).

139. Cai, Y.; Gaffney, S.H.; Goulding, P.N.; Haslam, E.; Lilley, T.H.; Liao, H.; Luck, G.; Martin, R.; Warminiski, E.E. Polyphenol complexation—a study in molecular recognition. *In*: Ho, C-T.; Lee, C.Y.; Huang, M-T. (eds.) Phenolic compounds in food and health. American Chemical Society Symposia Series 506, American Chemical Society, Washington DC. pp. 8–50, (1992).

140. Haslam, E. Tannins, polyphenols and molecular complexation. *Chemistry and Industry of Forest Products.* 12:1–24 (1992).

141. Cai, Y.; Haslam, E.; Liao, H. Polyphenol interactions. Part 6. Anthocyanins: co-pigmentation and the colour changes in red wines. *J. Sci. Food and Agric.* 59:299–305 (1992).

142. Haslam, E. Vegetable tannins and the durability of leather. *In*: Hallebeek, P.; Kite, M.; Calnan, C. (eds.) Conservation of leathercraft and related objects. ICOM, London, pp. 24–27 (1993).

143. Haslam, E. Polyphenol complexation. *In*: Scalbert, A. (ed.) Polyphenolic phenomena. INRA Publications, Paris, pp. 23–31 (1993).

144. Haslam, E. Polyphenols chameleons. *In*: van Beek, T.A.; Breteler, H. Phytochemistry and agriculture. Clarendon Press, Oxford, pp. 214–252 (1993).

145. Haslam, E. Nature's palette, *Chemistry in Britain* 29:875–878 (1993).

146. Haslam, E. Shikimic acid—metabolism and metabolites. John Wiley, New York, 387 p. (1993).

147. Haslam, E.; Lilley, T.H.; Murray, N.J.; Williamson, M.P. Study of the interaction between salivary proline-rich proteins and a polyphenol. *European J. Biochem.* 219:923–935 (1994).

148. Cai, Y.; Haslam, E. Plant polyphenols (vegetable tannins)—gallic acid metabolism. *Chem. Soc., Natural Product Reports* 11:41–66 (1994).

149. Grimmer, H.R.; Luck, G.; Haslam, E.; Liao, H.; Lilley, T.H.; Murray, N.J.; Warminski, E.E.; Williamson, M.P. Polyphenols, astringency and proline-rich proteins. *Phytochemistry* 37:357–371 (1994).

150. Haslam, E. Complexation and oxidative transformations of polyphenols. *In*: Brouillard, R.; Jay, M.; Scalbert, A. (eds.) Polyphenols 94, INRA Publications, Paris, pp. 46–55 (1995).

151. Haslam, E. Fruit and floral pigmentation. *Rev. Prog. Coloration* 25:18–28 (1995).

152. Haslam, E. Secondary metabolism—evolution and function: products or processes? *Chemoecology* 5/6:89–95 (1995).

153. Haslam, E. Natural polyphenols: complexation with peptides and proteins as a mechanism of action in medicine. *In*: Antus, S.; Gabor, M.; Vetschera, K. (eds.). Flavonoids and bioflavonoids 1995. Akademiai Kiado, Budapest, pp. 13–33 (1995).

154. Haslam, E. Natural polyphenols (vegetable tannins) as drugs: possible modes of action. *J. Natural Prod.* 40:205–215 (1996).

155. Brettle, R.; Cross, R.; Davies, G.M.; Frederickson, M.; Haslam, E. Synthesis of (–)-3(R)-Amino-4(R), 5(R)-dihydroxy-1-cyclohexene-1-carboxylic acid: The 3(R)-amino analogue of (–)-shikimic acid. *Biorganic and Medicinal Letters.* 6:291–294 (1996).

156. Brettle, R.; Cross, R.; Davies, G.M.; Frederickson, M.; Haslam, E.; MacBeath, F.S. Synthesis of (3R) and (3S)—fluoro analogues of shikimic acid. *Biorganic and Medicinal Letters* 6:1275–1278 (1996).

157. Adams, H.; Bailey, N.A.; Davies, G.M.; Frederickson, M.; Haslam, E.; MacBeath, F.S. On the stereochemical outcome of the reaction between (–)-chorismic acid and diazomethane: Proof of absolute stereochemistry of the major pyrazoline by x-ray crystallography of a cyclopropane based derivative. *J. Chem. Soc., Perkin Trans. 1* :1531–1533 (1996).

158. Adams, H.; Bailey, N.A.; Brettle, R.; Cross, R.; Davies, G.M.; Frederickson, M.; Haslam, E.; MacBeath, F.S. The shikimate pathway. Part 8. Synthesis of (–)-3(R)-amino-4(R), 5(R)-dihydroxy-1-cyclohexene-1-carboxylic acid: The 3(R)-amino analogue of (–)-shikimic acid. *Tetrahedron* 52:8565–8580 (1996).

159. Brettle, R.; Cross, R.; Davies, G.M.; Frederickson, M.; Haslam, E.; MacBeath, F.S. The shikimate pathway. Part 9. Halogenation at C-3 of the shikimate nucleus. *Tetrahedron* 52:10547–10556 (1996).

160. Baxter, N.J.; Charlton, A.J.; Haslam, E.; Lilly, T.H.; McDonald, C.J.; Williamson, M.P. Tannin interactions with a full length human salivary proline-rich protein display a stronger affinity than with single proline-rich repeats. *FEBS Letters* 382:289–292 (1996).

161. Baxter, N.J.; Haslam, E.; Lilly, T.H.; Williamson, M.P. Stacking interactions between caffeine and methyl gallate. *J. Chem. Soc., Faraday Trans.* 92:231–234 (1996).

162. Haslam, E. Aspects of the enzymology of the shikimate pathway. *In*: Hertz, W.; Kirby, G.W.; Moore, R.E.; Steglich, W.; Tamm, Ch. (eds.) Progress in the chemistry of organic natural products. Springer-Verlag, Vienna, 69:158–240 (1996).

163. Haslam, E. Vegetable tannage—where do the tannins go? *Journal of the Society of Leather Technologists and Chemists* 81:45–51 (1997).

164. Haslam, E. Practical polyphenolics—from structure to molecular recognition and physiological action. Cambridge University Press, Cambridge, 420 p. (1998).

CHE FARÒ SENZA POLIFENOLI?

Edwin Haslam

Department of Chemistry
University of Sheffield
Sheffield, S3 7HF
UNITED KINGDOM

1. INTRODUCTION

Many have spoken and written concerning the motivations which inspire a scientist in his work. For some there is an inherent curiosity about the workings of nature, and for others, the inspiration comes from the desire to make significant and lasting contributions to human knowledge. It would nevertheless be idle to deny that for a scientist the recognition of his work by his peers is an event which engenders the deepest personal satisfaction. I should therefore like to express to all those who have been responsible for this award and the decision to bestow it upon me, my sincere thanks. Much, much more importantly by honoring me you also honor my many colleagues who, over the years, have participated in this endeavor. It gives me enormous pleasure to record my gratitude to them.

When anyone is called upon to ride the rough torrents of an occasion such as this, there are a number of thoughts uppermost in his mind. Not the least of these is the topic for his address. The enigmatic title which I have selected doubtless reflects the uncertainty with which I have approached this task. Should this be a valedictory oration and call from the battlements? Should it be the epic tale of a scientific Odyssey, an adventurous journey, undertaken many years ago with the best of intentions? Or should it be a "Definitive State of the Art" commentary? I hope that you will find a judicious mixture of all three.

"It is not enough to have knowledge, one must use it" Goethe

Willingly or not we are all guilty of the modern compulsion to know more and more about less and less; we ride the information superhighway, consult our websites and surf the internet. But information does not equate with knowledge, nor does knowledge equate with wisdom. In recent years technological revolutions have made empirical research so fast and exhilarating that scientists often have

This lecture and chapter are dedicated to the memory of the late Larry G. Butler (1934–1997), Professor of Biochemistry at Purdue University.

Plant Polyphenols 2: Chemistry, Biology, Pharmacology, Ecology, Edited by
Gross et al. Kluwer Academic / Plenum Publishers, New York, 1999

15

little or no time for reflection. However, the relentless flood of data and information rarely brings with it the answer to the important question: *"What are we going to do with it all?"* Waist deep in data (and often none the wiser for it) disillusionment may set in. Indeed if the data get so far ahead of their assimilation into a conceptual framework, then the data itself may prove an encumbrance. There are evermore grants for producing data, but hardly any for standing back in contemplation. In our case we do not allow ourselves, sufficiently often, the apparent luxury of taking time out to reflect and ponder on what the chemistry and biochemistry of plant polyphenols are all about and where this most arcane of topics is headed. Polyphenol chemistry has surely reached the stage at which we can begin to stand back in active contemplation of the data before us and within the framework of ideas generated look beyond the present to many of the unanswered questions associated with the presence of polyphenols in plant materials. We are like the pioneers who settled this country. We have left the Eastern seaboard heading for the mountains of Oregon and the gates of the golden West; quite what is our position in the vast desert in between is something on which we can all speculate.

In what follows, however, I do not claim objectivity, indeed, I am instantly suspicious of any academic, writer or politician who lays claim to this rare and elusive quality. I am very much inclined to the view that the perception of any situation depends to a large extent on the background—the ambitions—the limitations—and the prejudices of the beholder. I hope that you will bear this in mind as I proceed to erect a few signposts to the future in this metaphorical desert, but I also hope that it provokes you to think, to debate, and doubtless to disagree not only with the inscriptions but also with the way in which they are directed.

The first significant research into the nature of vegetable tannins (plant polyphenols) was that of a quintet of distinguished German chemists—the Nobel laureate Emil Fischer and Karl Freudenberg, in the early years of this century, and later Otto Schmidt, Walter Mayer, and Klaus Weinges (all based in Heidelberg). The vastly expanded array of physical techniques for the separation and structure determination of natural products which began to be made available in the 1950's transformed the subject. Structural problems which previously took decades to solve now often yield a solution in a matter of weeks or months. In consequence isolation and structure determination, *by themselves*, no longer remain the principal goal; they are mere prologue. In this context my own attitudes and scientific aspirations were very accurately reflected in the words of that giant Californian redwood T.A. (Ted) Geissman made over 30 years ago:

"Certainly structures are important—but the determination of structure in itself is ceasing to be of much interest or importance, and often turns out to be an exercise in the manipulative skills of the investigator. Some syntheses are in the same class . . . My own tendency (in which, of course, I am not alone) is to look at biological relationships: taxonomy, phylogeny, biosynthesis and biotransformations . . . The future of phytochemistry is to use the chemical information as the starting point for inquiry into questions that lie in the realms of biology."

Future scientists may well think it odd that, over the past 30 years, we have learned so much about the structure of polyphenol molecules but yet so little about their intrinsic properties, about many of the important processes in which they become involved, both *in vivo* and *in vitro*, and about their function and the control of their synthesis in living systems. In this context aspects of the following areas will be considered:

(1) —**structural problems still awaiting resolution**

(2) —**the control (enzymic/chemical) of biosynthesis**

(3) —**the concept of molecular recognition intrinsic to polyphenol complexation 'and**

(4) —**the vexatious question of the role of plant polyphenols in plant metabolism**

2. STRUCTURE AND STRUCTURAL PROBLEMS

The word tannin has a long and well-established usage in the scientific literature. The importance of vegetable tannins to a range of scientific disciplines has been recognized for some time. However, a firm definition of what constitutes a vegetable tannin is not easy to give; probably the most acceptable, concise and simple definition is still that of Bate-Smith and Swain:[1]

"Water soluble phenolic compounds having molecular weights between 500 and 3,000 and, besides giving the usual phenolic reactions, they have special properties such as the ability to precipitate alkaloids, gelatin and other protein"

Many still prefer the term vegetable tannin, which they find valuable simply because of its lack of precision. Scientifically and terminologically, plant polyphenols is to be preferred as a descriptor for this class of higher plant secondary metabolites if serious attempts are to be made to interpret their diverse characteristics at the molecular level (however, see Postcript).

2.1 Properties and Classification

It is now possible to describe in broad terms the nature of plant polyphenols. They are secondary metabolites widely distributed in various sectors of the higher plant kingdom. They are distinguished by the following general features:

(a) *Water solubility*. Although when pure some plant polyphenols may be difficult to dissolve in water, in the natural state polyphenol—polyphenol interactions usually ensure some minimal solubility in aqueous media.

(b) *Molecular weights*. Natural polyphenols encompass a substantial molecular weight range from 500 to 3–4,000. Suggestions that polyphenolic metabolites occur which retain the ability to act as tannins but possess molecular weights up to 20,000 must be doubtful in view of the solubility proviso.

(c) *Structure and polyphenolic character*. Polyphenols, per 1,000 relative molecular mass, possess some 12–16 phenolic groups and 5–7 aromatic rings.

(d) *Intermolecular complexation*. Besides giving the usual phenolic reactions, they have the ability to precipitate some alkaloids, gelatin and other proteins from solution. These complexation reactions are not only of intrinsic scientific interest as studies in molecular recognition and possible biological function, but, as noted earlier, they have important and wide-ranging practical applications.

(e) *Structural characteristics*. Plant polyphenols are based upon two major and one minor structural theme, namely:

(1) ***condensed proanthocyanidins***. The fundamental structural unit in this group is the phenolic flavan-3-ol ("*catechin*") nucleus. Condensed proanthocyanidins exist as oligomers (soluble), containing two to five or six "*catechin*" units, and polymers (insoluble). The flavan-3-ol units are linked principally through the 4 and the 8 positions. In most plant tissues the polymers are of greatest quantitative significance but there is also usually found a range of soluble molecular species—monomers, dimers, trimers, *etc.*[2–5] The "monomer" units of procyanidins (R=H) and prodelphinidins (R=OH) are phenolic flavan-3-ols which are linked primarily through their 4 and 8 positions respectively. The stereochemistry at C-2 is most commonly encountered as the 2R configuration.

Oligomeric condensed proanthocyanidins have been held[6] to be most commonly responsible for the many distinctive properties of plants typically attributed to "condensed tannins". Mole[7] has however recently commented on some of this earlier work and has suggested that fewer plant families are characterised by the presence of "tannins" than heretofore thought. In the context of the later discussion on the polymeric proanthocyanidins it is pertinent to point out that, simply on the basis of solubility differences, Sir Robert and Lady Robinson[8] in the 1930's originally subdivided the leucoanthocyanins (condensed proanthocyanidins) into the three classes indicated below:

  (i)   *those that are insoluble in water and the usual organic solvents or give only colloidal solutions,*

  (ii)   *those readily soluble in water but not readily extracted therefrom by means of ethyl acetate, and*

  (iii)   *those capable of extraction from aqueous solution by ethyl acetate.*

Insofar as the total complement of condensed proanthocyanidins (procyanidins and prodelphinidins) found in plant tissues is concerned, the soluble oligomeric forms (monomers, dimers, trimers . . .) are in metabolic terms but the "tip of the iceberg". According to the Robinsons' classification they represent category (iii) above.

    For the generality of plants it is now quite clear that condensed proanthocyanidins which fall within the two other categories (i and ii) invariably strongly predominate over the more freely soluble forms. They are, metaphorically speaking, the base of the "metabolic iceberg". Indeed in the tissues of some plants such as ferns and fruit such as the persimmon (*Diospyros kaki*), there is an overwhelming preponderance of these forms. They are also of frequent occurrence in plant gums and exudates.

(2) **galloyl and hexahydroxydiphenoyl esters and their derivatives**. These metabolites are almost invariably found as multiple esters with D-glucose,[2,3,9–12] and a great many can be envisaged as derived from the key biosynthetic intermediate β-1,2,3,4,6-penta-O-galloyl-D-glucose. Derivatives of hexahydroxydiphenic acid are assumed to be formed by oxidative coupling of vicinal galloyl ester groups in a galloyl D-glucose ester;[12] dependent upon the positions on the D-glucopyranose ring between which oxidative coupling of the galloyl ester groups takes place then a particular chirality is induced in the twisted biphenyl of the resultant hexahydroxydiphenoyl ester.[2,3] Acid hydrolysis of hexahydroxydiphenoyl esters gives rise to the formation of the bis-lactone of hexahydroxydiphenic acid, the planar and virtually insoluble ellagic acid from which the characteristic nomenclature of ellagitannins is derived.

    Gallic acid is most frequently encountered in plants in ester form. These may be classified into several broad categories:

  (i)   Simple esters.

  (ii)   Depside metabolites (*syn* gallotannins).

  (iii)   Hexahydroxydiphenoyl and dehydrohexahydroxydiphenoyl esters (*syn* ellagitannins) based upon:

(a) $^4C_1$ conformation of D-Glucose.

(b) $^1C_4$ conformation of D-Glucose.

(c) "open-chain" derivatives of D-Glucose.

(iv) "Dimers" and "higher oligomers" formed by oxidative coupling of "monomers", principally those of class (iii) above.

bis-galloyl ester   hexahydroxydiphenoyl ester

ellagic acid

(3) **phlorotannins**. More recently a third, and relatively minor, class of natural polyphenol has been recognized—the phlorotannins. They are isolated from several genera of red-brown algae, and their structures are seen to be composed entirely of phloroglucinol sub-units linked by C—C and C—O chemical bonds. Presumably they are formed biosynthetically by oxidative C—C and C—O coupling of phloroglucinol nuclei. Chemical investigation of these polyphenols has revealed a variety of low, intermediate and high molecular weight compounds. The low molecular weight fraction can be separated into its individual components by chromatography; a typical low molecular weight phlorotannin is fucofureckol obtained from *Eisenia arborea*.[13]

fucofureckol

Although these low molecular weight polyphenols have received the greatest attention they often constitute only a minor proportion of the polyphenolic fraction from red algae; the remainder are composed of oligomers ($M_R \sim 10^3$–10^4) and polymers ($M_R > 10^4$) based similarly upon the phloroglucinol building unit.[14,15] Detailed chemical and spectroscopic analysis of the polymer from the marine brown alga *Fucus vesiculosus* suggests that it is highly branched with ~20 percent of the constituent units being at chain termini. There is no evidence for the presence in the polymer of large rings of phloroglucinol nuclei. The interior backbone consists predominantly of ether-linked phloroglucinol units. Based upon a comparative survey of analogous phloroglucinol polymers derived from other red-brown algae, McInnes and his colleagues concluded[15] that they are formed biosynthetically by a *non-random* assembly of subunits.

Although such statements are inevitably dangerous ones to make in science, it does not seem imprudent to conclude that in terms of structure we are fast approaching a steady state situation. Compared to 20 years ago the pace of discovery of new, original polyphenolic structures has diminished considerably. It is perhaps pertinent therefore to look back to many structural problems which have been discarded or ignored as "too difficult"; all are surely now amenable and worthy of renewed and more detailed study. Several such problems immediately spring to mind in the province of the flavan-3-ols and proanthocyanidins:

(i) *the insoluble polymeric proanthocyanidins,*[8]

(ii) *polymeric pigments of red wines,*

(iii) *thearubigins—black tea pigments,* and

(iii) *a stereospecific synthesis of phenolic flavan-3-ols.*

Mention should also be made of a series of problems which are intrinsically of a quite different order of magnitude of difficulty related to the *in vivo* and *in vitro* transformations undergone by many polyphenolics in the bark and heartwood of trees and which are the source of many commercially important extracts such as Mangrove and Quebracho.

(i) *the insoluble polymeric proanthocyanidins*

Sir Robert and Lady Robinson alluded to the first of these problems in their solubility studies of proanthocyanidins in the 1930s. Other workers confirmed their observations. Thus Hillis and Swain[16] noted that " *'leucoanthocyanins' of plum leaves were divisible into three classes, the first two being successively extractable with absolute, followed by aqueous methanol and the third remained in the residue, being non-extractable by these or other neutral solvents*". Similarly Bate-Smith[17] in a study of the proanthocyanidins of herbaceous Leguminosae observed that "*in most species examined, even in the most favorable conditions, that fraction which is extractable is only a very small proportion of the total proanthocyanidins present*". The same observations have been made time and again in the author's own laboratory. Plant debris which results from repeated extractions of fresh proanthocyanidin containing tissues with methanol, invariably retains far greater amounts of materials which give the various characteristic color tests for proanthocyanidins than that which has been extracted with methanol. Moreover further extraction of the plant debris with boiling DMSO or DMF (both powerful solvating reagents) effects no further extraction of proanthocyanidins. This feature is particularly common in the proanthocyanidins of ferns and the herbaceous Leguminosae. One interpretation of these observations is that these forms of plant proanthocyanidins (just as is the case of the lignins) are covalently bound to an insoluble carbohydrate (or other polymer) matrix within the plant cell.[18] The nature, structure and properties of the "insoluble" proanthocyanidins thus constitutes a major unresolved problem for phytochemists and ecologists in particular. Accepting the view of proanthocyanidin biogenesis—*namely the promiscuous encounter of quinone methide intermediates with phenolic flavan-3-ol metabolites derived in the same enzyme catalyzed reduction*—then it is possible to envisage the capture of the same quinone methide intermediate by other nucleophiles, for example, cellular polysaccharides. This would then render the final proanthocyanidin insolubilized by virtue of its attachment to a "terminal unit" which is the carbohydrate matrix (fig. 1).

A second series of unsolved structural problems concerns the polymeric structures which result from chemical transformation of flavan-3-ol and procyanidin precursors. Pre-eminent amongst these are the so-called polymeric wine pigments found in red wines and the thearubigins, red-brown pigments from black teas.

(ii) *polymeric pigments of red wines*

The polyphenolic procyanidins and the flavan-3-ols (−)-epicatechin and (+)-catechin play key roles in the processes associated with the maturation of red wines. These roles are manifested in several ways; the most important ones which have been identified to date are summarized in (fig. 2).

Whereas the brightness and purple tints of young red wines are due essentially to anthocyanins there is a progressive displacement of these monomeric pigments by more stable, darker "polymeric" forms during vinification. Somers[19] has

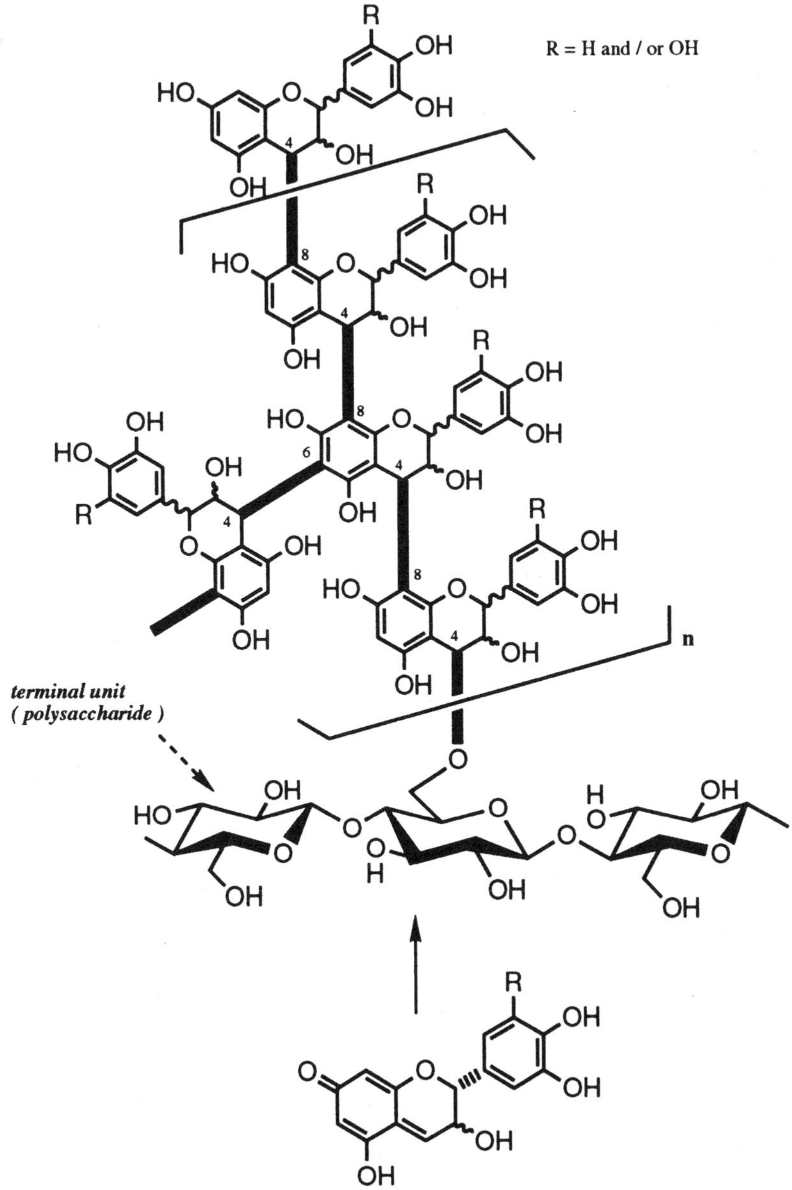

Figure 1. Capture of the quinone methide intermediate by a polysaccharide: attachment of the proanthocyanidin structure to an insoluble polysaccharide matrix.

estimated that, on average, these may constitute some 80 percent of the pigmentation of red wines after 10 years. Mean molecular weights of these polymeric pigments have been reported as rising from ~2,000 after 5 years to ~4,000 at 20 years. Somer's observations were first recorded some 25 years ago. Very probably they

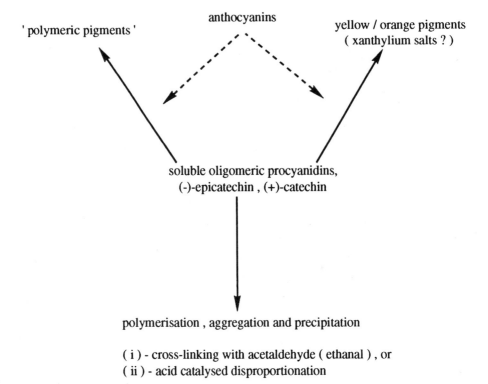

Figure 2. Reactions of polyphenolic procyanidins and the flavan-3-ols (–)-epicatechin and (+)-catechin during the maturation of red wines.

still remain scientifically the most significant ones in this area of oenology; it is therefore disappointing that precise structural data concerning these "polymeric pigments" are still totally lacking. Interpretations of the changes in *chemical* composition of the pigments as the red wine ages still therefore remain speculative and based upon reasoned chemical expectations.

Various hypotheses have been advanced; given the inherent complexity of red wines, it seems probable that more than one process is taking place. Most of these theories are based upon the original speculations of Somers[19] and were formulated by Somers as shown in figure 3. The first step is the nucleophilic attack at C-4 of oenin by a flavan-3-ol, (–)-epicatechin or (+)-catechin, or analogously by an oligomeric procyanidin; this is then followed by oxidation to regenerate a flavylium salt. Somers suggested that the effect of aryl-substitution at C-4 would be to stabilize the chromophore which, according to Somers, was best formulated as the quinonoidal anhydro-base. The oligomeric / polymeric nature of the pigments derives from the identity of the initial attacking nucleophile and additional "quinone-like" structures, arising from oxidation of the polymeric pigments, may well be formed and contribute to the browner tints of elderly wines.

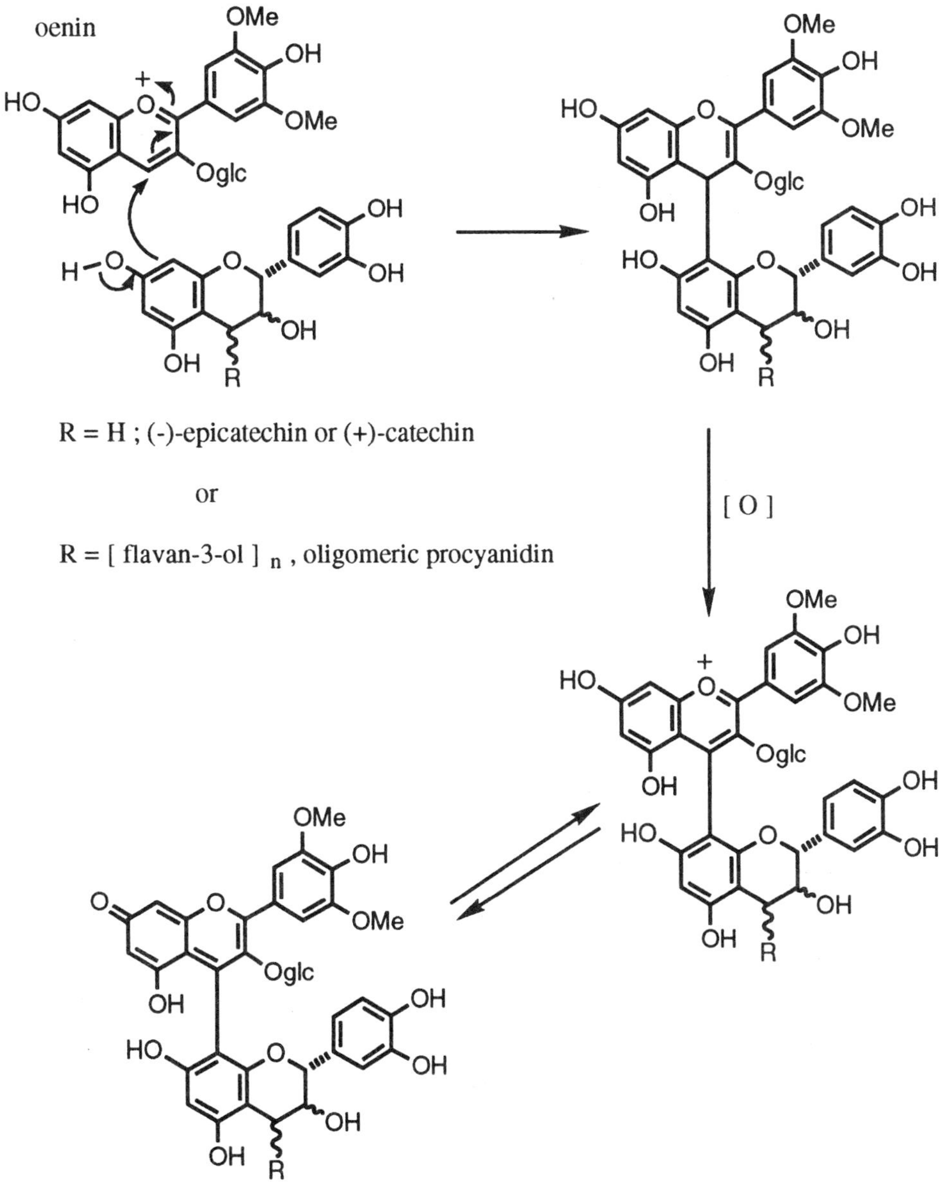
oenin
OMe
OH
OMe
Oglc
HO
HO
H
O
OH
OH
OH
OH
R
R = H ; (-)-epicatechin or (+)-catechin
or
R = [flavan-3-ol] n , oligomeric procyanidin
[O]
' polymeric ' proanthocyanidin pigments

Figure 3. Generation of oligomeric / polymeric pigments in red wines.[19]

(iii) *thearubigins—black tea pigments*

The outstanding characteristic of the chemical composition of the green tea flush is its very high concentration of polyphenolic metabolites—principally phenolic flavan-3-ols, which occur in the cytoplasmic vacuoles of leaf cells (fig. 4). Although amounts vary, dependent upon a number of factors such as variety and growing conditions, phenolic flavan-3-ols *usually* constitute some 20–30 percent of the dry matter of the leaf. Typically, the average content of the major phenolic flavan-3-ols in a Sri-Lankan tea shoot is shown in figure 4.

Since almost all the sensory characteristics of manufactured teas derive from the oxidative transformations of the green leaf phenolics, a very great deal of attention has been devoted to this topic. E.A.H. Roberts[20] made pioneering observations to this area and these have been seminal to its subsequent development. The enzymes of principal interest in the tea leaf are those intimately involved in the conversion of the tea flush to the commercially manufactured black and Oolong teas. Pre-eminent amongst these enzymes is tea-leaf polyphenoloxidase. It contains copper (0.32 percent) and exists as a mixture of several isoenzymes. Its main component has a relative molecular mass of ~140,000. Although various mechanisms (free radical, phenoxonium ion) are theoretically possible to initiate the polyphenol oxidase catalyzed oxidation, the primary step in tea fermentation is probably best envisaged as the overall conversion of the *ortho*-dihydroxy (*catechin*) and *ortho*-trihydroxy (*gallocatechin*)—phenyl 'B' rings of the substrate tea leaf polyphenolic flavan-3-ols to give the corresponding highly reactive orthoquinone derivatives. Thereafter these orthoquinones may react, under purely chemical

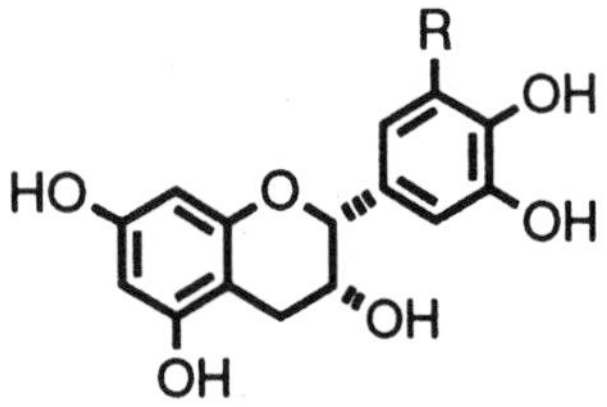

R = H : (-)-epicatechin ; 0.63 %

R = OH : (-)-epigallocatechin ; 2.35 %

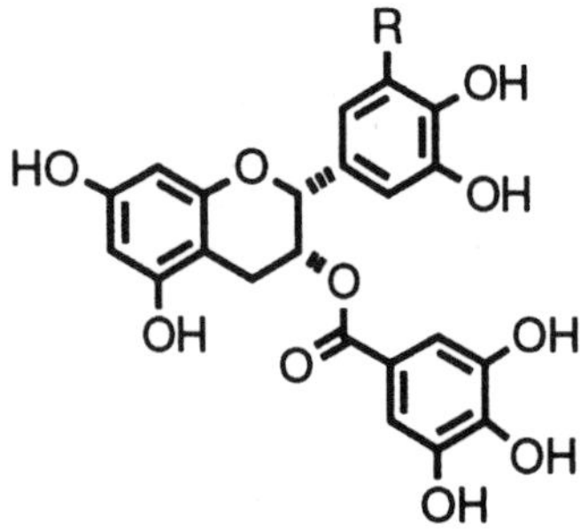

R = H : (-)-epicatechin-3-O-gallate ; 2.75 %

R = OH : (-)-epigallocatechin-3-O-gallate ; 10.55 %

Figure 4. Principal phenolic flavan-3-ols of a typical Sri-Lankan tea flush; percentage composition of dry weight of leaf.

control, to give a series of discrete yellow—orange compounds (theaflavins), or they may randomly polymerise, and very probably undergo other oxidative reactions such as aryl ring-fission, to give a complex, very poorly defined group of substances known, because of their red-brown color, as thearubigins. Although eventually some 75 percent of the substrate phenolic flavan-3-ols may be converted to the weakly acidic thearubigins they are the least well-characterized group of components of black tea notwithstanding their contribution to the color, mouthfeel and strength of the subsequent infusion.

It is generally assumed that the thearubigins are complex heterogeneous polymers formed by the oxidation (predominantly) of phenolic flavan-3-ols in the green tea leaf, all of which contain a "catechol" or "pyrogallol" 'B' ring and may in addition contain a gallate ester group. The first step in this enzymic oxidation is the conversion of some or all of these catechol and pyrogallol based nuclei to the respective *ortho*-quinones. Thereafter these intermediates partition to give either the theaflavins or the thearubigins (fig. 5a). The routes to the various theaflavins have been established and follow clearly defined chemical pathways. The routes to the thearubigins are ambiguous, but it is normally assumed that these involve oxidatively induced oligomerization and very probably further oxidation steps. The color of the thearubigins may thus derive from the presence of unreacted *ortho*-quinone groups in the final polymeric structures. Although model tea fermentation systems indicate that thearubigin type substances are formed by oxidation of any one of the green tea leaf phenolic flavan-3-ols, or a combination thereof, it seems highly probable that the principal substrates are (−)-epigallocatechin-3-O-gallate and, to a lesser extent, its parent flavan-3-ol. Such a conclusion seems inescapable for, as Roberts first pointed out, there are no other oxidizable substrates present in sufficient quantity to yield the amounts of thearubigins formed. Minor pathways doubtless are also in operation utilising not only the other green tea leaf phenolic flavan-3-ols but also the theaflavins (as they are formed) as substrates in direct or coupled oxidations.

Previous work on the thearubigins has concentrated almost exclusively on studies of the complex, ill-defined polymers themselves. In looking afresh at this problem it seems that one of the most promising ways forward would be to examine—both practically and theoretically—the oxidation of (−)-epigallocatechin-3-O-gallate and (−)-epigallocatechin and in particular the type of products which may derive from oxidative phenolic ring fission. Roberts[20] noted the acidic nature of the thearubigins and it seems most probable that this character derives from the oxidative breakdown of aromatic nuclei to give carboxylic acid functionalities. However, although the oxidative coupling of phenolic nuclei is a well-studied and well-documented phenomenon, pertinent information on the subsequent oxidative ring-opening of phenolic nuclei is by comparison poor. Without doubt in this context the most significant observations were made by Mayer and his colleagues[21] whilst studying the oxidation of methyl gallate. They showed (fig. 5b) that crystalline products bearing carboxylic acid functionalities could be derived by this means; on this basis there seem to be excellent grounds for similar model studies of the *in vitro* oxidation of (−)-epigallocatechin-3-O-gallate and/or (−)-epigallocatechin. Bearing in mind the analogies which flow from

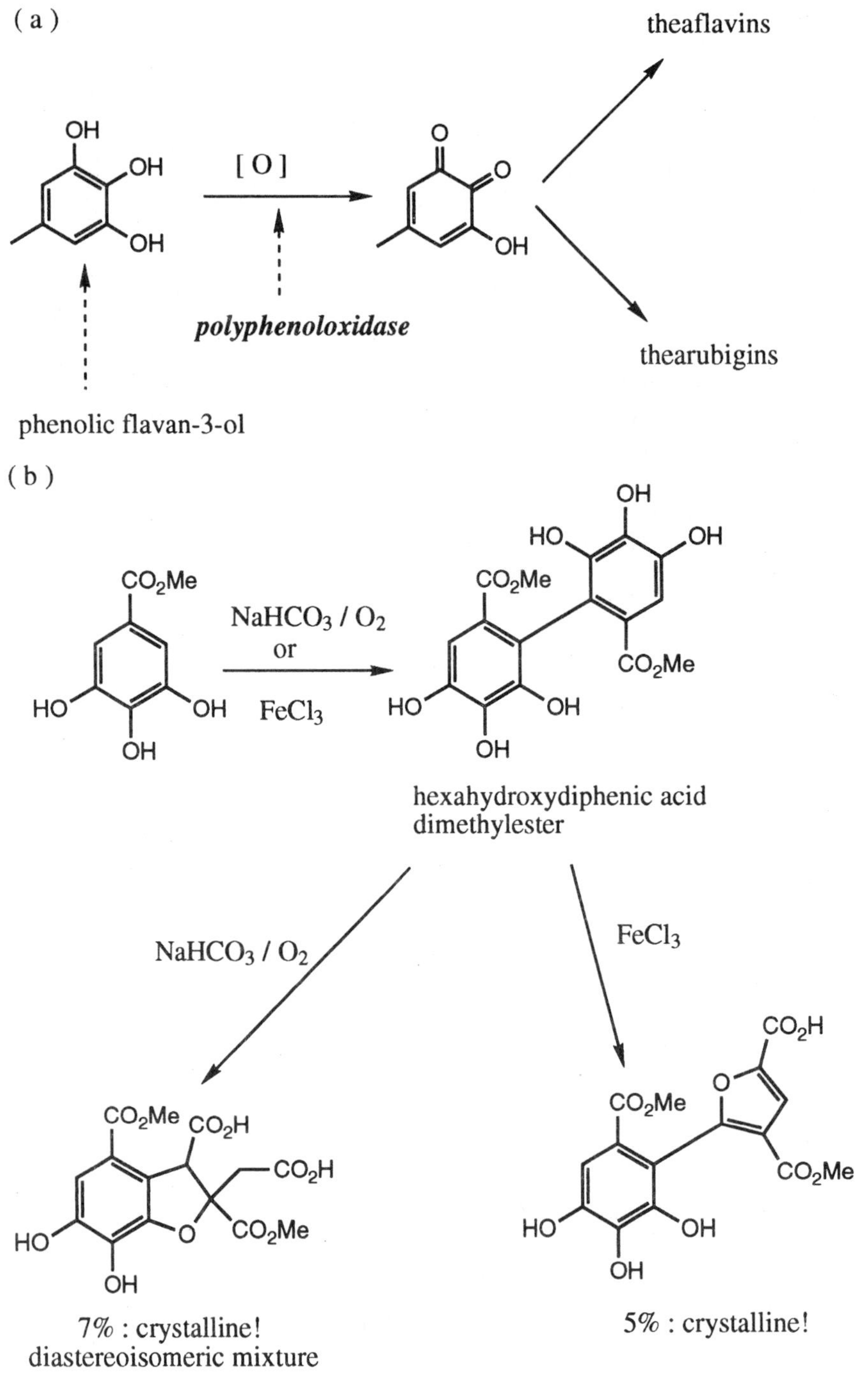

Figure 5. (a)—Principal pathways for the oxidative conversion of phenolic flavan-3-ols to theaflavins and thearubigins in tea fermentations: partitioning of the *ortho*-quinone intermediate; (b)—oxidation of methyl gallate.[21]

Mayer's work, it is also possible to speculate widely, if not wildly, on the probable structure of the thearubigins.[3]

(iv) *Stereospecific synthesis of phenolic flavan-3-ols*

Polyphenols constitute the important active principals of many medicinal plants and medicinal plant preparations. Notable studies in this area have been made by several groups worldwide. *In vitro* testing has identified a wide range of potentially significant biological activities, which are exhibited by natural polyphenols, but although these studies have revealed important differences in pharmacological activity between individual polyphenols and between classes of different polyphenol, overall they suggest selectivity rather than high specificity toward particular biological targets. Although proof of a cause and effect relationship does not yet exist substantial interest has thus recently been engendered by the epidemiological evidence which points to a reduced risk of certain degenerative diseases by the consumption of beverages containing polyphenols,[22] in particular green tea and red wines, both rich sources of polyphenols based on the flavan-3-ol carbon-oxygen skeleton. However, it is now also clear that the physiological and pharmacological actions of polyphenols very probably derive, at least in part, from some of the chemical features they have in common, *viz*—the various physical and chemical properties which are themselves associated principally with the possession of a concatenation of phenolic nuclei within the *same* molecule. On the basis of this proposition, it has been suggested that polyphenols exert certain of their roles in the medical treatment of diseased states by virtue of three distinctive *general* characteristics,[23] which they all possess to a greater or lesser degree, and which derive in essence from the properties of the simple phenolic nucleus itself, namely:

   (i) their complexation with metal ions (iron, manganese, vanadium, copper, aluminium, calcium, *etc.*),

  (ii) their antioxidant and radical scavenging activities, and

 (iii) their ability to complex with other molecules including macromolecules such as proteins and polysaccharides.

Nevertheless, it should also be pointed out that, as yet, it is not clear in humans how or if these complex polyphenolic substrates are absorbed from the gut, and this lack of precise knowledge on the fate of these compounds in the human body remains a major weakness in this area. What is urgently required at this juncture is evidence concerning their absorption, penetration and metabolism in the human body. Until such time, the evidence for their remedial effects remains based upon epidemiological evidence rather than scientific observation. In order to achieve this goal it is neccessary to be able to prepare *de novo* specifically radiochemically labelled [³H, ¹⁴C] samples of natural polyphenols to investigate their fate when ingested in the human body. Prime candidates for such a synthetic investigation would be the natural flavan-3-ols (–)-epicatechin and (–)-epigallocatechin because of their importance as anti-oxidant constituents of the widely consumed beverages—tea and wine. Parenthetically, it is also remarkable that these two molecules, although known for so many years, have yet to be synthesized, *de novo*, in their free phenolic forms.

3. BIOSYNTHESIS OF POLYPHENOLS

Organic chemists have long been familiar with the wealth of substances which adorn extracts of plants and micro-organisms but to which no useful purpose can be readily assigned. Polyphenols are just such a group of plant secondary metabolites. As the purely structural problems relating to these natural products were solved, attention switched to two closely interdependent questions: first to the manner in which they were synthesised *in vivo* and second to problems of a biological and physiological nature, in particular their role in the overall metabolism of the plant. The answers to this second question are of course of crucial importance in the context of the suggested ecological function of polyphenols as plant defense agents.

There is an inescapable logic to the argument that if metabolic pathways to a particular secondary metabolite or group of metabolites is to be fully defined it is essential to study the enzymes which catalyze each individual transformation. Only then will it be possible not only to state precisely how these compounds are formed but also to begin to address questions such as: *"how is their synthesis controlled?"*; *"what is the origin of their biosynthetic genes?"*; *"what is their function in the life of the organism?"*. For a great many years, probably because of the relative *biochemical* obscurity which secondary metabolites, such as polyphenols, have enjoyed and the perceived technical difficulties, the enzymology of secondary metabolism has remained in numerous cases, at best, poorly explored. Attention is drawn here in particular to the role which inbuilt chemical predisposition within the substrate appears to play in the biosynthetic transformations; this leads, in turn, to questions concerning the precise role and nature of the enzymes involved in these reactions.

3.1 Biosynthesis of Galloyl Esters

Gallic acid is unique amongst the various naturally occuring hydroxybenzoic acids[24,25] both in respect of its relative ubiquity in the plant kingdom and of the quantitative significance of its metabolism in many plants. Although some evidence exists to show that gallic acid may arise by oxidative degradation of L-phenylalanine,[26] the weight of experimental data favors the view[27,28,29] that it is nevertheless formed primarily *via* the dehydrogenation of 3-dehydroshikimate.

β-Glucogallin (β-1-O-galloyl-D-glucose), first isolated from Chinese rhubarb (*Rheum officinale*) in 1903, is, following the extensive studies of Gross,[30,31] considered to be the key intermediate in the biosynthesis of esters of gallic acid. Work with cell-free extracts from oak leaves and subsequently with the partially purified glucosyl transferase verified that β-glucogallin is generated by the reaction of gallic acid with UDP-glucose. Thereafter, β-glucogallin as substrate undergoes a series of further galloyl transfer reactions to yield ultimately β-1,2,3,4,6-pentagalloyl-D-glucose. It is very interesting to note that β-glucogallin acts as the principal galloyl group donor in these reactions. Thus, in the first of these reactions, a partially purified enzyme from young oak leaves (EC 2.3.1.90) catalyses the formation of β-1,6-digalloyl-D-glucose from two molecules of β-glucogallin. The sequence continues in analogous fashion, with β-glucogallin as prime galloyl donor, *via* β-1,2,6-trigalloyl-D-glucose and β-1,2,3,6-tetragalloyl-D-glucose, to give finally β-1,2,3,4,6-pentagalloyl-D-glucose. One of the most striking features of this bio-synthetic pathway is that the sequence of esterification steps with gallic acid:

$$1\text{-OH} > 6\text{-OH} > 2\text{-OH} > 3\text{-OH} > 4\text{-OH}$$

exactly parallels the sequence in the chemically mediated esterification of the hydroxyl groups of D-glucopyranose.[32]

This β-D-glucogallin dependent pathway is however by no means exclusive. Studies with enzyme extracts of sumach (*Rhus typhina*) have shown that in addition to β-D-glucogallin, β-1,6-digalloyl-D-glucose, β-1,2,6-trigalloyl-D-glucose and β-1,2,3,6-tetragalloyl-D-glucose may also act, although with progressively decreasing efficiency, as galloyl group donors (from the 1-position of the galloyl-ester).

Strong circumstantial evidence now exists to support the proposition[2,3,11] that the metabolite β-1,2,3,4,6-pentagalloyl-D-glucose then plays a pivotal role in the formation of the vast majority of gallotannins and ellagitannins which occur in many plants. Four principal pathways are then presumed to lead from β-1,2,3,4,6-pentagalloyl-D-glucose and to give, by appropriate chemical embellishment the various classes of metabolites (fig. 6—pathways a, b, c and d).

Important questions arise in the future study of the biosynthesis ellagitannins. Pre-eminent amongst these are:

(i) If the ellagitannins originate by processes initiated by phenolic oxida-tion,[3,11] what influences do enzymic control, on the one hand, and the intrinsic chemical reactivity of reaction intermediates (e.g., phenoxy radicals), on the other, exert upon the ultimate reaction profile?

(ii) —Why in the biogenesis of the ellagitannins do the *intramolecular* processes appear to proceed exclusively with the formation of new carbon—carbon linkages, but the *intermolecular* processes, namely those of oligomerization occur invariably with the generation of new carbon—oxygen linkages? Is this indicative of chemical or enzymic control?

Circumstantial evidence suggests that enzymes involved in the biosynthesis of ellagitannins do not act in the sense of a "Pauling enzyme", which lowers the activation energy by stabilizing transition states, but act by directing the very high energy intermediates along a particular pathway; *the chemistry is intrinsic to the intermediate*; the enzymes are thus permissive rather than catalytic ones.[33] It is a question worthy of fuller scrutiny.

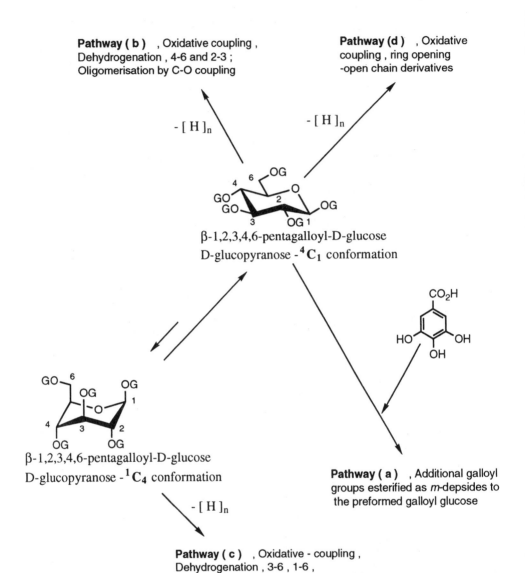

Figure 6. Biogenesis of gallotannins and ellagitannins; the metabolic embellishment of β-1,2,3,4,6-pentagalloyl-D-glucose, principal pathways.[3,11]

3.2 Proanthocyanidin Biosynthesis

An identical proposition also emerges strongly from any examination of the *status quo* of studies of the biosynthesis of plant proanthocyanidins. Current lines of thought suggest that the proanthocyanidins are formed as byproducts of the processes in which the parent flavan-3-ols are biosynthesized.[2,3] An impressive body of knowledge has been assembled concerning the general pathways of flavonoid biosynthesis,[34] but although Stafford and her colleagues have considerably clarified some steps in the biosynthesis of the phenolic flavan-3-ols[35,36,37] using cell suspension cultures of Douglas-fir needles (*Pseudotsuga menziesii*), the question of whether the final assembly of the proanthocyanidins is under enzymic control or not appears still undecided. So far it has not been possible to isolate a "proanthocyanidin condensing enzyme", or genes coding for such an enzyme and its regulation.

Current lines of thought suggest that the synthesis of oligomeric proanthocyanidins in plants is intimately involved with the formation of the phenolic flavan-3-ols[38,39,40] (fig. 7). It has been speculated[40] that the proanthocyanidins are formed in an oligomerization process in which the flavan-3-ol first forms a proanthocyanidin dimer by a stereospecific nucleophilic capture at C-4 of the putative quinone methide (or its protonated carbocation equivalent) intermediate in flavan-3-ol biosynthesis. The dimer then captures a further quinone methide (or its protonated carbocation equivalent) to generate a trimer and so progressively, by the capture of further quinone methide species, oligomers and finally polymers are formed (fig. 7). This biosynthetic reaction forms the basis of biogenetically patterned chemical syntheses of proanthocyanidins, which appear to mimic uniquely the proposed biosynthetic pathway.[38]

The extraordinarily facile nature of the biomimetic synthesis and the exceedingly close correspondence, in all aspects, between the *in vivo* and the *in vitro* proanthocyanidin syntheses raise the very important question whether the actual syntheses *in vivo* are under enzymic control or not. Once again, it is possible that the chemistry is inherent in the flavan-3-ol(s) and putative quinone methide intermediates; the enzyme(s) is permissive rather than catalytic. Alternatively, the formation of proanthocyanidins in plant tissues may simply reflect the fact that *reductase II*, is a 'leaky' or inefficient enzyme. In this scenario the proanthocyanidins would be formed by promiscuous capture of the end product (flavan-3-ol) by the highly reactive quinone methide intermediate leaking fom the enzyme active site, and the product profile would be determined both qualitatively and quantitatively by the characteristics of the chemical interaction between the flavan-3-ol(s) and putative quinone methide intermediate(s).

4. MOLECULAR RECOGNITION: ASTRINGENCY: CHEMICAL DEFENSE

In the case of plant polyphenols (*syn.* vegetable tannins) the surge in knowledge of a fundamental chemical and biochemical nature which has taken place over the past 40 years has now paved the way, as predicted by Geissman, to an interest in

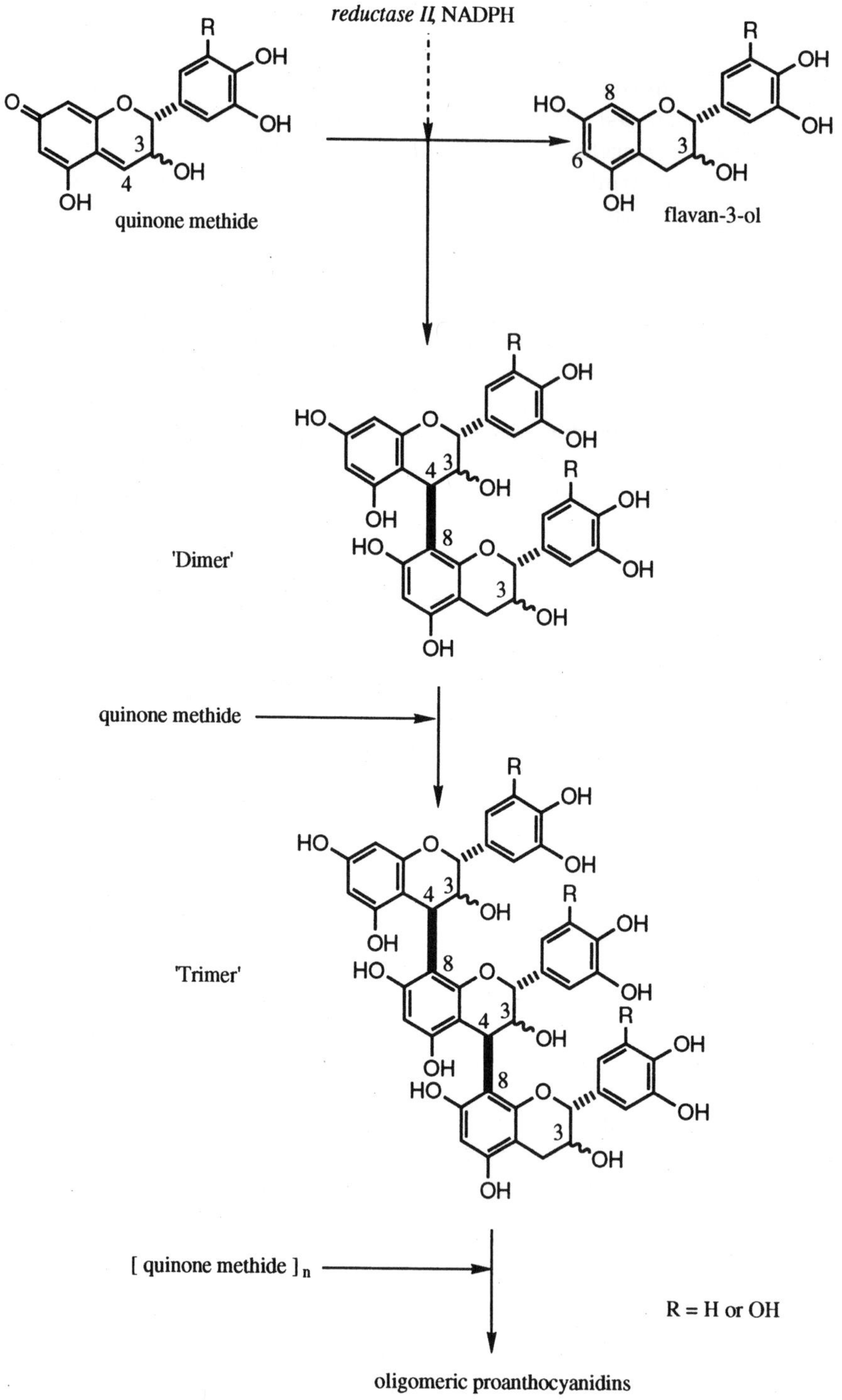

Figure 7. Suggested scheme of biogenesis of oligomeric proanthocyanidins.[40]

questions which lie in the realms of biology (not least the question of the purpose or function which these forms of metabolism serve in the organisms concerned). Equally it has given rise to a burgeoning concern with practical polyphenolic phenomena and an understanding of the importance of plant polyphenols in areas as diverse as agriculture, ecology and chemical defense in plants; foodstuffs, nutrition and beverages; fruit and floral pigmentation; natural glues, varnishes and exoskeletons; the efficacy of traditional and herbal medicines in the treatment of certain pathological conditions; traditional processes for the making of leather and the generation of important industrial chemicals. Whilst this change in emphasis might be considered timely, it also perfectly illustrates the seminal observation of Pasteur that the proper understanding of complex applied scientific problems invariably follows periods when substantial advances in fundamentals have first been accomplished. Underlying a great number of the practical phenomena in which polyphenols are intimately involved is their distinctive ability to form intermolecular complexes with each other and with other molecules, large and small— the property of *molecular recognition*. In the case of polyphenols such non-covalent interactions are not only important in their own right but are often an indispensable prelude to further chemical reactions in which new chemical bonds are formed. The fundamental basis of molecular recognition is provided by the potential energy surface that represents the interaction of two or more molecules in a cluster as a function of their mutual separation and orientation. The number of independent variables upon which this intermolecular energy depends increases as the molecular size increases; for molecules the size of typical polyphenols in an aqueous medium the number of variables is astronomical and rationalizations are, neccessarily, macroscopic and qualitative in their nature. Nevertheless, understanding how this facet of the physical and chemical behavior of polyphenols is achieved through their particular chemical structures has become a question of paramount importance and the area will be the one in which the major future advances in the field of polyphenols will undoubtedly occur.[3]

The view that secondary plant constituents are intimately concerned in the biochemical co-evolution of plants and herbivores was first put forward in the 1950s, notably by Fraenkel[41] in a paper entitled "The *Raison d'Etre* of Secondary Plant Substances" in which he drew on the views of the German botanist Stahl expressed some 70 years earlier. During the 1960s the topic developed experimentally with explosive effect and secondary metabolites became the cornerstone of a new theory of biochemical co-evolution. This work and the ideas of Feeny[42] and of Rhoades and Cates[43] led to a comprehensive view concerning the optimal patterns of chemical defense in plants. The theory had a great impact but its effect on other avenues of enquiry and intellectual endeavor was hypnotic. Some workers, for instance, advocated that the term secondary should now be suppressed on the basis of the clearly perceived biological significance of these compounds and the apparent misconceptions to which the term secondary now therefore gave rise. However it should also be noted that dissenting voices sought to put forward alternative theories to rationalize the occurrence of the extraordinarily diverse range of natural products (secondary metabolites) produced by plants, most notably Muller.[44]

The key element, the *sine qua non*, of the Darwinian view of nature is that the purposeful construction of living matter can be attributed to natural selection. Several assumptions—the *ifs*[45]—predicate that viewpoint. Most notably; *if* there are random variations in systems which can reproduce their kind; *if* such variations are inherited and confer some advantage to these reproducing systems— then these entities will have an enhanced chance of survival in any competitive situation. Whether or not they are the result initially of random variations the view that secondary plant metabolites (e.g., terpenoids, alkaloids and phenolics) may play a leading role in the ecological interactions between plants and herbivores fits intuitively into this intellectual framework. Because of their ubiquitous occurrence in plants more has probably been written (and speculated) about the roles of phenolic natural products in food selection and feeding deterrence than any other group of secondary metabolites. Indeed in the genesis of the general theory of biochemical co-evolution the roles of vegetable tannins (*syn.* plant polyphenols) were intensively investigated,[46,47] and considerable weight was ultimately placed on the results of these studies. Based on his own and other's work Bate-Smith[48] first succinctly stated the presumed role of vegetable tannins in plant defense:

"From the biological point of view the importance of tannins in plants lies in their effectiveness as repellants to predators, whether animal or microbial. In either case the relevant property is **astringency**, *rendering the tissues unpalatable by precipitating proteins or by immobilising enzymes, impeding invasion of the host by the parasite."*

The most prominent characteristic of polyphenols in this context, and as stated by Bate-Smith and Swain, was thus assumed to be their affinity for proteins which deleteriously affected tissue palatability and reduced the availability of food nitrogen to some herbivores.[47] Following the classic experiments of Feeny[46,47] on the detrimental effects of dietary tannins on the feeding of the winter moth (*Oerophtera brumata*) on oak leaf, the view was also espoused by Feeny that tannins constitute a unique quantitative defense for plants. Besides their strongly astringent taste[2,3,48] tannins also inhibit virtually every enzyme they are tested with *in vitro*. As a consequence, it is also widely assumed (but not proven) that they act as digestion inhibitors by the formation of immutable polyphenol-protein complexes.

Doubts and questions concerning the mode of action of polyphenols in mammalian diets and/or the idea of biochemical co-evolution have emerged from a purely physicochemical study of structure-activity relationships in protein-polyphenol complexation.[2,3,49] According to the Darwinian paradigm of molecular evolution the structure of a biochemically functional molecule is the result of the processes of selection, and presumes its preformation, either as a product of a biotic process or of a prebiotic one. Biotic preformation might include reactions catalyzed by enzymes and also non-enzymic processes. If the question is posed as to which factors have determined the molecular structure of a functional biomolecule then it is usual to argue that this is the result of selection processes directed towards the optimization of the molecule's biological function. In the case of plant polyphenols the question which must therefore be addressed is whether their structures represent the optimal ones predicted for their role as agents of chemical defense in plants. If their role in the chemical defense of plants is, as has been widely suggested, (*vide supra*), their complexation with proteins, then do their structures appear to have been fashioned in the crucible of evolution to maximize their potential to do so? According to Beart, Lilley and Haslam[49] the answer is certainly **not** an unequivocal yes! Likewise why do some plants which metabolize galloyl esters produce not one, not ten, but frequently a hundred or more closely related phenolic metabolites, if it is simply for the purposes of chemical defense. One may conclude, at this juncture, that either their mode of action is not, as has been widely suggested, to bind and complex with proteins **or** that the "purpose" of their synthesis is not directly related to plant chemical defense (and hence their ability to associate with proteins) and that the answers surrounding their origins and the circumstances of their metabolic generation may lie elsewhere.

Further factors which should be considered are illustrated by the example of the tea plant (*Camellia sinensis*). Many studies have shown that phenolic flavan-3-ol metabolites constitute ~20 percent and occasionally up to 30 percent of the dry weight of the fresh green tea shoot (fig. 4). These metabolites derive from the shikimate pathway which, in terms of the ATP equivalents required to derive just one molecule of the end-product amino acids L-phenylalanine or L-tyrosine, is the most energy demanding one of amino acid metabolism.[50] If the sole purpose of their synthesis is as agents of chemical defense then the costs to the tea plant of the biogenesis of these phenolic flavan-3-ol metabolites in such quantities is extraordinarily high.

Although it is a precept of modern chemical ecology that energy is not likely to be wasted in the production of secondary metabolites unless their is some compensating advantage to the organism in question, the structural diversity and the proliferation of plant polyphenols is best viewed as a result of there being very little selection pressure on their identity—the products bring no special advantage or disadvantage. An alternative proposition, and in the present situation perhaps more reasonable one, is that the presence of these polyphenolic metabolites does enhance, however marginally, the prospects for the plant's survival. Nevertheless, in making this proposal, it is not necessary to presume that this is the *raison d'etre* for their formation. Rather the presumption is made that their continued metabolism confers a sufficient advantage to balance the metabolic costs of their formation.

One of the alternative extant theories of secondary metabolism suggests that it provides organisms with a means of adjustment to changing circumstances. According to this view, the synthesis of enzymes designed to execute the processes of secondary metabolism thus permits the network of enzymes operative in primary / intermediary metabolism to continue to function until such time as conditions are propitious for renewed metabolic activity and growth.

5. POSTSCRIPT

We live in an age in all areas of artistic and scientific endeavor when presentation and packaging are often accorded a much greater importance than substance. Some 30 years ago when this adventure began, the author felt the terminology—*vegetable tannin*—was redolent of a bygone age and thereby inevitably consigned the chemistry (and biochemistry) of this group of substances to a rather murky, forgotten backwater of science. As time went by, he sought therefore to use the term as little as possible and to replace it with the name *plant polyphenols*. This, it was argued, was to be preferred as a descriptor for this class of higher plant secondary metabolites. Not only was it a more accurate description, but it was deemed essential if serious attempts were to be made to interpret the diverse characteristics of these substances at the molecular level within a modern scientific framework. The author is thus painfully aware of the inherent contradiction between these views and his acceptance of this third *tannin* award at the Third North American *Tannin* Conference. Likewise he is forced to reflect that the words *vegetable tannin* seem to possess everything of the aura of original sin—they decline to disappear. On the continent of Europe the "new" terminology has, however, been embraced with much greater enthusiasm, as the adoption by our French cousins of the name of their society—Groupe Polyphenols—readily testifies. Where Europe leads America must now surely follow.

REFERENCES

1. Bate-Smith, E.C.; Swain, T. *In*: Mason, H.S.; Florkin, A.M. (eds.). Comparative biochemistry, vol. 3. Academic Press, New York, p. 764 (1962).

2. Haslam, E. Plant polyphenols—vegetable tannins re-visited. Cambridge University Press, Cambridge, 230 pp. (1989).

3. Haslam, E. Practical polyphenolics: from structure to molecular recognition and physiological action. Cambridge University Press, Cambridge, 420 pp. (1998).

4. Porter, L.J. Flavans and proanthocyanidins. *In*: Harborne, J.B. (ed.). The flavanoids: Advances in research since 1980. Chapman and Hall, London, p. 21, (1988).

5. Porter, L.J. Tannins. *In*: Harborne, J.B. (ed.). Methods in plant biochemistry. Academic Press, London and New York, p. 389 (1989).

6. Bate-Smith, E.C.; Metcalfe, C.R. Leucoanthocyanins. 3. The nature and systematic distribution of tannins in dicotyledenous plants. *Biochem. J.* 55:669 (1957).

7. Mole, S. The systematic distribution of tannins in the leaves of angiosperms: a tool for ecological studies. *Biochemical Systematics and Ecology*, 21:833 (1993).

8. Robinson, R.; Robinson, G.M. A survey of anthocyanins. III. Notes on the distribution of leuco-anthocyanins. *Biochem. J.* 27:206 (1933).

9. Okuda, T.; Hatano, T.; Yoshida, T. New methods of analysing tannins. *J. Nat. Prod.* 52:1 (1989).

10. Okuda, T.; Hatano, T.; Yoshida, T. Oligomeric hydrolysable tannins, a new class of plant polyphenol. *Heterocycles*, 30:1195 (1990).

11. Haslam, E.; Cai, Y. Plant polyphenols (vegetable tannins): gallic acid metabolism. *Natural Products Reports (Roy. Soc. Chem.)* 11:44 (1994).

12. Schmidt, O.Th.; Mayer, W. Naturliche gerbstoffe. *Angew. Chem.* 68:103 (1956).

13. Glombitza, K.; Gerstberger; G. Phlorotannins with dibenzodioxin structural elements from the brown alga *Eisenia arborea*. *Phytochemistry* 24:543 (1985).

14. Ragan, M.A. The high molecular weight polyphloroglucinols of the marine brown alga *Fucus vesiculosus*. *Can. J. Chem.* 63:294 (1985).

15. McInnes, A.G.; Smith, D.G.; Walter, J.A.; Ragan, M.A. The high molecular weight polyphloroglucinols of the marine brown alga *Fucus vesiculosus*: ^{1}H and ^{13}C nuclear magnetic resonance spectroscopy. *Can. J. Chem.* 63:304 (1985).

16. Hillis, W.E.; Swain, T. The phenolic constituents of *Prunus domestica*. II. The analysis of tissues of the Victoria plum tree. *J. Sci. Food Agric.* 10:135 (1959).

17. Bate-Smith, E.C. Tannins of the herbaceous Leguminosae. *Phytochemistry* 12:1809 (1973).

18. Shen, Z.; Haslam, E.; Falshaw, C.P.; Begley, M.J. Procyanidins and polyphenols of *Larix gmelini* bark. *Phytochemistry* 25:2629 (1986).

19. Somers, T.S. The polymeric nature of wine pigments. *Phytochemistry* 10:2175 (1971).

20. Roberts, E.A.H. Economic importance of flavonoid substances: tea fermentation. *In*: Geissman, T.A. (ed.). The chemistry of flavanoid compounds, Pergamon Press, Oxford, London and New York, p. 468 (1962).

21. Mayer, W.; Hoffmann, E.H.; Losch, N.; Wolf, H.; Wolter, B.; Schilling, G. Dehydrierungsreaktionem mit gallusaureestern. *Liebig's Annalen de Chemie* :929–938 (1984).

22. Waterhouse, A.L. Wine and heart disease. *Chem. Ind.* p. 338 (1995).

23. Haslam, E. Natural polyphenols (vegetable tannins) as drugs: possible modes of action. *J. Nat. Prod.* 59:205 (1996).

24. Haslam, E. The metabolism of gallic and hexahydroxydiphenic acid in plants. *Fortschritt. Chem. Org. Naturstoffe* 41:1 (1982).

25. Haslam, E. Hydroxybenzoic acids and the enigma of gallic acid. *In*: Conn, E.E. (ed.). The shikimic acid pathway. *Recent Advances in Phytochemistry* 20:163 (1986).

26. Zenk, M.H. Zur frage de biosynthese von gallusaure. *Z. Naturforsch.* 19B:83 (1964).

27. Knowles, P.F.; Haslam, E.; Haworth, R.D. Gallotannins. Part 4. The biosynthesis of gallic acid. *J. Chem. Soc.* :1854 (1961).

28. Cornthwaite, D.C.; Haslam, E. Gallotannins. Part 9. The biosynthesis of gallic acid in *Rhus typhina*. *J. Chem. Soc.* :3008 (1965).

29. Dewick, P.M.; Haslam, E. Phenol biosynthesis in higher plants—gallic acid. *Biochem. J.* 113:537 (1969).

30. Gross, G.G.; Denzel, K.; Schilling, G. Biosynthesis of gallotannins. Enzymatic conversion of 1,6-digalloylglucose to 1,2,6-trigalloylglucose. *Planta* 176:135 (1988).

31. Gross, G.G.; Denzel, K. Biosynthesis of gallotannins. β-Glucogallin dependent galloylation of 1,6-digalloylglucose to 1,2,6-trigalloylglucose. *Z. Naturforsch.* 46C:389 (1991).

32. Richardson, A.C.; Williams, J.M. Selective acylation of pyranosides. I. Benzoylation of α-methylpyranosides of mannose, glucose and galactose. *Tetrahedron* 23:1369 (1967).

33. Haslam, E. Secondary metabolism—evolution and function; products or processes? *Chemoecology* 5/6:89 (1994/1995).

34. Hahlbrock, K.; Scheel, D. Physiology and molecular biology of phenylpropanoid metabolism. *Ann. Rev. Plant Physiol.* 40:347 (1989).

35. Stafford, H.A.; Lester, H.H. Flavan-3-ol biosynthesis. The conversion of (+)-dihydroquercetin and flavan-3,4-diol (leucocyanidin) to (+)-catechin by reductases extracted from cell suspension cultures of Douglas fir. *Plant Physiol.* 76:184 (1984).

36. Stafford, H.A.; Lester, H.H. Flavan-3-ol biosynthesis. The conversion of (+)-dihydromyricetin to its flavan-3,4-diol (leucodelphinidin) and (+)-gallocatechin by reductases extracted from cell suspension cultures of *Gingko biloba* and *Pseudostuga menziesii*. *Plant Physiol.* 76:184 (1985).

37. Stafford, H.A.; Lester, H.H.; Porter, L.J. Chemical and enzymic synthesis of monomeric procyanidins (leucocyanidins) from 2R,3R-dihydroquercetin. *Phytochemistry* 24:333 (1985).

38. Haslam, E. Biogenetically patterned synthesis of procyanidins. *J. Chem. Soc. Chemical Commun.* :594 (1974).

39. Hemingway, R.W.; Foo, L.Y. Quinone methide intermediates in procyanidin synthesis. *J. Chem. Soc. Chemical Commun.* :1035 (1983).

40. Haslam, E. Symmetry and promiscuity in proanthocyanidin biochemistry. *Phytochemistry* 16:1625 (1977).

41. Fraenkel, G. The raison d'etre of secondary plant substances. *Science* 129:1466 (1959).

42. Feeny, P. Plant apparency in chemical defence. *Rec. Adv. Phytochemistry* 10:1 (1976).

43. Rhoades, D.F.; Cates, R.G. Toward a general theory of plant herbivore chemistry. *Rec. Adv. Phytochemistry* 10:168 (1976).

44. Muller, C.H. Co-evolution. *Science* 165:415 (1969).

45. Cairns-Smith, A.G. Seven clues to the origin of life. Cambridge University Press, Cambridge (1985).

46. Feeny, P. Effects of oak leaf tannins on larval growth of the winter moth *Operophthera brumata*. *J. Insect Physiol.* 14:805 (1968).

47. Feeny, P. Seasonal changes in oak leaf tannins and nutrients as a cause of spring feeding by winter moth caterpillars. *Ecology* 51:565 (1970).

48. Bate-Smith, E.C. Haemanalysis of tannins—the concept of relative astringency. *Phytochemistry* 12:907 (1973).

49. Beart, J.E.; Haslam, E.; Lilley, T.E. Plant polyphenols—secondary metabolism and chemical defence. *Phytochemistry* 24:33 (1985).

50. Atkinson, D.E. Cellular energy metabolism and its regulation. Academic Press, New York (1977).

HYDROLYZABLE TANNINS

ANALYSIS OF GALLIC ACID BIOSYNTHESIS VIA QUANTITATIVE PREDICTION OF ISOTOPE LABELING PATTERNS

Ingo Werner, Adelbert Bacher, and Wolfgang Eisenreich

Department of Organic Chemistry and Biochemistry
Technische Universität München
D-85747 Garching
GERMANY

1. INTRODUCTION

Gallic acid (3,4,5-trihyroxybenzoic acid) serves as a basic building block of gallotannins and ellagitannins, which are abundant metabolites in plants and fungi. Tannins are produced by a wide variety of plants such as oak, acer, tea, and sumach (*R. typhina*). Several important functions such as stabilization of the cell wall and protection against herbivors have been attributed to tannins.[1,2] More recently, certain tannins were reported to have antineoplastic activity, possibly via antioxidant activity.

Studies on the biosynthesis of gallic acid were initiated four decades ago. The controversial results of earlier studies have been reviewed repeatedly.[3,4] Briefly, it has been proposed by several authors that gallic acid (**6**) is formed from the phenylpropanoid pathway via caffeic acid (**9**) or via trihydroxycinnamic acid (**10**) (fig. 1).[5–9] Alternatively, it has been reported that an early shikimate intermediate, such as 5-dehydroshikimic acid (**5**) or protocatechuic acid (**8**), could be directly converted to gallic acid.[10–17] As shown in fig. 1, the carboxylic atom of gallic acid would be derived from C-3 of phosphoenolpyruvate (indicated by □) via the phenylpropanoid pathway or alternatively from C-1 of phosphoenolpyruvate (indicated by *) via a more direct formation with an early shikimate derivative as the committed precursor.

2. THE EXPERIMENTAL APPROACH

Biosynthetic pathways can be studied by the retrobiosynthetic approach (fig. 2) which has recently been used in various biological systems[18–22] (for review see ref. 23). The use of stable isotopes in incorporation experiments is a general concept for elucidation of biosynthetic pathways. In the traditional approach, an

Plant Polyphenols 2: Chemistry, Biology, Pharmacology, Ecology, Edited by
Gross et al. Kluwer Academic / Plenum Publishers, New York, 1999

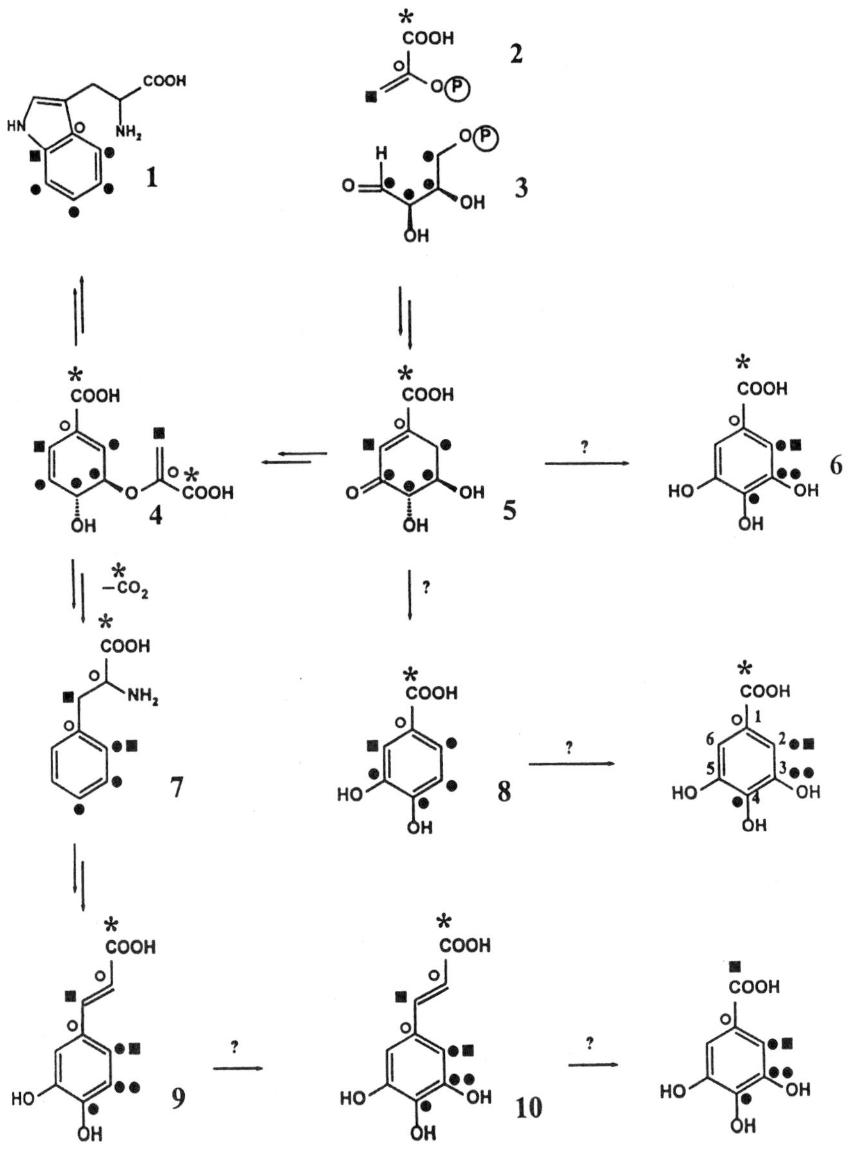

Figure 1. Hypothetical pathways for the biosynthesis of gallic acid (**6**).[5–17] Carbon atoms of aromatic amino acids biosynthetically equivalent to carbon atoms of phosphoenolpyruvate (**2**) and erythrose 4-phosphate (**3**) are indicated by symbols.

educated guess on the structure of a presumed precursor is derived from the structure of the metabolite under consideration of the plausibility of biochemical reactions leading from the presumed precursor to the target compound. The hypothetical precursor (compound Y, fig. 3A) is then proffered in isotope-labeled form to an organism or a cell culture. Incorporation of the isotope label into the downstream metabolite is then accepted as evidence for a more or less precursor

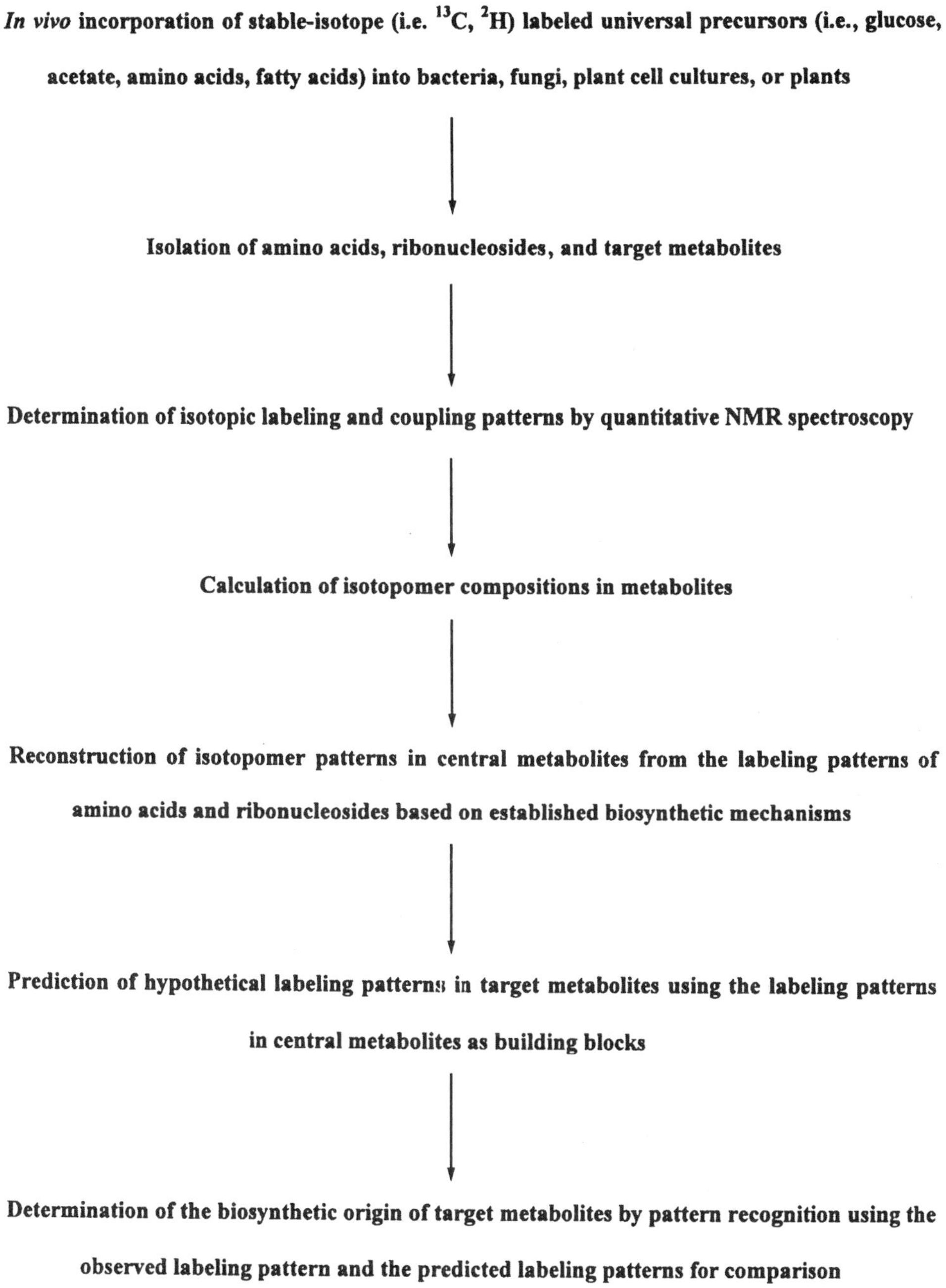

In vivo incorporation of stable-isotope (i.e. ^{13}C, ^{2}H) labeled universal precursors (i.e., glucose, acetate, amino acids, fatty acids) into bacteria, fungi, plant cell cultures, or plants

↓

Isolation of amino acids, ribonucleosides, and target metabolites

↓

Determination of isotopic labeling and coupling patterns by quantitative NMR spectroscopy

↓

Calculation of isotopomer compositions in metabolites

↓

Reconstruction of isotopomer patterns in central metabolites from the labeling patterns of amino acids and ribonucleosides based on established biosynthetic mechanisms

↓

Prediction of hypothetical labeling patterns in target metabolites using the labeling patterns in central metabolites as building blocks

↓

Determination of the biosynthetic origin of target metabolites by pattern recognition using the observed labeling pattern and the predicted labeling patterns for comparison

Figure 2. Experimental protocol for the elucidation of biosynthetic pathways by retrobiosynthetic NMR analysis.

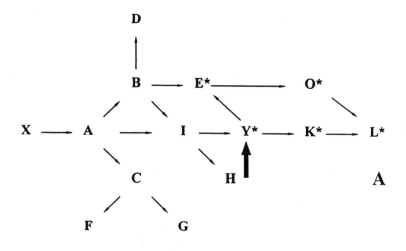

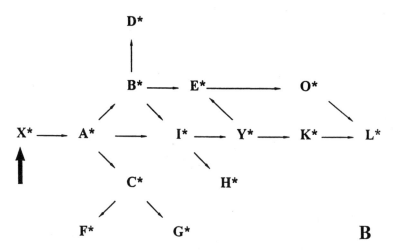

Figure 3. Strategies for *in vivo* labeling studies; A, by a conventional approach; the precursor Y is in close biogenetical proximity to the target metabolite, L; B, by the approach used in this study; the universal precursor X (i.e., glucose) is incorporated into a set of metabolites and yields a complex labeling pattern in the target molecule L. The symbol * indicates labeling of metabolites.

role of the proffered compound (from compound Y via intermediate K to the terminal product L in fig. 3A).

In cases with low incorporation rates, this assumption can be incorrect, since the precursor might also divert label to the general metabolism and the label might reach the target molecule via a completely different route (from compound Y via intermediates E and O to the terminal product L in fig. 3A).

An alternative approach is the use of general precursors, such as ^{13}C labeled glucose. It is obvious that in these cases the precursors serve as substrates for a variety of metabolic pathways. Consequently, the isotope label will be diverted to many if not to all cellular metabolites and complex labeling patterns will be generated in downstream metabolites (fig. 3B).

Starting from totally ^{13}C labeled glucose diluted by a large excess of unlabeled glucose, complex isotopomer mixtures are generated biosynthetically. Most importantly, they include isotopomers with adjacent carbon-13 atoms. Multiply ^{13}C labeled isotopomers can be diagnosed by one- and two-dimensional ^{13}C-NMR spectroscopy with high specifity and sensitivity. Specifically, ^{13}C atoms which were jointly transferred from the multiply ^{13}C labeled precursor appear as contiguous ^{13}C spin systems which can be diagnosed elegantly by modern NMR technology.

As an example, the signal of the symmetrical carbon pair C-3 and C-5 of gallic acid obtained after incorporation of [U-$^{13}C_6$]glucose (diluted with unlabeled glucose) into *P. blakesleeanus* gave no less than 25 lines[24] (fig. 4). By spectral simulation, deconvolution and coherence transfer methods (INADEQUATE, ^{13}C TOCSY), these lines were attributed to 5 multiply ^{13}C labeled isotopomers which are shown in figure 4. From the relative intensities of the signals, the isotopomers can be determined quantitatively. The data show that the totally carbon-13 labeled glucose contributed several intact blocks of labeled atoms to gallic acid with differing efficacies. Attributing these complex isotopomer patterns to a certain biosynthetic pathway is difficult if not impossible. However, the interpretation of those complex isotopomer compositions is possible by a pattern recognition approach described below.

It is obvious that all biosynthetic pathways utilize basic building blocks derived from central metabolic pathways, such as glycolysis, photosynthesis and others. Intermediates from those pathways are recruited for the biosynthesis of secondary metabolites via primary metabolites, such as amino acids or ribonucleosides, or via more direct routes (fig. 5). As a consequence, the isotope labeling patterns of downstream metabolites reflect the labeling patterns of the building blocks from which they were biosynthetically derived. More specifically, the labeling patterns of downstream metabolites can be interpreted as a mosaic of the patterns of their respective building blocks.

The labeling patterns of central intermediates can not be observed directly, since their steady state concentrations are low, and it is therefore not possible to isolate them in sufficient amounts for NMR analysis. However, on basis of established biosynthetic pathways, the labeling patterns of the elusive central metabolites can be reconstructed from the labeling data of amino acids and ribonucleosides, as shown in figures 6 and 7 for the fungus *P. blakesleeanus*. For example, the ^{13}C labeling pattern of ribose 5-phosphate (**13**) is reflected in histidine (**14**), the patterns of phosphoenolpyruvate (**2**) as well as the pattern of erythrose 4-phosphate (**3**) are reflected in the aromatic amino acids, and the pattern of acetyl-CoA (**11**) can be gleaned from leucine (**12**).

Based on the labeling patterns shown in figure 7, the labeling patterns of downstream metabolites can now be predicted in the forward biosynthetic direction via

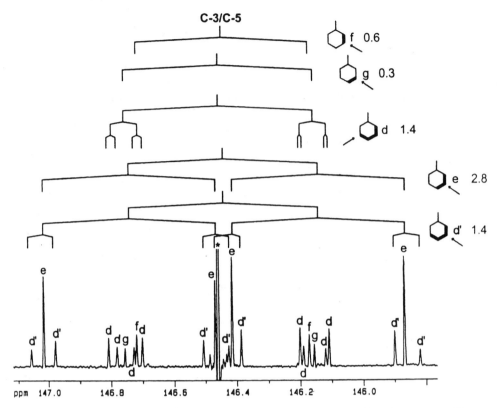

Figure 4. ^{1}H decoupled ^{13}C NMR signals of the symmetrical carbon atoms C3 and C5 of gallic acid from *P. blakesleeanus* after incorporation of [U-[13]C$_6$]glucose diluted with unlabeled glucose (1:25, w/w). ^{13}C-labeled isotopomers of gallic acid are shown schematically. Multiple ^{13}C labeling is indicated by bold bars. Arrows indicate the observed carbon atom in each respective isotopomer. The numbers indicate isotopomer contributions from [U-[13]C$_6$]glucose in mol percent.

different hypothetical pathways. A comparison between the observed labeling patterns and the predicted labeling patterns should then discriminate between alternative hypotheses. The application of the concept to the biosynthesis of gallic acid is illustrated below.

3. THE FUNGAL PATHWAY OF GALLIC ACID BIOSYNTHESIS

We will first consider experiments with [1-^{13}C]glucose. Shake cultures of *P. blakesleeanus* were grown at 25 °C for 6 days in medium containing [1-^{13}C]glucose (99.5% [13]C enrichment) as carbon source.[24] Gallic acid (**6**) was isolated from the culture medium and the cell mass.[24] Phenylalanine (**7**) and tyrosine (**15**) were isolated after acid hydrolysis of the biomass, and tryptophan (**1**) was isolated after alkaline hydrolysis of cell protein.[24] [13]C Enrichments were determined for each carbon atom in gallic acid and amino acids as described earlier (fig. 8).[18,24]

glycolysis

photosynthesis

gluconeogenesis

lipid catabolism

amino acid catabolism

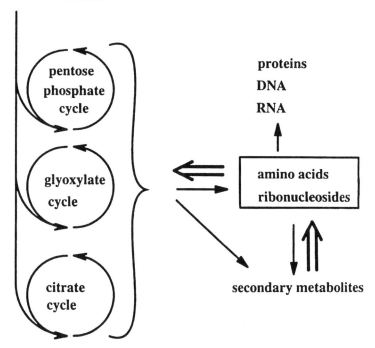

Figure 5. Biosynthesis of secondary metabolites from central metabolic intermediates and/or primary metabolites. Reconstruction of labeling patterns and biosynthetic pathways by pattern recognition is indicated by the retro-arrow.

Due to the symmetry of the aromatic rings of phenylalanine, tyrosine and gallic acid, the ring carbon atoms 2/6 and 3/5 yield averaged ^{13}C abundances, although they have different biosynthetic origins. Tryptophan reflects the original, i.e. non-symmetrical, labeling patterns of the shikimate ring system, and the labeling pattern of erythrose 4-phosphate (**3**) can thus be reconstructed by a retrobiosynthetic approach on basis of the tryptophan biosynthetic pathway (fig. 8). The labeling pattern of phosphoenolpyruvate (**2**) can be deduced from the labeling patterns of the side chains of phenylalanine and tyrosine.

The observed labeling patterns are explained on basis of carbohydrate metabolic pathways (for details see Ref. 24). Briefly, label from [1-^{13}C]glucose is diverted to C-3 of triose phosphate type metabolites by the glycolytic pathway. This results in the high ^{13}C enrichment of the β position of phenylalanine and tyrosine reflecting C-3 of phosphoenolpyruvate (22.3% ^{13}C). The observed ^{13}C labeling in the

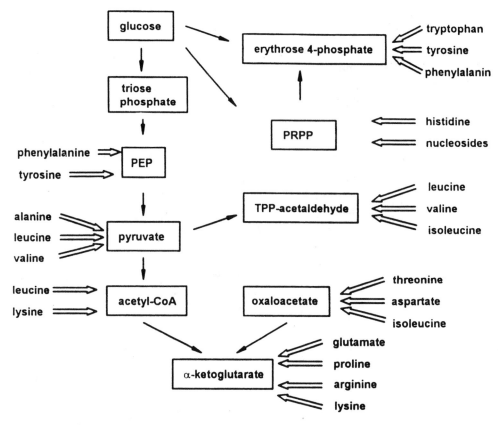

Figure 6. Reconstruction of labeling patterns of central metabolites from the labeling pattern of amino acids and ribonucleosides in the fungus *P. blakesleeanus* on the basis of established mechanisms.

4-position of erythrose 4-phosphate (17.4% ^{13}C) can be explained by futile cycling via the glycolytic/glucogenetic pathways and/or via the mannitol pathway. Specifically, regeneration of glucose from triose phosphate could divert ^{13}C label to the 6-position of glucose. Similarly, the reversible conversion of glucose to mannitol could also divert label to the 6-position of glucose. The resulting [6-^{13}C]glucose 6-phosphate could then be converted to [4-^{13}C]erythrose 4-phosphate via the pentose phosphate pathway with or without contribution of oxidative decarboxylation.

The labeling pattern of phosphoenolpyruvate (**2**), erythrose 4-phosphate (**3**), and phenylalanine (**7**) were then used to predict the labeling patterns of gallic acid (**6**) via the phenylpropanoid pathway and via an early shikimate derivative. As shown in figure 9, the observed labeling pattern of gallic acid closely agreed with the prediction via an early shikimate derivative, such as 5-dehydroshikimate (**16**), but not with the prediction via the phenylpropanoid pathway. Thus, a direct

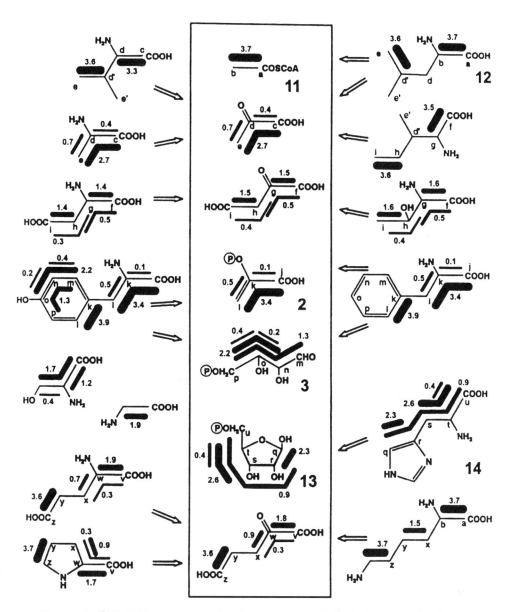

Figure 7. [13]C labeling patterns of amino acids from *P. blakesleeanus* after incorporation of [U-[13]C₆]glucose diluted with unlabeled glucose (1:25, w/w). Contiguous ^{13}C-labeled isotopomers are indicated by bold lines connecting ^{13}C atoms. The line widths reflect the relative contribution of the respective isotopomers. The numbers indicate isotopomer contributions from [U-[13]C₆]glucose in mol percent.

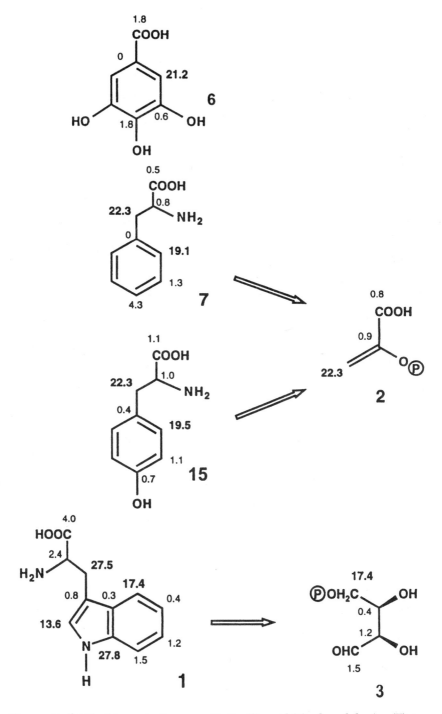

Figure 8. ^{13}C Enrichments (in percent) of gallic acid (**6**), phenylalanine (**7**), tyrosine (**15**), and tryptophan (**1**) from *P. blakesleeanus* cultured with [1-^{13}C]glucose (99.5% ^{13}C abundance). The labeling patterns of erythrose 4-phosphate (**3**) and phosphoenolpyruvate (**2**) were reconstructed from the amino acids based on conventional mechanisms of the shikimate biosynthetic pathway (for details see fig. 1).

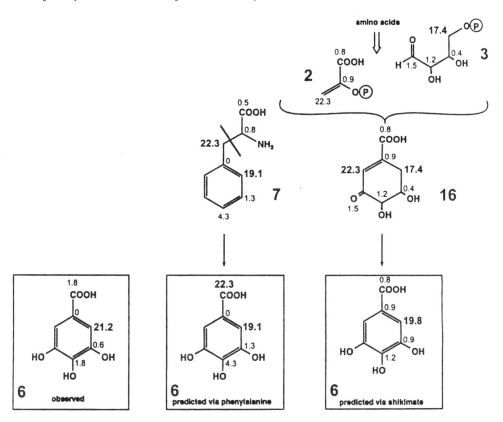

Figure 9. Observed and predicted labeling patterns for gallic acid from *P. blakesleeanus* supplied with [1-¹³C]glucose (for details see fig. 8).

conversion of 5-dehydroshikimate into gallic acid was in line with the experimental data.

Attempts to use the same experimental approach for the study of gallic acid biosynthesis in leaves of the tree, *R. typhina*, were unsuccessful. The amount of ¹³C label diverted to gallic acid and phenylalanine was so low (less than 2% ¹³C abundance in the most highly labeled carbon atoms) that it was impossible to draw firm conclusions. This was apparently due to the fact that exogeneous glucose supplied to the intact leaves via the leaf stem was poorly metabolized. The NMR spectra were therefore dominated by metabolic products which had been formed prior to glucose application. A much more sensitive technique was therefore required. As shown below, the sensitivity problem could be overcome by the use of a mixture of [U-¹³C₆]glucose and unlabeled glucose. This experimental approach will be described first with the fungal culture and then with plant material.

A shake flask culture of *P. blakesleeanus* was supplied with a mixture of [U-¹³C₆]glucose and unlabeled glucose at a ratio of 1:25.[24] The NMR signals of metabolites isolated from this culture are complex multiplets as a consequence of ¹³C¹³C coupling (for example, see figs. 4 and 10A).

The satellite patterns can be resolved by numerical simulation of the ^{13}C-^{13}C coupling patterns for each respective isotopomer in the mixture (as examples, see fig. 10B–D). As shown in figure 10, the chemical shift positions as well as the amplitude modulations due to higher order coupling show perfect agreement between the simulated and experimental spectra. It should be noted that the satellites arising from $^{13}C^{13}C$ coupling are shifted to high field by comparison with the singlets attributed to molecules with natural ^{13}C abundance. This is due to the heavy isotope shifts caused by ^{13}C atoms adjacent to the respective observed ^{13}C atom in the multiply labeled isotopomeric species. The heavy isotope shifts of individual nuclei in specific isotopomers as deduced from comparison of experimental signals with spectral simulations are summarized in figure 11. The various isotopomers in the biolabeled mixture are symbolized by bold lines indicating multiple

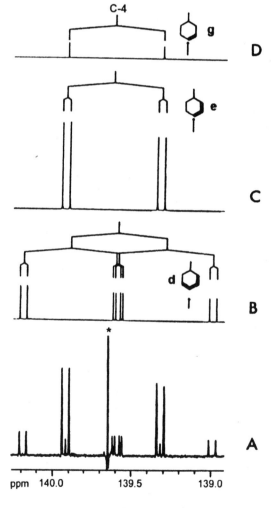

Figure 10. ^{13}C NMR signature of C4 of gallic acid from *P. blakesleeanus* cultured with [U-$^{13}C_6$]glucose diluted with unlabeled glucose (1:25, w/w). A, ^{1}H-decoupled spectrum; asterisk indicates natural abundance background. B–D, simulated signals of isotopomers based on NMR parameters shown in figure 11. The respective isotopomers are shown schematically by bold lines connecting contiguously adjacent ^{13}C atoms; and the observed carbon is marked by an arrow. The amplitudes of the simulated spectra were adjusted to reflect the relative abundance of each respective isotopomer in the mixture.

contiguous carbon atoms. A single adjacent ^{13}C atom typically results in a high field shift in the range of 2 Hz; two adjacent ^{13}C atoms are conducive to a high field shift of the observed carbon in the range of 4 Hz.

Single bond and multiple bond ^{13}C-^{13}C coupling constants are summarized in figure 11. It should be noted that the chemical shift degeneracy of the ring carbon atoms 3 and 5 is broken by the heavy isotope shift in isotopomer d where C-3 has a heavy isotope shift of –4.9 Hz and C-5 has a heavy isotope shift of –2.0 Hz. This results in higher order coupling between C-3 and C-5 which are no longer homotopic.

The fraction of each isotopomer in the mixture can be obtained by integration of the signal groups representing each individual isotopomer. Some isotopomers can be estimated independently from different spectral patterns. Thus, the abundance of isotopomer d can be diagnosed independently from the signatures of C-2, C-3, and C-4. The quantitative data obtained independently from different parts of the spectrum are in very good agreement.

The isotopomer composition of gallic acid (**6**) from the experiment with [U-^{13}C$_6$]glucose is summarized in figure 12. Based on the labeling data of phenylalanine (**7**), phosphoenolpyruvate and erythrose 4-phosphate (**2** and **3**, respectively, as reconstructed from the labeling data of aromatic amino acids), the coupling patterns of gallic acid were again predicted following the hypotheses via the phenylpropanoid pathway or via an early shikimate derivative (fig. 12). By comparison of the predicted patterns with the observed gallic acid labeling pattern, it is obvious that the bulk amount of gallic acid was synthesized in the fungus via an early shikimate intermediate and not via the phenylpropanoid pathway.

4. THE PLANT PATHWAY OF GALLIC ACID BIOSYNTHESIS

Initial experiments had shown that the incorporation of [1-^{13}C]glucose into leaves of *R. thyphina* is low. The poor utilization of exogenous nutrients had also hampered earlier studies in the literature and is probably the major reason for the conflicting reports on gallic acid biosynthesis. The preliminary experiments with fungal cultures indicated that the sensitivity problem could be addressed by the use of totally labeled glucose conducing to the formation of multiply labeled isotopomers. The signal contributions of the multiply labeled isotopomers are spread out in the frequency domain by ^{13}C-^{13}C coupling and are thus separated from the signals of molecules representing the high background of preformed gallic acid with natural ^{13}C abundance.

A large number of preliminary experiments (described in detail in Ref. 24) were required in order to find experimental conditions affording appropriate transfer of ^{13}C from exogenous glucose into plant metabolites. The optimized experimental conditions involved 150 young leaves of *R. typhina* (60 g) which were supplied with a solution (300 ml) containing 250 mg of [U-^{13}C$_6$]glucose (99.6% ^{13}C enrichment) and 6 g of natural abundance glucose.[24] The leaves were incubated in the dark at 18 °C and 60 percent humidity for 33 days. Gallic acid (50 mg) and amino acids were isolated by the standard procedure.

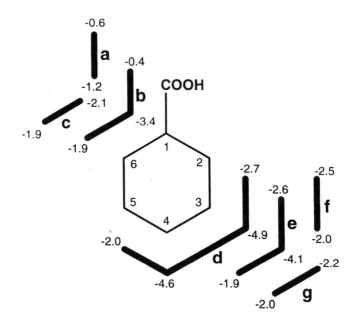

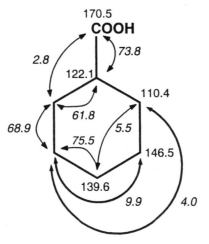

Figure 11. ^{13}C NMR parameters of gallic acid. The carbon connectivity of gallic acid is shown schematically. Top, heavy isotope shifts (Hz) in individual isotopomers (a–g) containing 2 or more ^{13}C atoms; contiguous ^{13}C labeled isotopomers are shown by bold lines. Bottom, ^{13}C chemical shifts (ppm) and [13]C ^{13}C coupling constants indicated by arrows (Hz, italic letters). Coupling between C-3 and C-5 occurs when the chemical shift degeneracy of these carbon atoms is broken by isotope shift effects.

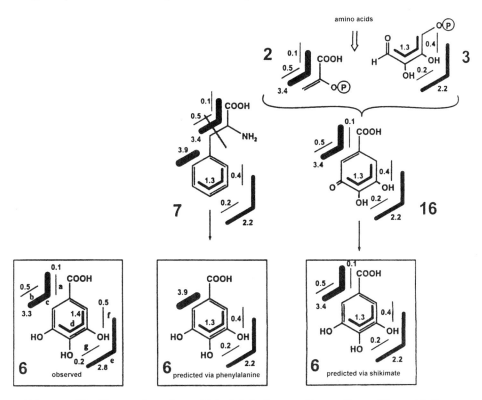

Figure 12. Observed and predicted labeling patterns for gallic acid from *P. blakesleanus* supplied with [U-¹³C₆]glucose diluted with unlabeled glucose (1:25, w/w). For details see figure 7.

The complex ¹³C NMR signals of the carboxylic carbon and ring atom C-1 of gallic acid are shown in figure 13. In contrast to the experiments with the fungal cultures (figs. 4 and 10), the relative intensity of the coupling patterns is low as compared to the contribution of isotopomers carrying a single ¹³C atom (marked by asterisks and truncated in fig. 13), which represent the contribution of natural abundance gallic acid. The large fraction of natural abundance material essentially represents gallic acid which had already been present at the beginning of glucose feeding.

A quantitative analysis of the biosynthetic gallic acid spectra yields the isotopomer composition shown in figure 14. Only isotopomers which were biosynthesized *de novo* from the proffered [U-¹³C₆]glucose are shown. The isotopomer composition of biosynthetic tyrosine (**15**) was determined by the same experimental approach and is also shown in figure 14. The isotopomer patterns of the aromatic rings of biosynthetic gallic acid (**16**) and tyrosine are similar and leave no doubt that the aromatic rings of both compounds originate from the shikimate pathway. Surprisingly, the incorporation of ¹³C-labeled glucose into the side chain of phenylalanine exceeds the incorporation into the ring by a factor of

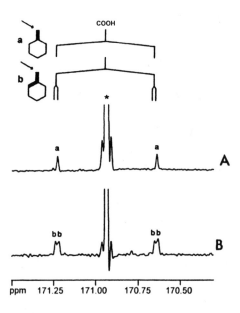

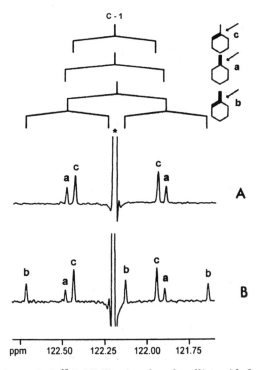

Figure 13. ^{1}H decoupled ^{13}C NMR signals of gallic acid from *R. typhina*. A, natural abundance sample; B, sample from *R. typhina* supplied with [U-^{13}C$_6$]glucose diluted with unlabeled glucose (1:25, w/w). The uncoupled signals are truncated. The high sensitivity and the good signal to noise ratio was achieved by 83,600 scans. For other details see legend to figure 4.

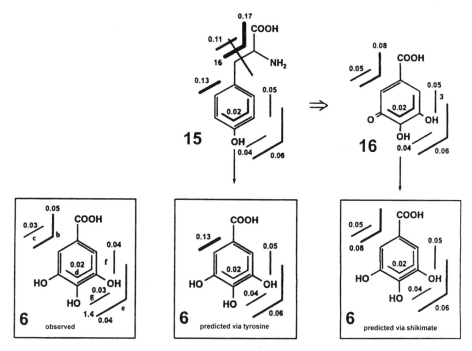

Figure 14. Observed and predicted labeling patterns of gallic acid from leaves of *R. typhina* supplied with [U-^{13}C$_6$]glucose diluted with unlabeled glucose (1:25, w/w). For details see figure 7.

2. Earlier studies had shown that plant tissues can contain substantial amounts of shikimate pathway intermediates such as shikimate and quinate.[25] We conclude from the labeling data that preformed early shikimate derivates were converted to chorismate and further to tyrosine during the [^{13}C]glucose feeding period. This would result in the formation of the aromatic amino acid from a preformed ring system at natural ^{13}C abundance by the addition of a enoylpyruvate side chain derived from the proffered ^{13}C labeled glucose mixture.

The prediction of gallic acid labeling patterns from [U-^{13}C$_6$]glucose via the phenylpropanoid pathway and via an early shikimate derivative is shown in figure 14. The observed labeling pattern in gallic acid agrees well with the latter hypothesis. Most notably, gallic acid showed a significant amount of the triple-labeled [carboxy, 1,6-^{13}C$_3$]isotopomer. In analogy with the *Phycomyces* experiment, it can be concluded that phenylalanine and tyrosine do not serve as major intermediates in the biosynthetic pathway of gallic acid in *R. typhina*.

5. CONCLUSIONS

The biosynthesis of gallic acid was studied in cultures of the fungus, *Phycomyces blakesleeanus*, and in leaves of the tree, *Rhus typhina*. Fungal cultures were grown with [1-^{13}C]glucose or with a mixture of unlabeled glucose and

[U-^{13}C$_6$]glucose. Young leaves of *R. typhina* were kept in an incubation chamber and were supplied with a solution containing a mixture of unlabeled glucose and [U-^{13}C$_6$]glucose via the leaf stem for 33 days. Gallic acid and amino acids were isolated and isotopomer compositions were determined by NMR spectroscopy.

From the ^{13}C labeling patterns of aromatic amino acids, the labeling patterns of their precursors, phosphoenolpyruvate and erythrose 4-phosphate, were reconstruced by a retrobiosynthetic approach. The labeling patterns of phosphoenolpyruvate and erythrose 4-phosphate were then used to predict labeling patterns of gallic acid biosynthesized (i) via the phenylpropanoid pathway and (ii) via an early shikimate derivative.

Comparison of observed and predicted labeling patterns showed that the bulk amount of gallic acid was biosynthesized in both experimental systems via an intermediate of the shikimate pathway prior to prephenate or arogenate, most probably 5-dehydroshikimate. Formation of gallic acid from phenylalanine via the lignin precursors, caffeic acid or 3,4,5-trihydroxycinammic acid, were ruled out as major pathways.

ACKNOWLEDGMENTS

The expert help of F. Wendling and A. Werner with the preparation of the manuscript is gratefully acknowledged. We thank Professor M.H. Zenk, Munich, for support and stimulating discussions and Professor P. Galland, Marburg, for a culture of *P. blakesleeanus*.

REFERENCES

1. Bu'Lock, J.D.; Powell, A.J. Secondary metabolism: An explanation in terms of induced enzyme mechanisms. *Experientia* 21:55–56 (1965).

2. Beart, J.E.; Lilley, T.H.; Haslam, E. Plant polyphenols—secondary metabolism and chemical defence: Some observations. *Phytochemistry* 24:33–38 (1985).

3. Gross, G.G. Enzymatic synthesis of gallotannins and related compounds. *In*: Stafford, H.A.; Ibrahim R.K. (eds.) Phenolic metabolism in plants. Plenum Press, New York, pp 297–324 (1992).

4. Haslam, E. Gallic acid and its metabolites. *Basic Life Sci.* 59:169–194 (1992).

5. Brucker, W. Zur Biogenese der gallus- und protocatechusäure durch *Phycomyces blakesleeanus. Planta* 48:627–630 (1957).

6. Brucker, W.; Drehmann, U. Zur phenolcarbonsäurebildung von phycomyces aus ^{14}C-markierten substraten. *Archiv f. Mikrobiol.* 30:396–408 (1958).

7. Brucker, W.; Hashem, M. Zur biogenese von shikimisäure und ihrer beziehung zur phenolkarbonsäure-bildung durch *Phycomyces. Ber. Dt. Bot. Gesell.* 75:3–13 (1962).

8. Zenk, M.H. Zur frage der biosynthese von gallussäure. *Z. Naturforsch.* 19b:83–84 (1964).

9. Neish, A.C.; Towers, G.H.N.; Chen, D.; El-Basyouni, S.Z.; Ibrahim, R.K. The biosynthesis of hydrobenzoic acids in higher plants. *Phytochemistry* 3:485–492 (1964).

10. Haslam, E.; Haworth, R.D.; Knowles, P.F. Gallotannins. IV. The biosynthesis of gallic acid. *J. Chem. Soc.* :1854–1859 (1961).

11. Dewick, P.M.; Haslam, E. Biosynthesis of gallic acid and caffeic acid. *J. Chem. Soc., Chem. Commun.* 673–675 (1968).

12. Cornthwaite, D.C.; Haslam, E. Gallotannins. IX. The biosynthesis of gallic acid in *Rhus typhina. J. Chem. Soc.* :3008–3011 (1965).

13. Conn, E.E.; Swain, T. Biosynthesis of gallic acid in higher plants. *Chem. Ind.* :592–593 (1961).

14. Marigo, G.; Alibert, G.; Boudet, A. Biosynthesis of phenols compounds in higher plants: origin of protocatechuic and gallic acids in *Quercus pedunculata. C. R. Acad. Sci. Ser. D* 269:1852–1854 (1969).

15. Dewick, P.M.; Haslam, E. Phenol biosynthesis in higher plants. *Biochem. J.* 113:537–542 (1969).

16. Ishikura, N.; Hayashida, S.; Tazaki, K. Biosynthesis of gallic and ellagic acids with ^{14}C-labeled compound in *Acer* and *Rhus* leaves. *Bot. Mag.* 97:355–367 (1984).

17. Haslam, E.; Cai, Y. Plant polyphenols (vegetable tannins): gallic acid metabolism. *Natural Product Reports* 11:41–66 (1994).

18. Strauß, G.; Eisenreich, W.; Bacher, A.; Fuchs, G. ^{13}C-NMR study of autotrophic CO_2 fixation pathways in the sulfur-reducing archaebacterium *Thermoproteus neutrophilus* and in the phototrophic eubacterium *Chloroflexus aurantiacus. Eur. J. Biochem.* 205:853–866 (1992).

19. Eisenreich, W.; Menhard, B.; Hylands, P.J.; Zenk, M.H.; Bacher, A. Studies on the biosynthesis of taxol: the taxane carbon skeleton is not of mevalonoid origin. *Proc. Natl. Acad. Sci. USA* 93:6431–6436 (1996).

20. Eisenreich, W.; Sagner, S.; Zenk, M.H.; Bacher A. Monoterpenoid essential oils are not of mevalonoid origin. *Tetrahedron Lett.* 38:3889–3892 (1997).

21. Rieder, C.; Strauß, G.; Fuchs, G.; Arigoni, D.; Bacher, A.; Eisenreich, W. Biosynthesis of the diterpene verrucosan-2β-ol in the phototrophic eubacterium *Chloroflexus aurantiacus.* A retrobiosynthetic NMR study. *J. Biol. Chem.* 273:18099–18108 (1998).

22. Rieder, C.; Eisenreich, W.; O'Brien, J.; Richter, G.; Götze, E.; Boyle, P.; Blanchard, S.; Bacher, A.; Simon, H. Rearrangement reactions in the biosynthesis of molybdopterin. An NMR study with multiply ^{13}C/^{15}N labelled precursors. *Eur. J. Biochem.* 255:24–36 (1998).

23. Bacher, A.; Rieder, C.; Eichinger, D.; Fuchs, G.; Arigoni, D.; Eisenreich, W. Elucidation of biosynthetic pathways and metabolic flux patterns via retrobiosynthetic NMR analysis. *FEMS Microbiol. Rev.*, 22:567–598 (1998).

24. Werner, I.; Bacher, A.; Eisenreich, W. Retrobiosynthetic NMR studies with ^{13}C-labeled glucose. Formation of gallic acid in plants and fungi. *J. Biol. Chem.* 272:25474–25482 (1997).

25. Gilchrist, D.G.; Koshuge, T. Aromatic amino acid biosynthesis and its regulation. *In:* The biochemistry of plants. Academic Press, San Diego, vol. 5/13, pp. 507–531 (1980).

BIOSYNTHESIS AND BIODEGRADATION OF COMPLEX GALLOTANNINS

Ruth Niemetz, Joerg U. Niehaus, and Georg G. Gross

Department of General Botany
University of Ulm
D-89069 Ulm
GERMANY

1. INTRODUCTION

According to an already classical definition formulated by Freudenberg in 1920,[1] plant tannins are divided into *condensed* tannins (nowadays often referred to as proanthocyanidins), which are derived from C–C linked flavan-3-ol units, and into *hydrolyzable* tannins. The latter are characterized by a central polyol moiety (β-D-glucose in most cases) whose hydroxy functions are partially or completely esterified with gallic acid (3,4,5-trihydroxybenzoic acid, **1**) (fig. 1) or also with more complex derivatives of this phenol. Stepwise substitution of glucose begins with the 1-O-galloyl derivative, β-glucogallin (**2**), and finally ends with 1,2,3,4,6-penta-O-galloyl-β-D-glucose (**3**), which is regarded as the immediate precursor of the two subclasses of hydrolyzable tannins, i.e., *gallotannins* and *ellagitannins*. Ellagitannins are thought to result from oxidative processes that form C–C linkages between adjacent galloyl residues of pentagalloylglucose to yield (R) or (S)-3,4,5,3',4',5'-hexahydroxydiphenoyl residues and that also lead to the subsequent formation of dimeric and oligomeric derivatives. Unfortunately, no experimental evidence is available to date for the biochemistry of these conversions.

Gallotannins, which are the subject of this chapter, originate by a quite different mechanism, i.e., by addition of further galloyl units to the pentagalloylglucose core, yielding digalloyl residues that are attached via so-called *meta*-depside bonds (**4**). Substitution degrees of as much as 10–12 galloyl residues have been reported for gallotannins from various sources.[2–5] It should be noted that evidence, based on NMR spectroscopy, was presented that gallotannins may be a mixture of *meta*- and *para*-depsides in nature.[3–5] However, as this view still awaits supporting experiments, only the traditional *meta*-bonds, known from the literature for decades, are used in this chapter for the illustration of gallotannin structures. To differentiate these polygalloylglucoses (gallotannins *sensu strictu*) from their

Plant Polyphenols 2: Chemistry, Biology, Pharmacology, Ecology, Edited by
Gross et al. Kluwer Academic / Plenum Publishers, New York, 1999

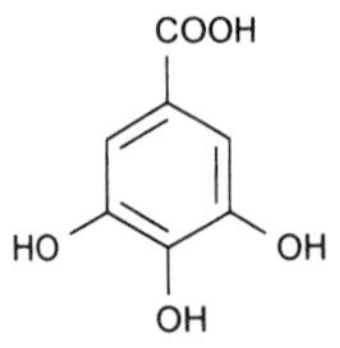

Figure 1. Structures of important gallotannin components. (**1**) Gallic acid (3,4,5-trihydroxybenzoic acid); (**2**) β-glucogallin(1-O-galloyl-β-D-glucose); (**3**) 1,2,3,4,6-pentagalloyl-β-D-glucose; (**4**) *meta*-digalloyl residue.

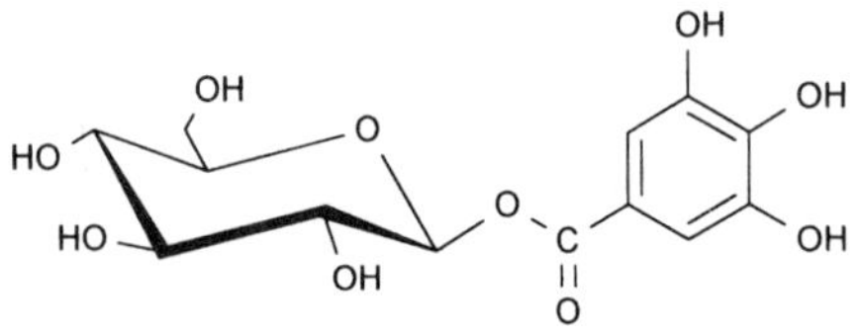

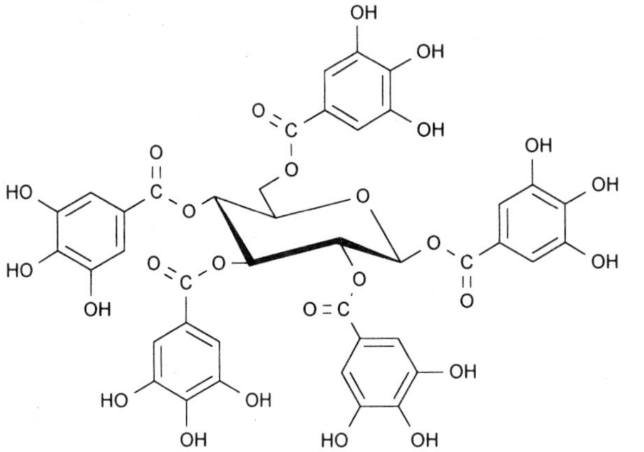

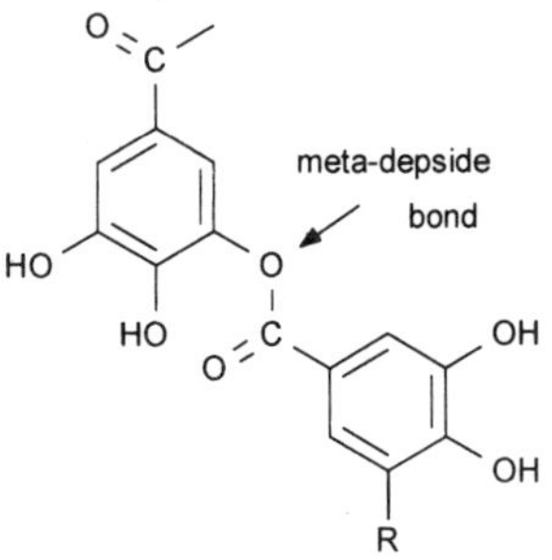

one- to fivefold substituted precursors (often called 'simple galloylglucose esters'), they are referred to as 'complex gallotannins' in this chapter. It must be emphasized that these terms say nothing about the tanning potential of these compounds; mono- and digalloylglucoses are practically inert, whereas the higher substituted analogs can act as tanning agents that display increasingly strong reactivities in relation to their increasing molecular weights.[6,7]

The above outlined gallotannin synthesizing reactions are summarized in figure 2, and it is evident that this biogenetic route can be subdivided into several sections that all comprise specific challenges, namely, (i) the origin of β-glucogallin as the first specific precursor of this pathway; (ii) the transformation of this monoester to pentagalloylglucose, including questions on the role of activated acyl donors and the exact structure of intermediates; and finally (iii) the galloylation reactions involved in the conversion of pentagalloylglucose to complex tannins. Until several years ago, no experimental evidence was available on the biochemical events of this pathway, as documented, for instance, by a review article published in 1985 on the biosynthesis of tannins.[8] It is the merit of intensive enzymatic investigations that were conducted over the past 15 years, and whose results are reported below, that this situation has changed considerably.

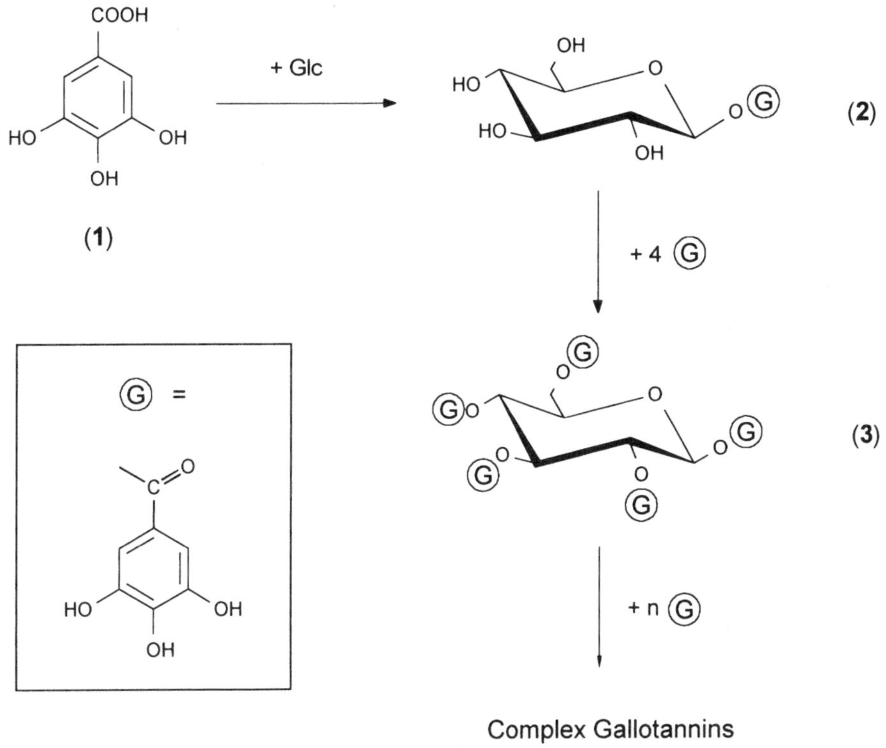

Figure 2. Principal steps in the pathway from gallic acid to complex gallotannins.

2. BIOSYNTHESIS OF β-GLUCOGALLIN

β-Glucogallin (1-O-galloyl-β-D-glucose) (2) was isolated as a natural constituent of Chinese rhubarb (*Rheum officinale*) in 1903 (cf.[9]) and has been regarded for many years as the primary metabolite in the pathway to gallotannins.[10] The biosynthesis of this compound depends, for thermodynamic reasons, on the presence of an 'activated' intermediate, i.e., a compound with a high group-transfer potential (reverse esterase reactions are an excellent tool for technical applications, but are inefficient in nature). This prerequisite can be fulfilled by reaction of an energy-rich galloyl derivative with free glucose, or by the participation of an activated glucose derivative (most likely the ubiquitous UDP-glucose) that combines with the free acid. Considering the well-documented role of acyl-CoA thioesters in the biosynthesis of phenolic esters (for references, see[11,12]), it appeared conceivable that galloyl-CoA represented the energy-rich metabolite required for the biosynthesis of β-glucogallin. This then unknown thioester was thus synthesized chemically via the N-hydroxysuccinimidyl derivative of 4-O-β-D-glucosidogallic acid;[13] however, *in vitro* studies with cell-free extracts from oak leaves revealed that galloyl-CoA was not involved in the biosynthesis of β-glucogallin.[14] Instead, the second of the above alternatives was found to be realized in

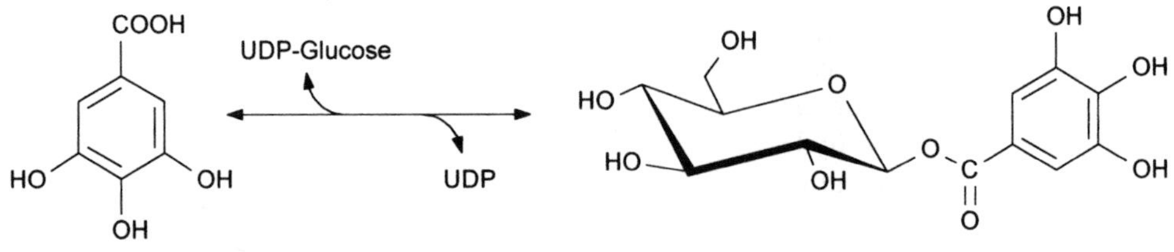

COOH

UDP-Glucose

UDP

(1)   **(2)**

Figure 3. Biosynthesis of β-glucogallin (**2**) by UDP-glucose: gallate 1-O-galloyltransferase. UDP, uridine-5′-diphosphate.

nature, i.e., the (reversible) reaction of free gallic acid with UDP-glucose, catalyzed by a glucosyltransferase that effected the biosynthesis of β-glucogallin (fig. 3), as well as of a variety of related 1-O-acyl-β-D-glucoses.[15–17]

In light of the presently available evidence, the existence of this glucosyltransferase is no longer surprising. Numerous enzymes of this type have since been identified from various plant sources that all catalyze the formation of phenolic 1-O-acylglucoses according to this reaction mechanism,[11] and it appears that UDP-glucose can be regarded now as the general activated donor required for the esterification of glucose with phenolic acids.

3. 'SIMPLE' GALLOYLGLUCOSE ESTERS—THE PATHWAY FROM β-GLUCOGALLIN TO PENTAGALLOYLGLUCOSE

3.1 1,6-Digalloylglucose

In the early phases of the investigations reported in this chapter, it was observed that cell-free extracts from young oak leaves were able to form digalloylglucose and trigalloylglucose in reaction mixtures that contained β-glucogallin (**2**) as sole substrate. Again, no evidence for the participation of galloyl-CoA was obtained. It was concluded that β-glucogallin must have played an unusual dual role, according to which it acted not only as an acceptor substrate, as expected, but apparently also as the acyl donor required for such a reaction.[14] From a thermodynamic view, the existence of such an enzyme activity was fairly surprising because the group-transfer potential of 1-O-acylglucoses was then considered to be compararatively low. It was known that the hemiacetal phosphate of glucose-1-phosphate, e.g., has a $\Delta G_o'$ of about $-21\,\mathrm{kJ\,mol^{-1}}$, and the rather inert ester linkage of isomeric glucose-6-phosphate has a $\Delta G_o'$ of only about -10.5 to $-12.5\,\mathrm{kJ\,mol^{-1}}$.[18] This question has been solved meanwhile by the finding that the related cinnamoyl ester, 1-O-sinapoyl-β-D-glucose, has an unexpectedly high group-transfer potential of $35.7\,\mathrm{kJ\,mol^{-1}}$.[19] This value is comparable to the well-known data for acyl-CoA thioesters (ca. $36\,\mathrm{kJ\,mol^{-1}}$), and it is reasonable to assume that the $\Delta G_o'$ of β-glucogallin (**2**) is in the same order of magnitude.

Subsequent investigations led to the isolation of an enzyme from oak leaves that catalyzed an unusual 'disproportionation' of β-glucogallin in which the galloyl moiety of the donor was transferred to the glucose-6-OH position of the acceptor molecule, thus forming 1,6-di-O-galloyl-β-D-glucose (**5**) under the concomitant liberation of 1 mol glucose as deacylated by-product (fig. 4).[20] Substrate specificity studies revealed that also the closely related esters 1-O-protocatechuoyl-β-D-glucose and 1-O-*p*-hydroxybenzoyl-β-D-glucose (structures depicted in fig. 4) could serve as substrates, with relative activities of 58 percent and 8 percent, respectively.[21]

These observations coincided with similar results from many other laboratories on the role of 1-O-acylglucoses as donors for the acylation of a wide variety of aliphatic and aromatic acceptors,[22] including the formation of chlorogenic acid via caffeoylglucose as an alternative to the long-established acyl-CoA dependent synthesis of this depside.[23] Also 'disproportionation' reactions of the above described type were found that catalyzed the biosynthesis of 1,2-disinapoylglucose in radish seedlings,[24,25] 3,5-dicaffeoylquinic acid in sweet potato,[26,27] and 1,3-di-isobutyroylglucose in wild tomato.[28] It must be concluded that 1-O-acylglucose esters, often regarded in the past as metabolically inert compounds, occupy a prominent position in the secondary metabolism of higher plants, which is comparable to that of the generally acknowledged role of acyl-CoA esters.

3.2 1,2,6-Trigalloylglucose

Initial enzyme studies had suggested that trigalloylglucose(s) might be formed by a mechanism similar to the formation of 1,6-digalloylglucose, i.e., by galloylation with β-glucogallin serving as the donor substrate.[14] This proposal was proven with an enzyme isolated from staghorn sumac (*Rhus typhina*) leaves that catalyzed the position-specific galloylation of the 2-hydroxyl of 1,6-digalloylglucose (**5**) to yield 1,2,6-tri-O-galloyl-β-D-glucose (**6**) (fig. 5).[29] Substrate specificity studies revealed that also the β-glucogallin (**2**) analogs, 1-O-*p*-hydroxybenzoylglucose and

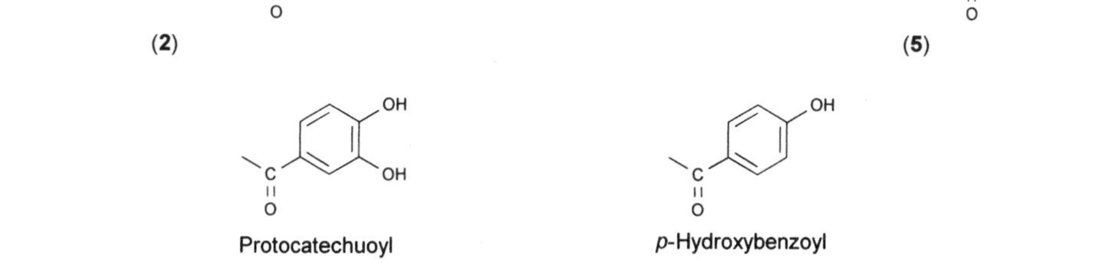

Figure 4. Enzymatic synthesis of 1,6-di-O-galloyl-β-D-glucose (**5**) by 'disproportionation' of two molecules β-glucogallin (**2**), catalyzed by β-glucogallin: β-glucogallin 6-O-galloyltransferase. The enzyme reacts also with the analogous esters, 1-O-protocatechuoyl and 1-O-*p*-hydroxybenzoyl-β-D-glucose.[21]

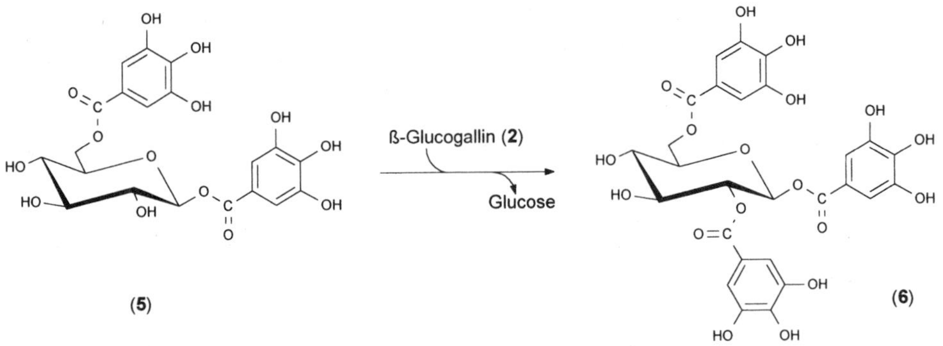

Figure 5. β-Glucogallin (**2**)-dependent acylation of 1,6-digalloylglucose (**5**) to 1,2,6-tri-O-galloyl-β-D-glucose (**6**) by 2-O-galloyltransferase. The enzyme reacts also with the 4-hydroxy and 3,4-dihydroxy analogs of gallic acid.

1-O-protocatechuoylglucose (cf. fig. 4) could act as potent acyl donors, exhibiting relative activities of 54 percent and 93 percent, respectively. Tri-O-protocatechuoylglucose was formed by this enzyme upon incubation of 1-O-protocatechuoylglucose in the presence of 1,6-di-O-protocatechuoylglucose.[30] (For an alternative β-glucogallin-'independent' synthesis of 1,2,6-trigalloylglucose, see section 3.5.)

3.3 1,2,3,6-Tetragalloylglucose

β-Glucogallin (**2**) was also found to function as galloyl donor in the subsequent acylation of 1,2,6-trigalloylglucose (**6**) to 1,2,3,6-tetra-O-galloyl-β-D-glucose (**7**) (fig. 6). The enzyme catalyzing this step was discovered in sumac leaves and

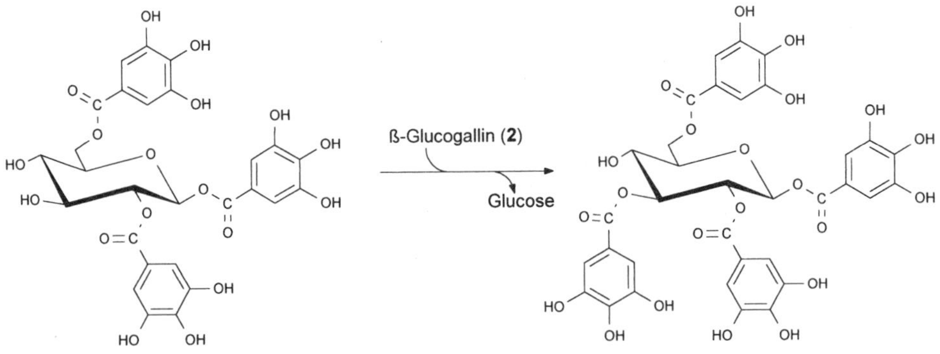

Figure 6. Enzymatic conversion of 1,2,6-trigalloylglucose (**6**) to 1,2,3,6-tetra-O-galloyl-β-D-glucose (**7**). 1,3,6-Trigalloylglucose is also accepted as substrate, yielding the same product.

purified from green acorns of pedunculate oak (*Quercus robur*, syn. *Q. pedunculata*). In addition to the natural acceptor substrate, 1,2,6-trigalloylglucose (**6**), its isomer 1,3,6-trigalloylglucose—which is no intermediate in the biosynthesis of hydrolyzable tannins in oak or sumac—was an extremely efficient acceptor molecule; in both cases, however, 1,2,3,6-tetragalloylglucose (**7**) was the sole reaction product.[31]

3.4 1,2,3,4,6-Pentagalloylglucose

The final enzyme of the pathway, catalyzing the formation of 1,2,3,4,6-penta-O-galloyl-β-D-glucose (**3**), was partially purified from young oak leaves and found to strictly depend on 1,2,3,6-tetragalloylglucose (**7**) as acceptor, whereas the 1,2,4,6-isomer was inactive (fig. 7).[32] The enzyme was recently purified more than a thousandfold to apparent homogeneity, allowing the preparation of antibodies against this pivotal catalyst in the metabolism of hydrolyzable tannins.[33]

3.5 Side Reactions

The above described studies give the impression of a straight, forward-directed pathway, depending on β-glucogallin (**2**) as sole and general acyl donor. This attractive picture was blurred by the discovery of additional and apparently β-glucogallin-*independent* sidereactions. Ambiguous results were encountered in the purification of the transferase catalyzing the formation of 1,2,6-trigalloylglucose (**6**) from 1,6-digalloylgluccose.[29] The initially unexplainable problems were clarified only after recognizing the unexpected existence of an interferring enzyme that effected the efficient formation of the same product, 1,2,6-trigalloylglucose (**6**), but evidently without any requirement for the established acyl donor, β-glucogallin (**2**).[34] As depicted in figure 8, this enzyme was found to represent another example of a 'disproportionation' reaction, in this instance, however, with the participation of two molecules 1,6-digalloylglucose (**5**), which were converted to 1,2,6-trigalloylglucose (**6**) and anomeric 6-galloylglucose as partially deacylated byproduct.

These findings stimulated ideas that, besides β-glucogallin, also higher substituted galloylglucoses could act as acyl donors, provided that they were bearing the energetically indispensable 1-O-galloyl group. This view was corroborated by a series of substrate specificity studies which, however, revealed that the reactivity of higher substituted analogs was diminished drastically, most likely because of steric hindrance as a consequence of increasing bulkiness of the molecule due to increasing substitution degrees. 1-Mono- and 1,6-di-esters function thus as the predominating galloyl donors, whereas the reactivities of higher analogs remain negligible.[34]

3.6 General Characteristics of the Pathway

The above reported enzyme studies have demonstrated the existence of a linear pathway from gallic acid to 1,2,3,4,6-pentagalloylglucose, and this sequence is almost identical to the metabolic route that had been postulated by Haslam and

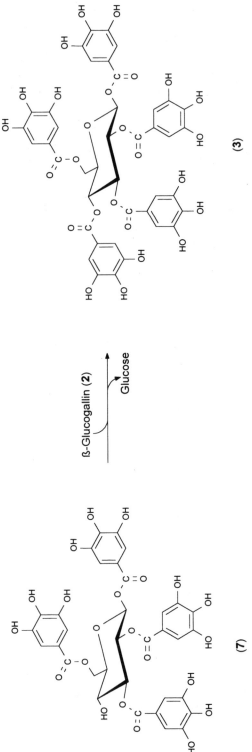

Figure 7. Galloylation of 1,2,3,6-tetragalloylglucose (**7**) to 1,2,3,4,6-penta-O-galloyl-β-D-glucose (**3**).

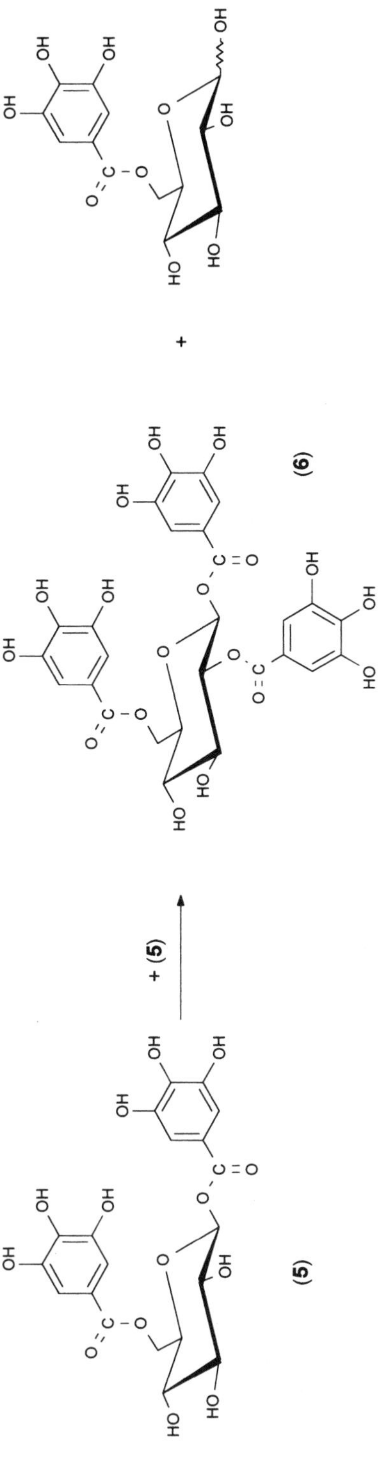

Figure 8. β-Glucogallin 'independent' enzymatic 'disproportionation' of two molecules 1,6-digalloylglucose (**5**) to 1,2,6-trigalloyl-glucose (**6**) and anomeric 6-O-galloylglucose.

coworkers in 1982 on the basis of the structures and natural distribution of numerous galloylglucose esters.[10] Only 1,2,4,6-tetragalloylglucose, discussed by these authors as a possible intermediate, was found to play no role in this biosynthetic sequence. One of the most striking findings was certainly the discovery of the surprisingly pronounced position-specificity of each of the individual enzyme-catalyzed galloylation steps in this sequence. Interestingly, an identical sequence of reactivities has been reported for the chemical esterification of the hydroxyl groups of glucose in studies with 1-benzyl- or 1-methyl-β-D-glucopyranose. Plausible arguments for the apparent reactivity differences were that, after the preferred semiacetal-OH at C-1, the primary 6-OH is more reactive than the residual secondary hydroxyls, the 2-OH among these being the most reactive one due to an activating effect of the neighboring anomeric center. The theoretically equivalent hydroxyls at C-3 and C-4 are discriminated by steric hindrance of the access to the 4-OH group adjacent to the already substituted bulky 6-position, resulting in a higher relative activity of the 3-OH.[35,36]

For detailed studies on this pathway, some of the above enzyme activities had been isolated from oak, whereas others were from sumac; however, supplementary experiments were carried out to ensure that their principal characteristics were identical in both plant species.[37] Generally, a pronounced uniformity of their properties was apparent, regarding their optimal reactivity and stability in slightly acidic media (pH 4-6), their temperature dependence (temperature optima at 40–50 °C, Q_{10} values of 1.8–2.0, activation energies of 30–50 kJ mol^{-1}), their common, unusual cold tolerance, as expressed by residudal reaction rates of 10–25 percent at 0 °C, and their pronounced trend to unusually high molecular weights of about 260,000–450,000 dalton. An exception was the 'β-glucogallin-independent' 1,6-digalloylglucose-'disproportionating' galloyltransferase[34] with a molecular weight of 56,000 dalton and a quite different Q_{10} value of 3.0, equivalent to an activation energy of 74.5 kJ mol^{-1}. This finding suggested that this enzyme did not belong to the central 'β-glucogallin-dependent' metabolic route to pentagalloylglucose.

This pathway has long been assumed to be unique. However, recent studies on the biosynthesis of various isobutyroyl-β-D-glucoses with enzyme extracts from wild tomato (*Lycopersicon penellii*) leaves have demonstrated the existence of a similar reaction pattern with striking similarity to the biosynthesis of galloylglucoses. A UDP-glucose-dependent glucosyltransferase catalyzed the conversion of free isobutyric acid to 1-O-isobutyroyl-β-D-glucopyranose (**8**), and this primary metabolite was found to serve as the energy-rich donor for a series of subsequent transacylation reactions yielding 1,3-di-, 1,2,3-tri- and 1,2,3,4-tetra-O-isobutyroylglucose (fig. 9).[28,38,39] In light of these exciting new results, it can be concluded that not only phenolic 1-O-acylglucoses occupy a central role as activated intermediates in secondary metabolism, but that this property must be ascribed now also to their aliphatic analogs, thus corroborating the presumed general importance of these ester acetals.

4. BIOSYNTHESIS OF COMPLEX GALLOTANNINS

Fortunately, the negative statement of a review article, published only a decade ago in 1989, that "virtually nothing is known about the formation of the

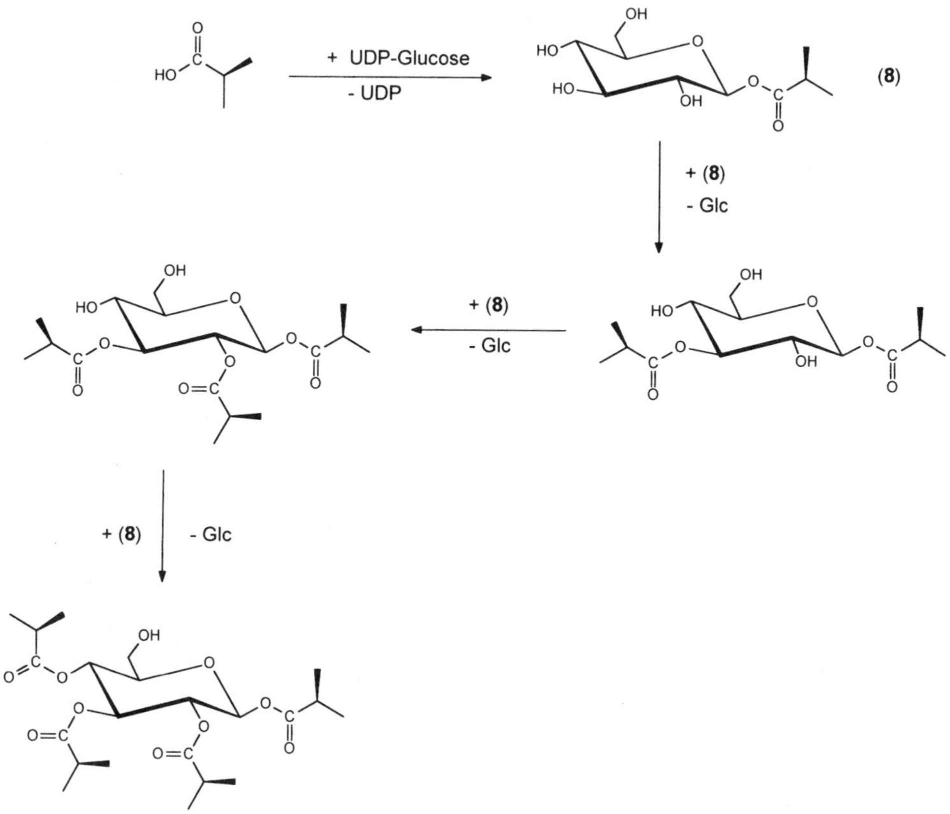

Figure 9. Enzymatic formation of 1-O-isobutyroyl-β-D-glucose (8) in wild tomato and utilization of this monoester as acyl donor in the biosynthesis of di-, tri- and tetra-O-isobutyroyl-β-D-glucoses.[39]

characteristic *meta*-depside bond"[11] of gallotannins has been superseded since. Only speculations were possible then about the nature of the activated galloyl derivative required in such reactions, i.e., whether, by analogy to the preceding steps, β-glucogallin (2) served also in these subsequent conversions as the energy-rich intermediate or whether other activated compounds were utilized, owing to the markedly differing nature of the phenolic OH-group to be substituted.

Shortly after the formulation of the above statement, it was discovered that cell-free extracts from sumac leaves catalyzed the galloylation of 1,2,3,4,6-pentagalloylglucose (3), affording a mixture of hexa-, hepta-, octa- and nonagalloylglucoses, together with traces of decagalloylglucose, and eventually even of undecagalloylglucose. Analogously to the preceding pathway to pentagalloylglucose, β-glucogallin (2) was again found to serve as galloyl donor in these conversions, thus demonstrating the predominating significance of this ester for the entire biogenetic route to complex gallotannins.

Evidence for the presumed gallotannin nature of the reaction products arose from degradation studies with purified *in vitro* reaction products. Treatment with fungal tannase resulted in their complete hydrolysis, leaving gallic acid (**1**) as sole phenolic component. Upon methanolysis, known to exclusively cleave the *meta*-depside bonds of gallotannins, *in vitro* formed hexa-, hepta- and octagalloylglucoses were degraded to 1,2,3,4,6-pentagalloylglucose (**3**) and methyl gallate in molar ratios of 1:1, 1:2, and 1:3, respectively. Hence, one, two, or three galloyl moieties, respectively, must have been depsidically attached to the 1,2,3,4,6-pentagalloylglucose core.[40]

Conclusive proof of the structures of these reaction products was obtained when three hexagalloylglucoses and four heptagalloylglucoses synthesized in scaled-up enzyme assays were isolated and analyzed by reversed-phase HPLC, ^{1}H and ^{13}C NMR spectroscopy, and comparison with previously published data on the occurrence and structure of these compounds in Chinese gallotannin from the related species *Rhus semialata*.[3] The structures of these polygalloylglucoses are depicted in figures 10 and 11.

It is important to note that the structures and, within certain limits, also the relative amounts of the reaction products obtained in the *in vitro* experiments with cell-free extracts from *Rhus typhina* were identical to those of the *in vivo* formed gallotannins from the related species *R. semialata*.[3] It appears to be a particular feature of this 'Chinese gallotannin' that the C-1 and C-6 positions generally

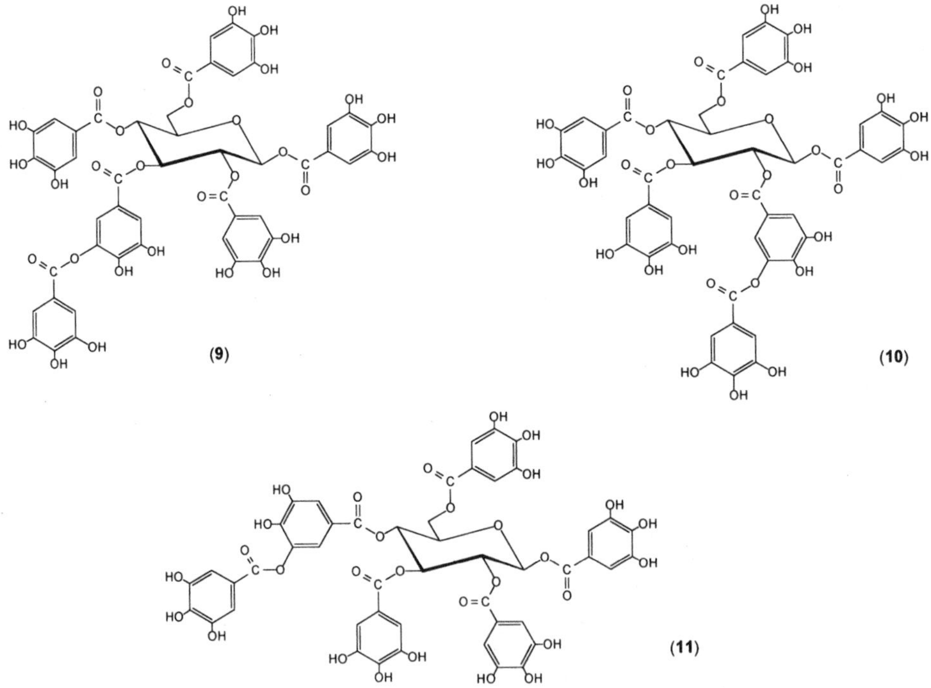

Figure 10. Structures of three hexagalloylglucoses (**9**)–(**11**) formed from 1,2,3,4,6-pentagalloylglucose (**3**) and the acyl donor β-glucogallin (**2**) by cell-free extracts from leaves of *Rhus typhina*.

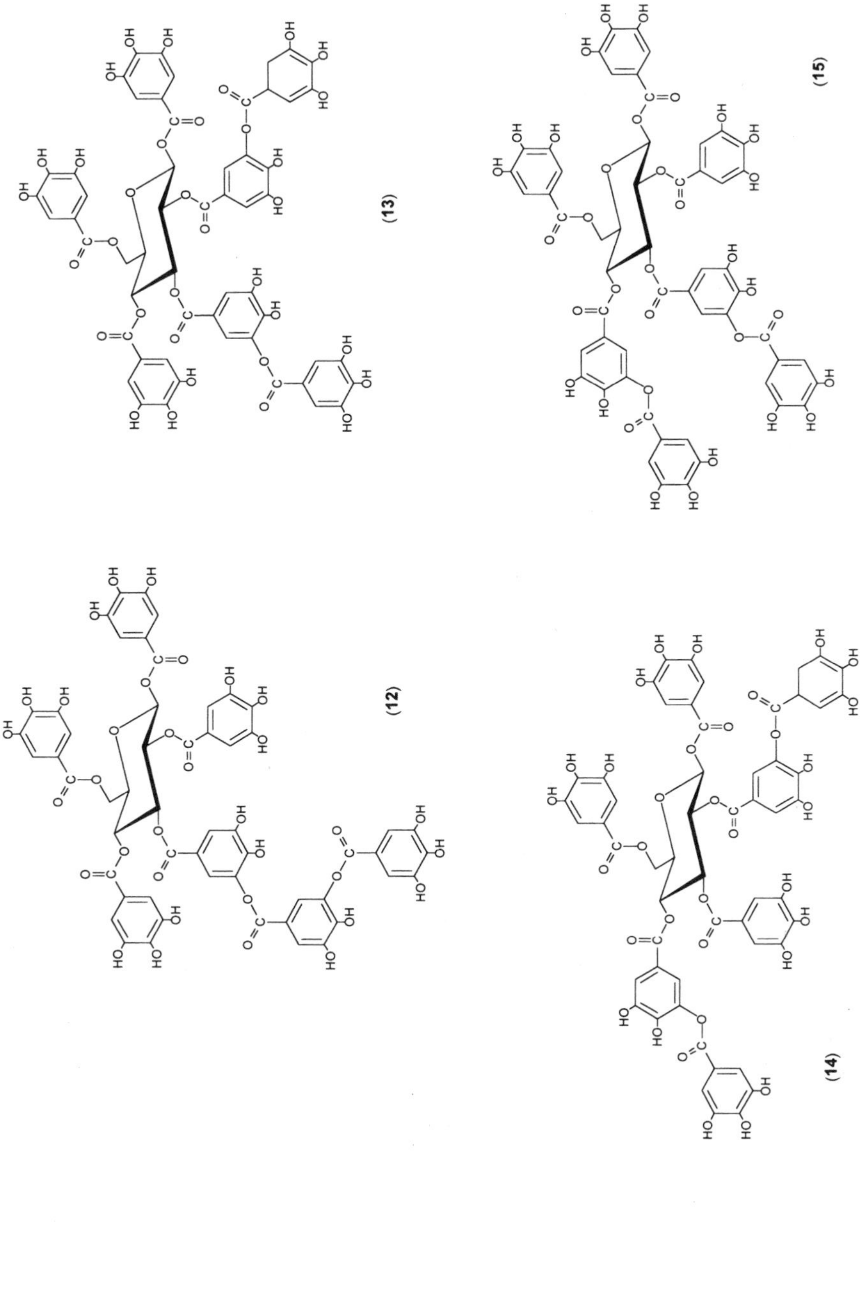

Figure 11. Structures of four heptagalloylglucoses (**12**)–(**15**) formed by enzymatical acylation of 1,2,3,4,6-pentagalloylglucose (**3**) with enzyme preparations from leaves of *Rhus typhina*.

remain free of depsidic substituents, in contrast to the gallotannins from *Quercus infectoria*[4] ('Turkish gallotannin') or *Paeonia lactiflora*[2,5] where depside bonds occur also at C-6 of the glucose core.

Detailed recent enzyme studies on the biosynthesis of complex gallotannins were conducted with cell-free extracts from sumac (*Rhus typhina*) leaves and revealed the existence of several isoenzymes that catalyzed the *in vitro* acylation of pentagalloylglucose (3) to higher substituted derivatives. Among these, three galloyltransferases were isolated and separated according to their different molecular weights of ca. 360,000, 260,000 and 170,000, respectively (transferases A, B, and C). Galloyltransferase C has been purified to apparent homogeneity and was found to consist of four identical subunits of M_r 42,000.[41] Substrate specificity studies showed that pentagalloylglucose (3) was the preferred acceptor, which was converted to hexa-, hepta- and octagalloylglucoses in a ratio of 30:10:1. Closer analysis revealed that 3-O-digalloyl-1,2,4,6-tetra-O-galloyl-β-D-glucose (9) was the predominantly formed hexagalloylglucose in this reaction. Further experiments in which the standard substrate, pentagalloylglucose, had been replaced by the hexagalloylglucoses (9)–(11) demonstrated that (9), the main product formed from pentagalloylglucose in the preceding step, was almost exclusively acylated to the heptagalloylglucose, 3-O-trigalloyl-1,2,4,6-tetra-O-galloyl-β-Dglucose (12). It is evident from these results that galloyltransferase C was specific in acylating penta- and hexagalloylglucoses in the 3-position of the glucose core (fig. 12).

The properties of the above-mentioned galloyltransferases A and B, characterized by molecular weights of 360,000 and 260,000, respectively, are currently investigated.[42] The principal properties of both enzymes were roughly similar to those of enzyme C, except for their significantly higher temperature stability. Differences between transferases A and B, apart from their molecular weights, were observed for their sensitivity against pH and elevated temperatures, enzyme A being less susceptible to these factors than enzyme B. Both enzymes were found to preferentially acylate the 4-position of pentagalloylglucose to hexagalloylglucose (11), followed by substitution of the 2-position of (11) to heptagalloylglucose (14) (fig. 13). A minor pathway with an inverse substitution sequence led to the series pentagalloylglucose (3) → hexagalloylglucose (10) → heptagalloylglucose (14), plus a trace activity toward heptagalloylglucose (13). In summary, it is evident that galloyltransferases A and B promoted galloylation at positions 2 and 4 of the galloylglucose core, in contrast to the above-reported transferase C with pronounced specificity toward the 3-position of the glucose core.

5. BIODEGRADATION OF GALLOYLGLUCOSES

In the above-reported investigations on the biosynthesis of galloylglucoses, pronounced hydrolysis of the standard acyl donor substrate, β-glucogallin, was encountered in the *in vitro* assays. It was finally discovered that this highly disturbing interference was due to the existence of a contaminating esterase that actively degraded both galloylglucose substrates and enzymatically formed reaction products.[21] This enzyme could be purified from leaves of *Q. robur*

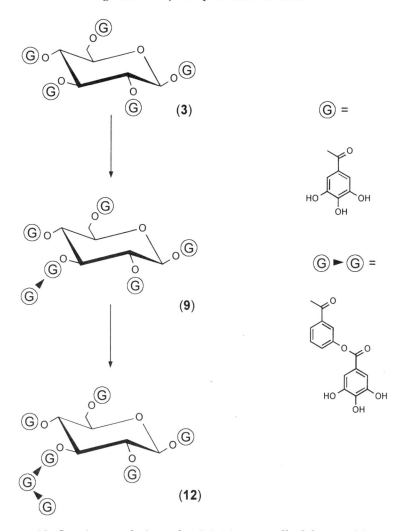

Figure 12. *In vitro* acylation of 1,2,3,4,6-pentagalloylglucose (**3**) to 3-O-digalloyl-(**9**) and 3-O-trigalloyl-1,2,4,6-tetra-O-galloyl-β-D-glucose (**12**) with purified galloytransferase C (cf. text) from sumac (*Rhus typhina*) leaves.

(pedunculate oak) more than nineteen-hundredfold to apparent homogeneity. Its molecular weight was determined by gel-filtration as 150,000 and 300,000 dalton, respectively. Two protein bands were also detected after native electrophoresis on polyacrylamide-gel, whereas denaturing electrophoresis in the presence of sodium dodecylsulfate revealed the existence of a single polypeptide band of M_r 75,000. It was concluded that the native enzyme preferentially existed as a tetramer of apparently four identical subunits in slightly acidic medium, whereas it dissociated partially or completely into still highly reactive dimers under the more alkaline conditions employed for gel-filtration or native PAGE.[43]

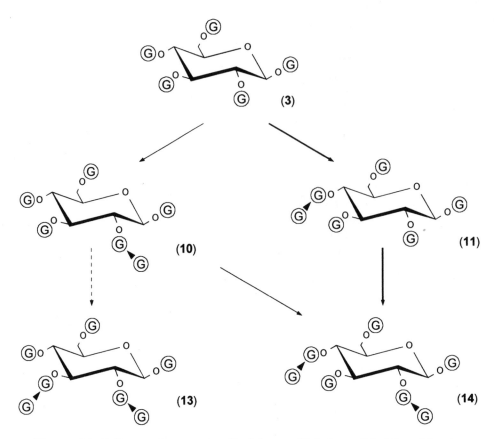

Figure 13. Pathways from pentagalloylglucose (**3**) to hexa-(**10**, **11**) and hepta-galloylglucoses (**13**, **14**), catalyzed by galloyltransferases A and B (cf. text) from leaves of sumac (*Rhus typhina*). Main reactions are symbolized by bold arrows, minor reactions by thin arrows; the dashed arrow symbolizes trace reactivity.

Detailed studies were carried out on the substrate-specificity of this enzyme with a wide array of pure compounds. It was found that this hydrolase was absolutely inactive with variously substituted methyl cinnamates, 1-O-sinapoyl-glucose, and substrates with nitro-substitution of the aromatic acid moiety. Hydrolysis occurred with simple galloyl esters (methyl, ethyl, propyl gallate), naphtyl acetate (but not with naphtyl propionate or butyrate), mono- to hexa-substituted galloyl-β-D-glucoses, variously ring-substituted 1-O-benzoyl-β-D-glucoses, and with depsides like *meta*-digallic acid or chlorogenic acid. It was evident that the enzyme exhibited both pronounced esterase and depsidase activities, i.e., properties that are typically ascribed to fungal tannases. Also, other characteristics of the enzyme from oak, e.g., pH and temperature optima or its already mentioned unusual high molecular weight, closely resembled those of microbial tannases (consult reference list[22,44]). By analogy to these enzymes, the esterase from oak leaves was thus classified as a new 'tannase' of plant origin.

There remains the question of the purpose of such an enzyme in green leaves. It has been recognized that condensed and hydrolyzable tannins contribute to the defense of plants against ruminants,[45,46] and the presence of tannin-degrading enzymes in green leaves would thus not make much sense. In contrast to herbivorous mammals, however, the situation with insects could be quite different. The feeding deterrent role traditionally ascribed to tannins due to their astringency, causing reduced palatability of plant parts, has been questioned; e.g., evidence has been presented for the hypothesis that the ellagitannin geraniin preferentially acted as protoxin that released insect growth inhibitors, particularly ellagic acid, upon hydrolytic cleavage.[47] According to this view, hydrolyzable tannins could play a dual protective role in plant-herbivore interactions, being active not only by direct protection against herbivorous animals, but also indirectly in the form of their degradation products. The occurrence of tannase in green leaves would thus significantly contribute to the latter process by release of harmful products after loss of cellular compartmentalization under the attack of an insect predator that brings this enzyme into contact with its tannin substrates. Such protoxin-based defense strategies are well documented;[48] given the correctness of this idea, the system tannin-'tannase' could be added to that list of chemical weapons in higher plants.

6. CONCLUSIONS

As discussed in this chapter, the mechanisms involved in the biosynthesis of pentagalloylglucose have been elucidated by consequent investigations on the enzyme level, and this applies also to the formation of complex gallotannins as one metabolic branch emerging from that pivotal intermediate. Future efforts must be devoted to unraveling the still highly enigmatical biosynthesis of the ellagitannin branch of hydrolyzable tannins; though our knowledge of the chemical synthesis of these compounds has been considerably augmented during the past few years, we still await a real breakthrough in the biochemistry of these widespread and important natural products.

ACKNOWLEDGMENTS

We thank the colleagues and coworkers who contributed to the research reported in this chapter. Generous financial support by the Deutsche Forschungsgemeinschaft, the Fonds der Chemischen Industrie, and by research grants of the University of Ulm is gratefully acknowledged.

REFERENCES

1. Freudenberg, K. Die Chemie der natürlichen Gerbstoffe. Springer, Berlin (1920).
2. Nishizawa, M.; Yamagishi, T.; Nonaka, G.; Nishioka, I. Structure of gallotannins in *Paeoniae radix*. *Chem. Pharm. Bull.* 28:2850 (1980).
3. Nishizawa, M.; Yamagishi, T.; Nonaka, G.; Nishioka, I. Tannins and related compounds. Part 5. Isolation and characterization of polygalloylglucoses from Chinese gallotannin. *J. Chem. Soc. Perkin Trans. I* :2963 (1982).

4. Nishizawa, M.; Yamagishi, T.; Nonaka, G.; Nishioka, I. Tannins and related compounds. Part 9. Characterization of polygalloylglucoses from Turkish galls. *J. Chem. Soc. Perkin Trans. I* :961 (1983).

5. Nishizawa, M.; Yamagishi, T.; Nonaka, G.; Nishioka, I.; Nagasawa, T.; Oura, H. Tannins and related compounds. XII. Isolation and characterization of galloylglucoses from Paeoniae radix and their effect on urea–nitrogen concentration in rat serum. *Chem. Pharm. Bull.* 31:2593 (1983).

6. Ozawa, T.; Lilley, T.H.; Haslam, E. Polyphenol interactions: astringency and the loss of astringency in ripening fruit. *Phytochemistry* 26:2937 (1987).

7. Kawamoto, H.; Nakatsubo, F.; Murakami, K. Stoichiometric studies on tannin–protein coprecipitation. *Phytochemistry* 41:1427 (1996).

8. Hillis, W.E. Biosynthesis of tannins. *In*: Higuchi, T. (ed.). Biosynthesis and biodegradation of wood components. Academic Press, Orlando, p. 325 (1985).

9. Haslam, E. Plant polyphenols. Vegetable tannins revisited. Cambridge University Press, Cambridge (1989).

10. Haddock, E.A.; Gupta, R.K.; Al-Shafi, M.K.; Layden, K.; Haslam, E.; Magnolato, D. The metabolism of gallic acid and hexahydroxydiphenic acid in plants: biogenetic and molecular taxonomic considerations. *Phytochemistry* 21:1049 (1982).

11. Gross, G.G. Enzymology of gallotannin biosynthesis. *In*: Lewis, N.G.; Paice, M.G. (eds.). Plant cell wall polymers: biogenesis and biological function. ACS Symp. Ser. Vol. 399, Washington, DC, p. 108 (1989).

12. Gross, G.G. Enzymatic synthesis of gallotannins and related compounds. *In*: Stafford, H.A.; Ibrahim, R.K. (eds.). Phenolic metabolism in plants. Recent advances in phytochemistry. Vol. 26, Plenum Press, New York, p. 297 (1992).

13. Gross, G.G. Synthesis of galloyl-coenzyme A thioester. *Z. Naturforsch.* 37c:778 (1982).

14. Gross, G. Synthesis of mono-, di- and trigalloyl-β-D-glucose by β-glucogallin-dependent galloyltransferase from oak leaves. *Z. Naturforsch.* 38c:519 (1983).

15. Gross, G.G. Synthesis of β-glucogallin from UDP-glucose and gallic acid by an enzyme preparation from oak leaves. *FEBS Lett.* 148:67 (1982).

16. Gross, G.G. Partial purification and properties of UDP-glucose: vanillate 1-O-glucosyl transferase from oak leaves. *Phytochemistry* 22:2179 (1983).

17. Weisemann, S.; Denzel, K.; Schilling, G.; Gross, G.G. Enzymatic synthesis of 1-O-phenylcarboxyl-β-D-glucose esters. *Bioorg. Chem.* 16:29 (1988).

18. Atkinson, M.L.; Morton, R.K. Free energy and the biosynthesis of phosphates. *In*: Florkin, M.; Mason, H.S. (eds.). Comparative biochemistry. Vol. II. Free energy and biological function. Academic Press, New York–London, p. 1 (1960).

19. Mock, H.P.; Strack, D. Energetics of the uridine 5'-diphosphoglucose: hydroxycinnamic acid acyl-glucosyltransferase reaction. *Phytochemistry* 32:575 (1993).

20. Schmidt, S.W.; Denzel, K.; Schilling, G.; Gross, G.G. Enzymatic synthesis of 1,6-digalloylglucose from β-glucogallin by β-glucogallin 6-O-galloyltransferase from oak leaves. *Z. Naturforsch.* 42c:87 (1987).

21. Gross, G.G.; Denzel, K.; Schilling, G. Enzymatic synthesis of di-O-phenylcarboxyl-β-D-glucose esters by an acyltransferase from oak leaves. *Z. Naturforsch.* 45c:37 (1990).

22. Gross, G.G. Biosynthesis of hydrolyzable tannins. *In*: Pinto, B.M. (ed.). Comprehensive natural products chemistry. Vol. 3. Carbohydrates and their derivatives including tannins, cellulose, and related lignins. Elsevier, Amsterdam, p. 799 (1999).

23. Stöckigt, J.; Zenk, M.H. Enzymatic synthesis of chlorogenic acid from caffeoyl coenzyme A and quinic acid. *FEBS Lett.* 42:131 (1974).

24. Dahlbender, B.; Strack, D. Enzymatic synthesis of 1,2-disinapoylglucose from 1-sinapoylglucose by a protein preparation from cotyledons of *Raphanus sativus* grown in the dark. *J. Plant Physiol.* 116:375 (1984).

25. Dahlbender, B.; Strack, D. Purification and properties of 1-(hydroxycinnamoyl)-glucose: hydroxycinnamoyltransferase from radish seedlings. *Phytochemistry* 25:1043 (1986).

26. Kojima, M.; Kondo, T. An enzyme in sweet potato root which catalyzes the conversion of chlorogenic acid, 3-caffeoylquinic acid, to isochlorogenic acid, 3,5-di-caffeoylquinic acid. *Agric. Biol. Chem.* 49:2467 (1985).

27. Villegas, R.J.A.; Shimokawa, T.; Okuyama, H.; Kojima, M. Purification and characterization of chlorogenic acid: chlorogenate caffeoyl transferase in sweet potato roots. *Phytochemistry* 26:1577 (1987).

28. Ghangas, G.S.; Steffens, J.C. 1-O-Acyl-β-D-glucoses as fatty acid donors in transacylation reactions. *Arch. Biochem. Biophys.* 316:370 (1995).

29. Denzel, K.; Schilling, G.; Gross, G.G. Biosynthesis of gallotannins. Enzymatic conversion of 1,6-digalloylglucose to 1,2,6-trigalloylglucose. *Planta* 176:135 (1988).

30. Gross, G.G.; Denzel, K. Biosynthesis of gallotannins. β-Glucogallin-dependent galloylation of 1,6-digalloylglucose to 1,2,6-trigalloylglucose. *Z. Naturforsch.* 46c:389 (1991).

31. Hagenah, S.; Gross, G.G. Biosynthesis of 1,2,3,6-tetra-O-galloyl-β-D-glucose. *Phytochemistry* 32:637 (1993).

32. Cammann, J.; Denzel, K.; Schilling, G.; Gross, G.G. Biosynthesis of gallotannins. β-Glucogallin-dependent formation of 1,2,3,4,6-pentagalloylglucose by enzymatic galloylation of 1,2,3,6-tetragalloylglucose. *Arch. Biochem. Biophys.* 273:58 (1989).

33. Grundhöfer, P.; Piffel, A.; Gross, G.G. Immunohistochemical studies on the localization of the pentagalloylglucose-forming enzyme in tannin producing tissues (manuscript in preparation).

34. Denzel, K.; Gross, G.G. Biosynthesis of gallotannins. Enzymatic 'disproportionation' of 1,6-digalloylglucose to 1,2,6-trigalloylglucose and 6-galloylglucose by an acyltransferase from leaves of *Rhus typhina* L. *Planta* 184:285 (1991).

35. Williams, J.M.; Richardson, A.C. Selective acylation of pyranosides. I. Benzoylation of methyl α-D-glycopyranosides of mannose, glucose and galactose. *Tetrahedron* 23:1369 (1967).

36. Reinefeld, E.; Ahrens, D. Der Einfluss der Konfiguration auf die partielle Veresterung von D-Glucopyranosiden. *Liebigs Ann. Chem.* 747:39 (1971).

37. Denzel, K.; Gross, G.G. (unpublished results).

38. Ghangas, G.S.; Steffens, J.C. UDPglucose: fatty acid transglucosylation and transacylation in triacylglucose biosynthesis. *Proc. Natl. Acad. Sci.; USA* 90:9911 (1993).

39. Kuai, J.-P.; Ghangas, G.S.; Steffens, J.C. Regulation of triacylglucose fatty acid composition. Uridine diphosphate glucose: fatty acid glucosyltransferase with overlapping chain-lenght specificity. *Plant Physiol.* 115:1581 (1997).

40. Hofmann, A.; Gross, G.G. Biosynthesis of gallotannins: formation of polygalloylglucoses by enzymatic acylation of 1,2,3,4,6-penta-O-galloylglucose. *Arch. Biochem. Biophys.* 283:530 (1990).

41. Niemetz, R.; Gross, G.G. Gallotannin biosynthesis: purification of β-glucogallin: 1,2,3,4,6-pentagalloyl-β-D-glucose galloyltransferase. *Phytochemistry* 49:327 (1998).

42. Niemetz, R.; Gross, G.G. Gallotannin biosynthesis: A new β-glucogallin-dependent galloyltransferase from sumac leaves acylating gallotannins at positions 2 and 4. *J. Plant Physiol* (in press).

43. Niehaus, J.U.; Gross, G.G. A gallotannin degrading esterase from leaves of pedunculate oak. *Phytochemistry* 45:1555 (1997).

44. Scalbert, A. Antimicrobial properties of tannins. *Phytochemistry* 30:3875 (1991).

45. Furstenburg, D.; Van Hofen, W. Condensed tannin as anti-defoliate agent against browsing by giraffe (*Giraffa camelopardalis*) in the Kruger National Park. *Comp. Biochem. Physiol.* 109A:425 (1994).

46. Robbins, C.T.; Hanley, T.A.; Hagerman, A.E.; Hjeljord, O.; Baker, D.L.; Schwartz, C.C.; Mautz, W.W. Role of tannins in defending plants against ruminants: reduction in protein availability. *Ecology* 68:98 (1987).

47. Klocke, J.A.; Van Wagenen, B.; Balandrin, M.F. The ellagitannin geraniin and its hydrolysis products isolated as insect growth inhibitors from semi-arid land plants. *Phytochemistry* 25:85 (1986).

48. Matile, P. Das toxische Kompartiment der Pflanzenzelle. *Naturwissenschaften* 71:18 (1984).

TOWARD UNDERSTANDING MONOMERIC ELLAGITANNIN BIOSYNTHESIS

Richard F. Helm,[a] Lei Zhentian,[a] Thilini Ranatunga,[a] Judith Jervis,[a] and Thomas Elder[b]

[a] Department of Wood Science and Forest Products
Virginia Tech University
Blacksburg, Virginia 24061-0346
USA

[b] Department of Forestry
Auburn University
Auburn, Alabama 36849
USA

1. INTRODUCTION

Ellagitannins are found in approximately 40 percent of all dicotyledenous plants ranging from raspberries to the oaks.[1-3] By definition, this class of compounds is composed of only glucose and biaryl-linked gallic acid. Though these two compounds are relatively simple in structure, the complexity of the compounds that can be formed is tremendous. Ellagitannins are seemingly a natural combinatorial library, with over 500 different compounds isolated and identified to date. Due to their relatively high oxidation potential[4] and ability to bind heavy metals[5] and proteins,[6] research efforts have been directed at understanding their roles in environment-plant interactions,[7] human health,[8] timber processing,[9] and spirits manufacture.[10]

Considering the relative abundance and economic importance of the plant species that contain these plant polyphenols, it is somewhat surprising that the biosynthetic pathway that leads to these compounds has not been elucidated. The primary objectives of this chapter, rather than provide a menu of structures or repeat the statements made in the more recent reviews on the subject,[11-13] are to: 1) provide a general overview of the proposed metabolic pathways that lead to the monomeric ellagitannins; 2) provide some insight into the conformational aspects of β-PGG; 3) discuss some of the more recent work performed in our

Plant Polyphenols 2: Chemistry, Biology, Pharmacology, Ecology, Edited by
Gross et al. Kluwer Academic / Plenum Publishers, New York, 1999

laboratories concerning the acyclic C-glycosidic ellagitannins castalagin and vescalagin.

2. BIOSYNTHETIC OVERVIEW

The enzyme-mediated reaction of gallic acid[14] and UDP-glucose provides β-glucogallin, which is sequentially acylated to form penta-galloyl β-D-glucopyranoside (β-PGG, fig. 1, **1**).[15] This compound is considered the precursor for all gallotannins and ellagitannins (the two classes of hydrolyzable tannins). Gallotannins (tannic acid) are biosynthesized by the formation of additional ester bonds (depside linkages) between gallic acid and phenolic hydroxyls of the galloyl moieties already esterified to D-glucose.[16] Monomeric ellagitannins, on the other hand, undergo a series of intramolecular reactions to provide compounds such as castalagin (**2**) and chebulagic acid (**3**, fig. 1). Castalagin is representative of the most common intramolecular couplings that occur between the 4,6 and 2,3 galloyl groups, whereas **3** is representative of the less common 3,6/2,4 coupling pathway.

A generalized schematic outlining the two routes to monomeric ellagitannins is shown in figure 2. All compounds are based on cyclization and modification of **1**, which can be envisioned as having two distinct conformations—the stable $^4C_1$ and the less stable $^1C_4$. Oxidation of **1**, or the loss of hydrogen atoms from two adjacent galloyl rings, followed by a biaryl coupling leads to the formation of a hexahydroxydiphenyl (HHDP) group. The orientation of the galloyl groups during bond formation is stereospecific and thought to be dictated by their location on the

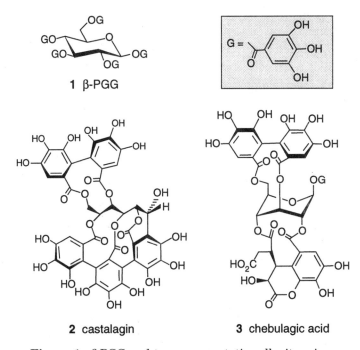

Figure 1. β-PGG and two representative ellagitannins.

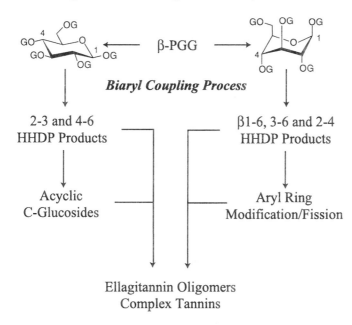

Figure 2. Generalized schematic of the ellagitannin biosynthetic pathway.

glucose ring.[12] This type of chirality is termed atropisomerism, where the chirality is due to the bulkiness of the substituents on the aryl rings, which prevent the rings from rotating past one another at room temperature.[17]

This is best exemplified by examining the structures of two of the simplest ellagitannins, tellimagrandin II (**4**) and punicafolin (**5**) (fig. 3). The galloyl groups would be oriented in such a way as to minimize the repulsive forces that would arise by the rings and/or substituents coming too close in space. This would allow for the stereoselective coupling affording the (*S*)-hexahydroxydiphenyl (HHDP) structure of **4** and the (*R*)-HHDP structure of **5**. While the exact mechanism of biaryl coupling is not known, the current working hypothesis (the so-called Haslam-Schmidt hypothesis) suggests that the stereospecific coupling is strictly due to stereoelectronic effects brought about by the glucose ring. However, a more "guided" enzymatic process involving a dirigent protein[18] can also be envisioned.

Further stereoselective couplings provide more highly condensed structures listed on the bottom of figure 4. Compounds that possess the 4,6- and 2,3-HHDP units result from oxidative coupling with the D-glucose moiety in the $^4C_1$ conformation and are almost always of the S-atropisomer. Ellagitannins with 1,6- and 3,6-HHDP units are typically envisioned as resulting from coupling while in the $^1C_4$ conformation. The 4,6- 2,3-HHDP units generally have an (*S*)-configuration, whereas the 3,6-HHDP type structures can be found in both forms.

3. β-PGG AND THE BIARYL COUPLING REACTIONS

Clearly, β-PGG plays a key role in the formation of ellagitannins, with the crucial biaryl coupling steps occurring in a regio- as well as stereo-specific manner.

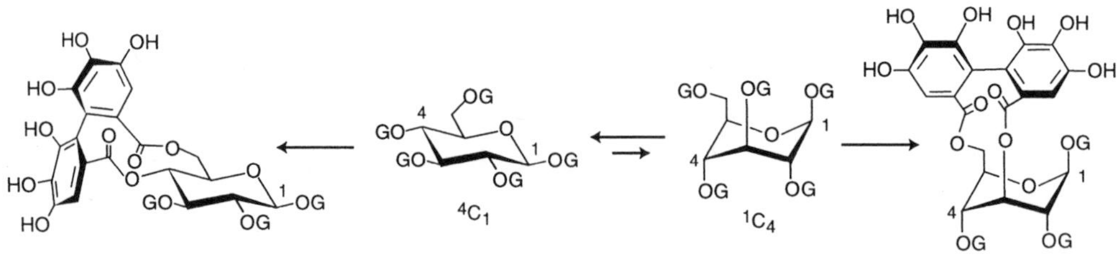

Figure 3. Ring conformation and biaryl coupling.

Since there have only been a few reports in the area of ellagitannin computer modeling,[11,19] we thought it useful to compare the structures of the two conformers. Figure 4 shows the lowest energy conformers of the two chair forms along with the distances from potential biaryl coupling sites and the stereochemical outcome if such a bond were to occur (without significant changes in the overall molecular shape prior to and during bond formation). Though this is a relatively simplistic method of analysis, it is somewhat gratifying to see that in the case of the $^4C_1$ conformer, the shortest distances lead to the most preferred conformer. Feldman and coworkers have applied the same type of analysis with less complex model structures as well as "true" $^4C_1$ ellagitannins.[11] In all cases, the structures obtained were similar to those found here, and a good agreement was found between the most energetically stable biaryl linkage site and those found in nature.

A more complex story emerges from analysis of the $^1C_4$ structure. The $\beta 1$–6 coupling distance of 4.87 Å would lead to the S-atropisomer observed for davidiin and helioscopinin B. The optimized structure has the 3-galloyl group oriented in such a way that only one of the potential carbons can be involved in the 3,6-HHDP bond. The distances between this carbon and the two potential carbons on the 6-galloyl group differ only slightly (6.83 and 6.95 Å), and each would provide a different atropisomer. Based on these distances, one could predict that either form may be found in nature, and indeed this is the case. However, if one were to invoke thermodynamics to justify such an occurrence, plants exhibiting such a biaryl linkage should have a mixture of each form present, not just the R- or S-atropisomer.

Another issue worth considering is whether the $^1C_4$ conformer can be observed experimentally. Is the energy barrier for the change too high at room temperature? In an effort to address this problem, a solution of synthetic [1-^{13}C]-$\mathbf{1}^{20}$ was dissolved in DMSO-d_6, and ^{1}H-NMR spectra were recorded at 20 and 80 °C, in order to see if it were possible to obtain NMR data to support such a change. The sample showed no measureable change in the $J_{1,2}$ (7.9 ± 0.1 Hz) or the $J_{C,H}$ of the anomeric carbon (170.3 Hz at 20 °C and 169.6 at 80 °C). If the $^1C_4$ conformer was present in significant amounts, one would have expected to see a decrease in the $J_{1,2}$ coupling constant as the percentage of the axial conformer increased.

The next experiment was to cool a solution of unlabeled **1** in acetone-d_6 to see if it is possible to "lock out" the individual conformers. Again, no significant change in the proton NMR spectrum occurred between 20° and –90 °C, leading one to

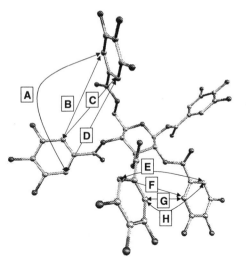

Distances/Sterochemical Outcome

A (4,6)	8.98 Å	R
B (4,6)	6.94 Å	S
C (4,6)	**6.18 Å**	S
D (4,6)	7.75 Å	R
E (2,3)	8.08 Å	R
F (2,3)	6.07 A	S
G (2,3)	**3.84 Å**	S
H (2,3)	6.07 Å	R

Distances/Sterochemical Outcome

A (3,6)	9.25 Å	R
B (3,6)	9.28 Å	S
C (3,6)	6.83 Å	R
D (3,6)	6.95 Å	S
E (β1,6)	5.61 Å	R
F (β1,6)	4.87 A	S
G (β1,6)	5.45 Å	S
H (β1,6)	6.84 Å	R

Compound	Galloyl	HHDP
Casuarictin	β1	(S, S); 2-3, 4-6
Cercidinin A	β1, 4, 6	(R); 2-3
Cercidinin B	β1	(R); 2-3
Gemin D	3	(S); 4-6
Nobotanin D	β1, 6	(S); 2-3
Pedunculagin	---	(S, S); 2-3; 4-6
Platycaryanin D	---	(R, R); 2-3; 4-6
Potentillin	α1	(S); 2-3; 4-6
Praecoxin B	4, 6	(S); 2-3
Pterocaryanin B	4	(S); 2-3
Pterocaryanin C	β1, 4, 6	(S); 2-3
Pterophyllin A	α1, 3	(S); 4-6
Roxin B	---	(S, S); β1-2; 4-6
Sanguin H-1	α1, 6	(S); 2-3
Sanguin H-4	α1	(S); 2-3
Strictinin	β1	(S); 4-6
Tellimagrandin I	2, 3	(S); 4-6
Tellimagrandin II	β1, 2, 3	(S); 4-6

Compound	Galloyl	HHDP
Corilagin	β1	(R); 3-6
Davidiin	2, 3, 4	(S); β1-6
Helioscopinin B	3	(S); β1-6
Macaranganin	β1, 2, 4	(S); 3-6
Nupharin A	α1, 2, 4	(S); 3-6
Nupharin B	α1, 2, 4	(R); 3-6
Punicafolin	β1, 2, 4	(R); 3-6
Tercatain I	β1, 4	(R); 3-6
Tercatain II	β1, 2	(R); 3-6

Figure 4. The minimized structures of the two conformers of β-PGG are shown at the top followed by the distances between potential coupling sites and the stereochemical result of the coupling. The known predominant monomeric ellagitannins with HHDP intramolecular coupling are listed at the bottom, categorized by conformer.

conclude that the conformer, if it is present, is not found in significant concentrations (<2 percent). This same experiment was also run on the α-anomer of **1** on the thought that the percentage of the least preferred conformer would increase if one galloyl group would be equatorial. However, no signals indicative of such a structure were observed.

Haslam and coworkers considered this issue several years ago.[21] The NMR data for the known 3–6 linked structures such as **3** exhibit coupling constants that are very similar to those of 1,6-anhydro β-D-glucopyranoside, commonly referred to as levoglucosan. This compound assumes a slightly skewed $^1C_4$ conformation with the hydroxyls oriented *quasi*-axial due to the internal bond. Acetylation tended to flatten the ring slightly. It was then hypothesized that the $^1C_4$-type ellagitannins had a similar overall shape. The rotational barrier for the R/S introconversion for a 4,6-HHDP ellagitannin model was recently determined[19] and was shown to be quite significant at room temperature.

4. ENZYMOLOGY OF THE BIARYL COUPLING REACTIONS

What enzymes would be involved in a biaryl coupling process? Naturally the first classes of enzymes that come to mind are those associated with lignification, the laccases and peroxidases, since lignin also contains biaryl interunit linkages. Since these enzymes are thought to work by a free radical coupling process, one could further hypothesize that ellagitannin biaryl bond formation also occurs via this route. At the onset, this may seem problematic—lignin is a racemic polymer, whereas the ellagitannins are stereospecific secondary metabolites. However, the discovery by Lewis and coworkers of dirigent proteins, compounds that aid in the production of the stereospecific lignans,[18] opens the door to the possibility that the free radical coupling process in ellagitannin formation may also be under such control.

The *in vitro* biaryl coupling of gallic acid or its simple ester derivatives via a peroxidase or laccase has been demonstrated, although yields are quite low.[13,15] In combination with HHDP-type structures, one also encounters significant over-oxidation, leading to destruction of the aromatic ring. In fact, the *in vitro* biaryl coupling of **1** has yet to be achieved with an isolated enzyme. Such results are almost certainly due to the relatively high oxidation potential of gallates. When these molecules are matched with oxidases with high redox potentials, the gallate structures break down rather than undergo biaryl coupling. A laccase or peroxidase with a lower redox potential may be employed *in vivo*.

5. POST COUPLING MODIFICATIONS, OLIGOMERS AND COMPLEX TANNINS

Compounds possessing the β1,6- and 3,6-HHDP units ($^1C_4$) also undergo a series of reactions that essentially destroy the aromaticity in one or more gallate rings. Typically, the modifications occur with the 2,4-HHDP units. There are about 150 compounds of this type identified to date, and these are limited in distribution to just a few plant families.[2] Little in the way of biochemistry has been done with these species, although callus and cell suspension cultures producing such compounds have been reported,[22-25] and geraniin has been chemically converted to chebulagic acid (**3**).[26] Considering the tools of biochemistry and molecular biology available today, use of callus and cell suspension cultures for investigating the pathway of their formation could prove to be quite rewarding.

The monomeric ellagitannins described above (both ${}^1C_4$ and ${}^4C_1$ based), can be further modified to form oligomers, and/or complex tannins (tannins that contain both gallotannin and condensed tannin moieties). The oligomers are typically formed by phenolic coupling processes, although there is a recent report of a biaryl linked dimer from *Rhoiptelea chiliantha*.[27] It has generally been thought that oligomer/complex tannin formation occurs subsequent to the assembly of the monomeric components.

6. ACYCLIC DERIVATIVES

While the ${}^1C_4$-type ellagitannins undergo a wide variety of post coupling modifications, a characteristic of the ${}^4C_1$-pathway is the opening of the glucose ring with formation of a C-glucosidic bond as well as a flavogallonyl moiety. These C-glycosidic compounds are the most abundant ellagitannins found in nature, being the primary tannins found in the heartwood of most oak, chestnut, and eucalyptus species. A simplified schematic starting with **1** is shown in figure 5. A 4,6-HHDP coupling is thought to precede 2,3-HHDP bond formation, leading to casuarictin (**6**), potentillin (**7**), and pedunculagin (**8**). This is followed by a series of transformations that include glucose ring opening, C-glucosidic bond formation with the galloyl group at the two position, placement of a galloyl group at the 5-position, and formation of the flavogallonyl structure. The order in which this intriguing biotransformation occurs (as well as the exact mechanism) is unclear. Since compounds **6–8** are often found along with **2** and **9**, Haslam has suggested both a redox mechanism directly from **7** as well as a stepwise pathway from **8**.[12] Tanaka has suggested a C-glucosidation reaction based upon nucleophilic attack of the anomeric aldehyde and has shown that treating pedunculagin (**8**) in phosphate buffer afforded an 11 percent yield of the pedunculagin C-glucosidation product, casuariin.[28] More recently, Vivas and Glories have suggested, based upon molecular modeling data, a mechanism where the first step is gallate transfer from the 1-position to the 5-position, exposing the pyranose ring. Sequential 4,6- and 2,3-HHDP coupling, flavogallonyl synthesis and finally C-glucoside formation would lead to castalagin (**2**) and vescalagin (**9**).[29] Higher oligomers and oxidized variants of these monomers are also known.[31-33]

7. METHODS FOR INVESTIGATING ELLAGITANNIN BIOSYNTHESIS

As can be seen from the above discussions, ellagitannin biosynthesis is an area where much fundamental work still needs to be done. A key aspect of performing this work is the availability of experimental protocols for isolating and

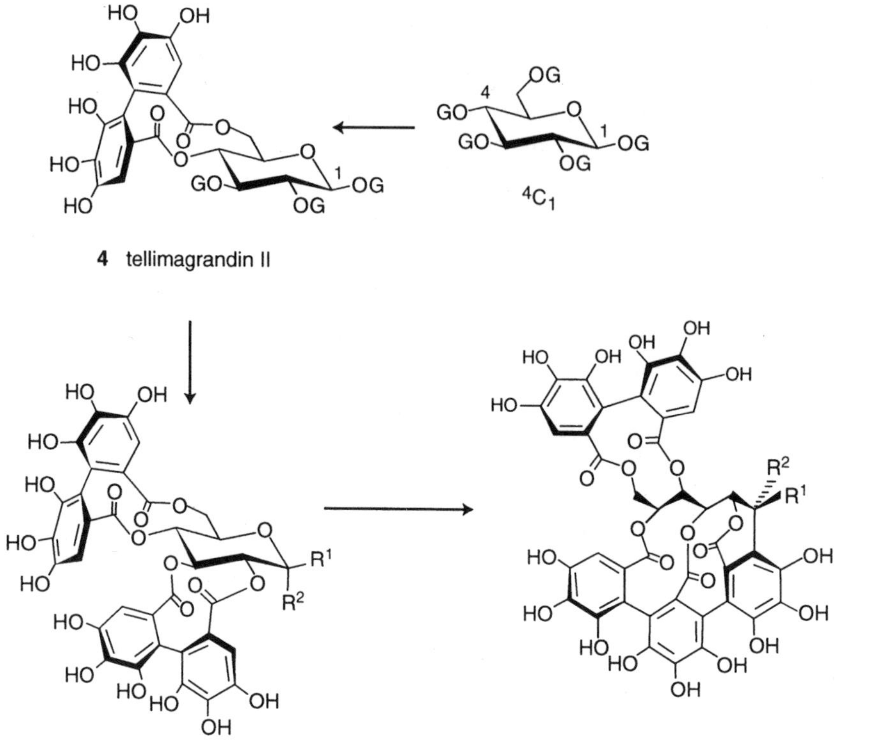

Figure 5. Generalized pathway to the C-glucosidic ellagitannins.

characterizing the ellagitannins as well as the enzymes and cofactors associated with their formation.

The simplest way to obtain ellagitannins is via extraction of the tannin-containing substrate with solvents such as methanol or acetone:water (7:3).[34-36] This of course requires preparation of the substrate to be extracted, and typically for wood, grinding with a Wiley mill is required. There are many variables involved in such processes, and these have been outlined in two excellent reviews.[34,36] It is imperative that sample preparation be accomplished as soon as possible after harvest to minimize tannin degradation, and efforts need to be made to minimize contact with metal ions and oxidants.

The quantification of tannins can be most conveniently accomplished in two ways, spectrophotometrically or by HPLC. In the past, most investigators relied on the rapid spectrophotometric methods.[34,36] These protocols typically involve degradation of a crude extract under acidic conditions in the presence of reagents that react with tannins/polyphenols to provide chromophores. These chromophores can then be quantified spectrophotometrically versus standards such as catechin or gallic acid. The most useful method for condensed tannins is vanillin/H_2SO_4, whereas degradation with nitrous acid is best for galloyl-containing tannins.[35] Total phenols are typically measured with the Folin-Ciocalteu reagent, which will detect almost all phenol-containing compounds.[34]

Although the above methods are quick and convenient, they are not absolute indicators of tannin content, nor do they provide any indication of what specific compounds are present. Investigators have therefore turned to HPLC to detect and quantify specific tannins. This is not as simple as one would think, especially for the ellagitannins, as the peak shapes and retention times tend to vary depending upon such seemingly minor issues as the water content of the injection solvent.[37] Nonetheless, the preferred routine method for monomeric ellagitannin separation and quantitation is via gradient C_{18} reverse-phase chromatography. In the near future, one can expect that capillary electrophoresis[38] as well as HPLC/MS[39] will become more routine for the quantification of ellagitannins.

The methods used for purification and characterization of the enzymes related to hydrolyzable tannin biosynthesis have been generated, almost exclusively, in the laboratory of Georg Gross (University of Ulm, Germany).[40–42] Buffered extracts are treated with insoluble polyvinylpyrrolidone (PVP) to eliminate inactivating phenolics and other unwanted contaminants, and the extract is then submitted to a series of chromatographic purifications, in combination with a treatment (protamine sulfate) to remove residual acids and an ammonium sulfate precipitation. These elegant efforts have led to the elucidation of the β-PGG biosynthetic pathway,[13,15] evidence of depside bond formation,[16] and the identification of galloyl-cleaving esterases from oak.[42]

Though these efforts have dramatically advanced our understanding of gallotannin biosynthesis, efforts at providing purified enzymes with *in vitro* activity for the biaryl coupling associated with ellagitannins have not been so successful. At present, only ellagic acid can be detected using purified protein extracts from native tissue isolates.[15] As stated in a recent review article concerning ellagitannin biosynthesis: *It is evident that sophisticated techniques and unconventional new strategies will be required for the eventual clarification of this challenging question.*[13]

If we are to advance our understanding of the processes associated with ellagitannin biosynthesis, we must think about new ways to systematically understand the process. If we wish to isolate and characterize the enzymes associated with the process, it will not be technically feasible to isolate such materials from the field. There would just be too much material to harvest and process in order to obtain enough protein. A simpler approach would be to generate cell cultures in the laboratory and ascertain what conditions are necessary to promote ellagitannin biosynthesis.

Several laboratories have embarked on such an endeavor.[22–25,44,45] Such tissues, if grown under appropriate conditions, would also be devoid of many of the toxic phenolics and contaminating photosynthetic pigments that have complicated previous efforts with leaf and tissue extracts. Furthermore, such tissues would be amenable to non-destructive NMR bioassays where callus tissue and/or cell suspension cultures could be placed in an NMR tube and supplemented with [1-^{13}C] β-PGG, the starting material for all ellagitannins. Such approaches have been highly successful with other biosynthetic pathways.[14,46] It is well established that the anomeric carbon of ellagitannins is extremely sensitive to the presence and absence of gallate and ellagate groups and is useful as a diagnostic tool for identifying specific ellagitannins.[47] This relatively simple bioassay could then be used to aid in the separation and purification of active fractions for further purification. If activity is lost, then one can go back and determine what step or process eliminated the activity.

8. STUDIES WITH *QUERCUS ALBA* CALLUS TISSUE

A longterm goal of our laboratory is to determine and then manipulate the ellagitannin biosynthetic pathway in oak in hopes of controlling heartwood formation. Since we desired a laboratory-based system to investigate the process, oak calli were derived from stem cutting explants from a healthy, young white oak tree (specimen is still living) growing in rural southwest Virginia. The surface sterilized explants were grown on Murashige-Skoog solid media supplemented with sucrose, napthaleneacetic acid (NAA), benzyladenine (BA), and Linsmaier/Skoog vitamins. After 5–6 weeks, the induced white/yellow calli were sectioned into 5 mm squares and subcultured onto fresh media. Subsequently, the calli were subcultured monthly.

While the media preparation employed was reported to be successful in producing ellagitannins with *Quercus acutissima*,[45] our initial callus extractions with either methanol or acetone:water (7:3) followed by HPLC analysis revealed only trace levels of 1, 2 and 9. We then began a series of experiments aimed at increasing ellagitannin production via media modification. Our first modification was to increase the copper concentration in the media tenfold. This modification was based on the following assumption. Compound 1 has a relatively high oxidation potential, thus the energy required to undergo biaryl coupling should be much lower than that of a lignin monomer. Since laccases are known to have a wide range of oxidation potentials,[48] we assumed that biaryl coupling may involve a laccase. If this is the case, perhaps there was not enough copper present in the media to prepare a fully operational laccase enzyme. Such a situation is not without precedent, as has been reported in studies of the laccases present in *Acer pseudoplatanus*.[49–52] In these studies, a fully functional laccase (one capable of oxidizing a lignin monomer) was only obtained when the media were supplemented with copper.

Callus cultures grown in both the regular and high copper media were removed from the plates, freeze-dried, and then ground to a powder with a mortar and pestle. The powdered material was extracted once with acetone:water (7:3) for 48 h. Processing provided an amber powder that was analyzed by HPLC using an end-capped C_{18} column as well as a gradient elution profile.[53] The resulting chromatogram for the high copper acetone:water extract is shown in figure 6 along with the heartwood extract of chestnut oak (*Quercus prinus*, a member of the white oak family). The figure clearly shows that the callus tissues produce ellagitannin profiles similar to that of a heartwood extract. The only compounds that are

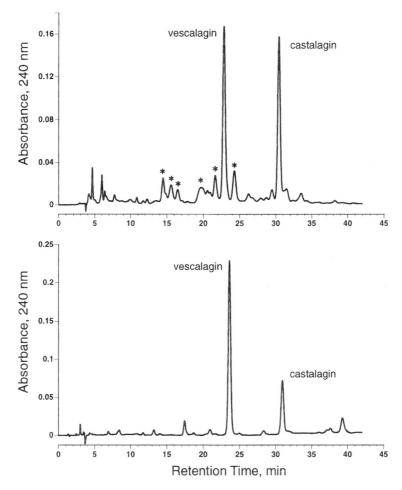

Figure 6. Gradient reverse-phase HPLC analyses of the acetone-water extracts of *Quercus prinus* (chestnut oak) heartwood (top) and *Quercus alba* (white oak) callus tissue (bottom). Structures have been confirmed in both isolates by purification and NMR analysis.

missing are the roburins and grandidin, which are the oligomer ellagitannins symbolized by asterisks on the *Q. prinus* chromatogram (top).[31,37] That the callus tissues do not produce high levels of dimers is not considered a detriment as this makes separation, analysis, and determination of the biochemical pathway to castalagin and vescalagin a bit easier. Subsequent studies with media with limited nitrogen have also shown the same ellagitannin profiles. Thus, we have two different media modifications that can produce ellagitannins from *Q. alba* callus tissue (table 1).[53]

The utility of a callus or cell suspension system is that they are amenable to a non-destructive NMR bioassay. Ellagitannin-producing tissues can be taken up

Table 1. Castalagin, vescalagin and insoluble ellagitannin contents of *Quercus alba* heartwood and callus tissues

Source	Vescalagin (percent)	Castalagin (percent)	Insoluble Ellagitannins (percent)
Heartwood	0.47	0.65	0.31
Copper Supplemented Callus[a]	1.46	1.25	0.70
Copper Supplememented Callus[a] minus Ammonium Nitrate	1.20	1.09	0.66

[a]Standard Murishage-Skoog media supplemented with cupric sulfate (250 mg/L), in the presence (regular MS) or absence (MS - NH_4NO_3) of supplemental nitrogen.

in a buffer and mixed with labeled **1**, with the samples analyzed by NMR at selected time points. Our initial NMR experiments involved grinding wet callus tissue in liquid nitrogen and taking this solid up in TRIS-acetate buffer. A portion of the suspension was placed in an NMR tube with [1-^{13}C]-**1**, and several spectra were recorded (unlocked) over a period of 48 hr. No changes were noted in the anomeric region, with one peak for the labeled position (arbitrarily given a value of 94 ppm). However, after flushing the NMR tube with oxygen, three peaks were observed at 96, 94, and 91.5 ppm (respective peak areas: 30, 50, and 20 percent, respectively). Half of the labeled starting material had been converted to other materials. This result supports our claim that the NMR bioassay method will be useful for ascertaining key elements of the ellagitannin biosynthetic process.

During normal callus tissue growth, small, straw-colored aqueous globules can be observed to grow on the surface of the tissues. This material was collected, and the protein concentration was determined to be 4.1 mg/mL (Sigma, bicinchoninic acid protein assay kit). This protein content is probably overestimated due to the presence of contaminating substances such as phenols. A portion of the exudate was mixed with a solution of [1-^{13}C]-**1**, which had been freshly dissolved in DMSO-d_6. This mixture was then added to a standard NMR tube and flushed with oxygen. NMR spectra were then obtained over the period of the next 2 weeks (25 min acquisition times), and the spectra are shown in figure 7.

The spectra clearly show the presence of several new compounds. After 2 weeks, 30 percent of the starting material had been converted to other labeled compounds. The sample was removed from the NMR tube and purified by separation on Toyopearl HW-40F followed by acetylation and preparative TLC.[54] Characterization of some of the enriched components (96 ppm and 91.5 ppm) led to the isolation of de-esterified variants of **1**. This de-galloylation suggests that there are esterases present that are capable of cleaving galloyl groups from **1**, the presence of which has recently been reported from extracts obtained from the leaves of pedunculate oak (*Quercus robur*, syn. *Quercus pedunculata*).[43] While this result suggests that additional fractionation will be required to provide preparations capable of providing biaryl coupling *in vitro*, the results demonstrate the applicability of the non-destructive NMR method for monitoring the modification of **1**. Though the conversion times are quite slow, it should be noted that all of the NMR

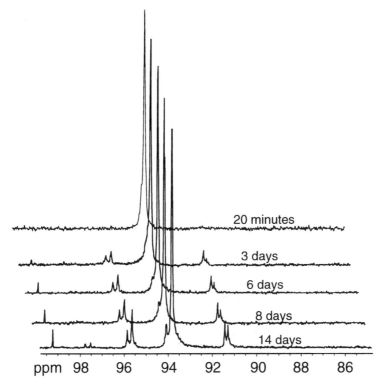

Figure 7. NMR spectra of a solution containing [1-^{13}C]-**1** and callus exudate. Spectra were recorded on a Bruker AM-360 at 27 °C using the standard proton-decoupled ^{13}C pulse program (1024 scans).

experiments that have been presented here have been accomplished with the calli grown on the regular copper media. Use of calli grown on the high copper and/or nitrogen-limited media should provide increased conversion rates. Conversion of these cells to suspension cultures is on the horizon as these systems would allow for much more quantitative control of growth and label addition as well as provide the ability to monitor the cells *in vivo*.

9. EXPERIMENTAL

The protocols for preparation of the wood samples and the protocols for preparation of the callus cultures were as described previously. [1-^{13}C]-β-PGG was prepared from perbenzylated gallic acid and [1-^{13}C]-D-glucose using a DMAP/DMAP·HCl/DCC system for acylation.[20] The perbenzylated β-anomer can be separated from its α-counterpart by silica gel chromatography; subsequent debenzylation provides [1-^{13}C]-**1** in about 90–95 percent purity. Final purification can be accomplished by preparative C$_{18}$ flash chromatography (LRP C$_{18}$, Whatman), eluting first with 20 percent MeOH/80 percent aq. TFA (0.2 percent) and then with

40 percent MeOH. Overall yields based on [1-^{13}C]-D-glucose are typically 50–52 percent.

The NMR bioassay experiments involved taking wet callus tissue (2.5 gm, wet weight) and grinding it to a powder in liquid nitrogen. This solid was then taken up in TRIS-acetate buffer (50 mM, pH 6.5, 10 mL), and a portion (0.5 mL) of the suspension was placed in an NMR tube with [1-^{13}C]-β-PGG. After 48 hr, the NMR tube was then flushed with oxygen, and spectra were recorded over a period of several weeks. The experiments using the exudate were performed in the same manner. A portion of the exudate (350 μL) was mixed with a solution of [1-^{13}C]-**1** (10 mg), which had been freshly dissolved in DMSO-d_6 (50 μL). This mixture was then added to a standard 5 mm NMR tube flushed with oxygen. NMR spectra were then obtained over the period of the next 2 weeks (25 min acquisition times).

All calculations centering on the modeling of **1** were performed at the PM3 semi-empirical level, with full geometry optimization, using Unichem under license from Cray Research. All molecular orbital calculations were executed on a Cray C90 supercomputer, operated by the Alabama Supercomputer Authority in Huntsville, Alabama, while visualizations were done through a Silicon-Graphics Indigo2 workstation, housed in the Department of Chemistry at Auburn University.

10. CONCLUSIONS

Despite years of study, the biosynthesis of ellagitannins remains a challenging problem. Although the initial compound leading to this class of compounds is almost certainly **1**, all steps subsequent to this remain obscure. The Haslam-Schmidt hypothesis states that the specifics of biaryl coupling are due to the conformation of the glucopyranose ring. The lack of NMR evidence for an all axial conformer with either **1** or its α-anomer, as well as the fact that ellagitannins are typically found as individual atropisomers rather than a mixture, would seem to suggest that more biological processes are involved. We have recently identified a *Quercus alba* callus system that produces significant quantities of the monomeric C-glycosidic ellagitannins castalagin and vescalagin without significant contamination with other polyphenols or ellagitannin oligomers. Although definitely a "work in progess," selective feeding studies with this system may prove invaluable to determining this challenging pathway.

ACKNOWLEDGMENTS

The authors would like to thank Professors Richard Veilleux and Gary Griffin (Virginia Tech University) for help in initially establishing our cell cultures and Professor Gross (University of Ulm) for his helpful correspondences. Portions of this work were financially supported by the USDA National Research Initiative Competitive Grants Program (Improved Utilization of Wood and Wood Fiber, Award Number 96-35103-3833) and the McIntire-Stennis Program.

REFERENCES

1. Haddock, E.A.; Gupta, R.K.; Al-Shafi, S.M.K.; Layden, K.; Haslam, E.; Magnolato, D. The metabolism of gallic acid and hexahydroxydiphenic acid in plants: biogenetic and molecular taxonomic considerations. *Phytochemistry.* 21:1049 (1982).

2. Okuda, T.; Yoshida, T.; Hatano, T. Classification of oligomeric hydrolysable tannins and specificity of their occurrence in plants. *Phytochemistry.* 32:507 (1993).

3. Gottlieb, O.R.; Borin, M.R.M.B.; Kaplan, M.A.C. Biosynthetic interdependence of lignins and secondary metabolites in angiosperms. *Phytochemistry.* 40:99 (1995).

4. Haslam, E. Natural polyphenols (vegetable tannins) as drugs: possible modes of action. *J. Nat. Prod.* 59:205 (1996).

5. Mila, I.; Scalbert, A.; Expert, D. Iron withholding by plant polyphenols and resistance to pathogens and rots. *Phytochemistry.* 42:1551 (1996).

6. Luck, G.; Liao, H.; Murray, N.J.; Grimmer, H.R.; Warminski, E.E.; Williamson, M.P.; Lilley, T.H.; Haslam, E. Polyphenols, astringency and proline-rich proteins, *Phytochemistry.* 37:357 (1994).

7. Appel, H. Phenolics in ecological interactions: the importance of oxidation. *J. Chem. Ecol.* 19:1521 (1993).

8. Okuda, T.; Yoshida, T.; Hatano, T. Ellagitannins as active constituents of medicinal plants. *Planta Medica.* 59:117 (1989).

9. Charrier, B.; Haluk, J.P.; Metche, M. Characterization of European oakwood constituents acting in the brown discoloration during kiln drying. *Holzforschung.* 49:168 (1995).

10. Vivas, N.; Glories, Y. Role of oak wood ellagitannins in the oxidation process of red wines during ageing. *Am. J. Enol. Vitic.* 47:103 (1996).

11. Quideau, S.; Feldman, K.S. Ellagitannin chemistry. *Accts. Chem. Res.* 96:475 (1996).

12. Haslam, E.; Cai, Y. Plant polyphenols (vegatable tannins): Gallic acid metabolism. *Nat. Prod. Rep.* 41 (1994).

13. Gross, G.G. Enzymatic synthesis of gallotannins and related compounds. *In:* Stafford, H.A.; Ibrahim, R.K. (eds.) Phenolic metabolism in plants. Plenum Press, New York. p. 297 (1992).

14. Werner, I.; Bacher, A.; Eisenreich, W. Retrobiosynthetic NMR studies with ^{13}C-labeled glucose. Formation of gallic acid in plants and fungi. *J. Biol. Chem.* 272:25474 (1997).

15. Gross, G.G. Enzymes in the biosynthesis of hydrolyzable tannins. *In:* Hemingway, R.W.; Laks, P.E. (eds.) Plant polyphenols—synthesis, properties, significance. Plenum Press, New York. p. 43 (1992).

16. Hofman, A.S.; Gross, G.G. Biosynthesis of gallotannins: Formation of polygalloylglucoses by enzymatic acylation of 1,2,3,4,6-penta-*O*-galloylglucose. *Arch. Biochem. Biophys.* 283:530 (1990).

17. Eliel, E.L.; Wilen, S.H. Stereochemistry of organic molecules. Wiley-Interscience, New York pp. 1119–1121 (1995).

18. Davin, L.B.; Wang; H.-B.; Crowell, A.L.; Bedgar, D.L.; Martin, D.M.; Sarkanen, S., Lewis, N.G. Stereoselective bimolecular phenoxy radical coupling by an auxiliary (dirigent) protein without an active center. *Science.* 275:362 (1997).

19. Feldman, K.S.; Kirchgessner, K.A. Probing ellagitannin stereochemistry: The rotational barrier of a model (4,6)-hexamethoxydiphenoyl glucopyranoside. *Tetrahedron Letters.* 38:7981 (1997).

20. Helm, R.F.; Ranatunga, T.D.; Chandra, M. Lignin-hydrolyzable tannin interactions in wood. *J. Agric. Food Chem.* 45:3100 (1997).

21. Haddock, E.A.; Gupta, R.K.; Haslam, E. The metabolism of gallic acid and hexahydroxydiphenic acid in plants. Part 3. Esters of (R)- and (S)-hexahydroxydiphenic acid and dehydrodiphenic acid with D-glucopyranose ($^1C_4$ and related conformations). *J. Chem. Soc., Perkin Trans. 1.* :2535 (1982).

22. Ishimaru, K.; Shimomura, K. Tannin production in hairy root culture *of Geranium thunbergii. Phytochemistry.* 30:825 (1991).

23. Neera, S.; Ishimaru, K. Tannin production in cell cultures of *Sapium sebiferum. Phytochemistry* 31:833 (1992).

24. Neera, S.; Arakawa, H.; Ishimaru, K. Tannin production in *Sapium sebiferum* callus cultures. *Phytochemistry.* 31:4143 (1992).

25. Taniguchi, S.; Yoshida, T.; Nakamura, N.; Nose, M.; Takeda, S.; Yabu-Uchi, R.; Ito, H.; Yazaki, K. Production of macrocyclic ellagitannin oligomers by *Oenothera laciniata* callus cultures. *Phytochemistry.* 48:981 (1998).

26. Tanaka, T.; Kouno, I.; Nonaka, G. Glutathione-mediated conversion of the ellagitannin geraniin into chebulagic acid. *Chem. Pharm. Bull.* 44:34 (1996).

27. Tanaka, T.; Jiang, Z.-H.; Kouno, I. Structures and biogenesis of rhiopteleanins, ellagitannins formed by stereospecific intermolecular C-C oxidative coupling, isolated from *Rhioptelea chiliantha. Chem. Pharm. Bull.* 45:1915 (1997).

28. Vivas, N.; Laguerre, M.; Glories, Y.; Bourgeois, G.; Vitry, C. Structure simulation of two ellagitannins from *Quercus robur* L. *Phytochemistry.* 39:1193 (1995).

29. Tanaka, T.; Kirihara, S.; Nonaka, G.; Nishioka, I. Tannins and related compounds. CXXIV. Five new ellagitannins, platycaryanins A, B, C, and D, and platycariin, and a new complex tannin, strobilanin, from the fruits and bark of *Platycarya strobilacea* (Sieb et Zucc.), and biomimetic synthesis of C-glycosidic ellagitannins from glucopyranose-based ellagitannins. *Chem. Pharm. Bull.* 41:1708 (1993).

30. Nonaka, G.; Sakai, T.; Tanaka, T.; Mihashi, K.; Nishioka, I. Tannins and related compounds. XCVII. Structure revision of C-glycosidic ellagitannins, castalagin, vescalagin casuarinin and stachyurin, and related hydrolyzable tannins. *Chem. Pharm. Bull.* 38:2151 (1990).

31. Hervé du Penhoat, C.L.M.; Michon, V.M.F.; Peng, S.; Viriot, C.; Scalbert, A.; Gage, D. Structural elucidation of new dimeric ellagitannins from *Quercus robur L.* Roburins A-E. *J. Chem. Soc. Perkin Trans. 1.* :1653 (1991).

32. Tanaka, T.; Ueda, N.; Shinohara, H.; Nonaka, G.; Fujioka, T.; Mihashi, K.; Kouno, I. C-Glycosidic ellagitannin metabolites in the heartwood of Japanese chestnut tree (*Castanea crenata* Sieb. et Zucc.). *Chem. Pharm. Bull.* 44:2236 (1996).

33. Tanaka, T.; Ueda, N.; Shinohara, H.; Nonaka, G.; Kouno I. Four new C-glycosidic ellagitannins, castacrenins D-G, from Japanese chestnut wood (*Castanea crenata Sieb. et Zucc.*). *Chem. Pharm. Bull.* 45:1751 (1997).

34. Waterman, P.G.; Mole, S. Analysis of phenolic plant metabolites. Blackwell Scientific Publications, Oxford. pp. 66–103 (1994).

35. Scalbert, A.; Monties B.; Janin, G. Tannins in woods: comparison of different estimation methods, *J. Agric. Food Chem.* 37:1324 (1989).

36. Scalbert, A., Quantitative methods for the estimation of tannins in plant tissues, *In*: Hemingway, R.W.; Laks, P.E. (eds.) Plant polyphenols—synthesis, properties, significance. Plenum Press, New York. p. 259 (1992).

37. Scalbert, A.; Duval, L.; Peng, S.; Monties, B.; du Penhoat, C. Polyphenols of *Quercus robur L.* II. Preparative isolation by low-pressure and high-pressure liquid chromatography of heartwood ellagitannins. *J. Chromatogr.* 502:107 (1990).

38. Fernandes, J.B.; Griffiths, D.W.; Bain, H. The evaluation of capillary zone electrophoresis and micellar electrokinetic capillary chromatographic techniques for the simultaneous determination of flavonoids, cinnamic acids and phenolic acids in Black currant (*Ribes nigrum*) bud extracts. *Phytochem. Anal.* 7:97 (1996).

39. Nawwar, M.A.M.; Marsouk, M.S.; Nigge, W.; Linsheid, M. High-performance liquid chromatographic/electrospray ionization mass spectrometric screening for polyphenolic compounds of *Epilobium hirsutum*—the structure of the unique ellagitannin epilobamide-A. *J. Mass Spectrom.* 32:645 (1997).

40. Gross, G.G. Synthesis of β-glucogallin from UDP-glucose and gallic acid by an enzyme preparation from oak leaves. *FEBS Letters*. 148:67 (1982).

41. Gross, G.G.; Schmidt S.W.; Denzel, K. β-Glucogallin-dependent acyltransferase from oak leaves. I. Partial purification and characterization. *J. Plant Physiol*. 126:173 (1986).

42. Cammann, J.; Denzel, K.; Schilling, G.; Gross, G.G. Biosynthesis of gallotannins: β-glucogallin-dependent formation of 1,2,3,4,6-pentagalloylglucose by enzymatic galloylation of 1,2,3,6-tetragalloylglucose. *Arch. Biochem. Biophys*. 273:58 (1989).

43. Neihaus, J.U.; Gross, G.G. A gallotannin degrading esterase from leaves of pedunculate oak. *Phytochemistry*. 45:1155 (1997).

44. Scalbert, A.; Monties, B.; Farve, J.-M. Polyphenols of *Quercus robur*: Adult tree and *in vitro* grown calli and shoots. *Phytochemistry*. 27:3483 (1988).

45. Tanaka, N.; Shimomura, K.; Ishimaru, K. Tannin production in callus cultures of *Quercus acutissima*. *Phytochemistry*. 40:1151 (1995).

46. Ford, Y.-Y.; Ratcliffe, R.G.; Robins, R.R. *In vivo* NMR analysis of tropane alkaloid metabolism in transformed root and de-differentiated cultures of *Datura stramonium*. *Phytochemistry*. 43:115 (1996).

47. Okuda, T.; Yoshida T.; Hatano, T. New methods of analyzing tannins. *J. Nat. Prod*. 52:1 (1989).

48. Xu, F.; Shin, W.; Brown, S.H.; Wahleithner, J.A.; Sundaram U.M.; Solomon, E.I. A study of a series of recombinant fungal laccases and bilirubin oxidase that exhibit differences in redox potential, substrate specificity and stability. *Biochim. Biophys. Acta*. 1292:303 (1996).

49. Bligny, R.; Gaillard J.; Douce, R. Excretion of laccase by sycamore (*Acer pseudoplatanus* L.) cells. Effects of a copper deficiency. *Biochem. J*. 237:583 (1986).

50. Driouich, A.; Lainé; A.C.; Vian, B.; Faye, L. Characterization and localization of laccase forms in stem and cell cultures of sycamore. *Plant J*. 2:13 (1992).

51. Sterjiades, R.; Dean, J.F.D.; Eriksson, K.-E.L. Laccase from sycamore maple *(Acer pseudoplatanus)* polymerizes monolignols. *Plant Physiol*. 99:1162 (1992).

52. Sterjiades, R.; Dean, J.F.D.; Gamble, G.; Himmelsbach, D.S., Eriksson, K.-E.L. Extracellular laccases and peroxidases from sycamore maple (*Acer pseudoplatanus*) cell-suspension cultures. Reactions with monolignols and lignin model compounds. *Planta*. 190:75 (1993).

53. Zhentian, L.; Jervis, J.; Helm, R.F. C-Glycosidic ellagitannins from white oak heartwood and callus tissues. *Phytochemistry* 51:751 (1999).

54. Viriot, C.; Scalbert, A.; Herve du Penhoat, C.L.M.; Rolando, C.; Moutounet, M. Methylation, acetylation and gel permeation of hydrolysable tannins. *J. Chromatogr. A*. 662:77 (1994).

PROSPECTS AND PROGRESS IN ELLAGITANNIN SYNTHESIS

Ken S. Feldman,[a] Kiran Sahasrabudhe,[a] Stéphane Quideau,[b]
Kendra L. Hunter,[a] and Michael D. Lawlor[a]

[a] Department of Chemistry
The Pennsylvania State University
University Park, Pennsylvania 16802
USA

[b] Laboratoire de Chimie des Substances Végétales
Institut du Pin—Université Bordeaux I
351 cours de la Libération
33405 Talance
FRANCE

1. INTRODUCTION

The ellagitannin family of secondary plant metabolites presents a rich array
of challenges for contemporary organic synthesis. The myriad intra- and intermolecular coupling modes available to galloyl rings appended to a glucose core,
along with a range of post-coupling modifications, define a vast matrix of
bond-forming possibilities, many of which are represented within the 500+ structurally characterized members.[1-4] A heightened interest in this class of natural
products is fueled by recent observations that several ellagitannins display
remarkable levels of activity in various anticancer and antiviral assays and hence
may serve as promising leads for development of novel therapeutics.[3,5-7] Progress toward that goal will benefit from access to significant quantities of natural
material as well as designed analogs for delineation of structure/activity profiles
and elucidation of biological mechanism-of-action. These latter goals, and possibly
the former goal as well, can only be met through a program of total chemical
synthesis.

2. BIOSYNTHESIS CONSIDERATIONS

The almost bewildering structural diversity among members of the ellagitannin family makes development of a comprehensive synthesis strategy a daunting

prospect. However, closer inspection through the lens of current biosynthesis dogma reveals some simplifying guidelines that might aid in the elaboration of a rational approach. Current understanding in the area of ellagitannin biosynthesis centers on the pivotal role of β-pentagalloylglucose (fig. 1; β-pgg; 1) as progenitor for each of the subfamilies via some permutation and/or combination of galloyl coupling.[1-4] Both C—C and C—O bond formation between galloyl esters are possible within this context, thus further expanding the repertoire of possible ellagitannin products. Imagination exceeds experiment at present, as direct evidence supporting even the most elementary steps in ellagitannin biosynthesis is not yet forthcoming. Nevertheless, the biosynthesis speculation has been extended to incorporate a second point of departure, tellimagrandin II (2), enroute to various other ellagitannin metabolites.[1-3] In this scenario, initial tellimagrandin II synthesis via O(4)/O(6) galloyl coupling within β-pgg 1 is followed by a host of "second-generation" coupling processes to deliver the representative species 3–6. Thus, subsequent O(2)/O(3) galloyl coupling within 2 might yield casuarictin (3), whereas oxidation of the O(6) galloyl ring within the hexahydroxydiphenoyl (HHDP) unit of 2 could afford the dehydrohexahydroxydiphenoyl (DHHDP) moiety of

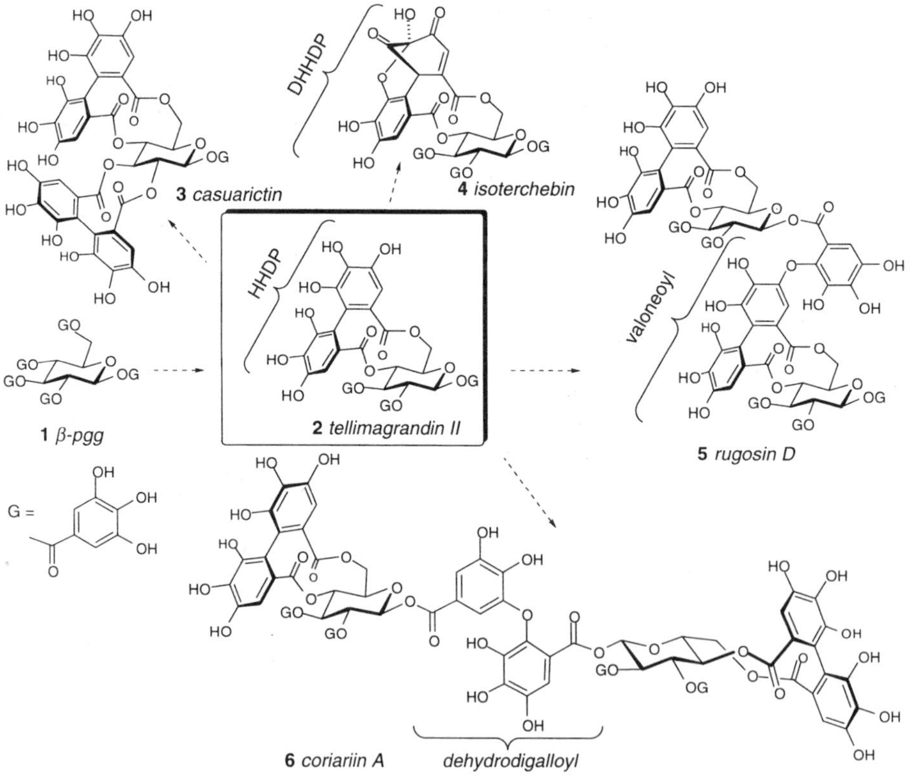

Figure 1. Possible biosynthesis of representative ellagitannins from a tellimagrandin II precursor.

isoterchebin (**4**). The dimeric ellagitannins rugosin D (**5**) and coriariin A (**6**) can be assembled by formal intermolecular C—O coupling of galloyls between two tellimagrandin II monomers. Oxidative C—O attachment between the O(1) galloyl in one molecule of **2** and the HHDP-bound O(6) galloyl in a second molecule of **2** will provide the valoneoyl ester component of **5**, whereas fashioning a C—O linkage between the anomeric galloyls of two tellimagrandin II monomers can yield the dehydrodigalloyl unit of coriariin A (**6**). It is interesting to note that tellimagrandin II is a co-isolate with each of the more complex ellagitannins **3–6**.[8–11] *This hierarchical perspective on ellagitannin biosynthesis can provide an instructive framework from which to design biomimetic chemical synthesis strategies for representative ellagitannins.*

Clarification of the intimate chemical details by which galloyl esters are oxidatively coupled during ellagitannin biosynthesis to form HHDP units remains for the future. Conjecture has, in general, fallen into one of two camps that differ in the level of oxidation experienced by each galloyl undergoing coupling (fig. 2).[12] The weight of historical precedence favors two classical one-electron oxidations (= phenolic hydrogen abstractions) to furnish an intermediate diradical **9**, which then can couple to produce the HHDP moiety. An alternate hypothesis cites a two-electron oxidation of one of the galloyl rings to furnish an electrophilic ortho-quinone intermediate (cf. **1** → **7**), which can be trapped by a proximal nucleophilic galloyl ring to complete HHDP formation. The electrophilicity of the orthoquinone can be enhanced, in principle, by protonation to generate the cationic species **7a/b**. This distinction is significant from the standpoint of synthesis design, as entirely different sets of chemistry must be employed to achieve either one- or two-electron galloyl oxidations.

The one-electron oxidation hypothesis seems most applicable to intramolecular cyclizations or perhaps even intermolecular homodimerizations under favorable concentration regimes. However, heterodimerization of distinct ellagitannin monomer radicals to generate either C—O or C—C linked ellagitannin dimers may

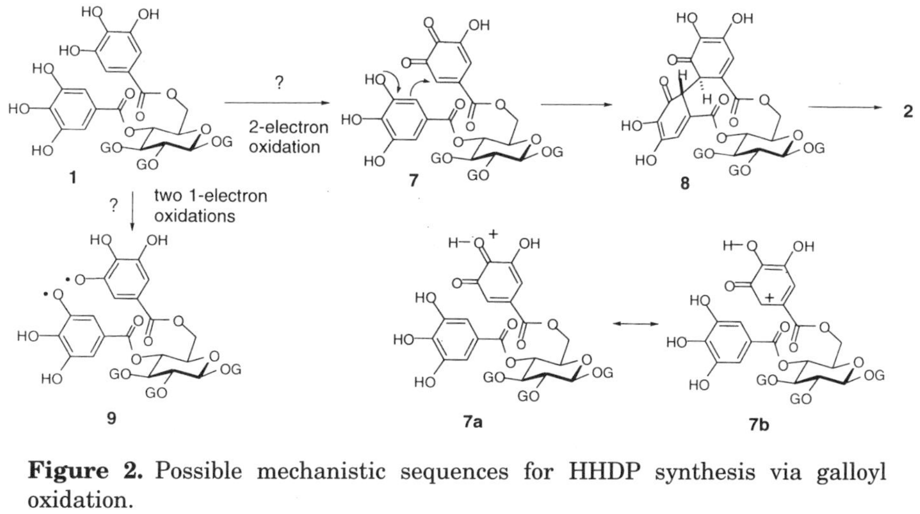

Figure 2. Possible mechanistic sequences for HHDP synthesis via galloyl oxidation.

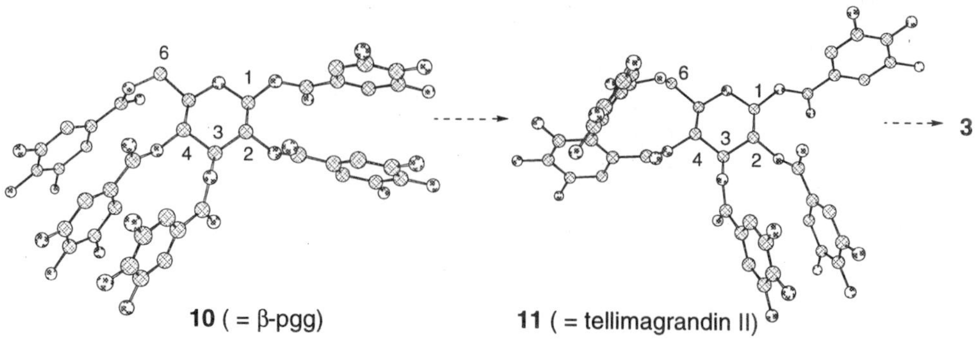

Figure 3. A proposed biosynthetic sequence from β-pgg **1** to casuarictin (**3**).

be problematic either *in vivo* or *in vitro*. Heteropolar galloyl coupling (i.e., **7** → **8**) doesn't suffer from this limitation. Thus, the orthoquinone-mediated nucleophile/electrophile pairing strategy seems to offer the most flexibility and therefore forms the basis for approaches to both C—C and C—O bond construction within the context of complex ellagitannin synthesis as described below.

The case for casuarictin biosynthesis via a tellimagrandin II intermediate in at least some plant species is strengthened by the results of an exhaustive search of the conformational space available to $^4C_1$ β-pgg **1** using molecular mechanics (MM) techniques[a] (fig. 3).[13]

The "global" minimum energy conformation **10** conveys a sense of the natural alignment of the pendant galloyl residues: the O(4) and O(6) rings are nicely juxtaposed for coupling, whereas the O(2) and O(3) galloyl units are splayed far apart. If ground state proximity translates into chemical reactivity in this system, then this modeling exercise would support the aforementioned biosynthesis hypothesis citing initial O(4)/O(6) coupling to yield **2** rather than reaction at the alternative O(2)/O(3) site to deliver the isomeric naturally occurring ellagitannin pterocaryanin C (not shown). A similar search of the conformational options accessible to the putative coupled product **2** indicates that, in the minimum energy conformer **11**, the O(2) galloyl has shifted position and now aligns itself with the O(3) galloyl ester as per expectations for casuarictin synthesis. However, this simple analysis does not tell the whole story. While casuarictin and tellimagrandin II are congeners in *Quercus phillyraeoides* A. GRAY,[8] *Melastroma malabathricum* features casuarictin and pterocaryanin C with no detectable tellimagrandin II.[14] Clearly, different plant species have evolved divergent biosynthetic pathways for the same target. The implications for developing a biomimetic synthesis strategy are evident. The two distinct coupling sequences implied by the isolation results provide permissive evidence that neither route suffers from insurmountable

[a]Conformational analyses were performed with the directed Monte Carlo algorithm of Macromodel 5.5 on a Silicon Graphics Indigo 2XZ computer equipped with an R8000 processor. Each molecule analyzed in this manner was subjected to either a 2,000-step search (3 rotatable bonds), a 5,000-step search (4 or 5 rotatable binds) or a 10,000-step search (6 rotatable bonds). In each case, the reported global minimum energy structure was found several times.

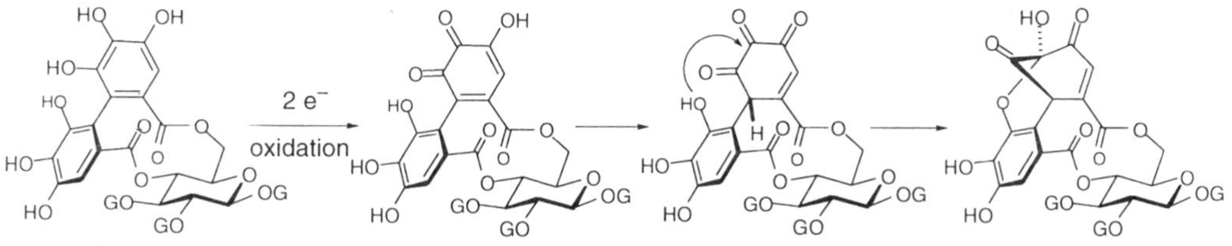

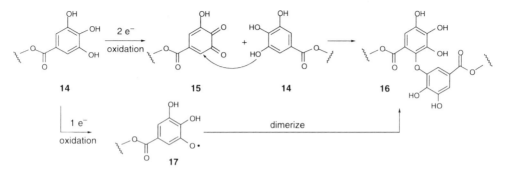

Figure 4. A proposed biosynthetic sequence from β-pgg **1** to isoterchebin (**4**).

obstacles to bond formation. However, the natural conformational preferences revealed by the MM calculation suggest that an "O(4)/O(6) first" approach to casuarictin might enjoy some kinetic advantages.

Isoterchebin (**4**) biosynthesis through a tellimagrandin II precursor is, in principle, more straightforward than casuarictin formation and requires no more than a formal 2-electron oxidation of the O(6) galloyl ring (fig. 4). The highly electrophilic central carbonyl of the trione tautomer **13** then can trap the neighboring nucleophilic phenolic hydroxyl to complete the isoterchebin skeleton. This pathway introduces an important theme in ellagitannin biosynthesis: galloyl ester oxidation leading to C—O rather than C—C bond formation. In this intramolecular case, geometric constraints dictate the regiochemistry of C—O bond closure. With these constraints removed, other C—O attachment modes between electrophilic galloyl-derived orthoquinones and phenolic hydroxyls are available.

Two of the many conceivable modes of tellimagrandin II dimerization are represented in the G6(O)→G1′(C) and G1(O)→G1′(C) connections of rugosin D (**5**) and coriariin A (**6**), respectively. As an initial iteration, it is possible to extend the two C—C coupling hypotheses discussed with figure 2 to the C—O bond-forming regime, figure 5. One plausible route features 2-electron oxidation and proceeds through an electrophilic orthoquinone **15** (or a protonated derivative) in partnership with a nucleophilic phenolic hydroxyl to formulate the key C—O bond via conjugate addition. An alternative sequence might rely on combination of

Figure 5. Possible mechanistic sequences for dehydrodigalloyl ether biosynthesis via galloyl oxidation.

two phenoxy radicals **17** to link the carbon of one galloyl ring with the oxygen of another ring.

Both of these intermolecular bond-forming processes raise selectivity issues that were not evident in the intramolecular cyclizations discussed earlier. Questions of site-specificity for both initial 2-electron oxidation and subsequent nucleophile (phenol) capture complicate the orthoquinone-based proposal. Similar uncertainties plague the diradical combination alternative, with the added burden of hetero- vs. homocoupling competition with dissimilar radical monomers. Finally, an overarching concern with both one- and two-electron oxidation-based biosynthetic hypotheses stems from the requirement for chemoselective C—O rather than C—C bond formation when the two reactive entities join. The role that enzymes play in steering the chemistry down productive channels still remains to be illuminated, and it is certainly within the realm of possibility that all of these concerns can be resolved favorably through enzyme-imposed alignment of reactive functionality. Such intrinsic control elements are not available to aid in the *in vitro* chemical assembly of these digalloyl ether linker units. Nevertheless, a biomimetic strategy for digalloyl ether synthesis that incorporates features of the putative orthoquinone-based biosynthesis proposal can be developed, as described below.

3. CHEMICAL SYNTHESIS CONSIDERATIONS

Synthesis of an HHDP unit from galloyl ester precursors must confront three issues: 1) bond formation itself, 2) control of regio- and atropselectivity, and 3) over-reaction (overoxidation) of the hexaphenolic product. A presumed biomimetic oxidative coupling strategy underlies numerous attempts to couple gallic acid or its derivatives over the past 130 years, with only limited success.[15-23] Invariably, one of two fates served to thwart high-yielding HHDP formation: Either ester hydrolysis under the coupling conditions discharged the HHDP unit from its glucose core (or from its simpler ester derivatives) to furnish ellagic acid (**19**), or an inadequate stop-message was built into the oxidation protocol and overoxidized products (e.g., benzotropolone **20**) resulted (fig. 6). Prior to 1993, the most encouraging example of oxidative HHDP formation was reported by Mayer, in which methyl gallate in the presence of horseradish peroxidase provided the HHDP derivative **21a** in 24 percent yield accompanied by substantial amounts of **19** (32%) and **20** (17%).[23] The opportunity for product oxidation can be removed by performing the galloyl coupling through a non-biomimetic *reductive* sequence with the bromogalloyl ester **22**.[24-26] This early Ullmann-type process (**22** → **21b**) presaged more recent chiral auxiliary-based attempts to impart absolute stereochemical control to the dihalide coupling, but as yet this approach has not been employed to deliver the free hexaphenoxy functionality required for ellagitannin synthesis.[25,26]

These early studies on unetherified gallate esters utilized both 1- and 2-electron oxidants with the former predominating as per the prevailing phenolic biosynthesis orthodoxy of the day. An alternative approach to galloyl oxidative coupling might exploit the capacity of substituents on the phenolic hydroxyls to

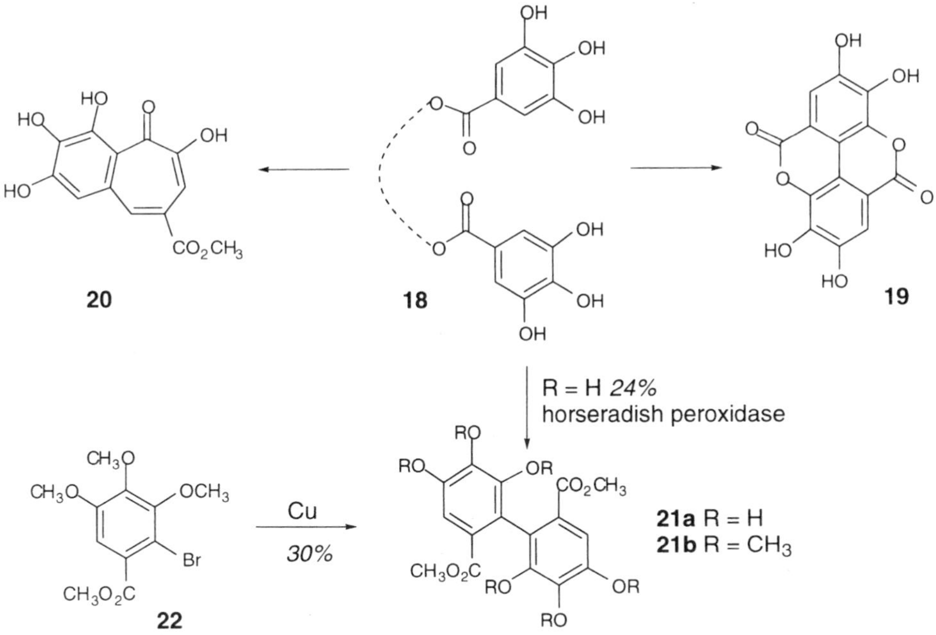

Figure 6. Early attempts at HHDP synthesis.

modulate the electrophilicity of a 2-electron oxidized intermediate (fig. 7). Thus, as an extension of the biosynthesis speculation indicated in figure 2, reactive electrophilic orthoquinone-like intermediates **24a/b**, whose reactivity can be tailored by design, may be available from functionalized galloyl precursors **23a/b**. In fact, much precedent for C-C bond formation between electrophilic orthoquinones and nucleophilic phenols can be found in the studies of benzotropolone synthesis (e.g., **20**).[27,28] The value of this "structure/reactivity" approach to formulating a galloyl coupling procedure lies in the potential to develop an oxidant/substrate pairing

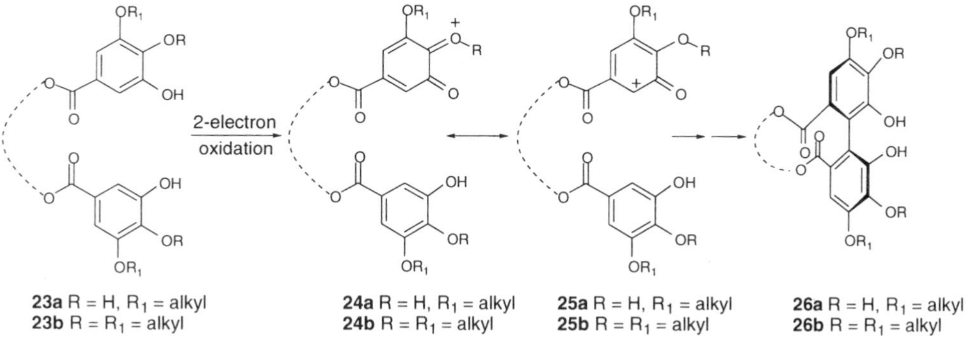

Figure 7. A two-electron oxidation approach to galloyl ester coupling.

that is just reactive enough to generate the key electrophilic intermediate, but not so reactive as to promote further oxidation of the HHDP product **26**. It is plausible that such selectivity can be imposed on the galloyl/HHDP system since each aryl ring in the HHDP product bears a relatively electron depleting sp^2 carbon substituent, a burden not shared by the parent galloyl rings in **23**. Thus, tuning the oxidant and galloyl ether substituents to permit incorporation of an oxidation "stop signal" can suppress overoxidation. In addition, modulation of the oxidation potential of individual galloyl rings may provide the means to impose regiochemical selectivity of galloyl coupling upon exposure of a polygalloylated glucose substrate to oxidant. Careful choice of phenol protecting groups should remove selected galloyl rings from participation in subsequent oxidative transformations and restrict oxidation to the galloyl rings of interest.

Atropisomer stereochemical control upon HHDP formation remains a pressing concern within the context of ellagitannin synthesis. The blueprint for a solution to this problem had been laid by Schmidt in the 1950's and later refined by Haslam in the 1980's.[22,29] These workers recognized that the strong structural bias for C—H/C=O eclipsing in secondary esters coupled with an energetically favorable antiparallel dipole alignment of adjacent galloyl ester carbonyls conspired to impose fairly strict conformational preferences upon pergalloylated glucopyranose species (e.g., β-pgg, **1**). They further speculated that these ground-state preferences translated into stereochemical imperatives upon galloyl oxidative coupling. However, the molecular-level transduction mechanism that connects substrate conformation with product stereochemistry remained unidentified.

More recent MM-based analysis of this system helps illuminate this relationship and provides a model with predictive value (fig. 8). Examination of the "global" minimum energy conformation **10** of β-pgg from a different perspective reveals that the O(6) and O(4) galloyls feature anti-parallel carbonyls and rest in offset parallel planes that have a relatively clockwise tilt, cf. **27a**. The indicated diastereotopic pairs of carbon atoms span distinctly different trans-ring distances. If reactivity scales with proximity, C—C bond formation between the closer carbon atom pair (3.64Å) will be favored. The product dione **28a** would result from this closure, and simple tautomerization within both cyclohexadienone rings unambiguously will deliver the natural (S)-HHDP isomer **29a**. The conformational search also identified another low energy isomer **27b** whose O(6) and O(4) galloyl rings have parallel carbonyl moieties and reside in planes tilted relatively counterclockwise with respect to the plane of the glucopyranose ring. Applying the proximity/reactivity model to **27b** (i.e., coupling between the carbon atoms separated by 4.18Å) would lead to the (R)-HHDP containing ellagitannin **29b**. The uniformly higher energies that attend the pro-(R)-channel (**27b** → **28b** → **29b**) are consistent with the prediction of kinetically favored reaction through the alternative less energetically costly pro-(S) sequence **27a** → **28a** → **29a**. These calculations augment the basic Schmidt-Haslam model by highlighting the importance of both galloyl "tilt", which defines the lowest energy route to HHDP formation, and carbon-pair proximity within a given "tilt" (conformation), which dictates the stereochemical outcome of coupling. Similar calculations/arguments can be made for rationalizing the stereochemical results of galloyl coupling across the O(2)/O(3)

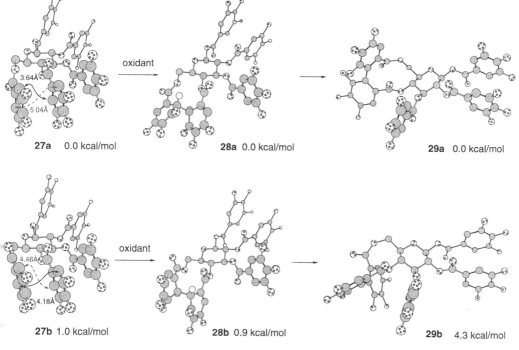

Figure 8. MM-based analysis of O(4)—O(6) galloyl coupling atropselectivity. Relative strain energies between the "**a**" and "**b**" series are shown with each pair of diastereomers.

and O(3)/O(6) gaps in β-pgg. Left unaddressed in this model, however, are questions of 1) thermodynamic vs. kinetic control of HHDP atropselectivity and 2) the basis for preferential (irreversible) oxidation at juxtaposed vs. isolated galloyls. Both of these topics will be probed through experiment as described below.

The development of a dehydrodigalloyl ether synthesis by adaptation of the proposed biosynthesis sequence described in figure 5 is likely to be less straightforward than the similar adaptation in the HHDP case. While C—C bond construction between an orthoquinone electrophile and a phenol nucleophile is well precedented, a survey of the literature reveals only a single example of diaryl ether synthesis from this pair of reactants, **30** + **31** → **32** (fig. 9).[30] The paucity of examples of C—O attachment in turn strengthens the case for enzymatic intervention in establishing the correct chemoselectivity between these reactive partners for dehydrodigalloyl ether assembly if the biosynthesis speculation remains valid. Since 1) such extrinsic organizing and orienting influences are currently beyond the scope of laboratory synthesis, and 2) direct phenolic hydroxyl addition to orthoquinones is not expected, an alternative to C—O bond-forming orthoquinone-phenol coupling must be devised. Such a procedure is likely to enjoy higher prospects for success if it is based on the normal C—O bond-forming chemical transformations of orthoquinones.

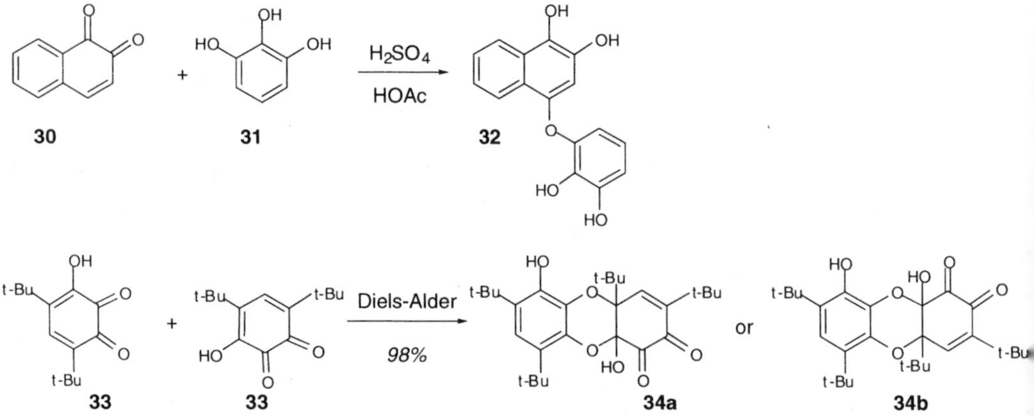

Figure 9. Uncommon and common C—O bond-forming reactions of orthoquinones.

One common C—O bond-forming reaction of orthoquinones is Diels-Alder dimerization, as exemplified by the conversion of **33** into **34a/b** (fig. 9).[31] Of course, a competition between C—O and C—C bond-forming modes will exist. However, this periselectivity can often be controlled by appropriate choice of substituent and/or Lewis acid additive.[31–35] Extension of this transformation to a galloyl-based substrate **35** relevant to dehydrodigalloyl synthesis is shown in figure 10. The benzodioxene structure **36** contains the desired C—O bond (indicated in bold) for eventual dehydrodigalloyl ether **37** synthesis, but it is embedded in an overly complex framework. Therefore, the challenge of dehydrodigalloyl ether synthesis from orthoquinone precursors now shifts to the interrelated questions of 1) selectively cleaving the indicated superfluous C—O bond, and 2) regenerating the aryl ring of **37**. The reduction of this Diels-Alder cycloaddition/fragmentation strategy to practice is described below in a novel approach to the dehydrodigalloyl-linking unit of dimeric ellagitannins.

Implementation of a biomimetic strategy for DHHDP ester synthesis from an HHDP-containing species requires chemoselective oxidation within a polygalloylated substrate (fig. 11).[36] The relative oxidation potentials of isolated galloyl esters

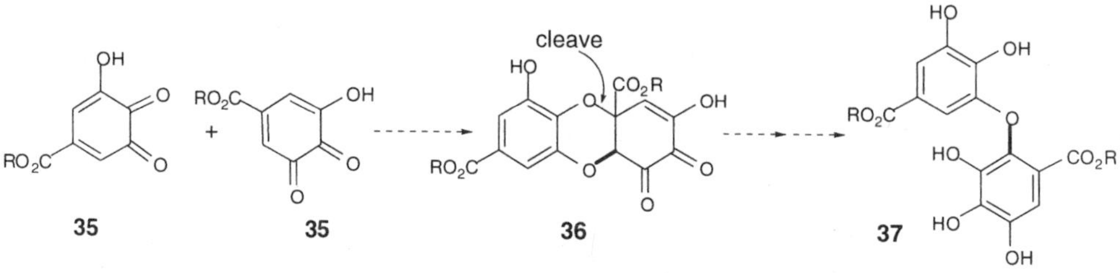

Figure 10. Application of a Diels-Alder orthoquinone dimerization to dehydrodigalloyl ether synthesis.

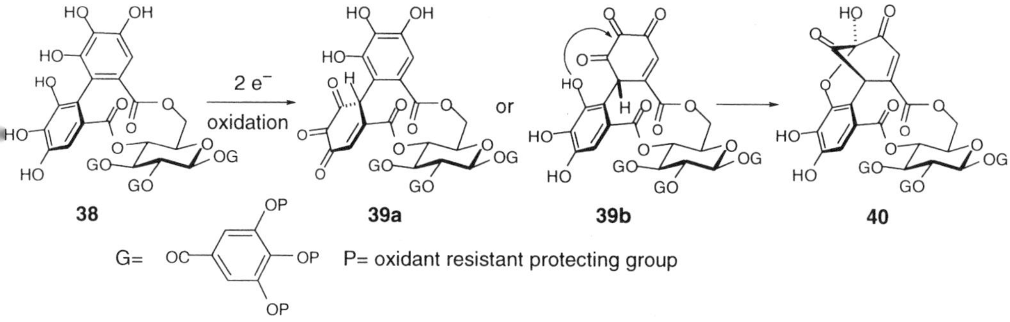

Figure 11. A biomimetic strategy for the oxidation of a protected version of tellimagrandin (**2**), **38**, to furnish a protected version of isoterchebin (**4**), **40**.

vs. an HHDP moiety vs. the product DHHDP unit are unreported, and so caution suggests that all ancillary galloyl esters on an HHDP-containing molecule should be appropriately masked prior to oxidation. The regiochemical selectivity of oxidation within an intact O(4)—O(6) HHDP unit is unsettled at present (cf. **39a** vs. **39b**). However, preliminary results presented later support a model for preferred 2-electron oxidation at an O(6) galloyl ester rather than its O(4) counterpart. If this selectivity can be exported to HHDP-bound galloyl rings as in **38**, then the desired O(6)-oxidized product **40** should emerge via the trione intermediate **39b**.

Alternative approaches to a trione of the type **39b** that don't proceed through a preformed HHDP unit can be imagined as well (fig. 12). These approaches utilize direct dimerization of galloyl-derived orthoquinone intermediates and are extensions of precedented chemistry within the orthoquinone literature.[27,37–39] Thus, two galloyl orthoquinones **41** might participate in conjugate addition (path *a*) to fashion the key inter-ring C—C bond and deliver the trione **42** directly. Similarly, Diels-Alder dimerization of two molecules of **41** could provide an intermediate bicyclo[2.2.2]octene adduct **43**, which is poised to fragment via retro-aldol cleavage to furnish **42** as well. The nature of the ether substituent "R" is likely to play a critical role in promoting (or suppressing) either reaction channel, as well as in influencing the mode of [4π + 2π] cycloaddition in the Diels-Alder route. The

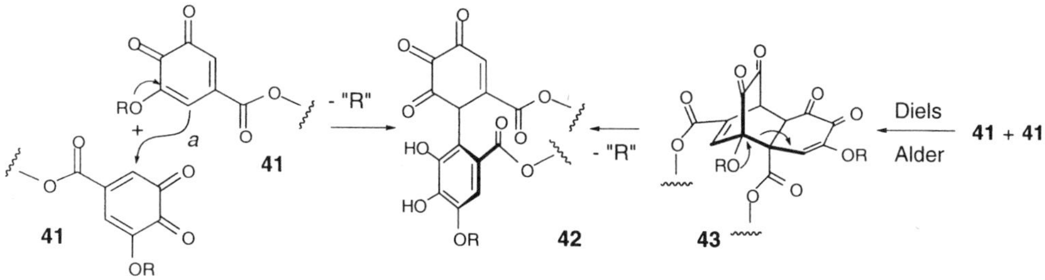

Figure 12. Direct dimerization of galloyl orthoquinones to afford a DHHDP unit.

exploration of each of these chemical processes in the context of DHHDP ester synthesis is detailed below.

4. HHDP SYNTHESIS: DEVELOPMENT AND APPLICATIONS

The HHDP unit remains the defining structural feature of the ellagitannins. Initial synthesis efforts therefore were focused on the efficient preparation of this biaryl moiety from readily available precursors. A biomimetic oxidative coupling of galloyl esters was adopted as the strategy of choice for reasons outlined earlier. Issues of chemical selectivity, e.g., 1) galloyl ester oxidation without subsequent HHDP-containing product oxidation, and 2) atropselectivity for the (*S*)-atropisomer, remain paramount and will ultimately determine the success or failure of this venture.

In formative studies, a panel of mono-, di-, and tri-*O*-methyl galloyl esters attached to the O(4) and O(6) positions of a glucopyranose-derived core was screened against a range of 2-electron oxidants as per the guiding hypothesis discussed with figure 7.[40,41] The initial efforts were frustrated by either isolation of unreacted starting material or observation of highly colored (overoxidized?) intractable product mixtures. Despite these discouraging beginnings, persistence eventually was rewarded by the discovery that the unsymmetrical tetra-*O*-methylated substrate **44** furnished two characterizable oxidation products upon exposure to Pb(OAc)$_4$ (Wessely oxidation[42]) (fig. 13). The major product(s) was an inseparable mixture of acetate adducts **47a** and **47b** (ratio unassigned) corresponding to OAc⁻ trapping of the putative electrophilic cyclohexadienonyl cation intermediate **45**. The minor product **46**, however, generated a good deal more excitement. This compound featured the desired (protected) (*S*)-HHDP moiety so long sought! Apparently, the oxidized intermediate modeled by **45** is quenched by

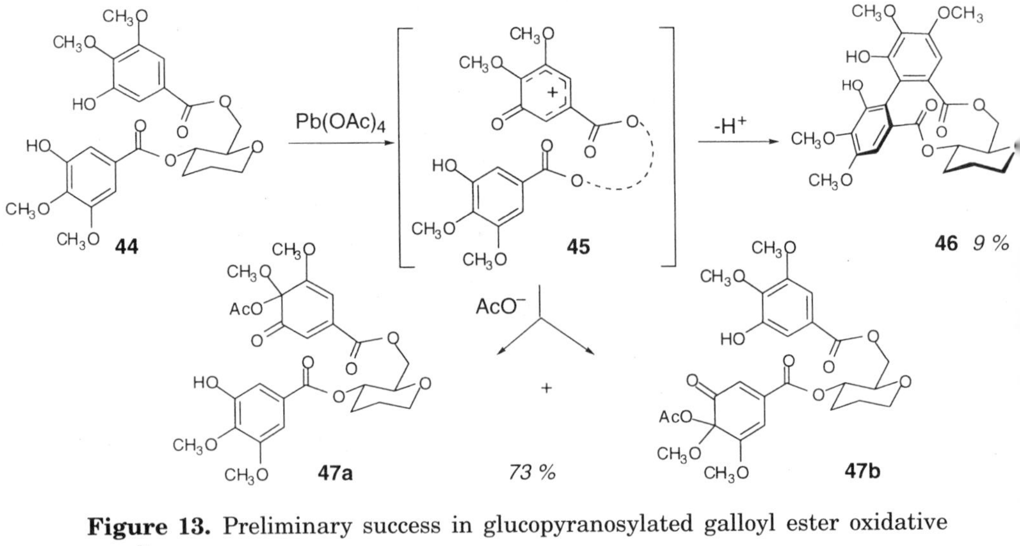

Figure 13. Preliminary success in glucopyranosylated galloyl ester oxidative coupling.

either the proximal, weakly nucleophilic galloyl ring to yield **46**, or by acetate provided by the lead reagent to deliver **47**.

Isolation of a single tetra-*O*-methyl-(*S*)-HHDP-containing diastereomer **46** is consistent with the Schmidt-Haslam hypothesis and the related MM calculations (*vide supra*). While these arguments cite a kinetic basis for product stereoselectivity, the strong thermodynamic preference for the (*S*)- over the (*R*)-HHDP isomer (cf. fig. 8) raises the possibility that atropselectivity is adjusted favorably through post-cyclization equilibration. This question was probed via the isomerization study indicated in figure 14, wherein the per-*O*-methyl-(*R*)-HHDP-containing glucopyranose derivative (*R*)-**48** (prepared by non-selective acylation chemistry[43,44]) was converted to the more stable isomer (*S*)-**48** upon heating.[45] Kinetic analysis through the Eyring equation allowed computation of an activation barrier for this isomerization (36.6 kcal/mol), which is likely to be insurmountable under the Pb(IV)-mediated oxidation conditions. Thus, atropselectivity upon galloyl ester coupling during oxidation of **44** is governed by kinetic and not thermodynamic control elements.

The stereochemical control issues related to HHDP formation were resolved favorably in this model system, but clearly attention to increasing product yield was warranted. Initial attempts at influencing the competition between the galloyl ring and acetate for putative cation **45** focused on Pb(IV) reagents with less nucleophilic counterions. These trials uniformly ended in failure, either as a consequence of unenhanced selectivity (Pb(OBz)$_4$) or of product destruction (Pb(OTFA)$_4$). Eventual exploitation of the different steric environments around the electrophilic atoms of note in **45** provided a solution to this problem. Replacement of the di-*O*-methyl ethers with a diphenyl ketal moiety afforded a Pb(OAc)$_4$ sensitive bis phenolic substrate **49** whose C(4) aryl carbons appear to be effectively blocked from acetate attack (fig. 15). Lead-mediated oxidation of **49** proceeded smoothly to furnish the biaryl product **50** in good yield as an inconsequential mixture of regioisomers all bearing the protected (*S*)-HHDP unit.[40,41] No evidence for acetate trapping products was forthcoming. A second advantage conferred by the diphenyl ketal moiety compared to simple methyl ethers can be appreciated by their ready removal under mild hydrogenolysis conditions to provide the desired free hexahydroxydiphenoyl-containing product **51** in excellent yield.

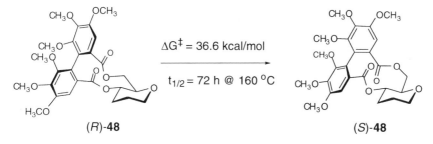

Figure 14. Isomerization of a glucopyranosyl-bound (*R*)-HHDP unit to its (*S*)-diastereomer.

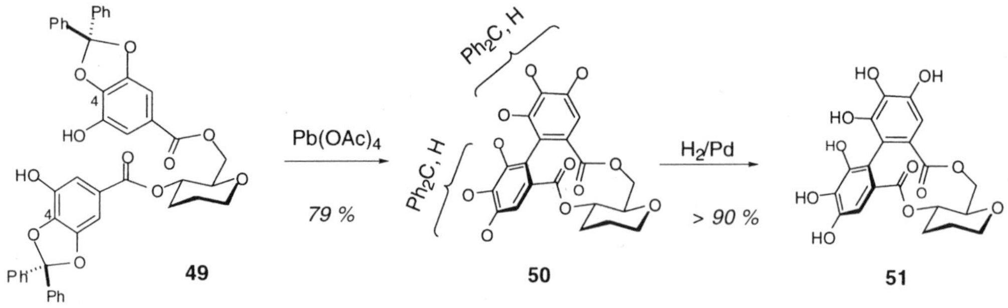

Figure 15. Efficient galloyl ester oxidative coupling on a glucopyranose core.

This methodology could be exported into the more demanding context of ellagitannin total synthesis. Thus, completely stereoselective and regioselective biomimetic assembly of the monomeric ellagitannin natural products tellima-grandin I (**52**),[46] tellimagrandin II (**2**),[47] sanguiin H-5 (**53**),[48] and pedunculagin (**54**)[49] have all been recorded (fig. 16). In each case, Pb(OAc)$_4$-mediated oxidative coupling of diphenyl ketal-protected galloyl esters defined the key transformation en route to product. β-Selective anomeric acylation chemistry was developed to accommodate the preparation of those ellagitannins (**2**, **53**) endowed with a galloyl ester at O(1).

The regioselectivity of galloyl coupling within a polygalloylated glucopyranose substrate was enforced by judicious choice of phenol protecting groups during these natural product syntheses. Only those galloyl esters destined for coupling were functionalized with the diphenyl ketal unit. The remaining spectator galloyl esters were fully O-benzylated to prevent Pb(IV)-based oxidation. However, one exception to this strategy emerged from a second-generation approach to tellimagrandin I (fig. 17).[46] In this chemistry, a glucopyranose substrate bearing oxidation sensitive phenolic galloyl esters at O(2), O(3), O(4), and O(6) was subjected to Pb(OAc)$_4$-mediated oxidative coupling to identify any intrinsic site selectivity for HHDP formation. Little preference was anticipated due to the irreversible nature of the

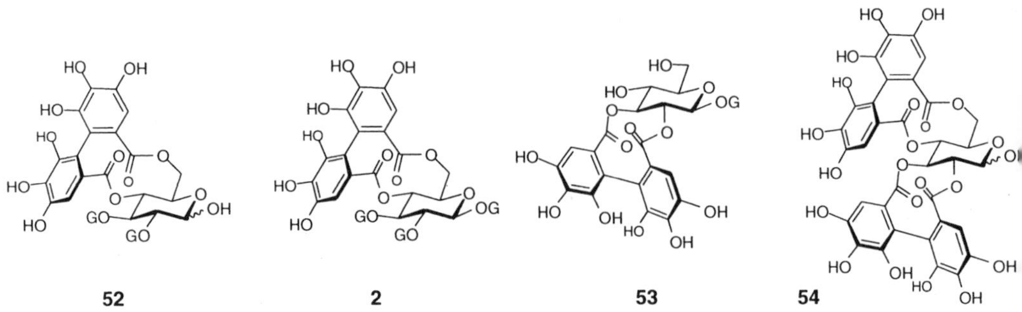

Figure 16. Ellagitannin natural products prepared by Pb(OAc)$_4$-mediated galloyl ester coupling.

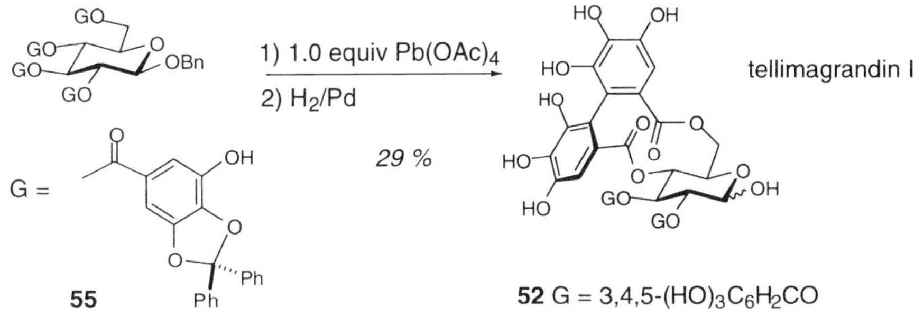

Figure 17. A facile synthesis of tellimagrandin I (**52**).

Pb(IV)-mediated oxidation, but in fact complete selectivity for O(4)/O(6) galloyl coupling was observed upon titration of **55** with 1.0 equivalent of oxidant.

The basis for this unexpected regioselectivity of oxidative coupling within **55** was probed by both MM calculations and by competitive oxidation experiments as described in figure 18.[50] Initial MM analysis of substrate conformation was achieved by examination of the simpler model compound β-methyl tetragalloyl-glucopyranoside. Inspection of the "global" minimum energy structure **56** reveals that, to the extent that a ground-state conformational model for predicting the course of galloyl oxidative coupling is applicable, the prospects for O(3)/O(4) galloyl union appear quite remote compared to the O(4)/O(6) or O(2)/O(3) alternatives.

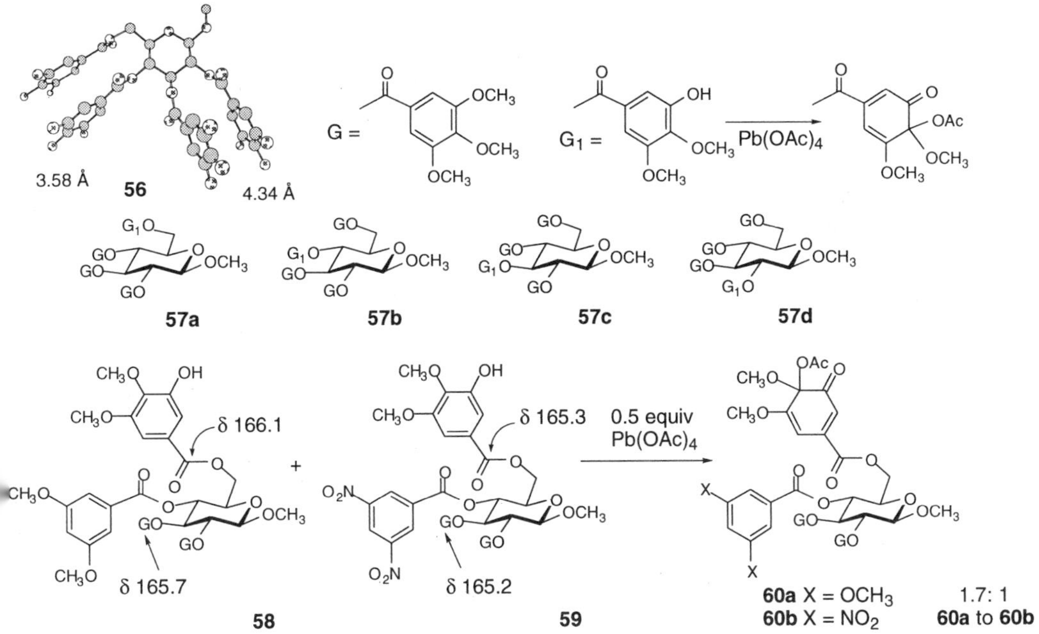

Figure 18. Probing the site selectivity of glucopyranosylated galloyl ester oxidation.

Greater proximity and a diminished energetic penalty[4] might conspire to favor aryl-aryl bond formation in the O(4)/O(6) case compared to the O(2)/O(3) option. Nevertheless, the relationship between these structural and energetic features and the factors that ultimately determine which galloyl ring is most rapidly oxidized are unclear.

A series of Pb(OAc)$_4$-mediated competitive oxidations of 3,4-di-*O*-methyl galloyl esters situated at different positions of the glucopyranose core was examined to help illuminate this connection. Thus, in each of three independent experiments, an equimolar mixture of the O(6)-phenoxy bearing substrate **57a** and either the O(4)-, O(3)-, or O(2)-regioisomers **57b**, **57c**, or **57d**, respectively, were subjected to a deficiency of Pb(OAc)$_4$ at low temperature. The ratio of oxidized products (e.g., orthoquinone monoketals) then reflects the intrinsic oxidation potential of the O(4)-, O(3)-, and O(2)-galloyl rings relative to the O(6)-ring on an authentic glucopyranose substrate. These studies revealed that the galloyl ring at O(6) was the most readily oxidized by lead, followed by the galloyls at O(2), O(4), and finally O(3): O(6)/O(2) = 1.3, O(6)/O(4) = 2.4, O(6)/O(3) = 2.7.

These results provide phenomenological support for the observation of exclusive O(4)/O(6) galloyl coupling with substrate **55**. Less clear, however, is the structural basis for this ranking of oxidation facility among the glucopyranose-bound galloyl esters. That the most sterically accessible O(2) and O(6) galloyl rings are preferentially oxidized should not go unnoticed. However, the possibility of an additional electronic influence cannot be discounted. Thus, the substrates **58** and **59** were pitted against each other in competition for limiting oxidant in an effort to unveil an electronic bias underlying this oxidation selectivity. A modest but real preference for oxidation of the overall more electron-rich species **58** was observed (**60a**:**60b** = 1.7:1). Disentangling ground state vs. transition state electronic effects remains a difficult challenge in this experiment. It is possible that ground-state elevation in **58** relative to **59**, perhaps as a consequence of either HOMO-HOMO mixing or of π/polar interactions[51,52] between the juxtaposed O(4) and O(6) galloyl rings, is responsible for the differential reactivity of these species. Perhaps more plausible is a scenario wherein anchimeric assistance at the electron deficient transition state, which will be enhanced with the more electron rich O(4) aryl ring of **58**, contributes to the preference for oxidation of the former substrate. The similarity in ^{13}C resonances for the O(4) and O(6) carbonyls noted in **58** and **59** suggest that through-bond inductive effects do not contribute significantly to the differential oxidation of the former substrate. Overall, these competitive oxidation experiments point to subtle but exploitable structural and electronic factors that contribute to selectivity upon polygalloylglucopyranose oxidative coupling. Whether these effects mediate selectivity during ellagitannin biosynthesis remains a matter of speculation.

5. DEHYDRODIGALLOYL ETHER SYNTHESIS: DEVELOPMENT AND APPLICATIONS

Development of a synthesis for the dehydrodigalloyl linker unit characteristic of coriariin A-type dimeric ellagitannins proceeded along two lines of inquiry.[53,54]

Both approaches utilized galloyl-derived orthoquinones (or related species) as electrophilically activated partners per the biosynthesis and chemical synthesis speculation accompanying figure 5 and figure 9, respectively. The initial route to diaryl ether synthesis relied on a heteropolar pairing of an orthoquinone **61** with a phenol or phenoxide nucleophile **62** (fig. 19). Unfortunately, no experimental conditions could be identified that promoted conjugate addition of **61** and **62** to furnish dehydrodigalloyl ethers. Parallel scouting studies of nucleophile addition to orthoquinone **61** revealed it to be an extremely selective electrophile that accepts only "soft" nucleophiles (e.g., thiols) in good yield. In contrast, the related orthoquinone monoketal **64**, derived in excellent yield from Pb(OAc)$_4$-mediated oxidation of methyl 3,5-di-*O*-methylgallate, did participate in C—O bond-forming union with the galloyl phenol **65** under Lewis acid catalysis. The product dehydrodigalloyl ether **66** was isolated in only modest yield despite extensive optimization studies, and therefore this strategy did not offer a significant advantage over the venerable but equally disappointing Ullmann coupling procedure (**67** + **65** → **68**).

The second thrust in this area followed the orthoquinone Diels-Alder dimerization strategy suggested in figure 10. Preliminary attempts to dimerize orthoquinone **69a** under mild heating led to complex product mixtures consisting of bicyclo[2.2.2]octene-based as well as benzodioxene-based structures. Apparently, little periselectivity accompanies [4π + 2π] cycloaddition, and equilibration among the adducts cannot be discounted. Various additives affected the course of the

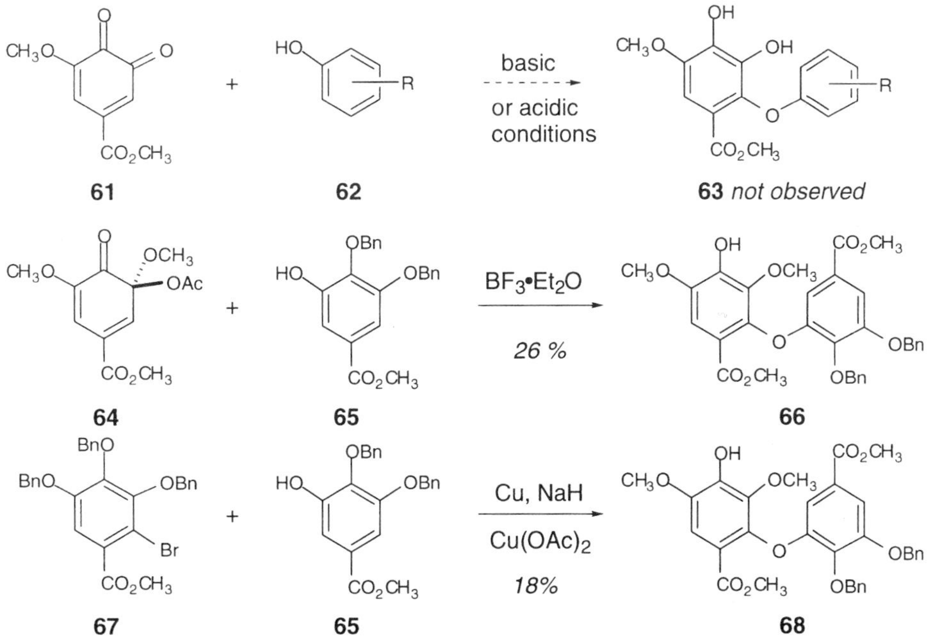

Figure 19. Preliminary attempts at dehydrodigalloyl ether synthesis by phenol/galloyl electrophile pairing.

dimerization: For example, warming **69a** in the presence of the mild Lewis acid Cu(OAc)$_2$ furnished a modest yield of a bicyclo[2.2.2]octene-containing Diels-Alder adduct.[54] Of more relevance to the goal of dehydrodigalloyl ether synthesis is the transformation of **69b** mediated by B(OAc)$_3$ (fig. 20). In this instance, good yields of the hetero Diels-Alder regioisomeric products **70a** and **70b** resulted to the exclusion of the bicyclooctene isomers. The regioisomer **70a** features the desired (emboldened) C—O bond that eventually will define the key linkage between galloyl rings. Implementation of this Diels-Alder dimerization process now required "deletion" of the other three new C—O bonds formed in this reaction. Two of these superfluous C—O bonds could be severed via base-mediated β-elimination of phenoxide to furnish a mixture of new orthoquinones **71a** and **71b**. Reduction of these species upon workup then delivered the desired dehydrodigalloyl ether **72a** along with its regioisomeric counterpart, ether **72b**. Fortunately, the undesired *para*-C—O linkage in **72b** could be completely converted to the requisite *meta*-C—O bond in **73** by mild base during benzylation via a Smiles rearrangement.[55] Thus, the discouraging regioisomeric mixture emerging from the Diels-Alder dimerization can be utilized in its entirety to provide ready access to protected versions of dehydrodigalloyl ethers, e.g., **73**, related to the dimeric ellagitannins.

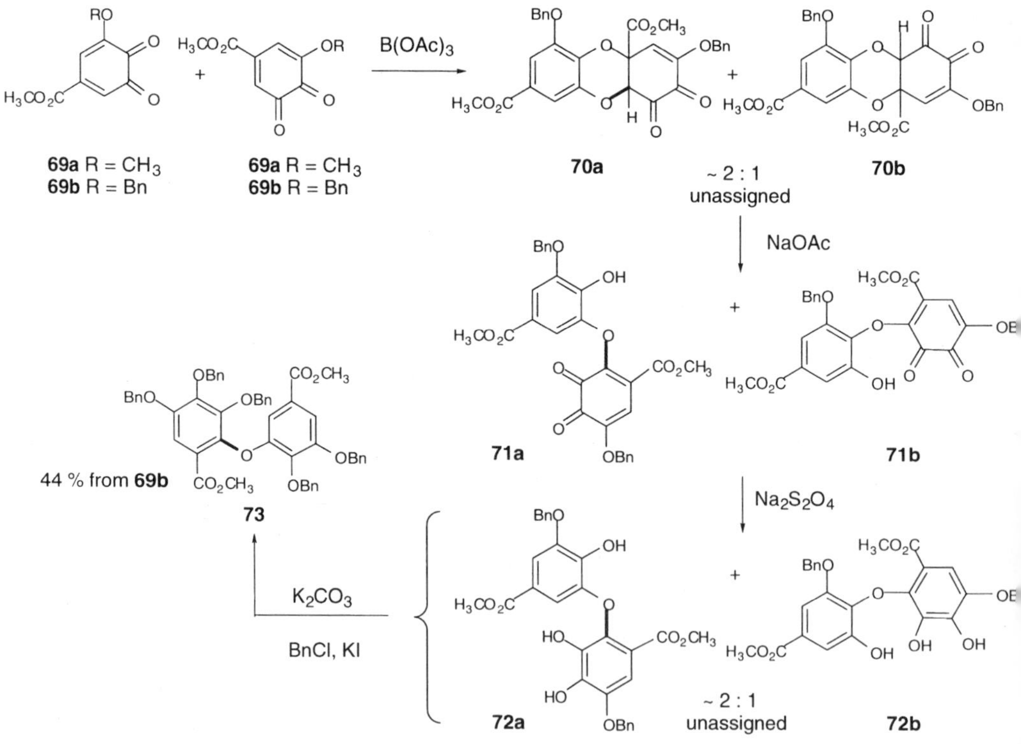

Figure 20. Dehydrodigalloyl synthesis via galloyl orthoquinone Diels-Alder dimerization.

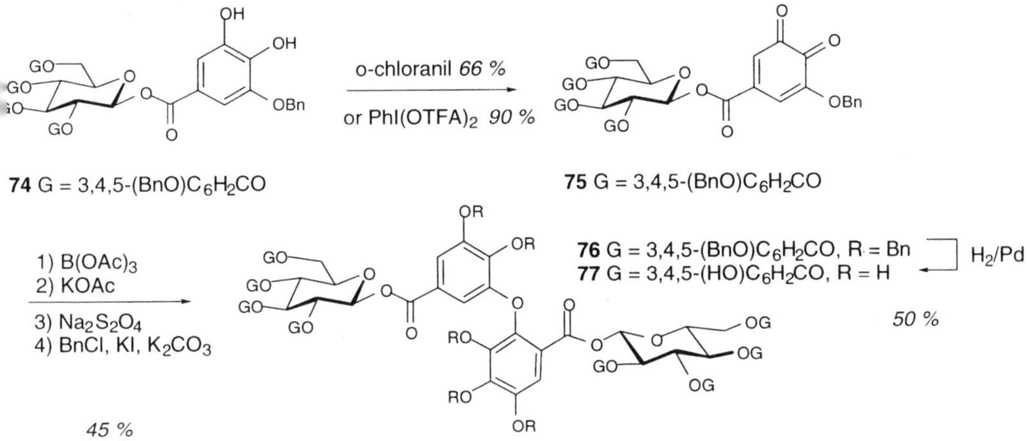

Figure 21. Synthesis of an ellagitannin-gallotannin hybrid.

Application of this orthoquinone-based dehydrodigalloyl ether synthesis to ellagitannin targets was preceded by work on the model system shown in figure 21.[47] Oxidation of the catecholic anomeric galloyl ester in **74** furnished the deep red orthoquinone **75** in good yield. Subjecting this galloyl orthoquinone to the hetero Diels-Alder dimerization sequence described in figure 20 reproducibly delivers the dehydrodigalloyl-containing gallotannin dimer **76** with satisfactory efficiency. Simple hydrogenolysis of all benzyl ethers in **76** provides the fully phenolic ellagitannin-gallotannin hybrid **77** as an off-white powder. Benzylation of the crude $Na_2S_2O_4$ reduction products is not necessary but aids in purification and characterization of the dehydrodigalloyl-containing species.

The facility of this procedure and the mildness of the reagents employed suggest that an analogous dimerization sequence cannot be discounted when evaluating biosynthesis options for the conversion of monomeric ellagitannins (e.g., tellimagrandin II (**2**)) into their dehydrodigalloyl-linked dimeric counterparts (e.g., coriariin A (**6**)).

6. DHHDP SYNTHESIS: DEVELOPMENT AND APPLICATIONS

The chemical conversion of a pre-existing HHDP unit into the more highly oxidized dehydrohexahydroxydiphenoyl (DHHDP) moiety was examined with the two substrates **78** and **79** and the oxidants previously employed to convert catecholic galloyls to orthoquinones (orthochloranil, PhI(OTFA)$_2$) (fig. 22). Suprisingly, all attempts at oxidation with these substrate/reagent combinations were unsuccessful at providing any evidence for DHHDP formation. Rather, starting material admixed with possibly polymeric species resulted. The implications for DHHDP biosynthesis are unsettling: perhaps biosynthesis proposals that cite HHDP precursors should be viewed with caution in the absence of definitive labeling experiments. However, it is equally plausible that the marked dissimilarity between

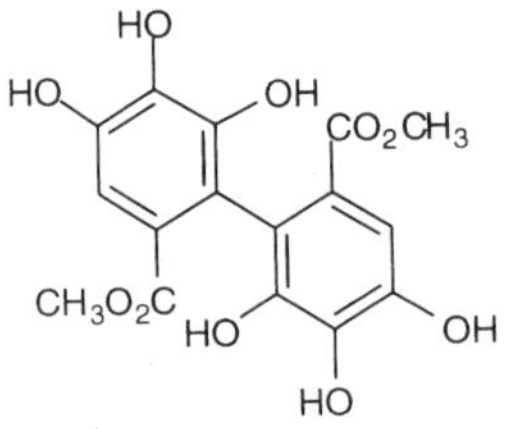

Figure 22. HHDP-containing substrates that did not lead to DHHDP formation upon oxidation.

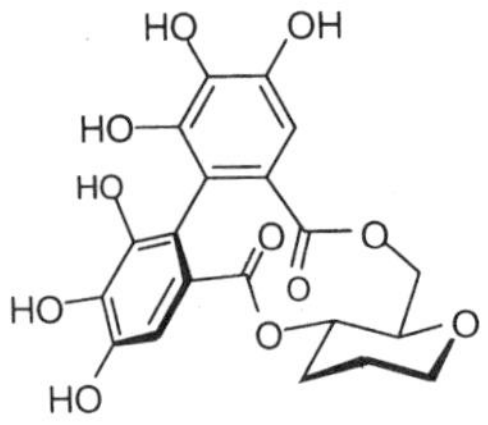

oxidation reagents *in vivo* and those examined herein account for the failure of this "biomimetic" chemistry.

Alternative approaches to DHHDP synthesis that do not pass through an HHDP precursor were examined as well.[56] As described in figure 12, initial galloyl oxidation ($\rightarrow$ **41**) can precede C—C coupling to deliver the DHHDP unit following internal ketalization of the first-formed trione **43**. Preliminary attempts at reducing this strategy to practice commenced with methyl gallate (**80**). Exposure of this simple unetherified gallic ester to orthochloranil under a carefully defined set of conditions led to the precipitation of a yellow solid (fig. 23). Extensive NMR analysis allowed the structural assignment **83** for this unstable species. The formation of the formal [5 + 2] adduct **83** can be explained most economically by initial methyl gallate oxidation to the unobserved orthoquinone **81**, which can dimerize through sequential Michael and then aldol additions. When redissolved in acetone-d_6, a slow conversion of this ether-insoluble dimer into an equilibrium mixture of the presumably more stable [5 + 5] adduct **84** and the DHHDP-containing species **85** (~ 1:2 $\rightarrow$ 1:4) was observed. These interconversions can be rationalized mechanistically by noting that an undetected trione intermediate **82** can precede all of the dimers **83**, **84**, and **85** via distinct C—C or C—O bond formation pathways.

These oxidation model studies provide evidence that the DHHDP substructure can be accessed directly from methyl gallate, in contrast to earlier reports wherein purpurgallin- or ellagic acid-type products were observed. The isomer mixture **83/84/85** obtained would complicate a total synthesis effort, but it is conceivable that a second-generation model system **86** featuring pyranose-bound gallic esters may have an inherent bias for the "natural" DHHDP connectivity in **85**. Unfortunately, oxidation of **86** under conditions successful with **80** (or many variations) went unrewarded with DHHDP formation. Apparently, the fortuitous precipitation of the [5 + 2] adduct **83** sequestered this sensitive species from further oxidant and thus enabled the chemistry described in figure 23 to be observed. No such "stop-message" could be built into the oxidation of **86**, and complex product mixtures that were consistent with overoxidation and then intermolecular reaction (oligomerization?) were the norm.

The results of these experiments suggested that a premium must be placed on developing reaction protocols that separated the oxidation from the cyclization steps. The Diels-Alder-based strategy described in figure 12 (**41** + **41** $\rightarrow$ **43** $\rightarrow$ **42**)

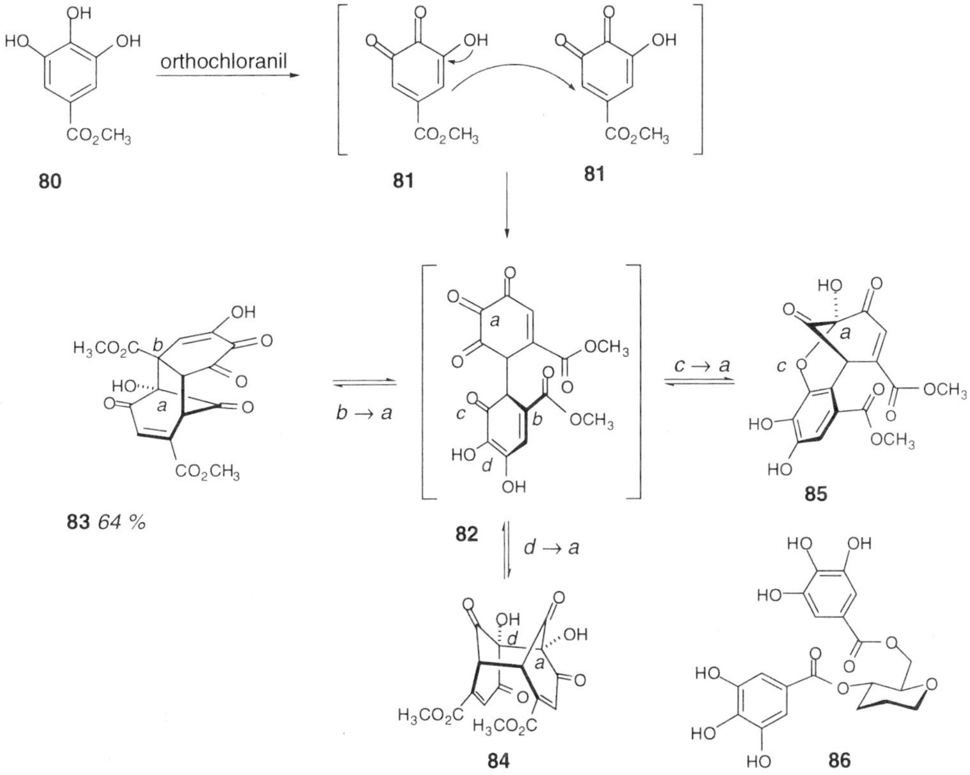

Figure 23. Synthesis of oxidized methyl gallate dimers, including DHHDP.

offered this separation, as oxidation of a monoetherified gallate derivative, in contrast to unetherified methyl gallate, can lead to an isolable orthoquinone Diels-Alder partner. Subsequent periselective $[4\pi + 2\pi]$ cycloaddition and adduct fragmentation could lead to the desired trione (cf. **42**) in a milieu free of oxidant. While earlier studies have documented the influence of B(OAc)$_3$ in steering the periselectivity of methyl ether-substituted galloyl orthoquinones toward hetero-Diels-Alder dimerization, companion experiments revealed that this preference could be overwhelmed by variations in structure. In this vein, simply warming the more reactive silyl ether-substituted orthoquinone **88** in Et$_2$O provided a good yield of the all-carbon $[4\pi + 2\pi]$ adduct **89**, isolated as a hydrate **90** following chromatography (fig. 24). The key C—C bond which unites the two galloyl residues is emboldened in **90**, whereas the remaining new C—C bond destined for scission is hatched. Efforts to selectively cleave this bond are underway.

7. CONCLUSIONS

Successful laboratory preparation of characteristic modules found in ellagitannins such as HHDP, DHHDP, and digalloyl ether functionalities sets the stage

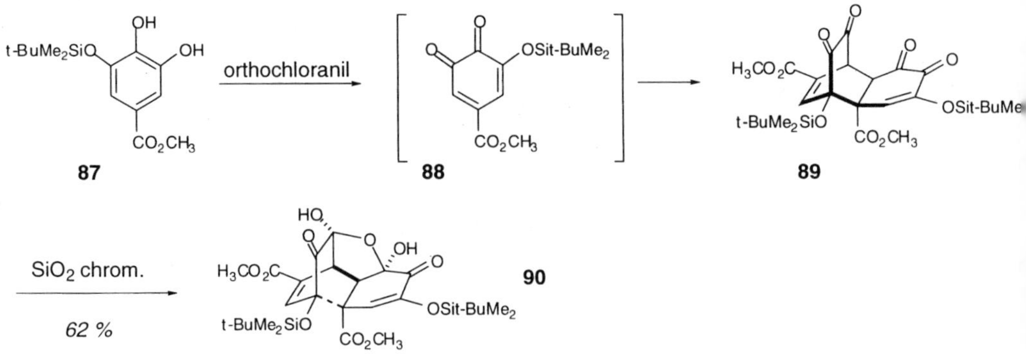

Figure 24. An attempt at DHHDP synthesis via Diels-Alder chemistry.

for continuing studies in complex ellagitannin total synthesis. Implementation of biomimetic strategies for simple ellagitannin synthesis has already yielded fruit; several HHDP-containing monomeric ellagitannins have been prepared. Further application of the chemistry described herein to more highly oxidized C—O connected ellagitannins represents an important future direction of these studies.

REFERENCES

1. Haslam, E. Plant polyphenols—vegetable tannins revisited. Cambridge University Press, Cambridge, (1989).

2. Haslam, E.; Cai, Y. Plant polyphenols (vegetable tannins*): gallic acid metabolism. *Nat. Prod. Reports* :41 (1994).

3. Berlinck, R.G.S.; Hatano, T.; Okuda, T.; Yoshida, T. Hydrolyzable tannins and related polyphenols. *In*: Herz, W.; Kirby, G.W.; Moore, R.E.; Steglich, W.; Tamm, Ch. (eds.). Progress in the chemistry of organic natural products. Springer-Verlag, New York, p.1 (1995).

4. Quideau, S.; Feldman, K.S. Ellagitannin chemistry. *Chem. Rev.* 96:475 (1996).

5. Okuda, T.; Yoshida, T.; Hatano, T. Chemistry and biological activity of tannins in medicinal plants. *In*: Wagner, H.; Farnsworth, N.R. (eds.) Economic and medicinal plant research, v.5. Plants and traditional medicines. Academic Press Ltd., London, p.129 (1991).

6. Okuda, T.; Yoshida, T.; Hatano, T. Polyphenols from Asian plants. *In*: Huang, M.-T., Ho; C.-T.; Lee, C.Y. (eds.) Phenolic compounds in food and their effects on health II. American Chemical Society, Washington, p. 160 (1992).

7. Okuda, T.; Yoshida, T.; Hatano, T. Pharmacologically active tannins isolated from medicinal plants. *In*: Hemingway, R.W.; Laks, P.E. (eds.) Plant polyphenols: synthesis, properties, significance. Plenum Press, New York, p. 539 (1992).

8. Nonaka, G-I.; Nakayama, S.; Nishioka, I. Tannins and related compounds. LXXXIII. Isolation and structures of hydrolyzable tannins, Phillyraeoidins A-E from *Quercus phillyraeoides. Chem. Pharm. Bull.* 37:2030 (1989).

9. Okuda, T.; Hatano, T.; Ogawa, N.; Kira, R.; Matsuda, M. Cornusiin A, a dimeric ellagitannin forming four tautomers, and accompanying new tannins in *Cornus officinalis. Chem. Pharm. Bull.* 32:4662 (1984).

10. Okuda, T.; Hatano, T.; Yazaki, K.; Ogawa, N. Rugosin A, B, C and praecoxin A, tannins having a valoneoyl group. *Chem. Pharm. Bull.* 30:4230 (1982).

11. Hatano, T.; Hattori, S.; Okuda, T. Tannins of *Coriaria japonica* A. GREY. I. Coriariins A and B, new dimeric and monomeric hydrolyzable tannins. *Chem. Pharm. Bull.* 34:4092 (1986).

12. McDonald, P.D.; Hamilton, G.A. Mechanisms of phenolic oxidative coupling reactions. *In*: Trahanovsky, W.S. (ed.) Oxidation in organic chemistry, v.5. Academic Press, New York, p.97 (1973).

13. Mohamadi, F.; Richards, N.G.J.; Guida, W.C.; Liskamp, R.; Lipton, M.; Caufield, C.; Chang, G.; Hendrickson, T.; Still, W.C. MacroModel—an integrated software system for modeling organic and bioorganic molecules using molecular mechanics. *J. Comput. Chem.* 11:440 (1990).

14. Yoshida, T.; Nakata, F.; Hosotani, K.; Nitta, A.; Okuda, T. Dimeric hydroysable tannins from *Melastoma malabathricum*. *Phytochemistry* 31:2829 (1992).

15. Löwe, J.F.F.F. Über die Bildung von Ellagsäure aus Gallussäure. *Z. Chem.* 4:603 (1868).

16. Grießsmayer, V. Über das Verhalten von Stärke und Dextrin gegen Jod und Gerbsäure. *Ann.* 160:40 (1871).

17. Perkin, A.G.; Nierenstein, M. Some oxidation products of the hydroxybenzoic acids and the constitution of ellagic acid. Part 1. *J. Chem. Soc.* 87:1412 (1905).

18. Erdtman, H. Phenoldehydrierungen VI. Dehydrierende kupplung einiger guajakolderivate. *Svensk. Kem. Tidskr.* 47:223 (1935).

19. Hathway, D.E. Autoxidation of polyphenols. Part 1. Autoxidation of methyl gallate and its *O*-methyl ethers in aqueous ammonia. *J. Chem. Soc.* :519 (1957).

20. Jurd, L. Plant polyphenols. VI Experiments on the synthesis of 3, 3′- and 4,4′-di-*O*-methylellagic acid. *J. Am. Chem. Soc.* 81:4606 (1959).

21. Szarkowski, J.W.; Golaszewski, T. Uber die Bildung der Ellagsäure durch UV-Strahlen. *Naturwissenschaften* 48:457 (1961).

22. Gupta, R.K.; Al-Shafi, S.M. K.; Layden, K.; Haslam, E. The metabolism of gallic acid and hexahydroxydiphenic acid in plants. Part 2. Esters of (*S*)-hexahydroxydiphenic acid with D-glucopyranose ($^4C_1$). *J. Chem. Soc., Perkin Trans. 1* :2525 (1982).

23. Mayer, W.; Hoffmann, E.H.; Losch, N.; Wolf, H.; Wolter, B.; Schilling, G. Dehydrierungsreaktionen mit Gallussäureestern. *Ann.* :929 (1984).

24. Grimshaw, J.; Haworth, R.D. Flavogallol. *J. Chem. Soc.* :4225 (1956).

25. Nelson, T.D.; Meyers, A.I. A rapid total synthesis of an ellagitannin. *J. Org. Chem.* 59:2577 (1994).

26. Lipshutz, B.H.; Liu, Z.-P.; Kayser, F. Cyanocuprate-mediated intramolecular biaryl couplings applied to an ellagitannin. Synthesis of (+)-*O*-permethyltellimagrandin II. *Tetrahedron Lett.* 35:5567 (1994).

27. Critchlow, A.; Haslam, E.; Haworth, R.D.; Tinker, P.B.; Waldron, N.M. The oxidation of some pyrogallol and purpurogallin derivatives. *Tetrahedron* 23:2829 (1967).

28. Horner, L.; Durckheimer, W.; Weber, K.-H.; Dolling, K. Synthese, struktur und eigenschaften von 1′.2′-dihydroxy-6.7-benzotropolonen. *Chem. Ber.* 97:312 (1964).

29. Schmidt, O.Th.; Mayer, W. Natürliche Gerbstoffe. *Angew. Chem.* 68:103 (1956).

30. Blumenfeld, S.; Friedlander, P. Allgemeine Reaktionen Aromatischer Chinone. *Chem. Ber.* 30:2563 (1897).

31. Pfundt, G.; Schenk, G.O. Cycloadditions to *o*-quinones, 1, 2-diketones and some of their derivatives. *In*: Hamer, J. (ed.) 1,4-Cycloaddition reactions. The Diels Alder reaction in heterocycle synthesis. Academic Press, New York, p. 345 (1967).

32. Ansell, M.F.; Gosden, A.F.; Leslie, V.J.; Murray, R.A. The reactions of *o*-benzoquinones with cyclopentadiene. *J. Chem. Soc.* (C). :1401 (1971).

33. Lee, J.; Snyder, J.K. Ultrasound promoted cycloadditions in the synthesis of *Salvia Militiorrhiza* abietanoid *o*-quinones. *J. Org. Chem.* 55:4995 (1990).

34. Waldron, N.M. The constitution of the dimers of 4,5-di-*t*-butyl-3-hydroxy-*o*-benzoquinone and 5-t-butyl-3-hydroxy-o-benzoquinone. *J. Chem. Soc.* (C) :1914 (1968).

35. Vol'era, V.B.; Khristyuk, A.L.; Zhorin, V.A.; Arhipov, I.L.; Stukan, R.A.; Ershov, V.V.; Enikolopyan, N.S. Conversion of 3,6-di-*tert*-butylorthobenzoquinone in Bridgeman presses. *Izv. Akad. Nauk SSSR, Ser. Khim.* 33:443 (1984) (Engl. Trans.)

36. Schmidt, O.Th.; Wieder, G. Zur Synthese der "Dehydro-hexahydroxy-diphensäure". *Ann.* 706:198 (1967).

37. Horner, L.; Dürckheimer, W. Die Konfiguration der dimeren *o*-Benzochinone und das Prinzip des geringsten Gesamtmoments für die Anordnung der Reaktionspartner in Orientierungskomplexen. *Chem. Ber.* 95:1219 (1962).

38. Magnusson, R.; Alder, E. Periodate oxidation of phenols. *Acta Chem. Scand.* 13:505 (1959).

39. Müller, E.; Bayer, O. Chinone, Teil II. *In*: Grundmann, C. (ed.) Methoden der Organischen Chemie (Houben-Weyl). Georg Thieme, Stuttgart, p. 139 (1979).

40. Feldman, K.S.; Ensel, S.M. Ellagitannin chemistry. The first example of biomimetic diastereoselective oxidative coupling of a glucose-derived digalloyl substrate. *J. Am. Chem. Soc.* 115:1162 (1993).

41. Feldman, K.S.; Ensel, S.M. Ellagitannin chemistry. Preparative and mechanistic studies of the biomimetic oxidative coupling of galloyl esters. *J. Am. Chem. Soc.* 116:3357 (1994).

42. Bubb, W.A.; Sternhell, S. The Wessely oxidation. *Tetrahedron Lett.* :4499 (1970).

43. Itoh, T.; Chika, J.-I.; Shirakami, S.; Ito, H.; Yoshida, T.; Kubo, Y.; Uenishi, J.-I. *J. Org. Chem.* 61:3700 (1996).

44. Khanbabaee, K.; Schulz, C.; Lotzerich, K. Synthesis of enantiomerically pure strictinin using a stereoselective esterification reaction. *Tetrahedron Lett.* 38:1367 (1998).

45. Feldman, K.S.; Kirchgessner, K.A. Probing ellagitannin stereochemistry: the rotational barrier of a model (4,6)-hexamethoxydiphenoyl glucopyranoside. *Tetrahedron Lett.* 38:7981 (1997).

46. Feldman, K.S.; Ensel, S.M.; Minard, R.M. Ellagitannin chemistry. The first total chemical synthesis of an ellagitannin natural product, tellimagrandin 1. *J. Am. Chem. Soc.* 116:1742 (1994).

47. Feldman, K.S.; Sahasrabudhe, K. Ellagitannin chemistry. Syntheses of tellimagrandin II and a dehydrodigalloyl ether-containing dimeric ellagitannin analog of coriariin A. *J. Org. Chem.* 63: (1999), in press.

48. Feldman, K.S.; Sambandam, A. Ellagitannin chemistry. The first total chemical synthesis of an *O*(2), *O*(3)-galloyl-coupled ellagitannin, sanguiin H-5. *J. Org. Chem.* 60:8171 (1995).

49. Feldman, K.S.; Smith, R.S. Ellagitannin chemistry. First total synthesis of the 2,3- and 4,6-coupled ellagitannin pedunculagin. *J. Org. Chem.* 61:2606 (1996).

50. Feldman, K.S.; Hunter, K.L. On the basis for regioselective oxidation within a tetragalloylpyranose substrate. *Tetrahedron Lett.* 39:8943 (1998).

51. Cozzi, F.; Cinquini, M.; Annunziata, R.; Dwyer, T.; Siegel, J.S. Polar/π interactions between stacked aryls in 1,8-diarylnaphthylenes. *J. Am. Chem. Soc.* 114:5729 (1992).

52. Cozzi, F.; Cinquini, M.; Annuziata, R.; Siegel, J.S. Dominance of polar/π over charge-transfer effects in stacked phenyl interactions. *J. Am. Chem. Soc.* 115:5330 (1993).

53. Quideau, S.; Feldman, K.S.; Appel, H.M. Chemistry of galloyl-derived *o*-quinones: reactivity toward nucleophiles. *J. Org. Chem.* 60:4982 (1995).

54. Feldman, K.S.; Quideau, S.; Appel, H.M. Galloyl-derived orthoquinones as reactive partners in nucleophilic additions and Diels-Alder dimerizations: a novel route to the dehydrodigalloyl linker unit of agrimoniin-type ellagitannins. *J. Org. Chem.* 61:6656 (1996).

55. Yoshida, T.; Maruyama, T.; Nitta, A.; Okuda, T. Eucalbanins A, B, and C, monomeric and dimeric hydrolyzable tannins from *Eucalyptus alba* REINW. *Chem. Pharm. Bull.* 40:1750 (1992).

56. Quideau, S.; Feldman, K.S. Ellagitannin chemistry. The first synthesis of dehydrohexahydroxydiphenoate esters from oxidative coupling of unetherified methyl gallate. *J. Org. Chem.* 62:8809 (1997).

HIGHLY OXIDIZED ELLAGITANNINS AND THEIR BIOLOGICAL ACTIVITY

Takashi Yoshida, Tsutomu Hatano, Hideyuki Ito, and Takuo Okuda

Faculty of Pharmaceutical Sciences
Okayama University
Tsushima, Okayama 700-8530
JAPAN

1. INTRODUCTION

Polyphenols such as flavonoids and tannins in medicinal plants, foods, and beverages are attracting considerable attention in human health care because of their biologically multifunctional properties. These properties include antioxidant activity due to radical scavenging action and anti-HIV, antitumor activities, and inhibitory effects on various enzymes.[1] Among them, the antioxidant potency of tannins and related polyphenols is thought especially promising for the prevention of human carcinogenesis and other aged-related diseases such as arteriosclerosis and cerebrovascular attack, which are implicated in accumulated cellular damage by active oxygen species or free radicals.

Potent antitumor or antioxidant activity in various *in vitro* and *in vivo* assays has been exhibited by ellagitannins with highly oxidized structures, e.g., dehydroellagitannins, macrocyclic dimers, and oligomeric hydrolyzable tannins having many hexahydroxydiphenoyl (HHDP) groups. For example, macrocyclic dimers represented by oenothein B,[2] woodfordin C,[3] and hirtellin B,[4] which are metabolites biogenetically produced by double C—O oxidative couplings between a galloyl and an HHDP group in each monomeric unit (tellimagrandin I or II), were revealed to have a remarkable host-mediated antitumor activity against Sarcoma-180 in mice.[5] Woodfordin C was also reported to exhibit an inhibitory effect against DNA topoisomerase-II, which was much stronger than that shown by adriamycin.[6] These findings suggest the significance of this type of ellagitannin to the exploration of structure-activity relationships and new bioactive substances in nature.

In this paper, we summarize the structural diversity of highly oxidized ellagitannin dimers currently found in the Euphorbiaceae and Myrtaceae and biological properties specific to the structures that might be associated with development of antitumor and anti-allergic drugs.

2. HIGHLY OXIDIZED ELLAGITANNINS FROM EUPHORBIACEOUS PLANTS

A number of hydrolyzable tannins are found in plants of the family Euphorbiaceae.[1] These compounds include chebulagic acid,[7] phyllanthusiins A-C,[8] mallotusinic acid,[9] and macarinins,[10] etc. Many of these compounds are regarded as oxidized metabolites of geraniin (1), which is frequently found as a major tannin in plants of this family. Geraniin itself is also classified as an oxidized ellagitannin (dehydroellagitannin) because it contains a dehydrohexahydroxydiphenoyl (DHDP) group, an oxidized form of hexahydroxydiphenoyl (HHDP) unit. Geraniin was first isolated as yellow crystalline compound in 1976 from Geranii herba (leaves of *Geranium thunbergii* Sieb. et. Zucc.),[11] a popular anti-diarrheic in Japan. It was later shown to be a ubiquitous tannin that is widely distributed in Euphorbiaceae,[12] Aceraceae,[13] Cercidiphyllaceae[13] and many other families[1] as well as Geraniaceae. The gross stereostructure of geraniin was elucidated as 1 by spectral and chemical methods.[13,14] Very recently, the structure of geraniin has been confirmed by means of X-ray crystallography.[15] Crystals of 1 suitable for X-ray analysis were obtained as prisms by recrystallization from an aqueous acetonitrile. Crystallographic data were collected by synchrotron radiation (wavelength 0.928 Å) at room temperature and low temperature (120 K), and those measured at 120 K are shown in figure 1.

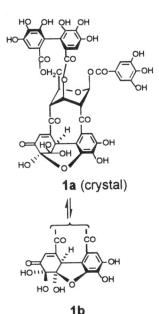

1a (crystal)

1b

X-ray Crystal Data

Crystal color	Yellow
Crystal size (mm)	0.4 x 0.1 x 0.004
Formula	$C_{41}H_{28}O_{27}\cdot 7H_2O$
Molecular weight	1078.5
Cell setting	Orthorhombic
Space group	$P2_12_12_1$
a (Å)	12.960 (2)
b	29.110 (2)
c	11.510 (2)
V (Å^3)	4342.3 (11)
Z	4
D (mg m^{-3})	1.650
Radiation type	Synchrotron
Wavelength (Å)	0.928
Temperature (K)	120
R-factor	0.0379

Figure 1. Structure and crystal data of geraniin (1).

The structure including seven moles of water was solved by a direct method with the DIRDIF program system, and refinement of the synchrotron data collected at 120 K allowed us to allocate almost all the H atoms including those of the seven water molecules. It should be noted that this was the first X-ray crystallographic elucidation of a natural ellagitannin, and it provided definite structural data including conformations of *1C* glucopyranose residue and hydrated cyclohexene trione ring, and also the inclined angle (60.6°) between two benzene rings in the (*R*)-HHDP moiety.

Geraniin is often found as an oxidized form in ellagitannin dimers such as euphorbins[16–19] which have been isolated from a number of species of Euphorbiaceae. These dimers are characterized by the presence of geraniin as a constituent monomer unit, and regarded as metabolites biogenetically produced by C—O oxidative coupling between an HHDP group of **1** and a galloyl group of tellimagrandin II or pentagalloylglucose as illustrated in figure 2.

The most recently discovered member of the euphorbin group of dimers, euphorbin I (**2**), was isolated from *Euphorbia watanabei*.[20] The structure of euphorbin I (**2**) was elucidated by spectral analyses and chemical degradations. Final confirmation of structure **2** including the orientation and absolute configuration of the linking unit (macaranoyl group) was provided by its chemical conversion into known euphorbin A (**4**) through Smiles rearrangement in phosphate buffer (pH 7.4). This reaction (room temperature; 8 hr) including an isomerization of the macaranoyl group to the valoneoyl group proceeded in a stepwise manner, and an intermediary product was found to be identical with euphorbin B by HPLC analysis. This finding was inconsistent with the structure of euphorbin B, which was previously proposed to be an isomer of euphorbin A concerning the orientation of the valoneoyl group.[19] Re-examination of euphorbin B led us to revise its structure to **3** possessing the tergalloyl group as the linking unit (fig. 3). Neither isomerization from **3** to **2** nor from **4** to **3** was observed, indicating that the rearrangement might be controlled by steric factor to give the least hindered valoneoyl group.

Euphorbin E (**5**), a notable dimer among the highly oxidized ellagitannins, was isolated from *Euphorbia hirta*.[21] This tannin was characterized by the presence of a dehydroeuphorbinoyl group and two DHHDP groups of 1'*R* and 1'*S* configuration, respectively. The ¹H and ¹³C NMR spectra of euphorbin E (**5**) were difficult to analyze because of the existence in an equilibrium mixture between six- and five-membered hemiacetal forms at the DHHDP moiety (fig. 4). The structure of euphorbin E was thus assigned on the basis of spectral analyses of its phenazine derivative (**5a**) similar to the case for geraniin[14] and many other dehydroellagitannins.[17–19]

Conclusive chemical evidence of structure **5** for euphorbin E was obtained by catalytic hydrogenation over Pd-C to a hexahydro-derivative (**6**) and a hydrogenolysis product (**7**) (fig. 5). These products were identified as dihydro-derivatives prepared from known euphorbin C (**8**)[17] and euphorbin F (**9**),[19] respectively (fig. 5). Production of compound **7** from **5** substantiated the presence of an allylic ether linkage in the dehydroeuphorbinoyl group.

Acalyphidin D₁ (**10**), a unique dimer possessing two DHHDP groups (fig. 6), was recently isolated from *Acalypha hispida* Burm. f., which has been used as a

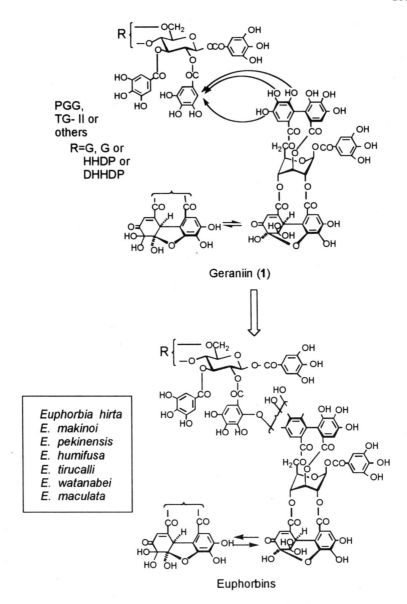

Figure 2. Biogenetic coupling modes of monomers leading to euphorbins.

remedy for thrush and boils in China and Indonesia. Isolation of acalyphidin D_1 was extremely difficult because of inevitable concomitance of a related dimer (excoecarianin) moving together on any column chromatography (Toyopearl HW-40, MCI-gel and/or Sephadex LH-20 etc.). Based on the ^{1}H NMR spectrum of crude acalyphidin D_1, it was assumed that this compound should have two DHHDP groups in the molecule. After several attempts at derivatization, acalyphidin D_1

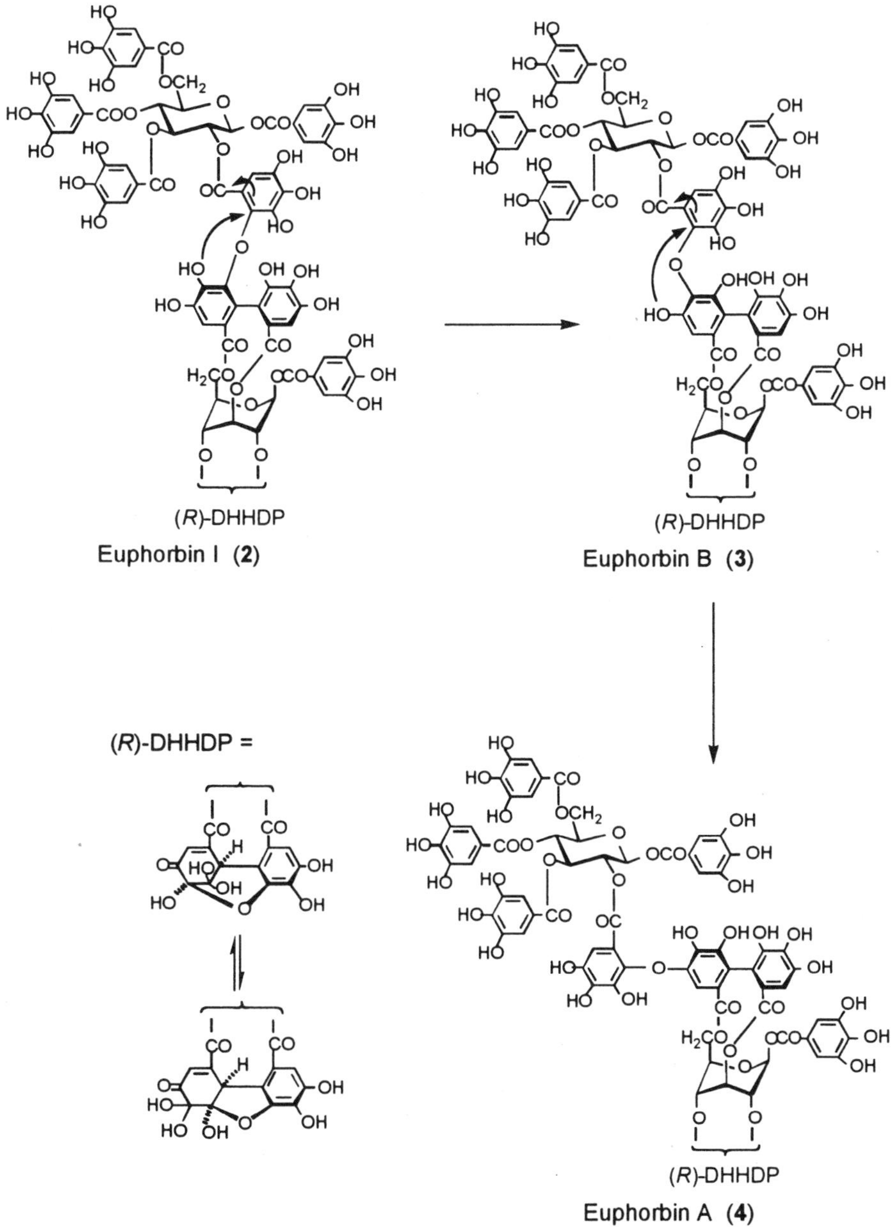

Figure 3. Smiles rearrangements from euphorbin I (2) to euphorbin A (4).

was finally purified as an acetonyl derivative (**10a**). The acetonyl derivative of dehydroellagitannin is easily prepared by treatment with ammonium formate in acetone[20,22] and can be used in place of the phenazine derivative to facilitate the purification and structural analysis of this class of tannins. In fact, the ^{1}H and ^{13}C

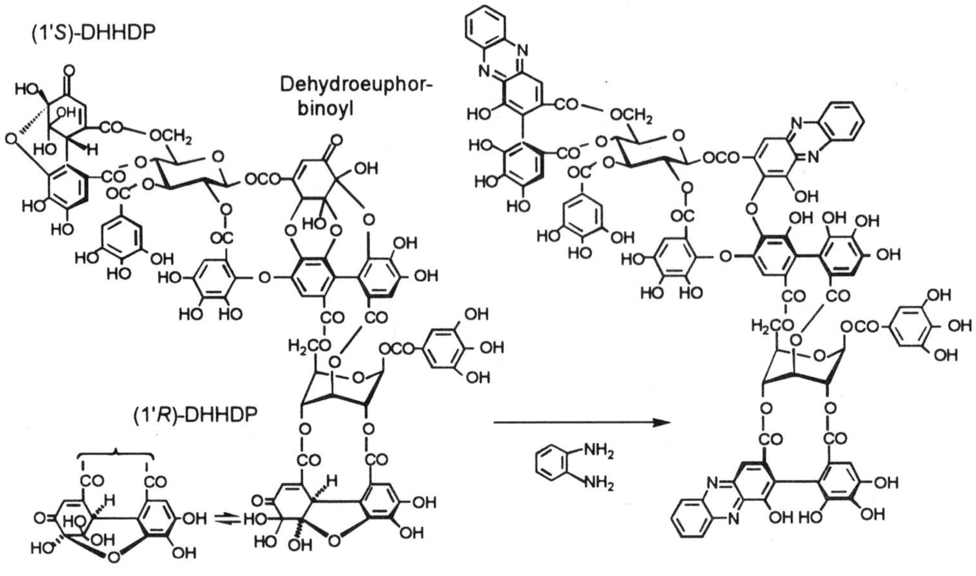

Figure 4. Structures of euphorbin E (**5**) and its phenazine derivative (**5a**).

NMR spectra of **10a** showed well-resolved sharp signals, unlike the complicated spectra of **10** due to an equilibrium mixture at the DHHDP groups. The spectra clearly indicated the presence of two acetonyl-DHHDP groups, an HHDP, a galloyl and a valoneoyl group as well as two *1C* glucopyranose residues. A comparison of the spectral data of **10a** with those of the acetonyl congener of **1** revealed aca-lyphidin D_1 to be a dimer composed of two moles of geraniin (**1**). The location of each acyl group including the orientation of the valoneoyl group in **10a** was unam-biguously determined from the HMBC spectrum, which showed connectivity between the aromatic proton and glucose proton through three-bond long-range correlations with the common ester carbonyl carbon as illustrated by arrows in the formula.

It is noteworthy that acalyphidin D_1 (**10**) is the first example of a dimer composed of two moles of geraniin (**1**), though many ellagitannin dimers having geraniin as a constituent monomer unit, such as euphorbins, are known.

3. MACROCYCLIC ELLAGITANNIN DIMERS FROM MYRTACEAE

Oenothein B (**11**) and related oligomers with a macrocyclic structure have been found in the genera *Oenothera*[2] and *Epilobium*[23] of Onagraceae (fig. 7). Woodfordin C (**12**) and cuphiin D_1 (**13**), which were characterized as gallates of **11**, were also obtained from species of Lythraceae.[3] These dimers were shown to possess *in vivo* antitumor activity or to have an *in vitro* inhibitory effect on DNA-topoisomerase II as described earlier. Accordingly, these compounds might be potential

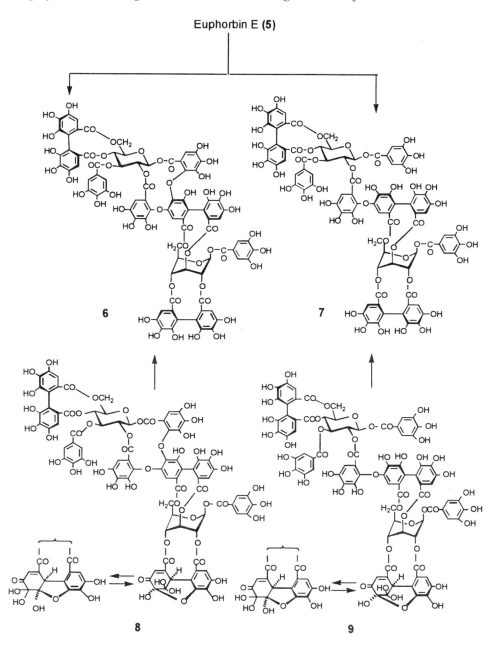

Figure 5. Hydrogenated derivatives of euphorbin E (**5**).

chemotherapeutic agents. Recently, oligomers of this type have been shown to be widely distributed in myrtaceous plants. Eugeniflorin D₁ (**14**) and D₂ (**15**) together with **11** were isolated from *Eugenia uniflora*.[24] Eugeninflorin D₁ (**14**) was assigned to an isomer of **12** concerning the configuration of the galloyloxy group at an

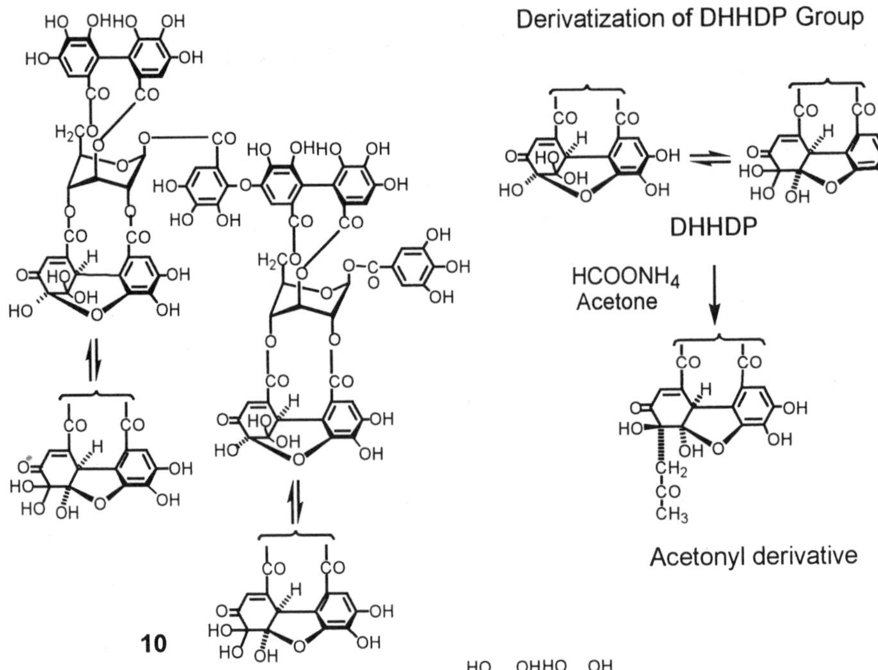

Derivatization of DHHDP Group

DHHDP

HCOONH$_4$
Acetone

Acetonyl derivative

10

Acetonyl acalyphidin D$_1$ (10a)

ESI-MS: *m/z* 2000 (M+NH$_4$)$^+$

[α]$_D$ -43° (MeOH)

CD (MeOH): [θ]$_{208}$ +3.2 x 10^4;

[θ]$_{236}$ -5.9 x 10^4;

[θ]$_{263}$ +2.3 x 10^4

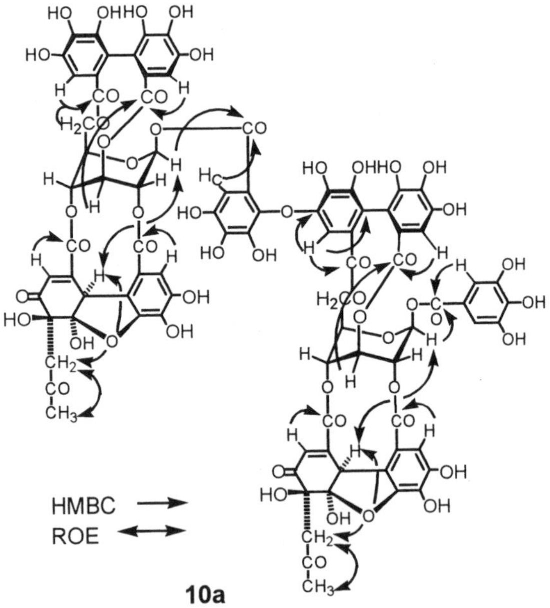

HMBC ⟶
ROE ⟷

10a

Figure 6. Structures of acalyphidin D$_1$ (**10**) from *Acalypha hispida* and its acetonyl derivative (**10a**).

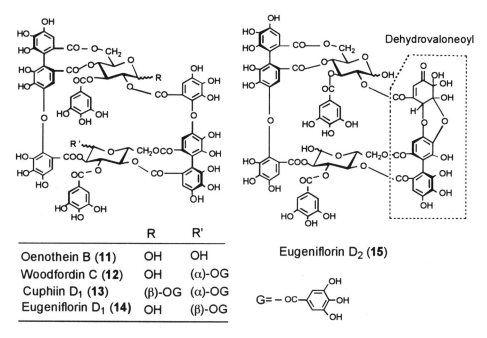

	R	R'
Oenothein B (11)	OH	OH
Woodfordin C (12)	OH	(α)-OG
Cuphiin D₁ (13)	(β)-OG	(α)-OG
Eugeniflorin D₁ (14)	OH	(β)-OG

Eugeniflorin D$_2$ (15)

Figure 7. Structures of macrocyclic ellagitannin dimers.

anomeric center. On the other hand, eugeniflorin D$_2$ (15) was characterized by the presence of a new acyl unit, a dehydrovaloneoyl group, as a linking unit. Its structure was elucidated based on the NMR spectral analyses including a ^{1}H-^{13}C long-range COSY spectrum. Conclusive evidence for the structure of 15 was obtained by chemical transformation into oenothein B (11) (hot water; 1 day). The transformation of the dehydrovaloneoyl group in 15 into the valoneoyl group could be interpreted in terms of intermolecular disproportionation reaction similarly to that reported for production of ellagic acid from the DHHDP group in dehydroellagitannin.[25] The stereochemistry at the methine carbon of the dehydrovaloneoyl group, however, remains to be defined.

More recently, a novel dimer, eurobustin C (16), along with eugeniflorin D$_2$ (15) and oenothein B (11), has been isolated from *Eucalyptus robusta* leaves (fig. 8). Eurobustin C (16) was characterized by the presence of a new acyl unit, named eurobustinoyl group, which is a plausible metabolite producible from the dehydrovaloneoyl group in 15. The ^{1}H NMR spectrum of 16 revealed structural similarity to 15. The main points of difference between 15 and 16 were a lack of an aromatic 1H-singlet and the presence of extra signals due to a methine and methylene protons coupled with each other in the latter. The structure of the linking unit (eurobustinoyl group) in 16 was characterized based on the comparison of the NMR spectral data of the cyclopentenone part with those of brevifolincarboxyl residue (fig. 8). Methylation of 16 followed by methanolysis gave trimethyl hexa-*O*-methyleurobustinoate (17) ([M+H]$^+$ 605), which showed a positive and a negative Cotton effect at 228 ($[\theta] + 1.1 \times 10^5$) and 254 nm ($[\theta] - 8.2 \times 10^4$) similar to

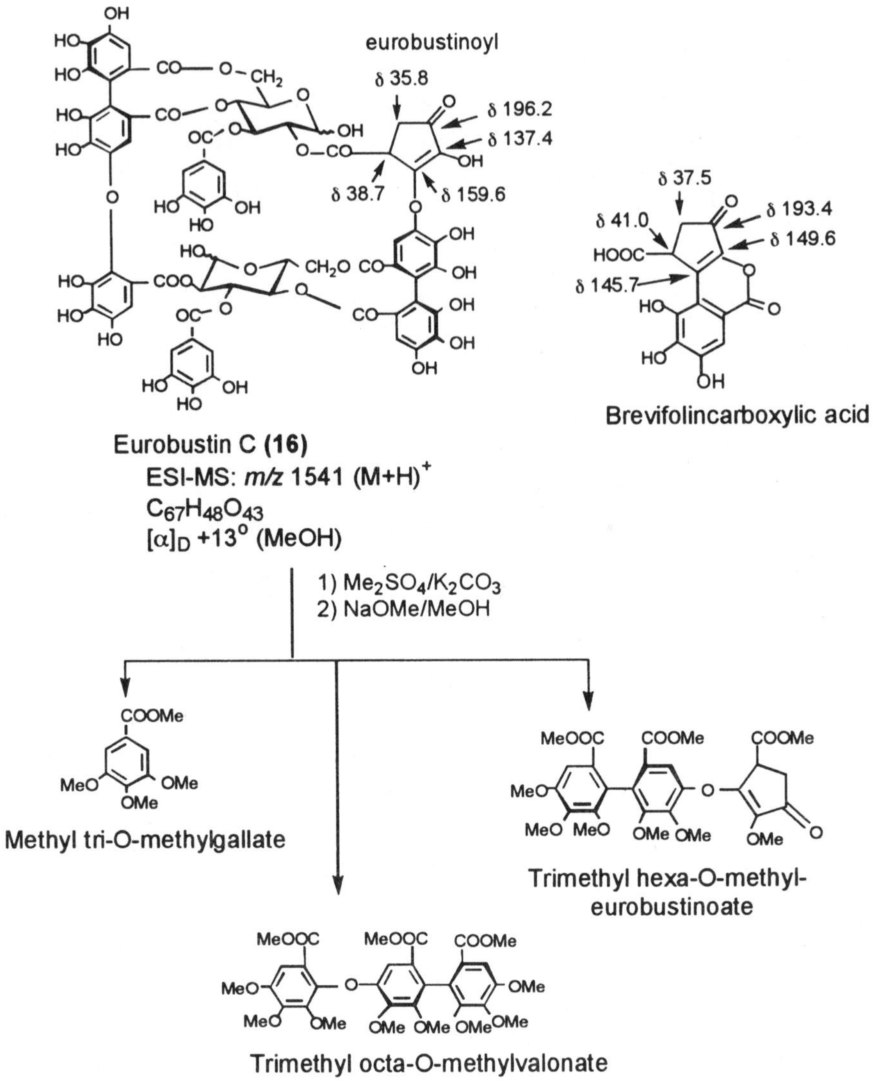

Figure 8. Structure elucidation of eurobustin C (**16**) from *Eucalyptus robusta*.

those of dimethyl (*S*)-hexamethoxydiphenate[26] in the CD spectrum. Atropisomerism of **17** was thus determined to be of (*S*)-series. The location of the eurobustinoyl group in **16** was ascertained by HMBC spectral analysis.

4. BIOLOGICAL ACTIVITY

Some oligomeric hydrolyzable tannins were shown to be promising candidates for new anticancer chemotherapeutic drugs in the investigation of the inhibitory effect of tannins on poly(ADP-ribose)glycohydrolase. The metabolism of poly(ADP-ribose) on chromosomal proteins in eukaryotic cells has been shown to be associated with gene activation, such as DNA repair, replication, and transcription.[27] During initiation of gene expression, DNA replication, and cell differentiation, degradation of poly(ADP-ribose) of specific chromosomal proteins was reported to be observed.[27] Therefore, it is likely that degradation of poly(ADP-ribose) is an important factor in the regulation of gene activation. Poly(ADP-ribose) on chromosomal proteins is hydrolyzed mainly by poly(ADP-ribose) glycohydrolase to give ADP-ribose and mono(ADP-ribosyl) proteins.[28] In a search for potent and specific inhibitors for poly(ADP-ribose) glycohydrolase purified from human placenta, we examined the inhibitory effect of tannins and related polyphenols on the enzyme.[29] The inhibitory activity was evaluated by the decrease in the radioactivity of [^{3}H]poly(ADP-ribose) upon addition of each tannin in various concentrations to the standard assay mixture containing $20\,\mu$M [^{3}H]poly(ADP-ribose) and an appropriate amount of enzyme at $37\,°$C for 10 min.

As shown in table 1, hyrolyzable tannins exhibited an appreciable inhibitory effect with IC_{50} values of 0.38–$31.8\,\mu$M, whereas procyanidin oligomers and their constituent monomers were not active even at $100\,\mu$M. Among the hydrolyzable tannins, the potency of the inhibitory activity increased in the order of gallotatannins < ellagitannin monomers <. dimers < trimers < tetramers. Structure-activity relationships revealed that the degree of effect was dependent on the oxidative score of the molecule, which was tentatively assumed based on dehydrogenation of the galloyl units (biogenetic consideration), i.e., the HHDP group producible by the C—C oxidative coupling between the two galloyl groups is scored as 1 and the valoneoyl group formed by C—O coupling between an HHDP and a galloyl group is scored as 2. Thus, the most potent inhibitory effect was exhibited by nobotanin K (fig. 9; **19**) ($0.3\,\mu$M), which had an oxidative score of 9.

It is noteworthy that the slope of the IC_{50} values versus the oxidative scores of oligomeric ellagitannins was hyperbolic, implying that highly oxidized structures with a rigid conformation compelled by the presence of HHDP and valoneoyl groups may be important in tight interaction with the glycohydrolase molecule (fig. 10). It should also be noted that the inhibitory effects of these oligomeric ellagitannins were more potent than those of the previously known inhibitors,[30,31] daunomycin and ethadridine (IC_{50} 50–100 μM), ADP-ribose (1–3 mM) and cAMP (5–10 mM).

To investigate the *in vivo* poly(ADP-ribose) glycohydrolase activity in intact cells, we examined the effect of a representative dimer, oenothein B (**11**), on glucocorticoid-sensitive mouse mammary tumor virus (MMTV) gene expression in 34I cells, because depoly(ADP-ribosyl)ation of chromosomal proteins was suggested to be involved in the initiation of MMTV transcription.[27] The induction of MMTV mRNA by dexamethasone (100 nM) was potently suppressed by pretreatment with oenothein B.[32] The inhibitory effect of oenothein B was dose-dependent, and 50 μM oenothin B showed about 90 percent inhibition of relative MMTV mRNA synthesis compared with control (without tannin). On the other hand, oenothein B showed no appreciable inhibition against NAD$^+$-poly(ADP-ribose) metabolizing enzymes such as poly(ADP-ribose) polymerase or NAD$^+$

Table 1. Inhibitory effects of tannins and related polyphenols on
poly(ADP-ribose)glycohydrolase[29]

Polyphenol		IC$_{50}$ (μM)	Oxid. Score	Number of HHDP	Val
Hydrolyzable Tannins					
Trigalloylglucose	Monomer	31.8	0	0	0
Tetragalloylglucose		24.2	0	0	0
Pentagalloylglucose		18.9	0	0	0
Tellimagrandin I		11.9	1	1	0
Casuarictin		11.7	2	2	0
Geraniin (**1**)		15.5	3	1 (+1)*	0
Cornusiin A	Dimer	7.1	3	1	1
Rugosin D		6.1	3	1	1
Coriariin A (**20**)		8.5	3	2 (+1)**	0
Nobotanin B (**18**)		4.4	4	2	1
Oenothein B (**11**)		4.8	4	0	2
Nobotanin E	Trimer	1.8	6	2	2
Nobotanin K (**19**)	Tetramer	0.3	9	3	3
Condensed Tannins					
ECG-dimer		>100			
ECG-trimer		>100			
ECG-tetramer		>100			
Flavan-3-ols					
(-)-epicatechin		>100			
(-)-Epicatechin gallate		>100			
(-)-Epigallocatechin		>100			
(-)-Epigallocatechin gallate		>100			

Oxid. score: HHDP = 1, valoneoyl (Val) = 2, DHHDP = 2, dehydrodigalloyl (DDG) = 1
* DHHDP
** DDG

glycohydrolase even at 0.5 mM. Oenothein B is thus a specific inhibitor of
poly(ADP-ribose) glycohydrolase.

On the other hand, we have demonstrated the antioxidant activity of tannins
and related polyphenols in various experimental models, including inhibitory
effects on lipid peroxidation induced by NADPH-ADP and ascorbic acid-ADP in
rat liver microsomes and mitochondria, respectively.[33] In order to clarify the mechanism of the inhibitory effect by tannins, a kinetic study on the autoxidation of
methyl linoleate as a model of lipid peroxidation was conducted.[34] The results
revealed that tannins terminated a radical chain-reaction of lipid oxidation
by donating a hydrogen radical. This mechanism was substantiated by *in situ*
ESR measurement showing transient ESR signal of tannin radicals. The

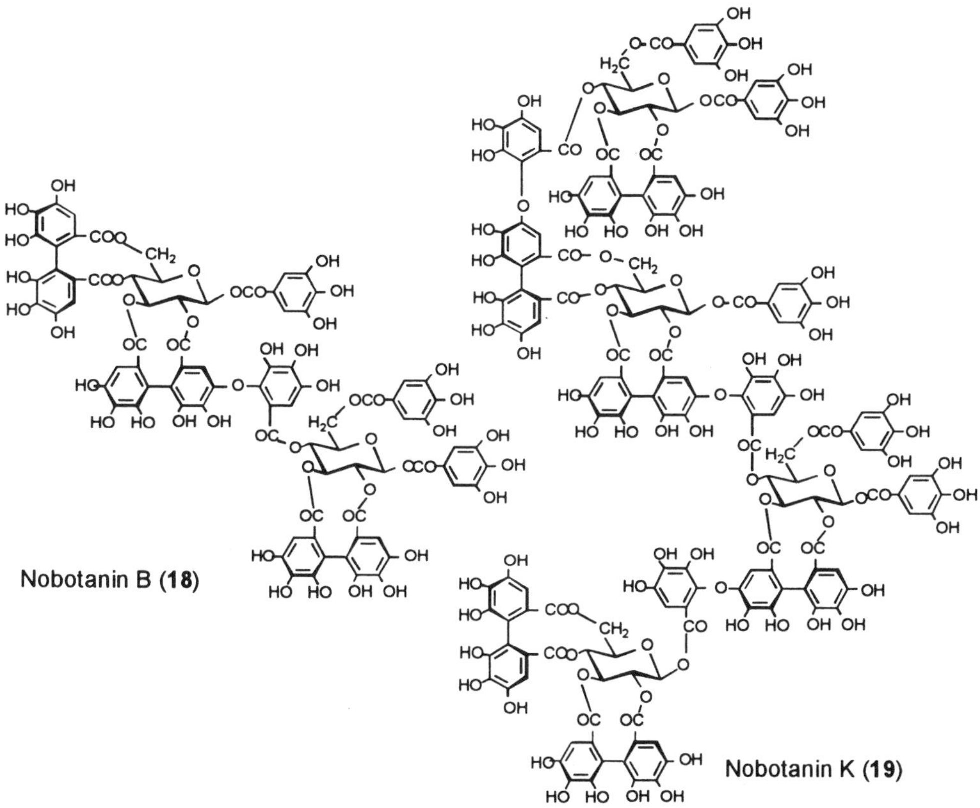

Figure 9. Structures of nobotanin B (**18**) and nobotanin K (**19**).

radical-scavenging effect of ellagitannins on 1,1-diphenyl-2-picrylhydrazyl (DPPH)[35] and on superoxide anion radical generated by the hypoxanthine-xanthine oxidase system[36] was stronger than those of galloylglucoses.

In ongoing studies on the antioxiant action of polyphenols, we have now examined the effects of tannins on potassium superoxide (KO_2)-induced histamine release from rat peritoneal mast cells. The mast cells (10^4 cells/ml) were preincubated with various concentrations of tannins (1–100 μM) at 37 °C for 15 min, and then KO_2 (10 mM) was added. The reaction mixture was kept for 20 min and chilled to terminate the reaction. The amounts of histamine in the supernatant and the residual mast cells were determined fluorometrically. As shown in table 2, the inhibitory effects of tannins and related polyphenols in this experimental system were largely dependent on the structure and molecular size of the polyphenols. Monomeric hydrolyzable tannins except for the dehydroellagitannin, geraniin (**1**), showed lower inhibitory activities (IC_{50} 55–83 μM)) than ellagitannin dimers (IC_{50} 0.68–48 μM). The most effective inhibition was exhibited by agrimoniin (fig. 11; **21**) and euphorbin C (**8**) with IC_{50} 0.68 and 0.80 μM, respectively. On the other hand, pretreatment with the antiallergic drugs, azelastine, astemizole, and ibudilast, had little inhibitory effect on the KO_2-induced histamine release. The potency of

IC$_{50}$ (µM)

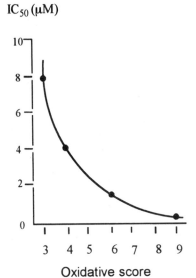

Figure 10. Plot of IC$_{50}$ values of tannins against poly(ADP-ribose) glycohydrolase versus oxidative score.

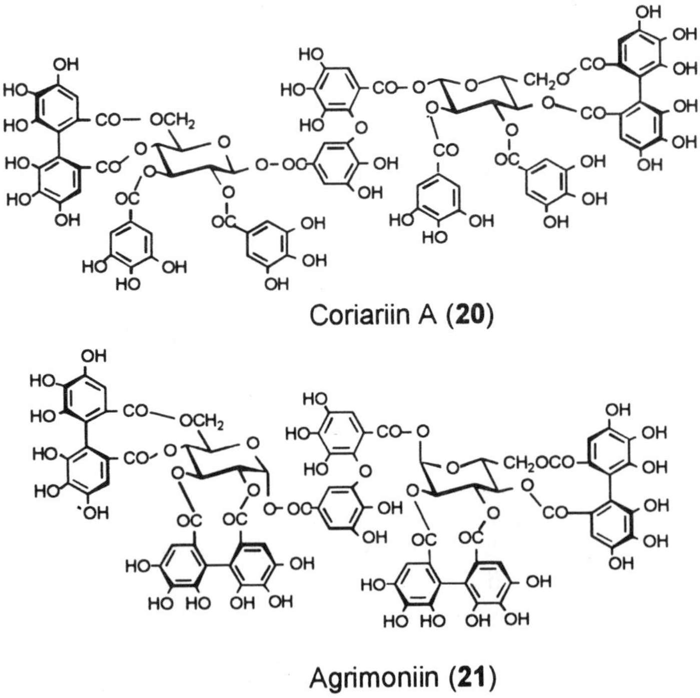

Oxidative score

Coriariin A (**20**)

Agrimoniin (**21**)

Figure 11. Structures of coriariin A (**20**) and agrimoniin (**21**).

Table 2. Inhibitory effects of tannins on histamine release (HR) induced by potassium superoxide and eompound 48/80

Tannin	Oxid. score	IC_{50} (µM)	
		KO_2-HR	48/80-HR
1,2,6-Trigalloylglucose	0	-	151.2
1,2,4,6-Tetragalloylglucose	0	80.4	89.2
Pentagalloylglucose	0	55.7	0.60
Pedunculagin	2	62.9	0.72
Geraniin (**1**)	3	37.5	0.76
Coriariin A (**20**)	3	43.1	2.97
Euphorbin A (**4**)	4	48.8	0.23
Nobotanin B (**18**)	4	18.7	0.63
Oenothein B (**11**)	4	4.1	0.65
Agrimoniin (**21**)	5	0.68	0.49
Euphorbin C (**8**)	6	0.80	0.50

inhibition by tannins seemed to increase with the number of HHDP and/or valoneoyl (euphorbinoyl) groups (oxidative score) in the molecule. This tendency was similar to that observed for radical scavenging activity of hydrolyzable tannins against superoxide anion radical.[36] Highly oxidized hydrolyzable tannins also showed potent inhibitory effects on the compound 48/80-induced histamine release (table 2).

However, the inhibitory effect on histamine-release caused by each of the two stimulants differed. These findings suggest that cell-membrane stabilizing action as well as the radical scavenging activity contribute to the inhibitory effects of ellagitannins. The inhalation of tannins may prevent airway hypersensitivity.

5. CONCLUSIONS

Recent progress in separation techniques and spectral analyses including 2D NMR correlation spectroscopy using high field magnets has allowed us to isolate and characterize a variety of hydrolyzable tannins in higher plants. In this paper, we focused on the highly oxidized oligomeric hydrolyzable tannins recently found in the Euphorbiaceae and the Myrtaceae and discussed their structural diversity. A tannin, acalyphidin D_1 (**10**), isolated from *Acalypha hispida* is the dimer that might be biosynthesized by intermolecular C—O oxidative coupling of two moles of geraniin. Eugeniflorin D_2 and eurobustin C constitute a new class of oligomer having a modified valoneoyl group as the linking unit of monomers. Discovery of these dimers as well as euphorbin E, which has three 'hydrated cyclohexene trione' moieties in the molecule, implied that diverse dehydrogenation processes specific to plants are involved in the gallic acid metabolism and that new hydrolyzable tannins with chemically interesting structures occur in nature. In addition to previous findings on the biological effects of oligomers, oenothein B (**11**) and nobotanins were found to be specific inhibitors against poly(ADP-ribose) glycohydrolase, which is involved in mammal tumor gene regulation. Oenothein B and other oligomers would thus be useful probes for studies on the physiological significance of poly(ADP-ribose) metabolism in regulation of nuclear functions, and will be candidates for new antiviral and anticancer chemotherapeutic drugs.

Upon investigation of the inhibitory effect of tannins and related polyphenols on KO_2-induced histamine release from rat peritoneal mast cells, higher molecular weight ellagitannins having large oxidative score such as agrimoniin and euphorbin C (IC_{50} less than $1\,\mu M$) showed stronger inhibitory activity than galloylglucoses and simple monomeric ellagitannin (IC_{50} more than $50\,\mu M$). Further study of these tannins may yield new antiallergic drugs.

ACKNOWLEDGMENTS

The authors are indebted to Professor P. Luger of Berlin University and Professor S. Kashino of Okayama University for crystallographic analysis, Professors Y.-K. Yen and L.-L. Yang of Taipei Medicinal College for supply of the plant materials, and Professor S. Tanuma of Science University of Tokyo, Professor H. Sakagami of Meikai University, and Professor M. Akagi of Tokushima Bunri University for biological assays. Financial support (in part) by a Grant-in-Aid for Scientific Research from the Ministry of Science, Education, Sports and Culture of Japan is gratefully acknowledged.

REFERENCES

1. Okuda, T.; Yoshida, T.; Hatano, T. Hydrolyzable tannins and related polyphenols. *Progr. Chem. Org. Nat. Prod.* 66:1 (1995).

2. Hatano, T.; Yasuhara, M.; Matsuda, K.; Yazaki, K.; Yoshida, T.; Okuda, T. Oenothein B, a dimeric hydrolyzable tannin with macrocyclic structure and accompanying tannins from *Oenothera erythrosepala*. *J. Chem. Soc., Perkin Trans. 1.* 2735 (1990).

3. Yoshida, T.; Chou, T.; Nitta, A.; Miyamoto, K.; Koshiura, R.; Okuda, T. Woodfordin C, a macro-ring hydrolyzable tannin dimer with antitumor activity and accompanying dimers from *Woodfordia fruticosa* flowers. *Chem. Pharm. Bull.* 38:1211 (1986).

4. Yoshida, T.; Hatano, T.; Ahmed, A.F.; Okonogi, A.; Okuda, T. Structures of isorugosin E and hirtellin B, dimeric hydrolyzable tannins having a trisgalloyl group. *Tetrahedron.* 47:3575 (1991).

5. Miyamoto, K.; Nomura, M.; Murayama, T.; Furukawa, T.; Hatano, T.; Yoshida, T.; Koshiura, R.; Okuda, T. Antitumor activities of ellagitannins against Sarcoma-180 in mice. *Biol. Pharm. Bull.* 16:379 (1993).

6. Motegi, A.K.; Kuramochi, H.; Kobayashi, F.; Ekimoto, H.; Takahashi, T.; Kadota, S.; Takamori, Y.; Kikuchi, T. Woodfruticosin (woodfordin C), a new inhibitor of DNA topoisomerase II. *Biochem. Pharmacol.* 44:1961 (1992).

7. Yoshida, T.; Fujii, R.; Okuda, T. Revised structures of chebulinic acid and chebulagic acid. *Chem. Pharm. Bull.* 28:3713 (1980).

8. Yoshida, T.; Itoh, H.; Matsunga, S.; Tanaka, R.; Okuda, T. Hydrolyzable tannins with $^1C_4$ glucose core from *Phyllanthus flexuosus* Muell. Arg. *Chem. Pharm. Bull.* 40:53 (1992).

9. Okuda, T.; Seno, K. Mallotusinic acid and mallotinic acid, new hydrolyzable tannins from *Mallotus japonicus. Tetrahedron Lett.* 39 (1978).

10. Lin, J.-H.; Ishimatsu, T.; Tanaka, T.; Nonaka, G.; Nishioka, I. Structures of macaranins and macarinins, new hydrolyzable tannins possessing macaranoyl and tergalloyl ester groups, from the leaves of *Macaranga sinensis* (Baill.) Muell.-Arg. *Chem. Pharm. Bull.* 38:1844 (1990).

11. Okuda, T.; Yoshida, T.; Nayeshiro, H. Geraniin, a new ellagitannin from *Geranium thunbergii. Tetrahedron Lett.* 3721 (1976).

12. Okuda, T.; Mori, K.; Hatano, T. The distribution of geraniin and mallotusinic acid in the order Geraniales. *Phytochemistry* 19:547 (1980).

13. Gupta, R.K.; Haddock, E.A.; Haslam, E. The metabolism of gallic acid and hexahydroxy-diphenic acid in plants. Part 3. Esters of (R) and (S)-hexahydroxydiphenic acid and dehy-drohexahydroxy-diphenic acid with D-glucopyranose ($^1C_4$) and related conformations). *J. Chem. Soc., Perkin Trans.1*: 2535 (1982).

14. Okuda, T.; Yoshida, T.; Hatano, T. Constituents of *Geranium thunbergii* Sieb. et. Zucc. Part 12. Hydrated stereostructure and equilibration of geraniin. *J. Chem. Soc., Perkin Trans.1.* 9 (1982).

15. Lugar, P.; Weber, M.; Kashino, S.; Amakura, Y.; Yoshida, T.; Okuda, T.; Beurskens, G.; Dauter, Z. Structure of the tannin geraniin based on conventional X-ray data at 295K and on synchrotron data at 293 and 120K. *Acta Crystallographica* B54:687 (1998).

16. Yoshida, T.; Chen, L.; Okuda, T. Euphorbins A and B, novel dimeric dehydroellagitannins from *Euphorbia hirta. Chem. Pharm. Bull.* 36:2940 (1988).

17. Yoshida, T.; Namba, O.; Chen, L.; Okuda, T. Euphorbin C, an equilibrated dimeric dehy-droellagitannin having a new tetrameric galloyl group. *Chem. Pharm. Bull.* 38:86 (1990).

18. Yoshida, T.; Namba, O.; Kurokawa, K.; Amakura, Y.; Liu, Y.-Z.; Okuda, T. Tannins and related polyphenols of euphorbiaceous plants. XII. Euphorbins G and H, new hydrolyzable tannins from *Euphorbia prostrata* and *Euphorbia makinoi. Chem. Pharm. Bull.* 42:2005 (1994).

19. Yoshida, T.; Yokoyama, K.; Namba, O.; Okuda, T. Tirucallins A and B and euphorbin F, monomeric and dimeric hydrolyzable tannins from *Euphorbia tirucalli. Chem. Pharm. Bull.* 39:1137 (1991).

20. Amakura, Y.; Yoshida, T. Euphorbin I, a new dimeric hydrolyzable tannins from *Euphorbia watanabeii. Chem. Pharm. Bull.* 44:1293 (1996).

21. Yoshida, T.; Namba, O.; Chen, L.; Okuda, T. Euphorbins E, a hydrolyzable tannin dimer of highly oxidized structure from *Euphorbia hirta. Chem. Pharm. Bull.* 38:1113 (1990).

22. Tanaka, T.; Fujisaki, H.; Nonaka, G.; Nishioka, I. Structures, preparation, high-performance liquid chromatography and some reactions of dehydroellagitannin-acetone condensates. *Chem. Pharm. Bull.* 40:2937 (1992).

23. Ducrey, B.; Marston, A.; Gohring, S.; Hartmann, R.W.; Hostettman, K. Inhibition of 5α-reductase and aromatase by the ellagitannins oenothein A and oenothein B from *Epilobium* species. *Planta Medica.* 63:111 (1997).

24. Lee, M.-H.; Nishimoto, S.; Yang, L.-L.; Yen, K.-Y.; Hatano, T.; Yoshida, T.; Okuda, T. Two macrocyclic hydrolyzable tannin dimers from *Eugenia uniflora. Phytochemistry.* 44:1343 (1997).

25. Schmidt, O.T.; Schanz, R.; Wurmb, R.; Groebke, W. Brevilagin 2. *Liebigs Ann. Chem.* 706:154 (1967).

26. Okuda, T.; Yoshida, T.; Hatano, T.; Koga, T.; Toh, N.; Kuriyama, K. Circular dichroism of hydrolyzable tannins. I. Ellagitannins and gallotannins. *Tetrahedron Lett.* 23, 3937 (1982).

27. Tanuma, S.; Johnson, L.D.; Johnson, G.S. ADP-ribosylation of chromosomal proteins and mouse mammary tumor virus gene expression. *J. Biol. Chem.* 58:15371 (1983).

28. Miwa, M.; Tanaka, M.; Matsushima, T.; Sugimura, T. Purification and properties of a glycohydrolase from calf thymus splitting ribose-ribose linkages of poly (adenosine diphosphate ribose). *J. Biol. Chem.* 249:3475 (1974).

29. Aoki, K.; Nishimura, K.; Abe, H.; Maruta, H.; Sakagami, H.; Hatano, T.; Okuda, T.; Yoshida T.; Tsai,Y.-J.; Uchiumi, F.; Tanuma, S. A Novel inhibitors of poly(ADP-ribose) glycohydrolase. *Biochim. Biophys. Acta.* 1158:251 (1993).

30. Tavassoli, M.; Hajihosaini, T.; Shall, S. Effect of DNA intercalators on poly(ADP-ribose) glycohydrolase activity. *Biochim. Biophys. Acta.* 827:228 (1985).

31. Uchida, K.; Suzuki, H.; Maruta, H.; Abe, H.; Aoki, K.; Niwa, M.; Tanuma, S. Preferential degradation of protein-bound (ADP-ribose)$_n$ by nuclear poly(ADP-ribose) glycohydrolase from human placenta. *J. Biol. Chem.* 268:3194 (1993).

32. Aoki, K.; Maruta, H.; Uchiumi, F.; Hatano, T.; Yoshida T.; Tanuma, S. A macrocircular ellagitannin, oenothein B, suppresses mouse mammary tumor gene expresiion *via* inhibition of poly(ADP-ribose)glycohydrolase. *Biochem. Biophys. Res. Commun.* 210:329 (1995).

33. Okuda, T., Kimura, Y., Yoshida, T., Hatano, T., Okuda, H., Arichi, S. Inhibitory effects on lipid peroxidation in mitochondria and microsomes of liver. *Chem. Pharm. Bull.* 31:1625 (1983).

34. Fujita, Y.; Komagoe, K.; Sasaki, Y.; Uehara, I.; Okuda, T.; Yoshida, T. Inhibition mechanism of tannins isolated from medicinal plants and related compounds on autoxidation of methyl linoleate. *Yakugaku Zasshi.* 108:528 (1988).

35. Yoshida, T.; Hatano, T.; Okumura, T.; Uehara, I.; Komogoe, K.; Fujita, Y.; Okuda, T. Radical-scavenging effects of tannins and related polyphenols on 1,1-diphenyl-2-picrylhydrazyl radical. *Chem. Pharm. Bull.* 37:1919 (1989).

36. Hatano, T.; Edamatsu, R.; Hiramatsu, M.; Mori, A.; Fujita, Y.; Yasuhara, T.; Yoshida, T.; Okuda, T. Effects of tannins and related polyphenols on superoxide anion radical and on 1,1-diphenyl-2-picrylhydrazyl radical. *Chem. Pharm. Bull.* 37:2016 (1989).

CONDENSED TANNINS AND RELATED COMPOUNDS

ENANTIOSELECTIVE SYNTHESIS OF FLAVONOIDS:

DIHYDROFLAVONOLS AND FLAVAN-3-OLS

Hendrik van Rensburg, Reinier J.J. Nel,
Pieter S. van Heerden, Barend B.C.B. Bezuidenhoudt,
and Daneel Ferreira

Department of Chemistry
University of the Orange Free State
P.O. Box 339
Bloemfontein 9300
SOUTH AFRICA

1. INTRODUCTION

The beneficial role of flavonoids in general and of the proanthocyanidins in particular to mankind is increasingly being recognized. In addition to the multitude of industrial applications, the oligo- and polymeric proanthocyanidins are now also credited for the profound health-promoting effects of tea, fruit juices, and red wine. This is mainly due to their *in vitro* antioxidant or radical scavenging properties, while the polyflavonoids in red wine have recently been implicated in protection against cardiovascular disorders, e.g., the "French paradox". These observations highlight the need for synthetic access to enantiopure flavonoid monomeric precursors exhibiting the aromatic oxygenation pattern of the constituent flavanyl units of the naturally occurring proanthocyanidins, hence paving the route to synthesize a representative series of these important biomolecules. This review accordingly focuses on methodologies aimed at the stereoselective syntheses of the different sets of the four diastereomers of both the dihydroflavonols and flavan-3-ols exhibiting characteristic "natural product" phenolic oxygenation patterns.

2. STEREOSELECTIVE SYNTHESIS OF DIHYDROFLAVONOLS

Only a limited number of dihydroflavonols with 2,3-*trans* configuration are "relatively readily available", providing of course that the researcher has access to the relevant natural source. Among these compounds are (+)-(2R,3R)-fustin (fig. 1, 1) (dihydrofisetin—commercially available), (−)-(2S,3S)-fustin (enantiomer

Plant Polyphenols 2: Chemistry, Biology, Pharmacology, Ecology, Edited by
Gross et al. Kluwer Academic / Plenum Publishers, New York, 1999

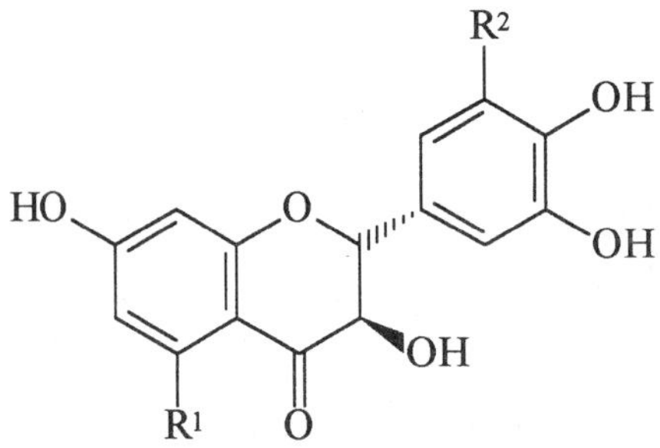

1 $R^1 = R^2 = H$
2 $R^1 = H, R^2 = OH$
3 $R^1 = OH, R^2 = H$
4 $R^1 = R^2 = OH$

Figure 1. Structures of dihydroflavonols 1–4.

of **1**—available from *Rhus typhina*[1]), (+)-(2R,3R)-dihydrorobinetin (fig. 1, **2**) (available from *Robinia pseudacacia* in partially racemized form[2]), (+)-(2R,3R)-dihydroquercetin (fig. 1, **3**) (taxifolin—available from *Pseudotsuga menziesii*— Douglas-fir outer bark[3]), and (+)-(2R,3R)-dihydromyricetin (fig. 1, **4**) [ampelopsin— available from the bark of *Pinus contorta*[4] (lodgepole pine)]. Notable is the conspicuous absence from this list of analogues possessing 2,3-*cis* configuration, thus clearly demonstrating the need for a synthetic protocol giving access to the full range of dihydroflavonol diastereomers with phenolic oxygenation patterns approximating those of the natural products.

Virtually all the synthetic efforts to synthesize enantiomerically enriched dihydroflavonols had hitherto focused on the asymmetric epoxidation of chalcones and the subsequent transformation of the chalcone epoxides into dihydroflavonols. The literature covering these developments up to 1990 was recently comprehensively reviewed.[5] Except for the reports from two different groups,[6–9] most of the literature protocols involved the "softer" but unsatisfactory options of selecting chalcones with the minimal oxygenation.[10–13] These deficiencies obviously reduce the synthetic utility, hence prompting us to continue with the focus on the transformation of chalcones exhibiting the hydroxylation patterns of the naturally occurring dihydroflavonols.

Another aspect that influences our thoughts regarding a future approach was our earlier observation[7] that the acid-catalyzed conversion of chalcone epoxide to dihydroflavonol, according to the method of Bognar and Stefanovsky,[14] proceeded with considerable "loss" of optical activity, presumably as a result of competition between protonation of the oxirane oxygen and hydrolysis of the protecting 2'-O-acetal functionality. This was in sharp contrast to the results of Onda and his coworkers[9] who reported a 100 percent ee for the formation of (2R,3R)-dihydromyricetin **4** and its (2S,3S)-enantiomer.

Arguably the most important recent development in this field was the use of the polyleucines instead of the polyalanines as the chiral catalysts inducing selectivity in the epoxidation step.[15,16] The [poly-(L)-leucine] catalyst could be prepared and used on a large (>200 g) scale and was found to be reusable, without detrimental effect on yield and enantiomeric excess.[17] We therefore revisited the chalcone → epoxide → dihydroflavonol protocol with a view to solving the problem of the "loss" of optical activity during the epoxide → dihydroflavonol transformation, and to assess the potential of the polyleucine catalysts to effect increased stereoselectivity in the epoxidation.

The requisite *trans*-(E)-chalcone methyl ethers **5–9**[7] were treated with hydrogen peroxide in the triphase system, aq. NaOH/poly-L- or D-alanine/CCl_4,[18–20] to afford the (–)-*trans*- **10a–14a** ($\alpha R,\beta S$) and (+)-*trans*-epoxides **10b–14b** ($\alpha S,\beta R$) ($J_{\alpha,\beta}$ = 1.5–2.2 Hz), respectively, in high yields (79–99 percent) (fig. 2, table 1).[21,22] These conversions proceeded slowly with reaction times varying from 36 to 96 hours. The (–)-chalcone oxiranes, *e.g.*, **10a**, exhibited higher optical purities (49–86 percent ee) than the (+)-isomers, *e.g.*, **10b** (47–75% ee) due to the higher purity of natural L-alanine $[\alpha]_D^{25} + 12.57$ (c 5.695 in 1M HCl) versus synthetic D-alanine $[\alpha]_D^{25} - 9.72$ (c 1.363 in 6M HCl, which was reflected in the optical purity of poly-L- $[\alpha]_D^{25} - 142.8$ (c 0.671 in CF_3CO_2H) and poly-D-alanine $[\alpha]_D^{25} + 102$ (c 0.314 in CF_3CO_2H) catalysts (table 1).

The enantiomeric purity of the epoxides was determined by ^{1}H NMR using Pr(hfc)$_3$ or Eu(hfc)$_3$ as chiral shift reagents and the absolute stereochemistry assigned by comparison of CD data with those of authentic samples.[7] The CD curves of the (–)-*trans*-chalcone oxiranes **10a–14a** exhibited high amplitude sequential negative and positive effects, respectively, in the 290–310 nm (n→π^* transition) and 240–264 nm (π→π^* transition) regions. The sign of these Cotton effects are reversed for the (+)-*trans*-epoxychalcones **10b–14b** (cf fig. 3 for comparison of epoxides **11a** and **11b**).

Our initial attempts toward cyclization of ($\alpha R,\beta S$)-chalcone epoxides to the corresponding (2R,3R)-2,3-*trans*-dihydroflavonols were hampered by low yields due to the formation of isoflavones *via* aroyl migration and by substantial loss of optical purity in the dihydroflavonols.[7] Lewis acids such as MgBr$_2$-OEt$_2$ and BF$_3$-OEt$_2$ smoothly cleave alkoxymethyl ethers and acetals under mild conditions.[23] It was thus anticipated that deprotection of the 2-O-methoxymethyl group with concomitant cyclization would enhance the preservation of optical integrity. Treatment of (–)-($\alpha R,\beta S$)-chalcone epoxide **11a** (86 percent ee) with MgBr$_2$-etherate indeed afforded (2R,3R)-2,3-*trans*-4',7-dimethoxy-dihydroflavonol **21a** with virtually no loss of optical purity but in a modest (20 percent) chemical yield, as well as 4',7-dimethoxyisoflavone (4 percent). Similar results were obtained with BF$_3$-OEt$_2$.[24]

In order to circumvent the problem of isoflavone formation, we investigated methods aimed at the initial nucleophilic opening of the oxirane functionality, followed by deprotection and cyclization. Owing to the excellent nuceophilic and nucleofugal properties of mercaptans, evaluation of thiols in the presence of Lewis acids resulted in the selection of the phenylmethanethiol (BnSH)-tin(IV) chloride (SnCl$_4$)[25] system as the reagent of choice for oxirane cleavage. Thus, treatment of the series of methoxymethylchalcone epoxides **10a/b–14a/b** with BnSH/SnCl$_4$ selectively cleaved the C$_\beta$—O bond of the oxirane functionality at –20 °C and effectively deprotected the methoxymethyl group at 0 °C to give the

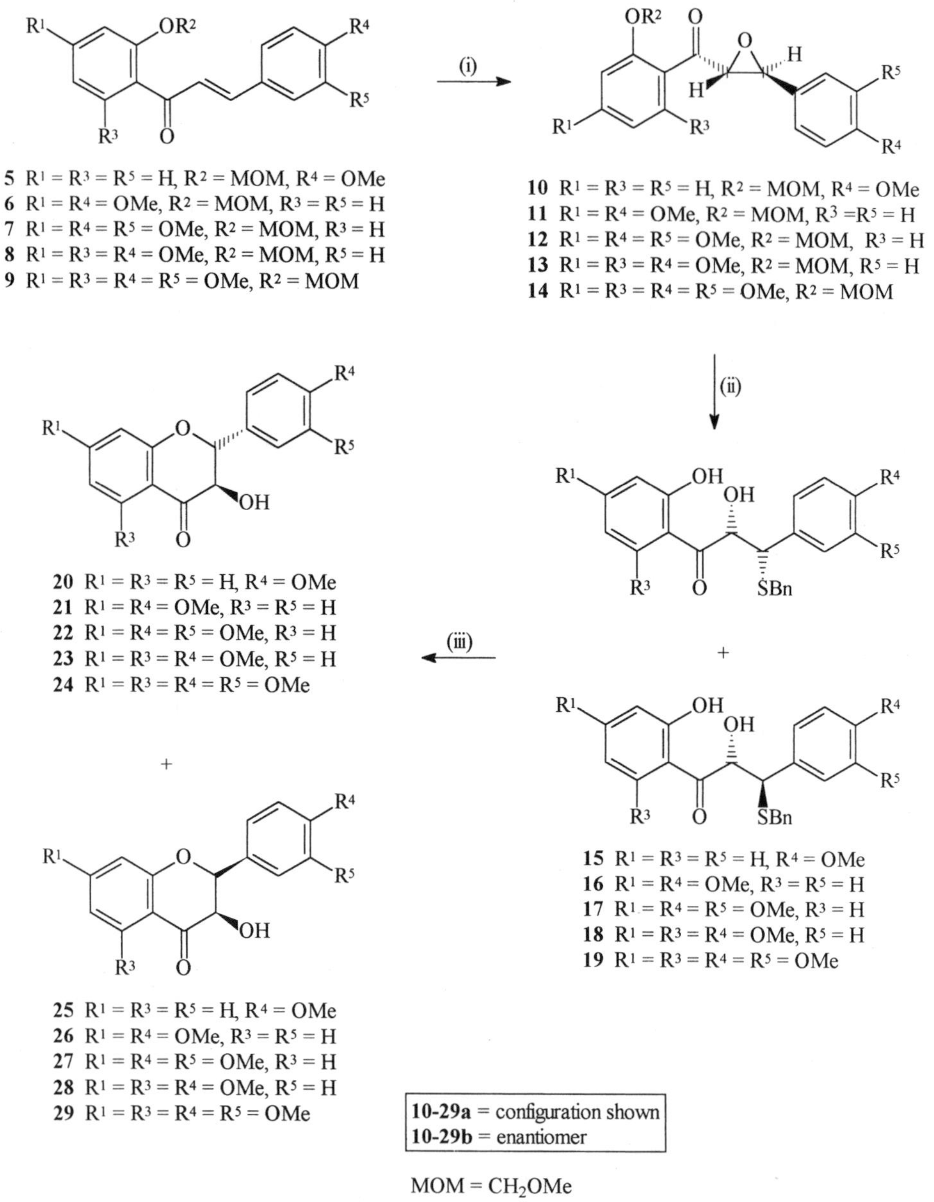

5   R¹ = R³ = R⁵ = H, R² = MOM, R⁴ = OMe
6   R¹ = R⁴ = OMe, R² = MOM, R³ = R⁵ = H
7   R¹ = R⁴ = R⁵ = OMe, R² = MOM, R³ = H
8   R¹ = R³ = R⁴ = OMe, R² = MOM, R⁵ = H
9   R¹ = R³ = R⁴ = R⁵ = OMe, R² = MOM

10   R¹ = R³ = R⁵ = H, R² = MOM, R⁴ = OMe
11   R¹ = R⁴ = OMe, R² = MOM, R³ = R⁵ = H
12   R¹ = R⁴ = R⁵ = OMe, R² = MOM, R³ = H
13   R¹ = R³ = R⁴ = OMe, R² = MOM, R⁵ = H
14   R¹ = R³ = R⁴ = R⁵ = OMe, R² = MOM

20   R¹ = R³ = R⁵ = H, R⁴ = OMe
21   R¹ = R⁴ = OMe, R³ = R⁵ = H
22   R¹ = R⁴ = R⁵ = OMe, R³ = H
23   R¹ = R³ = R⁴ = OMe, R⁵ = H
24   R¹ = R³ = R⁴ = R⁵ = OMe

15   R¹ = R³ = R⁵ = H, R⁴ = OMe
16   R¹ = R⁴ = OMe, R³ = R⁵ = H
17   R¹ = R⁴ = R⁵ = OMe, R³ = H
18   R¹ = R³ = R⁴ = OMe, R⁵ = H
19   R¹ = R³ = R⁴ = R⁵ = OMe

25   R¹ = R³ = R⁵ = H, R⁴ = OMe
26   R¹ = R⁴ = OMe, R³ = R⁵ = H
27   R¹ = R⁴ = R⁵ = OMe, R³ = H
28   R¹ = R³ = R⁴ = OMe, R⁵ = H
29   R¹ = R³ = R⁴ = R⁵ = OMe

10-29a = configuration shown
10-29b = enantiomer

MOM = CH₂OMe

Figure 2. Synthesis of dihydroflavonols. Reagents: i, 30% H_2O_2:6M NaOH 1: 0.32 (v/v), poly-*L*- or poly-*D*-alanine:chalcone 1:1 (m/m); ii, BnSH (4 equiv.), $SnCl_4$ (0.2 equiv.); iii, AgBF₄ (5 equiv.).

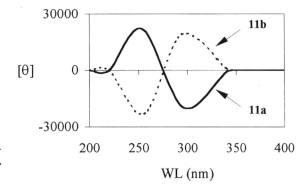

Figure 3. CD curves of the (–)-($\alpha R,\beta S$)-**11a** and (+)-($\alpha S,\beta R$)-**11b** chalcone epoxides.

corresponding $\alpha,2'$-dihydroxy-β-benzylsulfanyldihydrochalcones **15a/b–19a/b** as diastereomeric mixtures (*anti:syn ca.* 2.3:1) in 86–93 percent yield. The ^{1}H NMR data of the dihydrochalcones **15a/b–19a/b** were compatible with either a β-substituted α-hydroxy- or an α-substituted β-hydroxydihydrochalcone. Deuterium exchange, however, proved that the OH-group is on the carbon bearing the proton resonating at δ 5.12–5.40, i.e. the chemical shift of the α-hydrogen in α-hydroxydihydrochalcones.[8]

Treatment of α-hydroxy-β-bensylsulfanyldihydrochalcones **15a/b–19a/b** with the thiophilic Lewis acid, silver tetrafluoroborate (AgBF$_4$)[26] in CH$_2$Cl$_2$ at 0 °C, gave the 2,3-*trans*-dihydroflavonols **20–24** ($^3J_{2,3}$ 12.0 Hz) in good yields (48–80 percent) and, albeit in low proportions (5–15 percent), for the first time also the 2,3-*cis* analogues **25–29** (fig. 2) ($^3J_{2,3}$ 6.1 Hz) (table 1). Despite the fact that the

Table 1. Intermediate products in the conversion of chalcones **5–9** to dihydroflavonols **20–29**

Chalcone epoxide	Yield (%)	ee[a] (%)	Dihydro-chalcone	Yield (%)	Dihydro-flavonol	Yield (%)	ee[b] (%)	*trans:cis*
10a	99	84	**15a**	86	**20a/25a**	86	83	93:7
10b	98	69	**15b**	90	**20b/25b**	83	69	94:6
11a	98	86	**16a**	93	**21a/26a**	71	84	79:21
11b	98	74	**16b**	90	**21b/26b**	72	75	83:17
12a	99	67	**17a**	89	**22a/27a**	81	68	85:15
12b	98	58	**17b**	91	**22b/27b**	79	58	86:14
13a	97	70	**18a**	89	**23a/28a**	65	69	78:22
13b	96	53	**18b**	89	**23b/28b**	64	53	84:16
14a	79	49	**19a**	91	**24a/29a**	61	47	82:18
14b	76	49	**19b**	88	**24a/29b**	63	44	80:20

[a] Determined with Pr(hfc)$_3$ as chiral shift reagent.
[b] Determined with Eu(hfc)$_3$ as chiral shift reagent.

substituted dihydrochalcones **15–19** exhibited a high aptitude toward aroyl migration due to the *o*- and/or *o*-/*p*-hydroxylation of their A-rings, isoflavone formation (10–25 percent) was restricted to the transformations involving analogues **18** and **19**.

Assessment of optical purity of the dihydroflavonols was done by ¹H NMR using Eu(hfc)₃ as shift reagent. These results indicated that, within standard experimental deviation, the dihydroflavonols exhibited the same enantiopurities as the parent epoxides, thus unequivocally confirming the fact that optical integrity was being preserved in the transformation of the epoxides **10a/b–14a/b** to the corresponding *trans*-**20a/b–24a/b** and *cis*-dihydroflavonols **25a/b–29a/b** (table 1). The absolute stereochemistry of the 2,3-*trans*-derivatives **23a** and **24a** was assigned by comparison of their CD data with those of the same derivatives of authentic dihydroquercetin and dihydrokaempferol (see ref. 22 for references related to CD data of these compounds). The absolute configuration of the remaining *trans*-isomers **20a–22a** was then assigned assuming that the stereochemistry of the reactions leading to those compounds is the same as for products **23a** and **24a**. The mirror-image relationship of the two sets of *trans*-isomers (CD, fig. 4), confirmed the enantiomeric connection between these series and hence proof of the 2*S*,3*S* absolute configuration for analogues **20b–24b**. The (2*R*,3*R*)-2,3-*trans*-dihydroflavonols **20a–24a** exhibited positive and negative Cotton effects, respectively, in the 330–340 nm (n≈π* transition) and 285–315 nm (π→π* transition) regions, with these Cotton effects being reversed in the same regions for the (2*S*,3*S*)-2,3-*trans*-isomers **20b–24b** (*cf* fig. 2 for comparison of **21a** and **21b**). The *cis*-dihydroflavonols **25a/b–29a/b** accompanying the corresponding *trans*-isomers exhibited negative (340–345 nm) and positive (295–320 nm) Cotton effects for the (2*S*,3*R*)-2,3-*cis* analogues **25a–29a** and opposite sequential positive and negative Cotton effects for the (2*R*,3*S*)-2,3-*cis* isomers **25b–29b** (*cf* fig. 5 for comparison of **26a** and **26b**). Their absolute configurations then follow from the fact that optical integrity was preserved at C-3 in the transformation epoxide → dihydrochalcone → *cis*-dihydroflavonol.

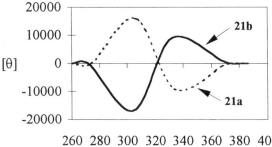

Figure 4. CD curves of the (2*R*,3*R*)-**21a** and (2*S*,3*S*)-**21b**-2,3-*trans*-dihydroflavonols.

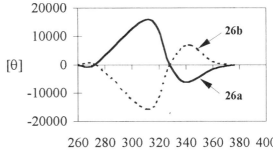

Figure 5. CD curves of the (2*S*,3*R*)-**26a** and (2*R*,3*S*)-**26b**-2,3-*cis*-dihydroflavonols.

The data for the 2,3-*cis*-dihydroflavonols represent the first systematic compilation of the circular dichroic properties of this rare class of C_6—C_3—C_6-type polyphenols and should hence contribute substantially toward elucidation of the absolute configuration of naturally occurring analogues. We have thus developed the first stereoselective route toward both *trans*- and *cis*-dihydroflavonols. These results demonstrate the considerable potential of this protocol especially in view of the amount of research[16] focusing on the development of improved asymmetric epoxidation of enones. These results should contribute significantly toward a general synthesis of oligomeric proanthocyanidins with 2,3-*trans* and also 2,3-*cis*-flavanoid chain extender units in order to assess the physical and chemical properties that determine their health-promoting properties in the human diet.

Although the Julia asymmetric epoxidation of chalcones has proven to be a reliable reaction affording epoxides in good yields and moderate to high enantiomeric excess, this protocol is not without limitations as reaction times are often unacceptably long (up to 3 days), reactions require continuous addition of oxidant and base, and degradation of the poly-amino acid often poses difficulties. We thus selected the adapted version recently developed by Roberts and his co-workers,[15,16] which involves a two-phase non-aqueous system consisting of oxidant, a non-nucleophilic base, immobilized poly (amino acid),[27] and an organic solvent.

Treatment of chalcones **5–7** with immobilized poly-(L)-leucine[15] (PLL), urea-hydrogen peroxide complex (UHP),[28] and 1,8-diazabicyclo[5.4.0]undec-7-ene (DBU) in dry THF afforded the (−)-(α*R*,β*S*)-*trans*-epoxychalcones **10a–12a** ($J_{\alpha,\beta}$ 1.5–2.2 Hz) in high yields (64–80 percent) and improved optical purity (85–95 percent ee) in comparison with the Julia procedure[6,22] (67–86 percent). The enantiomeric (+)-(α*S*,β*R*)-*trans*-epoxychalcones **30–32** ($J_{\alpha,\beta}$ 1.5–2.2 Hz) were similarly obtained by using immobilised poly-(D)-leucine (PDL) in the same two-phase system (61–76 percent yield, 81–90 percent ee) (58–74 percent ee for the Julia procedure)[6,22] (fig. 6, table 2). The enantiomeric purity of the epoxides was determined by ¹H NMR using Pr(hfc)₃ or Eu(hfc)₃ as shift reagents and the absolute stereochemistry assigned by comparison of CD data with those of authentic samples.[7]

We used these epoxides with superior enantiomeric excess to synthesize a series of poly-oxygenated β-hydroxydihydrochalcones.[29] The successful

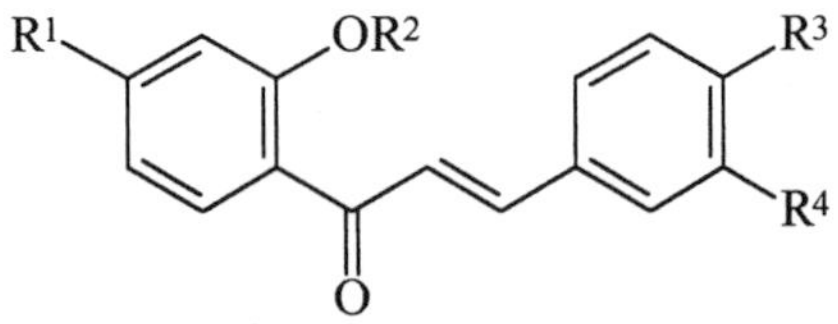

5 R¹ = R⁴ = H, R² = MOM, R³ = OMe
6 R⁴ = H, R² = MOM, R¹ = R³ = OMe
7 R² = MOM, R¹ = R³ = R⁴ = OMe

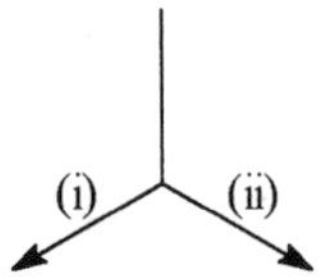

(i)   (ii)

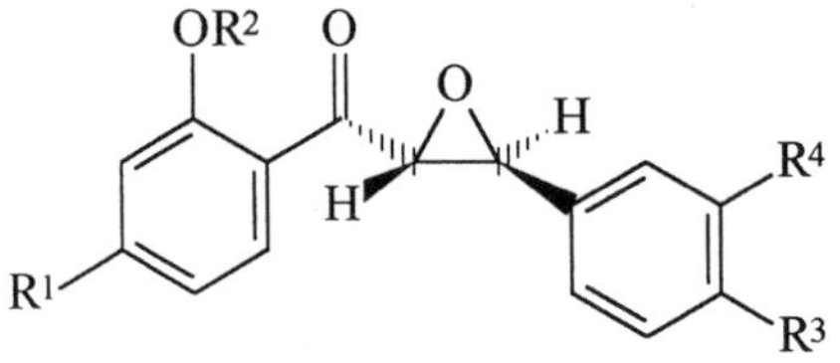

10 R¹ = R⁴ = H, R² = MOM, R³ = OMe
11 R⁴ = H, R² = MOM, R¹ = R³ = OMe
12 R² = MOM, R¹ = R³ = R⁴ = OMe

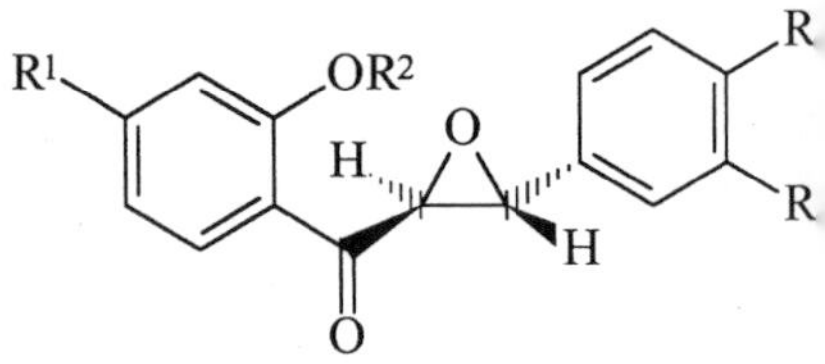

30 R¹ = R⁴ = H, R² = MOM, R³ = OMe
31 R⁴ = H, R² = MOM, R¹ = R³ = OMe
32 R² = MOM, R¹ = R³ = R⁴ = OMe

MOM = CH_2OMe

Figure 6. Epoxidation of chalcones using polyleucines. Reagents: i, poly-(L)-leucine, urea-hydrogen peroxide complex, DBU, ii, poly-(D)-leucine, urea-hydrogen peroxide complex, DBU.

Table 2. Intermediate products in the conversion of chalcones **5–7** to epoxides **10–12** and **30–32** using polyleucines

Chalcone	Poly-(amino acid)[a]	Chalcone epoxide	% Yield (% Conversion)	ee[b] (%)
5	PLL	**10a**	71	85
5	PDL	**30**	69	81
6	PLL	**11a**	80	95
6	PDL	**31**	76	90
7	PLL	**12a**	64 (72)	88
7	PDL	**32**	61 (71)	87

[a] PLL: poly-(L)-leucine; PDL; poly-(D)-leucine.
[b] Determined with Eu(hfc)₃ as chiral shift reagent.

transformation of epoxides to dihydroflavonols via the α-hydroxy-β-benzylsulfanyldihydrochalcones (*vide supra*) demonstrates the potential of this method and hence the synthetic utility of the chalcone epoxides synthesized with polyleucines as chiral catalysts.

3. STEREOSELECTIVE SYNTHESIS OF FLAVAN-3-OLS

Flavan-3-ols, the largest group of naturally occurring C_6—C_3—C_6 metabolites, serve as the nucleophilic entities in the semi-synthesis of oligomeric proanthocyanidins.[30,31] This semi-synthetic approach is often hampered by the limited availability of naturally occurring flavan-3-ols with 2,3-*trans* and especially 2,3-*cis* stereochemistry. Surprisingly, the only "synthetic access" to the stereochemically simple catechin, together with epicatechin the only flavan-3-ols that are commercially available, involves the cumbersome process of hydrogenation and reduction of the corresponding (2*R*,3*R*)-2,3-*trans*-dihydroquercetin tetra-*O*-benzyl ether.[32] In order to address the issue of stereocontrol at C-2 and C-3 of the flavan-3-ol molecular framework, we designed a concise protocol based upon the transformation of *retro*-chalcones into 1,3-diarylpropenes. These compounds are then subjected to asymmetric dihydroxylation to give poly-oxygenated diarylpropan-1,2-diols, which are used as chirons for essentially enantiopure flavan-3-ols.[33,34]

The series of (*E*)-*retro*-2-methoxymethylchalcone methyl ethers **33–37** ($J_{\alpha,\beta}$ 15.8–16.0 Hz) were prepared by base-catalyzed condensation of the appropriate oxygenated acetophenones and benzaldehydes. Compounds **33–37** were transformed by consecutive reduction (Pd-H_2 and $NaBH_4$) and elimination {$SOCl_2$ and 1,8-diazabicyclo[5.4.0]undec-7-ene (1,8-DBU)} of the ensuing alcohols **38–42**, exclusively affording the (*E*)-1,3-diarylpropenes **43–47** ($J_{1,2}$ 16 Hz) in good overall yield (65–73 percent) (fig. 7, table 3). Owing to the excellent results obtained by

Table 3. Intermediate products in the conversion of the chalcones **33–37** into the flavan-3-ols **53a/b–62a/b**

Propan-1-ols	Yield (%)	Propenes	Yield (%)	1,2-Diols	Yield (%)	ee[a] (%)	Flavan-3-ols	Yield (%)	ee[b] (%)	*trans : cis*
38	99	**43**	73	**48a**	82	99	**53a/58a**	87	99	1:0:33
				48b	84	99	**53b/58b**	88	99	1:0:31
39	98	**44**	74	**49a**	86	99	**54a/59a**	88	99	1:0:36
				49b	82	99	**54b/59b**	90	99	1:0:33
40	99	**45**	70	**50a**	85	99	**55a/60a**	82	99	1:0:32
				50b	83	99	**55b/60b**	80	99	1:0:30
41	98	**46**	68	**51a**	80	99	**56a/61a**	71	99	1:0:32
				51b	83	99	**56b/61b**	70	99	1:0:33
42	99	**47**	66	**52a**	80	99	**57a/62a**	66	99	1:0:34
				52b	87	99	**57b/62b**	65	99	1:0:35

[a] Determined by observing the spin systems of the corresponding mono- and/or bis-MTPA ester derivatives.
[b] Determined with Eu(hfc)₃ as chiral shift reagent.

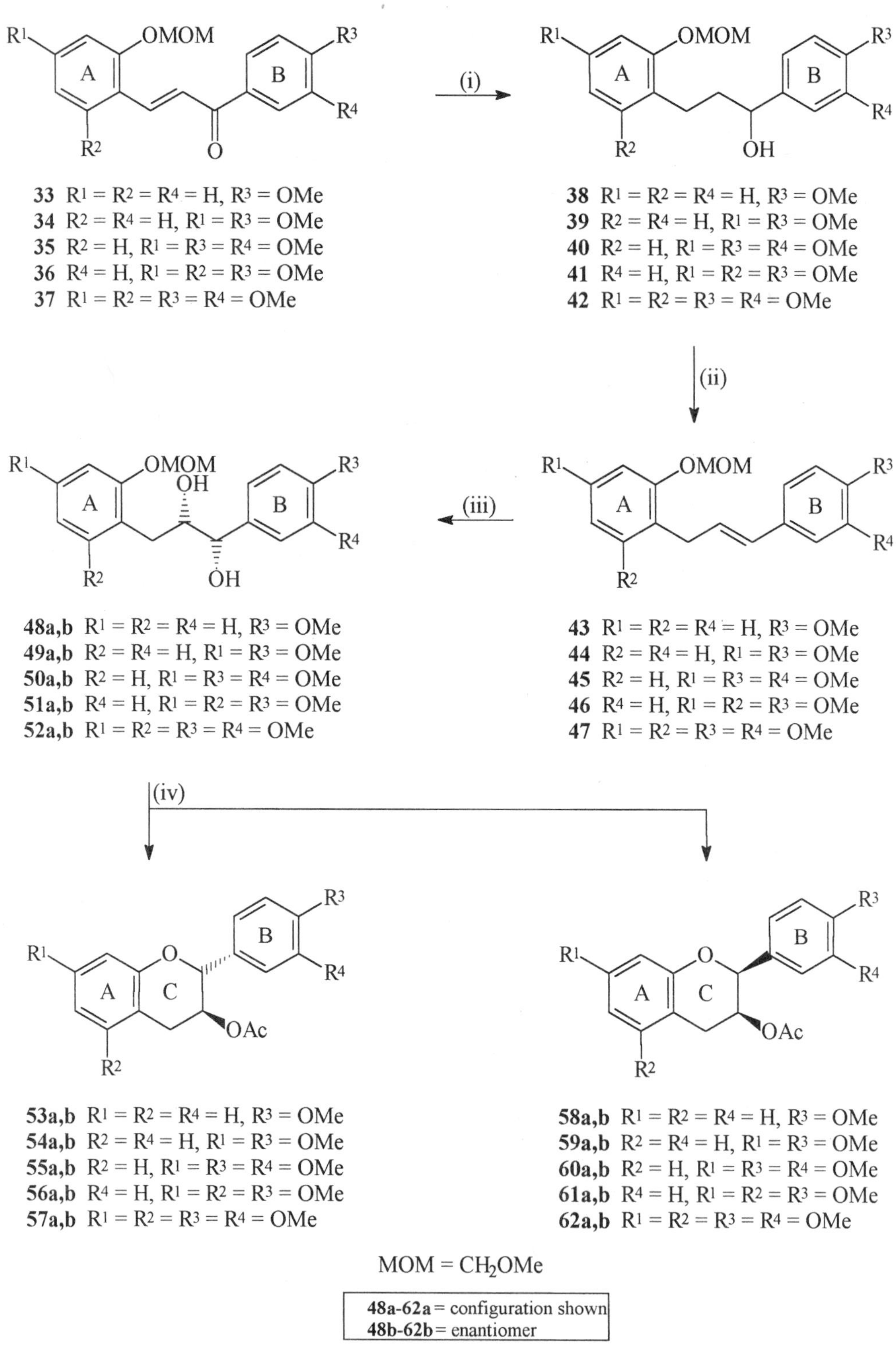

33 R¹ = R² = R⁴ = H, R³ = OMe
34 R² = R⁴ = H, R¹ = R³ = OMe
35 R² = H, R¹ = R³ = R⁴ = OMe
36 R⁴ = H, R¹ = R² = R³ = OMe
37 R¹ = R² = R³ = R⁴ = OMe

38 R¹ = R² = R⁴ = H, R³ = OMe
39 R² = R⁴ = H, R¹ = R³ = OMe
40 R² = H, R¹ = R³ = R⁴ = OMe
41 R⁴ = H, R¹ = R² = R³ = OMe
42 R¹ = R² = R³ = R⁴ = OMe

48a,b R¹ = R² = R⁴ = H, R³ = OMe
49a,b R² = R⁴ = H, R¹ = R³ = OMe
50a,b R² = H, R¹ = R³ = R⁴ = OMe
51a,b R⁴ = H, R¹ = R² = R³ = OMe
52a,b R¹ = R² = R³ = R⁴ = OMe

43 R¹ = R² = R⁴ = H, R³ = OMe
44 R² = R⁴ = H, R¹ = R³ = OMe
45 R² = H, R¹ = R³ = R⁴ = OMe
46 R⁴ = H, R¹ = R² = R³ = OMe
47 R¹ = R² = R³ = R⁴ = OMe

53a,b R¹ = R² = R⁴ = H, R³ = OMe
54a,b R² = R⁴ = H, R¹ = R³ = OMe
55a,b R² = H, R¹ = R³ = R⁴ = OMe
56a,b R⁴ = H, R¹ = R² = R³ = OMe
57a,b R¹ = R² = R³ = R⁴ = OMe

58a,b R¹ = R² = R⁴ = H, R³ = OMe
59a,b R² = R⁴ = H, R¹ = R³ = OMe
60a,b R² = H, R¹ = R³ = R⁴ = OMe
61a,b R⁴ = H, R¹ = R² = R³ = OMe
62a,b R¹ = R² = R³ = R⁴ = OMe

MOM = CH₂OMe

48a-62a = configuration shown	
48b-62b = enantiomer	

Figure 7. Synthesis of flavan-3-ols. Reagents: i, Pd-H₂, then NaBH₄; ii, SOCl₂, then 1,8-DBU; iii, AD-mix-α or AD-mix-β, MeSO₂NH; iv, 3M HCl, MeOH-H₂O, then Ac₂O/pyridine.

Sharpless *et al.*[35–38] during asymmetric dihydroxylation of olefins with AD-mix-α and AD-mix-β, these stereoselective catalysts were utilized for the introduction of chirality at C-2 and C-3 in the flavan-3-ol framework.

Thus, treatment of the protected (*E*)-propenes **43–47** at 0 °C with AD-mix-α in the two-phase system Bu^{t}OH:H$_2$O (1:1), afforded the (+)-(1*S*,2*S*)-*syn*-diols **48a–52a** (J$_{1,2}$ 5.8–6.5 Hz) in high yields (80–86 percent) and optical purity (99 percent ee). The (–)-(1*R*,2*R*)-*syn*-diols **48b–52b** (J$_{1,2}$ 5.8–6.5 Hz) were similarly obtained by using AD-mix-β in the same two-phase system (82–87 percent yield, 99 percent ee). These conversions proceeded slowly with reaction times in the range 12–48 h. The enantiomeric purity of the diols was determined by observing the H-1 and H-2 spin systems of the corresponding mono and/or bis-MTPA esters,[39] which show different chemical shifts in the ^{1}H NMR spectra of the two diastereoisomers. The absolute configuration was tentatively assigned according to the Sharpless model[35,36] for AD-mix (fig. 8). The successful Lewis acid-catalyzed phenylmethanethiol ring-opening and cyclization of chalcone epoxides in the synthesis of dihydroflavonols[21] prompted evaluation of this protocol in the cyclization of the diols **48a/b–52a/b**. Thus, the Lewis acid tin tetrachloride[40] in the presence of the powerful nucleophilic phenylmethanethiol (BnSH) was utilized for selective substitution of the benzylic C(1)-OH (–20 °C) and subsequent removal of the

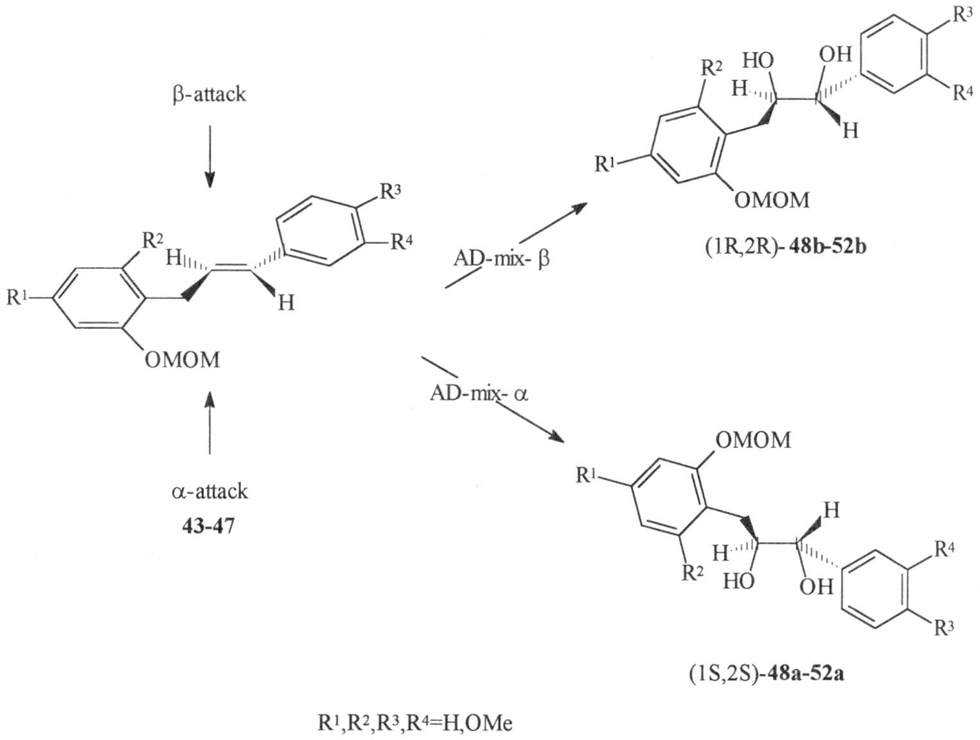

R^1,R^2,R^3,R^4=H,OMe

Figure 8. Sharpless model for diastereoselectivity in asymmetric dihydroxylation.

methoxymethyl group (0 °C) in **49a**, to give the benzylthio derivative **63** (70 percent, mixture of *syn* and *anti*) (fig. 9). Treatment of the benzylthio ether **63** with the thiophilic Lewis acid, silver tetrafluoroborate (AgBF$_4$),[26] in CH$_2$Cl$_2$ at 0 °C, however, resulted in slow (24 h) and low percentage conversion (10–20 percent) into the flavan-3-ols **54a–59a**.

In order to transform the diols effectively into the corresponding flavan-3-ols, methods aimed at selective removal of the 2′-*O*-methoxymethyl group and subsequent ring closure under mild acidic conditions, were explored. Simultaneous deprotection and cyclization of diols **48a/b–52a/b** in the presence of 3M HCl in MeOH, followed by acetylation, yielded 2,3-*trans*-(48–68 percent) **53a/b–57a/b,** and for the first time 2,3-*cis*-flavan-3-ol methyl ether acetate derivatives (17–22 percent) **58a/b–62a/b** in excellent enantiomeric excesses (>99 percent) (fig. 7, table 3). The optical purity was assessed by ¹H NMR using Eu(hfc)$_3$ as chiral shift reagent, which consistently indicated the presence of only one enantiomer, thus unequivocally confirming the fact that optical integrity was preserved in the transformation of the diols **48a/b–52a/b** into the corresponding *trans-* and *cis*-flavan-3-ols. The absolute stereochemistry of the naturally occurring *trans-***55a/b–57a/b** analogues was assigned by comparison of CD data with those of authentic samples in the catechin, fisetinidol, and afzelechin series of compounds from our collection of reference compounds. The absolute configuration of the *trans* **53a/b–59a/b** and *cis*-flavan-3-ols **58a/b** and **59a/b** was assigned assuming that the stereochemistry of the reactions leading to those compounds is the same as for the natural

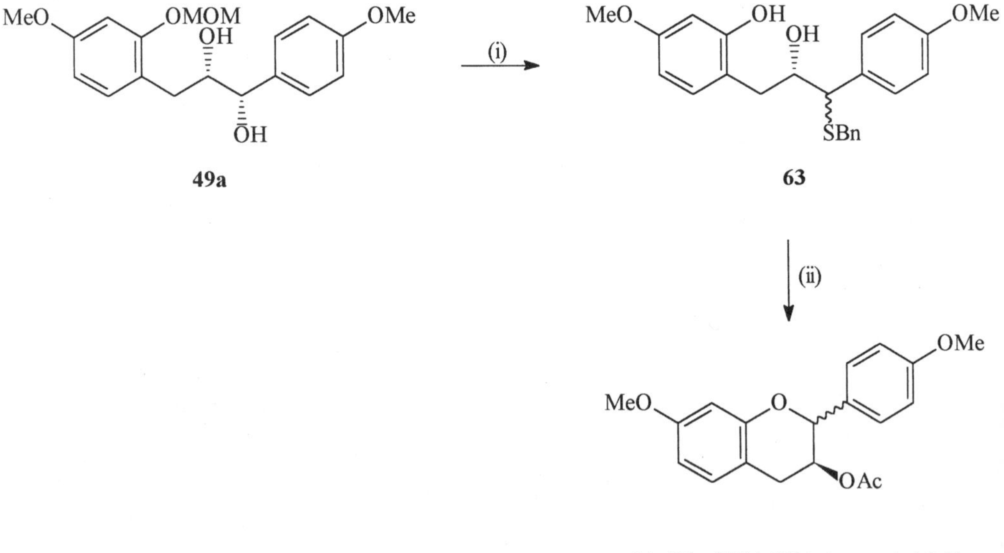

54a/59a (10%-20%, *trans:cis* 1:0.3)

Figure 9. Transformation of 1,3-diarylpropan-1,2-diols to flavan-3-ols. Reagents: i, BnSH (4 equiv.), SnCl$_4$ (0.2 equiv.), ii, AgBF$_4$ (5 equiv.), then Ac$_2$O/pyridine.

products. Thus, the stereochemistry of the flavan-3-ol acetates confirms the assigned configuration of the diols as derived from the Sharpless model.

Cyclization of the highly oxygenated diol **52a** led to the formation of two separable intermediate products **64** and **65** (fig. 10). Methanol thus initially acts as nucleophile in an S$_N$1-type acid-catalyzed solvolysis of the C-1 benzylic hydroxy group, affording a *syn:anti* mixture of isomers **64**. Prolonged treatment (*ca.* 2 h) eventually led to deprotection of the acetal functionality to give the intermediate **65**, which is susceptible to cyclization to afford both the 2,3-*trans*- and 2,3-*cis*-flavan-3-ols **57a** and **62a**. Formation of both these isomers is explicable in terms of the generation of an incipient C-1 carbocation by protonation of the benzylic methoxy functionality, and subsequent S$_N$1 cyclization that leads to predominant formation of the thermodynamically more stable *trans*-isomer. The aforementioned remarkable stability of the reputedly highly acid-labile methoxymethyl group

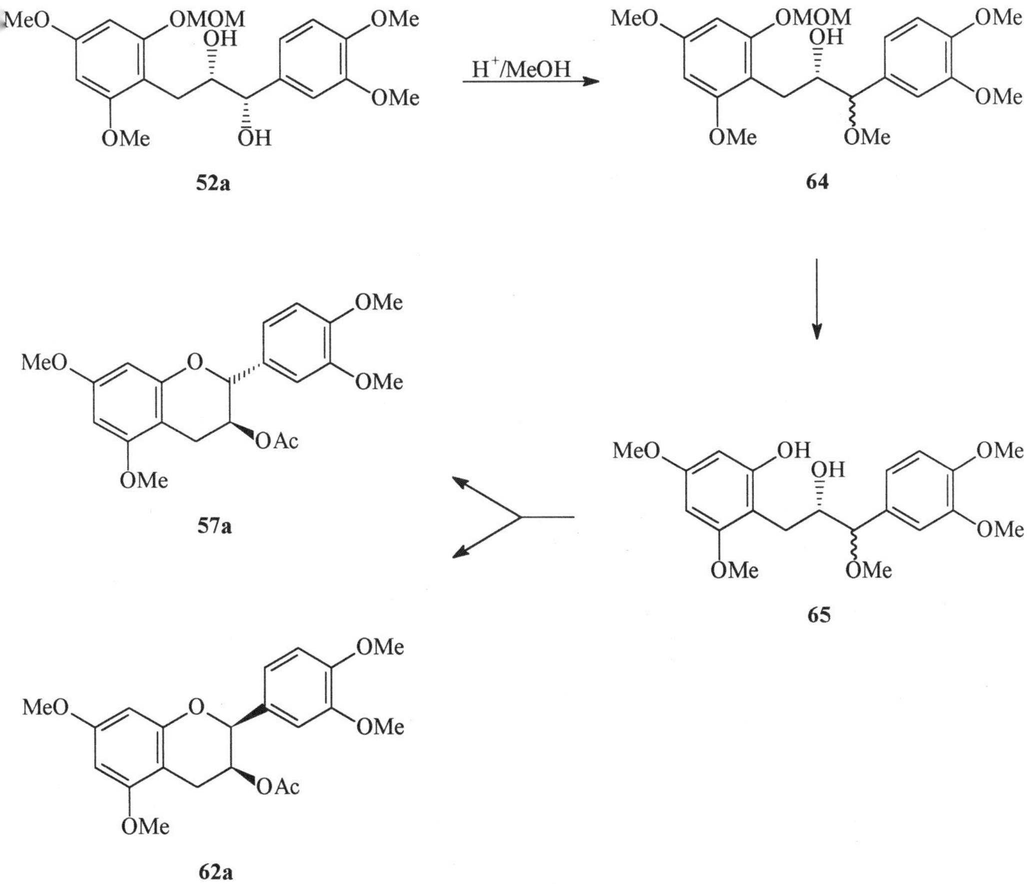

Figure 10. Intermediates in the acid-catalyzed 1,3-diarylpropan-1,2-diol → flavan-3-ol conversion.

toward acid hydrolysis was also observed during the enantioselective synthesis of isoflavans.[41,42]

We have thus developed the first and highly efficient synthesis of essentially enantiopure flavan-3-ols of both 2,3-*trans* and 2,3-*cis* configuration. These results abundantly demonstrate the utility of polyoxygenated diarylpropan-1,2-diols as chirons for naturally occurring flavanoids of this class. The potential of this protocol in the chemistry of the oligomeric proanthocyanidins and condensed tannins in general is evident, especially in view of its aptitude to the synthesis of free phenolic analogues. The latter analogues are as conveniently accessible by simply using labile protecting groups instead of *O*-methyl ethers. We are currently optimising the protocol that would lead to a general total synthesis of free phenolic flavan-3-ols.

4. CONCLUSIONS

In contrast to previous approaches, including our own, where stereoselective syntheses of flavonoid analogues possessing chirality at C-2, C-3 and C-4 were evaded, this review clearly demonstrates that these compounds are indeed readily accessible via utilization of the myriad of modern synthetic methods available to researchers in this field. It may be anticipated and hoped for that an increasing number of tannin chemists will focus more sharply on this much neglected area of the chemistry of oligomeric proanthocyanidins.

ACKNOWLEDGMENTS

We are grateful for the enthusiastic support of the research group in Bloemfontein. Financial support by the Foundation for Research Development, Pretoria, and the 'Sentrale Navorsingsfonds' of this University is gratefully acknowledged.

REFERENCES

1. Weinges, K. Stereochemical relations in the flavonoid group. *Liebigs Ann. Chem.* 229:627 (1959).

2. Freudenberg, K.; Hartmann, L. Inhaltsstoffe der *Robinia pseudacacia. Liebigs Ann. Chem.* 587:207 (1954).

3. Malan, J.C.S.; Chen, J.; Foo, L.Y.; Karchesy, J.J. Phlobaphene precursors in Douglas-fir outer bark. *In*: Hemingway, R.W.; Laks, P.E.; Branham, S.J. (eds.). Plant polyphenols: synthesis, properties, significance. Plenum Press, New York, p. 411 (1992).

4. Vercruysse, S.A.R.; Delcour, J.A,; Dondeyne, P. Isolation of quercetin, myricetin, and their respective dihydro compounds by Sephadex LH-20 chromatography. *J. Chromatogr.* 324:495 (1985).

5. Bezuidenhoudt, B.C.B.; Ferreira, D. Enantioselective synthesis of flavonoids. *In*: Hemingway, R.W.; Laks, P.E.; Branham, S.J. (eds.). Plant polyphenols: synthesis, properties, significance. Plenum Press, New York, p. 143 (1992).

6. Bezuidenhoudt, B.C.B.; Swanepoel, A.; Augustyn, J.A.N.; Ferreira, D. The first enantioselective synthesis of polyoxygenated α-hydroxydihydrochalcones and circular dichroic assessment of their absolute configuration. *Tetrahedron Lett.*, 28:4857 (1987).

7. Augustyn, J.A.N.; Bezuidenhoudt, B.C.B.; Ferreira, D. Enantioselective synthesis of flavonoids. Part 1. Polyoxygenated chalcone epoxides. *Tetrahedron* 46:2651 (1990).

8. Augustyn, J.A.N.; Bezuidenhoudt, B.C.B.; Swanepoel, A.; Ferreira, D. Enantioselective synthesis of flavonoids. Part 2. Polyoxygenated α-hydroxydihydrochalcones and circular dichroic assessment of their absolute configuration. *Tetrahedron* 46:4429 (1990).

9. Takahashi, H.; Kubota, Y.; Miyazaki, H.; Onda, M. Heterocycles. XV. Enantioselective synthesis of chiral flavanonols and flavan-3,4-diols. *Chem. Pharm. Bull.* 32:4582 (1984).

10. Enders, D.; Zhu, J.; Raabe, G. Asymmetric epoxidation of enones with oxygen in the presence of diethylzinc and (R,R)-N-methylpseudoephedrine. *Angew. Chem. Int. Ed. Engl.* 35:1725 (1996).

11. Elston, C.L.; Jackson, R.F.W.; MacDonald, S.J.F.; Murray, P.J. Asymmetric epoxidation of chalcones with chirally modified lithium and magnesium *tert*-butyl peroxides. *Angew. Chem. Int. Ed. Engl.* 36:410 (1997).

12. Bougauchi, M.; Watanabe, S.; Arai, T.; Sasai, H.; Shibasaki, M. Catalytic asymmetric epoxidation of α,β-unsaturated ketones promoted by lanthanoid complexes. *J. Am. Chem. Soc.* 119:2329 (1997).

13. Lygo, B.; Wainwright, P.G. Asymmetric phase-transfer mediated epoxidation of α,β-unsaturated ketones using catalysts derived from Cinchona alkaloids. *Tetrahedron Lett.* 39:1599 (1998).

14. Bognar, R.; Stefanovsky, J. Flavonoids. Part VI. The preparation and reactions of the epoxides of 2′-hydroxychalcone derivatives. *Tetrahedron* 18:143 (1962).

15. Lasterra-Sánchez, M.E.; Felfer, U.; Mayon, P.; Roberts, S.M.; Thornton, S.R.; Todd, C.J. Development of the Julia asymmetric epoxidation reaction. Part 1. Application of the oxidation to enones other than chalcones. *J. Chem. Soc., Perkin Trans. 1* :343 (1996).

16. Bentley, P.A.; Bergeron, S.; Cappi, M.W.; Hibbs, D.E.; Hursthouse, M.B.; Nugent, T.C.; Pulindo, R.; Roberts, S.M.; Wu, L.E. Asymmetric epoxidation of enones employing polymeric α-amino acids in non-aqueous media. *J. Chem. Soc., Chem. Commun.*: 739 (1997).

17. Flisak, J.R.; Gombatz, K.J.; Holmes, M.M.; Jarmas, A.A.; Lantos, I.; Mendelsohn, W.L.; Novack, V.J.; Remich, J.J.; Snyder, L. A practical, enantioselective synthesis of SK- and -F 104354. *J. Org. Chem.* 58:6274 (1993).

18. Julia, S.; Masana, J.; Vega, J.C. 'Synthetic enzymes'. Highly stereoselective epoxidation of chalcone in a triphasic toluene-water-poly-(S)-alanine system. *Angew. Chem. Int. Ed. Engl.* 19:929 (1980).

19. Julia, S.; Guixer, J.; Masana, J.; Rocas, J.; Colonna, S.; Annuziata, R.; Molinari, H. Synthetic enzymes. Part 2. Catalytic asymmetric epoxidation by means of polyaminoacids in a triphase system. *J. Chem. Soc., Perkin Trans. 1* :1317 (1982).

20. Banfi, S.; Colonna, S.; Molinari, H.; Julia, S.; Guixer, J. Asymmetric epoxidation of electron-poor olefins. Part 5. Influence on stereoselectivity of the structure of polyaminoacids used as catalysts. *Tetrahedron* 40:5207 (1984).

21. Van Rensburg, H.; Van Heerden, P.S.; Bezuidenhoudt, B.C.B.; Ferreira, D. The first enantioselective synthesis of *trans*- and *cis*-dihydroflavonols. *J. Chem. Soc., Chem. Commun.*: 2747 (1996).

22. Van Rensburg, H.; Van Heerden, P.S.; Bezuidenhoudt, B.C.B.; Ferreira, D. Stereoselective synthesis of flavonoids. Part 4. *Trans*- and *cis*-dihydroflavonols. *Tetrahedron* 53:14141 (1997).

23. Kim, S.; Park, Y.H.; Kee, I.S. Mild deprotection of methoxymethyl, methylthiomethyl, methoxyethoxymethyl, and β-(trimethylsilyl)ethoxymethyl ethers with magnesium bromide in ether. *Tetrahedron Lett.*, 32:3099 (1991).

24. Moberg, C.; Rakos, L.; Tottie, L. Stereospecific Lewis acid catalyzed methanolysis of styrene oxide. *Tetrahedron Lett.* 33:2191 (1992).

25. Chini, M.; Crotti, P.; Giovani, E.; Machia, F.; Pineschi, M. Metal salt-promoted alcoholysis of 1,2-epoxides. *Synlett.*: 673 (1992).

26. Barrett, A.G.M.; Bezuidenhoudt, B.C.B.; Howell, A.R.; Lee, A.C.; Russel, M.A. Redox glycosidation *via* thionoester intermediates. *J. Org. Chem.* 54:2275 (1989).

27. Itsuno, S.; Sakakura, M.; Ito, K. Polymer-supported poly(amino acids) as new asymmetric epoxidation catalyst of α,β-unsaturated ketones. *J. Org. Chem.* 55:6047 (1990).

28. Heaney, H. Novel organic peroxygen reagents for use in organic synthesis. *Top Curr. Chem* 164 (Organic Peroxygen Chemistry): 1 (1993).

29. Nel, R.J.J.; Van Heerden, P.S.; Van Rensburg, H.; Ferreira, D. Enantioselective synthesis of flavonoids. Part 5. Poly-oxygenated β-hydroxydihydrochalcones. *Tetrahedron Lett.* 39:5623 (1998).

30. Botha, J.J.; Ferreira, D.; Roux, D.G. Synthesis of condensed tannins. Part 4. A direct biomimetic approach to [4,6]- and [4,8]-biflavanoids. *J. Chem. Soc., Perkin Trans. 1* :1235 (1981).

31. Delcour, J.A.; Ferreira, D.; Roux, D.G. Synthesis of condensed tannins. Part 9. The condensation sequence of leucocyanidin with (+)-catechin and with the resultant procyanidins. *J. Chem. Soc., Perkin Trans. 1* :1711 (1983).

32. Weinges, K. Über catechine und ihre herstellung aus leuko-anthocyanidin-hydraten. *Annalen* 615:203 (1958); Freudenberg, K.; Weinges, K. Leuco- und pseudoverbindungen der anthocyanidine. *Annalen* 613:61 (1958).

33. Van Rensburg, H.; Van Heerden, P.S.; Bezuidenhoudt, B.C.B.; Ferreira, D. Enantioselective synthesis of the four catechin diastereomer derivatives. *Tetrahedron Lett.* 38:3089 (1997).

34. Van Rensburg, H.; Van Heerden, P.S.; Ferreira, D. Enantioselective synthesis of flavonoids. Part 3. *Trans*- and *cis*-flavan-3-ol methyl ether acetates. *J. Chem. Soc., Perkin Trans. 1* :3415 (1997).

35. Sharpless, K.B.; Amberg, W.; Bennani, Y.L.; Crispino, G.A.; Hartung, J.; Joeng, K.; Kwong, H.; Morikawa, K.; Wang, Z.; Xu, D.; Zhang, X. The osmium-catalyzed asymmetric dihydroxylation—a new ligand class and a process improvement. *J. Org. Chem.* 57:2768 (1992).

36. Joeng, K.; Sjo, P.; Sharpless, K.B. Asymmetric dihydroxylation of enynes. *Tetrahedron Lett.* 33:3833 (1992).

37. Wang, Z.; Zhang, X.; Sharpless, K.B. Asymmetric dihydroxylation of aryl allyl ethers. *Tetrahedron Lett.* 34:2267 (1993).

38. Kolb, H.C.; Van Nieuwenhze, M.S.; Sharpless, K.B. Catalytic asymmetric dihydroxylation. *Chem. Rev.* 94:2483 (1994).

39. Dale, J.A.; Mosher, H.S. Nuclear magnetic resonance enantiomer reagents. Configurational correlations *via* nuclear magnetic resonance chemical shifts of diastereomeric mandelate, *O*-methylmandelate, and α-methoxy-α-trifluoromethylphenylacetate (MTPA) esters. *J. Am. Chem. Soc.* 95:512 (1973).

40. Pereyre, M.; Quitard, J.P.; Rahm, A. *In*: Tin in organic synthesis, Butterworth, London: p. 4 (1987).

41. Versteeg, M.; Bezuidenhoudt, B.C.B.; Ferreira, D.; Swart, K.J. The first enantioselective synthesis of isoflavonoids: (*R*)- and (*S*)-isoflavans. *J. Chem. Soc., Chem. Commun.* 1317 (1995).

42. Versteeg, M.; Bezuidenhoudt, B.C.B.; Ferreira, D. Stereoselective synthesis of isoflavonoids: (*R*)- and (*S*)-isoflavans. *Tetrahedron Memorial Issue* (the late Sir Derek Barton), (in press).

MANIPULATION OF SOME CRUCIAL BONDS IN THE MOLECULAR BACKBONE OF PROANTHOCYANIDINS AND RELATED COMPOUNDS

Daneel Ferreira, E. Vincent Brandt, Hendrik van Rensburg, Riaan Bekker, and Reinier J.J. Nel

Department of Chemistry
University of the Orange Free State
P.O. Box 339
Bloemfontein, 9300
SOUTH AFRICA

1. INTRODUCTION

Recent years have witnessed considerable development toward an understanding of the intricate principles governing the chemistry of oligomeric proanthocyanidins. These developments were initiated by the growing realization of the industrial importance of the proanthocyanidins, which is based upon their interactions with key molecules in the biological domain. Despite the surprisingly limited array of carbon-carbon and carbon-oxygen bonds in the heterocycles of proanthocyanidin oligomers, a rich variety of ingenious protocols aimed at manuevering these bonds in a controlled mode has emerged over the past 10–15 years. These studies blend with the heavy focus on the formation and rupture of the interflavanyl bond(s), which played such a decisive role in the structural elucidation of proanthocyanidin oligomers during the seventies and early eighties. Methodologies to chemically exploit the heterocyclic bonds are highly relevant to progress in the chemistry of these important and intriguing biomolecules and constitute the subject of this review.

2. CLEAVAGE OF THE INTERFLAVANYL BOND

The readily occurring cleavage of the interflavanyl bond in proanthocyanidins exhibiting C(5) oxygenation of the A-ring of their chain-extender units with sulfur[1,2] and aromatic carbon nucleophiles[3] under acid catalysis has played a key role in the structure elucidation of this complex group of natural products. In the

Plant Polyphenols 2: Chemistry, Biology, Pharmacology, Ecology, Edited by
Gross et al. Kluwer Academic / Plenum Publishers, New York, 1999

5-deoxy (A-ring) series of compounds, e.g., the fisetinidol-(4→8)- and -(4→6)-cate-chin profisetinidins **1**, **4** and **6** (fig. 1), and the analogous prorobinetinidins **2** and **7** (fig. 1) from the commercially important bark of *Acacia mearnsii* (black wattle),[4,5] this $C(sp^3)$—$C(sp^2)$ bond is remarkably stable under a variety of conditions[6,7] and has hitherto resisted all efforts at cleavage in a controllable manner. Such a stable interflavanyl bond has adversely affected both the structure investigation of the polyflavanoid tannins in black wattle bark and of those from other commercial sources, e.g., *Schinopsis* spp. (quebracho) as well as the establishment of the absolute configuration of the chain-terminating flavan-3-ol moiety in the 5-deoxy-oligoflavanoids. We have therefore investigated methods to cleave the interflavanyl bond in profisetinidins efficiently under conditions sufficiently mild to allow the isolation and identification of the constituent flavanyl units.[8,9]

Treatment of the fisetinidol-(4α→8)-catechin **1**,[7] representing a typical tannin unit of commercial wattle extract, with sodium cyanoborohydride [Na(CN)BH₃][10] in trifluoroacetic acid (TFA) for 6h at 0 °C gave conversion comprising the start-ing material **1** (24 percent recovery), catechin **10** (15 percent) and the (2*R*)-1-(2,4-dihydroxyphenyl)-3-(3,4-dihydroxyphenyl)propan-2-ol **13** (16 percent) (54 percent recovery of material) (fig. 2). Similar treatment of the fisetinidol-(4β→8)- and -(4α→6)-catechins **4**[7] and **6**[7] with their respective more and less labile interflavanyl bonds compared with the C(4)—C(8) bond in compound **1** under acidic conditions[11] also afforded a mixture consisting of starting material (**4**, **6**; 16, 12 percent recovery, respectively), catechin **10** (17, 4 percent, respectively), and the (2*R*)-1,3-

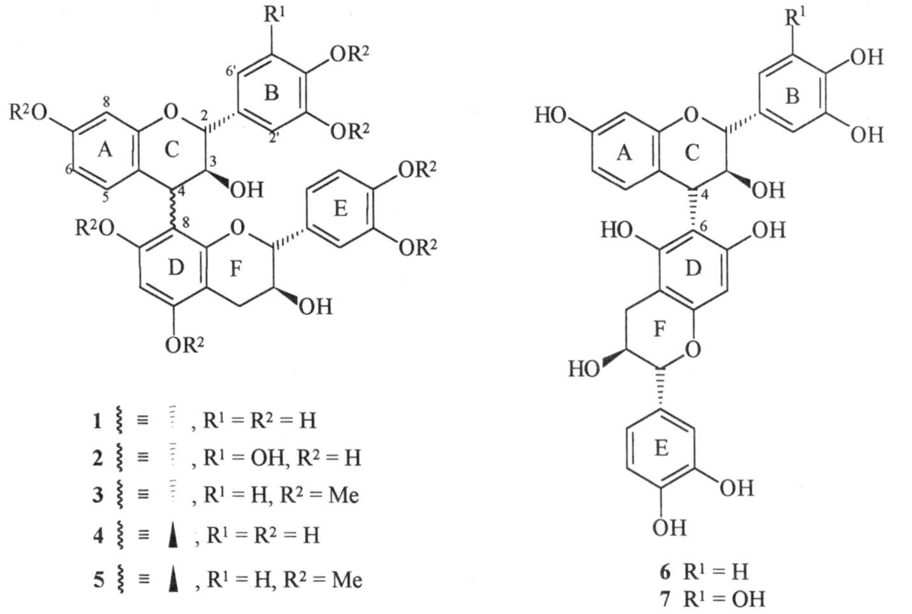

Figure 1. Structures of dimeric profisetinidins and prorobinetinidins based on catechin.

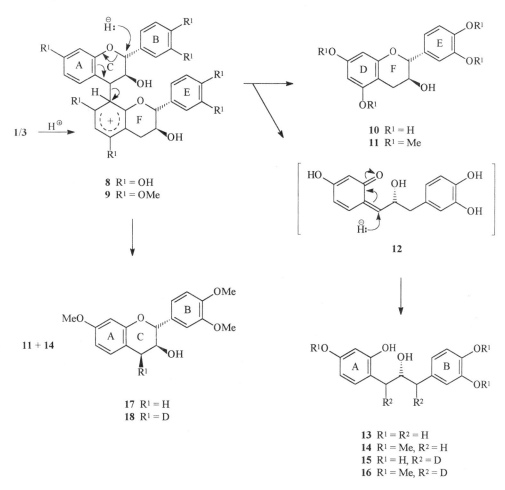

Figure 2. Proposed route to the cleavage of the interflavanyl bond and of the C-ring in profisetinidin **1**.

diarylpropan-2-ol (**13**, **18**, 4 percent, respectively) (50 and 20 percent recovery of material).

Similar conditions also effected cleavage of the interflavanyl bond in the fisetinidol-(4α→8)-catechin permethylaryl ether **3** to afford tetra-*O*-methylcatechin **11** (21 percent), the 1,3-diarylpropan-2-ol **14** (12 percent), and tri-*O*-methylfisetinidol **17** (12 percent). Such a rupture of the interflavanyl bond in the permethylaryl ether **3** introduced an important dimension to these cleavages in relation to the chemistry of the 5-deoxyoligoflavanoids where the additional chromatographic steps involved with derivatization are often prerequisites for sample purity. The 'liberation' of the chain-terminating flavan-3-ol unit **10** irrespective of whether the phenol **1** or methyl ether **3** was employed, provided a powerful probe toward addressing the hitherto unsolved problem of defining the absolute configuration at the stereocenters of this moiety in naturally occurring proanthocyanidins that are often synthetically inaccessible due to the unavailability of the flavan-3,4-diol and/or flavan-3-ol precursors.

The mild conditions effecting simple cleavage of the strong interflavanyl bond in the profisetinidins **1**, **4**, and **6** prompted application of the same protocol to the procyanidins B-1 **19** and B-3 **21** and their respective permethylaryl ethers **20** and **22** (fig. 3) with less rigid C(4)—C(8) linkages compared with those in the profisetinidins **1** and **4**. Treatment of procyanidin B-1 **19** with Na(CN)BH$_3$ in TFA for 1 h at 0 °C gave a mixture comprising the starting material **19** (14 percent recovery), catechin **10** (20 percent) and epicatechin [C(3) epimer of **10**] (21 percent). Under identical conditions, procyanidin B-3 **21** afforded catechin **10** (35 percent) and a residue (15 percent) of starting material. The permethylaryl ethers **20** and **22** gave, within 30 min, respectively, tetra-*O*-methylcatechin **11** (31 percent), tetra-*O*-methylepicatechin (33 percent), and starting material **20** (10 percent recovery), and tetra-*O*-methylcatechin **11** (56 percent) and starting material **22** (12 percent recovery).

Whereas the heterocyclic ring of the catechin DEF moiety invariably remained intact during the reductive process, cleavage of both the (4→6)- and (4→8)-interflavanyl bonds in the free phenolic profisetinidins **1**, **4**, and **6** is apparently associated with the simultaneous opening of the C-ring of the chain-extender unit. Protonation of the electron-rich phloroglucinol D-ring[12,13] in profisetinidin **1** (fig. 2), and concomitant delivery of the equivalent of a hydride ion at C(2) (C-ring) of intermediate **8** effects the concurrent rupture of the pyran C-ring and of the C(4)—C(8) bond. This gives catechin **10** and the *o*-quinone methide intermediate **12**, which is subsequently reduced to the 1,3-diarylpropan-2-ol **13**. Such an interdependence of the cleavage of the O(1)—C(2) and C(4)—C(8) bonds was demonstrated by the inability of the reagent to effect rupture of the heterocycle of catechin **10**.

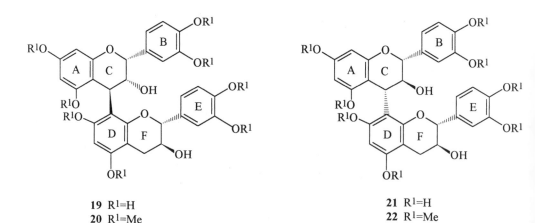

19 R^1=H
20 R^1=Me

21 R^1=H
22 R^1=Me

Figure 3. Structures of procyanidin B-1 **19** and B-3 **21**, and their permethylaryl ethers **20** and **22**.

The selective cleavage of the interflavanyl bonds in procyanidins B-1 **19** and B-3 **21** and their permethylaryl ethers **20** and **22** presumably results from the relative lability of this bond imposing a high degree of S_N1 character to the processes of protonation and delivery of hydride ion.

The mechanism for cleavage of the interflavanyl bond in the profisetinidin biflavanoids (fig. 2) was corroborated using sodium cyanotrideuterioborohydride [Na(CN)BD$_3$] in TFA. Under these conditions, the fisetinidol-(4α→8)-catechin **1** was converted into catechin **10** (26 percent) and the (2R)-1,3-dideuterio-1,3-diarylpropan-2-ol **15** (25 percent). The permethylaryl ether **3** and the fisetinidol-(4β→8)-catechin hepta-O-methyl ether **5** both gave tetra-O-methylcatechin **11** (12, 32 percent, respectively), the dideuterio-1,3-diarylpropan-2-ol tri-O-methyl ether **16** (14, 16 percent, respectively) and the 4β-deuteriofisetinidol derivative **18** (12, 14 percent, respectively). Formation of the deuteriated 1,3-diarylpropan-2-ols **15** and **16** (mixtures of diastereomers) thus confirmed the conjecture regarding the genesis of the propan-2-ols via reduction of the o-quinone methide **12**.

The protonated species **8** presumably also served as precursor to the 4β-deuteriotri-O-methylfisetinidol **18** via delivery of hydride ion from the β-face in a predominant S_N2 mode. Compound **18** persistently formed also when fisetinidol-(4α→8)- and (4β→8)-catechin hepta-O-methyl ethers **3** and **5** were treated with Na(CN)BD$_3$ in TFA. This observation prompted an investigation of the structural features of the substrates that direct the stereochemistry of the delivery of hydride ion at C(4) in intermediates of type **9**. Treatment of the epifisetinidol-(4β→8)-catechin hepta-O-methyl ether **23** (fig. 4) with Na(CN)BD$_3$ afforded the 4β-deuteriotri-O-methylepifisetinidol **25** (18.5 percent), tetra-O-methylcatechin **11** (32 percent) and the (2S)-1,3-dideuterio-1,3-diarylpropan-2-ol (6 percent, enantiomer of compound **16**). The *ent*-fisetinidol-(4β→8)-catechin hepta-O-methyl ether **24** (fig. 4) gave 4α-deuteriotri-O-methyl-*ent*-fisetinidol (13 percent, the enantiomer of compound **18**), tetra-O-methylcatechin **11** (24 percent) and the (2S)-1,3-dideuterio-1,3-diarylpropan-2-ol (12 percent, enantiomer of **16**).

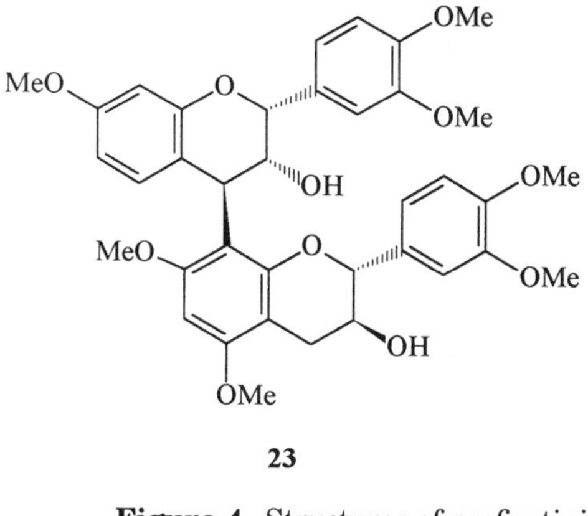

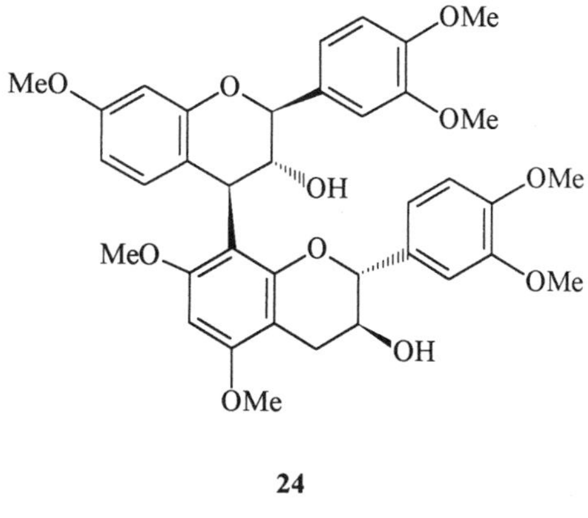

Figure 4. Structures of profisetinidin permethylaryl ethers **23** and **24**.

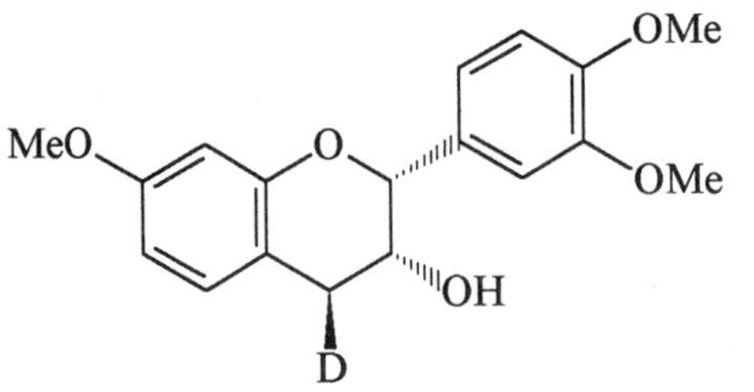

Figure 5. Structure of the 4β-deuterio-epifisetinidol derivative 25.

25

Thus, the formation of the 4β-deuteriofisetinidol- and epifisetinidol derivatives **18** (fig. 2) and **25** (fig. 5) from the reduction of the profisetinidin permethylaryl ethers **3**, **5**, and **23** with Na(CN)BD₃ in TFA, and of the enantiomer of compound **18** during reduction of the *ent*-fisetinidol-(4β→8)-catechin derivative **24**, indicated that the deuterium ion is consistently delivered at C(4) of a protonated species of type **8** from the side opposite to the 2-aryl group of the C-ring. This presumably indicates that delivery of hydride ion occurs from a complex between the reducing agent and the C-ring heterocyclic oxygen lone pair *trans* to the 2-aryl group, such transfer being most readily facilitated in an A-conformer[14] of type **26** (fig. 6).

The potential of this development toward the structural elucidation of the proanthocyanidin condensed tannins, especially the 5-deoxy analogues, from important commercial sources is clear. In addition, the method facilitates the ready definition of the absolute configuration of the chain-terminating flavan-3-ol moiety in 5-deoxyoligoflavanoids, especially in view of the demonstration that these units may also comprise *ent*-catechin and *ent*-epicatechin.[15,16]

The double interflavanyl linkage in A-type proanthocyanidins introduces a high degree of conformational stability that culminates in high-quality and unequivocal NMR spectra, conspicuously free of the effects of dynamic rotational isomerism

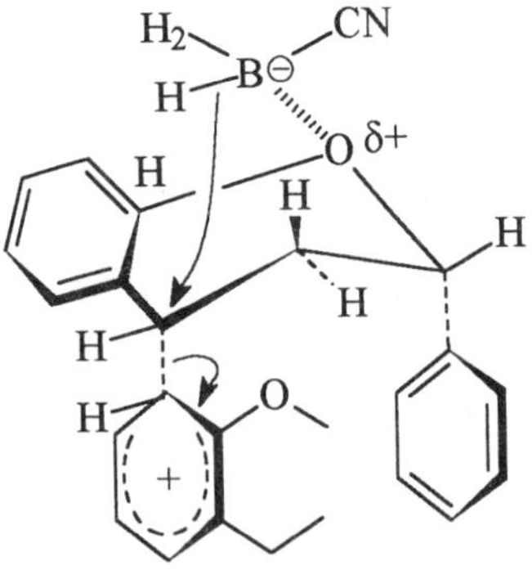

Figure 6. Complex permitting selective delivery of hydride ion.

26

at the dimeric level. Compounds of this class are readily recognizable from the characteristic AB-doublet ($^3J_{3,4}$ = 3–4 Hz) of C-ring protons in the heterocyclic region of their ^{1}H NMR spectra,[17] and may possess either (2α,4α)- or (2β,4β)-double interflavanyl bonds. Two fundamental structural problems, i.e., establishment of the mode of linkage of the C- to the D-ring and assignment of the absolute configuration at the stereocentres of the F-ring, have however limited progress in this field. These and related problems have hitherto been approached via exotic spectroscopic methods[18–22] that prompted us to search for a simple and more general chemical method based upon the reductive cleavage of the acetal functionality of A-type proanthocyanidins. The potential to address these problems by reduction of either of the C—O acetal bonds was demonstrated[23] for the known procyanidins A-1 **27** and A-2 **28**, available from the skins of mature peanuts (*Arachus hypogea*),[24] using Na(CN)BH$_3$ in TFA. The readily accessible hepta-*O*-methyl ethers **29** and **30** were selected as model compounds with a view to using the *O*-substituents of the D-ring as probes for anticipated much simplified ^{1}H NMR studies.

Separate treatment of the hepta-*O*-methylprocyanidins A-1 **29** and A-2 **30** with Na(CN)BH$_3$ in TFA for 1.5 h at 0 °C (fig. 7) gave conversion into mixtures comprising the starting materials (**29**, 3.4 percent and **30**, 4.4 percent, respectively), and as anticipated from cleavage "a" the tetrahydropyrano[2,3-*f*]chromene derivatives (**31**, 5.2 percent and **32**, 7 percent, respectively). The envisaged B-type procyanidin biflavanoids **33** and **34** from the "b" pathway were not obtained but instead, the respective monomeric units, i.e., tetra-*O*-methyl-*ent*-catechin **35** (4 percent) and tri-*O*-methylcatechin **37** (3.4 percent) from the A-1 derivative **29**, and tetra-*O*-methyl-*ent*-catechin **35** (3 percent) and. tri-*O*-methylepicatechin **36** (1.3 percent) from the A-2 derivative **30** were isolated. These compounds were accompanied by some related analogues[23] that have no relevance to the problem under discussion and will not be further dealt with.

Both the carbon-oxygen bonds of the acetal functionality in the procyanidin A-1 **29** and A-2 **30** derivatives are thus susceptible to reductive cleavage under acidic conditions. This process is presumably triggered by the random protonation of the acetal oxygens and concomitant delivery of the equivalent of hydride ion at the antibonding (σ*) orbitals of the carbon-oxygen bonds in a predominant S$_N$2 manner. Such a transfer of hydride ion apparently occurs from a complex between the reducing agent and the axial C(3) (C-ring) oxygen lone pair, the proximity of the boron-hydrogen bonds to the backside of the acetal carbon being a prerequisite for reduction of either one of the acetal bonds. Reduction thus leads to "inversion" of configuration at C(2)(C) of both B-type procyanidin intermediates **33** and **34** and of the tetrahydropyrano[2,3-*f*]chromene derivatives **31** and **32**. The chemistry and the unequivocal structure elucidation including assessment of absolute configuration at all the stereocenters of the latter class of compounds are well understood[25–28] and facilitated confirmation of the absolute stereochemistry of ring F in the natural product derivatives **29** and **30**.

Biflavanoids **33** and **34** are prone to facile cleavage of their interflavanyl bonds via protonation of the electron-rich phloroglucinol D-ring[12,13] and attack of hydride ion at C(4)(C)[8,9] to give the *ent*-catechin derivative **35** from the ABC-unit

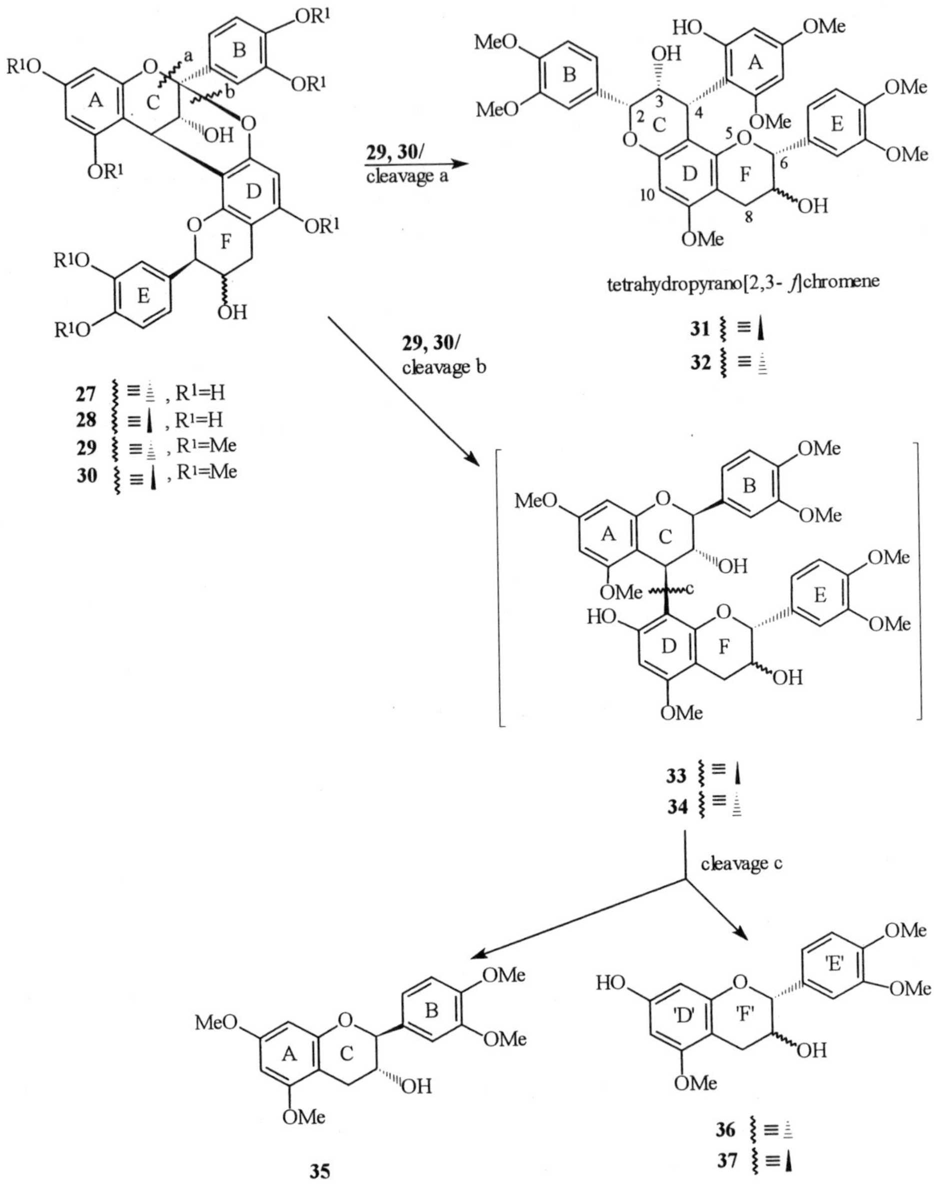

Figure 7. Cleavage of the acetal functionality of proanthocyanidin A-1 and A-2 permethylaryl ethers **29** and **30** with Na(CN)BH$_3$ in TFA.

and, respectively, the epicatechin and catechin derivatives **36** and **37** from the DEF-moieties. The 'liberation' of the latter two chain terminating flavan-3-ol units unambiguously defines the D-ring oxygen that is involved in the acetal functionality of the parent compounds **29** and **30**. It furthermore provides a powerful probe toward addressing the hitherto unsolved problem of establishing the absolute configuration at the stereocenters of this moiety in naturally occurring A-type proanthocyanidins. The flavan-3-ol unit **35**, albeit with inversed C(2) configuration, should facilitate the assignment of the absolute configuration at C(3) (C-ring) of the parent compounds **29** and **30**, especially in view of the inability to differentiate between 3,4-*cis*- and 3,4-*trans*-configuration in these compounds on the basis of $^3J_{HH}$ values.[18] The mode of the C—C linkage between the constituent flavan-3-ol units in the A-type procyanidin, *e.g.* (4→6) or (4→8) is defined by the nature of tetrahydropyranochromene,[25,26] *i.e.* [2,3-*f*], [3,2-*g*] or [2,3-*h*] that is formed via reductive cleavage "a".

The protocol described here should thus contribute substantially toward a straightforward chemically orientated structural definition of proanthocyanidins of the A-class.

3. INTERFLAVANYL BOND FORMATION IN PROCYANIDINS UNDER NEUTRAL CONDITIONS

Progress in the structure elucidation of natural proanthocyanidin oligomers with phloroglucinol-type A-rings, e.g., the procyanidins, is still impeded by the absence of methodology to facilitate the progressive construction of the interflavanyl bonds. The existing semi-synthetic methods involve coupling of electrophilic C(4)-substituted flavan-3-ol units with nucleophilic flavan-3-ols under either acidic or basic conditions.[29] Under these conditions, however, the interflavanyl bond is labile, which invariably leads to an equilibrium between substrates and products. Such a labile bond and an apparent preference of the electrophile for the di- and trimeric products once condensation is initiated furthermore give poor control regarding the level of oligomerization. We thus investigated the effectiveness of the thiophilic Lewis acids, dimethyl(methylthio)-sulfonium tetrafluoroborate (DMTSF),[30,31] and silver tetrafluoroborate (AgBF₄),[32] to activate the C(4)—S bond in the 4-thioethers of flavan-3-ols toward carbon nucleophiles, and hence to generate the interflavanyl bond of proanthocyanidins under neutral conditions.[33]

The feasibility of the procedure to establish the C(sp^3)—C(sp^2) interflavanyl bond in procyanidin oligomers under neutral conditions was assessed using the model reaction between the 4-thiobenzyl ether of a flavan-3-ol and phloroglucinol as a potent capture nuceophile (fig. 8). Treatment of a mixture of 4β-benzylsulfanylepicatechin **38**, available via acid-catalyzed thiolysis of loblolly pine purified tannin polymer,[34] representing the (2*R*,3*R*)-2,3-*cis*-flavan-3-ol chain extender unit of the procyanidins, and phloroglucinol in THF with a fresh batch of DMTSF (1:4:1 molar ratio) for 1 h at 0 °C gave epicatechin-(4β→2)-phloroglucinol **39**[35] (28 percent yield) and the analogous bis-epicatechin-(4β→2:4β→4) phloroglucinol **40**[35] (19 percent).

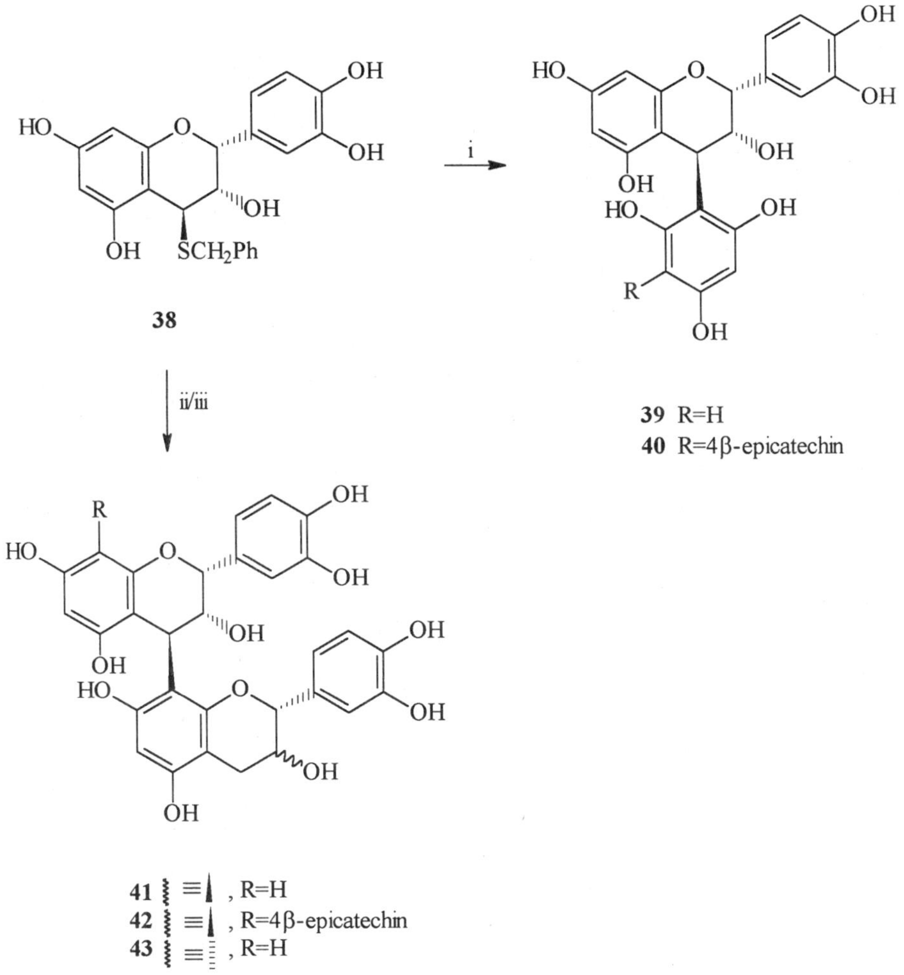

Figure 8. Procyanidin synthesis using Lewis acid activation of 4β-benzylsul-fanylepicatechin **38**; *Reagents and conditions*: i, Phloroglucinol/DMTSF in THF, 0 °C, 1.25 h; ii, catechin/DMTSF in THF, –15 °C, 2 h or catechin/AgBF₄ in THF, 0 °C, 1 h; iii, epicatechin/AgBF₄ in THF, 0 °C, 1 h.

The aqueous workup afforded an acidic reaction mixture (pH ~ 2.5 by Merck ACILIT® indicator, pH 0–6), hence indicating that the coupling process might have occurred under acid catalysis. Workup in a pH 6.85 buffer solution, however, gave an identical product distribution. The stability of the C(4)—S bond in the 4β-benzylsulfanylepicatechin **38** in acidic medium was additionally ascertained by treatment of a mixture of thioether **38** and phloroglucinol in 0.1 M HCl solution for 3.5 h at 20 °C which afforded only the starting materials. The C(sp³)—C(sp²) bond formation to give the 4-arylflavan-3-ol **39** and the trimeric analogue **40** thus occurred under neutral conditions via activation of the C(4)—S bond in the thioether **38** by DMTSF.

The potential of DMTSF to facilitate the aforementioned bond formation prompted an attempt at generating the interflavanyl bond of the ubiquitous procyanidin B-1 **41**.[36,37] A mixture of 4β-benzylsulfanylepicatechin **38** and catechin in THF treated with DMTSF (1:10:1 molar ratio) at −15 °C for 2h indeed afforded procyanidin B-1[2] **41** (22 percent) and the analogous trimeric procyanidin **42**[38] (19 percent) (fig. 8). In spite of the persistent formation of the triflavanoid **42**, formation of regiomeric dimers or higher oligomers could not be detected in several subsequent runs. This protocol thus compares favorably with the classical acid-catalyzed condensation of catechin-4α-ol and catechin,[39,40] which gave a mixture of procyanidins B-3 **46** and B-6, the trimeric procyanidin C(2) **47** and its 4,6-regioisomer, and the presumed all-*trans*-[4,8]-linked tetraflavanoid analogue in the proportions 10:1:12:1:3 and in *ca* 45 percent overall yield.

Problems have been experienced with the moisture sensitivity of DMTSF. In view of this and the successful activation of the benzylic carbon-sulphur bond of α-hydroxy-β-benzylsulfanyl-dihydrochalcones toward the intramolecular formation of a carbon-oxygen bond in stereoselective synthesis of dihydroflavonols using AgBF$_4$,[41,42] the potential of this thiophilic Lewis acid to effect interflavanyl bond formation in procyanidins was next assessed. Treatment of a solution of 4β-benzylsulfanylepicatechin **38** and catechin (10 *eq.*) in THF with AgBF$_4$ (2.5 *eq.*) for 1h at 0 °C (fig. 8) indeed afforded procyanidin B-1[2] **41** in improved yield (38 vs. 22 percent for DMTSF as activator) as the only product formed in meaningful quantities. When this procedure was repeated with epicatechin as the capture nucleophile, procyanidin B-2[2] **43** was formed in 37 percent yield.

In order to broaden the scope of this thiophilic Lewis acid-catalyzed interflavanyl bond formation under neutral conditions, 4-benzylsulfanylcatechin **44**, representing the (2R,3S)-2,3-*trans* chain extender unit of the procyanidins was next used as the source of the flavan-3-ol C(4) electrophilic moiety (fig. 9). Compound **44** was synthesized as a mixture of the 4β- and 4α-epimers (4:1) via reduction of (2R,3R)-2,3-*trans*-dihydroquercetin with sodiumborohydride in ethanol[39] and acid-catalyzed thiolysis of the unstable catechin-4-ol intermediate using phenylmethanethiol-acetic acid. Both the 4β- and 4α-benzylsulfanylcatechin epimers **44** will be converted to the same transient C(4) carbocationic intermediate by AgBF$_4$, hence both will display the same stereochemical course of interflavanyl coupling. Separate treatment of a mixture of the epimeric 4-benzylsulfanylcatechins **44** and epicatechin and catechin in THF with AgBF$_4$ (2.5 *eq.*) for 1h at 0 °C, afforded procyanidin B-4[2] **45** (35 percent) and B-3[2] **46** (51 percent), respectively. The preferences for the formation of 4β- and 4α-interflavanyl bonds using the epicatechin- and catechin-4-thiobenzyl ethers **38** and **44**, respectively, and for the (4→8)-interflavanyl linkages were anticipated.[40] The thiobenzyl ethers are converted by the Lewis acids into relatively stable intermediates permitting both the regioselective attack of the nucleophile via C(8) where the HOMO displays maximum amplitude and stereoselectivity by approach from the sterically least-hindered side.

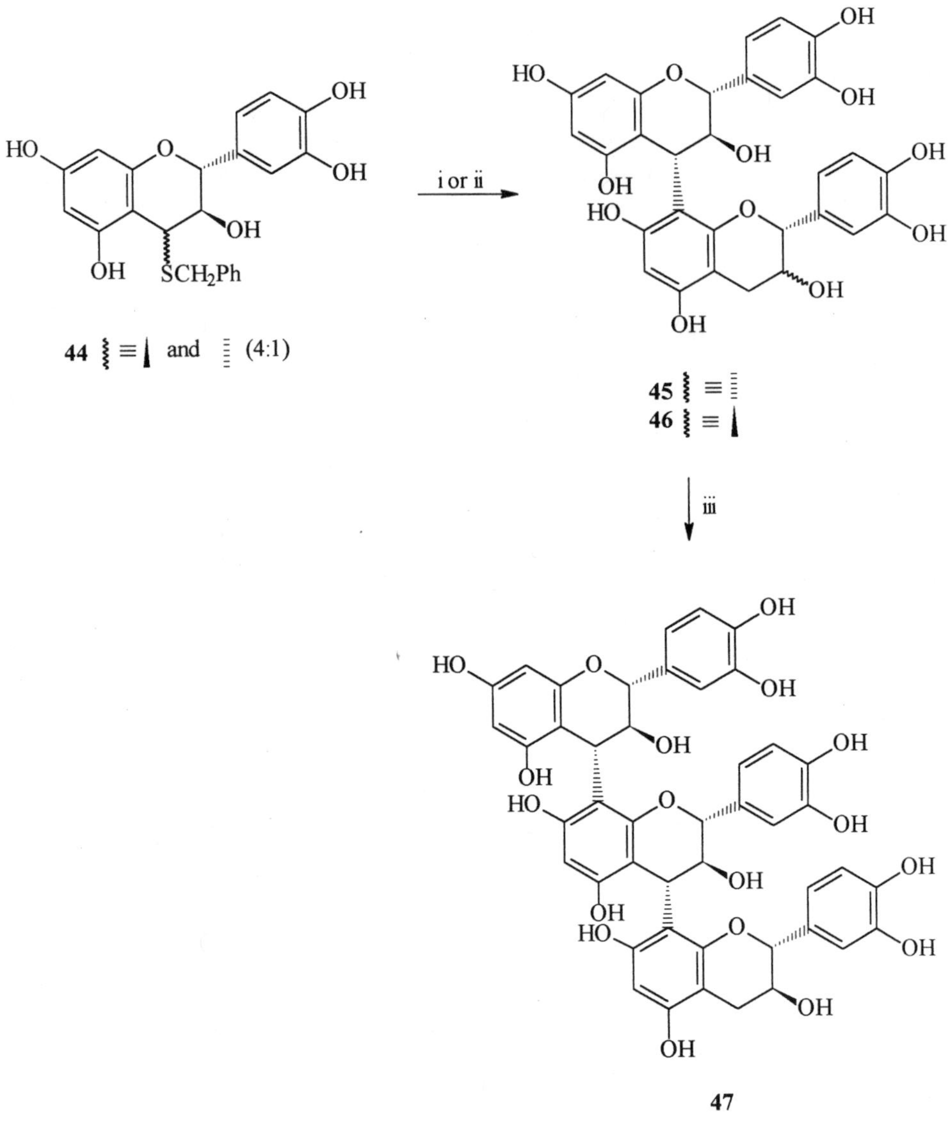

Figure 9. Procyanidin synthesis using Lewis acid activation of a mixture (4:1) of 4β- and β-benzyl-sulphanylcatechin **44**; *Reagents and conditions*: i, epicatechin/AgBF$_4$ in THF, 0 °C, 1 h; ii, catechin/AgBF$_4$ in THF, 0 °C, 1 h; iii, **44**/AgBF$_4$ in THF, 0 °C, 1 h.

Finally procyanidin B-3 **46**, and the 4-benzylsulfanylcatechin epimers **44** were coupled using AgBF$_4$ in THF to afford the trimeric procyanidin C(2)[40] **47** (26 percent) as the only isolable product (fig. 9). The sequences **38** → procyanidin B-1 **41** and B-2 **43**, and **44** → procyanidin B-4 **45**, B-3 **46** and procyanidin C(2) **47** using AgBF$_4$ as the thiophilic Lewis acid, no doubt offer advantages as far as control over the level of oligomerization, reversibility and 'scattering' of the interflavanyl bond(s) are concerned. This can be compared with the formation of these products via catalysis of the coupling of catechin-4-ol and catechin by protic acids.[39,40]

4. ABSOLUTE CONFIGURATION OF FLAVANONE-BENZOFURANONE FLAVANOIDS AND 2-BENZYL-2-HYDROXYBENZOFURANONES

Although the biflavonoids represent a major group of phenolic natural products,[43] the assessment of the absolute configuration of analogues possessing stereocenters other than those originating from atropisomerism of the interflavonoid bond has hitherto been unsuccessful. This also precluded stereochemical assignment of zeyherin, the first and thus far sole entry with a benzofuranoid constituent unit, which was isolated some 27 years ago from the heartwood of the 'red ivory' (*Berchemia zeyheri* Sond.).[44]

Continued investigation of this source recently[45,46] led to the purification of two zeyherin epimers **48** and **50**, thus posing the challenge to develop a protocol that would permit assessment of the absolute configuration of the three stereocenters of these unique flavanone-2-benzylbenzofuranone biflavonoids (fig. 10). The O(1)—C(2) and C(3)—C(4) bonds in the permethylaryl ethers **49** and **51** of the flavanone-benzofuranone biflavonoids **48** and **50** were susceptible to facile reductive cleavage with sodium cyanoborohydride [Na(CN)BH$_3$] in trifluoroacetic acid (TFA) at 0 °C leading to the formation of the 7-(4-methoxyphenethyl)tetra-*O*-methylmaesopsin enantiomers **52** and **53**. Conformational information available from computational data in conjunction with ^{1}H NMR and CD spectroscopic observations of the biflavonoid derivatives **49** and **51** and their degradation products

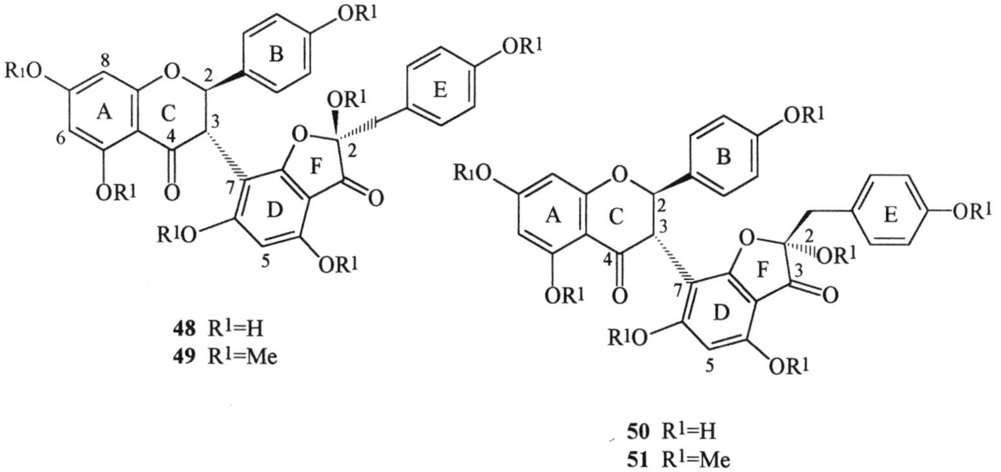

48 R^1=H
49 R^1=Me

50 R^1=H
51 R^1=Me

Figure 10. Structures of flavanone-benzylbenzofuranones **48** and **50**, and derivatives **49** and **51**.

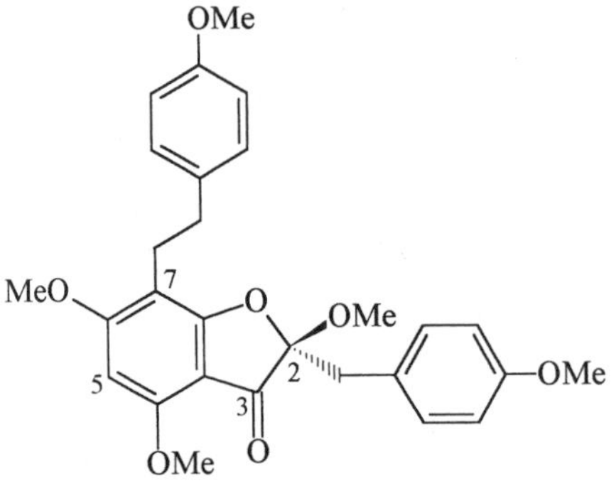

52

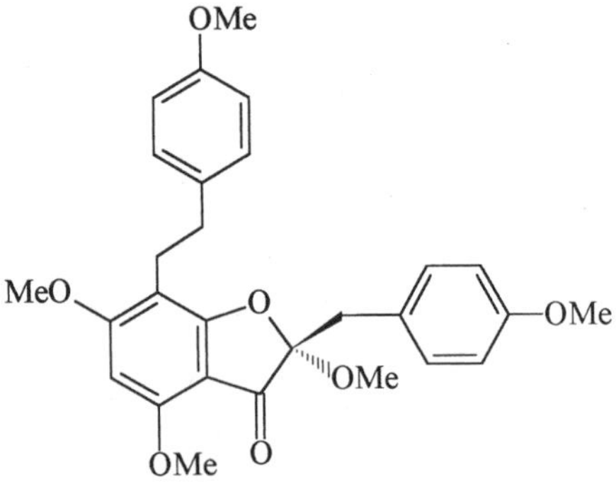

53

Figure 11. Structures of 7-(4-methoxyphenethyl)tetra-*O*-methylmaesopsin enantiomers **52** and **53**.

52 and **53**, permitted assignment of absolute configuration to the biflavonoids and the 2-benzyl-2-hydroxy-1-benzofuran-3(2*H*)-one group of natural products. A detailed report of these results, as well as the structure elucidation of the first two isoflavanone-benzofuranone biflavonoids **54** and **55** (fig. 12) are covered later in this book.

5. ENANTIOSELECTIVE SYNTHESIS OF FLAVONOID MONOMERS

Progress in the study of chemical and biological properties of the oligo- and polymeric proanthocyanidins depends to a large extent on the availability of the monomeric precursors that are required for their synthesis. This prompted investigations into the enantioselective synthesis of dihydroflavonols, i.e., the precursors to the electrophilic flavan-3,4-diol chain extender units, and flavan-3-ols, i.e, the source of the nucleophilic chain terminating units.

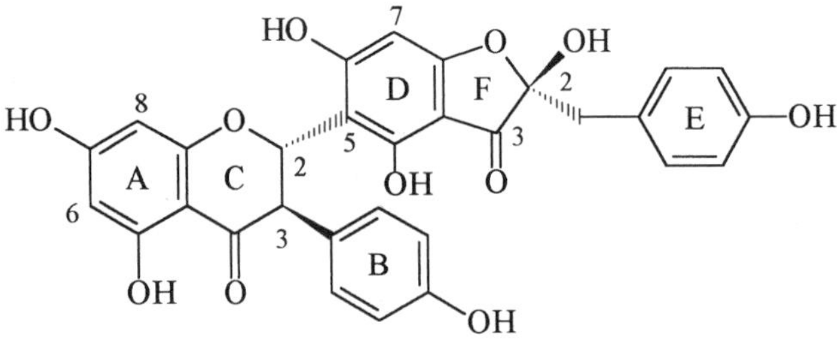

54
55 C-2(F) epimer

Figure 12. Structures of isoflavanone-benzofuranones **54** and **55**.

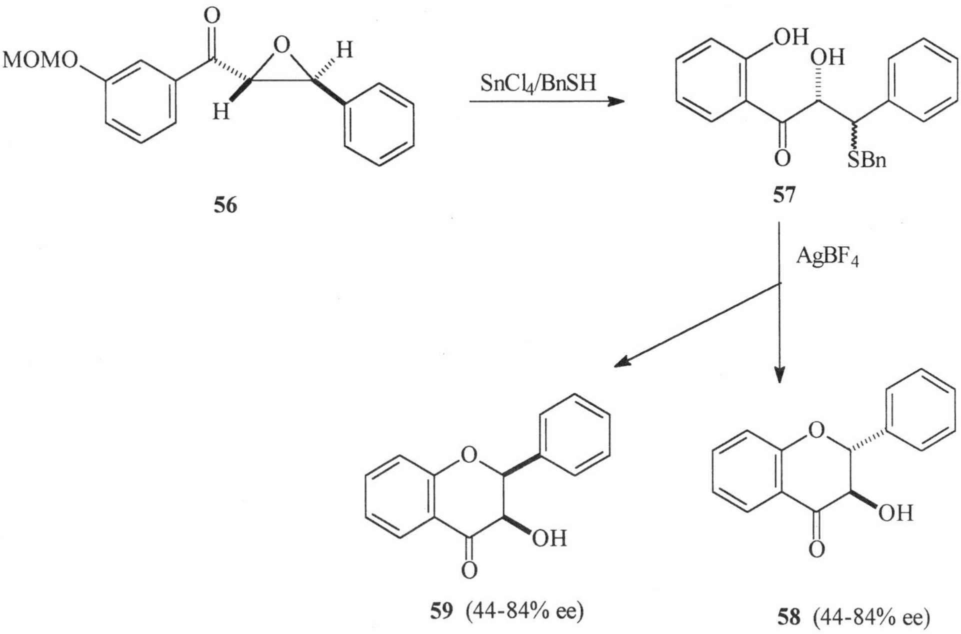

Figure 13. Enantioselective synthesis of 2,3-*trans*- and 2,3-*cis*-dihydroflavonols.

Enantiomerically enriched chalcone epoxides **56** were selected as precursors to both 2,3-*trans*- and 2,3-*cis*-dihydroflavonols **58** and **59** via α,2′-dihydroxy-β-benzylsulfanyldihydrochalcones, e.g., **57** (fig. 13).[41,42] Both 2,3-*trans*- and 2,3-*cis*-flavan-3-ols **62** and **63** were accessible *via* cyclization of 1,3-diaryl-1,2-*syn*-diols of type **61** which were obtained by asymmetric dihydroxylation of the appropriate 1,3-diarylpropenes, e.g., **60** (fig. 14).[48,49] These results are discussed in detail in the previous chapter. They clearly demonstrate a meaningful extension of the number of C—C and C—O bonds of the flavonoid heterocycle that may be stereoselectively manipulated.

6. REARRANGEMENT OF THE PYRAN HETEROCYCLE

Profisetinidin condensed tannins are arguably the most important polyflavanoids of commerce, making up the major constituents of wattle and quebracho tannins. The profisetinidin biflavanoids are susceptible to facile C-ring rearrangement under mild basic conditions, i.e., at pH *ca* 10.0 and temperatures below 50 °C. The mechanisms explaining the intricate genesis of the pyran-ring rearranged analogues were recently reviewed[29,50] and are summarized in figure 15 for the fisetinidol-(4α→8)- and (4β→8)-catechin mono-*O*-methyl ethers **64** and **65**.[1]

---

[1] These compounds are selectively protected at 4-OH (E-ring) to avoid side reactions associated with and E-ring quinone methide.

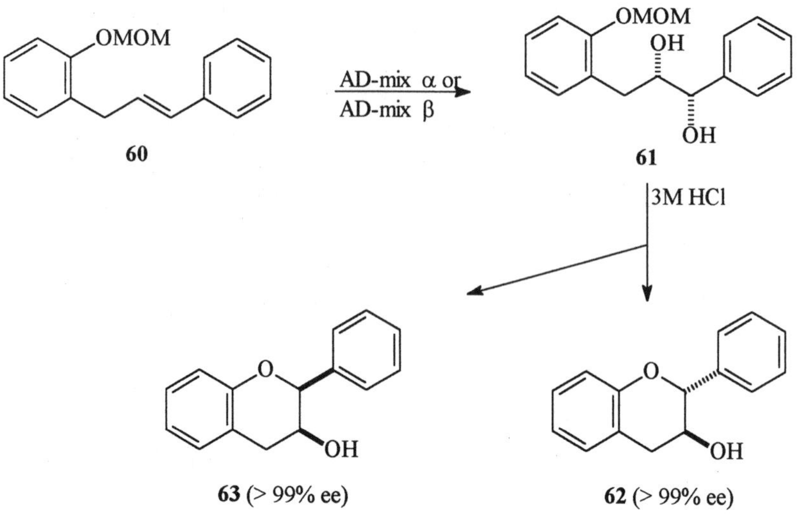

Figure 14. Enantioselective synthesis of flavan-3-ol diastereomers.

Both precursors are transformed by base into the B-ring quinone methides **66** and **67**, the former of which is converted highly stereoselectively into the tetrahydropyrano[2,3-*f*]chromene **68**. Besides conversion into the tetrahydropyrano[2,3-*f*]chromenes **69** and **70**, the intermediate quinone methide **67** with a 4β-flavanyl substituent is susceptible to 1,3-flavanyl migration of the DEF moiety leading to an A-ring quinone methide **71**. Stereoselective cyclization involving 7-OH(D) and C(4) then gives the tetrahydropyrano[2,3-*f*]chromenes **72** and **73** with "interchanged" A- and B-rings and inverted absolute configuration at the stereocenters of the C-ring compared to the C-ring constitution/-configuration in analogues **69** and **70**.

The same principles also ·govern the base-catalyzed conversions of the profisetinidin triflavanoids as is demonstrated in figure 16 for the bis-fisetinidol-(4α→6: 4α→8)-catechin mono-*O*-methyl ether **74**.[51] Here, the intermediate bifunctional quinone methide **75** served as the precursor to the hexahydropyranochromene **76** as well as the 'isomerization-intermediates' **77** and **78**, their permethylaryl ether triacetates being identical to the corresponding derivatives of the natural products.[51]

Continued investigations of the heartwood extractives of *Baikiaea plurijuga*, *Colophospermum mopane*, and *Guibourtia coleosperma* revealed the presence of a series of profisetinidin triflavanoid related compounds with rearranged C-rings. This prompted extension of the base-catalyzed pyran ring rearrangements of the bis-fisetinidol-(4β→6: 4α→8)-catechin,[52] bis-fisetinidol-(4β→6: 4β→8)-catechin,[53] bis-fisetinidol-(4α→6: 4β→8)-catechin,[54] and to the as yet unknown bis-fisetinidol-(4α→6: 4α→8)-epicatechin.[55] In contrast to the limited number of compounds obtained when the all-*trans* triflavanoid **74** was treated with base (see fig. 16), the susceptibility of analogues with 3,4-*cis* fisetinidol constituent units to secondary

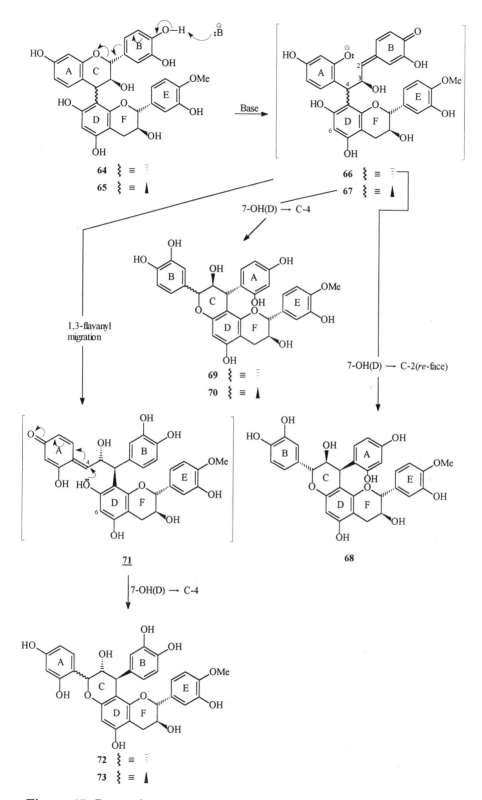

Figure 15. Proposed route to the formation of C-ring rearranged products related to fisetinidol-(4→8)-catechin profisetinidins.

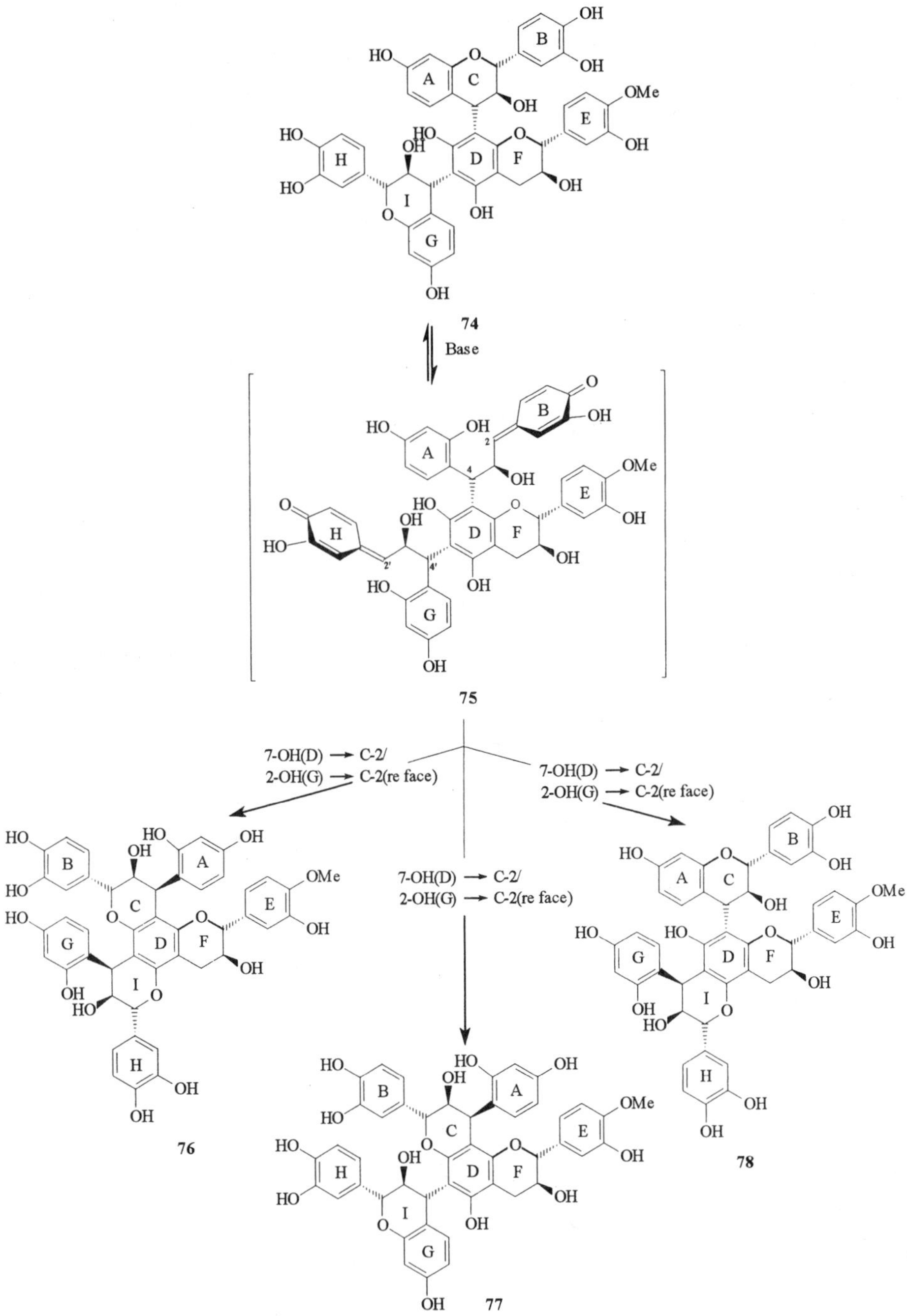

Figure 16. Proposed route to the formation of hexahydropyranochromene **76** and isomerization-intermediates **77** and **78**.

rearrangements led to the formation of an unacceptable large number of compounds, e.g., sixteen when the bis-fisetinidol-(4β→6: 4β→8)-catechin was subjected to base treatment.[53]

The complex structures of the economically important group of profisetinidin triflavanoid related phlobatannins were, however, synthetically accessible[56] in a controlled biomimetic fashion via the repetitive formation of the interflavanyl bond and pyran ring rearrangement of the chain extender unit under mild alkaline conditions. The unrestrained course of the base-catalyzed C-ring rearrangement reactions of profisetinidin triflavanoids possessing 2,3-*trans*-3,4-*cis* flavanyl constituent units thus results in exceptionally complex reaction mixtures. The stepwise construction of the dipyranochromene framework via sequential interflavanyl bond formation and pyran ring rearrangement, however, permitted concise synthetic access to phlobatannins at the trimeric level.

The susceptibility of the constituent flavanyl units of proanthocyanidins to intramolecular rearrangement via B-ring quinone methides under basic conditions was also demonstrated[57] in an unusual dimerization-rearrangement reaction of catechin **10** at pH 12 and 40 °C (fig. 17). Deprotonation of the 4'-OH of catechin **10** leads to an intermediate B-ring quinone methide **79**, which is intermolecularly trapped by the powerful nucleophilic C(8) of catechin in a highly stereoselective fashion to give the 'dimeric' intermediate **80**. This compound is transformed by base into the E-ring quinone methide **81**, which then stereoselectively recyclizes to compound **82**, i.e., the 1-(3,4-dihydroxyphenyl)-3-(2,4,6-trihydroxyphenyl)propan-2-ol substituted at C(1) with a catechinic acid moiety.

It was recently reported[58] that the application of thiolytic cleavage to study the condensed tannins from tree barks and nutshells did not provide quantitative yields of monomeric flavanyl cleavage products. Application[59] of thiolytic cleavage of the polymeric proanthocyanidins from pecan nut pith, known to comprise epigallocatechin, gallocatechin, and epicatechin chain extender units in the approximate ratios of 5:2:1 with either catechin or gallocatechin as terminal units,[60] consistently afforded significant amounts of phloroglucinol and a mixture of 1,3-dithiobenzyl-2,4,5,6-tetrahydroxyindane diastereomers **89**.

Such a conversion is demonstrated in figure 18 for a typical prodelphinidin **83** with 2,3-*cis* configuration of the chain extender units. Thiolytic cleavage of **83** gives the 4β-benzylsulfanylepigallocatechin **84**, which is protonated at the electron-rich phloroglucinol A-ring to afford intermediate **85** with a labile C(4)—C(10) bond. This bond then ruptures under the influence of the electron-releasing benzylsulfanyl group. It is a unique process representing the equivalent of the cleaving of the interflavanyl bond under acidic conditions, but under the influence of an external sulfur nucleophile. Rearrangement of the intermediate sulfonium ion **86** leads to the formation of the indane diastereomeric mixture **87** with its labile benzylic ether linkage, which is cleaved, with the release of phloroglucinol, to carbocation **88**. Reaction of the latter with the capture nucleophile, phenylmethane thiol, affords the mixture of 1,3-dibenzylsulfanyl-2,4,5,6-tetrahydroxyindane diastereomers **89**. These results clearly invalidate the use of extended thiolysis to provide meaningful estimates of the molecular weight of

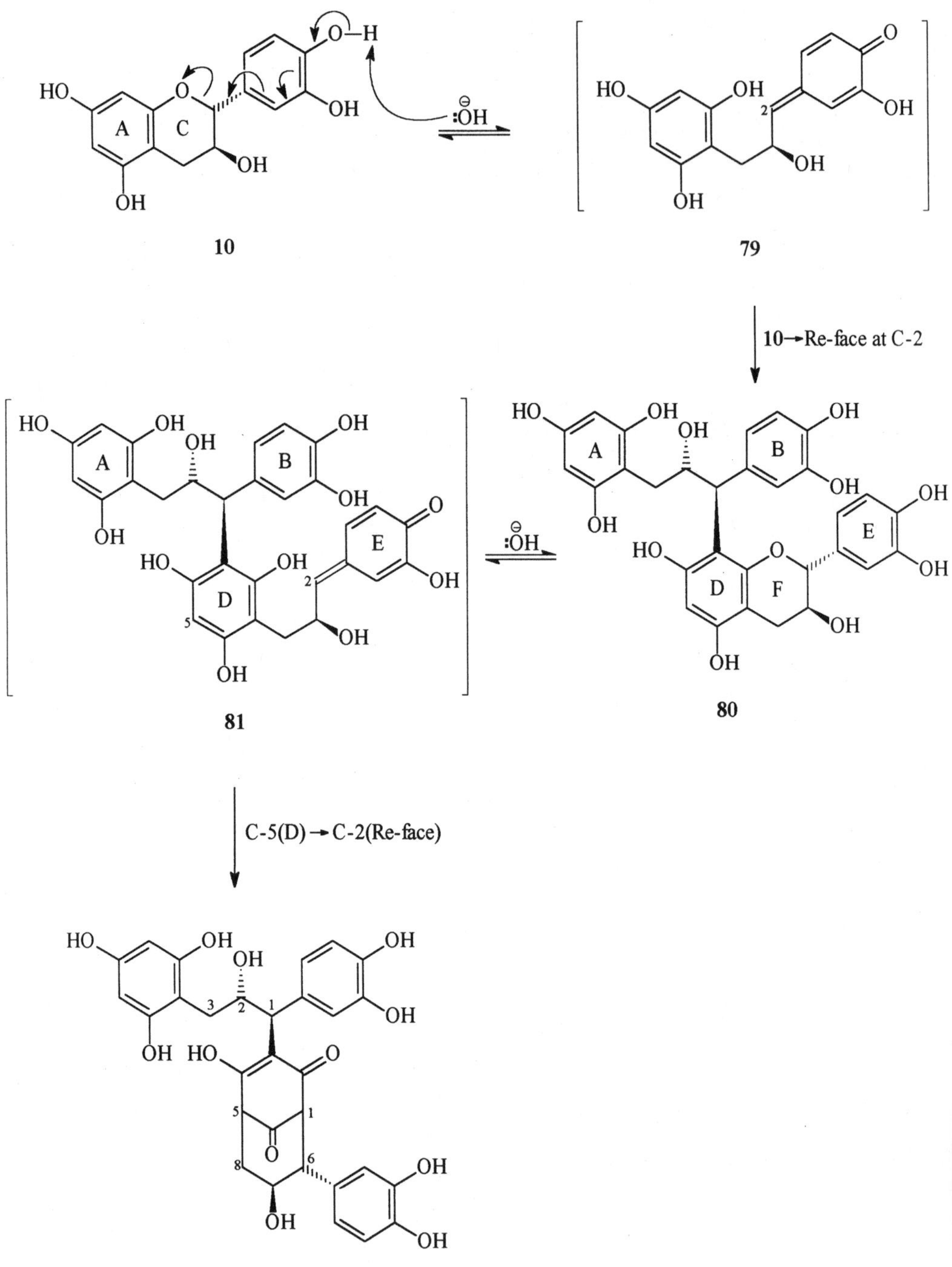

Figure 17. Proposed route to the dimerization-rearrangement of catechin **10** at pH 12 and 40 °C.

polymeric proanthocyanidins. It also calls into question the use of thiolysis as a means of obtaining 'quantitative' information on the composition of mixed proanthocyanidin polymers.

The aforementioned failures to obtain quantitative yields of 4-benzylsulfanylflavan-3-ols or the corresponding phloroglucinol adducts from acetic acid-catalyzed cleavage of condensed tannins, prompted further study[61] of the products of acid-catalyzed cleavage of these tannins. Particular emphasis was put on defining the rearrangements of the 4-phloroglucinol adducts that might occur when reactions at 100 °C are extended to 24 h or more.

Pecan (*Caraya illinoensis*) nut pith tannins were thus reacted with phloroglucinol and acetic acid at 100 °C for 24 or 48 h. Following methylation and acetylation of the products, a series of proanthocyanidin derivatives were isolated. These included the known methyl ether acetate derivatives epicatechin-(4β→2)-phloroglucinol **90**, epigallocataechin-(4β→2)-phloroglucinol **91**, catechin-(4α→2)-phloroglucinol **92**, gallocatechin-(4α→2)-phloroglucinol **93**, and gallocatechin-(4β→2)-phloroglucinol **94**. The methyl ether acetate derivatives of catechin **95** and gallocatechin **96** were also isolated. Acid-catalyzed thiolysis and acetylation afforded the peracetate derivatives of 4α-benzylsulfanylcatechin **99**, 4α-benzylsulfanylgallocatechin **100**, 4β-benzylsulfanylepicatechin **101**, and 4β-benzylsulfanylepigallocatechin **102**. Formation of the two epigallocatechin-phloroglucinol adducts **91** and **103** were not anticipated (fig. 19). The formation of the 2*S* all-*cis* product **103** was explained by invoking the dynamic equilibrium between adducts of 2,3-*trans*-3,4-*cis* and 2,3-*trans*-3,4-*trans* configuration via a B-ring quinone methide as is depicted in figure 20.

Besides the compounds indicated in figure 19, four [1]benzofuro[2,3-*c*] chromenes, identified as the methyl ether acetate derivatives **104–107**, were also isolated. The formation of the unique ring system of these compounds suggested an S_N2 attack of the phloroglucinol hydroxyl function at C(3)(C), thus inverting the stereochemistry of the C(3) oxygen function as is indicated in figure 21. Such a mechanism has a precedent in the formation of the dihydrobenzofuran ring system of gambiriin B1–B3.[62]

A primary objective of the investigation of the rearrangements of the 4-arylflavan-3-ols under relatively drastic conditions was to establish whether cleavage of the C(4)—C(10) bond (see section 6.3) would also occur in these compounds that have traditionally served as good models for the 5,7-dihydroxyprocyanidins and prodelphinidins. The isolation of the two products, also with novel ring structures, i.e., the spirobi[2,3-dihydro[1]benzofurans] **110** and **111**, indeed confirmed the lability of the C(4)—C(10) bond when the proanthocyanidins are subjected to prolonged heating in acetic acid (fig. 22). Such a bond cleavage leads to the formation of a D-ring quinone methide, which is susceptible to rearrangement via donation of electrons from the 3-OH group and a concomitant 1,2-hydride shift. Attack of the A-ring, probably directed to the side opposite the B-ring, on the protonated carbonyl intermediate then forms the dihydrofuran C-ring. Displacement of the protonated benzylic hydroxyl group by phloroglucinol (with retention of configuration of the oxygen) results in formation of the dihydrofuran E-ring favouring this stereochemistry. Considering the similarity of the C(4)—C(10) and

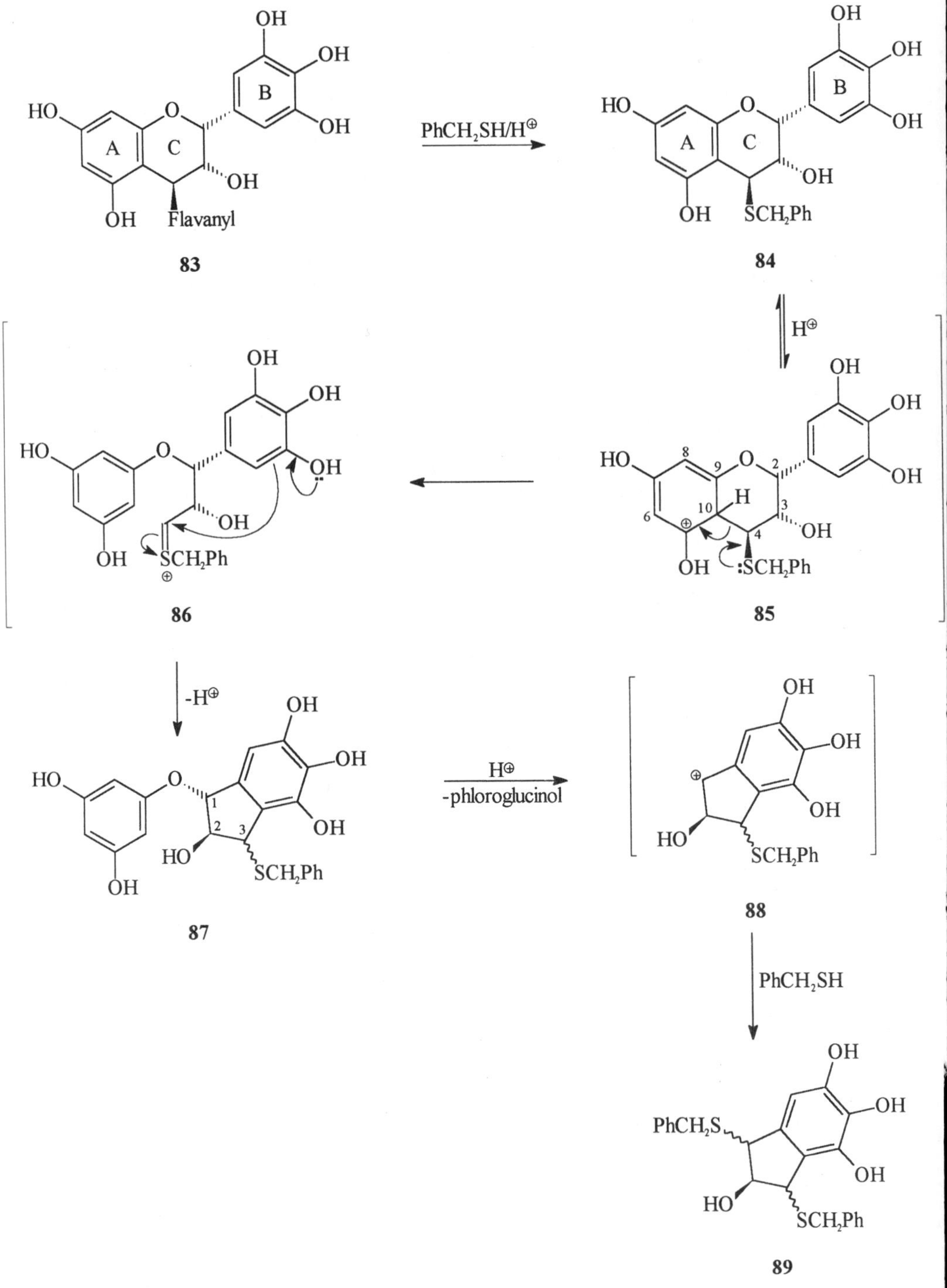

Figure 18. Proposed route to the formation of phloroglucinol and indane diastereomers **88** during thiolysis of a prodelphinidin.

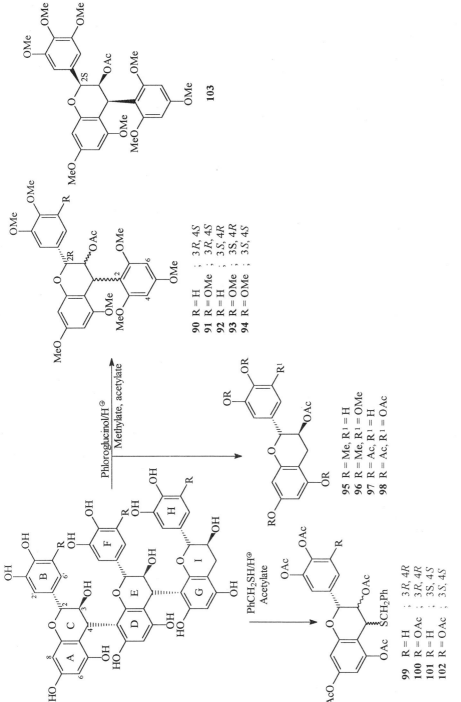

Figure 19. Products of acid-catalyzed cleavage of pecan nut pith tannins using phloroglucinol and phenylmethane thiol as capture nucleophiles.

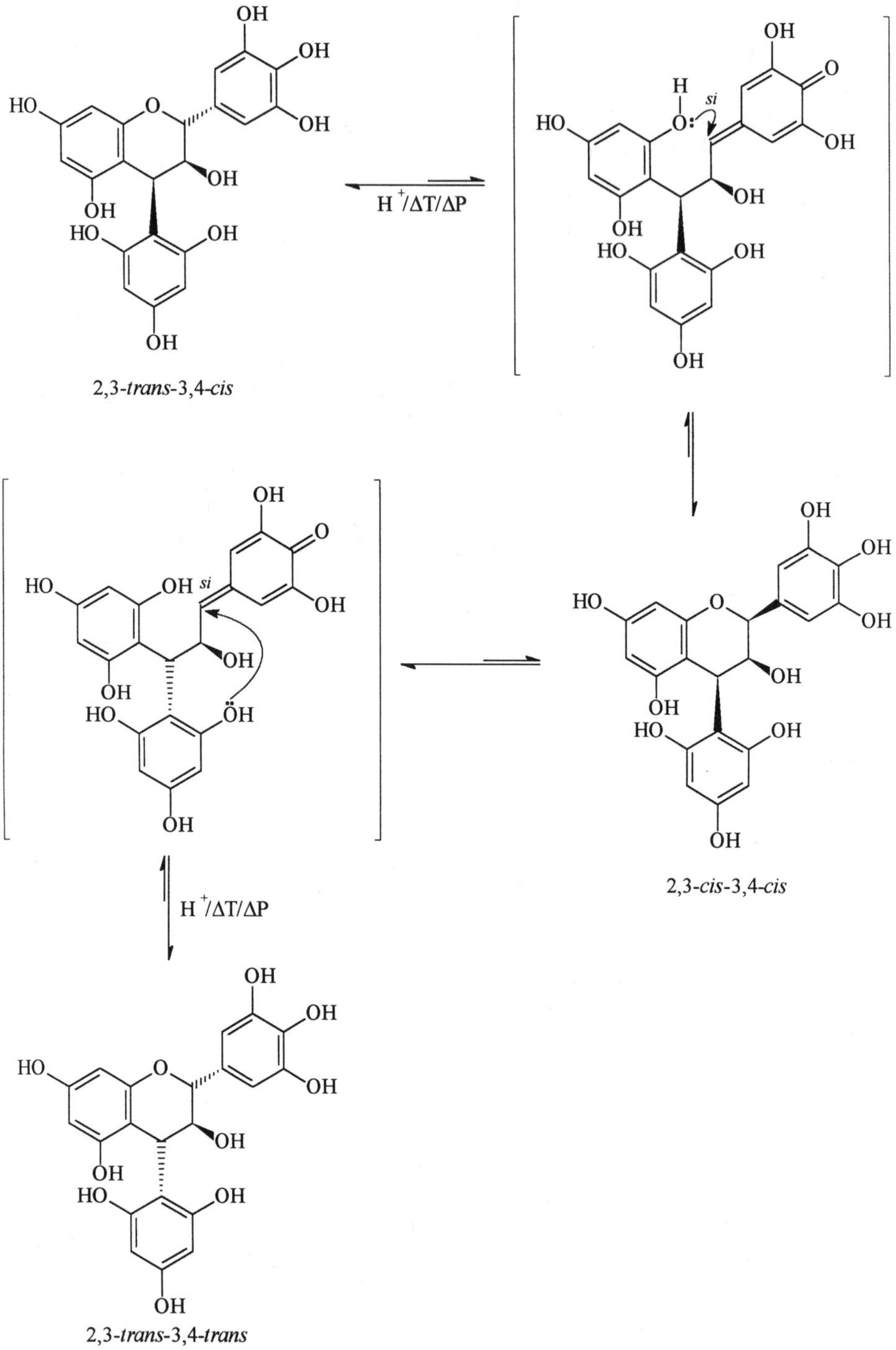

Figure 20. Proposed routes to the formation of the 2*S* all-*cis*-4-arylflavan-3-ol.

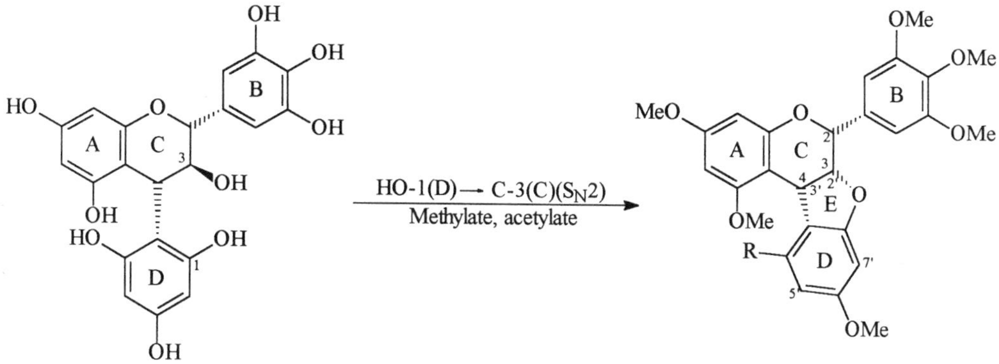

109

104 R = OMe
105 R = OAc

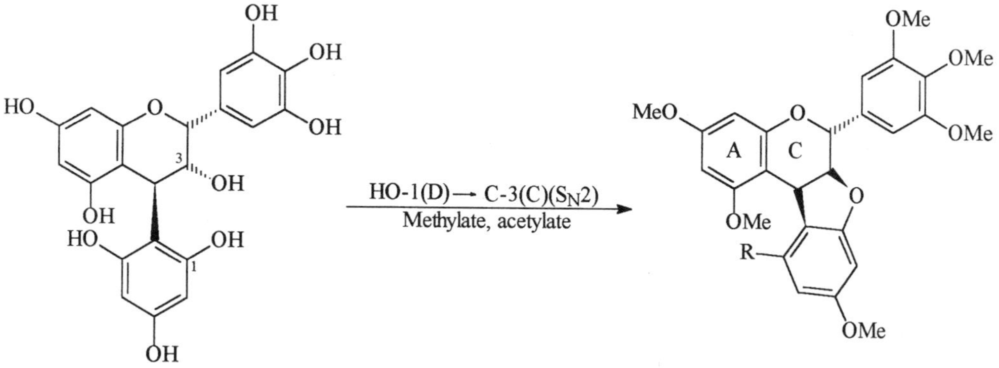

108

106 R = OMe
107 R = OAc

Figure 21. Proposed route to the formation of the [1]benzofuro[2,3-*c*] chromenes.

C(4)—C(2) (D-ring) bonds of the phloroglucinol adducts **90–94** (fig. 19), it seems logical that this reaction should readily occur in acid-catalyzed cleavage of these compounds.

7. CONCLUSIONS

This review clearly demonstrates the considerable progress that has been achieved in the quest to manipulate the carbon-carbon and carbon-oxygen bonds in the heterocyclic rings of the proanthocyanidin oligomers. The results should contribute substantially toward a better understanding of the complex principles that govern their chemistry as well as their industrial applications.

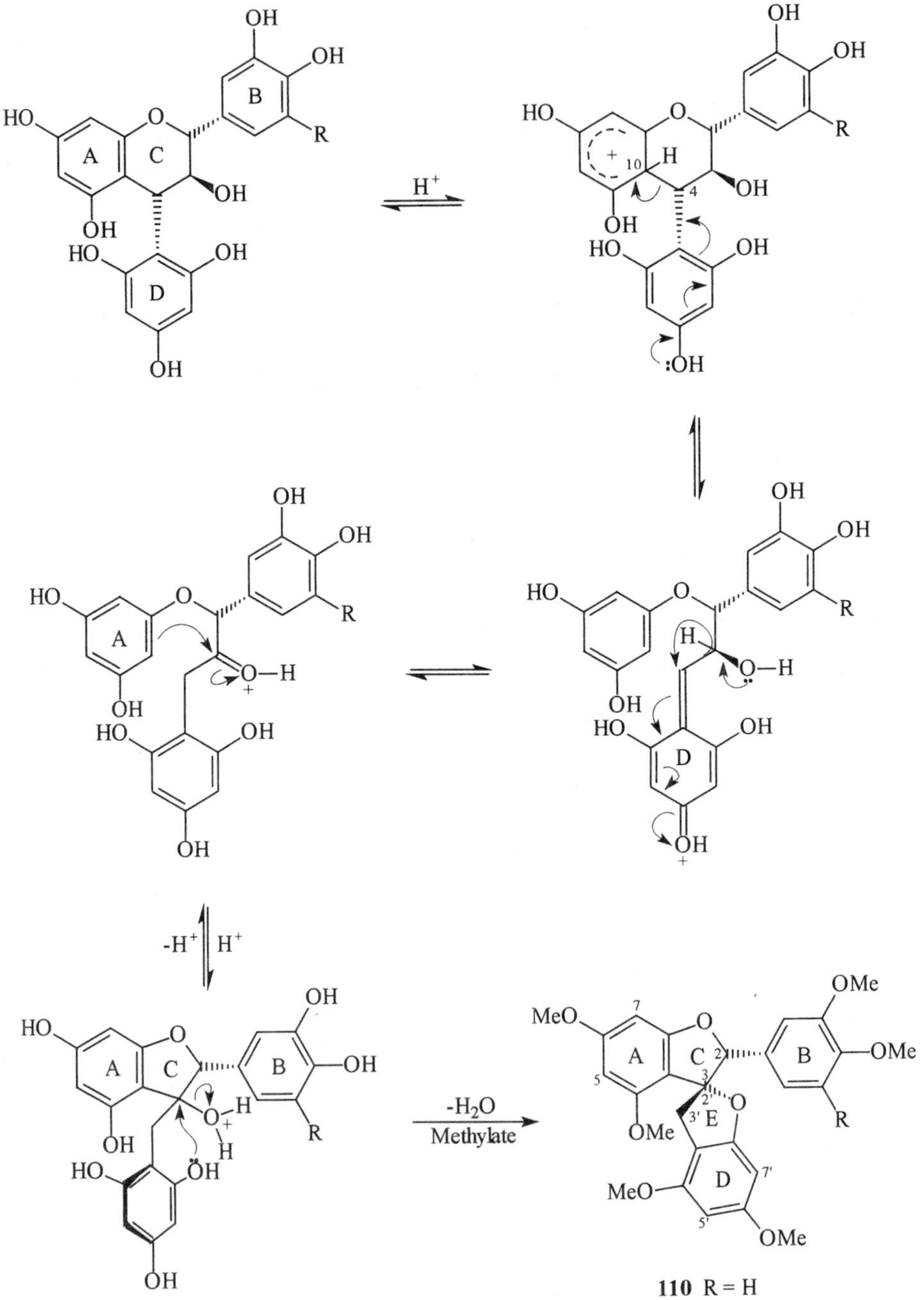

110 R = H
111 R = OMe

Figure 22. Proposed route to the formation of the spirobi[2,3]dihydro[1]benzofurans **110** and **111**.

ACKNOWLEDGMENTS

We are grateful for the enthusiastic support of our coworkers, Pieter S. van Heerden, Barend C.B. Bezuidenhoudt, Susan L. Bonnet, Catharina M. Saunders, Jan P. Steynberg, and Petrus J. Steynberg. Financial support by the Foundation for Research Development, Pretoria, the 'Sentrale Navorsingsfonds' of this University, and the Marketing Committee, Wattle Bark Industry of South Africa, Pietermaritzburg, is gratefullly acknowledged.

REFERENCES

1. Betts, M.J.; Brown, B.R.; Brown, P.E.; Pike, W.T. Degradation of condensed tannins: Structure of the tannin from common heather. *Chem. Commun.* :1110 (1967).

2. Thompson, R.S.; Jacques, D.; Haslam, E.; Tanner, R.J.N. Plant proanthocyanidins. Part 1. Introduction; the isolation, structure, and distribution in nature of plant procyanidins. *J. Chem. Soc., Perkin Trans. 1* :1387 (1972).

3. Foo, L.Y.; Porter, L.J. Prodelphinidin polymers: definition of structural units. *J. Chem. Soc., Perkin Trans. 1* :1186 (1978).

4. Drewes, S.E.; Roux, D.G.; Saayman, H.M.; Feeney, J.; Eggers, S.H. The stereochemistry of biflavanoids from black wattle bark: leucofisetinidin-(+)-catechin, leucorobinetinidin-(+)-catechin and leucorobinetinidin-(+)-gallocatechin. *Chem. Commun.* :370 (1966).

5. Drewes, S.E.; Roux, D.G.; Saayman, H.M.; Eggers, S.H.; Feeney, J. Some stereochemically identical biflavonols from the bark tannins of *Acacia mearnsii*. *J. Chem. Soc.* (C). :1302 (1967).

6. Roux, D.G.; Paulus, E. Polymeric leuco-fisetinidin tannins from the heartwood of *Acacia mearnsii*. *Biochem. J.* :320 (1962).

7. Young, D.A.; Cronjé, A.; Botes, A.L.; Ferreira, D.; Roux, D.G. Synthesis of condensed tannins. Part 14. Biflavanoid profisetinidins as synthons. The acid-induced 'phlobaphene' reaction. *J. Chem. Soc., Perkin Trans. 1.* :2521 (1985).

8. Steynberg, P.J.; Steynberg, J.P.; Bezuidenhoudt, B.C.B.; Ferreira, D. Cleavage of the interflavanyl bond in 5-deoxy (A-ring) proanthocyanidins. *J. Chem. Soc., Chem. Commun.* :31 (1994).

9. Steynberg, P.J.; Steynberg, J.P.; Bezuidenhoudt, B.C.B.; Ferreira, D. Oligomeric flavanoids. Part 19. Reductive cleavage of the interflavanyl bond in proanthocyanidins. *J. Chem. Soc., Perkin Trans. 1.* :3005 (1995).

10. Lane, C.F. Sodium cyanoborohydride, a highly selective reducing agent for organic functional groups. *Synthesis* :135 (1975).

11. McGraw, G.W.; Hemingway, R.W. Electrophilic aromatic substitution of catechin. Bromination and benzylation. *J. Chem. Soc., Perkin Trans. 1* :973 (1982).

12. Brown, A.G.; Eyton, W.B.; Holmes, A.; Ollis, W.D. Identification of the thearubigenins as polymeric proanthocyanidins. *Phytochemistry* 8:2333 (1969).

13. Beart, J.E.; Lilley, T.H.; Haslam, E. Polyphenol interactions. Part 2. Covalent binding of procyanidins to proteins during acid-catalyzed decomposition; observations on some polymeric proanthocyanidins. *J. Chem. Soc., Perkin Trans. 2* :1439 (1985).

14. Porter, L.J.; Wong, R.Y.; Benson, M.; Chan, B.G.; Vishwanadhan, V.N.; Gandour, R.D.; Mattice, W.L. Conformational analysis of flavans. ^{1}H NMR and molecular mechanical (MM2) studies of the benzopyran ring of 3',4',5,7-tetrahydroxyflavan-3-ols: the crystal and molecular structure of the procyanidin: (2R,3S,4R)-3',4',5,7-tetramethoxy-4-(2,4,6-trimethoxyphenyl)-flavan-3-ol. *J. Chem. Res. M* :830 (1986); *S* :86 (1986).

15. Steynberg, P.J.; Burger, J.F.W.; Bezuidenhoudt, B.C.B.; Steynberg, J.P.; Van Dyk, M.S.; Ferreira, D. The first natural condensed tannins with (–)-catechin 'terminal' units. *Tetrahedron Lett.* 31:2059 (1990).

16. Delle Monache, F.; Ferrari, F.; Poce-Tucci, A.; Marini-Bettollo, G.B. Catechins with (+)-epi-configuration in nature. *Phytochemistry* 11:2333 (1972).

17. Jacques, D.; Haslam, E.; Bedford, G.R.; Greatbanks, D. Plant proanthocyanidins. Part II. Proanthocyanidin-A2 and its derivatives. *J. Chem. Soc., Perkin Trans. 1* :2663 (1974).

18. Cronjé, A.; Burger, J.F.W.; Brandt, E.V.; Kolodziej, H.; Ferreira, D. Assessment of 3,4-*trans* and 3,4-*cis* relative configuration in the A-series of (4,8)-linked proanthocyanidins. *Tetrahedron Lett.* 31:3789 (1990).

19. Kolodziej, H.; Sakar, M.J.; Burger, J.F.W.; Engelshowe, R.; Ferreira, D. A-Type proantho-cyanidins from *Prunus spinosa*. *Phytochemistry* 30:2042 (1991).

20. Gonzalez, A.G.; Irizar, A.C.; Ravelo, A.G.; Fernandez, M.F. Type-A proanthocyanidins from *Prunus spinosa*. *Phytochemistry* 31:1432 (1992).

21. Balde, A.M.; De Bruyne, T.; Pieters, L.; Kolodziej, H.; Van den Berghe, D.; Claeys, M.; Vlietinck, A. Oligomeric proanthocyanidins possessing a doubly linked structure from *Pavetta owariensis*. *Phytochemistry* 38:719 (1995).

22. Balde, A.M.; De Bruyne, T.; Pieters, L.; Kolodziej, H.; Van den Berghe, D.; Claeys, M.; Vlietinck, A. Tetrameric proanthocyanidins containing a double interflavanoid (A-type) linkage from *Pavetta owariensis*. *Phytochemistry* 40:933 (1995).

23. Steynberg, P.J.; Cronjé, A.; Steynberg, J.P.; Bezuidenhoudt, B.C.B.; Brandt, E.V.; Ferreira, D. Oligomeric flavanoids. Part 25. Cleavage of the acetal functionality in A-type proantho-cyanidins. *Tetrahedron* 53:2591 (1997).

24. Karchesy, J.J.; Hemingway, R.W. Condensed tannins: $(4\beta{\approx}8; 2\beta{\approx}O{\rightarrow}7)$-linked procyanidins in *Arachis hypogea* L. *J. Agric. Food Chem.* 34:966 (1986).

25. Steynberg, J.P.; Burger, J.F.W.; Young, D.A.; Brandt, E.V.; Steenkamp, J.A.; Ferreira, D. Oligomeric flavanoids. Part 3. Structure and synthesis of phlobatannins related to (–)-fisetinidol-$(4\alpha,6)$- and $(4\alpha,8)$-(+)-catechin profisetinidins. *J. Chem. Soc., Perkin Trans. 1* :3323 (1988).

26. Steynberg, J.P.; Burger, J.F.W.; Young, D.A.; Brandt, E.V.; Steenkamp, J.A.; Ferreira, D. Oligomeric flavanoids. Part 4. Base-catalyzed conversions of (–)-fisetinidol-(+)-catechin profisetinidins with 2,3-*trans*-3,4-*cis*-flavan-3-ol constituent units. *J. Chem. Soc., Perkin Trans. 1* :3331 (1988).

27. Steynberg, J.P.; Burger, J.F.W.; Young, D.A.; Brandt, E.V.; Ferreira, D. Oligomeric flavanoids. Part 6. Evidence supporting the inversion of absolute configuration at 3-C asso-ciated with base-catalyzed A-/B-ring interchange of precursors having 2,3-*trans*-3,4-*cis*-flavan-3-ol constituent units. *Heterocycles* 28:923 (1988).

28. Steynberg, J.P.; Bezuidenhoudt, B.C.B.; Burger, J.F.W.; Young, D.A.; Ferreira, D. Oligomeric flavanoids. Part 7. Novel base-catalyzed pyran rearrangements of procyanidins. *J. Chem. Soc., Perkin Trans. 1* :2203 (1990).

29. Ferreira, D.; Steynberg, J.P.; Roux, D.G.; Brandt, E.V. Diversity of structure and function in oligomeric flavanoids. *Tetrahedron* 48:1473 (1992).

30. Trost, B.M.; Murayama, E. Dimethyl(methylthio)sulfonium fluoroborate. A chemoselective initiator for thionium ion induced cyclizations. *J. Am. Chem. Soc.* 103:6529 (1981).

31. Trost, B.M.; Sato, T. Dimethyl(methylthio)sulfonium tetrafluoroborate initiated organometallic additions to and macrocyclizations of thioketals. *J. Am. Chem. Soc.* 107:719 (1985).

32. Barrett, A.G.M.; Bezuidenhoudt, B.C.B.; Howell, A.R.; Lee, A.C.; Russell, M.A. Redox gly-cosidation *via* thionoester intermediates. *J. Org. Chem.* 54:2275 (1989).

33. Steynberg, P.J.; Nel, R.J.J.; Van Rensburg, H.; Bezuidenhoudt, B.C.B.; Ferreira, D. Oligomeric flavanoids. Part 27. Interflavanyl bond formation in procyanidins under neutral conditions. *Tetrahedron* 54:8153 (1998).

34. Hemingway, R.W.; Karchesy, J.J.; McGraw, G.W.; Wielesek, R.A. Heterogeneity of interflavanoid bond location in loblolly pine bark proanthocyanidins. *Phytochemistry* 22:275 (1983).

35. Van Zyl, P.W.; Steynberg, J.P.; Brandt, E.V.; Ferreira, D. Spectroscopic properties of free phenolic 4-arylflavan-3-ols as models for natural condensed tannins. *Magn. Reson. in Chem.* 31:1057 (1993).

36. Porter, L.J. Flavans and proanthocyanidins. *In*: Harborne, J.B. (ed.) The flavonoids advances in research since 1980, Chapman and Hall, London, p. 21 (1988).

37. Porter, L.J. Flavans and proanthocyanidins. *In*: Harborne, J.B. (ed.) The flavonoids advances in research since 1986, Chapman and Hall, London, p. 23 (1994).

38. Hemingway, R.W.; Foo, L.Y.; Porter, L.J. Linkage isomerism in trimeric and polymeric 2,3-*cis* procyanidins. *J. Chem. Soc., Perkin Trans. 1* :1209 (1982).

39. Botha, J.J.; Ferreira, D.; Roux, D.G. Synthesis of condensed tannins. Part 4. A direct biomimetic approach to [4,6]- and [4,8]-biflavanoids. *J. Chem. Soc., Perkin Trans. 1* :1235 (1981).

40. Delcour, J.A.; Ferreira, D.; Roux, D.G. Synthesis of condensed tannins. Part 9. The condensation sequence of leucocyanidin with (+)-catechin and with the resultant procyanidins. *J. Chem. Soc., Perkin Trans. 1* :1711 (1983).

41. Van Rensburg, H.; Van Heerden, P.S.; Bezuidenhoudt, B.C.B.; Ferreira, D. The first enantioselective synthesis of *trans*- and *cis*-dihydroflavonols. *J. Chem. Soc., Chem. Commun.* :2747 (1996).

42. Van Rensburg, H.; Van Heerden, P.S.; Bezuidenhoudt, B.C.B.; Ferreira, D. Stereoselective synthesis of flavonoids. Part 4. *Trans*- and *cis*-dihydroflavonols. *Tetrahedron* 53:14141 (1997).

43. Geiger, H. Biflavonoids and triflavonoids. *In*: Harborne, J.B. (ed.) The flavonoids advances in research since 1986, Chapman and Hall, London, p. 95 (1994).

44. Volsteedt, F. du R.; Roux, D.G. Zeyherin, a natural 3,8-coumaranonylflavanone from *Phyllogeiton zeyheri* Sond., *Tetrahedron Lett.* 20:1647 (1971).

45. Bekker, R.; Brandt, E.V.; Ferreira, D. The absolute configuration of biflavonoids and 2-benzyl-2-hydroxybenzofuranones. *J. Chem. Soc., Chem., Commun.* :957 (1996).

46. Bekker, R.; Brandt, E.V.; Ferreira, D. Absolute configuration of flavanone-benzofuranone-type biflavonoids and 2-benzyl-2-hydroxybenzofuranones. *J. Chem. Soc., Perkin Trans. 1* :2535 (1996).

47. Bekker, R.; Brandt, E.V.; Ferreira, D. Structure and stereochemistry of the first isoflavanone-benzofuranone biflavonoids. *Tetrahedron Lett.* 39:6407 (1998).

48. Van Rensburg, H.; Van Heerden, P.S.; Bezuidenhoudt, B.C.B.; Ferreira, D. Enantioselective synthesis of the four catechin diastereomer derivatives. *Tetrahedron Lett.* 38:3089 (1997).

49. Van Rensburg, H.; Van Heerden, P.S.; Ferreira, D. Enantioselective synthesis of flavonoids. Part 3. *Trans*- and *cis*-flavan-3-ol methyl ether acetates. *J. Chem. Soc., Perkin Trans. 1* :3415 (1997).

50. Ferreira, D.; Steynberg, J.P.; Burger, J.F.W.; Bezuidenhoudt, B.C.B. Oxidation and rearrangement reactions of condensed tannins. *In*: Hemingway, R.W.; Laks, P.E.; Branham, S.J. (eds.) Plant polyphenols: synthesis, properties, significance, Plenum Press, New York, p. 349 (1992).

51. Steynberg, J.P.; Steenkamp, J.A.; Burger, J.F.W.; Young, D.A.; Ferreira, D. Oligomeric flavanoids. Part 11. Structure and synthesis of the first phlobatannins related to (4α,6:4α,8)-bis-fisetinidol-(+)-catechin profisetinidin triflavanoids. *J. Chem. Soc., Perkin Trans. 1* :235 (1990).

52. Bonnet, S.L.; Steynberg, J.P.; Bezuidenhoudt, B.C.B.; Saunders, C.M.; Ferreira, D. Oligomeric flavanoids. Part 20. Structure and synthesis of phlobatannins related to the (4β,6:4α,8)-bis-fisetinidol-catechin profisetinidin. *Phytochemistry* 43:215 (1996).

53. Bonnet, S.L.; Steynberg, J.P.; Bezuidenhoudt, B.C.B.; Saunders, C.M.; Ferreira, D. Oligomeric flavanoids. Part 21. Structure and synthesis of phlobatannins related to the (4β,6:4β,8)-bis-fisetinidol-catechin profisetinidin triflavanoid. *Phytochemistry* 43:229 (1996).

54. Bonnet, S.L.; Steynberg, J.P.; Bezuidenhoudt, B.C.B.; Saunders, C.M.; Ferreira, D. Oligomeric flavanoids. Part 22. Structure and synthesis of phlobatannins related to the (4α,6:4β,8)-bis-fisetinidol-catechin profisetinidin triflavanoid. *Phytochemistry* 43:241 (1996).

55. Bonnet, S.L.; Steynberg, J.P.; Bezuidenhoudt, B.C.B.; Saunders, C.M.; Ferreira, D. Oligomeric flavanoids. Part 23. Structure and synthesis of phlobatannins related to bis-fisetinidol-epicatechin profisetinidin triflavanoid. *Phytochemistry* 43:253 (1996).

56. Saunders, C.M.; Bonnet, S.L.; Steynberg, J.P.; Ferreira, D. Oligomeric flavanoids. Part 24. Controlled biomimetic synthesis of profisetinidin triflavanoid related phlobatannins. *Tetrahedron* 52:6003 (1996).

57. Ohara, S.; Hemingway, R.W. Condensed tannins: the formation of a diarylpropan-2-ol-catechinic acid dimer from base-catalysed reactions of (+)-catechin. *J. Wood. Chem. Technol.* 11:195 (1991).

58. Hemingway, R.W. Reactions at the interflavanoid bond of proanthocyanidins. *In*: Hemingway, R.W.; Karchesy, J.J.; Branham, S.J. (eds.) Chemistry and significance of condensed tannins, Plenum Press, New York (1989).

59. McGraw, G.W.; Steynberg, J.P.; Hemingway, R.W. Condensed tannins: a novel rearrangement of procyanidins and prodelphinidins in thiolytic cleavage. *Tetrahedron Lett.* 34:987 (1993).

60. McGraw, G.W.; Rials, T.G.; Steynberg, J.P.; Hemingway, R.W. Chemistry of pecan tannins and analysis of cure of pecan tannin-based, cold-setting adhesives with a DMA 'micro-beam' test. *In*: Hemingway, R.W.; Laks, P.E.; Branham, S.J. (eds.) Plant polyphenols: synthesis, properties, significance, Plenum Press, New York, p. 979 (1992).

61. Steynberg, P.J.; Steynberg, J.P.; Hemingway, R.W.; Ferreira, D.; McGraw, G.W. Acid-catalyzed rearrangements of flavan-4-phloroglucinol derivatives to novel 6-hydroxyphenyl-6a,11b-dihydro-6*H*-[1]-benzofuro[2,3-*c*]chromenes   and   hydroxyphenyl-3,2′-spirobi-dihydro[1]benzofurans. *J. Chem. Soc., Perkin Trans. 1* :2395 (1997).

62. Nonaka, G-I., Nishioka, I. Novel biflavonoids, chalcan-flavan dimers from Gambir. *Chem. Pharm. Bull.* 28:3145 (1980).

NMR CHARACTERIZATION AND BIOLOGICAL EVALUATION OF PROANTHOCYANIDINS: A SYSTEMATIC APPROACH

Tess De Bruyne,[a] Luc Pieters,[a] Roger Dommisse,[b]
Herbert Kolodziej,[c] Victor Wray,[d] Dirk Vanden Berghe,[a]
and Arnold Vlietinck[a]

[a] Department of Pharmaceutical Sciences
University of Antwerp (UIA)
B-2610 Antwerp
BELGIUM

[b] Department of Chemistry
University of Antwerp (RUCA)
B-2020 Antwerp
BELGIUM

[c] Fachbereich Pharmazie
Institut für Pharmazeutische Biologie (WE02)
D-14195 Berlin
GERMANY

[d] Gesellschaft für Biotechnologische Forschung
D-38124 Braunschweig
GERMANY

1. INTRODUCTION

Condensed tannins or proanthocyanidins constitute an important group of secondary plant metabolites and are in many cases the active compounds of the medicinal plants from which they are isolated. Reports of several *in vitro* assays demonstrate potentially significant interactions with biological systems, such as antiviral, antibacterial, molluscicidal, enzyme-inhibiting, antioxidant, and radical scavenging properties.[1] Their anticipated interference with biological systems is, at least in part, due to their characteristic ability to form complexes with macromolecules, combined with their polyphenolic skeleton.[1,2]

Plant Polyphenols 2: Chemistry, Biology, Pharmacology, Ecology, Edited by
Gross et al. Kluwer Academic / Plenum Publishers, New York, 1999

NMR spectroscopy is used as a standard technique for the identification and structure elucidation of proanthocyanidins, together with parameters obtained from mass spectrometry and circular dichroism. However, structural analogy of different procyanidin isomers and the sterically hindered rotation between the composing moieties are reflected in similar and very complex NMR spectra. Rotational isomerism often results in the duplication of NMR signals. Structure elucidation is therefore usually accomplished by recording NMR spectra of suitably derivatized procyanidins at higher temperatures to avoid the problem of rotational isomerism. Since derivatization is often cumbersome, and certainly undesirable for bio-guided isolations, another approach was sought for the structure elucidation of proanthocyanidins. In this work, two-dimensional NMR techniques are systematically used to obtain complete assignments for the free phenolic forms of the different isomers of B-series procyanidins and for two doubly linked dimers. The general strategy followed in the assignment and combination of data of the different 2D spectra available has been outlined for procyanidin B_3 as an example.[3] In parallel, the properties of this B-series of procyanidins, including procyanidin B_1 or epicatechin-$(4\beta\rightarrow8)$-catechin, procyanidin B_2 or epicatechin-$(4\beta\rightarrow8)$-epicatechin, procyanidin B_3 or catechin-$(4\alpha\rightarrow8)$-catechin, procyanidin B_4 or catechin-$(4\alpha\rightarrow8)$-epicatechin, procyanidin B_5 or epicatechin-$(4\beta\rightarrow6)$-epicatechin, procyanidin B_6 or catechin-$(4\alpha\rightarrow6)$-catechin and procyanidin B_8 or catechin-$(4\alpha\rightarrow6)$-epicatechin were investigated. In addition, representatives of the doubly linked A-series (proanthocyanidin A_1 or epicatechin-$(4\beta\rightarrow8,2\beta\rightarrow O\rightarrow7)$-catechin and proanthocyanidin A_2 or epicatechin-$(4\beta\rightarrow8,2\beta\rightarrow O\rightarrow7)$-epicatechin), as well as a trimeric procyanidin C_1 or epicatechin-$(4\beta\rightarrow8)$-epicatechin-$(4\beta\rightarrow8)$-epicatechin and some related polyphenols as (−)-epicatechin, (+)-catechin, epigallocatechin, (+)-taxifolin, and gallocatechin-$(4'\rightarrow O\rightarrow7)$-epigallocatechin were evaluated in a panel of biological screening assays such as antibacterial, antiviral (including anti-HIV), complement modulation, and radical scavenging test systems.[4]

For this comparative investigation, the procyanidins B_3 and B_6 were isolated from the biomimetic condensation reaction of (+)-catechin with the $NaBH_4$-reduction product of (+)-taxifolin, whereas the procyanidins B_4 and B_8 were formed in the semisynthetic reaction between (−)-epicatechin and the reduced (+)-taxifolin.[3,5] Procyanidins B_2 and C_1 were isolated from a *Theobroma cacao* extract,[4,5] procyanidin B_5 from *Aesculus hippocastanum*,[5] procyanidin B_1[4] and the biflavonoid gallocatechin-$(4'\rightarrow O\rightarrow7)$-epigallocatechin[6] from *Bridelia ferruginea*, and the proanthocyanidins A_1 and A_2 from *Vaccinium vitis-ideae*.[5]

2. NMR INVESTIGATIONS

The interpretation and subsequent deduction of spectral parameters from NMR spectra of proanthocyanidins are hampered by several spectroscopic limitations. First of all, the rotational isomerism, which is solvent dependent, is an important complicating factor, resulting in duplication or broadening of NMR resonances, depending on the presence of either (+)-catechin or (−)-epicatechin upper units. Therefore, derivatization as peracetates or methyl ether acetates and temperature

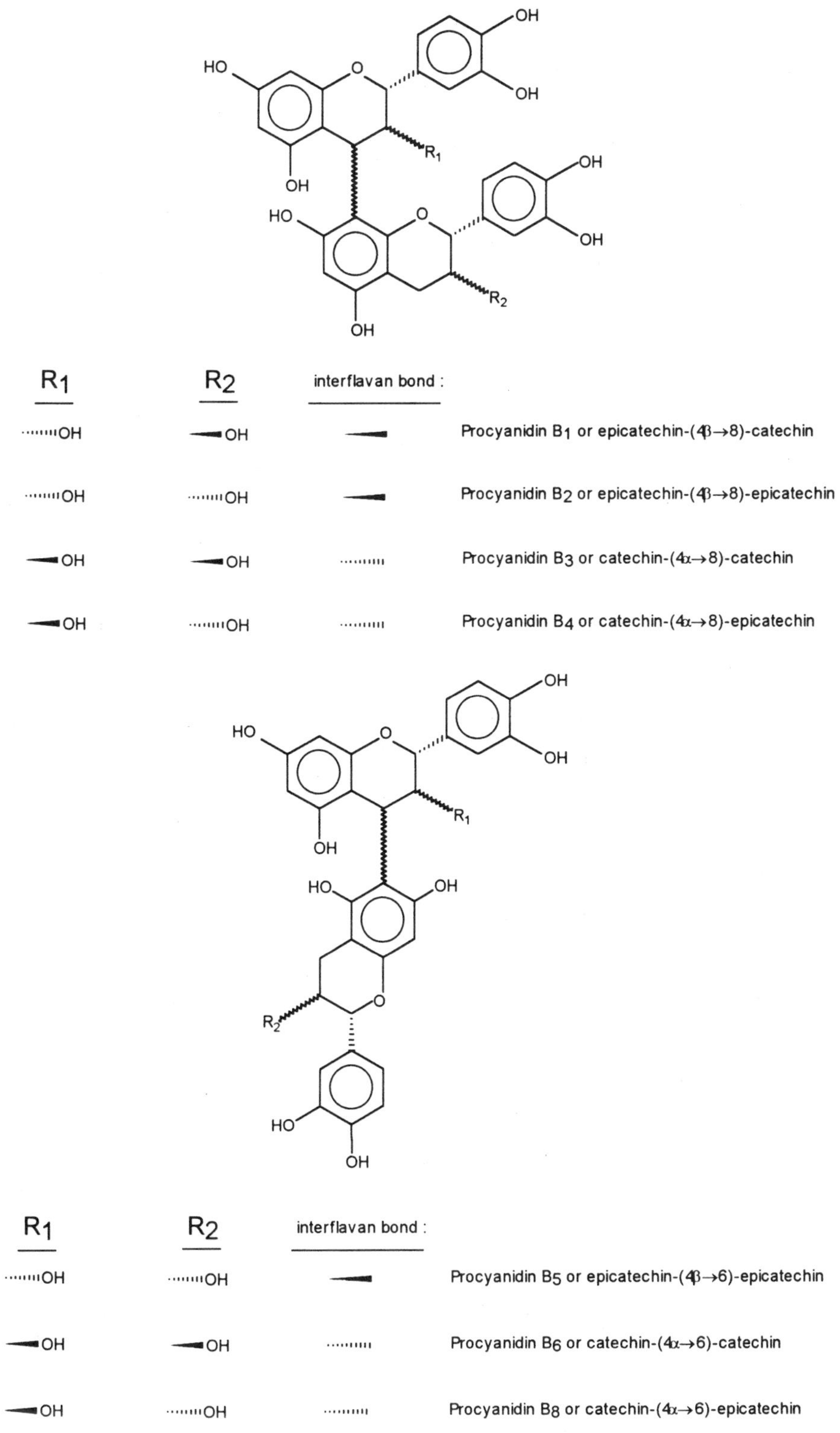

R_1	R_2	interflavan bond :	
········ΙΙOH	━━ OH	━━	Procyanidin B$_1$ or epicatechin-(4β→8)-catechin
········ΙΙOH	········ΙΙOH	━━	Procyanidin B$_2$ or epicatechin-(4β→8)-epicatechin
━━ OH	━━ OH	········ΙΙΙ	Procyanidin B$_3$ or catechin-(4α→8)-catechin
━━ OH	········ΙΙOH	········ΙΙΙ	Procyanidin B$_4$ or catechin-(4α→8)-epicatechin

R_1	R_2	interflavan bond :	
········ΙΙOH	········ΙΙOH	━━	Procyanidin B$_5$ or epicatechin-(4β→6)-epicatechin
━━ OH	━━ OH	········ΙΙΙ	Procyanidin B$_6$ or catechin-(4α→6)-catechin
━━ OH	········ΙΙOH	········ΙΙΙ	Procyanidin B$_8$ or catechin-(4α→6)-epicatechin

Figure 1. Structures of proanthocyanidin dimers and related polyphenols.

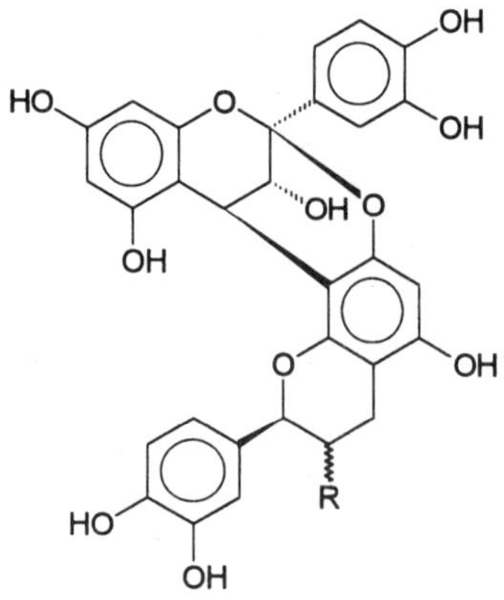

R

·······'''OH   Proanthocyanidin A$_1$ or epicatechin-(4β→8, 2β→O→7)-catechin

▬OH   Proanthocyanidin A$_2$ or epicatechin-(4β→8, 2β→O→7)-epicatechin

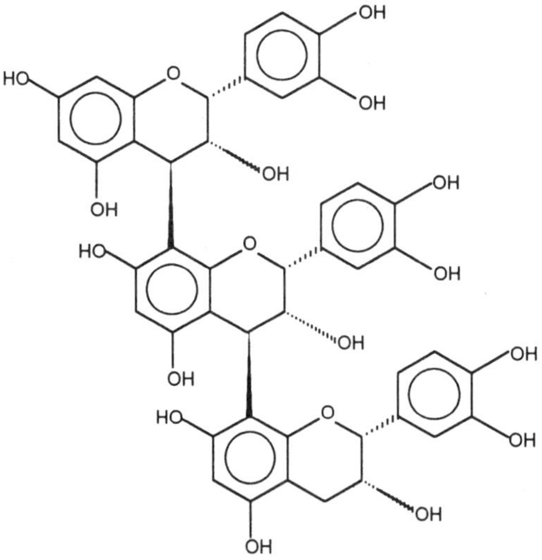

Proanthocyanidin C$_1$ or epicatechin-(4β→8)-epicatechin-(4β→8)-epicatechin

Figure 1. *Continued.*

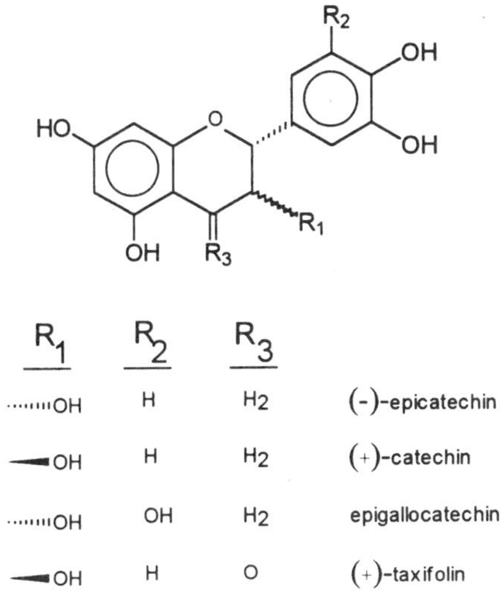

R₁	R₂	R₃	
·······ᴵᴵᴵOH	H	H2	(-)-epicatechin
◄━OH	H	H2	(+)-catechin
·······ᴵᴵᴵOH	OH	H2	epigallocatechin
◄━OH	H	O	(+)-taxifolin

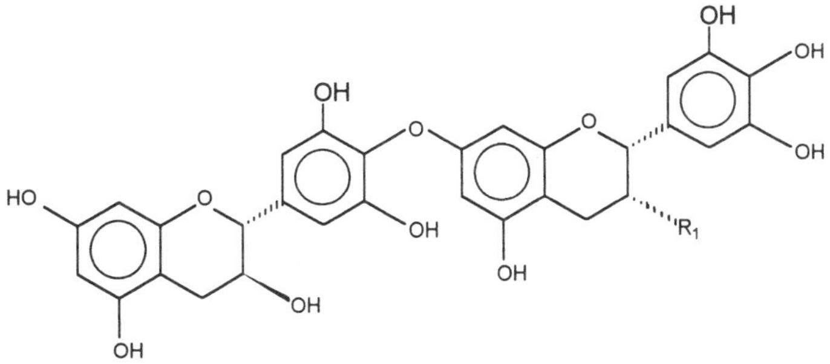

gallocatechin-(4'→O→7)-epigallocatechin

Figure 1. *Continued.*

elevation during NMR analysis was considered necessary to obtain spectra that are suitable for unambiguous interpretation.

Moreover, there is a severe ambiguity in literature about the reports of the 6/8 assignments for flavonoids, and attributions of 6/8 resonances of proanthocyanidins are often based on flavonoid data, resulting sometimes in publication of contradictory chemical shifts for those signals.[7] Therefore, H-6 and H-8 frequencies were determined unequivocally after addition of a trace of cadmium nitrate to the NMR sample in acetone-d_6. The cadmium nitrate reduces the exchange rate of the OH-protons to such an extent that phenolic protons appear as sharp lines instead of the broad OH-peaks in the low field region of the spectrum. Assignments of 6- and 8-carbons are then obtained from their long-range heteronuclear correlations in long-range HETCOR or HMBC spectra to the 5-OH and 7-OH protons.[3]

Combining two-dimensional NMR techniques like homo- and heteronuclear correlation experiments, either with direct or inverse detection (COSY, long-range COSY, HETCOR or HMQC, and long-range HETCOR or HMBC), not only afforded a tool for distinguishing the different isomers, but also provided a complete spectral assignment for the free phenolic forms of the different isomers of B-series procyanidins and for two doubly linked dimers without the need for derivatization. Assignments could be made for both rotameric isomers whenever present. A 4→8 interflavanoid bond was proven by the simultaneous HMBC-correlations from C-8a*t* to H-2*t* and H-4*u*. Evidence for a 4→6 linkage was provided by the HMBC correlations C-4a*u* to H-4*u* and C-6*t* to H-4*u*, whereas C-8*t* showed only a correlation to H-2t.[3,5] ^{1}H- and ^{13}C-NMR data for proanthocyanidin dimers are summarized in tables 1 and 2, respectively.

3. BIOLOGICAL EVALUATION

Pharmacological modulation of the complement system, one of the major effector pathways in inflammation processes, is of potential interest for the treatment and control of immune-pathological reactions linked with complement activation. The screening system used to evaluate complement modulation was based on the haemolytical properties of the complement system and measured spectrophotometrically the amount of hemoglobin released at λ 414 nm. Modulation of this effect after addition of the test compound is represented by the 50 percent inhibitory concentrations. Within one assay, each value was the mean of two independent samples; the IC_{50}'s were overall results of two assays. Briefly, in the classical pathway assay, several dilutions of the test compound were incubated with human pooled serum (HPS as a source for the complement factors) at 37 °C. Sheep red blood cells (SRBC) sensibilized with monoclonal anti-SRBC antibodies were added, and after further incubation and subsequent centrifugation, the absorption of the supernatans was read in a multiscan spectrophotometer at $\lambda = 414$ nm. In the alternative pathway assay, on the other hand, after incubation of test compound dilutions with HPS, rabbit red blood cells were added, and the assay was completed as for the classical system. The exact experimental conditions were described earlier.[8,9,10] The 50 percent inhibitory concentrations on the classical and alternative pathways of the complement system are represented in table 3.

The results of inhibition on the classical pathway showed clearly that the dimers tested were more active than the monomeric flavonoids. The trimer procyanidin C_1 had still a more pronounced activity. The flavanoid epigallocatechin was an important exception, having a IC_{50} value of 19.6 μM. Comparing the dimers with single interflavanoid linkage, differences were not very pronounced, although a tendency could be observed that (4→6)-dimers had a better inhibitory activity than their respective (4→8)-counterparts. Procyanidins B_6 ($CH_{50} = 18.5\,\mu$M) and B_8 ($CH_{50} = 19.7\,\mu$M) proved the most active. The doubly linked proanthocyanidin A_2 established a very important inhibitory action on classical complement ($IC_{50} = 11.6$

μM). The *Bridelia ferruginea* biflavanoid gallocatechin-(4′→O→7)-epigallocatechin also showed an interesting IC_{50}-value. The *ortho*-trihydroxyl group is probably responsible for the potent complement inhibition as also demonstrated for other interactions of procyanidins and related compounds on biological systems such as radical scavenging.

Table 1. ¹H-NMR assignments for free phenolic proanthocyanidin dimers

¹H Assignments	B₂	B₃	B₄	A₁	A₂
6'u	6.75 d 7.0Hz	6.51 dd 8.4 ; 2.0Hz	6.87 dd 8.4 ; 2.0Hz	7.06 dd 8.4 ; 2.0Hz	7.06 dd 8.4 ; 2.4Hz
6'u min		6.84 dd 8.4 ; 2.0Hz	6.56 dd 8.4 ; 2.0Hz		
6't	6.90 br s	6.28 dd 8.0 ; 2.0Hz	6.93 dd 8.4 ; 2.0Hz	6.88 dd 8.4 ; 1.6Hz	7.05 dd 8.4 ; 2.4Hz
6't min		6.88 dd 8.0 ; 2.0Hz	6.38 dd 8.4 ; 2.0Hz		
2'u	7.00 d 1.2Hz	6.78 d 2.0Hz	7.02 d 2.0Hz	7.18 d 2.0Hz	7.19 d 2.4Hz
2'u min		6.99 d 2.0Hz	6.78 d 2.0Hz		
5'u	6.76 d 1.8Hz	6.679 d 8.0Hz	6.80 d 8.4Hz	6.83 d 8.4Hz	6.83 d 8.4Hz
5'u min		6.79 d 8.0Hz	6.66 d 8.4Hz		
2't	6.76 d 1.8Hz	6.64 d 2.0Hz	7.20 d 2.0Hz	7.01 d 1.6Hz	7.29 d 2.4Hz
2't min		7.01 d 2.0Hz	6.71 d 2.0Hz		
5't	7.13 br s	6.683 d 8.0Hz	6.68 d 8.4Hz	6.85 d 8.4Hz	6.84 d 8.4Hz
5't min		6.76 d 8.0Hz	6.70 d 8.4Hz		
6u	6.05 br s	5.93 d 2.4Hz	5.85 d 2.4Hz	5.96 d 2.4Hz	5.97 d 2.4Hz
6u min		5.85 d 2.4Hz	5.94 d 2.4Hz		
8u	6.02 br s	5.82 d 2.4Hz	5.88 d 2.4Hz	6.06 d 2.4Hz	6.06 d 2.4Hz
8u min		5.86 d 2.4Hz	5.95 d 2.4Hz		
6t	6.00 s	6.16 s	6.04 s	6.15 s	6.13 s
6t min		6.04 s	6.20 s		
2u	5.12 br s	4.30 d 9.2Hz	4.46 d 9.6Hz	–	–
2u min		4.39 d 8.8Hz	4.40 d 9.6Hz		
2t	4.99 br s	4.59 d 7.2Hz	5.00 br s	4.71 d 8.8Hz	4.96 br s
2t min		4.74 d 7.6Hz	4.88 br s		
3u	4.04 br s	4.42 m	4.56 m	4.14 m	4.14 d 3.6Hz
3u min		4.55 m	4.37 m		
3t	4.34 br s	3.83 m	4.26 br s	4.19 m	4.33 d 3.6Hz
3t min		4.08 m	4.11 br s		
4u	4.75 br s	4.47 d 7.6Hz	4.70 d 7.6Hz	4.21 d 3.2Hz	4.33 d 3.6Hz
4u min		4.57 d 7.6Hz	4.52 d 7.6Hz		
4t α	2.77 dd 16.4 ; 4.4Hz	2.79 dd 16.4 ; 5.6Hz	2.94 dd 16.8 ; 4.4Hz	3.04 dd 16.4 ; 6.0Hz	2.95 dd 17.2 ; 4.8Hz
β	2.93 dd 16.4 ; 3.5Hz	2.54 dd 16.4 ; 8.0Hz	2.82 dd 16.8 ; 2.4Hz	2.60 dd 16.4 ; 8.8Hz	2.80 dd 17.2 ; 2.4Hz
4t min α		2.89 dd 16.0 ; 5.6Hz	2.85 dd 16.8 ; 4.4Hz		
β		2.62 dd 16.0 ; 8.0Hz	2.73 dd 16.8 ; 2.4Hz		

400 MHz, acetone-d_6 for B₃ and B₄ additionally at 600 MHz

[a...e] assignments may be mutually interchanged
u: upper unit   t: terminal unit
min: minor rotamer

Table 1 *Continued*

¹H Assignments	B₅	B₆	¹H Assignments	B₈
6'u	6,72 dd (8.0;2.0 Hz)	6.84 dd 8.1;1.9Hz	6'uv	6.82 m
6'u min		6.71 dd 8.1;1.9Hz	6'uw	6.86 m
6't	6,84 dd (8.0;2.0 Hz)	6.74 dd 8.1;1.9Hz	6'tx	6.79 dd (8.0;1.9 Hz)
6't min		6.75 dd 8.1;1.9Hz	6'ty	6.68 dd (8.3;1.9 Hz)
2'u	6,98 d (2.0 Hz)	6.99 d 1.9Hz	2',5' ty	6.77
2'u min		6.86 d 1.9Hz	5'uw	6.78^d
5'u	6,77^c d (8.0 Hz)	6.80^b d 8.1Hz		6.76^d
5'u min		6.79^a d 8.1Hz	tx	6.66
2't	7,06 d (2.0 Hz)	6.89 d 1.9Hz	2'uw	7.01 d (1.7 Hz)
2't min		6.91 d 1.9Hz		6.82 m
5't	6,78^c d (8.0 Hz)	6.78^b d 8.1Hz	tx	6.96 d (*)
5't min		6.77^a d 8.1Hz		
6/8u	6.05 br s	5.93 d 2.3Hz	6u,8u,8t	5.96
6/8u min		6.01 d 2.0Hz		5.88/5.89
8/6u	6.09 br s	6.02 d 2.3Hz		5.88/5.89
8/6u min		5.88 d 2.0Hz		5.92
8t	6.10 s	5.97 s		5.89
8t min		6.05 s		6.07
2u	4.97 br s	4.76 d 4.6Hz	2uv	4.98 d
2u min		5.08 d 5.9Hz	2uw	4.46 d
2t	4.84 br s	4.53 d 7.8Hz	2ty	4.61 d (10 Hz)
2t min		4.56 d 7.8Hz	2tx	4.82 d (7 Hz)
3u	4.08 m	4.35 m	3uv	4.33 m
3u min		4.35^e m	3uw	4.41 m
3t	4.17 m	3.99 m	3ty	4.22 m
3t min		3.99^e m	3tx	4.19 m
4u	4.66 d (1.6 Hz)	4.43 br s	4uw	4.38 m
4u min		4.58 br s	4uv	4.52 m
4t α	2.80 dd (16.8;3.2 Hz)	2.91 dd (16.2;5.4Hz)	4ty	2.90 dd (14.5;10 Hz)
β	2.65 dd (16.8; * Hz)	2.53 dd (16.2;8.4Hz)		2.50 dd (14.5;10 Hz)
4t min		2.99 m	4tx	2.79 m
		2.60 m		2.61 m

400 MHz, acetone-d_6 for B₆ and B₈ 400 and 600 MHz, methanol-d_4

$^{a...e}$ assignments may be mutually interchanged

u: upper unit   t: terminal unit   u,v: upper units of both rotamers for **(6)**

min: minor rotamer   x,y: terminal units of both rotamers for **(6)**

*: spectrum insufficiently resolved to determine small J

Table 2. ^{13}C-NMR assignments for free phenolic proanthocyanidin dimers

^{13}C Assignments	B$_2$	B$_3$	B$_4$	A$_1$	A$_2$	^{13}C Ass.	B$_5$	^{13}C Ass.	B$_6$	^{13}C Ass.	B$_8$
8au	157.62[a]	158.47	158.37	154.05	154.06[e]	8au	159.35[f]	u	158.59	5,7,8a	158.76
8au min		158.36	158.19					u	158.59		157.91
7u	158.36	157.02	157.57	157.99	158.02	7u	158.90		158.26		157.76
7u min		157.30	157.12					t min	157.65		156.38
5u	158.32[a]	156.89	157.24	156.61	156.99[e]	5u	157.94[f]		157.65		155.81
5u min		157.10	157.17					5/7 t min	157.13		155.67
8at	155.96[b]	154.69	155.26	151.24	151.77	8at	155.53		156.86		155.25
8at min		154.59	154.93					7/8a t	155.36		
7t	155.85[b]	154.90	155.38	151.93	151.99	7t	155.19	5/7 t	155.36		
7t min		155.29	155.38					7/8a t min	155.10		
5t	155.85[b]	155.41	156.09	155.66	156.17	5t	155.97	7/8a	154.78		
5t min		155.45	156.11					5/8a u min	154.78		
3'u	145.31	144.97	145.04	145.13	145.17	3'u	145.39	3',4'	145.47	3',4'	146.37
3'u min		145.43	145.17						145.55		146.14
4'u	145.11	145.40	145.76	146.27	146.14	4'u	145.28		145.82		146.05
4'u min		145.68	145.56						145.64		145.88
3't	145.25	144.85	145.49[d]	146.46	146.31	3't	145.39				145.71
3't min		145.65	145.48[d]								145.65
4't	145.41	145.19	145.39	145.96	145.71	4't	145.36				
4't min		145.55	145.11								
1'u	131.74	132.44	132.41	132.33	132.56	1'u	132.35	1'u	131.85	1'	132.42
1'u min		132.29	132.35					1'u min	131.73		132.31
1't	132.31	131.43	131.94	130.25	130.98	1't	132.19	1't	132.00		132.19
1't min		131.73	131.47					1't min	132.04		129.82
6'u	119.30	120.45	120.82	119.81	119.84	6'u	119.28	6'u	120.74	6'uv	119.37
6'u min		120.69	120.55					6'u min	120.02	6'uw	120.91
6't	119.12	119.52	119.26	120.94	120.72	6't	119.41	6't	120.06	6'tx	119.37
6't min		120.00	119.71					6't min	120.15	6'ty	120.42
2'u	115.27	116.07	116.04	115.66	115.77	2'u	115.19	2'u	115.72	2',5' ty	116.73
2'u min		115.87	116.13					2'u min	115.24	5'uw	116.05
5'u	115.56[c]	115.55	115.49	115.37	115.38	5'u	115.57[g]	5'u	115.67		115.94
5'u min		115.51	115.51					5'u min	115.37	tx	115.86
2't	115.54[c]	115.14	115.26	116.07	116.30	2't	115.35	2't	115.28		115.32
2't min		114.97	114.41					2't min	115.81		
5't	114.98	115.71	115.49	116.12	115.72	5't	115.53[g]	5't	115.85		
5't min		115.67	115.49					5't min	115.89		

[a] ... [g] assignments may be mutually interchanged

u: upper unit   t: terminal unit

min: minor rotamer

u,v: upper units of both rotamers for (6)

x,y: terminal units of both rotamers for (6)

Table 2 *Continued*

^{13}C Assignments	B$_2$	B$_3$	B$_4$	A$_1$	A$_2$	^{13}C Ass.	B$_5$	^{13}C Ass.	B$_6$	^{13}C Ass.	B$_8$
8t	107.08	107.65	107.75	106.77	107.09	6t	107.65	6t	109.57	4au,4at,6t	106.23
8t min		107.54	107.18					6t min	109.53		101.08
4au	100.63	106.46	105.91	103.9	104.11	4au	99.42	4au	102.32		100.55
4au min		106.08	106.44					4au min	102.35		
4at	100.63	102.23	99.91	103.29	102.43	4at	100.75	4at	104.40		
4at min		100.55	101.47					4at min	105.50		
6u	96.53	96.98	97.54	98.13	98.29	6/8u	96.72	6/8u	97.29	6u,8u,8t	96.45
6u min		97.22	97.54					6/8u min	97.77		95.92
8u	96.00	96.67	96.41	96.40	96.37	8/6u	96.16	8/6u	95.85		
8u min		96.12	97.02					8/6u min	95.45		
6t	97.17	96.09	97.90	96.78	96.66	8t	96.47	8t	96.54		
6t min		97.47	96.83					8t min	96.16		
2u	76.94	83.54	83.49	100.07	100.02	2u	77.12	2u	82.47	2uv	79.75
2u min		83.68	83.58					2u min	82.66	2uw	84.35
2t	79.26	82.07	80.08	84.71	81.77	2t	79.33	2t	84.05	2ty	77.92
2t min		82.89	79.32					2t min	83.90	2tx	79.60
3u	72.90	73.23	73.38	67.51	67.75	3u	72.18	3u	74.55	3uv	74.05
3u min		73.02	73.28					3u min	74.55	3uw	73.92
3t	66.40	68.31	66.89	67.64	66.13	3t	66.89	3t	68.33	3ty	67.50
3t min		68.10	67.09					3t min	68.26	3tx	67.50
4u	36.88	38.09	38.31	28.94	28.96	4u	37.33	4u	39.09	4uw	39.3
4u min		37.97	38.58					4u min	38.27	4uv	38.5
4t	29.03	28.82	29.80	29.60	29.99	4t	29.59	4t	28.74	4ty	+
4t min		28.64	29.42					4t min	28.98	4tx	+

[a...g] assignments may be mutually interchanged

+ : masked by residual acetone

u: upper unit   t: terminal unit

min: minor rotamer

u,v: upper units of both rotamers for (6)

x,y: terminal units of both rotamers for (6)

An increasing number of processes has been associated with enhanced free radical production including inflammation, radiation damage, anticancer reaction, immunity, arteriosclerosis, myocardial ischaemia, and aging. The superoxide scavenging activities were measured by the nitrite method after generation of the superoxide anions by a hypoxanthine-xanthine oxidase system.[11] Briefly, test solutions were prepared by mixing hypoxanthine (0.2 mM), hydroxylamine (0.2 mM), EDTA (0.1 mM) with the test compound (or without test compound as controls). The reaction was started by the addition of xanthine oxidase (final concentration: 2.5 mU/mL) in a buffered solution (final concentrations: 20.8 mM KH_2PO_4 and 15.6 mM $Na_2B_4O_7$). After incubation (30 min, 37 °C), the coloring reagent (final concentration: 300 μg/mL sulfanilic acid, 5 μg/mL N-1-naphtylethylenediamine, and 16.7

Table 3. Superoxide radical scavenging, complement modulation, and anti-HIV
activities for dimeric proanthocyanidins and related polyphenols

	Superoxide scavenging activity IC_{50} (μM) ± SD	Complement classical pathway IC_{50} (μM)	Complement alternative pathway IC_{50} (μM)	Anti-HIV		
				IC_{50} (μg/mL)	CC_{50} (μg/mL)	SI
(-)-epicatechin	18.29 ± 0.69	655.5	n.a.	34.08	136.18	4.0
(+)-catechin	13.06 ± 0.09	647.2	n.a.	>86.81	86.81	<1
epigallocatechin	3.40 ± 0.26	19.6	179.4	>16.09	16.09	<1
(+)-taxifolin	24.16 ± 1.51	850.3	n.a.	>25.44	25.44	<1
procyanidin B_1	–	31.3	72.0	–	–	–
procyanidin B_2	10.52 ± 0.04	58.0	n.a.	17.18	135.06	8.0
procyanidin B_3	41.35 ± 5.44	37.7	n.a.	27.05	117.79	5.0
procyanidin B_4	29.97 ± 1.96	45.5	n.a.	>0.31	0.31	<1
procyanidin B_5	11.36 ± 0.89	51.7	n.a.	8.32	53.49	8.7
procyanidin B_6	27.34 ± 1.81	18.5	n.a.	29.86	141.18	5.0
procyanidin B_8	32.95 ± 2.94	19.7	83.7	19.57	126.58	8.0
procyanidin C_1	10.69 ± 0.53	6.0	85.5	4.26	16.55	5.0
procyanidin A_1	14.70 ± 0.99	57.1	105.0	14.05	144.26	10.5
procyanidin A_2	17.83 ± 0.67	11.6	112.8	5.81	137.63	24.0
gallocatechin-(4'→O→7)- epigallocatechin	2.01 ± 0.03	14.6	86.0	–	–	–

- : not tested   n.a.: not active   SI : selectivity index SI=CC_{50}/IC_{50}

percent acetic acid) was added. The mixtures were allowed to stand at room temperature for 30 minutes, and the absorbances were measured at 550 nm. The scavenging activities were measured with different concentrations of the test compounds, and the 50 percent inhibitory concentrations (IC_{50}) were calculated by regression analysis. The IC_{50} values are listed in table 3.

Comparing the results of the flavonoids (–)-epicatechin, (+)-catechin, epigallocatechin, and (+)-taxifolin, conclusions similar to those found in a previous study on flavonoids were obtained.[12] Epigallocatechin had a very interesting activity. Three B-ring hydroxyl groups, a hydroxyl at C-3, a saturated C2–C3, bond and the absence of C-4 carbonyls contributed to a good superoxide scavenging activity. The best results for dimeric proanthocyanidins were obtained for the procyanidins B_2 and B_5. The procyanidins B_3, B_4, B_6, and B_8 were about three times less active. For the dimeric procyanidins with a single linkage, those with two epicatechin units (B_2 and B_5) give thus better antioxidant activity than products containing catechin. The type of linkage had no consistent significant influence. Procyanidin C_1 produced no higher scavenger properties than its dimeric analogue B_2. The biflavanoid gallocatechin-(4'→O→7)-epigallocatechin displayed a very prominent activity. The doubly linked proanthocyanidins A_1 and A_2 showed interesting activities, although they were not exactly as potent as B_2 or B_5. Superoxide radical scavenging properties observed were in line with the observations of Ricardo da Silva (1991), who stated for the first time that it was not the degree of polymerization as such, but the number of hydroxyl groups (preferably *ortho*-trihydroxyl) (also increasing with the degree of polymerization) that was important for the activity.[13] A significant $4\rightarrow6/4\rightarrow8$ difference was, however, not apparent from our results.

Condensed tannins exhibit antiviral activities as shown by different research groups.[14] However, no structure-activity differences within an isomeric group of proanthocyanidins were investigated before. The *in vitro* antiviral screening method estimated the inhibition of the cytopathic effects (CPE) on a host cell monolayer (VERO cells) infected with *Herpes simplex* virus after incubation in the presence of the test compound.[15] Antiviral activity is expressed as a reduction factor (RF), being the ratio of the viral titres in the virus control and in the presence of the maximal nontoxic dose (MNTD) of test substance. A RF of 10^3 to 10^4 and more indicates a pronounced antiviral activity and is suitable as a selection criterion for further investigation. In the virucidal test system, the test compounds are pre-incubated with the virus suspension; after which the residual infectious virus is titrated. Results of the antiviral and virucidal assays on *Herpes simplex* virus are presented in table 4.

Table 4. Antiviral and virucidal activities on *Herpes simplex* virus for dimeric proanthocyanidins and related polyphenols

	Antiviral		Virucidal	
	HSV		HSV	
	conc (μg/ml)	RF	conc (μg/ml)	RF
(+)-taxifolin	100-1	1	-	
epigallocatechin	100	10	200-10	1
	50	1		
procyanidin B$_2$	100	10	200-10	1
	50	1		
procyanidin B$_3$	100-1	1	200-10	1
procyanidin B$_4$	100-1	1		
procyanidin B$_5$	100	10^3	200-10	1
	50	1		
procyanidin B$_6$	100	10^2	-	
	50	1		
procyanidin B$_8$	100	10	200-10	1
	50	1		
procyanidin C$_1$	100	10^3	200-10	1
	50	1		
procyanidin A$_1$	100	10^4	200-10	1
	50	10		
	25	1		
procyanidin A$_2$	100-10	T	-	
	1	1		

- : actvity not tested   RF: reduction factor   T: cytotoxic

The antiviral assay demonstrated antiherpetic activity for the majority of the compounds tested. Antiviral activity was most prominent for the doubly linked proanthocyanidins A_1 and A_2, for the trimeric procyanidin C_1, and for the $(4{\to}6)$-coupled dimers procyanidins B_5 and B_6. Comparing the $4{\to}8/4{\to}6$ pairs procyanidin B_2/B_5 and B_3/B_6, a gain of 10^2 in reduction factor (at $100\,\mu g/mL$) is noted for the $(4{\to}6)$-coupled dimers, with the two epicatechin B_5 dimer as most active isomer. In the pair procyanidin B_4/B_8, only a tenfold difference could be observed. Further elongation of procyanidin B_2 to procyanidin C_1 increased the reduction factor 10^2 at $100\,\mu g/mL$. Epigallocatechin had a weak antiherpetic activity, though still better than some dimers (B_3,B_4), confirming that *ortho*-trihydroxylgroups potentiated activity. Other potentiating factors were polymerization and mode of linkage, double linkage, or $(4{\to}6)$-linkage being the better ones. In the extracellular (virucidal) assay, no activities could be observed at concentrations up to $200\,\mu g/mL$ against *Herpes simplex* (table 4).

A microtray assay evaluated the protection of a drug against the cytopathic effects of the HIV-virus.[16,17] This colorimetric assay monitors the exclusive ability of metabolically active cells to take up and reduce 3-(4,5-dimethylthiazol-2-yl)-2,5-diphenyl-tetrazolium bromide (MTT) to a blue formazan product, which is measured spectrophotometrically at $540\,nm$. Determination of cell viability has the advantage that both antiviral and cytotoxic activities can be determined in parallel. Comparison of effects on HIV-infected and mock-infected MT-4 cells allows calculation of a selectivity index, which is listed in table 3.

The most interesting activity was seen for the doubly linked proanthocyanidin A_2 with a selectivity index of 24. For its analogue with a (+)-catechin terminal unit proanthocyanidin A_1, the SI was more than halved, but was still larger than for the B-type procyanidins with single linkages. Tested flavonoid-monomers showed no mentionable activity except for (−)-epicatechin with a selectivity index of 4. Looking at the procyanidin dimers with a single linkage, a general tendency of higher anti-HIV activity for (−)-epicatechin containing products became apparent. No significant difference was observed for the different types of interflavanoid linkage. Procyanidin B_4, however, was an exception to both considerations. Further elongation to form procyanidin C_1 did not increase the SI more. Although the IC_{50} value was rather small, also the 50 percent cytotoxic concentration was very low. Since with a higher degree of polymerization both cytotoxicity and HIV-inhibition appeared to increase, there might be an optimum polymerization extent with maximum selectivity index.

These compounds were also tested for antibacterial activity. In the dilution methods for the estimation of activity, samples being tested were mixed with a suitable medium, which had previously been inoculated with the test organism. After incubation, growth of the microorganism was determined by direct visual comparison of the test culture with a control culture without the test compound. A series of dilutions of the sample in culture medium was made. After incubation, the endpoint of the test was taken at the highest dilution that just prevented perceptible growth of the test organism (minimum inhibitory concentration: MIC-value). The minimum bactericidal concentration (MBC) was determined by plating out samples of completely inhibited dilution cultures on solid or liquid media containing no antibiotic.[15] The microbial battery included representatives of most human pathogenic bacteria as well as those causing food infections and intoxications. The test was limited to microorganisms only requiring a standard growth medium. *Neisseria gonorrhoea* and *Campylobacter fetus* were thus excluded from the procedure applied.

None of the compounds tested was active against *Escherichia coli, Pseudomonas aeruginosa, Salmonella paratyphi, Enterobacter cloaca, Mycobacterium fortuitum, Staphylococcus aureus,* or *Candida albicans* (MIC and MBC > 100 μg/mL). The results revealed a moderate antibacterial activity for certain compounds on *Streptococcus pyogenes, Bacillus cereus, Klebsiella pneumoniae,* and *Proteus vulgaris* at concentrations ≤100 μg/mL (table 5). Since active concentrations were very close to the maximum concentrations tested, absence of activity in higher concentrations against other representatives of the same groups could not be excluded. Whenever growth-inhibition was observed, it was due to bactericidal action at the same or double concentration. For activities of proanthocyanidin A_1 against *Proteus vulgaris,* or of procyanidin A_2 against *Bacillus cereus,* the maximum concentration tested of 100 μg/mL failed to confirm the bactericidal mechanism in those cases. *Streptococcus pyogenes* proved most susceptible to several of the tested compounds. Several dimers and the one tested trimer were active. Against *Streptococcus pyogenes,* no monomers showed any activity, although the ortho-trihydroxyl compound epigallocatechin did inhibit the gram-negatives *Proteus vulgaris* and *Klebsiella pneumoniae.* Gram-positive inhibition worked a little better for the doubly linked proanthocyanidin A_2 and for procyanidin B_4 (catechin-$(4\alpha\rightarrow8)$-epicatechin). The $(4\rightarrow6)$-linked counterpart of B_4, procyanidin B_8 was significantly less active. However, this difference was not observed in the pairs B_3/B_6 or B_2/B_5. Procyanidin B_2 showed exactly the same MIC- and MBC-values as did procyanidin B_5. Addition of another epicatechin unit to form procyanidin C_1 did not alter the activity.

4. CONCLUSIONS

A series of dimeric procyanidins, obtained either by isolation or by a biomimetic condensation reaction, were used as model compounds for the more complex oligomers in an comparative investigation for their spectroscopic (NMR) characteristics and for various biological activities in order to obtain structure-activity related indications. Two-dimensional NMR techniques were used systematically to obtain complete assignments for the free phenolic forms of the different isomers of B-series procyanidins and for two doubly linked dimers. The combination of several two-dimensional NMR techniques provides a powerful instrument for distinction between dimeric procyanidin isomers, and, moreover, affords complete assignments, even for both rotameric forms when present, without the need for derivatization.

Antiviral (*Herpes simplex* virus and human immunodeficiency virus), antibacterial, superoxide radical scavenging and complement modulating properties were determined for proanthocyanidin dimers and related polyphenols. Although

Table 5. Antibacterial activity of dimeric proanthocyanidins and related polyphenols (MIC- and MBC-values in μg/mL)

		Bacillus cereus	Klebsiella pneumoniae	Proteus vulgaris	Streptococcus pyogene
(+)-taxifolin	MIC	> 100	> 100	> 100	> 100
	MBC	> 100	> 100	> 100	> 100
epigallocatechin	MIC	> 100	100	50	> 100
	MBC	> 100	100	100	> 100
procyanidin B$_2$	MIC	> 100	> 100	> 100	100
	MBC	> 100	> 100	> 100	100
procyanidin B$_3$	MIC	> 100	> 100	> 100	> 100
	MBC	> 100	> 100	> 100	> 100
procyanidin B$_4$	MIC	> 100	> 100	> 100	25
	MBC	> 100	> 100	> 100	50
procyanidin B$_5$	MIC	> 100	> 100	> 100	100
	MBC	> 100	> 100	> 100	100
procyanidin B$_6$	MIC	> 100	> 100	> 100	100
	MBC	> 100	> 100	> 100	100
procyanidin B$_8$	MIC	> 100	> 100	> 100	> 100
	MBC	> 100	> 100	> 100	> 100
procyanidin C$_1$	MIC	> 100	> 100	> 100	100
	MBC	> 100	> 100	> 100	100
procyanidin A$_1$	MIC	> 100	> 100	50	> 100
	MBC	> 100	> 100	> 100	> 100
procyanidin A$_2$	MIC	100	> 100	> 100	50
	MBC	> 100	> 100	> 100	50

> 100: not active up to 100 μg/mL

vegetable tannins all have intrinsically the same chemical tools for interference with biological systems, the effective manifestation of a physiological effect is strongly dependent on the chemical surroundings and conformation of polyphenol and target. In general, tendencies toward more pronounced activities were seen with epicatechin-containing dimers for anti-HSV, anti-HIV, and radical scavenging effects, whereas the presence of *ortho*-trihydroxyl groups in the B-ring was important in anti-HSV and radical scavenging effects and complement classical pathway inhibition. Double interflavan linkages gave rise to interesting antiviral effects (HSV and HIV) and complement inhibition. Influences of the degree of poly-merization or of the type of interflavan linkage (4→6 or 4→8) differed in the different biological systems evaluated. Only minor or moderate antibacterial effects were observed. Variabilities observed can thus be approached as some selectivity rather than specificity toward a specific target system.

ACKNOWLEDGMENTS

Our gratitude is directed toward Professor E. de Clercq and Dr. M. Witvrouw (REGA Institute, Catholic University of Louvain (KUL), Louvain, Belgium) for the determination of the anti-HIV properties. T.D.B. is a postdoctoral fellow of the Fund for Scientific Research (FWO; Flanders—Belgium). This work was financially supported by the Flemish Government (Concerted Action no. 92/94-09) and the Fund for Scientific Research (FWO, Flanders—Belgium; grant no. G.0119.96).

REFERENCES

1. De Bruyne, T.; Pieters, L.; Deelstra, H.; Vlietinck, A. Condensed vegetable tannins: biodiversity in structure and biological activities. *Biochem. Sys. Ecol.* 27:445–459 (1999).

2. Haslam, E. Natural polyphenols (vegetable tannins) as drugs: possible modes of action *J. Nat. Prod.*, 59:205–215 (1996).

3. De Bruyne, T.; Pieters, L.; Dommisse, R.; Kolodziej, H.; Wray, V.; Domke, T.; Vlietinck, A. Unambiguous assignments for free dimeric proanthocyanidin phenols from 2D NMR, *Phytochemistry*, 43(1):265–272 (1996).

4. De Bruyne, T.; Pieters, L.; Witvrouw, M.; De Clercq, E.; Vanden Berghe, D.; Vlietinck, A. Biological evaluation of proanthocyanidin dimers and related polyphenols *J. Nat. Prod.* 62(7):954–958 (1999).

5. De Bruyne, T.; Pieters, L.; Dommisse, R.; Kolodziej, H.; Wray, V.; Vlietinck, A. Comparative investigation of proanthocyanidins. Part I. NMR characterization of proanthocyanidin dimers. *J. Nat. Prod.* (submitted).

6. De Bruyne, T.; Cimanga, K.; Pieters, L.; Claeys, M.; Dommisse, R.; Vlietinck, A. Gallocatechin-(4′→O→7)-epigallocatechin, a new biflavonoid isolated from *Bridelia ferruginea Nat. Prod. Lett.* 11:47–52 (1997).

7. Van Loo, P.; De Bruyne, A.; Budesinsky, M. Reinvestigation of the structural assignment of signals in the ^{1}H and ^{13}C spectra of the flavone apigenin. *Magn. Res. Chem.* 24:879 (1986)

8. De Bruyne, T.; Van Driessen, G.; Van Poel, B.; Laekeman, G.; Vlietinck, A. Inhibiting properties of (+)-taxifolin on complement classical pathway. *Life Sci. Adv.* 11:63–66 (1992).

9. Lasure, A.; Van Poel, B.; Cimanga, K.; Pieters, L.; Vanden Berghe, D.; Vlietinck, A. Modulation of the complement system by flavonoids. *Pharm. Pharmacol. Lett.* 4:32–35 (1994).

10. Huang, Y.; De Bruyne, T.; Apers, S.; Ma, Y.; Claeys, M.; Vanden Berghe, D.; Pieters, L.; Vlietinck, A. Complement-inhibiting cucurbitacin glycosides from *Picria fel-terrae*. *J. Nat. Prod.* 61:757–761 (1998).

11. Hu, J.; Calomme, M.; Lasure, A.; De Bruyne, T.; Pieters, L.; Vlietinck, A.; Vanden Berghe, D. Structure-activity relationship of flavonoids with superoxide scavenging activity. *Biol. Trace Elem. Res.* 47:327–331 (1995).

12. Cos, P.; Ying, L.; Calomme, M.; Hu, J.; Cimanga, K.; Van Poel, B.; Pieters, L.; Vlietinck, A.; Vanden Berghe, D. Structure-activity relationship and classification of flavonoids as

inhibitors of xanthine oxidase and superoxide scavengers. *J. Nat. Prod.* 61(1):71–76 (1998).

13. Ricardo da Silva, J.; Darmon, N.; Fernandez, Y.; Mitjavila, S. Oxygen free radical scavenger capacity in aqueous models of different procyanidins from grape seeds. *J. Agric. Food Chem.* 39:1549–1552 (1991).

14. Baldé, A.M.; Van Hoof, L.; Pieters, L.; Vanden Berghe, D.; Vlietinck, A. Plant antiviral agents. VII. Antiviral and antibacterial procyanidins from the bark of *Pavetta owariensis.* *Phytotherapy Res.* 4(5):182–188 (1990).

15. Vanden Berghe, D.; Vlietinck, A. Screening methods for antibacterial and antiviral agents from higher plants. *In:* Dey, P.M.; Harborne, J.B. (eds.) Methods in plant biochemistry, Vol. 6: Assays for bioactivity, Academic Press, London, pp. 47–69 (1991).

16. Pauwels, R.; Balzarini, J.; Baba, M.; Snoeck, R.; Schols, D.; Herdewijn, P.; Desmyter, J.; De Clercq, E. Rapid and automated tetrazolium-based colorimetric assay for the detection of anti-HIV compounds. *J. Virol. Meth.* 16:171–185 (1987).

17. Pauwels, R. Development of new agents against the human immunodeficiency virus (HIV). Evaluation methods, structure-activity relationships and mechanism of action. Ph. D. Thesis, Louvain (1990).

METHODS FOR DETERMINING THE DEGREE OF POLYMERIZATION OF CONDENSED TANNINS: A NEW ^{1}H-NMR PROCEDURE APPLIED TO CIDER APPLE PROCYANIDINS

Sylvain Guyot,[a] Christine Le Guernevé,[b] Nathalie Marnet,[a] and Jean-François Drilleau[a]

[a] Laboratoire de Recherches Cidricoles, Biotransformation des Fruits et Légumes
INRA, F-35650 Le Rheu
FRANCE

[b] Unité de Recherches des Biopolymères et des Arômes
IPV, INRA
F-34060 Montpellier
FRANCE

1. INTRODUCTION

The degree of polymerization (DP) that corresponds to the number of flavan-3-ol units is one of the most important features that characterize condensed tannins (proanthocyanidins) because of its direct link to the various properties of this kind of phenolic compound. In their definition of vegetable tanning substances, Bate-Smith and Swain[1] referred to the molecular weight that must range from 500 to 3,000. By considering their extraction, their biological activities, their sensory effects, condensed tannins often behave according to their molecular weight, although this single feature is quite insufficient to show evidence of all their properties. On the whole, the molecular weight of condensed tannins is related to their ability to associate with proteins and polysaccharides; this "tanning capacity" varies in an increasing order with the DP.[2–6] This property is also related to other applications. For instance, the work of Lea and Arnold[7] pointed out the influence of the DP of procyanidins in relation to bitterness and astringency of cider. Proanthocyanidins are also partly involved in haze formation in beers: the capacity of beer tannins to precipitate proteins increases with the DPn.[4,5,8] Many studies dealing with the biological activities of proanthocyanidins also show that antioxi-

Plant Polyphenols 2: Chemistry, Biology, Pharmacology, Ecology, Edited by Gross et al. Kluwer Academic / Plenum Publishers, New York, 1999

dant,[9] antifungal,[10] anti-enzymic,[11] antisecretory,[12] or antitumor[13] activities may correlate with the DP.

The main difficulty in the studies dealing with condensed tannins is probably to obtain them in an individual molecular form. In most cases, the complete purification of proanthocyanidins having a DP above five is almost impossible. Therefore, only mixtures containing more or less polymerized structures are considered for the studies of their structures and their properties. These mixtures may be characterized by their average degree of polymerization (DPn) or their weight average (Mw) and number average (Mn) molecular weight the ratio of which (Mw/Mn) can provide an index of polydispersity.

The first part of this chapter is a review of the various methods used to characterize condensed tannins on a molecular weight basis. Following that discussion, our recent results on the development of a new ^{1}H-NMR method for the determination of the DPn of homogeneous cider apple procyanidins are presented.

2. ESTIMATION OF NUMBER AVERAGE MOLECULAR WEIGHT (Mn) AND AVERAGE DEGREE OF POLYMERIZATION (DPn) OF CONDENSED TANNINS

Several methods have been used to determinate the DPn of the condensed tannins. They can be divided into two categories: chemical methods mainly based on depolymerization and physical methods that are non-degradative.

Chemical methods based on color measurement were the first methods used in estimation of the molecular weight of these compounds. In the effort to obtain information on the polymerization state, Goldstein and Swain[14] have compared several ripe and unripe fruit tannin extracts on the basis of three chemical assays: acid degradation (LA), vanillin reaction (V), and Folin reaction (FD). The acid degradation of condensed tannins leads to the rupture of the interflavanic linkage and yields to the formation of anthocyanidin structures corresponding to the extension units of the initial polymers (fig. 1). Vanillin (V) apparently only reacts with one end unit of the polymers when these reactions are performed in glacial acetic acid,[15] and Folin-Denis gives a positive reaction with all phenolic groups. V/LA and V/FD ratio can be taken, therefore, as a measure of polymerization.[14,16] In most cases, V/LA was preferred because both reagents react only with flavanols, whereas FD reagent reacts also with other simple phenolic compounds that may be present in fruit extracts. Imprecision of V/LA method corresponds to the fact that the acid degradation does not take into account the lower units of the polymers.[17] Moreover, secondary reactions such as phlobaphene formations may affect the degradation yields in acid media.[18] In other studies, 4-dimethylaminocinnamaldehyde was preferred to vanillin because of its better sensitivity.[19,20] These methods are easy to use; however, they do not allow the calculation of the true value for the degree of polymerization. Only a relative polymerization index for the comparison between several tannin extracts can be obtained.[14,16,21]

Acid-catalyzed cleavage of the interflavanyl linkages (IFL) affords the flavanol extender units their carbocations, whereas the lower units are liberated as monomeric flavanols. In the presence of a competing nucleophile, the carbocations immediately combine leading to the formation of ether adducts the structures of

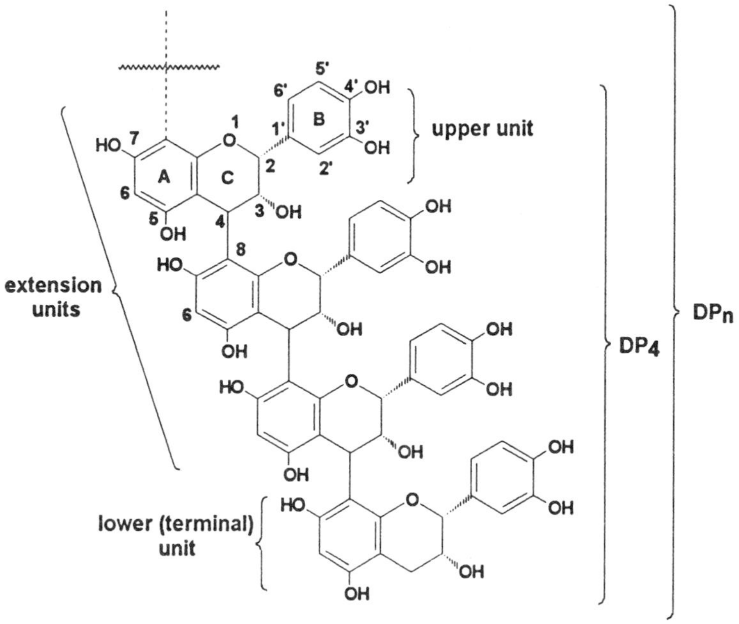

Figure 1. Structure of condensed tannins—example of a (−)-epicatechin based structure.

which directly derive from that of extension units in the initial molecule.[22] The lower units and the extension units are then separately assayed in the reaction media. The method provides access to useful information on the nature of the constitutive flavanol units of proanthocyanidin oligomers, and the molar ratio of extension and terminal units allows the calculation of DPn of condensed tannin extracts.[23–28] Because the calculation of DPn is based on the quantitative estimation of the reaction products, the reaction must be performed with a good yield. Nucleophilic reagents are generally thiol compounds, the most commonly used being benzylmercaptan (toluene-α-thiol). Some investigators have used 2-sulfanylethanol[29] or phloroglucinol.[30–33] Parameters such as temperature,[34] solvent,[35] concentration, and nature of the acid[34,35] and of the nucleophile[33,35] may influence the reaction kinetics and yields. By comparing several sulphur nucleophiles, Brown and Shaw[35] have observed that maximum yields were obtained with a minimum of acidity and a minimum of reaction time. Ethanol and methanol are the most common solvents used; presence of water in the reaction medium significantly lowers the yields.[33] Yields obtained with benzylmercaptan are generally higher than those with phloroglucinol.[23,33] On the whole, two kinds of conditions are encountered: Many researchers conduct the reaction during several hours in a mild acid medium (acetic acid) by heating with reflux,[22,24,36] whereas others prefer a small proportion of strong acid (hydrochloric or sulfuric acids) at room temperature and for a short reaction time.[6] Yields may vary according to the nature of condensed tannins in the extracts. Therefore, an optimization of the conditions should be performed when the reaction is applied to extracts from very different origins. Other factors such as the presence of impurities, the occurrence of side reactions,[35,37] and the instability of some reaction products[33] may alter the yields. The separations and assays of the reaction products also contribute to the validity of these methods. HPLC[12,24–26,28,32–33] or GC[33] techniques have replaced the weighing of the reaction products,[23] paper chromatography,[34] or TLC,[39] which were initially used. Reversed phase HPLC appears as a sensitive and accurate method provided that the reaction products have been reliably identified and calibrated in the HPLC system. Moreover, it can be applied on micro-quantities by direct injection of the reaction medium in the HPLC system.[6,38] Recent works have shown the benefit of these methods for the quantification of partly purified or insoluble condensed tannins[33,40] in plant extracts.

A ^{13}C-NMR method was first proposed by Czochanska et al.[41] for the characterization of a tannin extract of *Ribes sanguineum*. It was then applied to tannins from various origins.[42–44] It is based on the distinction of the chemical shifts and relative signal areas of C-3 of the lower flavanol units from C-3 of the extension units. The former are located between 65 and 68 ppm, whereas the latter are between 72 and 73 ppm.[41,42,10] DPn is directly obtained by the ratio of the signal areas. Nevertheless, the method is limited by the accurate observation of the C-3 signal of the terminal units, which is all the less intense as DPn is higher. Therefore, polymers presenting molecular weight above 8,000 cannot be precisely characterized according to this procedure.[41] In comparison with chemical degradation, the ^{13}C-NMR method has the advantage of being nondestructive. Moreover, it gives precise information on the stereochemistry of the heterocycles of the constitutive units and also on the hydroxylation pattern of the B nuclei (procyanidins/prodelphinidins ratio).[42,44] In some favorable cases, it may also provide information on the proportion of C4-C6 interflavanic linkages and on the existence of chain branching in the polymers.[45]

Gel permeation chromatography (GPC) is another method used in the field of condensed tannins characterization.[46–48,32,12] It allows the estimation of the number average molecular weight (Mn) of tannin fractions and also provides useful information on the polydispersity, since an estimation of the distribution (Mw) can also be obtained.[47] Two main disadvantages can be pointed out about these methods. First, the column calibration is often performed by using polystyrene standards, which may have a different chromatographic behavior than that of polymeric proanthocyanidins. Second, derivatization of the molecules (i.e., methylation or acetylation) is needed before the chromatographic separation. These chemical reactions are rarely quantitative and may produce unknown products by side reactions. A method that does not require derivatization has been described; the separation was performed with dimethyformamide as elution solvent.[49]

Some other methods are more rarely encountered. Some articles refer to the use of vapor pressure osmometry (VPO)[46,50] or elevation of boiling point[51] for condensed tannin analysis. VPO is a classical method for Mn estimation of polymeric structures. This analytical technique is based on the Raoult's law and allows a direct analysis without degradation of the molecules. A good correlation of VPO measurements with ^{13}C-NMR method has been obtained for the calculation of DPn of homogeneous condensed tannins.[50]

By combining thiolytic degradation and ¹H-NMR, Cai et al.[52] have proposed another procedure for DPn measurements. The H-2 signals of the two classes of reaction products (i.e., flavan-3-ol and thioether derivatives corresponding, respectively, to terminal and extension units) were well separated and identified on the ¹H-NMR spectra, and the DPn could be estimated.

Mass spectrometry (MS) constitutes a technique of real benefit in the field of condensed tannin analysis and for other polyphenolic oligomers and polymers.[53] It is discussed in another chapter of this book (see chapter 3.5). However, until now, it does not allow measurement of DPn of procyanidin fractions, but has been largely employed to obtain accurate assessment of molecular size of purified procyanidins. It also showed evidence of a series of homogeneous oligomers with various DP in a plant extract.[28] At present, electrospray ionization seems to give the best results in its application to high molecular weight polyphenolic compounds.[53]

3. CIDER APPLE PROCYANIDINS. THIOLYSIS, HPLC, AND A NEW ¹H-NMR METHOD FOR DPn DETERMINATION

Freeze-dried cider apple (*Malus domestica*, var. Kermerrien) parenchyma was successively extracted with methanol and aqueous-acetone. The methanolic extract contained mainly simple sugars, organic acids, and alcohol soluble phenolics (hydroxycinnamic acids, dihydrochalcones, monomeric catechins, and oligomeric procyanidins), whereas the aqueous-acetone extract contained polymeric condensed tannins.[40] Non-phenolic substances were removed from the dry methanol and aqueous-acetone extracts according to the following procedure: each dry extract was dissolved in diluted acetic acid (2.5 percent v/v), and the solution was eluted on C18 Sep-Pak cartridges, which retain phenolic compounds. Polyphenols were further recovered by elution with a mixture of diluted acetic acid and methanol (1:1 v/v). Dry extracts were then obtained by evaporation and freeze-drying.

Dry methanol soluble polyphenols were fractionated on a normal phase semi-preparative HPLC in order to separate oligomeric procyanidins according to their molecular weight as already described by Rigaud et al.[48] Simple phenolics (hydroxycinnamic acids, dihydrochalcones and monomeric catechins) were eluted at first (FM1, fig. 2), whereas oligomeric procyanidins were successively eluted according to their increasing degree of polymerization (fig. 2). Seven procyanidin fractions, numbered from FM2 to FM8+, were collected and evaporated to dryness.

Dry aqueous-acetone soluble polymeric procyanidins were fractionated by reversed phase HPLC at a semi-preparative scale and five fractions, numbered from FA1 to FA5, were collected according to the increasing elution time (fig. 2) and evaporated to a reduced volume and freeze-dried.

The dry collected fractions were submitted to thiolysis, and reaction media were directly analyzed by reversed-phase HPLC coupled with diode array detection according to the method described by Guyot et al. (1998).[40] Three main reaction products were formed under thiolysis. (–)-Epicatechin, which appeared as its

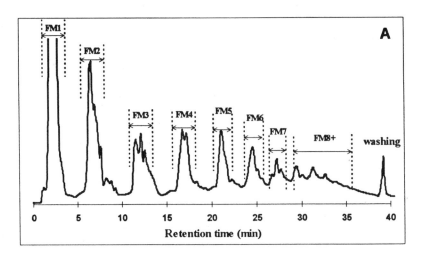

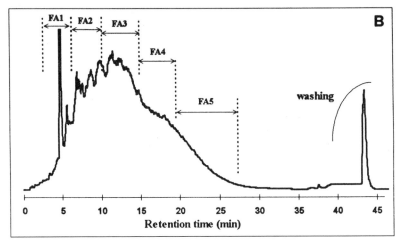

Figure 2. Chromatographic fractionation of Kermerrien cider apple procyanidins. (A) normal phase (SI) chromatogram (280 nm) of the methanol extract; (B) reversed-phase (C18) chromatogram (280 nm) of the aqueous-acetone extract.

benzylthioether derivative, was the only extension unit in the constitution of cider apple procyanidin fractions, whereas terminal units were mainly (−)-epicatechin and also (+)-catechin in a very low proportion. The results of thiolysis-HPLC are presented in table 1.

Thiolysis yields were wholly homogeneous with a value around 75 percent, which indicates that the reaction was performed efficiently for all fractions. As a whole, (−)-epicatechin represented more than 95 percent of the constitutive units in all fractions, which showed the great homogeneity of procyanidins for this cider apple variety.

Table 1. Thiolysis-HPLC characterization of *Kermerrien* cider apple procyanidin fractions according to their constitutive units and their average degree of polymerization (DPn)

		FM2	FM3	FM4	FM5	FM6	FM7	FM8	FA2	FA3	FA4	FA5
Terminal units (%)	CAT[a]	4	4	3	2	2	2	2	1	1	<1	<1
	EC[b]	45	29	22	18	14	13	11	13	7	6	4
Extension units (%)	EC[b]	51	67	75	80	84	85	87	86	92	93	95
DPn		2.0	3.1	4.1	5.0	6.0	6.8	7.9	6.7	11.8	14.9	22.1
Thiolysis yield (%)		71	77	83	78	78	78	78	69	73	77	72

[a]CAT : (+)-catechin ; [b]EC : (-)-epicatechin. Presented values correspond to the mean of three replicates. FA1 values are not presented because not enough product was available and the fraction was contaminated by non procyanidin compounds (mainly chlorogenic acid).

The calculated DPn (table 1) confirmed the normal phase separation of oligomeric procyanidins according to their molecular weight as was already observed for cider,[51] grape seeds,[48] and cacao[48] procyanidins. Polymeric procyanidins (FA2 to FA5) were also separated in an increasing DPn order on reversed-phase. Such a chromatographic behavior, which does not constitute a rule in proanthocyanidin separation, was, however, observed on reversed phase for sorghum procyanidins[20] and may be due to the homogeneous structure of the polymerized molecules in these particular cases.

In a critical review, Kolodziej[54] has summarized the advances in ^{1}H-NMR interpretation of procyanidins and the limits of ^{1}H parameters when applied to oligomers. Thus, ^{1}H parameters have been used to give information on the C-ring configuration and on the position of the interflavanic linkage. In most cases, ^{1}H-NMR data have been obtained for permethylated or peracetylated proanthocyanidins and, as far as we know, they have never been used for the estimation of DPn of oligomeric and polymeric procyanidins extracts.

Our NMR analyses were performed on several fractions without any prior derivatization of the molecules. Dry procyanidin fractions were dissolved in a mixture of acetone-d_6 and D_2O 1:1 (v/v). Care was taken not to let the samples sit around in D_2O a long time because of the facile exchange of the H-6 and H-8 protons of the A-ring. Proton NMR spectra were recorded at 30 °C in 3 mm sample tubes on a Varian Unity-Inova-500 spectrometer at 500 MHz. In all samples, 128 transients were acquired, and a recycle time of 20 seconds was used (i.e., at least 5T1 for all protons). A 8,000 Hz spectral width was used with 32 k data points. Chemical shifts are reported relative to internal tetramethylsilane. Baseline correction was applied before integration.

The ^{1}H-NMR spectra of three fractions (FM3, FM6, and FA3) with significantly different DPn are presented in figure 3. The signals are globally broadened as commonly observed for ^{1}H-NMR spectra of oligomeric procyanidins when acquired at ambient temperature.[54] However, spectra were comparable on the basis of four

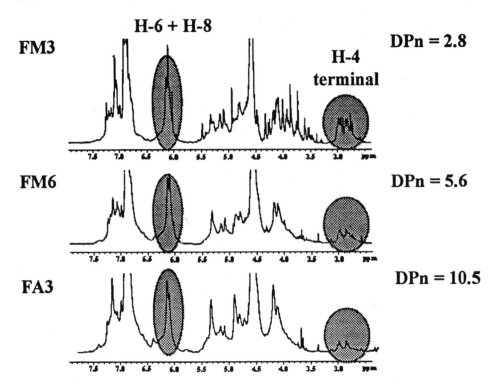

Figure 3. ¹H-NMR spectra of three apple cider procyanidin fractions with significantly different Dpn.

zones regarding chemical shifts. Signals between 6.5 and 7.5 ppm correspond to B-ring protons, those located between 3.2 and 5.6 ppm are attributed to C-ring protons except for the H-4 of terminal units with signals between 2.4 and 3.0 ppm. Resonances of the latter were observed at distinctly higher field than signals of the H-4 of extension units because they do not suffer the deshielding effect of the flavanyl substitution. Moreover, these signals were not masked or overlapped by acetate proton signals as was observed when analyses were performed with peracetylated procyanidins.[55] The zone of the spectra between 5.8 and 6.5 ppm corresponds to resonances of H-6 and H-8 protons of the A-rings.

As already described by Czochanska et al.[41] for DPn estimation with ¹³C-NMR, the ¹H-NMR method was based on the ratio of a resonance area attributed to all flavanol units to a resonance area that is specific to the terminal units. In this case, as seen in figure 3, the integration ratio (H-6 + H-8)/(H-4*terminal*) increases with DPn according to the following formula:

$$DP_n = 2 \left(\frac{(H\text{-}6 + H\text{-}8)\ \text{signal area}}{(H\text{-}4_{\text{terminal}})\ \text{signal area}} \right) - 1$$

Table 2. Comparison of DPn calculated to ${}^1H$-NMR and HPLC following thiolysis

	FM3	FM4	FM6	FM8	FA3
DPn (${}^1H$-NMR)	2.8	3.9	5.6	7.0	10.5
DPn (thiolysis-HPLC)	3.1	4.1	6.0	7.9	11.8

The DPn that have been estimated by HPLC following thiolysis and by the ${}^1H$-NMR method are compared in table 2. A good equivalency between the two methods was obtained for the calculation of DPn.

4. CONCLUSIONS

Unidimensional ${}^1H$-NMR might constitute a valuable method to determine the average degree of polymerization of homogeneous procyanidin fractions. This approach offers several interesting features compared to other techniques used until now for such a purpose. First, this non-destructive method could be applied while conserving molecules in their natural state. A consequence is that there is no risk of undesired reactions in the sample. Thus it might be used to confirm DPn calculations obtained by other techniques such as thiolysis following HPLC. Additionally, although ${}^{13}C$-NMR spectroscopy allows better resolved spectra to be obtained, ${}^1H$-NMR has the great advantage of being less time consuming. Moreover, the lower resolution of ${}^1H$ spectra does not seem to affect the quality of integration. Further work is in progress to validate the method with a more extensive series of oligomeric and polymeric procyanidin fractions.

REFERENCES

1. Bate-Smith, E.C.; Swain, T. *In*: Masson, H.S.; Florkin, A.M. (eds.) Comparative biochemistry. Vol. III, Academic Press, New York, (1962).

2. Haslam, E. Polyphenol-protein interactions. *Biochem. J.* 139:285 (1974).

3. Arnold, R.A.; Noble, A.C.; Singleton, V.L. Bitterness and astringency of phenolic fractions in wine. *J. Agric. Food Chem.* 28:675 (1980).

4. Porter, L.J.; Woodruffe, J. Haemanalysis: the relative astringency of proanthocyanidin polymer. *Phytochemistry* 23:1255 (1984).

5. Asano, K.; Ohtsu, K.; Shinagawa, K.; Hashimoto, H. Affinity of proanthocyanidins and their oxidation products for haze-forming proteins of beer and the formation of chill haze. *Agric. Biol. Chem.* 48:1139 (1984).

6. Cheynier, V.; Rigaud, J.; Ricardo da Silva, J.M. Structure of procyanidin oligomers isolated from grape seeds in relation to some of their chemical properties. *In*: Hemingway, R.W.; Laks, P.E. (eds.) Plant polyphenols—synthesis, properties, significance. Plenum Press, New York, p. 281 (1992).

7. Lea, A.G.H.; Arnold, G.M. The phenolics of ciders: bitterness and astringency. *J. Agric. Food Chem.* 29:478 (1978).

8. Delcour, J.A.; Schoeters, M.M.; Meysman, E.W.; Dondeyne, P. The intrisic influence of cat-echins and procyanidins on beer haze formation. *J. Inst. Brew.* 90:381 (1984).

9. Vennat, B.; Bos, M.-A.; Pourrat, A.; Bastide, P. Procyanidins from tormentil: fractionation and study of the anti-radical activity towards superoxide anion. *Biol. Pharm. Bull.* 17:1613 (1994).

10. Eberhard, T.L.; Young, R.A. Conifer seed cone proanthocyanidin polymers: characterization by 13C NMR spectroscopy and determination of antifungal activities. *J. Agric. Food Chem.* 42:1704 (1994).

11. Bos, M.-A.; Vennat, B.; Meunier, M.T.; Pouget, M.-P.; Pourrat, A.; Fialip, J. Procyanidins from tormentil: antioxidant properties towards lipoperoxidation and anti-elastase activity. *Biol. Pharm. Bull.* 19:146 (1996).

12. Hor, M.; Heinrich, M.; Rimpler, H. Proanthocyanidin polymers with antisecretory activity and proanthocyanidin oligomers from *Guazuma ulmifolia* bark. *Phytochemistry* 42:109 (1996).

13. Gali, H.U.; Perchellet, E.M.; Gao, X.M.; Karchesy, J.J.; Perchellet, J.P. Comparison of the inhibitory effects of monomeric, dimeric and trimeric procyanidins on the biochemical markers of skin tumor promotion in mouse epidermis *in vivo. Planta Med.* 60:235 (1994).

14. Goldstein, J.L.; Swain, T. Changes in tannins in ripening fruits. *Phytochemistry* 2:371 (1963).

15. Butler, L.G.; Price, L.; Brotherton, J.E. Vanillin assay for proanthocyanidins (condensed tannins) modification of the solvent for estimation of the degree of polymerization. *J. Agric. Food Chem.* 30:1087 (1982).

16. Ribéreau-Gayon, P.; Stonestreet, E. Dosage des tanins du vin rouge et détermination de leur structure. *Chim. Anal.* 48:188 (1966).

17. Scalbert, A. Quantitative methods for the estimation of tannins in plant tissues. *In:* Hem-ingway, R.W.; Laks, P.E. (eds.) Plant polyphenols—synthesis, properties, significance. Plenum Press, New York, p. 259 (1992).

18. Swain T.; Hillis, W.E. The phenolic constituents of *Prunus domestica*. I—The quantitative analysis of phenolic constituents. *J. Sci. Food Agric.* 10:63 (1959).

19. McMurrough I.; McDowell, J. Chromatographic separation and automated analysis of flavanols. *Analytical Biochemistry* 91:92 (1978)

20. Putman, L.J.; Butler, L.G. Separation of high molecular weight sorghum procyanidins by high performance liquid chromatography. *J. Agric. Food Chem.* 37:943 (1989).

21. Butler, L.G. Relative degree of polymerization of sorghum tannin during seed development and maturation. *J. Agric. Food Chem.* 30:1090 (1982).

22. Thompson, R.S.; Jacques, D.; Haslam, E.; Tanner, R.N.J. Plant proanthocyanidins. Part I. Introduction; the isolation, structure, and distribution in nature of plant procyanidins. *J. Chem. Soc., Perkin Trans. I* :1387 (1972).

23. Gupta, R.K.; Haslam, E. Plant proanthocyanidins. Part 5. Sorghum polyphenols. *J. Chem. Soc., Perkin Trans. I* :892 (1978).

24. Shen, Z.; Haslam, E.; Falshaw, C.P.; Begley, M.J. Procyanidins and polyphenols of *Larix gmelini* bark. *Phytochemistry* 25:2629 (1986).

25. Prieur, C.; Rigaud, J.; Cheynier, V.; Moutounet, M. Oligomeric and polymeric procyanidins from grape seeds. *Phytochemistry* 36:781 (1994).

26. Souquet, J.M.; Cheynier, V.; Brossaud, F.; Moutounet, M. Polymeric proanthocyanidins from grape skins. *Phytochemistry* 43:509 (1996).

27. Matthews, S.; Mila, I.; Scalbert, A.; Donnelly, D.M.X. Extractible and non-extractible proan-thocyanidins in barks. *Phytochemistry* 45:405 (1997).

28. Guyot, S.; Doco, T.; Souquet, J.-M.; Moutounet, M.; Drilleau, J.F. Characterization of highly polymerized procyanidins in cider apple (*Malus silvestris var. Kermerrien*) skin and pulp. *Phytochemistry* 44:351 (1997).

29. Tanaka, T.; Takahashi, R.; Kouno, I.; Nonaka, G.I. Chemical evidence for de-astringency (insolubilization of tannins) of persimmon fruit. *J. Chem. Soc., Perkin Trans. I* :3013 (1994).

30. Foo, L.Y.; Porter, L.J. Prodelphinidin polymers: definition of structural units. *J. Chem. Soc., Perkin Trans. I* :1186 (1978).

31. Matsuo, T.; Tamaru, K.; Saburo, I. Chemical degradation of condensed tannin with phloroglucinol in acidic solvents. *Agric. Biol. Chem.* 48:1199 (1984).

32. Koupai-Abyazani, M.R.; Muir, A.D.; Bohm, B.A.; Towers, G.H.N.; Gruber, M.Y. The proanthocyanidin polymers in some species of *Onobrychis*. *Phytochemistry* 34:113 (1993).

33. Matthews, S.; Mila, I.; Scalbert, A.; Pollet, B.; Lapierre, C., Hervè du Penhoat, C.L.M.; Rolando, C.; Donnelly, D.M.X. Method for estimation of proanthocyanidins based on their acid depolymerization in the presence of nucleophiles. *J. Agric. Food Chem.* 45:1195 (1997).

34. Hemingway, R.W.; McGraw, G.W. Kinetics of acid-catalysed cleavage of procyanidins. *J. Wood Chem. Tech.* 3:421 (1983).

35. Brown, B.R.; Shaw, M.R. Reactions of flavonoids and condensed tannins with sulphur nucleophiles. *J. Chem. Soc., Perkin Trans I* :2036 (1974).

36. Hsu, F.L.; Nonaka, G.I.; Nishioka, I. Tannins and related compounds. XXXIII. Isolation and characterizationof procyanidin in *Dioscorea cirrhosa* Lour. *Chem. Pharm. Bull.* 33:3293 (1985).

37. McGraw, G.W.; Steynberg, J.P.; Hemingway, R.W. Condensed tannins: a novel rearrangement of procyanidins and prodelphinidins in thiolytic cleavage. *Tetrahedron Letts.* 34:987 (1993).

38. Rigaud, J.; Perez-Ilzarbe, J.; Ricardo da Silva, J.M.; Cheynier, V. Micromethod for identification of proanthocyanidin using thiolysis monitored by high-performance liquid chromatography. *J. Chromatogr.* 540:401 (1991).

39. Lea, A.G.H.; Bridle, P.; Timberlake, C.F.; Singleton, V.L. The procyanidins of white grapes and wines. *Am. J. Enol. Vitic.* 30:289 (1979).

40. Guyot, S.; Marnet, N.; Laraba, D.; Sanoner, P.; Drilleau, J.-F. Reversed-phase HPLC following thiolysis for quantitative estimation and characterization of the four main classes of phenolic compounds in different tissue zones of a French cider apple variety (*Malus domestica* var. Kermerrien). *J. Agric. Food Chem.* 46:1698 (1998).

41. Czochanska, Z.; Foo, L.Y.; Newman, R.H.; Porter, L.J.; Thomas, W.A.; Jones, W.T. Direct proof of a homogeneous polyflavan-3-ol structure for polymeric proanthocyanidins. *J. Chem. Soc. Chem. Commun.* :375 (1979).

42. Czochanska, Z.; Foo, L.Y.; Newman, R.H.; Porter, L.J. Polymeric proanthocyanidins. Stereochemistry, structural units and molecular weight. *J. Chem. Soc., Perkin Trans. I* :2278 (1980).

43. Foo, L.Y.; Porter, L.J. The phytochemistry of proanthocyanidin polymers. *Phytochemistry* 19:1747 (1980).

44. Porter, L.J.; Newman, R.H.; Foo, L.Y.; Wong, H.; Hemingway, R.W. Polymeric proanthocyanidins. ^{13}C N.M.R. studies of procyanidins. *J. Chem. Soc., Perkin Trans I* :1217 (1982).

45. Newman, R.H.; Porter, L.J.; Foo, L.Y.; Johns, S.R.; Willing, R.I. High-resolution ^{13}C NMR studies of proanthocyanidin polymers (condensed tannins). *Magn. Reson. Chem.* 25:118 (1987).

46. Karchesy, J.J.; Hemingway, R.W. Loblolly pine bark polyflavonoids. *J. Agric. Food Chem.* 28:222 (1980).

47. Williams, V.M.; Porter, L.J.; Hemingway, R.W. Molecular weight profiles of proanthocyanidin polymers. *Phytochemistry* 22:569 (1983).

48. Rigaud, J.; Escribano-Bailon, M.T.; Prieur, C.; Souquet, J.M.; Cheynier, V. Normal-phase high-performance liquid chromatographic separation of procyanidins from cacao beans and grape seeds. *J. Chromatogr. A*, 654:255 (1993).

49. Bae, Y.S.; Foo, L.Y.; Karchesy, J.J. GPC of natural procyanidin oligomers and polymers. *Holzforschung* 48:4 (1994).

50. Porter, L.J. Number- and weight-average molecular weights for some proanthocyanidin polymers (condensed tannins). *Aust. J. Chem.* 39:557 (1986).

51. Lea, A.G.H. The phenolics of ciders: oligomeric and polymeric procyanidins. *J Sci. Food Agric.* 29:471 (1978).

52. Cai, Y.; Evans, F.J.; Phillipson, J.D.; Zenk, M.H.; Gleba, Y.Y. Polyphenolic compounds from *Croton lechleri. Phytochemistry* 30:2033 (1991).

53. Cheynier, V.; Doco, T.; Fulcrand, H.; Guyot, S.; Le Roux, E.; Souquet, J.M.; Rigaud, J.; Moutounet, M. ESI-MS analysis of polyphenolic oligomers and polymers. *Analusis* 28:32 (1997).

54. Kolodziej, H. ^{1}H NMR spectral studies of procyanidin derivatives: Diagnostic ^{1}H NMR parameters applicable to the structural elucidation of oligomeric procyanidins. *In*: Hemingway, R.W.; Laks, P.E. (eds.) Plant polyphenols—synthesis, properties, significance. Plenum Press, New York, p. 295 (1992).

55. Balas, L.; Vercauteren, J.; Laguerre, M. 2D NMR Structure elucidation of proanthocyanidins: the special case of the catechin-$(4\alpha\text{-}8)$-catechin-$(4\alpha\text{-}6)$-catechin trimer. *Magn. Reson. Chem.* 33:85 (1995).

ELECTROSPRAY CONTRIBUTION TO STRUCTURAL ANALYSIS OF CONDENSED TANNIN OLIGOMERS AND POLYMERS

Hélène Fulcrand,[a] Sylvain Guyot,[b] Erwan Le Roux,[a] Sophie Remy,[a] Jean-Marc Souquet,[a] Thierry Doco,[a] and Véronique Cheynier[a]

[a] Unité de Recherches des Polymères et des Techniques Physico-chimiques
Institut des Produits de la Vigne, INRA—ENSAM
34060 Montpellier
FRANCE

[b] Station de Recherche Cidricole
Biotransformation des Fruits et Légumes, INRA
35650 Le Rheu
FRANCE

1. INTRODUCTION

Phenolic compounds have attracted considerable interest because of their ubiquitous occurrence within the plant kingdom and numerous important properties. In particular, they are essential components of plant-derived foods and beverages, responsible for some major organoleptic features, including color and astringency. They are also receiving increasing attention as natural antioxidants and potential health-promoting agents.

Polyphenols show a great diversity of structures from rather simple molecules (monomers, oligomers) to polymers. The latter are usually designated by the term "tannins," referring to their ability to complex with proteins, which were originally used in the formation of leather from hide. Among them, condensed tannins are particularly important with respect to their wide distribution in plants and contribution to major food qualities.[1,2] These phenolic compounds are highly unstable and are rapidly transformed into various reaction products when the plant cells are damaged, for instance, during food processing. Transformation products of the genuine phenolic compounds may bind to proteins and thus be regarded as tannins.[3] The occurrence of specific tannins arising from oxidation (i.e., thearubi-

Plant Polyphenols 2: Chemistry, Biology, Pharmacology, Ecology, Edited by
Gross et al. Kluwer Academic / Plenum Publishers, New York, 1999

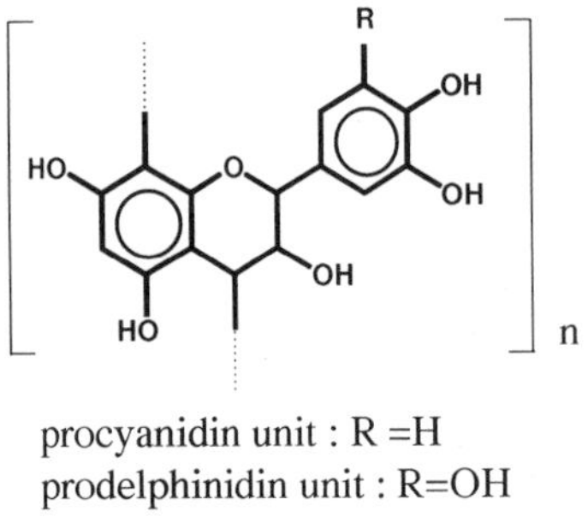

procyanidin unit : R =H
prodelphinidin unit : R=OH   **Figure 1.** Structure of proanthocyanidins.

gins and theaflavins) is well documented in black tea.[4] As well, grape phenolics
are known to proceed to various polymeric species during winemaking and
aging.[5-7] Although the influence of such reactions on color and taste changes has
been suspected for many years, the structure of some resulting products[8-12] and
their presence in wine[13,14] have been established only recently.

Condensed tannins are oligomeric and polymeric flavan-3-ols, also called
proanthocyanidins, as they release the red anthocyanidin pigments when heated
in acidic media. Several classes can be distinguished on the basis of the hydroxylation pattern of the constitutive units (e.g., procyanidins, prodelphinidins, fig. 1).

Monomeric units may be linked by C—C (C4'→C6 and/or C4'→C8) bonds
(B-type) or doubly linked, with an additional ether (C2→O→C7) linkage (A-type)
(fig. 2). Besides, flavan-3-ol units may be encountered as 3-*O*-esters, in particular,
of gallic acid, or as glycosides.[15,16] Finally, the degree of polymerization (DP) may
vary greatly as proanthocyanidins have been described up to 10,000 in molecular
weight.[17]

The organoleptic properties (e.g., bitterness, astringency) and the reactivity of
tannins including radical scavenging effects and protein-binding ability largely
depend on the nature and number of their constitutive units. In particular, the
number of active sites (phenolic and especially *o*-diphenolic groups)[18] increases
with the degree of polymerization and with gallate esterification. Therefore, determination of tannin structure is a prerequisite to study their chemical properties
and eventual health effects.

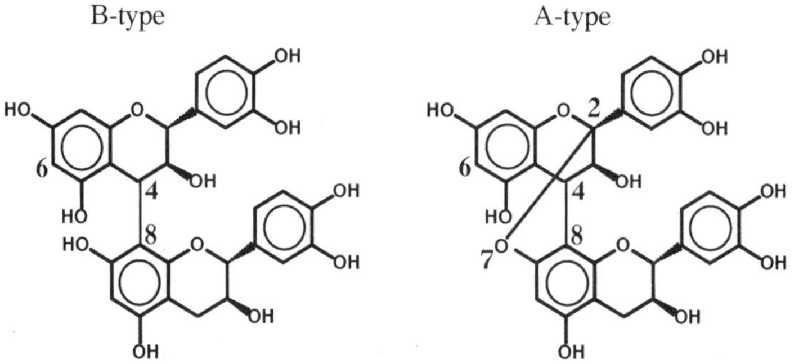

Figure 2. Types of interflavanic linkages.

Formal identification of proanthocyanidins is based on spectroscopic methods including in particular NMR and MS analyses, generally restricted to pure compounds. It thus implies tedious isolation procedures, which become more and more difficult as the molecular weight increases, owing to the larger number of possible isomers, smaller amounts of each individual compound, and poorer resolution of the chromatographic profiles. This is especially true in the case of grape products, which contain a large diversity of tannin structures based on several monomers, whereas some other plants synthesize essentially one series, e.g., (–)-epicatechin derivatives in the case of apple[19] or cacao.[20]

Several methods have been developed to analyze oligomeric and polymeric tannins in complex solutions. Some of them, based on acid-catalyzed cleavage in the presence of nucleophilic agents (e.g., toluene-α-thiol, phloroglucinol),[21–23] allow determination of the monomeric composition and mean degree of polymerization (DP) of tannin extracts or fractions. However, they only give access to an average composition and provide no information on polymer size distribution.

Other methods, including counter-current chromatography,[24] chromatography on Fractogel TSK HW-40[25,26] or Sephadex LH20[27] columns, normal phase TLC,[28] or HPLC,[29–30] aim to separate oligomeric and polymeric proanthocyanidins on a chain-length basis. These methods are poorly resolutive for a given DP, so that separation of oligomers is usually better achieved by reversed-phase liquid chromatography. Detection is classically performed by UV-visible spectrophotometry, either directly or after derivatization with DMCA,[31–33] but structural similarities of molecules based on the same constitutive units lead to identical UV-visible spectra. Alternatively, mass spectrometry allows one to distinguish related oligomers and polymers and provides important information on their constitutive moieties.

Classical ionization sources such as chemical ionization or electronic impact are poorly suited to the analysis of most phenolic compounds because of their thermosensitivity and non-volatility in the native (non-derivatized) form. These techniques require rather tedious preliminary derivatization (permethylation, peracetylation, trimethylsilylation) to render the molecules volatile. In addition, they usually yield only fragment ions that provide no information on the molecular weight of the original polymer. In contrast, fast atom bombardment mass spectrometry (FAB-MS) gives access to the molecular weight and some specific fragments. Therefore, it has been for many years the method of choice for mass spectrometry analysis of proanthocyanidin oligomers.[34,35]

Nevertheless, since the review of FAB-MS applications in the elucidation of proanthocyanidin structures published by D.G. Barofsky in the first edition of the Tannin Conference Proceedings,[36] technology of mass spectrometers has evolved well. In particular, ionization sources have been considerably developed in accordance with the popular use of high-performance liquid chromatography (HPLC) with reversed-phase columns. As detectors proved to be limited for analyses of biological samples, investigations were undertaken to combine liquid chromatography with mass spectrometry. The difficulties of such a concept were related to ionization process, usually consisting of creating ions from a gas phase. In this case, the process should occur starting from a condensed phase (LC effluent) at atmospheric pressure. Hence, several interfaces such as thermospray, particle beam, continuous-flow fast atom bombardment (CF-FAB) have emerged.[37] Electrospray, plasma desorption, and UV laser desorption appeared more recently, allowing analyses of very large molecules like biopolymers. Principles and strategies of interfacing a liquid chromatographic system to a mass spectrometer are reviewed in the book edited by Niessen and van der Greef.[38]

Regarding flavonoid analysis, mass coupling systems most often cited in recent literature are thermospray, FAB, LSIMS, and electrospray. Thermospray (TSP) ionization consists in vaporization of LC effluent by heating.[39] Consequently, this technique is suitable for non-thermolabile compounds. Some criticisms are generally leveled at it:

- poor (or lack of) sensitivity particularly for large and non-volatile compounds

- rare or no fragmentations

- restricted chromatographic conditions

However, good results have been obtained in the identification of flavones, xanthones, and flavonol glycosides extracted from plant material by TSP LC-MS in combination with UV, UV-shifted, and MS/MS methods.[40,41] In contrast, thermospray is not appropriate for analysing non-volatile, polar, and thermolabile molecules such as tannins, given their polymeric nature.

Electrospray (ESI)[42] is one of the softest of ionization techniques and is well adapted for high molecular weight compounds. Numerous books and reviews on this technique and its applications have recently been published and compiled in web sites.[43]

Historically, ESI-MS grew in popularity with protein analyses. Thus, the possiblilty of creating multiple-charged ions detectable in the mass range of a quadrupole allowed analyses of such large molecules. Their mass spectrum appears as an ion peak series, corresponding to the statistical distribution of charges that the protein is able to bear. Then, the mass is reconstructed from the ion peak series by a mathematical deconvolution process.

After a brief presentation of principles, signal acquisition, and interpretation of proanthocyanidin spectra obtained by ESI-MS, applications carried out in our laboratory to analyze tannins will be described. This section will be divided in two main parts: analysis of native proanthocyanidins originating from various fruits and investigation of polyphenol reactions in the course of food processing.

2. PRINCIPLES OF ESI-MS

In the ESI source, positive and negative multiple-charged ions are produced by applying a high voltage at the liquid exit and a lower voltage at the analyzer entrance.[44] The sign of the voltage applied determines the selection between positive and negative ions. The resulting field gradient allows electrostatic nebulization of the solution containing analyte ions. Droplets are thus formed in a dry bath gas at atmospheric pressure. As neutral solvent evaporates, the size of droplets reduces until the charge repulsion overcomes the cohesive droplet forces, leading to a "Coulombic explosion". In the model proposed by Iribarne and Thomson,[45,46] the smaller droplets continue to evaporate and the process repeats as long as the droplet surface curvature becomes sufficiently high to permit the field-assisted evaporation of charged solutes (fig. 3).

Details of the electrospray ionization mechanism are still a matter of debate.[47] What is clear is that molecular ions are formed from a liquid solution under mild conditions by electrospray ionization. These ions usually occur by addition of proton, alkali cation, or ammonium ion in positive ion formation and by abstraction of proton or cation in negative ion formation. Besides, ESI process can be pneumatically assisted by a high-velocity flow of gas (referred to as nebulizer gas), surrounding the capillary injection. This technique called ionspray[48] allows introduction of higher amounts of solvent. Typically, an ionspray probe tolerates 20–200 μl/min, whereas the real electrospray capacity is 1–10 μl/min.

Sample introduction into the electrospray ion source requires a flowing liquid stream. According to the type of sample and the complexity of the matrix, samples can be introduced directly with no separation, either by flow injection through a loop injection or by pumping the sample continuously through a microliter syringe installed in a syringe infusion pump. When chromatography is required, samples are introduced through the loop injection and separated on reversed-phase column before entering in the ion source. Then, the delicate point of the ion source held at atmospheric pressure is the efficient sampling and transport of ions into the mass spectrometer at operating vacuum (e.g., >10–5 torr for a quadrupole mass filter). One of the oldest models (but also the simplest) uses a direct transition between atmospheric pressure from the ion source and high vacuum region into the mass analyzer. This transition is ensured by a single laser-drilled pinhole orifice (100 to 130 μm) and a cryopumping, capable of very high pumping speeds.

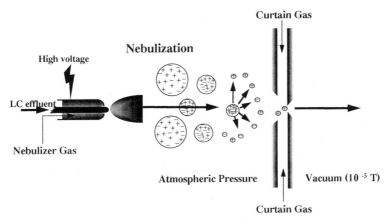

Figure 3. Ionspray—Ion formation by liquid phase ionization.

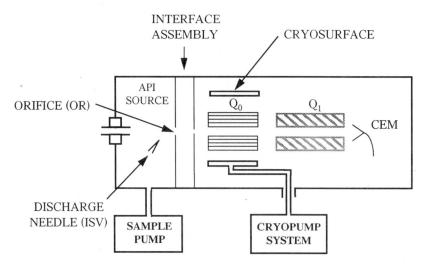

Figure 4. Adapted from the functional scheme of the main assembly of an API *I* mass spectrometer (Sciex-Perkin Elmer, Toronto).

Sample ions are drawn by electric fields through the interface assembly into the vacuum chamber (fig. 4).

The interface assembly uses nitrogen gas to form a physical barrier between atmospheric pressure in the source and the very low pressure of the vacuum chamber. This "curtain" gas also serves to dry the droplets and decluster the ions.

Alternatively, other instruments based on differentially pumped vacuum technology have been described by Fenn and coworkers.[49] The major difference between these two types of technology seems to result from the possible use of higher field gradients (15 kV) in the latter, allowing coupling with magnetic sector analyzer enabling high resolution.

Several operations are necessary prior to analyzing a sample. First, the analyzer should be calibrated using a commercial polymer standard (usually a **P**oly**P**ropylene**G**lycol). This consists of obtaining the exact values of PPG ion peaks with a signal response and a resolution defined by the manufacturer. Then, a product chemically close to the sample to be analyzed is used to optimize the signal. Calibration and optimization are performed by tuning iteratively the following parameters: voltages of the ionspray (ISV) and the orifice (OR), flows of nebulizer gas and curtain gas (N_2), and position of the discharge needle (fig. 4). Orifice voltage is also used to induce cluster dissociations and fragmentations in the course of sample analyses.

3. INTERPRETATION OF PROANTHOCYANIDIN SPECTRA OBTAINED BY ESI-MS

Multiply charged ions created by electrospray ionization can be detected either in the positive ion mode as $(M + zH)^{z+}/z$ or in the negative ion mode as

$(M - zH)^{z-}/z$. The sign and the number (z) of charges depend on the chemical nature of the molecule. Thus, small molecules are detected as singly charged ions, whereas larger ones are found as multiply charged ions. Nevertheless, the potential number of charges can be modulated by environmental factors (pH, presence of ions and organic solvent).

Proanthocyanidin molecules are better detected in the negative ion mode than in the positive ion mode due to the relative acidity of phenolic protons. They are also more negatively charged as the chain length increases. This means that tannin oligomers are detected as singly charged or multiply charged ions up to a DP of 8 and only as multiply charged species beyond this polymerization degree, due to the mass range limits of a quadrupole (2,400 amu). The degree of purity and the complexity of the sample determine the mode of introduction in the ion source: either by LC-MS coupling or by continuous flow injection (CFI). The use of mass spectrometry detection after HPLC separation allows one to determine the mass of each eluted compound and to detect eventual contaminations that may not be apparent on the UV spectrum. The continuous flow injection is usually preferred in the case of larger molecules (beyond DP5), which cannot be properly separated and often elute as unresolved humps at the end of the chromatograms because of the large number of isomers present. Consequently, using this technique, isomers add up in a single signal, resulting in a gain of sensitivity.

In addition to molecular ion peaks, proanthocyanidins can be detected as non-covalent adducts (clusters), leading to possible artifacts when interpreting ESI-MS data. In particular, formation of non-covalent adducts with solvent molecules like formic acid may give additional signals. As well, non-covalent dimers formed by staking of two phenolic molecules (one neutral and one charged) are frequently detected at high concentrations. The mass clusters thus obtained can help for identification of the molecular ion peak[50] but may lead to a complication of the spectrum and an intensity decrease due to signal division. Their formation may be limited by increasing the orifice voltage.

Although mass spectrometry is not sufficient to achieve formal identification, characteristic fragments can be obtained by increasing the orifice voltage and often prove extremely helpful for identification purposes. In particular, the fragmentations observed earlier by FAB-MS[34–36] such as Retro-Diels-Alder fission, interflavanic bond cleavage and substituent loss (e.g., glycosyl, galloyl or benzylthioether moities) are retrieved by ESI-MS.

A necessary step in the interpretation of proanthocyanidin mass spectra is to distinguish between singly charged and multiply charged signals. Distinction is provided by examination of the distance between isotopic peaks. Basically, the peak corresponding to the mass (M + 1) represents the mass of a molecule containing one ^{13}C and appears just after the molecular ion signal. Since the value measured by the mass analyzer is the mass to the charge ratio (m/z), the molecular ion peak of a singly charged ion (z = 1) is detected at $m/z = (M - H)^-$ in the negative ion mode. The species containing one ^{13}C appears at $m/z = (M + 1 - H)^-$. For a doubly charged ion (z = 2), the molecular ion peak is detected at $m/z = (M - 2H)^{2-}/2$ and that containing one ^{13}C at $m/z = (M + 1 - 2H)^{2-}/2$. The difference between these two signals $[(M + 1 - 2H)^{2-}/2] - [(M - 2H)^{2-}/2]$ is equal to 1/z = 1/2 amu. Thus, isotopic peaks are separated by 1 amu in the case of singly charged species and 0.5 amu for doubly charged ions. Examples of a procyanidin tetramer detected as singly charged ion and a procyanidin heptamer detected as doubly charged ion are presented in figure 5.

Following the same reasoning, spacing between isotopic ions for a triply charged ion (z = 3) is then 1/3 amu and finally for a z multiply charged ions 1/z amu. Consequently, the distance between isotopic peaks becomes smaller as the number of charges increases. Considering the quadrupole resolution, ions bearing more than four charges cannot be differentiated by the isotopic peak distance. However, calculation software is able to determine the charge of all peaks belonging to an ion peak series. Owing to the natural abundance of carbon 13 (1.1 percent), the intensity of the ion peak corresponding to M + 1 (mass of a compound with one ^{13}C) increases with the number of carbons within the molecule. This ion peak becomes the predominant one when the molecule reaches 90 carbons. As well, M + 2 (compound with two ^{13}C) corresponds to the most intense peak beyond 180 carbons. In a general way, the mass is incremented by one mass unit every 90 carbons. Applied to procyanidin polymers, the mass takes one additionnal mass unit approximatively every six monomer units. However, software calculating exact isotopic distribution facilitates signal attribution (indicating signal apex and consequently molecular size).

Regarding polymers, a particular problem for interpreting mass spectra is due to overlap of ion peak signals. This results from the fact that polymers are never available as pure compounds but rather as mixtures. In the case of tannins, singly charged DPn (m/z = m), doubly charged DP2n (m/z = 2 m/2), and triply charged DP3n (m/z = 3 m/3) ion peaks have very close values that overlap. Signal simulation based on a singly charged DP4, a doubly charged DP8, and a triply charged DP12 procyanidin polymer is presented in figure 6 to illustrate the problem. The isotopic distribution of each ion was calculated from nominal mass (entire part of the mass) and divided by its respective charge number, giving m/z values (fig. 6A).

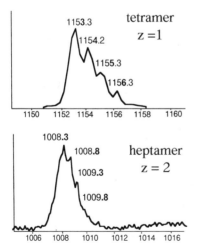

Figure 5. Comparison of isotopic distance between a singly charged and doubly charged ions.

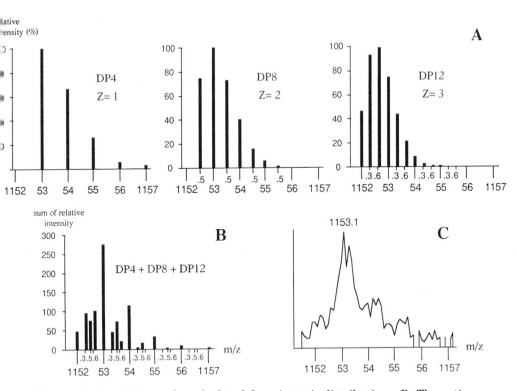

Figure 6. A: m/z ion peaks calculated from isotopic distributions. B: Theoretical signal obtained by addition of the calculated profiles (A) in ratio 1:1:1. C:Real signal containing DP4, DP8, and DP12 in unknown ratio.

The sum of the three calculated m/z profiles (fig. 6A) results in signal enlargement as can be also observed in the real signal (fig. 6B). Besides, the theorical and real spectra match pretty well, although the former shows the case in which all these ions occur in the same proportion. It is important to note that in fact the shape of the signal varies with the relative amounts of each species.

The simultaneous presence of DP4, DP8, and DP12 in the real signal detected at 1153.1 amu should be checked by finding the ion peaks corresponding to intermediate DPs (between DP5 to DP11), in particular those with prime number DPs.

Applications of ESI-MS analysis to the characterization of proanthocyanidin extracts from various plant sources are presented below. The examples selected (i.e., tannins from cider apple,[19] litchi pericarp,[51] grape seeds, and grape skins are increasingly complex as they contain, respectively, only B-type epicatechin-based procyanidins, A- and B-type epicatechin-based procyanidins, partly galloylated procyanidins, and heterogenous proanthocyanidins consisting of both (epi)gallocatechin and (epi)catechin units partly galloylated.

Analysis of a cider apple fraction (with an average degree of polymerization of 4) by LC-ESI-MS in the negative ion mode showed the presence of B-type procyanidin oligomers from dimer (DP2) to tetramer (DP4) detected as singly charged ions [M – H]⁻ (fig. 7) along with other apple constituents such as chlorogenic acid and phloretin derivatives.

Continuous flow injection (CFI) of another apple fraction (DP > 4)[19] into the electrospray ionization source (fig. 8A) yielded ion peaks at m/z 1441.0, 1729.5, 2017.5, and 2305.5 in addition to those observed above.

These signals can be interpreted as [M – H]⁻ ion peaks of pentameric, hexameric, heptameric, and octameric procyanidins, respectively, but they may also correspond to the doubly charged ions [M – 2H]²⁻ of DP10, DP12, DP14, and DP16 species. Existence of doubly charged ions was proven by the presence of additional signals that can be unambiguously attributed to the doubly charged ions [M – 2H]²⁻ of uneven DPs, starting from DP5. Triply charged ions of the DP10, DP11, DP13, DP14, DP16, and DP17 procyanidins were also detected, indicating that the number of charges increases, as expected, with the polymer length. In fact, procyanidins between DP9 and DP14 were detected both as doubly charged and triply charged ions, whereas polymers above DP15 can only appear as triply charged ions, given the detection limit of the ESI-MS apparatus. This suggests that the signals attributed to singly charged DPn are contaminated with those of doubly charged DP2n and triply charged DP3n, as confirmed by the multiplicity of mass values within the corresponding peak and variable spacing between them.

No multiply charged species beyond the triply charged ones were detected, presumably because of the lower concentration of larger tannin molecules. However,

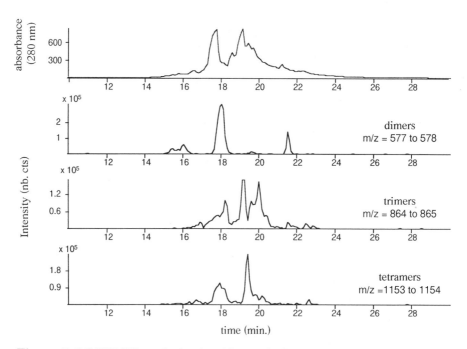

Figure 7. LC-ESI-MS analysis of a cider apple fraction. Mass chromatograms have been extracted from the total ion current.

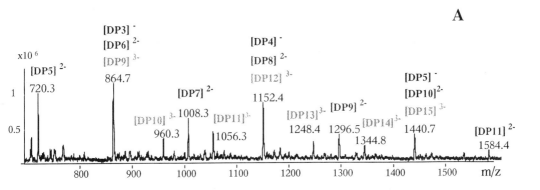

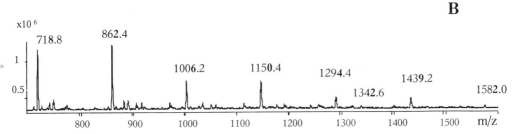

Figure 8. Mass spectra analyzed by CFI-ESI-MS: A) Cider apple extract; B) Litchi pericarp extract.

the apparent decrease of polymer concentration as the molecular weight increases may also be due to an increased dispersion of signals among variously charged ions, including large ones that cannot be detected.

The major peaks of a litchi pericarp tannin extract detected by LC-ESI-MS gave mass signals corresponding to those of B-type procyanidins minus 2 amu, which can be explained by the presence of one additional linkage (–2H) in each oligomer. This result is consistent with that of thiolysis analysis indicating that litchi tannins are based on epicatechin units linked by both B-type (cleaved by thiolysis) and A-type (resistant to thiolysis) bonds (fig. 2).[51] The presence of A-type linkage and the sequence of some oligomers were also confirmed by fragmentation.[13] Procyanidin A2 (A-type linked procyanidin dimer) present in the litchi tannin extract gave the same fragmentation pattern as in FAB-MS,[36] with one ion peak corresponding to the Retro-Diels Alder fission ([M—H—$C_8H_8O_3$]$^-$ at m/z = 423) and two fragments generated by breakage of the A-type interflavanic linkage (at m/z = 285 and 289). The different fragments obtained from two trimeric species detected in this extract allowed us to determine their sequence, in agreement with the results of thiolysis analyses. Similarly, characteristic fragment ions at 863 and 573 indicated that the tetramer sequence was: epicatechin—procyanidin A2—epicatechin, so that the only remaining ambiguity concerns the position of the B-type linkages (C4→C6 or C4→C8).

ESI-MS analysis of the litchi tannin extract by continuous flow injection (fig. 8B) showed numerous peaks in addition to those arising from the oligomeric species. Among them, mass signals at m/z 1,439.2, 1,752.6, and 2,014.2 can be interpreted as the pentameric, hexameric, and heptameric procyanidins containing one A-type interflavanoid linkage. Signals corresponding to doubly charged ions of procyanidins containing one A-type bound with uneven DPs from 5 to 13 were detected at m/z 718.8, 1,006.2, 1,294.4, 1,582.0, and 1,872.0, respectively, whereas other ones corresponding to triply charged ions of DP 14, 16, 17, 19, 20, and 22 were found at m/z 1,342.6, 1,534.4, 1,630.4, 1,822.2, 1,917.2, and 2,109.2, respectively. To our knowledge, these are the largest species ever detected, although the occurrence of larger molecular weight tannins (up to DP 80) have been demonstrated by thiolysis in apple and in grape.[52]

As described above, coexistence of singly charged DPn, doubly charged DP2n, and triply charged DP3n in a given mass peak resulted in enlargement of the mass range and poorer resolution of the mass values within the signal. In addition, each signal attributed to a specific polymer length also consisted of several distinct peaks corresponding to various numbers of A-type linkages within the molecule. For instance, DP11 procyanidins (fig. 9, table 1), which can be at least detected as doubly charged ions owing to the detection limit (2,400 amu) of the ESI-MS apparatus, appeared highly heterogeneous, with one to five A-bonds.

The mass spectra obtained with grape tannin extracts were more complex than those of apple and litchi samples because of the larger diversity of constitutive monomeric units present in the tannin structures, but could be interpreted in the same way.

Thus, grape seed extracts were shown to contain B-type procyanidin oligomers from DP2 to DP9, yielding the same signals as in cider apple extract, but also the

doubly charged DP11

Figure 9. Heterogenity in the number of A-type linkages within DP11 signal. The average number of A-type bonds increased with the chain length, the predominant DP11 species presenting three A-type linkages.

Table 1. Calculated m/z values for a doubly charged DP11
containing an increasing number of A-type linkages

Number of A-type linkages	[a](M-2H)$^{2-}$/2
1A	1583.99
2A	1582.98
3A	1581.97
4A	1580.97
5A	1579.96
6A	1578.95

[a]highest ion peak in the isotopic distribution

corresponding monogalloylated and digalloylated species (fig. 10) as expected from earlier works.[26,30,53]

Although mass signals of doubly charged digalloylated DP2n species overlapped with those of singly charged monogalloylated DPn, values at m/z 872.7, 1,160.8, and 1,449.4 could be unambiguously interpreted, respectively, as the [M − 2H]$^{2-}$ molecular ions of digalloylated pentameric, heptameric, and nonameric procyanidins.[13] Signals corresponding to the doubly charged trigalloylated hexamer, heptamer, and octamer, respectively, were also detected, indicating that the number of galloyl substituents increases with the chain length, in agreement with earlier studies.[30]

Regarding LC-ESI-MS analysis, several ion peaks corresponding to dimers, galloylated dimers, trimers, and galloylated trimers were eluted all along the HPLC profile, due to the presence of isomers based on different sequences of catechin, epicatechin, and epicatechin-3-O-gallate units.

Analysis of a grape skin extract yielded essentially the same mass data as the seeds, but additional series of peaks were also present. The 16 mass unit difference between these new compounds and procyanidins indicated the substitution of one hydrogen atom by one hydroxyl group and could thus be attributed to the presence of (epi)gallocatechin units in the oligomer structure. Thus, the ion peak series (fig. 11) between 1,153.8 and 1,201.8 amu could be interpreted as singly charged ions [M − H]$^{-}$ of proanthocyanidin tetramers containing from one to three (epi)-gallocatechin units.

As observed for procyanidins, larger oligomers gave multiply charged ions. For instance, doubly charged ion peaks of pentamers with zero to three trihydroxylated units appeared in the spectrum simultaneously with the corresponding

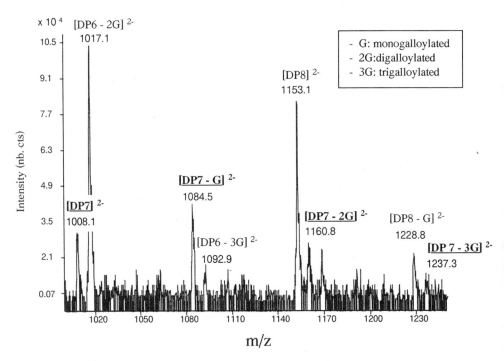

Figure 10. Part of the mass spectrum obtained by CFI analysis of a grape seed extract showing a heptamer ion series containing from 0 to 3 galloyl substituents.

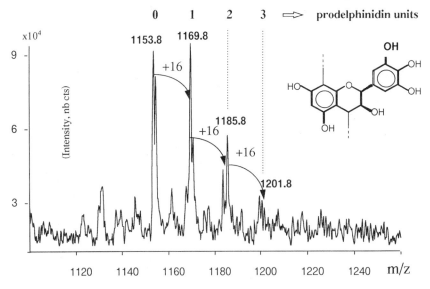

Figure 11. Part of a grape skin mass spectrum showing a tetramer series detected by CFI-ESI-MS.

singly charged ions in trace amounts as well as some doubly charged ions of hexa- and heptamers

The presence of oligomeric proanthocyanidins based on both (epi)catechin and (epi)gallocatechin units had been previously demonstrated in grapes,[5,53] and more recently that of epigallocatechin, along with trace amounts of gallocatechin and epigallocatechin gallate, in grape skin tannins[52] has been established. However, a dimer based on both types of units (gallocatechin-(4→8)-catechin) had previously been isolated from barley.[54]

Several mass spectra corresponding to galloylated derivatives were also found. In particular, detection of a signal attributed to a tetramer containing one (epi)gallocatechin unit and one galloyl substituent confirmed that these moieties can coexist in a single proanthocyanidin oligomer.

4. STUDY OF PHENOLIC REACTIONS

Phenolic compounds are highly reactive species that undergo various types of reactions in the course of food processing. Among these reactions, oxidation processes and addition reactions, both leading to various adducts and eventually tannin-like polymeric compounds, are particularly important. LC-ESI-MS has been successfully applied, on one hand, for the study of these reactions in model solutions, and, on the other hand, for the detection of the products arising from them in processed foods and beverages. Examples of such applications are presented below.

Enzymatic oxidation mechanisms, initiated by polyphenol oxidase (PPO) action, have been thoroughly studied as they are responsible for discoloration phenomena referred to as enzymatic browning. The first step of the reaction process is the enzymically catalyzed transformation of *o*-diphenolic (and monophenolic) substrates to *o*-quinones. The resulting oxidized species are highly unstable, being both strong electrophiles and powerful oxidants,[55] and rapidly proceed to various reaction products.

In particular, polyphenol oxidase-catalyzed oxidation of (+)-catechin yielded several products, including both colorless compounds and yellow pigments.[56,57] The major colorless oxidation product showed the same UV spectrum as (+)-catechin and procyanidins and gave a molecular ion peak at m/z 577, in the negative ion mode, corresponding to the mass of procyanidin dimers. It was also coeluted with procyanidin B3 on the reverse-phase and normal phase HPLC columns used. However, both compounds could be distinguished on the basis of the fragment ion at m/z 289 (formed by rupture of the interflavanic linkage), by comparing its relative intensity to that of the molecular ion and of fragment ions yielded by retro-Diels-Alder fission of the pyran flavanol rings. NMR analysis showed that this product was a catechin dimer in which the two units were linked by a C8→C6′ biphenyl bond. In contrast, the fragmentation pattern of other oxidation products, consisting of two catechin moieties linked by C3′→O→C8 or C4′→O→C8 ether bonds, was similar to that of procyanidins. Note that although these molecules are not tannins *stricto sensu*, they were shown to have the same enzymatic inhibition properties as procyanidin dimers,[58] and can thus be regarded as tannins with respect to their ability to interact with proteins.

Another important mechanism leading to tannin-derived species is the acetaldehyde-mediated reaction of flavanols, generating ethyl-linked flavanol polymers. In the presence of anthocyanins, flavanol-aldehyde-anthocyanin adducts are formed in addition to the previous ones. The reaction involving only flavanols is believed to be implied in the loss of astringency observed during ripening of persimmon,[59] whereas the latter has been reported to participate in color and taste changes taking place in red wine aging.[5,6,60]

Acetaldehyde-induced reactions have been studied in model solutions and monitored by HPLC.[61–65] Some of the lower molecular weight reaction products were thus separated, whereas the larger species were eluted as an unresolved hump at the end of the chromatographic profile. Flavanol polymers can be distinguished from anthocyanin-flavanol adducts on the basis of their UV-visible spectra, but this provides no information on the number of units in the different molecules. By contrast, ESI-MS seems to be the method of choice to solve this problem and has been successfully applied to investigate these reactions.[8,9,11,12]

Monitoring of acetaldehyde-induced polymerization of flavanols by LC-ESI-MS[8] showed the presence of ethyl-linked dimers, trimers, and tetramers detected as the [M-H]⁻ ions at m/z 605, 921, and 1,237. Intermediate ethanol adducts on the monomer, dimer, and trimer were also detected, confirming the reaction mechanism postulated by Timberlake and Bridle.[66] This process starts with the protonation of the aldehyde in the acidic medium, followed by nucleophilic attack of the resulting carbocation by a flavanol unit. The intermediate ethanol adduct thus formed loses a water molecule to give a new carbocation that suffers a nucleophilic attack by another flavanol molecule. The C6 and C8 nucleophilic positions of the flavanol seem equally reactive since four dimers (two C6→C8 (R and S), one C6→C6, one C8→C8), were obtained starting from each monomer.[8,12] Polymerization continued following the same mechanism, yielding an increasing number of species at each reaction step, and thus resulting in poorer resolution in the chromatogram. Direct ESI-MS analysis of the sample enabled the detection of ethyl-linked oligomers up to the hexamer. The ethyl-linked pentamer was observed both as [M – H]⁻ and as [M – 2H]²⁻ ions.

LC-ESI-MS analysis of solutions containing malvidin-3-glucoside, acetaldehyde, and (epi)catechin was performed in the positive ion mode to take advantage of the positive charge in the anthocyanin structure.[67] It showed the presence of both ethyl-linked flavanol oligomers, detected as [M + H]⁺ ions, and anthocyanin-ethyl-flavanol adducts, detected as flavylium cations M⁺, meaning that the anthocyanin competed with the flavanol in the addition process (fig. 12).

Among the latter series, dimeric, trimeric, and tetrameric species containing one anthocyanin, and one, two, or three flavanol units, were detected. A tetrameric species containing two anthocyanin and two flavanol units was also found as the doubly charged M²⁺ ion. However, it seems that the reaction stops when both ends are occupied by an anthocyanin moiety. Equivalent dimeric and trimeric adducts were also formed when malvidin-3-glucoside was replaced with cyanidin-3-glucoside.[11] As well, carboxyl methine ("ethanoic")-linked adducts were formed in

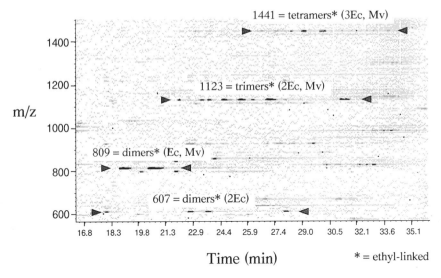

Figure 12. 3D-map of LC/MS analysis showing some products of the acetalde-hyde induced reaction in a malvidin 3-*O*-glucoside (Mv)–flavanol (Fl) solution.

the presence of glyoxylic acid, arising from oxidation of tartaric acid, which is a major wine component.[10] This suggests that a great diversity of products can be generated during wine aging, their respective levels depending on the nature and proportions of the various precursors present.

LC-ESI-MS analysis of a red wine confirmed that it contained several reaction products in addition to the different proanthocyanidin species detected in grape. Thus LC-ESI-MS allowed to be unambiguously demonstrated the presence of ethyl-linked catechin and trimers, of ethyl-linked malvidin-3-glucoside (epi)cate-chin adducts[13,14] as well as the presence of "ethanoic" linked dimers as shown in figure 13.

5. CONCLUSIONS

Mass spectrometry is a complementary analytical technique to NMR, IR, UV, and XRF spectroscopies in formal identification of structures. In particular, ESI-MS proved extremely efficient in determining the molecular weight of labile com-pounds such as proanthocyanidins and their derivatives, giving access to the nature and number of their constitutive units and substituents. This is particu-larly true in the case of oligomeric compounds (up to tetramers) that yield essen-tially molecular ions ([M + H]+ or [M − H]−). ESI-MS also produces characteristic fragments of proanthocyanidins like FAB-MS, being extremely helpful for identification purposes.

Moreover, the possibility of coupling ESI-MS with liquid chromatography systems (e.g., HPLC, capillary electrophoresis) makes this technique a highly selective detector, enabling on-line characterization of each individual compound.

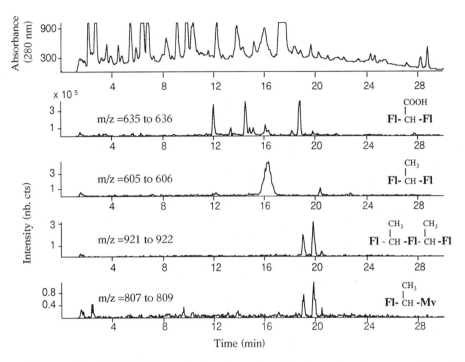

Figure 13. LC-ESI-MS analysis of a wine fraction showing some derived tannins. Fl = flavanol, i.e., epicatechin or catechin.

In particular, it enables the determination of the constitutive units of the various species present in a tannin extract, which cannot be distinguished, for instance, on the basis of their UV-visible or fluorescence spectra.

However, the application of ESI-MS suffers from the limitations described in the discussion that follows. For a given compound, the signal intensity is proportional to the concentration, but it depends on the capacity of the molecule to become ionized, which is greatly influenced both by the structure and by the environment. In particular, proanthocyanidins are more easily charged as the degree of polymerization increases. Consequently, the use of ESI-MS for quantitative purposes requires a calibration curve to be established for each compound under the acquisition conditions used for the analysis.

Although polymeric species up to DP22 were detected, interpretation became increasingly difficult as the molecular weight increased. This problem is partly due to the limited mass range imposed by the quadrupole analyzer and may be solved to some extent by the development of new techniques permitting soft ionization on a wider mass range. An interesting alternative may be the use of electrospray ionization coupled to a Time Of Flight analyzer, combining the advantages of ESI in ion production and TOF (wide range of masses).

Another difficulty also arises from the larger number of charges generated in the ESI source as the tannin length increases. For example, under our experimental conditions (negative ion mode, solvent: methanol containing 0.5 percent of formic acid), oligomers from pentamers to octamers were found as doubly charged ions, although masses of the corresponding singly charged ions fell within the quadrupole detection limits. In fact, despite its obvious advantages with the quadrupole analyzer, the easy generation of multiple ions for larger molecules leading to dispersion of peak signals and frequent overlapping constitutes one of the major limitations of ESI.

Matrix Assisted Laser Desorption Ionization (MALDI) may be a good ion source to limit the production of multiply charged species. Thus, Maldi-Tof has allowed detection of apple procyanidins up to the decamer level as the singly charged molecular ions.[68] However, its applications have so far been limited by strong interferences of polyphenolic compounds with the matrix. It would be interesting to perform assays without matrix since proanthocyanidins have similar properties.

Therefore, further research should focus on methods associating soft ionization processes yielding essentially monocharged ions with analyzers enabling detection of larger molecular weight species.

REFERENCES

1. Haslam, E. Symmetry and promiscuity in procyanidin biochemistry. *Phytochemistry*, 16:1625 (1977).

2. Haslam, E. Vegetable tannins. *Rec. Adv. Phytochem.* 12:475 (1979).

3. Haslam, E.; Lilley, T.H. Natural astringency in foodstuffs. A molecular interpretation. *C.R.C. Crit. Rev. Food Sci. Nutr.* 27:1 (1988).

4. Harbowy, M.E.; Balentine D.A. Tea chemistry. *Crit. Rev. Plant Sci.* 16:415 (1997).

5. Haslam, E. In vino veritas: oligomeric procyanidins and the ageing of red wines. *Phytochemistry* 19:2577 (1980).

6. Somers, T.C. The polymeric nature of wine pigments. *Phytochemistry* 10:2175 (1971).

7. Ribéreau-Gayon, P. The anthocyanins of grapes and wines. *In*: Markakis P. (ed.). Anthocyanins as food colors. Academic Press, New York, p. 209 (1982).

8. Fulcrand, H.; Doco, T.; Es-Safi, N.; Cheynier, V.; Moutounet, M. Study of the acetaldehyde induced polymerisation of flavan-3-ols by liquid chromatography ion spray mass spectrometry. *J. Chromatogr. A* 752:85 (1996).

9. Fulcrand, H.; Es-Safi, N.; Doco, T.; Cheynier, V.; Moutounet, M. LC-MS study of acetaldehyde induced polymerisation of flavan-3-ols. *Polyphenol Communications* 96:202 (1996).

10. Fulcrand, H.; Cheynier, V.; Oszmianski, J.; Moutounet, M. An oxidized tartaric acid residue as a new bridge potentially competing with acetaldehyde in flavan-3-ols condensation. *Phytochemistry* 46:223 (1997).

11. Guerra, C.; Saucier, C.; Bourgeois, G.; Vitry, C.; Busto, O.; Glories, Y. Partial characterization of coloured polymers of flavan-3-ols-anthocyanins by mass spectrometry. Proceedings of the First Symposium In vino Analytica Scientia, Analytical chemistry for wine, brandy and spirits, 124 (1997).

12. Saucier, C.; Guerra, C.; Pianet, I.; Laguerre, M.; Glories, Y. (+)-Catechin-acetaldehyde condensation products in relation to wine ageing. *Phytochemistry* 46:229 (1997).

13. Cheynier, V.; Doco, T.; Fulcrand, H.; Guyot, S.; Le Roux, E.; Souquet, J.M.; Rigaud, J.; Moutounet, M. ESI-MS analysis of polyphenolic oligomers and polymers. *Analusis* 25:M32 (1997).

14. Saucier, C.; Little, D.; Glories Y. First evidence of acetaldehyde-flavanol condensation products in red wine. *Am. J. Enol. Vitic*, 48:370 (1997).

15. Cheynier, V.; Rigaud, J.; Ricardo-da-Silva, J.M. Structure of procyanidins oligomers isolated from grape seeds in relation to some of their chemical properties. *In*: Hemingway, R.W.; Laks, P.E. (eds.). Plant polyphenols: synthesis, properties, significance. Plenum Press, New York, p. 281 (1992).

16. Bae, Y-S.; Burger, J.F.W.; Steynberg, J.P.; Ferreira, D.; Hemingway, R.W. Flavan and procyanidin glycosides from the bark of blackjack oak. *Phytochemistry* 35:473 (1994).

17. Hör, M.; Heinrich, M.; Rimpler, H. Proanthocyanidin polymers with antisecretory activity and proanthocyanidin oligomers from *Guazuma ulmifolia* bark. *Phytochemistry* 42:109 (1996).

18. Haslam, E. Polyphenol-protein interactions. *Biochem. J.* 139:285 (1974).

19. Guyot, S.; Doco, T.; Souquet, J.M.; Moutounet, M.; Drilleau, J.F. Characterization of highly polymerized procyanidins in cider apple *(Malus sylvestris* var. Kermerrien) skin and pulp. *Phytochemistry* 44:351 (1997).

20. Porter, L.J.; Ma, Z.; Chan, B.G. Cacao procyanidins: major flavanoids and identification of some minor metabolites. *Phytochemistry* 30:1657 (1991).

21. Thompson, J.; Jacques, D.; Haslam, E.; Tanner, R.J.N. Plant proanthocyanidins. Part. I. Introduction: the isolation, structure, and distribution in nature of plant procyanidins. *J. Chem. Soc., Perkin Trans.* 1:1387 (1972).

22. Foo, L.Y.; Porter, L.J. Prodelphinidin polymers: definition of structural units. *J. Chem. Soc., Perkin Trans I.* :1186 (1978).

23. Hemingway, R.W.; McGraw, G.W. Kinetics of acid-catalyzed cleavage of procyanidins. *J. Wood Chem. Technol.* 3:421 (1983).

24. Putman, L.J.; Butler, L.G. Fractionation of condensed tannins by counter-current chromatography. *J. Chromatogr.* 318:85 (1985).

25. Derdelinckx, G.; Jerumanis, J. Separation of malt hop proanthocyanidins on Fractogel TSK HW-40 (S). *J. Chromatogr.* 285:231 (1984).

26. Ricardo-da-Silva, J.M.; Rigaud, J.; Cheynier, V.; Cheminat, A.; Moutounet, M. Procyanidin dimers and trimers from grape seeds. *Phytochemistry* 30:1259 (1991).

27. Lea; A.G.H.; Timberlake, C.F. The phenolics of ciders.1. Procyanidins. *J. Sci. Food Agric.* 25:1537 (1974).

28. Lea, A.G.H. The phenolics of cider: oligomeric and polymeric procyanidins. *J. Sci. Food Agric.* 29:471 (1978).

29. Rigaud, J.; Escribano-Bailon, M.T.; Prieur, C.; Souquet, J-M.; Cheynier, V. Normal-phase high-performance liquid chromatographic separation of procyanidins from cacao beans and grape seeds. *J. Chromatogr.* 654:55 (1993).

30. Prieur, C.; Rigaud, J.; Cheynier, V.; Moutounet, M. Oligomeric and polymeric procyanidins from grape seeds *(Vitis vinifera). Phytochemistry* 36:781 (1994).

31. Treutter, D. Chemical reaction detection of catechins and proanthocyanidins with 4-dimethylaminocinnamaldehyde. *J. Chromatogr.* 467:185 (1989).

32. Treutter, D.; Santos-Buelga, C.; Gutman, M.; Kolodziej, H. Identification of flavan-3-ol and procyanidins by HPLC and chemical reaction detection. *J. Chromatogr. A.* 667:290 (1994).

33. De Pascual-Teresa, S.; Treutter, D.; Santos-Buelga, C.; Rivas-Gonzalo, J.C. Screening of plant foodstuffs and beverages for flavan-3-ols. *In*: Armado, R.; Andersson, H.; Bardocz, S.; Serra, F. (eds.). COST 916-Polyphenols in food, Office for Official Publications of the European Communities, Luxembourg, p. 63 (1998).

34. Self, R.; Eagles, J.; Galetti, G.C.; Mueller-Harvey, I. Fast atom bombardment mass spectrometry of polyphenols (syn. vegetable tannins). *Biomed. Environ. Mass Spectrom.* 13:449 (1986).

35. Karchesy, J.J.; Hemingway, R.W.; Foo, Y.L.; Barofsky, E.; Barofsky, D.F. Sequencing procyanidin oligomers by fast atom bombardment mass spectrometry. *Anal. Chem.* 58:2563 (1986).

36. Barofsky, D.F. FAB-MS applications in the elucidation of proanthocyanidin structures. *In*: Hemingway, R.W.; Karchesy, J.J. (eds.). Chemistry and significance of condensed tannins. Plenum Press, New York and London, p. 175 (1989).

37. McCloskey, J.A. Methods in enzymology, Volume 193: Mass spectrometry. Academic Press, San Diego, (1990).

38. Niessen, W.M.A.; van der Greef, J. Liquid chromatography-mass spectrometry: principles and applications. Chromatography Science Series, Vol. 58; Marcel Dekker: New York, NY, (1992).

39. Vestal, M.L. Liquid chromatography-mass spectrometry. *In*: McCloskey, J.A. (ed.). Methods in enzymology, Volume 193: Mass spectrometry. Academic Press, San Diego, p. 107 (1990).

40. Wolfender, J.-L.; Maillard, M.P.; Hostettmann, K. Thermospray liquid chromatography-Mass spectrometry in phytochemical analysis. *Phytochemical Analusis* 5:153 (1994).

41. Claeys, M.; Li, Q.M.; Heuvel, H.V.D.; Dillen, L. Mass spectrometric studies on flavonoid glycosides. *In*: Russell, T.J.W.; Newton, P. (eds.). Application of modern mass spectrometry in plant science research. Proc. Phytochem. Soc. Europe (40). Clarendon Press, Oxford, p. 182 (1995).

42. Cole, R.B. Electrospray ionization mass spectrometry: fundamentals, instrumentation, and applications. Wiley, New York, (1997).

43. http://www.mpi-muelheim.mpg.de/stoecki/esi_refs.html http://www.lcms.com/books.htm

44. Edmonds, C.G.; Smith, R.D. Electrospray ionization mass spectrometry. *In*: McCloskey, J.A. (ed.). Methods in enzymology, Volume 193: Mass spectrometry. Academic Press, San Diego, p. 413 (1990).

45. Iribarne, J.V.; Thomson, B.A. On the evaporation of small ions from charged droplets. *J. Chem. Phys.* 64:2287 (1976).

46. Thomson, B.A.; Iribarne, J.V. Field induced ion evaporation from liquid surfaces at atmospheric presure. *J. Chem. Phys.* 71:4451 (1979).

47. Kebarle, P.; Tang, L. From ions in solution to ions in the gas phase: The mechanism of electrospray mass spectrometry. *Anal. Chem.* 65:972A (1993).

48. Bruins, A.P.; Weidolf, L.O.G.; Henion, J.D. Determination of sulfonated azo-dyes by liquid chromatography/atmospheric pressure ionization mass spectrometry. *Anal. Chem.* 59:2647 (1987).

49. Yamashita, M.; Fenn, J.B. Electrospray ionization mass spectrometry-electrospray ion source. Another variation on the free-jet theme. *J. Phys. Chem.* 88:4451 (1984).

50. Fulcrand, H.; Remy, S.; Souquet, J-M.; Cheynier, V.; Moutounet, M. Identification of wine tannin oligomers by on line liquid chromatography electrospray ionisation mass spectrometry. *J. Agric. Food Chem.* (1999, in press).

51. Le Roux, E.; Doco, T.; Sarni-Manchado, P.; Lozano, Y.; Cheynier, V. A-type proanthocyanidins from pericarp of *Litchi chinensis*. *Phytochemistry* 48:1251 (1998).

52. Souquet, J.-M.; Cheynier, V.; Brossaud, F.; Moutounet, M. Polymeric proanthocyanidins from grape skins. *Phytochemistry* 43:509 (1996).

53. Czochanska, Z.; Foo, L.; Porter, L. Compositional changes in lower molecular weight flavans during grape maturation. *Phytochemistry* 18:1819 (1979).

54. Outtrup, H. Structure of prodelphinidins in barley. European Convention Brewery Congress, proceeding of the 18th congress, Copenhagen, p. 323 (1981).

55. Pierpoint, W.S. The enzymic oxidation of chlorogenic acid and some reactions of the quinones produced. *Biochem. J.* 98:567 (1966).

56. Guyot, S.; Cheynier, V.; Souquet, J-M.; Moutounet, M. Influence of the pH on the enzymatic oxidation of (+)-catechin in model systems. *J. Agric. Food Chem.* 43:2458 (1995).

57. Guyot, S.; Vercauteren, J.; Cheynier, V. Colourless and yellow dimers resulting from (+)-catechin oxidative coupling catalysed by grape polyphenoloxidase. *Phytochemistry* 42:1279 (1996).

58. Guyot, S.; Pellerin, P.; Brillouet, J-M.; Cheynier, V. Inhibition of β-glucosidase (*Amygdalae dulces*) by (+)-catechin oxidation products and procyanidin dimers. *Biosci. Biotech. Biochem.* 60:1131 (1996).

59. Tanaka, T.; Takahashi, R.; Kouno, I.; Nonaka, G.-I. Chemical evidence for the deastringency (insolubilisation of tannins) of persimmon fruit. *J. Chem. Soc., Perkin Trans.* 1:3013 (1994).

60. Ribéreau-Gayon, P. Connaissance de la nature des combinaisons de l'anhydride sulfureux dans les vins. *Bull. O.I.V.* 507:406 (1973).

62. Dallas, C.; Ricardo-da-Silva, J.M.; Laureano, O. Degradation of oligomeric procyanidins and anthocyanins in a tinta roriz red wine during maturation. *Vitis.* 34:51 (1995).

62. Rivas-Gonzalo, J.C.; Bravo-Haro, S.; Santos-Buelga, C. Detection of compounds formed through the reaction of malvidin-3-monoglucoside in the presence of acetaldehyde. *J. Agric. Food Chem.* 43:1444 (1995).

63. Dallas, C.; Ricardo-da-Silva, J.M.; Laureano, O. Interactions of oligomeric procyanidins in model wine solutions containing malvidin-3-glucoside and acetaldehyde. *J. Sci. Food Agric.* 70:493 (1996).

64. Dallas, C.; Ricardo-da-Silva, J.M.; Laureano, O. Products formed in model wine solutions involving anthocyanins, procyanidin B2 and acetaldehyde. *J. Agric. Food Chem.* 44:2402 (1996).

65. Saucier, C.; Pianet, I.; Busto, O.; Little, D.; Bourgeois, G.; Vitry, C.; Glories, Y. LC-MS study and structural identification of acetaldehyde catechin dimers in red wine. Proceedings of the First Symposium In vino Analytica Scientia, Analytical chemistry for wine, brandy and spirits, p. 173 (1997).

66. Timberlake, C.F.; Bridle, P. Interactions between anthocyanins, phenolic compounds, and acetaldehyde and their significance in red wines. *Am. J. Enol. Vitic.* 27:97 (1976).

67. Es-Safi, N.; Fulcrand, H.; Cheynier, V.; Moutounet, M.; Hmamouchi, M.; Essassi, E.M. Kinetic studies of acetaldehyde-induced condensation of flavan-3-ols and malvidin-3-glucoside in model solution systems. *Polyphenol Communications*, p. 279 (1996).

68. Ohnishi-Kameyama, M.; Yanagida, A.; Kanda, T.; Nagata, T. Identification of catechin oligomers from apple (*Malus pumila cv. Fuji*) in matrix-assisted laser desorption/ionization time-of-flight mass spectrometry and fast-atom bombardment mass spectrometry. *Rapid Comm. Mass Spectr.* 11:31 (1997).

STRUCTURE AND ABSOLUTE CONFIGURATION OF

BIFLAVONOIDS WITH BENZOFURANOID CONSTITUENT UNITS

E. Vincent Brandt, Riaan Bekker, and Daneel Ferreira

Department of Chemistry
University of the Orange Free State
P.O. Box 339
Bloemfontein 9300
SOUTH AFRICA

1. INTRODUCTION

Notwithstanding the somewhat arbitrary differentiation between oligomeric flavonoids and oligomeric flavans (proanthocyanidins), the former, which are considered to originate via phenol oxidation, represent a less common but major group of phenolic natural products.[1] Unlike proanthocyanidins, however, assessment of the absolute configuration of oligomeric flavonoids possessing stereocenters other than those originating from atropisomerism of the interflavonoid bond has hitherto met with very little success. In effect, this prevented the stereochemical assignment to zeyherin,[2] the first (and thus far sole) entry with a benzofuranoid constituent unit, which was isolated some 27 years ago from the heartwood of "red ivory" (*Berchemia zeyheri*) Sond.[3] The collective utilization of chemical degradation and ¹H NMR-, CD- and computational data now permits estimation of the absolute stereochemistry of such zeyherin analogues as well as related structures from the same source,[4,5,6] and for the first time also of 2-benzyl-2-hydroxybenzo[1]furan-3(2*H*)-one enantiomers, a small but biosynthetically significant group of aurone derivatives.[7]

2. DIMERIC BENZOFURANOIDS AND THEIR ISOLATION

Our first reports[4,5] on the phenolic constituents of the red heartwood of *Berchemia zeyheri* (fig. 1) revealed the presence of two flavanone-benzofuranoid oligomers, (2*R*, 3*S*)-naringenin-(3α→7)-(2*R*)-maesopsin (**1**) and its 2*S*(F)-epimer (**2**). Continuation of the investigation has since also illustrated the existence of a unique pair of "rearranged" epimers, (2*S*, 3*R*)-dihydrogenistein-(2α→7)-(2*R*)-

Plant Polyphenols 2: Chemistry, Biology, Pharmacology, Ecology, Edited by
Gross et al. Kluwer Academic / Plenum Publishers, New York, 1999

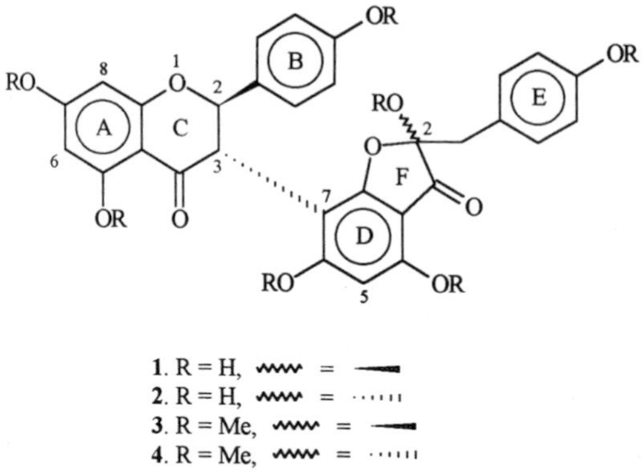

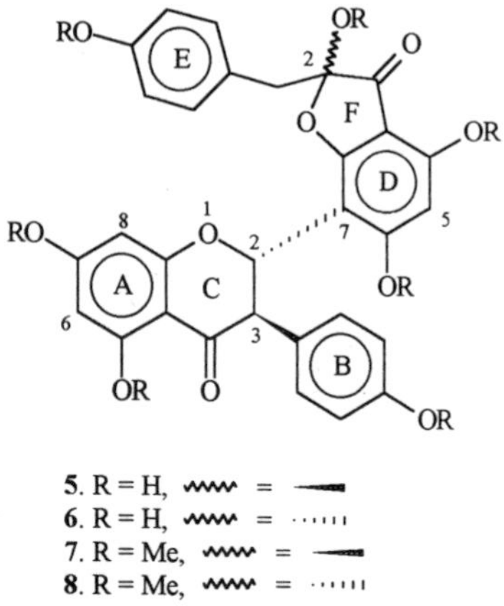

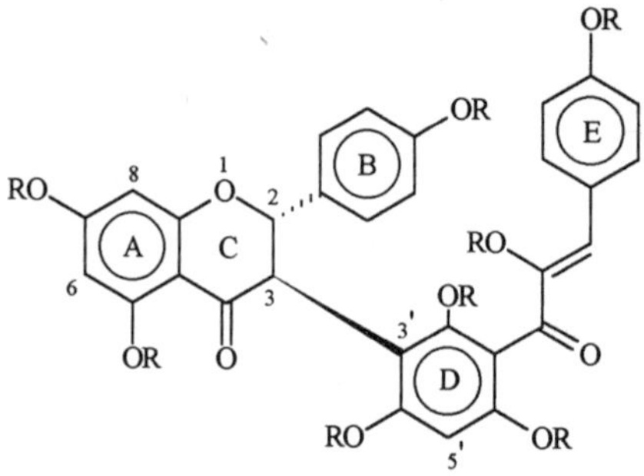

1. R = H, ∿ = ▬
2. R = H, ∿ = ·····ıı
3. R = Me, ∿ = ▬
4. R = Me, ∿ = ·····ıı

5. R = H, ∿ = ▬
6. R = H, ∿ = ·····ıı
7. R = Me, ∿ = ▬
8. R = Me, ∿ = ·····ıı

9. R = H
10. R = Me

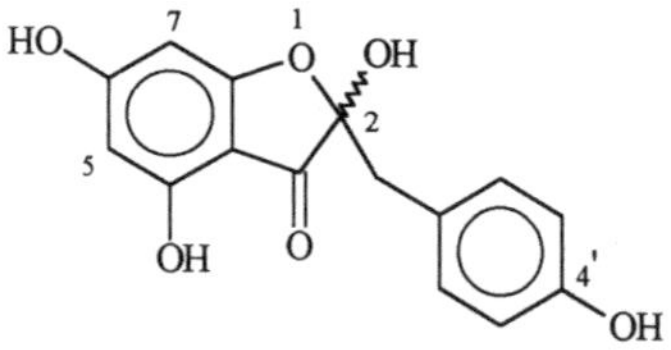

11

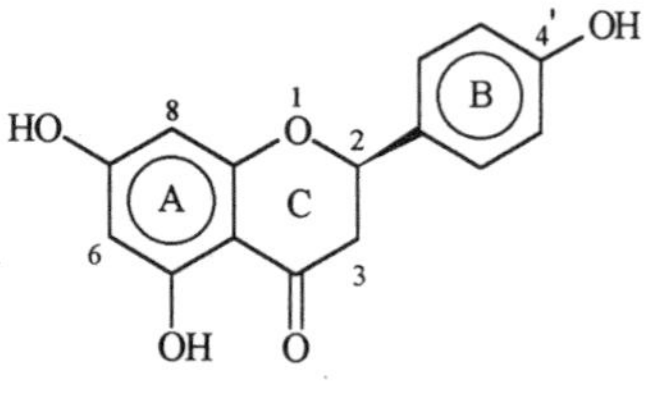

12

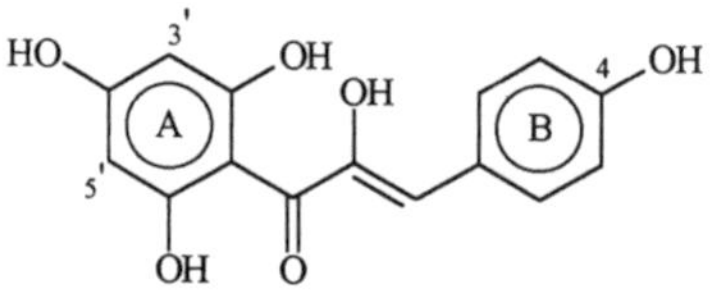

13

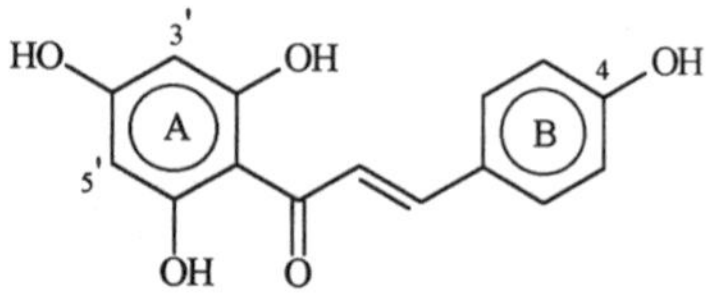

14

Figure 1. Phenolic constituents (**1**) to (**14**) from the heartwood of *Berchemia zeyheri*.

maesopsin (**5**) and (2*S*, 3*R*)-dihydrogenistein-(2α→7)-(2*S*)-maesopsin (**6**), representative of the first natural isoflavanone-benzofuranoid oligomers.[6] These and the closely related analogue (**9**), with an $\alpha,2'4,4',6'$-pentahydroxychalcone constituent unit, coexist in *B. zeyheri* with the predominant benzofuranone, (±)-maesopsin (**11**)[2,8] (7 percent of the total extract), (2*R*)-4',5,7-trihydroxyflavanone (**12**) (naringenin),[7] the $\alpha,2',4,4',6'$-pentahydroxychalcone (**13**),[8] 2',4,4',6'-tetrahydroxychalcone (**14**)[7], and a variety of monomeric[9] and oligomeric flavonoids.

The extract of heartwood drillings of *B. zeyheri*, obtained by extraction with Me_2CO/H_2O (8:2), was fractionated by repetitive countercurrent distribution in H_2O/sec-BuOH/*n*-hexane (5:3.5:1.5) (20 transfers) and in H_2O/sec-BuOH/*n*-hexane (5:4:1) (103 transfers). The combined last 10 tubes of the latter were subjected to column chromatography on Sephadex LH-20 with EtOH. Owing to the complexity of the resulting 25 fractions and having established the absence of natural methoxy groups by ^{1}H NMR, fraction 15 was methylated with dimethylsulphate under anhydrous conditions. This allowed further purification by flash column chromatography in *n*-hexane/C_6H_6/Me_2CO/MeOH (40:40:15:5) and repetitive PLC to yield the polyphenols (**1**), (**2**), (**5**), (**6**), and (**9**) as their permethyl ethers (**3**), (**4**), (**7**), (**8**), and (**10**).

3. STRUCTURAL ELUCIDATION OF OLIGOMERIC BENZOFURANOIDS

The close structural relationships between the permethyl ethers (**3**), (**4**), (**7**), and (**8**) are evident from the conspicuous resemblance of their ^{1}H NMR spectra (table 1), notably free of the effects of dynamic rotational isomerism about the interflavonoid bond, in conjunction with COSY- and NOESY-data. All exhibit an AM-spinsystem emanating from H-2 (δ 5.60–5.82) and H-3 (δ 4.43–4.66) of the heterocyclic C-ring with 2,3-*trans* relative configuration ($J_{2,3}$ = 10.5–12.0 Hz) and an AB-spinsystem (J_{AB} = 2.0 Hz) associated with H-6 and H-8 of the A-ring. Presence of the elements of a tetra-*O*-methylmaesopsin unit substituted at C-5 or C-7 of the D-ring is revealed by the residual singlet (δ 5.79–5.95), the presence of the shielded 2(F)-OMe (δ 2.99–3.24) together with the 2(F)-methylene (J_{AB} = 13.5–14.0 Hz). Differentiation between the two very similar AA'BB'-spin systems, reminiscent of the B- and E-ring protons, are effected via COSY spectra using the benzylic 2(F)-methylene and the H-2(C) [for flavanone analogues (**3**) and (**4**)] or H-3(C) [for isoflavanone analogues (**7**) and (**8**)] resonances as reference signals. NOE-association of the residual singlet with two methoxy groups [4(D)- and 6(D)-OMe] for all derivatives, (**3**), (**4**), (**7**), and (**8**) defines them as 5(D)-linked dimers. Long-range COSY- and NOESY-experiments allow definition of compounds based on a 3-substituted flavanone by association of H-2'(B) and H-2/6'(B) with H-2(C), as displayed by the 3α(C)/7(D)-linked zeyherins (**3**) and (**4**). This contrasts notably with the analogues (**7**) and (**8**) where coupling of H-2'(B) and H-6'(B) occurs with H-3(C), indicative of a 2(C)→3(C)-rearrangement of the B-ring and therefore of 2α(C)/7(D)-linked oligomers based on a 2α-substituted isoflavanone.

The above structural assignments correlate with the EI mass spectral fragmentation data of the derivatives (**3**), (**4**), (**7**), and (**8**), which apart from the

Table 1. ¹H NMR data for biflavonoid derivatives (**3**), (**4**), (**7**), (**8**), and (**10**) and phenethylmaesopsin enantiomers (**15**), (**16**), and (**17**). Chemical shifts in PPM (d), multiplicities and coupling constants (J) in Hz

Compound	(3)	(4)	(7)	(8)	(10)	(15) - (17)
H-2,6(B)	7.27 (d, 8.5)	7.07 (d, 8.5)	7.32 (d, 8.5)	7.38 (d, 8.5)	7.26 (d, 8.5)	7.07 (d, 8.5)
H-2,6(E)	7.12 (d, 8.5)	7.04 (d, 8.0)	7.01 (d, 8.5)	7.15 (d, 8.5)	7.57 (d, 8.5)	7.13 (d, 8.5)
H-3,5(B)	6.76 (d, 8.5)	6.69 (d, 8.5)	6.84 (d, 8.5)	6.82 (d, 8.5)	6.76 (d, 8.5)	6.79 (d, 8.5)
H-3,5(E)	6.65 (d, 8.5)	6.67 (d, 8.0)	6.65 (d, 8.5)	6.62 (d, 8.5)	6.85 (d, 8.5)	6.69 (d, 8.5)
H-6 (A)[a]	6.11 (d, 2.0)	6.13 (d, 2.0)	5.99 (d, 2.0)	6.05 (d, 2.0)	6.16 (d, 2.0)	
H-8 (A)[a]	6.14 (d, 2.0)	6.16 (d, 2.0)	5.98 (d, 2.0)	6.00 (d, 2.0)	6.13 (d, 2.0)	
H-5(D)	5.87 (s)	5.79 (s)	5.93 (s.)	5.95 (s.)	6.18 (s)	5.87 (s)
H-2 (C)	5.82 (d, 12.0)	5.60 (d, 12.0)	5.61 (d, 10.5)	5.62 (d, 10.5)	5.85 (d, 12.0)	
H-3 (C)	4.65 (d, 12.0)	4.43 (d, 12.0)	4.66 (d, 10.5)	4.62 (d, 10.5)	4.58 (d, 12.0)	
2(F)-CH₂	2.99 (d, 13.5)	2.98 (d, 13.5)	3.13 (d, 14.0)	3.19 (d, 14.0)		
	2.76 (d, 13.5)	3.05 (d, 13.5)	3.05 (d, 14.0)	3.06 (d, 14.0)		
H-β					6.25 (s)	2.69 (d, 8.0)
H-2 [b]						2.75 (d, 8.0)
H-3 [b]						2.71 - 2.63 (m)
2-CH₂						2.81 - 2.73 (m)
3-CH₂						
2(F)-OMe	3.11 (s)	2.99 (s)	3.24 (s.)	3.19 (s.)	3.67 (s, α-OMe)	3.18 (s)
Av-OMe	3.89 (s)	3.86 (s)	3.89 (s.)	3.92 (s.)	3.87 (s)	3.88 (s)
	3.86 (s)	3.84 (s)	3.87 (s.)	3.88 (2 x s.)	3.82 (2 x s)	3.79 (s)
	3.80 (2 x s)	3.79 (s)	3.86 (s.)	3.82 (s.)	3.73 (s)	3.76 (s)
	3.72 (s)	3.82 (s)	3.79 (s.)	3.79 (s.)	3.68 (s)	3.69 (s)
	3.71 (s)	3.69 (2 x s)	3.78 (s.)	3.71 (s.)	3.67 (s)	
			3.73 (s.)		3.57 (s)	

[a] May be interchanged ; [b] For the dideuterio analogue (**16**).

molecular ion M^+, m/z 656, all display prominent peaks at m/z 535, 476, 355, 313, 312, 181, and 121. The latter fragment reflects the loss of a 4-methoxybenzyl radical involving the E-ring hence affording the m/z 535 ion, whereas that at m/z 476 results from the equivalent of an RDA fragmentation of the flavanone- (**3**) and (**4**) or isoflavanone moiety (**7**) and (**8**) (ABC-unit).

The ^{1}H NMR spectrum (table 1) of the octa-*O*-methylether derivative (**10**) of the natural product (**9**), the first biflavonoid with an α-hydroxychalcone constituent unit, resembling the flavanone-chalcone dimer that was obtained from *Brackenridgea zanguebarica*,[10] exhibits the typical effects of dynamic rotational isomerism about the interflavonoid bond at ambient temperature. At elevated temperature (70 °C), the data are reminiscent of an $\alpha,2',4,4',6'$-pentamethoxychalcone moiety (vinylic H-β singlet, δ 6.25) coupled via C-3'(D) [residual H-5'(D) singlet, δ 6.18] to C-3 of tri-*O*-methylnaringenin, which expectedly displays doublets for H-2(C) (δ 5.85) and H-3(C) (δ 4.58) with 2,3-*trans* relative configuration ($J_{2,3}$ = 12.0 Hz). H-6(A) and -8(A) resonate as the familiar AB-spin system ($J_{6,8}$ = 2.0 Hz), and differentiation of the AA'BB'-spin systems of the B- and E-rings is again effected by a long-range COSY spectrum using the H-2(C) and the vinylic H-β resonances, respectively, as reference signals. NOE associations of H-5'(D) with two methoxy groups [4'- and 6'-OMe(D)] permitted differentiation between H-5'(D) and the vinylic H-β, which shows a conspicuous absence of an NOE association with the enolic α-OMe resonance (δ 3.76). Since *E*-2'-hydroxy-$\alpha,4,4',6'$-tetramethoxychalcone is known to display this association prominently,[11] a *Z*-configuration is inferred for the chalcone double bond. The mass spectrum confirms the molecular formula, $C_{38}H_{38}O_{11}$ (M^+, m/z 670) and is dominated by an RDA fragmentation involving the C-ring, thus leading to the base peak, m/z 490, and the A-ring fragment, m/z 181.

4. REDUCTIVE CLEAVAGE OF BENZOFURANOID-TYPE BIFLAVONOIDS

Owing to the complexity imposed by the three stereocenters, the well-defined CD spectra (fig. 2b), per se, did not permit stereochemical assignment to either of the derivatives (**3**) and (**4**). Thus, in the absence of direct stereochemical levers that allow the use of CD data in conjunction with NMR techniques, the problem was addressed by chemical degradation in an attempt to reduce the number of stereocenters. Whereas the interflavonoid bond in epimers (**3**) and (**4**) is stable toward acid-catalyzed thiolytic cleavage,[12,13] treatment of these compounds with sodium cyanoborohydride,[14] respectively, led to the formation of the 7-(4-methoxyphenethyl)-tetra-*O*-methylmaesopsin enantiomers (**15**) and (**17**) (fig. 3). The procedure involves the addition of Na(CN)BH$_3$ (12 molar excess) in portions over a period of 30 min to the zeyherin derivatives (**3**) and (**4**) dissolved in TFA under N$_2$ at 0 °C, followed by agitation for an additional 90 min.[15] The reaction is quenched by the careful addition of H$_2$O and adjustment of the pH to ~6.9 (Merck special indicator, pH 4.0–7.0) with 2 percent *aq*. NaHCO$_3$. The reaction mixture is subsequently extracted with EtOAc, dried over Na$_2$SO$_4$, and evaporated to dryness. Purification by PLC yields the product (**15**) or (**17**). The structures become evident

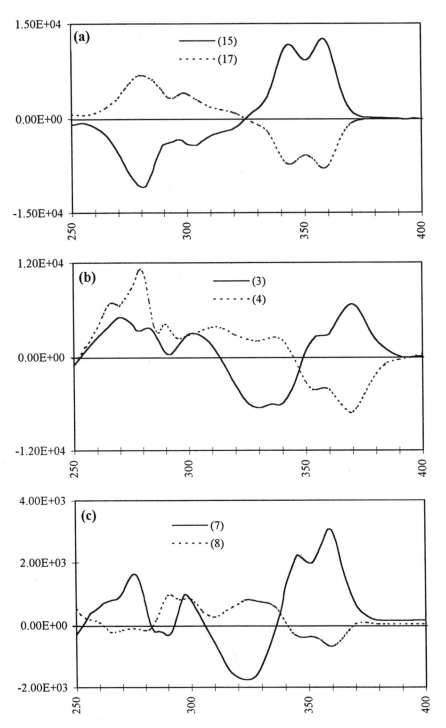

Figure 2. CD data for phenethylmaesopsin fragments (**15**) and (**17**) and biflavonoid derivatives (**3**), (**4**), (**7**), and (**8**).

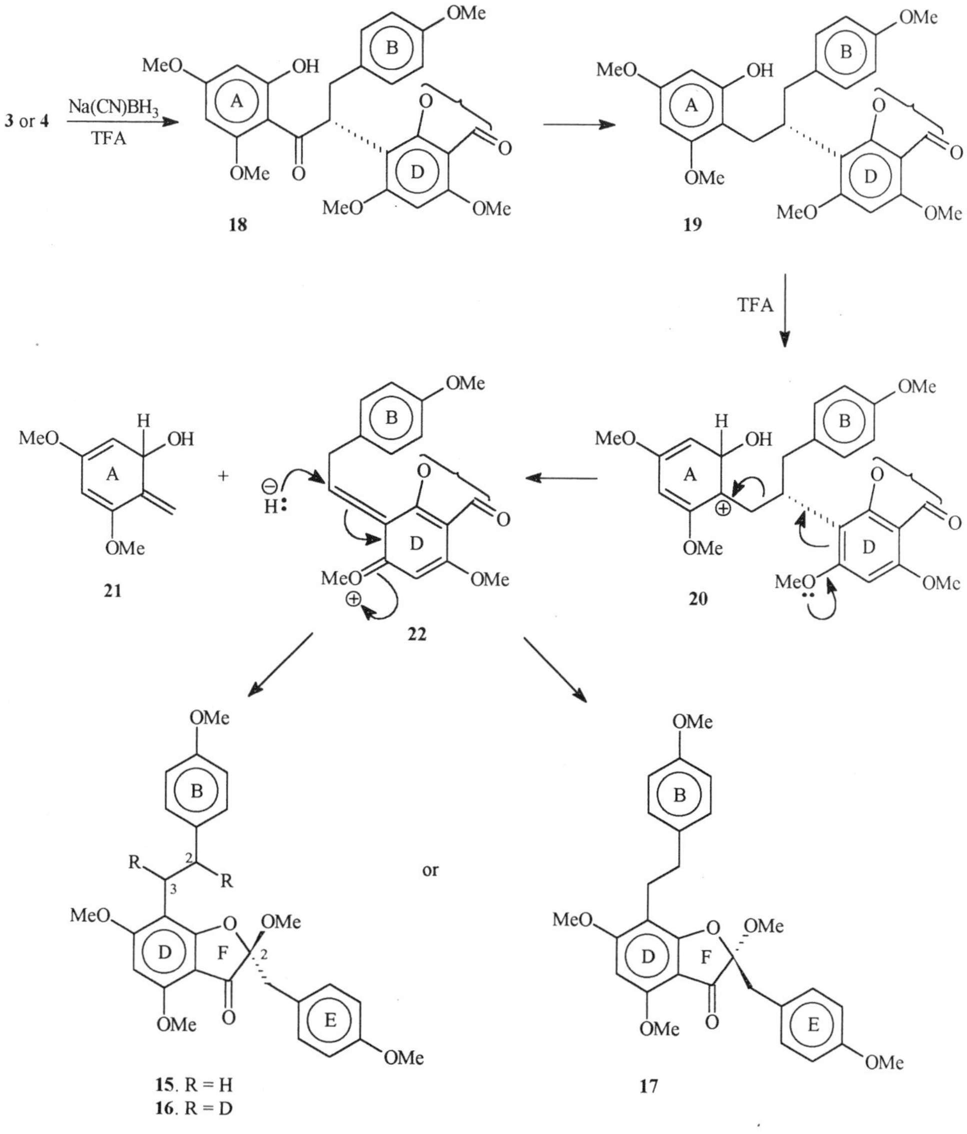

Figure 3. Degradation of the zeyherin epimers (**3**) and (**4**) with sodium cyanoborohydride in TFA.

by comparison of the ^{1}H NMR spectra (table 1) with those of the substrates and clearly illustrate retention of the maesopsin moiety and the B-ring but replacement of the elements of the A- and C-rings by two vicinal methylene groups in both compounds.

The genesis, respectively, of the two fragments (**15**) and (**17**) requires the breaking of both the O-1/C-2 and C-3/C-4 bonds of the C-rings of the epimers (**3**) and

(**4**). The initial step presumably involves reductive cleavage of the O-1/C-2 bond by the protonation of O-1 followed by hydride tranfer to C-2, hence leading to the formation of a 2,3-diarylpropiophenone (**18**), which is subsequently reduced to the 1,2,3-triarylpropane (**19**).[16,17] Protonation of the electron-rich phloroglucinol-type A-ring presumably imposes lability to the equivalent of the C-3/C-4 bond, which then breaks under the influence of the electron-releasing D-ring in the interme-diate (**20**). Aromatization of the A-ring fragment (**21**) via [1,3]-sigmatropic rearrangement results in the formation of 2-hydroxy-4,6-dimethoxytoluene, whereas reduction of the *o*-quinone methide type intermediate (**22**) affords the 2*R*(F)- or 2*S*(F)-7-(4-methoxyphenethyl)-tetra-*O*-methylmaesopsin (**15**) and (**17**). This mechanism was confirmed by utilizing sodium cyanotrideuterioborohydride under the same conditions for reduction of the zeyherin epimer (**3**), which resulted in the formation of the dideuteriomaesopsin analogue (**16**). The ^{1}H NMR data of the latter compound are given in table 1.

5. CONFORMATIONAL ANALYSIS OF THE BENZOFURANOID MOIETIES

The CD spectra (fig. 2a) of the maesopsin analogues (**15**) and (**17**) may in prin-ciple be used to define the absolute configuration at C-2(F) providing that the pre-ferred conformations of their benzofuranone moieties are known. Estimation of the latter by semi-empirical methods[18] (fig. 4) indicates that the oxacyclopentenone ring preferentially adopts a β-O-1-envelope conformation (**23**) with the heteroatom projecting above the plane (dihedral angle C$\underline{O}$/C-3/C-2/O-1 = −178.8°) of the enone ring system in the 2*R*(F)-enantiomer (**15**) and for the 2*S*(F)-enantiomer (**17**) an α-O-1-envelope conformation (**24**) with the heteroatom projecting below the plane (dihedral angle C$\underline{O}$/C-3/C-2/O-1 = 179.5°) of the enone ring system. The calcula-tions were performed by MOPAC 93[18] using the AM1 Hamiltonian with gradient minimization and GNORM = 0.1. Criteria for terminating optimizations were increased by a factor 10 (PRECISE).

To account, however, for the dependence of the CD-curves on the total ensem-ble of conformers significantly populated at ambient temperature rather than a single preferred conformer, a global search routine[18] was employed to explore the

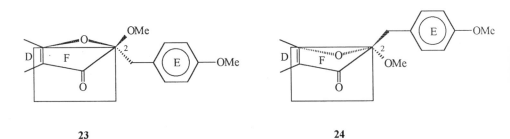

23   24

Figure 4. Calculated conformations (**23**) and (**24**) for the cyclopentenone F-ring of the 2*R*(F) and 2*S*(F)-phenylethylmaesopsins.

Table 2. GMMX data for the conformational analysis of compounds (**15**), (**17**), (**25**), and (**26**).

Structure	Total conformers[a]	Unique conformers[b]	Final ensemble[c]	Dihedral angle[d]	E_{min}[e]
(**15**)	7464	2993	1961	-168 to -179°	52.09
(**17**)	6547	2616	1582	168 to 179	52.04
(**25**)	8989	2804	613[f]	100 to 110°	77.70
(**26**)	3183	1098	331[f]	50 to 60°	76.02

[a] Total number of conformers considered during search; [b] Number of unique conformers (carbon skeleton only) within $3 kcal mol^{-1}$ of the most stable conformer; [d] Dihedral CO/C-3/C-2/O-1 for (15) and (17) and dihedral C-2(C)/C-3(C)/C-7(D)/C-6(D) for (25) and (26); [e] E_{min} of the most stable conformer in the final ensemble ($kcal mol^{-1}$); [f] Resulting conformers further refined by four additional cycles.

potential energy surface (PES) of both enantiomers (**15** and **17**). The search routine, GMMX version 1.0, based on the MMX forcefield of PC-MODEL,[18] was used to explore conformational space. Input was prepared with PC-MODEL's graphical user interface treating aromatic carbons as type 40 and the search run via the statistical option, alternating between internal (bonds) and external (cartesian) coordinates. The hydrogen bond function was activated and a dielectric constant ($\varepsilon = 1.5$) employed. Searches were allowed to run until the default cutoff criteria were reached and the Boltzman populations calculated at 25 °C for the final ensemble of conformers within $3 kcal mol^{-1}$ of the minimum energy. The results (table 2) conform to those of the initial calculations, indicating conformers (Boltzman population, 99.72 percent) with a β-O-1-envelope conformation (**23**) for the 2R(F)-enantiomer (**15**) and conformers (Boltzman population, 99.77 percent) with an α-O-1-envelope conformation (**24**) for the 2S(F)-enantiomer (**17**) within $3 kcal mol^{-1}$ energy window of the minimum.

6. ABSOLUTE CONFIGURATION OF BENZOFURANOIDS

According to Snatzke's chirality rule for cyclopentenones,[19] positive and negative Cotton effects for the n→π* transition in the 330–365 nm region of their CD spectra are in accord with β-O-1- and α-O-1-envelope conformations, respectively. Having established, therefore, the association of β-O-1- and α-O-1-envelope conformations with 2R(F) and 2S(F) absolute configurations for the maesopsin analogues (**15**) and (**17**) by conformational analyses, application of the rule to these compounds permits allocation of a 2R(F) absolute configuration to the enantiomer (**15**), displaying a positive Cotton effect $\{[\theta]_{359} + 3.04 \times 10^3\}$ and 2S(F)- to the enantiomer (**17**) with a negative Cotton effect $\{[\theta]_{361} - 3.14 \times 10^3\}$ (fig. 2a). The same configurations, 2R(F) and 2S(F), are consequently implied for the substrates (**3**) and (**4**) prior to degradation. Respective positive and negative Cotton effects in the 330–365 nm region of the CD spectra of these derivatives (**3**) and (**4**) (fig. 2b) confirm the conclusion. The CD curves (fig. 2a) for derivatives (**15**) and (**17**) should additionally also permit the unambiguous assessment of the absolute stereochemistry of the 2-benzyl-2-hydroxybenzofuranone group of naturally occurring flavonoids in general.

The CD data of the epimers (**3**) and (**4**) subsequently also permitted tentative assignment of a 2*R*(C) configuration to both compounds via high-amplitude positive Cotton effects $\{[\theta]_{274.3} + 4.8 \times 10^3$ and $[\theta]_{278.3} + 1.4 \times 10^4$, respectively} for the $\pi \rightarrow \pi^*$ transitions in their CD spectra[20] (fig. 2b). A global conformational search routine (GMMX 1.0),[18] utilizing the conditions mentioned above, was used to estimate the dihedral angle [C-2(C)/C-3(C)/C-7(D)/C-6(D)] about the C-3(C)/C-7(D) bond for conformers of zeyherins (**25**) and (**26**) (fig. 5) significantly populated within 3 kcal mol^{-1} of the minimum energy (table 2). The results indicated preferred E-conformers (C-ring) for both (**25**) and (**26**) with interflavonoid dihedral angles of 100–110° (Boltzman population, 99.98 percent) and 50–60° (Boltzman population, 99.90 percent), respectively. When taken in conjunction with the observed NOE association (0.7 percent) of 2-OMe(F) (δ 3.11) with 2-H(C) (δ 5.82) in conformer (**25**) of derivative (**3**) with the 2*R*-benzofuranoid constituent unit and its conspicuous absence in conformer (**26**) of derivative (**4**) with 2*S*(F) configuration, the 2*R*(C) absolute configuration of the zeyherins (**1**) and (**2**) is unequivocally confirmed. The 3*S*(C) configuration of both these natural products is then evident from their ^{1}H NMR coupling constants ($^3J_{2,3}$ 12.0 Hz), establishing a 2,3-*trans* (C) relative configuration. Although the same basic skeleton of the epimers (**1**) and (**2**) was proposed for zeyherin by Volsteedt and Roux,[2] the 60 MHz ^{1}H NMR data of their methylated derivative differ notably from the 300 MHz data of the permethyl ethers (**3**) and (**4**). This may suggest that their compound could have been a different diastereomer.

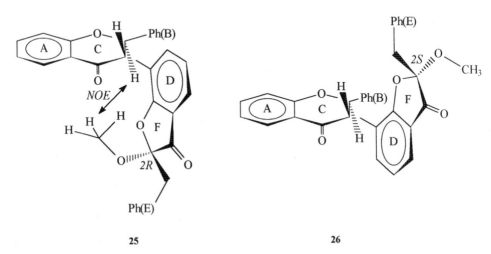

25   26

Figure 5. Preferred conformations about the 3α(C),7(D)-interflavonoid bond (**25**) and (**26**) for the zeyherin epimer derivatives (**3**) and (**4**) (substituents have been omited for clarity).

Although similar positive and negative Cotton effects in the 360 nm region $\{[\phi]_{359} + 3.06 \times 10^3$ and $[\phi]_{360} - 3.50 \times 10^2\}$ allow the allocation of $2R(F)$- and $2S(F)$-configurations, respectively, to the compounds (7) and (8) (fig. 2c), Cotton effects originating from the isoflavanone moieties are not comparable with those of flavanones. Isoflavanones display a distinctive Cotton effect in the 310–330 nm region for their n→π* transitions, a positive effect reflecting a $3R$-configuration and a negative effect a $3S$-configuration.[21,22] The strong positive effect observed in this region for compound (7) is therefore reminiscent of a $3R$-configuration, the high amplitude being the result of the cumulative effect of the chirality at C-3(C) and that at C-2(F), which would make a positive contribution to the effect [$2R(F)$-configuration]. The much weaker and negative Cotton effect displayed by the isomer (8) would then likewise be in accord with a $3R$(C)-configuration, since the positive effect originating from the chirality at C-3(F) and the negative effect associated with a $2S(F)$-configuration would now be opposing each other, the latter being slightly dominant (fig. 2c). When taken in conjunction with the 2,3-*trans* relative configuration for the the C-ring protons ($J_{2,3}$ = 10.5 Hz), a $2S(F)$-configuration is implied for both compounds and therefore permits assignment of the absolute configuration of these novel biflavonoids as $(2S, 3R)$-dihydrogenistein-$(2\alpha{\to}7)$-$(2R)$-maesopsin (5) and its epimer, $(2S, 3R)$-dihydrogenistein-$(2\alpha{\to}7)$-$(2S)$-maesopsin (6). The configurations are confirmed by the expected biogenesis of these compounds (see below).

The CD data of the octamethyl ether (10) of the flavanone-$(3,3')$-α-hydroxychalcone biflavonoid (9) resemble those of dihydrokaempferol [$(2R,3R)$-2,3-*trans*-$3',4',5,7$-tetrahydroxyflavanone]. Thus, the high-amplitude positive Cotton effect $\{[\theta]_{326.1}\ 3.2 \times 10^3\}$ for the n→π* transition and the negative Cotton-effect $\{[\theta]_{275} - 3.6 \times 10^3\}$ for the π→π* transition, in conjunction with ^{1}H NMR data indicating the 2,3-*trans* relative configuration of the C-ring, are reminiscent of $2S,3R$ absolute stereochemistry[20] for the biflavonoid derivative (10). Owing to the fact that the absolute stereochemistries of biflavonoids (1), (2), (5), (6), and (9) are now established, we choose to name these compounds according to the proposals of Hemingway et al.[23] for the proanthocyanidins, e.g. (1), as $(2R,3S)$-naringenin-$(3\alpha{\to}7)$-$(2R)$-maesopsin.

7. BIOSYNTHETIC SIGNIFICANCE

The biosynthetic significance of the enantiomeric relationship of the C-ring stereocenters of compounds (1) and (2) as compared to those of the flavanone-α-hydroxychalcone (9), as well as the enantiomerism of the maesopsin constituent units of the zeyherins (1) and (2) in contrast to the racemic nature of the parent maesopsin (11) in *B. zeyheri*, is not clear. It appears reasonable to propose (fig. 6) that a quinone methide radical (27) originating by oxidation of $2',4,4',6$-tetrahydroxychalcone[24] (14) effects electrophilic substitution[25] to the phloroglucinol-type ring of either maesopsin (11), presumably a specific enantiomer, or the α-hydroxychalcone (13). The facile interconversion,[26] 2-benzyl-2-hydroxybenzofuranone ↔ α-hydroxychalcone and flavanone ↔ chalcone, may feature prominently during the aging process thus explaining the array of enantiomerism depicted

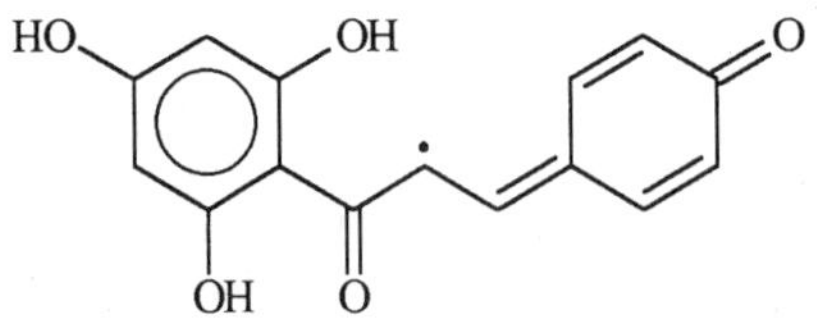

27

Figure 6. Proposed quinone methide radical (**27**).

above. The former interconversion also precludes an unambiguous claim of the flavanone-α-hydroxychalcone being a natural product since such a conversion may feasibly be induced by the process of *in vitro* methylation.

The absolute configuration of the isoflavanone moieties (2*S*, 3*R*) in (**5**) and (**6**), relative to those of the flavanone moieties (2*R*, 3*S*) in (**3**) and (**4**), is in agreement with the proposed biogenetic origin of the biflavonoids (**5**) and (**6**) via initial 2,3-phenyl migration of the (2*R*)-4′,5,7-trihydroxyflavanone (naringenin) (**12**), which coexists in *B. zeyheri*. Irrespective of the particular mechanism involved, *i.e.*, ionic or radical, several of which have been proposed,[27,28,29] such a migration always proceeds in a *suprafacial* fashion. (2*R*)-Naringenin would therefore experience phenyl migration from C-2β to C-3β, leaving C-2 electron deficient and prone to nucleophylic attack from the less hindered α-face by the phloroglucinol-type D-ring of maesopsin. Since the latter coexists as the racemate in *B. zeyheri*, the two epimers (2*S*, 3*R*)-dihydrogenistein-(2α→7)-(2*R*)-maesopsin (**5**) and (2*S*, 3*R*)-dihydrogenistein-(2α→7)-(2*S*)-maesopsin (**6**) originate.

8. CONCLUSIONS

The phenolic content of the heartwood of *Berchemia zeyheri* ("red ivory") includes a range of unique flavanone- or isoflavanone-benzofuranoid-type biflavonoids comprising (2*R*, 3*S*)-naringenin-(3α→7)-(2*R*)-maesopsin (**1**) and its epimer, (2*R*, 3*S*)-naringenin-(3α→7)-(2*S*)-maesopsin (**2**). These are accompanied by a pair of unique "rearranged" epimers, (2*S*, 3*R*)-dihydrogenistein-(2α→7)-(2*R*)-maesopsin (**5**) and (2*S*, 3*R*)-dihydrogenistein-(2α→7)-(2*S*)-maesopsin (**6**), representative of the first natural isoflavanone-benzofuranoid oligomers.

The O-1/C-2 and C-3/C-4 bonds of the C-ring in the flavanone-benzofuranone-type biflavonoids are subject to cleavage with sodium cyanoborohydride in trifluoroacetic acid at 0 °C. Comparison of the CD data of the resulting degradation products with those of the biflavonoids, when taken in conjunction with conformational analysis by molecular modeling and ^{1}H NMR, allows assignment of the absolute configuration to the biflavonoids and the 2-benzyl-2-hydroxybenzo[1]furan-3(2*H*)-one group of natural products. Thus the absolute configuration of at least some of the biflavonoids possessing stereocenters is indeed assessable when chemical degradation and appropriate physical data are used collectively. It additionally led to the generation of CD data facilitating the determination of the absolute configuration of the 2-benzyl-2-hydroxybenzofuranones for the first time.

ACKNOWLEDGMENTS

Financial support by the Foundation of Research Development, Pretoria, the Sentrale Navorsingsfonds of this University and the Marketing Committee, Wattle Bark Industry of South Africa, Pietermaritzburg, is gratefully acknowledged.

REFERENCES

1. Geiger, H. Biflavonoids and triflavonoids. *In*: Harborne, J.B. (ed.) The flavonoids–advances in research since 1986. Chapman and Hall, London, p. 96 (1994).

2. Volsteedt, F. du R.; Roux, D.G. Zeyherin, a natural 3,8-coumaranonylflavanone from *Phyllogeiten Zeyheri* sond. *Tetrahedron Letts.* :1647 (1971).

3. Palgrave, K.C. Dicotyledons. *In*: Moll, J. (ed.) Trees of southern Africa. C. Struik Publishers, Cape Town, p. 553 (1983).

4. Bekker, R.; Brandt, E.V.; Ferreira, D. The absolute configuration of biflavonoids and 2-benzyl-2-hydroxybenzofuranones. *J. Chem. Soc., Chem. Commun.* :957 (1996).

5. Bekker, R.; Brandt, E.V.; Ferreira, D. Absolute configuration of flavanone-benzofuranone-type biflavonoids and 2-benzyl-2-hydroxybenzofuranones. *J. Chem. Soc., Perkin Trans. 1* :2535 (1996).

6. Bekker, R.; Brandt, E.V.; Ferreira, D. Structure and stereochemistry of the first isoflavanone-benzofuranone biflavonoids. *Tetrahedron Letts.* (in press).

7. Bohm, B.A. Flavones and flavonols. *In*: Harborne, J.B. (ed.) The flavonoids—advances in research since 1986. Chapman and Hall, London, p. 329 (1988).

8. Volsteedt, F. du R.; Rall, G.J.H.; Roux, D.G. *Cis-trans* isomerism of a new α-hydroxychalcone from *Berchemia zeyheri* sond. (red ivory). *Tetrahedron Letts.* :1001 (1973).

9. Bekker, R.; Smit, R.S.; Brandt, E.V.; Ferreira, D. Benzofuranoids wih carbon frameworks reminiscent of products of benzylic acid rearrangement. *Phytochemistry* 43:673 (1996).

10. Drewes, S.E.; Hudson, N.A.; Bates, R.B.; Linz, G.S. Medicinal plants of Southern Africa. Part 1. Dimeric chalcone based pigments from *Brackenridgea zanguebarica*. *J. Chem. Soc., Perkin Trans. 1* :2809 (1987).

11. Brandt, E.V.; Ferreira, D. Unpublished results.

12. Betts, M.J.; Brown, B.R.; Brown, P.E.; Pike, W.T. Degradation of condensed tannins: structure of the tannin from common heather. *J. Chem. Soc., Chem. Commun.* :1110 (1967).

13. Thompson, R.S.; Jacques, D.; Haslam, E.; Tanner, R.J.N. Plant proanthocyanidins. Part 1. Introduction; the isolation, structure and distribution in nature of plant procyanidins. *J. Chem. Soc., Perkin Trans. 1* :1387 (1972).

14. Lane, C.F. Sodium cyanoborohydride, a highly selective reducing agent for organic functional groups. *Synthesis* :135 (1975).

15. Steynberg, P.J.; Steynberg, J.P.; Bezuidenhoudt, B.C.B.; Ferreira, D. Oligomeric flavanoids. Part 19. Reductive cleavage of the interflavanyl bond in proanthocyanidins. *J. Chem. Soc., Perkin Trans. 1* :3005 (1995).

16. Elliger, C.A. Deoxygenation of aldehydes and ketones with sodium cyanoborohydride. *Synth. Commun.* 15:1315 (1985).

17. Lewin, G.; Bert, M.; Dlaugnet, J.-C.; Schaeffer, C.; Guinamant, J.-L.; Volland, J.-P. Reduction de flavanones par le cyanoborohydrure de sodium dans l'acide trifluoroacetique-I/. *Tetrahedron Letts.* 30:7049 (1989).

18. GMMX, Version 1.0; PC MODEL, Version 3.0. Serena Software, P.O. Box 3076, Bloomington, IN 474-3076. MOPAC, Version 93.00. Stewart, J.J.P., Fujitsu Ltd., Tokyo, Japan.

19. Snatzke, G. Circulardichoismus VIII. Modifizierung der octantenregel für α,β-ungesättigte ketone: theorie. Circular dichoismus IX: Modifizierung der octantenregel für α,β-ungesättigte ketone: transoide enone. *Tetrahedron* 21:413, 421 (1965).

20. Gaffield, W. Circular dichroism, optical rotatory dispersion and absolute configuration of flavanones, 3-hydroxyflavanones and their glycosides. *Tetrahedron* 26:4093 (1970).

21. Kurosawa, K.; Ollis, W.D.; Redman, B.T.; Sutherland, I.O.; Alves, H.M.; Gottlieb, O.R. Absolute configurations of isoflavans. *Phytochemistry* 17:1423 (1978).

22. Hatano, T.; Kagawa, H.; Yasuhara, T.; Okuda, T. Two new flavonoids and other constituents in licorice root: their relative astringency and radical scavenging effects. *Chem. Pharm. Bull.* 36:2090 (1988).

23. Hemingway, R.W.; Foo, L.Y.; Porter, L.J. Linkage isomerism in trimeric and polymeric 2,3-cis-procyanidins. *J. Chem. Soc., Perkin Trans. 1* :1209 (1982).

24. Jackson, B.; Locksley, H.D.; Scheinmann, F.; Wolstenholme, W.A. Extractives from *Guttiferae*. Part XXII. The isolation and structure of four novel biflavanones from the heartwoods of *Garcinia buchananii* Baker and *G. eugeniifolia* Wall. *J. Chem. Soc.* (C) :3791 (1971).

25. Molyneaux, R.J.; Waiss, A.C.; Haddon, W.F. Oxidative coupling of apigenin. *Tetrahedron* 26:1409 (1970).

26. Ferreira, D.; Brandt, E.V.; Volsteedt, F. du R.; Roux, D.G. Parameters regulating the α- and β-cyclization of chalcones. *J. Chem. Soc., Perkin Trans. 1* :1437 (1975).

27. Dewick, P.M. Isoflavonoids. *In*: Harborne, J.B. (ed.) The flavonoids—advances in research since 1980. Chapman and Hall, London, p. 199 (1988).

28. Dewick, P.M. Isoflavonoids. *In*: Harborne, J.B. (ed.) The flavonoids—advances in research since 1986. Chapman and Hall, London, p. 203 (1994).

29. Crombie, L.; Holden, I.; Van Bruggen, N.; Whiting, D.A. Pre-isoflavonoid stages in the biosynthesis of amorphigenin: Ring-D formation and ring-A migration. *J. Chem. Soc., Chem. Commun.* :1063 (1986).

NEW NMR STRUCTURE DETERMINATION METHODS FOR PRENYLATED PHENOLS

Toshio Fukai and Taro Nomura

Faculty of Pharmaceutical Sciences
Toho University
Funabashi, Chiba 274-8510
JAPAN

1. INTRODUCTION

Recently, the occurrences of isoprenoid-substituted phenols from natural sources have become fairly familiar. Over the past 20 years, there have been increasing reports of isoprenylated compounds with structural, biological, and pharmacological interest.[1] Some of them have interesting bioactivities, e.g., anti-tumor promoting activity,[2] hypotensive effect,[3] antagonism for bombesin receptor,[4] inhibitory effects for some enzymes,[5,6] because they have both hydrophilic and hydrophobic groups in the molecule. It would not seem that use of modern NMR techniques (e.g., 2D NMR measurements) could result in the proposing of an incorrect structure. On the other hand, many structures of isoprenoid-substituted phenols had been reported without unambiguous evidence before 2D NMR spectrometry became routine work. Some structures were revised in reinvestigation of the plant source or in conflicting of a structure for other compounds. Nevertheless, most compounds have not been thoroughly described when a significant bioactivity is not found in the sources.[7] We wanted to get new techniques for the inspection of proposed structure by using earlier reported data. In the course of our study on phenolic compounds from medicinal plants, we found two new NMR methods for structure determination of 3-methyl-2-butenyl (prenyl) or (E)-3,7-dimethyl-2,6-octadienyl (geranyl) phenols. These methods also suit our objective of reinspecting previously reported structures. The first method involves classification with variation in the chemical shift of methylene carbon of the prenyl group and the second method is based on variation in the chemical shift of the 5-hydroxy proton of prenylated flavonoids. In this chapter, we describe these methods and a third method of classification of isoprenoid-substituted flavones.

Plant Polyphenols 2: Chemistry, Biology, Pharmacology, Ecology, Edited by
Gross et al. Kluwer Academic / Plenum Publishers, New York, 1999

2. VARIATIONS IN THE CHEMICAL SHIFT OF THE METHYLENE CARBON OF A PRENYL GROUP

The benzylic methylene carbon signal of the prenyl or geranyl group (C1) of phenols appears between δ 20–30. In the course of our work on prenylated phenols, we noticed that the chemical shift of the C1 signal depends on the nature of the substituents located at the adjacent positions.[8] To generalize the observation, we examined the ^{13}C NMR data of prenylated and/or geranylated phenols reported in the literature along with the phenols isolated or synthesized by our group (257 data sets including use of more than two prenyl groups in a compound and measurements in different solvents). On this basis, the prenyl (geranyl) groups could be classified into six types depending on the substituents located at the adjacent positions as follows (see fig. 1): The prenyl group of type 1 exists between two oxygen functions. The prenyl group of type 2 is substituted at C-3 position of the flavone. Such 3-prenylflavones have been isolated only from Moraceous plants.[3,9,10] In type 3 phenols, an oxygen function is located at one of the positions *ortho* to the prenyl group and the other *ortho*-position is substituted with a carbon function (an alkyl or alkenyl group). In compounds of type 4, one of the positions *ortho* to the prenyl group is unsubstituted, and an oxygen function is located at the other *ortho*-position. In type 5 compounds, both positions *ortho* to the prenyl group are replaced by carbon functions (alkyl and/or alkenyl groups). Type 6 prenyl groups are found in xanthone derivatives in which the prenyl group is attached to C-8 of 7-oxygenated xanthones (located between carbonyl and hydroxyl groups).

The chemical shifts of the C1 signals of these six types of prenyl groups are observed in restricted ranges specific to each type except for compounds of types

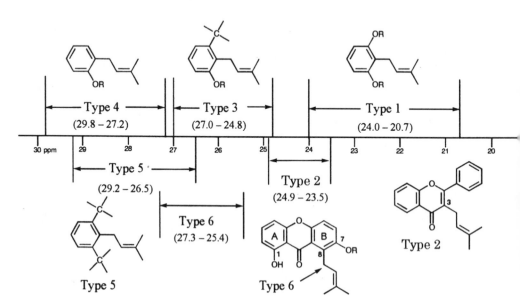

Figure 1. Variations in the chemical shift of C1 of the prenyl (geranyl) group (earlier version).

4 and 5 that overlap as shown in figure 1. To differentiate between compounds of type 4 and type 5, it is necessary to determine whether there is spin-spin coupling between the methylene carbon and an aromatic proton (^3J = ca. 4 Hz). Distinguishing of a prenyl group of type 3 from that of type 5 is difficult when the C1 signal is observed between δ 26.5–27.0. However, prenyl groups of type 5 are rarely found in natural compounds. Generally, such phenolic compounds have two isoprenoid substituents in one ring, and these groups are adjacent to each other as in broussoflavonol E (**1**),[8] kazinols C (**2**), and E (**3**)[11] (fig. 2).

This method is useful for the characterization of complex phenolic compounds. For example, the structures of broussoflavonols C and D isolated from *Broussonetia papyrifera* had been assigned to formulae **4'** and **5'** (fig. 3), respectively, by the observations of NOE between the methyl signal of 1,1-dimethylallyl group and H-5', a fragment ion at *m/z* 165 in EI mass spectra (assigned to **4'a**), and by chemical shifts of methoxy methyl carbons of their methyl ethers. In the course of the research of the C1 chemical shift, we found that the C1 chemical shifts of these compounds do not fit the above rule. The C1 signals of prenyl groups of broussoflavonol C appear at δ 26.2 and 29.2 in CDCl$_3$. If the structure of the compound is **4'**, one of the C1 signals of the prenyl groups should have occurred between δ 20.7–24.0 (8-prenyl group, type 1). The C1 signals of the compound indicated that the *ortho*-position to one of the prenyl groups is type 4 (*ortho*-oxygen function and *ortho*-unsubstituted) or type 5 (di*ortho*-carbon functions) (see fig. 1). From the above rule and the data reported earlier, the structure of broussoflavonol C must be formula **4**. Thus, we reinvestigated the structure of the compound and revised the structure to formula **4** using ^{13}C-NMR experiments including ^{1}H-^{13}C longrange correlation (LSPD-technique). The structure of broussoflavonol D was also revised to formula **5** from **5'** by the same manner. From Dreiding models of formulae **4** and **5**, it was indicated that a nearest distance between the methyl protons of the 1,1-dimethylallyl group and H-6' is 1.4 Å.[8]

This method is useful for the structure determination of prenyl phenols, but the importance of this technique decreased with the development of new NMR techniques such as HMBC experiment. However, erroneous structures have been proposed by reliance on HMBC results due to artificial cross peaks. The structures were revised with our rule and remeasurements of the NMR spectra.[7,12,13] Thus, we reinvestigated the C1 signals of the prenyl group, and revised figure 1 with the

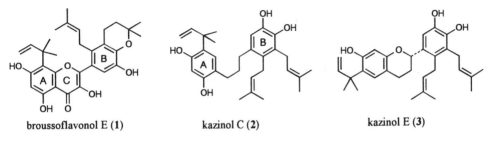

broussoflavonol E (**1**)   kazinol C (**2**)   kazinol E (**3**)

Figure 2. Structures of **1**–**3** having two isoprenoid groups in B ring.

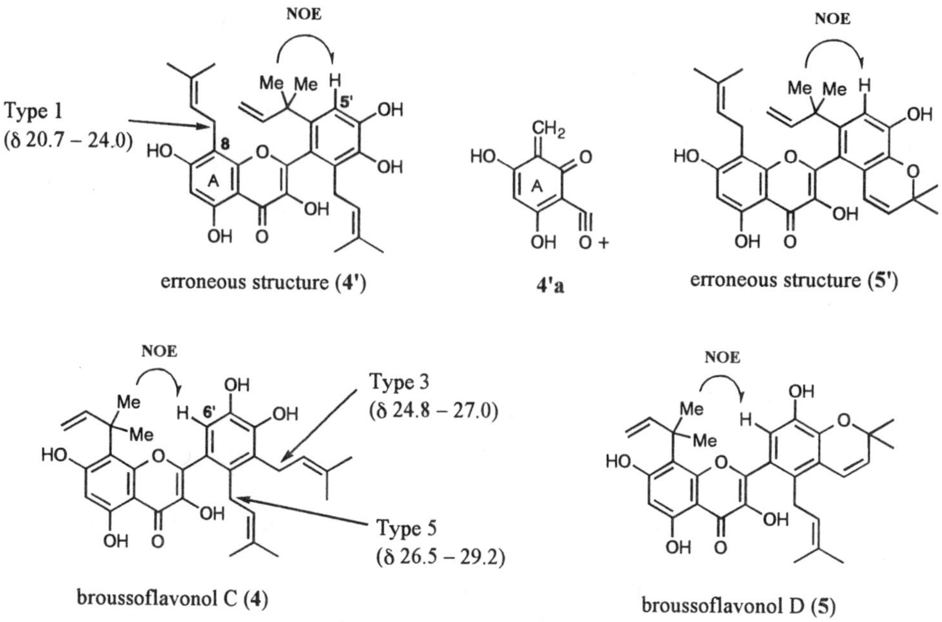

Figure 3. Structures of broussoflavonols C (**4**), D (**5**), and their erroneous structures (**4′** and **5′**).

addition of new data obtained through 1994.[14–16] Further revision is shown in table 1 (fig. 4). The table was revised with additional data up to the middle of 1998. In the above classification of chemical shifts, the boundaries of each group are somewhat ambiguous.[14–16] Nevertheless, the boundaries are sharp when the chemical shifts are compared in the same solvent (fig. 5). Furthermore, among compounds with the same carbon skeleton, the methylene signals of each group occur within narrower regions. In table 1, the prenylphenols are divided into seven groups depending upon the similarity of the skeleton as shown in figure 4.

It is noteworthy that the di*ortho*-substituent effect for the chemical shift of C1 of prenyl group is different from a general substituent effect. Similar unusual chemical shifts are observed for methoxyl carbons: di*ortho*-substituted methoxyl carbons appear more downfield (δ ca. 60) than mono*ortho*-substituted or di*ortho*-unsubstituted methoxyl carbons (δ ca. 55).[17] Generally, di*ortho*-substitution causes an upfield shift of the methylene carbon (known as steric compression). For example, the α-methylene carbon of gancaonin R (**6**, δ 32.8 in acetone-d_6) appears more upfield than that of gancaonin S (**7**, δ 36.2 in acetone-d_6) (fig. 6).[18]

The above new NMR method may be a useful inspection as a final stage of the structure determination. In a review of published assignments, some compounds did not fit our rule. Either the assignments for the structure and/or the NMR chemical shift assignments need to be reinvestigated. The same method could probably be useful for structure determination of phenolic alkaloids containing prenyl group residue, and the application was reported briefly by Furukawa et al.[19]

Table 1. The chemical shifts of methylene carbons of the prenyl (geranyl) groups of phenolic compounds (δ in ppm).

Group 1[a]

Type			
Type 1	21.6 – 22.7 (A ring) 23.2 – 24.8 (B ring)	21.0 – 22.7 (A ring) ————	20.8 – 22.0 (A ring) 22.2 – 22.6 (B ring)
Type 2	24.4 – 25.3	24.1 – 24.4	23.4 – 24.1
Type 3	25.8 – 27.3	25.8 – 26.2	25.2
Type 4	28.6 – 29.5	28.9 – 29.2	27.4 – 28.2
Type 5	————	28.0 – 29.2	27.3 – 28.4
solvent	acetone-d_6	CDCl$_3$	DMSO-d_6

Group 2

Type			
Type 1	21.4 – 22.9 (A ring) 22.9 – 24.6 (B ring)	20.8 – 22.8 (A ring) ————	20.5 – 22.3 (A ring) 23.0 – 23.3 (B ring)
Type 3	24.9 – 25.6		————
Type 4	27.6 – 29.4	27.6 – 29.9	26.9 – 29.1
solvent	acetone-d_6	CDCl$_3$	DMSO-d_6

Group 3

Type			
Type 1	22.6 – 22.8 (catechins) 22.9 – 23.6	22.0 – 22.7 (catechins) 22.0 – 23.2	———— ————
Type 3	————	25.1 – 25.6	————
Type 4	28.5 – 28.9	27.9 – 29.7	
solvent	acetone-d_6	CDCl$_3$	DMSO-d_6

Group 4

Type			
Type 1	22.0 – 24.0	21.6 – 23.5	21.0 – 22.3
Type 3	26.3	25.8 – 27.3	————
Type 4	————	28.4	27.8
solvent	acetone-d_6	CDCl$_3$	DMSO-d_6

[a] See Figure 4.

3. VARIATIONS IN THE CHEMICAL SHIFT OF THE 5-HYDROXYL PROTON OF 6- OR 8-ISOPRENYL FLAVONOIDS

Many 5,7-dihydroxyflavonoids with isoprenyl groups (prenyl, geranyl, or lavandulyl) at the C-6 or C-8 position have been isolated from natural sources.[1,3,9,15,20,21] Various methods exist for determining the position of substitution, such as the Gibbs test for flavonoids having no hydroxyl group at C-2' or C-3', the anomalous aluminum induced UV shift, cyclization reactions (6-prenylflavonoids afford two

Table 1. *Continued*

Group 5

Type			
Type 1	21.6 – 22.8 (A ring)	21.2 – 23.3 (A ring)	20.7 – 21.2 (A ring)
	23.0 – 25.6 (B ring)	22.6 (B ring)	————
Type 3	————	————	24.1
Type 4	28.8 – 29.0	27.4 – 28.1	28.3 – 28.6
Type 5	————	————	27.9
Type 6	26.0 – 29.4	25.6 – 26.8	25.3 – 25.8
solvent	acetone-d_6	CDCl$_3$	DMSO-d_6

Group 6

Type			
Type 1	21.5 – 23.2	21.4 – 23.4	21.0 – 21.8
Type 3	25.5 – 26.2	24.4 – 26.1	———
Type 4	28.2 – 29.2	28.2 – 30.0	———
Type 5	———	26.2 – 27.8	———
solvent	acetone-d_6	CDCl$_3$	DMSO-d_6

Group 7

Type			
Type 1	23.7	23.6 – 24.6	———
Type 2	25.3 – 25.9	25.5 – 25.9	———
Type 4	28.3	26.5 – 29.9	———
solvent	acetone-d_6	CDCl$_3$	DMSO-d_6

products because of cyclization to 5-OH or 7-OH, 8-prenylflavonoids afford only one product by cyclized to 7-OH), shifts of chromene protons after acetylation of a 5-hydroxyl group (for this opportunity for a cyclization reaction must exist), chemical shifts of C-6 and C-8 carbons, etc.[22–26] These methods are not always applicable and may therefore lead to erroneous structures. The examples of unsuitable methods for structure determination have been listed elsewhere.[15]

An unambiguous method for the structure determination of such prenylated flavonoids is the observation of spin-spin coupling between the 5-hydroxyl proton and C-6 ($^3J = 6$ Hz) by a ^{13}C NMR measurement such as an undecoupled spectrum, HMBC spectrum, etc.[27,28] This method needs relatively large amounts of a sample, but the following ^{1}H NMR method requires only a small amount of the flavonoid (less than 200 μg for routine measurements).

The hydrogen-bonded hydroxyl proton on C-5 (5-OH) of flavonoids appears between δ 12.1–12.5 in acetone-d_6 (fig. 7). The 5-OH chemical shift depends to some degree on the frequency of the spectrometer used for the measurement. When the spectrum is measured by a 100 MHz spectrometer, the signal appears further upfield (about 0.1 ppm) than when a 400 MHz instrument is used. The chemical shifts of the 5-OH signals reported in this section were observed at 400 or 500 MHz

Group 1

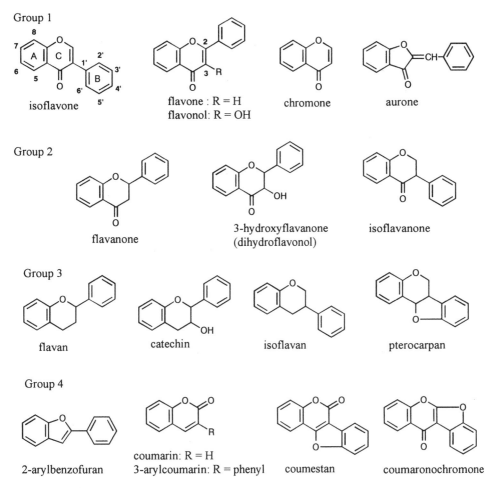

isoflavone

flavone : R = H
flavonol: R = OH

chromone

aurone

Group 2

flavanone

3-hydroxyflavanone
(dihydroflavonol)

isoflavanone

Group 3

flavan

catechin

isoflavan

pterocarpan

Group 4

2-arylbenzofuran

coumarin: R = H
3-arylcoumarin: R = phenyl

coumestan

coumaronochromone

Group 5 [xanthone]

Group 6 [chalcone, dihydrochalcone, acetophenone, benzophenone, benzoic acid, coumaric acid,
1,3-diphenylpropane, stilbene, dihydrostilbene, dibenzoylmethane, etc.]

chalcone

stilbene

dibenzoylmethane

Group 7 [phenol, dihydrophenancerene]

Figure 4. Grouping of prenylphenols in table 1.

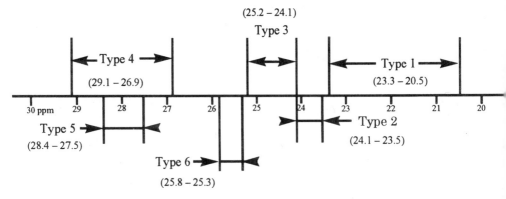

Figure 5. Variations in the chemical shifts of C1 of the prenyl (geranyl) group measured in DMSO-d_6 (new version).

in acetone-d_6, unless otherwise stated. The chemical shift of the signal also shows a slight but small upfield shift as temperature is increased (flavone, isoflavone, and flavanone: 0.0022 ppm per degree, flavonol: 0.0033 ppm per degree).[29] The shift with change in concentration is negligible and is not changed over a concentration range of 0.03–2 percent.

In the course of our work on the flavonoids, we noticed that the chemical shift of the 5-OH of flavonoids is shifted by prenylation at C-6 or C-8.[30] On examining the ^{1}H NMR data of 50 isoprenoid-substituted flavanones reported in the literature and the flavanones isolated or synthesized by our group[31] where the hydrogen-bonded hydroxyl proton (5-OH) appears between δ 12.0–12.6 (in acetone-d_6) and δ 11.9–12.4 (in CDCl$_3$), one finds that the 5-OH of 6-prenylated flavanones appears further downfield than that of 6-unsubstituted flavanones having the same B ring. In contrast, the 5-OH signal of an 8-prenylated flavanone is shifted slightly upfield compared with that of a flavanone having the same B ring and no side chain. The same regularities were observed in similar flavanones containing other isoprenoid groups such as geranyl, isoamyl, 3-hydroxy-3-methylbutyl, and lavandulyl residues. The downfield shift of 6-prenylated flavanones is about

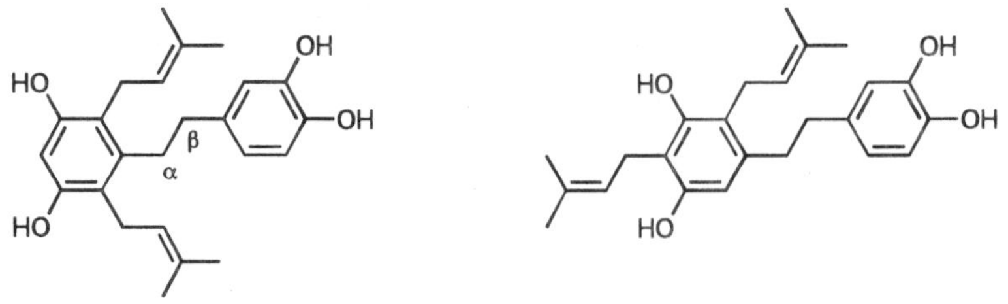

gancaonin R (6)   gancaonin S (7)

Figure 6. Structures of gancaonins R (**6**) and S (**7**).

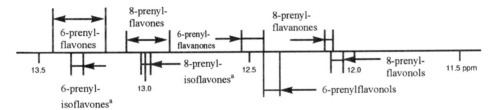

Figure 7. Chemical shifts of the hydrogen-bonded hydroxyl group (5-OH) of flavonoids in acetone-d_6 at 25 °C (400 and 500 MHz).

0.3 ppm, and the upfield shift of 8-prenylated flavanones is about 0.05 ppm when the ^{1}H NMR spectra are measured in acetone-d_6. The downfield shift is 0.35 ppm when measured in $CDCl_3$.[31,32]

The difference of the chemical shift of the 5-OH of 6- and 8-prenylated eriodictyols isolated from *Wyethia helenioides* had been reported by Bohlmann, et al.[33] They pointed out a possibility for structure determination based on that chemical shift. Unfortunately, their structures were wrong. We synthesized 6- and 8-prenylated eriodictyol (**8** and **9**, fig. 8) by an unambiguous method and revised the formulae proposed previously for the natural flavanones.[31] In examining the literature, we found that the above generalization was not in accord. Recently, two of these flavanones and some analogs were re-isolated or synthesized.[32,34,35] From these results, their chemical shifts now followed our rule.

The 5-OH signal of 3-hydroxyflavanone (dihydroflavonol) appears further upfield than that of a flavanone having the same A and B rings ($\Delta\delta$ 0.46 ± 0.01 ppm). Substituent effects on 3-hydroxyflavanones are almost the same as those on the flavanones. Prenylation effects at C-6 or C-8 are also useful for the structure determination of prenylated (geranylated) 3-hydroxyflavanones.[36] The parameters for flavanone are summarized with the addition of the data of 3-hydroxyflavanone derivatives in table 2.

The 5-OH signals of flavones appear at δ 12.8–13.4 in acetone-d_6.[37] The effects of substitution on the hydroxyl signal are shown in table 2. Parameter (a) of

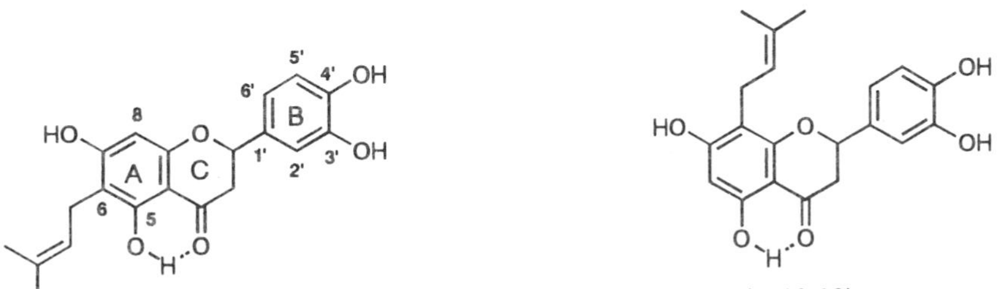

(δ 12.48)

6-prenylated eriodictyol (**8**)

(δ 12.12)

8-prenylated eriodictyol (**9**)

Figure 8. The chemical shifts of 5-OH of 6- and 8-prenylated eriodictyols (**8** and **9**) in acetone-d_6.

Table 2. Substituent parameter (ppm) on the 5-hydroxyl group of flavonoids (in acetone-d_6).

Substituents	Flavanones (3-Hydroxyflavanones)	Flavones (Flavonols)
(*a*) Prenylation at C-6	+ 0.29 ± 0.01	+ 0.25 ± 0.02
(Prenylation at both C-6 and C-8)		
(*b*) Prenylation at C-8	− 0.05 ± 0.03	− 0.06 ± 0.01
(*c*) Hydroxylation at C-3	− 0.46 ± 0.01	− 0.83 ± 0.01
(*d*) Methylation of 7-OH		
1) 6-Prenylflavonoid	− 0.16 ± 0.03	
2) 8-Prenylflavonoid	+ 0.10	+ 0.14 ± 0.04
3) 6,8-Unsubstituted flavonoid	− 0.04	
(*e*) Hydroxylation at C-2'		
1) 3'-Unsubstituted flavonoid	+ 0.03 ± 0.02	+ 0.12 ± 0.02
2) 3'- Alkylated flavonoid	+ 0.02	+ 0.07
(*f*) Hydroxylation at C-3'	+ 0.01	− 0.01
(*g*) Hydroxylation at C-4'	+ 0.02 ~ + 0.03	0 ± 0.01
(*h*, Methylation of 3-OH)		+ 0.09 ± 0.02

Substituents	Isoflavones	Isoflavanones
(*a*) Prenylation at C-6	+ 0.27 ± 0.03	+ 0.29 ± 0.01
(Prenylation at both C-6 and C-8)		
(*b*) Prenylation at C-8	− 0.06 ± 0.02	− 0.06 ± 0.02
(*d*) Methylation of 7-OH		
1) 6-Prenylflavonoid	− 0.11 ± 0.01	− 0.15 ± 0.01
2) 8-Prenylflavonoid	+ 0.09	
3) 6,8-Unsubstituted flavonoid	− 0.05	
(*e*) Hydroxylation at C-2'		
1) 3'-Unsubstituted flavonoid	− 0.28 ± 0.03	+ 0.12
2) 3'-Alkylated flavonoid	− 0.48 ± 0.05	
(*i*) Cyclization between 2'-OH and 3'-prenyl group	− 0.51 ± 0.05	+ 0.03

flavones (prenylation at C-6) is smaller than that in flavanones. The effect of hydroxylation at C-2' of a flavone (parameter *e*) is of considerable value in structure determination as it results in a downfield shift of 0.12 ± 0.02 ppm. The effect of the hydroxylation at C-3 (flavonol) is an upfield shift of 0.83 ± 0.01 ppm. The 5-OH signals of prenylated flavonols are observed between δ 12.0 and 12.4 in acetone-d_6 when the 3-hydroxyl group is free.[38] The substituent parameters of flavonols are almost the same as those of flavones. Thus, the effects of substitution are shown along with the parameters of flavone in table 2. It should be noted that the 5-OH signal of flavonols is sometimes shifted due to chelation with metals. When a flavonol fluoresces under irradiation with UV light,[39] the sample needs further purification. The 5-OH signals of flavonols are shifted downfield shift by 0.64–0.67 ppm with methylation of the 3-OH. However, the 5-OH signals of 3-*O*-methylflavonols appear still further upfield (about 0.2 ppm) than those of flavones having the same A and B rings.

The 5-OH signals of isoflavones appear between δ 12.5 and 13.5 in acetone-d_6. The changes in the chemical shift of the 5-OH signal upon hydroxylation at C-2' of isoflavones as well as the effect of prenylation at C-6 or C-8 were reported briefly by us[30] and in detail later by Tahara et al.[40] Hydroxylation at C-2' of a flavone or flavonol causes a downfield shift of the 5-OH signal (table 2); however, the 5-OH signal of 2'-hydroxyisoflavones appears farther upfield than that of 2'-unsubstituted isoflavones or 2'-*O*-alkyl(alkenyl)isoflavones (0.25–0.56 ppm, table 2). The 5-OH signals of prenylated isoflavanones appear at δ 12.0–12.7 in acetone-d_6. The substituent parameters of isoflavanones are shown in table 2, but these parameters are incomplete due to the small number of examples in the literature.[41] The usefulness of the parameters in table 2 for structure determination was described in our earlier papers.[15,16,29–31,36–38,41]

The prenylation effect of a 7-*O*-methyl or 7-*O*-prenyl ether of a flavanone is different from that of 7-hydroxyflavanones. The 5-OH of a 7-*O*-methyl or prenyl ether of a 6-prenylflavanone appears further downfield than that of a 6-unsubstituted flavanone having the same B ring (0.13–0.14 ppm in acetone-d_6, 0.03–0.05 ppm in CDCl$_3$). The 5-OH signal of an 8-prenylated 7-*O*-methyl or 7-*O*-prenyl ether of a flavanone is also downfield compared with that of an 8-unsubstituted flavanone having the same B ring (0.10–0.11 ppm in acetone-d_6 or CDCl$_3$).[32] It should be noted that the signals of the 5-OH of 6- or 8-prenylated 7-*O*-methylflavanones appeared in a relatively narrow range (δ 12.20–12.24 in acetone-d_6, 12.04–12.12 in CDCl$_3$). Thus, comparison of the 5-OH signal of a 7-*O*-methyl or prenyl ether requires measurement under same conditions, frequency of instrument and use of CDCl$_3$.

A hypothesis concerning the substituent effects on the 5-OH signal has been presented.[15] Significant differences in electron density (at the 4-carbon, 4-oxygen, 5-carbon, 5-oxygen, and 5-hydrogen) were not found among the models of 6- or 8-prenylated and non-prenylated flavanones **10**, **11** and **12** (fig. 9). The distances between the 5-OH hydrogen and the oxygen of the carbonyl group of compounds **10**, **11**, and **12** are almost the same. Since the electron densities near the hydrogen bond of **10**, **11**, and **12** are about the same, and because the contribution of chemical exchange (saturation transfer) is small, the following is the most reasonable cause for the downfield shift of the 5-OH signal on prenylation at the 6-position: The 6-prenyl group interferes with rotation of the bond between the C-5 carbon and the C-5 hydroxyl group and/or distorts the molecular orbital at the 5-OH hydrogen. Thus, the 5-OH signal of 6-prenylated flavonoids may appear further downfield than that of flavonoids having no substituent at C-6.

This hypothesis may be supported by the substituent effect of 2-α,α-dimethylallyl group on 1-OH signal of xanthone.[42] The substituent effect of the 2-prenyl

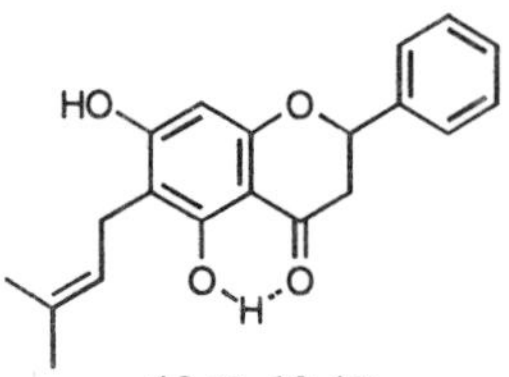

10 (δ 12.43) 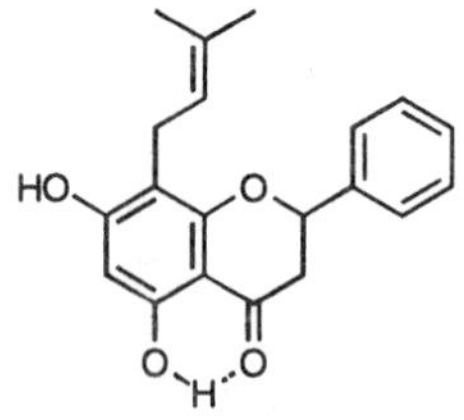   11 (δ 12.10) 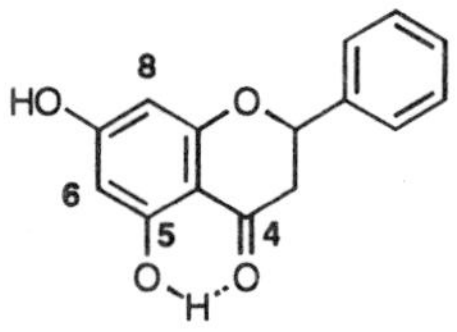   12 (δ 12.14)

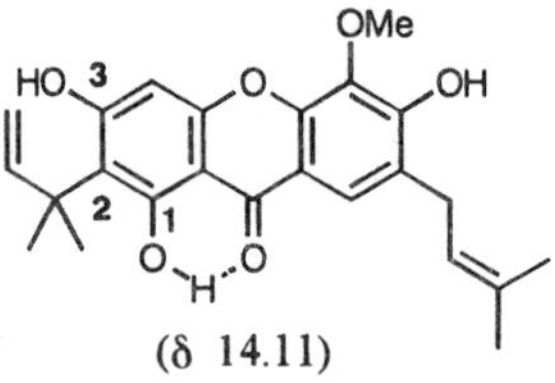

(δ 14.11)   (δ 13.41)   (δ 13.80)

cudraxanthone E (13)   cudraxanthone F (14)   paxanthonin (15)

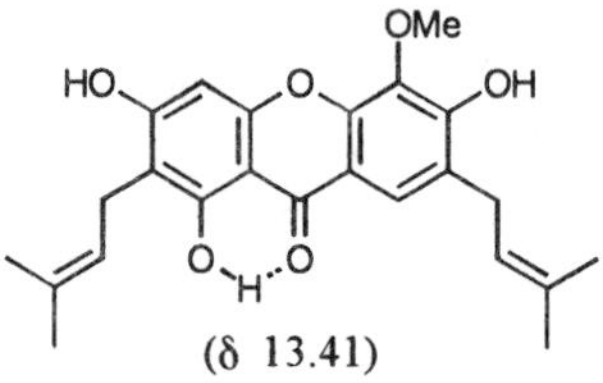

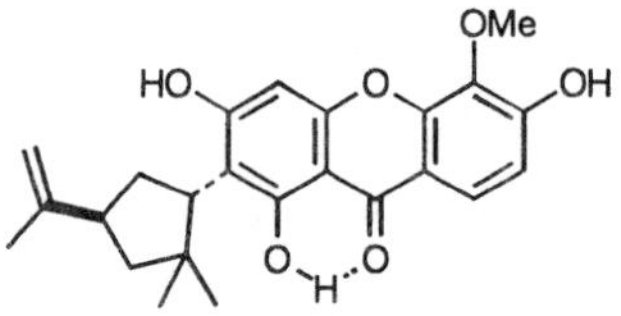

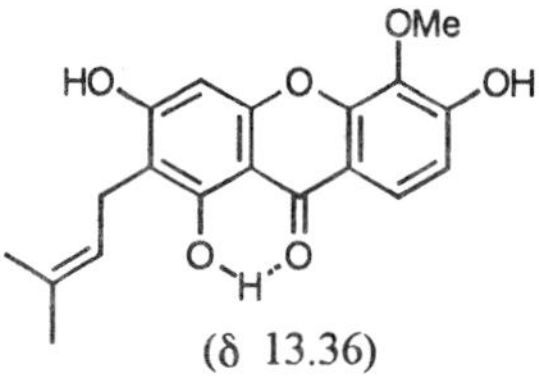

(δ 13.36)   17 R = H   (δ 13.24, 300 MHz)

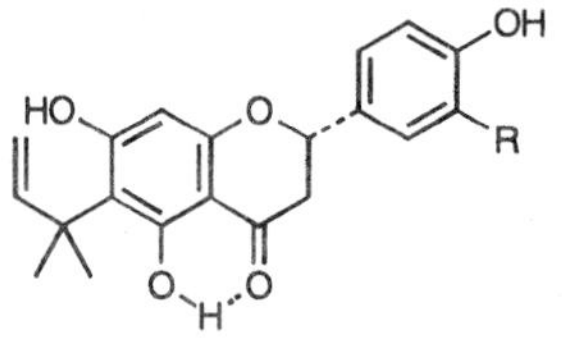

morusignin D (16)   18 R = OH   (δ 13.23, 300 MHz)
                    19 R = OMe (δ 13.24, 300 MHz)   20 R = OH (δ 12.48, 300 MHz)
                                                     21 R = OMe (δ 12.49, 300 MHz)

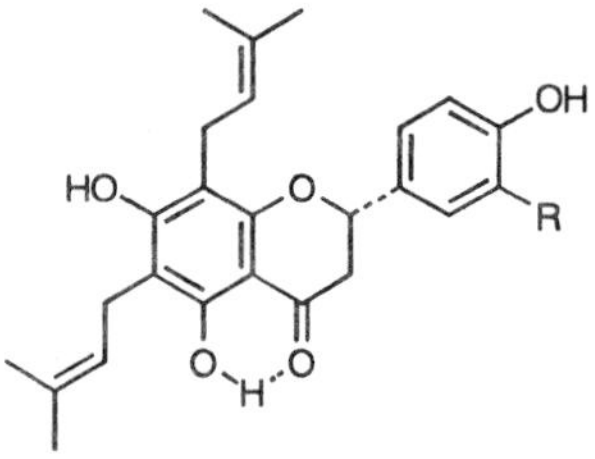

Figure 9. The chemical shifts of 5-OH of flavanones (**10–12, 17–21**[a]), 1-OH of xanthones (**15, 16**)[b] in acetone-d_6 and cudraxanthones E (**13**) and F (**14**) in CDCl$_3$. [a]Data from reference 45. [b]Data from reference 43.

group on the 1-OH of 1,3-dihydroxyxanthone derivative is a downfield shift of about 0.3 ppm (unpublished data). The hydrogen-bonded hydroxyl proton (1-OH) of cudraxanthone E (**13**, fig. 9) having a 2-α,α-dimethylallyl group appeared further downfield (0.70 ppm) than that of cudraxanthone F (**14**) with a 2-prenyl group.[42] A similar downfield shift (0.44 ppm) has also been observed for a monoterpene-substituted xanthone, paxanthonin (**15**, fig. 9).[43] The downfield shift of **13** and **15** could be explained with the above hypothesis. The rotation of the bond between the C-1 and the 1-OH group receives further interference with more bulky groups. Thus, the 1-OH signal of cudraflavone E (**13**) and paxanthonin (**15**) may appear further downfield than that of 2-prenylxanthone (**14**) and morusignin D (**16**),[44] respectively. Recently, some 6-(α,α-dimethylallyl)-flavanones (**17–19**, fig. 9) were isolated from *Monotes engleri*.[45] The 5-OH signals of the flavanones (**18** and

19) also appear further downfield (0.75 ppm) than that of 6,8-diprenylated eriodictyol (**20**) or its derivative (**21**), respectively.

The 5-OH signal of a linear pyranoisoflavone such as alpinumisoflavone (**22**) appears further downfield (0.41 ppm) than that of the comparable isoflavone with no substituent at the C-6 and C-8 positions (genistein, **23**) in acetone-d_6 (see fig. 10). In the angular pyranoisoflavone derrone (**24**), the 5-OH signal is also downfield from that of genistein (0.11 ppm).[40] The notable difference between the chemical shifts of linear and angular isoflavones (0.3 ppm) is a useful tool for structure determination. A similar shift difference is also observed in pyranoflavones as shown in figure 10. The difference between the chemical shifts of the 5-OH signals in the linear dihydropyranoflavanone (**28**) and its angular type isomer (**29**) is large (0.66 ppm, fig. 10). It is also useful for the structure determination of pyranoflavanones.

In examining the literature, we found that the above generalizations were not in accord with the chemical shifts reported for some isoprenoid-substituted flavonoids. We reinvestigated the structure of six synthesized flavonoids as well

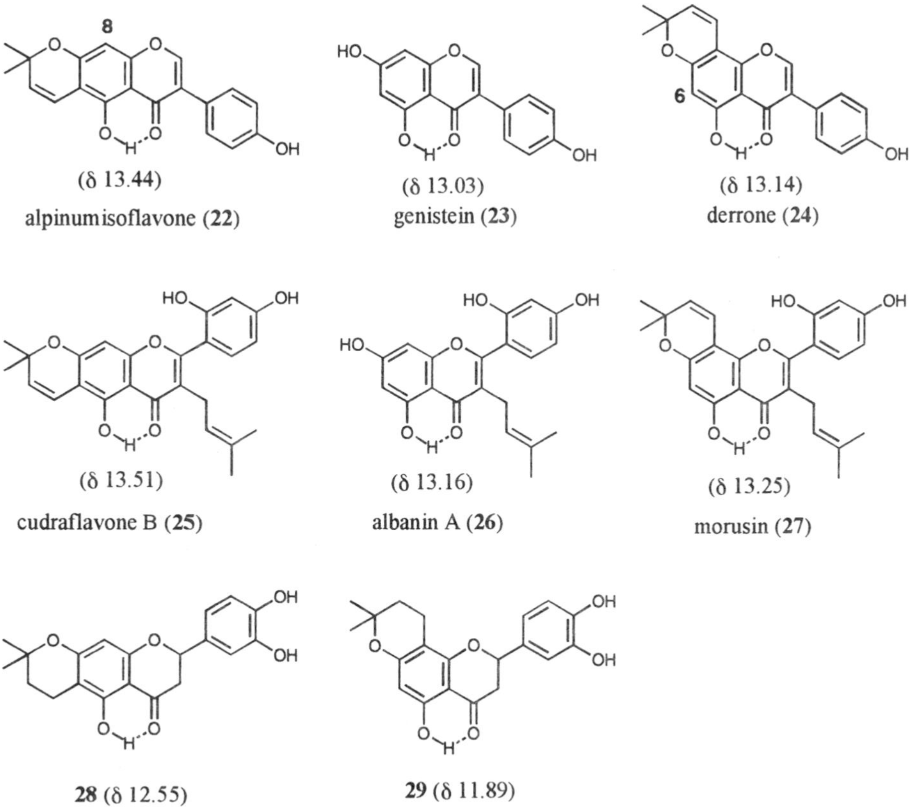

Figure 10. Chemical shifts of 5-OH of pyranoflavonoids and dihydropyranoflavanones (**28** and **29**) in acetone-d_6.

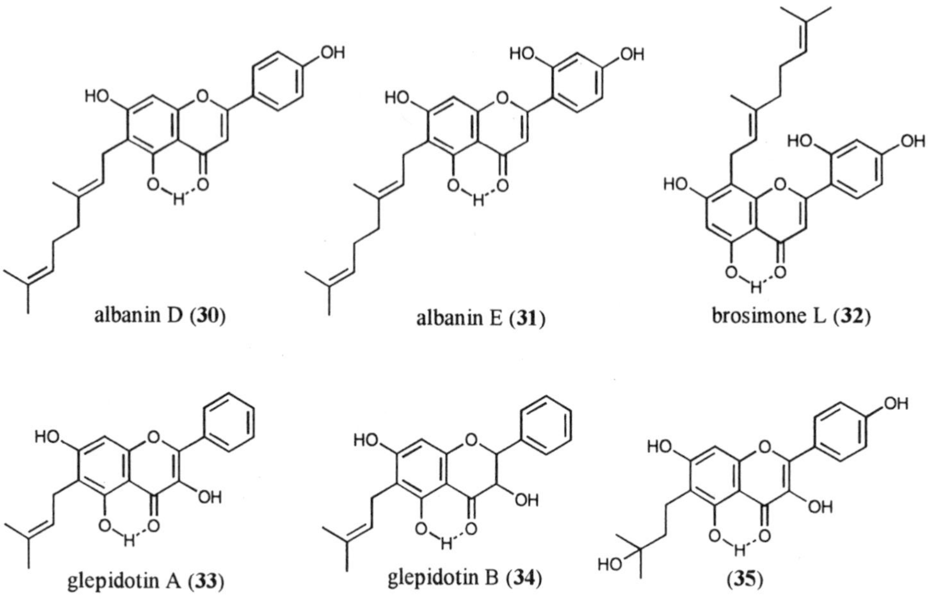

Figure 11. Revised structures with synthetic method.

as 6-prenyleriodictyol (**8**) and 8-prenyleriodictyol (**9**) in figure 11. The structures of the isoprenoid-substituted flavonoids albanins D (**30**) and E (**31**) from *Morus alba*,[37,46,47] brosimone L (**32**) from *Brosimopsis oblongifolia*,[37,48,49] glepidotins A (**33**) and B (**34**) from *Glycyrrhiza lepidota*[38,50] and 6-(γ-hydroxyisoamyl)-kaempferol (**35**) from *Bursera leptophloeos* were revised.[38,51]

4. VARIATION IN THE CHEMICAL SHIFT OF METHYLENE AND A-RING PROTONS OF FLAVONOIDS

The third method is classification based on variation in the chemical shift of benzylic methylene protons of the prenyl (geranyl) group and the A ring proton of 6- and 8-prenylated flavonoids. The method may be useful for confirmation of the structure when the 5-OH chemical shift has not been reported or possibly reported but with a strange chemical shift that might be due to a solvent impurity. The signals of H-6 and H-8 of 8- and 6-prenylated flavanones appear at almost the same region as well as methylene protons (H$_2$-1) of the prenyl (geranyl) groups. On the other hand, the chemical shifts of those of the 6- and 8-prenylated flavone, isoflavone, and flavonols are somewhat different as shown in table 3. In the course of the research, we found that the structure of artobilochromen (**36**, fig. 12) isolated from *Artocarpus nobilis*[52] is doubtful from the chemical shift of the H$_2$-1.[53] The physical and spectral data of the compound are almost the same as those of artonin E (KB-3) (**37**) isolated from *Artocarpus communis*.[54,55] The classification with the chemical shifts of A ring proton and H$_2$-1 might be useful for structure determination, but the method must be as an auxiliary method.

Table 3. Chemical shifts (δ) of methylene protons of prenyl (geranyl) group, and H-6 and H-8 of 6- or 8-prenylated (geranylated) flavonoids (in acetone-d_6)

	H-8 of 6-Prenylated flavonoids	H-6 of 8-Prenylated flavonoids	CH$_2$ of 6-prenyl group	CH$_2$ of 8-Prenyl group
Isoflavone	6.36 – 6.43	6.49 – 6.57	3.36 – 3.38	3.45 – 3.46
Flavone	6.33 – 6.35	6.57 – 6.63[a]	3.35 – 3.37	3.56 – 3.58[b]
Flavonol	6.26 – 6.39	6.53 – 6.65	3.36 – 3.37	3.52 – 3.58

CH$_2$ of Prenyl group on B ring (*ortho*-OH and -H):		3.30 – 3.48
(di*ortho*-OH):		3.44 – 3.45
CH$_2$ of 3-Prenyl group of flavone:		3.08 – 3.16

[a] Except 3-prenylflavone; 3,8-diprenylflavone, δ 6.32.

[b] Except 3-prenylflavone; 3,8-diprenylflavone, δ 3.36.

5. CONCLUSIONS

We found new structure determination methods for prenylated phenols with the NMR technique. The first method is variations in the chemical shift of the methylene carbon of a prenyl group. This method is useful for complex prenylated phenols. HMBC spectrometry is also a useful tool for these compounds but requires hydrogen atom(s) that exist at unambiguous positions. Furthermore, sometimes we do not know whether weak cross peaks are artificial or not when the dihedral angle cannot be calculated. The combination use of these methods may be the best way. The details of the variation of methylene carbon of the prenyl group were described elsewhere.[14,15] The second method is based on variations in the chemical shift of the hydrogen-bonded hydroxyl proton (5-OH) of prenylated flavonoids. This technique requires only a small amount of the sample. This method is especially useful for 6- or 8-prenylflavanones and isoflavanones. The other unambiguous method for these compounds is the observation of spin-spin coupling between 5-OH and C-6, but this ^{13}C NMR method requires a large amount of sample. The

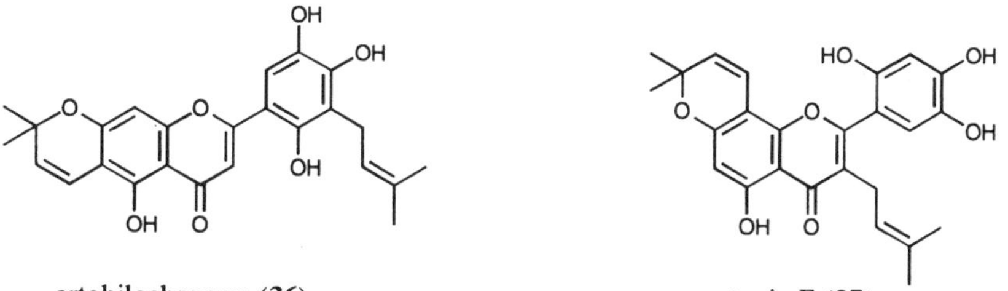

artobilochromen (**36**)   artonin E (**37**)

Figure 12. Structures of artobilochromen (**36**) and artonin E (**37**).

signal of 5-OH of flavonoids appears further downfield than δ 11, and the other peaks are not observed in the range. Thus, this method will be a useful check to see if new prenylflavonoids are present in semi-purified fractions. The third method is based on the variation in the chemical shift of methylene and the A-ring protons of prenylated flavones, flavonols, and isoflavones. This method is an auxiliary method. We have not found any exceptions in these new rules. Thus, using a combination of these methods is a useful tool for structure determination of prenylflavonoids.

REFERENCES

1. Barron, D.; Ibrahim, R.K. Isoprenylated flavonoids—a survey. *Phytochemistry* 43:921 (1996).

2. Yoshizawa, S.; Suganuma, M.; Fujiki, H.; Fukai, T., Nomura, T.; Sugimura, T. Morusin, isolated from root bark of *Morus alba* L., inhibits tumour promotion of teleocidin. *Phytotherapy Res.* 3:193 (1989).

3. Nomura, T. Phenolic compounds of the mulberry tree and related plants. *In*: Herz, W.; Grisebach, H.; Kirby, G.W.; Tamm, Ch. (eds.) Progress in the chemistry of organic natural products, Vol. 53, Springer-Verlag, Wien, p. 87 (1988).

4. Mihara, S.; Hara, M.; Nakamura, M.; Sakurawi, K.; Tokura, K.; Fujimoto, M.; Fukai, T.; Nomura, T. Non-peptide bombesin receptor antagonists, kuwanon G and H, isolated from mulberry. *Biochem. Biophys. Res. Commun.* 213:594 (1995).

5. Yanagisawa, T.; Sato, T.; Chin, M. (Chen, Z.); Mitsuhashi, H.; Fukai, T.; Hano, Y.; Nomura, T. Testosterone 5α-reductase inhibitors of *Morus* flavonoids. *In*: Das, N.P. (ed.) Flavonoids in biology and medicine III, current issues in flavonoids research. National University of Singapore, Singapore, p. 557 (1990).

6. Reddy, G.R.; Ueda, N.; Hada, T.; Sackeyfio, A.C.; Yamamoto, S.; Hano, Y.; Aida, M.; Nomura, T. A prenylflavone, artonin E, as arachidonate 5-lipoxygenase inhibitor. *Biochem. Phar.* 41:115 (1991).

7. Tahara, S. Structural diversity in isoflavonoids and erroneously proposed structures (in Japanese). *Kagaku to Seibutsu (Chem. Biol.; J. Jpn. Agrical. Chem. Soc.)* 29:493 (1991); *Chem. Abstr.* 115:203246a (1991).

8. Fukai, T.; Nomura, T. NMR spectra of isoprenoid substituted phenols. 1. Constituent of the Moraceae plants. 5. Revised structures of broussoflavonols C and D, and the structure of broussoflavonol E. *Heterocycles* 29:2379 (1989).

9. Nomura, T.; Hano, Y. Isoprenoid-substituted phenolic compounds of Moraceous plants. *Nat. Prod. Rep.* 11:205 (1994).

10. Nomura, T.; Hano, Y.; Aida, M. Isoprenoid-substituted flavonoids from *Artocarpus* plants (Moraceae). *Heterocycles* 47:1179 (1998).

11. Ikuta, J.; Hano, Y.; Nomura, T.; Kawakami, Y.; Sato, T. Constituents of the cultivated mulberry tree. 32. Components of *Broussonetia kazinoki* Sieb. 1. Structures of two new isoprenylated flavans and five new isoprenylated 1,3-diphenylpropane derivatives. *Chem. Pharm. Bull.* 34:1968 (1986).

12. Tahara, S.; Moriyama, M.; Ingham, J.L.; Mizutani, J. Structure revision of piscidone, a major isoflavonoid in the root bark of *Piscidia erythrina*. *Phytochemistry* 31:679 (1992).

13. Lin, C.-H.; Chiu, P.-N.; Fang, S.-C.; Shieh, B.-J.; Wu, R.-R. Revised structure of broussoflavonol G and the 2D NMR spectra of some related prenylflavonoids. *Phytochemistry* 41:1215 (1996).

14. Fukai, T.; Nomura, T. Variations in the chemical shift of benzylic methylene carbon of prenyl group on heterocyclic prenylphenols, *Heterocycles* 42:911 (1996).

15. Nomura, T.; Fukai, T. Phenolic constituents of licorice (*Glycyrrhiza* species). *In*: Herz, W.; Kirby, G.W.; Moore, R.E.; Steglich, W.; Tamm, Ch. (eds.) Progress in the chemistry of organic natural products, Vol. 73. Springer-Verlag, Wien, p. 1 (1998).

16. Nomura, T.; Fukai, T. Novel methods of structure determination of prenylated phenols with ¹H- and ¹³C-NMR spectra. *In*: Ageta, H.; Aimi, N.; Ebizuka, Y.; Fujita, T.; Honda, G. (eds.) Towards natural medicine research in 21st century, proceedings of the international symposium on natural medicines. Elsevier, Amsterdam, p. 561 (1998).

17. Dhami, K.; Stothers, S. ¹³C NMR studies. 8. ¹³C spectra of some substituted anisoles. *Can. J. Chem.* 44:2855 (1966).

18. Fukai, T.; Wang, Q.-H.; Nomura, T. Phenolic constituents of *Glycyrrhiza* species. 6. Six prenylated phenols from *Glycyrrhiza uralensis*. *Phytochemistry* 30:1245 (1991).

19. Furukawa, H.; Yogo, M.; Wu, T.-S. Acridone alkaloids. 10. ¹³C-nuclear magnetic resonance spectra of acridone alkaloids. *Chem. Pharm. Bull.* 31:3084 (1983).

20. Harborne, J.B. (ed.). The flavonoids: advances in research since 1980. Chapman and Hall, London (1988).

21. Harborne, J.B. (ed.). The flavonoids: advances in research since 1986. Chapman and Hall, London (1994).

22. Feigl, F.; Anger, V. Spot tests in organic analysis, 7th ed. (Engl. trans. by Oesper, R.E.), Elsevier, Amsterdam, p. 185 (1966).

23. Sherif, E.A.; Gupta, R.K.; Krishnamurti, M. Anomalous AlCl₃ induced U.V. shift of *C*-alkylated polyphenols. *Tetrahedron Lett.* 21:641 (1980).

24. Jain, A.C.; Gupta, R.C.; Sarpal, P.D. Synthesis of (±) lupinifolin, di-*O*-methyl xanthohumol and isoxanthohohumol and related compounds. *Tetrahedron* 34:3563 (1978).

25. Arnone, A.G., Cardillo, A.G., Merlini, L., Mondelli, R. NMR effects of acetylation and long-range coupling as a tool for structural elucidation of hydroxychromenes. *Tetrahedron Lett.* :4201 (1967).

26. Chari, V.M.; Ahmad, S.; Österdahl, B.-G. ¹³C NMR Spectra of chromeno- and prenylated flavones structure revision of mulberrin, mulberrochromene, cyclomulberrin and cyclomulberrochromene. *Z. Naturforsch.* 33b:1547 (1978).

27. Wehrli, F.W. Proton-coupled ¹³C nuclear magnetic resonance spectra involving ¹³C-¹H spin-spin coupling to hydroxyl-protons, a complementary assignment aid. *J. Chem. Soc., Chem. Commun.* :663 (1975).

28. Shirataki, Y.; Yokoe, I.; Endo, M.; Komatsu, M. Determination of C-6 or C-8 substituted flavanone using ¹³C-¹H long range coupling and the revised structures of some flavanones. *Chem. Pharm. Bull.* 33:444 (1985).

29. Fukai, T.; Nishizawa, J.; Nomura, T. Phenolic constituents of *Glycyrrhiza* species. 14. Variations in the chemical shift of the 5-hydroxyl proton of isoflavones; two isoflavones from licorice. *Phytochemistry* 36:225 (1994).

30. Fukai, T.; Wang, Q.-T.; Takayama, M.; Nomura, T. Phenolic constituents of *Glycyrrhiza* species. 4. Structures of five new prenylated flavonoids, gancaonins L, M, N, O, and P from aerial parts of *Glycyrrhiza uralensis*. *Heterocycles* 31:373 (1990).

31. Fukai, T.; Nomura, T. NMR spectra of isoprenoid substituted phenols. 3. Structure of 6- or 8-isoprenoid substituted flavanone: chemical shift of the hydrogen-bonded hydroxyl group. *Heterocycles* 31:1861 (1990).

32. Fukai, T.; Nomura, T. NMR spectra of isoprenoid substituted phenols. 9. Variations in the chemical shift of the 5-hydroxyl proton of 7-*O*-prenylated flavanones. *Heterocycles* 43:1361 (1996).

33. Bohlmann, F.; Zdero, C.; Robinson, H.; King, R.M. Naturally occurring terpene derivatives. 357. A diterpene, a sesquiterpene quinone and flavanones from *Wyethia helenioides*. *Phytochemistry* 20:2245 (1981).

34. Wu, L.-J.; Miyase, T.; Ueno, A.; Kuroyanagi, M.; Noro, T.; Fukushima, S. Studies on the constituents of *Sophora flavescens*. 2. *Chem. Pharm. Bull.* 33:3231 (1985).

35. Iinuma, M.; Yokoyama, J.; Ohyama, M.; Tanaka, T.; Mizuno, M.; Ruangrungsi, N. Seven phenolic compounds in the roots of *Sophora exigua*. *Phytochemistry* 33:203 (1993).

36. Fukai, T.; Pei, Y.-H.; Nomura, T.; Xu, C.-Q.; Wu, L.-J.; Chen, Y.-J. Constituents of Moraceous plants. 29. Components of the root bark of *Morus cathayana*. 2. Isoprenylated flavanones from *Morus cathayana*. *Phytochemistry* 47:273 (1998).

37. Fukai, T.; Nomura, T. NMR spectra of isoprenoid substituted phenols. 4. Revised structures of albanins D and E, geranylated flavones from *Morus alba*. *Heterocycles* 32:499 (1991).

38. Fukai, T.; Nomura, T. NMR spectra of isoprenoid substituted phenols. 5. ^{1}H-NMR chemical shift of the flavonol 5-hydroxy proton as a characterization of 6- or 8-isoprenoid substitution. *Heterocycles* 34:1213 (1992).

39. Feigl, F. Spot tests in inorganic analysis, 5th ed. (Engl. trans. by Oesper, R.E.), Elsevier, Amsterdam, p. 182 (1958).

40. Tahara, S.; Ingham, J.L.; Hanawa, F.; Mizutani, J. ^{1}H NMR chemical shift value of the isoflavone 5-hydroxyl proton as a convenient indicator of 6-substitution or 2'-hydroxylation. *Phytochemistry* 30:1683 (1991).

41. Fukai, T.; Tantai, L.; Nomura, T. NMR spectra of isoprenoid substituted phenols. 8. ^{1}H NMR chemical shift of the isoflavanone 5-hydroxyl proton as a characterization of 6- or 8-prenyl group. *Heterocycles* 37:1819 (1994).

42. Hano, Y.; Matsumoto, Y.; Sun, J.-Y.; Nomura, T. Constituents of the Moraceae plants. 1. Components of root bark of *Cudrania tricuspidata*. 4. Structures of three new isoprenylated xanthones, cudraxanthones E, F, and G. *Planta Med.* 56:399 (1990).

43. Ishiguro, K.; Nakajima, M.; Fukumoto, H.; Isoi, K. Xanthones in cell suspension cultures of *Hypericum patulum*. 3. A xanthone substituted with an irregular monoterpene in cell suspension cultures of *Hypericum patulum*. *Phytochemistry* 39:903 (1995).

44. Hano, Y.; Okamoto, T.; Nomura, T.; Momose, Y. Constituents of the Moraceae plants. 11. Components of the root bark of *Morus insignis* BUR. 1. Structures of four new isoprenylated xanthones, morusignins A, B, C, and D. *Heterocycles* 31:1345 (1990).

45. Seo, E.-K.; Silva, G.L.; Chai, H.-B.; Chagwedera, T.E.; Farnsworth, N.R.; Cordell, G.A.; Pezzuto, J.M.; Kinghorn, A.D. Cytotoxic prenylated flavanones from *Monotes engleri*. *Phytochemistry* 45:509 (1997).

46. Takasugi, M.; Ishikawa, S.; Masamune, T.; Shirata, A.; Takahashi, K. Anti-bacterial compounds in the branch bark of mulberry tree. 42nd Annual Meeting of the Chemical Society of Japan, Abstract Paper, Sendai, p. 352 (1980).

47. Shirata, A.; Takahashi, K.; Takasugi, M.; Nagao, S.; Ishikawa, S.; Ueno, S.; Muñoz, L.; Masamune, T. Antimicrobial spectra of the compounds from mulberry tree (in Japanese). Sanshi Shikenjo Hokoku (Bull. Sercul. Exp. Sta.) 28:793 (1983).

48. Ferrari, F.; Messana, I.; Do Carmo Mesquita de Araujo, M. Constituents of *Brosimopsis oblongifolia*. 2. Structures of three new flavones, brosimones G, H, and I form *Brosimopsis oblongifolia*. *Planta Med.* 55:70 (1989).

49. Ferrari, F.; Nomura, T. Letter to editor. *Planta Med.* 58:116 (1992).

50. Mitscher, L.A.; Raghavrao, G.S.; Khanna, I.; Veysoglu, T.; Drake, S. Antimicrobial agents from higher plants: prenylated flavonoids and other phenols from *Glycyrrhiza lepidota*. *Phytochemistry* 22:573 (1983).

51. Souza, M.P.; Machado, M.I.L.; Braz-Filho, R. Six flavonoids from *Bursera leptophloeos*. *Phytochemistry* 28:2467 (1989).

52. Kumar, N.S.; Pavanasasivam, G.; Sultanbawa, U.S.; Mageswaran, R. Chemical investigation of Ceylonese plants. 24. New chromenoflavonoids from the bark of *Artocarpus nobilis* Thw. (Moraceae). *J. Chem. Soc. Perkin Trans. 1* :1243 (1977).

53. Fukai, T.; Nomura, T. NMR spectra of isoprenoid substituted phenols. 6. ^{1}H NMR spectra of prenylated flavonoids and pyranoflavonoids. *Heterocycles* 36:329 (1993).

54. Hano, Y.; Yamagami, Y.; Kobayashi, M.; Isohata, R.; Nomura, T. Constituents of the Moraceae plants. 8. Artonins E and F, two new prenylflavones from the bark of *Artocarpus communis* Forst. *Heterocycles* 31:877 (1990).

55. Fujimoto, Y.; Zhang, X.-X.; Kirisawa, M.; Uzawa, J.; Sumatra, M. New flavones from *Artocarpus communis* Forst. *Chem. Pharm. Bull.* 38:1787 (1990).

CHEMISTRY, BIOSYNTHESIS, AND BIOLOGICAL ACTIVITY OF NATURAL DIELS-ALDER TYPE ADDUCTS FROM MORACEOUS PLANTS

Taro Nomura and Yoshio Hano

Faculty of Pharmaceutical Sciences
Toho University
Funabashi, Chiba 274-8510
JAPAN

1. INTRODUCTION

The mulberry tree, a typical plant of the genus *Morus*, has been widely culti-vated in China and Japan. Its leaves are indispensable as food for silkworms. On the other hand, the root bark of the mulberry tree, *Morus alba* L. and other plants of the genus *Morus*, has been used as an antiphlogistic, diuretic, expectorant, and laxative in the Chinese herbal medicine called "Sang-Bai-Pi" ("Sohakuhi" in Japanese).[1] In the pharmacological field, several researchers have reported on the hypotensive effect of the extract. Considering these reports, it was suggested that the hypotensive constituents were made up of a mixture of many phenolic con-stituents of the mulberry tree and related plants. About 70 kinds of new phenolic compounds could be isolated from the Japanese cultivated mulberry tree and Chinese crude drug "Sang-Bai-Pi". Among them, a hypotensive compound, kuwanon G (**1**), was isolated in 0.2 percent yield from the root bark.[2] Furthermore, kuwanon G is considered to be formed through an enzymatic Diels-Alder type reac-tion of a chalcone (**2**) and dehydrokuwanon C (**3**) or its equivalent (fig. 1). Subse-quently, about 40 kinds of Diels-Alder type adducts have been isolated from moraceous plants.[3–5]

2. STRUCTURES OF DIELS-ALDER TYPE ADDUCTS FROM MORACEOUS PLANTS

The Diels-Alder reaction is well known as a [4 + 2]cycloaddition of a diene and a dienophile to form a six-membered ring. In the case of kuwanon G (**1**), dehy-drokuwanon C (**3**) can be regarded as the diene, and the chalcone derivative (**2**)

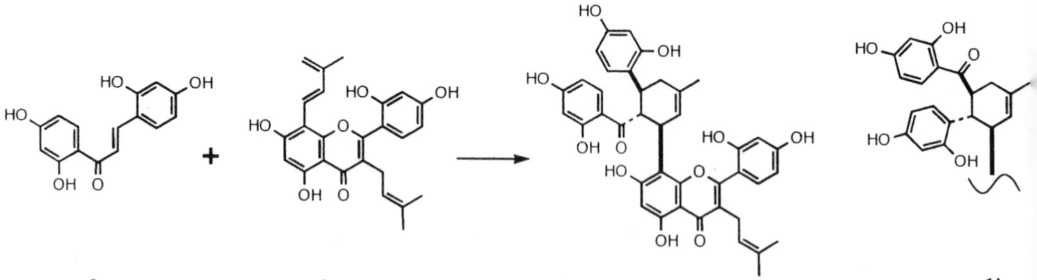

Figure 1. Formation of kuwanon G (**1**) from **2** and **3** through the Diels-Alder reaction and the structure of different isomers (**1'**).

the dienophile. From the chemical and spectroscopic evidence, two possible plane structures were reported for kuwanon G as structures **1** and **1'**.[2-3] These two structures are isomers due to different regioselectivity of the Diels-Alder reaction. To confirm the regioselectivity of the reaction, a Diels-Alder reaction of the *trans*-chalcone (**4**) and 3-methyl-1-phenylbutadiene (**5**) gave two cycloproducts, one of which is all-*trans* (**6**) in relative configuration, whereas the other is the *cis-trans* type adduct (**7**) in relative configuration (fig. 2).[6] The structure of kuwanon G was confirmed by the following results.[6]

Kuwanon G octamethyl ether (**1a**) was pyrolyzed to give *trans*-chalcone (**2a**) and dehydrokuwanon C tetramethyl ether (**3a**). These structures were confirmed by chemical evidence. The Diels-Alder reaction of the fragmentation compounds, **2a** and **3a**, gave two [4 + 2]cycloadducts (fig. 3), one of which was identical with

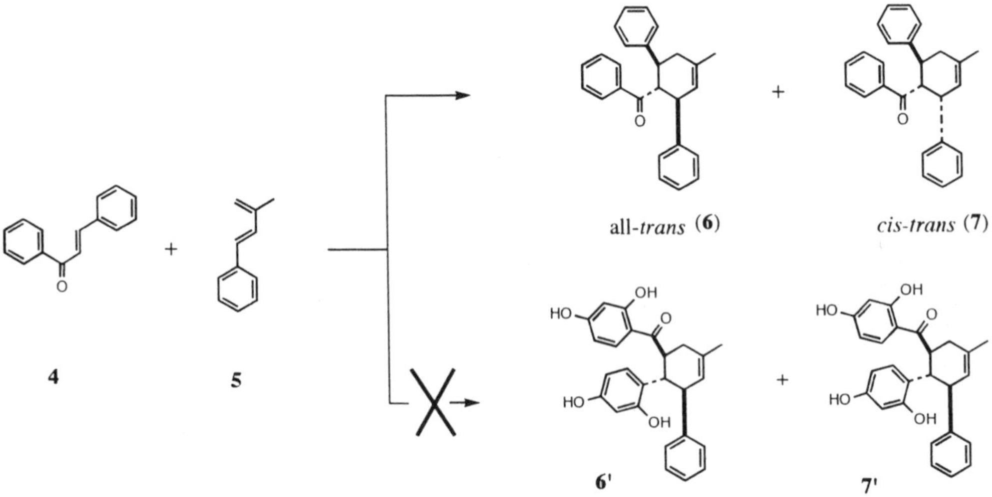

Figure 2. The Diels-Alder reaction of model compounds (**4** and **5**).

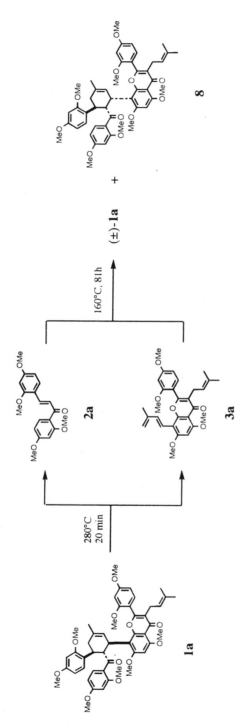

Figure 3. Pyrolysis of kuwanon G octamethyl ether (**1a**) and the reconstruction of the ether (±-**1a**) by the Diels-Alder reaction.

(±)-kuwanon G octamethyl ether (**1′a**), whereas the other was the *cis-trans* type cycloadduct (**8**). The structure of kuwanon G has thus been confirmed as depicted in figure 1.

About 40 kinds of optically active Diels-Alder type adducts have been isolated from *Morus* root bark. It is interesting that pairs of isomers such as kuwanons I (**9**)[7] and J (**10**)[8] coexist in the *Morus* root bark. Taking these findings into account, the mulberry Diels-Alder type adducts are regarded biogenetically as [4 + 2]cycloadduct of a chalcone and dehydroprenylphenol. These Diels-Alder type adducts are the most characteristic components of the plants of *Morus* species.[3,4] Recently, from the other species belonging to Moraceae, about 10 kinds of Diels-Alder type adducts were isolated by our group and Italian researchers.[4]

One of our coworkers, the late Dr. Ueda of Kyoto University, obtained pigment-producing callus tissues of *Morus alba* L. The callus tissues induced from seedlings were cultured under specified conditions and subjected to selection over a period of 9 years, giving rise to cell strains having a high-pigment productivity.[8] From the extract of the callus tissues, six Diels-Alder type adducts, kuwanons J (**10**), Q (**11**), R (**12**), V (**13**), mulberrofuran E (**14**), and chalcomoracin (**15**), were isolated along with morachalcone A (**16**), isobauachalcone B (**17**), and moracin C (**18**) (fig. 4).[8–10] If we call morachalcone A (**16**) as A, isobauachalcone B (**17**) as B, moracin C (**18**) as C, kuwanon J (**10**) is an adduct of dehydro-A with A (AA type), kuwanon Q (**11**) is an adduct of dehydro-A with B (AB type), kuwanon R (**12**) is an adduct of dehydro-B with A (BA type), and kuwanon V (**13**) is an adduct of dehydro-B with B (BB type). Furthermore, chalcomoracin (**15**) is an adduct of dehydro-C with A (CA type), and

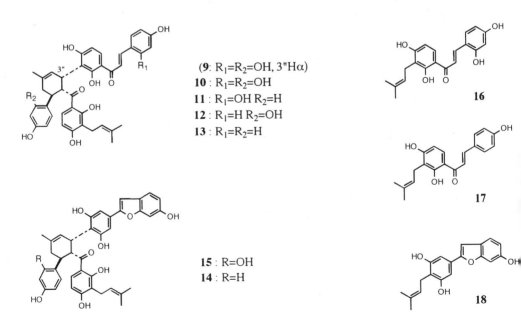

(**9**: R$_1$=R$_2$=OH, 3″Hα)
10 : R$_1$=R$_2$=OH
11 : R$_1$=OH R$_2$=H
12 : R$_1$=H R$_2$=OH
13 : R$_1$=R$_2$=H

15 : R=OH
14 : R=H

Figure 4. Phenolic components of *Morus alba* callus tissues.

mulberrofuran E (**14**) is an adduct of dehydro-C with B (CB type). It is interesting that all possible combinations of these monomers (**16, 17, 18**) could be isolated from *Morus* callus tissues. These results strongly suggest that kuwanons J (**10**), Q (**11**), R (**12**), V (**13**), mulberrofuran E (**14**), and chalcomoracin (**15**) isolated from the callus cultures are naturally occurring Diels-Alder type adducts.

The mulberry Diels-Alder type adducts could be divided into two groups, one of which is an all-*trans* type adduct, and the other is a *cis-trans* type adduct. The all-*trans* type adducts seem to be formed by *exo*-addition in the Diels-Alder reaction of a chalcone and a dehydroprenylphenol, whereas the *cis-trans* type adducts seem to be formed by *endo*-addition. Absolute stereochemistries of mulberry Diels-Alder type adducts were confirmed by circular dichroism (CD) spectroscopic evidence and by an X-ray crystallographic analysis.

Mulberrofuran C (**19**) is one of the hypotensive Diels-Alder type adducts and seems to be an adduct of dehydroprenyl-2-arylbenzofuran and chalcone derivative.[11] The absolute configuration of **19** was confirmed from the following results. The ketalized Diels-Alder type adduct, mulberrofuran G (**20**) isolated from *Morus* root bark, could be derived stereospecifically from the original adduct, mulberrofuran C (**19**), under acidic conditions described in figure 5.[12] The relative configurations of the four chiral centers of the pentamethyl ether of mulberrofuran G (**20a**) were confirmed by X-ray crystallographic analysis.[13] Monobromomulberrofuran G pentamethyl ether (**20b**) was derived from **20a** by treatment with NBS, and **20b** was converted to an aromatized compound (**20c**) through dehydrogenation by DDQ (fig. 5). The X-ray crystallographic analysis of **20c** revealed that the absolute configuration of the chiral center at C-8″ is *R*. As the correlation between **20c** and mulberrofuran C (**19**) through the **20a** was confirmed, the absolute stereochemistry of **19** was determined to be 3″*S*, 4″*R*, 5″*S*. Furthermore, as the stereochemistry of mulberrofuran J (**21**) at the C-3″ position was determined to be antipodal to that of mulberrofuran C (**19**) by the CD spectra of related compounds, the absolute configuration of **21** was then expressed as 3″*R*, 4″*R*, 5″*S* (fig. 6).[14,15] Regarding optical rotation values of mulberrofuran C (**19**, $[\alpha]_D$ + 153°) and mulberrafuran J (**21**, $[\alpha]_D$ – 341°), the sign of optical rotation depends on the stereochemistry of the C-3″ position. Meanwhile, other all-*trans* type adducts showed minus values as in **21**, while *cis-trans* type adducts showed plus values as in **19**.[15] Absolute configurations of a series of the mulberry Diels-Alder type adducts were thus determined, and the adducts having all-*trans* relative configuration are *exo*-addition products in the Diels-Alder reaction, whereas the *cis-trans* type adducts are formed through *endo*-addition (fig. 7).

3. BIOSYNTHESIS OF MULBERRY DIELS-ALDER ADDUCTS KUWANON J AND CHALCOMORACIN

Examination of the biosynthesis of the mulberry Diels-Alder type adducts is very important because of no evidence for the biological intermolecular Diels-Alder reaction exists to date. Some cell strains of *Morus alba* callus tissues induced from the seedlings or the leaves have a high productivity of the mulberry Diels-Alder type adducts. The yields of major adducts are about 100–1,000 times more than

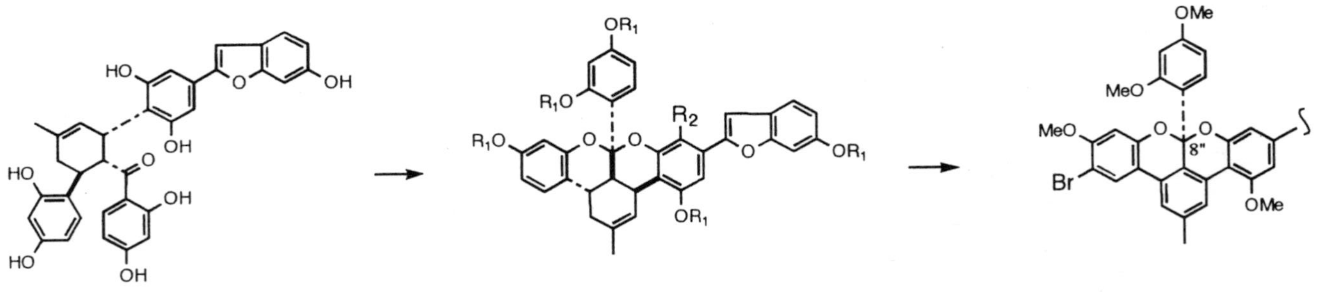

19

20: $R_1=R_2=H$
20a: $R_1=Me$ $R_2=H$
20b: $R_1=Me$ $R_2=Br$

20c

Figure 5. Leading process from mulberrofuran C (**19**) to an aromatized compound (**20c**) by way of mulberrofuran G (**20**).

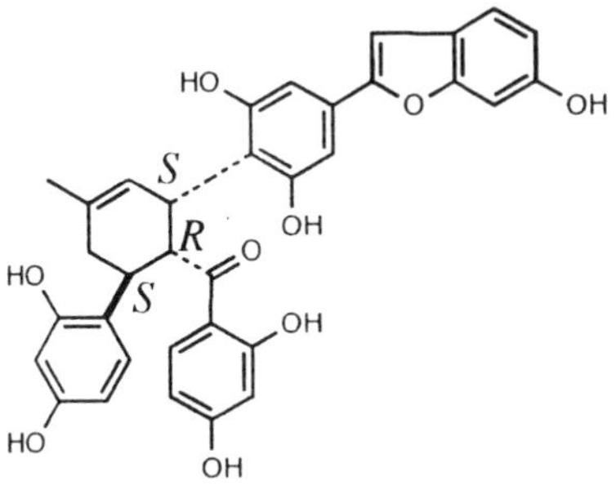

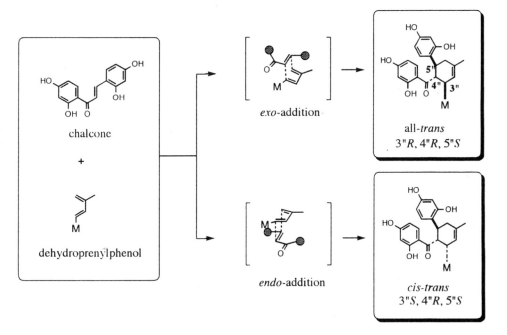

Figure 6. Absolute stereochemistries of mulberrofurans C (**19**) and J (**21**).

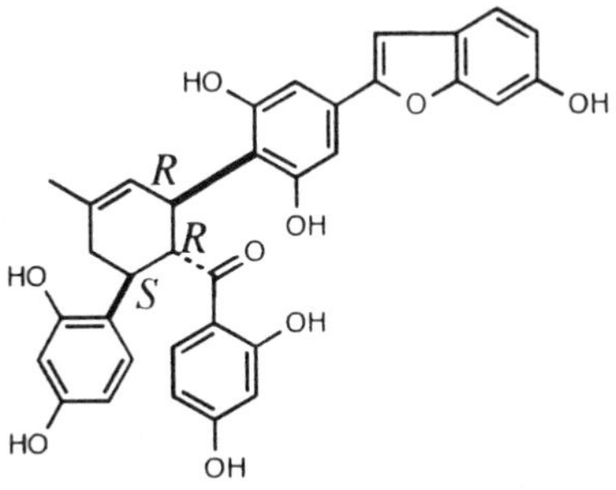

those of the intact plant (fig. 4).[8–10] The biosynthesis of the mulberry Diels-Alder type adducts has been studied with the aid of the cell strains.

Administration experiments of [1-¹³C]- and [2-¹³C]acetates to the cell cultures revealed that both kuwanon J (**10**) and chalcomoracin (**15**) are composed of two molecules of cinnamoylpolyketide skeletons.[16] Namely, kuwanon J (**10**) is regarded as a dimer of isoprenylated chalcone, whereas chalcomoracin (**15**) is composed of isoprenylated 2-arylbenzofuran and isoprenylchalcone (fig. 8). From the labeling

Figure 7. Absolute configuration of mulberry Diels-Alder type adducts.

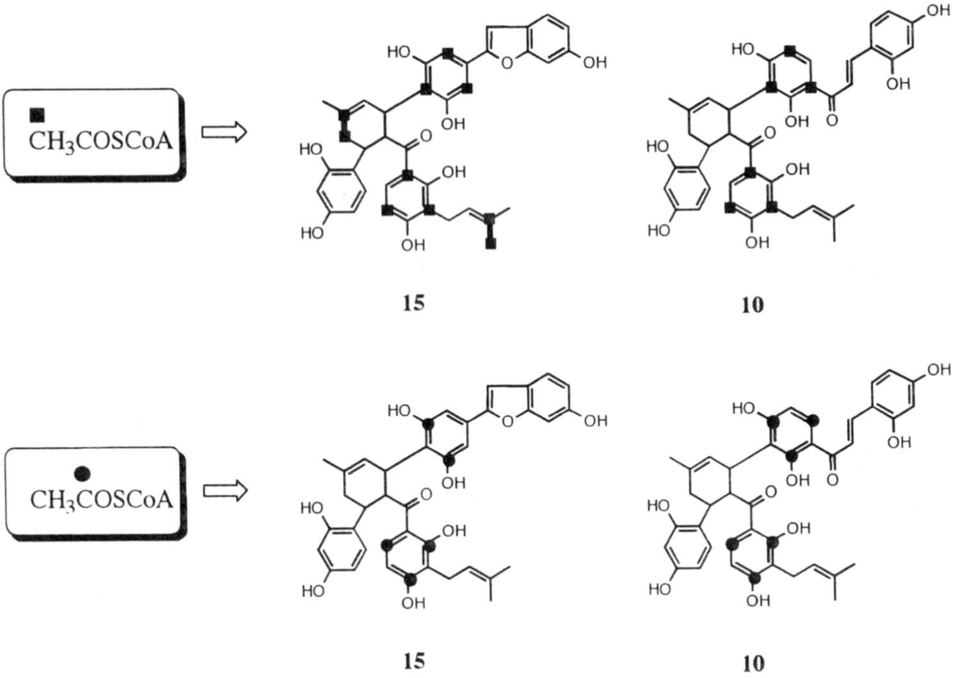

Figure 8. The ^{13}C-labeling patterns of kuwanon J (**10**) and chalcomoracin (**15**).

patterns, the chalcone skeleton seems to be originated through the Claisen type condensation of the cinnamoylpolyketide and the 2-arylbenzofuran skeleton through the aldol type condensation (fig. 9).

Administration of ^{13}C-labeled acetate to the cell cultures resulted in the highly ^{13}C enriched aromatic carbons of chalcomoracin (**15**) (about 17 percent), whereas two isoprenyl units of **15** were labeled to a lesser extent (about 0.4 percent). In the case of feeding experiments with [2-^{13}C]acetate, contiguous ^{13}C-labeled C2 units were incorporated into the isoprenyl units.[16] Moreover, the incorporated positions were only those corresponding to the starter acetate unit of the mevalonate biosynthesis. On the other hand, incorporation of [1-^{13}C]acetate was not found in the isoprenyl units.[16]

These findings suggest the participation of the tricarboxylic acid (TCA) cycle to the biosynthesis of the isoprenyl units of chalcomoracin (**15**). In the experiment with [2-^{13}C]acetate, the contiguous ^{13}C atoms can be derived from the two methyl groups of the intact acetate administered by way of at least two passages through the TCA cycle. Accordingly, the acetate incorporated into the isoprenyl units of chalcomoracin (**15**) was not the intact acetate administered, but [1,2-[13]C$_2$]acetate reorganized from the methyl group of the intact acetate through the TCA cycle (fig. 10).[16] This hypothesis was reinforced by the administration experiment with [2-^{13}C]acetate in a pulsed manner.[17] This result enabled us to disclose the satellite peaks based on the [13]C-^{13}C spin-spin coupling between the carbons at C-25″ and

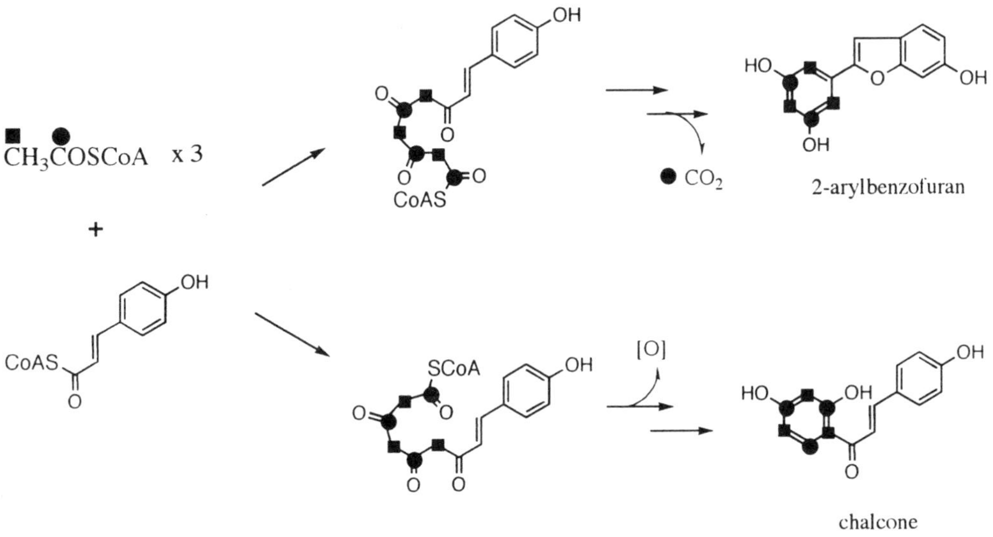

Figure 9. Formation of the 2-arylbenzofuran and chalcone skeletons in the *Morus alba* cell cultures.

23″ as well as that between the carbons at C-7″ and C-1″, in addition to the ^{13}C-^{13}C spin-spin coupling between C-23″ and C-24″ and that between C-6″ and C-1″. No satellite peaks were observed at C-22″ or C-2″ (fig. 11). This result suggested that the three carbons were contiguously enriched with ^{13}C atoms. However, this assumption was ruled out from the coupling pattern of the central carbons at the C-1″ and C-23″. The central carbons at the C-1″ and C-23″ appeared as doublet signals. If the ^{13}C-labelings continuously related in the sequence, the central carbons must appear as the doublet of doublet signals. The appearance of the doublet signal indicates that the central carbon is independently coupled with the two adjacent methyl carbons. The independent ^{13}C-labeling pattern at the iso-prenyl group might be explained as transfer of ^{13}C-labeling from *cis*-methyl to *trans*-methyl through a diene formation. Furthermore, the phenomenon of the ^{13}C enrichment of both third acetate units found at C-7″ and C-25″, in spite of the lack of ^{13}C-labeling at both second acetate units, can be explained by the isomerization between the two 3,3-dimethylallyl and 3-methyl-butadienyl groups (fig. 11). It is noteworthy that the isomerization takes place not only at the prenyl group par-ticipating in the intermolecular Diels-Alder type reaction, but also in the other prenyl group that remains intact. In conclusion, this finding gives confirmative evidence on the formation of diene structure at the prenyl moieties of the chalcone for the Diels-Alder type cyclization.

Final confirmation of the biosynthesis of the mulberry Diels-Alder type adducts was obtained by an administration experiment with *O*-methylchalcone derivative to the *Morus alba* cell cultures.[18] Administration of *O*-methylated chalcone (**22**) to the cell cultures yielded the metabolites (**23**), (**24**), (**25**), (**26**), and (**27**) (fig. 12). The metabolite (**23**) is an interesting product because the formation of the chalcone

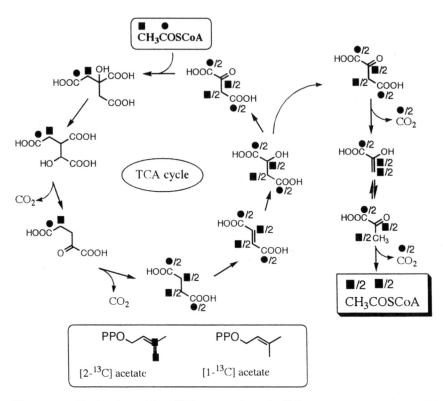

Figure 10. Formation of [1,2-^{13}C]acetate from [2-^{13}C]acetate through the TCA cycle and the labeling patterns of the isoprenyl units of **15**.

(**23**) in the tissue cultures indicated that an isoprenylation takes place after the completion of chalcone skeleton. The metabolites (**24**), (**25**), (**26**), and (**27**) revealed that the precursory chalcone (**22**) was incorporated intact into the Diels-Alder adducts. An analogous experiment employing tri-*O*-methylated chalcone (**28**) gave the Diels-Alder type metabolite (**29**). These results suggest that one molecule of isoprenylated chalcone is recognized as a dienophile at the α,β-double bond, while another molecule of the chalcone acts as a diene at the isoprenyl portion. Furthermore, the Diels-Alder metabolites from the precursory chalcones (**22**), (**23**), and (**28**) were all optically active, having the same stereochemistries as those of kuwanon J (**10**) and chalcomoracin (**15**). These results revealed that kuwanon J (**10**) and chalcomoracin (**15**) have thus been proved to be enzymatic Diels-Alder reaction products.

One of the Diels-Alder type adducts isolated from the *Artocarpus* spp. plants (Moraceae), artonin I (**30**) was considered to be formed through the Diels-Alder reaction of a chalcone derivative, morachalcone A (**16**) and artocarpesin (**31**), as precursors.[19] Both artonin I (**30**) and artocarpesin (**31**) are not inevitably detected in *Morus alba* cell cultures. We attempted the synthesis of natural Diels-Alder type adduct, artonin I (**30**), with the aid of an enzyme system of *Morus alba* cell

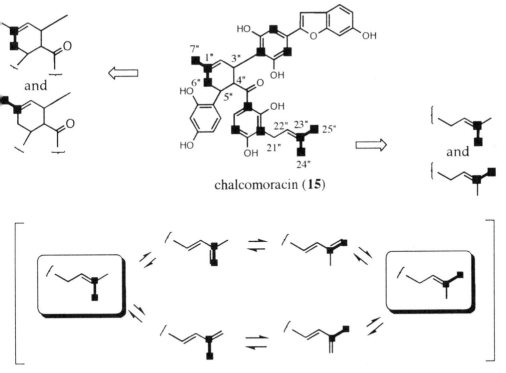

chalcomoracin (**15**)

Figure 11. Two independent ^{13}C-labeling patterns at the isoprenyl units of **15** and the transfer of the ^{13}C-labeling from *cis*-methyl carbon to *trans*-methyl carbon through the diene formation.

cultures. Artocarpesin (**31**) was added to the *Morus alba* cell cultures, which were treated in the usual way. The metabolite isolated from the cells was identical with artonin I (**30**) by direct comparisons (fig. 13).[20] This is the first example of the elucidation of the structure of an organic natural product by application of an enzymatic synthesis of the target substance with the aid of the cell cultures of the related plants.

To confirm the stage of the Diels-Alder reaction in the cell cultures, the hydroxylation process of the adducts was studied.[21] The incorporation of [2-^{13}C]acetate into chalcomoracin (**15**) (about 17 percent) is higher than that into kuwanon J (**10**) (about 4 percent). While chalcomoracin (**15**) has the same chalcone part as kuwanon J (**10**), the ^{13}C-enrichment of the two compounds is very different. We examined the ^{13}C-enrichment of the Diels-Alder type adducts in the cell cultures by the feeding experiment as described in figure 14. Considering these results, kuwanon J (**10**) consists of two molecules of morachalcone A (**16**) with 4 percent of the ^{13}C-enrichment. Similarly, kuwanon V (**13**) is composed of two molecules of isobauachalcone B (**17**) with 24 percent of the ^{13}C-enrichment. In the case of kuwanon R (**12**), the upper unit must be the chalcone part (**17**) with 24 percent of the ^{13}C-enrichment and the lower unit must have the chalcone (**16**) with 4 percent

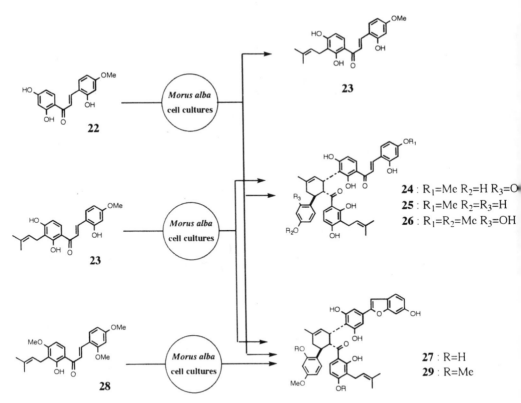

Figure 12. Aberrant metabolism of *O*-methylated precursory chalcones in the *Morus alba* cell cultures.

of the ^{13}C-enrichment. However, the actual ^{13}C-enrichment of kuwanon R (**12**) was about 14 percent, and the ^{13}C-enrichment of the upper unit of **12** is the same as that of the lower unit. If kuwanon R (**12**) is biosynthesized through the respective Diels-Alder type reaction of molecules of the chalcone derivatives (**16** and **17**), the agreement of the ^{13}C-enrichment between the upper and lower units of **12** seems to be unlikely (fig. 15). From the results, it could be indicated that the successive increases of the hydroxyl groups diminish the ^{13}C-enrichment. If lesser hydroxylated adduct, kuwanon V (**13**), is initially biosynthesized followed by successive hydroxylation reaction to form **12** and then **10**, the ^{13}C-enrichment of the two chalcone parts must always be the same number in every adduct. The major adducts, kuwanon J (**10**) and chalcomoracin (**15**), in the cell cultures are presumably derived from kuwanon V (**13**) and mulberrofuran E (**14**) as "primer" of the Diels-Alder type adducts, respectively (fig. 16).

4. BIOLOGICAL ACTIVITY OF DIELS-ALDER ADDUCTS FROM MORACEAE

The methanol extract, 1 mg–20 mg, showed a dose-dependent decrease in arterial blood pressure in a pentobarbital-anesthetized rabbit. From the extract,

16
30
Morus alba callus
endogenous
precursory chalcone (16)
artonin I (30, 9 mg)
[α]D +91°
31 (50 mg)

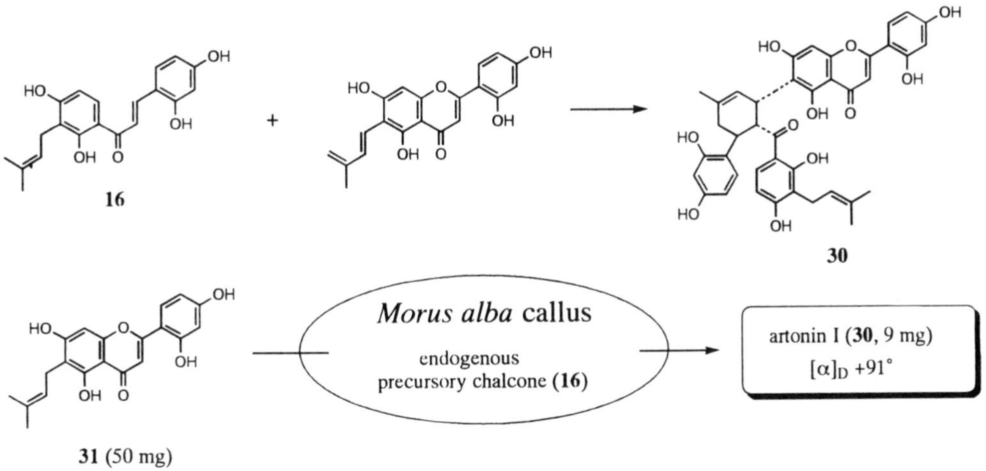

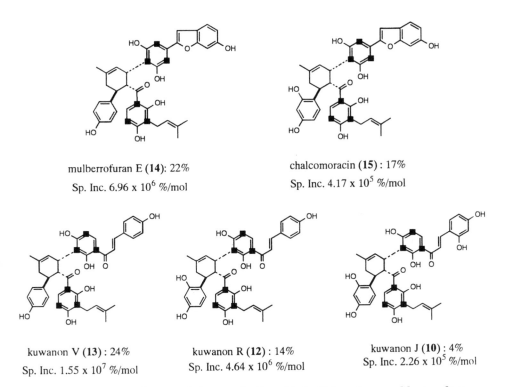

Figure 13. Formation of artonin I (**30**) by Diels-Alder reaction and bioconversion of artocarpesin (**31**) to artonin I (**30**) with the aid of the *Morus alba* cell cultures.

mulberrofuran E (14): 22%
Sp. Inc. 6.96 x 10^6 %/mol
chalcomoracin (15) : 17%
Sp. Inc. 4.17 x 10^5 %/mol
kuwanon V (13) : 24%
Sp. Inc. 1.55 x 10^7 %/mol
kuwanon R (12) : 14%
Sp. Inc. 4.64 x 10^6 %/mol
kuwanon J (10) : 4%
Sp. Inc. 2.26 x 10^5 %/mol

Figure 14. Enrichment factors of the Diels-Alder type adducts from [2-^{13}C]acetate.

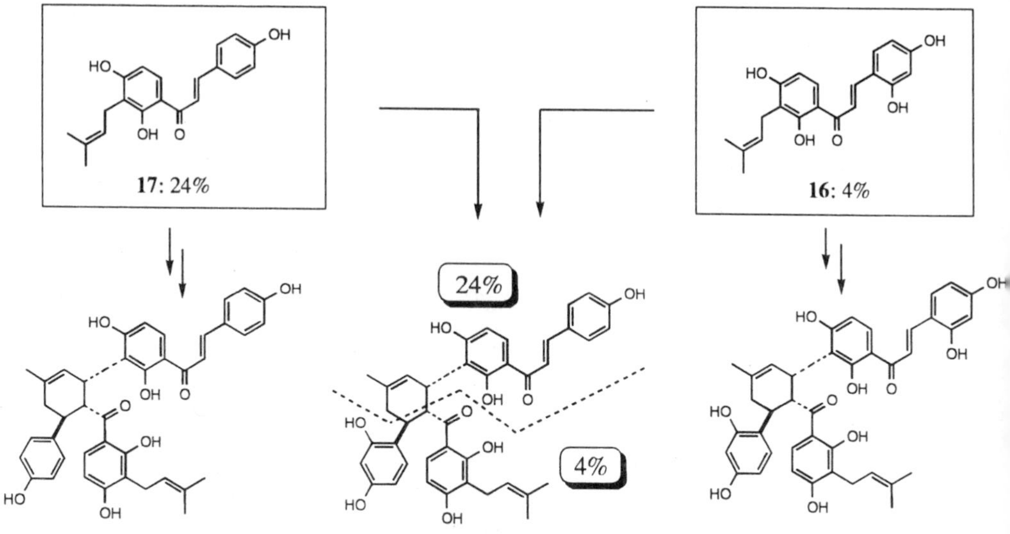

Figure 15. Expected ^{13}C-enrichment for kuwanon R (**12**).

kuwanons G (**1**) and H (**32**) could be isolated in 0.2 percent and 0.13 percent yields, respectively. Both compounds (**1** and **32**) almost equally caused a decrease in arterial blood pressure in a dose-dependent and reversible manner at doses between 0.1 and 3 mg/kg i.v. in pentobarbital-anesthetized as well as in unanesthetized, gallamine-immobilized rabbits (figs. 17, 18). The hypotensive action of kuwanons G (**1**) and H (**32**) were not modified by atropine nor eserine, suggesting a non-cholinergic natural origin. Furthermore, neither propranol nor diphenylhydramine affected arterial blood pressure. They produced no significant changes in either ECG (electrocardiogram) or respiration when administered intravenously in rabbits. The hypotensive effects of kuwanons G (**1**) and H (**32**) were not accompanied by a heart rate change. In pentobarbital-anesthetized pithed dogs, kuwanons G (**1**) and H (**32**) also significantly decreased femoral arterial blood pressure. These effects suggested that the mechanism of hypotensive effects of kuwanons G (**1**) and H (**32**) was mediated through the peripheral system.[3,22]

Mulberrofuran C (**19**) is thought to be formed by a Diels-Alder type of enzymatic reaction process of a chalcone derivative and dehydroprenylmoracin C (**18**) or its equivalent.[11] Furthermore, mulberrofurans F (**34**) and G (**20**) seem to be Diels-Alder type adducts derived from chalcomoracin (**15**) and mulberrofuran C (**19**), respectively, by the intramolecular ketalization reaction of the carbonyl group with the two adjoining hydroxyl groups.[3,12] Intravenous injection of mulberrofuran C (**19**, 1 mg/kg) produced a significant hypotension in rabbits.[11] Single intravenous injection of mulberrofurans F (**34**) and G (**20**) (both, 0.1 mg/kg) caused a marked depressor effect in rabbits by 26 mmHg and 16 mmHg, respectively.[12]

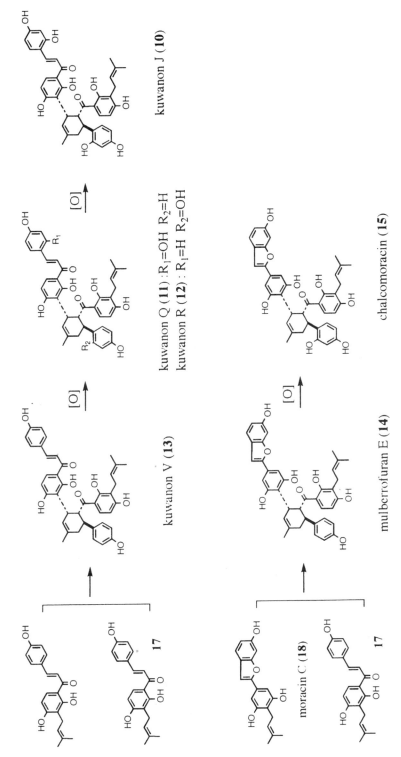

Figure 16. Successive hydroxylation of the primary adduct **13** (or **14**) resulting in the formation of **10** (or **15**) in the *Morus alba* cell cultures.

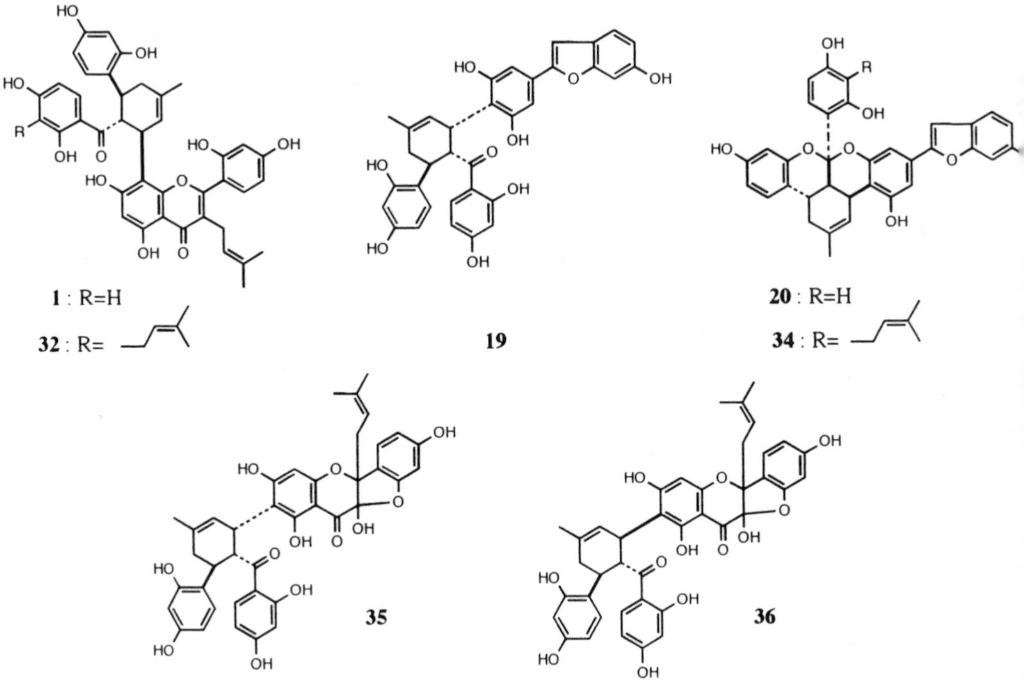

Figure 17. Structures of hypotensive active Diels-Alder type adducts from Moraceae plants.

In Japan, "Sang-Bai-Pi" (the root bark of Chinese mulberry tree) imported from China has been used as an herbal medicine, hence a study of the components of the crude drug purchased in the Japanese market was undertaken. Sanggenons C (**35**) and D (**36**), the Diels-Alder type adducts from "Sang-Bai-Pi", showed hypotensive effects as follows: Sanggenon C (**35**) caused a transient 15 mmHg decrease in arterial blood pressure at the doses of 1 mg/kg in pentobarbital-anesthetized rabbits. At doses of 5 mg/kg, the compound (**35**) caused a transient decrease of 100 mmHg which changed over more than 1 hour to 15 mmHg.[23] Sanggenon D (**36**) caused a transient decrease at the doses of 1 mg/kg in pentobarbital and urethane-anesthetized male Wister strain rats by 35 mmHg, whereas the compound (**36**) caused a decrease by 80 mm Hg at the doses of 1 mg/kg in SHR.[24]

A series of observations for bombesin and its mammalian counterparts indicate that the mammalian bombesin-like peptides may act as autocrine growth factors in human small cell lung carcinoma (SCLC) and other cancers. First, many human SCLC cell lines have been shown to express bombesin-like peptides. Second, peptide bombesin receptor antagonism or anti-bombesin antibodies inhibit SCLC cell growth *in vitro* and *in vivo*. These data suggested that the bombesin receptor antagonists may be useful for the treatment of some kinds of SCLC and other cancers. Dr. Fujimoto's group (Shionogi Research Laboratories, Shionogi Co. Ltd, Osaka, Japan) screened 400 plant extract samples to search for non-peptide

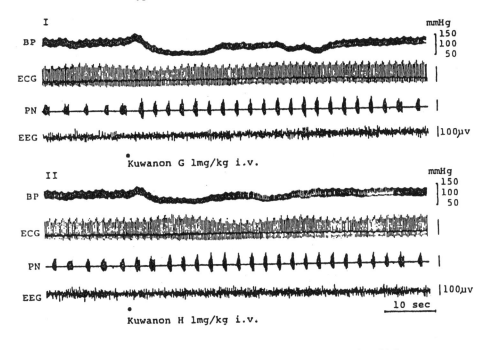

ECG, PN and EEG in a gallamine-immobilized rabbit.
BP: systemic blood pressure
ECG: electrocardiogram
PN: phrenic nerve discharge
EEG: electroencepharogram

Figure 18. Hypotensive effects of kuwanons G (**1**) and H (**32**).

bombesin receptor antagonists. The methanol extract of the underground part of cultivated mulberry tree, *Morus bombycis*, was found to potently inhibit [^{125}I]gastrin-releasing peptide (GRP) binding to Swiss 3T3 cells. Further work was carried out in cooperation with our group and Dr. Fujimoto's group. Bioassay-directed fractionation led to the isolation of kuwanons G (**1**) and H (**32**).

Kuwanons G (**1**) and H (**32**) inhibited specific binding of [^{125}I]GRP to GRP-preferring receptors in murin Swiss 3T3 fibroblasts with K1 values = 470 ± 60 nM and 290 ± 50 nM, respectively. Kuwanon H (**32**) was one order of magnitude less potent for inhibiting [^{125}I]bombesin binding to neuromedin B (NMB)-preferring receptors in rat esophagus membranes. Kuwanon H (**32**) antagonized bombesin-induced increases in the cytosolic free calcium concentration and GRP-induced DNA synthesis in Swiss 3T3 cells. Kuwanon H (**32**) was the most potent of non-peptide bombesin receptor antagonists that had been reported at that time. However, its affinity may still be too low to determine whether the non-peptide antagonist is effective against human lung cancers. Kuwanon H (**32**) and possibly kuwanon G (**1**) also can serve as lead compounds for more rational drug design in the synthesis of more potent antagonists.

5. CONCLUSIONS

The optically active intermolecular Diels-Alder adducts comprising two molecules of isoprenyl phenols were characeristic components of Moraceous plants. The stereochemistries of the adducts were consistent with those of ones in the Diels-Alder reaction involving *exo*- and *endo*-addition. The biosynthetic study of the mulberry Diels-Alder adducts with the aid of *Morus alba* L. cells and suspension cultures demonstrated participation of an enzyme in the formation of the adduct. Some Diels-Alder type adducts including kuwanon G (**1**) showed interesting biological activities, hypotensive effect in rodents, and inhibitory effect against specific binding of [^{125}I]GRP to GRP-preferring receptors in murin Swiss 3T3 fibroblasts. Thus, the mulberry Diels-Alder adducts are attractive compounds not only in the phytochemistry, but also in the biological point of view.

REFERENCES

1. Jiangsu New Medical College (ed.). The encyclopedia of Chinese materia medica. Shanghai People's Publishing House, Shanghai (1977).

2. Nomura, T.; Fukai, T. Kuwanon G, a new flavone derivative from the root bark of the cultivated mulberry tree (*Morus alba* L.). *Chem. Pharm. Bull.* 28:2548 (1980).

3. Nomura, T. Phenolic compounds of the mulberry tree and related plants. *In*: Herz, W.; Grisebach, H.; Kirby, G.W.; Tamm, Ch. (eds.) Fortschritte der chemie organischer naturstoffe. Springer-Verlag, Wien, 53:87 (1988).

4. Nomura, T.; Hano, Y. Isoprenoid-substituted phenolic compounds of moraceous plants. *Nat. Prod. Reports*: :205 (1994).

5. Nomura, T.; Hano, Y.; Ueda, S. Chemistry and biosynthesis of natural Diels-Alder type adducts from moraceous plants. *In*: Atta-ur-Rahman (ed.) Studies in natural products chemistry. Elsevier, Amsterdam. 17:451 (1995).

6. Nomura, T.; Fukai, T.; Narita, T.; Terada, S.; Uzawa, J.; Iitaka, Y.; Takasugi, M.; Ishikawa, S.; Nagao, S.; Masamune, T. Confirmation of the structure of kuwanons G and H (albanins F and G) by partial synthesis. *Tetrahedron Lett.* 22:2195 (1981).

7. Nomura, T.; Fukai, T.; Matsumoto, J.; Imashimizu, A.; Terada, S.; Hama, M. Structure of kuwanon I, a new natural Diels-Alder adduct from the root bark of *Morus alba*. *Planta Med.* 46:167 (1982).

8. Ueda, S.; Nomura, T.; Fukai, T.; Matsumoto, J. Kuwanon J, a new Diels-Alder adduct and chalcomoracin from callus culture of *Morus alba* L. *Chem. Pharm. Bull.* 30:3042 (1982).

9. Ueda, S.; Matsumoto, J.; Nomura, T. Four new natural Diels-Alder type adducts, mulberrofuran E, kuwanons Q, R, and V from callus culture of *Morus alba* L. *Chem. Pharm. Bull.* 32:350 (1984).

10. Ikuta (nee Matsumoto), J.; Fukai, T.; Nomura, T.; Ueda, S. Constituents of the cultivated mulberry tree 35. Constituents of *Morus alba* L. cell cultures. 1. Structures of four new natural Diels-Alder type adducts, kuwanons J, Q, R, and V. *Chem. Pharm. Bull.* 34:2471 (1986).

11. Nomura, T.; Fukai, T.; Matsumoto, J.; Fukushima, K.; Momose, Y. Structure of mulberrofuran C, a natural hypotensive Diels-Alder type adduct from root bark of the cultivated mulberry tree (*Morus bombycis* Koidzumi). *Heterocycles* 16:759 (1981).

12. Fukai, T.; Hano, Y.; Hirakura, K.; Nomura, T.; Uzawa, J.; Fukushima, K. Structures of mulberrofurans F and G, two natural hypotensive Diels-Alder type adducts from the cultivated mulberry tree (*Morus lhou* (Ser.) Koidz.). *Heterocycles* 22:474 (1984).

13. Rama Rao, A.V.; Deshpande, V.H.; Shastri, R.K.; Tavale, S.S.; Dhaneshwar, N.N. Structures of albanols A and B, two novel phenols from *Morus alba* bark. *Tetrahedron Lett.* 24:3013 (1983).

14. Hano, Y.; Suzuki, S.; Kohno, H.; Nomura, T. Absolute configuration of kuwanon L, a natural Diels-Alder type adduct from the *Morus* root bark. *Heterocycles* 27:75 (1988).

15. Hano, Y.; Suzuki, S.; Nomura, T.; Iitaka, Y. Absolute configuration of natural Diels-Alder type adduct from the *Morus* root bark. *Heterocycles* 27:2315 (1988).

16. Hano, Y.; Nomura, T.; Ueda, S. Biosynthesis of chalcomoracin and kuwanon J, the Diels-Alder type adducts, in *Morus alba* cell cultures. *Chem. Pharm. Bull.* 37:554 (1989).

17. Hano, Y.; Ayukawa, A.; Nomura, T.; Ueda, S. Dynamic participation of primary metabolites in the biosynthesis of chalcomoracin and b-sitosterol in *Morus alba* cell cultures. *Naturwissenschaften* 79:180 (1992).

18. Hano, Y.; Nomura, T.; Ueda, S. Biosynthesis of optically active Diels-Alder type adducts revealed by an aberrant metabolism of O-methylated precursors in *Morus alba* cell cultures. *J. Chem. Soc., Chem. Commun.* :610 (1990).

19. Nomura, T.; Hano, Y.; Aida, M. Isoprenoid-substituted flavonoids from *Artocarpus* plants (Moraceae). *Heterocycles* 47:1179 (1998).

20. Hano, Y.; Aida, M.; Nomura, T.; Ueda, S. A novel way of determining the structure of artonin I, an optically active Diels-Alder type adduct, with the aid of an enzyme system of *Morus alba* cell cultures. *J. Chem. Soc., Chem. Commun.* :1177 (1992).

21. Hano, Y.; Nomura, T.; Ueda, S. Late stage of intermolecular Diels-Alder type adducts in *Morus alba* L. cell cultures. *Heterocycles* 51:231 (1999).

22. Nomura, T.; Fukai, T.; Momose, Y.; Takeda, R. Hypotensive constituents of the root bark of the mulberry tree (*Morus alba* L.) and the mechanism of their actions. Third symposium on the development and application of naturally occurring drug materials. Symposium paper (Tokyo) p. 13 (1980).

23. Nomura, T.; Fukai, T.; Hano, Y.; Uzawa, J. Structure of sanggenon C, a natural hypotensive Diels-Alder adduct from Chinese crude drug "Sang-Bai-Pi" (*Morus* root barks). *Heterocycles* 16:2141 (1981).

24. Nomura, T.; Fukai, T., Hano, Y.; Uzawa, J. Structure of sanggenon D, a natural hypotensive Diels-Alder adduct from Chinese crude drug "Sang-Bai-Pi" (*Morus* root barks). *Heterocycles* 17(special issue) :381 (1982).

25. Mihara, S.; Hara, M.; Nakamura, M.; Sakurai, K.; Tokura, K.; Fujimoto, M.; Fukai, T.; Nomura, T. Non-peptide bombesin receptor antagonists, kuwanon G and H, isolated from mulberry. *Bioch. Biophys. Res. Commun.* 213:594 (1995).

BIOTECHNOLOGY

DESIGNER TANNINS: USING GENETIC ENGINEERING TO MODIFY LEVELS AND STRUCTURES OF CONDENSED TANNINS IN *LOTUS CORNICULATUS*

Mark P. Robbins, Adrian D. Bavage, and Phillip Morris

Institute of Grassland and Environmental Research
Aberystwyth Research Center
Aberystwyth, Ceredigion SY23 3EB
UNITED KINGDOM

1. INTRODUCTION

In terms of the genetic improvement of forage legumes, a number of targets have been identified.[1] However, it is interesting to note that when one discusses the modification of chemical composition of these species, condensed tannins have been identified as being of critical importance by a number of independent research groups.

The reasons for the widespread interest in condensed tannins relate to two different classes of forage species. In the case of temperate herbage legumes, principally white clover (*Trifolium repens*) and lucerne (*Medicago sativa*), these species do not contain tannins in vegetative tissues. Therefore, while these species are highly proteinaceous and suitable for a wide range of European, American, and Australasian agricultural systems, improvements are still possible. Basically, problems with the effective use of these crops relate to processes of microbial deamination of proteins within the rumen. Grazing ruminants, whether they are sheep or cattle, lose a significant portion of ingested protein in the rumen. The efficiency of agricultural systems, which include white clover or lucerne, could be very significantly increased by reducing levels of protein degradation. The introduction of condensed tannins into the foliar tissues of temperate forage legumes should decrease rates of protein deamination in the rumen, and this offers an opportunity for increasing levels of bypass protein, which can then be absorbed as amino acids lower down in the intestinal tract.[2]

Another benefit that would accrue from the introduction of tannins into white clover and lucerne relates to pasture bloat. Rapid digestion of proteins in the rumen results in stable protein foam which, unless treated, can result in the death

Plant Polyphenols 2: Chemistry, Biology, Pharmacology, Ecology, Edited by
Gross et al. Kluwer Academic / Plenum Publishers, New York, 1999

of livestock. The syndrome of pasture bloat currently causes significant economic losses both in the United States and Australasia. It has been estimated that cattle deaths due to bloat cost the US \$180M and Canada \$45M per annum, respectively.[3] Additionally, pasture bloat has been reported as a significant problem in New Zealand where death rates in an individual herd can reach 15 percent.[4] Interestingly, this is a longstanding agricultural problem, and graphic descriptions of bloat are even included in nineteenth century English literature.[5] Condensed tannins precipitate proteins and collapse protein foams[6] and would therefore be of clear value in future white clover and lucerne varieties.

For the above reasons, a number of groups are striving to introduce tannins into the foliage of white clover or lucerne using modern biotechnologies. Methods being pursued for the engineering of condensed tannin biosynthesis include the introduction of the terminal steps of this pathway into foliar tissues or the attempted transactivation of this pathway in vegetative organs. Nevertheless, an important question remains: how can one determine the optimal levels and structures of condensed tannins in new transgenic plants from the perspective of protein protection and bloat safety? Analyses can be carried out using forages with differing levels of condensed tannins. However, a more elegant solution might be to analyse transgenic *Lotus* lines that have an isogenic background but have specific modifications in condensed tannin biosynthesis. With this information, it should be a great deal easier to assess optimal tannin structures and levels in future transgenic varieties of white clover and lucerne.

There is also a second class of forage crops for which condensed tannins are of critical importance. These are tropical and semiarid herbage legumes that contain very high levels of condensed tannins within their foliage. In these cases, levels of tannins can be extremely high; for example, *Desmodium* has foliage that can contain up to 15 percent DW in the form of condensed tannin polymers.[7] Such crops are either unpalatable or of significantly reduced nutritive value. These issues are also of importance in temperate agriculture as a number of alternative forage species may well accumulate deleterious levels of condensed tannins under some environmental conditions.[8] Once again, the assessment of optimal tannin levels and structures may be of value both for conventional plant breeders and also for biotechnologists who may be able to dramatically reduce condensed tannin levels in these crop species using modern methods.

In this chapter, we will discuss some of the first experiments aimed at altering the condensed tannin pathway in higher plants. Strategies for altering flux and end product accumulation are discussed as are some future opportunities for modifying the pathways of condensed tannin biosynthesis in herbage legumes and other crop species.

2. A MODEL TRANSGENIC SYSTEM

Few transformable model systems biosynthesize condensed tannin polymers. For this reason, we have chosen to work on *Lotus corniculatus*, which is both a bonafide crop species and also an excellent laboratory organism with good tissue culture properties. *Lotus corniculatus* is readily transformable with *Agrobacterium rhizogenes*,[9] a microorganism that naturally introduces genes into the genomes of higher plant species. Our efforts enable us to readily introduce genes into this species and allow us to produce root cultures and also regenerated plants that look substantially similar to non-transformed plants.[10] We have developed, in collaboration with Dr Judith Webb at IGER, a range of clonal transformable genotypes of *Lotus corniculatus* cv. Leo and have based our experimental work on this resource. A brief description of IGER genotypes for use in genetic manipulation experiments is presented in table 1.

Our basic experimental approach can be summarized as follows. Gene constructs are built in conventional binary transformation vectors, transferred into *Agrobacterium rhizogenes,* and then *A. rhizogenes* strains that harbor unrearranged construct sequences are then used for *Lotus* transformation. "Hairy roots" are produced when recipient genotypes are wounded with a scalpel loaded with *A. rhizogenes* harboring a binary vector and after decontamination with cefotaxime, these independently derived "hairy root" lines can then be used directly as experimental material. After selection on kanamycin or alternative selectable markers, we have found that co-transformation rates of 60–70 percent are typical. The identity of construct-positive lines can then be confirmed by PCR, Southern analysis, RT-PCR, and Northern blot analysis.

In a typical experiment, control (empty vector) and construct-positive lines are produced in a given recipient genotype. For experiments where we have analyzed co-transformed root cultures, three different genotypes have been used; S33 (low tannin levels in root cultures), S50 (medium levels), or S41 (high levels). Pairwise comparison of independent construct-positive lines with a control population then

Table 1. Tannin content of IGER clonal genotypes of
Lotus corniculatus var. Leo.

Line	Derived root culture (mg/g FW)	Leaves (percent DW)	Stems (percent DW)	Roots[a] (percent DW)
S33	0.3-0.45	0.4-1.3	0.9-2.1	1.3-4.2
S50	0.4-0.7	0.5-1.1	0.7-2.0	1.0-3.7
S41	2.0-3.8	2.2-6.9	1.1-2.3	1.4-4.8

[a] For root cultures, values refer to the range of tannin content noted with control lines grown for 3 weeks under standard conditions in liquid medium.[19] Leaf, stem, and root values refer to the range of tannin levels noted in plants grown under a range of environmental conditions.[8]

permits the identification of novel chemical and developmental phenotypes. Such comparisons can be made using transgenic root cultures and also with plants regenerated from these root cultures, which can be grown up for analysis either in growth rooms or in transgenic glasshouse facilities.

3. CONDENSED TANNIN BIOSYNTHESIS IN *LOTUS CORNICULATUS*

In figure 1, we show a representation of the condensed tannin biosynthetic pathway in *Lotus corniculatus*. It is a common assumption that all of enzymatic

Coumaroyl CoA + 3 x malonyl CoA

↓**CHS gene family**

Naringenin chalcone

↓**CHI**	**F3'OH**		
Naringenin	→	Eriodictyol	
↓**F3OH**	**F3'OH** ↓	**F3'5'OH**	
Dihydrokaempferol →	Dihydroquercetin →	Dihydromyricetin	**Common with**
↓**DKR**	↓**DQR**	↓**DMR**	**anthocyanins**
Afzelechin 4-ol	Catechin 4-ol	Gallocatechin	**and flavonoids**

....

↓**FDR ± E**	↓**FDR ± E**	↓**FDR ± E**	**CT-specific**
Afzelechin	Catechin, epi-catechin	Gallocatechin, epi-gallocatechin	
[Propelargonidin units]	[Procyanidin units]	[Prodelphinidin units]	

↓**CE**

Dimers

↓**PE**

Condensed tannin polymers

Figure 1. Proposed pathway to condensed tannins in the leaves of *Lotus corniculatus*. CHS, chalcone synthase; CHI, chalcone isomerase, F3OH, flavanone 3-hydroxylase; F3'OH, flavonoid 3' hydroxylase; F3'5'OH, flavonoid 3'5' hydroxylase; DKR, dihydrokaempferol reductase; DQR, dihydroquercetin reductase; DMR, dihydromyricetin reductase; FDR, flavan-3,4-diol reductase *syn* leucoanthocyanidin reductase; E, epimerase; CE, condensing enzyme; PE, polymerising enzyme.

steps comprising the pathway are expressed in the "tannin cells" where condensed tannin (CT) polymers accumulate in root, leaf, stem, and petal tissues of *Lotus*. The location of tannin-containing cells in *Lotus corniculatus* is illustrated in figure 2. However, of course, it is possible that some of the early steps occur in other cell types and that intermediates are then transported to tannin cells where the terminal and polymerization steps are carried out.

We have attempted to immunolocalize dihydroflavonol reductase (DFR) in root tissues of *Lotus corniculatus* using antibodies supplied by Dr. Anne Depicker (University of Ghent). Unfortunately, results were inconclusive (Robbins, Bavage, and Thomas, unpublished). However, one striking observation was that strong signals were noted in vascular tissues, and this may relate to the strong homology of DFR with cinnamyl CoA reductase (CCR), a stereospecific reductase involved in the biosynthesis of lignin polymers.[11]

The current position with regard to the condensed tannin pathway is that genes encoding a number of these enzymes have been and are continuing to be cloned from plants that biosynthesize anthocyanins. However, recently workers have started to clone genes from tanniferous plants that appear to encode for enzymes directly involved in the condensed tannin pathway. Chalcone synthase (CHS) and DFR sequences have been cloned from *Onobrychis viciifolia*,[12] flavanone

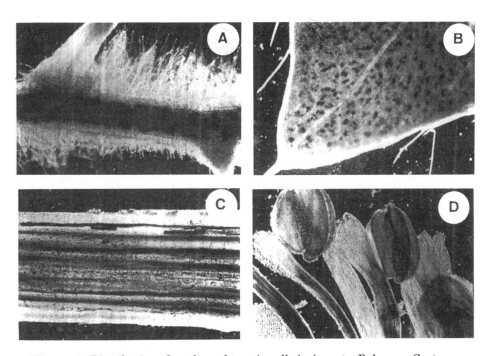

Figure 2. Distribution of condensed tannin cells in A; roots, B; leaves, C; stems, and D; floral tissue of *Lotus corniculatus*. (Reproduced with permission from "Biotechnology and the improvement of forage legumes", McKersie and Brown (eds), CAB International, 1997, figure 7.1).

3-hydroxylase (F3OH) and DFR from barley[13] and DFR sequences from *Lotus corniculatus*[14] and strawberry.[15]

4. GENETIC MANIPULATION STRATEGIES UTILIZED IN *LOTUS CORNICULATUS*: MODIFICATION OF TANNIN LEVELS

A variety of strategies and genes have been used to modify levels and structures of condensed tannins in *Lotus corniculatus*. One surprising strategy for increasing the levels of condensed tannins in *Lotus* relates to the antisense expression of one of the genes that codes for an initial step in this biosynthetic pathway. We introduced and expressed an antisense construct containing a CHS gene from French bean (*Phaseolus vulgaris*).[16] The predicted chemical phenotype from this intervention was a reduction in tannin levels, but when we screened root cultures, it became apparent that unexpected results had been obtained. In figure 3, we show results from screening independently derived control and antisense chalcone synthase (AS-CHS) lines in an S50 genetic background. We analyzed

A. Control lines

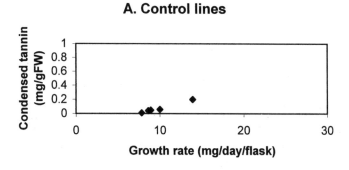

B. Antisense CHS lines

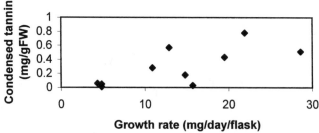

Figure 3. Condensed tannin content and growth rate of *Lotus corniculatus* root cultures derived from S50 recipient genotype. A, control transformants; B, lines harbouring an antisense chalcone synthase construct. Note that in these experiments lines were grown in small flasks (10 mL medium in 50 mL flasks), hence control tannin levels are lower than indicated for S50 control root cultures in Table 1 (50 mL medium in 250 mL flasks).

lines over five successive subcultures in liquid medium and measured condensed tannin content (mg/g Fresh Weight) and average growth rate over a period of 3 weeks after subculture. It became apparent that control lines and some ineffective antisense events clustered together but that some antisense lines were identified with a higher condensed tannin content than controls. In some of the lines, increased tannin content was accompanied by an increase in root culture growth rate. High tannin phenotypes were also noted in the S33 genotype at early subcultures, but no up-regulated lines were noted in the S41 (high tannin) genetic background.

A detailed analysis was performed on S50 AS-CHS lines, and it was confirmed that some of the antisense lines had elevated tannin levels and that this was accompanied by an increase in steady-state levels of endogenous CHS transcripts as measured using a strand-specific DNA probe in Northern blot experiments.[16] It is difficult to find a simple explanation for such counter-intuitive observations. Our current hypothesis is that this antisense construct has down-regulated one or more members of the CHS gene family in *Lotus* and that genetic over-compensation by other member(s) of the gene family has resulted in an overall increase in CHS transcript and condensed tannin end products. Genetic compensation has also been noted by other workers who have studied transgenic plants.[17] Additionally, it is perhaps informative to compare these antisense CHS results with the data of Todd and Vodkin who have analyzed some classical CHS mutants that control proanthocyanidin accumulation in the seedcoat tissues of soybean.[18] Todd and Vodkin found a dominant *I* allele, which includes a CHS sequence that suppresses pigment accumulation and also report that deletion events within this member of the CHS gene family restored higher levels of CHS transcript and led to seedcoat pigmentation.

In conclusion, from our results, there appear to be opportunities to increase flux into the tannin biosynthetic pathway using manipulations involving antisense CHS. It would be of considerable interest to find out whether this up-regulatory phenomenon is restricted to *Lotus corniculatus* and other tanniferous legumes or is of a more general nature. We are currently analyzing AS-CHS plants, and initial results indicate that a number of plants may have reduced shoot biomass and a reduction in plant width and numbers of growing points (Buckridge and Robbins, unpublished). We are currently trying to determine whether AS-CHS plants have an increase in tannin content in leaf, stem, and other tissues in regenerated plants.

With reference to reducing the levels of condensed tannins in *Lotus* root cultures, we have carried out experiments using an antisense *Antirrhinum majus* DFR gene.[19] Antisense dihydroflavonol reductase (AS-DFR) constructs were introduced into S33, S50, and S41 genotypes. A number of S33 AS-DFR lines showed reductions in tannin levels relative to controls, and down-regulation was also noted in some S50 antisense lines, but no phenotypes were seen in S41 lines.

We have also analyzed some AS-DFR lines as regenerated plants,[20] and a summary of results relating to S33 AS-DFR transgenics is shown in table 2. From these results, we can conclude that low-tannin phenotypes in root cultures do not necessarily correspond to low-tannin phenotypes in leaves and stems of plants

Table 2. Chemical phenotypes of *Lotus corniculatus* S33
transformants harboring an antisense DFR construct[a]

Line	Root culture	Leaves	Stems
RFD4	-	-	-
RFD7	-	↓	↓
RFD8	↓	↓	↓
RFD19	↓	-	-
RFD38	↓	-	-
RFD49	na	-	-
RFD50	na	↓	↓

[a] -, not sig. different from S33 control lines; ↓, reduced levels relative to S33 controls; na, not analyzed.

regenerated from such root cultures. For example, RFD8 shows a phenotype in root cultures and also in foliar tissues, RFD7 has a phenotype in leaves and stems but not in root cultures, whereas RFD19 and RFD38 have reduced tannin levels in root cultures but no reductions in shoot tissues. We can thus conclude that suppression effects are tissue-specific and that differing patterns of tissue-specific reductions can be noted in independently derived transformation events. Additionally, one can make a conclusion that the type (or class) of phenotype noted in root cultures is reflected in whole plant regenerants but not necessarily in the same lines.

Similar effects were noted when S50 AS-DFR lines were analyzed. Some lines showed statistically significant reductions in condensed tannin content in leaf, stem, and root tissues or combinations thereof.[20] Interestingly, one AS-DFR construct comprising the first half of the *A. majus* DFR gene gave rise to lines with reduced tannin levels and also some lines with above control levels of condensed tannin. These results may reflect genetic over-compensation events within the DFR multigene family with this construct (pMAJ2) in this genotype (S50).

5. GENETIC MANIPULATION STRATEGIES UTILIZED IN *LOTUS CORNICULATUS*: MODIFICATION OF TANNIN POLYMER STRUCTURE

We have also reported changes in polymer hydroxylation in AS-DFR lines. Briefly, our analyses indicated that decreases in condensed tannin levels were accompanied by a decrease in levels of polymer hydroxylation.[9,19] These data were obtained by scanning TLC plates of butanol-HCl hydrolysates. We have now developed more accurate HPLC methods for determining the monomer composition of condensed tannin polymers, and in the future, we would like to reanalyze the hydroxylation status of tannins from AS-DFR lines.

One conclusion from antisense dihydroflavonol reductase experiments is that results can be interpreted to suggest that this enzymatic step can be rate-limiting for the quantity of condensed tannins in *Lotus*. Therefore, we decided to investigate whether DFR could be involved in the control of hydroxylation and hence carried out sense experiments aimed at over expressing heterologous DFR sequences. Initially, we chose to introduce and express a full-length *A. majus* DFR cDNA.[21] The substrate specificity of DFR genes has been analyzed in some detail, and it has been proposed that the *A. majus* DFR corresponds to a dihydrokaempferol reductase activity. Therefore, a predicted phenotype for over-expressing this enzyme is the production of condensed tannin polymers, which contain propelargonidin units. Of course this assumes that dihydrokaempferol is a metabolic intermediate in *Lotus corniculatus* and that enzymes subsequent to DFR are able to use afzelechin 4-ol and its derivatives.

In fact, a range of root culture phenotypes resulted from sense *A. majus* DFR experiments, and a summary of data relating to selected lines is shown in table 3. One line (ADFR10) was identified with enhanced tannin levels and a high propelargonidin content relative to control lines. This chemical phenotype is exactly as predicted for the over-expression of an *A. majus* DFR sequence in *Lotus*. Additionally, two lines were identified with normal tannin levels but with distorted procyanidin: prodelphinidin ratios, and this is an unexpected result. ADFR01 has a high prodelphinidin content, whereas by contrast ADFR11 has a high procyanidin content in tannin hydrolysates. Our explanation for these novel results is that there may be at least two DFR genes expressed in *Lotus* root cultures and that these genes encode DFR enzymes with different substrate specificities. Therefore, ADFR01 may correspond to the suppression of a dihydroquercetin reductase, whereas ADFR11 may be an event where dihydromyricetin reductase has been suppressed. While we know that DFR is a small gene family in *Lotus*,[21] we have no evidence for enzymes with different substrate specificities and no direct

Table 3. Subunit composition of condensed tannins from Selected S50 *Lotus corniculatus* root cultures harboring a sense *A. majus* DFR construct

Line	CT level	PC(%)	PD(%)	PP(%)	PC:PD:PP relative to controls
Controls	Normal	80	14	4	
ADFR10	Enhanced	73	13	14	***
ADFR01	Normal	68	29	2	***
ADFR11	Normal	91	3	6	**
ADFR02	Reduced	78	18	4	
ADFR07	Reduced	83	12	4	

evidence for a dihydromyricetin reductase. However, recent reports by Paulocci, Damiani, and Arcioni (see chapter in this volume) indicate that more than one class of DFR cDNAs can be isolated from *Lotus corniculatus*, and it would be of value to determine whether these cDNA sequences have different substrate specificities when expressed *in vitro* or in heterologous expression systems. Finally, two other sense DFR lines were found (ADFR02 and ADFR07) that have reduced tannin content but normal procyanidin: prodelphinidin: propelargonidin ratios. It can be hypothesized that these are lines where there has been suppression both of dihydroquercetin reductase and dihydromyricetin reductase genes. So in summary, the range of chemical phenotypes noted in this experiment may actually modify our view of the enzymatic control of condensed tannin biosynthesis by dihydroflavonol reductase.

In addition, we have also carried out tannin cell counts in *Lotus* root cultures expressing sense *A. majus* DFR, and some typical results are shown in figure 4. High tannin lines such as ADFR10 have a greater number of detectable tannin cells both in root tips and in mature root sections. By contrast, putatively co-suppressed lines such as ADFR02 and ADFR07 have lower numbers of tannin cells than controls, and therefore, this may be a method for rapidly screening transgenic root cultures for quantitative changes in tannin levels.

We have pursued two additional approaches for specifically increasing the hydroxylation of condensed tannin polymers. In one set of experiments, we have introduced and expressed a full-length DFR sequence from *Petunia hybrida*, which has been reported to encode dihydroquercetin reductase activity. We have started to analyze data from root cultures, and initial analyses indicate that in addition to high prodelphinidin lines, other lines have been produced with unusual procyanidin: prodelphinidin: propelargonidin ratios (Bavage, Robbins and Morris, unpublished).

An alternative approach for increasing tannin hydroxylation relates to the introduction and expression of enzyme activities that directly hydroxylate dihydroflavonol intermediates. To this end, in collaboration with Dr. Paul Beuselinck (USDA, ARS), we are attempting to introduce a flavonoid 3'5'hydroxylase sequence from *Lisianthus* into *Lotus corniculatus* with the aim of increasing tannin polymer hydroxylation patterns in a direct and predictable manner.

6. CONCLUSIONS

Condensed tannins are critically important with reference to forage quality in herbage legumes because of their ability to protect dietary protein. We have developed *Lotus corniculatus* (bird's foot trefoil) as an experimental system for the genetic manipulation of this pathway in higher plants. In this article, we have described recent work aimed at altering the levels and the structures of condensed tannin polymers using genes cloned from other plant species. Opportunities now exist for the predictable genetic modification of tannin molecules, and in the future there will be opportunities and possible options for the cloning of the terminal steps of the pathway.

Clearly there are major research opportunities for work that relates to the terminal steps of condensed tannin biosynthesis, but unfortunately, details

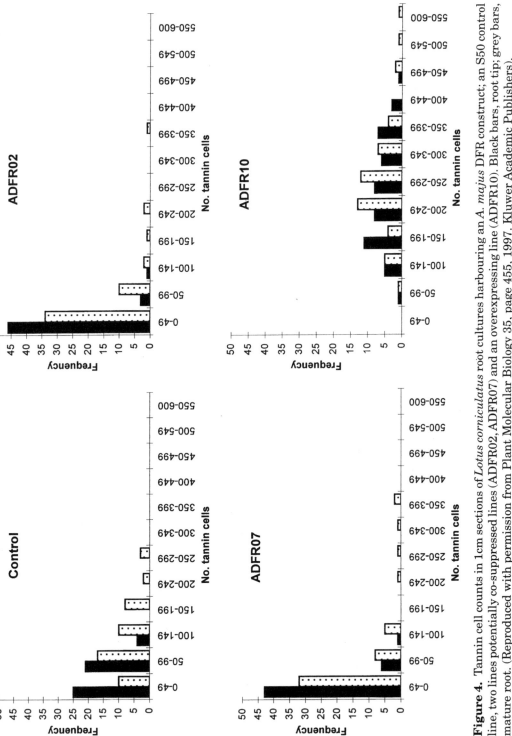

Figure 4. Tannin cell counts in 1cm sections of *Lotus corniculatus* root cultures harbouring an *A. majus* DFR construct; an S50 control line, two lines potentially co-suppressed lines (ADFR02, ADFR07) and an overexpressing line (ADFR10). Black bars, root tip; grey bars, mature root. (Reproduced with permission from Plant Molecular Biology 35, page 455, 1997. Kluwer Academic Publishers).

regarding this sector of polyphenolic metabolism are not well understood. In addition to flavan-3,4-diol reductase (leucoanthocyanidin reductase), other enzymatic steps exist. It is reasonable to suggest the presence of an epimerase in this pathway. Evidence from barley mutants suggests that other steps in the pathway are catalyzed by a condensing enzyme that initiates polymer formation and a polymerizing enzyme that is involved in polymer extension.[22,23] However, it is important to point out that Dr. Greg Tanner (CSIRO) has developed an interesting hypothesis that instead of corresponding to soluble enzymatic activities these steps may in fact correspond to vacuolar transporters responsible for the import and polymerization of catechins and catechin-4-ols, respectively.

To our knowledge, with the possible exception of flavan-3,4-diol reductase, none of the genes coding for these steps has been cloned and identified from higher plants. The cloning of these genes is a considerable technical challenge. However, a new approach may offer new options for cloning and then exploiting these terminal genes of the condensed tannin pathway. In pioneering work, Damiani and colleagues[24] have introduced Sn, a maize anthocyanin regulatory gene, into *Lotus corniculatus*. The consequence of this genetic intervention appears to be the suppression of tannin biosynthesis in leaf tissues and a decrease in DFR and FDR enzyme activities. Therefore, this intervention appears to be suppressing the expression of terminal pathway genes, which are presumably normally regulated by a Sn homologue. Subtractive cloning of Sn-suppressed vs non-suppressed lines in an isogenic background may yield sequences that correspond to these as yet uncloned genes. Functional assessment of these sequences could be carried out in *Lotus* by expressing sequences in short sense or in antisense, and for relevant genes, lines should be produced with modified tannin molecular weights and stereo-structure and possibly with altered crosslinking within polymer molecules.

In conclusion, the genetic modification of lignin structure is now well established, and there is no reason to believe that condensed tannins cannot be engineered in a similar manner. The consequences of tannin design may include the use of forages and other crops as sources of specialty tannin polymers, and in view of the taste-related properties of tannins, there may also be applications in the food and beverage industries.[25]

ACKNOWLEDGMENTS

This work at the Institute of Grassland and Environmental Research was funded by competitive strategic grants from the Biotechnology and Biological Sciences Research Council. Adrian Bavage was supported by a BBSRC Plant Molecular Biology Grant (phase 2). We acknowledge Teri Davies, Ian Davies, and other members of the laboratory for technical assistance. Also, we would like to thank colleagues for making sequences, vectors, and constructs available to us: Cathy Martin for *A. majus* DFR , Rick Dixon for *P. vulgaris* CHS and Kevin Davies for DFR and F3'5'OHase constructs. In addition, we thank our international collaborators, particularly Greg Tanner and Phil Larkin, for discussions on the later steps in the tannin pathway and Francesco Damiani, Francesco Paulocci, and Sergio

Arcioni relating to collaborative work on Sn-suppression phenomena in *Lotus corniculatus*.

REFERENCES

1. McKersie, B.D.; Brown, D.C.W. Biotechnology and the improvement of forage legumes. CAB International, Oxford (1997).

2. Barry, T.N.; Duncan, S.J. The role of condensed tannins in the nutritional value of *Lotus pedunculatus* for sheep. 1. Voluntary intake. *Br. J. Nut.* 51:485 (1984).

3. Lees, G.L. Condensed tannins in some forage legumes: their role in the prevention of ruminant pasture bloat. *In*: Hemingway, R.W.; Laks, P.E. (eds.). Plant polyphenols—synthesis, properties, significance. Plenum Press, New York p. 915 (1992).

4. Reid, C.S.W. Bloat in New Zealand cattle. *Bovine Practitioner* 1:24 (1976).

5. Hardy, T. Far from the madding crowd. McMillan, London (1874).

6. Tanner, G.J.; Moate, P.J.; Davis, L.H.; Laby, R.H.; Yuguang, L.; Larkin, P.A. Proanthocyanidins (condensed tannin) destabilise plant protein foams in a dose dependent manner. *Aust. J. Agric. Res.* 46:1101 (1995).

7. Reed, J.D. Nutritional toxicology of tannins and related polyphenols in forage legumes. *J. Anim. Sci.* 34:82 (1994).

8. Carter, E.B.; Theodorou, M.K.; Morris, P. Responses of *Lotus corniculatus* to environmental change. 2. Effect of elevated CO_2, temperature and drought on tissue digestion in relation to tannin and carbohydrate accumulation. *J. Sci. Food Agric.* (in press).

9. Robbins, M.P.; Carron, T.R.; Morris, P. Transgenic *Lotus corniculatus*: a model system for modification and genetic manipulation of condensed tannin biosynthesis. *In*: Hemingway, R.W.; Laks, P.E. (eds.). Plant polyphenols—synthesis, properties, significance. Plenum Press, New York, p. 111 (1992).

10. Webb, K.J.; Jones, S.; Robbins, M.P.; Minchin, F.R. Characterisation of transgenic root cultures of *Trifolium repens*, *Trifolium pratense* and *Lotus corniculatus* and transgenic plants of *Lotus corniculatus*. *Plant Sci.* 70:243 (1990).

11. Lacombe, E.; Hawkins, S.; Van Doorsselaere, J.; Piquemal, J.; Goffner, D.; Poeydomenge, O.; Boudet, A.M.; Grima-Pettenati, J. Cinnamoyl CoA reductase, the first committed enzyme of the lignin branch biosynthetic pathway: cloning, expression and phylogenetic relationships. *Plant J.* 11:429 (1997).

12. Joseph, R.; Tanner, G.; Larkin, P. Proanthocyanidin synthesis in the forage legume *Onobrychis viciifolia*. A study of chalcone synthase, dihydroflavonol 4-reductase and leucoanthocyanidin 4-reductase in developing leaves, *Aust. J. Plant Physiol.* 25:271 (1998).

13. Meldgaard, M. Expression of chalcone synthase, dihydroflavonol reductase and flavanone 3-hydroxylase in mutants of barley deficient in anthocyanin and proanthocyanidin biosynthesis. *Theor. Appl. Genet.* 83:695 (1992).

14. Bavage, A.D.; Robbins, M.P. Dihydroflavonol reductase, a *Lotus corniculatus* L. tannin biosynthesis gene: isolation of a partial gene clone by PCR. *Lotus Newsletter* 25:37 (1994).

15. Moyano, E.; Portero-Robles, I.; Medina-Escobar, N.; Valpuesta, V.; Munoz-Blanco, J.; Caballero, J.L. A fruit-specific putative dihydroflavonol 4-reductase gene is differentially expressed in strawberry during the ripening process. *Plant Physiol.* 117:711 (1998).

16. Colliver, S.P.; Morris, P.; Robbins, M.P. Differential modification of flavonoid and isoflavonoid biosynthesis with an antisense chalcone synthase construct in transgenic *Lotus corniculatus*. *Plant Mol. Biol.* 35:509 (1997).

17. Beffa, R.S.; Neuhaus J.-M.; Meins F. Physiological compensation in antisense transformants: Specific induction of an "ersatz" glucan endo-1,3-glucosidase in plants infected with necrotizing viruses. *Proc. Natl. Acad. Sci. USA* 90:8792 (1993).

18. Todd, J.J.; Vodkin, L.O. Duplications that suppress and deletions that restore expression from a chalcone synthase multigene family. *Plant Cell* 8:687 (1996).

19. Carron, T.R.; Robbins, M.P.; Morris, P. Genetic modification of condensed tannin biosynthesis in *Lotus corniculatus*. I. Heterologous antisense dihydroflavonol reductase down-regulates tannin accumulation in "hairy root" cultures. *Theor. Appl. Genet.* 87:1006 (1994).

20. Robbins, M.P.; Bavage, A.D.; Strudwicke, C.; Morris, P. Genetic manipulation of condensed tannins in higher plants. II. Analysis of birdsfoot trefoil plants harbouring antisense dihydroflavonol reductase constructs. *Plant Physiol.* 116:1133 (1998).

21. Bavage, A.D.; Davies, I.G.; Robbins, M.P.; Morris, P. Expression of an *Antirrhinum* dihydroflavonol reductase gene results in changes in condensed tannin structure and accumulation in root cultures of *Lotus corniculatus* (bird's foot trefoil). *Plant Mol. Biol.* 35:443 (1997).

22. Jende-Strid, B. Gene-enzyme relations in the pathway of flavonoid biosynthesis in barley. *Theor. Appl. Genet.* 81:668 (1991).

23. Jende-Strid, B. Genetic control of flavonoid biosynthesis in barley. *Hereditas* 119:187 (1993).

24. Damiani, F.; Paulocci, F.; Cluster, P.D.; Arcioni, S.; Tanner, G.J.; Joseph, R.J.; Yi, Y.G.; Demajnik, J.; Larkin, P.J. Tissue-specific up- and down- regulation of tannin synthesis in transgenic *Lotus corniculatus* plants. *In*: Vercauteren, J.; Cheze, C.; Dumon, M.C.; Weber, J.F. (eds.) Polyphenols communications 96. Groupe Polyphenols, Bordeaux (France). 219 (1996).

25. Robbins, M.P.; Bavage, A.D.; Morris, P. Options for the genetic manipulation of astringent and antinutritional metabolites in fruit and vegetables. *In*: Tomas-Barberan, F.A.; Robins, R.J. (eds.). Phytochemistry of fruit and vegetables. Clarendon Press, Oxford, p. 251 (1997).

GENETIC SYSTEMS FOR CONDENSED TANNIN BIOTECHNOLOGY

Margaret Y. Gruber,[a] Heather Ray,[a] Patricia Auser,[a]
Birgitte Skadhauge,[b] Jon Falk,[c] Karl K. Thomsen,[b] Jens
Stougaard,[d] Alister Muir,[a] Garry Lees,[a] Bruce Coulman,[a]
Bryan McKersie,[e] Steve Bowley,[e] and Diter von Wettstein[f]

[a] Agriculture and Agri-Food Canada
Saskatoon Research Centre
Saskatoon, Saskatchewan S7N 0X2
CANADA

[b] Carlsberg Research Centre
DK-2500 Copenhagen
DENMARK

[c] Institute of Botany
University of Köln
Köln 50931
GERMANY

[d] Department of Molecular Biology
University of Aarhus
DK-8000 Aarhus C
DENMARK

[e] Department of Plant Agriculture
University of Guelph
Guelph, Ontario N1G 2W1
CANADA

[f] Department of Soil and Crop Science
Washington State University
Pullman, Washington 99164-4420
USA

Plant Polyphenols 2: Chemistry, Biology, Pharmacology, Ecology, Edited by
Gross et al. Kluwer Academic / Plenum Publishers, New York, 1999

1. INTRODUCTION

Condensed tannins (proanthocyanidins) are plant phenolic polymers with protein-binding, carbohydrate-binding, and antioxidant properties. Dietary condensed tannins deter some insects from feeding on crops, disrupt insect digestion and growth,[1] and deter larger browsing and foraging animals.[2] As well, the very high levels of condensed tannins found in tropical plant species likely protect these plants from the damaging effects of strong sunlight. Patterns of expression of condensed tannins vary widely within tissues and among plant species, but could be substantially improved in some plant species to suit agricultural and industrial applications.

Condensed tannins have been shown to disrupt the lethal foam, called bloat, which results in the rumen from rapid initial digestion of several major temperate forage legumes.[3,4] Tannins also reduce N loss in ruminants by improving ruminal escape protein and by inhibiting gastrointestinal parasites.[5-8] Historically, forage breeders have been unsuccessful in expressing even small amounts of condensed tannins in the leaves of alfalfa and white clover. The introduction of moderate amounts is expected to greatly improve forage quality and productivity for the beef, dairy, and sheep industries. In addition, tannins prevent the buildup of spoilage microorganisms in silage and will likely result in insect resistance if introduced into major crops.

Conversely, reduction of tannins in some temperate and tropical plant species would improve feed quality and industrial potential. For example, health and digestion of poultry and swine, which have sensitive monogastric digestive systems, can be negatively affected by even low quantities of dietary condensed tannins. Barley lines that are free of condensed tannins in the seeds have already been developed and exhibit improved rates of weight in chickens.[9] Chemical stabilizers are not required to malt and brew with these lines, since removal of tannins has prevented the formation of beer haze.[10,11] A similar approach to breeding apple lines could reduce the tannin-protein haze of apple juice and improve the shelf-life of stored fruit juice.

Many plant species accumulate condensed tannins in their vegetative, floral, and seed tissues.[12] Legumes are a particularly rich source of these compounds. With the exception of barley and sorghum seedcoats[10,13] and one report in rice,[14] the major cereal crops do not express condensed tannins in any tissues. Several other important crops such as alfalfa and white clover and the oilseed *Brassicas* do not express condensed tannins in any tissues other than in seedcoats. Moreover, these crop species have highly complex genomes, a feature that, in the case of alfalfa, has frustrated efforts to find or develop suitable germplasm that expresses floral or vegetative tannins.[15] Hence, plant breeders have turned to molecular biology in recent years in the expectation that the isolation of genes for condensed tannin biosynthesis will lead to the successful manipulation of tannins in crop species.

In this chapter, we review the present knowledge of condensed tannin biosynthesis, as well as efforts to modify condensed tannins in forage legumes. We also present new information on genetic systems that are now in place to facilitate the isolation of novel and recalcitrant condensed tannin genes. It is through these systems that researchers expect to effect changes in the expression of condensed tannins in major crop species.

2. CONDENSED TANNIN BIOSYNTHESIS IN LEGUMES AND OTHER PLANT SPECIES

Condensed tannins are synthesized as end products of the flavonoid pathway (fig. 1). They share several biochemical steps with anthocyanin biosynthesis, after which the pathways diverge.[16] The common steps include 1) condensation of coumaroyl CoA with malonyl CoA by chalcone synthase (CHS) to form chalcones, 2) closure of the chalcone ring structure by chalcone isomerase (CHI) to form the flavanone, naringenin, 3) 3′- and 5′-hydroxylations within the B-ring (boxed area in fig. 1), 4) hydroxylation of carbon 3 within the C-ring (F3H), and 5) reduction of the C-ring ketone by dihydroflavonol reductase (DFR) to form 3,4-*cis*-leucoanthocyanidins. Several of these steps can also lead to phlobaphenes, flavonols, and flavones.

Genes involved in the common steps of flavonoid biosynthesis have been isolated and characterized in a variety of plant species in relation to anthocyanin biosynthesis, since mutants for color are easily recognized. In addition to those genes and mutations indicated for maize, barley, petunia, snapdragon and *Arabidopsis* (table 1), many of the genes have been cloned or identified by mutations in other species, for example, the soybean chalcone synthase (CHS) gene (*I*), the carnation chalcone isomerase (CHI) biochemical step (*i*), and the *Gentiana triflora* flavonoid 3′(5′) hydroxylase gene.[16–24] A number of regulatory genes, which control the content and tissue distribution of anthocyanins, flavonols and phlobaphenes, have also been identified, cloned, and well characterized.

Conversely, enzymes and genes that catalyze or regulate the steps that are unique to condensed tannin biosynthesis (after leucoanthocyanidin formation) have been largely ignored, although mutations that affect these steps have been identified in barley.[22] These include the formation of flavan-3-ol monomers, such as catechin (3′,4′ OH on the B-ring) and gallocatechin (3′,4′,5′ OH on the B-ring) by leucoanthocyanidin reductases (LARs), and the subsequent formation of condensed tannin polymers (proanthocyanidins) (fig. 1). The nature and extent of additional stereospecific enzymes required to synthesize different polymer isomers are also largely unknown at this time.

Leucocyanidin reductase (LCR), the best studied LAR, has been characterized in extracts made from developing alfalfa seeds (*Medicago sativa cv* Beaver), barley seeds (*Hordeum vulgare*), and developing leaves and flowers of sainfoin (*Onobrychis viciifolia*), sulla (*Hedysarum sulfurescens*), black locust (*Robinia pseudoacacia*), big trefoil (*Lotus uliginosis*), and *Lotus japonicus* (fig. 1).[25–28] Another LAR, leucodelphinidin reductase (LDR), has also been demonstrated using sainfoin leaf extracts.[29] Both LCR and LDR activities are under developmental regulation in legume tissues.[25,29–31] For example, the sainfoin LCR is expressed in very young leaf tissue and the alfalfa LCR is expressed in very young seeds when procyanidin subunits are being synthesized, but the activity in both these tissues declines dramatically very early in development.[25,28] LDR activity peaks somewhat later

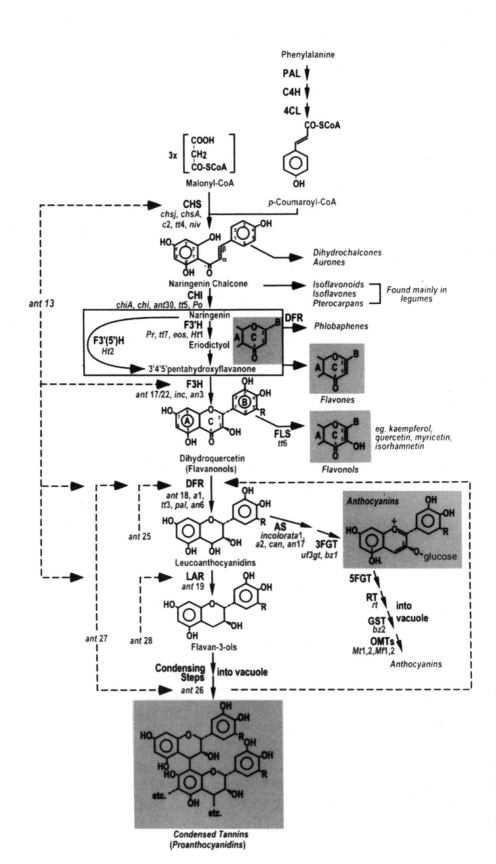

Phenylalanine

PAL

C4H

4CL

CO-SCoA

3x [COOH / CH2 / CO-SCoA]

Malonyl-CoA

CHS
chsj, chsA, c2, tt4, niv

p-Coumaroyl-CoA

Dihydrochalcones
Aurones

Naringenin Chalcone

CHI
chiA, chi, ant30, tt5, Po

Isoflavonoids
Isoflavones   *Found mainly in*
Pterocarpans   *legumes*

Naringenin

F3'H
Pr, tt7, eos, Ht1

DFR   Phlobaphenes

F3'(5')H
Ht2

Eriodictyol

3'4'5'pentahydroxyflavanone

F3H
ant 17/22, inc, an3

Flavones

FLS
tt6

*eg. kaempferol,
quercetin, myricetin,
isorhamnetin*

Flavonols

Dihydroquercetin
(Flavanonols)

DFR
ant 18, a1, tt3, pal, an6

AS
incolorata1, a2, can, an17

Anthocyanins

3FGT
uf3gt, bz1

glucose

Leucoanthocyanidins

LAR
ant 19

5FGT

RT
rt   into
vacuole

GST
bz2

OMTs
Mt1,2,Mf1,2

Anthocyanins

Flavan-3-ols

**Condensing
Steps**   into vacuole
ant 26

ant 13

ant 25

ant 27   *ant 28*

etc.   etc.

*Condensed Tannins
(Proanthocyanidins)*

during sainfoin leaf development, concomitant with the rise in the proportion of prodelphinidin subunits.[30,31] The two activities suggest that LAR comprises a class of stereospecific enzymes, each of which could contribute, along with other enzymes, to the distinct polymer composition that can change in a tissue as tannins accumulate. In sainfoin, it is unknown whether each of these activities is specific to only one of the two structurally different types of cells that accumulate condensed tannins in leaves (fig. 2).[32] The dynamic nature of condensed tannin biosynthesis and accumulation can also be observed in developing sorghum seed, in Douglas-fir bark and barley seedcoats.[13,33–36]

LAR enzymes are difficult to purify since many of them are unstable during purification and exhibit nonuniform patterns of expression during plant development (Gruber et al., unpublished).[25] They are easily solubilized, NADPH-dependent enzymes, however, the purification regimes differ for each plant species that has been examined. Tissues with high enzyme activity have been identified for several plant species, and have resulted in purified enzyme preparations (Gruber et al., unpublished; Larkin et al., personal communication).[37] Technical difficulties in LAR protein characterization are responsible for the current inability to substantiate a LAR gene by reverse genetics.

A mutation in barley, termed *ant26*, suggests that a protein-mediated "condensing" step occurs in condensed tannin biosynthesis after LAR *in planta* (table 1).[22] Polymer synthesis can also occur *in vitro* without enzyme preparations, when allowed to proceed under the moderately acidic conditions found in plant vacuoles.[38] This step is considered to occur by the sequential addition of a leucoanthocyanidin in a C4-C8 linkage onto a flavan-3-ol base to form a linear polymer (fig. 1).[16] Alternatively, the addition can occur to form a C4-C6 linkage. Our efforts to measure enzymatic activity at this step with plant extracts,[29] together with the

---

Figure 1. Flavonoid pathway illustrating steps leading to condensed tannins and other end products. Biosynthetic mutant lines described for maize, barley, petunia, snapdragon, and *Arabidopsis* (italics) are indicated at the affected biochemical steps (solid arrows). The effects of four condensed tannin regulatory gene mutant lines (italics) of barley (*ant*13,25,27,28) and the putative condensing step mutant line *ant26* on biochemical steps are indicated (dotted arrows). A,B,C, indicate ring structures as designated in dihydroquercetin. R indicates H or OH in the B-ring. The ring numbering system of dihydroquercetin applies to naringenin and to all subsequent structures, including condensed tannins. Biochemical steps include: PAL, phenylalanine ammonia lyase; C4H, 4-cinnamate hydroxylase; 4CL, 4-coumaroyl CoA ligase; CHS, chalcone synthase; CHI, chalcone isomerase; F3H, flavanone 3â-hydroxylase; F3′H, flavonoid 3′-hydroxylase; F3′(5′)H, flavanoid 3′(5′) hydroxylase; DFR, 4-dihydroflavanol reductase; LAR, leucoanthocyanidin reductases; FLS, flavonol synthase; AS, anthocyanin synthase [leucoanthocyanidin dioxygenase (LDOX) or anthocyanidin hydroxylase]; 3FGT, flavanol-UDP-3-O-glucosyl transferases; 5FGT, flavanol-UDP-5-O-glucosyl transferases; RT, flavanol-UDP-O-glycosyl-O-rhamnosyl transferases; GST, glutathione-S-transferases; OMT, O-methyl transferases.

Table 1. Biochemical genes and genetic loci in condensed tannin and anthocyanin biosynthesis

*Gene	Plant Species and their [5]Cloned Genes and Genetic Loci				
	Zea mays (maize)	*Antirrhinum majus* (snapdragon)	*Petunia hybrida* (petunia)	*Arabidopsis thaliana* (cress)	*Hordeum vulgare* (barley)
Shared Biochemical Steps					
CHS	*c2, wph* (both cloned)	*nivea* (*niv*) (cloned)	(cloned)	tt4 (cloned)	(cloned)
CHI	(cloned)	(cloned)	*po* (cloned)	tt5 (cloned)	ant30
F3H		*incolorata* (*inc*) (cloned)	an3 (cloned)	(cloned)	ant17,22 (17 cloned)
F3'H	*Pr*	*eosina* (*eos*)	Ht1,2 (cloned, but not Ht locus)	tt7	
F3'(5')H			Hf1,2 (both cloned)		
DFR	*a1* (cloned)	*pallida* (*pal*) (cloned)	an6 (cloned)	tt3 (cloned)	ant18 (cloned)
Anthocyanin-Specific Branch					
AS (LDOX)	*a2* (cloned)	*candi* (*can*) (cloned)	an17 (cloned)	(cloned)	ant1,2,5?
3FGT	*bronze 1* (*bz1*) (cloned)	(cloned)	(cloned)		(cloned)
5FGT				(cloned)	

protein-precipitating nature of condensed tannins, underscore the importance of whole, stable, condensed tannin vacuoles or tannin vacuolar vesicles to this area of biochemistry. Polymer formation is likely to be directly linked with polymer packaging, requiring tannin-binding proteins or an intact vacuolar structure and direct deposition into the vacuole. Hence, demonstration of the biochemical nature of this step awaits the development of preparations enriched in tannin cells, such as those illustrated from sainfoin (fig. 2). Tannin cell preparations will be useful sources from which to isolate whole tannin vacuoles, "condensing step" proteins, and their respective cDNAs.

Table 1 *Continued*

RT			*rt* (cloned)		
GST	*bronze* 2 (*bz*2) (cloned)		*an*13 (cloned)	(cloned)	
OMT	(lignin OMT cloned)		*Mt, Mf* (cloned)	(phenolic acids OMT cloned)	
Condensed Tannin-Specific Branch					
LAR					*ant*19
Polymer Step(s)					*ant*26
Vacuole Import					ant*25-28*?

*see fig. 1 for full biochemical name

§sequences found in EMBL and GenBank DNA databases at time of print

3. HISTOCHEMISTRY OF CONDENSED TANNIN ACCUMULATION

Studies of *Pseudotsuga menziesii, Pinus taeda, Aesculus hippocastanum*, and *Populus euramericana* have supported the widely held hypothesis that condensed tannin accumulation occurs initially in small vesicles of endoplasmic reticulum (ER) and that these vesicles subsequently associate with, and deposit tannins into, the main plant vacuole.[16,39-41] In contrast, condensed tannins appear to be deposited directly into the central vacuole in sainfoin leaves, since associations between ER vesicles and condensed tannins, or between ER and the central vacuole, could not be found even after examination of serial sections of tannin cells.[42] In addition, the initial accumulation of tannin in very young leaf cells of this species can be observed by osmium staining as deposits adjacent to the internal surface of the vacuolar membrane and as thin strands deep within the vacuole matrix.[42,43] Membrane vesicles have been found directly associated with these strands.[43] These findings have led us to the hypothesis that thin invaginations of vacuolar membrane can form deep within the vacuole in a leaf pre-expansion phase of vacuole development. If so, then such invaginations could serve to increase the surface area for tannin-specific enzymes and allow a faster flow of tannin precursors into the vacuole.

Accumulation of condensed tannins in tissues follows distinct patterns that vary among plant species. This is well illustrated in some legume species, in which condensed tannins completely fill the vacuoles of specialized cells, whereas in closely related species, a different type of cell within the same tissue might synthesize these compounds. For example, condensed tannins of *L. uliginosis* flowers are located in a three-dimensional lattice network of cells that completely spans the petal, whereas condensed tannins of *L. japonicus* petals are located in surface papillae of the petal epidermis (fig. 2).[25] Furthermore, differential staining of

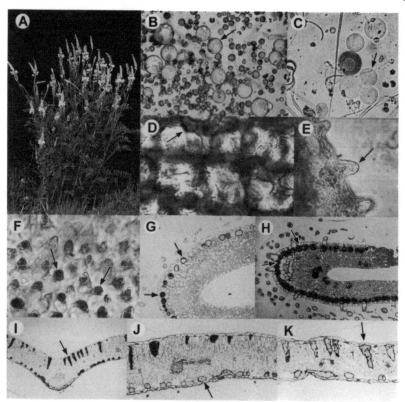

Figure 2. Accumulation of condensed tannins in forage legumes. A. Sainfoin plant. B. Unstained sainfoin leaf protoplasts, illustrating clear, colorless tannin cells (arrow) mixed with small, green mesophyll cells. C. Unstained preparation enriched for sainfoin tannin protoplasts (arrow). D. Internal structure of *L. uliginosis* petals stained with butanol:HCl to detect condensed tannins (arrow) in a 3-dimensional network of cells. E. Edge of fresh *L. japonicus* petal illustrating papilla containing clear, colorless tannin vacuoles (arrow). F. *L. japonicus* petal stained with butanol:HCl to illustrate tannins in papilla vacuoles (arrows). G-K. Cross section of osmium-stained sainfoin leaves illustrating the developmental rhythm of condensed tannin accumulation (arrows). G.H. Tannins are initially accumulated in network cells on the abaxial surface of immature, folded leaves. I. Tannins subsequently accumulate in large "sacs" on the adaxial surface. J. Tannins in network cells are depleted first during leaf maturity. K. Tannins in adaxial "sacs" are depleted during senescence (arrow in K). A color representation of this figure can be found following page 326.

several legume species indicates that tannins and anthocyanins are not usually found within the same cells (Gruber, unpublished).[25] Together, these results indicate a highly regulated, spatial deposition of tannin-specific cells.

Further evidence that condensed tannins accumulate in a precise, but dynamic and developmentally regulated pattern, is revealed by histochemical and osmiophilic staining of cells in the leaves of sainfoin, *H. sulfurescens*, and *L. uliginosis*. For example, in sainfoin, a flat network of tannin-accumulating cells appears in the subepidermal layer of the abaxial surface of very young, folded, immature leaves (fig. 2).[32,42] These cells continue to fill with condensed tannins until close to the leaf unfolding stage. At this stage, the abaxial cells begin to empty, and tannins start to accumulate in very large, distinct subepidermal cells scattered throughout the mesophyll cells on the adaxial leaf surface. By the time the leaves begin to senesce, the condensed tannin content has started to decline in both cell types, and the cells eventually empty completely. This dynamic profile strongly suggests that condensed tannins are recycled by the sainfoin plant when no longer required. Development-regulated accumulation of condensed tannins has also been observed in most tissues of *L. corniculatus*, *L. japonicus*, *Robinia pseudacacia*, *R.. tortuoso*, and sainfoin, regardless of whether they are expressed in specialized cells.[25] As well, condensed tannins are under developmental regulation in the testa layers of barley and alfalfa seedcoats,[25,35,36] and the content in plants can be increased by stress conditions, such as heat and poor nutrition.[44,45]

Legume leaf and flower vacuoles that contain condensed tannins are normally clear and colorless such as those in *L. japonicus* flowers (fig. 2). In seedcoats, however, phenolic oxidation or dehydration-associated chemical changes usually occur, leading to a brown color and reduced tannin extractibility in some species. It is not clear whether such oxidized phenolics are further metabolized by the plant. In the alfalfa seedcoat, the testa layer still contains clear, colorless, extractable condensed tannins, even though other layers are brown. However, in *Brassica* species, the brown seedcoat has very little extractable condensed tannin, although it can be detected histochemically (Gruber and Marles, unpublished results).

4. GENE ISOLATION FROM CONDENSED TANNIN *ANT* MUTANTS OF BARLEY

The *ant* lines of barley comprise a large collection of mutations that were chemically induced in the anthocyanin and condensed tannin pathways (table 1, fig. 1).[22,36,46] Barley normally accumulates anthocyanins in the cross cell layer of the seedcoat as well as in several other vegetative and reproductive tissues, but condensed tannins accumulate only in the testa layer of the seedcoat. Some of the *ant* mutant lines lack condensed tannins in the testa layer,[35,36] and, therefore, fine differences in gene expression between the *ant* mutants and their parental lines can be exploited to isolate unique or difficult-to-isolate genes for condensed tannin biosynthesis using recently developed differential mRNA screening techniques. One line from which it would be exciting to isolate new genes is *ant*26–485, originally proposed to have a mutation in the seedcoat "a condensing step" and recently shown to have reduced LAR activity (fig. 1).[26,37] Other lines include *ant*25–264, which has reduced DFR and LAR activity, *ant*27–489, which has altered DFR and LAR activity and polymer formation, and *ant*28–484, which has reduced LAR activity.[26,37] Since these mutations affect the expression of more than one enzyme, they may, in fact, represent uncharacterized vacuolar transport proteins or regulatory proteins. However, it is not known whether these mutations arose from a change in the transcript level or from a post-transcriptional event. The latter factor would compromise the outcome of these mutants in differential mRNA screening.

*Ant*13–152 appeared to be a good candidate mutant to analyze for differences in expressed tannin genes, since it regulates the transcription of several genes. Transcripts of chalcone synthase (CHS), flavanone 3â-hydroxylase (F3H), and dihydroflavanol reductase (DFR) were not detected in the testa layer of *ant*13, nor were LCR activity nor condensed tannins.[22,47] As a result, cDNA representational different analysis (RDA) was applied to mRNA isolated from dissected seedcoat testa/pericarp layers of *ant*13 and parental wildtype barley lines. RDA is a subtractive hybridization technique that uses PCR to selectively amplify genes expressed in one line and not in a related line.[48] However, mRNAs present at low levels in the mutant line rather than altogether absent will not be selectively amplified by RDA. The RDA technique had the potential with *ant*13 to amplify wildtype sequences for the unknown "condensing step" gene, LAR genes, and regulatory genes, in addition to common flavonoid genes.

Using the RDA technique, two unique RDA cDNAs were amplified and cloned, then subsequently mapped to two barley genomic clones. The two RDAs accumulate at low levels in testa/pericarp tissue of wildtype barley lines, but are not detected in wildtype barley leaf tissue, nor in testa/pericarp tissue of *ant*13 or *ant*27 lines using sensitive reverse transcriptase PCR assays (fig. 3).

In addition, the RDAs are not homologous to any other gene or protein sequences. These data open the possibility that either unknown tissue-specific regulatory genes that activate tannin biosynthesis in the wildtype line or difficult-to-isolate tannin biochemical genes have been cloned. The low number of amplified sequences has led to the speculation that the *ant*13 mutation is leaky, and may, in fact, express low levels of mRNA for other flavonoid genes (CHS, etc.) not detectable using Northern blot analysis.

We anticipate using these RDAs as transgenes to genetically complement *ant*13 and other mutant lines, in order to determine the biological function of these genes. If their activity controls tannin biosynthesis, they may be used to manipulate tannins in important agricultural and industrial crops.

5. *LOTUS JAPONICUS* TRANSPOSON MUTANTS

Barley is the only plant species in which a full complement of mutants for condensed tannin biosynthesis has been developed. However, widespread, multiple mutations, which are induced in lines derived from chemically-mutagenized seed, have the potential to confound or limit the ability of researchers to isolate genes from the *ant* mutants by some methods.[49,50] Transposon and *T-DNA* insertions are now being used in several plant species to create new types of lines with a limited number of gene-tagged mutations.[51,52] Such mutants expand the repertoire or genetic tools available with which to clone difficult-to-isolate genes and greatly facilitate the dissection of biochemical and developmental pathways. Recently, the *Ac/Ds* transposon system from maize was introduced on a *T-DNA* plasmid into the

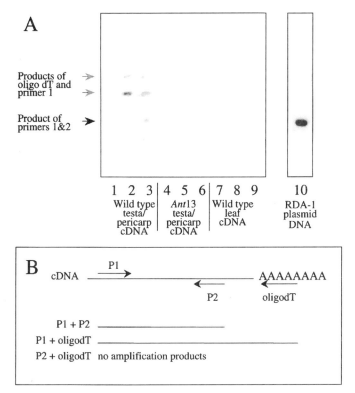

Figure 3. Expression of *ant*13 RDA-1 in wildtype barley testa. Panel A. Southern hybridization of RDA-1 to agarose gel-separated products amplified in a reverse transcript (RT) PCR assay from cDNA of *ant*13 and wildtype tissues. Lanes 1–3, wildtype testa/pericarp cDNA; Lanes 4–6, *ant*13 testa/pericarp cDNA; Lanes 7–9, wildtype leaf cDNA; Lane 10, control reaction using primers 1 and 2 with cloned RDA-1 plasmid DNA. Lanes 1,4,7, RDA-1 primer 2 control reactions; Lanes 2,5,8, RDA-1 primer 1 control reactions; Lanes 3,6,9, primers 1&2 complete reactions. NB: all reactions contained oligodT primers from cDNA synthesis. Panel B. Schematic illustration of primer binding and products which accumulate in the RT-PCR reaction in Panel A. Different sized amplicons (indicated by lines) are produced in reactions using different primer combinations. Expression of RDA-2 had a similar pattern to RDA-1 (data not shown).

genome of *Lotus japonicus*,[53] a fast-growing, diploid legume that synthesizes large amounts of condensed tannin in the flowers, but none in leaves.[25]

Two types of transposon tagging systems have been developed in transgenic *L. japonicus*. In the one-element system, the *T-DNA* carries the autonomous maize transposable element, *activator* (*Ac*) subcloned between a spectinomycin resistance gene and its promoter (fig. 4, insert A). *Ac* comprises a gene for a DNA transposition enzyme that is located between two DNA inverted repeat sequences. The enzyme recognizes the inverted repeats and catalyzes element self-excision, usually at meiosis. In this way, the complete element can move to a new location

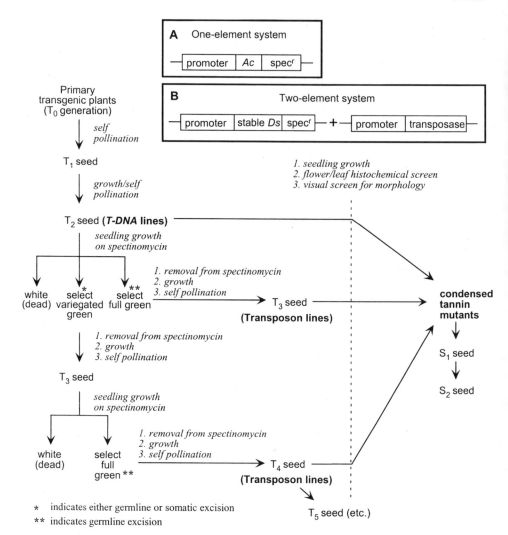

Figure 4. Schematic of the procedure to select gene tagged tannin mutants. Insert A. Portion of the inserted *T-DNA* in a plant line with autonomous (self) excision of *Ac*. Insert B. Portion of the inserted *T-DNA* in each of the two lines required for two-element excision of *Ds*. T_0, primary transformed plant. T_{1-5}, generations 1–5 of selfed progeny derived from primary transgenic plants. $S_{1,2}$, generations 1 and 2 of selfed transgenic plants after condensed tannin selection.

in the genome, generating a new set of mutant phenotypes with each generation. The second system is based on two stable *Ac* derivative lines (fig. 4, insert B), one called *dissociator* (*Ds*) containing both inverted repeats but a disabled transposase gene, and the other called *activator*, containing one or neither of the inverted repeats, but able to direct the synthesis of the transposase. Excision and reinsertion of *Ds* can only occur in crosspollinated progeny of the two derivative lines. In

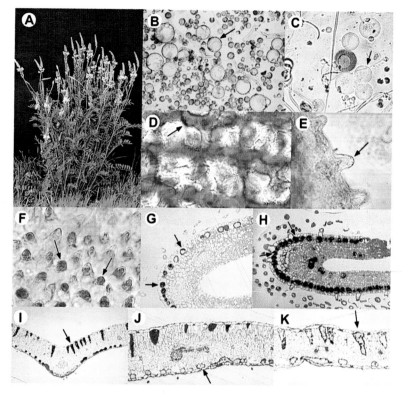

Figure 2.

Figure 5.

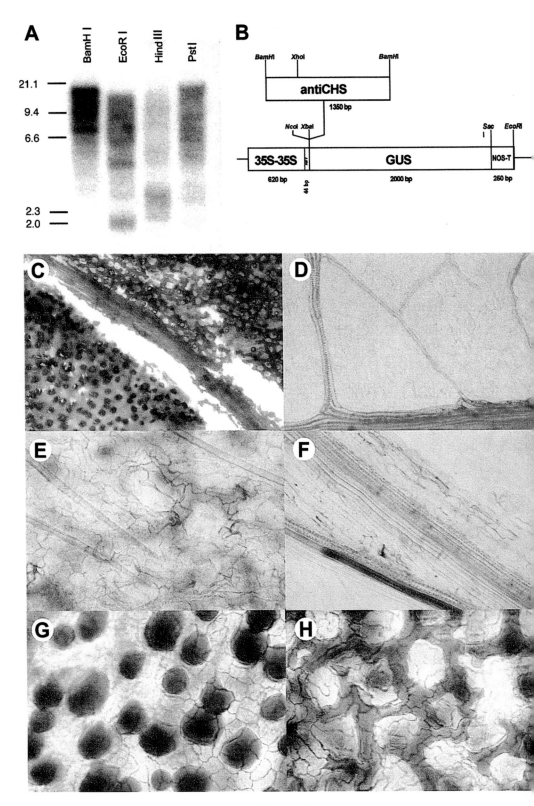

Figure 7.

both systems, when the transposon leaves its site, expression of spectinomycin resistance can begin. As a result, excision lines in which transposition of *Ac* has occurred in germinal cells can be selected from *T-DNA* insertion lines by full green seedling growth in the presence of spectinomycin (fig. 4).

Both *T-DNA* and transposon mutants are equally valuable resources for isolating tagged genes responsible for selectable phenotypes. The major advantage to transposon mutagenesis is that new mutants can be generated simply by undertaking a seed increase, in contrast to the continuous rounds of plant transformation required for *T-DNA* insertion mutagenesis. In addition, the physical separation of transposase and inverted repeats in the two-element system gives greater control over the generation and selection of stable mutations than in the one-element system. In *L. japonicus,* the potential for new mutants in progeny seed is 10–15 percent, based on the rate of recovery of full-green spectinomycin resistant plants in the one-element system.

L. japonicus transposon plant lines were screened under greenhouse and field conditions in Saskatoon, SK, Canada (52° 07′ N latitude, 106° 38′ W longitude) for the presence of leaf tannins and flower petal tannins using butanol: HCl hydrolysis (fig. 5).[54] The field trial was conducted on Sutherland clay-loam brown soil under mist irrigation.

Condensed tannin content in field-grown flowers was generally much higher and more variable than in greenhouse grown flowers. Plant growth varied between the field plots and was more difficult to control compared with the greenhouse, due to small differences in drainage capacity between the plots. Hence, selection of flower mutations with subtle genetic differences in tannin expression was impossible under field conditions. However, suitable methodology was developed to facilitate recovery of precious seed in the field and to accomodate the large-scale seed increases necessary to generate new populations of mutants.

No flower mutants have been recovered from our screening efforts with *L. japonicus* to date. However, we have recovered 21 "gain-of-function" mutants containing leaf tannins.[37] Several of these mutants (*tan1-tan6*) are being classified genetically and biochemically. With the exception of *tan1*, all *tan* leaf mutants had large flowers, large smooth leaves, and long internodes, resembling the morphology of untransformed *L. japonicus.* Conversely, *tan1* plants resembled the morphology of *Lotus angustissimus*, a related species with small trichome-covered leaves and stems, short internodes, and small flowers. The leaf monomeric and polymeric flavonoid profiles of *tan1* and its S_2 progeny also appeared more like *L. angustissimus* than like untransformed *L. japonicus* (fig. 6).

Tan1 appears to be a pleiotropic mutant, affecting more than one character. The correlation of tannins and hairs resembles the linkage between anthocyanins and trichomes observed in *Arabidopsis,* in which a mutant line, *ttg* (*transparent testa, glauca*), missing these characters could be complemented with the maize *Lc* and *C1* anthocyanin regulatory genes.[55]

Leucocyanidin reductase was detected in leaf extracts of the mutant *L. japonicus* genotypes, and the activity correlated well with the level of leaf tannin (table 2). The relatively low expression of LCR in *tan2* compared with *tan1* may reflect different sites of integration of the transgene in these mutants. In general,

Figure 5. Greenhouse and field trial conditions and methodology used to increase seed and to select tannin mutations in *L. japonicus* gene-tagged lines. A. Seedlings growing on greenhouse bench covered with porous cloth to prevent pollen escape. Greenhouse seedlings were screened for leaf mutants at 6–7 weeks after germination. B. Transplanting mutant seedlings into small field plots through fibre-reinforced plastic sheeting. The sheeting minimized seed loss from pod shattering, reduced plot management by preventing weed growth, increased the length of the growing season by maintaining soil temperature, and enabled plot development and maintenance in a clay-based soil under wet early spring conditions or heavy rain. Valuable seeds from shattered pods could be recovered from the sheeting using a small hand-held vacuum equipped with disposable filters. C. Mist sprinkler system to maintain *L. japonicus* mutants in hot, dry conditions in Saskatoon and to reduce pod shattering in a continuously flowering species. D. Small plot illustrating 6″-spaced flowering plants, grown in five 6″-spaced rows per line. Disposable row crop covers were held in place by scrap metal pipes and were used to prevent the spreading of transgenic pollen by insects to other lines and to other nearby *Lotus* species. Crop covers also prevented the spread of insect pests that infested two of the lines to other plots. A color representation of this figure can be found following page 326.

tannin accumulation was stable in two subsequent generations of *tan*1 after self pollination (S_1 and S_2), although 7 percent of S_1 progeny of *tan*1 do not accumulate leaf tannin (table 2). This reversion is consistent with the rate of *Ac*-mediated transposition that occurs in active transposase lines of *L. japonicus*.[53] Curiously, tannin did not accumulate in most of the S_1 progeny of *tan*2, but could be found

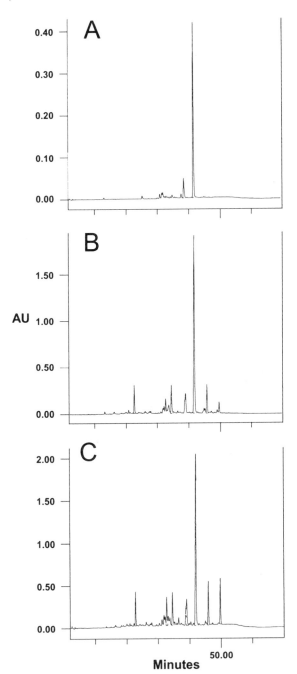

Figure 6. Leaf flavonoids in a *Lotus japonicus* leaf mutant line and in *L. angustissimus*. HPLC trace of ethyl acetate extractable fraction separated by a Waters C18 column. A. *L. japonicus* wild type leaves. B. *L. angustissimus* wild type leaves. C. *L. japonicus tan*1 mutant leaves.

Table 2. Leaf tannin biochemistry and genetics in *Lotus japonicus* wild type and mutant lines

S_0 Genotype (*T-DNA* or *Tn* generation)	Tannin Content in Selected Leaf Mutants (ug·mg⁻¹ FW)			Leaf LCR[a] Activity	% Mutant Progeny Recovered with Leaf Tannin	
	Flowers	Leaves	Stems		S_1[b]	S_2[b]
Untransformed	188	ND[c]	6	ND[c]	-	-
*tan*1 (T$_{3,\,etc.}$)	137	118	5	16.8	93.2	100
*tan*2 (T$_2$)	222	59	9	9.3	8.5	11.3

[a]Leucocyanidin reductase, ³H-(+)catechin (DPM·10⁻³·mg FW⁻¹·h⁻¹) detected by radioHPLC after incubation of [C$_4$-³H]-2R,3S-*trans*-3S,4S-*cis*-leucocyanidin (2·10⁶ DPM) with leaf extracts.

[b]leaf tannin segregating generations

[c]ND, not detected

in a small, but significantly greater proportion of the S_2 progeny (table 2). The repression of tannin accumulation in *tan*2 S_1 progeny was not due to transposition of the stable *Ds* element through the activity of a native *L. japonicus* transposase (Skadhauge et al., unpublished). Although as yet untested, the repression of the mutant phenotype may involve any of the mechanisms underlying transgene silencing, for example, DNA methylation.[56] Partial release from repression of tannin accumulation in S_2 progeny is less easily explained but has been observed in other transgenic plant species.[57,58]

The mutants we selected appear to be mutated in regulatory genes that induce the expression of leaf tannins. At present, we are cross pollinating these mutants to classify them into allelic groups. We intend to complete the analysis of their leaf phenolic biochemistry and to use the mutants to isolate legume-specific leaf regulatory genes. At a later date, more *Tn* lines could be screened in continuing efforts to isolate LAR and condensing step genes.

6. ANTISENSE GENES CONTROLLING CONDENSED TANNIN EXPRESSION

The development of strategies to express flavonoid genes in forage plant species as a means to alter condensed tannin content and composition, has been driven by the potential to develop improved crop species. These strategies have been facilitated largely by the similarities that exist among plant species between the genes that code for the common steps of anthocyanin and condensed tannin biosynthesis. Hence, the potential of heterologous antisense technology has been explored for reducing tannin gene expression in outcross pollinating tetraploid species from which it is particularly difficult to isolate mutants. Lines with a range of tannin content in these types of forage plant species would be useful to determine optimum tannin concentration *in planta* for improved rumen digestion and plant resistance to insects. Antisense may reduce expression of several members of a multigene family simultaneously, an advantage when manipulating flavonoid genes. Several tannin transgenic systems expressing antisense versions of heterologous anthocyanin genes have been developed. The *Lotus corniculatus* transgenic systems, expressing either *Antirrhinum majus* CHS or DFR genes, are described in detail in another chapter of this book.[59-63]

Since the sainfoin CHS gene is sufficiently similar to the petunia CHS gene, a petunia antisense CHS gene was expressed from an enhanced 35S promoter to develop transgenic sainfoin plants (fig. 7).[64,65] The population expressed a range of tannin content in mature sainfoin leaves and included plants in which condensed tannins were totally eliminated in both types of tannin cells. Unfortunately, the transgene was unstable in sainfoin. Approximately 15 months later, the mature plants began to accumulate condensed tannins (fig. 7), and we could no longer detect the CHS insert. Our findings supported gene elimination as the main mechanism for reversion to the untransformed phenotype. Phenotypic reversion has also been observed with petunia plants transformed with the antisense CHS gene.[66] However, the mechanism for the petunia reversion does not involve gene elimination and is distinct from cosuppression using sense CHS genes,[67] since equivalent amounts of antisense CHS mRNA were detectable in both white and colored petunia petal sectors.

Naturally occuring antisense CHS genes were recently shown to affect seedcoat color in nontransformed soybean. The soybean *I* gene is a 10 kb region encoding three chalcone synthase genes, one of which is situated in antisense orientation between the others.[68] Molecular analysis of this DNA segment and analysis of lines of recessive soybean mutations with restored dark seedcoat color suggest that the inverted CHS gene acts as a repressor, silencing the adjacent genes in the uncolored seedcoat.

7. EFFECT OF INTRODUCED ANTHOCYANIN REGULATORY GENES ON ANTHOCYANIN AND CONDENSED TANNIN BIOSYNTHESIS

Genes that regulate condensed tannin biosynthesis are largely uncharacterized at this time. Hence, the strategy has emerged to introduce well-characterized anthocyanin regulatory genes from maize into forage legume species in transgenic strategies to manipulate condensed tannin content and improve forage quality. Little is known about the capacity of anthocyanin regulatory genes to control the condensed tannin-specific branch of the biosynthetic pathway. While it seems real-

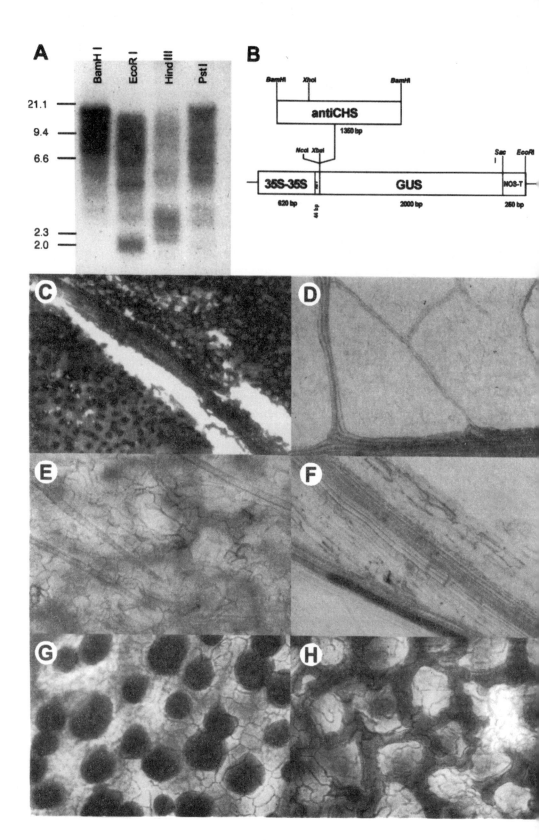

istic to expect increased flux through the flavonoid pathway with these maize transgenes, in actuality, the final product will depend, in part, on how much of the tannin biochemical pathway is present in a recipient plant species or plant tissue.

The anthocyanin regulatory genes that have been introduced into plants, including forage legumes, fall into two categories of gene families known as *myb* and *myc* genes.[69-77] These classifications are based on their structural similarities to transcriptional regulatory genes and proto-oncogenes found in other kingdoms. *Myb* genes encode DNA-binding proteins that affect transcription. Well-characterized anthocyanin-related myb genes include the *colorless* gene, *C1*, that affects transcription of most of the major maize seed anthocyanin biochemical genes (CHS, DFR, AS, 3FGT, GST),[72,73] the light-independent *purple leaf* gene, *Pl*, which controls an equivalent set of genes in the maize plant body,[74] and the *an2* gene which is a *C1* homologue controlling anthocyanin accumulation and vacuolar pH in the petunia flower.[78] *Myb* genes are generally (but not always) quite different in their acidic carboxy-terminal regions, but related *myb* genes show strong homology and similar DNA sequence binding specificity in the amino-terminal regions.[76,77,79,80] The acidic region appears important to the function of *myb* genes, since a mutation in this region in the maize mutant, *C1-I*, changed the *C1* gene from a transcriptional activator into a dominant repressor.[81]

Myc proteins contain the basic helix-loop-helix (*bHLH*) motif often identified in protein-protein interactions. They affect anthocyanin expression by binding to *myb* proteins. However, the requirements for *myc* proteins to activate specific *myb* proteins is not all that well understood.[21,82] The maize seed genes, *B, R*, and *Sn (bol3)*, and the maize leaf gene, *Lc (leafy)*, are the best characterized of the isolated anthocyanin *myc* genes and serve as functional equivalents in different seed tissues and leaves, affecting the same profile of biosynthetic genes as *myb* genes.[83-88] Alleles of some of these genes affect subsets of anthocyanin genes. For example, *R-S* affects CHS, DFR, GST and anthocyanin expression in aleurone tissue, and R-P affects DFR, GST, and anthocyanin expression in coleoptiles and anthers.

Figure 7. Condensed tannins in transformed and nontransformed sainfoin. A. Southern blot of restriction enzyme digested sainfoin leaf DNA hybridized with the 1.35 kb petunia CHS gene. B. Schematic of vector used to develop transgenic sainfoin with a petunia antisense CHS gene. C—H. stained with butanol:HCl to detect condensed tannins as dark-red deposits. C. Leaf structure of mature untransformed plant; *left*, adaxial leaf surface with stained tannin cells (arrow); *right*, abaxial leaf surface with network of stained tannin cells (arrow). D. Closeup of stained subepidermal tannin cells ("sacs") in mature leaf from 13-month old transgenic plant containing no tannins. D. Closeup of stained subepidermal network of tannin cells (arrow) in mature leaf of 13-month old transgenic plant containing no tannins. F. stem from transformed genotype containing no tannins except in a small number of cells. G. As in D but accumulating condensed tannins (arrow) 2 months later. H. As in E but accumulating condensed tannins (arrow) 2 months later. A color representation of this figure can be found following page 326.

Each plant species is likely to have many *myb* and *myc* regulatory genes, each varying somewhat in structure. Together with other novel regulatory sequences, they lend exquisite sensitivity to the regulation of flavonoid gene expression and uniqueness to the phenotype of different plant species. For example, the presence of an active *myb* gene, *P*, regulates CHS and DFR genes to divert flavonoid intermediates into C-glycosylflavones, 3-deoxyanthocyanins, flavan-4-ols and the reddish-brown phlobaphenes in maize pericarp, cob and floral organs, as distinct from *C1* which directs CHS and DFR into anthocyanin biosynthesis.[89,90] Normally, expression of one member of both the *myb* and *myc* family is required to activate or induce the expression of anthocyanin biosynthetic genes.

The behavior of regulatory genes and their ability to stimulate tannin gene expression on their own may be different in a heterologous plant species or *in vitro* than in the native species, depending on their interactions with the structural features and range of genes with which they associate in the recipient species. This has been the case for the effects of *myb* and *myc* transgenes on anthocyanin and lignin biosynthetic genes. For example, the petunia *myb* gene, *an2*, paired with the petunia *myc*-like *jaf13* gene, normally stimulates DFR expression in petunia but not CHS or F3H. However, the same pair of genes activated the CHS gene in a maize transient expression system, while the *an2* gene alone fully recovered anthocyanin accumulation in the maize *pl* mutant.[21,76,77] *Jaf13* and the snapdragon *myc*-like gene, *delila*, also activated heterologous anthocyanin genes on their own *in trans*.[77,91–93] Variation in the ability to induce anthocyanin expression occurred within a series of studies, in which the maize *C1* gene and combinations of *myc* genes were constitutively expressed in maize, *Arabidopsis*, chysanthemum, tomato, petunia, and oats.[55,72,73,94,95] Finally, *myb305* and *myb340* from snapdragon activated PAL, CI and F3H promoters in a yeast expression system, but bound the CHS promoter *in vitro*. When introduced into tobacco, the snapdragon genes induced phenolic acids and lignin and had no effect on CHS, DFR or AS.[77,80,96,97]

Our lack of detailed understanding of flavonoid gene regulation complicates our ability to select suitable genes which will induce optimal levels of condensed tannins in leaves of forage legume species. For example, the introduction of the maize *Sn* gene boosted normally low root tannin levels to high levels in *Lotus corniculatus*, while either completely suppressing tannin genes (DFR, LCR), tannins and *Sn* expression in leaf mesophyll and epidermal tissue or having no effect on leaf tannin at all.[98,99] Post-transcriptional degradation of leaf *Sn* RNA was observed in the suppressed plants,[99] a factor which has been observed elsewhere to result from high gene copy number.[100,101] In contrast, the maize *B-Peru* and *C1* genes inserted into white clover and peas induced anthocyanins in leaves of both species, although expression was very low in white clover.[102,103] Transgenic strategies to elevate tannins through insertion of regulatory genes should be more predictable in the future when factors that regulate tannin genes are better characterized and when species-specific and tissue-specific promoters are more common. The development of tannins in plant species that appear incapable of inducing condensed tannins in leaves, such as white clover, also awaits the isolation of the elusive LCR gene to divert the flow of flavonoids toward condensed tannins.

8. STRATEGIES TO MANIPULATE CONDENSED TANNIN EXPRESSION IN ALFALFA

Alfalfa accumulates condensed tannins in the seedcoat, but not in leaves nor in other tissues.[25,30] Historically, the development of alfalfa germplasm with leaf tannins has suffered from difficulties to select rare mutants accumulating leaf anthocyanins or condensed tannins in a tetraploid, outcross-pollinating species.[15] As well, it has been difficult to overcome the genome instabilities which occur from attempting to introduce tannin genes through intergeneric hybridization and somaclonal variation (Lees, unpublished results; Larkin, personal communication).[104] Hence, we have introduced several different genes into alfalfa in an effort to stimulate anthocyanin and tannin biosynthesis. These genes include sense and antisense petunia and alfalfa CHS cDNAs,[64,105] a genomic alfalfa F3H (cv. Beaver), and the maize *B-Peru*, *Lc* and *C1* regulatory genes.[55,72,83,88] Several sets of these transgenic alfalfa plants are now being grown under field conditions to evaluate their flavonoid profile. Winter hardiness and disease resistance are also being evaluated, since flavonoids are induced during cold temperatures and have been shown to be involved in plant-microbe interactions.[106]

We suspect that an antisense CHS may improve the usually poor expression of CHS that exists in alfalfa leaves.[105,107] Our hypothesis is that alfalfa may normally transcribe such a high level of CHS mRNA, as a result of its large multigene family, that CHS expression is suppressed. If this is true, then lowering the level of CHS mRNA with an antisense gene may actually improve the accumulation of chalcone synthase protein and increase flux through the alfalfa leaf flavonoid pathway. This may have been the case for *L. corniculatus*, in which an antisense CHS transgene stimulated higher levels of native CHS transcripts and resulted in higher condensed tannin levels in transgenic root cultures.[63]

None of the transgenic alfalfa plants are expected to have leaf tannins, particularly since related species and rare alfalfa plants have small red sectors in the leaves and stems rather than tannins. At the very least, we expect a profile of leaf flavonoids closer in structure to condensed tannins than the profile normally found in alfalfa leaves, in addition to altered seed tannin content, altered flower color, and accumulation of leaf anthocyanins if we are inserting the right combination of regulatory genes. The progeny derived from crosses of CHS- and F3H-transgenic alfalfa should also produce leucoanthocyanins, provided that a portion of the DFR mRNA normally expressed strongly in alfalfa leaves can be co-opted from its current unknown function.[107] These progeny lines would then be good test recipients to determine whether expression of an LCR gene will shift alfalfa leaf flavonoid synthesis toward tannins. Unfortunately, anthocyanins did not develop as predicted in another transgenic alfalfa system using *B-Peru* and *C1* maize regulatory genes (Larkin et al., personal communication). At the present time, it is unclear whether our transgenic plants with regulatory genes will also have similar expression problems.

9. CONCLUSIONS

Condensed tannins can be manipulated by genetically transferring anthocyanin biosynthetic or regulatory genes into crop plants, but the success of this approach is species-specific. Consequently, it is important to characterize the remaining, poorly defined steps of tannin biosynthesis, as well as the factors that regulate this pathway. The isolation of genes for these steps has been difficult using biochemical approaches, and several new genetic systems have recently been developed to facilitate this task. These include the application of representational difference analysis to a barley *ant* mutant and the isolation of new *L. japonicus* transposon-tagged lines with ectopic expression of leaf tannins. Another approach that could advance this area of research is the isolation of regulatory genes from tannin-accumulating plant species based on homologies to known anthocyanin regulatory genes.

ACKNOWLEDGMENTS

Ms. Maureen Woods and Mr. Ralph Underwood are acknowledged for the development of the sainfoin transgenic plants and for assistance with graphics, respectively. This is AAFC Publication No. 1296.

REFERENCES

1. Manuwoto, S.; Scriber, J.M. Effects of hydrolyzable and condensed tannin on growth and development of two species of polyphagous lepidoptera: *Spodoptera eridania* and *Callosamia promethea*. *Oecologia (Berlin)* 69:225 (1986).

2. Furstenburg, D.; van Hoven, W. Condensed tannin as anti-defoliate agent against browsing by giraffe (*Giraffa camelopardalis*) in the Kruger National Park. Comp. Biochem. Physiol. 107A:425 (1994).

3. Howarth, R.E.; Chaplin, R.K.; Cheng, K.-J.; Goplen, B.P.; Hall, J.W.; Hironaka, R.; Majak, W.; Radostits, O.M. Bloat in cattle. Agriculture Canada Publication 1858/E. Communications Branch. Agriculture and Agri-Food Canada, Ottawa (1991).

4. Tanner, G.J.; Moate, P.; Dailey, L.; Laby, R.; Larkin, P.J. Proanthocyanidins (condensed tannins) destabilise plant protein foams in a dose dependent manner. *Aust. J. Agric. Res.* 46:101 (1995).

5. Jones, G.A.; McAllister, T.A.; Muir, A.D.; Cheng, K.-D. Effects of sainfoin (*Onobrychis viciifolia* Scop.) condensed tannins on growth and proteolysis by four strains of ruminal bacteria. *Appl. Environ. Microbiol.* 60:1374 (1994).

6. Tanner, G.J.; Moore, A.E.; Larkin, P.J. Proanthocyanidins inhibit hydrolysis of leaf proteins by rumen microflora *in vitro*. *Brit. J. Nutr.* 71:47 (1994).

7. Min, B.R.; Barry, T.N.; McNabb, W.C.; Kamp, P.D. Effect of condensed tannins on the production of wool and on its processing characteristics in sheep grazing *Lotus corniculatus*. *Aust. J. Agric. Res.* 49:597 (1998).

8. Niezen, K.E.; Waghorn, T.S.; Charleston, W.A.G.; Waghorn, G.C. Growth and gastrointestinal nematode parasitism in lambs grazing either lucerne (*Medicago sativa*) or sulla (*Hedysarum coronarium*) which contains condensed tannins. *J. Agric. Sci. (Cambridge)* 125:81 (1995).

9. Newman, R.K.; Newman, C.W.; El-Negoumy, A.M.; Aastrup, S. Nutritive quality of proanthocyanidin-free barley. *Nutrition Reports Int'l.* 30:809 (1984).

10. Erdal, K. Proanthocyanidin-free barley. *J. Inst. Brewing* 92:220 (1986).

11. Von Wettstein, D.; Jende-Strid, B.; Ahrenst-Larsen, B.; Sørensen, J.A. Biochemical mutant in bartley renders chemical stabilization of beer superfluous. *Carlsberg Res. Commun.* 42:341 (1979).

12. Porter, L.J. Flavans and proanthocyanidins. *In*: Harborne, J.B. (ed.) The flavonoids. advances in research since 1980. Chapman and Hall, New York. pp. 21 (1988).

13. Butler, L.G. Relative degree of polymerization of sorghum tannin during seed development and maturation. *J. Agric. Food Chem.* 30:090 (1982).

14. Reddy, V.S.; Dash, S.; Reddy, A.R. Anthocyanin pathway in rice (*Orza sativa* L.): identification of a mutant showing dominant inhibition of anthocyanins in leaf and accumulation of proanthocyanidins in pericarp. *Theor. Appl. Genet.* 91:301 (1995).

15. Goplen, B.P.; Howarth, R.E.; Sarkar, S.K.; Lesins. K. A search for condensed tannins in annual and perennial species of *Medicago, Trigonella,* and *Onobrychis. Crop Sci.* 20:801(1980).

16. Stafford, H. Flavonoid metabolism. CRC Press, Boca Raton, FL (1990).

17. Dooner, H.K.; Robbins, T.P. Genetic and developmental control of anthocyanin biosynthesis. *Ann. Rev. Genet.* 25:173 (1991).

18. Martin, C.; Prescott, A.; Mackay, S.; Bartlett, J.; Vrijlandt, E. Control of anthocyanin biosynthesis in flowers of *Antirrhinum majus. Plant J.* 1:37 (1991).

19. Holton, T.A.; Cornish, E.C. Genetics and biochemistry of anthocyanin biosynthesis. *Plant Cell* 7:1071 (1995).

20. Shirley, B.W.; Kubaske, W.L.; Storz, G.; Bruggemann, E.; Koornneef, M.; Ausubel, F.M.; Goodman, H.M. Analysis of *Arabidopsis* mutants deficient in flavonoid biosynthesis. *Plant J.* 8:65 (1996).

21. Mol. J.; Grotewold, E.; Does, R. How genes paint flowers and seeds. *Trends in Plant Sci.* 3:212 (1998).

22. Jende-Strid, B. Genetic control of flavonoid biosynthesis in barley. *Hereditas* 119:187 (1993).

23. Koorneef, M. Mutations affecting the testa colour in *Arabidopsis. Arabid. Inf. Service* 27:1 (1990).

24. Reuber, S.; Jende-Strid, B.; Wray, V.; Weissenbock, G. Accumulation of the chalcone isosalipurposide in primary leaves of barley flavonoid mutants indicates a defective chalcone isomerase. *Physiol. Plant.* 101:827 (1997).

25. Skadhauge, B.; Gruber, M.Y.; Thomsen, K.K.; von Wettstein, D. Leucocyanidin reductase activity and accumulation of proanthocyanidins in developing legume tissue. *Am. J. Botany* 84:494 (1997).

26. Tanner, G.J.; Kristiansen, K.N.; Jende-Strid, B. Biosynthesis of proanthocyanidins (condensed tannins) in barley. *Proc. XVI Int. Conf. Groupe Polyphenols*, Portugal (1992).

27. Tanner, G.J.; Kristiansen, K.N. Synthesis of ^{3}H-3,4-*cis*-leucocyanidin and enzymatic reduction to catechin. *Anal. Biochem.* 209:274 (1993).

28. Joseph, R.; Tanner, G.; Larkin, P. Proanthocyanidin synthesis in the forage legume *Onobrychis viciifolia*. A study of chalcone synthase, dihydroflavonol 4-reductase and leucoanthocyanidin 4-reductase in developing leaves. *Aust. J. Plant Physiol.* 25:27 (1998).

29. Singh, S.; McCallum, J.; Gruber, M.Y.; Towers, G.H.N.; Muir, A.D.; Bohm, B.A.; Koupai-Abazani, M.R.; Glass, A.D.M. Biosynthesis of flavan-3-ols by leaf extracts of *Onobrychis viciifolia. Phytochemistry* 44:425 (1997).

30. Koupai-Abyazani, M.R.; McCallum, J.; Muir, A.D.; Lees, G.L.; Bohm, B.A.; Towers, G.H.N.; Gruber, M.Y. Purification and characterization of a proanthocyanidin polymer from seed of alfalfa (*Medicago sativa* cv. Beaver). *J. Agric. Food Chem.* 41:565 (1993).

31. Koupai-Abyazani, M.R.; McCallum, J.; Muir, A.D.; Bohm, B.A.; Towers, G.H.N.; Gruber, M.Y. Developmental changes in the composition of proanthocyanidins from leaves of sainfoin (*Onobrychis viciifolia* Scop.) as determined by HPLC analysis. *J. Agr. Food Chem.* 41:1066 (1993).

32. Lees, G.L.; Suttill, N.H.; Gruber, M.Y. Condensed tannins in sainfoin. 1. A histological and cytological survey of plant tissues. *Can. J. Bot.* 71:1147 (1993).

33. Stafford, H.A.; Smith, E.C.; Weider, R.M. The development of proanthocyanidins (condensed tannins) and other phenolics in bark of *Pseudotsuga menziessii. Can. J. Bot.* 67:1111 (1989).

34. Brandon, M.J.; Foo, L.Y.; Porter, L.J.; Meredith X. Proanthocyanidins of barley and sorghum: composition as a function of maturity of barley ears. *Phytochemistry* 21:2953 (1982).

35. Kristiansen, K.N. Biosynthesis of proanthocyanidins in barley: Genetic control of the conversion of dihydroquercetin to catechin and procyanidins. *Carlsberg Res. Commun.* 49:503 (1984).

36. Skadhauge, B.; Thomsen, K.K.; von Wettstein, D. The role of the barley testa layer and its flavonoid content in resistance to *Fusarium* infections. *Hereditas* 126:147 (1997).

37. Skadhauge, B. Genetics and biochemistry of proanthocyanidin biosynthesis and their biological significance in crop plants. *PhD Thesis*. The Royal Veterinary and Agriculture University, Copenhagen, Denmark (1996).

38. Delcour, J.A.; Ferreira, D.; Roux, D.G. Synthesis of condensed tannins. Part 9. The condensation sequence of leucocyanidin with (+)-catechin and with the resultant procyanidins. *J. Chem. Soc., Perkin Trans.* 1:1711 (1983).

39. Parham, R.A.; Kaustinen, H.M. Differential staining of tannin in sections of epoxy-embedded plant cells. *Stain Technol.* 51:237 (1976).

40. Parham, R.A.; Kaustinen, H.M. On the site of tannin synthesis in plant cells. *Bot. Gaz.* 138:465 (1983).

41. Rao, K.S. Fine structural details of tannin accumulations in nondividing cambial cells. *Ann. Bot.* 62:575 (1988).

42. Lees, G.L.; Gruber, M.Y.; Suttill, N.H. Condensed tannins in sainfoin. 2. Occurrence and changes during leaf development. *Can. J. Bot.* 73:1540 (1995).

43. Lees, G.L. Condensed tannins in some forage legumes: their role in the prevention of ruminant pasture bloat. *In*: Hemingway, R.W.; Laks, P.E. (eds.) Plant polyphenols: synthesis, properties, significance. Plenum Press, New York. pp 914 (1992).

44. Lees, G.L.; Hinks, C.F.; Suttill, N.H. Effect of high temperature on condensed tannin accumulation in leaf tissues of big trefoil (*Lotus uliginosis* Schkuhr). *J. Sci. Food Agric.* 65:415 (1994).

45. Barry, T.N.; Forss, D.A. The condensed tannin content of vegetative *Lotus pedunculatus*, its regulation by fertilizer application, and effect upon protein solubility. *J. Sci. Food Agric.* 34:1047 (1983).

46. Jende-Strid, B.; Møller, B.L. Analysis of proanthocyanidins in wild type and mutant barley (*Hordeum vulgare* L.). *Carlsberg Res. Commun.* 46:53 (1981).

47. Meldgaard, M. Expression of chalcone synthase, dihydroflavonol reductase and flavanone-3-hydroxylase in mutants of barley deficient in anthocyanin and proanthocyanidin biosynthesis. *Theor. Appl. Genet.* 83:695 (1992).

48. Hubank, M.; Schatz, D.F. Identifying differences in the mRNA expression by representation difference analysis of cDNA. *Nucl. Acids Res.* 22:5640 (1994).

49. Olson, O.; Wang, X.; von Wettstein, D. Sodium azide mutagenesis: Preferential generation of A:T-G:C transitions in the barley *Ant*18 gene. *Proc. Nat'l. Acad. Sci. USA* 90:8043 (1993).

50. Wang, X.; Olsen, O.; Knudsen, S. Expression of the dihydroflavonol reductase gene in an anthocyanin-free barley mutant. *Hereditas* 119:67 (1993).

51. Gierl, A.; Saedler, H. Plant-transposable elements and gene tagging. *Plant Mol. Biol.* 19:39 (1992).

52. Jones, D.G.; Jones, D.; Bishop, G.J.; Harrison, K.; Carroll, B.J.; Scofield, S.R. Use of the maize transposon *Activator* and *Dissociation* to show phosphinothricin and spectinomycin

resistance genes act non-cell-autonomously in tobacco and tomato seedlings. *Transgen. Res.* 2:63 (1993).

53. Thykjaer, T.; Stiller, J.; Handberg, K.; Jones, J.; Stougaard, J. The maize transposable element *Ac* is mobile in the legume *Lotus japonicus*. *Plant Mol. Biol.* 27:981 (1995).

54. Watterson, J.J.; Butler, L.G. Occurrence of an unusual leucoanthocyanidin and absence of proanthocyanidins in *Sorghum* leaves. *J. Agric. Food Chem.* 31:41 (1983).

55. Lloyd, A.M.; Walbot, V.; Davis, R.W. *Arabidopsis* and *Nicotiana* anthocyanin production activated by maize regulators *R* and *C1*. *Science* 258:1773 (1992).

56. Kilby, N.J.; Leyser, H.M.O.; Furner, I.J. Promoter methylation and progressive transgene inactivation in *Arabidopsis*. *Plant Mol. Biol.* 20:103 (1992).

57. Scheid, M.O.; Paszkowski, J.; Potrykus, I. Reversible inactivation of a transgene in *Arabidopsis thaliana*. *Mol. Gen. Genet.* 228:104 (1991).

58. Dehio, C., Schell, J. Identification of plant genetic loci involved in a posttranscriptional mechanism for meitoically reversible transgene silencing. *Proc. Nat'l. Acad. Sci. (USA)* 91:5538 (1994).

59. Morris, P.; Robbins, M.P. Condensed tannin formation by *Agrobacterium rhizogenes* transformed root and shoot organ cultures of *Lotus corniculatus*. *J. Exp. Bot.* 43:221 (1992).

60. Carron, T.R.; Robbins, M.P.; Morris, P. Genetic modification of condensed tannin biosynthesis in *Lotus corniculatus*. 1. Heterologous antisense dihydroflavonol reductase downregulates tannin accumulation in "hairy root" cultures. *Theor. Appl. Genet.* 87:1006 (1994).

61. Robbins, M.P.; Bavage, A.D.; Strudwicke, C.; Morris, P. Genetic manipulation of condensed tannins in higher plants. II. Analysis of birdsfoot trefoil plants harboring antisense dihydroflavonol reductase constructs. *Plant Physiol.* 116:1133 (1997).

62. Bavage, A.D.; Davies, I.G.; Robbins, M.P.; Morris, P. Expression of an *Antirrhinum* dihydroflavonol reductase gene results in changes in condensed tannin structure and accumulation in root cultures of *Lotus corniculatus* (birdsfoot trefoil). *Plant Mol. Biol.* 35:443 (1997).

63. Colliver, S.P.; Morris, P.; Robbins, M.P. Differential modification of flavonoid and isoflavonoid biosynthesis with an antisense chalcone synthase construct in transgenic *Lotus corniculatus*. *Plant Mol. Biol.* 35:509 (1997).

64. van der Krol, A.R.; Lenting, P.E.; Veenstra, J.; van der Meer, I.M.; Koes, R.E.; Gerats, A.G.M.; Mol, J.N.M.; Stuitje, A.R. An anti-sense chalcone synthase gene in transgenic plants inhibits flower pigmentation. *Nature* 333:866 (1988).

65. Datla, R.S.S.; Bekkaoui, F.; Hammerlindo, J.K.; Pilate, G.; Dunstan, D.I.; Crosby, W.L. Improved high-level constitutive foreign gene expression in plants using an AMV RNA4 untranslated leader sequence. *Plant Sci.* 94:139 (1993).

66. van der Meer, I.M.; Stam, M.E.; van Tunen, A.J.; Mol, J.N.M.; Stuitje, A.R. Antisense inhibition of flavonoid biosynthesis in petunia anthers results in male sterility. *Plant Cell* 4:253 (1992).

67. Que, Q.D.; Wang, H.Y.; English, J.J.; Jorgensen, R.A. The frequency and degree of cosuppression by sense chalcone synthase transgenes are dependent on transgene promoter strength and are reduced by premature nonsense codons in the transgene coding sequence. *Plant Cell* 9:1357 (1997).

68. Todd, J.J.; Vodkin, L.O. Duplications that suppress and deletions that restore expression from a chalcone synthase multigene family. *Plant Cell* 8:687 (1996).

69. van der Meer, I.M.; Stuitje, A.R.; Mol, J.N.M. Regulation of general phenyl-propanoid and flavonoid gene expression. *In*: Verma, D.P.S. (ed.) Control of plant gene expression. CRC Press, Boca Raton, FL pp. 125 (1993).

70. Mol, J.; Jenkins, G.; Schafer, E.; Weiss, D. Signal perception, transduction, and gene expression involved with anthocyanin biosynthesis. *Critical Rev. Plant Sci.* 15:525 (1996).

71. Styles, E.D.; Ceska, O. The genetic control of flavonoid synthesis in maize. *Can. J. Genet. Cytol.* 19:289 (1997).

72. Cone, K.C.; Burr, F.A.; Burr, B. Molecular analysis of the maize anthocyanin regulatory locus *C1*. *Proc. Nat'l. Acad. Sci. (USA)* 83:9631 (1986).

73. Paz-Arez, J.; Ghosal, D.; Weinard, U.; Peterson, P.; Saedler, H. The regulatory *C1* locus of *Zea mays* encodes a protein with homology to myb proto-oncogene products and with structural similarities to transcriptional activators. *EMBO J.* 6:3553 (1987).

74. Cone, K.C.; Coccioloni, S.M.; Burr, F.A.; Burr, B. Maize anthocyanin regulatory gene *pl* is a duplicate of *c1* that functions in the plant. *Plant Cell* 5:1795 (1993).

75. Cone, K.C.; Cocciolone, S.M.; Moehlenkamp, C.A.; Weber, T.; Drummond, B.J.; Tagliani, L.A.; Bowen, B.A.; Perrot, G.H. Role of the regulatory gene *Pl* in the photocontrol of maize and anthocyanin pigmentation. *Plant Cell* 5:1807 (1993).

76. Quattrocchio, F.; Wing, J.F.; Leppen, H.T.C.; Mol, J.N.M.; Koes, R.E. Regulatory genes controlling anthocyanin pigmentation are functionally conserved among plant species and have distinct sets of target genes. *Plant Cell* 5:1497 (1993).

77. Quattrocchio, R.; Wing, J.F.; van der Woude, K.; Mol, J.N.M.; Koes, R. Analysis of *bHLH* and *myb* domain proteins: species specific regulatory differences are caused by divergent evolution of target anthocyanin genes. *Plant J.* 13:475 (1998).

78. Chuck, G.; Robbins, T.; Nijjar, C.; Ralston, E.; Courtney-Gutterson, N.; Dooner, H.K. Tagging and cloning of a petunia flower color gene with the maize transposable element Activator. *Plant Cell* 5:371 (1993).

79. Avila, J.; Nietao, C.; Canas, L.; Benito, J.M.; Paz-Ares, J. *Petunia hybrida* genes related to the maize regulatory *C1* and to animal *myb* proto-oncogenes. *Plant J.* 3:553 (1993).

80. Moyano, E.; Martinez-Garcia, M.F.; Martin, C. Apparent redundancy in *myb* gene function provides gearing for the control of flavonoid biosynthesis in *Antirrhinum* flowers. *Plant Cell* 8:1519 (1996).

81. Goff, S.A.; Cone, K.C.; Fromm, M.E. Identification of functional domains in the maize transcriptional activator C1: comparison of wildtype and dominant inhibitor proteins. *Genes and Development* 5:298 (1991).

82. Goff, S.A.; Cone, K.C.; Chandler, V.L. Functional analysis of the transcriptional activator encoded by the maize *B* gene: evidence for a direct functional interaction between two classes of regulatory proteins. *Genes and Development* 6:864 (1992).

83. Chandler, V.L.; Radicella, J.P.; Robbins, T.P.; Chen. J.; Turks, D. Two regulatory genes of the maize anthocyanin pathway are homologous: Isolation of the *B* utilizing *R* genomic sequences. *Plant Cell* 1:1175 (1989).

84. Consonni, G.; Geuna, F.; Gavazzi, G.; Tonelli, C. Molecular homology among members of the *R* gene family in maize. *Plant J.* 3:335 (1993).

85. Gerats, A.G.; Bussard, J.; Coe, E.H.Fr.; Larson, R. Influence of *B* and *R* on UDPG:flavonoid-3-0-glucosyltransferase in *Zea mays*. *L. Biochem. Genet.* 22:1161 (1984).

86. Tonelli, C.; Consonni, G.; Donfini, S.F.; Dellaporta, S.L.; Viotti, A.; Gavazzi, G. Genetic and molecular analysis of *Sn*, a light-inducible tissue specific regulatory gene in maize. *Mol. Gen. Genet.* 225:401 (1991).

87. Gavazzi, G.; Mereghetti, M.; Consonni, G.; Tonelli, C. *Sn*, a light-dependent and tissue specific gene of maize: the genetic basis of its instability. *Genetics* 125:193 (1990).

88. Ludwig, S.R.; Habera, L.F.; Dellaporta, S.L.; Wessler, S.R. *Lc*, a member of the maize *R* gene family responsible for tissue-specific anthocyanin production encodes a protein similar to anthocyanin transcriptional activators and contains the *myc*-homology region. *Proc. Nat'l. Acad. Sci. (USA)* 86:7092 (1989).

89. Grotewold, E.; Drummond, B.J.; Bowen, B.; Peterson, T. The *myb*-homologous *P* gene controls phlobaphene pigmentation in maize floral organs by directly activating a flavonoid gene subset. *Cell* 76:543 (1994).

90. Styles, E.D.; Ceska, O. The genetic control of flavonoid synthesis in maize. *Can. J. Genet. & Cytol.* 19:289 (1977).

91. Goodrich, J.; Carpenter, R.; Coen, E.S. A common gene regulates pigmentation pattern in diverse plant species. *Cell* 68:955 (1992).

92. Quattrocchio, F.M. Regulatory genes controlling flower pigmentation in *Petunia hybrida*. *PhD Dissertation*. Vrije Universiteit, Amsterdam, The Netherlands (1994).

93. Mooney, M.; Desnos, T.; Harrison, K.; Jones, J.; Carpenter, R.; Coen, E. Altered regulation of tomato and tobacco pigmentation genes caused by the *delila* gene of *Antirrhinum*. *Plant J.* 7:333 (1995).

94. Wong, J.R.; Walker, L.S.; Drikeilis, H.; Klein, T.M. Anthocyanin regulatory genes from maize *B-Peru* and *C1* activate the anthocyanin pathway in wheat, barley and oat cells. *J. Cell Biochem. Suppl.* 0(15 part A):159 (1991).

95. Bradley, J.M.; Davies, K.M.; Deroles, S.C.; Bloor, S.J.; Lewis, D.H. The maize *Lc* regulatory gene up-regulates the flavonoid biosynthetic pathway of petunia. *Plant J.* 13:381 (1998).

96. Tamagnone, L.; Merida, A.; Parr, A.; Mackay, S.; Culliznez-Macia, F.A.; Roberta, K.; Martin, C. The AmMYB308 and AmMYB330 transcription factors from *Antirrhinum* regulate phenylpropanoid and lignin biosynthesis in transgenic tobacco. *Plant Cell* 10:135 (1998).

97. Sablowski, R.W.M.; Moyano, E.; Cullianezmacia, F.A.,; Schuch, W.; Martin, C.; Beven, M. A flower-specific MYB protein activates transcription of phenylpropanoid biosynthetic genes. *EMBO J.* 13:128 (1994).

98. Damiani, F.; Paolocci, F.; Consonni, G.; Crea, F.; Tonelli, C.; Arcioni, S. A maize anthocyanin transactivator induces pigmentation in hairy roots of dicotyledenous species. *Plant Cell Rep.* 17:339 (1998).

99. Damiani, F.; Paolocci, F.; Cluster, P.D.; Arcioni, S.; Tanner, G.J.; Joseph, R.G.; Li, Y.G.; deMajnik, J.; Larkin, P.J. The maize transcription factor *Sn* alters proanthocyanidin synthesis in transgenic *Lotus corniculatus* plants. *Aust. J. Plant Phys.* 26: in press (1999).

100. English, J.J.; Mueller, E.; Baulcombe, D.C. Suppression of virus accumulation in transgenic plants exhibiting silencing of nuclear genes. Plant Cell 8:179 (1996).

101. Dougherty, W.; Lindbo, Smith, H.; Parks, T.; Swaney, S.; Proebsting, W. RNA-mediated virus resistance in transgenic plants: Exploitation of a cellular pathway involved in RNA degradation. *Mol. Plant-Microbe Interact.* 7:544 (1994).

102. de Majnik, J.; Joseph, R.G.; Tanner, G.J.; Larkin, P.J.; Dmordjevic, M.A.; Rolfe, B.G.; Weinman, J.J. A convenient set of vectors for expression of multiple gene combinations in plants. *Plant Mol. Biol. Rep.* 15:134 (1997).

103. de Majnik, J.; Tanner, G.J.; Joseph, R.G.; Larkin, P.J.; Weinman, J.J.; Djordjevic, M.A.; Rolfe, B.G. Transient expression of maize anthocyanin regulatory genes influences anthocyanin production in white clover and peas. *Aust. J. Plant Physiol.* 25:335 (1998).

104. Larkin, P.J.; Yuguang, L.; Tanner, G.J.; Banks, P.M. Using alien genes—translocations, transfusions and transgressions. *In*: Focused plant improvement. Towards responsible and sustainable Agriculture. *Proc. Tenth Australian Plant Breeding Conference*. Gold Coast, Australia (April) (1993).

105. Junghans, H.; Dalkin, K.; Dixon, R.A. Stress responses in alfalfa (*Medicago sativa* L.). Part 15. Characterization and expression patterns of members of a subset of the chalcone synthase multigene family. *Plant Mol. Biol.* 22:239 (1993).

106. Shirley, B.W. Flavonoid synthesis: "new functions" for an "old pathway." *Trends in Plant Sci.* 1:377 (1996).

107. Charrier, B.; Coronado, C.; Kondorosi, A.; Ratet, P. Molecular characterization and expression of alfalfa (*Medicago sativa* L.) flavanone-3-hydroxylase and dihydroflavonol-4- reductase encoding genes. *Plant Mol. Biol.* 29:773 (1995).

BIRDSFOOT TREFOIL: A MODEL FOR STUDYING THE SYNTHESIS OF CONDENSED TANNINS

Francesco Paolocci, Raffaella Capucci, Sergio Arcioni, and Francesco Damiani

Istituto di Ricerche sul Miglioramento
Genetico delle Piante Foraggere
Consiglio Nazionale delle Ricerche
Perugia
ITALY

1. INTRODUCTION

Forage legumes are essential components in the diet of cattle and sheep unless they are fed with rendered animal proteins, which may have consequent serious problems for animal and human health. The most cultivated forage species are lucerne and clovers. However, the large amounts of proteins available from these species are poorly exploited, since they are degraded extensively both by plant proteases and rumen microorganisms where N may be lost through NH_3 formation. These degradative phenomena are the basis of a range of different and serious agricultural problems: 1) bloating with consequent risks for the health and the life of animals;[1] 2) loss of nutritive value and reduced growth of the animals;[2] 3) pollution of the environment; in fact, excreted ammonia-rich manure is a pollution factor that alters ecological equilibrium and the environment.[3]

For these reasons, the most cultivated forage legumes cannot be freshly consumed but necessitate a pretreatment (silage, hay, or dehydration), with additional costs and/or loss of nutrients. Alternatively, they have to be consumed in mixtures with grasses. As it is often not possible to use these species as pasture, many lands are not utilized for animal feeding purposes and are left abandoned.

Bloat (ruminal tympanites) is a common digestive disorder of ruminants. Its typical symptoms are the over-distension of the rumen reticulum resulting from the formation of a stable, proteinaceous foam in the rumen, which causes mortality or morbidity.[4] The foam formation is closely related to the ingestion of fresh forage such as lucerne and clover, which are rich in soluble proteins.[5] Bloat is the most serious problem in the utilization of lucerne and clover as grazing forage when they are not mixed with grasses. The etiology of bloating involves animal,

Plant Polyphenols 2: Chemistry, Biology, Pharmacology, Ecology, Edited by
Gross et al. Kluwer Academic / Plenum Publishers, New York, 1999

plant, microbial, and environmental factors. Climatic factors such as changes in temperature and shading and agronomic factors such as application of fertilizers affect the chemical composition of plants and as consequences their bloating propensity. A number of plant factors have been implicated in the etiology of bloat: pectins, saponins, water-soluble proteins, lipids, lipoproteins, polysaccharides, enzymes, phenolics (especially condensed tannins), and cell wall structures related to the mechanical rupture of the cells and the initial rate of digestion. But so far, only the presence of condensed tannins (CT), the cell wall strength, saponins, and the rate of protein degradation have been correlated with bloat.

Some forage legumes do not cause bloating symptoms, and these lesser culti-vated species contain a certain amount of CT. The role of CT in preventing bloat-ing has been extensively examined. They destabilize and prevent formation of the proteinaceous foam and protect dietary protein from rumen degradation.[6]

These functions are due to the ability of CT to form reversible pH-dependent complex with proteins, the formation of these protein-tannin complexes making protein unavailable for ruminal microbial deamination. Moreover, CT inhibit the activity of extracellular microbial enzyme and activate other mechanisms like deprivation of metal ions and inhibition of oxidative phosphorylation, which also result in the prevention of foam formation and bloating.

However, owing to their ability to inhibit enzymatic activities, high concentra-tions of CT (as reported for several temperate and tropical forages) can be detri-mental to the nutritive value and prevent the intake and digestibility of the forage by the animal.[7] For these reasons, a number of well-adapted and well-growing high-tannin species are poorly exploited for animal feeding. The desirable con-centration of CT for ruminants should represent a balance between positive and negative effects, and therefore the ability to modulate CT concentration at the optimal levels in each species would enlarge the pool of species to be utilized in animal feeding.

Condensed tannins are secondary metabolites originating from a branch of the anthocyanin pathway. Their biosynthetic pathway is well characterized, and the genes coding for the different enzymes have been cloned in different species (*Antir-rhinum*,[8] Petunia,[9] maize,[10] *Vitis*[11]) and are regulated by genes belonging to the *myc* and *myb* families,[12] which interact with promoter regions of the structural genes and initiate their transcriptions. The biosynthesis of anthocyanidin and CT share the same sequence of reactions to the last common intermediate, leucoan-thocyanidin, the formation of which is catalyzed by the enzyme dihydroflavanol reductase (DFR). Less is known about the molecular biology of the branch of the flavonoid pathway leading from leucoanthocyanidin to CT. The NADPH-dependent reduction of labeled leucoanthocyanidin to catechin has been demonstrated in a number of species, including *L. corniculatus*[13] (Tanner, unpublished). This activ-ity is due to a leucoanthocyanidin reductase (LAR). The nature of the subsequent condensation steps leading to CT is unknown. The characterization of the enzy-matic activities and the cloning of the relative genes are in progress in different labs with the aim of introducing these genes in species when absent or, when present, to regulate their activity in such a way as to obtain a palatable, non-bloating, protein-rich forage.

Different approaches are being pursued toward this objective: a) the purification of the LAR enzyme from CT-rich species (e.g., *Onobrychiis viciifolia*) and the subsequent cloning of the gene on the basis of the deduced nucleotide sequence (Tanner, personal communication); b) the gene-tagging strategies through transformation of *Lotus japonicus*, a leaf negative tannin species, with transposons;[14] c) the production of *Lotus corniculatus* mutants through genetic transformation with sense and antisense sequences of structural genes of the CT pathway;[15,16,17] d) the production of *L. corniculatus* mutants through genetic transformation with heterologous regulatory genes of the flavonoid biosynthetic pathway.[18]

This last approach proved particularly useful and, through a proposed co-suppression mechanism, CT negative mutants were obtained. The characterization of these mutants indicated that the introduced transgene (*Sn*[19]) was silent in CT negative tissues, and that silencing of transgenes was paralleled by silencing of DFR genes and by the absence of LAR enzymatic activity.

This paper discusses the novel experiences of transformation with the maize *Sn* gene on three *Lotus* genotypes (a gift of Mark Robbins and Phil Morris from IGER, Aberystwyth UK) that are well characterized for different CT levels.

2. EFFECT OF GENETIC TRANSFORMATION OF SELECTED GENOTYPES

Hairy root transformation was performed in each genotype utilizing a binary *Agrobacterium rhizogenes* vector harboring the wild type Ri plasmid 1855 and the p121.Sn plasmid[20] derived from pBI121.1[21] where uidA gene is replaced by the maize *Sn* cDNA. As a control, the binary vector harboring pBI121.1 plasmid was utilized.

Several *Sn* transgenics were obtained from each starting genotype applying the protocol described by Robbins and colleagues.[22] Small alterations of plant morphology due to the effect of rol genes present in the Ri plasmid were observed,[23] but the plant structures and habits of growth were substantially unaffected.

Sn is a transactivator gene belonging to the *myc* family. The genes of this family are widely distributed all over species and genera and control several aspects of plant physiology and morphology. All these features derive from their influence in regulating transcription of structural genes of the flavonoid pathway, from which a large number of important compounds originates such as anthocyanins and tannins. Genetic transformations with these regulatory genes produced modifications of: pattern pigmentation,[24,20] CT levels,[18] and thricome development[25] in various species.

In our transformed material, apart from CT levels discussed below, no relevant alterations in plant morphology and pigmentation were observed. However, some plants showed some red pigmentation in leaves and stems in certain stages of growth. Three genotypes belonging to the cv. Leo named S50, S33, S41 and characterized by low, medium-low, and high CT levels in leaves, respectively, were used in this study. These CT data derive from analyses performed in Perugia and are slightly different from those provided by IGER, where S33 showed lower CT content respective to S50 when the hairy root were utilized for the assay.[16]

This confirms that CT synthesis is subject to environmental and tissue-specific mechanisms. For this reason, the starting genotypes and the transgenic control and *Sn* transformed plants were repeatedly assayed for CT levels in leaves by using a test based on the ability of 4-dimethyl-aminocinnamaldehyde (DMACA) to produce blue coloration when reacted with proanthocyanidins under acidified conditions.[26] After staining, the plants were visually scored (1–7) for the intensity of pigmentation.

The average value of all transgenic plants is reported in the histograms relative to each population (fig. 1) and compared to the relative starting genotype. The comparison between mother plants and the control-transformed plants (last two columns of each histogram) indicates that no effect on CT levels is attributable to the Ri T-DNA, since no differences between these two mean values were observed in any genotype.

Data relative to the *Sn* transgenics show a variable distribution of CT levels in relation to the starting genotype. S41 transgenics were practically unaffected

CT levels in transgenic plants

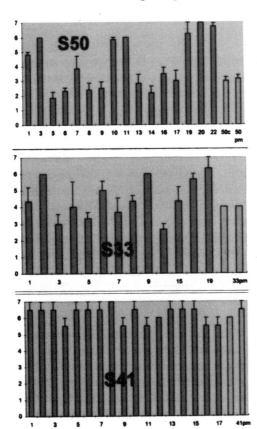

Figure 1. Histograms of the CT level within each transgenic population. Each bar represents the average values for CT score attributed to each plant in six assays performed from spring to autumn in 1997.

as regards CT levels by *Sn* transformation, all plants showing a high score for CT not different from S41.

Increasing variability is observed in transgenics derived from low CT genotype. As a matter of fact, S33 derived transformants showed both higher and lower CT scores with respect to the starting plants, and no completely suppressed genotypes were reported. In the case of S50-derived transgenics, the variability among transformed lines increased, and both almost completely CT-suppressed and highly CT-expressing individuals were observed.

The differences observed in response to transformation suggest some differences among the starting genotypes in the mechanism of induction of CT synthesis. To investigate this aspect, the original genotypes and several derived *Sn*-transgenics were *in vitro* micropropagated[27] and repeatedly assayed for CT levels. For these last analyses, plantlets were grown in a growth chamber at the following conditions: 12 h photoperiod, 22 °C, 150 μE/s light intensity. After 4 weeks of these treatments, plants were evaluated for CT levels, and light intensity was increased to 330 μE/s. After 4 more weeks, plants were assayed for CT levels and moved in the greenhouse (photoperiod around 14 hours, temperature in the range 12–27 °C, light intensity >1,000 μE/s). After 4 more weeks, the CT test was repeated.

This experiment indicates that a different mechanism of interaction with light is involved among genotypes. The S41 starting genotype and its *Sn* transgenic lines were CT positive at any light intensity, and the increase of light intensity resulted in an increase of CT level. Conversely, S33 and S50 starting genotypes were both negative for CT at low light intensity but differed with each other at higher light intensity. At 330 μE/s, both faintly stained for CT, but the further increase of light did not affect S50. CT content of S33 was further increased to the plateau level of the genotype (fig. 2).

In short, it seems that S41 has both a light-independent and a light-dependent mechanism for inducing CT synthesis. Alternatively, a low light input of less than 150 μE/s, is enough for the induction of CT biosynthesis, as observed in leaves of sainfoin (unpublished results).

Figure 2. Effect of light intensity on CT levels in the three starting plants. From left to right 150, 330, >1,000 μE/s. From top to bottom S41, S33, S50 plant.

On the contrary, in S33 and S50, the CT synthesis is strictly related to the quantum of light received. They differ for the threshold level of CT that is very low in S50 and medium low in S33.

Transgenic plants derived from S41 were never different from the starting genotype either in level or in timing of CT synthesis. S33 CT-depressed transgenics showed a similar trend as the mother plant with respect to light induction but always with lower CT levels. S33 with high CT levels such as S33-19 were conversely not different from S41 untransformed plant. S50 transgenic CT negative plants remained always negative, whereas S50 plants with increased levels of CT such as S50-11 differed from the starting plant only at elevated light intensity.

The final aim of the work is the production of stable CT-contrasting isogenic lines to apply strategies of cDNA subtraction or Reverse Transcription Differential Display. For this purpose, several micropropagated individuals of some selected transgenics were analyzed again for CT level in the second year.

From this viewpoint, a different behavior was observed in the three populations, strictly related with the degree of variability induced by *Sn* transformation. S41 transformants were very stable, in fact all maintained the same levels of CT in the second year. Within population S33, only one plant (S33-1) showed a significant increase in CT level in the second year respective to the results of first year analysis, but mean values of the 2 years were positively correlated. In S50 transgenics, there was the tendency for extreme values to revert toward the average value of the population (fig. 3).

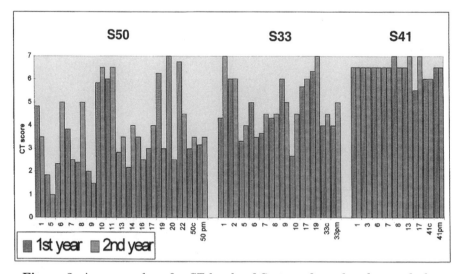

Figure 3. Average values for CT levels of *Sn* transformed and control plants recorded in two subsequent years. From left to right S50, S33, and S41 transgenic and control plants.

One plant in particular (S50-20) that in the first year had elevated levels of CT and was not distinguishable from S41 plants, in the second year showed a level of CT less than the average value of the S50 transgenic population (fig. 4). Some other plants displayed a similar trend but with smaller shift. As a consequence, the correlation between CT levels in the 2 years were not significant.

However, some stable transgenic lines were also found in the S50 population, plants 5 and 9 were stable suppressed and plants 10 and 11 overaccumulated CT with respect to the starting genotype (fig. 3). On these plants, it will be possible to apply subtractive strategies for isolating unknown sequences involved in CT synthesis.

3. MOLECULAR ANALYSIS

Transgenic plants were scored through southern blot analysis for the presence and number of copies of the transgene. For this purpose, DNA was restricted with *Hind*III enzyme electrophoresed on 1 percent agarose gel and hybridized with a *Sn* sequence generated through PCR.[18] Some hybridisation pattterns of S50 transgenics are reported in figure 5.

Results of this investigation indicated that the copy number of the transgene ranged from one to six but no direct relation between number of copies and CT levels was observed. However, the S50 plants selected for the stable low level of CT (such as plant 50–9) had a elevated number of copies of *Sn*, whereas plants 10 and 11 (not shown), which tested stable in over-expressing CT, had only one copy of the transgene. Moreover, plant S20, which was very unstable, showed four copies of the transgene (not shown). The elevated number of *Sn* copies in this plant offers a possible explanation for its unstable behavior, in fact, it has been observed in *Lotus* that 35S driven transgenes are very sensible to methylation when multiple

Figure 4. Instability for CT levels in transgenic plant S50/20. DMACA stain was performed on open air grown plants in summertime. Left 1st year, right 2nd year.

S50 transgenic plants:

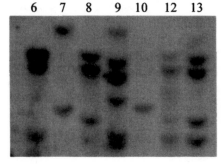

Figure 5. Southern blot analysis of some *Sn* transgenic plants derived from S50 genotype.

copies (more than three) are present.[28] It is thus conceivable that, during the cycle of micropropagation, methylation of the transgene occurred with the consequent silencing of the transgene and, at the end, with a direct effect on CT levels.

This hypothesis, which was confirmed by analysing the DNA methylation through the restriction with suitable enzymes, is supported by the observation that individuals with contrasting CT phenotypes have different levels of expression of the transgene.

For this purpose, analysis of transgene expression was performed in some individuals and, because previous results indicated that a strict correlation between DFR activity and CT synthesis occurred, also the expression of DFR was assayed. RNA was isolated in duplicate lots from transformants and mother plants according to Chang.[29]

Northern analysis for *Sn* was performed on total RNA of the starting mother genotypes, and the absence of hybridizing bands indicated that a gene with strong homology to *Sn* was not expressed. When the same blot was hybridized with a DFR *Lotus* probe, cloning and sequence of which will be discussed later, only S41 genotype showed a band. This supports the previous observation that CT is strongly and quantitatively related to DFR activity.[18] Of the transgenic plants, we tested only some contrasting S50 transgenics and only one, with high levels of CT, resulted also positive for both DFR and *Sn* hybridization (fig. 6). This indicates that *Sn* acts on the CT pathway inducing DFR synthesis and finally CT production. This was also observed in a previous experiment[18] where *Sn* transformed plants resulted suppressed for CT synthesis in leaves but showed increased levels of CT in roots. This tissue-specific behavior was paralleled by the *Sn* expression; *Sn* transcripts were not detected in leaves but were present in roots.

Through Northern analysis, we can discriminate between strong and low or null activity of the tested genes, but we cannot exclude that negative signals are obtained not for the silencing of the genes tested but merely because of the very low abundance of its transcripts. For a more powerful analysis, RT-PCR was performed. For such analysis, $5\,\mu g$ of DNA-free total RNA were reverse transcribed in the presence of SuperScript II Rnase H-Reverse Transcriptase by using random hexamers in a final volume of $20\,\mu l$. Four μl of cDNA were used for amplification with *Sn* primers as previously described.[18] Twenty microliters of PCR products

Northern analysis

S50 transgenic: Untrans.:
6   14   20   S50 S41

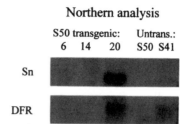

Sn

DFR

Figure 6. Northern blot analysis for *Sn* and DFR expression in some transformed and control plants.

were run on 1 percent agarose gel, blotted and hybridized with *Sn* probes. As control for DNA contamination, non reverse transcribed RNA were also amplified and run on the gel. As internal control for RNA loading $5\,\mu l$ of cDNA were also amplified in the presence of two partially degenerated primers, EF-1α forward and reverse derived by Rosati et al.[30] to amplify a *L. corniculatus* elongation factor 1α fragment. The PCR conditions were as for *Sn*, but the amplification cycles were reduced from 30 to 20.

The RT-PCR analysis proved to be very interesting for the differences observed in *Sn* transgenic plants derived from different starting genotypes. In fact, S33 and S50 derived transgenic lines behaved similarly: CT negative or depressed plants did not show *Sn* transcription; the transgenics with increased levels of CT, on the contrary, showed that *Sn* mRNA was produced. S41 transgenes showed neither differences with respect to the starting plant for CT level, nor for *Sn* expression.

This observation confirms that in plants with a low amount of leaf CT, the expression of *Sn* can increase the synthesis of CT, however, for a possible co-suppression mechanism,[31] the transformation with *Sn* can also decrease the CT levels. In this case, no *Sn* transcripts are observed. In S41 genetic background *Sn* does not produce any effect on CT levels and it is not expressed. Two hypotheses can be proposed: a) a pre-transcriptional mechanism of gene silencing has occurred and inhibits *Sn* transcription with no effect on CT accumulation; b) the interaction between *Sn* and an endogenous *Sn*-like gene has caused a post-transcriptional mutual silencing. In this last case, the silencing of the endogenous gene does not seem to affect CT levels and, as reported by Robbins et al.[32] (who found increased levels of CT in some S50 plants transformed with a heterologous DFR gene in anti-sense orientation), the unexpected CT production may be due to an over-compensation of the silenced gene by some other genes of the same family.

4. GENE CLONING

The final aim of the study is to isolate *Lotus* genes involved in the CT synthesis pathway. While waiting for the selection of suitable material for starting RT-PCR Differential Display analysis,[33] the cloning of *Lotus* sequences utilizing information from genes already sequenced in other species was attempted.

For this purpose, a cDNA library from unexpanded leaves of plants of the cv Leo was produced. A preliminary screening of this library for its representativeness and quality displayed the presence of transcripts ranging from 500 bp to 2.5 kbp. Moreover, by random selection of single phage plaques, we have been able to clone birdsfoot transcripts related to highly and poorly expressed genes, multi- or single-copy genes (unpublished results).

Despite this, the screening of the library with *Sn* fragments did not yield useful results. This was in agreement with Southern analysis of untransformed Leo plants with *Sn* probes spanning the most conserved amino acid domains. No hybridizing signals, even at low stringency conditions, were detected. This indicates that, although *Sn* is able to induce or suppress CT synthesis in *Lotus*, its sequence is poorly conserved in this species.

For a faster isolation of Lotus genes, RT-PCR was applied. In order to clone a *Lotus corniculatus* DFR cDNA fragment, two degenerated oligonucleotides, Dfrfrw1 (5′-cctgagaatgaaa/gta/gatcaagcc-3′) and Dfrbackw1 (5′-taatgagct/ctca/gccagt-3′), were designed by aligning the DFR sequences cloned in lucerne, petunia, and grapevine using the Clustal program. Total leaf RNA was isolated from *L. corniculatus* plants cv Leo, and poly A + RNA was prepared using the Poly ATtract mRNA system. About 0.5 µg of mRNA were reverse transcribed and amplified using the SuperScript One-Step RT-PCR according to the supplier's instructions. The PCR product, about 300 bp, was directly sequenced by using Big Dye Terminator Cycle Sequencing Ready Reaction Kit in a 310 ABI Prism automatic sequence analyzer. The sequence analysis displayed the presence of more than one PCR product. Therefore, the fragments resulting from the PCR amplification were cloned into pGEM-T-easy vector and sequenced. Four clones showing single-point mutations were obtained. Sequence comparisons on NCBI databank displayed high homology with another cDNA fragment recently cloned in *L. corniculatus*.[17] All these DFR partial sequences named DFR1, DFR2, DFR3, DFR4 have been deposited in Gene Bank under the following accession numbers respectively: AF117261, AF117262, AF117263, AF117264. Many of these single point mutations led to amino acid changes. Figure 7 shows the amino acid sequence alignment of the five partial cDNA fragments cloned in birdsfoot trefoil. Further experiments are in progress to clone the entire DFR genes and to assess whether these DFR sequences are different allelic forms or different genes, which differ for substrate and/or tissue specificity.

With a similar technique, a *myc*-like sequence putatively related to the flavonoid pathway has been rescued. Although we do not yet have any proof that this sequence is related to the mechanism of induction/suppression of CT synthesis in leaves, it is interesting to go further in its study because it showed a DNA polymorphism among the three starting plants of the experiment. This has been demonstrated with a Southern analysis where DNA from S33, S50, S41, and S52, one of the plants of the cv. Leo we utilized for the preparation of the cDNA library, were restricted with two enzymes independently (*Hind*III and *Kpn*I) and probed with an internal sequence of this gene (fig. 8). The analyses of cross progeny derived by S41 and S50 plants could demonstrate whether this gene is somehow linked with the polymorphic accumulation of CT observed in these two plants.

The utilization of degenerated primers deduced from genes cloned in other species[30] allowed us to isolate and clone an internal sequence of a gene coding for the "1α elongation factor" that is useful to standardize molecular analyses.

Sequence alignment of *L.corniculatus* DFR proteins

```
DFR1    NEVIKPTINGVLDIMKACQKAKTVQRLVFTSSAGTLNAVEH    41
DFR2    NEVIKPTINGVLDIMKACQKAKTVRRLVFTSSAGTLDAVEH    41
DFRA    NEVIKPTINGVLDIMKACQKAKTVRRLVFTSSAGTLNAVEH    41
DFR3    NEVIKPTINGVLDIMKACQKAKTVRRLVFTSSAGTLNVIEH    41
DFR4    NEVIKPTINGVLDIMKACQKAKTVRRLVFTSSAGTLNVIEH    41
        *************************.***********...**


DFR1    QKQMYDESCWSDVEFCRRVKMTGWMYFVSKTLAEQEAWKFA    82
DFR2    QKQMFDESCWSDVEFCRRVKMTGWMYFVSKTLAEQEAWKFA    82
DFRA    QKQMFDESCWGDVEFCRRVKMTGWMSLGSKTLAEQEAWKFA    82
DFR3    QKQMFDESCWSDVEFCRRVKMTGWMYFVSKTLAEQEAWKFA    82
DFR4    QKQMFDESCWSDVEFCRRVKMTGWMYFVSKTLAGQEAWKFA    82
        ****.*****.**************** .*****.*******


DFR1    QEHDIDLITIIPSLVVGSFLMPTMP---    107
DFR2    KEHDIDLITTIPSLVVGSFLMPTMP---    107
DFRA    QEHDIDFITIIPSLVVGSFLMPTMP---    107
DFR3    KEHGIDFITIIPPLVVGSFLMPTMP---    107
DFR4    KEHGIDFITIIPPLVVGSFLMPTMP---    107
        .**.**.**.**.************
```

Figure 7. Partial amino acid sequence of *L. corniculatus* DFR proteins deduced from DNA cloned sequences DFR1, DFR2, DFR3, DFR4 compared with the *Lotus* sequence DFRA.[17]

5. CONCLUSIONS

Among forages, *L. corniculatus* is a particularly amenable species to work with. In fact, along with its suitability to tissue culture, genetic transformation and regeneration, this species shows production and CT accumulation in stems, veins, and flowers, whereas roots and leaves present a large variability for this trait. The presence of genotypes displaying different levels of leaf tannins makes this species

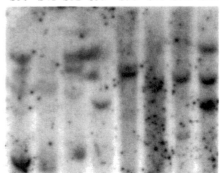

Figure 8. RFLP analysis of four plants of the cv. Leo hybridized with a myc-like sequence isolated from *L. corniculatus*.

of remarkable interest in understanding the molecular and physiological regulation of the phenylpropanoid pathway.

The transformation of three *Lotus corniculatus* genotypes characterized by different levels of CT with the maize gene *Sn* produced modifications on the levels of CT respective to the mother plants. A large variability was reported for transgenic plants derived from a genotype featured by low levels of CT (S50) and in lesser extent, from a genotype with medium low levels of CT (S33). In these populations of transgenics, some over-expressed tannins whereas others reduced CT levels respective to the initial plant. This fact strongly suggests that a modification of the genotypic specific regulation mechanism occurred both by transactivation of regulator and structural genes and/or silencing functional homologous genes. On the contrary, none of the S41 transgenic lines display CT levels significantly different from mother plants, suggesting that CT control is difficult to modulate by the presence of a transgene when this trait is strongly expressed.

The effect of light on CT accumulation in leaves, even in association with other environmental parameters such as temperature, differs among the genotypes tested and strongly suggests the presence of a different mechanism of CT pathway activation. The different interactions between CT synthesis and light intensity suggest that regulator genes controlling the pathway are polymorphic and interact differently with *Sn*. The R family of anthocyanin transactivators, to which *Sn* belongs, includes loci and alleles with a different sensitivity to light induction. The fact that S41 genotype, differently from S50 and S33, accumulates CT even at very low light intensity may suggest the presence in the first genotype of R-like allelic forms highly sensitive even to low input of light and able to compensate any exogenous perturbation by a transgene.

Although it is difficult to draw conclusions from these preliminary studies about physiological and molecular control of CT synthesis in *L. corniculatus* genotypes, the results reported confirm the possibility of regulating this complex pathway by transformation with an exogenous transactivator gene.

The availability of stable CT-suppressed and CT-elicited transgenic plants derived from the same starting genotype, i.e., S50/10 and S50/11 vs. S50/5 and S50/9, allows the use of these isogenic mutants as valuable material for subtractive strategies in order to clone both regulatory and structural genes involved in tannin synthesis and accumulation.

It is worth noting that in *L. japonicus*, a forage species that does not accumulate tannin into the leaves, mutants that do express tannin in leaves have been produced through a gene-tagging approach. The presence of a repressor "tagged" gene has been hypothesized to explain the development of leaf CT accumulating mutants as well as the block of flavonoid leaf synthesis in lucerne and white clover.[14] Despite the differences in controlling CT in leaves, mutants of *L. corniculatus and L. japonicus* are equally valuable resources to isolate genes responsible for tannins.

ACKNOWLEDGMENTS

The authors wish to thank Professor C. Tonelli and Dr. G. Consonni of University of Milan for providing the *Sn* gene; Dr. P. Larkin and Dr. G. Tanner of

CSIRO Canberra for introducing the problems of bloating, for protocols and useful discussion; Dr. M. Robbins and Dr. P. Morris of IGER (Aberystwyth, UK) for providing selected genotypes of *Lotus* and helpful suggestions and discussion. We also thank Mr. A. Bolletta for photographic plates and Ms. G. Labozzetta for the excellent technical assistance.

REFERENCES

1. Clark, R.T.J.; Reid, C.S.W. Foamy bloat of cattle. A review. *Journal of Dairy Sci.* 57:753–785 (1974).

2. Barry, T.N.; Manley, T.T. The role of condensed tannins in the nutritional value of *Lotus pedunculatus* for sheep. 2. Quantitative digestion of carbohydrates and proteins. *British J. of Nutr.* 51:493–504 (1984).

3. Jarvis, S.C. The pollution potential and flows of nitrogen to waters and the atmosphere from grassland under grazing. *In*: Dewi, I.A.; Axford, R.F.E.; Marai, I.F.M.; Omed, H.M. (eds.) Pollution in livestock production systems. CAB International, Wallingford pp. 227–239 (1994).

4. Lees, G.L. Condensed tannins in some forage legumes: their role in the prevention of ruminant pasture bloat. *In*: Hemingway, R.W.; Laks, P.E. (eds.) Plant polyphenols: synthesis, properties, significance. Plenum Press, New York, pp. 915–934 (1992).

5. Jones, W.T.; Lyttleton, J.W. Bloat in cattle. XXXIV. A survey of forage legumes that do not produce bloat. *NZ J. Agric. Res.* 13:149–156 (1971).

6. Tanner, G.J.; Moate, P.; Dailey, L.; Laby, R.; Larkin P.J. Proanthocyanidins (condensed tannins) destabilise plant protein foams in a dose dependent manner. *Aust. J. Agric. Res.* 46:1101–1109 (1995).

7. Kumar, R.; Singh, M. Tannins, their adverse role in ruminant nutrition. *J. Agric. Food Chemistry* 32:447–453 (1984).

8. Sommer, H.; Saedler, H. Structure of the chalcone synthase gene in of *Antirrhinum majus*. *Mol. Gen. Genet.* 202:429–434 (1986).

9. Koes, R.E.; Spelt, C.E.; van den Elzen, P.J.M.; Mol, J.N.M. Cloning and molecular characterization of the chalcone synthase multigene family of petunia. *Gene* 81:245–257 (1989).

10. Wienenad, U.; Weydemann, U.; Niesbach-Kloesgen, U.; Peterson, P.A.; Saedler, H. Molecular cloning of C2 locus of *Zea mays*, the gene coding for chalcone synthase. *Mol. Gen. Genet.* 203:202–207 (1986).

11. Sparvoli, F.; Martin, C.; Scienza, A.; Gavazzi, G.; Tonelli, C. Cloning and molecular analysis of structural genes involved in flavonoid and stilbene biosynthesis in grape (*Vitis vinifera* L.) *Plant Mol. Biol.* 24:743–755 (1994).

12. van der Meer, I.M.; Stuitje, A.R.; Mol, J.N.M. Regulation of general phenyl-propanoid and flavonoid gene expression. *In*: Verma, D.P.S. (ed.) Control of plant gene expression, CRC Press, Boca Raton, FL, pp. 125–155 (1993).

13. Tanner, G.J.; Kristiansen, K.N. Synthesis of 3,4-cis-leucocyanidin and enzymatic reduction to catechin. *Anal. Biochem.* 209:274–277 (1993).

14. Gruber, M.Y.; Skadhauge, B.; Stougaard, J. Condensed tannin mutation in *Lotus japonicus*. *Polyphenols Actualites. Lettre d'information du Groupe polyphénols*. Février 1998, pp. 4–8 (1998).

15. Carron, T.R.; Robbins, M.P.; Morris, P. Genetic modifications of condensed tannin biosynthesis in *Lotus corniculatus*. 1. Heterologous antisense dihydroflavonol reductase downregulates tannin accumulation in "hairy roots" cultures. *Theor. Appl. Genet.* 87:1006–1015 (1994).

16. Colliver, S.P.; Morris, P.; Robbins, M.P. Differential modification of flavonoid and isoflavonoid biosynthesis with an antisense chalcone synthase construct in transgenic *Lotus corniculatus*. *Plant Mol. Biol.* 35:509–522 (1997).

17. Bavage, A.D.; Davies, I.G.; Robbins, M.P.; Morris, P. Expression of an *Antirrhinum* dihydroflavanol reductase gene results in changes in condensed tannin structure and accumulation in root cultures of *Lotus corniculatus*. *Plant Mol. Biol.* 35:443–458 (1997).

18. Damiani, F.; Paolocci, F.; Cluster, P.D.; Arcioni, S.; Tanner, G.J.; Joseph, R.G.; Li, Y.G.; deMajnik, J.; Larkin, P.J. The maize transcription factor *Sn* alters proanthocyanidin synthesis in transgenic *Lotus corniculatus* plants. *Aust. J. of Plant Physiol.* 26:159–169 (1999).

19. Tonelli, C.; Consonni, G.; Dolfini, S.F.; Dellaporta, S.L.; Viotti, A.; Gavazzi, G. Genetic and molecular analysis of *Sn*, a light-inducible tissue-specific regulatory gene in maize. *Mol. Gen. Genet.* 225:401–410 (1991).

20. Damiani, F.; Paolocci, F.; Consonni, G.; Crea, F.; Tonelli, C.; Arcioni, S. A maize anthocyanin transactivator induces pigmentation in several transgenic dycotiledonous species. *Plant Cell Report* 17:339–344 (1998).

21. Jefferson, R.A.; Kavanagh, T.A.; Bevan, M.W. Gus fusion: β-glucuronidase as a sensitive and versatile gene fusion marker in higher plants. *EMBO J.* 6:3901–3907(1987).

22. Robbins, M.P.; Carron, T.R.; Morris, P. Transgenic *Lotus corniculatus*: a model system for modification and genetic manipulation of condensed tannin biosynthesis. *In*: Hemingway, R.W.; Laks, P.E. (eds.) Plant polyphenols: synthesis, properties, significance. Plenum Press, New York, pp. 111–131 (1992).

23. Damiani, F.; Nenz, E.; Paolocci, F.; Arcioni, S. Introduction of hygromycin resistance in *Lotus* spp. through *Agrobacterium rhizogenes* transformation. *Transgenic Research* 2:330–335 (1993).

24. Lloyd, A.M.; Walbot, W.; Davis, R.W. *Arabidopsis* and *Nicotiana* anthocyanin production activated by maize regulators R and C1. *Science* 258:1773–1775 (1992).

25. Larkin, J.C.; Oppenheimer, D.G.; Lloyd, A.M.; Paparozzi, E.T.; Marks, M.D. Roles of the GLABROUS1 and TRANSPARENT TESTA genes in *Arabidopsis* trichome development. *Plant Cell* 6:1065–1076 (1994).

26. Li, Y.G.; Tanner, G.J.; Larkin, P.J. The DMACA-HCl protocol and the threshold proanthocyanidin content for bloat safety in forage legumes. *J. Sci. Food Agric.* 70:89–101 (1996).

27. Pupilli, F.; Damiani, F.; Nenz, E.; Arcioni, S. *In vitro* propagation of *Medicago* and *Lotus* species by node culture. *In vitro Cell. and Dev. Biol.* 28:167–171 (1992).

28. Bellucci, M.; Alpini, A.; Paolocci, F.; Damiani, F.; Arcioni, S. Transcription of a maize cDNA in *Lotus corniculatus* is regulated by T-DNA methylation and transgene copy number. *Theor. Appl. Genet.* 98:257–264 (1999).

29. Chang, S.; Puryear, J.; Cairney, J. A simple and efficient method for isolating RNA from pine trees. *Plant Mol. Biol. Rep.* 11:113–116 (1993).

30. Rosati, C.; Cadic, A.; Duron, M.; Renou, J.P.; Simoneau, P. Molecular cloning and expression analysis of dihydroflavanol 4-reductase gene in flower organs of *Forsytia x intermedia*. *Plant Mol. Biol.* 35:303–311 (1997).

31. Metzlaff, M.; O'Dell, M.; Cluster, P.D.; Flavell, R.B. RNA-mediated RNA degradation and chalcone synthase A silencing in petunia. *Cell* 88:845–854 (1997).

32. Robbins, M.P.; Bavage, A.D.; Strudwicke, C.; Morris, P. Genetic manipulation of condensed tannins in higher plants. II analysis of *Lotus corniculatus* plants harboring antisense dihydroflavonol reductase constructs. *Plant Physiol.* 116:1133–1144 (1998).

33. Colonna-Romano, S.; Leone, A.; Maresca, B. Differential-display reverse transcription PCR (DDRT-PCR). Springer-Verlag (1998).

ISOTOPIC LABELLING OF DIETARY POLYPHENOLS FOR BIOAVAILABILITY STUDIES

Stéphanie Déprez[a] and Augustin Scalbert[b]

[a] Laboratoire de Chimie Biologique
(INRA), INA-PG
78850 Thiverval-Grignon
FRANCE

[b] Unité des Maladies Métaboliques et Micronutriments
INRA de Clermont-Ferrand/Theix
63122 St. Genès-Champarelle
FRANCE

1. INTRODUCTION

People are increasingly concerned by the effect of their diet on health. Excessive consumption of some dietary components (fat, sugar) may have negative effects, whereas the consumption of other components (fiber, micronutrients, etc.) is encouraged. Epidemiology studies have suggested a protective role for different constituents of our diet against diseases. The reduced prevalence of various cancers by the consumption of fruit and vegetables is now well established.[1] Other associations, such as the protective role of wine against cardiovascular diseases[2] or of soya against breast cancer[3] have been suggested. Many researchers try today to identify the molecules in food responsible for these protective effects.

Polyphenols are widespread in edible fruit and vegetables.[4,5] We ingest about 1 gram of polyphenols every day mainly from fruit and vegetables or beverages (tea, coffee, wine, cider, fruit juices, etc.).[6] A large number of experimental studies on animals and cell cultures have shown that polyphenols may well contribute to the protective effects of fruit and vegetables and derived food products against cancers and cardiovascular diseases. However, it is not possible to extrapolate these data to humans due to our poor knowledge of the bioavailability (the proportion of a nutrient capable of being absorbed and available for use and storage)[7] of polyphenols and on their concentration in our tissues. Bioavailability of polyphenols also depends on their metabolism, either by the microflora in the gut or by organs such as the liver. Because their metabolites may be more or less active

Plant Polyphenols 2: Chemistry, Biology, Pharmacology, Ecology, Edited by
Gross et al. Kluwer Academic / Plenum Publishers, New York, 1999

357

than the parent polyphenols, it is important to identify these metabolites, determine their metabolic origin, and concentration in our body.

Isotopically labelled molecules have been widely used in bioavailability studies. The use of radioactive isotopes such as ^{3}H and ^{14}C greatly facilitates the detection and positive identification of polyphenols present in trace amounts. The advantage of using radiolabelled polyphenols is well illustrated in a study of the absorption and metabolism of condensed tannins (CTs) fed to sheep.[8] Application of the classic butanol/HCl assay for CTs showed an apparent disappearance of 85 percent condensed tannins in feces suggesting that they had been absorbed. However, this was ruled out by the administration of ^{14}C—CTs, which showed that none of the radiolabelled CTs were absorbed. Using radiolabelled molecules, the whole administered dose can be traced in the body, whatever the chemical structure of the parent compound and its metabolites. The absence of other reliable and sensitive methods for CT analysis and estimation makes the use of radioisotopes particularly valuable. Polyphenols labelled with stable isotopes have also been used particularly in human studies because of their relative innocuousness. The development of mass spectrometry techniques for the analysis and estimation of the labelled compounds has reinforced the interest in stable isotope labelled molecules.

Labelled polyphenols then differ according to the nature of the isotope label, stable or radioactive, the position and number of labels in the molecule, and the proportion of labelled molecules (specific activity for radiolabelled molecules). The characteristics of labelled polyphenols, and hence labelling method, are determined by the nature of the nutrition study to be undertaken. The purpose of this paper is to briefly review applications of isotopically labelled polyphenols in nutrition studies and to describe the methods that have been used for the preparation of these labelled polyphenols.

2. ISOTOPICALLY LABELLED POLYPHENOLS IN NUTRITION STUDIES

Nutrition studies are ideally carried out on human volunteers. However, technical and safety requirements make such studies often difficult. Animal models or human cell lines can then be used. They are precious tools to study the mechanisms governing bioavailability. A few examples are given below to illustrate the requirements on labelled polyphenols.

Very few studies on polyphenol bioavailability have used radiolabelled compounds in humans. In one study, oral doses of [U-^{14}C]-(+)-catechin (2 g, 70 μCi) were administered to volunteers. This allowed identification of the main metabolites in urine.[9] The use of radioactive substances in human studies is often made difficult owing to too low sensitivity or too high a radiation hazard. A low bioavailability of the studied compound or a high body burden may require for estimation of the compound's concentration in plasma, the administration of radioactivity doses exceeding those possible either by the specific activity of the compound administered or the safety limits. Administration of a stable isotopically labelled form and determination of stable isotopic ratio by mass spectroscopy may then provide a safe alternative.[10] Apart from these considerations of sensitivity or safety, stable isotopically labelled compounds have been particularly useful in the area of pharmacology for the following applications:[11]

- *Identification of a drug and metabolites through the "twin-peak" method.* A mixture of labelled and unlabelled forms of the drug is administered, and metabolites are recognized by the presence of characteristic doublet ions in their mass spectra.

- *Measurement of amounts of a drug or its metabolite in plasma by stable isotope dilution.* When the circulating level of the drug or metabolite is at or below the microgram per liter, their estimation is greatly facilitated by the addition of molecules labelled with stable isotopes as internal standard in ten- to hundredfold excess.

- *Pharmacokinetic studies.* Labelled and unlabelled forms of a drug are coadministered by either two different routes or in two different formulations, and their concentrations in tissues are followed simultaneously by mass spectroscopy. To facilitate their detection by mass spectrometry, the labelled and unlabelled compounds should differ by two or more mass units.

So far, few authors have applied these methods to the study of polyphenol bioavailability with human volunteers. Deuterated phytoestrogenic lignans and isoflavones[12,13] and deuterated tyramines have been used as internal standards for the estimation of trace amounts of these compounds as low as 0.4 ng/ml in urine or 1 ng/100 g stool. The "twin peak" method has only been used in animal experiments for the detection of rutin[14] and genistein[15] metabolites.

Radioisotopes such as tritium and carbon-14 have commonly been used in *in vivo* studies on animals to trace polyphenols and their metabolites in the different tissues. Radiolabelled polyphenols are detected by liquid scintillation or autoradiography. Applications to the study of bioavailability of catechin,[16] quercetin,[17,18] hesperetin,[19] and diosmin[20] in rats, of proanthocyanidins in rats,[21] mice,[22] sheep,[8] and chickens[23] have been reported.

The success of such studies depends on the amount of radioactivity administered, on the extent of absorption of the labelled polyphenols, and on the size of the samples to be analysed. With rats, an average oral dose of 2–5 μCi is usually enough to detect and estimate a polyphenol and its metabolites in plasma and urine, at least for compounds largely absorbed. For compounds of moderate or low absorption, the radioactivity administered must be increased. The radioactivity administered also depends on the study objectives. The dose required to simply trace excretion routes of radiolabelled compounds will be low compared to the dose needed to isolate and identify metabolites formed from the parent polyphenol in the biological fluids.

In pharmacokinetic studies, another limiting factor is the volume of available plasma. The following experiment illustrates the requirements for the specific activity of the parent polyphenol. [U-^{14}C]-Proanthocyanidins (2 μCi, 4 mg) were given orally to mice and 50 µl blood aliquots sampled at various time points. Radioactivity levels reached a maximum 45 minutes after ingestion and then decreased to levels close to the detection limit of scintillation counters (300 dpm for 50 µl blood samples) 6 hours later.[22] The dose of proanthocyanidins given to each animal corresponds on a weight basis to doses of about 4 g for humans. It exceeds by at least one order of magnitude the daily dose typically consumed by humans. However, administration of this high dose was necessary to detect proanthocyanidins and their metabolites in the blood. This example underlines the need to obtain the highest specific activities when preparing radiolabelled polyphenols in order to work at doses as close as possible to those characteristic of our diet.

Metabolism of nutriments by the gut microflora or inner tissues and the mechanisms controlling their intestinal absorption are commonly studied with animal explants, human cell lines, and microflora grown *in vitro*. Radiolabelled compounds can be particularly useful.[24,25] Here again, experiments should preferentially be carried out at physiological concentrations. This imposes requirements on the specific activity of the parent molecules. Barnes et al. used sections of duodenum to study the intestinal uptake and biliary excretion of isoflavone estrogens.[25] They used genistein concentration of 42 mg/l close to that expected in the gut of volunteers having consumed soy products containing 80 mg genistein. This required ^{14}C-labelled genistein with specific activities of 634 µCi/mmol for identification of the main metabolites.

Higher specific activities are needed for metabolic studies of polyphenols in inner tissues due to their low plasmatic concentrations compared to those found in the gut. Barnes et al.[15] studied the metabolism of genistein by mammary epithelial cells at a 1.0 mg/l concentration (3.7 µM), close to that measured in the plasma of volunteers having ingested soy proteins containing 80 mg genistein (0.24 mg/l).[26] They were able to detect radiolabelled metabolites by mass spectroscopy by using [4-^{14}C]-genistein with specific activity of 23,000 µCi/mmol. This specific activity corresponds to a preparation in which half of the molecules contain one carbon-14 isotope.

The same considerations can be applied to *in vitro* studies of proanthocyanidin metabolism by gut flora or of absorption by intestinal epithelial cell lines. Proanthocyanidin daily ingestion can be roughly estimated as 0.5 g (1 g flavonoids consumed per day,[6] proanthocyanidin dimers accounting for half of flavonoids). This would result in a 3.4 mM concentration expressed as catechin unit equivalents in a 0.5-liter intestinal volume. One can calculate the minimal specific activity required for detecting a metabolite. It will be 3 µCi/mmol if we assume a total volume of 2 ml, a 1 percent yield from the parent compound and a detection limit of 500 dpm. Similarly, the same value of 3 µCi/mmol can be determined for an absorption experiment using an intestinal epithelial cell monolayer grown on a filter with 2-ml compartment volumes and a 1 percent absorption of the parent compound.

3. BIOLABELLING OF POLYPHENOLS

The success of a study on the bioavailability of labelled polyphenols relies on the characteristics of the labelled molecule: a) nature of the label isotope;

b) stability of the label in the molecule; c) level of isotopic enrichment (specific activity for radiolabelled compounds); d) purity of the labelled compound and e) the available quantity. All these characteristics depend on the labelling method.

Two routes lead to isotopically labelled polyphenols: the synthetic route and the biosynthetic route.[27] Both start from a preferentially cheap precursor with a simple chemical structure. The synthesis is then either carried out by the chemist in the laboratory or by a plant selected for its ability to accumulate the compound in high yield. The choice between one or the other routes depends on the required characteristics of the labelled compound (table 1).

The synthetic route gives access to isotopically labelled molecules in high yields, and the method is usually easily scaled up. Up to 100 percent isotopic enrichment can be achieved. Multistep synthetic sequences have been developed. However, the time needed to set up the synthesis is often long, and complex molecules are often beyond reach. For instance, 3-monodeutero-flavanols have been prepared in a four-step scheme starting from the parent compound, and the same method may also lead to dideuterated dimers.[27] Another synthesis of monodeuterated dimer B3 in five steps starting from (+)-catechin and dihydroquercetin has recently been published.[28] However, deuteration of oligomers of high molecular weight using similar schemes will not be easily achieved.

Deuterated polyphenols can also be prepared in one-step synthesis by exchange of aromatic hydrogens. The difficulty is to find the right experimental conditions allowing exchange of stable hydrogens without degrading the molecule. Various isoflavonoids and lignans were deuterated using PB_3 or NaOD in D_2O.[29] More recently, the same authors reported the full exchange of aromatic protons in B-ring of isoflavonoids using D_3PO_4/BF_3 as reagent.[30] This method has also been applied to the deuteration of lignans and a coumestan.[31]

The synthesis of 2',5',6'-trideuterorutin as a mixture of mono-, di- and trideuterorutin by base-catalyzed exchange in D_2O has been reported.[14] The same experimental conditions when applied to quercetin failed to label the B-ring. Rolando and his collegues found that these conditions also could not be applied to labelling of (+)-catechin because it was largely degraded.

Table 1. A comparison of chemical synthesis and biosynthesis for isotopic labelling

Methods	Advantages	Limitations
Chemical synthesis	High specific activity	Limited access to complex molecules
	Definite location of the label isotope and high isotopic purity	Long setup of synthetic schemes
Biosynthesis	Labelling of complex molecules	Limited specific activities
	No setup of synthetic scheme	Long purification process to ensure the removal of radioactive contaminants

Alternatively, the biosynthetic approach allows isotopic labelling of complex molecules as they occur in plants. No setup of a synthetic scheme is then needed. The choice of a plant material in which the biosynthesis of the compound to label is high is the primary determinant for the success of biolabelling. Other parameters such as the choice of the precursor, the administration method, and the growth conditions also influence the results of the biolabelling experiments. These factors are discussed below.

The selection of a plant material for labelling experiments depends on: a) the nature and content of polyphenols present in the tissues indicative of an active biosynthesis of the polyphenols of interest; b) the availability of the plant material throughout the year; and c) the survival duration of the plant material during labelling. A plant containing high amounts of the specific molecule is first selected. Das and Griffiths[32] selected *Uncaria gambir* for its high content in (+)-catechin (9.4 percent of leaf fresh weight). In our case, we looked for a plant suitable for labelling a series of procyanidin dimer, trimer, oligomers, and related flavanol monomer.[27,33] The male catkins of willow trees (*Salix caprea*) were reported to contain (+)-catechin, B3 dimer and C2 trimer. However, their availability is restricted to 2 to 3 weeks at the end of the winter. Furthermore, the use of catkins for radiolabelling experiments was not compatible with the confinement of radioactivity because of the easy loss of hair at maturity. We therefore selected the leaves because they are available over a longer period of the year and were found to contain similar concentrations of the same procyanidins.

It is important to examine the polyphenols present in the tissue over a whole season. Variations in the nature and content of polyphenols in leaves according to their state of maturity have commonly been reported.[34–39] Variations of the biosynthetic activity of the different polyphenols with time are thus expected. The proanthocyanidin content in leaves of *Salix caprea* increased from 0.5 to 1.1 percent dry wt. from April to July (fully matured leaves) and the incorporation of [1-^{14}C]-acetate into proanthocyanidins was highest (3 percent) in early July. Incorporation dropped to 0.05 percent in August.[33]

We could extend the period over the year when willow shoots are available for labelling experiments by growing young trees in the greenhouse. The proanthocyanidin content in the leaves of these shoots remained constant from February to August. High incorporation rates (around 3 percent) were observed from May to July. However, flavanols present in leaves were found to differ in several respects from those of adult trees grown outdoors. Leaves of outdoor adult trees were found to contain high amounts of dimer B3, trimer C2 and procyanidin polymers (total: 1.9 percent dry wt.), whereas leaves from willow plants grown in the greenhouse contained no dimer and trimer but large amounts (2.9 percent dry wt.) of procyanidin and prodelphinidin polymers. Shoots of adult trees were thus used to label mono-, di- and trimers and those of the plants grown in the greenhouse for labelling polymers.

Survival of the plant or explant during the labelling experiment is essential as the yield of incorporation of a precursor increases in proportion to the duration of labelling. Just like flowers in a bouquet, an excised shoot will resist fading differently according to species. A *Gingko biloba* shoot may resist wilting for up to a month,[40] whereas willow shoots fade 2 to 8 days after the stem has been cut. They were kept in the labelling chamber after administration of the labelled precursor until they were fully dried. Plant metabolism can be altered to enhance the biosynthesis of a particular polyphenol. Glyphosate was added to *Rhus typhina* shoots to stimulate biosynthesis of gallotannins. $^{14}CO_2$ was incorporated into labelled gallotannins in 1.5 percent yield.[41]

Apart from shoot explants, cell cultures have also been used to label polyphenols. They present several advantages: they are available all year long; a precursor is easily administered as no translocation through the plant is needed; polyphenols are often freed in the medium and recovery of the products is thus made easier; long labelling duration is often possible. Fry has administered [^{14}C]-cinnamate to spinach cells in order to label cell wall phenolic acids.[42] After only 1 min., 85 percent of the radioactivity supplied was taken up by the cells, and after 33 days, 38 percent was found in the cell wall.

So far, however, cell cultures and labelled precursors have principally been used in biosynthesis studies[43,44] and only recently to obtain labelled polyphenols, such as anthocyanins in *Vitis vinifera*.[45] The main limits of cell cultures are that they are not always easily established and maintained and that the nature and quantities of the polyphenols produced are often not predictable and different from the compounds present in the parent tissues.

Either hydrogen or carbon atoms can be labelled in polyphenols. Higher specific activities are obtained with tritium (20,980 Ci/mol) than with carbon-14 (45 Ci/mol). So far, only carbon isotopes (^{14}C or ^{13}C) have been used to prepare labelled polyphenols with plants because it is often easier to trace metabolites (including CO_2) by labelling the carbon skeleton rather than hydrogens which can be exchanged or lost.

The choice of a precursor depends on: a) the effectiveness of its transformation into the polyphenol of interest; b) its toxicity to the plant; c) the ease of handling (a chamber will be required for administration of gaseous precursors such as $^{14}CO_2$) and d) its price.

A few authors have compared incorporation yields and specific activities of labelled polyphenols obtained with different precursors. $^{14}CO_2$ was three to five times less efficient than ^{14}C-acetate to label (+)-catechin in *Uncaria gambir*.[32] $^{14}CO_2$ was also less effective than ^{14}C-phenylalanine when fed to sorghum to label condensed tannins.[23] Shikimic acid was 20 times more effective than L-phenylalanine in the labelling of gallic acid.[46]

Polyphenols are rarely labelled on their own. A varying amount of the precursor is diverted to non-phenolic compounds such as sugars.[23,47] With willow shoots, labelled sugars accounted for 4 to 80 percent of the radioactivity in the leaf extract depending on the choice of the precursor, acetate or phenylalanine, and the labelling duration that varied between two and six days.[33] Labelled sugars were two to seven times more abundant than labelled flavanols. Similar conversion of phenylalanine to sugar was demonstrated in *Eucalyptus*.[48] However, no incorporation into cell wall glycosyl residues was observed when radiolabelled cinnamate was administered to cell cultures of spinach over 11 days.[42]

Ideally, the precursor should be as close as possible to the product in the biosynthetic pathway to limit its incorporation into other metabolites. This is why phenylalanine is sometimes preferred to acetate to label polyphenols, but its use is limited by its higher cost. The use of such a precursor does not eliminate incorporation into non-phenolic products, particularly when long periods of incorporation are used. This is due to the turnover of the precursor or that of the polyphenols themselves.[48,49] A major fraction (80 percent) of ^{14}C-catechins when administered to tea shoots through their cut internodes was catabolized into CO_2 within 70 hours.[50]

When selecting a precursor, it is also essential to consider the position of the isotopic labels in the precursor molecule and to ensure isotopes will be incorporated in the final polyphenol molecule. Some hydrogen isotopes may be exchanged or lost during biosynthesis. A shift of 50 percent of the ^{3}H label from aromatic position 4 of cinnamic acid to aromatic position 3 of 4-coumaric acid occurs in the course of the hydroxylation reaction.[51] Biolabelling of flavan-3-ols using [2-^{3}H]-cinnamic acid resulted in 90 percent loss of the tritium label.[43]

Appropriate labelling of some particular carbon(s) or hydrogen(s) in a polyphenol allows one to trace them in human or animals and clarify their metabolism. Some polyphenols such as flavonoids have mixed biosynthesis with carbons originating from different biosynthetic pathways. The choice of different precursors allows different parts of the phenolic molecule to be preferentially labelled. Administration of [1-^{14}C]-acetate to leafy shoots of *Uncaria gambir* led to the exclusive labelling of the A ring in (+)-catechin, as established by alkaline degradation of the flavanoid skeleton.[32] The metabolism of this labelled (+)-catechin could be compared to that of (+)-catechin labelled in both A and B ring obtained by incorporation of [14]CO_2.

As stressed above, not only the nature of the labelled polyphenols but also the extent of labelling (specific activity for radiolabelled compounds) is important. Incorporation yields and specific activities usually increase with the total radioactivity fed with the precursor.[33,52] However, when too high an amount of precursor is applied, some toxicity to the plant may result. Feeding 1 mCi of sodium ^{14}C-acetate (100μmol; 0.1 M) to a willow shoot induced blackening of the leaf periphery, premature wilting, low absorption rates of the precursor through the stem section, and a fourfold decrease in the incorporation into proanthocyanidins.[33] Toxicity was only partly alleviated by addition of hydrochloric acid to neutralize the alkaline solution of the precursor. Toxic effects of acetate were even more drastic with buckwheat leaves rich in rutin: administration of 25μmol sodium acetate (neutralized solution) to a shoot induced leaf wilting within 24 hrs. Toxicity could also be reduced by applying successive low doses of precursor. Administration of a precursor in a few minutes or continuously over a longer period of time did not affect its incorporation into saponifiable phenols from the cell wall, such as coumarate and ferulate in spinach cell culture.[42]

Methods for feeding labelled precursors into plants should have limited effects on polyphenol metabolism, maintain the plant alive as long as possible, and direct the precursor preferentially to the tissues showing the highest polyphenol biosynthesis. Ideal methods use intact plants since polyphenol metabolism can be affected by wounding and isolation of tissues from the rest of the plant. The precursor can be given as a gas or as a solution. Gaseous precursors, either [13]CO_2[53] or [14]CO_2,[52] can be fed during several weeks or even months to obtain high incorporation yields in plants grown in pots. However, when the plant is too big to be introduced in the airtight chamber, cuttings can be used. This is suitable for short duration labelling with $^{14}CO_2$ (usually less than a week).[32,41] Precursors in solution such as phenylalanine can be fed through the roots of intact plants grown hydroponically.[54] Solutions of precursors can also be injected with a syringe in stems of leafy shoots[40] or fruit-bearing branches.[55]

However, most workers use explants, either stems bearing leaves or fruits, leaves, leaf discs or fruits. The choice of the nature of the explant is important because it effects the metabolism of the precursor into polyphenols or contaminants as well as the yield of incorporation. Incorporation of labelled phenylalanine into sugar contaminants was much lower when the precursor was fed to whole leaves rather than to discs.[48] The incorporation yield of polyphenols varies in proportion to the duration of metabolism, and thus with the survival of the explant after it has been cut away from the plant. We found it difficult to control the survival of willow explants that varied between 2 to 7 days after cutting.[33]

When selecting an explant, one has also to make sure that a sufficient fraction of the precursor effectively reaches the tissues showing active synthesis of the polyphenol(s) of interest. Solutes such as ^{13}C-phenylalanine were easily translocated through the roots of *Fagus grandifoliia*, whereas ^{13}C-ferulic acid did not reach the aerial parts of the plant.[54] When ^{14}C-phenylalanine was fed to willow cuttings, 78 to 86 percent of the radioactivity was found in the stem and did not reach the leaves.[33] Similarly, administration of ^{14}C-phenylalanine to a sorghum panicle showed that 50 percent of the radioactivity remained in the stem while only 6 percent was incorporated into the seeds.[56]

Results of biolabelling experiments carried out with different plants show wide variations in the yields of incorporation and the specific activities obtained (table 2).

As expected, the incorporation yields and specific activities vary considerably with the nature of the plant material, the precursors, and the duration of incorporation. Yields of incorporation never exceed 5 percent. A major part of the precursor is either not translocated to the right tissue (precursor brought as a solution), not assimilated (CO_2), or incorporated in sugars and other non-phenolic metabolites.

There is no link between the yields of incorporation and the specific activities of the purified products. Specific activities also depend on the dilution of the labelled product with pre-existing polyphenols. The low activities of epicatechin and dimer A2 labelled in *Aesculus hippocastaneum* are explained by a high dilution in the 42 fruits, which gave 1.3 kg of seed shells. The highest specific activities or enrichments are thus achieved with long durations of incorporation, which increase the yields of incorporation and with plant materials showing a rapid *de novo* synthesis of polyphenols as in the *Vitis vinifera* cell cultures showing a very efficient conversion of the precursor into anthocyanins (higher than 50 percent) and a low dilution with polyphenols in preexisting biomass. The specific activities obtained by the different investigators are well above some of the requirements detailed above. For higher activities, synthetic routes will be the method of choice.

Table 2. Biolabelling experiments of plant polyphenols with carbon-14 and carbon-13, yields of incorporation and specific activities of the purified products

Plant species	Plant material	Labelled precursor	Quantity administered (μCi)	Duration of incorporation (day)	Labelled polyphenols	Yield of incorporation (%)	Quantity purified (mg)	Specific activity (μCi/mmol)	Reference
Geranium endressi	leaf	D-[U-^{14}C]-glucose	5	1	^{14}C-gallic acid	0.01	20		57
Rhus typhina	shoot	$^{14}CO_2$	5,000	7	^{14}C-pentagalloylglucose			3500	41
Viola tricolor	entire plant	$^{14}CO_2$		8	^{14}C-rutin			1275	52
Aesculus hippocastaneum	branch with fruits	[1-^{14}C]-acetate	1,700	14	^{14}C-epicatechin ^{14}C-dimer A2	0.4 0.1	1500 630	1.6 1.1	55
Uncaria gambir	shoot	$^{14}CO_2$ [1-^{14}C]-acetate	1,000 500	0.5 1	^{14}C-catechin ^{14}C-catechin	1.5 5		10 35	32
Sorghum bicolor	panicle	$^{14}CO_2$	80	0.3	^{14}C-proanthocyanidins			0.78^b	23
		[U-^{14}C]-phenylalanine	5	2	^{14}C-proanthocyanidins	0.5			56
Vitis vinifera	entire plant with grapes	$^{14}CO_2$		3	^{14}C-proanthocyanidins			290^b	21
Salix caprea	shoot	[1-^{14}C]-acetate	400	3	^{14}C-catechin ^{14}C-proanthocyanidins - dimer B3 - trimer C2 - polymers	1.2 0.6 0.5 3.2	9 11 11 70	157 135 171 52^b	33
		[U-^{14}C]-phenylalanine	200	3	^{14}C-proanthocyanidins	2.7	18	87^b	
Populus euramericana x	entire plant	$^{13}CO_2$	10^a	90	^{13}C-lignins			6^a	53
Vitis vinifera	cell culture	[1-^{13}C]-phenylalanine		6	[4-^{13}C]-anthocyanins	52		60^a	45

^{a} Isotopic enrichment (%)
^{b} Specific activities of polymers calculated on a catechin unit basis (M=288)

4. CONCLUSIONS

Biolabelling polyphenols in plants remains the only method suitable to obtain labelled complex molecules such as tannins that are not easily synthesized. It is also used for the rapid preparation of labelled phenolic compounds of lower molecular weight. Yields of incorporation are maximized by choosing the right plant material (species, tissue, state of maturity), precursor, and culture conditions.

The range of possibilities offered by biolabelling with stable or radioactive isotopes differs widely due to the requirements needed for detecting the products. Detection of polyphenols labelled with stable isotopes is carried out by mass spectrometry. Detection of new metabolites by the "twin-peak" method requires high isotopic enrichments and long incorporations with costly precursors. Synthetic routes are thus usually preferred when they are available, and it is expected that biolabelling with stable isotopes will be mainly applied for labelling molecules of a too high structural complexity to be easily obtained by synthesis. Lower levels of enrichment with stable isotopes will otherwise be acceptable to determine the fate of polyphenols in our body and their transformation into previously identified metabolites.

On the other hand, the low yields of incorporation of precursors in plants, 5 percent or lower, are usually sufficient to trace radiolabelled polyphenols and their metabolites in animals or cell culture models. Specific activities of $100\,\mu$Ci/mmol or higher are commonly obtained and are suitable for the study of their intestinal absorption and excretion and for the detection of the main metabolites by liquid scintillation or autoradiography.

ACKNOWLEDGMENTS

We wish to thank Daniel Massimino and Pierre Chagvardieff (CEA, Cadarache) for helpful comments on the manuscript and the European Community for financial support (FAIR CT contract No 95-0653).

REFERENCES

1. Steinmetz, K.A.; Potter, J.D. Vegetables, fruit, and cancer prevention: a review. *J. Am. Diet. Assoc.* 96:1027 (1996).

2. St. Leger, A.S.; Cochrane, A.L.; Moore, F. Factors associated with cardiac mortality in developed countries with particular reference to the consumption of wine. *Lancet* :1017 (1979).

3. Adlercreutz, H.; Mazur, W. Phyto-oestrogens and Western diseases. *Ann. Med.* 29:95 (1997).

4. Swain, T. Economic importance of flavonoid compounds: foodstuffs. *In*: Geissman, T.A. (ed.). The chemistry of flavonoid compounds. Pergamon Press, Oxford p. 513 (1962).

5. Shahidi, F.; Naczk, M. Food phenolics, sources, chemistry, effects, applications. Technomic Publishing Co. Inc., Lancaster, (1995).

6. Kühnau, J. The flavonoids. A class of semi-essential food components: their role in human nutrition. *World Rev. Nutr. Diet.* 24:117 (1976).

7. Bender, A.E. Nutritional significance of bioavailability. *In*: Southgate, D.; Johnson, I.; Fenwik, G.R. (eds.). Nutrient availability: Chemical and biological aspects. Royal Society of Chemistry, Cambridge, p. 3 (1989).

8. Terrill, T.H.; Waghorn, G.C.; Woolley, D.J.; McNabb, W.C.; Barry, T.N. Assay and digestion of ^{14}C-labelled condensed tannins in the gastrointestinal tract of sheep. *Brit. J. Nutr.* 72:467 (1994).

9. Hackett, A.M.; Griffiths, L.A.; Broillet, A.; Wermeille, M. The metabolism and excretion of (+)-[^{14}C]cyanidol-3 in man following oral administration. *Xenobiotica* 13:279 (1983).

10. Klein, P.D.; Hachey, D.L.; Kreek, M.J.; Shoeller, D.A. Stable isotopes: essential tools in biological and medical research. *In*: Baillie, T.A. (ed.). Stable isotopes—Applications in pharmacology, toxicology and clinical research. Macmillan Press Ltd., London, p. 3 (1978).

11. Draffan, G.H. Stable isotopes in human drug metabolism studies. *In*: Baillie, T.A. (ed.). Stable isotopes—Applications in pharmacology, toxicology and clinical research. Macmillan Press Ltd., London, p. 27 (1978).

12. Adlercreutz, H.; Fotsis, T.; Bannwart, C.; Wähälä, K.; Brunow, G.; Hase, T. Isotope dilution gas chromatographic-mass spectrometric method for the determination of lignans and isoflavonoids in human urine, including identification of genistein. *Clin. Chem. Acta* 199:263 (1991).

13. Adlercreutz, H.; Fotsis, T.; Kurzer, M.S.; Wähälä, K.; Makela, T.; Hase, T. Isotope dilution gas chromatographic mass spectrometric method for the determination of unconjugated lignans and isoflavonoids in human feces, with preliminary results in omnivorous and vegetarian women. *Anal. Biochem.* 225:101 (1995).

14. Baba, S.; Furuta, T.; Fujioka, M.; Goromaru, T. Studies on drug metabolism by use of isotopes. XXVII. Urinary metabolites of rutin in rats and the role of intestinal microflora in the metabolism of rutin. *J. Pharm. Sci.* 72:1155 (1983).

15. Peterson, T.G.; Coward, L.; Kirk, M.; Falany, C.N.; Barnes, S. The role of metabolism in mammary epithelial cell growth inhibition by the isoflavones genistein and biochanin A. *Carcinogenesis* 17:1861 (1996).

16. Hackett, A.M.; Shaw, I.C.; Griffiths, L.A. 3'-O-Methyl-(+)-catechin glucuronide and 3'-O-methyl-(+)-catechin sulphate: new urinary metabolites of (+)-catechin in the rat and the marmoset. *Experientia* 38:538 (1982).

17. Petrakis, P.L.; Kallianos, A.G.; Wender, S.H.; Shetlar, M.R. Metabolic studies of quercetin labeled with C[14]. *Arch. Biochem. Biophys.* 85:264 (1959).

18. Ueno, I.; Nakano, N.; Hirono, I. Metabolic fate of [^{14}C]quercetin in the ACI rat. *Jap. J. Exp. Med.* 53:41 (1983).

19. Honohan, T.; Hale, R.L.; Brown, J.P.; Wingard, R.E. Synthesis and metabolic fate of hesperetin-3-[14]C. *J. Agric. Food Chem.* 24:906 (1976).

20. Oustrin, J.; Fauran, M.J.; Commanay, L. A pharmacokinetic study of ^{3}H-diosmine. *Arzneim. Forsch.* 27 (II):1688 (1977).

21. Harmand, M.F.; Blanquet, P. The fate of total flavanolic oligomers (OFT) extracted from *Vitis vinifera* L. in the rat. *Eur. J. Drug Metabol. Pharmacokinet.* 3:15 (1978).

22. Laparra, J.; Michaud, J.; Lesca, M.F.; Blanquet, P.; Masquelier, J. Etude pharmacocinétique des oligomères procyanidoliques totaux du raisin. *Acta Ther.* 4:233 (1978).

23. Jimenez-Ramsey, L.M.; Rogler, J.C.; Housley, T.L.; Butler, L.G.; Elkin, R.G. Absorption and distribution of ^{14}C-labelled condensed tannins and related sorghum phenolics in chickens. *J. Agric. Food Chem.* 42:963 (1994).

24. Peterson, T.G.; Coward, L.; Kirk, M.; Falany, C.N.; Barnes, S. Isoflavones and breast epithelial cell growth: the importance of genistein biochanin A metabolism in the breast. *Carcinogenesis* 17:101 (1996).

25. Sfakianos, J.; Coward, L.; Kirk, M.; Barnes, S. Intestinal uptake and biliary excretion of the isoflavone genistein in rats. *J. Nutr.* 127:1260 (1997).

26. Gooderham, M.J.; Adlercreutz, H.; Ojala, S.T.; Wähälä, K.; Holub, B.J. A soy protein isolate rich in genistein and daidzein and its effects on plasma isoflavone concentrations, platelet

aggregation, blood lipids and fatty acid composition of plasma phospholipid in normal men. *J. Nutr.* 126:2000 (1996).

27. Déprez, S.; Buffnoir, S.; Scalbert, A.; Rolando, C. Isotopic labelling of proanthocyanidins. *Analusis* 25:M43 (1997).

28. Pierre, M.C.; Chèze, C.; Vercauteren, J. Deuterium labeled procyanidin syntheses. *Tetrahedron Lett.* 38:5639 (1997).

29. Wähälä, K.; Mäkelä, T.; Bäckström, R.; Brunow, G.; Hase, T. Synthesis of the [²H]-labelled urinary lignans, enterolactone and enterodiol, and the phytoestrogen daidzein and its metabolites equol and O-demetylangolensin. *J. Chem. Soc., Perkin Trans. 1* :95 (1986).

30. Wähälä, K.; Rasku, S. Synthesis of D-4-genistein, a stable deutero labeled isoflavone, by a predeuteration—selective dedeuteration approach. *Tetrahedron. Lett.* 38:7287 (1997).

31. Mazur, W.; Fotsis, T.; Wähälä, K.; Ojala, S.; Salakka, A.; Adlercreutz, H. Isotope dilution gas chromatographic mass spectrometric method for the determination of isoflavonoids, coumestrol, and lignans in food samples. *Anal. Biochem.* 233:169 (1996).

32. Das, N.P.; Griffiths, L.A. Studies on flavonoid metabolism. Biosynthesis of (+)-[¹⁴C]catechin by the plant *Uncaria gambir* Roxb. *Biochem. J.* 105:73 (1967).

33. Déprez, S.; Scalbert, A. Carbon-14 biolabelling of (+)-catechin and proanthocyanidin oligomers in willow-tree cuttings. *J. Agric. Food Chem.* (in press) (1999).

34. Scalbert, A.; Haslam, E. Polyphenols and chemical defence of the leaves of *Quercus robur*. *Phytochemistry* 26:3191 (1987).

35. Lister, C.E.; Lancaster, J.E.; Sutton, K.H.; Walker, R.L. Developmental changes in the concentration and composition of flavanoids in skin of a red and a green apple cultivar. *J. Sci. Food Agric.* 64:155 (1994).

36. Bohm, B.A.; Singh, S.; Koupai-Abyazani, M.R. Biosynthesis and turnover of tannins in sainfoin (*Onobrychis viciifolia*). *Polyphénols Actualités* n°11:18 (1994).

37. Feeny, P.P.; Bostock, H. Seasonal changes in the tannin content of oak leaves. *Phytochemistry* 7:871 (1968).

38. Tissut, M. Etude du cycle annuel des dérivés phénoliques de la feuille de hêtre, *Fagus sylvatica* L. *Physiol. Vég.* 6:351 (1968).

39. Koupai-Abyazani, M.R.; McCallum, J.; Muir, A.D.; Böhm, B.A.; Towers, G.H.N.; Gruber, M.Y. Developmental changes in the composition of proanthocyanidins from leaves of sainfoin (*Onobrychis viciifolia* Scop.) as determined by HPLC analysis. *J. Agric. Food Chem.* 41:1066 (1993).

40. Fukushima, K. Biolabelling of polyphenols in woody plants. I. Feeding methods. *Polyphénols Actualités* n°16:11 (1997).

41. Rausch, H.; Gross, G.G. Preparation of [¹⁴C]-labelled 1,2,3,4,6-penta-O-galloyl-β-D-glucose and related gallotannins. *Z. Naturforsch.* 51:473 (1996).

42. Fry, S.C. Incorporation of ¹⁴C-cinnamate into hydrolase-resistant components of the primary cell wall of spinach. *Phytochemistry* 23:59 (1984).

43. Jacques, D.; Opie, C.T.; Porter, L.J.; Haslam, E. Plant proanthocyanidins. Part 4. Biosynthesis of procyanidins and observations on the metabolism of cyanidin in plants. *J. Chem. Soc., Perkin Trans. 1* :1637 (1977).

44. Stafford, H.A.; Shimamoto, M.; Lester, H.H. Incorporation of (¹⁴C)phenylalanine into flavan-3-ols and procyanidins in cell suspension cultures of Douglas fir. *Plant Physiol.* 69:1055 (1982).

45. Krisa, S.; Waffo Teguo, P.; Vitrac, X.; Vercauteren, J.; Deffieux, G.; Mérillon, J.M. Production de polyphénols marqués en ¹³C à l'aide de culture cellulaire. *Proc. 19th International Conference on Polyphenols*, Lille, France, p. 245 (1998).

46. Zaprometov, M.N.; Bukhlaeva, V.Y. Two pathways of biosynthesis of gallic acid. *Biochemistry* 33:317 (1968).

47. Hillis, W.E.; Hasegawa, M. The formation of polyphenols in trees-I: Administration of ^{14}C glucose and subsequent distribution of radioactivity. *Phytochemistry* 2:195 (1963).

48. Hillis, W.E.; Isoi, K. The biosynthesis of polyphenols in *Eucalyptus* species. *Phytochemistry* 4:905 (1965).

49. Ellis, B.E. Degradation of aromatic compounds in plants. *Lloydia* 37:168 (1974).

50. Zaprometov, M.N. *Dokl. Akad. Nauk SSSR* 125:1359 (1959).

51. Reed, D.J.; Vimmerstedt, J.; Jerina, D.M.; Daly, J.W. Formation of phenols from aromatic substrates by plant and animal mono-oxygenases: the effect of adjacent deuteriums on the magnitude of the NIH shift of tritium. *Arch. Biochem. Biophys.* 154:642 (1973).

52. Michaud, J.; Laparra, J.; Lesca, M.F.; Harmand, M.F.; Masquelier, J.; Blanquet, P. Phytosynthèse de rutoside marqué au [14]C. *Bull. Soc. Pharm. Bordeaux* 108:133 (1969).

53. Lapierre, C.; Gaudillère, J.P.; Monties, B.; Guittet, E.; Rolando, C.; Lallemand, J.Y. Enrichissement photosynthetique en carbone 13 de peuplier: caracterisation préliminaire par acidolyse et RMN [13]C. *Holzforschung* 37:217 (1983).

54. Lewis, N.G.; Inciong, M.E.; Razal, R.A.; Yamamoto, E.; Davin, L.B. Monolignol biogenesis, lignin structure and biodegradation. *Proc. 15émes Journées Internationales du Groupe Polyphénols*, Strasbourg, p. 365 (1990).

55. Eastmond, R.; Gardner, R.J. ^{14}C-Epicatechin and ^{14}C-procyanidins from seed shells of *Aesculus hippocastaneum*. *Phytochemistry* 13:1477 (1974).

56. Reddy, V.; Butler, L.G. Incorporation of ^{14}C from ^{14}C-phenylalanine into condensed tannin of sorghum grain. *J. Agric. Food Chem.* 37:383 (1989).

57. Bernays, E.A.; Woodhead, S. Incorporation of dietary phenols into the cuticle in the tree locust *Anacridium melanorhodon*. *J. Insect Physiol.* 28:601 (1982).

METABOLIC ACTIVITY OF URIDINE 5'-DIPHOSPHOGLUCOSE: CINNAMYL ALCOHOL GLUCOSYLTRANSFERASE AS AN INTRINSIC INDICATOR OF CAMBIAL GROWTH IN CONIFERS

Hartmut Förster,[a] Ulrich Pommer,[a] and Rodney A. Savidge[b]

[a] Faculty of Biology and Pharmacy
Friedrich-Schiller-University of Jena
99189 Erfurt-Kühnhausen
GERMANY

[b] Faculty of Forestry and Environmental Management
University of New Brunswick
Fredericton NB, E3B 6C2
CANADA

1. INTRODUCTION

In terms of both frequency and magnitude of plant biomass accumulation, the transfer of the glycosyl moiety is one of the most important biochemical reactions known. Two-thirds of the biosphere's carbon existing as carbohydrates is due to this kind of reaction.[1] Transfer of the glycosyl moiety from sugar nucleotides to a large number and molecular diversity of acceptors is catalyzed by glycosyltransferases.

Glucosyltransferases as one subclass of glycosyltransferases are involved in the glucosylation of phenolics,[2-7] steroids,[8-12] polysaccharides,[13,14] hydroxamic acids,[15] thiohydroximates,[16,17] xenobiotics[18,19] as well as cinnamates[20,21] and benzoic acids.[22-24] In secondary metabolism, glucosyltransferases occur as both soluble[5,6,8] and membrane-bound[10,13] proteins, with both forms in close proximity having been reported to catalyze glucosylation of hydroxysteroids in oat seedlings.[11] Glucosyl-transferases are eminent as constituents of protein complexes responsible for cell wall biogenesis.[25]

UDPG:coniferyl alcohol glucosyl transferase (UDPG:CAGT), an enzyme catalyzing glucosylation of the monolignol coniferyl alcohol and also acting on other cinnamyl alcohols, has been investigated with special regard to the formation of coniferin (i.e., the 4-O-β-D-glucopyranoside of *trans*-coniferyl alcohol) in trees.[26-30] Coniferin is an established lignin precursor and is also considered to be a

Plant Polyphenols 2: Chemistry, Biology, Pharmacology, Ecology, Edited by
Gross et al. Kluwer Academic / Plenum Publishers, New York, 1999

371

biochemical indicator of impending tracheid differentiation in conifers.[31,32] Parallel variation in seasonal oscillations of coniferin and cambial growth has been clearly demonstrated in a number of conifer species.[29–33] The presence and significance to lignification of glucosylated cinnamyl alcohols in the cambium of woody plants were emphasized several decades ago by Freudenberg,[34,35] and UDPG:CAGT was immunolocalized to vascular bundles as well as to the epidermal and subepidermal layers in young spruce seedlings.[28] Thus, the view that coniferin participates in lignification has merit.

Reports on monolignol glucosides have been directed toward their perceived participation in lignification,[36–39] and their possible roles in other aspects of metabolism have remained little investigated. For example, knowledge about intracellular storage and intercellular transport of monolignols and their corresponding glucosides is extremely limited. Thus, conclusions about the physiological role of monolignol glucosides tend to be narrow, insufficiently supported by objective experimentation.[40] Gross,[41] for example, clearly demonstrated the enzymatic potential of lignifying tissues to synthesize monomeric lignin units independently of cinnamyl glucosides, although others had considered the glucosides essential for lignification.[42–44] Moreover, non-lignifying phloem as well as developing xylem has been found to contain phenolic glucosides, including syringin.[45] Alternative interpretations of the physiological meaning for monolignol glucosides, beyond their perceived role as lignin precursors, have been discussed but deserve greater attention.[46]

As a tissue that can be readily obtained from quite large as well as small trees, differentiating xylem can be a rich source of enzymes catalyzing not only lignification but various other aspects of wood formation.[47,48] The cambial zone, where cell division occurs but cellular differentiation is normally absent, can also yield cells in abundance.[48] Despite these advantages, opportunities for progress in molecular physiology and plant biochemistry with the cambial meristem and its derivative cells have not yet been fully appreciated. In addition, knowledge about the regulation of seasonal cambial activity and xylogenesis at both genetic and biochemical levels remains fragmentary and needing of much more effort to explain how diameter growth occurs in woody plants.

Knowing that coniferin is not present in the cambium except during cambial growth,[31,32,48] we investigated coniferin formation, as catalyzed *in vitro* by UDPG:CAGT, in relation to the annual cycle of cambial activity and dormancy. Tissues of cambium and developing xylem were harvested from stems of *Pinus banksiana* Lamb. and *Pinus strobus* L. from March to November in order to explore the relationship between UDPG:CAGT activity and developmental stages of wood formation.

2. CATALYTIC FORMATION OF MONOLIGNOL GLUCOSIDES IN CAMBIAL ZONE AND DEVELOPING XYLEM OF *PINUS BANKSIANA* AND *PINUS STROBUS*

Figure 1 shows the theoretical succession of compounds emerging from the shikimate pathway into the stage of phenylpropanoid metabolites, monolignols

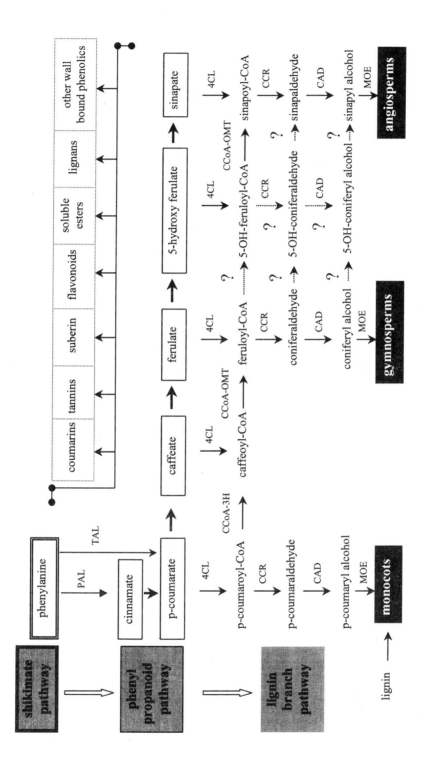

such as coniferyl alcohol, and lignins derived from those monolignols. UDP-glucose:cinnamyl alcohol glucosyltransferase catalysis takes place at the level of cinnamyl alcohols (i.e., of monolignols). Thus, the biosynthesis of coniferin requires the availability of precusors descended from the shikimate pathway (fig. 1), as well as products derived from the oxidative pentose phosphate and glycolysis pathways.

Coniferin derived from coniferyl alcohol is the predominant glucoside in gymnosperm stems, and some angiosperms accumulate either coniferin or syringin, or both, due to endogenous coniferyl and sinapyl alcohols being produced and glucosylated.[26,37,41,42,49] Lignin in monocots and in compression wood of conifers has significant amounts of *p*-coumaryl alcohol,[38,49-52] indicating the possible presence of the corresponding glucoside, however, only Freudenberg[42] has isolated that molecule to date.[53]

Interestingly, although lignin is the second most abundant organic polymer after cellulose on earth,[36,54] monolignol glucosides are frequently found only in small quantities, and significantly fewer precursor monolignols are present.[31,32,53,55,56] However, in contrast to angiosperms, conifers can accumulate coniferin in considerable amounts.[27,32,32,34,44,45,57] For example, Schmid and Griesebach[27] found coniferin from cambial sap of *Picea abies* to be at a concentration of 3.7 mmol/L. Cambium and developing xylem of *Pinus banksiana* exhibited 2 mmol/L and 10 mmol/L coniferin, respectively.[29,30]

Figure 2 shows the seasonal activity of uridine 5′-diphosphoglucose:coniferyl alcohol glucosyltransferase (UDPG:CAGT, E.C. 2.4.1.111), the enzyme catalyzing synthesis of coniferin from coniferyl alcohol and uridine 5′-diphosphoglucose) throughout an annual cycle of cambial growth and dormancy in stems of *Pinus banksiana*. Our *in vitro* UDPG:CAGT kinetic investigations were done in conjunction with microscopical investigations of the corresponding tissues.[30]

The UDPG:CAGT activity obtained from seasonal harvests of cambium and developing xylem appeared synchronously with resumption of cambial cell-division activity (fig. 2). UDPG:CAGT activity in the cambial zone appeared first on April 22, exactly when the initiation of springtime cell-division activity first became evident. UDPG:CAGT activity increased rapidly over the next 2 weeks, accompanied by repeated cell divisions in fusiform cambial cells to more than double the number of cells per radial file in the cambial zone. Following this widening of the cambial zone, the first primary walled radially enlarging (non-dividing) cells appeared on the inner margin on the cambial zone as the earliest stage in formation of earlywood tracheids. This developing xylem could readily be separated from cells of the cambial zone by bark peeling, and it yielded higher UDPG:CAGT activity than cambial cells (fig. 2).

A dramatic increase over earlier dates of UDPG:CAGT activity in both developing xylem and cambium was found on May 17 (fig. 2). In the cambial zone, UDPG:CAGT activity exceeded the April 22 rate by 15 times, and in the developing xylem, a fourfold increase had occurred. The markedly increased UDPG:CAGT activity coincided with production of an increased number of primary walled, radially enlarged cambial derivative cells and, as a new stage of wood development, with the first appearance of secondary wall formation and lignification in the

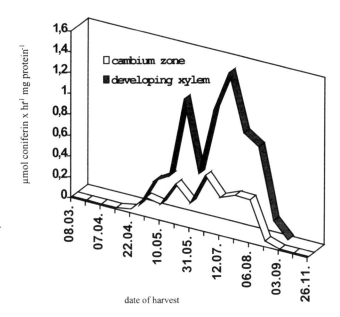

Figure 2. Seasonal activity of the UDP glucose: coniferyl alcohol glucosyltransferase in cambium and developing xylem of *Pinus banksiana*.

developing earlywood. UDPG:CAGT activity remained readily detectable and particularly strong in developing xylem throughout the summer period of cambial growth.

On August 13, microscopy indicated cell-division activity in the cambial zone had ceased or was ceasing. However, wood formation, in terms of secondary wall polysaccharide deposition, lignification, and protoplasmic autolysis, was still occurring on August 13. UDPG:CAGT activity was still detectable in both tissues at that time, but at low levels. The UDPG:CAGT activity in cambium and developing xylem fell to non-detectability after September 3, coinciding with the completion of annual ring formation and onset of cambial dormancy.

The pronounced variation in UDPG:CAGT activity with the changing seasons can be seen as an adjustable metabolic process obeying developmental requirements. Oscillations in metabolic processes are well known but nevertheless poorly understood.[45,56] The seasonal variation observed in UDPG:CAGT activity evidently is more than a mere wavelike oscillation, however. It appears to be a qualitative on-off switching of gene expression that underlies either synthesis or activation of UDPG:CAGT. Other than the recognition that it is related to the changing seasons, exactly what mechanism acts to control the switching remains unknown.

UDPG:CAGT from *Pinus* cambium and developing xylem may be involved in the regulation of lignification. Conceivably, provision of coniferyl alcohol for polymerization into lignin in walls of lignifying cells requires genetic and biochemical coordination between coniferin-hydrolyzing β-glucosidase and UDPG:CAGT activity. To date, no experimental data about seasonal behavior of β-glucosidase(s) specific for coniferin hydrolysis have been published, but past observations indicated that coniferin-specific β-glucosidase activity was weakly (if at all) present in dormant cambium and most active during cambial growth in developing xylem (R.A. Savidge and V. Leinhos, unpublished data).

The time course of the UDPG:CAGT activity over the growing season (late April to late August, fig. 2) clearly shows its dependence on the stage of cambial growth. Although UDPG:CAGT activity in developing xylem was higher than that in the cambial zone, control of UDPG:CAGT activity evidently was by the cambium. This is indicated by 1) the initiation of UDPG:CAGT activity in the cambium in early spring just when cell-division activity began and well before any developing xylem was present, 2) the continuing presence of UDPG:CAGT activity in both cambial zone and developing xylem throughout the growing season, and 3) the disappearance of UDPG:CAGT activity from the cambial zone in early autumn, in synchrony with the cessation of cell-division activity, both at a time when late-wood formation was still occurring. These findings point to the unexpected conclusion that gene expression for UDPG:coniferyl alcohol glucosyl transferase in *Pinus* species is linked to the cambium's seasonal competence for mitosis and/or cytokinesis.

Coniferin is an inert and reasonably stable water-soluble compound; therefore, it appears well suited to serve as a storage metabolite as well as a transportable molecule. However, the absence of coniferin from the cambial region during dormancy and the exclusive appearance of coniferin in the cambial region during cambial growth[29,30,31,32,33,48] suggest it is a genuine storage compound in gymnosperms. Whether high coniferin content influences equilibrium conditions regarding precursor flow from phenylalanine to lignin, as shown in figure 1, has to be clarified by further research.

3. CHARACTERIZATION OF THE UDP-GLUCOSE:CONIFERYL ALCOHOL GLUCOSYLTRANSFERASE

The coniferyl alcohol glucosyltransferase extracted from cambial zone and developing xylem of *Pinus* was partially purified by low- and high-speed centrifugation to remove cell debris, organelles, and microsomes, precipitated by saturated ammonium sulfate solution between 40 to 85 percent, and subsequently fractionated by desalting followed by gel filtration through a column of Sephadex G-25. The Sephadex G-25 step separated both salt and endogenous coniferin from the desired protein. There was a double peak of protein elution from the column; however, only the first peak exhibited distinct activity of the UDPG:CAGT.

No UDPG:CAGT activity was found in the microsomal fraction of *Pinus* cambium or developing xylem. This confirmed the soluble nature of the enzyme, in agreement with findings published earlier for UDP-glucose:coniferyl alcohol glucosyl transferases in plants.[26,27,28] Consequently, the step of removing the microsomal fraction was inserted into the purification scheme. Compared to the activity of the transferase in the crude extract, the enzyme as thus processed had reached a purification factor of 25. A further purification was carried out by preparative IEF to achieve a purification factor of 42; however, for the following described experiments, the enzyme was investigated after Sephadex G-25.

Preparative IEF was done overnight over the range from pH 4.0 to 6.5. A distinct optimum of the UDPG:CAGT activity was localized to p*I* 4.9 (fig. 3). The bulk of proteins present in the enzyme preparation were near p*I* 5.2 and showed no UDPG:CAGT activity. The highest UDPG:CAGT activity was at pH 7.6 and 40 °C, and the optimal concentration of protein was below 100 μg/assay. However, the UDPG:CAGT activity was promoted by addition of coniferin, the optimal coniferin concentration for maximal UDPG:CAGT activity being 0.1 mmol/L.

Recently, purification of specific coniferin-hydrolyzing β-glucosidases from developing xylem of jackpine[58] and lodgepole pine[59] was reported. This enzyme acts

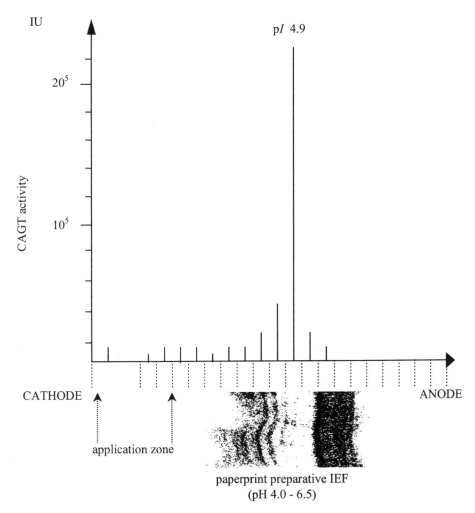

Figure 3. Preparative isoelectric focusing (IEF) of partially purified protein extracted from developing xylem of jackpine.

as the counterpart of UDPG:CAGT and is considered to be involved in the lignification of woody plants.[42,58-63] Therefore, the possibility that the UDPG:CAGT may function as a glucosidase was investigated. The pH optimum for glucosylation was 7.6 in phosphate buffer and 8.0 in TRIS/HCL-buffer, in agreement with results reported elsewhere. The coniferin specific β-glucosidase had been characterized as having a more acidic pH optimum, around 5.8.[58,59] When the enzyme was incubated with coniferin and uridine 5'-diphosphate, hydrolytic activity yielding coniferyl alcohol was about 3 percent at pH 7.6 and about 17 percent at pH 5.8, compared to UDPG:CAGT activity. It is possible, therefore, that maintenance of a coniferin pool is a function of pH; however, we did not exclude the possibility of a coniferin-specific β-glucosidase being present in the protein preparation as an impurity.

The investigated UDPG:coniferyl alcohol glucosyltransferase had a pronounced substrate specificity for UDPglucose as donor and for coniferyl alcohol as acceptor. Additionally, the enzyme preparation exhibited a high specificity toward sinapyl alcohol (table 1). The ability of the enzyme to transfer glucose to both coniferyl and sinapyl alcohols was previously reported by Freudenberg and Neish.[42] Our results, supported by combined gas chromatography-mass spectrometric (electron impact, 70 eV) analysis, confirmed formation of syringin with high efficiency when sinapyl alcohol was used as substrate. The high specificity of the UDPG:CAGT for 4-hydroxy-3-methoxyphenylpropenols and UDPG was determined by testing a number of diverse phenolic compounds such as cinnamates, phenylpropanates, and benzoic acid derivatives as well as other sugar nucleotides. Excepting sinapyl alcohol, none of the phenolics could replace coniferyl alcohol, nor could any of the sugar nucleotides substitute for UDPG at all effectively.

Following preparative electrophoresis (fig. 3), gel sectors expressing the highest UDPG:CAGT activity were pooled, transferred to Centriprep 10 (Amicon) tubes, concentrated, and again subjected to a second preparative IEF. The purity of proteins eluted from this run was assessed by native polyacrylamide gel electrophoresis (PAGE, fig. 4A) and analytical IEF (fig. 4B), revealing a singular band of UDPG:CAGT activity.

The resulting protein had a pI of 5.0 and a relative molecular weight of 43 kD, in good agreement with published data on other purified plant glucosyltransferases. Glucosyltransferases from different species have been found to be relatively conserved in terms of molecular weight and pI, being within a range of 43 to 56 kD[7,10,21] with corresponding pIs of 4.4 to 5.4.[4,6,15,16,18,20,22] Only a few exceptions regarding the optimum pH have been reported. Whereas glucosyltransferases in general show a pH optimum around 7.5, a UDPG:sinapic acid glucosyltransferase[20] and a O-glucosyltransferase specific for phenolics[6] exhibited their optimal turnover at pH 5.8 and 9.0, respectively.

As analyzed by non-denaturating polyacrylamide gel electrophoresis (native PAGE, 4–15 percent), one major band was obtained catalyzing the glucosidation of coniferyl alcohol and sinapyl alcohol (table 1, fig. 4). Following SDS-PAGE (12.5 percent homogeneous), the resulting bands were electro-transferred in Tris/glycine buffer (pH 8.3, 25 mM/192 mM, respectively) in 20 percent methanol to a

Table 1. Substrate specificity for different phenolic acceptors, cinnamyl alcohols and sugar nucleotides of the partially purified coniferyl alcohol glucosyltransferase

Substrate	Relative activity %	K_m [mM]
Sugar nucleotides		
UDP glucose	100	0.56
TDP glucose	0	
CDP glucose	0	
ADP glucose	0	
GDP glucose	0	
UDP mannose	0	
Cinnamyl alcohols		
coniferyl alcohol	100	0.83
sinapyl alcohol	96	
p-coumaryl alcohol	0	
Phenolic compounds		
p-coumaric acid	0	
cinnamic acid	0	
ferulic acid	0	
caffeic acid	0	
dihydroxyacetonphosphate	0	
salicylic acid	0	
mandelic acid	0	
vanilic acid	0	
vanillin	0	

polyvinylidene difluoride membrane, stained with 0.1 percent Coomassie Blue, destained, and subjected to N-terminal sequencing (table 2). Although IEF and native PAGE had indicated purification to homogeneity, SDS-PAGE yielded several bands, and the major one was found during sequencing to yield major and minor signals, indicating the presence of two proteins.

The major signal (65 percent of total) was new, without any evident homology to known proteins (table 2). The N-terminal of the second signal (25 percent of total) yielded 100 percent homology to the internal sequence of a precursor to a protein known as UDP glucose-starch glucosyltransferase[64] (EC 2.4.1.11) that has been isolated from *Zea mays* chloroplasts (accession no. SP P04713 in the Swiss Prot database). The obtained sequence has also been described as part of "granule-bound glycogen (starch) synthase precursor" and as amyloplast-specific transit

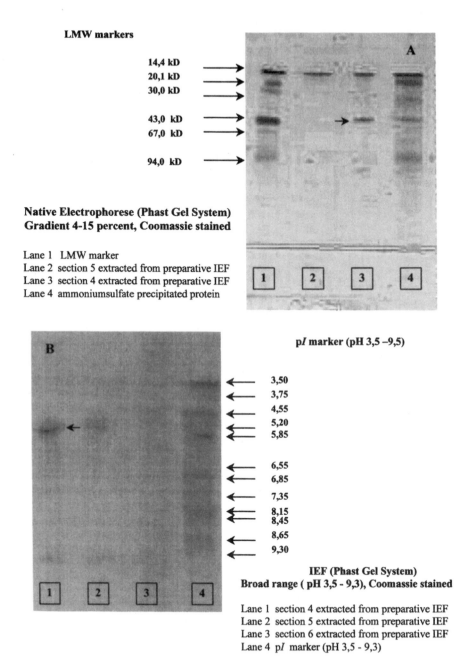

LMW markers

14,4 kD
20,1 kD
30,0 kD

43,0 kD
67,0 kD

94,0 kD

Native Electrophorese (Phast Gel System)
Gradient 4-15 percent, Coomassie stained

Lane 1 LMW marker
Lane 2 section 5 extracted from preparative IEF
Lane 3 section 4 extracted from preparative IEF
Lane 4 ammoniumsulfate precipitated protein

p*I* marker (pH 3,5 –9,5)

3,50
3,75
4,55
5,20
5,85

6,55
6,85

7,35

8,15
8,45

8,65
9,30

IEF (Phast Gel System)
Broad range (pH 3,5 - 9,3), Coomassie stained

Lane 1 section 4 extracted from preparative IEF
Lane 2 section 5 extracted from preparative IEF
Lane 3 section 6 extracted from preparative IEF
Lane 4 p*I* marker (pH 3,5 - 9,3)

Figure 4. Native PAGE gradient and analytical IEF electrophoresis of the purified protein obtained after preparative IEF.

Table 2. N-terminal sequence of the UDP-glucose:coniferyl alcohol glucosyltransferase purified from developing xylem of *Pinus strobus*

---

43 000 Dalton band (25 percent of total signal)

NH_2 – **Asn – Tyr – Gln – Ser – His – Gly – Ile – Tyr –**

43 000 Dalton band (co-migrating protein, 65 percent of total signal)

NH_2 – **X – Glu – Lys – Thr – Val – Trp – Pro – Met – Glu – Lys -**

---

protein from *Zea mays*, and has also been reported in *Triticum aestivum*, *Hordeum vulgare*, *Sorghum bicolor*, and *Musa acuminata*.

Excepting *p*-coumarate-3-hydroxylase and the monolignol glucosyltransferases, all other enzymes and their corresponding genes involved in general phenylpropanoid pathway and lignin-specific pathway have been characterized.[40,65,66] As a result of this gap in our current knowledge, full sequence information for any higher plant UDPG:cinnamyl alcohol glucosyltransferases are not yet available, hence further research remains to be done in order to get unambiguous evidence. The sequence data for our UDPG:CAGT appear to support it being similar to known glucosyltransferases; however, we have not ruled out the possibility that the dominant comigrating band, containing a novel protein, of our SDS-PAGE was the real UDPG:CAGT. Further research is obviously needed.

4. FORMATION OF CINNAMYL ALDEHYDE GLUCOSIDES AS CATALYZED BY THE PARTIALLY PURIFIED UDPG:CONIFERYL ALCOHOL GLUCOSYLTRANSFERASE

Cinnamyl aldehyde has been used to investigate substrate specifity for coniferin production as catalyzed by glucosyltransferases isolated from suspension cultures of Paul's scarlet rose[67] and from cambial sap of *Picea abies*.[27] Although the reported effectiveness, as compared to the corresponding cinnamyl alcohols, of 46 percent[67] and 63 percent[27] for coniferyl aldehyde and 48 percent for sinapyl aldehyde[27] have indicated an interesting type of catalysis, the enzymatic products generated from cinnamyl aldehydes and UDPG have never actually been confirmed.

Additionally, we confirmed that dihydroconiferyl alcohol, a monomer isolated from developing xylem of *Pinus contorta*[69] but usually not regarded as an important component of the lignin biosynthetic pathway (although accounting for around 30 percent per unit of the lignin in a mutant loblolly pine),[70] had also been converted into the 4-*O*-β-D-glucopyranoside.[68]

It has long been known that cinnamyl aldehydes must be present in lignin because the phloroglucinol-HCl lignin specific reaction depends on the presence of

Table 3. Analytical HPLC separation, spectrophotometric data and structure-confirming results[68] of cinnamyl alcohols, aldehydes and corresponding glucosides

compound	Retention time (R_t) [min]	UV maximum (UV_{max}) [nm]	compound confirmed by
coniferyl alcohol	28.1	268	^{1}H-NMR, COSY, GC-MS
coniferyl aldehyde	33.7	344	^{1}H-NMR, COSY
coniferin	21.3	260	^{1}H-NMR, COSY, GC-MS
coniferyl aldehyde glucoside	24.1	332	^{1}H-NMR, COSY
sinapyl alcohol	29.6	276	^{1}H-NMR, COSY, GC-MS
sinapyl aldehyde	34.8	348	^{1}H-NMR, COSY
syringin	23.4	268	^{1}H-NMR, COSY, GC-MS
sinapyl aldehyde glucoside	26.4	320	^{1}H-NMR, COSY

aldehyde groups.[41,42] Over the last decade, investigation of lignin structure has provided new insights due to more sensible and reliable methods.[71,72] There can be no doubt that in addition to the three main constituents forming lignin, viz. p-coumaryl, coniferyl and sinapyl alcohols, the cinnamyl aldehydes serve as normal structural elements and should not be regarded as abnormal units within the lignin biopolymer.[46,70,71]

The ongoing revolution in molecular biology has opened up paramount opportunities in genetic engineering of the lignin manipulation's underlying modification of wood.[47,73] Besides O-methyltransferases,[47,74,75] much attention has been paid to other presumably key enzymes of lignification such as hydroxycin-namate:coenzyme A ligase (4CL),[76,77] cinnamyl alcohol dehydrogenase (CAD)[75] as well as monolignol oxydizing enzymes.[78] By down-regulating CAD,[79,80,81] increased incorporation of cinnamyl aldehyde units into lignin of transgenic plants has been observed. The lignin composition of a mutant loblolly pine deficient in CAD[82] showed dramatic modifications, including increased coniferylaldehyde units, indicating the inherent opportunity for the polymer's modulation.

Matsui et al.[83] demonstrated the conversion of coniferin to syringin via their monolignols[84] in shoots of *Magnolia kobus*. The question of where in the general phenylpropanoid and lignin-specific pathway the conversion from guiacyl to syringyl units occurs thus remains uncertain. Plausibly, interconversions also occur between the corresponding glucosides of cinnamyl aldehydes to influence the composition of lignin and/or other metabolic proceedings. These possibilities remain to be explored.

5. INVOLVEMENT OF THE LIGNIFICATION PATHWAY'S CONCOMITANT GLUCOSIDES IN METABOLIC EVENTS OF PLANTS

Cinnamyl alcohol glucosides have been frequently isolated from numerous plant families,[85] implicating the genetic expression of the corresponding glucosyl-transferases. Cell suspensions of *Linum flavum* can accumulate endogenous coniferin up to 12.4 percent on a dry-weight basis.[86] Addition of coniferin stimulated the production of podophyllotoxin, a cytotoxic lignan occurring in several plant species, more than fivefold compared to coniferyl alcohol.[87] Furthermore, coniferin as well as syringin could be isolated in ample amounts from *Paulownia tomentosa* bark[88] and cambial tissues of various angiosperm woods.[89] Publications about novel phenolic glucosides[90] in plants, including the very unusual compound syringin-4-*O*-β-glucoside discovered recently,[91] continue to appear. Nevertheless, monolignols are not abundant in their free form in lignifying tissues, and the corresponding glucosides have been considered to be limited to gymnosperms and lower developed angiosperms.[92] However, it seems obvious that these compounds are part of the tight regulation of lignin biosynthesis since they arise from and must influence the functioning of the shikimate, phenylpropanoid, and lignin-branch pathways.

Coniferin derived from the cambial region of *Tsuga heterophylla* was considered to be a precursor of lignans.[93] Interestingly, bark of beech (*Fagus grandiflora*) contains only Z-monolignols and their glucosides, not the corresponding E-isomers abundant in conifers and normally present in other plant species.[53,94,95] Consequently, the glucosyltransferase probed from *Fagus* bark exhibited a marked substrate specificity for Z and not E monomers.[95] In contrast, the exclusive association of E-monolignols (or corresponding glucosides) with lignification[96] must be the metabolic scheme intrinsic to production of beechwood lignin,[97] as there has been no evidence for incorporation of Z-constituents reported.

Since lignin and its derivatives share common pathways with a wide variety of secondary metabolites such as flavonoids, suberin, coumarins, stilbenes, and lignans,[55] joint points of contact between the various branches must occur. Coniferin is known to be an inhibitor of both pine CAD[50,98] and 4CL,[99] actions that must influence the substrate flux from phenylalanine to lignin. In addition, coniferin is described as a inducer of *vir*-gene inducing activities in *Agrobacterium tumefaciens* causing crown gall disease,[100,101,102] whereas coniferyl alcohol and a related β-galactopyranoside had negligible inducer activities.[102] Acetosyringone, coumaryl, and sinapyl alcohol are well known *vir*-inducers in dicots, but in monocots their presence and characteristics are not well established. Yet recently, *vir*-gene inducers of *Agrobacterium* have been isolated from *Dendrobium* orchids.[103] Coniferyl alcohol as well as "protocorm-like bodies (PLBs)" being identified as an aryl beta-glucoside similar to coniferin increase the inducer production significantly.

The concentration of Z-coniferin was lowered after the combined attack by beech scale (*Cryptococcus fagisuda*) and the bark canker fungus (*Nectria coccinea*) causing beech bark disease. In contrast, the concentration of *Z-iso*-coniferin and Z-syringin remained unchanged.[104] Likewise, an accumulation of lignans and a decrease of cinnamyl alcohol glucosides were observed after hatching of gall larvae in *Picea glauca*.[105] The cell division promoting activity of dihydrodiconiferyl alcohol glucosides initiating rapidly growing crown gall tumors in *Vinca rosea* is also note-worthy.[106,107] In conclusion, the participation of glucosides derived from the lignin-branch pathway in defense mechanisms of plants is evident. In this context, the immediate formation of lignin or suberin after pathogen attack[108] is of interest, since the afflicted surface not uncommonly is situated near the cambial zone, and the responding tissues can be sites of cambial activity.

6. CONCLUSIONS

Since Ibrahim and Grisebach[67] first reported the isolation and characterization of a UDPG:coniferyl alcohol glucosyltransferase in 1976, scarce attention has been paid to the enzyme's occurrence, participation, and presumable physiological role in primary and secondary metabolism of plants. The resultant glucosides of the cinnamyl alcohols have merely been regarded as "storage reserves" in support of lignification. Nevertheless, cinnamyl glucosides as well as their glucosyltrans-ferase have been found throughout the plant kingdom[85] including bryophytes,[26,109] which generally are regarded to have no lignin, although Savidge[48] has called this perception into question. In addition, coniferin has found to be effective in inhibiting cinnamyl alcohol dehydrogenase, an enzyme active in non-lignifying tissues[98,110] as well as developing xylem[79] of conifers.

The glucosides of cinnamyl alcohols apparently have limited distribution in vascular plants, being present in conifers and mainly restricted to the *Magnoli-aceae* and *Oleaceae* families in angiosperms.[60,111] If the monolignol glucosides con-tribute merely to lignification, one might expect a more general occurrence in woody plants. This problem has always been explained in terms of glucosidases, such as the coniferin specific β-glucosidase, being more or less efficient in cat-alyzing cinnamyl alcohol glucosides in different species. However, so far no con-vincing evidence has been provided in support of this assumption. On the contrary, β-glucosidase isolated from lignifying tissues has been found only in low abun-dance,[43,58,59] and so far there has been little research addressing the presumed reg-ulation of glucoside levels by glucosyltransferase/β-glucosidase.

UDPG:CAGT activity disappeared when the cambial zone entered dormancy in late August, prior to completion of lignification in last differentiating latewood tracheids in early September, but in springtime UDPG:CAGT activity could be readily detected well before lignification commenced (fig. 2).[30] Typically, a full

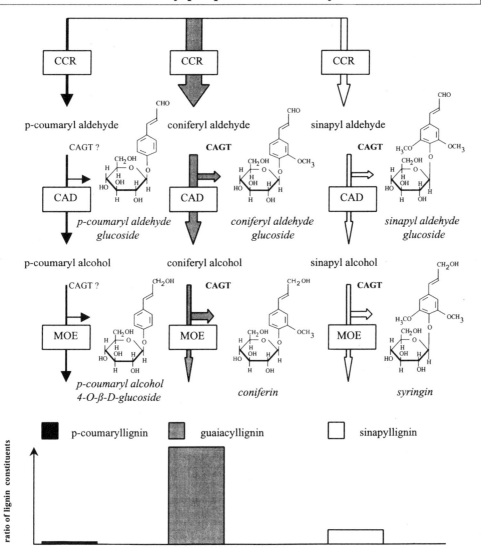

Figure 5. Phenolic glucosides arise during the lignin-specific pathways in conifers UDPG:CAGT, in catalyzing formation of cinnamyl alcohol glucosides, evidently competes with CAD and oxidases/peroxidases for coniferyl alcohol, the former by reverse activity generating the corresponding aldehyde, and the latter enabling the monolignol to be polymerized into lignin.

month elapses from when springtime cytokinetic activity resumes to when initiation of lignification becomes evident.[48] Being cambial produced before the onset of lignification and vanishing when cambial growth enters dormancy, it follows that UDPG:CAGT activity is not specifically linked to lignification but rather specifically associated with the annual cycle of growth and dormancy that occurs in perennial plants.

The participation of isotopically labelled monolignol glucosides in lignification has been demonstrated by Terashima et al.[112,113] Nevertheless, it is clear that the polymerization mechanism finally leading to lignin requires that the cinnamyl alcohol (or aldehyde) bears a free hydroxyl group.[30,41] When para position of the aromatic ring is shielded by a sugar moiety, the quinone methide radical cannot be produced, hence the monolignol is not susceptible to oxidative polymerization. Thus, UDPG:CAGT activity, in conferring the glucose to the corresponding monolignol, prevents rather than facilitates lignin synthesis. This may be part of the biochemical and genetic mechanism for maintaining the cambial zone as a meristem, preventing it from differentiating into lignified tissue.

Our data indicate that UDPG:CAGT must also compete with CAD for available cinnamyl aldehydes, forming the corresponding aldehydic glucosides. As illustrated in figure 5, the competition for substrate that exists between UDPG:CAGT and other enzymes in the lignification pathway very probably diminishes precursor flow into lignification, which could well be an important factor bearing on the normally slow rate at which lignification proceeds during xylem cell differentiation. For this reason, we consider that both glucosyltransferases and β-glucosidases are suitable target enzymes when the aim is to manipulate lignin content and composition. To our current knowledge, there are no single tissues or certain species reported where the entire pathway of lignification has been characterized biochemically and/or genetically.[114] Recent results[83,84] indicated that hydroxylation/methoxylation controlling the guaiacyl/syringyl ratio may occur further downstream in the lignin branch pathway than previously thought,[38,40] and the participation of monolignols and their glucosides in modulating the ratio deserves consideration.

ACKNOWLEDGMENTS

This research was supported by the Ministry of Science, Research and Culture of Thuringia. R.A.S. acknowledges NSERC research fundings. The authors wish to thank Valerie Steeves and Larry Calhoun (University of New Brunswick, Department of Chemistry) for providing structure data of cinnamyl aldehyde glucosides by ¹H-NMR and COSY (table 3). We want to acknowledge the skillful technical assistance of Claire Deslongchamps.

REFERENCES

1. Sinott, M.L. Catalytic mechanism of enzymic glycosyl transfer. *Chem. Rev.* 90:1171 (1990).

2. Cheng, G.W.; Malencik, D.A.; Breen, P.J. UDP-glucose:flavonoid *O*-glucosyl-transferase from strawberry fruit. *Phytochemistry* 35:1435 (1994).

3. Ishikura, N.; Mato, M. Partial purification and some properties of flavonol 3-*O*-glucosyltransferases from seedlings of *Vigna mungo*, with special reference to the formation of kaempferol 3-*O*-galactoside and 3-*O*-glucoside. *Plant Cell Physiol.* 34:329 (1993).

4. Ishikura, N.; Yang, Z-Q. Multiple forms of flavonol *O*-glucosyltransferases in young leaves of *Euonymus alatus* F. *ciliato-dentatus*. *Phytochemistry* 36:1139 (1994).

5. Lutterbach, R.; Ruyter, C.M.; Stöckigt, J. Isolation and characterization of an UDPG-dependent glucosyltransferase activity from *Rauwolfia serpentina* Benth. cells suspension cultures. *Can. J. Chem.* 72:51 (1993).

6. Parry, A.D.; Edwards, R. Characterization of *O*-glucosyltransferases with activities toward phenolic substrates in alfalfa. *Phytochemistry* 37:655 (1994).

7. Vellekoop, P.; Lugones, L.; van Brederode, J. Purification of an UDP-glucose: flavone, 7-*O*-glucosyltransferase, from *Silene latifolia* using a specific interaction between the enzyme and phenyl-sepharose. *FEBS Letters* 330:36 (1993).

8. Kreis, W.; May, U., Reinhard, E. UDP-glucose:digitoxin 16'-*O*-glucosyltransferase from suspension-cultured *Digitalis lanata* cells. *Plant Cell Reports* 5:442 (1986).

9. Ullmann, P.; Ury, A.; Rimmele, D.; Benveniste, P.; Bouvier-Navé, P. UDP-glucose sterol β-D-glucosyltransferase, a plasma membrane-bound enzyme of plants: enzymatic properties and lipid dependence. *Biochimie* 75:713 (1993).

10. Warnecke, D.C.; Heinz, E. Purification of a membrane-bound UDP-glucose:sterol β-D-glucosyltransferase based on ist solubility in diethyl ether. *Plant Physiol.* 105:1067 (1994).

11. Kalinowska, M. Glucosylation of C_{19}- and C_{21}-hydroxysteroids by soluble and membrane-bound glucosyltransferase from oat (*Avena sativa*) seedlings. *Phytochemistry* 36:617 (1994).

12. Paczkowski, C.; Wojciechowski, Z.A. Glucosylation and galactosylation of diosgenin and solasodine by soluble glucosyltransferase(s) from *Solanum melongena* leaves *Phytochemistry* 35:1429 (1994).

13. White, A.R.; Xin, Y.; Pezeshk, V. Xyloglucan glucosyltransferase in golgi membranes from *Pisum sativum* (pea). *Biochem. J.* 294:231 (1993).

14. Xin, Y.; White, A.R.; Pezeshk, V. Solubilization and partial purification of a xyloglucan glucosyltransferase from pea golgi membranes. *Plant Phys.* 102:90 (1993).

15. Leighton, V.; Niemeyer, H.M.; Jonsson, L.M.V. Substrate specifity of a glucosyl-transferase and an N-hydroxylase involved in the biosynthesis of cyclic hydroxamic acids in gramineae. *Phytochemistry* 36:887 (1994).

16. Reed, D.W.; Davin, L.; Jain, J.C.; Deluca, V.; Nelson, L.; Underhill, E.W. Purification and properties of UDP-glucose:thiohydroximate glucosyltransferase from *Brassica napus* L. seedlings. *Arch. Biochem. Biophys.* 305:526 (1993).

17. Grootwassink, J.W.D.; Reed, D.W.; Kolenovsky, A.D. Immunopurification and immunocharacterization of the glucosinolate biosynthetic enzyme thiohydroximate S-glucosyltransferase. *Plant Physiol.* 105:425 (1994).

18. Wetzel, A.; Sandermann Jr., H. Plant biochemistry of xenobiotics: isolation and characterization of a soybean *O*-glucosyltransferase of DDT metabolism. *Arch. Biochem. Biophys.* 314:323 (1994).

19. Gallandt, E.R.; Balke, N.E. Purification of soybean UDP-glucose:6-hydroxybentazon glucosyltransferase. *Plant Physiol.* 102:48 (1993).

20. Nurmann, G.; Strack, D. Formation of 1-sinapoylglucose by UDP-glucose:sinapic acid glucosyltransferase from cotyledons of *Raphanus sativus*. *Z. Pflanzenphysiol.* 102:11 (1981).

21. Shimizu, T.; Kojima, M. Partial purification and characterization of UDPG:t-cinnamate glucosyltransferase in the root of sweet potato, *Ipomoea batatas* Lam. *J. Biochem.* 95:205 (1984).

22. Yalpani, N.; Schulz, M.; Davis, M.P.; Balke, N.E. Partial purification and properties of an inducible uridine 5'-diphosphate-glucose:salicylic acid glucosyltransferase from oat roots. *Plant Physiol.* 100:457 (1992).

23. Yalpani, N.; Balke, N.E.; Schulz, M. Induction of UDP-glucose:salicylic acid glucosyltrans-ferase in oat roots. *Plant Physiol.* 100:1114 (1992).

24. Enyedi, A.J.; Raskin, I. Induction of UDP-glucose:salicylic acid glucosyltransferase activity in tobacco mosaic virus-inoculated tobacco (*Nicotiana tabacum*) leaves. *Plant Physiol.* 101:1375 (1993).

25. Northcote, D.H. Control of plant cell wall biogenesis. *In:* Lewis, N.G.; Paice, M.G. (eds.). ACS Symposium Series 399, Plant cell wall polymers—biogenesis and biodegradation. American Chemical Society, Washington, DC, p. 1 (1989).

26. Ibrahim, R.K. Glucosylation of lignin precursors by uridine diphosphate glucose: coniferyl alcohol glucosyltransferase in higher plants. *Z. Pflanzenphysiol.* 85:253 (1977).

27. Schmid, G.; Grisebach, H. Enzymic synthesis of lignin precursors. Purification and prop-erties of UDPglucose:coniferyl-alcohol glucosyltransferase from cambial sap of spruce (*Picea abies* L.). *Eur. J. Biochem.* 123:363 (1982).

28. Schmid, G.; Hammer, D.K.; Ritterbusch, A.; Grisebach, H. Appearance and immuno-histochemical localization of UDP-glucose:coniferyl alcohol glucosyltransferase in spruce (*Picea abies* (L.) Karst.) seedlings. *Planta* 156:207 (1982).

29. Förster, H.; Savidge, R.A. Characterization and seasonal activity of UDPG: coniferyl alcohol glucosyl transferase linked to cambial growth in jackpine. Proc. Pl. Gr. Reg. Soc. Am., 22nd annual meeting, 402 (1995).

30. Savidge, R.A.; Förster, H. Seasonal activity of uridine 5′-diphosphoglucose: coniferyl alcohol glucosyltransferase in relation to cambial growth and dormancy in conifers. *Can. J. Bot.* 76:486 (1998).

31. Savidge, R.A. A biochemical indicator of commitment to tracheid differentiation in *Pinus contorta. Can. J. Bot.* 66:2009 (1988).

32. Savidge, R.A. Coniferin, a biochemical indicator of commitment to tracheid differentiation in conifers. *Can. J. Bot.* 67:2663 (1989).

33. Förster, H.; Savidge, R.A. Seasonal occurence of coniferin in the cambium of *Pinus banksiana* Lamb. linked to enzymatic activity of UDPG: coniferyl alcohol glucosyl trans-ferase. *In:* Vercauteren, J.; Chéze, C.; Dumon, M.J.; Weber, J.F. (eds.). Polyphenols Com-munications 96, p. 527 (1996).

34. Freudenberg, K.; Harkin, J.M. The glucosides of cambial sap of spruce. *Phytochemistry* 2:189 (1963).

35. Freudenberg, K.; Torres-Serres, J. Umwandlung des phenylalanins in lignin-bildende glu-coside. *Liebigs Ann. Chem.* 703:225 (1967).

36. Grisebach, H. Biochemistry of lignification. *Naturwissenschaften* 64:619 (1977).

37. Grisebach, H. Lignins. *In:* Conn, E.E. (ed.). The biochemistry of plants, Vol 7, Academic Press, New York, p. 457 (1981).

38. Boudet, A.M.; Grima-Pettenati, J. Lignin genetic engineering. *Molecular Breeding* 2:25 (1996)

39. Barceló, A.R. Lignification in plant cell walls. *In:* Jeon, K.W. (ed.). International review of cytology, Academic Press, Inc., San Diego, London, p. 87 (1997).

40. Boudet, A.M.; Goffner, D.P.; Grima-Pettenati, J. Lignins and lignification: recent biochem-ical and biotechnological developments. C.R. Acad. Sci. Paris, Life sciences, plant biology and pathology 319:317 (1996).

41. Gross, G.G. The biochemistry of lignification. *In:* Woolhouse, H.W. (ed.). Advances in bio-chemical research. Academic Press, New York, p. 25 (1980).

42. Freudenberg, K.; Neish, A.C. Constitution and biosynthesis of lignin. *Mol. Biol. Biochem. Biophys. Series* 2:1 (1968).

43. Sarkanen, K.V. Precursors and their polymerisation. *In:* Sarkanen, K.V.; Ludwig, C.H. (eds.). Lignins. Wiley-Interscience, New York, p. 95 (1971).

44. Marcinowski, S.; Grisebach, H. Turnover of coniferin in pine seedlings. *Phytochemistry* 16:1665 (1977).

45. Terazawa, M.; Miyake, M. Phenolic compounds in living tissue of woods II. Seasonal variations of phenolic glycosides in the cambial sap of woods. *Mokuzai Gakkaishi* 30:329 (1984).

46. Barber, M.S.; Mitchell, H.J. Regulation of phenylpropanoid metabolism in relation to lignin biosynthesis in plants. *In:* Jeon, K.W. (ed.). International review of cytology. Academic Press, Inc., San Diego, London, p. 243 (1997).

47. Sederoff, R.; Campbell, M.; O'Malley, D.; Whetten, R. Genetic regulation of lignin biosynthesis and the potential modification of wood by genetic engineering in loblolly pine. *In:* Ellis, B.E.; Kuroki, G.W.; Stafford, M.A. (eds.) Recent advances in phytochemistry. Vol. 28, Genetic engineering of plant secondary metabolism, Plenum Press, New York—London, p. 313 (1994).

48. Savidge, R.A. Xylogenesis, genetic and environmental regulation—a review. *IAWA Journal* 17:269 (1996).

49. Shimada, M. Biochemical studies on bamboo lignin and methoxylation in hardwood and softwood lignins. *Wood Res.* 53:19 (1972).

50. Campbell, M.M. The biochemistry and molecular biology of lignification: problems, progress and prospects. *In:* Scalbert, A. (ed.) Polyphenolic phenomena, INRA Editions Paris 1993, p. 99 (1993).

51. Campbell, M.M.; Ellis, B.E. Fungal elicitor-mediated responses in pine cell cultures I. Induction of phenylpropanoid metabolism. *Planta* 186:409 (1992).

52. Higuchi, T. Biosynthesis of lignin. *In:* Higuchi, T. (ed.) Biosynthesis and biodegradation of wood components. Academic Press, New York, p. 141 (1985).

53. Lewis, N.G.; Inciong, M.E.J.; Dhara, K.P.; Yamamoto, E. High-performance liquid chromatographic separation of E- and Z-monolignols and their glucosides. *Journal of Chromatography* 479:345 (1989).

54. Brown, A. Review of lignin in biomass. *J. Appl. Biochem.* 7:371 (1985).

55. Sederoff, R.; Chang, H-M. Lignin biosynthesis. *In:* Lewin, M.; Goldstein, I.S. (eds.) Wood structure and composition. Marcel Dekker, Inc., New York, Basel, Hong Kong, p. 263 (1991).

56. Savidge, R.A. Seasonal cambial activity in *Larix lariciana* saplings in relation to endogenous indol-3-ylacetic acid, sucrose and coniferin. *Forest Sci.* 37:953 (1991).

57. Leinhos, V.; Savidge, R.A. Isolation of protoplasts from developing xylem of *Pinus banksiana* and *Pinus strobus*. *Can. J. For. Res.* 23:343 (1993).

58. Leinhos, V.; Udagama-Randeniya, P.V.; Savidge, R.A. Purification of an acidic coniferin-hydrolysing β-glucosidase from developing xylem of *Pinus banksiana*. *Phytochemistry* 37:311 (1994).

59. Dharmawardhana, D.P.; Ellis, B.E.; Carlson, J.E. A β-glucosidase from lodgepole pine xylem specific for the lignin precursor coniferin. *Plant Physiol.* 107:331 (1995).

60. Marcinowski, S.; Grisebach, H. Enzymology of lignification. Cell-wall-bound β-glucosidase for coniferin from spruce *(Picea abies)* seedlings. *Eur. J. Biochem.* 87:37 (1978).

61. Burmeister, G.; Hösel, W. Immunohistochemical localisation of β-glucosidases in lignin and isoflavone metabolism in *Cicer arietinum* L. seedlings. *Planta* 152:578 (1981).

62. Hösel, W.; Todenhagen, R. Characterization of a β-glucosidase from *Glycine max* which hydrolyses coniferin and syringin. *Phytochemistry* 19:1349 (1980).

63. Hösel, W.; Fiedler-Preiss, A.; Borgmann, E., Relationship of coniferin β-glucosidase to lignification in various plant cell suspension cultures. *Plant Cell Org. Cult.* 1:137 (1982).

64. Klösgen, R.B.; Gierl, A.; Schwarz-Sommer, Z.; Saedler, H. Molecular analysis of the waxy locus of *Zea mays*. *Mol. Gen. Genet.* 203:237 (1986).

65. Whetten, R.; Sederoff, R.R. Lignin biosynthesis. *The Plant Cell* 7:1001 (1995).

66. Boudet, A.M.; Lapierre, C.; Grima-Pettenati, J. Biochemistry and molecular biology of lignification. *New Phytol.* 129:203 (1995).

67. Ibrahim, R.K.; Grisebach, H. Purification and properties of UDP-glucose:coniferyl alcohol glucosyltransferase from suspension cultures of Paul's scarlet rose. *Arch. Biochem. Biophys.* 176:700 (1976).

68. Förster, H.; Steeves, V.; Calhoun, L.; Savidge, R.A. (unpublished results, manuscript in preparation).

69. Savidge, R.A. Dihydroconiferyl alcohol in developing xylem of *Pinus contorta. Phytochemistry* 26:93 (1987).

70. Ralph, J.; MacKay, J.J.; Hatfield, R.D.; O'Malley, D.M.; Whetten, R.W.; Sederoff, R.R. Abnormal lignin in a loblolly pine mutant. *Science* 277:235 (1997).

71. Monties, B. Recent advances on lignin inhomogeneity. *In*: Van Sumere, C.F.; Lea, P.J. (eds.). The biochemistry of plant phenolics. Annual proceedings of the Phytochemical Society of Europe. Vol. 25, Academic Press, London, p. 161 (1985).

72. Lapierre, C. Applications of new methods for the investigation of lignin structure. *In*: Jung, H.G.; Buxton, D.R.; Hatfield, R.D.; Ralph, J. (eds.). Forage cell wall structure and digestibility. American Society of Agronomy, Madison, WI, p. 133 (1993).

73. Timmis, R.; Trotter, P.C. Biological and economic feasibility of genetically engineered trees for lignin properties and carbon allocation. *In*: Dhawan, V. (ed.). Applications of biotechnology in forestry and agriculture. Plenum Press, New York, London, p. 349 (1989).

74. Dixon, R.A.; Maxwell, C.A.; Ni, W.; Oommen, A.; Paiva, N.L. Genetic manipulation of lignin and phenylpropanoid compounds involved in interactions with microorganisms. *In*: Ellis, B.E.; Kuroki, G.W.; Stafford, M.A. (eds.). Recent advances in phytochemistry. Vol. 28, Genetic engineering of plant secondary metabolism. Plenum Press, New York—London, p. 153 (1994).

75. Atanassova, R.; Favet, N.; Martz, F.; Chabbert, B.; Tollier, M-T.; Monties, B.; Fritig, B.; Legrand, M. Altered lignin composition in transgenic tobacco expressing *O*-methyltransferase sequences in sense and antisense orientation. *The Plant Journal* 8:465 (1995).

76. Kajita, S.; Katayama, Y.; Omori, S. Alterations in the biosynthesis of lignin in transgenic plants with chimeric genes for 4-coumarate:coenzyme A ligase. *Plant Cell Physiol.* 37:957 (1996).

77. Lee, D.; Meyer, K.; Chapple, C.; Douglas, C.J. Antisense suppression of 4-coumarate:coenzyme A ligase activity in *Arabidopsis* leads to altered lignin subunit composition. *The Plant Cell* 9:1985 (1997).

78. Savidge, R.A.; Udagama-Randeniya, P.V.; Xu, Y.; Leinhos, V.; Förster, H. Coniferyl alcohol oxidase: a new enzyme involved with lignification. *In*: Lewis, N.G.; Sarkanen, S. (eds.). ACS Symposium Series 697, Lignin and lignan biosynthesis. American Chemical Society, Washington, DC, p. 109 (1998)

79. Baucher, M.; Chabbert, B.; Pilate, G.; Van Doorsselaere, J.; Tollier, M-T.; Petit-Conil, M.; Cornu, D.; Monties, B.; Van Montagu, M.; Inzè, D.; Jouanin, L.; Boerjan, W. Red xylem and higher lignin extractability by down-regulating a cinnamyl alcohol dehydrogenase in poplar. *Plant Physiol.* 112:1479 (1996).

80. Stewart, D.; Yahiaoui, N.; McDougall, G.J.; Myton, K.; Marque, C.; Boudet, A.M.; Haigh, J. Fourier-transform infrared and Raman spectroscopic evidence for the incorporation of cinnamaldehydes into the lignin of transgenic tobacco (*Nicotiana tabacum* L.) plants with reduced expression of cinnamyl alcohol dehydrogenase. *Planta* 201:311 (1997).

81. Hibino, T.; Takabe, K.; Kawazu, T.; Shibata, D.; Higuchi, T. Increase of cinnamaldehyde groups in lignin of transgenic tobacco plants carrying an antisense gene for cinnamyl alcohol dehydrogenase. *Biosci. Biotech. Biochem.* 59:929 (1995).

82. Mackay, J.J.; O'Malley, D.M.; Presnell, T.; Booker, F.L.; Campbell, M.M.; Whetten, R.W.; Sederoff, R.R. Inheritance, gene expression, and lignin characterization in a mutant pine deficient in cinnamyl alcohol dehydrogenase. *Proc. Natl. Acad. Sci. USA* 94:8255 (1997).

83. Matsui, N.; Fukushima, K.; Yasuda, S.; Terashima, N. On the behavior of monolignol glucosides in lignin biosynthesis. III. Synthesis of variously labeled coniferin and incorporation of the label into syringin in the shoot of *Magnolia kobus*. *Holzforschung* 50:408 (1996).

84. Matsui, N.; Fukushima, K.; Yasuda, S. On the behavior of monolignol glucosides in lignin biosynthesis. IV. Incorporation of the aglycons of 4-*O*-β-D-glucosides of caffeyl alcohol and 5-hydroxyconiferyl alcohol into shoots of *Magnolia kobus*. *Mokuzai Gakkaishi* 43:663 (1997).

85. Kremers, R.E. The ocurrence of lignin precursors. *TAPPI* 40:262 (1957).

86. Van Uden, W.; Pras, N.; Batterman, S.; Visser, J.F.; Malingré, T.M. The accumulation and isolation of coniferin from a high-producing cell suspension of *Linum flavum* L. *Planta* 183:25 (1990).

87. Woerdenbag, H.J.; Van Uden, W.; Frijlink, H.W.; Lerk, C.F.; Pras, N.; Malingrè, T.M. Increased podophyllotoxin production in *Podophyllum hexandrum* cell suspension cultures after feeding coniferyl alcohol as a β-cyclodextrin complex. *Plant Cell Rep.* 9:97 (1990).

88. Sticher, O.; Lahloub, M.F. Phenolic glycosides of *Paulownia tomentosa* bark. *Planta Medica* 46:145 (1982).

89. Terazawa, M.; Okuyama, H.; Miyake, M. Isolation of coniferin and syringin from the cambial tissue and inner-bark of some angiospermous woods. *Mokuzai Gakkaishi* 30:409 (1984).

90. Sugiyama, M.; Nagayama, E.; Kikuchi, M. Lignan and phenylpropanoid glycosides from *Osmanthus asiaticus*. *Phytochemistry* 33:1215 (1993).

91. Park, H-J. Syringin 4-O-β-glucoside, a new phenylpropanoid glycoside, and costunolide, a nitric oxide synthase inhibitor, from the stem bark of *Magnolia sieboldii*. *J. Nat. Prod.* 59:1128 (1996).

92. Gross, G.G. Biosynthesis and metabolism of phenolic acids and monolignols. *In:* Higuchi, T. (ed.). Biosynthesis and biodegradation of wood components. Academic Press, New York, p. 229 (1985).

93. Krahmer, R.L.; Hemingway, R.W.; Hillis, W.E. The cellular distribution of lignans in *Tsuga heterophylla* wood. *Wood Science and Technology* 4:122 (1970).

94. Lewis, N.G.; Inciong, M.E.J.; Ohashi, H.; Towers, G.H.N.; Yamamoto, E. Exclusive accumulation of Z-isomers of monolignols and their glucosides in bark of *Fagus grandifolia*. *Phytochemistry* 27:2119 (1988).

95. Yamamoto, E.; Inciong, M.E.J.; Davin, L.B.; Lewis, N.G. Formation of cis-coniferin in cell-free extracts of *Fagus grandifolia* Ehrh bark. *Plant Physiol.* 94:209 (1990).

96. Eberhardt, T.L.; Bernards, M.A.; He, L.; Davin, L.B.; Wooten, J.B.; Lewis, N.G. Lignification in cell suspension cultures of *Pinus taeda*. In situ characterization of a gymnosperm lignin. *The Journal of Biological Chemistry* 268:21088 (1993).

97. Nimz, H. Das Lignin der Buche—Entwurf eines konstitutionsschemas. *Angew. Chem.* 86:336 (1974).

98. O'Malley, D.M.; Porter, S.; Sederoff, R.R. Purification, characterization, and cloning of cinnamyl alcohol dehydrogenase in loblolly pine (*Pinus taeda* L.). *Plant Physiol.* 98:1364 (1992).

99. Voo, K.S.; Whetten, R.W.; O'Malley, D.M.; Sederoff, R.R. 4-coumarate-coenzyme—A ligase from loblolly pine xylem—isolation, characterization, and complementary—DNA cloning. *Plant Physiol.* 108:85 (1995).

100. Morris, J.W.; Morris, R.O. Identification of an *Agrobacterium tumefaciens* virulence gene inducer from the pinaceous gymnosperm *Pseudotsuga menziesii*. Proc. Natl. Acad. Sci. USA 87:3614 (1990).

101. Castle, L.A.; Smith, K.D.; Morris, R.O. Cloning and sequencing of an *Agrobacterium tumefaciens* beta-glucosidase gene involved in modifying a *vir*-inducing plant signal molecule. *J. Bacteriol.* 174:1478 (1992).

102. Delay, D.; Dyé, F.; Wisniewski, J-P.; Delmotte, F. Synthesis and *Agrobacterium* inducing activities of coniferyl alcohol β-glucosides. *Phytochemistry* 36:289 (1994).

103. Nan, G.L.; Tang, C.S.; Kuehnle, A.R.; Kado, C.I. *Dendrobium* orchids contain an inducer of *Agrobacterium* virulence genes. *Physiological and Molecular Plant Pathology* 51:391 (1997).

104. Dübeler, A.; Voltmer, G.; Gora, V.; Lunderstädt, J.; Zeeck, A. Phenols from *Fagus sylvatica* and their role in defence against *Cryptococcus fagisuda*. *Phytochemistry* 45:51 (1997).

105. Kraus, C.; Spiteller, G. Comparison of phenolic compounds from galls and shoots of *Picea glauca*. *Phytochemistry* 44:59 (1997).

106. Binns, A.N.; Chen, R.H.; Wood, H.N.; Lynn, D.G. Cell division promoting activity of naturally ocurring dehydrodiconiferyl glucosides: do cell wall components control cell division? Proc. Natl. Acad. Sci. USA 84:980 (1987).

107. Lynn, D.G.; Chen, R.H.; Manning, K.S.; Wood, H.N. The structural characterization of endogenous factors from *Vinca rosea* crown gall tumors that promote cell division of tobacco cells. *Proc. Natl. Acad. Sci. USA* 84:615 (1987).

108. Pearce, R.B. Cell wall alterations and antimicrobial defense in perennial plants. *In*: Lewis, N.G.; Paice, M.G. (eds.). ACS Symposium Series 399, Plant cell wall polymers—biogenesis and biodegradation. American Chemical Society, Washington, DC, p. 346 (1989).

109. Erickson, M.; Miksche, G.E. On the occurrence of lignin or polyphenols in some mosses and liverworts. *Phytochemistry* 13:2295 (1974).

110. Bergmann, F.; Vornam, B.; Hosius, B. Coniferyl alcohol dehydrogenase, a multifunctional isozyme-gene-system in Norway spruce, affects the *Armillaria* resistance of young trees. *Silvae Genetica* 45:256 (1996).

111. Yamamoto, E.; Bokelman, G.H.; Lewis, N.G. Phenylpropanoid metabolism in cell walls. *In*: Lewis, N.G.; Paice, M.G. (eds.). ACS Symposium Series 399, Plant cell wall polymers—biogenesis and biodegradation. American Chemical Society, Washington, DC, p. 68 (1989).

112. Terashima, N.; Fukushima, K.; He, L.; Takabe, K. Comprehensive model of the lignified plant cell wall. *In*: Jung, H.G.; Buxton, R.D.; Hatfield, R.D.; Ralph, J. (eds.). Forage cell wall structure and digestibility. *Am. Soc. Agr.*, Madison, WI, USA, p. 247 (1993).

113. Terashima, N.; Atalla, R.H.; Ralph, S.A.; Landucci, L.L.; Lapierre, C.; Monties, B. New preparations of lignin polymer models under conditions that approximate cell wall lignification. I. Synthesis of novel lignin polymer models and their structural characterization by ^{13}C NMR. *Holzforschung* 49:521 (1995).

114. Campbell, M.M.; Sederoff, R.R. Variation in lignin content and composition. Mechanisms of control and implications for the genetic improvement of plants. *Plant Physiol.* 110:3 (1996).

LOCALIZATION OF DIRIGENT PROTEIN INVOLVED IN LIGNAN BIOSYNTHESIS: IMPLICATIONS FOR LIGNIFICATION AT THE TISSUE AND SUBCELLULAR LEVEL

Mi Kwon, Vincent Burlat, Laurence B. Davin,
and Norman G. Lewis

Institute of Biological Chemistry
Washington State University
Pullman, Washington 99164-6340
USA

1. INTRODUCTION

Lignans and lignins are both derived from the phenylpropanoid pathway and are formed through phenolic oxidative coupling processes.[1] However, the metabolic fates of both classes of natural products are quite distinct, since they have different physiological roles during plant growth and development. Lignins have several functions in their role as integral cell wall components. These include mechanical support, as well as imparting properties to tracheary elements enabling them to be suitable for long-distance water conduction. Roles have been also proposed in plant defense via formation of "impenetrable" barriers, these being generated following attack, for example, by encroaching pathogens and herbivores.

Lignins are major components of cell walls where they are distributed mainly throughout the secondary xylem.[2] Lignans, on the other hand, often accumulate in the heartwood of many woody species,[3,4] although they are also present in various organs including leaves, fruits, seeds, roots, and petioles.[5-7] Unlike the lignin biopolymers, lignans are usually dimeric phenylpropanoids although higher oligomeric forms also exist.[8] Because of their antioxidant, antifungal, antibacterial, cytotoxic, and antiviral properties, their roles appear to be primarily in plant defense.[8,9]

Lignin macromolecules are formed by the dehydrogenative polymerization of the three monolignols, E-p-coumaryl (1), E-coniferyl (2), and E-sinapyl (3) alcohols,

Plant Polyphenols 2: Chemistry, Biology, Pharmacology, Ecology, Edited by
Gross et al. Kluwer Academic / Plenum Publishers, New York, 1999

which differ only in the degree of methoxylation of the aromatic ring.[2,10] Dehydrogenative polymerization occurs in such a manner as to give rise, in the lignin biopolymers, to a finite variety of substitution patterns and intermolecular linkages,[2,9] which are only now becoming fully established, e.g., the recent discovery of 5–5' dibenzodioxocin substructure (4) in native lignin (fig. 1).[11]

In early models of lignin biosynthesis, it was envisaged that monolignols were transported into the cell wall where they would then undergo one electron-oxidation, this being catalyzed by putative oxidases and peroxidases to generate the corresponding free-radical forms.[10,12] Based on comparisons of artificial 'lignin' preparations obtained *in vitro* by the action of, for example, peroxidases and H_2O_2, it was initially considered that lignin biopolymer assembly occurred randomly.[10,12] However, this hypothesis could not explain various features of lignin biosynthesis and lignification *in vivo*. For example, the random coupling model based on *in vitro* studies using monolignols such as *E*-coniferyl alcohol (2) gave preparations with very high frequencies of 8–5' and 8–8' linkages, whereas the 8–*O*–4' linked moieties were present in very low amounts.[13] On the other hand, the frequency of the 8–*O*–4' intermolecular linkage in natural lignins is estimated to be very high (50–70 percent). By contrast, the dehydrodiconiferyl alcohol (8) (8–5' linkage, 9–12 percent) and other substructures, such as 5–5' dibenzodioxocin (4) (18–20 percent) are present at lower frequencies.[2,11,14] In addition, the random coupling hypothesis did not explain the formation of optically active lignans. For example, (+)-pinoresinol (5a), (–)-secoisolariciresinol (6), and (–)-matairesinol (7) are found in *Forsythia intermedia* (fig. 1).[1,8] Moreover, lignans are considered to mainly arise from stereoselective bimolecular phenoxy radical coupling of *E*-coniferyl alcohol (2). These are often linked through 8–8' bonds between the two phenylpropanoid units; however, other inter-unit linkages also exist, e.g., 8–5', 8–*O*–4', and 8–1'.

Significantly, lignin contents and monomeric compositions as well as their intermolecular linkage patterns vary predictably between different tissue types, cell wall layers, and species.[15-19] For example, in gymnosperms, lignins are mainly derived from coniferyl alcohol (2) together with a small amount of *p*-coumaryl alcohol (1).[2] However, lignins in angiosperms are typically derived from coniferyl alcohol (2) and sinapyl alcohol (3) moieties in roughly equal proportion.[2] Additionally, in the angiosperms, e.g., birch (*Betula verrucosa*), sinapyl moieties are mainly present in the secondary cell wall of fibers, whereas coniferyl alcohol (2) subunits are found in the middle lamella of vessels.[17] Specific lignin patterning, in terms of monomers, has also been noted in specialized reinforced tissues, called compression wood, which is characterized by a higher lignin content in its cell wall and a preferential increase in its *p*-coumaryl alcohol (1) amount.[20] Thus, more and more evidence is gathering that reveals that lignin formation is under full biochemical control, in harmony with formation of all other known biopolymers.[9]

Although the phenylpropanoid pathway leading to formation of three monolignols is relatively well defined, the subsequent and obligatory monolignol coupling and post-coupling modifications in both lignin and lignan biosynthesis are

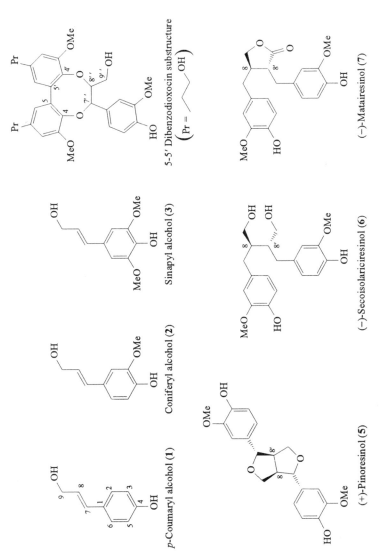

Figure 1. The three monolignols, *E-p*-coumaryl alcohol (**1**), *E*-coniferyl alcohol (**2**), and *E*-sinapyl alcohol (**3**), the 5–5′ dibenzodioxocin substructure (**4**) and examples of lignans present in *Forsythia intermedia*, (+)-pinoresinol (**5a**), (–)-secoisolariciresinol (**6**), and (–)-matairesinol (**7**).

not well understood. There are several enzymes proposed as possible candidates in lignin polymerization, i.e., H_2O_2-dependent peroxidases,[10,12,21-23] laccases,[10,12,24-27] peroxidases and laccases,[10,12,21,28] coniferyl alcohol oxidases,[29,30] (poly)phenol oxidases,[31,32] and cytochrome oxidases,[33] etc. However, there has been no direct demonstration showing an essential involvement for any of these enzymes in lignin biosynthesis. Instead, correlation with lignification was largely dependent on their capabilities to oxidize monolignols *in vitro* and/or to have a presumed association with vascular tissues.

Recently, the initial step in the 8–8' linked lignan biosynthetic pathway was clarified in *F. intermedia*.[1] This involved stereoselective monolignol coupling of two molecules of coniferyl alcohol (**2**) in the presence of a 78 kDa dirigent protein and a one-electron oxidase (such as laccase). The one-electron oxidant is considered only to provide oxidative capacity, with the dirigent protein binding, orientating, and coupling the free-radical forms and releasing (+)-pinoresinol (**5a**). The dirigent protein was purified from *F. intermedia* stem tissue and its encoding gene cloned, the analysis of which reveals no homology to any protein of known function.[34] The native protein (~78 kDa) is highly glycosylated and apparently exists as a homotrimer of ~27 kDa subunits. The gene encoding the dirigent protein, however, only codes an ~18 kDa polypeptide with the remainder of the 27 kDa subunit being added via glycosylation.

Its overall mode of action can be described as follows: When coniferyl alcohol (**2**) is incubated with laccase alone, its one-electron oxidation gives free-radical intermediates that couple to afford only racemic products such as (±)-dehydrodiconiferyl alcohols (**8**), (±)-pinoresinols (**5**), and (±)-*erythro*/*threo* guaiacylglycerol 8–*O*–4' coniferyl alcohol ethers (**9**) (fig. 2). When the dirigent protein and laccase are combined together in judicious amounts, however, only (+)-pinoresinol (**5a**) formation is engendered (fig. 2). Indeed, the dirigent protein is the only one of its type found thus far in nature capable of engendering stereoselective coupling of two phenolic molecules.[1] Significantly, the corresponding free-radical intermediates of *p*-coumaryl (**1**) and sinapyl (**3**) alcohols did not serve as substrates for the dirigent protein, although they only differ in the degree of the methoxylation on the aromatic ring;[1] this suggests that the binding site is able to distinguish between the various monolignol-derived radicals.

The dirigent protein cDNA (psd-Fi1) was transferred to an eukaryotic expression system, i.e., *Spodoptera frugiperda*/baculovirus: this produced a functional recombinant dirigent protein,[34,35] thereby demonstrating the authenticity of the gene cloned. The cDNA (psd-Fi1) was next used as a probe to screen several other species that accumulate lignans in specific tissue types, e.g., western-red-cedar (*Thuja plicata*), western hemlock (*Tsuga heterophylla*), and flax (*Linum usitatissimum*). This resulted in the isolation of several dirigent protein homologs.[34,35] Indeed, comparable dirigent proteins were also obtained from various other species (e.g., *Eucommia ulmoides*) suggesting that dirigent proteins may be ubiquitous in plant species. In addition, since different plant species can produce various lignans with coupling modes other than 8–8', this suggests that a family of dirigent proteins exists that can catalyze distinct modes of coupling in different species.

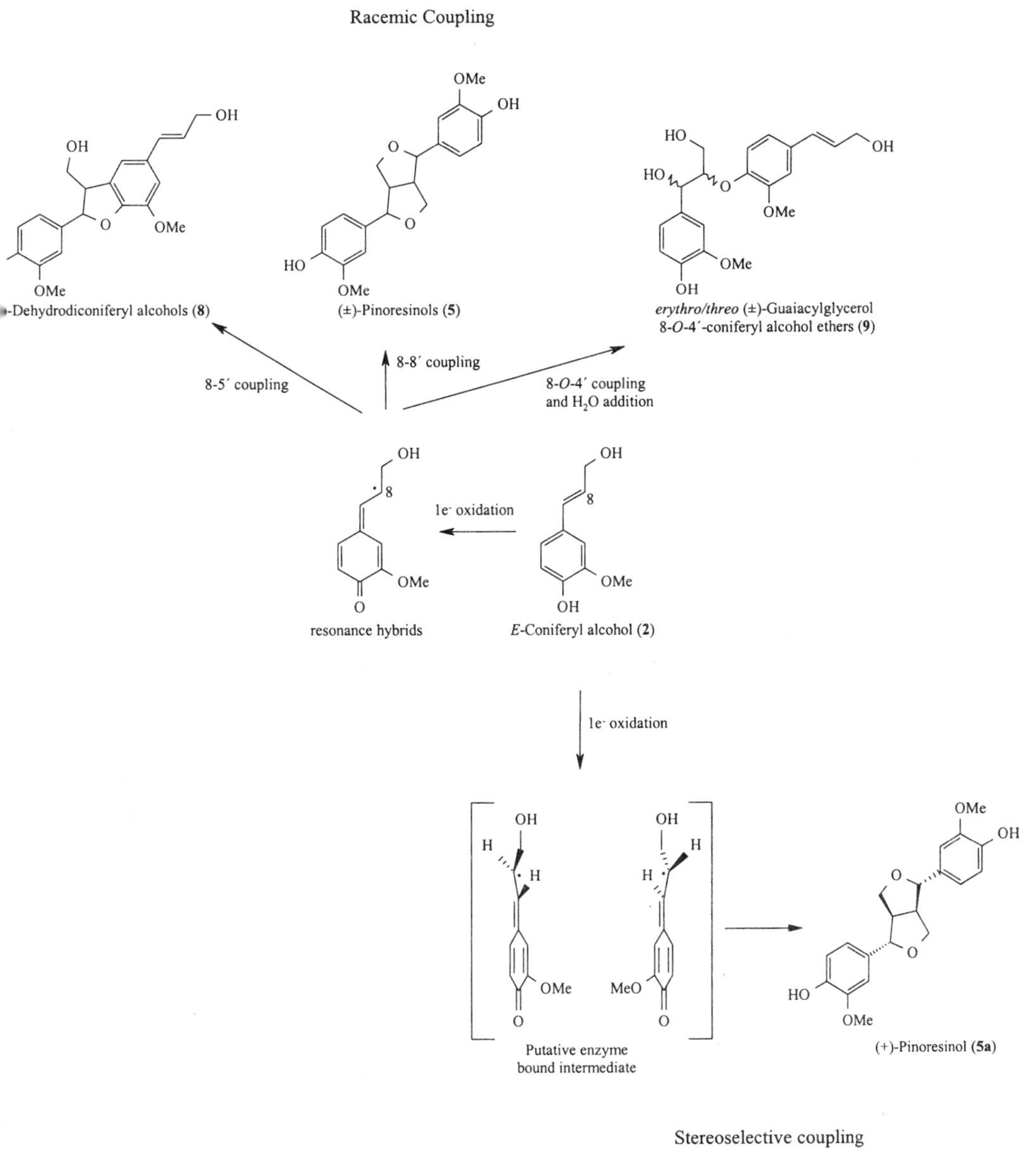

Figure 2. Racemic coupling versus stereoselective coupling. Random: *E*-coniferyl alcohol (**2**) molecules are converted to (±)-dehydrodiconiferyl alcohols (**8**), (±)-pinoresinols (**5**), and (±)-*erythro/threo* guaiacylglycerol 8–*O*–4′ coniferyl alcohol ethers (**9**) in the presence of an oxidant, e.g., laccase. Stereoselective: in the presence of both oxidant and dirigent protein, stereoselective coupling occurs to give rise only to (+)-pinoresinol (**5a**) in *F. intermedia*. (Redrawn from Davin et al.[1])

2. TISSUE-SPECIFIC DIRIGENT PROTEIN mRNA EXPRESSION IN *F. INTERMEDIA* STUDIED BY TISSUE PRINT AND *IN SITU* HYBRIDIZATION

To study the tissue specificity of dirigent protein gene expression in *F. intermedia*, total RNA (20 μg) was isolated from various parts of the plant. However, it was of interest to first establish the presumed sites of lignin deposition for comparative purposes. This was achieved by staining, with phloroglucinol-HCl, various stem internode sections as well as the roots and petioles. As can be seen (fig. 3a–3h), secondary xylem development and lignin deposition were clearly visualized following formation of the second stem internode (fig. 3c, 3d), this becoming more intense with internode maturation (fig. 3e–3h). Additionally, partial cross-sections of phloroglucinol–HCl stained petioles (fig. 3j) and roots (fig. 3l) also revealed lignin deposition in the xylem tissue of the vascular bundle and vascular cylinder, respectively.

Next, RNA gel blot analysis of these different tissues (fig. 4) showed high dirigent protein mRNA expression in the nearly unlignified meristematic region (first internode) of the stem, although it was still strongly expressed in all internodes undergoing lignification and secondary xylem development. On the other hand, root and petiole tissues displayed a lower expression level, whereas the leaf tissue gave a prominent signal.

These results were confirmed further by the use of the tissue print hybridization technique,[36] which also afforded a quantitative estimation of relative dirigent protein mRNA expression in the different stem internode, petiole, and root tissues. Thus, as can be seen in figure 5, the labeling with the dirigent protein antisense probes on tissue prints was highest in the first and second internodes in the vascular cambial region, whereas the levels steadily declined with increasing maturation and formation of the fully lignified secondary xylem. The negative control experiments using the sense probe, on the other hand, gave essentially no specific signals. Thus, dirigent protein mRNA expression was associated with the differentiating vascular cambial region of the stems. In addition to the vascular cambium, however, the epidermis of the young stem (first and second internode) also showed low levels of dirigent protein mRNA expression. As can be seen in figure 5, dirigent protein mRNA was also associated with the meristematic pericycle regions in the roots and vascular tissues of petioles. As before, the sense

→

Figure 3. Comparison of morphology and presumed lignin deposition in different developmental stages of stem, petioles, and roots in *F. intermedia*. Panels (a), (c), (e), (g), (i), and (k): partial cross-sections of first, second, fifth, and tenth internodes as well as petioles and roots; panels (b), (d), (f), (h), (j), and (l): visualization of lignin in vascular elements by staining with phloroglucinol-HCl. Thick sections (8 μm) obtained from paraffin embedded blocks were stained by methylene blue for visualization of morphology. For lignin staining, sections were stained with phloroglucinol-HCl after removing paraffin in Hemo-D. Abbreviations: co, cortex; vc, vascular cambium; p, pith; dx, developing xylem; sx, secondary xylem; pf, phloem fiber; x, xylem; ca, cambium; en, endodermis; pe, pericycle. Bar = 50 μm.

Morphology Lignin localization

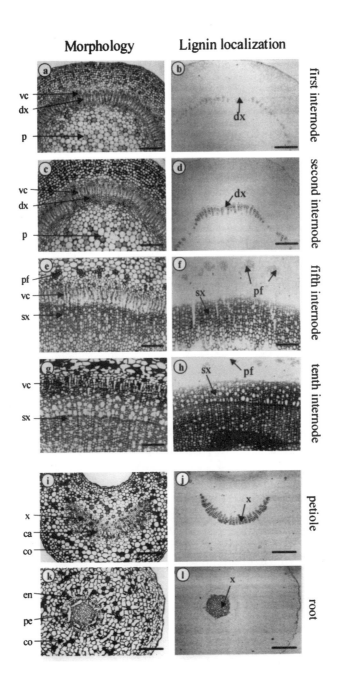

first internode

second internode

fifth internode

tenth internode

petiole

root

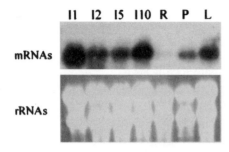

Figure 4. RNA gel blot analysis of transcripts of dirigent protein in *F. intermedia*. I1, I2, I5, and I10 correspond to the first, second, fifth, and tenth internode of the stem, respectively. R, P, and L correspond to the root, petiole, and leaf, respectively. Each lane contains 20 µg of total RNA.

(control) showed no labeling. Thus, in every instance, dirigent protein mRNA expression was observed predominantly in meristematic regions, i.e., cambial regions in vascular tissues.

Although the tissue print hybridization technique established specific dirigent protein mRNA expression in the vascular cambial regions, it did not, however, permit identification of the exact cell types involved due to the low resolving power of this technique. Accordingly, it was instructive to conduct *in situ* hybridization experiments in order to attempt to identify the precise cellular localization of dirigent protein mRNA expression in the samples studied. The approach used for *in situ* hybridization utilized digoxigenin (digoxigenin-11-uridine-5'-triphosphate) labeled antisense riboprobes (which specifically recognize dirigent protein mRNA) and sense riboprobes (as negative control) in both stem second internode and petiole sections. The second stem internode was selected for study on the basis that development of secondary xylem had been initiated as clearly revealed through lignin histochemical staining (fig. 3d). In every case, visualization of hybridized probes was achieved through the employment of alkaline phosphatase (AP) conjugated anti-digoxigenin (anti-dig) antibodies: 5-bromo-4-chloro-3-indolyl-phosphate (BCIP) and nitroblue tetrazolium choride (NBT) when hydrolyzed by the reporter enzyme (AP) gave characteristic purple chromophores (fig. 6a, 6d, 6g) in the vascular regious. For control purposes, complementary sense probes were synthesized, but these gave no specific signals since they did not hybridize with mRNA of interest (fig. 6b, 6e, 6h). Note that dirigent protein mRNA was mainly associated with the vascular cambial region, and ray cells (fig. 6a, 6d). However, tracheary elements actively expanding and differentiating also expressed the dirigent protein mRNA in both stem (fig. 6a, 6d) and weakly in petiole vascular cambium. (fig. 6g) Expression was also observed in differentiating tracheary elements (mainly vessels) of the secondary xylem. On the other hand, no evidence of expression was obtained with fully developed tracheary elements of the secondary xylem, this being expected since these cells were now presumably dead. As previously noted with tissue print hybridization, the dirigent protein mRNA expression was mainly in the vascular cambial and ray cells adjacent to differentiating tracheary elements in developing secondary xylem; a much lower level of expression was also observed in both the epidermis and pith parenchyma (fig. 6a, 6d) as well as leaves (fig. 4 lane L).

The expression of dirigent protein mRNAs in the ray cells, epidermis (younger stems) and leaves could be associated with the involvement of the dirigent protein

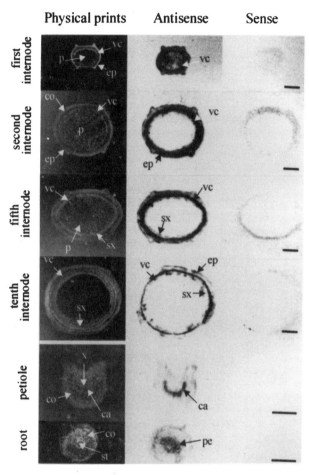

Figure 5. Localization of dirigent protein mRNA in *F. intermedia* by tissue print hybridization. Physical prints for anatomical details were obtained by pressing freshly cut cross-sections onto sobo glue (polyvinyl acetate) coated microscopic slides. The resulting tissues were next printed onto a nylon membrane (gene screen) with blots hybridized either with antisense or sense ^{35}S-labeled probes; they were obtained from the dirigent protein cDNA (psd-Fi1) inserted into the pBluescript vector containing T7/T3 RNA polymerase promoters. For the antisense RNA probe, cDNA was cleaved using BamH1 and transcribed by T7 RNA polymerase. For the sense probe, cDNA was linearized by Xho1 and transcribed by T3 RNA polymerase. Hybridization was performed overnight at 68 °C with a final wash in $0.1 \times$ SSPE / 1% SDS for 3 hr at 68 °C. The autoradiogram was obtained following 5 days exposure at room temperature using Kodak scientific imaging film. Abbreviations: vc, vascular cambium; p, pith; ep, epidermis; co, cortex; sx, secondary xylem; x, xylem; ca, cambium; st, stele; pe, pericycle. Bar = 1.0 mm.

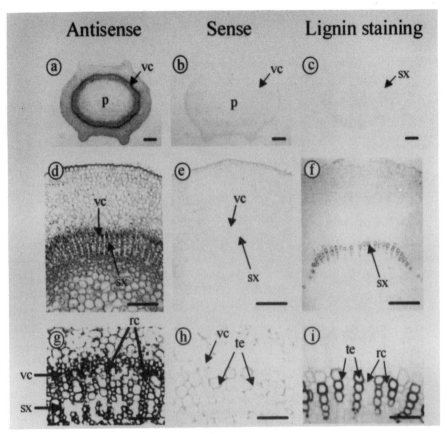

Figure 6. *In situ* hybridization of dirigent protein mRNA with digoxigenin-labeled riboprobes, (a), (b), (d), (e), (g) and (h) and patterns of lignin deposition in both tissues (c), (f), and (i). Panels (a), (d), and (g); sections were labeled with antisense probes. Panels in second stem internodes, (b), of *F. intermedia*, (e) and (h); sections were labeled with the sense probes as negative control. Visualization used an alkaline phosphatase conjugated anti-digoxigenin antibodies, with BCIP and NBT as chromophore substrates without counter-staining. Panels (c), (f), and (i) revealed regions of presumed lignin deposition through staining with phloroglucinol-HCl. Sections (8 μm thick) were prepared on silane (3-amino-propylsilane) treated microscopic slides from paraffin-embedded blocks fixed in FAA (50 percent formalin, 5 percent glacial acetic acid and 10 percent alcohol) solution. The sections were digested with proteinase K (2 μg/ml) in TE buffer (50 mM Tris, 5 mM EDTA, pH 7.4) for 20 min at 37 °C, washed twice in TE buffer (50 mM Tris, 5 mM EDTA, pH 7.4) and DEPC (diethyl pyrocarbonate)-treated water and then subjected to *in situ* hybridization. Incorporation of digoxigenin and specific activity of probes were evaluated as described by Pechoux and Morel.[37] Digoxigenin-labeled riboprobes were hydrolyzed into 150–200 bp fragments. Hybridization was carried out in a moist chamber at 37 °C overnight with a final probe concentration of 0.1 μg/ml. Immunodetection was carried out as described by Bochenek and Hirsch.[38]

Abbreviations: vc, vascular cambium; sx, secondary xylem; p, pith; te, tracheary elements; rc, ray cells. Scale bar = 100 μm (panels a, b and c), 25 μm (d), (e) and (f) and 10 μm (g), (h), and (i).

in lignan biosynthesis in agreement with the occurrence of lignans and other extractives in, for example, the ray cells of heartwood.[39,40] On the other hand, the dirigent protein mRNA detection in the cambial cells and actively expanding, differentiating, tracheary elements might be more closely correlated to a possible role in lignin biosynthesis. If this proves correct, these data may be consistent with the known formation and occurrence of discrete lignin initiation sites within the cell wall[41] as well as with Sarkanen et al. findings[42,43] that suggest that lignification occurs via a template replication process. It could therefore be hypothesized that dirigent protein sites or arrays are involved in stipulating formation of a primary lignin chain, which is then further replicated.[44]

3. TISSUE SPECIFIC AND SUBCELLULAR LOCALIZATION OF DIRIGENT PROTEIN EPITOPES IN *F. INTERMEDIA* STEMS BY IMMUNOLOCALIZATION

As described above, dirigent protein gene expression gave preliminary indications that dirigent protein mRNA expression was possibly involved in both lignan and lignin biosynthesis, respectively. Accordingly, complementary efforts were next directed to determine the actual localization of dirigent protein (epitopes). In this respect, polyclonal antibodies were first raised in rabbits against the 78 kDa recombinant *F. intermedia* dirigent protein expressed in the *Spodoptera*/baculovirus system.[34] The immune serum so obtained specifically cross-reacted with both recombinant and native forms of the *F. intermedia* dirigent protein, with characterization of the immune serum performed using western-immunoblots and competitive inhibition ELISA (indirect enzyme linked immunosorbent assays).[45] These antibodies were then used to determine precisely the tissue-specific and subcellular localization of the dirigent protein involved in (+)-pinoresinol (**5a**) biosynthesis. However, it was also considered that the dirigent protein epitopes might also be part of arrays related to lignin biosynthesis initiation; indeed, in another study by Donaldson,[41] it had been observed that lignification was correlated with precise lignin initiation sites in the cell wall.

Thus, samples of *F. intermedia* stem (second internode) and petiole were next processed for paraffin embedding in the same way as for the *in situ* hybridization experiments. Serial thick sections were deparaffinized and immunolabeling was performed using an alkaline phosphatase-coupled secondary antibody. Using the dirigent protein immune serum (fig. 7a, 7c, 7e), these experiments revealed a very specific dark purple signal as compared to the pre-immune serum control (fig. 7b, 7d, 7f). Significantly, this labeling was concentrated in the cambial region and developing xylem of stem (fig. 7a, 7c) as well as in the petiole (fig. 7e), this being in very good agreement with the pattern of dirigent protein gene expression observed at the same developmental stages (fig. 6).

Next, in order to obtain better resolution, samples were fixed with a mixture of aldehydes and then embedded in resin.[34,35] Immunolabeling of *F. intermedia* stem serial cross-sections, at the light microscopy level, revealed a very specific black signal with the dirigent protein immune serum (fig. 8a) as compared to that of the pre-immune serum control (fig. 8b). Significantly, this labeling was

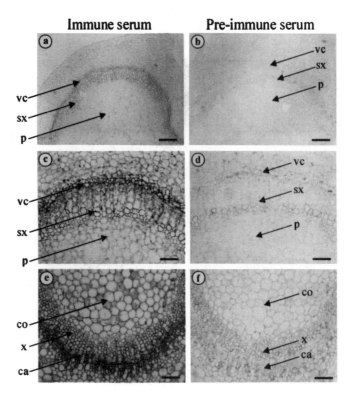

Figure 7. Immunocytochemical labeling of dirigent protein epitoper in *F. intermedia* second stem internode (a), to (d) and petiole (e), and (f) embedded in paraffin, using an alkaline phosphatase-coupled secondary antibody. Panels (a), (c), and (e), labeling with the immune serum. Panels (b), (d), and (f), control with the pre-immune serum performed in the same conditions on serial sections. Note the dark purple labeling in the cambial region and secondary xylem of stem and petiole. Abbreviations: vc, vascular cambium; sx, secondary xylem; p, pith; co, cortex; x, xylem; ca, cambium. Scale bar = 50 μm (a), and (b), 25 μm (c), and (d), and 10 μm (e), and (f).

distributed within the cell walls of lignified secondary xylem (vessels and ray cells) and phloem fibers (fig. 8a). This excellent correlation between lignin occurrence and detection of dirigent protein epitopes was also observed at different developmental stages of the stem, as well as in petioles and root tissues.[45]

On the other hand, the relatively faint labeling observed in the vascular cambium of the resin embedded sections, as compared to that from paraffin treatment, may be due to at least two technical limitations of the method. Firstly, during aldehyde fixation (necessary for ultrastructural preservation), a relatively higher loss of protein antigenicity might occur with the soluble dirigent proteins in the cytoplasm as compared to that of the dirigent protein epitopes localized in the lignified cell wall, since the former are presumed to be more accessible to the fixative. Secondly, since the paraffin thick sections (10 μm) were deparaffinized before immuno-labeling, this might be expected to result in an overestimation of any soluble dirigent protein labeling in the cytosol of cambial cells as compared to that immobilized in the lignified cell wall regions. However, with the resin embedded sections, a more representative surface reaction presumably occurs with the dirigent protein epitopes in both the cytoplasm and the lignified cell wall.

Transmission electron microscopy was next used in order to detect the precise subcellular localization of the dirigent protein epitopes. In this regard, it was found that the S1 sublayer of the secondary wall, and, to a lesser extent, the middle lamella, were predominantly labeled with the dirigent protein immune serum (fig. 8c, 8e), as compared to controls with pre-immune serum performed on serial sections (fig. 8d). Significantly, this labeling was coincident with known sites of lignin initiation as previously demonstrated by Donaldson,[41] thus reinforcing the concept of dirigent protein epitope arrays involved in the formation of the initial lignin template(s).[34,43,46] Indeed, this observation may provide a basis for explaining the predominance of 8–*O*–4' linkages within the natural (heterogenous) lignin biopolymers,[14] an observation that contrasts markedly with the hypothesis of random coupling of monolignol-derived phenoxy radicals during lignin biosynthesis as developed by Freudenberg in the 1950s.[10]

Attempts to implicate various proteins in the control of lignin deposition have also been made with proline-rich proteins[47–49] and glycine-rich proteins.[49–53] These are putatively also correlated, both temporally and spatially, with lignin deposition, although no precise biochemical role has ever been given as to their function. On the other hand, the first apparent evidence for a biochemical control of lignin assembly is now obtained through the temporal and spatial localization of arrays of dirigent protein epitopes with the onset of lignification, i.e., via the existence of an excellent match between dirigent protein gene-specific expression in the cambial and ray cells and the subcellular localization of the dirigent protein epitopes, and, on the other hand, the initiation sites of lignification in the outer part of the cell wall (S1 sublayer and compound middle lamella).[41,54,55] The dirigent protein epitopes could then be responsible for the formation of the initial lignin templates being further replicated by a template polymerization mechanism. Moreover, Sarkanen's group recently showed that *in vitro* enzyme-catalyzed dehydropolymerization of monolignols led to higher molecular weight species when lignin macromolecules were added to the reaction mixture.[42,43] This suggested that macromolecular lignin chains could be formed *in vivo* by such a direct template polymerization mechanism: this may, in turn, provide an explanation as to how the 8–*O*–4' linkages in plant lignins predominate.

Finally, as regards the role of dirigent proteins during lignan biosynthesis [e.g., (+)-pinoresinol (**5a**)], it should be noted that little is yet known about lignan subcellular localization. Nevertheless, related studies on extractive- and lignan-rich heartwood have strongly indicated that heartwood 'extractives' were formed and transported along ray parenchyma cells.[39,40] Moreover, these 'extractive' metabolites accumulate in a very specific manner within different cell types to the extent that even one pure substance (i.e., the lignan matairesinol) is present in a single heartwood cell, whereas a different lignan, e.g., 7'-hydroxymatairesinol occurs in a neighboring cell.[4,40,56,57] Moreover, an earlier model by Chattaway[39] illustrated

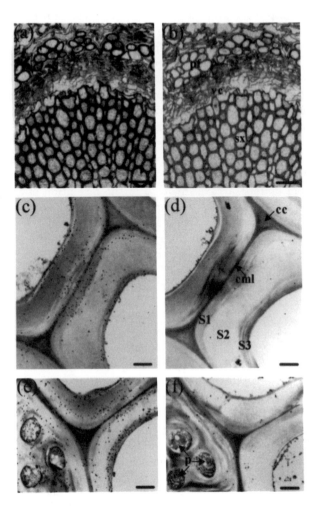

Figure 8. Immuno-gold labeling of *F. intermedia* mature stems using polyclonal antiserum raised against the recombinant dirigent protein at the light microscopy (a) and (b) and transmission electron microscopy (c) to (f) levels. Panels (a), (c), and (e), labeling with the immune-serum. Panels (b), (d), and (f) control with the pre-immune serum on the same zones of tissues on serial sections. Note the black labeling due to the silver-gold particles in the lignified cell wall of secondary xylem cells and phloem fibers (a), and the black dots in the compound middle lamella, the S1 and S3 sublayers of the secondary wall (c) and (d), and around the bordered pits (e). Abbreviations: pf, phloem fibers; vc, vascular cambium; sx, secondary xylem; cc, cell corner; cml, compound middle lamella (middle lamella and primary wall); S1, S2, S3, sublayers of the secondary wall from the outer part (near the middle lamella) to in the inner part (near the lumen); p, bordered pit. Bar: 20 μm (a) and (b); 1 μm (c) to (f).

the secretion of heartwood constituents by ray parenchyma cells into the lumen of neighboring cells through the pit apertures (see Gang et al.[35]). In this context, it is interesting to note that, besides the labeling of the S1 sublayer and middle lamella, the S3 sublayer of the secondary wall was also specifically labeled for the dirigent protein, as well as the cell wall around the pits (fig. 8e). This immunolabeling (at least that closest to the lumen/cytoplasm and to the pit aperture) might therefore be associated with lignan formation. Indeed, this could explain the occurrence of lignans in these areas. In an analogous manner, lignans may also be biosynthesized, albeit to a lesser extent, in other tissues such as the leaves and epidermis of younger stems where they are involved in additional defense functions.

4. CONCLUSIONS

Dirigent protein gene expression and protein deposition were studied in *F. intermedia* by a combination of tissue print hybridization, *in situ* hybridization, and immunolabeling at both light and transmission electron microscopy levels. These procedures were carried out to determine the significance of dirigent protein mRNA expression and protein occurrence relative to both lignan biosynthesis and possibly to lignification. Dirigent protein mRNA was extensively localized in the cambial region of the stem, petioles, and roots as well as in ray cells as revealed from the *in situ* hybridization of the young stems. Labeling was also noted in the young epidermis and leaf tissue in harmony with the role of lignans in defense. The dirigent protein mRNA expression in ray cells suggested the involvement of dirigent proteins in lignan biosynthesis (such as pinoresinol) even though little is yet known about lignan subcellular localization; indeed, the involvement of ray cells in lignan biosynthesis has long been contemplated based on previous studies on extractive- and lignan-rich heartwood. Thus, this study provides the first direct evidence implicating ray cells in lignan biosynthesis.

Dirigent protein epitopes were also detected, using specific polyclonal immune serum, in the vascular cambium and lignified cell walls of secondary xylem and phloem fibers. These data suggest a post-translational transport system into the cell wall, this being presumably consistent with the presence of the signal peptide within the dirigent protein amino acid sequence.[35] Subcellular localization of dirigent protein epitopes further revealed its presence mainly in the S1 secondary wall regions and, to a lesser extent, in the S3 layer as well as in the compound middle lamella. Significantly, in the lignified cells, the dirigent protein epitopes in S1 sublayer and middle lamella were coincident with known sites of initiation of lignification, this being consistent with the proposed involvement of contiguous arrays of dirigent protein sites controlling the formation of primary lignin chains. On the other hand, the detection of dirigent protein epitopes in the S3 sublayer was provisionally viewed as consistent with lignan formation along or within the inner part of the cell walls. Indeed, this latter expression may involve lignan biosynthesis just prior to cell death. Based upon these observations, dirigent proteins are thus proposed to have complementary functions, i.e., in the control of lignin polymerization and lignan biosynthesis, respectively.

ACKNOWLEDGMENTS

The authors thank Dr. V.R. Franceschi, Christine M. Davitt, and Valerie J. Lynch-Holm of the Electron Microscopy Center (Washington State University) for use of their facilities and helpful technical assistance. Thanks are extended to the United States Department of Energy (DE-FG03-97ER20259), the National Science Foundation (MCB09631980), the National Aeronautics and Space Administration (NAG100164), the United States Department of Agriculture (9603622), McIntire-Stennis, the Arthur M. and Kate Eisig Tode Foundation, and the Lewis B. and Dorothy Cullman and G. Thomas Hargrove Center for Land Plant Adaptation Studies for generous support of this study.

REFERENCES

1. Davin, L.B.; Wang, H.-B.; Crowell, A.L.; Bedgar, D.L.; Martin, D.M.; Sarkanen, S.; Lewis, N.G. Stereoselective bimolecular phenoxy radical coupling by an auxiliary (dirigent) protein without an active center. *Science* 275:362–366 (1997).

2. Lewis, N.G.; Yamamoto, E. Lignin: occurrence, biogenesis and biodegradation. *Annu. Rev. Plant Physiol. Plant Mol. Biol.* 41:455–496 (1990).

3. Swan, E.P.; Jiang, K.S.; Gardner, J.A.F. The lignans of *Thuja plicata* and the sapwood-heartwood transformation. *Phytochemistry* 8:345–351 (1969).

4. Krahmer, R.L.; Hemingway, R.W.; Hillis, W.E. The cellular distribution of lignans in *Tsuga heterophylla* wood. *Wood Sci. Technol.* 4:122–139 (1970).

5. San Feliciano, A.; Miguel del Corral, J.M.; Lopez, J.L.; de Pascual-Teresa, B. Lignans from polar extracts of *Juniperus thurifera*. *Phytochemistry* 31:267–270 (1992).

6. Fang, J.-M.; Liu, M.-Y.; Cheng, Y.-S. Lignans from wood of *Calocedrus formosana*. *Phytochemistry* 29:3048–3049 (1990).

7. MacRae, W.D.; Towers, G.H.N. Biological activities of lignans. *Phytochemistry* 23:1207–1220 (1984).

8. Lewis, N.G.; Davin, L.B. Lignans and neolignans: a phytochemical treasure. *Polyphénols Actualités* 10:17–19 (1994).

9. Lewis, N.G.; Davin, L.B.; Sarkanen, S. The nature and function of lignins. *In*: Barton, Sir D.H.R.; Nakanishi, K.; Meth-Cohn, O. (eds.). Comprehensive natural products chemistry. Vol. 1. Elsevier, London, pp. 618–739 (1999).

10. Freudenberg, K. Biosynthesis and constitution of lignin. *Nature* 183:1152–1155 (1959).

11. Brunow, G.; Kilpeläinen, I.; Sipilä, J.; Syrjänen, K.; Karhunen, P.; Setälä, H.; Rummakko, P. Oxidative coupling of phenols and the biosynthesis of lignin. *In*: Lewis, N.G.; Sarkanen, S. (eds.). Lignin and lignan biosynthesis. Vol. 697. ACS symposium series, Washington, DC, pp. 131–147 (1998).

12. Freudenberg, K. The constitution and biosynthesis of lignin. *In*: Freudenberg, K.; Neish, A.C. (eds.). Constitution and biosynthesis of lignin. Springer-Verlag, New York, NY, pp. 47–122 (1968).

13. Chen, C.-L. Characterization of milled wood lignins and dehydrogenative polymerisates from monolignols by carbon-13 NMR spectroscopy. *In*: Lewis, N.G.; Sarkanen, S. (eds.). Lignin and lignan biosynthesis. Vol. 697. ACS symposium series, Washington, DC, pp. 255–275 (1998).

14. Adler, E. Lignin chemistry—past, present and future. *Wood Sci. Technol.* 11:169–218 (1977).

15. Fergus, B.J.; Goring, D.A.I. The distribution of lignin in birch wood as determined by ultra-violet microscopy. *Holzforschung* 24:118–124 (1970).

16. Saka, S.; Thomas, R.J. Evaluation of the quantitative assay of lignin distribution by SEM-EDXA technique. *Wood Sci. Technol* 16:1–18 (1982).

17. Saka, S.; Goring, D.A.I. Localization of lignins in wood cell walls. *In*: Higuchi, T. (ed.). Biosynthesis and biodegradation of wood components. Academic Press, Orlando, FL, pp. 51–61 (1985).

18. Terashima, N.; Fukushima, K. Biogenesis and structure of macromolecular lignin in the cell wall of tree xylem as studied by microautoradiography. *In*: Lewis, N.G.; Paice, M.G. (eds.). Plant cell wall polymers: biogenesis and biodegradation. Vol. 399. ACS symposium series, Washington, DC, pp. 160–168 (1989).

19. Burlat, V.; Ambert, K.; Ruel, K.; Joseleau, J.-P. Relationship between the nature of lignin and the morphology of degradation performed by white-rot fungi. *Plant Physiol. Biochem.* 35:645–654 (1997).

20. Timell, T.E. Compression wood in gymnosperms. Springer-Verlag, Berlin 2150 p. (1986).

21. Higuchi, T. Biochemical studies of lignin formation. I. *Physiol. Plant.* 10:356–372 (1957).

22. Ros Barceló, A. Peroxidase and not laccase is the enzyme responsible for cell wall lignification in the secondary thickening of xylem vessels in *Lupinus*. *Protoplasma* 186:41–44 (1995).

23. Nose, M.; Bernards, M.A.; Furlan, M.; Zajicek, J.; Eberhardt, T.L.; Lewis, N.G. Towards the specification of consecutive steps in macromolecular lignin assembly. *Phytochemistry* 39:71–79 (1995).

24. Higuchi, T. Further studies on phenol oxidase related to the lignin biosynthesis. *J. Biochem.* 45:515–528 (1958).

25. Driouich, A.; Lainé, A.-C.; Vian, B.; Faye, L. Characterization and localization of laccase forms in stem and cell cultures of sycamore. *Plant J.* 2:13–24 (1992).

26. Sterjiades, R.; Dean, J.F.D.; Eriksson, K.-E.L. Laccase from sycamore maple (*Acer pseudoplatanus*) polymerizes monolignols. *Plant Physiol.* 99:1162–1168 (1992).

27. Bao, W.; O'Malley, D.M.; Whetten, R.; Sederoff, R.R. A laccase associated with lignification in loblolly pine xylem. *Science* 260:672–674 (1993).

28. Sterjiades, R.; Dean, J.F.D.; Gamble, G.; Himmelsbach, D.S.; Eriksson, K.-E.L. Extracellular laccases and peroxidases from sycamore maple (*Acer pseudoplatanus*) cell suspension cultures. *Planta* 190:75–87 (1993).

29. Savidge, R.A.; Udagama-Randeniya, P. Cell wall-bound coniferyl alcohol oxidase associated with lignification in conifers. *Phytochemistry* 31:2959–2966 (1992).

30. Udagama-Randeniya, P.; Savidge, R. Electrophoretic analysis of coniferyl alcohol oxidase and related laccases. *Electrophoresis* 15:1072–1077 (1994).

31. Mason, H.S.; Cronyn, M. On the role of polyphenoloxidase in lignin biosynthesis. *J. Amer. Chem. Soc.* 77:491 (1955).

32. Chabanet, A.; Goldberg, R.; Catesson, A.-M.; Quinet-Szély, M.; Delaunay, A.-M.; Faye, L. Characterization and localization of a phenoloxidase in mung bean hypocotyl cell walls. *Plant Physiol.* 106:1095–1102 (1994).

33. Koblitz, H.; Koblitz, D. Participation of cytochrome oxidase in lignification. *Nature* 204:199–200 (1964).

34. Gang, D.R.; Costa, M.A.; Fujita, M.; Dinkova-Kostova, A.T.; Wang, H.-B.; Burlat, V.; Martin, W.; Sarkanen, S.; Davin, L.B.; Lewis, N.G. Regiochemical control of monolignol radical coupling: a new paradigm for lignin and lignan biosynthesis. *Chemistry and Biology* 6:143–151 (1999).

35. Gang, D.R.; Fujita, M.; Davin, L.B.; Lewis, N.G. The 'abnormal lignins': mapping heartwood formation through the lignan biosynthetic pathway. *In*: Lewis, N.G.; Sarkanen, S. (eds.). Lignin and lignan biosynthesis. Vol. 697. ACS symposium series, Washington, DC, pp. 389–421 (1998).

36. Ye, Z.-H.; Song, Y.-R.; Varner, J.E. Gene expression in plants. *In*: Reid, P.D.; Pont-Lezica, R.F. (eds.). Tissue printing. Tools for the study of anatomy, histochemistry, and gene expression. Academic Press, San Diego, CA, pp. 95–123 (1992).

37. Pechoux, C.; Morel, G. *In situ* hybridization in semithin sections. *In*: Morel, G. (ed.). Hybridization techniques for electron microscopy. CRC press, Boca Raton, FL, pp. 99–138 (1993).

38. Bochenek, B.; Hirsch, A.M. *In situ* hybridization of nodulin mRNAs in root nodules using non-radioactive probes. *Plant Mol. Biol. Rep.* 8:237–248 (1990).

39. Chattaway, M.M. The sapwood-heartwood transition. *Aust. For.* 16:25–34 (1952).

40. Hergert, H.L. Secondary lignification in conifer trees. *In*: Arthur, J.C. (ed.). Cellulose chemistry and technology. Vol. 48. ACS symposium series, Washington, DC, pp. 227–243 (1977).

41. Donaldson, L.A. Mechanical constraints on lignin deposition during lignification. *Wood Sci. Technol.* 28:111–118 (1994).

42. Guan, S.-Y.; Mlynár, J.; Sarkanen, S. Dehydrogenative polymerization of coniferyl alcohol on macromolecular lignin templates. *Phytochemistry* 45:911–918 (1997).

43. Sarkanen, S. Template polymerization in lignin biosynthesis. *In*: Lewis, N.G.; Sarkanen, S. (eds.). Lignin and lignan biosynthesis. Vol. 697. ACS symposium series, Washington, DC, pp. 194–208 (1998).

44. Lewis, N.G.; Davin, L.B. Lignans: biosynthesis and function. *In*: Barton, Sir D.H.R.; Nakanishi, K.; Meth-Cohn, O. (eds.). Comprehensive natural products chemistry. Vol. 1. Elsevier, London, pp. 639–712 (1999).

45. Burlat, V.; Kwon, M.; Davin, L.B.; Lewis, N.G. Proposed control of lignan and lignin biosynthesis by dirigent proteins in *Forsythia intermedia*: an ultrastructural study. (submitted).

46. Lewis, N.G.; Davin, L.B. The biochemical control of monolignol coupling and structure during lignan and lignin biosynthesis. *In*: Lewis, N.G.; Sarkanen, S. (eds.). Lignin and lignan biosynthesis. Vol. 697. ACS symposium series, Washington, DC, pp. 334–361 (1998).

47. Marcus, A.; Greenberg, J.; Averyhart-Fullard, V. Repetitive proline-rich proteins in the extracellular matrix of the plant cell. *Physiol. Plant.* 81:273–279 (1991).

48. Müsel, G.; Schindler, T.; Bergfeld, R.; Ruel, K.; Jacquet, G.; Lapierre, C.; Speth, V.; Schopfer, P. Structure and distribution of lignin in primary and secondary cell walls of maize coleoptiles analyzed by chemical and immunological probes. *Planta* 201:146–159 (1997).

49. Ye, Z.-H.; Song, Y.-R.; Marcus, A.; Varner, J.E. Comparative localization of three classes of cell wall proteins. *Plant J.* 1:175–183 (1991).

50. Ryser, U.; Schorderet, M.; Zhao, G.-F.; Studer, D.; Ruel, K.; Hauf, G.; Keller, B. Structural cell-wall proteins in protoxylem development: evidence for a repair process mediated by a glycine-rich protein. *Plant J.* 12:97–111 (1997).

51. Keller, B.; Sauer, N.; Lamb, C.J. Glycine-rich cell wall proteins in beans: gene structure and association of the protein with the vascular system. *EMBO J.* 7:3625–3633 (1988).

52. Keller, B.; Templeton, M.D.; Lamb, C.J. Specific localization of a plant cell wall glycine-rich protein in protoxylem cells of the vascular system. *Proc. Natl. Acad. Sci., USA* 86:1529–1533 (1989).

53. Ye, Z.-H.; Varner, J.E. Tissue-specific expression of cell wall proteins in developing soybean tissues. *Plant Cell* 3:23–37 (1991).

54. Terashima, N.; Fukishima, K.; Takabe, K. Heterogeneity in formation of lignin. VIII. An autoradiographic study on the formation of guaiacyl and syringyl lignin in *Magnolia kobus* DC. *Holzforschung* 40 (Suppl.):101–105 (1986).

55. Wardrop, A.B.; Bland, D.E. The process of lignification in woody plants. *In*: Kratzl, K.; Billek, G. (eds.). Biochemistry of wood. Pergamon Press, New York, pp. 92–116 (1959).

56. Hergert, H.L. Infrared spectra of lignin and related compounds. II. Conifer lignin and model compounds. *J. Org. Chem.* 25:405–413 (1960).

57. Hergert, H.L. Infrared spectra. *In*: Sarkanen, K.V.; Ludwig, C.H. (eds.). Lignins, occurrence, formation, structure and reactions. Wiley—Interscience, New York, pp. 267–297 (1971).

ANTIOXIDANT PROPERTIES AND HEART DISEASE

HUMAN METABOLISM OF DIETARY QUERCETIN GLYCOSIDES

Andrea J. Day and Gary Williamson

Biochemistry Department
Institute of Food Research
Norwich Research Park
Colney, Norwich, NR4 7UA
ENGLAND

1. INTRODUCTION

Flavonoids, along with other phytochemicals, are thought to have a role in reducing the risk of chronic diseases such as coronary heart disease and cancers.[1,2] Fruits and vegetables undoubtedly afford some protection against these diseases,[3] but the effects of individual compounds that may play a role is unclear. Evidence has been accumulating from both *in vivo* and *in vitro* studies that flavonoids have biological activity that may be beneficial. They are good antioxidants and can act by scavenging free radicals, chelation of metal ions or inhibiting lipid peroxidation,[4] and they can inhibit platelet aggregation.[5] Flavonols also have the ability to induce phase II detoxification enzymes[6] and have been shown to inhibit the growth of certain human cancer cells such as those of the colon,[7] ovary,[8] and gastrointestinal tract.[9]

The biological activity is highly dependent on the structure of the flavonol.[10] Figure 1 shows the general structure for flavonols and the related flavonoids. The unsaturated C ring, with the 4-carbonyl, provides the basis for a delocalized structure, which is important for action as a stable free radical acceptor. The presence of a catechol group in the B-ring and a free 3- or 5-hydroxyl in combination with the 4-carbonyl, provides the basis for metal ion chelation. Flavonols, unlike the majority of the flavonoids lacking a 3-hydroxyl group, are generally glycosylated at the 3 position in the C-ring. Occasionally, as with onion, glycosylation will occur in the 4'- or 7-position.[11] The presence of the conjugate may reduce or even abolish the biological activity associated with the aglycone (see chapter by Williamson).

Total flavonoid glycoside consumption in the United States was estimated by Kühnau to be about 1 g/day, with the 4-oxo-flavonoid intake estimated at 110 mg/day (aglycone basis).[12] Hertog et al. used more specific analysis to show that the average Dutch intake of flavonols was about 22 mg/day, with the flavones contributing about 1 mg/day.[13] Although the flavonol content in plants can vary

Plant Polyphenols 2: Chemistry, Biology, Pharmacology, Ecology, Edited by
Gross et al. Kluwer Academic / Plenum Publishers, New York, 1999

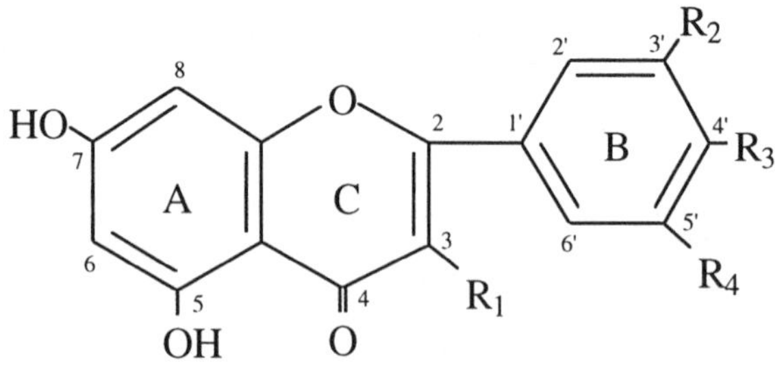

	R_1	R_2	R_3	R_4
Flavonol				
Quercetin	OH	OH	OH	H
Kaempferol	OH	H	OH	H
Myricetin	OH	OH	OH	OH
Flavone				
Apigenin	H	H	OH	H
Luteolin	H	OH	OH	H

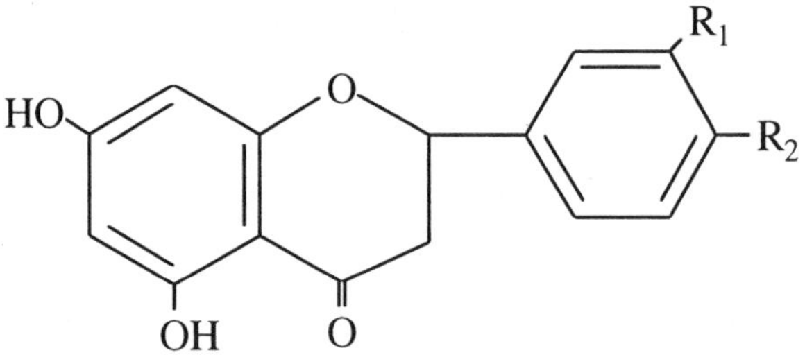

	R_1	R_2
Flavanone		
Naringenin	H	OH
Eriodictyol	OH	OH

Figure 1. Structure of various flavonoids.

considerably depending on the variety, maturity, growth conditions, exposure to UV, time of storage, and post-harvest processing,[14] 110 mg/day in the American diet is likely to be greatly overestimated. However diets rich in red wine, fruits, and vegetables, such as those of the southern Mediterranean, may well achieve this level. In addition, populations consuming high levels of tea, such as the Japanese, have a flavonol content estimated at 70 mg/day.[15]

The flavonol-rich foods are represented in table 1. In the Dutch diet, the main contributors of flavonols include, onions (29 percent), black tea (48 percent), apples

(7 percent), french beans and kale (7 percent), and red wine (1 percent).[13] Flavones are only found in a few foods, and their main contribution in the diet will arise from use of herbs such as parsley. The flavonol glycoside composition of many foods has not been investigated, but table 2 shows the range of flavonol glycosides present in six varieties of commercially available black tea.[17]

Only a few epidemiological studies have investigated the relative risk of coronary heart disease or cancers and flavonol intake. These studies are summarized in table 3. Flavonol intake and cancer were not found to be related in any of the studies, and there are inconsistent data on the relationship with coronary heart disease. The relative risk of mortality from coronary heart disease was reduced in the Zutphen Elderly Study.[18] The Finland Study also found a decreased relative risk although this was only statistically significant for women and not men.[19] If the antioxidant vitamins were removed as confounding factors in the epidemiology, the relationship of flavonoids with coronary heart disease for women was weakened to non-significant (before adjustment relative risk was 0.54, 0.33–0.87). The Zutphen Study on middle-aged men also found a low relative risk with flavonoid consumption and incidence of stroke.[20] The U.S. Male Health Professional Study did not find any relationship for coronary heart disease and flavonoid intake, although a reduced risk of mortality was observed in men with prevalent coronary heart disease.[21] The main conflicting study is the Caerphilly Study

Table 1. Flavonol content of commonly consumed foods

Food	Level (mg/kg fresh edible wt)	
	Quercetin	Kaempferol
Apple	36	<2
Apricot	25	<2
broad bean	5.5	7
Broccoli	30	72
Endive	<1.3	46
french bean	39	12
grape, black	15	<2
grape, white	12	<2
Kale	110	211
Leek	<1	30
Lettuce	14	<2
Onion	347	<2
red currant	13	<2
Strawberry	8.6	12
sweet cherry	15	<2
turnip tops	7.3	48

Modifed from Hertog et al, 1992,[16] expressed as aglycone.
Myricetin was found in broad bean, black and white grapes (26, 4.5, 4.5 mg/kg). Apigenin was found at a high level in celery (108 mg/kg) and luteolin in red bell peppers (11 mg/kg).

Table 2. Flavonol glycoside composition of six varieties of black tea

Quercetin glycoside	Level (mg/L)[a]	Kaempferol glycoside	Level (mg/L)[a]
3-O-gal-rha-glu	1-6	3-O-gal-rha-glu	0-6
3-O-glu-rha-glu	2-17	3-O-glu-rha-glu	0-6
3-O-glu-rha	0-15	3-O-glu-rha	0-7
3-O-gal	2-13	3-O-gal	1-3
3-O-glu	2-12	3-O-glu	1-9
3-O-rha	0-1		
3-O-ara	0-1		
Quercetin aglycone	0	kaempferol aglycone	0
Total quercetin	13-38	total kaempferol	7-19

Adapted from Price et al, 1998.[17]
[a]expressed as aglycone; gal = galactose, glu = glucose, rha = rhamnose, ara = arabinose Major brands of tea were analyzed. 3 g of tea was brewed with 200 ml of boiling water and left to stand for 4 min before extraction.

carried out in Wales on middle-aged men.[22] This actually showed an increase in risk from coronary heart disease and flavonoid intake. In this study, tea comprised 82 percent of the flavonol intake, a larger contribution than from the other studies. The authors suggest that the addition of milk in tea, typical of British culture, may reduce the absorption of flavonols by protein binding, and therefore the relative risk of the high flavonol intake group versus the low intake group is artificial. However, the absorption of both catechin and quercetin from green or black tea has recently been shown to be unaffected by the addition of milk.[24] It is more likely that the positive association found in the Caerphilly study is a result of confounding factors that could not be accounted for, such as those arising from

Table 3. Summary of epidemiological studies on flavonol intake and risk of cardiovascular disease

Study	Number in study	Age (yr)	Relative risk of mortality from coronary heart disease[a]	Ref
Zutphen (Netherlands)	805	65-84	0.32 (0.15-0.71)[b]	18
			0.47 (0.27-0.82)[c]	23
Finland	5133	30-69	0.73 (0.41-1.32)[d]	19
			0.67 (0.44-1.00)[e]	19
US Male Health Professional	34 789	40-75	1.31 (0.42-3.05)[f]	21
Caerphilly (Wales)	1900	49-59	1.6 (0.9-2.9)	22
Zutphen (Netherlands)	552	50-69	0.27 (0.11-0.7)[g]	20

[a] after adjustment of known risk factors [b]5-year follow-up, [c]10-year followup, [d]female, [e]male, [f]prevalent CHD at baseline gives a relative risk of mortality of 0.63 (0.33-1.2), [g]incidence of stroke.

socioeconomic class and manual labor. Additionally, tea intake was associated with a less healthy lifestyle such as higher intakes of fat and energy, along with less alcohol and increased smoking.

The usefulness of epidemiology relies on (i) accurate food consumption data and (ii) assessed compounds being absorbed relative to their concentration in the diet. The epidemiology on relative risk of coronary heart disease, stroke, and cancers after high or low consumption of flavonols was calculated based on the amounts of flavonols from foods analyzed after hydrolysis to the aglycone.[13] This provides a useful and rapid indicator of flavonol concentrations in the diet. However, flavonols exist in foods mainly as glycosides and the type and positioning of the glycoside may substantially alter the bioavailability from the diet, thus making the epidemiology far more complex. Data may be required on the individual availability of various glycosides in the diet before the relationship of flavonol intake and chronic disease can be assessed. For example, the rhamnoglucoside of both quercetin (rutin) and kaempferol are important glycosides in tea,[17] but from absorption studies rutin is not absorbed as efficiently as other flavonol glycosides. In the Caerphilly study, 82 percent of the flavonols came from tea, which greatly increases the proportion of rutin to other flavonol glycosides, thus biasing the flavonol intake data toward less bioavailable flavonols.

The biological activity of flavonols has generally been assessed by experimentation using the aglycone. Not only may the nature and position of the glycosidic substitution affect this activity, but so may the metabolic products of the flavonols, likely to be sulphate or glucuronide conjugates. The ability of the flavonoids to act as antioxidants is through the availability of the hydroxyl groups.[10] Once the compound is conjugated, the antioxidant potential and other biological activities may be greatly reduced. It is, therefore, important to know the relative bioavailability of the flavonol glycosides consumed from the diet and to what extent the metabolites retain biological activity.

2. FLAVONOL GLYCOSIDE BIOAVAILABILITY

Although the flavonols are found predominately glycosylated in plants, whether they are consumed in this form is potentially important to their mechanism of absorption. Processing of plant foods may affect the flavonol glycoside content or change the ratio of the glycosides present, thus affecting content in the diet. Processes that may be responsible for changes to the flavonol glycoside content are shown by figure 2 and include: (a) mechanical removal, (b) solubility, (c) heat treatment, and (d) enzyme activity.

Flavonols are often produced by plants as a response to UV light and as such are concentrated mainly in the skin or outer layers of the fruit and vegetables, with the notable exception of onions.[14] Of the flavonol glycosides remaining, substantial quantities may be leached into the cooking water. This has been demonstrated by boiling of broccoli,[25] boiling of red- and brown-skinned onions,[26] and canning of green beans.[27] An average of 82 percent for broccoli florets, 22 percent for green beans, 14 and 22 percent for red- and brown-skinned onions, were leached into the cooking water during boiling for 15 minutes. Total recovery of the

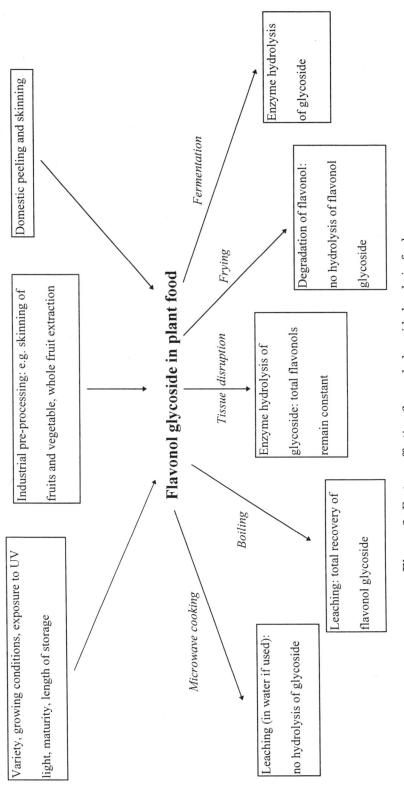

Figure 2. Factors affecting flavonol glycoside levels in food.

flavonol glycoside from the tissue and cooking water was greater than 98 percent for broccoli and beans. Larger losses were observed for red- and brown-skinned onions, 25 and 14 percent, respectively, although it was thought that a large proportion of this may be held on the muslin cloth used to constrain the onion. Free aglycone was not found in any of the tissues or the cooking water. Crozier et al. also measured the amount of flavonol (after hydrolysis of the glycoside) remaining in sliced onion after boiling, but unfortunately did not measure the content leached into the water.[28] The authors also measured the content of flavonols in onion after microwave cooking and demonstrated a substantial loss. However, water was added to the onions when cooking, and again the flavonol content leached out was not analyzed; therefore, the degree of overall degradation could not be calculated. Price (personal communication, 1998) investigated the degradation of flavonol glycosides in broccoli after microwave cooking in the absence of added water and showed that there was no loss even after 15 minutes.

In all these studies, the flavonols are stable on heat treatment and are not deglycosylated. Although excessive frying (15 min) of onions did result in a loss of 23–29 percent flavonol, there was not a corresponding increase in the aglycone. This implies that although some flavonol glycosides may be degraded, the glycoside is still the compound that will be consumed. In contrast to this, enzymes released when plant tissue is disrupted may act specifically on the flavonol glycoside. For example, onions have a high level of β-glucosidase which will become active when the onion is chopped.[29] Price and Rhodes observed that quercetin 3,4'-glucoside was converted initially to quercetin 4'-glucoside and then to quercetin over the course of 24 hours, with no overall loss of total flavonol.[30] It is possible that similar activity may occur during production of unpasteurized fruit juices where enzymes have not been heat inactivated, although further work would be needed to verify this. Wine is also a potential source of flavonol aglycone since fermentation of the grape will result in deglycosylation of the flavonol glycoside. McDonald et al. showed that up to 67 percent of the flavonol content of red wine was in the free form.[31] Red wine, therefore, probably represents the greatest source of flavonol aglycone in the diet.

Early investigations on the absorption of flavonols were mainly carried out using rats as a model, often after administeration of large doses of quercetin. Previously, the glycosides were not considered important as no β-glucosidase is secreted into the intestine during digestion. Hence, absorption would only occur in the large intestine where the flavonol glycoside is subjected to extensive metabolism by the colon microflora. Hydrolysis to the aglycone and products of ring fission would result. Although these studies are important and will be discussed later, attention will first be focused on recent investigations of flavonol glycoside absorption in man at more typical dietary concentrations. The results of these are summarized in table 4.

The first study of this nature was carried out by Hollman et al. on the absorption of flavonols in healthy ileostomy volunteers.[33] Trials involving ileostomy subjects have the advantage of being able to look at absorption from the small intestine without complications of microbial degradation in the colon. Onions were used as a rich source of flavonol glycosides and compared to the absorption from

Table 4. Quercetin concentrations in plasma or urine after dietary supplementation of flavonol glycosides in humans

Food	flavonol glycoside	Number of subjects	Amount ingested (mg)[a]	Max concentration in plasma (μM)	Time to reach max (hr)	% excreted	Ref
onion	quercetin glucosides	9	89	n/m	-	0.3	33
supplement	quercetin aglycone	9	100	n/m	-	0.12	33
supplement	quercetin rutinoside	9	100	n/m	-	0.07	33
onion	quercetin glucosides	2	64	0.6	2.9	n/m	35
onion	quercetin glucosides	9	68	0.74	<0.7	1.39	36
apple	quercetin glycosides	9	98	0.3	2.5	0.44	36
supplement	quercetin rutinoside	9	100	0.3	9.3	0.35	36
supplement	quercetin 4'-glucoside	9	100	3.2	<0.6	n/m	32
supplement	quercetin rutinoside	9	100	0.23	5.6	n/m	32
onion	quercetin glucosides	24	114	1.48[b]	-	-	38
tea	quercetin glycosides	15	49	0.1[c]	-	0.5	37
tea	kaempferol glycosides	15	27	0.05[c]	-	2.5	37
onion	quercetin glucosides	15	13	0.07[c]	-	1.1	37
broccoli	kaempferol glycosides	2	12.5	n/m	-	0.88	56
blackcurrant/ apple juice	quercetin glycosides	5	4.8, 6.4, 9.6	-	-	0.47[d]	40
parsley	apigenin glycosides	14	50	n/m	-	0.58	39
parsley	apigenin glycosides	24	84	n/d	-	n/d	38
flavonol-rich meal	quercetin glycosides	10	87	0.37[c]	-	n/m	62

n/m- not measured; n/d - not detected. [a]aglycone equivalent.
[b]blood sampled 90 min after intake on day 7 of intervention, [c]daily portion split over three occasions for a 7-day intervention period. Blood was sampled approx. 4 hr after first portion, [d]average percent excretion on day 7 of intervention (steady-state), on day 1 the percent excretion of quercetin was 0.29 percent, [e]blood sampled 3 hrs after meal.

supplements of either quercetin aglycone, or quercetin 3-rhamnoglucoside (rutin). Absorption was calculated as the amount of quercetin ingested minus the amount remaining in the ileostomy effluent after adjustment for degradation in the ileostomy bag.

Hollman showed that 52 percent of the flavonol glycosides from onions were absorbed, whereas only 24 percent and 17 percent of quercetin and rutin were absorbed, respectively. The excretion of quercetin in hydrolyzed samples was measured in the urine over a 13-hour period and expressed as a percentage of the amount ingested. Only 0.3 percent of total quercetin from onion flavonol glycosides and 0.1 percent of total quercetin or rutin supplements were excreted in this time. Post-column derivatization of the flavonol with aluminum provides a sensitive method of detection with fluorescent. However, a free 3-hydroxyl group is required for this method and so conjugates cannot be detected.[34] Although 3'- or 4'-methylquercetin can be detected by this method, the values were not reported.

Hollman et al. continued this work by looking at the absorption and disposition of the flavonol glycosides from foods in human subjects.[35,36] Appearance of quercetin in the plasma was monitored and 24 h recovery of the flavonol in the urine calculated. The authors showed that the flavonol glucosides from onions are rapidly absorbed across the intestinal wall (<0.7 h), but other flavonol glycosides, such as rutin, have a delayed absorption (9.3 h), suggesting that they require prior hydrolysis by colon microorganisms. Apple flavonol glycosides appeared to be absorbed at an intermediate rate (2.5 h), which may be related to the various glycoside conjugates in apples being absorbed at differential rates.

De Vries et al. measured the concentration of kaempferol and quercetin in the plasma and urine of individuals after a 7-day randomized crossover study on the consumption of tea and onions.[37] Quercetin was excreted at an average of 1.1 percent of intake after onion consumption and 0.5 percent after tea, suggesting that quercetin from onions is more bioavailable than from tea. Kaempferol, only found in the tea, was excreted at a level of 2.5 percent of the amount ingested, which suggests that either kaempferol is absorbed better than quercetin, or that quercetin is more extensively modified and has not been detected by the methodology used.

Apigenin, a flavone of related structure to kaempferol, was also investigated. Janseen et al. did not detect any apigenin in the plasma after 7 days of feeding 5 g of dried parsley (84 mg apigenin) to individuals using methods capable of detecting apigenin at 1.1 μmol/l.[38] Nielson et al., however, found apigenin excreted by 14 volunteers after 7-day consumption of 24 g of fresh parsley (50 mg apigenin).[39] An average of 0.5 percent per day was excreted in the urine, but unlike quercetin over a 7-day period[40] a large inter-individual variation was observed. Ten subjects excreted an average of 0.1–0.5 percent, whereas four subjects excreted 0.8–4 percent. On one of the days, one subject excreted up to 7.4 percent of the ingested dose. These results indicate that absorption of apigenin glycosides from parsley is highly dependent on the individual. Microbial metabolism may be required for hydrolysis of the glycoside before absorption can occur and this may account for the large intra-individual variation observed. Levels of such bacterial populations may be affected by other constituents of the diet.

The evidence clearly indicates that flavonols and flavones are absorbed from the diet, and can appear in the plasma within a very short time period (<0.7 hr). This suggests absorption from the small intestine is an important mechanism, but it does not rule out additional absorption from the colon. The ability of some flavonol glycosides to be absorbed more rapidly than others indicates a potential role of the nature of the glycoside to affect the rate of absorption. The onion glycosides, quercetin 4'-glucoside and quercetin 3, 4'-diglucoside appear to be absorbed better than either tea flavonol glycosides, apple flavonol glycosides, or rutin from a supplement. It may be possible for flavonol glucosides to be actively transported via the sodium-dependent glucose transporter in the small intestine,[41] whereas flavonols with other sugar conjugates may not be recognised and transported.

The amount of hydrolyzable flavonol and flavone conjugates excreted in the urine is very low compared to the amount ingested, typically less than 3 percent. Quercetin excretion remained relatively constant after apple and blackcurrant juice were consumed over a period of 7 days.[40] The quercetin content of the juice consumed ranged from 4.8–9.6 mg, with the amount excreted averaging 0.47 percent of the dose by day 7 of the study (a steady-state was reached on day 4). This implies that urinary quercetin would be a good biomarker for quercetin consumption, at least at low concentrations. These data also agree well with urinary excretion of quercetin after tea consumption for 7 days, providing 0.5 percent of the daily dose[37] and 0.44 percent excreted after a single dose of apples.[36] However, after onions were consumed for 7 days, the average percentage excretion of quercetin was 1.1 percent. This again implies that the type of flavonol glycoside may affect absorption.

3. METABOLISM OF FLAVONOLS

The first stage of metabolism is likely to be deglycosylation of the flavonol glycoside. None of the studies mentioned thus far have determined whether or not the flavonol glycoside remains intact in the circulating plasma. One study detected flavonol glycosides, based on the retention time and UV spectrum, in the plasma of two subjects.[42] The identification of these compounds is somewhat dubious as conjugated metabolites of quercetin likely to be formed have similar retention times and UV-spectrum to the parent glycoside. Further evidence such as mass spectrometry data is required before it can be concluded that the flavonol glycoside remains intact on absorption.

Deglycosylation does occur in the intestine by the action of gut microflora. Most of this microbial activity occurs in the colon, although a small population of bacteria exists throughout the intestinal tract.[43] Release of the aglycone is also accompanied by degradation of the flavonol by ring fission, and so although some of the flavonol may passively diffuse through the colon wall, much of the flavonol will be further metabolized to phenolic acids.[44] These will also be absorbed and metabolized further and have in themselves potential biological activity. It is interesting to note that rats actually have a high level of colonic microflora in their stomach, potentially leading to a higher level of microbial degradation prior to absorption in the small intestine.

The relatively small bacterial population in the duodenum does not explain the rapid increase in plasma quercetin levels observed after quercetin 4′-glucoside ingestion but not after rutin ingestion in humans. The bacterial β-glucosidase may be more specific for the 4′-glucoside, rather than the 3-rutinoside. However, Gee et al. have shown that there is an interaction of some flavonol glycosides, such as quercetin 3-glucoside, quercetin 4′-glucoside and quercetin 3,4′-glucoside but not rutin, with the sodium-dependent glucose transporter (SGLT1) of the rat small intestine.[41] Although this experiment does not directly show transport across the intestine, the fact that an interaction occurs means that the potential for transport of the flavonols exists. Synthetic compounds such as p-nitrophenol glucoside are also transported across the intestine by SGLT1,[45] increasing the likelihood that this mechanism can operate for flavonoid glycosides.

Day et al. have shown that a cytosolic β-glucosidase present in cell-free extracts of human small intestine and liver is capable of hydrolyzing various flavonoid and isoflavone glycosides.[46] The enzyme has a high affinity for quercetin 4′-glucoside and genistein 7-glucoside with a K_m of 32 and 14 μM, respectively. The enzyme was not active on rutin or naringin, a rhamnoglucoside of naringenin. Only the small intestine had activity toward quercetin 3-glucoside, and this was at a much slower rate. Effects of known inhibitors of the cytosolic broad-specificity β-glucosidase previously identified by Daniels et al.[47] indicate that this enzyme is responsible for the activity demonstrated on the flavonol substrates. However, an additional β-glucosidase exists in the brush-border membrane of the small intestine, lactase phlorizin hydrolase (LPH).[48] This enzyme complex is capable of hydrolyzing phlorizin to the aglycone phloretin. Since phlorizin is closely related to the flavonoid glycosides, it is possible that LPH also has similar activity on the flavonol glycosides, and it may be responsible for the hydrolysis of quercetin 3-glucoside observed in the small intestine cell-free extract.

If LPH hydrolyzes certain flavonol glycosides, then it could act on the apical side of the brush-border, releasing the aglycone into the intestinal lumen. The close proximity of the aglycone to the membrane may increase the ability of the flavonol to passively diffuse into the enterocyte. Thus active transport of the flavonol glycoside may not be the only mechanism of absorption in the small intestine.

It is clear that, regardless of transport mechanisms in the small intestine, colon microflora play the major role in glycoside hydrolysis. However, once deglycosylated the flavonol will be subjected to extensive breakdown. Ring fission products of flavonoids have been detected by many researchers after oral administration of flavonoids to both rats and humans.[49–53] In general, phenolic acids arise from the B-ring and hence the type of metabolite depends on the hydroxylation pattern. The product from the A-ring, 2,4,6-trihydroxybenzoic acid, is thought to be oxidized to CO_2, which has been measured in some rat experiments using radiolabelled flavonols.[54]

Oral administration of rutin to humans led to the detection of 3,4-dihydroxytoluene, 3-hydroxyphenylacetic acid, 3,4-dihydroxyphenylacetic acid, and

3-methoxy-4-hydroxy-phenylacetic acid in the plasma after 4 h.[53] Maximum detection of all compounds was between 8 and 12 h, decreasing to baseline at 20–35 h. The total urinary excretion in 48 h accounted for 50.5 percent of the dose. Baba et al. also determined the same urinary metabolites after ingestion of deuteriated rutin by humans[49] and rats.[50] The authors also detected β-m-hydroxy-phenyl-hydracyrylic acid in humans, although not with rats. Rats, however, produced 3,4-dihydroxyphenyl propionic acid. Rutin was not detected in either rats or humans. None of these metabolites were found after intraperitoneal administeration in bile-duct cannulated rats or after administration of the antibiotic neomycin for 4 days prior to orally administering rutin to rats. The phenolic acids were detected in both the free and conjugated forms in the plasma and urine of both rats and humans.

Once absorbed, flavonols will be subjected to metabolism involving phase I or phase II enzymes. Hydroxylation and demethylation are examples of cytochrome P450 mono-oxygenase dependent activities that may be involved in the metabolism of flavonols. Nielsen showed that microsomes prepared from normal and Aroclor-1254 induced rats were capable of hydroxylating certain flavonols and flavones.[55] The requirement for this metabolic activity was that either a single or no hydroxyl groups were present on the B-ring. Two or more hydroxyl groups in the B-ring prevented further hydroxylation. Demethylation in the same system was only observed when a methyl group was present in the 4′ position, but not in the 3′ position. The P450 activity was suggested to arise through action of CYP1A and not CYP3A, both induced by Araclor-1254, as inhibition of CYP3A by trole-andomycin did not result in a reduction of biotransformation of kaempferol or 4′-methylquercetin to quercetin.

Nielsen also described the metabolites of various flavonols and flavones in female rats after administration of 100 mg aglycone for 14 days.[39] Interestingly, the 4′-hydroxylated flavonoids were recovered in the urine in an unchanged form, suggesting that P450 metabolism is not specifically required for these compounds as indicated by the above *in vitro* data. The exception was naringenin, a flavanone of related structure to kaempferol, where very small amounts of the 3′-hydroxy-lated naringenin (eriodictyol) were detected in urine using LC-MS, although the concentration was too low to be quantified by HPLC (<0.17 percent). Lack of P450 hydroxylation for the flavonols in rats agrees with two human feeding trials carried out by the same author. In the first study, urinary excretion of kaempferol was monitored after broccoli was consumed for 7 days, but no quercetin could be detected.[56] This suggests that if hydroxylation occurred, the quercetin produced is rapidly converted to more complex metabolites other than methyl-quercetin, which are not detected under the present methods. In the second study, apigenin but not luteolin was also excreted after parsley consumption.[39] Human studies involving naringenin metabolism have not described the appearance of eriodictyol in the urine after consumption of pure naringin (the rhamnoglucoside present in grape-fruit juice) or grapefruit juice.[57]

Several investigations of quercetin metabolism in rats have found 3′- and 4′-methyquercetin in the bile and/or urine.[39,54,58–61] The 3′-position (relative to 4′) for methylation appears to be the preferential site of catechol *O*-methyltransferase activity with a ratio of > 2:1.[39,59,60] The 4′-methylquercetin declines more rapidly than the 3′-methylquercetin with little detectable after 12 hours. In fact, 3′-methylquercetin was the only metabolite identified in the urine or bile of rats after prolonged feeding (10 days).[58] Nielsen et al. suggests that active demethylation of 4′-methylquercetin by P450 enzymes may be responsible for the accumulation of 3′-methylquercetin observed over time.[39] There is no evidence for methylation at any other hydroxyl groups of the flavonoids apart from the catechol grouping on the B-ring.

In humans, 3′-methylquercetin but not 4′-methylquercetin was reported by Hollman, although details were not given.[32] Manach et al. also reported only the 3′-methylquercetin in the plasma of 3 out of 10 human subjects consuming a meal rich in quercetin.[62] These subjects were also those having the highest concentration of quercetin in the plasma. The 3′-methylquercetin concentration was 0.1–0.2 μM after 3h, which although significant, accounted for only 20–30 percent of the total quercetin content in plasma. This is in contrast to the data presented on rats, where methylated quercetin represented most of the recovered metabolites. Species variation may account for the relative importance of methylation; however, as suggested by Manach, the high doses of quercetin fed to rats may direct the metabolism of quercetin through a methylation pathway. At the lower concentrations of quercetin from a flavonol-rich meal, the potential threshold level of methylation may not have been reached.

Conjugation of flavonols with glucuronide or sulphate are the mostly likely metabolic routes. Conjugation is a common detoxification reaction leading to increased solubility of compounds and a higher molecular weight, which is important for excretion particularly in the bile.[63] Sulphation is likely to be a major pathway of metabolism at low concentrations for many xenobiotics, although this pathway can easily become saturable, whereas glucuronidation is particularly important for increasing molecular weight. It should be noted that using animal models for determining conjugation is highly dependent on the species, e.g., pigs are not capable of forming sulphate conjugates, and cats cannot glucuronidate.[64] Free and conjugated flavonols have been detected in rats after large doses, but Manach did not detect any free flavonols in humans after a flavonol-rich meal.[62] This suggests that the glucuronide pathway can become saturated at high levels. Saturation of preferred pathways is well known for metabolism of many drugs.

The major route for excretion of flavonol conjugates is probably through the bile, although this evidence comes largely from rat studies, since it is very difficult to measure biliary excretion in humans. Sfakianos et al. showed that 75 percent of genistein 7-glucuronide was excreted in the bile after 4 hours infusion of genistein into the intestine.[65] Bile is excreted back into the small intestine through the gallbladder, although the glucuronide is unlikely to be able to diffuse across the intestine as a charged molecule. Hence the flavonol glucuronide will reach the colon where it is metabolized by microflora. The aglycone released may be reabsorbed, producing enterohepatic circulation, or undergo ring fission to the phenolic acids.

Many researchers have incubated metabolites isolated form the urine, plasma, or bile with β-glucuronidase or aryl-sulphatase. Information can be gained on the type of conjugation from analysis before and after enzymatic treatment. No information on the position of conjugation is obtained, which will be of significance in determining potential biological activity. One study on the metabolism of diosmetin, 4'-methyl luteolin, showed that the 3' and 7,3'- glucuronides were the major metabolites in rat.[66] Genistein, an isoflavone with a 4'-hydroxyl in the B-ring, has also been shown to have a glucuronide conjugated at the 7- position when fed to rats.[65] There is no conclusive evidence at present for conjugation in the 3-position of flavonols, although *in vitro* incubations of human liver cell-free extracts with UDP-glucuronic acid and quercetin produced a metabolite of the same retention time and UV-spectrum as quercetin 3-glucuronide (unpublished data). The preferential sequence of conjugation has not been determined for flavonols, but for flavones it would appear the order is 3'- > 7- > 5-; the 4'- position is available for conjugation, but the preference for this site has not been determined.[66] Additionally, the 5-position appears to be glucuronidated if it is the only hydroxyl available.

The liver is generally believed to play the major role in the metabolism of xenobiotics, with the contribution by the small intestine and kidney often overlooked. These other organs also have the ability to perform metabolic transformations since they express the cytochrome monooxygenase enzymes, methyl-transferase and the conjugating enzymes. However, only a few researchers have studied the effect of the intestinal metabolism of the flavonoids. Mizuma et al. demonstrated that naphthyl glucoside was metabolized to the corresponding glucuronide after passing through the intestinal mucosa in a rat everted sac.[67] They also showed the conversion of *p*-nitrophenol glucoside to the glucuronide in similar experiments.[45,68] Day et al. have also demonstrated that the human intestinal mucosa is capable of deglycosylating certain flavonoid glycosides, at a rate potentially higher than that of the liver.[46]

Sfakianos et al. collected the portal vein blood in rats fed genistein.[65] They showed that genistein 7-glucuronide was the predominant compound in the portal vein, indicating that most of genistein metabolism had occurred prior to the liver. The liver performed identical metabolism when genistein was directly administered to the portal vein. Piskula and Terao also proposed that glucuronidation of epicatechin is likely to occur in the small intestine since this tissue demonstrates a higher UDP-glucuronosyl transferase activity than the liver, kidney, or lung toward this flavonoid.[69] Crespy et al. found that most of absorbed quercetin was conjugated in the portal vein of rats after perfusion of quercetin through the intestinal lumen.[70] The authors showed approximately 50 percent of the absorbed quercetin was directly conjugated and excreted back into the intestine, resulting in a lower net absorption of quercetin. Analysis of the effluent also indicated that 5 percent of the quercetin had been methylated, consistent with methyl transferase activity in the intestinal mucosa.

4. CONCLUSIONS

Figure 3 shows a schematic representation of the potential mechanisms involved in absorption of flavonol glycosides. Flavonol glycosides are not readily

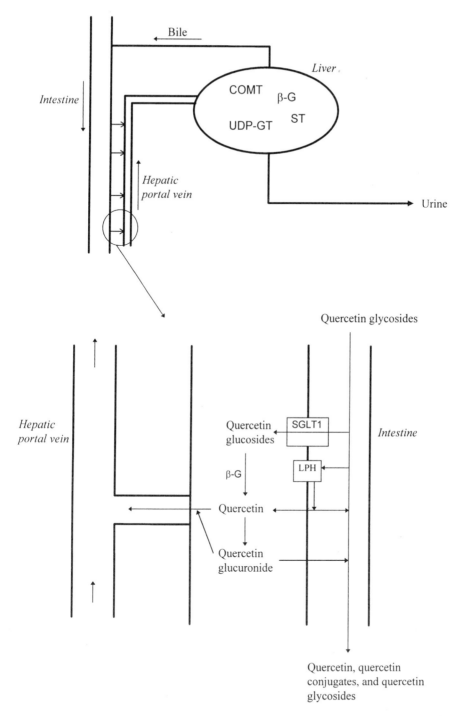

Figure 3. Scheme of enterohepatic circulation and proposed mechanism of flavonol glycoside absorption. COMT, catechol o-methyltransferase; ST, sulphur transferase; UDPGT, UDP-glucuronsyl transferase; β-G, broad specificity β-glucosidase; SGLT1, sodium dependent glucose transporter; LPH, lactase phlorizin hydrolase.

deglycosylated during domestic food processing, and thus, the glycoside conjugates are the major form in which flavonols are consumed. It may be possible for the flavonol glycoside to be actively transported across the brush-border by the sodium-dependent glucose transporter, but once inside the enterocyte, the glycosides are likely to be deglycosylated by a cytosolic β-glucosidase. Lactase phlorizin hydrolase may also be capable of deglycosylating certain flavonol glycosides with the released aglycone diffusing across the intestine. Additionally, the gut microorganisms will release the flavonol aglycone. The intestinal mucosa is active in conjugating the flavonol; the resulting metabolites are then either excreted back into the intestinal lumen or transported to the liver for further processing. The biotransformation of the flavonol within the enterocyte would help to maintain a favorable concentration gradient for absorption.

The liver, after any additional metabolism, will excrete the flavonol either into the bile or the blood for excretion by the kidneys. Bilary excretion will result in enterohepatic circulation, although deconjugation of the glucuronide is more likely to occur in the colon where ring fission may be predominant.

Quercetin metabolites are present in the plasma at concentrations likely to range from 0.3–$0.6\,\mu M$ after a flavonol-rich meal. Steady-state concentrations at such levels are attainable with repeated dietary intakes. The rate of uptake and maximum concentration reached appears to depend on the nature and position of the sugar in the conjugate. Such information on the bioavailability of flavonol glycosides is still fragmentary and requires further research and evaluation. The plasma concentration of quercetin commonly observed is similar to that found for other antioxidant vitamins, such as β-carotene,[71] and so potentially can have biological activity. However, the effects of flavonol metabolites should be further investigated.

ACKNOWLEDGMENTS

We would like to thank the Biotechnology and Biological Sciences Research Council for funding, M.R.A. Morgan for support, and K.R.Price for help with the manuscript.

REFERENCES

1. Middleton, E.; Kandaswami, C. The impact of plant flavonoids on mammalian biology: implications for immunity, inflammation and cancer. *In*: Harborne, J.B. (ed.) The flavonoids: advances in research since 1986. Chapman and Hall, London p. 619 (1994).

2. Huang, M.-T.; Ferraro, T. Phenolic compounds in food and cancer prevention. *In*: Huang, M.-T.; Ho, C.-T.; Lee, C.Y. (eds.) Phenolic compounds in food and their effects on health II. American Chemical Society, Washington, DC. p. 8 (1992).

3. Block, G.; Patterson, B.; Subar, A. Fruit, vegetables and cancer prevention: a review of the epidemiological evidence. *Nutr. Cancer* 18:1 (1992).

4. Cotelle, N.; Bernier, J.-L.; Catteau, J.-P.; Pommery, J.; Wallet, J.-C.; Gaydou, E.M. Antioxidant properties of hydroxy-flavones. *Free Rad. Biol. Med.* 20:35 (1996).

5. Tzeng, S.H.; Ko, W.C.; Ko, F.N.; Teng, C.M. Inhibition of platelet aggregation by some flavonoids. *Thromb. Res.* 64:91 (1991).

6. Uda, Y.; Price, K.R.; Williamson, G.; Rhodes, M.J.C. Induction of the anticarcinogenic marker enzyme quinone reductase in murine hepatoma cells *in vitro* by flavonoids. *Cancer Lett.* 120:213 (1997).

7. Ranelletti, F.O.; Ricci, R.; Larocca, L.M. Growth inhibitory effects of quercetin and presence of estrogen binding sites in human colon cancer cell lines and primary colorectal tumours. *Int. J. Cancer* 50:486 (1992).

8. Scambia, G.; Ranelletti, F.O; Panici, P.B.; Piantelli, M.; Bonanno, G.; Devincenzo, R.; Ferrandina, G.; Rumi, C.; Larocca, L.M.; Mancuso, S. Inhibitory effects of quercetin on OVCA-433 cells and presence of type II oestrogen binding sites in primary ovarian tumours and cultured cells. *Brit. J. Cancer* 62:942 (1990).

9. Yoshida, M.; Sakai, T.; Hosokawa, N.; Marui, N.; Matsumoto, K.; Fujioka, A.; Nishino, H.; Aoika, A. The effect of quercetin on cell cycle progression and growth of human gastric cancer cells. *FEBS Lett.* 260:10 (1990).

10. Rice-Evans, C.A.; Miller, N.J.; Paganga, G. Structure—antioxidant activity relationships of flavonoids and phenolic acids. *Free Rad. Biol. Med.* 20:933 (1996).

11. Fossen, T.; Pedersen, A.T.; Andersen, O.M. Flavonoids from red onion (Allium Capa). *Phytochemistry* 47:281 (1998).

12. Kühnau, J. The flavonoids: a class of semi-essential food components: their role in human nutrition. *Wld. Rev. Nutr. Diet* 24:117 (1976).

13. Hertog, M.G.L.; Hollman, P.C.H.; Katan, M.B.; Kromhout, D. Intake of potentially anticarcinogenic flavonoids and their determinants in adults in The Netherlands. *Nutr. Cancer* 20:21 (1993).

14. Herrmann, K. Flavonols and flavones in food plant: a review. *J. Food Tech.* 11:433 (1976).

15. Hertog, M.G.L.; Kromhout, D.; Aravanis, C.; Blackburn, H.; Buzina, R.; Fidanza, F.; Giampaoli, S.; Jansen, A.; Menotti, A.; Nedeljkovic, S.; Pekkarinen, M.; Simic, B.S.; Toshima, H.; Feskens, E.J.M.; Hollman, P.C.H.; Katan, M.B. Flavonoid intake and long-term risk of coronary heart disease and cancer in the seven countries study. *Arch. Intern. Med.* 155:381 (1995).

16. Hertog, M.G.L.; Hollman, P.C.H.; Katan, M.B. Content of potentially anticarcinogenic flavonoids of 28 vegetables and 9 fruits commonly consumed in The Netherlands. *J. Agric. Food Chem.* 40:2379 (1992).

17. Price, K.R.; Rhodes, M.J.C.; Barnes, K.A. Flavonol glycoside content and composition of tea infusions made from commercially available teas and tea products. *J. Agric. Food Chem.* 46:2517 (1998).

18. Hertog, M.G.; Feskens, E.J.M.; Hollman, P.C.H.; Katan, M.B.; Kromhout, D. Dietary antioxidant flavonoids and risk of coronary heart disease: The Zutphen Elderly Study. *The Lancet* 342:1007 (1993).

19. Knekt, P.; Jarvinen, R.; Reunanen, A.; Maatela, J. Flavonoid intake and coronary mortality in Finland: a cohort study. *BMJ* 312:478 (1996).

20. Keli, S.O.; Hertog, M.G.L.; Feskens, E.J.M.; Kromhout, D. Dietary flavonoids, antioxidant vitamins, and incidence of stroke. *Arch. Intern. Med.* 156:637 (1996).

21. Rimm, E.B.; Katan, M.B.; Ascherio, A.; Stampfer, M.J.; Willett, W.C. Relation between intake of flavonoids and risk for coronary heart disease in male health professionals. *Ann. Intern. Med.* 125:384 (1996).

22. Hertog, M.G.L.; Sweetnam, P.M.; Fehily, A.M.; Elwood, P.C.; Kromhout, D. Antioxidant flavonols and ischemic heart disease in a Welsh population of men: the Caerphilly study. *Am. J. Clin. Nutr.* 65:1489 (1997).

23. Hertog, G.L.; Feskens, E.J.M.; Kromhout, D. Antioxidant flavonols and coronary heart disease risk. *The Lancet* 349:699 (1997).

24. van het Hof, K.H.; Kivits, G.A.A.; Weststrate, J.A.; Tijburg, L.B.M. Bioavailability of catechins from tea: the effect of milk. *Eur. J. Clin. Nutr.* 52:356 (1998).

25. Price, K.R.; Casuscelli, F.; Colquhoun, I.J.; Rhodes, M.J.C. Composition and content of flavonol glycosides in broccoli florets (*Brassica olearacea*) and their fate during cooking. *J. Sci. Food Agric.* 77:468 (1998).

26. Price, K.R.; Bacon, J.R.; Rhodes, M.J.C. Effect of storage and domestic processing on the content and composition of flavonol glucosides in onion (*Allium cepa*). *J. Agric. Food Chem.* 45:938 (1997).

27. Price, K.R.; Colquhoun, I.J.; Barnes, K.A.; Rhodes, M.J.C. The composition and content of flavonol glycosides in green beans and their fate during processing. *J. Agric. Food Chem.* 48:4898 (1998).

28. Crozier, A.; Lean, M.E.J.; McDonald, M.S.; Black, C. Quantitative analysis of the flavonoid content of commercial tomatoes, onions, lettuce, and celery. *J. Agric. Food Chem.* 45:590 (1997).

29. Tsushida, T.; Suzuki, M. Content of flavonoid glucosides and some properties of enzymes metabolizing the glucoside in onion. *Nippon Shokuhin Kagaku Kaishi* 43:642 (1996).

30. Price, K.R.; Rhodes, M.J.C. Analysis of the major flavonol glycosides present in four varieties of onion and changes in composition resulting from autolysis. *J. Sci. Food Agric.* 74:331 (1997).

31. McDonald, M.S.; Hughes, M.; Burns, J.; Lean, M.E.J.; Matthews, D.; Crozier, A. Survey of the free and conjugated myricetin and quercetin content of red wines of different geographical origins. *J. Agric. Food Chem.* 46:368 (1998).

32. Hollman, P.C.H. Determinants of the absorption of the dietary flavonoid quercetin in man. Ph.D Thesis, Wageningen Agricultural University, The Netherlands (1997)

33. Hollman, P.C.H.; de Vries, J.H.M.; van Leeuwen, S.D.; Mengelers, M.J.B.; Katan, M.B. Absorption of dietary quercetin glycosides and quercetin in healthy ileostomy volunteers. *Am. J. Clin. Nutr.* 62:1276 (1995).

34. Hollman, P.C.H.; van Trijp, J.M.P.; Buysman, M.N.C.P. Fluorescence detection of flavonols in HPLC by postcolumn chelation with aluminium. *Anal. Chem.* 68:3511 (1996).

35. Hollman, P.C.H.; Gaag, M.V.D.; Mengelers, M.J.B.; van Trijp, J.M.P.; de Vries, J.H.M.; Katan, M.B. Absorption and disposition kinetics of the dietary antioxidant quercetin in man. *Free Rad. Biol. Med.* 21:703 (1996).

36. Hollman, P.C.H.; Van Trijp, J.M.P.; Buysman, M.N.C.P.; Gaag, M.S.; Mengelers, M.J.B.; de Vries, J.H.M.; Katan, M.B. Relative bioavailablity of the antioxidant quercetin from various foods in man. *FEBS Lett.* 418:152 (1997).

37. de Vries, J.H.M.; Hollman, P.C.H.; Meyboom, S.; Buysman, M.N.C.P.; Zock, P.L.; van Staveren, W.A.; Katan, M.B. Plasma concentrations and urinary excretion of the antioxidant flavonols quercetin and kaempferol as biomarkers for dietary intake. *Am. J. Clin. Nutr.* 68:60 (1998).

38. Janseen, P.L.T.M.; Mensink, R.P.; Cox, F.J.J.; Harryvan, J.L.; Hovenier, R.; Hollman, P.C.H.; Katan, M.B. Effects of the flavonoids quercetin and apigenin on hemostasis in healthy volunteers: results from an *in vitro* and a dietary supplement study. *Am. J. Clin. Nutr.* 67:255 (1998).

39. Nielsen, S.E. Metabolism and biomarker studies of dietary flavonoids. Ph.D thesis, Danish Veterinary and Food Administration, Denmark (1998)

40. Young, J.F.; Nielsen, S.E.; Haraldsdottir, J.; Daneshvar, B.; Lauridsen, S.T.; Knuthsen, P.; Crozier, A.; Sandstrom, B.; Dragsted, L.O. Effect of fruit juice intake on urinary quercetin excretion and biomarkers of antioxidant status. *Am. J. Clin. Nutr.* 69:87 (1999).

41. Gee, J.M.; DuPont, M.S.; Rhodes, M.J.C.; Johnson, I.T. Quercetin glucosides interact with the intestinal glucose transport pathway. *Free Rad. Biol. Med.* 25:19 (1998).

42. Paganga, G.; Rice-Evans, C. The identification of flavonoids as glycosides in human plasma. *FEBS Lett.* 401:78 (1997).

43. Frank, H.K. Dictionary of food microbiology. Technonomic Publishing Company, Lancaster, USA (1992).

44. Hackett, A.M. The metabolism of flavonoid compounds in mammals. *In*: Cody, V.; Middleton, E.; Harborne, J.B. (eds.) Plant flavonoids in biology and medicine: biochemical, pharmological and structure-activity relationships. Alan Liss Inc., New York p. 177 (1986).

45. Mizuma, T.; Ohta, K.; Hayashi, M.; Awazu, S. Comparitive study of active absorption by the intestine and disposition of anomers of sugar-conjugated compounds. *Biochem. Pharm.* 45:1520 (1993).

46. Day, A.J.; DuPont, M.S.; Ridley, S.; Rhodes, M.; Rhodes, M.J.C.; Morgan, M.R.A.; Williamson, G. Deglycosylation of flavonoid and isoflavonoid glycosides by human small intestine and liver β-glucosidase activity. *FEBS Lett.* 436:71 (1998).

47. Daniels, L.B.; Coyle, P.J.; Chiao, Y.-B.; Glew, R.H. Purification and characterization of a cytosolic broad specificity beta-glucosidase from human liver. *J. Biol. Chem.* 256:13004 (1981).

48. Leese, H.J.; Semenza, G. On the identity between the small intestinal enzymes phlorizin hydrolase and glycosylceramidase. *J. Biol. Chem.* 248:8170 (1973).

49. Baba, S.; Furuta, T.; Horie, M.; Nakagawa, H. Studies on drug metabolism by use of isotopes XXVI: Determination of urinary metaboiltes of rutin in humans. *J. Pharm. Sci.* 70:780 (1981).

50. Baba, S.; Furuta, T.; Fujioka, M.; Goromaru, T. Studies in drug metabolism by use of isotopes XXVII. Urinary metabolites of rutin in rats and the role of intestinal microflora in the metabolism of rutin. *J. Pharm. Sci.* 72:1155 (1983).

51. Booth, A.N.; Murray, C.W.; Jones, F.T.; DeEds, F. The metabolic fate of rutin and quercetin in the animal body. *J. Biol. Chem.* 223:251 (1956).

52. Petrakis, P.L.; Kallianos, A.G.; Wender, S.H.; Shetlar, M.R. Metabolic studies of quercetin labeled with ¹⁴C. *Arch. Biochem. Biophys.* 85:264 (1959).

53. Sawai, Y.; Kohsaka, K.; Nishiyama, Y.; Ando, K. Serum concentrations of rutinoside metabolites after oral administration of a rutoside formulation to humans. *Drug Res.* 37:729 (1987).

54. Ueno, I.; Nakano, N.; Hirono, I. Metabolic fate of ¹⁴C quercetin in the A.C.I. rat. *Jpn. J. Exp. Med.* 53:41 (1983).

55. Nielsen, S.E.; Breinholt, V.; Justesen, U.; Cornett, C.; Dragsted, L.O. *In vitro* biotransformation of flavonoids by rat liver microsomes. *Xenobiotic* 28:389 (1998).

56. Nielsen, S.E.; Kall, M.; Justesen, U.; Schou, A.; Dragsted, L.O. Human absorption and excretion of flavonoids after broccoli consumption. *Cancer Lett.* 114:173 (1997).

57. Ameer, B.; Weintraub, R.A.; Johnson, J.V.; Yost, R.A.; Rouseff, R.L. Flavanone absorption after naringin, hesperidin, and citrus administration. *Clin. Pharm. Ther.* 60:34 (1996).

58. Manach, C.; Morand, C.; Demigne, C.; Texier, O.; Regerat, F.; Remesy, C. Bioavailability of rutin and quercetin in rats. *FEBS Lett.* 409:12 (1997).

59. Manach, C.; Morand, C.; Texier, O.; Favier, M.L.; Agullo, G.; Demigune, C.; Regerat, F.; Remesy, C. Quercetin metabolites in plasma of rats fed diets containing rutin or quercetin. *J. Nutr.* 125:1911 (1995).

60. Manach, C.; Texier, O.; Regerat, F.; Agullo, G.; Demigne, C.; Remesy, C. Dietary quercetin is recovered in rat plasma as conjugated derivates of isorhamnetin and quercetin. *Nutr. Biochem.* 7:375 (1996).

61. Zhu, B.T.; Ezell, E.L.; Liehr, J.G. Catechol-o-methyltransferase catalyzed rapid o-methylation of mutagenic flavonoids. *J. Biol. Chem.* 269:292 (1994)

62. Manach, C.; Morand, C.; Crespy, V.; Demigne, C.; Texier, O.; Regerat, F.; Remesy, C. Quercetin is recovered in human plasma as conjugated derivatives which retain antioxidant properties. *FEBS Letters* 426:331 (1998).

63. Smith, R.L. The excretory function of bile. Chapman and Hall, London (1973).

64. Gibson, G.G.; Skett, P. (eds.) Factors affecting drug metabolism: internal factors. *In:* Introduction to drug metabolism. Blackie Academic and Professional, London p.107 (1994).

65. Sfakianos, J.; Coward, L.; Kirk, M.; Barnes, S. Intestinal uptake and bilary excretion of the isoflavone genistein in rats. *J. Nutr.* 127:1260 (1997).

66. Boutin, J.A.; Meunier, F.; Lambert, P.-H.; Hennig, P.; Bertin, D.; Serkiz, B.; Volland, J.-P. *In vivo* and *in vitro* glucuronidation of the flavonoid diosmetin in rats. *Drug Met. Disposition* 21:1157 (1993).

67. Mizuma, T.; Ohta, K.; Awazu, S. The β-anomeric and glucose preferences of glucose transport carrier for intestinal active absorption of monosaccharide conjugates. *Biochim. Biophys. Acta.* 1200:117 (1994).

68. Mizuma, T.; Awazu, S. Intestinal Na$^+$/glucose cotransporter mediated transport of glucose conjugate formed from disaccharide conjugate. *Biochim. Biophys. Acta.* 1379:1 (1998).

69. Piskula, M.K.; Terao, J. Accumulation of (–)-epicatechin metabolites in rat plasma after oral administration and distribution of conjugated enzymes in rat tissues. *J. Nutr.* 128:1172 (1998).

70. Crespy, V.; Manach, C.; Morand, C.; Besson, C.; Demigne, C.; Remesy, C. Intestinal absorption and metabolism of quercetin. *In:* Polyphenol communications 98, XIXth International conference on polyphenols. Lille, France p. 75 (1998).

71. Stocker, R.; Frei, B. Endogenous antioxidant defenses in human blood plasma. *In:* Sies, H. (ed.) Oxidative stress: oxidants and antioxidants, Academic Press, London p. 213 (1991).

SPIN STABILIZING APPROACH TO RADICAL CHARACTERIZATION OF PHENYLPROPANOID ANTIOXIDANTS: AN ESR STUDY OF CHLOROGENIC ACID OXIDATION IN THE HORSERADISH PEROXIDASE, TYROSINASE, AND FERRYLMYOGLOBIN PROTEIN RADICAL SYSTEMS

Stephen C. Grace,[a] Hideo Yamasaki,[b] and William A. Pryor[a]

[a] Biodynamics Institute
Louisiana State University
Baton Rouge, Louisiana 70803-1800
USA

[b] Laboratory of Cell and Functional Biology
Faculty of Science
University of the Ryukyus
Nishihara, Okinawa 903-0213
JAPAN

1. INTRODUCTION

Polyphenols are widespread in the environment, especially as constituents of edible plants. Over the last decade there has been a growing interest in the antioxidant properties of dietary polyphenols and their potential benefits for human health. Although some of these properties were first recognized over 60 years ago by Szent-Györgyi and colleagues,[1] only in recent years has significant progress been made toward understanding the molecular details of the antioxidant mechanisms.[2-4]

Plant polyphenols are synthesized from metabolic precursors formed in the phenylpropanoid biosynthetic pathway. An important but often overlooked group of products are the phenylpropanoid esters that can accumulate to high levels in some plants under certain conditions. Phytochemical surveys indicate that chlorogenic acid (CGA), an ester of caffeic and quinic acid (fig. 1), is one of the most abundant phenylpropanoid esters in vascular plants, both taxonomically[5] and

Plant Polyphenols 2: Chemistry, Biology, Pharmacology, Ecology, Edited by
Gross et al. Kluwer Academic / Plenum Publishers, New York, 1999

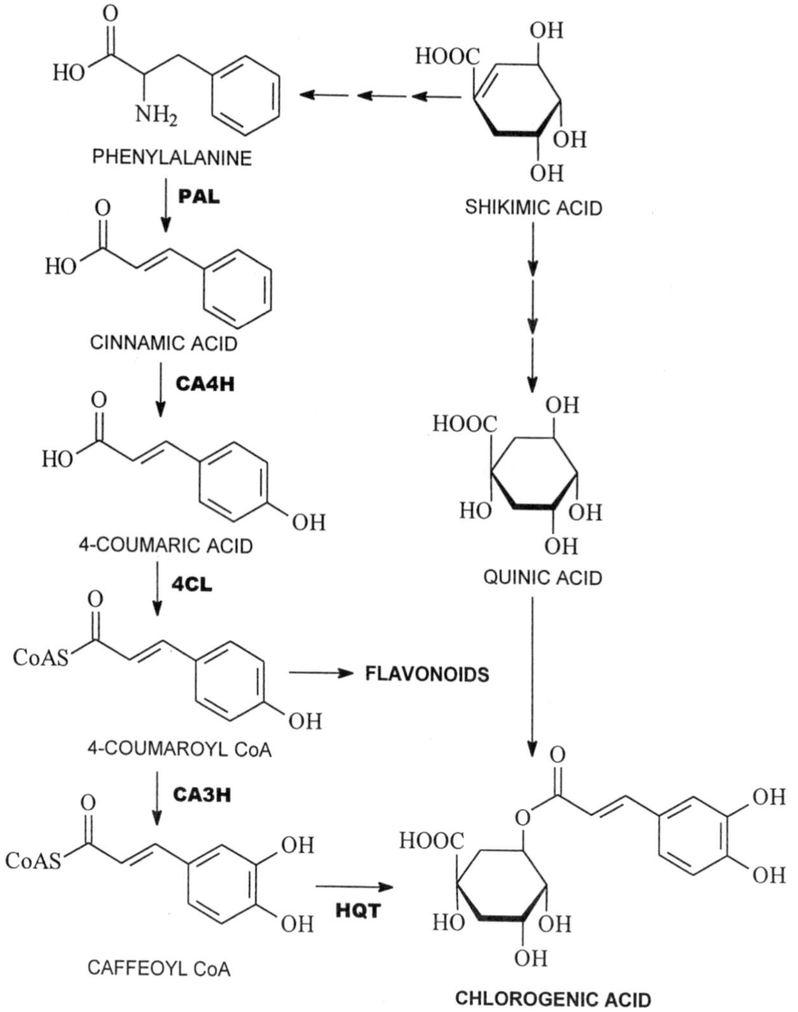

Figure 1. Biosynthetic pathway for chlorogenic acid in plants. The enzymes involved are phenylalanine-ammonia lyase (PAL), cinnamate 4-hydroxylase (CA4H), 4-coumaroyl:CoA ligase (4CL), coumarate 3-hydroxylase (CA3H), and hydroxycinnamoyl CoA 5-quinate: hydroxycinnamyl transferase (HQT).

compositionally.[6] Chlorogenic acid is a potent scavenger of oxygen free radicals as well as other reactive species such as peroxynitrite[7-10] and may have a similar function in plant tissues, particularly under conditions of environmental stress.[11] It is also a natural phenolic substrate for various oxidative enzymes found in plant and animal tissues, such as horseradish peroxidase, tyrosinase, and human myeloperoxidase.[10,12]

Although polyphenols have been recognized largely as beneficial antioxidants, there have been a number of reports in the literature suggesting that dietary catechols and their oxidized derivatives exhibit both prooxidant and cytotoxic properties in living systems.[13-17] The catechol group found in many polyphenolic structures is readily oxidized by free radicals, transition metals, and oxidative enzymes, reflecting the fact that these compounds are universally good hydrogen atom donors. However, *o*-semiquinones, the one-electron oxidation products, also have the potential to *enhance* free radical production through redox cycling and related mechanisms.[18] For example, many phenolic antitumor drugs (e.g., etoposide) and environmental carcinogens (e.g., benzene, cigarette tar) generate free radicals via a semiquinone intermediate, and this process has been associated with their genotoxic and in some cases mutagenic effects.[19-21] Since dietary polyphenols also have the potential to act as prooxidants under certain conditions, understanding the properties and reactivities of their free radical intermediates is essential in any effort to assess their putative health benefits.

Electron spin resonance (ESR) spectroscopy is an ideal technique for obtaining information about semiquinones and other types of organic radicals produced during autoxidations and in reactions with metabolizing enzymes. Analysis of the hyperfine splitting constants (hfsc) and the *g*-factor in an ESR spectrum provides information on the interactions of the unpaired electron with neighboring atoms and the identity of the radical species under investigation. Although the ESR properties of polyphenols can be easily studied under extreme alkaline conditions (pH > 10) due to the relative stability of the semiquinones,[22-24] in the biological pH range where most enzyme reactions occur (pH 5–8), *o*-semiquinones are unstable intermediates that rapidly decay by disproportionation to yield quinones and other non-radical products.[25] Therefore, it is generally not possible to detect these species at physiological pH without the aid of "spin stabilizing" metal ions (e.g., Zn^{2+}, Mg^{2+}) that increase the lifetimes of semiquinones through the formation of stable complexes.[26] Although the spin stabilization technique has been applied extensively to investigate the ESR properties of catecholamines,[26,27] there have been few studies of plant polyphenols using this approach. Here, we report the use of spin stabilization for the detection and analysis of radicals derived from chlorogenic acid in three enzyme systems of biological interest: horseradish peroxidase, tyrosinase, and the ferrylmyoglobin protein radical.

2. MATERIALS AND METHODS

Chlorogenic acid, horseradish peroxidase (Type VI-A), mushroom tyrosinase, and horse heart myoglobin were purchased from Sigma (St. Louis, Missouri, USA). The concentration of horseradish peroxidase was determined from the absorbance at 402 nm ($\varepsilon_{402} = 2 \times 10^5 \, M^{-1} cm^{-1}$).[27] Metmyoglobin ($Mb^{III}$) was prepared by treatment of 1 mM myoglobin with 2 mM potassium ferricyanide, followed by purification on a Sephadex G-25 gel exclusion column. The concentration of metmyoglobin was determined from the absorbance at 632 nm ($\varepsilon_{632} = 3,770 \, M^{-1} cm^{-1}$).[28]

For spin stabilization studies, chlorogenic acid was dissolved in aqueous buffer containing 50 mM 2-[N-morpholino]ethanesulfonic acid (MES)-NaOH and 200 mM zinc sulfate and the final pH adjusted to 5.6. The use of higher pH values was not

possible because of zinc insolubility. The high concentration of zinc required to optimize the ESR signal has previously been shown to be without marked effect on
horseradish peroxidase activity.[26] Solutions were transferred to an aqueous quartz
flat cell, and ESR measurements were recorded at room temperature using a Varian
109 ESR spectrometer operating at X band (9.5 GHz) and employing 100 kHz field
modulation. The operating conditions were as follows: microwave power 10–20 mW,
modulation amplitude 0.4 gauss, receiver gain 4×10^3, time constant 250 ms.

Computer-based simulations of ESR spectra were performed using software
provided by the National Institute of Environmental Health Sciences (Research
Triangle Park, North Carolina, USA) available at the NIEHS website at
http://epr.niehs.nih.gov.

3. RESULTS AND DISCUSSION

Chlorogenic acid is rapidly oxidized by horseradish peroxidase in the presence
of H_2O_2.[9] Addition of Zn^{2+} to the reaction mixture as a spin stabilizing metal ion
produces a strong ESR signal (fig. 2, *upper spectrum*), confirming that radicals are
formed in the peroxidase reaction and identifying the catechol group is the primary
radical site. The *g* factor for this species is 2.0044, similar to other zinc-complexed
semiquinones,[26,27,29] as well as the uncomplexed semiquinone found in cigarette
tar.[21] No ESR signals are observed in the absence of either enzyme or H_2O_2 (not
shown). When the reaction is carried out in the absence of a complexing metal ion,
a weak ESR signal is observed (fig. 2, *lower spectrum*) that decays within minutes
to ESR silent species (not shown). Thus, Zn^{2+} or other divalent metals (e.g., Mg^{2+},
Cd^{2+})[30] can stabilize the CGA semiquinone anion to allow its detection under static
ESR conditions, as previously shown for several catecholamine radicals.[27] The
major pathway for decay of the semiquinone in the absence of chelating metals is
by second-order disproportionation.[26]

With prolonged incubation, a secondary radical species appears in the horseradish peroxidase reaction coinciding with the loss of the primary *o*-semiquinone
(fig. 3). The intensity of this new, secondary radical increases as the reaction proceeds and reaches a stable value after 20 min (fig. 3B). The rate constant, k_{obs}, for
decay of the primary radical is approximately 0.92 min^{-1}, whereas the formation
of the secondary radical is considerably slower ($k_{obs} = 0.40$ min^{-1}), indicating that
the secondary radical is not the *direct* product of the reaction responsible for decay
of the primary radical. A mechanism for secondary radical formation will be discussed below.

We also employed the ESR spin stabilization method to study the oxidation of
CGA in the tyrosinase reaction. Tyrosinase is a bifunctional copper-containing
enzyme that catalyzes the *o*-hydroxylation of phenols to their corresponding
catechols as well as the two-electron oxidation of catechols to their corresponding
o-quinones. The enzyme has been implicated in several metabolic processes,
including the bioactivation of phenolic antitumor drugs against human cancer
cells[31] and the browning reaction observed upon wounding of plant tissues.[32]

In aerated solutions containing tyrosinase and Zn^{2+}, the CGA semiquinone is
detected as a transient species (fig. 4). However, in contrast to the horseradish

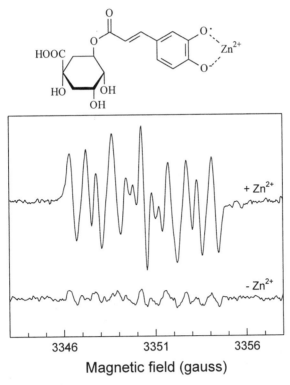

Figure 2. ESR spectrum of the chlorogenic acid *o*-semiquinone radical in the presence and absence of zinc ions. The reaction contained 10 mM CGA, 14 mM H_2O_2 and 0.1 nM HRP in 50 mM MES-NaOH buffer, pH 5.6 in the presence (*upper spectrum*) or absence (*lower spectrum*) of 200 mM $ZnSO_4$. The ESR spectra were recorded approximately 1 minute after addition of H_2O_2. Zinc ions stabilize the semiquinone anion (pK_a 3–4)[29] by forming a complex, as shown at the top. The ESR instrument conditions were as follows: microwave power, 10 mW; field modulation, 100 kHz; modulation amplitude, 0.4 gauss; time constant, 0.25 s; receiver gain, 4×10^2; sweep time, 20 gauss min^{-1}.

peroxidase reaction, in which a stable secondary radical species appears, the semiquinone formed in the tyrosinase reaction decays to an ESR silent species within several minutes, and no other radicals are observed. This unique behavior can be explained by the well-known two-electron mechanism of catechol oxidation by tyrosinase.

As shown in figure 5, the initial oxidation product in the tyrosinase reaction is the CGA *o*-quinone. This species can subsequently react with the parent catechol by reverse disproportionation to generate the *o*-semiquinone. Such a mechanism was previously proposed to explain the formation of catecholamine *o*-semiquinones in the tyrosinase reaction.[33] As observed in the horseradish peroxidase reaction, the CGA semiquinone is a short-lived species that decays at a rate that is dependent on the concentration of enzyme added (fig. 4B). Under

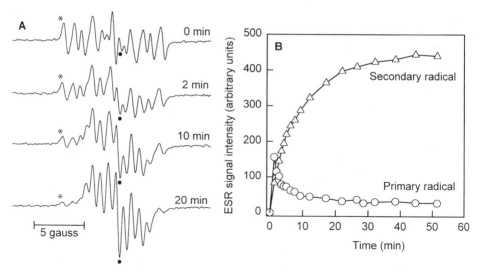

Figure 3. Formation of a secondary CGA radical species in the horseradish per-oxidase-H_2O_2 reaction. (**A**) Time-dependent changes in the ESR spectrum. (**B**) Kinetics of radical transition; ($\circ$) primary radical, ($\triangle$) secondary radical. The changes in the ESR signal intensities were determined at g $\approx$ 2.0073 for the primary radical (*marked* *) and g $\approx$ 2.0039 for the secondary radical (*marked* •) where the spectral overlap is minimal. All conditions were as in figure 2, except that the HRP concentration was 0.4 nM.

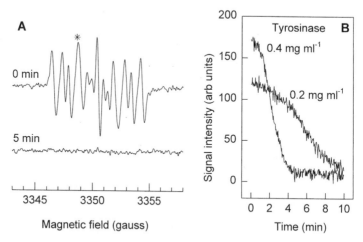

Figure 4. Detection of CGA radicals in the tyrosinase reaction. (**A**) ESR spectra recorded 30 s and 5 min after addition of 0.4 mg ml^{-1} enzyme. (**B**) Kinetics of the reaction monitored from the changes in the g $\approx$ 2.0056 peak (*marked* *).

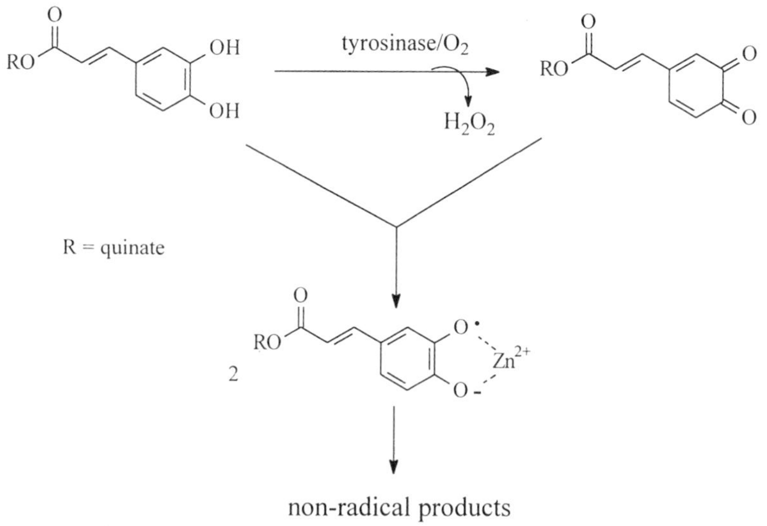

Figure 5. Mechanism for CGA *o*-semiquinone formation in the tyrosinase reaction. Tyrosinase oxidizes CGA to the *o*-quinone in the presence of molecular oxygen. The quinone reacts with the parent catechol by reverse disproportionation to form the semiquinone which subsequently decays to nonradical products.

these experimental conditions, CGA oxidation presumably becomes limited by the availability of molecular oxygen in the solution and thus, as quinone formation ceases, the ESR signal decays as the semiquinone is consumed by secondary reactions.

Myoglobin is the major heme-containing protein of cardiac tissue.[34,35] The ferric form of the protein (Mb^{III}) can react with hydrogen peroxide to produce ferrylmyoglobin, a two-electron oxidation product containing the oxoferryl species, $Fe^{IV} = O$, and a transient protein radical. Ferrylmyoglobin is a strong one-electron oxidant that has been implicated in the oxidation of low density lipoprotein (LDL), a process thought to be involved in atherosclerosis and other human disorders.[36] A variety of antioxidants can reduce ferrylmyoglobin to Mb^{III}, with the concomitant production of antioxidant radicals.[34] Dietary polyphenols such as CGA and caffeic acid also catalyze the conversion of ferrylmyoglobin to metmyoglobin,[37] and in doing so inhibit ferrylmyoglobin-mediated oxidative modification of LDL.[38]

Ferrylmyoglobin has two target sites for redox reactions, the oxoferryl heme moiety and an amino acid radical located at the protein surface.[28] Both sites can undergo electron-transfer reactions with reducing agents such as ascorbate and Trolox C, the water soluble analog of vitamin E.[34] Laranjinha et al.[37] postulated that CGA and other dietary phenylpropanoids can also react with both sites, but they were unable to detect radical products due to the instability of the semiquinones.

In the presence of Zn^{2+} as a spin stabilizing ion, we were able to detect the formation of radicals in the reaction of CGA with ferrylmyoglobin (fig. 6A). Addition

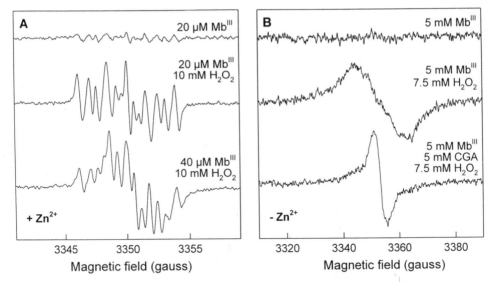

Figure 6. Detection of CGA radicals in the ferrylmyoglobin reaction. (**A**) Reactions contained 10 mM CGA, 50 mM MES-NaOH (pH 5.6), 200 mM ZnSO$_4$ and, where indicated, 20 μM or 40 μM myoglobin (Mb^{III}). Ferrylmyoglobin was generated by the addition of 10 mM H$_2$O$_2$. (**B**) Reactions contained 5 mM Mb^{III} in PBS buffer (10 mM sodium phosphate, 150 mM NaCl, pH 7.4) and, where indicated, 7.5 mM H$_2$O$_2$ and 5 mM CGA. The instrument conditions in panel A were as in figure 2 and in panel B were as follows: 20 mW power, 2 gauss modulation amplitude, 1 $\times$ 10^4 receiver gain, 0.25 s time constant, 50 G min^{-1} sweep time. Note the difference in scale for the field sweeps in the two panels.

of H$_2$O$_2$ to solutions containing CGA and Mb^{III} generates a radical with an ESR spectrum identical to that observed in the HRP-H$_2$O$_2$ reaction, confirming that the CGA semiquinone is produced during the reduction of ferrylmyoglobin to Mb^{III}. At higher concentrations of Mb^{III}, the ESR spectrum of the secondary radical species appears, indicating that oxidation of CGA by horseradish peroxidase and ferrylmyoglobin follows similar pathways and yields similar products. A weak ESR signal is observed in the presence of Mb^{III} even without the addition of hydrogen peroxide (fig. 6A), possibly due to a slow reduction of the ferric heme or autoxidation of the polyphenol.

In the absence of Zn^{2+} and at higher concentrations of Mb^{III} (5 mM), redox transitions occurring within the protein itself can be studied directly. Although ferrylmyoglobin contains several radical species with different lifetimes,[28] we detected only a broad, featureless signal with a linewidth of ~20 gauss and a *g* value of 2.0048 in the static ESR system (fig. 6B). This signal has been interpreted to originate from an immobilized aromatic residue in the protein, i.e., histidine or phenylalanine,[28] or tryptophan.[35] In the presence of CGA, the broad signal is replaced by a narrower (~5 gauss) signal with a similar *g* value. Although the hyperfine structure of this radical could not be resolved, the line width and g value are consistent with the formation of the secondary CGA radical.

Figure 7 shows the dependence of the CGA semiquinone ESR intensity on oxidant concentration in the horseradish peroxidase and ferrylmyoglobin reactions. In the presence of H_2O_2, the CGA semiquinone is detected at HRP concentrations as low as 10^{-11} M whereas at least 10^{-6} M myoglobin is required to achieve a similar signal intensity, indicating that chlorogenic acid is a markedly better substrate in the HRP reaction as compared to the ferrylmyoglobin reaction.

Information about the structure of CGA radicals was obtained from a computer-based simulation of their ESR spectra and analysis of the hyperfine splitting constants (hfsc). The simulated spectra were in excellent agreement with the experimental spectra for both radical species (fig. 8). The primary radical shows coupling between the unpaired electron and four inequivalent protons with hfsc values of $a_H^2 = 0.87$ G, $a_H^5 = 1.36$ G and $a_H^6 = 3.00$ G and $a_H^\beta = 2.52$ G. The assignment of coupling constants to specific protons is based on a comparison with Zn^{2+}-stabilized catecholamine radicals.[27,29,30] We did not observe hyperfine interaction between the unpaired electron and the side chain proton at position 8 ($a_H^\gamma \approx 0$), consistent with previous studies of p-substituted benzosemiquinones.[30] These data suggest an "open-chain" structure for the Zn^{2+}-complexed primary CGA o-semiquinone as depicted in figure 8.

To analyze the structure of the secondary CGA radical, the ESR spectrum of the primary radical was subtracted from the superimposed spectra after normalizing the relative intensity of the outer lines. The result is a symmetrical ESR spectrum, suggestive of a single radical species (fig. 8). However, unlike

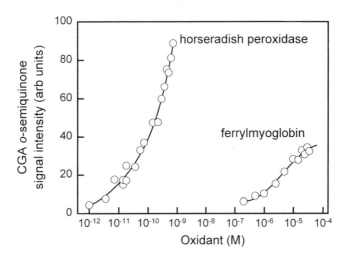

Figure 7. Oxidant dependence of CGA o-semiquinone formation in the horseradish peroxidase and ferrylmyoglobin reactions. The concentration of CGA radicals was estimated from the height of the low-field peak at $g \approx 2.0073$ Experimental conditions were as in figure 2.

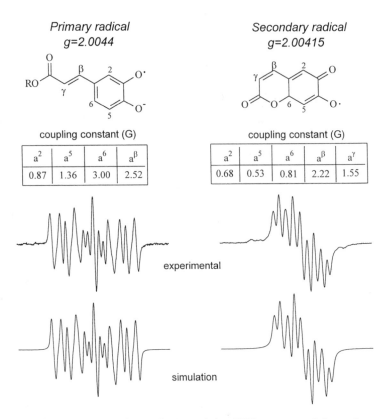

Figure 8. Computer-based simulation of the ESR spectra of the primary and seconday CGA radicals. Values for the hyperfine splitting constants of each radical species are also shown. See text for details.

the primary radical, hyperfine interaction is observed between the unpaired electron and a total of *five* protons with hfsc values of $a_H^2 = 0.68$, $a_H^5 = 0.53\,G$, $a_H^6 = 0.81\,G$, $a_H^\beta = 2.22\,G$, and $a_H^\gamma = 1.55\,G$. The larger number of proton couplings in this species is consistent with a cyclized structure in which the unpaired electron is delocalized over a more extended π-conjugated system in comparison to the primary radical. A putative structure for the CGA secondary radical is shown in figure 8.

We have shown that the secondary CGA radical is formed in the HRP and ferrylmyoglobin reactions, both one-electron oxidizing systems, but not in the tyrosinase reaction, a two-electron oxidizing system. Furthermore, the kinetics of primary radical decay and secondary radical formation are markedly dissimilar (fig. 3), indicating that there is not a simple interconversion of one radical species to another in the peroxidase system. We suggest that the stable CGA secondary radical arises from intramolecular cyclization of the primary radical followed by a second peroxidase-catalyzed oxidation reaction. A mechanism for decay of the CGA semiquinone to an ESR- silent lactone followed by the formation of a stable secondary radical species in the peroxidase reaction, but not in the tyrosinase reaction, is presented in figure 9.

First, consider the structure of the Zn^{2+}-stabilized CGA o-semiquinone: phenoxyl radicals contain a delocalized π-electron capable of interacting with neighboring atoms. The magnitude of this interaction depends on the coupling constant, which is proportional to the spin density on the atom to which the hydrogen is bonded.[39] Theoretical and empirical analyses of a range of alkyl-substituted o-semiquinones have shown that the highest spin density occurs on carbon atoms located *para* to the diphenolic oxygens, i.e., at ring positions 1 and 6.[30,39] Thus, the largest coupling constant in the CGA primary radical (3.00 G) is assigned to the carbon at position 6 (fig. 8). As will be discussed below, the concentration of spin density at this carbon would be favorable for an intramolecular cyclization reaction.

An important feature of hydroxycinnamates such as CGA is the side-chain double bond that allows these compounds to exist as *cis* (Z) and *trans* (E) stereoisomers. Although in the native state hydroxycinnamates are predominantly *trans*,[3] in aqueous solvents they isomerize to the less stable *cis* form, particularly under the influence of light.[40] As shown in figure 9, hydrogen abstraction by horseradish peroxidase or other oxidative enzymes would generate a radical, **III**, where spin density is brought into the vicinity of side-chain carbonyl oxygen through resonance stabilization. This species can undergo intramolecular cyclization as a result of nucleophilic attack of the unpaired electron on the carbonyl oxygen, generating radical **IV**. Cyclization represents an equilibrium process in which the radical

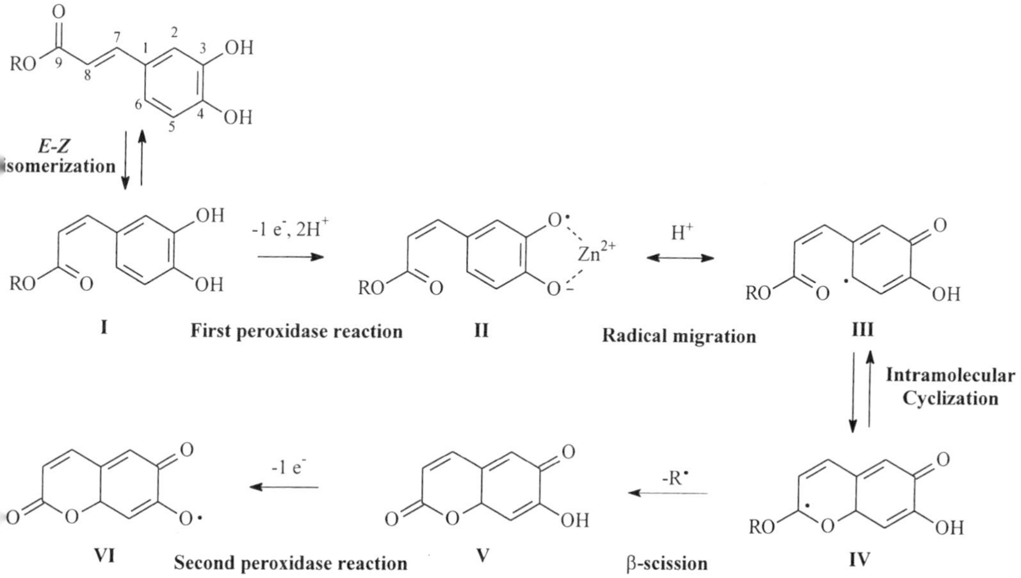

Figure 9. Proposed mechanism for formation of a secondary CGA radical species in peroxidase-catalyzed reactions. See text for details.

character of the molecule is retained (see below), and therefore does not explain the decay of the primary radical species in the tyrosinase system.

We postulate that decay of the primary CGA radical involves a β-scission reaction within the cyclized intermediate **IV** to yield a carbon-centered quinyl radical and a nonradical lactone as products. As shown in figure 10, if homolysis occurs at the C-6 carbon of the phenyl ring, radical **III** is regenerated, which, as alluded to above, is in equilibrium with the primary semiquinone. In contrast, if homolysis occurs at the 5-O-quinyl linkage, the cleavage reaction would be essentially irreversible and would generate 3-oxo-4-hydroxycoumarin (**V**) and the quinyl radical as products. Like most alkyl radicals, the quinyl radical would be expected to decompose rapidly, either by reaction with dioxygen or through self-termination reactions such as dimerization. Therefore, it is unlikely that ESR signals arising from this species can be detected under these conditions. However, preliminary studies using the spin trap DMPO have provided strong evidence for the formation of a carbon-centered radical adduct in the CGA-peroxidase reaction (not shown). The identity of this adduct is currently under investigation.

According to our proposed scheme, the cyclized derivative **V** retains a single phenolic group that would be subject to a second oxidation in the presence of HRP or ferrylmyoglobin, but would not be readily oxidized by tyrosinase. Although less active than catechols, monophenols (e.g., p-coumaric acid, tyrosine, etoposide, vitamin E) are also oxidized by heme peroxidases in the presence of H_2O_2.[41-44] Thus, we propose that the secondary radical arises from a second peroxidase reaction involving oxidation of the cyclized CGA derivative (fig. 9). This species exhibits hyperfine coupling at the side-chain protons at positions 7 and 8, consistent with greater delocalization of the π-radical character and unlike the "open-chain" structure of the primary radical. Although the mechanism outlined in figure 9 provides a consistent explanation for all of the ESR results presented thus far, conclusive

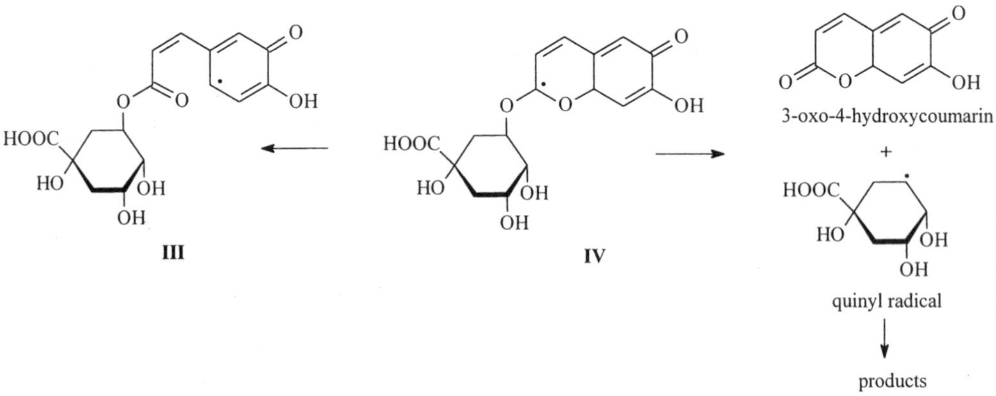

Figure 10. β-scission reaction of the ester bond of the CGA radical. Homolysis at the C-6 carbon of the phenyl ring regenerates radical **III**, whereas homolysis at the 9-O-quinyl linkage generates 3-oxo-4-hydroxycoumarin and the quinyl radical as products.

evidence awaits a detailed analysis of the products, which is beyond the scope of the present study.

4. CONCLUSIONS

Chlorogenic acid is one of the major products of phenylpropanoid metabolism in vascular plants. Although its physiological role is unclear, several studies have shown that CGA can act as both a powerful hydrogen donating antioxidant and a natural phenolic substrate for heme peroxidases such as horseradish peroxidase.[7-12] The one-electron oxidation of CGA by both radical scavenging and enzymatic mechanisms generates a semiquinone radical that may be involved in secondary reactions in biological systems. However, there has been little detailed characterization of this intermediate since o-semiquinones rapidly decay by disproportionation at physiological pH. To overcome this problem, we have used the ESR spin stabilization method first applied to catecholamines[26,27] to study and characterize CGA radicals formed in several oxidative enzyme systems. The Zn^{2+}-complexed CGA o-semiquinone is a relatively short-lived species ($t\frac{1}{2} = 178$ s) that gives an ESR multiline signal at $g = 2.0044$. A stable secondary radical is also formed in the HRP and ferrylmyoglobin reactions that we have tentatively identified as a cyclized coumarin-like derivative based on analysis of its hyperfine structure. In contrast to these one-electron oxidizing systems, the transient CGA semiquinone is the only radical species observed during the two-electron oxidation of CGA by tyrosinase.

A functional role for chlorogenic acid as an electron donor in several types of peroxidase reactions could well have physiological implications in both plants and animals. In plants, cellular H_2O_2 levels increase under conditions of high stress or rapid growth, factors also associated with the accumulation of phenolic compounds.[11] Yamasaki et al.[45] have proposed that vacuolar phenolics participate in the "delocalized detoxification" of H_2O_2 by acting as electron donors to heme peroxidases, which also are located in the vacuole. Monitoring the appearance of phenoxyl radicals after stress events by ESR could be used to test this hypothesis.

The health benefits of dietary polyphenols are thought to arise, in part, from their potent antioxidant properties. Among many possible roles in mammalian cellular metabolism, chlorogenic acid and related compounds may have antioxidant activity in cardiac tissue by converting ferrylmyoglobin, a species that has been implicated in the early events of atherosclerosis, to native metmyoglobin.[38] Chlorogenic acid might also interact with the heme enzyme myeloperoxidase produced by phagocytic cells to scavenge reactive peroxides such as peroxynitrite generated at localized sites of infection and inflammation.[10] Thus, the study of polyphenols as natural peroxidase donors is of considerable interest for its functional significance in several areas of biology.

ACKNOWLEDGMENTS

This work was supported by Grant 97-35100-5144 from the U.S. Department of Agriculture (SCG) and Grant ES-06754 from the National Institute of Environmental Health Sciences (WAP).

DEDICATION

This chapter is dedicated to the memory of Autumn Mir Grace, who passed away on August 18, 1997.

REFERENCES

1. Bentsath, A.; Rusznyak, S.; Szent-Györgyi, A. Vitamin P. *Nature* 139:326 (1937).

2. Jovanovic, S.V.; Steenken, S.; Tosic, M.; Marjanovic, B.; Simic, M.G. Flavonoids as antioxidants. *J. Am. Chem. Soc.* 116:4846 (1994).

3. Rice-Evans, C.A.; Miller, N.J.; Paganga, G. Structure-antioxidant activity relationships of flavonoids and phenolic acids. *Free Rad. Biol. Med.* 20:933 (1997).

4. van Acker, S.A.; van den Berg, D.J.; Tromp, M.N.; Griffioen, D.H.; van Bennekom, W.P.; van der Vijgh, W.J.; Bast, A. Structural aspects of antioxidant activity of flavonoids. *Free Rad. Biol. Med.* 20:331 (1996).

5. Mølgaard, P.; Ravn, H. Evolutionary aspects of caffeoyl ester distribution in dicotyledons. *Phytochemistry* 27:2411 (1988).

6. Macheix, J.J.; Fleuriet, A. Phenolic acids in fruits. *In*: Rice-Evans, C.; Packer, L. (eds.). Flavonoids in health and disease. Marcel Dekker, New York, p. 35 (1998).

7. Ohnishi, M.; Morishita, H.; Iwahashi, H.; Toda, S.; Shirataki, Y.; Kimura, M.; Kido, R. Inhibitory effects of chlorogenic acids on linoleic acid peroxidation and haemolysis. *Phytochemistry* 36:579 (1994).

8. Castelluccio, C.; Paganga, G.; Melikian, N.; Bolwell, G.P.; Pridham, J.; Sampson, J.; Rice-Evans, C. Antioxidant potential of intermediates in phenylpropanoid metabolism in higher plants. *FEBS Lett.* 368:188 (1995).

9. Kono, Y.; Kobayashi, K.; Tagawa, S.; Adachi, K.; Ueda, A.; Sawa, Y.; Shibata, H. Antioxidant activity of polyphenolics in diets. Rate constants of reactions of chlorogenic acid and caffeic acid with reactive species of oxygen and nitrogen. *Biochim. Biophys. Acta* 1335:335 (1997).

10. Grace, S.C.; Salgo, M.G.; Pryor, W.A. Scavenging of peroxynitrite by a phenolic-peroxidase system prevents oxidative damage to DNA. *FEBS Lett.* 426:24 (1998).

11. Grace, S.C.; Logan, B.A.; Adams, W.W. III. Seasonal differences in foliar content of chlorogenic acid, a phenylpropanoid antioxidant, in *Mahonia repens*. *Plant Cell Environ.* 21:513 (1998).

12. Yamasaki, H.; Grace, S.C. EPR detection of phytophenoxyl radicals stabilized by zinc ions: evidence for the redox-coupling of plant phenolics with ascorbate in the H_2O_2-peroxidase system. *FEBS Lett.* 422:377 (1998).

13. Hodnick, W.F.; Kung, F.S.; Roettger, W.J.; Bohmont, C.W.; Pardini, R.S. Inhibition of mitochondrial respiration and production of toxic oxygen radicals by flavonoids. A structure-activity study. *Biochem. Pharmacol.* 35:2345 (1986).

14. Laughton, M.J.; Halliwell, B.; Evans, P.J.; Hoult, J.R.S. Antioxidant and prooxidant actions of the plant phenolics quercetin, gossypol and myricetin. *Biochem. Pharmacol.* 38:2859 (1989).

15. Morgan, J.F.; Klucas, R.V.; Grayer, R.J.; Abian, J.; Becana, M. Complexes of iron with phenolic compounds from soybean nodules and other legume tissues: prooxidant and antioxidant properties. *Free Rad. Biol. Med.* 22:861 (1997).

16. Yamanaka, N.; Oda, O.; Nagao, S. Prooxidant activity of caffeic acid, dietary non-flavonoid phenolic acid, on Cu^{2+}-induced low density lipoprotein oxidation. *FEBS Lett.* 405:186 (1997).

17. Yang, G.; Liao, J.; Kim, K.; Yurkow, E.J.; Yang, C.S. Inhibition of growth and induction of apoptosis in human cancer cell lines by tea polyphenols. *Carcinogenesis* 19:611 (1998).

18. Flowers, L.; Ohnishi, T.; Penning, T.M. DNA strand scission by polycyclic aromatic hydrocarbons *o*-quinones: Role of reactive oxygen species, Cu(II)/Cu(I) redox cycling, and *o*-semiquinone anion radicals. *Biochemistry* 36:8640 (1997).

19. Kalyanaraman, B.; Nemec, J.; Sinha, B.K. Characterization of free radicals produced during oxidation of etoposide (VP-16) and its catechol and quinone derivatives. An ESR study. *Biochemistry* 28:4839 (1989).

20. Zhang, L.; Robertson, M.L.; Kolachana, P.; Davison, A.J.; Smith, M.T. Benzene metabolite, 1,2,4-benzenetriol, induces micronuclei and oxidative DNA damage in human lymphocytes and HL60 cells. *Environ. Mol. Mutagen.* 21:339 (1993).

21. Pryor, W.A.; Stone, K.; Zang, L.Y.; Bermúdez, E. Fractionation of aqueous cigarette tar extracts: Fractions that contain the tar radical cause DNA damage. *Chem. Res. Toxicol.* 11:441 (1998).

22. Kuhnle, J.A.; Windle, J.J.; Waiss, A.C. EPR spectra of flavonoid anion-radicals. *J. Chem. Soc. B* 613 (1969).

23. Jensen, O.N.; Pederson, J.A. The oxidative transformation of (+) catechin and (–)-epicatechin as studied by ESR. *Tetrahedron Lett.* 39:1609 (1983).

24. van Acker, S.A.; de Groot, M.J.; van den Berg, D.J.; Tromp, M.N.; den Kelder, D.O.; van der Vijgh, W.J.; Bast, A. A quantum chemical explanation of the antioxidant activity of flavonoids. *Chem. Res. Toxicol.* 9:1305 (1996).

25. Bindoli, A.; Rigobello, M.P.; Deeble, D.J. Biochemical and toxiciological properties of the oxidation products of catecholamines. *Free Rad. Biol. Med.* 13:391 (1992).

26. Kalyanaraman, B.; Sealy, R.C., Electron spin resonance-spin stabilization in enzymatic systems: Detection of semiquinones produced during peroxidatic oxidation of catechols and catecholamines. *Biochem. Biophys. Res. Comm.* 106:1119 (1982).

27. Kalyanaraman, B.; Felix, C.C.; Sealy, R.C. Electron spin resonance-spin stabilization of semiquinones produced during oxidation of epinephrine and its analogues. *J. Biol. Chem.* 259:354 (1984).

28. Giulivi, C.; Cadenas, E. Heme protein radicals: formation, fate, and biological consequences. *Free Rad. Biol. Med.* 24:269 (1998).

29. Felix, C.C.; Sealy, R.C. Photolysis of melanin precursors: formation of semiquinone radicals and their complexation with diamagnetic metal ions. *Photochem. Photobiol.* 34:423 (1981).

30. Felix, C.C.; Sealy, R.C. Electron spin resonance characterization of radicals from 3,4-dihydroxy-phenylalanine: Semiquinone anions and the metal chelates. *J. Am. Chem. Soc.* 103:2831 (1981).

31. Bolton, J.L.; Pisha, E.; Shen, L.; Krol, E.S.; Iverson, S.L.; Huang, Z.; van Breeman, R.B.; Pezzuto, J.M. The reactivity of *o*-quinones which do not isomerize to quinone methides correlates with alkylcatechol-induced toxicity in human melanoma cells. *Chem.-Biol. Interact.* 106:133 (1997).

32. Richard-Forget, F.C.; Gauillard, F.A. Oxidation of chlorogenic acid, catechins, and 4-methylcatechol in model solutions by combinations of pear (*Pyrus communis* cv. Williams) polyphenol oxidase and peroxidase: a possible involvement of peroxidase in enzymatic browning. *J. Agric. Food Chem.* 45:2472 (1997).

33. Korytowski, W.; Sarna, T.; Kalyanaraman, B.; Sealy, R.C. Tyrosinase-catalyzed oxidation of dopa and related catechol(amine)s: a kinetic electron spin resonance investigation using spin stabilization and spin label oximetry. *Biochim. Biophys. Acta* 924:383 (1987).

34. Giulivi, C.C.; Cadenas, E. Ferrylmyoglobin: Formation and chemical reactivity toward electron donating compounds. *Meth. Enzymol.* 233:189 (1994).

35. DeGray, J.A.; Gunther, M.R.; Tschirret-Guth, R.; Ortiz de Montellano, P.R.; Mason, R.P. Peroxidation of a specific tryptophan of metmyoglobin by hydrogen peroxide. *J. Biol. Chem.* 272:2359 (1997).

36. Leake, D.S. Effects of flavonoids on the oxidation of low-density lipoproteins. *In*: Rice-Evans, C.; Packer, L. (eds.). Flavonoids in health and disease. Marcel Dekker, New York, p. 253 (1998).

37. Laranjinha, J.; Almeida, L.; Madeira, V. Reduction of ferrylmyoglobin by dietary phenolic acid derivatives of cinnamic acid. *Free Rad. Biol. Med.* 19:329 (1995).

38. Laranjinha, J.; Vieira, O.; Almeida, L.; Madeira, V. Inhibition of metmyoglobin/H_2O_2-dependent low density lipoprotein lipid peroxidation by naturally occurring phenolic acids. *Biochem. Pharmacol.* 51:395 (1996).

39. Depew, M.C.; Wan, J.K.S. Quinhydrones and semiquinones. *In*: Patai, S.; Rappoport, Z. (eds.) The chemistry of quinonoid compounds, Vol. II. John Wiley and Sons Ltd., New York, p. 963 (1988).

40. Ibrahim, R.; Barron, D. Phenylpropanoids. *In*: Dey, P.M.; Harborne, J.B. (eds.). Methods in plant biochemistry, vol. 1. Plant phenolics. Academic Press, London, p. 75 (1989).

41. Nakamura, M. One-electron oxidation of trolox C and vitamin E by peroxidases. *J. Biochem.* 110:595 (1991).

42. Kagan. V.E.; Yalowich, J.C.; Day, B.W.; Goldman, R.; Gantchev, T.G.; Stoyanovsky, D.A. Ascorbate is the primary reductant of the phenoxyl radical of etoposide in the presence of thiols both in cell homogenates and in model systems. *Biochemistry* 33:9651 (1994).

43. Michon, T.; Chenu, M.; Kellershon, N.; Desmadril, M.; Guéguen, J. Horseradish peroxidase oxidation of tyrosine-containing peptides and their subsequent polymerization. A kinetic study. *Biochemistry* 36:8504 (1997).

44. Witting, P.K.; Upston, J.M.; Stocker, R. Role of α-tocopheroxyl radical in the initiation of lipid peroxidation in human low-density lipoprotein exposed to horseradish peroxidase. *Biochemistry* 36:1251 (1997).

45. Yamasaki, H.; Sakihama, Y.; Ikehara, N. Flavonoid-peroxidase reaction as a detoxification mechanism of plant cells against H_2O_2. *Plant Physiol.* 115:1405 (1997).

POLYPHENOLS INCREASE ADHESION BETWEEN LIPID BILAYERS BY FORMING INTERBILAYER BRIDGES

Thomas J. McIntosh,[a] Michael P. Pollastri,[b] Ned A. Porter,[b] and Sidney A. Simon[c]

[a] Departments of Cell Biology and [c] Neurobiology
Duke University Medical Center
Durham, North Carolina 27710
USA

[b] Department of Chemistry
Duke University
Durham, North Carolina 27706
USA

1. INTRODUCTION

Tannins are naturally occurring polyphenolic compounds that are present in the leaves, bark, and fruit of many higher plants. Tannic acid (TA) is a hydrolyzable tannin, meaning that it is an ester of digallic acid and D-glucose. TA has been shown to aggregate or precipitate a number of macromolecules or macromolecular assemblies, including various polymers,[1] peptides,[2] proteins,[3–7] and phospholipid vesicles.[8] Several physicochemical interactions are involved in the TA-induced aggregation of polymers and proteins. Two of the most important are hydrophobic interactions,[5,6] arising from the free energy decrease associated with removing the digallic acid rings from water and multidentate bond formation between the hydrogen-bond donating phenolic groups on digallic acid and the hydrogen-bond accepting groups on the polymers or proteins.[1–4,6,7] Because of the complexity of the interactions between polyphenols and proteins, it has been difficult to formulate a molecular picture of the interactions between these compounds. However, there has been a recent publication characterizing at a molecular level the interaction of a variety of polyphenols with a 19 amino acid residue repeating unit found in salivary proline-rich proteins.[2] It was found that the more hydrophobic the polyphenol, the greater the binding.

Plant Polyphenols 2: Chemistry, Biology, Pharmacology, Ecology, Edited by Gross et al. Kluwer Academic / Plenum Publishers, New York, 1999

Larger polyphenols can interact with several prolines thus increasing the binding constant. It was also suggested that the interaction includes the cooperative stacking of the galloyl rings and the proline residues.[2]

There is a long history of the interaction of polyphenols in general and TA in particular with biological membranes and lipid bilayers, primarily because of TA's utility as a fixative for electron microscopic studies of biological membranes and lipid bilayers.[9] Few details are known about the mechanisms of TA-induced aggregation of phospholipid vesicles. It has been shown, however, that the interaction of tannic acid is stronger with vesicles composed of phosphatidylcholine than those containing other phospholipids such as phosphatidylinositol or phosphatidylserine.[9] The involvement of the choline groups of the phosphatidylcholine (PC) molecules in the stabilization of the TA-PC complex was first noted in the work of Kalina and Pease[9] who found that aqueous dispersions of TA formed a stable "complex" with PC. Kalina and Pease[9] also reported (no data shown) that Compound V (fig. 1) and hexagalloylglucose were about as effective as TA in producing dispersions that are reactive to OsO_4. They did not, however, propose explanations based on physical principles as to why these compounds reduced the fluid space between bilayers.

This paper summarizes our work characterizing, at the molecular level, the interaction of TA and related polyphenols with lipid bilayer vesicles. The reasons to use lipid bilayers to investigate the molecular details of polyphenol interactions with macromolecules are that bilayers are chemically pure, that any of their groups can be chemically altered so that ideas regarding their interaction can be critically tested, and that they can be analyzed by a battery of physiochemical methods. Our specific goals have been to determine: 1) the location of polyphenols in the bilayer, 2) their effect on bilayer structure, and 3) the mechanism(s) by which TA and TA analogs varying in the size, shape, and number and position of gallic acid moieties induce bilayer aggregation. In particular, we have analyzed how TA and TA analogs modify the repulsive and attractive interactions between bilayer surfaces, thereby reducing the interbilayer fluid spacing and causing vesicles to adhere. To accomplish these goals, we have used a variety of techniques, including absorbance measurements, electron microscopy, and X-ray diffraction to monitor lipid aggregation in the presence and absence of polyphenols. We have also used dipole potential measurements of lipid monolayers and direct binding measurements to assess polyphenol binding, and surface pressure (Π)-area isotherms, X-ray diffraction, and micropippette aspiration to determine the changes in bilayer structure caused by polyphenol binding.

We find that TA binds to lipid vesicles by inserting its digallic acid residues deep into the bilayer interfacial region, thereby increasing the area of the vesicle surface. The addition of TA brings apposing vesicles (originally at a separation of 1.5 nm) to within 0.5 nm of each other, thus collapsing the interbilayer fluid space. The collapse of the fluid space occurs abruptly (cooperatively) at a TA to lipid ratio of about 5:1. Experiments with TA analogs indicate that in order to aggregate vesicles, the molecule must have a central core structure with at least two covalently bound gallic acids. The greater the number of gallic acid residues covalently bound to the D-glucose, the lower the concentration needed to precipitate lipid vesicles. However, we have found that in addition to TA, only pentagalloyl–

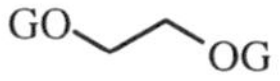

I. 1, 2-ethanediol-bis-(3,4,5-trihydroxy)benzoate

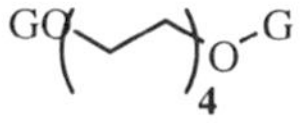

II. tetraethyleneglycol-bis-(4,4,5-trihydroxy)benzoate

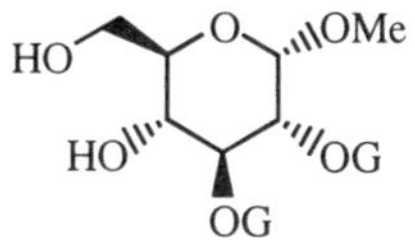

III. methyl 2,3-digalloyl-α-glucopyranoside

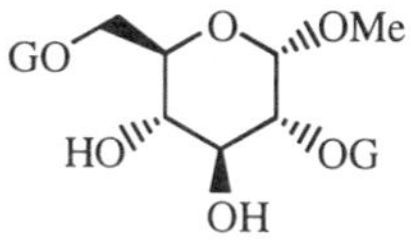

IV. methyl 2,6-digalloyl-α-glucopyranoside

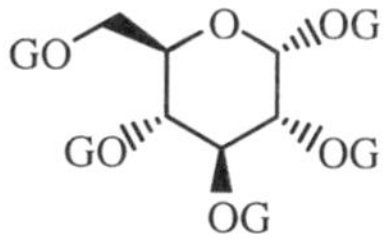

V. pentagalloyl-α-D-glucopyranose

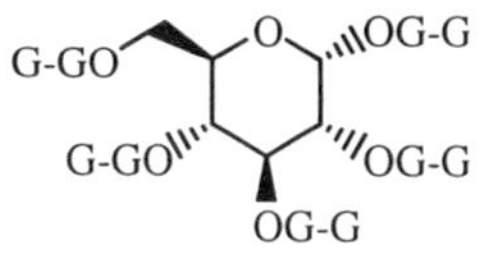

TA = tannic acid

G = gallic acid

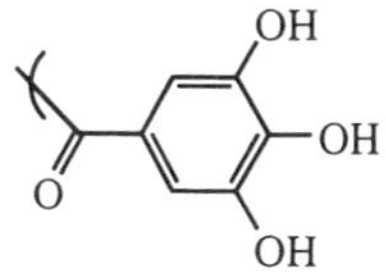

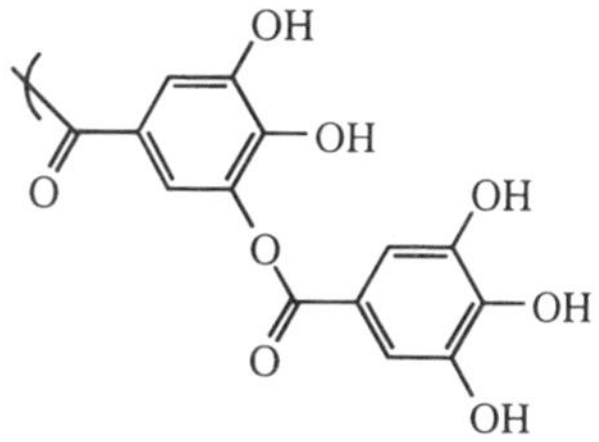

G-G = digallic acid

Figure 1. Chemical structures of tannic acid and several synthesized tannic acid analogs (Compounds I–V). Also included are tannic acid analog (TA), gallic acid (G), and digallic acid (G-G).

α-D-glucopyranose (Compound V, fig. 1) can fully collapse the fluid spacing. Based on this information and other data presented below, we argue that tannic acid and certain tannic acid analogs aggregate vesicles by forming molecular bridges between adjacent bilayers.

2. MATERIALS AND METHODS

Phospholipids including phosphatidylcholine isolated from egg yolks (EPC) and phosphatidylethanolamine isolated from bacterial membranes (BPE) were purchased from Avanti Polar Lipids (Alabastar, Alabama) and tannic acid was purchased from Fluka Chemical Co. (Bućhs, Switzerland) and recrystallized.[10] Cholesterol and the hydrophilic polymer polyvinylpyrrolidone (PVP) were purchased from Sigma Chemical Co. (St. Louis, Missouri). Tannic acid analogs (fig. 1) were synthesized by methods given in Huh et al.[10] for Compounds I, II, III, and V and in Pollastri[11] for Compound IV. Synthetic compounds were characterized by proton nuclear magnetic resonance and mass spectroscopy.[10]

Multilamellar liposomes (MLVs) were made by standard methods.[12] Lipids were co-dissolved in chloroform:methanol (2:1 v:v), and the solvent was removed by rotary evaporation. The lipids were then hydrated in 2 mM Hepes buffer and extensively vortexed. Large unilamellar vesicles (LUVS) were formed from MLVs using the freeze-thaw high-pressure extrusion method.[13] MLVs, at concentrations of 5–15 mg/mL, were frozen and thawed 3 times and extruded 20 times through a 0.1 mm polycarbonate filter with a LiposoFast lipid extruder (Avestin, Ottawa, Canada).

Absorbance measurements were performed with a dual beam Shimadzu spectrophotometer at a wavelength of 600 nm. Neither TA nor the lipid dispersions absorb appreciably at this wavelength. The final lipid concentration was 0.14 mg/mL. Aliquots of stock solutions containing TA, TA analogs, or PVP were added to the dispersion to reach the concentrations indicated in the figures.

Negative staining of LUVs of EPC in the presence and absence of tannic acid was accomplished by placing a drop of the dispersion (2 mg/mL) on a carbon-formvar grid (400 mesh), blotting with filter paper, adding a drop of a 2-percent uranyl acetate, and blotting again with filter paper. Specimens were studied with a JOEL 1200 EX electron microscope.

Surface pressure area isotherms were performed on an automated Langmuir film balance of the torsionhead/floating barrier type.[14] The trough consisted of a thermostatted rectangular trough of dimensions $14 \times 80 \times 0.5$ cm. The compression rate was 17Å^2 molecule^{-1} min^{-1}. The subphase contained tannic acid in 0.01 M NaCl and 2 mM HEPES buffer whose pH was adjusted to 7.0 with NaOH. EPC was dehydrated under vacuum, carefully weighed, and then dissolved in purified chloroform:methanol (2:1 v:v). The lipid was applied to the monolayer surface with an Alga microsyringe and prior to taking Π-A isotherms, the solvent was permitted to evaporate.

Measurements of dipole potential (V) were made on EPC monolayers formed by spreading 10 mL of 25 mg/mL lipid in chloroform on an aqueous subphase in a trough having a surface area of about 20 cm^2. The subphase was stirred using a small Teflon coated magnet. Under these conditions, liposomes are in equilibrium with the surface monolayer.[15] The dipole potential was measured between a Ag/AgCl electrode in the subphase and a Americium electrode in air connected to a Keithley electrometer. The dipole potential represents the difference in potential in the presence and absence of the monolayer. Changes in dipole potential (ΔV) due to the presence of tannic acid, tannic acid analogs, or PVP were obtained by injecting into the subphase under the monolayer aliquots of concentrated solutions of these molecules.

For micropipet experiments, giant (30 μm to 40 μm) unilamellar lipid vesicles were made by gentle rehydration, with 180 mOsm sucrose, from dried lipid films deposited on a teflon substrate as described by Needham and Nunn.[16] Micropipets were filled with equiosmotic NaCl solution, connected to a water-filled manometer system, and mounted in a deFonbrune micromanipulator operated by a joystick (Research Instruments Inc., Durham, North Carolina). Individual single walled lipid vesicles were selected and held in the pipet at a base suction pressure of 5,000 dyn/cm^2. With a transfer pipet,[17] vesicles were transferred into a chamber containing tannic acid solutions. With the measuring pipet and aspirated vesicle in focused view, the transfer pipet was pulled back, exposing the test vesicle to tannic acid solution. Once exposed to the new bathing medium, the change in projection length of the vesicle membrane in the pipet was recorded on videotape. Equiosmotic solutions were used in order to keep vesicle volume constant. Upon exposure to the TA solution, the projection length of the vesicle changed in the micropipet due to lipid bilayer area change. In this manner, small changes in membrane area could be readily measured from pipet and vesicle geometry[16] because the micropipet acts as a very sensitive transducer of area change.

For X-ray diffraction experiments, both LUVs and MLVs of EPC in the presence of tannic acid or tannic acid analogs were pelleted with a bench centrifuge, sealed in thin-walled X-ray capillary tubes, and mounted in a point collimation X-ray camera. X-ray films were processed by standard techniques and densitometered with a Joyce-Loebl microdensitometer as described previously.[18–21] After background subtraction, integrated intensities, I(h) were obtained for each order h by measuring the area under each diffraction peak, and the structure amplitude F(h) was set equal to $\{h^2 I(h)\}^{1/2}$.[22,23] One-dimensional electron density profiles were calculated from

$$\rho(x) = (2/d)\sum \exp\{i\phi(h)\} \cdot F(h) \cdot \cos(2\pi x h/d)$$

where $\rho(x)$ is the electron density on a relative density scale in a direction perpendicular to the bilayer surface, x is the distance from the center of the bilayer, d is the lamellar repeat period, $\phi(h)$ is the phase angle for order h, and the sum is over h. Phase angles were determined using the sampling theorem[24] as described in detail previously.[18,19] Electron density profiles presented in this paper are at a resolution of $d/2h_{max}$ = 6 to 8 Å.

The number of polyphenols bound to EPC vesicles that resulted in a significant increase in absorbance was determined by obtaining the total lipid concentration and the free polyphenol concentration. The total phosphate concentration was obtained by a modified Bartlett procedure. The free polyphenol concentration was obtained by first forming LUVs of EPC at 1 mg/mL as described above. Polyphenols at various concentrations were then added, the suspension vortexed and centrifuged through ultracentrifuge filter units with a 0.22 μm pore size (PGC Scientific Gaithersberg, Maryland). In separate experiments without lipid present, we found that the polyphenols passed completely through the filter. The filtrates were washed with CHCl$_3$ to remove the small amount of lipid that passed through the filter. The concentrations of TA and PGG in the filtrates were then obtained by UV spectrometry.

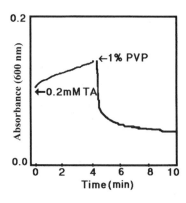

Figure 2. The change in absorbance at 600 nm of LUVs of EPC upon the addition of 0.2 mM tannic acid at time zero minutes. After steady state was achieved, solutions of PVP, a polymer that precipitates TA, were added to give a final concentration of 1% PVP. Figure is from Simon et al.[26] and used with permission of the *Biophysical Journal*.

3. RESULTS

The effect of TA on the aggregation of lipid vesicles was monitored by absorbance measurements. Figure 2 shows that the addition of tannic acid to preparations of either multilamellar vesicles (MLVs) or large unilamellar vesicles (LUVs) rapidly increased the absorbance at 600 nm. This TA-induced increase in optical density was partially reversed by the addition of PVP, a polymer that precipitates TA by having an H-bond acceptor for the H-bond donating phenolic groups on TA.[1,6]

The observed TA-induced increases in absorbance (fig. 2) were undoubtedly due to the aggregation of vesicles, as shown by the electron micrographs of figure 3. In the absence of TA, the LUVs appeared as individual structures with average

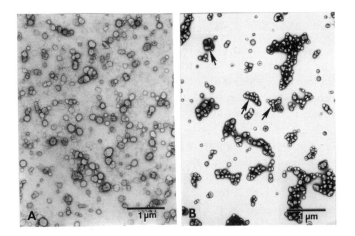

Figure 3. Electron micrographs of negatively stained large unilamellar vesicles of EPC in the absence (A) and presence (B) of 1×10^{-4} M tannic acid. In the absence of TA individual vesicles and small clusters of vesicles are present. The addition of TA causes the formation of larger aggregates, including elongated structures consisting of stacks of flattened vesicles (arrow in B). Figure is from Simon et al.[26] and used with permission of the *Biophysical Journal*.

diameters of about 0.1 μm or else as small aggregates containing a few spherical vesicles (fig. 3A). The addition of TA caused these LUVs to adhere and aggregate, often forming large aggregates containing many flattened vesicles (fig. 3B).

To characterize the association of TA to lipid films, we used several methods, including dipole potential and surface pressure (Π)—area measurements on lipid monolayers and pipet aspiration and X-ray diffraction measurements on lipid vesicles. Figure 4 shows the TA-induced changes in dipole potential (ΔV) for monolayers of EPC, equimolar EPC:cholesterol, BPE, and equimolar BPE:cholesterol. PE monolayers were chosen to compare with PC monolayers because they differ in the H-bonding profile.[25] For each lipid, the injection under the monolayer of increasing concentrations of TA produced decreases in the dipole potentials that can be described by a Langmuir isotherm $\Delta V = \Delta V_{max}/(1 + K_{1/2}/(TA.))$. In this analysis, $K_{1/2}$ is the half maximal concentration, and ΔV_{max} is the maximum decrease in dipole potential. The maximum decrease in dipole potential (ΔV_{max}) depended on the lipid headgroup, being $-245\,mV$ for EPC, and about $-70\,mV$ for BPE. The incorporation of equimolar cholesterol changed ΔV_{max} to $-125\,mV$ for EPC and to $-40\,mV$ for BPE monolayers (fig. 4).

The data in figure 5 show that the presence of TA in the subphase markedly changed the surface pressure (Π)-area isotherms. In particular, at $\Pi = 35\,dyn/cm$, the presence of $1 \times 10^{-4}\,M$ TA increased the area per lipid molecule by about $25\,\text{Å}^2$. Since the interaction of TA monolayers and bilayers may not be the same (see Conclusions), we also determined whether TA increased the area per molecule of lipid bilayers. The change in surface area of large, spherical, single walled bilayers was measured with micropipets.[26] Figure 6 shows that the addition of $1 \times 10^{-6}\,M$, $1 \times 10^{-5}\,M$, $5 \times 10^{-5}\,M$, and $1 \times 10^{-4}\,M$ TA produced systematically increasing changes in vesicle area. In the absence of TA, there was no perceivable change in

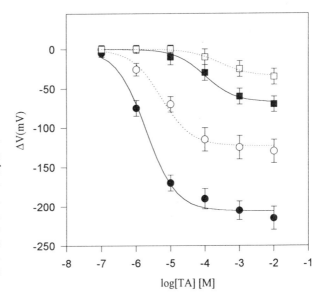

Figure 4. Tannic acid-induced changes in dipole potential for monolayers of EPC (solid circles), equimolar EPC:cholesterol (open circles), BPE (solid squares), and equimolar BPE:cholesterol (open squares). The dotted and solid lines represent Langmuir isotherms fit to the data.

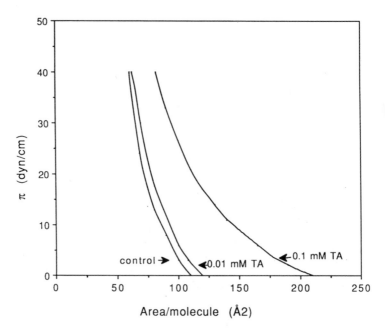

Figure 5. Surface pressure-area isotherms of EPC monolayers over subphases containing buffered solutions of 0, 1×10^{-5} M, and 1×10^{-4} M tannic acid at pH 7.0.

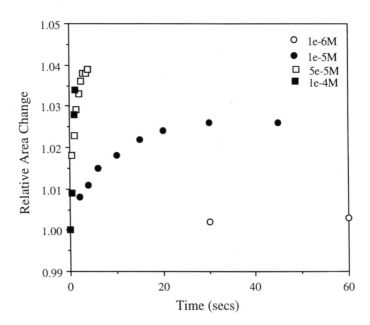

Figure 6. Micropipet measurements of the relative change in vesicle area (A/A_o) upon the transfer into equiosmolar solutions containing 1×10^{-6} M, 1×10^{-5} M, 5×10^{-5} M, and 1×10^{-4} M tannic acid. Figure is from Simon et al.[26] and used with permission of the *Biophysical Journal*.

area over a period of 10 minutes. The rate of absorption increased linearly with TA concentration, showing that the TA adsorption was diffusion limited. At the higher TA concentrations of 5×10^{-5} M and 1×10^{-4} M, the vesicles rapidly expanded and broke at average critical area changes of 3.2 ± 0.8 percent and 2.5 ± 0.3 percent, respectively. Thus, TA increased the area per lipid molecule in both monolayers and bilayer vesicles, but the percentage changes were very different. The area changes in bilayers seen with TA were limited because the rate that TA partitions into the outer monolayer is greater than that which it can be transferred to the inner monolayer. Consequently, the vesicle developed a tension that was greater than the lysis tension (about 4 dynes/cm) and therefore ruptured.[26]

Information on the change in both the bilayer structure and the interbilayer distance in the absence and presence of TA was provided by X-ray diffraction analysis. Figure 7 shows the repeat periods of multilamellar vesicles (MLVs) of EPC formed in the presence of increasing concentrations of TA. In the absence of TA, the repeat period was 63 Å, and at 10^{-4} M TA, it decreased slightly to about 61 Å. A TA concentration of 6×10^{-4} M, produced a diffraction pattern with two repeat periods, at 61 Å and 51 Å. Higher TA concentrations gave patterns with a single repeat period at 51 Å. Similar diffraction patterns were recorded from EPC LUVs that were precipitated by the addition of TA.

Electron density profiles for EPC bilayers in the presence and absence of TA are shown in figure 8. Each profile contains two unit cells with two bilayers and the fluid space between these adjacent bilayers. For each profile, the bilayer on the left is centered at the origin, so that the low-density trough at 0 Å corresponds

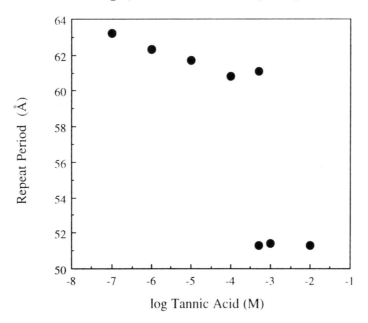

Figure 7. Repeat periods of EPC liposomes plotted versus TA concentration. Figure is from Simon et al.[26] and used with permission of the *Biophysical Journal*.

to the location of the terminal methyl groups of the lipid hydrocarbon chains. The high-density peaks centered at $\pm 19 \text{Å}$ in the upper profile and $\pm 18 \text{Å}$ in the lower profile correspond to the lipid head groups of the bilayer, and the medium-density regions between the headgroup peaks and the terminal methyl dip correspond to the lipid hydrocarbon chains. The medium-density regions, centered at 32Å in the top profile and 26Å in the bottom profile, correspond to the middle of the fluid layer between adjacent bilayers. This figure shows that TA had four major effects on the electron density profile of EPC: 1) the distance between head group peaks across the bilayer was reduced by about 2Å from 38Å to 36Å, 2) the terminal methyl trough in the center of each bilayer was broadened, 3) the distance between head groups across the fluid space between adjacent bilayers was reduced, and 4) the electron density in the interbilayer fluid space was increased.

These profiles can be used to estimate the fluid spacing between adjacent bilayers.[18,20,21] For profiles of EPC bilayers at this resolution ($d/2h_{max} \approx 6 - 8\text{Å}$), the high electron density head group peaks are located between the phosphate moiety and the glycerol backbone.[27,28] Thus, the edge of the bilayer lies about 5Å outward from the center of the high-density peaks in the electron density profiles.[18,20] Using this definition for fluid spacing, we estimated that 10^{-4}M TA reduced the fluid spacing between bilayers from about 15Å to about 5Å.[26]

TA contains five digallic acid moieties covalently attached to a D-glucose core (fig. 1). To determine what part(s) of the TA molecule are essential for binding and precipitating lipid vesicles, a variety of TA analogs (fig. 1) were synthesized and

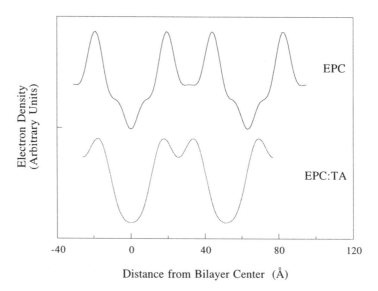

Figure 8. Electron density profiles of EPC in water (top) and in $1 \times 10^{-4} \text{M}$ TA (bottom). In each profiles two unit cells are shown, containing two apposing bilayers and the fluid space between bilayers. Figure is from Simon et al.[26] and used with permission of the *Biophysical Journal*.

tested.[10] These analogs had different cores (ethylene glycol in Compound I, tetraethylene glycol in Compound II, and D-glucose in Compounds III, IV, and V). These compounds also had different numbers and positions of gallic acids (two gallic acids in Compounds I and II on the terminals of the ethylene glycol and tetraethylene glycol, respectively, two gallic acids in Compounds III and IV with different attachment sites to the glucose core, and five gallic acids in Compound V around the D-glucose core). We[10] also performed experiments with methyl gallate to determine if the gallate moiety by itself was able to aggregate bilayers.

The steady state absorbance changes in LUVs of EPC produced by the addition of tannic acid, methyl gallate, and Compounds III and V are shown in figure 9. Methyl gallate did not have a marked effect on absorbance, whereas both Compound V and TA produced large changes in absorbances, with the maximum increases occurring at about $10\,\mu$M. Compound III (fig. 9) and Compound IV (data not shown) had similar effects and also produced large changes in absorbance, but required about a hundredfold higher concentration than either TA or Compound V.[10] The absorbance data for Compounds I and II (data not shown) tracked those of Compound III up to 10^{-3}M. However, at higher concentrations, Compounds I and II aggregated in solution, making absorbance measurements impossible at those concentrations because of light scattering.

The number of EPC molecules per TA or Compound V molecule were obtained at concentrations that produced large increases in absorbance (fig. 9). The mole ratios of EPC:TA and EPC:Compound V were 153:1 and 177:1 at 10^{-5}M, 20.4:1 and 20.2:1 at 10^{-4}M, and 4.8:1 and 4.6:1 at 5×10^{-4}M.[10] That is, at the bulk concentration that produced the greatest increase in absorbance, the mole ratio of EPC to bound TA or Compound V was about 5:1.[10]

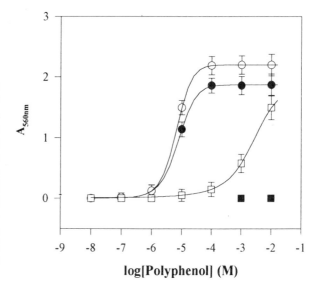

Figure 9. Absorbance changes in LUVs of EPC produced by the addition of TA and TA analogs. The steady state changes in absorbance (at 560 nm) are plotted versus the concentration of tannic acid (solid circles), methyl gallate (squares), Compound III (open square), and Compound V (open circles). Figure is taken from Huh et al.[10] and used with permission of the *Biophysical Journal.*

The changes in dipole potential (ΔV) of EPC monolayers produced by the addition of TA, methyl gallate, and Compounds II, III, and V are shown in figure 10. The solid and dotted lines correspond to Langmuir isotherms of the form $\Delta V = \Delta V_{max}/(1 + K_{1/2}/(\text{Conc.}))$ where ΔV_{max} is the maximum change in dipole potential, and $K_{1/2}$ is the concentration where $\Delta V = \Delta V max/2$. The maximum changes in dipole potential were 0 mV for methyl gallate, –70 mV for Compounds I, II, and III, –245 mV for TA, and –347 mV for Compound V.[10] The $K_{1/2}$'s were about 1 μM for TA and Compound V, 100 μM for Compound III, and 35 μM for Compounds l and II.[10]

X-ray diffraction showed that Compound I and II had very small effects on lamellar repeat periods; the repeat periods of EPC in 10^{-2}M solutions of Compounds I and II were 61 to 62 Å, compared to d = 63 Å for EPC in buffer.[10] EPC had a repeat period of 56 to 57 Å in 10^{-2}M Compounds III and IV. However, with 10^{-3} to 10^{-2}M Compound V, the repeat period decreased to 50 Å.[10] Under these conditions, the mole ratio of Compound V to EPC was about 5:1.[10]

Electron density profiles were used to calculate the widths of the bilayer and fluid spacing caused by the incorporation of Compounds III and V. Profiles of EPC in water and in 10^{-3}M compound V are shown in figure 11. In each profile, two unit cells are shown with two apposing bilayers and the fluid space between bilayers. Similar to the case of TA (fig. 8), Compound V reduced both the thickness of the bilayer and the width of the interbilayer fluid spacing. Figure 12 shows a plot of the interbilayer fluid spacing for EPC MLVs as a function of the aqueous concentration of TA or compound V. In a similar manner to that of TA, Compound V decreased the fluid spacing between bilayers from about 15 Å to about 5 Å.

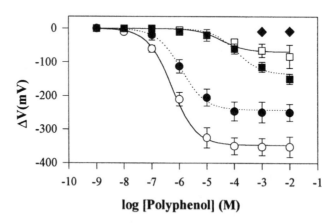

Figure 10. Changes in dipole potential produced by various concentration of: tannic acid (solid circles), methyl gallate (diamonds), Compound II (open squares), Compound III (closed squares), and Compound V (open circles). The lines represent Langmuir isotherms fit to the data. Figure is taken from Huh et al.[10] and used with permission of the *Biophysical Journal*.

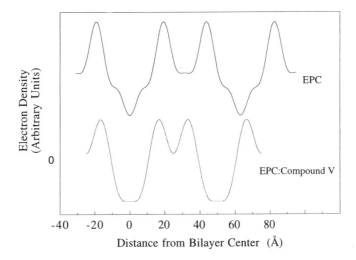

Figure 11. Electron density profiles of EPC (top) and EPC in 10^{-3} M compound V (bottom). Figure is taken from Huh et al.[10] and used with permission of the *Biophysical Journal*.

4. CONCLUSIONS

These studies provide information on the binding of TA and selected analogs to lipid bilayers and their effects on both bilayer structure and the width of the interbilayer fluid space. The experiments with TA analogs show the critical features of the TA molecule that are necessary to aggregate lipid vesicles.

As determined by absorbance measurements (figs. 2 and 9), electron microscopy (fig. 3), and X-ray diffraction, TA aggregates lipid vesicles. Absorbance is an effective method of monitoring aggregation as the formation of vesicle aggregates results in larger particles that scatter more light.[29] The electron micrographs directly show the vesicle aggregation and indicate that large columns of flattened vesicles are formed by the addition of TA (fig. 3). X-ray diffraction provides another direct indication that TA aggregates and flattens vesicles. In the absence of TA, EPC large unilamellar vesicles give no sharp diffraction reflections as there is a large and variable spacing between vesicles. However, the addition of TA produces X-ray patterns with sharp lamellar reflections, indicating that the vesicles had aggregated, flattened, and stacked together in regular repeating units.

Of the TA analogs that were synthesized and tested (fig. 1), only Compound V was as effective as TA in aggregating vesicles (fig. 9) and in reducing the repeat period and fluid space between bilayers (fig. 12). Concentrations of D-glucose up to 0.01 M did not alter the absorbance, and hence aggregation, of LUVs of EPC. This is an important result because at high concentrations D-glucose has been shown to decrease the fluid space between bilayers by decreasing the van der Waals interaction.[30] Similarly, up to 0.01 M methylgallate did not aggregate vesicles, showing that merely having high concentrations of gallic acid in the interbilayer space is insufficient to increase the adhesion between bilayers. In

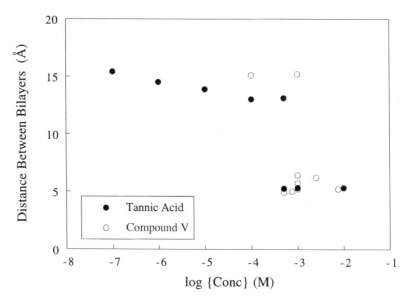

Figure 12. Fluid spacing between EPC bilayers plotted against the concentration of tannic acid (solid circles) and compound V (open circles). Figure is taken from Huh et al.[10] and used with permission of the *Biophysical Journal*.

contrast, Compounds I–IV, which contained two gallic acid moieties per molecule, did aggregate vesicles, but required about a hundredfold higher concentration than Compound V (fig. 9). Compounds III and IV reduced the repeat period of EPC vesicles more than Compounds I and II, indicating that the D-glucose core is important orienting and/or stabilizing the conformation of the gallic acids in the aggregation/precipitation process. Compound V is similar to TA in that it has gallic acids entirely surrounding the central D-glucose core. However, Compound V is simpler than TA in that it contains gallic acid rather than digallic acid moieties, (fig. 1). Thus, our experiments show that the presence of digallic acid moieties is not required for the aggregation properties of TA.

The dipole potential measurements (figs. 4 and 10) provide several important pieces of information regarding the interaction of TA with lipids. They show that TA has a larger effect on ΔVmax with phosphatidylcholine (PC) than with phosphatidylethanolamine (PE) monolayers. This preference for PC may be due, at least in part, to electrostatic interactions between the π electrons in the polyphenol rings and the choline groups in the PC headgroups.[10] In this regard, each choline moiety represents a potential binding site for each digallic acid in TA. It follows then that reducing the number of binding sites by the addition of equimolar cholesterol (which increases the distance between phospholipid headgroups in the plane of the bilayer[21]) decreases the dipole potential by about half. A similar phenomenon is observed for BPE monolayers (fig. 4).

The data in figure 10 also show that TA and Compound V bind much more strongly to EPC than do Compounds I–IV. This can be partially rationalized by the greater number of polyphenol resides per molecule (10 for TA, 5 for Compound V, and 2 for Compounds I–IV). In other words, as was previously found in other biological systems,[2] we found the greater the number of gallic acid moieties, the better the binding.

The Π-A isotherms, micropipet manipulation, and X-ray diffraction experiments gave complementary information on the changes in lipid organization induced by the addition of TA. That is, they all showed that TA increased the molecular area. The changes in area observed in the Π-A curves are difficult to relate to changes in area seen in bilayers. This is because monolayers do not break like vesicles do in the presence of exogenously added compounds (fig. 6) and also because there is controversy as to what monolayer surface pressure best represents half a bilayer. In the micropipet experiments, where TA was added to the outside of preformed vesicles, changes in area per vesicle of 3 to 4 percent were measured upon TA adsorption (fig. 6). For TA concentrations greater than 5×10^{-5} M, the bilayers become unstable and rupture (fig. 6). Previous measurements of bilayer stability[16] showed that PC bilayers fail at critical area expansions on the order of 3 to 4 percent. While the outer monolayer is expanded due to TA adsorption, the inner monolayer cannot expand more than a critical area of about 3 to 4 percent. Therefore, if neither lipid nor TA can "flip-flop" from one monolayer to the other rapidly enough, the integrity of the whole bilayer is compromised, and the bilayer will break. Thus, the micropipet experiments indicate that TA partitions into the lipid bilayer increasing the area per lipid molecule by at least 3 to 4 percent.

A molecule such as TA that increases the area/molecule (fig. 6) and partitions into the bilayer interfacial region would be expected to decrease the bilayer thickness.[31] That is, if the density of the acyl chains remained constant, and if TA partitioned only into the interfacial region, then $\Delta A/A_o \approx -(\Delta L/L_o)(1 + \Delta A/A_o)$, where A_o and L_o are the area/molecule and hydrocarbon thickness in the absence of TA, respectively.[10] For EPC bilayers, $\Delta A/A_o = 0.04$ (fig. 6) and $L_o \approx 30$ Å,[18] and thus from the above equation, $\Delta L \approx -1.2$ Å. Thickness decreases of this magnitude were indeed observed in our X-ray diffraction experiments (figs. 8 and 11). Moreover, the increase in area/molecule would also affect the packing of the chains in the center of the bilayer, causing increased disorder and a delocalization of the terminal methyl groups. Such delocalization of the terminal methyls was observed in the electron density profiles of EPC in the presence of TA (fig. 8) or Compound V (fig. 11).

Other amphiphilic compounds, such as tetracaine[32] and benzyl alcohol[31] also decrease the bilayer thickness. Since tetracaine and benzyl alcohol do not cause vesicle aggregation, the ability of TA to cause vesicle aggregation does not lie with the relatively small changes in bilayer structure that ensue upon its binding.

TA and Compound V both dramatically reduced the fluid spaces between adjacent bilayers (fig. 12). As the TA or Compound V concentration was increased, the reduction in fluid space was discontinuous (not gradual), and only two equilibrium fluid spacings were observed, one at $d_f \approx 15$ Å for low TA concentrations and the other at $d_f \approx 5$ Å for high TA concentrations or Compound V (fig. 12). Since

$d_f \approx 5\,\text{Å}$ corresponds to the fluid spacing where adjacent EPC bilayers come into steric contact,[20,21] TA and Compound V both essentially collapsed the fluid spacing between apposing bilayers. These two equilibrium fluid spaces can be represented in a potential energy diagram by two distinct minima representing the different adhesion energies in these states.[33] The smaller minimum, at $d_f \approx 15\,\text{Å}$, found at zero and low TA or Compound V concentrations, has an adhesion energy of -0.013 erg/cm^2.[34,35] This energy is determined by a balance between the attractive and repulsive pressures between the bilayers.[30,34,36] For electrically neutral EPC bilayers in excess buffer, the attractive pressure is the van der Waals pressure,[30,34,36] and the repulsive pressures include a short-range hydration pressure arising from the reorientation of the water interacting with the bilayer surface[18,30,36–39] and an entropic undulation pressure that arises from thermally-induced out-of-plane undulations (fluctuations) of the bilayer.[39–43] In the presence of TA or Compound V, the energy minimum at $5\,\text{Å}$ represents a larger (because the surfaces are closer), but unknown, adhesion energy. Previous analysis has shown that for EPC bilayers, the work to decrease d_f from $15\,\text{Å}$ to $5\,\text{Å}$ is about $0.5\,\text{erg/cm}^2$.[18] These two energy minima must be separated by an energy barrier that arises from the work required to overcome the repulsive undulation and hydration pressures ($0.5\,\text{erg/cm}^2$) as well as any steric pressures arising from the intercalated TA or Compound V molecules. Thus, once a critical concentration of TA or compound V is bound (about one TA for five lipids), the free energy gain is sufficiently large to overcome the barrier so that the bilayers fall into the very deep and narrow potential well at a fluid spacing of $5\,\text{Å}$. For this to occur, TA or Compound V must either increase the attractive pressure(s) or decrease the repulsive pressures between bilayers.

We now consider the possibilities that TA or compound V increases the van der Waals pressure or decreases one of the repulsive pressures (hydration or undulation) in a manner consistent with it decreasing the fluid spacing over a narrow concentration range (fig. 12). The possibility that the polyphenols and the lipids compete for the same limited supply of water is eliminated because absorbance and X-ray measurements were performed on dispersions in excess water. Therefore, the reduction in the fluid spacing must result from the direct interaction of the polyphenols with the bilayers. The possibility that TA or Compound V caused the fluid space to decrease by decreasing the undulation pressure is unlikely for two reasons. First, in liquid-crystalline EPC bilayers, both TA and Compound V decreased the bilayer thickness (figs. 8 and 11), which should decrease the bending modulus and increase the undulation pressure.[44] Second, in gel bilayers, where the undulation pressure should be very small,[39,43] both TA and Compound V also collapsed the fluid spacing.[10,26]

The possibility that the aggregation process is due to TA or Compound V increasing the van der Waals pressure can be eliminated. If TA or Compound V were *only* increasing the van der Waals pressure, then the repeat period should decrease continuously as a function of the volume fraction of Compound V, rather than with the abrupt decrease that is experimentally observed (fig. 12). Of course, as the bilayers become closer together as a consequence of polyphenol binding, the van der Waals pressure will increase.

We next consider how the polyphenols could affect the repulsive hydration pressure (P_h). This pressure has been shown to decay exponentially as $P_h = P_o$ exp.($-d_f/\lambda$), where P_o is its magnitude and λ is its decay length.[18,30,36] Experimental studies have correlated P_o with V^2, where V is the dipole potential.[45,46] Since both TA and Compound V markedly reduced the dipole potential (figs. 4 and 10), these molecules would be expected to decrease the hydration pressure. However, the argument against using the decrease in dipole potential as the primary cause for decrease in fluid spacing is that the dipole potential decreased monotonically from 10^{-8}M to 10^{-5}M for both TA and Compound V (figs. 4 and 10), whereas the fluid space decreased abruptly at a concentration of about 10^{-3}M TA or Compound V (fig. 12).

Therefore, since neither an increased van der Waals nor decreased repulsive hydration or undulation pressures can completely explain the observed sharp decrease in fluid spacing caused by TA or Compound V (fig. 12), we postulate the existence of an additional attractive pressure. We argue that TA, or Compound V, simultaneously binds to apposing bilayers, forming a molecular interbilayer bridge.[26] In support of this model are several observations. First, the micropipet and X-ray data indicate that TA and Compound V both partition into the interfacial regions of the bilayers, changing the area per molecule and thickness. Second, as noted above, upon the addition of increasing concentrations of TA or Compound V, only two equilibrium fluid spaces are observed: $d_f \approx 15$Å for TA or Compound V $< 10^{-3}$M, and $d_f \approx 5$Å for TA or Compound V $> 10^{-3}$M (fig. 12). The fact that only two stable states are observed is consistent with a bridging mechanism since membranes convert to the more stable state only when a sufficient number of bridges are formed to overcome the hydration and steric pressures. Third, our experiments with Compounds I–V indicate that to collapse the fluid space, a molecule must contain between three and five gallic acids. A critical point in this argument for molecular bridging is that when the fluid spacing was decreased to 5Å, the mole ratio of EPC to Compound V was about 5:1.[10] Since each Compound V molecule contains five gallic acids (fig. 1), this means that under conditions where the fluid space was collapsed each PC headgroup could be in contact with a gallic acid residue. *Thus, the collapse of the fluid space occurred only when the entire surface was covered with polyphenols.* Another factor that might be important to the proposed bridging mechanism is the arrangement of gallic acids in Compound V. Molecular space filling models show that anhydrous Compound V has a "globular" shape with its five gallic acids projecting in different directions from the central D-glucose (fig. 9). The molecular size of compound V ("diameter" ≈ 16Å) and this distribution of gallic acids means that compound V can span the interbilayer space and have gallic acids in contact with PC molecules from apposing bilayers. That is, since the width of each headgroup region of the bilayer is about 10Å and the interbilayer space is about 5Å for EPC:Compound V, it follows that gallic acid residues on opposite sides of compound V could simultaneously penetrate into the interfacial regions of apposing bilayers if Compound V spanned the interbilayer space. Even though Compounds I–IV interact with PC and are sufficiently long to form bridges, because of their relatively low affinity (fig. 10), the surface concentrations of these polyphenols must remain below what is necessary to form a sufficient number of bridges.

One similarity between Compound V and TA is that they both have gallic acid or digallic residues extending in different directions from a central D-glucose core. Our data suggest that Compound V and TA collapse the fluid space at concentration when each PC headgroup is in contact with gallic or digallic acid. The similarities in the results shown in figure 12 indicate that the additional five gallic acids present on TA are not necessary to collapse the fluid spacing. However, since TA is a larger molecule than Compound V, the observed similarity in fluid spacing implies that TA has to penetrate further into the bilayer interfacial region. In this regard, quadrupole splitting revealed that TA binding changes the orientation of the PC headgroup and also perturbs the order of the first few methylene groups in the bilayer.[26]

Thus, selected polyphenols such as TA and Compound V can induce PC bilayers to undergo a discontinuous transition in the fluid space. Although such discontinuous transitions have been observed with negatively charged lipids upon the addition of cations,[29,38,47] we believe that these polyphenols are the only small molecules known to collapse the fluid space upon binding to neutral bilayers.

ACKNOWLEDGMENTS

We thank Drs. V. Borovyagin (deceased), E. A. Disalvo, K. Gawrisch, N-W. Huh, D. Needham, and E. Toone for their important scientific contributions to these studies. This work was supported by grants GM27278 and HL17921 from the National Institutes of Health.

REFERENCES

1. Nash, T.; Allison, A.C.; Harington, J.S. Physico-chemical properties of silica in relation to its toxicity. *Nature* 210:259 (1966).

2. Baxter, N.J.; Lilley, T.H.; Haslam, E.; Williamson, M.P. Multiple interactions between polyphenols and a salivary proline-rich protein repeat in complexation and precipitation. *Biochemistry* 36:5566 (1997).

3. Goldstein, J.L.; Swain, T. The inhibition of of enzymes by tannins. *Photochemistry (Oxf.)* 4:185 (1965).

4. Haslam, E. Polyphenol-protein interactions. *Biochem. J.* 139:285 (1974).

5. Mehansho, H.; Butler, L.G.; Carlson, D.M. Dietary tannins and salivary proline-rich proteins: interactions, induction, and defense mechanisms. *Ann. Rev. Nutrition* 7:423 (1987).

6. Oh, H.I.; Hoff, J.E.; Armstrong, G.S.; Haff, L.A. Hydrophobic interaction in tannin-protein complexes. *J. Agric. Food Chem.* 28:394 (1980).

7. Luck, G.; Liao, H.; Murray, N.J.; Grimmer, H.R.; Warminski, E.E.; Williamson, M.P.; Lilley, T.H.; Haslam, E. Polyphenols, astringency and proline-rich proteins. *Phytochemistry* 37:357 (1994).

8. Schrijvers, A.H.G.J.; Frederik, P.M.; Stuart, M.C.A.; Burger, K.N.J.; Heijnen, V.V.T.; Van der Vusse, G.J.; Reneman, R.S. Formation of multilamellar vesicles by addition of tannic

acid to phosphatidylcholine-containing small unilamellar vesicles. *J. Histochem. Cytochem.* 37:1635 (1989).

9. Kalina, M.; Pease, D.C. The preservation of ultrastructure in saturated phosphatidyl-cholines by tannic acid in model systems and type II pneumocytes. *J. Cell Biology* 74:726 (1977).

10. Huh, N.-W.; Porter, N.A.; McIntosh, T.J.; Simon, S.A. The interaction of polyphenols with bilayers: conditions for increasing bilayer adhesion. *Biophys. J.* 71:3261 (1996).

11. Pollastri, M.P. Synthesis and study of tannic acid analogs in tannin-phopsholipid adhesion. Master's Thesis, Department of Chemistry, Duke University (1998).

12. New, R.R.C. Liposomes: a practical approach. IRL Press, New York (1990).

13. Hope, M.J.; Bally, M.B.; Webb, G.; Cullis, P.R. Production of large unilamellar vesicles by a rapid extrusion procedure. *Biochim. Biophys. Acta* 812:55 (1985).

14. Harvey, N.; Mirajovsky, D.; Rose, P.L.; Verbiar, R.; Arnett, E.M. Molecular recognition in chiral monolayers of steroylserine methyl ester. *J. Am. Chem. Soc.* 111:1115 (1989).

15. MacDonald, R.C.; Simon, S.A. Lipid monolayer states and their relationship to bilayers. *Proc. Nat. Acad. Sci. USA* 84:4089 (1987).

16. Needham, D.; Nunn, R.S. Elastic deformation and failure of lipid bilayer membranes containing cholesterol. *Biophys. J.* 58:997 (1990).

17. Needham, D. Measurement of interbilayer adhesion energy. *In:* Duzgunes, N. (ed.) Membrane fusion techniques. Academic Press, New York, 79 (1992).

18. McIntosh, T.J.; Simon, S.A. The hydration force and bilayer deformation: a reevaluation. *Biochemistry* 25:4058 (1986).

19. McIntosh, T.J.; Holloway, P.W. Determination of the depth of bromine atoms in bilayers formed from bromolipid probes. *Biochemistry* 26:1783 (1987).

20. McIntosh, T.J.; Magid, A.D.; Simon, S.A. Steric repulsion between phosphatidylcholine bilayers. *Biochemistry* 26:7325 (1987).

21. McIntosh, T.J.; Magid, A.D.; Simon, S.A. Cholesterol modifies the short-range repulsive interactions between phosphatidylcholine membranes. *Biochemistry* 28:17 (1989).

22. Blaurock, A.E.; Worthington, C.R. Treatment of low angle X-ray data from planar and c oncentric multilayered structures. *Biophys. J.* 6:305 (1966).

23. Herbette, L.; Marquardt, J.; Scarpa, A.; Blasie, J.K. A direct analysis of lamellar x-ray dif-fraction from hydrated oriented multilayers of fully functional sarcoplasmic reticulum. *Biophys. J.* 20:245 (1977).

24. Shannon, C.E. Communication in the presence of noise. *Proc. Inst. Radio Engrs., N. Y.* 37:10 (1949).

25. Zhou, F.; Schulten, K. Molecular dynamics study of a membrane-water interface. *J. Phys. Chem.* 99:2194 (1995).

26. Simon, S.A.; Disalvo, E.A.; Gawrisch, K.; Borovyagin, V.; Toone, E.; Schiffman, S.S.; Needham, D.; McIntosh, T.J. Increased adhesion between neutral lipid bilayers: inter-bilayer bridges fromed by tannic acid. *Biophys. J.* 66:1943 (1994).

27. Lesslauer, W.; Cain, J.E.; Blasie, J.K. X-ray diffraction studies of lecithin bimolecular leaflets with Incorporated fluorescent probes. *Proc. Nat. Acad. Sci. USA* 69:1499 (1972).

28. Hitchcock, P.B.; Mason, R.; Thomas, K.M.; Shipley, G.G. Structural chemistry of 1,2 dilau-royl-DL-phosphatidylethanolamine:molecular conformation and intermolecular packing of phospholipids. *Proc. Nat. Acad. Sci. USA* 71:3036 (1974).

29. Day, E.P.; Kwok, Y.W.; Hark, S.K.; Ho, J.T.; Vail, W.J.; Bentz, J.; Nir, S.; Papahadjopoulos, D. Reversibility of sodium-induced aggregation of sonicated phosphatidylcholine vesicles. *Proc. Natl. Acad. Sci. USA* 77:4206 (1980).

30. LeNeveu, D.M.; Rand, R.P.; Parsegian, V.A.; Gingell, D. Measurement and modification of forces between lecithin bilayers. *Biophys. J.* 18:209 (1977).

31. Ebihara, L.E.; (?) H.J.; MacDonald, R.C.; McIntosh, T.J.; Simon, S.A. The effect of benzyl alcohol on lipid bilayers: a comparison of bilayer systems. *Biophys. J.* 28:185 (1979).

32. Boulanger, Y.; Schier, S.; Smith, I.C.P. Molecular details of anesthetic-lipid interaction as seen by deuterium and phosphorous-31 nuclear magnetic resonance. *Biochemistry* 20:6824 (1981).

33. Lipowsky, R. Discontinuous unbinding transitions of flexible membranes. *J. Phys. 2 France* :1755 (1994).

34. Evans, E.; Metcalfe, M. Free energy potential for aggregation of giant, neutral lipid bilayer vesicles by van der Waals attraction. *Biophys. J.* 46:423 (1984).

35. Evans, E.; Needham, D. Physical properties of surfactant bilayer membranes: thermal transitions, elasticity, rigidity, cohesion, and colloidal interactions. *J. Phys. Chem.* 91:4219 (1987).

36. Parsegian, V.A.; Fuller, N.; Rand, R.P. Measured work of deformation and repulsion of lecithin bilayers. *Proc. Nat. Acad. Sci. USA* 76:2750 (1979).

37. Marcelja, S.; Radic, N. Repulsion of interfaces due to boundary water. *Chem. Phys. Lett.* 42:129 (1976).

38. Rand, R.P.; Parsegian, V.A. Hydration forces between phospholipid bilayers. *Biochim. Biophys. Acta* 988:351 (1989).

39. McIntosh, T.J.; Simon, S.A. Contribution of hydration and steric (entropic) pressures to the interaction between phosphatidylcholine bilayers: experiments with the subgel phase. *Biochemistry* 32:8374 (1993).

40. Helfrich, W. Elastic properties of lipid bilayers: theory and possilbe experiments. *Z. Naturforsch.* 28C:693 (1973).

41. Helfrich, W.; Servuss, R.-M. Undulations, steric interactions and cohesion of fluid membranes. *Il Nuovo Cimento* 3:137 (1984).

42. Evans, E.A.; Parsegian, V.A. Thermal-mechanical fluctuations enhance repulsion between bimolecular layers. *Proc. Nat. Acad. Sci. USA* 83:7132 (1986).

43. Evans, E. Entropy-driven tension in vesicle membranes and unbinding of adherent vesicles. *Langmuir* 7:1900 (1991).

44. Simon, S.A.; Advani, S.; McIntosh, T.J. Temperature dependence of the repulsive pressure between phosphatidylcholine bilayers. *Biophys. J.* 69:1473 (1995).

45. Simon, S.A.; McIntosh, T.J. Magnitude of the solvation pressure depends on dipole potential. *Proc. Nat. Acad. Sci. USA* 86:9263 (1989).

46. Simon, S.A.; McIntosh, T.J.; Magid, A.D.; Needham, D. Modulation of the interbilayer hydration pressure by the addition of dipoles at the hydrocarbon/water interface. *Biophys. J.* 61:786 (1992).

47. Cevc, G.; Marsh, D. Phospholipid bilayers. Physical principles and models; John Wiley and Sons, New York, (1987).

ANTIATHEROGENIC EFFECTS OF TEA POLYPHENOLS

(FLAVAN-3-OLS) IN HUMANS AND ApoE-DEFICIENT MICE

Takako Tomita,[a] Yukiko Miura,[b] Tsuyoshi Chiba,[b]
Eiji Kawai,[a] Keizo Umegaki,[c] Shinji Miura,[b]
Haruko Koizumi,[d] Masahiko Ikeda,[a] and Isao Tomita[b]

[a] Graduate School of Health Sciences
University of Shizuoka
52-1 Yada, Shizuoka 422-8526
JAPAN

[b] School of Pharmaceutical Sciences
University of Shizuoka
52-1 Yada, Shizuoka 422-8526
JAPAN

[c] National Institute of Nutrition and Health
1-23-1 Toyama, Shinjuku-ku, Tokyo 162-8636
JAPAN

[d] Central Pharmaceutical Research Institute
Japan Tobacco Inc
23 Nakogi, Hatano, Kanagawa, 257-0024
JAPAN

1. INTRODUCTION

In the subendothelial space, low-density lipoproteins (LDL) are converted to oxidized forms through contact with macrophages, endothelial and smooth muscle cells.[1] Oxidatively modified LDL have chemotactic properties and recruit blood monocytes developing into tissue macrophages. In addition, modified LDL are taken up into macrophages through scavenger receptors to convert to lipid-laden foam cells. Thus, the prevention of LDL oxidation is assumed to be one of the initial and critical measures for antiatherosclerosis.

Naturally occurring antioxidants in foods include ascorbic acid, tocopherols, carotenoides, and various kind of flavonoids. Oxidative modification of LDL takes

Plant Polyphenols 2: Chemistry, Biology, Pharmacology, Ecology, Edited by
Gross et al. Kluwer Academic / Plenum Publishers, New York, 1999

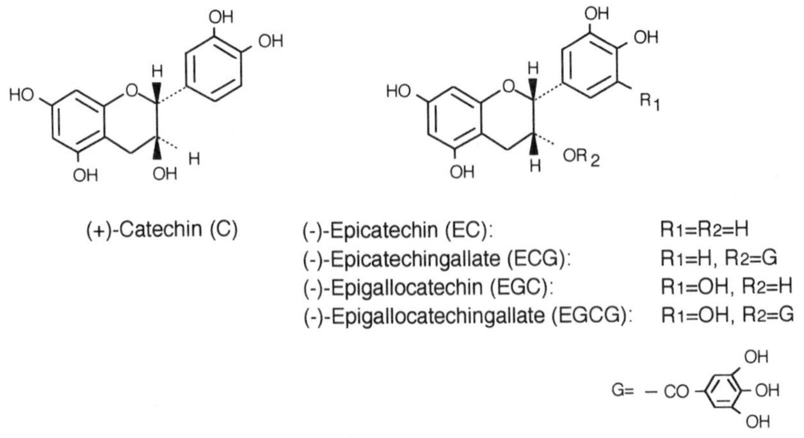

(+)-Catechin (C)

(-)-Epicatechin (EC):   R1=R2=H
(-)-Epicatechingallate (ECG):   R1=H, R2=G
(-)-Epigallocatechin (EGC):   R1=OH, R2=H
(-)-Epigallocatechingallate (EGCG): R1=OH, R2=G

Figure 1. Chemical structures of flavan-3-ol derivatives in tea.

place when antioxidants present in LDL are largely consumed.[2] An epidemiological study shows that European populations higher in the plasma levels of the natural antioxidants, ascorbic acid and α-tocopherol, have a low incidence of coronary heart disease.[3] Hertog et al.,[4] and Kneckt et al.[5] report that flavonoid intake is inversely associated with coronary heart disease mortality.

The leaves of tea (*Camellia sinensis*) contain antioxidative polyphenols, which consist of various flavan 3-ols (fig. 1): (+)-catechin (C), (–)-epicatechin (EC), (–)-epicatechin gallate (ECG), (–)-epigallocatechin (EGC) and (–)-epigallocatechin gallate (EGCG).[1] Among them, EGCG with hydroxy group at R1 and galloyl group at R2 is a principal component. We have already shown that these polyphenols have a variety of pharmacological effects; antioxidative,[7] antimutagenic,[8] anticancer promoting,[9] and hypolipidemic effects.[7] More recently, we have reported that these polyphenols exert potent inhibitory effects on Cu^{2+}-mediated oxidative modification of LDL.[10,11] In this article, we describe our recent findings on *in vitro* antioxidative effects of LDL and antiplatelet aggregating activity by each pure component of tea polyphenols, and *exo vivo* and *in vivo* antiatherogenic effects observed when tea polyphenols were ingested by humans and mice, respectively.

2. *IN VITRO* ANTIOXIDATIVE EFFECTS ON LDL

The compositional and structural changes accompanying LDL oxidation[1,10] include an increase in electrophoretic mobility and lipid hydroperoxide contents, fragmentation of apolipoprotein B (apoB)-100, a decrease in phosphatidylcholine with increasing lysophosphatidylcholine, derivation of lysine amino groups, and the generation of fluorescent adducts due to the covalent binding of lipid peroxidation products to apoB. These changes give the protein moiety of LDL an increased negative charge and lead to recognition by the scavenger receptors on macrophages.

For this study, LDL (d = 1.019–1.063) was isolated from porcine serum by discontinuous density gradient ultracentrifugation in the presence of 0.01 percent EDTA. LDL subfraction was purified by FPLC system (Pharmacia Co., Sweden) using a Sephacryl S400 HR column.

LDL free of EDTA was oxidized by incubating with $5\,\mu$M $CuSO_4$ in the presence of tea polyphenols or reference compounds. The oxidation of LDL was monitored either by an increase in absorbance at 234 nm due to the conjugated diene formation or the formation of hydroperoxides[12,13] and thiobarbituric acid reactive substances (TBARS).[14] Figure 2 shows the inhibitory effects of tea polyphenols on diene formation when LDL (0.1 mg protein/mL) was incubated with $5\,\mu$M $CuSO_4$ in the presence of $0.5\,\mu$M purified polyphenols at 37 °C. As the oxidation of LDL proceeds, absorbance at 234 nm starts to increase. The lag time before the onset of diene formation is proportional to the antioxidative potency present in the medium. Without any antioxidants, lag time was 15 min. Lag time of t-butylated hydroxy toluene (BHT), a food additive antioxidant, was 45 min. That of EGC, C, and EC was 100, 123, and 143 min, respectively. ECG and EGCG, both possessing galloyl moiety at 3 position of C ring have rather long lag time, 161 and 188 min, respectively, indicating that the galloyl group at 3 position is important to exert antioxidant activity. Table 1 shows IC_{50} of tea polyphenols in Cu^{2+}-mediated formation of lipid hydroperoxides and TBARS. The result also shows that gallate esters of tea polyphenols inhibited lipid peroxidation and TBARS formation more efficiently than their free forms, EC, C, and EGC. ApoB fragmentation due to LDL oxidation was also inhibited by tea polyphenols, especially by EGCG.

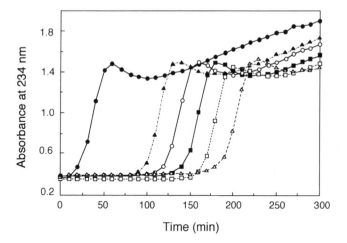

Figure 2. Inhibitory effects of tea flavan-3-ols on Cu^{2+}-mediated conjugated diene formation in low-density lipoproteins. LDL was prepared from porcine plasma by repeated discontinuous density gradient ultracentrifugation. LDL (0.1 mg protein/mL) was incubated with $5\,\mu$M $CuSO_4$ at 37 °C in the presence and absence of $0.5\,\mu$M flavan 3-ols or t-butylated hydroxytoluene (BHT) as a reference compound. Lag time (min): Control 15.0, EGC 100, C 122.5, EC 142.5, ECG 160.5, EGCG 167.5, BHT 45.0.

Table 1. IC_{50} of flavan 3-ols in Cu^{2+}-mediated LOOH and TBARS formation in LDL[a]

Flavan 3-ols	IC_{50} (μM)	
	LOOH	TBARS
ECG	0.90±0.05	0.95±0.07
EGCG	1.05±0.09	1.03±0.07
EC	1.23±0.05	1.13±0.04
C	1.41±0.04	1.36±0.02
EGC	2.38±0.20	2.74±0.05
BHT[b]	2.69±0.03	3.09±0.02

a) LDL (0.2 mg protein/ml) was incubated with 5 μM $CuSO_4$ for 2 hr at 37 °C in the presence or absence of the flavan 3-ols. BHT was used as a standard compound. LOOH was measured as described in 12, 13 and the results are expressed as nmol equivalents of linoleic acid 13-hydroperoxide. TBARS was measured as described in 14, and the results are expressed as nmol equivalents of malon dialdehyde.

b) IC_{50} of BHT was 2.69±0.03 (LOOH) and 3.09±0.02 (TBARS), respectively. Each value represents mean±SD of triplicate determinations.

LOOH: lipid hydroperoxides, TBARS: thiobarbituric acid reactive substance, BHT: *t*-butylated hydroxytoluene,

3. *IN VITRO* INHIBITORY EFFECTS ON PLATELET AGGREGATION

Platelets also play important roles in the initiation mechanism of atherosclerosis by causing damage to endothelial cells and by releasing various vasoactive factors such as platelet-derived growth factors (PDGF), thromboxane A_2 (TXA_2), and serotonin. Compounds with antioxidative properties often possess antiplatelet activities as observed in various flavonoids. Therefore, tea polyphenols were examined on their inhibitory effects of platelet aggregation with rabbit washed platelets, which show properties similar to those from humans. When rabbit washed platelets were stimulated with thrombin at the concentration to induce approximately 60 percent aggregation in the presence of varying concentrations of tea polyphenols, EGCG and ECG exerted stronger inhibitory effect than aspirin, which was used as a reference compound. The inhibitory effect of EC, EGC, and C was weaker than that of aspirin. IC_{50} of aspirin, EGCG, and ECG was 300, 71.4, and 57.7 μM, respectively. Thus potency ratio to aspirin was approximately fourfold in EGCG and fivefold in ECG.

4. *EXO VIVO* ANTIOXIDATIVE EFFECTS ON LDL IN HUMANS

The two *in vitro* results shown above suggested that ordinary tea consumption might prevent *in vivo* lipid peroxidation and platelet aggregation. We recruited

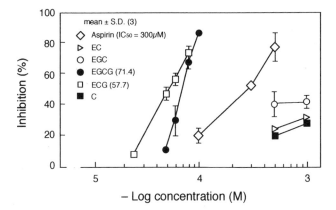

Figure 3. Inhibitory effects of tea flavan 3-ols on washed platelet aggregation. Washed platelets were prepared from rabbit platelet-rich plasma by centrifuging for 10 min at 1,500 g, and they were stimulated with thrombin at a concentration to induce approx. 60 percent aggregation following 2 min preincubation with varying concentrations of flavan-3-ols or aspirin at 37 °C. Aggregation was monitored by turbidimetry.

young volunteers to participate in our *exo vivo* study. Twenty-two male volunteers at an average age of 25 years had meals under the same dietary regimen during the 2 weeks. They were restricted to drinking green and black tea and fruit and vegetable juice during the experimental period. After a 1 week washout period, they were equally divided into two groups (control and tea group) in respect to body mass index (BMI) and serum lipid levels. The tea group only had 300 mg of the tea extract twice daily just before breakfast and dinner for 1 week. Blood was withdrawn at a 9-hr fast before breakfast and 1 hr after the tea extract was ingested before breakfast at the end of the experiment.

The tea extract used for this experiment contained 58.4 percent (–)-EGCG, 11.7 percent (–)-EGC, 6.6 percent (–)-EC, 1.6 percent (+)-GCG: gallocatechin gallate, 0.5 percent (–)-ECG, and 0.4 percent caffeine. Ingestion of the tea extract did not significantly change serum cholesterol nor triglycerides levels. Plasma concentration of EGCG, the principal component of the tea extract, was analyzed by HPLC at the end of the experiment. Plasma from the tea group contained 29.2 ± 9.0 nM in free EGCG and 22.0 ± 18.0 nM (mean ± SE for 7) in conjugated form. Total concentration of EGCG was 56.0 nM on an average. EGCG was neither detected in plasma before the experiment nor in plasma from the control group.

LDL contains various endogenous antioxidants, including α- and γ-tocopherol, β-carotene, lycopene, and retinyl stearate. If the oxidative modification of LDL occurs, these antioxidants are consumed at first. When the LDL is depleted from its endogenous lipophilic antioxidants, lipid peroxidation can enter into a propagating chain reaction, and the formation of various aldehydes is observed.[2] However, when water-soluble antioxidants such as urate and ascorbate are present outside LDL particles, these antioxidants could retard the destruction of the endogenous antioxidants in LDL. Figure 4 shows changes in plasma antioxidant

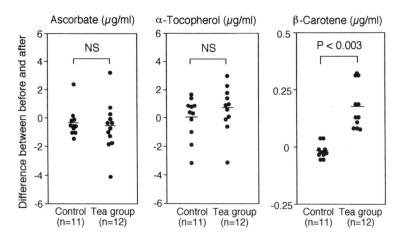

Figure 4. Changes in plasma antioxidant vitamins by ingestion of the tea extract. Blood was withdrawn before breakfast at the start and the end of the experimental period. Plasma ascorbate was measured by the formation of dinitrophenyl hydrazone. α-Tocopherol and β-carotene were extracted by n-hexane and analyzed by HPLC using nucleosil 5NH₂ column for α-tocopherol, and Inertsil ODS-2 reversed-phase column for β-carotene. Each point indicates differences between values before and after the experimental period in each person.

levels by ingestion of tea extract. Ascorbate level was not changed in both groups. α-Tocopherol level tended to increase in the tea group, whereas β-carotene level became significantly higher in the tea group. These results suggest that the presence of tea polyphenols with lipid and water-soluble (dual) properties in plasma might prevent the *in vivo* consumption of the lipid soluble antioxidants.

The LDL of each person was prepared from the plasma collected before and at the end of the experiment by discontinuous density gradient ultracentrifugation. After dialysis, LDL (0.1 mg protein/mL) was subjected to Cu^{2+} (5 μM)-mediated oxidation at 37 °C. Progress of oxidation was monitored by absorbance at 234 nm. Figure 5 shows typical oxidation curves of LDL prepared from the plasma of the tea group before and at the end of experiment. The curve shifted toward the right after a week's ingestion of tea extract. Thus, the lag time was significantly prolonged by 15 min compared with the lag time before the experiment (before: 64.6 ± 2.4 min vs after: 79.6 ± 5.9 min, mean ± SE [11 samples], P < 0.02 by paired t-test). In contrast, the lag time in the control group was unchanged before and after the experiment (before: 68.1 ± 2.9 min vs after: 68.3 ± 2.0 min, mean ± SE [11 samples]). Propagation rate ($min^{-1}mg^{-1}$ LDL protein) was also not altered in both groups (control: 2.27 ± 0.21 vs tea group: 2.21 ± 0.21, mean ± SE [11 samples]).

In an intervention study of α-tocopherol,[15] approximately 15 min prolongation of lag time was observed in the group whose plasma α-tocopherol levels were 1.5 to 1.8 times higher than the normal level. For this elevation, 200–400 mg daily supplementation of α-tocopherol seems to be needed. Ingestion of 300 mg of the tea extract twice daily is equivalent to drinking seven to eight Japanese-size tea

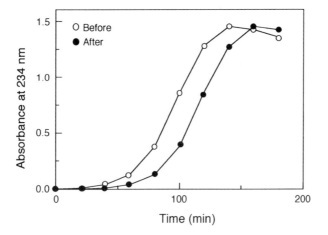

Figure 5. Changes in lag time of Cu^{2+}-mediated LDL oxidation by ingestion of tea extract. LDL was prepared from plasma of each subject by discontinuous density gradient ultracentrifugation. LDL preparation (0.1 mg protein/mL) was incubated in the presence of $5\,\mu M$ $CuSO_4$ at $37\,°C$. Before (○) and after (●) 1-week tea ingestion. Lag time (min): Before 64.6 ± 2.4, After 79.6 ± 5.9 (P < 0.02 by paired t-test).

cups of green tea a day, and also equivalent to drinking two to three cups of green tea in an American-size coffee mug. These results imply that daily drinking of ordinary amounts of green tea may exert preventive effects on *in vivo* LDL oxidation.

5. *IN VIVO* ANTIATHEROGENIC EFFECTS IN ApoE-DEFICIENT MICE

As we have confirmed the beneficial effects of the tea extract *in vitro* and *exo vivo* experiments mentioned above, we undertook an *in vivo* experiment to examine whether a long-term ingestion of tea exerts antiatherogenic effects in apolipoprotein E (apoE)-deficient mice. ApoE is a glycoprotein with a molecular size of approximately 34 kDa and plays a central role in maintaining plasma cholesterol homeostasis. One of the most important functions is to serve as a high-affinity ligand for the apoB and apoE (LDL) receptor and for the chylomicron remnant receptor, thereby allowing the specific uptake of apoE containing particles by the liver.[16] In addition, apoE in HDL facilitates the reuptake of cholesterol from peripheral cells.[17] Mainly, the lack of these two functions of apoE leads to elevation of these lipoprotein levels in the circulation and progression of atherosclerosis in the deficient mice. ApoE-deficient mice were created by Zhang et al.[18] and Plump et al.[19] in 1992. Since then, the mice have been proven to be useful animal models prone to atherosclerosis.

In our experiment, 10-week old male apoE-deficient mice C57BL/6J (–) were fed for 14 weeks an atherogenic diet containing 1.25 percent cholesterol, 0.5 percent sodium cholate, 12.5 percent cocoa butter in a regular chow. In the tea group, the tea extract was given through drinking water supplemented with

0.8 mg/ml of the tea extract (the same tea extract used for the human study) during a 14-week experimental period. Mice in the tea group thus ingested approximately 3 mg tea extract/day. Blood samples were withdrawn from retro-orbital puncture every 4 weeks under light anesthesia with ether and from the abdominal aorta at the end of the experiment.

Figure 6 shows lipoprotein profiles of plasma from apoE (+)- and apoE-deficient mice. ApoE (+) mice show a high HDL peak and small VLDL and LDL peaks (fig. 6, upper), while apoE (−) mice show a VLDL peak only even at week 0, and VLDL peak became higher with the time on the atherogenic diet (fig. 6, lower).

Plasma cholesterol level elevates with time and plateaus (approximately 3,000 mg/dl) at 4 weeks. There was no difference in serum cholesterol levels between control and tea group through the whole period. For the purpose of estimating the oxidative state of plasma lipoprotein, TBARS levels in plasma and Cu^{2+}-oxidized β-VLDL were examined (table 2). β-VLDL was prepared from pooled plasma collected at the fourth week and incubated for 3 hrs with 5 μM $CuSO_4$. TBARS level in β-VLDL was significantly lower in the tea group than in the control group. Additionally, TBARS level in plasma collected at the eighth week was also lower in the tea group. These results indicate that plasma lipids from the tea group are less susceptible to *in vivo* and *in vitro* oxidation.

The aorta from the arch to the femoral bifurcation was dissected at the end of the experiment. The weight of the cleaned aorta from the tea group was reduced by 33 percent compared with that from the control group (P < 0.01) as shown in figure 7 (left). Atherosclerotic area was analyzed by a NIH image. In the control group, 36 percent of the aortic area was covered with fatty plaque, whereas

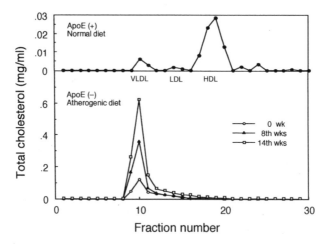

Figure 6. Lipoprotein profile in apoE-deficient mice fed an atherogenic diet. Upper: 10-week old male C57/BL6J (+) mice were maintained for 14 weeks on a regular mice chow (Oriental CRF-1). Lower: 10-week old male C57/BL6J (−) mice (apoE-deficient) were maintained for 8 and 14 weeks on an atherogenic diet. Blood was withdrawn from the abdominal aorta, and plasma lipoprotein profile was analyzed by FPLC. Cholesterol in each fraction was analyzed by an enzymatic method.

Table 2. Changes to *in vitro* and *in vivo* oxidative states of plasma lipoproteins by tea ingestion in ApoE-deficient mice.

Oxidation	Source	TBARS (nmole MDA/ml)[c]	
		Control	Tea
in vitro	Cu^{2+}-mediated oxidation of β-VLDL[a]	21.5±0.36 (3)	17.1±0.93 (3)[d]
in vivo	Plasma[b]	30.3±4.14 (5)	20.0±4.90 (5)[d]

a) β-VLDL were prepared from pooled plasma of 3 apoE deficient mice fed an atherogenic diet with and without tea extract for 4 weeks. Three samples of VLDL fraction (0.2 mg protein/ml) were incubated with 5 μM $CuSO_4$ for 3 hr at 37 °C.

b) Plasma was obtained from apoE deficient mice fed an atherogenic diet with and without tea extract for 8 weeks.

c) TBARS was measured in samples from 1) and 2) as an indicator of oxidative state. Each value indicates mean±SE for the numbers in the parentheses.

d) Significance **$P<0.01$ vs Control. TBARS: thiobarbituric acid reactive substance, MDA: malon dialdehyde

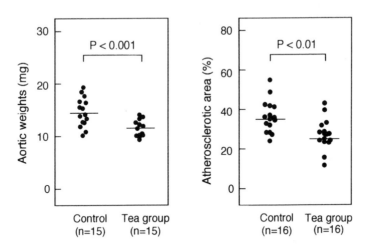

Figure 7. Reduction by tea ingestion of aortic weights and atherosclerotic lesions in apoE-deficient mice fed an atherogenic diet. Ten-week old male apoE deficient mice were maintained for 14 weeks on an atherogenic diet with and without the tea extract. The aorta was dissected from the arch to the femoral bifurcation. The cleaned aorta was longitudinally cut, and atherosclerotic lesions were analyzed by a NIH image.

that area was only 27 percent in the tea group. There was a significant 27 percent reduction of atherosclerotic area by tea ingestion (fig. 7, right). Aortic lipids were extracted, and total cholesterol and triglycerides were enzymatically determined (fig. 8). The cholesterol content in the tea group was 25 percent lower than in the control group (P < 0.01). Reduction of triglyceride content was more marked than cholesterol content; there was a 50 percent reduction in the tea group compared with the control group. These results clearly indicate that long-term tea consumption prevents the development of atherosclerosis in apoE-deficient mice.

ApoE-deficient mice develop atherosclerotic lesions that contain epitopes formed during oxidative modification of lipoproteins, and they demonstrate high titers of circulating autoantibodies against such epitopes.[20] Tangirala et al.[21] treated apoE-deficient mice for 6 months with an antioxidant N,N'-diphenyl 1,4-phenylene diamine (DPPD) to examine whether lipoprotein oxidation contributes to lesion formation. They found that atherosclerosis was significantly reduced in the treated mice. Hayek et al.[22] fed red wine, or its polyphenols quercetin or catechin, to apoE-deficient mice maintained on a regular diet. They observed reduced susceptibility to oxidation of LDL and attenuation in the development of the atherosclerotic lesion in the aortic arch area in mice fed red wine or quercetin, and, to a lesser extent, in mice fed catechin. Quercetin or catechin was given with alcohol in their experiment. Our experiment was carried out in more advanced atherosclerotic stages than that by Hayek et al.[22] Nevertheless, we demonstrated that tea polyphenols prevented *in vivo* lipoprotein peroxidation and development of atherosclerosis.

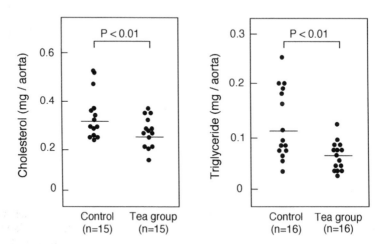

Figure 8. Reduction by tea ingestion of arotic lipid contents in apoE-deficient mice fed an atherogenic diet. ApoE-deficient mice were treated as described in the legend to figure 7. Total cholesterol and triglycerides in lipid extracts were measured by enzymatic methods.

6. CONCLUSIONS

It was concluded from our study that

1. All tea polyphenols, especially EGCG, strongly prevented copper-mediated oxidation of LDL *in vitro*. Their potency was greater than that of BHT, α-tocopherol, and vitamin C.

2. ECG and EGCG potently inhibited platelet aggregation. Their potency was fivefold and fourfold that of aspirin, respectively.

3. Ingestion of the tea extract for 1 week increased plasma EGCG level and significantly prolonged the lag time of copper-mediated oxidation of LDL in young volunteers.

4. Ingestion of the tea extract rendered plasma lipids less susceptible to oxidation and prevented the development of atherosclerosis in apoE-deficient mice.

These results strongly suggest that long-term consumption of ordinary amounts of tea (7–8 cups) exerts beneficial effects on the prevention of atherosclerosis. The beneficial effects may result mainly from antioxidative and antiplatelet properties.

ACKNOWLEDGMENTS

This work was supported in part by Research Promotion Project Grant (1997–1998) of Shizuoka Prefectural Government.

REFERENCES

1. Steinberg, D.; Parthasarathy, S.; Carew, T.E.; Khoo, J.C.; Witztum, J.L. Beyond cholesterol.—Modification of low-density lipoprotein that increases its atherogenicity. *N. Engl. J. Med.* 320:915 (1989).

2. Esterbauer, H.; Jurgens, G.; Quehenberger, O.; Koller, E. Autoxidation of human low density lipoprotein: loss of polyunsaturated fatty acids and vitamin E and generation of aldehydes. *J. Lipid Res.* 28:495 (1987).

3. Gey, L.F.; Brubacher, G.B.; Stahelin, H.B. Plasma levels of antioxidant vitamins in relation to ischemic heart disease and cancer. *Am. J. Clin. Nutr.* 45:1368 (1987).

4. Hertog, M.G.L.; Feskens, E.J.M.; Hollman, P.C.H.; Karan, M.B.; Kromhout, D. Dietary antioxidant flavonoids and risk of coronary heart disease: the Zutphen Elderly Study. *Lancet* 342:1007 (1993).

5. Hertog, M.G.L.; Feskens, E.J.M.; Kromhout, D. Antioxidant flavonols and coronary heart disease risk. *Lancet* 349:699 (1997).

6. Knekt, P.; Jarvinen, R.; Reunanen, A.; Maatela, J. Flavonoid intake and coronary mortality in Finland: a cohort study. *BMJ* 312:478 (1996).

7. Yoshino, K.; Tomita, I.; Sano, M.; Oguni, I.; Hara, Y.; Nakano, M. Effects of long-term dietary supplement of tea polyphenols on lipid peroxide levels in rats. *Age* 17:79 (1994).

8. Jain, A.J.; Shimoi, K.; Nakamura, Y.; Kada, T.; Hara, Y.; Tomita, I. Crude tea extracts decrease the mutagenic activity of N-methyl-N'-nitro-N-nitrosoguanidine *in vitro* and in intragastric tract of rats. *Mutation. Res.* 210:1 (1989).

9. Nakamura, Y.; Kawase, I.; Harada, S.; Matsuda, M.; Honma, T.; Tomita, I. Antitumor promoting effects of tea aqueous non-dialysates in mouse epidermal JB6 cells. *In*: Ohigashi, H. et al. (eds.). Food factors for cancer prevention. Springer Verlag, Tokyo, p. 138 (1997).

10. Miura, S.; Watanabe, J.; Tomita, T.; Sano, M.; Tomita, I. The inhibitory effects of tea polyphenols (flavan-3-ol derivatives) on Cu^{2+} mediated oxidative modification of low density lipoprotein. *Biol. Pharm. Bull.* 17:1567 (1994).

11. Miura, S.; Watanabe, J.; Sano, M.; Tomita, T.; Osawa, T.; Hara, Y.; Tomita, I. Effects of various natural antioxidants on the Cu^{2+}-mediated oxidative modification of low density lipoprotein. *Biol. Pharm. Bull.* 18:1 (1995).

12. Asaka, K.; Suzuki, T.; Ohrui, H.; Meguro, H. Study on aromatic phosphines for novel fluorometry of hydroperoxides (II). The determination of lipid hydroperoxides with diphenyl-1-pyrenylphosphine. *Anal. Lett.* 20:797 (1987).

13. Asaka, K.; Suzuki, T.; Ohrui, H.; Meguro, H. Study on aromatic phosphines for novel fluorometry of hydroperoxides (I). Synthesis and spectral properties of diphenyl aryl phosphines and their oxides. *Anal. Lett.* 20:731 (1987).

14. Princen, H.M.G.; van Duyvenvoorde, W.; Buytenhek, R.; van der Laarse, A.; van Poppel, G.; Leuven, J.A.G.; van Hinsbergh, V.W.M. Supplementation with low doses of vitamin E protects LDL from lipid peroxidation in men and women. *Arterioscler. Thromb. Vasc. Biol.* 15:325 (1995).

15. Jialal, I.; Fuller, C.J.; Huet, B.A. The effect of a-tocopherol supplementation on LDL oxidation. A dose-response study. *Arterioscler. Thromb. Vasc. Biol.* 15:190 (1995).

16. Mahley, R.W. Apolipoprotein E: cholesterol transport protein with expanding role in cell biology. *Science* 240:622 (1988).

17. Koo, C.; Innerarity, T.L.; Mahley, R.W. Obligatory role of cholesterol and apolipoprotein E in the formation of large cholesterol enriched and receptor active high density lipoproteins. *J. Biol. Chem.* 260:11934 (1985).

18. Zhang, S.H.; Reddick, R.L.; Piedrahita, J.A.; Maeda, N. Spontaneous hypercholesterolemia and arterial lesions in mice lacking apolipoprotein E. *Science* 258:468 (1992).

19. Plump, A.S.; Smith, J.D.; Hayek, T.; Aalto-Aetälä, K.; Walsh, A.; Verstuyft, J.G.; Rubin, E.M.; Breslow, J.L. Severe hypercholesterolemia and atherosclerosis in apolipoprotein E-deficient mice created by homologous recombination in ES cells. *Cell* 71:343 (1992).

20. Palinski, W.; Ord, V.; Plump, A.S.; Breslow, J.L.; Steinberg, D.; Witztum, J.L. ApoE-dificient mice are a model of lipoprotein oxidation in atherogenesis: demonstration of oxidation-specific epitopes in lesions and high titers of autoantibodies to malondialdehyde-lysine in serum. *Arterioscler. Thromb.* 14:605 (1994).

21. Tangirala, R.K.; Casanada, F.; Miller, E.; Witztum, J.L.; Steinberg, D.; Palinski, W. Effect of the antioxidant *N,N'*-diphenyl 1,4-phenylenediamine (DPPD) on atherosclerosis in apoE-deficient mice. *Arterioscler. Thromb. Vasc. Biol.* 15:1625 (1995).

22. Hayek, T.; Fuhrman, B.; Vaya, J.; Rosenblat, M.; Belinky, P.; Coleman, R.; Elis, A.; Aviram, M. Reduced progression of atherosclerosis in apolipoprotein E-deficient mice following consumption of red wine, or its polyphenols quercetin or catechin, is associated with reduced susceptibility of LDL to oxidation and aggregation. *Arterioscler. Thromb. Vasc. Biol.* 17:2744 (1997).

GLYCOSYLATION, ESTERIFICATION AND POLYMERIZATION OF FLAVONOIDS AND HYDROXYCINNAMATES: EFFECTS ON ANTIOXIDANT PROPERTIES

Gary Williamson, Geoff W. Plumb, and Maria T. Garcia-Conesa

Biochemistry Department
Institute of Food Research
Norwich Research Park
Colney, Norwich, NR4 7UA
ENGLAND

1. INTRODUCTION

Dietary flavonoids and hydroxycinnamates are effective antioxidants that may affect health and are also important for food preservation. Of the flavonoids, quercetin is a common representative and is found in many plant foods, especially onions, apples, tea, and broccoli. Quercetin is glycosylated in most plants, and the position and the nature of substitution of the sugar are species specific. Catechins are a well-studied group of flavonoids found at high levels in tea. Hydroxycinnamates are also found at exceptionally high levels in many foods including coffee and cereal brans and include ferulic, sinapic, p-coumaric, and caffeic acids. These compounds are commonly ester-linked to sugars or organic acids. This chapter reviews the action of flavonoids and hydroxycinnamates in two antioxidant assays: direct scavenging of the ABTS radical in the aqueous phase[1] and inhibition of iron/ascorbate-induced lipid peroxidation of phosphatidylcholine liposomes.[2]

2. WHY IS THE ANTIOXIDANT ACTIVITY OF PHENOLICS IMPORTANT?

Flavonoids and hydroxycinnamates are a broad class of (poly)phenolic compounds that are synthesized by plants as defenses against stresses such as invading organisms, UV light, and tissue injury. These secondary metabolites can occur at high levels in the diet of humans, and several studies have estimated the dietary intakes of phenolics.[3,4] In general, the most common sources of flavonoids

Plant Polyphenols 2: Chemistry, Biology, Pharmacology, Ecology, Edited by
Gross et al. Kluwer Academic / Plenum Publishers, New York, 1999

483

in the Western diet are onions, wine, tea, broccoli, and apples, whereas hydroxy-cinnamates are found at high levels in coffee, cereals (especially the bran of wheat and corn), and several fruits.

The structure of flavonoids is based around two aromatic rings (A and B) and a third non-aromatic ring (C) (fig. 1). Most flavonoids occur in plants and foods as conjugates, with sugars, organic acids, and other molecules. Flavonols such as quercetin are linked by a glycosidic bond to a range of sugars such as glucose, rhamnose, glucuronic acid, and arabinose. The position of attachment is species-dependent but is usually at the C(3) and/or C(4') atoms, and the substitution may consist of up to three sugars. The glycoside substitution positions are shown in figure 1. Flavanols such as catechin are less commonly glycosylated but are often galloylated or polymerized. Typical structures are shown in figures 1 and 2.

Hydroxycinnamic acids in the diet are most commonly ferulic, caffeic, sinapic, or *p*-coumaric acids. These are rarely found in the free form in the plant or in foods, but are usually ester-linked to sugars, organic acids, or lipids, and further, may be dimerized particularly in the plant cell wall of many cereals. The structures of some derivatives are shown in figures 3 and 4.

Antioxidant activity is a chemical process that depends on the redox, partitioning, chelating, hydrogen-donating and radical scavenging properties of a compound. Ideally, all of these properties would be measured for each compound to assess total antioxidant activity. In practice, many assays have been developed to combine the properties into an estimate of antioxidant capacity under certain

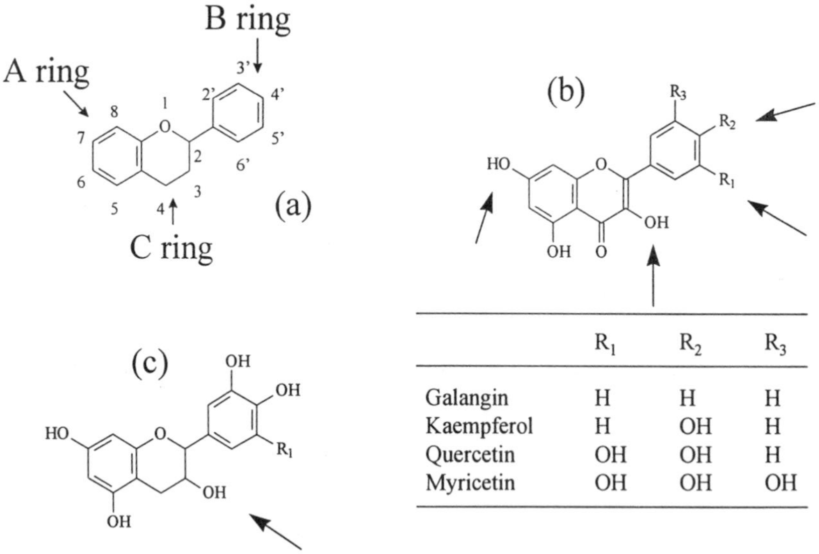

	R_1	R_2	R_3
Galangin	H	H	H
Kaempferol	H	OH	H
Quercetin	OH	OH	H
Myricetin	OH	OH	OH

Figure 1. (a) General flavonoid structure; (b) flavonols: galangin, kaempferol, quercetin, and myricetin. The positions of glycosylation are indicated by arrows; (c) flavanols: epicatechin and epigallocatechin (R_1 = OH). The position of galloylation (if present) is indicated by an arrow.

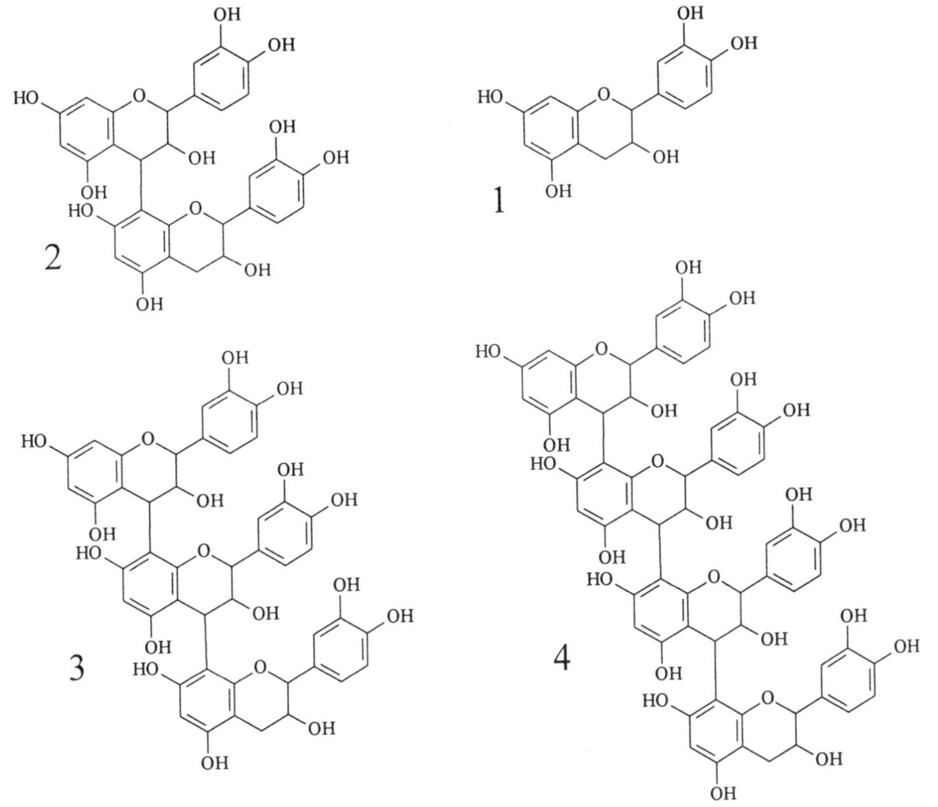

Figure 2. Oligomers of (epi)-catechin. The catechin monomer (**1**) is shown, in addition to a dimer (**2**) of catechin-[(+)-epicatechin], a trimer (**3**) of [(+)-epicatechin]₂-catechin and a tetramer (**4**) of [(+)-epicatechin]₃-catechin.

conditions. Each of these assays has advantages and disadvantages, and no assay system is perfect. In principle, however, many assays assess the ability of the test compound to scavenge a radical, although conditions vary widely and may emphasize other aspects such as lipid solubility or chelation of transition metal ions.

The most widely used assay (although under many conditions) is the inhibition of lipid peroxidation. This was developed by food chemists many decades ago to assess the ability of test molecules to prevent rancidity in fats. The method has been through several modifications and sophistications, but essentially the basic principle is unaltered and is shown in figure 5. An antioxidant, AH, interacts with a lipid peroxyl radical (LOO·) and forms a second radical (A·). The most effective antioxidants form a stable radical; this increases the chance of a termination reaction, in which two antioxidant radicals interact to form a putative dimer. AH can also chelate transition metal ions (M^{III}). The inhibition of the peroxidation reaction is also influenced by the partition coefficient, which determines the

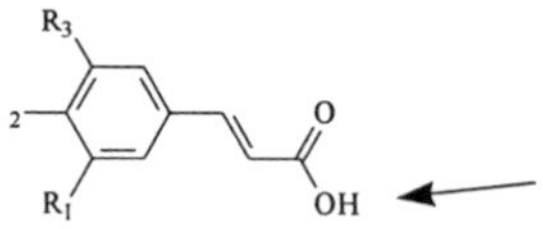

	R_1	R_2	R_3
Ferulic	OCH_3	OH	H
Sinapic	OCH_3	OH	OCH_3
p-Coumaric	H	OH	H
Caffeic	OH	OH	H

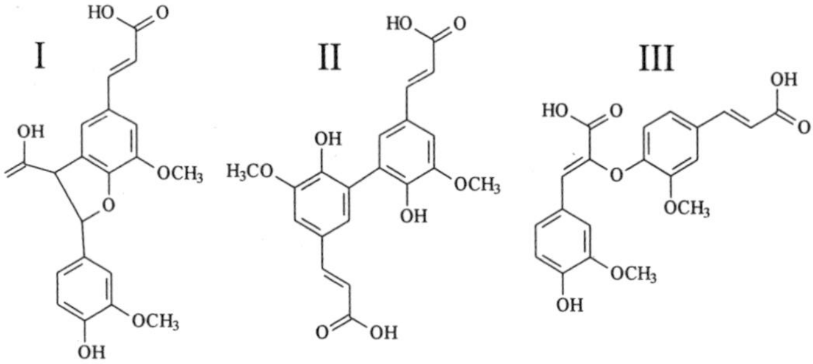

Figure 3. Unsubstituted hydroxycinnamates may be ester-linked at the carboxyl group (indicated by an arrow); dimers of ferulic acid are shown: I, *trans*-5-[(E)-2-carboxyvinyl]-2-(4-hydroxy-3-methoxy-phenyl)-7-methoxy-2,3-dihydrobenzofuran-3-carboxylic acid (5,8'-benzofuran diferulic acid); II, (E,E)-4,4'-dihydroxy-5,5'-dimethoxy-3,3'-bicinnamic acid (5-5' diferulic acid); III, (Z)-β-{4-[(E)-2-carboxyvinyl]-2-methoxy-phenoxy}-4-hydroxy-3-methoxycinnamic acid (8-O-4 diferulic acid).

chance that AH could encounter a lipid hydroperoxyl radical. Several variations of this type of assay system have been used by many researchers to measure the antioxidant properties of some flavonoids: OH· scavenging;[5] iron-induced lipid peroxidation of rat liver microsomes;[6] linoleic acid peroxidation in cetyl trimethyl-ammonium bromide micelles;[7] Fe^{2+}-induced linoleate peroxidation;[8] autoxidation of rat cerebral membranes;[8] peroxidation of red blood cell membranes;[9] oxidation of low-density lipoprotein;[10,11] linoleic acid peroxidation in sodium dodecyl sulphate micelles.[12] Many other methods have been developed that have been reviewed.[13]

Another commonly used assay is to assess the ability of a test compound to scavenge an organic radical. Examples include: ORAC (oxygen radical absorbance capacity) using OH·, peroxyl radicals or Cu^{2+};[14] inhibition of peroxynitrite-mediated tyrosine nitration;[15] and the TEAC (Trolox equivalent antioxidant capacity) assay (fig. 6).

It should be emphasized that these assays give information on the *chemical* nature of the compounds tested. Biological effects generally cannot be extrapolated from the data. These assays should be considered as valuable tools to assess

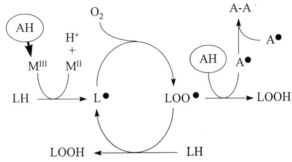

	R₁	R₂
1,2'disinapoyl-2-feruloyl gentiobiose	feruloyl	sinapoyl
1-sinapoyl-2-feruloyl gentiobiose	feruloyl	H
1,2,2'-trisinapoyl gentiobiose	sinapoyl	sinapoyl
1,2-disinapoyl gentiobiose	sinapoyl	H

Figure 4. Soluble, esterified hydroxycinnamates isolated from broccoli.

structure-function relationships of molecules, especially natural phytochemicals such as flavonoids and hydroxycinnamates, which, unlike vitamins, exist naturally as a huge number of structures. Once certain compounds are identified as most active in *in vitro* assays, then these compounds can be assessed in further *in vivo* and *ex vivo* trials.

Dietary phenolics may play a role in reducing the risk of heart disease, cancer, and cataract. A detailed discussion of the evidence is beyond the scope of this article, but epidemiological studies generally, but not always, indicate a reduced risk of chronic disease with high flavonoid intake.[3,4,16,17,18,19] This is supported by several lines of evidence from *in vitro* and animal studies, which have been reviewed.[20,21,22,23] The possible systemic benefits depend on the rate of uptake,

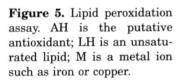

Figure 5. Lipid peroxidation assay. AH is the putative antioxidant; LH is an unsaturated lipid; M is a metal ion such as iron or copper.

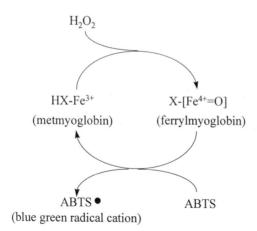

H_2O_2

HX-Fe^{3+} X-[Fe^{4+}=O]
(metmyoglobin)   (ferrylmyoglobin)

ABTS •   ABTS
(blue green radical cation)

Figure 6. TEAC (Trolox equivalent antioxidant capacity) assay. ABTS is 2,2'-azinobis(3-ethyl-benzothiazoline-6-sulfonate).

absorption, and metabolism in humans, and this is reviewed by Day et al. in this book.

3. ANTIOXIDANT ACTIVITY OF FLAVONOIDS AND HYDROXYCINNAMATES

Several publications have described a structure-function relationship for the antioxidant activity of flavonoids using a variety of different assay methods. Despite the use of different assay methods, the conclusions are generally very similar. The 3',4'dihydroxy structure is important for Cu^{2+} complexation[24] and for activity in a variety of antioxidant assays.[25,26,27,28] Both quercetin and catechin contain this structure, whereas substitutions at either position, or loss of one of the hydroxyl groups as in kaempferol, decreases the antioxidant activity. The double bond between C(2) and C(3) is also important in most assays for antioxidant activity[26,27,28] as is the carbonyl moiety at C(4).[26,27,28] There is often increased antioxidant activity with increased number of hydroxyl groups (depending on position) on the molecule.[9,25,26] Table 1 summarizes antioxidant data using two methods of analysis, the TEAC assay and inhibition of lipid peroxidation. Results are presented for flavonols, flavan-3-ols, and hydroxycinnamates, together with some representatives from other classes for comparison. Most of the data in table 1 derive from our laboratory. It supports and agrees with data on the influence of flavonoid structure on antioxidant activity as described above, but it extends these observations to show the effects of the nature of substitutions and of oligomerization.

From the data in table 1, several structure-function relationships can be derived.

1. Most of the free flavonols are efficient antioxidants in both the aqueous and lipid phases, but increased hydroxyl groups on the B ring tend to increase activity.

2. Glycosylation of flavonols decreases the antioxidant activity, but the effect is much more marked when the substitution is in the B ring (as a 4'-substitution). The antioxidant activity tends to decrease with increasing number of sugar moieties on one position. The substitution of quercetin with glucose, glucose-rhamnose, or glucose-rhamnose-glucose decreases the antioxidant activity in both the TEAC and lipid peroxidation assays, but the difference between one and two sugars is very small. However, when the B ring does not contain an *ortho*-dihydroxy group, as in kaempferol, the effect of substituting different numbers of sugars is different with quercetin, presumably because kaempferol aglycone is a much poorer antioxidant than quercetin.

3. Dimerization and trimerization of (epi)-catechin increased antioxidant activity, but the tetramer showed decreased activity.

4. Galloylation of flavanols increased aqueous phase antioxidant activity but decreased lipid phase activity.

5. Esterification of hydroxycinnamates decreases antioxidant activity. The hydroxycinnamate esters contain more than one hydroxycinnamate group per molecule, but the TEAC value is not increased proportionately. Further, dimethyl esterification of 8,5-benzofurandiferulic acid causes a pronounced decrease in the antioxidant action.

6. Dimerization of hydroxycinnamates affects antioxidant activity, but the exact effect depends on the nature of the linkage between the hydroxycinnamates. There are several diferulate structures that have been identified in plant cell walls, and each of these exhibits different antioxidant activities. The chain termination properties of ferulic acid and diferulate derivatives depend on the total number of phenolic hydroxyl groups, the presence of the alkyl chains of the hydroxycinnamic acid, and also on the number of sites that can accommodate an unpaired electron (i.e., the extent of resonance stabilization).

A large and growing number of reports show effects of flavonoids and hydroxycinnamates on mammalian cells, some of which are (probably) independent of the antioxidant action. Some examples are given below.

Quercetin inhibits the cytochrome P450-catalyzed metabolism of ethoxycoumarin and of ethylresorufin by HepG2 cells at concentrations as low as $0.1\,\mu$M.[29] Myricetin inhibits phosphatidylinositol 3-kinase α with an IC$_{50}$ value of $1.8\,\mu$M, and the inhibition is increased by the presence of hydroxyl groups in the B ring.[30] Methyl 4'-O-quercetin inhibits xanthine oxidase with an IC$_{50}$ value of $2.5\,\mu$M,[31] and quercetin inhibits human phenolsulfotransferase with K$_d$ = $0.4\,\mu$M.[32] Quercetin is a very potent inhibitor of Maloney mouse leukemia virus reverse transcriptase with an IC$_{50}$ value of $0.2\,\mu$M, and inhibits HIV reverse transcriptase with K$_i$ = $0.52\,\mu$M.[21] Quercetin suppresses α-tumour necrosis factor-mediated stimulation of interleukin-8 expression partly by inhibiting the activation of the

Table 1. Antioxidant activities of purified compounds. IC_{50} values are from
ascorbate/iron-induced peroxidation of phospholipid liposomes[2]

Compound	TEAC value (mM)	IC_{50} value (μM)	Reference
Flavonol/glycosides			
Galangin	1.22 ± 0.02	20.7 ± 0.8	
Kaempferol (K)	1.35 ± 0.02	12.6 ± 0.06	
	1.34 ± 0.08		27
K 3-O-glu	1.06 ± 0.01	>100	
K 3-O-glu-rha	1.05 ± 0.02	79 ± 1	
K 3-O-glu-glu	1.03 ± 0.02	>100	
K 3-O-glu-rha-glu	0.96 ± 0.01	37 ± 1	
K 3-O-gal-rha-glu	0.96 ± 0.02	33 ± 2	
K 3-O-glu(xyl)rha	1.19 ± 0.03	91 ± 4	
K 3-O-galA	1.23 ± 0.06	> 100	
Quercetin (Q)	4.43 ± 0.02	7.7 ± 0.3	
	4.7 ± 0.1		27
Q 3'-O-methyl	1.22 ± 0.02	11.2 ± 0.6	
Q 4'-O-methyl	1.97 ± 0.03	6.6 ± 0.4	
Q 3-O-glu	2.15 ± 0.09	5.1 ± 0.4	
Q 3-O-gal	2.22 ± 0.07	4.5 ± 0.1	
Q 3-O-rha	2.23 ± 0.04	5.3 ± 0.2	
Q 4'-O-glc	0.91 ± 0.07	> 100	
Q 3-O-glu-rha	2.09 ± 0.07	5.7 ± 0.2	
Q 3-O-glu-glu	1.45 ± 0.02	9.2 ± 0.3	
Q 3,4'-(glu)$_2$	0.58 ± 0.01	> 100	
Q 3-O-glu-rha-glu	1.24 ± 0.05	19 ± 1	
Q 3-O-gal-rha-glu	1.25 ± 0.02	14 ± 1	
Q 3-O-gluA	2.02 ± 0.02	5.1 ± 0.2	
Myricetin	2.9 ± 0.1	7.8 ± 0.5	
	3.1 ± 0.3		27
Flavan-3-ols			
Catechin (cat)	2.47 ± 0.02	3.4 ± 0.5	
	2.2 ± 0.05		27
(+)-epicatechin (Ec)	2.23 ± 0.02	5.3 ± 0.3	
	2.5 ± 0.02		27
(+)-epicatechin gallate (EcG)	5.03 ± 0.05	25 ± 4	
	4.9 ± 0.02		14
(+)-epigallocatechin gallate	4.73 ± 0.05	27 ± 3	
	4.75 ± 0.06		10

transcription factor, NF-kB.[33] The galloylated flavan-3-ol, epigallocatechin gallate,
inhibits AP-1 activation.[33] However, because of the design of the experiments, it is
difficult to assess whether the results are physiologically relevant. If more than
10μM is required for an effect, it is difficult to extrapolate to a significant effect
in vivo. The effect of some phenolics on a variety of receptors has been examined,[23]
and the tightest binding was observed with the β-adrenergic receptor, the 5HT1
receptor, and the opiate receptor.

Table 1. *Continued*

flavan-3-ol dimers			
Ec 4-8 cat	4.73 ± 0.03	4.1 ± 1.0	
Ec 4-8 Ec	4.39 ± 0.04	3.2 ± 1.2	
cat 4-8 Ec	4.61 ± 0.02	4.3 ± 1.1	
Ec 4-6 Ec	4.47 ± 0.02	4.6 ± 1.7	
Ec 4-6 cat	4.52 ± 0.04	4.8 ± 1.1	
Ec 4-8 2-O-7 Ec	3.79 ± 0.04	4.1 ± 0.4	
cat-EcG	5.16 ± 0.04	13 ± 1	
EcG-cat	5.34 ± 0.08	16 ± 0.1	
flavan-3-ol trimer			
Ec-Ec-cat	4.87 ± 0.04	6.5 ± 1.0	
flavan-3-ol tetramer			
Ec-Ec-Ec-cat	3.02 ± 0.03	40 ± 2	
other compounds			
gallic acid	2.91 ± 0.04	69 ± 6	
	3.01 ± 0.05		10
butylated hydroxytoluene	1.49 ± 0.02	5.0 ± 0.2	
trolox C (standard)	1	12.6 ± 1.4	10
Phenethylisothiocyanate	0.28 ± 0.05	>100	
Benzylisothiocyanate	0.17 ± 0.02	>100	
diallyl sulphide	0.35 ± 0.01	>100	
diallyl disulphide	0.21 ± 0.02	>100	
vitamin E	0.97 ± 0.01	6.2 ± 0.4	27
vitamin C	0.99 ± 0.04	-	27
glutathione (GSH)	0.90 ± 0.03	> 100	27
Urate	1.02 ± 0.06	64 ± 1	27
Hydroxycinnamates			
p-coumaric acid	2.12 ± 0.09	52 ± 1	
	2.2 ± 0.06		10
caffeic acid	1.33 ± 0.02	3.9 ± 0.1	
	1.26 ± 0.01		36
ferulic acid	1.96 ± 0.01	26 ± 0.4	
	1.90 ± 0.02		36
sinapic acid	1.84 ± 0.01	12 ± 1	
Hydroxycinnamate dimers			
5,5'-diferulic acid	2.19 ± 0.01	6.9 ± 0.7	
8-O-4-diferulic acid	2.60 ± 0.10	14 ± 1	
8-8'-diferulic acid	4.00 ± 0.20	14 ± 1	
8,5-benzofurandiferulic acid	1.49 ± 0.02	9.1 ± 0.5	
Hydroxycinnamate esters			
dimethyl-8,5- benzofurandiferulic acid	0.33 ± 0.1		
1,2-disinapoyl gentiobiose	1.67 ± 0.02	15 ± 1	
1-sinapoyl-2-feruloyl gentiobiose	1.79 ± 0.03	24 ± 3	
1,2,2'-trisinapoyl gentiobiose	2.15 ± 0.02	14 ± 1	
1,2'-disinapoyl-2-feruloyl gentiobiose	2.26 ± 0.01	13 ± 1	

4. CONCLUSIONS

Interactions with LDL (low-density lipoprotein) and biological membranes depend partly on the antioxidant activity and partly on the partition coefficient. Quercetin inhibited Cu^{2+}-induced oxidation of LDL with an IC_{50} value of $0.22\,\mu M$, and inhibited peroxidation of rat liver microsomes with $IC_{50} < 1.5\,\mu M$.[6] Quercetin may also reduce the *in vivo* oxidizability of LDL in mice when fed as part of the diet[34] and quercetin metabolites in human plasma may show antioxidant activity after consumption of quercetin-rich foods.[35] Quercetin also interacts with liposomes to reduce the T_m (melting temperature).[8]

These results demonstrate that polyphenols are biologically active and that flavonoids are particularly potent. However, the lower biological activity and antioxidant action of hydroxycinnamates may be compensated for by the higher amounts in the diet of these compounds relative to flavonoids. Whether polyphenolics improve human health remains to be seen, but any effects must be accompanied by chemical and mechanistic studies. Hopefully the results summarized here will provide the groundwork for future research into the health effects of polyphenolics.

ACKNOWLEDGMENTS

We thank the Biotechnology and Biological Sciences Research Council, UK, and the European Union (FAIR-CT95-0653 and FAIR-CT96-1099) for funding.

REFERENCES

1. Miller, N.; Rice-Evans, C. Spectrophotometric determination of antioxidant activity. *Redox. Rep.* 2:161–171 (1996).

2. Williamson, G.; Plumb, G.W.; Uda, Y.; Price, K.R.; Rhodes, M.J.C. Dietary quercetin glycosides: antioxidant activity and induction of the anticarcinogenic phase II marker enzyme quinone reductase in Hepalclc7 cells. *Carcinogenesis* 17:2385–2387 (1996).

3. Hertog, M.G.L.; Hollman, P.C.H.; Katan, M.B.; Kromhout, D. Intake on potentially anticarcinogenic flavonoids and their determinants in adults in the Netherlands. *Nutr. Cancer* :20–29 (1993).

4. Hertog, M.G.L.; Kromhout, D.; Aravanis, C.; Blackburn, H.; Buzina, R.; Fidanza, F.; Giampaoli, S.; Jansen, A.; Menotti, A.; Nedeljkovic, S.; Pekkarinen, M.; Simic, B.S.; Toshima, H.; Feskens, E.J.M.; Hollman, P.C.H.; Katan, M.B. Flavonoid intake and long-term risk of coronary heart disease and cancer in the seven countries study. *Arch. Intern. Med.*, 155:381–386 (1995).

5. Hanasaki, Y.; Ogawa, S.; Fukui, S. The correlation between active oxygen scavenging and antioxidant effects of flavonoids. *Free Rad. Biol. Med.* 16:845–850 (1994).

6. Laughton, M.J.; Halliwell, B.; Evans, P.J.; Hoult, J.R. Antioxidant and pro-oxidant actions of the plant phenolics quercetin, gossypol and myricetin. *Biochem. Pharmacol.* 38:2859–2865 (1989).

7. Wang, P.-F.; Zheng, R.-L. Inhibitions of autoxidation of linoleic acid by flavonoids in micelles. *Chem. Phys. Lipids* 63:37–40 (1992).

8. Saija, A.; Scalese, M.; Lanza, M.; Marzullo, D.; Bonina, F.; Castelli, F. Flavonoids as antioxidant agents: importance of their interaction with biomembranes. *Free Rad. Biol. Med.* 19:481–486 (1995).

9. Chen, Z.Y.; Chan, P.T.; Ho, K.Y.; Fung, K.P.; Wang, J. Antioxidant activity of natural flavonoids is governed by number and location of their aromatic hydroxyl groups. *Chem. Phys. Lipids* 79:157–163 (1996).

10. Salah, N.; Miller, N.J.; Paganga, G.; Tijburg, L.; Bolwell, G.P.; Rice-Evans, C. Polyphenolic flavanols as scavengers of aqueous phase radicals and as chain-breaking antioxidants. *Arch. Biochem. Biophys.* 322:339–346 (1995).

11. Vinson, J.A.; Dabbagh, Y.A.; Serry, M.M.; Jang, J.H. Plant flavonoids, especially tea flavonols, are powerful antioxidants using an *in vitro* oxidation model for heart disease. *J. Agr. Food Chem.* 43:2800–2802 (1995).

12. Foti, M.; Piattelli, M.; Baratta, M.T.; Ruberto, G. Flavonoids, coumarins and cinnamic acids as antioxidants in a micellar system. Structure-activity relationship. *J. Agric. Food Chem.* 44:497–501 (1996).

13. Aruoma, O.I. Antioxidant methodology: *in vivo* and *in vitro* concepts. AOCS Press, Champaign, IL, (1997).

14. Cao, G.; Sofic, E.; Prior, R.L. Antioxidant and pro-oxidant behaviour of flavonoids: structure-activity relationships. *Free Radical Biol. Med.* 22:749–760 (1997).

15. Pannala, A.; Rice-Evans, C.A.; Halliwell, B.; Singh, S. Inhibition of peroxynitrite-mediated tyrosine nitration by catechin polyphenols. *Biochem. Biophys. Res. Commun.* 232:164–168 (1997).

16. Hertog, M.G.L.; Sweetman, P.M.; Fehily, A.M.; Elwood, P.C.; Kromhout, D. Antioxidant flavonols and ischemic heart disease in a Welsh population of men: the Caerphilly study. *Am. J. Clin. Nutr.* 65:1489–1494 (1997).

17. Dorant, E.; Van den Brandt, P.A.; Goldbohm, R.A.; Sturmans, F. Consumption of onions and a reduced risk of stomach carcinoma. *Gastroenterology* 110:12–20 (1996).

18. Knekt, P.; Jarvinen, R.; Seppanen, R.; Heliovaara, M.; Teppo, L.; Pukkala, E.; Aromaa, A. Dietary flavonoids and the risk of lung cancer and other malignant neoplasms. *Am. J. Epidemiol.* 146:223–230 (1997).

19. Rimm, E.B.; Katan, M.B.; Ascherio, A.; Stampfer, M.J.; Walter, M.D.; Willett, M.D. Relation between intake of flavonoids and risk of coronary heart disease in male health professionals. *Ann. Intern. Med.* 125:384–389 (1996).

20. Mitscher, L.A.; Jung, M.; Shankel, D.; Dou, J.H.; Steele, L.; Pillai, S.P. Chemoprotection: A review of the potential therapeutic antioxidant properties of green tea (*Camellia sinensis*) and certain of its constituents. *Med. Res. Rev.* 17:327–365 (1997).

21. Formica, J.V.; Regelson, W. Review of the biology of quercetin and related bioflavonoids. *Food Chem. Toxicol.* 33:1061–1080 (1995).

22. Das, A.; Wang, J.H.; Lien, E.J. Carcinogenicity, mutagenicity and cancer preventing activities of flavonoids: a structure-system-activity relationship (SSAR) analysis. *Progress Drug Res.* 42:166 (1994).

23. Zhu, M.; Phillipson, J.D.; Greengrass, P.M.; Bowery, N.E.; Cai, Y. Plant polyphenols: Biologically active compounds or non-selective binders to protein? *Phytochemistry* 44:441–447 (1997).

24. Brown, J.; Khodr, H.; Hider, R.C.; Rice-Evans, C. Structural dependence of flavonoid interactions with Cu^{2+} ions: implications for their antioxidant properties. *Biochem. J.* 330:1173–1178 (1998).

25. Yokozawa, T.; Dong, E.; Liu, Z.W.; Shimizu, M. Antioxidative activity of flavones and flavonols *in vitro*. *Phytother. Res.* 11:446–449 (1997).

26. Cook, N.C.; Samman, S. Flavonoids—Chemistry, metabolism, cardioprotective effects, and dietary sources. *J. Nutr. Biochem.* 7:66–76 (1996).

27. Rice-Evans, C.A.; Miller, N.J.; Bolwell, P.G.; Bramley, P.M.; Pridham, J.B. The relative-antioxidant activities of plant-derived polyphenolic flavonoids. *Free Rad. Res.* 22:375–383 (1995).

28. Rice-Evans, C.A.; Miller, N.J.; Paganga, G. Structure-antioxidant activity relationships of flavonoids and phenolic acids. *Free Rad. Biol. Med.* 20:933–956 (1996).

29. Musonda, C.A.; Helsby, N.; Chipman, J.K. Effects of quercetin on drug metabolizing enzymes and oxidation of 2′,7-dichlorofluorescin in HepG2 cells. *Hum. Exp. Toxicol.* 16:700–708 (1997).

30. Agullo, G.; Gamet-Payrastre, L.; Manenti, S.; Viala, C.; Remesy, C.; Chap, H.; Payrastre, B. Relationship between flavonoid structure and inhibition of phosphatidylinositol 3-kinase: A comparison with tyrosine kinase and protein kinase C inhibition. *Biochem. Pharmacol.* 53:1649–1657 (1997).

31. Cos, P.; Ying, L.; Calomme, M.; Hu, J.P.; Cimanga, K.; Van Poel, B.; Pieters, L.; Vlietinck, A.J.; Van den Berghe, D. Structure-activity relationship and classification of flavonoids as inhibitors of xanthine oxidase and superoxide scavengers. *J Nat. Prod.* 61:71–76 (1998).

32. Walle, T.; Eaton, E.A.; Walle, U.K. Quercetin, a potent and specific inhibitor of the human p-form phenolsulfotransferase. *Biochem. Pharmacol.* 50:731–734 (1995).

33. Sato, M.; Miyazaki, T.; Kambe, F.; Maeda, K.; Seo, H. Quercetin, a bioflavonoid, inhibits the induction of interleukin 8 and monocyte chemoattractant protein-1 expression by tumor necrosis factor-alpha in cultured human synovial cells. *J. Rheumatol.* 24:1680–1684 (1997).

34. Dong, Z.G.; Ma, W.Y.; Huang, C.S.; Yang, C.S. Inhibition of tumor promoter-induced activator protein 1 activation and cell transformation by tea polyphenols, (–)-epigallocatechin gallate, and theaflavins. *Cancer Res.* 57:4414–4419 (1997).

35. Hayek, T.; Fuhrman, B.; Vaya, J.; Rosenblat, M.; Belinky, P.; Coleman, R.; Elis, A.; Aviram, M. Reduced progression of atherosclerosis in apolipoprotein E-deficient mice following consumption of red wine, or its polyphenols quercetin or catechin, is associated with reduced susceptibility of LDL to oxidation and aggregation. *Arterioscler. Thromb. Vasc. Biol.* 17:2744–2752 (1997).

36. Manach, C.; Morand, C.; Crespy, V.; Demigne, C.; Texier, O.; Regerat, F.; Remesy, C. Quercetin is recovered in human plasma as conjugated derivatives which retain antioxidant properties. *FEBS lett.* 426:331–336 (1998).

37. Miller, N.J.; Diplock, A.T.; Rice-Evans, C.A. Evaluation of the total antioxidant activity as a marker of the deterioration of apple juice oil storage. *J. Agr. Food Chem.* 43:1794–1801 (1995).

TANNINS AS BIOLOGICAL ANTIOXIDANTS

Ann E. Hagerman, Ken M. Riedl, and Robyn E. Rice

Department of Chemistry and Biochemistry
Miami University
Oxford, Ohio 45056
USA

1. INTRODUCTION

The consumption of fruits and vegetables is correlated with reduced risk for cancer and cardiovascular disease.[1,2] Diets rich in fruits and vegetables contain several known nutritive factors that have been linked to diminished risk of disease such as ascorbic acid, tocopherols, and carotenoids.[3,4,5] However, it is clear that known nutritive factors are not entirely responsible for the beneficial properties of fruit- and vegetable-containing diets. It is widely believed that non-nutritive components of the diets are in part responsible for these benefits. That theory is supported by studies in which more cancers are produced by chemical carcinogens if animals are fed a complete synthetic diet than if animals are fed a diet based on plant foods.[6]

A wide array of non-nutritive constituents is found in plant foods, including the organosulfur compounds, the isothiocyanates, and the phenolics.[7] Since the most beneficial diets contain a wide variety of fruits and vegetables, it is likely that the beneficial components are widely distributed among plant foods. However, many phytochemicals have rather restricted distributions among foods. For example, organosulfur compounds are found in *Allium spp.* (onion and garlic), while isothiocyanates are found in cruciferous vegetables (e.g., broccoli). The phenolics are widely distributed plant metabolites found in virtually all plant foods.[8] Many low molecular weight phenolics appear to have specific roles in maintaining health; for example, resveratrol inhibits stages of tumor initiation, promotion, and progression.[9] The potential benefits of consuming the naturally occurring polyphenolics known as tannins have not been investigated.

The term "phenolic" includes any compound with an hydroxyl-substituted benzene ring and includes primary metabolites such as the antioxidant vitamin α-tocopherol and the amino acid tyrosine. Secondary plant phenolics include phenolic acids, esters, and glycosides; flavonoids; cell-wall bound phenolics; and

Plant Polyphenols 2: Chemistry, Biology, Pharmacology, Ecology, Edited by
Gross et al. Kluwer Academic / Plenum Publishers, New York, 1999

tannins.[10] The focus of this chapter is the subset of polymeric plant phenolics known as tannins (or polyphenols). Tannins are plant phenolics with relatively high molecular weights (>1,000), which precipitate protein from solution.[11]

Beneficial effects have been noted in animal models for cancer or cardiovascular diseases when polyphenolics such as ellagic acid,[12] tannic acid,[13,14] or food-derived polyphenols [15,16,17] are orally administered to animals. Other routes of administration reveal a broad spectrum of activities for polyphenolics. For example, dermal application of some polyphenols inhibits promotion of skin cancer by TPA.[18] Only those tannins containing tellimagrandin II subunits were able to inhibit sarcoma cell growth in mice when administered by intraperitoneal injection.[19] It is not surprising that tannin activity is structure-dependent, since tannins with different structures are likely to have different lifetimes and different chemical and biological properties. Unfortunately, many studies of the physiological activities of tannins have been performed with poorly defined mixtures of polyphenolics instead of with well-characterized compounds. Little attempt has been made to differentiate between the role of low molecular weight phenolics and tannins.

So far, there has been little success in establishing biologically plausible disease-preventive mechanisms of action for tannins. In general, tannins are chemically stable polymers and are difficult to degrade to monomers. As a result of their size, polarity, and stability, the metabolic fate of the tannins is unusual; they are probably not readily absorbed from the alimentary tract but persist in the tract during digestion.[20,21] Small phenolics are readily absorbed and excreted after ingestion,[22] but tannins are large (MW > 1,000) so that direct absorption is unlikely.[20,21,23] The chemical stability of condensed tannins, which remain intact for months at mild temperatures and pH[24] and decompose only at high temperatures and extreme pH,[25] suggests that chemical depolymerization in the digestive tract is unlikely. Enzymatic degradation of the condensed tannins has not been reported. The complex hydrolyzable tannins found in foods (ellagitannins) are also recalcitrant to degradation, so that only about 10 percent of the ellagic acid comprising the parent tannin casuarictin was released under conditions simulating digestion.[26] Metabolic stability appears to be the rule for dietary tannins, although simple gallotannins such as tannic acid are readily hydrolyzed to yield absorbable phenolic acids.[27]

If tannins are unlikely to be absorbed and transported to target cells, then activities such as the ability of tannin to induce Phase II detoxification in the liver[28] or to inhibit oxidation of LDL[29] are unlikely to be physiologically significant. In model systems, tannins can break DNA in a sequence-specific fashion[30] and can alter gene expression.[31] However, intercellular targets such as DNA are unlikely to be reached by tannins *in vivo*. We believe that tannins are likely to have an impact extracellularly and in the alimentary tract.

2. BIOLOGICAL ANTIOXIDANTS

A variety of pathologies including cancer and cardiovascular disease have been linked to formation of reactive oxygen species (ROS).[32,33] By reducing ROS to harmless forms, antioxidants may prevent oxidative damage to biomolecules including lipids, proteins, and nucleic acids. A variety of types of antioxidants have been identified in living systems, including enzymes (e.g., superoxide dismutase); endogenous protein (e.g., albumin) and nonprotein (e.g., uric acid) species; and dietary antioxidants (tocopherols, carotenoids, ascorbic acid). Epidemiological studies have shown that in humans low levels of antioxidants in blood plasma are correlated with elevated risk of disease.[5,34,35] Although levels of plasma antioxidants are related to risks, increased dietary intake of antioxidant vitamins does not necessarily reduce risk.[36,37] Studies of the role of dietary antioxidants in maintaining health are complicated by the fact that levels of plasma antioxidants are a function not only of intake of nutrient antioxidants but also of oxidative demand of the organism.[38,39]

Either exogenous or endogenous reactions can generate ROS and increase demand for antioxidants. For example, levels of plasma ascorbic acid are generally low in smokers, perhaps because oxidative metabolism of smoke byproducts consumes the antioxidant.[40] Similarly, the digestive tract is directly exposed to exogenous ROS when the diet is rich in polyunsaturated fats, which are susceptible to oxidation.[41] Endogenous ROS are produced in the digestive tract at sites of inflammation or mechanical irritation[33] or by Fenton-type reactions involving dietary iron or other metals.[42]

The level of ROS in the digestive tract could influence the demand for dietary antioxidants such as ascorbic acid. If nutritive antioxidants are depleted by reaction with exogenously or endogenously formed ROS during digestion, they will be unavailable for uptake and transport to the serum. We suggest that one potential benefit of dietary tannins is protection of mobile antioxidants from destructive oxidative reactions during digestion. If our hypothesis is correct, then consumption of a tannin-rich diet will result in elevated levels of nutritive antioxidants in the plasma.

3. REDOX CHEMISTRY OF TANNINS

Our recent efforts have been directed toward understanding the oxidative chemistry of tannins. At the molecular level, the antioxidant activity of phenolics may be due to their favorable redox potential,[43] their ability to form stable radicals,[44] or their ability to chelate metal ions.[45] We used *in vitro* methods to demonstrate that tannins are very effective antioxidants under physiological conditions.

We established that the electrochemical redox potentials of tannins are similar to the potentials of related small phenolics.[46] The redox potentials of various tannins at pH 6–8 are substantially below 1,000 mV, and thus these compounds are reducing agents for the peroxyl ($E^-1,000$ mV) and hydroxyl ($E^-2,300$ mV) radical. Using the deoxyribose method,[47] we established that the rate of reaction between a representative gallotannin and OH$^\bullet$ ($k = 3.1 \times 10^{11} M^{-1}s^{-1}$) was two orders of magnitude higher than the rate of reaction with mannitol[46] and was comparable to the rate of reaction of other macromolecules with OH$^\bullet$.[48] Small phenolics often exhibit pro-oxidant activities in the deoxyribose method, enhancing oxidative damage rather than providing protection,[49] but we have found that *in vitro*

Table 1. Ability of tannins and small
phenolics to quench radicals in the
metmyoglobin method[a]

Compound	Trolox ratio
sorghum procyanidin	28.4
polygalloyl glucose	15.1
epigallocatechin gallate	11.3
oenothein B	6.0
theaflavin[b]	5.5
catechin	2.6
methyl gallate	2.6

[a] Triplicate determinations were fit with linear
regression to calculate the concentration for
50 per cent inhibition. The average SD for
these determinations was 1.9 per cent of the
mean. Trolox ratio is the concentration of
Trolox to give 50 per cent inhibition divided
by the concentration of test substance to give
50 per cent inhibition when the metmyoglobin
method is run for a fixed time of three min.
[b] Assuming a molecular weight of 656 g/mol
for theaflavin.

most tannins do not behave as pro-oxidants. However, at pH 7.4, condensed tannin
had mixed behavior, inhibiting TBARS formation at high concentrations but stimulating it at low concentrations. The deoxyribose method is limited in its utility
for phenolics because complexation and redox reactions with iron can interfere
with the reactivity of a putative antioxidant with OH[•].

Using the metmyoglobin (carbon radical scavenging) method,[50] we found that
five chemically distinct tannins were all antioxidants with potencies up to an order
of magnitude more effective than Trolox (table 1). Based on these data, if tannin
consumption is about 1 g/day,[51] then on a daily basis tannin provides about a hundredfold more antioxidant activity than is provided by ascorbic acid (RDA 60 mg).
This is convincing evidence that tannins are potent antioxidants and suggests that
further work to test their potential to spare nutritive antioxidants *in vivo* may be
fruitful.

4. TANNINS AS DIETARY ANTIOXIDANTS

In order to test the effectiveness of tannins as dietary antioxidants, we are conducting feeding trials with lab animals. Because we hypothesize that the tannins
may act in the digestive tract to protect absorbable nutritive antioxidants, in our
initial experiments we are assessing the levels of antioxidants in the serum during

the feeding trials. If that hypothesis is unsupported by our data, we will alter our analytical approach to directly assess the redox activity of tannins in the digestive tract.

In a pilot feeding trial, the condensed tannin from sorghum grain was used to supplement diets fed to laboratory rats. Many common foods including fruits, legumes, and beverages contain condensed tannins,[51] so the results obtained with sorghum tannin were of particular interest. Furthermore, the sorghum tannin is free of low molecular weight nontannin phenolics,[24] so effects of non-tannin phenolics such as flavonoids[52] were eliminated. Sorghum tannin can be added to the diets of rats at high levels (2–5 percent by weight) without acute tox-icity, although protein metabolism and growth rates of the animals are altered by those levels of tannin;[53] there is little effect on protein metabolism when tannin is added at the low levels (~0.2 percent) appropriate for assessing tannins as dietary antioxidants.

Groups of six rats were fed control diets or tannin-containing diets. The base diets were either nutritionally complete rat chow or antioxidant-free rat chow [Harlan Teklad formula TD 97104; prepared with stripped corn oil and standard levels of minerals (AIN-93-G), and standard amounts of the other vitamins (AIN-76-A) except for tocopherol and retinoids]. Tannin was mixed with dry powdered feed (final tannin level 0.16 percent by weight) and animals were provided with 15 g of food per day for 4 weeks. Animals were weighed on a weekly basis, and blood (200–400 μL) was collected from the tail vein into EDTA-coated Microtainer tubes. The blood was immediately transferred to the laboratory, spun down, and plasma collected and frozen at –80 in the dark. Total antioxidants in plasma samples were determined with the ORAC method.[54] The ORAC assay provides a measure of radical quenching capacity of biological samples and has been used successfully to measure total antioxidants in plasma. The reaction between the peroxyl radical (generated with AAPH [2,2'-azobis-(2-amidinopropane) dihy-drochloride]) and the indicator protein β-phycoerythrin is monitored fluorimetri-cally. Antioxidants prolong the lifetime of the indicator protein by preventing its oxidation. An advantage of ORAC is its ability to measure all of the antioxidants present, including water-soluble and lipid-soluble nutritive antioxidants, and the endogenous antioxidants (serum albumin, uric acid, etc.).

Both the weight data and the plasma antioxidant data indicated that the treat-ments had no effect on the rats (table 2). Feeding the animals an antioxidant-free diet for 4 weeks did not deplete the levels of total antioxidants in the plasma, so it is not surprising that tannin amendment of the diets did not affect antioxidant status of the animals. It is reasonable to conclude that for rats, at least during short-term laboratory trials, endogenously synthesized ascorbic acid[55] substitutes for other dietary antioxidants when necessary. Rats are thus an inappropriate model system since they differ from humans in their ability to synthesize ascor-bic acid.

In a more recent feeding trial we have used guinea pigs, which are similar to humans in their requirement for ascorbic acid.[55] Groups of six weanling animals were fed tannin-free or tannin-containing diets for 2 weeks. For this trial, a tannic acid preparation (MCB #CB792) comprising 15 percent tetragalloyl glucose, 30

Table 2. Body mass and serum antioxidant level of rats fed antioxidant-depleted and tannin-supplemented diets[a]

	day 0		day 26	
	body mass (g)	serum antioxidants	body mass (g)	serum antioxidants
		(Trolox equivalents)[b]		(Trolox equivalents)[b]
nutritionally complete	215 ± 21	4.7 ± 1.4	232 ± 9	3.5 ± 1.2
nutritionally complete + tannin	214 ± 12	4.7 ± 1.4	231 ± 8	3.8 ± 1.0
antioxidant-free	213 ± 10	4.7 ± 1.4	233 ± 9	4.0 ± 0.8
antioxidant-free + tannin	216 ± 10	4.7 ± 1.4	240 ± 16	4.5 ± 1.1

[a]Rats (six in each group) were fed nutritionally complete diets with or without 0.16 per cent condensed tannin; or antioxidant-free diets with or without 0.16 per cent condensed tannin; for 4 weeks.
[b]Serum antioxidants were determined using the ORAC[54] method. Values are the mean ± SD. Values for day 0 were averaged for all animals in the experiment so the same value is reported for each group of animals. None of the values are significantly different (t-test).

percent pentagalloyl glucose, 25 percent hexagalloyl glucose, and 12 percent heptagalloyl glucose was used rather than the purified sorghum condensed tannin. Choice of tannin was dictated by our *in vitro* data (which suggested that gallotannins were more effective antioxidants than condensed tannins) and their commercial availability.

The base diet contained low levels of ascorbic acid (33 mg/kg) and of tocopherol (15 mg/kg) (Harlan Teklad formula TD 98116; modified from formula 170679) since guinea pigs have high mortality if fed ascorbic acid-free diets. Unamended base diet was fed to one group of six weanling animals, and diet amended with 0.3 percent gallotannin was fed to another group of six animals. The control group of animals was fed base diet supplemented with ascorbic acid (467 mg/kg) and tocopherol acetate (85 mg/kg). The animals were accommodated to diet by slowly introducing the nutritionally complete diet to conventional chow over the course of 1 week and were then provided the experimental diets ad lib for 2 weeks. Animals were weighed daily, and at the end of the trial, blood was collected into EDTA-coated tubes from lightly anesthetized animals by cardiac puncture. Blood was immediately transported to the laboratory, centrifuged, and plasma frozen at −40. Subsamples were made to 1 mM dithioerythritol to preserve ascorbic acid before freezing.

The weight data showed that the vitamin-deficient diet was not tolerated by the animals. Although initial masses of the animals in the three groups were identical, by the eighth day of the trial, the animals consuming the vitamin-deficient diet weighed significantly less than those in the nutritionally complete group. Tannin-supplementation improved performance of the animals on the vitamin-deficient diet, although animals still weighed less than those on the nutritionally complete diet (fig. 1). Differences in food consumption did not appear to explain

Figure 1. Body mass of guinea pigs fed antioxidant-depleted or tannin-supplemented diets. Guinea pigs were fed a nutritionally complete diet (n = 6) (Δ); vitamin C and E-depleted diet (n = 5) (○); or vitamin-depleted diet supplemented with 0.3 percent gallotannin (n = 6) (■); for 2 weeks. The masses of the animals on the vitamin-depleted diet are significantly lower than those of the animals on the nutritionally complete diet after the eighth day of the trial (t-test, p = 0.05).

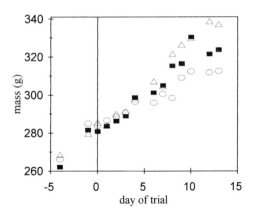

the differences in weight, although because guinea pigs spill and soil their food, exact quantitation of consumption is difficult.

Complete analysis of blood antioxidant levels will establish whether gallotannin supplementation protected nutritive antioxidants and thus improved performance of the animals on the low ascorbic acid/tocopherol diet. Serum analyses that will be done include total antioxidants (ORAC),[54] ascorbic acid,[56] and levels of oxidative damage to proteins[57] and lipids.[58]

5. FUTURE DIRECTIONS

Determining whether changes in the antioxidant status of an animal are meaningful is dependent upon proper choice of both the animal system and the analytical method. For example, using rats as an experimental model for studying non-nutritive dietary antioxidants such as tannins does not appear to be useful because in rats, endogenously synthesized ascorbic acid overwhelms the effects of dietary antioxidants. Even in animals with a requirement for dietary ascorbic acid, it may be necessary to impose oxidative stress such as exercise[59] on the animals before collecting serum in order to see significant differences in levels of dietary antioxidants.

The profusion of analytical methods that have been devised for determining total antioxidants reflects the difficulty of devising a method that is universally acceptable. Methods that are widely used include the ORAC method;[54] the metmyoglobin method;[50] and the FRAP method.[60] Since each of these methods provides a slightly different perspective, total antioxidants in plasma samples should probably be assessed by several of these methods to obtain a complete assessment of antioxidant level. The potential of tannins to interfere with each of these methods must be considered. For example, the intrinsic fluorescence of tannins may interfere with the ORAC method; the tendency to form complexes with iron and protein may interfere with the FRAP and metmyoglobin methods, respectively. Some of these problems will be overcome by specific measurement of the antioxidants of interest such as ascorbic acid.[56] In order to fully assess the effectiveness of potential dietary antioxidants, levels of oxidative damage and of pro-oxidants should be measured. Experimental artifacts including post-sampling oxidative damage to samples are associated with measurement of protein carbonyls[57] and lipid peroxidation.[58] Measurement of other metabolic products of oxidation such as prostacyclins[61] may overcome these problems.

Our studies to date have focused on the hypothesis that dietary tannin may "spare" nutritive antioxidants during the digestive process, and thus may exert an indirect effect on antioxidant status in the serum. Absorption of tannins or tannin-derived metabolic products may contribute more directly to antioxidant status. For example, in our experiment in which gallotannins were used as the dietary supplement, it is possible that small, absorbable phenolics such as gallic acid and its simple esters may have been produced during digestion. We believe that one important direction for future research on the effects of dietary tannins is understanding the metabolic fate of ingested tannins. We are developing appropriate methodology for specific determination of phenolics in blood plasma; other groups are focusing efforts on producing high specific radioactivity tannins that can be used in metabolic experiments.[23] There are also other potential roles for dietary tannin in disease prevention; tannins may bind carcinogens and prevent their absorption;[62] they may minimize nitrosamine formation;[63] or they may induce intestinal Phase I/Phase II metabolism.[64] The large quantities of tannin consumed as a constituent of fruit- and vegetable-rich diets suggest that tannin is a phytochemical worthy of careful consideration as a disease-preventative agent.

ACKNOWLEDGMENTS

This work was supported by grants from the Research Corporation, the National Science Foundation, the Howard Hughes Medical Research Foundation, and the Miami University Undergraduate Summer Scholars Program.

REFERENCES

1. Hansson, L.E.; Nyren, O.; Bergstrom, R.; Wolk, A.; Lindgren, A.; Baron, J.; Adami, H.O. Nutrients and gastric cancer risk. A population-based case-control study in Sweden. *Int. J. Cancer* 57:638 (1994).

2. Appel, L.J.; Moore, T.J.; Obarzanek, E.; Vollmer, W.M.; Svetkey, L.P.; Sacks, F.M.; Bray, G.A.; Vogt, T.M.; Cutler, J.A.; Windhauser, M.M.; Lin, P.H.; Karanja, N. A clinical trial of the effects of dietary patterns on blood pressure. *New Engl. J. Med.* 336:1117 (1997).

3. Block, G. Epidemiologic evidence regarding vitamin C and cancer. *Am. J. Clin. Nut.* 54:1310S (1991).

4. Ziegler, R.G. Vegetables, fruits, and carotenoids and the risk of cancer. *Am. J. Clin. Nut.* 53:251S (1991).

5. Gey, K.F. Prospects for the prevention of free radical disease, regarding cancer and cardiovascular disease. *Brit. Med. Bull.* 49:679 (1993).

6. Hecht, S.S.; Morse, M.A.; Amin, S.; Stoner, G.D.; Jordan, K.G.; Choi, C.; Chung, F. Rapid single-dose model for lung tumor induction in A/J mice by 4-(methylnitrosamino)-1-(3-pyridyl)-1-butanone and the effect of diet. *Carcinogenesis* 10:1901 (1989).

7. Huang, M.-T.; Ferraro, T.; Ho, C.-T. Cancer chemoprevention by phytochemicals in fruits and vegetables. *In*: Huang, M.-T.; Osawa, T.; Ho, C.-T.; Rosen, R.T. (eds.). Food phytochemicals for cancer prevention I. American Chemical Society, Washington DC, p. 2 (1994).

8. Ho, C.-T. Phenolic compounds in food: An overview. *In*: Ho, C.-T.; Lee, C.-Y.; Huang, M.-T. (eds.). Phenolic compounds in food and their effects on health. American Chemical Society, Washington DC, p. 2 (1992).

9. Jan, M.; Cai, L.; Udeani, G.O.; Slowing, K.V.; Thomas, C.F.; Beecher, C.W.W.; Fond, H.H.S.; Farnsworth, N.R.; Kinghorn, A.D.; Mehta, R.G.; Moon, R.C.; Pezzuto, J.M. Cancer chemopreventive activity of resveratrol, a natural product derived from grapes. *Science* 275:218 (1997).

10. Robinson, T. The organic constituents of higher plants. Cordus Press, North Amherst, MA (1980).

11. Haslam, E. Plant polyphenols—vegetable tannins revisited. Cambridge University Press, Cambridge (1989).

12. Daniel, E.M.; Stoner, G.D. The effects of ellagic acid and 13-cis-retinoic acid on N-nitrosobenzylmethylamine-induced esophageal tumorigenesis in rats. *Cancer Lett.* 56:117 (1991).

13. Athar, M.; Khan, W.A.; Mukhtar, H. Effect of dietary tannic acid on epidermal, lung, and forestomach polycyclic aromatic hydrocarbon metabolism and tumorigenicity in Sencar mice. *Cancer Res.* 49:5784 (1989).

14. Yugarani, T.; Tan, B.K.; Das, N.P. The effects of tannic acid on serum and liver lipids of RAIF and RICO rats fed on high fat diet. *Comp. Biochem. & Physiol.* 104:339 (1993).

15. Yang, C.S.; Wang, Z.-Y.; Hong, J.-Y. Inhibition of tumorigenesis by chemicals from garlic and tea. *In*: Jacobs, M.M. (ed.). Diet and cancer. Plenum Press, New York, p. 113 (1994).

16. Uchida, S.; Ohta, H.; Niwa, M.; Mori, A.; Nonaka, G.; Nishioka, I.; Ozaki, M. Prolongation of life span of stroke-prone spontaneously hypertensive rats (SHRSP) ingesting persimmon tannin. *Chem. Pharm. Bull.* 38:1049 (1990).

17. Uchida, S.; Ozaki, M.; Akashi, T.; Yamashita, K.; Niwa, M.; Taniyama, K. Effects of (–)-epigallocatechin-3-O-gallate (green tea tannin) on the life span of stroke-prone spontaneously hypertensive rats. *Clin. Exp. Pharm. Physiol.* 22:S302 (1995).

18. Perchellet, J.P.; Gali, H.U.; Perchellet, E.M.; Laks, P.E.; Bottari, V.; Hemingway, R.W.; Scalbert, A. Antitumor-promoting effects of gallotannins, ellagitannins and flavonoids in mouse skin *in vivo*. *In*: Huang, M-T.; Osawa, T.; Ho, C-T.; Rosen, R.T. (eds.). Food phytochemicals for cancer prevention I. American Chemical Society, Washington DC, p. 303 (1994).

19. Miyamoto, K.; Nomura, M.; Murayama, T.; Furukawa, T.; Hatano, T.; Yoshida, T.; Koshiura, R.; Okuda, T. Antitumor activities of ellagitannins against Sarcoma-180 in mice. *Biol. Pharm. Bull.* 16:379 (1993).

20. Robbins, C.T.; Hagerman, A.E.; Austin, P.J.; McArthur, C.; Hanley, T.A. Variation in mammalian physiological responses to a condensed tannin and its ecological implications. *J. Mammal.* 72:480 (1991).

21. Jimenez-Ramsey, L.M.; Rogler, J.C.; Housley, T.L.; Butler, L.G.; Elkin, R.G. Absorption and distribution of 14C-labeled condensed tannins and related sorghum phenolics in chickens. *J. Agric. Food Chem.* 42:963 (1994).

22. Hollman, P.C.H.; Gaag, M.V.D.; Mengelers, M.J.B.; vanTrijp, J.M.P.; deVries, J.H.M.; Katan, M.B. Absorption and disposition kinetics of the dietary antioxidant quercetin in man. *Free Rad. Biol. Med.* 21:703 (1996).

23. Deprez, S.; Mila, I.; Scalbert, A.; Huneau, J.-F.; Tome, D. C-14 biolabeling of proanthocyanidins in willow shoots for bioavailability studies (this volume) (1999).

24. Schofield, J.A.; Hagerman, A.E.; Harold, A. Loss of tannins and other phenolics from willow leaf litter. *J. Chem. Ecol.* 24:1409 (1998).

25. Hemingway, R.W. Reactions at the interflavanoid bond of proanthocyanidins. *In*: Hemingway, R.W.; Karchesy, J.J. (eds.). Chemistry and significance of condensed tannins. Plenum Press, New York, p. 265 (1989).

26. Daniel, E.M.; Ratnayake, S.; Kinstle, T.; Stoner, G.D. The effects of pH and rat intestinal contents on the liberation of ellagic acid from purified and crude ellagitannins. *J. Nat. Prod. (Lloydia)* 54:946 (1991).

27. Booth, A.N.; Masri, M.S.; Robbins, D.J.; Emerson, O.H.; Jones, F.T.; DeEds, F. The metabolic fate of gallic acid and related compounds. *J. Biol. Chem.* 234:3014 (1959).

28. Prochaska, H.J.; Talalay, P. Phenolic antioxidants as inducers of anticarcinogenic enzymes. *In*: Ho, C.-T.; Lee, C.-Y.; Huang, M.-T. (eds.). Phenolic compounds in food and their effects on health. American Chemical Society, Washington DC, p. 150 (1992).

29. Frankel, E.N.; Kanner, J.; German, J.B.; Parks, E.; Kinsella, J.E. Inhibition of oxidation of human low-density lipoprotein by phenolic substances in red wine. *Lancet* 341:454 (1993).

30. Bhat, R.; Hadi, S.M. DNA breakage by tannic acid and Cu(II): Sequence specificity of the reaction and involvement of active oxygen species. *Mut. Res.* 313:39 (1994).

31. Tsai, Y.J.; Aoki, T.; Maruta, H.; Abe, H.; Sakagami, H.; Hatano, T.; Okuda, T.; Tanuma, S. Mouse mammary tumor virus gene expression is suppressed by oligomeric ellagitannin, novel inhibitors of poly (ADP-ribose) glycohydrolase. *J. Biol. Chem.* 267:14436 (1992).

32. Kehrer, J.P. Free radicals as mediators of tissue injury and disease. *Crit. Rev. Toxicol.* 23:21 (1993).

33. Janssen, Y.M.W.; Van Houten, B.; Borm, P.J.A.; Mossman, B.T. Cell and tissue responses to oxidative damage. *Lab. Invest.* 69:261 (1993).

34. Ames, B.N.; Shigenaga, M.K.; Hagen, T.M. Oxidants, antioxidants, and the degenerative diseases of aging. *Proc. Nat. Acad. Sci. USA* 90:7915 (1993).

35. Byers, T.; Perry, G. Dietary carotenes, vitamin C, and vitamin E as protective antioxidants in human cancers. *Ann. Rev. Nut.* 12:139 (1992).

36. Albanes, D.; Heinonen, O.P.; Huttunen, J.K.; Taylor, P.R.; Virtamo, J.; Edwards, B.K.; Haapakoski, J.; Rautalahti, M.; Palmgren, J. Effects of alpha-tocopherol and beta-carotene supplements on cancer incidence in the Alpha-Tocopherol Beta-Carotene Cancer Prevention Study. *Am. J. Clin. Nut.* 62:1427S (1995).

37. Omenn, G.S.; Goodman, G.E.; Thronquist, M.D.; Cullen, M.R.; Glass, A.; Keogh, J.P.; Meyskens, F.L.; Valanis, B.; Williams, J.H.; Barnhart, S.; Hammar, S. Effects of a combination of beta carotene and vitamin A on lung cancer and cardiovascular disease. *New Engl. J. Med.* 334:1150 (1996).

38. Peng, Y.M.; Peng, Y.S.; Lin, Y.; Moon, T.; Roe, D.J.; Ritenbaugh, C. Concentrations and plasma-tissue-diet relationships of carotenoids, retinoids, and tocopherols in humans. *Nut. Cancer* 23:233 (1995).

39. Gey, K.F. Cardiovascular disease and vitamins. Concurrent correction of 'suboptimal' plasma antioxidant levels may, as an important part of 'optimal' nutrition, help to prevent early stages of cardiovascular disease and cancer, respectively. *Bibliotheca Nutritio et Dieta* 52:75 (1995).

40. Ross, M.A.; Crosley, L.K.; Brown, K.M.; Duthie, S.J.; Collins, A.C.; Arthur, J.R.; Duthie, G.G. Plasma concentrations of carotenoids and antioxidant vitamins in Scottish males: influences of smoking. *Eur. J. Clin. Nut.* 49:861 (1995).

41. Kubow, S. Routes of formation and toxic consequences of lipid oxidation products in foods. *Free Rad. Biol. Med.* 12:63–81 (1992).

42. Ryan, T.P.; Aust, S.D. The role of iron in oxygen-mediated toxicities. *Crit. Rev. Toxicol.* 22:119 (1992).

43. Simic, M.G.; Jovanovic, S.V. Inactivation of oxygen radicals by dietary phenolic compounds in anticarcinogenesis. *In*: Ho, C.-T.; Osawa, T.; Huang, M.-T.; Rosen, R.T. (eds.). Food phytochemicals for cancer prevention II. American Chemical Society, Washington, DC, p. 20 (1994).

44. Bors, W.; Heller, W.; Michel, C.; Saran, M. Flavonoids as antioxidants: Determination of radical-scavenging efficiencies. *Meth. Enzymol.* 186:343 (1990).

45. Slabbert, N. Complexation of condensed tannins with metal ions. *In*: Hemingway, R.W.; Laks, P.E. (eds). Plant polyphenols—synthesis, properties, significance. Plenum Press, New York, p. 421 (1992).

46. Hagerman, A.E.; Riedl, K.M.; Jones, G.A.; Sovik, K.N.; Ritchard, N.T.; Hartzfeld, P.W.; Riechel, T.L. High molecular weight plant polphenolics (tannins) as biological antioxidants. *J. Agric. Food Chem.* 46:1887 (1998).

47. Halliwell, B.; Gutteridge, J.M.C.; Aruoma, O.I. The deoxyribose method: A simple "test-tube" assay for determination of rate constants for reactions of hydroxyl radicals. *Analyt. Biochem.* 165:215 (1987).

48. Smith, C.; Halliwell, B.; Aruoma, O.I. Protection by albumin against the pro-oxidant actions of phenolic dietary components. *Food Chem. Toxicol.* 30:483 (1994).

49. Laughton, M.J.; Halliwell, B.; Evans, P.J.; Hoult, J.R.S. Antioxidant and pro-oxidant actions of the plant phenolics quercetin, gossypol and myricetin. *Biochem. Pharmacol.* 38:2859 (1989).

50. Miller, N.J.; Rice-Evans, C.A.; Davies, M.J.; Gopinathtan, V.; Milner, A. A novel method for measuring antioxidant capacity and its application to monitoring the antioxidant status in premature neonates. *Clin. Sci.* 84:407 (1993).

51. Pierpoint, W.S. Flavonoids in the human diet. *In*: Cody, V.; Middleton, E.; Harborne, J.B. (eds.). Plant flavonoids in biology and medicine—biochemical, pharmacological, and structure-activity relationships. Alan R. Liss Inc. New York, p. 125 (1986).

52. Bors, W.; Saran, M. Radical scavenging by flavonoid antioxidants. *Free Rad. Res. Comm.* 2:289 (1987).

53. Mole, S.; Rogler, J.C.; Morell, C.J.; Butler, L.G. Herbivore growth reduction by tannins: Use of Waldbauer ratio techniques and manipulation of salivary protein production to elucidate mechanisms of action. *Biochem. System. Ecol.* 18:183 (1990).

54. Cao, G.; Alessio, H.M.; Cutler, R.G. Oxygen-radical absorbance capacity assay for antioxidants. *Free Rad. Biol. Med.* 14:303 (1993).

55. Tuffery, A.A. Laboratory animals. Wiley, New York (1996).

56. Barja, G.; Hernanz, A. Vitamin C, dehydroascorbate, and uric acid in tissues and serum: High-performance liquid chromatography. *Meth. Enz.* 234:331 (1994).

57. Levine, R.L.; Garland, D.; Oliver, D.N.; Amici, A.; Climent, I.; Lenz, A.; Ahn, B.; Shaltiel, S.; Statman, E.R. Determination of carbonyl content in oxidatively modified proteins. *Meth. Enz.* 187:100 (1990).

58. Marshall, T.A.; Roberts, R.J. *In vitro* and *in vivo* assessment of lipid peroxidation of infant nutrient preparations: Effect of nutrition on oxygen toxicity. *J. Am. Coll. Nutr.* 9:190 (1990).

59. Sen, C.K. Oxidants and antioxidants in exercise. *J. Appl. Physiol.* 79:675 (1995).

60. Benzie, I.F.F.; Strain, J.J. The ferric reducing ability of plasma (FRAP) as a measure of "antioxidant power". The FRAP assay. *Analyt. Biochem.* 239:70 (1996).

61. Pratico, D.; Barry, O.P.; Lawson, J.A.; Adiyaman, M.; Hwang, S.W.; Khanapure, S.P.; Iuliano, L.; Rokach, J.; FirtGerald, G.A. IPF2alpha-I: an index of lipid peroxidation in humans. *Proc. Nat. Acad. Sci. USA* 95:3449 (1998).

62. Stavric, B.; Matula, T.I.; Klassen, R.; Downie, R.H.; Wood, R.J. Effect of flavonoids on mutagenicity and bioavailability of xenobiotics in foods. *In*: Huang, M-T.; Ho, C-T.; Lee, C-Y. (eds.). Phenolic compounds in food and their effects on health. American Chemical Society, Washington, DC, p. 239 (1992).

63. S'Dhenoy, N.R.; Choughuley, A.S.U. Effect of certain plant phenolics on nitrosamine formation. *J. Agric. Food Chem.* 37:721 (1989).

64. Koster, A.S.; Richter, E.; Lauterbach, F.; Hartmann, F. Intestinal metabolism of xenobiotics. Springer-Verlag, New York (1989).

CONFORMATION, COMPLEXATION, AND ANTIMICROBIAL PROPERTIES

INTERACTION OF FLAVANOIDS WITH PEPTIDES AND PROTEINS AND CONFORMATIONS OF DIMERIC FLAVANOIDS IN SOLUTION

Tsutomu Hatano,[a] Takashi Yoshida,[a]
and Richard W. Hemingway[b]

[a]Faculty of Pharmaceutical Sciences
Okayama University
Tsushima, Okayama 700-8530
JAPAN

[b]Southern Research Station
USDA Forest Service
Pineville, Louisiana 71360
USA

1. INTRODUCTION

Although the physiological roles of tannins and related polyphenols in plants have not yet been clarified, their ability to form complexes with proteins or related biopolymers has been correlated with some protection of the plants from predators such as animals, insects, and microbes.[1] Similarly, commercial uses of tannins, especially in the leather and brewing industries, are also based on their binding with proteins.[2] Pharmacological properties of tannins have been investigated based on recent advances in the structural study of tannins in medicinal plants, and various actions of tannins including antitumor and antiviral effects have been revealed.[3] These effects are attributed to interactions with certain biomolecules in organisms, too.

Most of these effects are dependent on the chemical structures or molecular shapes of tannins.[3] Therefore, clarification of the molecular conformations of tannins and related polyphenols is requisite to understanding the process of molecular interaction of tannins with the biomolecules. Here we summarize our reports on the conformational analyses of dimeric flavanoids related to condensed tannins and interaction of these flavanoids with peptides. Because molecular

Plant Polyphenols 2: Chemistry, Biology, Pharmacology, Ecology, Edited by
Gross et al. Kluwer Academic / Plenum Publishers, New York, 1999

conformations of oligomeric flavanoids[4] in the free phenolic form in solution with the biologically significant solvent (water) had not yet been characterized, we focused our efforts on definition of the conformations of the natural free phenols in aqueous solutions.

2. CONFORMATIONS IN ORGANIC SOLVENTS

Among various proanthocyanidins, we used catechin-(4α→8)-catechin and catechin-(4α→8)-epicatechin for the conformational analysis, since the presence of high rotational barriers around their interflavan linkages[5] resulted in sharp NMR spectra for the two conformers seen for each of the two compounds. However, complete assignment of these spectra had not been attempted. These dimeric procyanidins were synthesized as shown in figure 1.[6] The 2α,3β-dihydroquercetin, prepared by Karchesy at Oregon State University, was treated with sodium borohydride and the resulting hydroxyl group at C-4 of the reduction product was removed in acidic conditions to give a cation. Nucleophilic A-rings of either (+)-catechin or (−)-epicatechin then attacked the cation to form the dimeric procyanidins. These two dimers, which have (+)-catechin as the upper unit and either (+)-catechin or (−)-epicatechin as the lower unit, were then available without structural ambiguity.

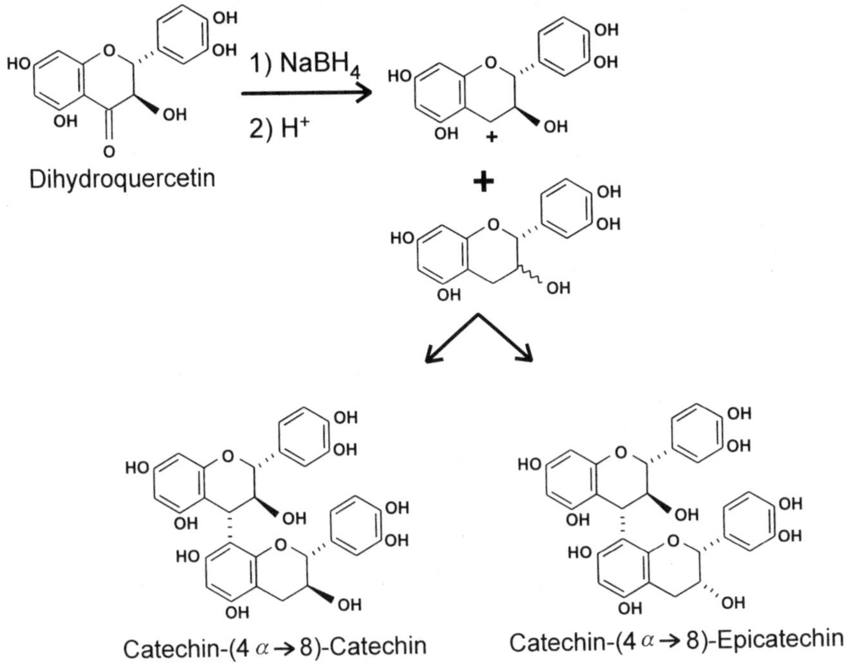

Figure 1. Syntheses of catechin-(4α→8)-catechin and catechin-(4α→8)-epicatechin.

The ^{1}H NMR spectrum of catechin-($4\alpha\rightarrow8$)-epicatechin in acetone-d_6 is shown in figure 2. Steric hindrance minimizes rotation around the interflavan bond, and two relatively stable conformational forms were observed in the NMR spectra. An expanded spectrum shows the signals in the aromatic region at around 6 ppm are A-ring and D-ring protons. Signals at lower field of the aromatic region are composed of four sets of three protons each forming the ABC 3-spin system characteristic of catechol rings. These are attributable to protons of B-ring and E-ring of this dimeric procyanidin. Therefore, the aromatic protons indicate the presence of two conformers, one of which shows signals due to the minor conformer with significant upfield shifts relative to those of the major conformer. The upfield shift of these B- and E-ring protons indicates that the minor conformer adopts a conformation where the two aromatic rings are close to each other enabling these anisotropic effects.

To establish the spatial positions of the constituent flavan units around the interflavan bonds of the dimeric procyanidins, definition of the conformations of heterocyclic C- and F-rings that affect locations of B- and E-rings in each dimer is required. Coupling constants of heterocyclic C- and F-ring protons in the ^{1}H NMR spectra of the two dimers, catechin-($4\alpha\rightarrow8$)-catechin and catechin-($4\alpha\rightarrow8$)-catechin, were analyzed with the aid of spectral simulation. The coupling constants for the upper units of both dimers were similar to those for the half-chair form of the heterocyclic ring of the monomeric catechin in which the B-ring is equatorial. However, the coupling constants for the lower units are different from the constituent monomers.

For the lower residue of catechin-($4\alpha\rightarrow8$)-epicatechin, estimation of the $J_{2,3}$ coupling constant is difficult, since H-2 appears as a broad singlet and H-3 proton

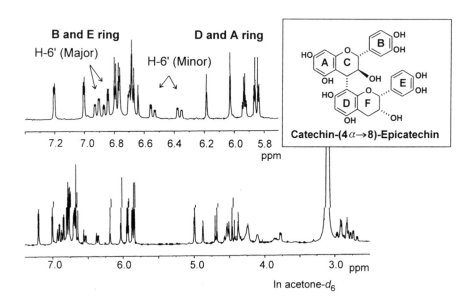

Figure 2. ^{1}H NMR spectrum of catechin-($4\alpha\rightarrow8$)-epicatechin.

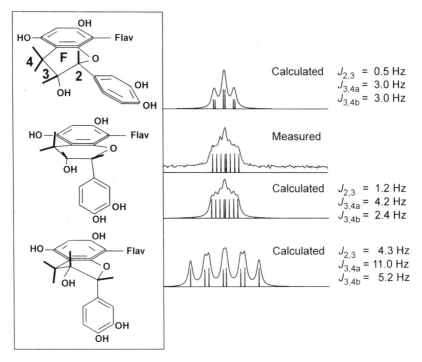

Figure 3. Conformations and coupling patterns for F-ring H-3 of catechin-($4\alpha{\rightarrow}8$)-epicatechin in acetone-d_6 containing D_2O.

shows a multiplet signal due to coupling with the two H-4 protons as shown in figure 3. However, spectral simulation using PCPMR[7] led to the assignment that the coupling constant between H-2 and H-3 is 1.2 Hz, and $J_{3,4a}$ and $J_{3,4b}$ are 4.2 and 2.4, respectively, when recorded in acetone-d_6. These F-ring coupling constants are consistent with neither the half-chair conformation with an equatorial phenyl ring nor the reverse half-chair conformation with an axial phenyl ring. An energetically unfavorable skewed-boat conformation was thus suggested for the F-ring, even though the observed coupling may be the result of an "averaged" coupling over the time frame of the NMR experiment[8] for reasons described below.

In D_2O, catechin-($4\alpha{\rightarrow}8$)-epicatechin showed a coupling pattern (fig. 4) somewhat different from that in acetone-d_6 as shown in figure 3. Obviously, the observed signal pattern of F-ring H-3 is far from that for the energetically preferable half-chair or reverse half-chair conformation, and the simulated signal patterns indicated that the F-ring adopts the C3 sofa conformation.

On the other hand, the dimer catechin-($4\alpha{\rightarrow}8$)-catechin showed a signal pattern of F-ring H-3 as shown in figure 5. The observed signal pattern is closer to the calculated signal pattern for the half-chair heterocyclic ring with an equatorial phenyl group. The calculated signal pattern for the reverse chair form with an axial phenyl group is far from the measured one, indicating that this form does not participate strongly in the conformation of this compound.

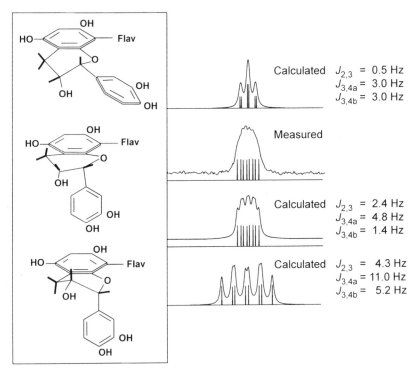

Figure 4. Conformations and coupling patterns for F-ring H-3 of catechin-$(4\alpha\rightarrow8)$-epicatechin in D_2O.

In addition, the molecular shapes of proanthocyanidins are strongly affected by steric hindrance around their interflavan bonds, so it is necessary to define the dihedral angles between the two flavan units of dimeric procyanidins. Spatial positions of the two halves in each of the two dimers were analyzed based on ^1H-^1H long-rang COSY[9] and NOESY measurements.

The ^1H-^1H long-range COSY spectrum of catechin-$(4\alpha\rightarrow8)$-epicatechin shown in figure 6 enabled the assignments of the B- and E-ring protons to each conformer by cross peaks due to the coupling between C-ring H-2 and B-ring H-2′ and H-6′, and between F-ring H-2 and E ring H-2′ and H-6′. In addition, this measurement permitted assignment of the orientation of D-ring relative to the upper half of the dimer in the following way. The ^1H-^1H long-range COSY spectrum showed strong cross peaks between C-ring H-4 and H-6 and H-8 of A-ring where the A-ring plane is at 90 degrees to the C4-H4 bond. The cross peak between D-ring H-6 and C-ring H-4 of the major conformer therefore indicates that the bond C4-H4 and the D-ring plane form an angle of about 90 degrees. On the other hand, the minor conformer showed a much weaker cross peak for the coupling between the C-ring H-4 and D-ring H-6. Therefore, the dihedral angle between C4-H4 and the D-ring plane in the minor conformer is larger or smaller than 90 degrees tending to either 180 or 0 degrees when the cross peak is absent.

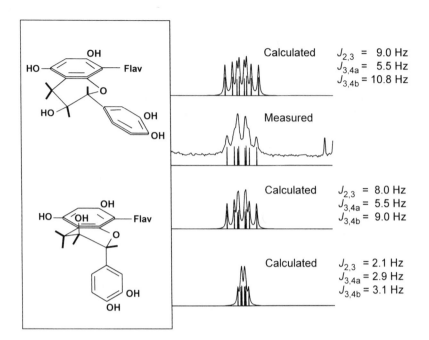

Figure 5. Conformations and coupling patterns for F-ring H-3 of catechin-(4α→8)-catechin in D₂O.

In support of the above conclusions, the NOESY spectrum of catechin-(4α→8)-epicatechin shown in figure 7 indicates that the C- and E-rings of this molecule are spatially close to each other. A cross peak due to NOE correlation between C-ring H-4 and E-ring H-2′ of the major conformer indicated that the major conformer of this dimeric procyanidin adopts a conformation in which the α-oriented E-ring of the lower unit locates near the β-oriented H-4 of the upper unit. Although the spectrum also showed many correlations between the protons of one conformer and the corresponding protons of the other conformer, these cross peaks are not due to NOE correlations but to the conformational exchange. For example, the H-2′ of the lower half of one conformer shows a correlation with H-2′ of the lower half of another conformer.

The presence of the NOE between C-ring H-4 and E-ring H-2′ in the molecule of catechin-(4α→8)-epicatechin was confirmed by measurement of a NOE difference spectrum (fig. 8). Irradiation of H-2′ of the lower residue of the major conformer caused increase of the peak area of C-ring H-4 along with that of F-ring H-2. Therefore, the orientation of the lower residue is that shown in figure 7, where the E-ring of the lower residue locates on the same side as that of H-4 of the upper residue. Irradiation of H-2′ of the lower unit of the major conformer also caused a negative peak for the corresponding proton of the minor conformer. The appearance of the negative peak of the corresponding proton of another conformer is explained by conformational exchange accompanied by saturation transfer.

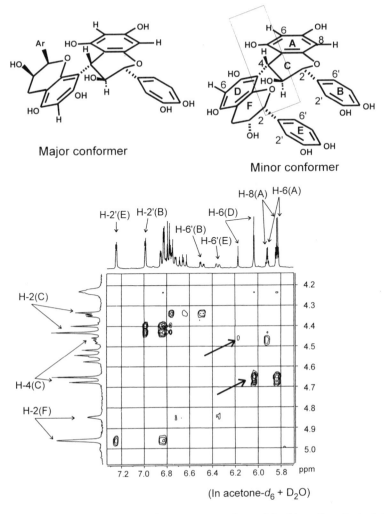

Figure 6. ^{1}H-^{1}H Long-range COSY spectrum of catechin-(4α→8)-epicatechin in acetone-d_6 containing D$_2$O.

Analogous spatial correlations between the protons of C- and E-rings in one conformer of the dimeric procyanidin, catechin-(4α→8)-catechin were also observed in the NOESY and NOE difference spectra.

3. CONFORMATIONS IN WATER

The ^{1}H NMR spectra of catechin-(4α→8)-epicatechin in three different solvents, acetone-d_6, dioxane-d_8 and D$_2$O, differed as shown in figure 9. The chemical shifts of several aromatic protons in the spectrum in dioxane-d_8 varied from those in the spectrum in acetone-d_6, whereas relative abundances of the two conformers in

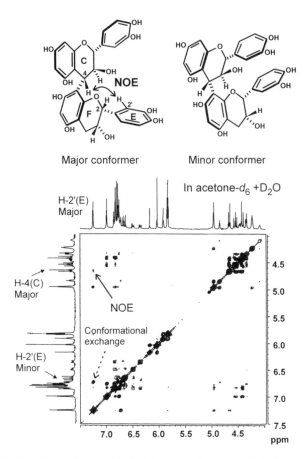

Figure 7. NOESY Spectrum of catechin-(4α→8)-epicatechin in acetone-d_6 containing D_2O.

these two solvents were almost the same. In D_2O, the signals of the major conformer in the organic solvents almost disappeared, and only the signals of another conformer were clearly observed. The aromatic protons of the major conformer in D_2O (i.e., the minor conformer in the organic solvents) shifted upfield noticeably relative to those of the other conformer. These changes are attributable to the anisotropic effects between B- and E-rings where the two aromatic rings overlap each other.

The composition of the two conformers of catechin-(4α→8)-epicatechin was dependent on the water content in the organic solvent. As shown in figure 10, the abundance of the major conformer in the organic solvents increases in the presence of a small amount of D_2O and then rapidly decreases upon further addition of D_2O. The relative proportion of the two conformers was estimated from the peak area in the ${}^1H$ NMR spectra. This change in rotamer population can be explained in the following way. The more "extended" conformer (i.e., the major conformer in

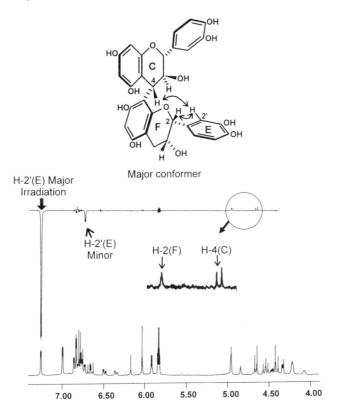

Figure 8. NOE difference spectrum of catechin-(4α→8)-epicatechin in acetone-d_6 containing D_2O.

acetone-d_6) is considered to be energetically preferable in the organic solvents. On the other hand, in D_2O, the dimer prefers the more "compact" conformation that is stabilized by hydrophobic or π-π interactions. These two conformations, which have been discussed for catechin-(4α→8)-epicatechin, are represented by those shown in figure 11.[10]

Interestingly, for catechin-(4α→8)-catechin, the relative abundance of the more "extended" conformer where the C- and E-rings are close to each other is smaller than that of the more "compact" conformer even in organic solvents such as acetone-d_6 and dioxane-d_8. In D_2O, this dimer exclusively adopts the more "compact" conformation.

4. INTERACTIONS BETWEEN POLYPHENOLS AND PEPTIDES CONTAINING PROLINE RESIDUES

Utilization of tannins for leather-tanning is based on the affinity of tannins for collagen, which is rich in proline and hydroxyproline residues.[2] The abundant prolyl residues in salivary proteins also are considered to contribute substantially

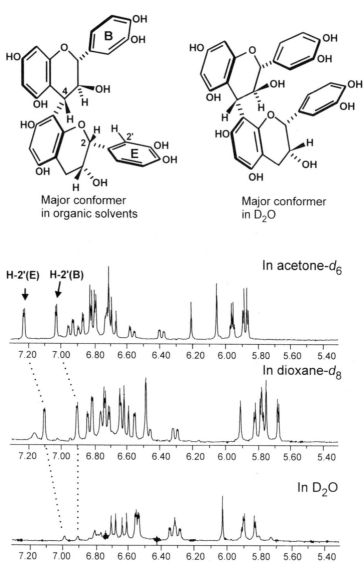

Figure 9. Solvent dependent changes of the ^{1}H NMR spectrum of catechin-($4\alpha{\rightarrow}8$)-epicatechin.

to preferential complexation with tannins resulting in the astringency so important to the flavor of fine red wines.[2,11] The benefits of traditional uses of medicinal plants containing high levels of tannins on such problems as skin diseases or digestive disorders may also be attributable to the interaction of tannins with proline-rich proteins. Participation of proline residues in the complexation between pentagalloylglucose and proline-rich peptides has been shown.[12]

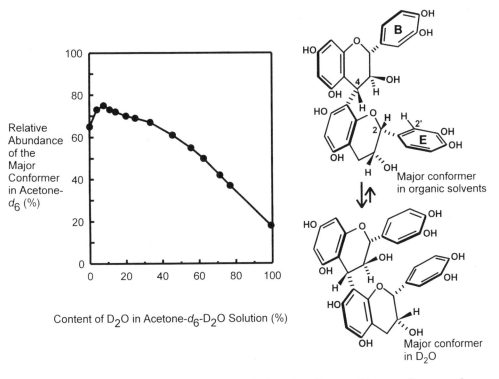

Figure 10. Effects of D_2O content on relative abundances of two conformers of catechin-$(4\alpha\rightarrow8)$-epicatechin.

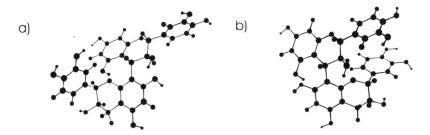

Figure 11. Conformations of catechin-$(4\alpha\rightarrow8)$-epicatechin in solutions a) More "extended", major conformation in acetone-d_6 and dioxane-d_8. b) More "compact", minor conformation in D_2O.

Our studies started with the analysis of the simplest combination, catechin and proline (or hydroxyproline), using NOESY experiments as a tool for clarifying the positions in each of the two molecules involved in molecular interaction. Catechin has been regarded as too small to interact to form precipitates with proteins.[13] However, when even a small amount of catechin was added to an aqueous solution (containing 10 percent of CD_3OD) of poly-L-proline (m.w. 10,000–30,000)

(9 mM for catechin and 120 mM for a monomeric proline residue), a white precipitate was formed. The interaction of catechin with a local portion of the polypeptide chain was proposed as the mechanism of this interaction.[14] NMR experiments on mixtures of catechin and proline in H_2O (containing 10 percent CD_3OD) with the NOESYHG pulse sequence[9] showed cross peaks of intermolecular NOEs between catechin B-ring protons and the proline C_γ protons (fig. 12). The C_β and C_δ protons of proline showed weaker interactions with aromatic protons of both the A- and B-rings of catechin. On the other hand, the NOESY spectrum of a mixture of catechin and *cis*-4-hydroxy-L-proline showed significant cross peaks between the C_δ protons of hydroxyproline and B-ring protons of catechin. The spectrum also showed correlations between a hydroxyproline C_δ proton (α-oriented

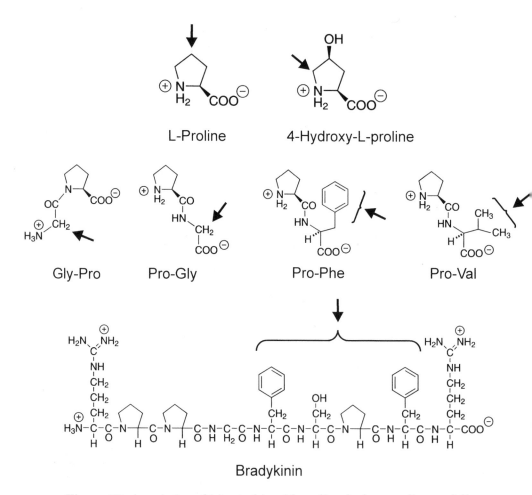

Figure 12. Association of (+)-catechin with proline, hydroxyproline, and dipeptides containing proline. The arrows indicate the sites that showed NOE interaction with catechin protons.

proton) to catechin H-6 and H-8. Importantly, noticeable changes in the ^{1}H or ^{13}C chemical shifts, which would suggest the presence of strong hydrogen bonding effects, were not observed. Therefore, hydrophobic interaction seems to be the dominant feature of these intermolecular associations. The aromatic rings of catechin and the aliphatic regions of proline or hydroxyproline come in close contact, but the complex is not so tight that we see significant changes in chemical shifts.

As mentioned above, polyphenols are believed to preferentially interact with proline residues in peptides. However, catechin showed intermolecular NOEs with the other amino acid residues rather than proline when dipeptides containing a proline residue, shown in figure 12, were used as targets for intermolecular recognition. An aqueous solution (containing 10 percent CD$_3$OD) of a mixture of Pro-Gly and (+)-catechin showed intermolecular NOE between the glycine C$_\alpha$ protons and the (+)-catechin H-8 proton. Pro-Val showed intermolecular NOE with catechin between methyl signals of the valine residue and the catechin H-8 as well as the H-2' proton. Pro-Phe showed intermolecular NOE with catechin between the phenyl ring of the peptide and the catechin B-ring. Gly-Pro showed two sets of ^{1}H and ^{13}C signals, which are attributable to the isomers concerning the amino (imino) nitrogen of proline. The major isomer, which showed a 2H singlet-like signal for glycine C$_\alpha$ protons, was assigned the *trans*-configuration. The minor isomer, which showed two doublets of the corresponding protons whose chemical shifts are separated due to the effect of the neighboring proline carbonyl carbon, was assigned the *cis*-configuration. The NOESY spectrum of a mixture of catechin and Gly-Pro showed significant cross peaks between the glycine protons of the *cis*-isomer and the catechin H-6 and H-8. On the other hand, the *trans*-isomer showed only very weak cross peaks with the B-ring protons of catechin. These results indicate that two aromatic rings of the catechin molecule preferentially associate with hydrophobic moieties in peptides other than the proline residues. Self-association of (+)-catechin in water was also shown by an NOE correlation of H-2' with H-8 of another molecule of catechin, even in competition with these peptides.

NOESY experiments on the interactions of polyphenols with two oligopeptides containing proline residues suggested the importance of the molecular conformations of both the polyphenol and peptide in addition to the hydrophobicity, to the specificity of their complexation. Bradykinin is a nonapeptide, Arg-Pro-Pro-Gly-Phe-Ser-Pro-Phe-Arg, containing three prolyl residues. This peptide exhibits various physiological effects including the regulation of blood pressure, and conformational studies on it in various solvents have been conducted.[15-17] Complexation of this peptide with polyphenols has also been investigated.[18] Our NOESY study on the interaction of bradykinin and (+)-catechin showed NOEs of the proline C$_\delta$ protons, the Ser C$_\beta$ methylene protons, and the phenyl protons of the phenylalanyl residues with the catechin A-ring protons. The phenylalanyl residue also showed correlations with catechin B-ring protons. The amino acid residues in bradykinin showing these NOEs are contained in the sequence. Phe$_5$-Ser$_6$-Pro$_7$-Phe$_8$. The interaction of this same sequence in bradykinin with micelles has also been reported.[15]

Gly-Pro-Gly-Gly is a tetrapeptide with inhibitory effects on the activity of dipeptidyl peptidase IV. This compound has been reported to inhibit the entry of

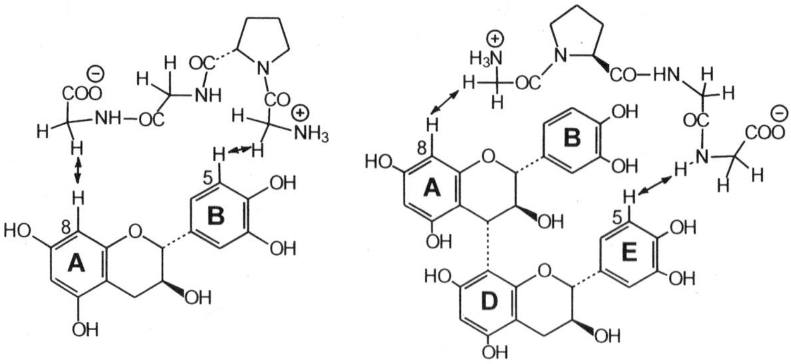

Gly-Pro-Gly-Gly
+
(+)-Catechin

Gly-Pro-Gly-Gly
+
Catechin-(4 α → 8)-catechin

Figure 13. Observed NOE interaction of flavanoids with Gly-Pro-Gly-Gly in water.

HIV into cells.[19] Although the *cis*-isomer of the proline nitrogen is present, its content is low, so the mixture can be treated practically as *trans*-isomer. Differentiation of the three sets of methylene protons of the glycine residues in the peptide was substantiated by the ^{1}H-^{1}H and ^{1}H-^{13}C long-range COSY analyses.

The NOESY spectrum of the mixture of (+)-catechin and Gly-Pro-Gly-Gly in water (containing 10 percent CD$_3$OD) showed correlations of catechin H-6 and H-8 with the C$_\alpha$ methylene protons of the *C*-terminal glycine residue, and of the catechin H-5 proton with methylene protons of the *N*-terminal glycine residue, suggesting that the catechin A-ring preferentially associates with the *C*-terminal and the B-ring with the *N*-terminal glycine residues.[20] Intermolecular NOEs, which suggest participation of the proline residue in the complexation, were not observed. Proline probably plays a role in the restriction of the relative spatial positions of the two glycine residues.

Intermolecular NOEs were also observed for the combination of Gly-Pro-Gly-Gly and catechin-(4α→8)-catechin. The association sites in the molecules of the polyphenol and peptide are, however, different from those observed in catechin. The NOESY spectrum showed intermolecular NOE between H-8$_A$ (the upper unit of the procyanidin dimer) and the methylene protons of the *N*-terminal glycine residue and between the E-ring protons, H-5$_E$ and H-2$_E$ of the dimer with the amide proton of the *C*-terminal glycine residue. These cross peaks suggest an association as exemplified by figure 13.

5. CD SPECTRAL ANALYSIS OF POLYPEPTIDE CONFORMATION CHANGES

The NMR results described above indicated that complexation is directed to conformationally accessible hydrophobic regions; therefore, the molecular shapes of both the polyphenol and polypeptide are important to selectivity. The NMR measurements of mixtures of flavans and bradykinin did not show significant changes in the chemical shifts of proton signals. This fact suggested interaction without conformational change of either the polyphenol or the peptide. Because circular dichroism (CD) is known to be sensitive to the conformational changes of peptides,[21] we used CD to verify whether the conformational changes in the peptide molecule occur with addition of polyphenols or not. When bradykinin was used as the peptide for the complexation, CD spectral measurements did not show any significant spectral changes upon the addition of either catechin or catechin-($4\alpha\rightarrow8$)-catechin.

However, the CD spectrum of cytochrome *c* showed a noticeable change in the visible region when either catechin or catechin-($4\alpha\rightarrow8$)-catechin was added to the solution of cytochrome *c* (fig. 14). This CD spectral change can, however, be attributed to the reduction of Fe^{3+} to Fe^{2+} of the heme prosthetic group based on the change in the visible spectrum (fig. 14). Most proteins in aqueous media, including cytochrome *c*, adopt conformations where the hydrophobic and hydrophilic side chains of the constituent amino acid residues are inside and outside, respectively. The reduction of the heme group by the polyphenols must be explained in terms of the interaction at the hydrophobic "inside" region in the cytochrome *c* molecule. Further study on the interaction of flavans with cytochrome *c* is now in progress.

6. CONCLUSIONS

Complete assignment of the ^{1}H and ^{13}C NMR spectra of the two conformational isomers for each of the dimeric proanthocyanidins catechin-($4\alpha\rightarrow8$)-catechin and catechin-($4\alpha\rightarrow8$)-epicatechin, in the natural free phenolic form and in the biologically significant solvent water, has been accomplished.[10] Knowledge of the assignments of the proton spectra, together with the application of various two-dimensional NMR experiments, has permitted the definition of the shapes of the heterocyclic C and F rings in the upper and lower units, the dihedral angles between the upper and lower flavan units, and the effect of changes in solvent composition on the relative proportions and shapes of the two conformers of these dimeric proanthocyanidins in organic and water solvents.[10] In organic solvents, the more extended conformation is favored for catechin-($4\alpha\rightarrow8$)-epicatechin; and a similar extended conformation, although not the major isomer, is present in nearly equal proportion to a more compact conformer for catechin-($4\alpha\rightarrow8$)-catechin. However, in water the more compact conformers dominate. These "unfavorable" conformations seem to be stabilized by hydrophobic (or π-π) interactions between the upper and lower unit catechol rings.

The importance of hydrophobic interactions was further highlighted in studies of the interaction of proanthocyanidins with oligopeptides. Even though trimeric and higher molecular weight proanthocyanidins are commonly assumed to be responsible for the precipitation of proteins, when even low mole ratios of the monomeric flavan-3-ol (+)-catechin was combined with polyproline, a white precipitate was obtained.[20] Evidence for hydrophobic interaction was seen in NOESYHG experiments, but no noticeable changes in ^{1}H or ^{13}C chemical shifts

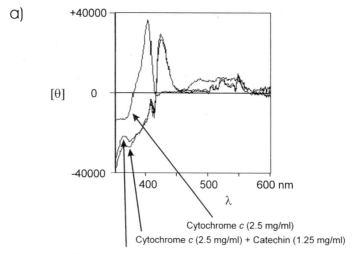

a)

Cytochrome *c* (2.5 mg/ml)
Cytochrome *c* (2.5 mg/ml) + Catechin (1.25 mg/ml)
Cytochrome *c* (2.5 mg/ml) + Catechin-(4α→8)-catechin (1.25 mg/ml)

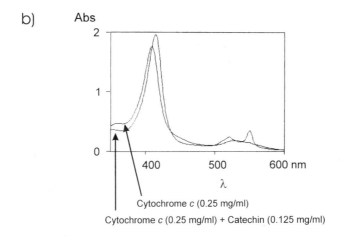

b)

Cytochrome *c* (0.25 mg/ml)
Cytochrome *c* (0.25 mg/ml) + Catechin (0.125 mg/ml)

Figure 14. Spectral changes of cytochrome *c* upon the addition of flavans a) CD spectra (in H₂O-CH₃OH, 9:1, v/v). b) Visible spectra (in H₂O-CH₃OH, 9:1, v/v).

were seen that would suggest strong hydrogen bonding in mixtures of (+)-catechin with L-proline or 4-hydroxy-L-proline. Studies of a series of dimeric peptides with various amino acid residues coupled with proline once more highlighted the significance of hydrophobic interaction. In association of catehin with dimeric peptides, stronger cross peaks were seen between either the A- or B-ring protons with the glycine, valine, or phenylalanine protons than to the prolyl unit protons. Once more, no measurable changes in chemical shifts suggesting strong hydrogen bonding were observed. It is important to note that intermolecular self-association of catechin also resulted in strong cross peaks in these experiments.

Addition of catechin or its dimer to Gly-Pro-Gly-Gly did not show association with the prolyl residue but rather to the conformationally accessible methylenes of glycine.[20] No significant changes in NMR chemical shifts or CD spectra that would suggest a "tight" binding or strong hydrogen bonding effects were noted. Studies of the interaction of catechin and dimeric procyanidins with bradykinin and cytochrome *c* also showed no significant change in the conformation of the peptides. Rather, association of the polyphenol with the presumed biologically active center of bradykinin and reduction of the heme group that resides within the interior of the cytochrome *c* polymer suggested that hydrophobic interaction dominates in the association of polyphenols with proteins in water. No significant changes in NMR chemical shifts that might suggest strong hydrogen bonding were observed. The evidence of all these experiments suggests strong interaction of polyphenols with hydrophobic centers where both self-association and preference for accessible hydrophobic regions of oligomeric peptides dominate.

ACKNOWLEDGMENTS

This work was funded by USDA NRICGP Grant No. 94-03395.

REFERENCES

1. Schultz, J.C.; Hunter, M.D.; Appel, M. Antimicrobial activity of polyphenols mediates plant-herbivore interactions. *In*: Hemingway, R.W., Laks, P.E. (eds.). Plant polyphenols—synthesis, properties, significance. Plenum Press, New York, p. 621 (1992).

2. Haslam, E. Practical polyphenolics: from structure to molecular recognition and physiological action. Cambridge University Press, Cambridge (1998).

3. Okuda, T.; Yoshida, T.; Hatano, T. Hydrolyzable tannins and related polyphenols. *Fortschr. Chem. Org. Naturst.* 66:1 (1995).

4. Steynberg, J.P.; Vincent Brandt, E.; Ferreira, D.; Helfer, C.A.; Mattice, L.; Gornik, D.; Hemingway, R.W. Conformational analysis of oligomeric flavanoids 2. Methyl ether acetate derivatives of profisetinidins. *Magn. Res. Chem.* 33:611 (1995).

5. Haslam, E. Symmetry and promiscuity in proanthocyanidin biochemisry. *Phytochemistry* 16:1625 (1977).

6. Botha, J.J.; Ferreira, D.; Roux, D.G. Synthesis of condensed tannins. Part 4. A direct biomimetic approach to [4,6]- and [4,8]-biflavanoids. *J. Chem. Soc. Perkin Trans. 1.* :1235 (1981).

7. PCPMR Spectral Simulation Program, Version 1. Serena Software, Bloomington, IN.

8. Tobiason, F.L.; Kelley, S.; Midland, M.M.; Hemingway, R.W. Temperature dependence of (+)-catechin pyran ring proton coupling constants as measured by NMR and modeled using GMMX search methodology. *Tetrahedron Letts.* 38(6):985 (1997).

9. Ziegler, P. A practical guide to 1-D and 2-D experiments for AC/AM Systems, Bruker Instruments, Billerica, MA (1991).

10. Hatano, T.; Hemingway, R.W. Conformational isomerism of phenolic procyanidins: preferred conformations in organic solvents and water. *J. Chem. Soc., Perkin Trans. 2.* :1035 (1997).

11. Hagerman, A.; Butler, L.G. The specificity of proanthocyanidin-protein interactions. *J. Biol. Chem.* 256:4494 (1981).

12. Murray, N.J.; Williamson, M.P.; Lilley, T.H.; Haslam, E. Study of the interaction between salivary proline-rich proteins and a polyphenol by ^{1}H-NMR spectroscopy. *Eur. J. Biochem.* 219:923 (1994).

13. Bate-Smith, E.C. Haemanalysis of tannins: The concept of relative astringency. *Phytochemistry* 12:907 (1973).

14. Tilstra, L.F.; Cho, D.; Bergmann, W.R.; Mattice, W.L. Interaction of condensed tannins with biopolymers. *In*: Hemingway, R.W.; Karchesy, J.J. (eds.). Chemistry and significance of condensed tannins. Plenum Press, New York, p. 335 (1989).

15. Denys, L.; Bothner-By, A.A.; Fisher, G.H.; Ryan, J.W. Conformational diversity of bradykinin in aqueous solution. *Biochemistry* 21:6531 (1982).

16. Young, J.K.; Hicks, R.P. NMR and molecular modeling investigations of the neuropeptide bradykinin in three different solvent systems: DMSO, 9:1 dioxane/water, and in the presence of 7.4 mM lyso phosphatidylcholine micelles. *Biopolymers* 34:611 (1994).

17. Cann, J.R.; Liu, Z.; Stewart, J.M.; Gera, L.; Kotovych, G. A CD and an NMR study of multiple bradykinin conformations in aqueous trifluoroethanol solutions. *Biopolymers*. 34:869 (1994).

18. Haslam, E. Natural polyphenols (vegetable tannins) as drugs: possible modes of action. *J. Nat. Prod.* 59:205 (1996).

19. Callebaut, C.; Krust, B.; Jacotot, E.; Hovamessian, A.G. T cell activation antigen, CD26, as a cofactor for entry of HIV in CD4$^+$ cells. *Science* 262:2045 (1993).

20. Hatano, T.; Hemingway, R.W. Association of (+)-catechin and catechin-(4α→8)-catechin with oligopeptides containing proline residues. *J. Chem. Soc., Chem. Commun.* :2537 (1996).

21. Das, T.K.; Mazumdar, S.; Mitra, S. Heme CD as a probe for monitoring local structural changes in heme-proteins: alkaline transition in hemeproteins. *Proc. Indian Acad. Sci.* 107:497 (1995).

MODELING THE CONFORMATION OF POLYPHENOLS AND THEIR COMPLEXATION WITH POLYPEPTIDES: SELF-ASSOCIATION OF CATECHIN AND ITS COMPLEXATION WITH L-PROLINE GLYCINE OLIGOMERS

Fred L. Tobiason,[a] Richard W. Hemingway,[b]
and Gérard Vergoten[c]

[a] Department of Chemistry
Pacific Lutheran University
Tacoma, Washington 98447
USA

[b] Southern Research Station
USDA Forest Service
Pineville, Louisiana 71360
USA

[c] CRESIMM
Université des Sciences et Technologies de Lille
UFR de Chemie, Bât C8
59655 Villeneuve d'Ascq
FRANCE

1. INTRODUCTION

Over the past 10 years, several scientific thrusts have come together in the study of flavanoids that make it possible to move forward into the study of complexation between polyphenols and polypeptides. Enhanced understanding of the conformational properties of flavanoid monomers and polyflavanoids through molecular modeling, combined with the detailed NMR experimental data now in the literature, provide the foundation.[1-13] Recent work using conformational searching techniques with the GMMX[6-8] protocol has shown additional detail about the distribution of pseudo equatorial and pseudo low-energy axial conformers in the ensemble, as shown in figure 1. This leads to information about the relationship

Plant Polyphenols 2: Chemistry, Biology, Pharmacology, Ecology, Edited by
Gross et al. Kluwer Academic / Plenum Publishers, New York, 1999

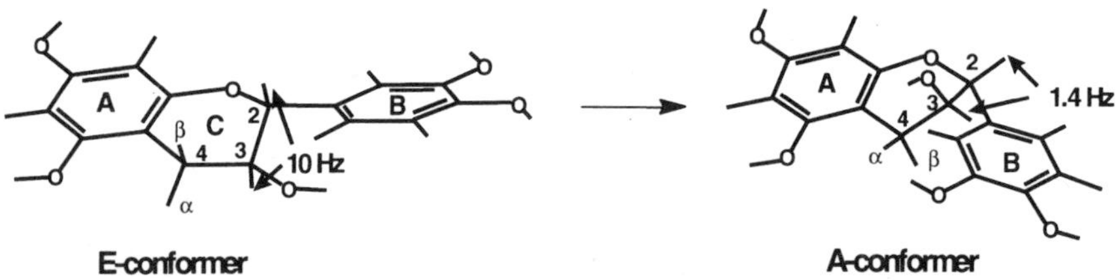

Figure 1. The pseudo equatorial conformer to pseudo axial conformer is illustrated. The structure-sensitive $J_{2,3}$ coupling constant varies between the E-conformer and A-conformer from 10.0 Hz to 1.4 Hz, respectively, as determined by MacroModel with a GB/SA water solvent model.

between the conformer ensemble and the Boltzmann averaged NMR proton coupling constants that one would expect to observe in a solution. Figure 1 also illustrates the pseudo equatorial to axial transformation that takes place in all catechin or (+)-catechin-($4\alpha{\rightarrow}8$)-(+)-catechin (B3) dimer complexes during the conformer searches and which would also be expected to occur in solution. Interest continues to further understand the details about this conformer distribution as well as in the prediction of complexation of tannins with metal ions and proteins. Although the GMMX software has given many interesting results, it is limited in handling cases that require systematic conformational searching of molecules combined in a complex. In addition, there are no solvent model options.

Recent NMR studies on procyanidin dimers[14] and NOE results of the complexation of L-proline-glycine compounds with (+)-catechin and polyflavanoid dimers[15] have given data to help guide computational studies. Couple this with the improved molecular computational software available,[16–19] and it becomes possible to explore complexation searching conformational space through Monte Carlo and molecular dynamics protocols using water as a solvent. The importance of this is highlighted by the renewed interest in its pharmacological characteristics such as the antiviral and antitumor behavior of tannins and other polyphenols[20] as well as reported interaction of polyphenols with proteins in aqueous solutions.[21–23] In this chapter, we explore computational models for molecules such as L-proline-glycine and glycl-L-prolyl-glycl-glycine ion (GPGG ion) interacting with (+)-catechin and (+)-catechin-($4\alpha{\rightarrow}8$)-(+)-catechin (B3) to form complexes. These results are compared to the close-contact positions obtained from NOE NMR experiments in aqueous solution. The complex structures found using conformational search methods are discussed in terms of the specific hydrophobic and hydrophilic interactions observed.

2. COMPUTATIONAL METHODS

The two protocols applied in this study for searching conformational space are MacroModel[16] version 5.5 and HyperChem 5.1.[17] MacroModel was applied in the Monte Carlo conformational multiple minimum structure searching mode using the Amber and Merck (MMFF 94s) force fields.[24] Typically, 5000 MC steps were used within a 16 kJ/mol energy window using the GB/SA[25] water solvent model. Sometimes a 30 kJ/mol energy window was used for the collection of conformer structures over a wider range of energies. The molecules involved in the complex were related to each other during the searches through the MOLS command which selects a molecule in the complex and defines an axis system for independent molecular rotation and translation for complexation. A distance constraint of between 2 to 8 Å was applied between selected atoms. The distance constraint keeps the molecules from flying apart by discarding all structures generated outside of the constraint limits. Since no quantitative NOE distances were available, the FIXD command to keep the molecules at some fixed distance between NOE contacts was not used. All flexible torsion angles for both molecules were rotated and, except for the amide linkage, bond angles were allowed to vary between 0° and ±180°. The catechin pyran torsion angles were varied, but the proline ring was not opened, and the configuration was left *trans*. The number of combinations for bonds rotated at any given MC step was randomly selected between 2 and a maximum of 20 for two catechin molecules. The searching procedure was started with many different initial structure complex combinations; parallel and anti-parallel (+)-catechin molecular forms, for example.

HyperChem was applied using the molecular dynamics protocol with the low-energy complexation structure found from the MacroModel search studies. The MD studies were run for 10 ps at 1 fs steps. The heating from 0 K and cooling to 0 K were done in 0.3 ps each. The constant temperature bath was examined at 300 and 350 K. Structures were examined by sampling the stored structure files accumulated at every 5 fs. Ensembles were filtered for computation of distances between selected atoms.

3. CATECHIN/CATECHIN SELF-ASSOCIATION

NOE results showed considerable cross-peak correlation, indicating that self-association was occurring. The structures illustrated below are from among the low-energy conformers in the ensemble found from the molecular searches. Figure 2 shows the lowest energy structure found for the self-association catechin/catechin complex with the MMFF force field. The B-rings are nearly parallel and overlap each other. The hydrogen bonding in the MMFF force field comes from the natural electrostatic and van der Waals interactions. A measure of the complexation binding energy can be obtained from taking the difference in energy between the complex formed and the two molecules separated by a large distance, e.g., 12 Å. This energy difference, ($E_{complex} - E_{isolated}$) for catechin/catechin, is -30.0 kJ/mol for the complex shown in figure 2. Figure 3 shows the second lowest-energy structure found from a number of Monte Carlo MacroModel searches.

The third lowest-energy structure found for catechin self-association using the MMFF force field and the water solvent was found at an energy of 1.5 kJ/mol and is shown in figure 4. This structure is also in parallel form, but translated slightly from that shown in figure 3.

The perspective of the complex formed in figure 2 shows that B-rings for molecules I and II are parallel and aligned, but OH groups are opposed. In that figure,

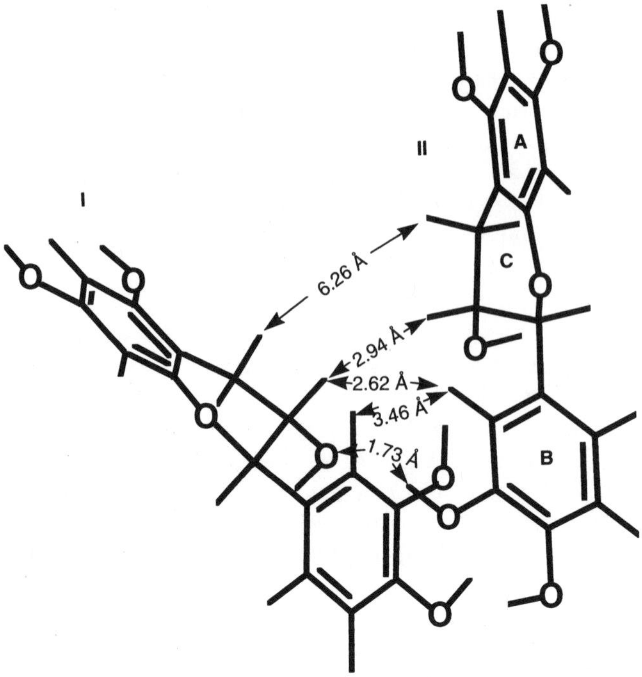

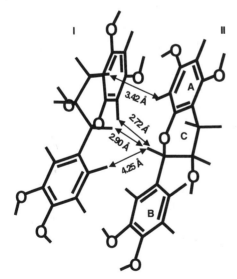

Figure 2. The lowest energy catechin/catechin complex found in a typical MacroModel MMFF force field search, having an energy of 122.9 kJ/mol (0.0 kJ/mol relative energy). The B-rings are nearly parallel and overlap, with the hydroxyl groups opposed. The distances listed show the closest hydrogen atom contact points.

Figure 3. The second energy of the lowest three catechin/catechin complexes found in a typical MacroModel MMFF force field search has a relative energy of 0.3 kJ/mol. This structure and that in figure 2 are nearly parallel, but molecule II is rotated approximately 180° around the long axis relative to molecule I. The distances listed show the closest hydrogen atom contact points.

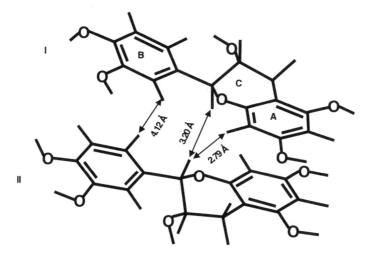

Figure 4. The third lowest-energy complex found from the MacroModel MC MMFF force field search for catechin/catechin complexes. The conformer relative energy is 1.5 kJ/mol. The molecules are rotated nearly 180° relative to each other.

the B-ring for molecule II is oriented in the plane of the page. The A-rings are oriented out of the page plane. There is a strong hydrogen bond formed between I (OH)-3_C and II (OH)-3_B. This structure suggests that strong interaction should appear between II H-3_C and I C-ring protons along with some weak interaction between B-rings, for example, H-2_B and H-2_B. Examination of the figure 3 complex shows that the two catechin molecules are parallel to each other, but have the OH-3_C groups rotated around the long axis by nearly 180° to each other. The contact points are clearly close enough to cause NOE behavior, especially the 2.90 Å between protons II H-2_C and I H-2_C, and the 2.72 Å between protons II H-2_C and I H-8_A. There is also contact between the I H-$(4\alpha)_C$ proton on catechin I and the H-8_A proton on catechin II. These structures are crossed over at the pyran ring.

In the figure 4 structure, notice that there is NOE contact between molecules I and II through the H-2_C protons at 3.20 Å, and across from the H-2_C proton to the H-8_A proton at 2.79 Å. In the conformational searches, it was common to observe that the C-ring protons are involved as contact points in these catechin self-associated complexes.

Figure 5 illustrates the HyperChem molecular dynamics low-energy structure as determined from a self-association simulation in a water-box containing about 1665 water molecules. Simulations were run at a constant temperature of 300 K. The structure is in the anti-parallel form with the A-ring and B-rings paired. The complex structure actually has a cross-shape. There is contact between B- and A-ring protons H-2_B to H-8_A of 3.64 Å, as predicted by NOE results. However, again it would appear that the H-2_C protons should show strong NOE behavior with a

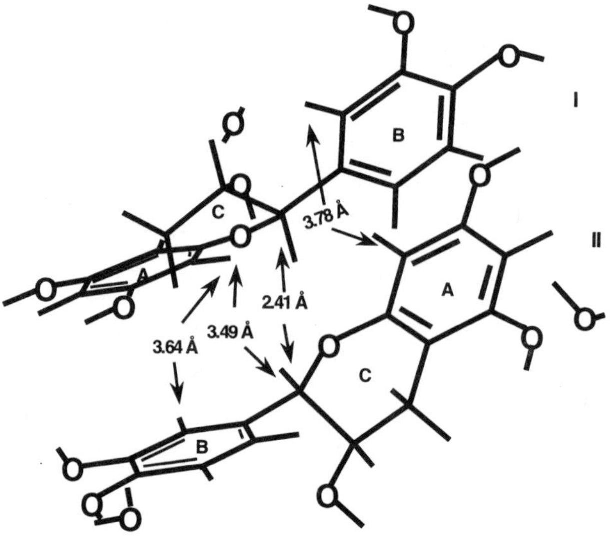

Figure 5. A low-energy complex form stabilized during the MD HyperChem run in a water solvent box. The structure shows the anti-parallel form. The closest contact distance is 2.41 Å between the H-2_C protons between molecules I and II. For clarity, all but a few water molecules have been removed.

separation of 2.41 Å. This has not been observed by NMR experiments.[15] In figure 5, the water molecules have been cut away with only several neighbors retained. There are problems in trying to achieve the minimization of energy for a structure in a water box, since it is mainly the individual water molecules that are moved. If the water molecules are removed, minimization can substantially change the interaction, allowing the individual molecules of the complex in some cases to drift apart. One of the benefits of the GB/SA water model when used in molecular dynamics simulations is that explicit water molecules are not present. Molecular dynamics experiments were run by MacroModel to check some of the structures.

One example of the interaction between catechin molecules when one is in the axial configuration is shown in figure 6. This particular MacroModel complex conformer has a relative energy of 8.6 kJ/mol above the lowest one found (fig. 2) and so would contribute only slightly to the Boltzmann distribution. However, this axial/equatorial anti-parallel form does give a picture of one type of structure that may be needed to explain coupling constant values. A lower relative energy value is needed, however, for there to be a significant contribution to the Boltzmann properties.

Figure 7 shows the minimum energy complex structure found in a run with a modification of the electrostatic hydrogen bonding function in the MMFF force field with a Lennard-Jones 6–12 potential. This might be the lowest energy [119 kJ/mol] catechin/catechin complex. The next higher one in this series is only 1.0 kJ/mol

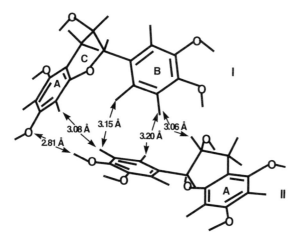

Figure 6. The complexation of an axial catechin conformer I with the equatorial conformer II in an anti-parallel form showing the interaction of the B-ring protons with the A-ring protons. This higher energy complex is 8.6 kJ/mol above the global minimum found so far in MacroModel MC conformational searching.

higher than the one shown in figure 2. It is also a cross-like structure with good interaction between the A- and B-rings, as shown in figure 8.

According to the work by Hatano and Hemingway,[15] the cross-peak interactions in the catechin solutions in water were observed between the molecule I H-2_B proton and the molecule II H-8_A proton. In the lowest energy conformers found so far by MC conformational searching, there is considerable B-ring/B-ring

Figure 7. This is the lowest energy structure found in a modified Monte Carlo MacroModel run with the hydrogen bonding function changed to a Lennard-Jones 6–12 potential. The energy of 119 kJ/mol is lower than that shown in figure 1, but this may be due to H-bonding function changes.

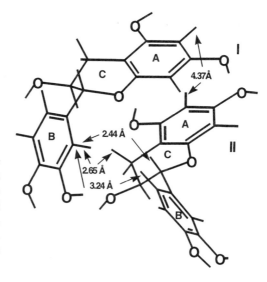

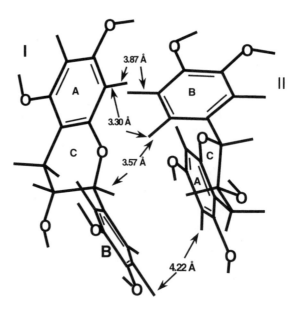

Figure 8. The second highest-energy structure from this run with the hydrogen bonding set to a different function. The hydrogen bonding parameter in the MacroModel program was set to run a Lennard-Jones 6–12 potential. The structures cross over at the pyran rings. The ensemble shows a number of good NOE contacts (hydrophobic interactions) between the A- and B-rings.

association with definite interaction between the H-4_C proton and the H-3_B proton; see figures 7 and 8. These results suggest that further work should examine the NMR spectra for indication of C-ring proton involvement in the complexes through hydrophilic and hydrophobic interaction. It is possible that it might not appear because of relaxation properties.

In the MacroModel study, several of the above structures were found by using complexes started in the anti-parallel structural form. However, there did not seem to be a propensity for the complexes to end with the B-ring of one molecule interacting directly (that is, overlaid) with the A-ring of another moiety. Specifically, the protons of the B-ring need to be oriented directly into the A-ring as would be required by Hatano's NMR results.[15] A number of higher energy conformers found in MacroModel ensembles appear to have properties that fit with the NOE NMR results.

To better examine the hydrophobic interactions of the ensemble, filtered data were studied. The distance data given below were filtered from several Macro-Model runs. The interatomic distance filter was set between 2.0 and 4.0 Å with contact numbers collected on structures within this distance window. Although a Boltzmann average was not done, a couple of examples of percentages are given here showing interactions between molecules I and II. Run A, with 788 complexes in the ensemble (50 kJ/mol energy window), had 6.35 percent H-2_B to H-6_B, and

4.57 percent H-2$_B$ to H-2$_B$. Run B, 388 complexes in an energy window of 30 kJmol, had 13.4 percent H-8$_A$ to H-6$_B$, 12.3 percent H-6$_A$ to H-5$_B$, and 7.73 percent H-2$_B$ to H-2$_B$. Run C, 325 complexes, 30 k/mol energy window, had 12.9 percent H-2$_C$ to H-2$_C$, 9.5 percent H-2$_B$ to H-6$_B$, and 8.9 percent H-8$_A$ to H-2$_B$.

Although the lowest energy conformer found might not show all of the experimental NOEs, it is important to examine the higher energy structures. Intermolecular distances were filtered from the 111 conformer complexes found over a 16 kJ/mol window for structures from the search shown in figures 7 and 8. The following percentage contact was noted for an interatomic distance window between 2.0 Å and 4.0 Å: I H-2$_B$ to II H-8$_A$ found a 28.2 percent population; and II H-2$_B$ to I H-8$_A$, 16.2 percent. This shows a large number of potential NOE contacts that agree with the noted experimental NMR results.

Searches made with NOE distance constraints on those conformers allowed in the ensemble were not performed. Besides not having the quantitative data needed, it is also possible that the relaxation time for some nuclei in the complexes are very fast and may or may not contribute to the NOEs in expected ways.[26] Therefore, some close interactions may not lead to observed NOEs.

4. CATECHIN/L-PROLINE-GLYCINE OLIGOMER COMPLEXES

Although many of the complexes between catechin and L-proline and proline-glycine complexes have been examined, just three complexes will be discussed here. These are catechin/L-prolyl-glycine, the catechin/glycyl-L-prolyl-glycyl-glycine tetramer zwitterion, and the dimer B3/glycyl-L-prolyl-glycyl-glycine tetramer ion complex. Searches were started by placing the L-proline derivative near the catechin or the B3 dimer [(+)-catechin-(4α→8)-(+)-catechin] in a position guided by the NMR NOE data. However, a number of different starting structures were examined to allow many possible complexes over the thousands of MC steps in the search routine. Since the *trans* form of proline was predominant in the mixtures studied,[15] except for glycyl-L-proline, only the *trans* form of the proline was considered in this study.

The NMR NOE results for catechin/L-prolyl-glycine show strong cross-peaks from both the H-6$_A$ and the H-8$_A$ to the glycine protons. In addition, the *trans* form showed very weak cross-peaks with the B-ring protons. The catechin/L-prolyl-glycine structure complex given in figure 9 shows that the proline ring settles over the C-ring in a planar fashion. The glycine unit shows hydrogen bonding between the NH$_2$ and carboxylate group. This compound was modeled in the neutral state. Examination of the figure shows interaction between the methylene protons and H-2$_B$ of 4.1 Å, and the proline (C$_\delta$H$_2$) and the catechin H-(4β)$_C$ proton of 2.6 Å.

In this low-energy catechin/L-proline-glycine complex, and in nearly all modeled cases, there has been a lot of interaction between the proline ring protons (γ and δ) to H-4$_C$ as well as to the H-2$_C$ and H-3$_C$ protons from catechin. In these studies, there has been indication that the C-ring protons from catechin should be involved in NOEs for the complexes. It is not clear yet why these have not been observed.

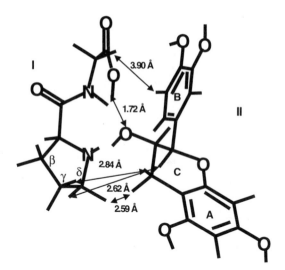

Figure 9. The lowest-energy catechin/L-proline-glycine complex. The proline-glycine is *trans*, and the amide structure is locked in the planar position. In this case, the glycine C-OH was allowed to rotate, and it shows a strong H-bond (1.72 Å) to the catechin. Considerable interaction is noted from H-2$_B$ and H-3$_B$ to the γ and δ L-proline ring protons, as well as from the glycine methylene proton to the H-5$_B$ proton. The structure is shown here in the neutral form.

First, the structural characteristics of the GPGG ion tetramer were examined. This was accomplished by carrying out MacroModel searches on the GPGG ion using the water solvent model, GB/SA. After starting the structure in many different conformers, the lowest energy forms found always showed a β-turn structure. An example of the lowest energy structure is given in figure 10. The β-turn structure is confirmed by a number of studies, including NMR and Raman spectroscopy as well as with modeling.[27,28]

There appears to be some controversy about the nature of this tetramer structure. Perly et al. predicted a II β-turn structure.[27] Although we came up with a β-turn structure, not all of the torsion angles agreed with Perly's NMR paper. In our study, the torsion angle ϕ_4 agreed exactly with the NMR study done in DMSO, but for ϕ_2, ψ_2 Perly found –60°, 120°. Our structure is more like a I β-turn structure (fig. 10); ϕ_2, ψ_2, ϕ_3, and ψ_3 values are –60°, –30°, –90°, and 0°, respectively.[28,29] The NH $\cdots$ O=C hydrogen bonding across the glycine arms is predicted satisfactorily. This structure became the starting point for forming complex structures with the catechin molecule and with the B3 dimer. In each search case, all non-ring torsion angles were allowed to vary. Figure 11 shows one of the lower-energy catechin/GPGG ion complexes found with a MacroModel search and the MMFF force field. Here we see the arms of the GPGG ion structure interacting through hydrogen bonding with the (OH)-7$_A$. The methylene of the N-terminal glycine interacts hydrophobically with the H-6$_A$ (4.62 Å) proton of catechin and the C-terminal (CH$_2$)$_\alpha$ interacts with H-8$_A$ proton (3.46 Å). The proline γ and δ CH$_2$

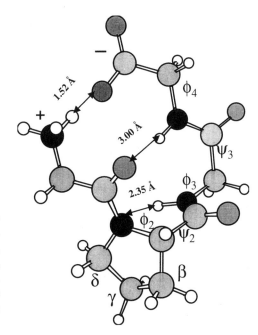

Figure 10. The low-energy conformer for GPGG zwitterion from the Macro-Model search shows a β-turn structure. The ϕ_2, ψ_2, ϕ_3, ψ_3, and ϕ_4 values are $-80°$, $-29°$, $-98°$, $20°$, and $169°$, respectively. Some similar form of β-turn structure is found with all of the low-energy catechin complexes.

hydrogen atoms also interact with the catechin H-8_A hydrogen atom. According to work done by Hatano and Hemingway,[15] there is strong association only between the C-terminal glycine methylene group and the H-8_A proton and the other N-terminal methylene group and the H-5_B proton. Their work might suggest a preferred extended structure for the polypeptide oligomer, but the observations could arise by averaging NOEs from several low energy complexes, for example; see figure 12. It should be noted here that the energy difference ($E_{complex} - E_{isolated}$) between the complexed molecules in figure 11 and the isolated molecules minimized by MMFF and in water was -32.0 kJ/mol. This again supports the favorable energy lowering by molecular association.

In figure 12, the next higher-energy conformer in this series (1.7 kJ/mol) shows the gylcine arms interacting primarily with the pyran ring oxygen atom. There is close contact between the H-8_A proton of catechin and the NH_3^+-CH_2 group (3.08 Å) and proline $(CH_2)_\delta$ (3.41 Å). The proline $(CH_2)_\gamma$ hydrogen atoms interact directly with H-2_C, and H-2_B, and the H-5_B and H-2_B catechin hydrogen atoms have close contact with C-terminal $(CH_2)_\alpha$ protons (3.68 Å, 2.92 Å). This again shows the hydrophilic interaction helping to tie the molecules together coupled with hydrophobic interactions as well. This study indicates that proline methylene hydrogen atoms are involved in many hydrophobic interactions. Conformer complexes arise with the catechin situated in the axial form, too. Figure 13 shows that the (OH)-3_C is axial, and that it is in good position to interact through hydrogen bonding with the C-terminal and N-terminal glycine arms. The energies of these complexes are typically in the order of 8 kJ/mol higher and consequently not low

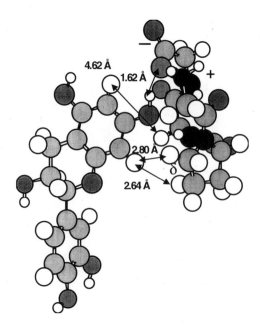

Figure 11. NMR experiments show NOE between H-8$_A$ of catechin and the methylene of the C-terminal glycine as well as between the H-6$_B$ proton of catechin and the C$_\alpha$ methylene of the C-terminal glycine unit. Molecular search results show preference for both hydrophobic and hydrophilic interactions. The numbers of favorable interactions are large, and relative binding energies are low. The arms of the β-turn tetramer interact through hydrogen bonding with the (OH)-7$_A$ group. Methylene protons from NH$_3^+$-CH$_2$ and the C-terminal (CH$_2$)$_\alpha$ methylene group interact with H-6$_A$ and H-8$_A$, respectively. The γ and δ hydrogen atoms on the proline ring interact hydrophobically with the H-8$_A$ proton at 2.64 Å and 2.80 Å, respectively.

enough to play a significant role in Boltzmann summations, e.g., in NMR coupling constants and the weighting of NOE data. This relative energy dropped substanially with the Amber force field.

Searches were also carried out with the Amber force field. One of the low-energy conformers is shown in figure 14. Primary contacts are through the β-turns interacting with both the (OH)-3$_B$ and (OH)-4$_B$ hydroxyl groups, and with the proline methylene hydrogen atoms interacting directly with the C-ring hydrogen atoms. The Amber force field favored A-conformer catechin low energy conformers.

5. CATECHIN-(4α→8)-CATECHIN/GPGG ION COMPLEX

Only a few searches were done on this complex structure, and figure 15 illustrates one of the low-energy structures found using the MacroModel protocol. Several low-energy structures were found that showed the glycine arms expanded between the upper and lower units of catechin. The lowest-energy axial conformer

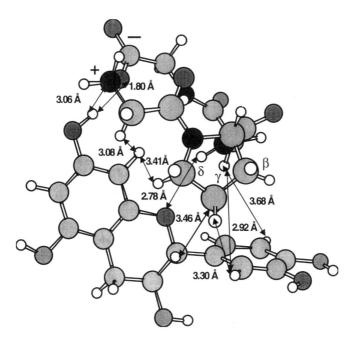

Figure 12. This figure illustrates the diversity of interactions with the combination of hydrophilic and hydrophobic contacts. Here, the β-turn arms are interacting directly with the pyran oxygen atom. The C-terminal α-methylene group shows interaction to the H-8$_A$ proton (3.46 Å) whereas the proline methylene hydrogen atoms (γ and δ) interact through hydrophobic contact with the H-2$_C$, H-2$_B$ and H-5$_B$ protons.

complex found is higher in energy by nearly 8 kJ/mol from the equatorial counterpart. One structure, illustrated in figure 15, shows the β-turn arms interacting with the upper A-ring and lower E-ring. There is also sufficient hydrophobic interaction between the proline hydrogen atoms and the A-ring. Specifically, the NH$_3^+$-CH$_2$ is 2.70 Å from H-6$_A$ and 5.38 Å from H-8$_A$, with the C$_{terminal}$-(CH$_2$)$_1$, 2.59 Å from H-8$_A$. The proline (C$_\gamma$H$_2$) is 4.17 Å from H-8$_A$ and the(C$_\delta$H$_2$) is 2.86 Å from H-6$_A$. Several close hydrogen-bonding contacts are present; for example, NH to (OH)-7$_A$, 2.36 Å. The terminal CO$^-$ to (OH)-4$_E$ shows an unrealistically short hydrogen bonding distance of 1.3 Å. This is probably what causes the β-turn to rotate into a more open configuration for the outer glycine torsion angles. This result comes from two 3000-step MC MacroModel searches, so other potentially low-energy conformer complexes are possible. However, this shows the propensity for interaction between the upper and lower units as was observed in the NOE experiments.[15]

6. CONCLUSIONS

Molecular modeling has been used to search conformational space for the self-association of catechin and the complexation of catechin with L-proline

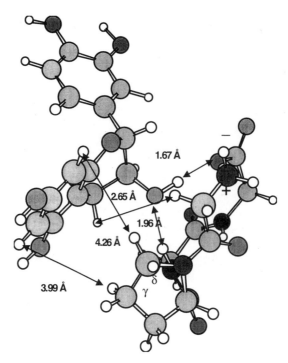

Figure 13. A low-energy search conformation found with interaction between
H-4$_C$ of catechin and glycine in GPGG ion. Catechin in an A-conformation is
always seen in the ensemble of low-energy complexes. The relative energy here
is 3.0 kJ/mol. The ^{1}H NMR spectrum shows $J_{2,3} = 8$ Hz, indicating some axial-
conformation. There is commonly H-H contact between the H-4$_C$, H-8$_A$, and
methylenes of glycine. Close contact is seen between H-8$_A$ and proline C$_\delta$ hydrogens (2.80 Å) in this conformer.

glycine oligomeric peptides, including the glycyl-L-prolyl-glycyl-glycine zwitterion
tetramer (GPGG). Monte Carlo MacroModel conformational searching was applied
using the Merck force field (MMFF) with the GB/SA water solvent model. Interatomic contact distances between the two moieties in the complexes were compared with results from NOE NMR experiments. Searching for polyphenol
complexes looks very promising for giving insight into ways in which catechin
might self-associate, and how polyphenols interact with proline-based molecules.
Binding energies calculated from the difference in energy between the complex
and the isolated molecules show significant association. The MMFF force field in
its native state, however, seems to overweight hydrogen bonding even when using
a water solvent model. Catechin/catechin self-association has been established by
NMR NOE experiments and verified by MacroModel molecular complex conformational search methods. A series of loosely-bound low-energy catechin/catechin
conformer complexes that show both hydrophilic and hydrophobic interactions

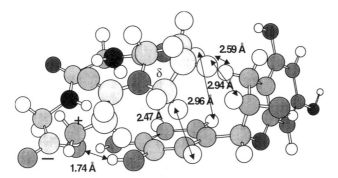

Figure 14. This –1,067 kJ/mol GPGG ion/catechin conformer (Amber force field), with a relative energy of 1.7 kJ/mol, shows the interaction of the β-turn arms with the B-ring hydroxyl groups. There is considerable hydrophobic interaction between the proline H-atoms and H-2_C and the glycine methylene groups and H-2_B and H-5_B. The catechin is favored in A-conformer form.

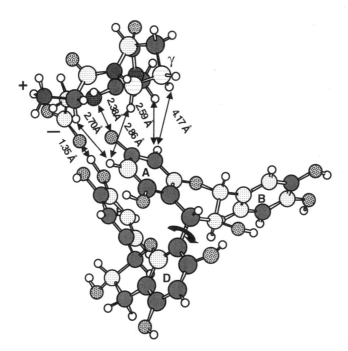

Figure 15. This shows the lowest energy structure found so far in the search in the interaction of the GPGG ion with the B3 dimer [(+)-catechin-($4\alpha \rightarrow 8$)-(+)-catechin]. There is typically interaction found between the glycine arms and the upper and lower units of B3. For example, NH_3^+-CH_2 to H-6_A is 2.70 Å.

were found. The low-energy structures were B-ring aligned (parallel) and crossed at the pyran rings. These show NOE contact properties like those found in the NMR NOE experiments.

Catechin/GPGG ion complexes have been shown to exist by molecular conformational analysis using MacroModel and the MMFF and Amber force fields. It has been shown that the GPGG zwitterion occurs in a β-turn structure, and that the glycl arms interact with the various OH groups of catechin through hydrogen bonding. The methylene groups and proline hydrogen atoms show hydrophobic contact with the catechin A- and B-rings. Further work is needed with some of the other hydrogen-bonding functions that are available. It appears that the strong hydrogen bonding for the ions may be over-weighted. Finally, this work shows that the combination of the interplay between molecular modeling experiments and NMR experiments is important to guide the direction of modeling structural studies. Complexes formed with the B3 dimer show interactions between the GPGG ion and the upper A-ring and the lower E-ring. Other low-energy conformers for the dimer structure are possible.

ACKNOWLEDGMENTS

Fred L. Tobiason wishes to give special thanks to the University of Science and Technology, Lille 1, Lille, France for supporting conformational studies there; to the USDA for continuing financial support; and to the Pacific Lutheran University media services personnel for their help. Financial support from USDA NRICGP Project No. 94-03395 is greatly appreciated.

REFERENCES

1. Hemingway, R.W.; Karchesy, J.J. (eds.) Chemistry and significance of condensed tannins. Plenum Press, New York (1989).

2. Hemingway, R.W.; Laks, P.E. (eds.) Plant polyphenols: synthesis, properties, significance. Plenum Press, New York (1992).

3 Fronczek, F.R.; Gannuch, G.; Tobiason, F.L.; Broeker, J.L.; Hemingway, R.W.; Mattice, W. L. Dipole moment, solution conformation and solid state structure of (−)-epicatechin, a monomer of procyanidin polymers. *J. Chem. Soc., Perkin Trans. 2* :1611 (1984).

4. Fronczek, F.R.; Gannuch, G.; Mattice, W.L.; Hemingway, R.W.; Chiari, G.; Tobiason, F.L.; Houglum, K.; Shanafelt, A. Preference for occupancy of axial positions by substituents bonded to the heterocyclic ring in penta-*O*-acetyl-(+)-catechin in the crystal state. *J. Chem. Soc., Perkin Trans. 2* :1385 (1985).

5. Tobiason, F.L.; Fronczek, F.R.; Steynberg, J.P.; Steynberg, E.C.; Hemingway, R.W. Crystal structure, conformational analysis, and charge density distributions for *ent*-epifisetinidol: an explanation for regiospecific electrophilic aromatic substitution of 5-deoxyflavans. *Tetrahedron* 49(2):5927 (1993).

6. Tobiason, F.L.; Hemingway, R.W. Predicting heterocyclic ring coupling constants through a conformational search of tetra-*O*-methyl-(+)-catechin. *Tetrahedron Lett.* 35(14):2137 (1994).

7. Hemingway, R.W.; Tobiason, F.L.; McGraw, G.W.; Steynberg, J.P. Conformation and complexation of tannins: NMR spectra and molecular search modeling of flavan-3-ols. *Magn. Reson. in Chem.* 34:424 (1996).

8. Tobiason, F.L.; Kelley, S.; Midland, M.M.; Hemingway, R.W. Temperature dependence of (+)-catechin pyran ring proton coupling constants as measured by NMR and modeled using GMMX search methodology. *Tetrahedron Lett.* 38(6):985 (1997).

9. Vercauteren, J.; Cheze, C.; Dumon, M.C.; Weber, J.F. Polyphenols Communications 96, Vols. 1 and 2, Groupe Polyphenols, Université Bordeaux 2, Bordeaux, Cedex, FR (1996).

10. Bergmann, W.R.; Barkley, M.D.; Hemingway, R.W.; Mattice, W.L. Heterogeneous fluorescence decay of 4→6 and 4→8 linked dimers of (+)-catechin and (−)-epicatechin as a result of rotational isomerism. *J. Am. Chem. Soc.* 109:6614 (1987).

11. Viswanadhan, V.N.; Mattice, W.L. Preferred conformations of the sixteen (4,6) and (4,8) β-linked dimers of (+)-catechin and (−)-epicatechin with axial or equatorial dihydroxy-phenyl rings at C(2). *J. Chem. Soc., Perkin Trans. 2* :739 (1987).

12. Steynberg, J.P.; Brandt, E.V.; Hoffman, M.J.H.; Hemingway, R.W.; Ferreira, D. Conformations of proanthocyanidins. *In*: Hemingway, R.W.; Laks, P.E. Plant polyphenols: synthesis, properties, significance. Plenum Press, New York, pp. 501–520 (1992).

13. Steynberg, J.P.; Brandt, E.V.; Ferreira, D.; Gornik, D.; Hemingway, R.W. Definition of the conformations of proanthocyanidins: methylether acetate derivatives of profisetinidins. *Magn. Reson. in Chem.* 33:611 (1995).

14. Hatano, T.; Hemingway, R.W. Conformational isomerism of phenolic procyanidins. Preferred conformations in organic solvents and water. *J. Chem. Soc., Perkin Trans. 2* :1035 (1997).

15. Hatano, T.; Hemingway, R.W. Association of (+)-catechin and catechin-(4α → 8)-catechin with oligopeptides containing proline residues. *J. Chem. Soc., Chem. Commun.* :2537 (1996).

16. MacroModel, Interactive Modeling Program, Stille, M.C., Version 5.5, Department of Chemistry, Columbia University, New York, NY 10027, USA. (1997).

17. HyperChem, Version 5.1, HyperCube, Inc., 1115 NW 4th St., Gainesville, FL 32601, USA. (1997).

18. PCModel Version 7.0, Serena Software, Bloomington, IN, 47402-3076 USA. (1998).

19. Spartan PC and Spartan (5.5), Wavefunction, Inc., 18401 von Karman Ave, Suite 370, Irvine, CA 92612, U.S.A (1998).

20. Murray, N.J.; Williamson, M.P.; Lilley, T.H.; Haslam, E. Study of the interaction between salivary proline-rich proteins and a polyphenol by 1H-NMR spectroscopy. *Eur. J. Biochem.* 219:923 (1994).

21. Haslam, E. Practical polyphenolics: from structure to molecular recognition and physiological action. Chapter 3, Cambridge University Press, Cambridge (1998).

22. Charlton, A.J.; Baxter, N.J.; Lilley, T.H.; Haslam, E.; McDonald, C.J.; Williamson, M.P. Tannin interactions with a full-length human salivary proline-rich protein display a stronger affinity than with single proline-rich repeats. *FEBS Lett.* 382:289 (1996).

23. Baxter, N.J.; Lilley, T.H.; Haslam, E.; Williamson, M.P. Multiple interactions between polyphenols and a salivary proline-rich protein repeat result in complexation and precipitation. *Biochem.* 36:5566 (1997).

24. Halgren, T.A. Merckmolecular force field. I. Basis, form, scope, parameterization, and performance of MMFF94; II. MMFF94 van der Waals and electrostatic parameters for intermolecular interactions. *J. Comput. Chem.* 17:490–519 through 616–641 (1996).

25. Still, W.C.; Tempczyk, A.; Hawley, R.C.; Hendrickson, T. Semianalytical treatment of solvation for molecular mechanics and dynamics. *J. Am. Chem. Soc.*, 112:6127 (1990).

26. Derome, A.E., Modern NMR techniques for chemistry research. Pergamon Press, Oxford, Chapter 5 (1988).

27. Perly, B.; Helbecque, N.; Forchioni, A.; Loucheux-Lefebvre, M.H. Glycyl-L-prolylglycyl-glycine in β-turn-supporting environment. *Biopolymers* 22:1853 (1983).

28. Lagant, P.; Vergoten, G.; Fleury, G.; Loucheux-Lefebvre, M.H. Raman spectroscopic evidence for β-turn conformation. *J. Raman Spect.* 15:421 (1984).

29. Venkatachalam, C.M. Stereochemical criteria for polypeptides and proteins. V. Conformation of a system of three linked peptide units. *Biopolymers* 6:1425 (1968).

POLYPHENOLS, METAL ION COMPLEXATION AND BIOLOGICAL CONSEQUENCES

Augustin Scalbert,[a] Isabelle Mila,[b] Dominique Expert,[c]
Frank Marmolle,[d] Anne-Marie Albrecht,[d] Richard Hurrell,[e]
Jean-François Huneau,[f] and Daniel Tomé[f]

[a] Unité des Maladies Métaboligues et Micronutriments
INRA de Clermont Feurand / Theix
FRANCE

[b] Laboratoire de Chimie Biologique (INRA)
INA-PG
F-78850 Thiverval-Grignon
FRANCE

[c] Laboratoire de Pathologie Végétale (INRA)
INA-PG, 16 rue Claude Bernard
75231 Paris cedex 05
FRANCE

[d] Laboratoire de Physico-Chimie Bioinorganique
UMR 7512 CNRS
Faculté de Chimie, 1 rue Blaise Pascal
F-67000 Strasbourg
FRANCE

[e] Labor für Humanernährung
ETH, Seestrasse 72
Postfach 474
CH-8803 Ruschlikon
SWITZERLAND

[f] Unité de Nutrition Humaine et Physiologie Intestinale (INRA)
INA-PG, 16 rue Claude Bernard
F-75231 Paris cedex 05
FRANCE

Plant Polyphenols 2: Chemistry, Biology, Pharmacology, Ecology, Edited by
Gross et al. Kluwer Academic / Plenum Publishers, New York, 1999

1. INTRODUCTION

Tannins, together with lignins, are the most widespread and abundant polyphenols in plants. Proanthocyanidins (PA) appeared early in the evolution of terrestrial plants and are commonly found in fern allies, ferns, gymnosperms and angiosperms.[1,2] Gallotannins (GT) and ellagitannins (ET) (hydrolyzable tannins) are found in a large number of angiosperm families[3] and often co-occur with proanthocyanidins in plant tissues.

The involvement of tannins in the protection of plants against pathogens, decay fungi,[4-6] and herbivores[7] is well established. Most investigators explain these properties by a common ability of tannins to interact with proteins. They could inhibit microbial or herbivore enzymes or form complexes with the plant protein substrates, thereby reducing the feeding value of the plant. The same interactions with proteins, either from saliva or from mucosa, are assumed to be involved in the perception of astringency by mammals[8] and contribute to deter herbivores from feeding on tannin-rich plants.

Tannin interactions with metal ions may be as significant as interactions with proteins in plant defense and animal nutrition. Virtually all tannins found in plants have catechol or pyrogallol residues rather than simple phenol residues. This makes them excellent chelators of various metal ions. These properties have formed the basis of various technological applications of tannins. Tannins have been used as a component for iron dyes[9] and writing inks,[10] as anticorrosive primers for steel[11] and non-ferrous metals,[12] modifiers of rheological properties of minerals and clays,[13] fixing agents of metal micronutrients in foliar sprays,[14] agents for metal ion recovery from waste waters,[15] as a component for the tannin/aluminium tannage, or as copper-fixing agents in wood preservation.[16,17]

In contrast, very limited attention has been paid to the significance of metal chelation by tannins in biology. It is remarkable to observe that tannins lacking catechol or pyrogallol chelating groups such as propelargonidins are exceedingly rare in plants even in the more primitive ferns.[18] Furthermore, experiments on synthetic proanthocyanidins have shown that tannins such as propelargonidins are just as efficient to precipitate bovine serum albumin as procyanidins or prodelphinidins containing catechol and pyrogallol rings, respectively.[19] In view of some results recently obtained, we would like to emphasize the importance of iron chelation by tannins in plant defense and in human nutrition.

2. IRON COMPLEXATION BY TANNINS IN PLANT DEFENSE

Most living cells have an absolute requirement for iron, as it is involved in various essential cellular processes.[20] However, the concentration of free iron in the environment is often too low to fulfill the needs of an organism unless the organism possesses a high-affinity carrier system. This low concentration is explained by the formation of insoluble ferric hydroxides in aerobic environments and by the chelation of iron by various ligands produced by other living organisms such as microbial siderophores or some animal proteins.

The high affinity of tannins for iron may well limit iron availability to pathogens or decay microorganisms. We compared the relative affinity of tannins and lower molecular weight phenolic ligands by using the chrome azurol S (CAS) assay, commonly used to detect high affinity ligands such as siderophores.[21] Tannins such as vescalagin, castalagin, or penta-*O*-galloyl-*β*-D-glucose were far more efficient than gallic acid, protocatechuic acid, (+)-catechin, or gallocatechin having a single catechol or pyrogallol group to displace iron(III) from the yellow iron/CAS complex (fig. 1). Tannins thus appear to be very efficient in displacing iron from other ligands.

The effect of iron-binding by tannins on the growth of various strains of *Erwinia chrysanthemi,* a pectinolytic bacterium causing soft rot on a wide range of plants, was studied.[21] The growth on tannin-rich media of three mutants altered in their siderophore-mediated iron transport pathway was compared to that of the wild-type strain 3937. The mutants were either unable to synthesize chrysobactin, the catechol-type siderophore of *E. chrysanthemi* [3937 *cbs29* (Cbs⁻)], or to synthesize both chrysobactin and its receptor [3937 *fct18* (Fct⁻Cbs⁻)]. A third mutant, 3937 *fct34acsA12* (Fct⁻Acs⁻), carries a transposon insertion in the *fct* gene and a second one in the *acs* gene coding for the second siderophore of *E. chrysanthemi* of yet unknown structure.[22]

The growth of all mutants, and more particularly that of the one altered in both siderophore systems, was more inhibited than that of the wild strain at various concentrations of a commercial chestnut extract rich in vescalagin, castalagin, and other ellagitannins. A gallotannin such as penta-*O*-galloyl-*β*-D-glucose was far more effective than the lower molecular weight phenols in inhibiting growth of the two mutants, showing that they may well inhibit growth of the mutants by iron deprivation (fig. 2).

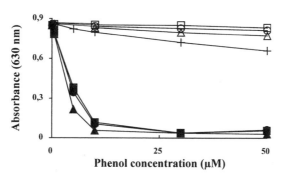

Figure 1. Removal of iron(III) from its chrome azurol S complex (absorbing at 630 nm) by hydrolysable tannins and low molecular weight phenols. (●) vescalagin; (■) castalagin; (▲) penta-*O*-galloyl-*β*-D-glucose; (△) gallic acid; (○) protocatechuic acid; (+) (+)-catechin; (□) salicylic acid (reprinted from reference 21 with permission of Elsevier Science).

The Cbs⁻ mutant was also grown in agar plates containing minimum inhibitory concentrations of castalagin, vescalagin, penta-*O*-galloyl-β-D-glucose, and 1,2,3,6-tetra-*O*-galloyl-β-D-glucose (250 or 500 μM). Filter disks containing 10, 100, and 1,000 nmol of iron(III) chloride were added on top of the agar. Colonies were observed in the vicinity of the disks containing 100 and 1,000 nmol iron showing that iron was the limiting factor (fig. 3).

Tannins also inhibited growth of the wild strain (fig. 2 C), but in that case, levels of inhibition were not much different from those of other phenolic compounds of lower molecular weight. This suggests that other mechanisms of growth inhibition may also operate when the microorganisms are able to produce siderophores and to efficiently compete with tannins for assimilating iron.

Plants can be compared to animals with respect to their defence strategies against microorganisms based on iron deprivation. Plants accumulate tannins to

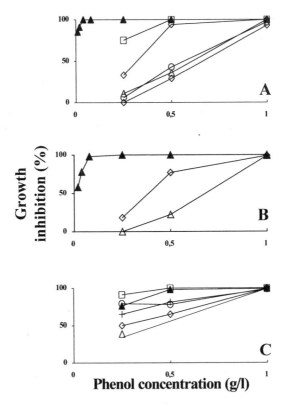

Figure 2. Growth inhibition by penta-*O*-galloyl-β-D-glucose and low molecular weight phenols of Cbs⁻ (A) and Fct⁻ Cbs⁻ (B) mutants altered in their siderophore-mediated iron transport pathway and of the wild-type strain 3937 (C) of *Erwinia chrysanthemi*. (▲), Penta-*O*-galloyl-β-D-glucose; (△), gallic acid; (○), protocatechuic acid; (☆), chlorogenic acid; (+), (+)-catechin; (■), salicylic acid (reprinted from reference 21 with permission of Elsevier Science).

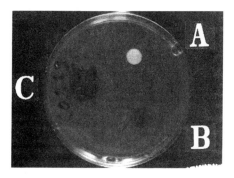

Figure 3. Cross-feeding of *Erwinia chrysanthemi* cbs 29, grown in agar plate containing castalagin with various quantities of iron chloride. A, 10 nmol; B, 100 nmol; C, 1,000 nmol. Note the halo of colonies around disks B and C.

efficiently fix iron, whereas animals accumulate iron-binding proteins such as lactoferrin and conalbumin.[23] Lactoferrin found in various external secretions and secreted by leukocytes in inflamed tissues protects the organism against infection.[24] Conalbumin, which accounts for 10 percent of the egg white dry weight, prevents microorganisms from reaching the embryo in the yolk.[25] Like conalbumin, tannins accumulate in high concentrations in protective tissues such as bark or fruit periderms. The tannin content in barks may exceed 50 percent of the dry weight.[26] These high concentrations likely contribute to limit penetration of microorganisms in the underlying living tissues.

3. IRON COMPLEXATION BY TANNINS IN HUMAN NUTRITION

Polyphenol compounds are widely present in the human diet as components of fruits, vegetables, spices, pulses, and cereals, and they are especially high in tea, coffee, red wine, cocoa, and the different herb teas. In the United States, consumption of polyphenols is estimated at 1 g/day,[27] and in the United Kingdom as much as 0.5 g polyphenol/day is ingested from tea alone.[28] The consumption of black tea and coffee has been shown to strongly inhibit Fe absorption from composite meals[29–31] with coffee having about half the inhibitory effect of tea. Other beverages such as red wine,[32,33] herb teas,[34] and cocoa,[35] as well as various vegetables with a high polyphenol content,[36,37] have been reported to inhibit iron absorption. Red wine polyphenols would appear less inhibitory than the phenolics of tea or coffee. They reduced iron absorption from a simple bread roll meal,[33] but had little effect on iron absorption from more complex composite meals.[30] On the other hand, a serving of yod kratin, a widely consumed vegetable in Thailand, reduced iron absorption from a composite meal of rice, fish, and vegetables by almost 90 percent. Iron absorption from other vegetables rich in polyphenols such as spinach and aubergine has also been observed to be low, and significant negative correlation has been reported between the total polyphenol content of vegetable foods and their iron absorption in man.[36]

Using model compounds, Brune et al.[37,38] demonstrated that tannic acid, a mixture of gallotannins, inhibited iron absorption in human subjects from a bread meal in a dose-dependent manner. Two low molecular weight phenolic compounds, gallic acid and chlorogenic acid, similarly inhibited iron absorption, whereas catechin showed no effect.

The mechanism for reduced iron absorption is usually assumed to be the formation of unabsorbable iron-polyphenol complexes. In order to clarify the mechanisms involved, a study of iron transport through a Caco-2 human intestinal epithelial cell line grown *in vitro* in the presence of various polyphenols was undertaken. Procyanidins (the main polyphenols in many fruits, wine, or cocoa) with an average polymerization degree of 7 at $100\,\mu M$ (calculated as catechin equivalent) concentration, totally inhibited the transport of iron-59(III) through the cell monolayer.[39] The structurally related (+)-catechin and procyanidin dimer B1 at the same concentration also inhibited iron absorption but were twofold less effective than the procyanidin polymer.

The studies with the polyphenol-containing foods mentioned above have shown that all major types of polyphenols inhibit non-heme iron absorption. The inhibitory compounds would appear to include phenolic acids such as chlorogenic acid from coffee, monomeric flavonoids such as found in herb teas, and the complex polymerization products found in black tea and cocoa. In relation to public health, it would appear appropriate to advise those population groups most susceptible to developing iron deficiency (infants, children, pregnant women) to avoid the consumption of coffee, tea, herb teas, or cocoa with their meals. On the other hand, the consumption of black tea with meals has been suggested as a strategy for reducing iron absorption in patients with iron overload disorders.[40]

4. TANNIN/IRON COMPLEXES

In order to provide a better understanding of the biological properties of tannins related to iron deprivation, the coordination properties of various gallotannins (fig. 4) were recently investigated by applying a combination of electrospray mass spectrometry and classical methods (UV-visible absorption spectrophotometry, potentiometry).[41,42] A spectrophotometric (450–800 nm) titration of β-glucogallin with iron(III) versus pH showed the presence of three different ferric complexes depending on the pH (fig. 5). The stoichiometry of these chelates was determined by ESMS and shown to be monoferric. Numerical treatment[43,44] of the spectrophotometric data provided the values of the corresponding stability constants. Under acidic conditions, Fe(III) is coordinated by a single ligand, while it is tetracoordinated by two molecules of β-glucogallin in media close to neutrality (pH 6.3). Our results also showed that esterification of gallic acid to a glucose moiety such as in β-glucogallin increases the affinity for iron.

For 1,2,6-tri-*O*-galloyl-β-D-glucose and penta-*O*-galloyl-β-D-glucose, the formation of a single complex with two ligands and three or five ferric cations, respectively, was observed. When the concentration of iron(III) was in excess, insoluble polymeric species were formed.

It was observed in a slow process that penta-*O*-galloyl-β-D-glucose was able to remove iron(III) from its stable complex with EDTA at pH 6.3 in methanol/water

β-Glucogallin 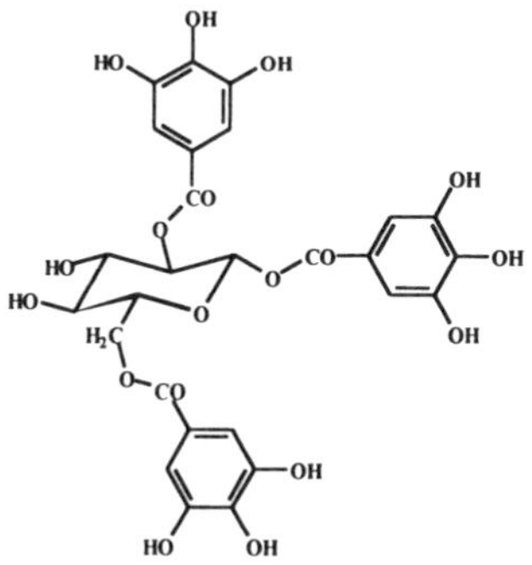   1,2,6-Tri-*O*-galloyl-β-D-glucose

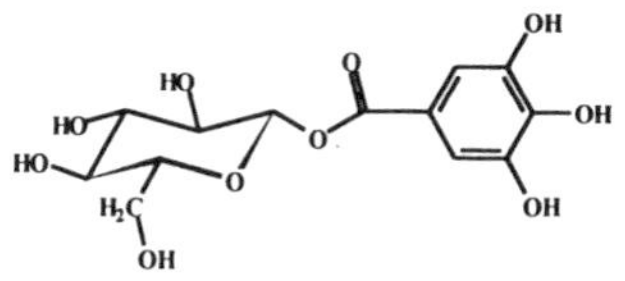

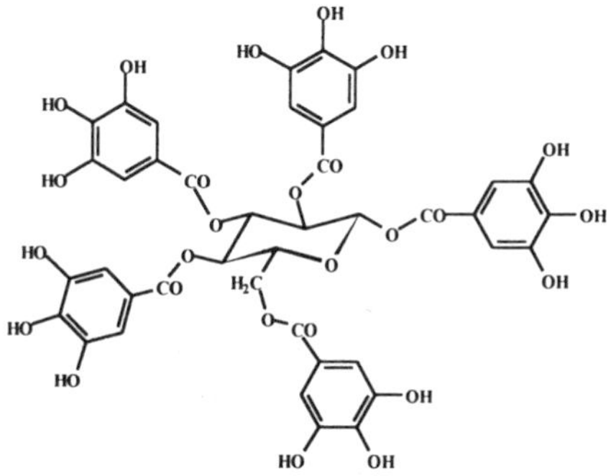

Penta-*O*-galloyl-β-D-glucose

Figure 4. Three galloyl esters considered in the present study of tannin coordination properties.

(80/20) and for equivalent concentrations of the ferric species and of the scavenger (Marmolle, F., Mila, I., Scalbert, A., Albrecht-Gary, A.M., unpublished results).

5. CONCLUSIONS

The results of the above-reported experiments demonstrate the importance of iron chelation by tannin in plant biology and human nutrition. Without denying the role of tannin/protein interaction in explaining the biological role of tannins, it is evident that greater attention should be paid to these chelation properties.

Further studies are being carried out in our laboratories to determine which characteristics of the complexes are essential to explain iron deprivation, whether their stability or their solubility/insolubility represent the decisive factor. Alternatively, inhibition of iron transport through binding of tannins to the membrane

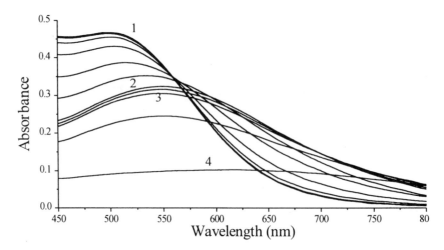

Figure 5. Spectrophotometric titration of ferric β-glucogallin complexes versus pH. [β-glucogallin]$_{tot}$ = 4 × [Fe(III)]$_{tot}$ = 2.9 × 10^{-4} M. Solvent: water, I = 0.01 M, l = 1 cm, T = (25.0 ± 0.2) °C. Spectrum (1): pH: 10.0, (2): 6.1, (3) 5.0, (4) 3.7.

iron transporter cannot be excluded. A better understanding of the mechanisms of iron deprivation should help to clarify the structural features in polyphenols controlling iron bioavailability.

Tannins, and plant polyphenols in general, do not only chelate iron(III). They also form complexes with other metals such as aluminium and copper. These complexes, although less stable than those formed with iron(III), may affect mineral absorption in the gut. Their structures, stability, and nutritional significance are presently being studied.

REFERENCES

1. Gottlieb, O.R. Plant phenolics as expressions of biological diversity. *In*: Hemingway, R.W.; Laks, P.E. (eds.) Plant polyphenols: synthesis, properties, significance. Plenum Press, New York, p. 523 (1992).

2. Markham, K.R. Distribution of the flavonoid in the lower plants and its evolutionary significance. *In*: Harborne, J.B. (ed.) The flavonoids—advances in research since 1980. Chapman and Hall Ltd, London, p. 427 (1988).

3. Bate-Smith, E.C. Systematic distribution of ellagitannins in relation to the phylogeny and classification of the angiosperms. *In*: Bendz, G.; Santesson, J. (eds.) Chemistry and botanical classification. Academic Press, London, p. 93 (1973).

4. Jersch, S.; Scherer, C.; Huth, G.; Schlösser, E. Proanthocyanidins as basis for quiescence of *Botrytis cinerea* in immature strawberry fruits. *Z. Pflanzenkr. Pflanzenschutz* 96:365 (1989).

5. Scalbert, A. Antimicrobial properties of tannins. *Phytochemistry* 30:3875 (1991).

6. Scalbert, A. Tannins in woods and their contribution to microbial decay prevention. *In*: Hemingway, R.W.; Laks, P.E. (eds.) Plant polyphenols: synthesis, properties, significance. Plenum Press, New York, p. 935 (1992).

7. Harborne, J.B.; Grayer, R.J. Flavonoids and insects. *In*: Harborne, J.B. (ed.) The flavonoids—advances in research since 1986. Chapman and Hall, London, p. 589 (1994).

8. Breslin, P.A.S.; Gilmore, M.M.; Beauchamp, G.K.; Green, B.G. Psychophysical evidence that oral astringency is a tactile sensation. *Chem. Senses* 18:405 (1993).

9. Cardon, D.; Du Chatenet, G. Guide des teintures naturelles. Delachaux and Niestlé, Neuchatel (1990).

10. Grimshaw, J. Phenolic aralkylamines, monohydric alcohols, monocarbaldehydes, monoketones and monocarboxylic acids. *In*: Coffey, S. (ed.) Rodd's chemistry, 2nd edition. Elsevier, Amsterdam, p. 141 (1976).

10. Seavell, A.J. Anticorrosive properties of mimosa (wattle) tannin. *J. Oil Col. Chem. Assoc.* 61:439 (1978).

11. Sampat, S.S.; Vora, J.C. Influence of colloids on the corrosion of 3S aluminium in low flow velocity water. *Indian J. Technol.* 13:476 (1975).

13. Chang, C.W.; Anderson, J.U. Flocculation of clays and soils by organic compounds. *Soil Sci. Soc. Am. Proc.* 32:23 (1968).

14. Durkee, G.E. Micronutrient foliar sprays. *Agrichem. West* 1:17 (1965).

15. Randall, J.M.; Bermann, R.L.; Garrett, V.; Waiss, A.C.J. Use of bark to remove heavy metal ions from waste solutions. *Forest Prod. J.* 24:80 (1974).

16. Laks, P.E.; Mc Kaig, P.A.; Hemingway, R.W. Flavonoid biocides: wood preservatives based on condensed tannins. *Holzforschung* 42:299 (1988).

17. Scalbert, A.; Cahill, D.; Dirol, D.; Navarrete, M.-A.; de Troya, M.-T.; Van Leemput, M. A tannin/copper preservation treatment for wood. *Holzforschung* 52:133 (1997).

18. 18. Porter, L.J. Flavans and proanthocyanidins. *In*: Harborne, J.B. (ed.) The flavonoids—advances in research since 1980. Chapman and Hall Ltd, London, p. 21 (1988).

19. Kawamoto, H.; Nakatsubo, F.; Murakami, K. Relationship between the B-ring hydroxylation pattern of condensed tannins and their protein-precipitating capacity. *J. Wood Chem. Technol.* 10:401 (1990).

20. Weinberg, E.D. Cellular regulation of iron assimilation. *Quart. Rev. Biol.* 64:261 (1989).

21. Mila, I.; Scalbert, A.; Expert, D. Iron withholding by plant polyphenols and resistance to pathogens and rots. *Phytochemistry* 42:1551 (1996).

22. Mahé, B.; Masclaux, C.; Rauscher, L.; Enard, C.; Expert, D. Differential expression of two siderophore-dependent iron-acquisition pathways in *Erwinia chrysanthemi* 3937: characterization of a novel ferrisiderophore permease of the ABC transporter family. *Molec. Microbiol.* 18:33 (1995).

23. Weinberg, E.D. Iron withholding: a defense against infection and neoplasia. *Physiol. Rev.* 64:65 (1984).

24. Van Snick, J.L.; Masson, P.L.; Heremans, J.F. The involvement of lactoferrin in the hyposideremia of acute inflammation. *J. Exp. Med.* 140:1068 (1974).

25. Alderton, G.; Ward, W.H.; Fevold, H.L. Identification of the bacteria-inhibiting iron-binding protein in egg white as conalbumin. *Arch. Biochem.* 11:9 (1946).

26. Hathway, D.E. The condensed tannins. *In*: Hillis, W.E. (ed.) Wood extractives and their significance to the pulp and paper industries. Academic Press, New York, p. 191 (1962).

27. Kühnau, J. The flavonoids: a class of semi-essential food components: their role in human nutrition. *World Rev. Nutr. Diet.* 24:117 (1976).

28. Stagg, G.V.; Millin, D.J. The nutritional and therapeutic value of tea—a review. *J. Sci. Food Agric.* 26:1439 (1975).

29. Disler, P.B.; Lynch, S.R.; Charlton, R.W.; Torrance, J.D.; Bothwell, T.H.; Walker, R.B.; Mayet, F. The effect of tea on iron absorption. *Gut* 16:193 (1975).

30. Hallberg, L.; Rossander, L. Effect of different drinks on the absorption of non-heme iron from composite meals. *Human Nutr. Appl. Nutr.* 36A:116 (1982).

31. Morck, T.A.; Lynch, S.R.; Cook, J.D. Inhibition of food iron absorption by coffee. *Am. J. Clin. Nutr.* 37:416 (1983).

32. Beswoda, W.R.; Torrance, J.D.; Bothwell, T.H.; McPhail, A.P.; Graham, B.; Mills, W. Iron absorption from red and white wines. *Scand. J. Haematol.* 34:121 (1985).

33. Cook, J.D.; Reddy, M.B.; Hurrell, R.F. The effect of red and white wines on nonheme-iron absorption in humans. *Amer. J. Clin. Nutr.* 61:800 (1995).

34. Hurrell, R.F.; Reddy, M.; Cook, J.D. Inhibition of non-heme iron absorption in man by herb teas, coffee and black tea. *Brit. J. Nutr.* (in press).

35. Gillooly, M.; Bothwell, T.H.; Charlton, R.W.; Torrance, J.D.; Bezwoda, W.R.; MacPhail, A.P.; Derman, D.P.; Novelli, L.; Morrall, P.; Mayet, F. Factors affecting the absorption of iron from cereals. *Brit. J. Nutr.* 51:37 (1984).

36. Gillooly, M.; Bothwell, T.H.; Torrance, J.D.; MacPhail, A.P.; Derman, D.P.; Bezwoda, W.R.; Mills, W.; Charlton, R.W.; Mayet, F. The effects of organic acids, phytates and polyphenols on the absorption of iron from vegetables. *Brit. J. Nutr.* 49:331 (1983).

37. Tuntawiroon, M.; Sritongkul, N.; Brune, M.; Rossander-Hulten, L.; Pleehachinda, R.; Suwanik, R.; Hallberg, L. Dose-dependent inhibitory effect of phenolic compounds in foods on nonheme-iron absorption in men. *Amer. J. Clin. Nutr.* 53:554 (1991).

38. Brune, M.; Rossander, L.; Hallberg, L. Iron absorption and phenolic compounds: importance of different phenolic structures. *Eur. J. Clin. Nutr.* 43:547 (1989).

39. Mila, I.; Deprez, S.; Huneau, J.-F.; Tomé, D.; Scalbert, A., Proanthocyanidins inhibit nutrient absorption in a human intestinal epithelial cell line. Symposium papers of the 19th International Conference on Polyphenols, Lille, France, p. 121 (1998).

40. de Alareon, P.A.; Donovan, M.E.; Forbes, G.B.; Landau, S.; Stockman, J.A. Iron absorption in thalassemic syndromes and its inhibition by tea. *New Engl. J. Med.* 300:5 (1979).

41. Jacquinot, M.; Leize, E.; Potier, N.; Albrecht, A.M.; Shanzer, A.; Van Dorsselaer, A. Charcterization of non-covalent complexes by electrospray mass spectrometry. *Tetrahedron Lett.* 34:2771 (1993).

42. Marmolle, F.; Leize, E.; Mila, I.; Van Dorsselaer, A.; Scalbert, A.; Albrecht-Gary, A.M. Polyphenol metallic complexes: characterization by electrospray mass spectrometric and spectrophotometric methods. *Analusis* 25:M53 (1997).

43. Maeder, M.; Zuberbühler, A.D. Nonlinear least-squares fitting of multivariate absorption data. *Anal. Chem.* 62:2220 (1990).

44. Sillen, L.G.; Warnqvist, B. High-speed computers as a supplement to graphical methods. 10. Application of Letagrop to spectrophotometric data, for testing models and adjusting equilibrium constants. *Arkiv Kemi* 31:377 (1968).

ANTI-CARIES ACTIVITY OF BARK PROANTHOCYANIDINS

Tohru Mitsunaga

Department of Forest Products
Faculty of Bioresources
Mie University
Kamihama, Tsu 514-8507
JAPAN

1. INTRODUCTION

Present-day accumulation of woody bark reaches 5.6 million tons per year in Japan. The distribution of its utilization is shown in figure 1. Taking into account that about 75 percent of the bark goes to fuel and waste matter, it can be said that bark is not effectively utilized. Furthermore, it is regarded as a troublesome product in the lumber industry. However, bark is without question a very important tissue for the protection of a tree from external invasions by wood-rotting fungi or insects, and polyphenols (proanthocyanidins) contained in the bark are believed to play an important role in these defensive actions.[1] This is based on the properties of phenols that denature proteins or inhibit enzymes. We attempted to examine how to effectively utilize bark by considering these characteristics of the polyphenols.

In regard to the enzyme inhibition caused by polyphenols, the inhibitory effects on glucosyltransferase (GTase) by the polyphenols contained in green and oolong tea extracts have been reported recently.[2,3,4] Dental caries are developed by the degradation of water-insoluble glucans synthesized from sucrose by GTase, which is derived from *Streptococcus sobrinus* or *S. mutans*.[5] Application of these tea polyphenols as a preventive for dental caries in foods and sanitary supplies is utilized in some consumer products in Japan. The effective polyphenols in green tea for GTase inhibition are mainly monomeric polyphenols having galloyl groups, and those in oolong tea are enzymatically oxidized compounds[6] produced by semifermentation of green tea polyphenols. On the other hand, bark polyphenols are mainly proanthocyanidins (PAC) consisting of polyhydroxyflavan-3-ol repeating units linked through C4—C6 or C4—C8 bonds.[7] In this chapter, the GTase inhibition ability of some bark proanthocyanidins and the correlation between the chemical structure of PAC and their inhibitory activity are discussed.

Plant Polyphenols 2: Chemistry, Biology, Pharmacology, Ecology, Edited by
Gross et al. Kluwer Academic / Plenum Publishers, New York, 1999

555

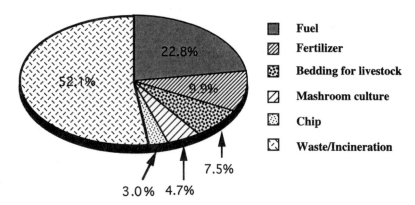

Figure 1. Utilization of bark in Japan.

2. BACKGROUND OF DENTAL CARIES PREVENTION

Dental caries are highly prevalent affecting more than 90 percent of 5- to 15-year-old children because exposure to foods and beverages containing sugar has been increasing with rising consumerism. Therefore, food and food materials resistant to dental caries have been a subject of much research.

Dental caries are a bacterially caused disease produced by mutans streptococci, *Streprococcus mutans* and *S. sobrinus*. These bacteria produce glucosyltransferase (GTase), which synthesizes the insoluble and adhesive glucans from sucrose. The bacterially synthesized glucans attach to tooth surfaces and combine with bacteria in the mouth to form plaque on the teeth. Then, organic acids produced by the degradation of plaque by another bacteria cause the pH on the teeth to be too low and, consequently, dental caries occur.

Thus far, the findings of GTase inhibitors by polyphenols have been made mainly on foodstuffs and medicinal plants, and some active polyphenols from teas and traditional medicines have been identified. Several polyphenols present in green tea (*Camellia sinensis*) extracts inhibit the growth of mutans streptococci,[8] and glucan synthesis from sucrose by GTase of *S. sobrinus*[2] and epigalocatechin gallate (**1**) particularly shows high GTase inhibitory effect. In oolong tea extracts (semifermented tea leaves of *Camellia sinensis*), polymeric polyphenolic compounds that had a molecular weight of approximately 2,000 have reduced caries development and plaque accumulation as well as inhibited GTase activity in humans and experimental animals.[9] To investigate the effective compound in oolong tea, enzymic reaction of (+)-catechin by peroxidase was carried out *in vitro*; consequently, dehydrodicatechin-A (**2**), an oxidative coupling product between the A and B aromatic ring, was obtained.[10] Therefore, it was concluded that oligomers of the coupling product were effective compounds for GTase inhibition in oolong tea. A similar investigation has been undertaken on hydrolyzable tannins from crude drugs, and penta- (**3**) and hexagalloylglucose (**4**) have turned out to be the most potent inhibitors of GTase. The inhibitory potency of gallotannins decreased with the decrease of the number of galloyl residues.[11]

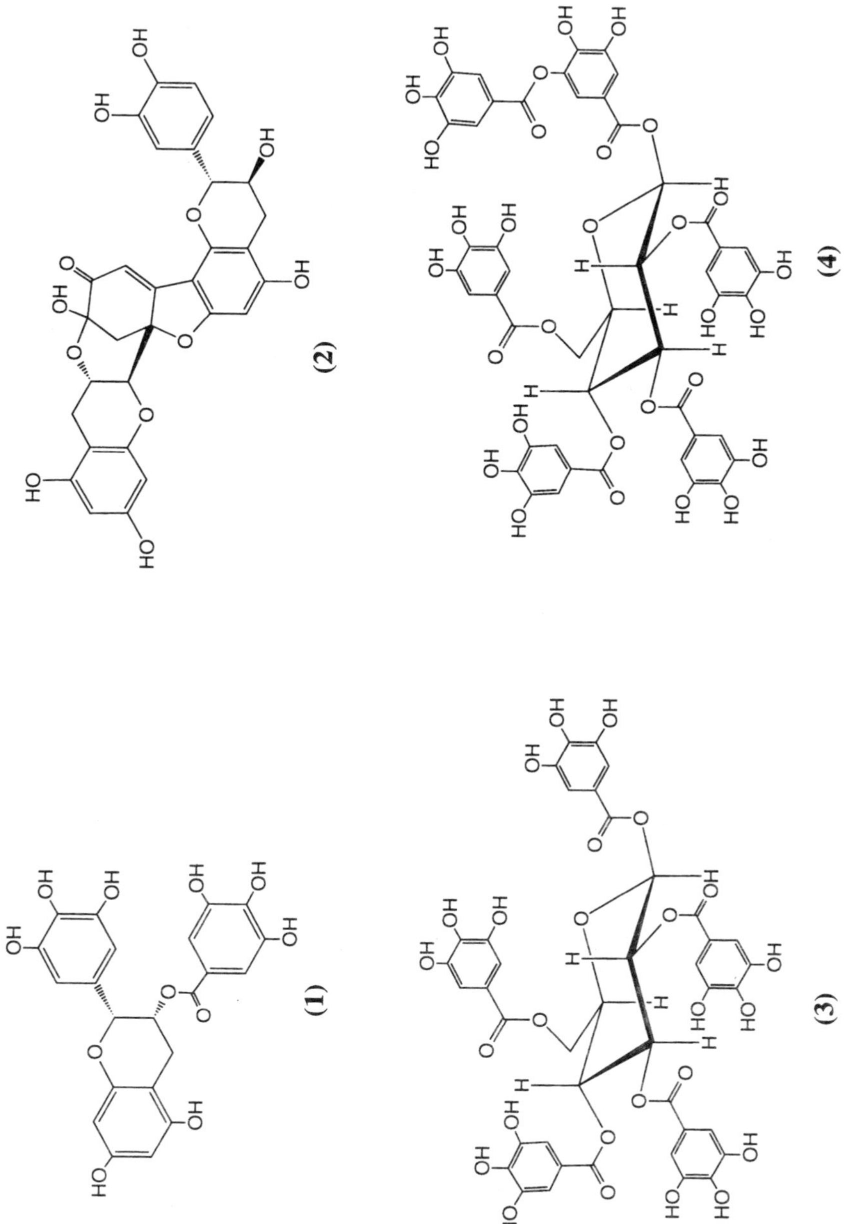

Figure 2. Structures of some known polyphenols inhibiting GTase.

3. INHIBITORY EFFECT OF BARK PROANTHOCYANIDINS ON GTASE

Green tea extracts (GTE), oolong tea extracts (OTE), commercially available Quebracho (*Schinopsis lorentzii*) extracts (QE), bark extracts of Acacia (ABE), Karamatsu (KBE), Sugi (SBE), Hinoki (HBE), and Kuromatsu (KUBE) were prepared by using 70-percent aqueous acetone. ABE and KBE were fractionated by LH-20 gel column chromatography using ethanol, methanol, and 70-percent aqueous acetone as eluents. These eluted fractions were termed EE, ME, and AE, respectively. The polyphenol analyses were conducted by using the following specific color reactions: the Folin-Denis method[12] for the determination of total phenol content and vanillin-HCl method[13,14] to determine proanthocyanidin content. Furthermore, tannin content was determined by the Lowenthal method[15] using a redox reaction. Table 1 shows the data obtained by these analysis methods and IC_{50} values, the concentration giving 50 percent of inhibition, on GTase of 70-percent aqueous acetone extracts from teas and barks. Flavanol contents of GTE and OTE are very similar, but their PAC contents are quite different. Because GTE contains mainly monomeric polyphenols, it does not exhibit tanning

Table 1. Polyphenol analyses and GTase inhibition of teas and barks extracted by 70-percent aqueous acetone

Extracts	Analysis methods [a]			GTase inhibition [b] IC_{50} (µg/ml)
	Folin-Denis	Vanillin-HCl	Lowenthal	
GTE	40.3	49.5	5.2	200
OTE	42.5	42.2	29.5	40
ABE	66.5	57.3	45.5	11
KBE	70.2	66.1	53.9	5
SBE	52.0	29.3	15.2	16
HBE	40.5	25.6	10.3	23
KUBE	42.3	18.4	9.5	30

GTE : Green tea extracts   OTE : Oolong tea extracts
ABE : Acacia bark extracts   KBE : Karamatsu bark extracts
SBE : Sugi bark extracts   HBE : Hinoki bark extracts
KUBE: Kuromatsu bark extract

a) These figures show the weight % of the amounts caluculated by the calibration curves using (+)-catechin to that of the extracts used.

b) The concentration giving 50% inhibition of GTase

characteristics. The Lowenthal analysis of ABE and KBE represents the concentration of PAC among the bark extracts. Because the values show about 80 percent of that provided by the vanillin-HCl method, most of the polyphenols in ABE and KBE would consist of PAC. Both bark extracts exhibited higher inhibition than GTE or OTE, particularly, KBE showed very high inhibition, and its IC_{50} value was about $5.0\,\mu g/mL$. Judging from these IC_{50} values, the inhibitory effect of KBE was about 10 times higher than that of OTE and 40 times that of GTE. Therefore, the inhibitory activity of these extracts was presumed to relate to the contents and structures of proanthocyanidins.

4. INFLUENCE OF MOLECULAR WEIGHT AND HYDROXYLATION PATTERNS OF BARK PAC ON INHIBITORY ACTIVITY OF GTASE

The nucleus exchange reaction (NER) is a method designed for the analysis of phenolic nuclei constituting lignin[16] and condensed tannins.[17] In the case of condensed tannins, phloroglucinol or resorcinol from the A-rings and catechol and pyrogallol from the B-rings are liberated in large amounts. This means that the NER method gives more information with regard to the phenolic nuclei constituting condensed tannins than other degradation methods such as butanol-hydrochloric acid[18] or thiolysis.[19,7,20,21,22,23] Thiolysis has been used as one of the efficient methods for the structural analysis of natural condensed tannins. This method gives information not only about the types of flavan-3-ol constituting condensed tannins, but also of the stereochemistry of sp^3 carbons constructing the pyran-ring. However, this information does not represent the whole analysis of materials used because the yields of the degradation products are often very low and insignificant for the 5-deoxyprofisetinidins, for example. On the other hand, the yield of products obtained by NER has proved to be extremely high, so this method has been applied to the analysis of almost all classes of proanthocyanidins constituting bark polyphenols.

NER is the nuclear substitution reaction by excess phenol in the presence of borontrifluoride (BF_3). The mechanism of phenolation of PAC is described in our previous published work[24,25,26] and was proposed by using (+)-catechin as a model. That is as follows: a lone pair of electrons of oxygen on the pyran ring coordinates to BF_3 first, and then the ether bond is cleaved at the same time as a lone pair of electrons of a hydroxyl group at the C-3 position makes a nucleophilic attack on the C-2 carbon to form an epoxide intermediate. Successively, a phenol existing in excess makes a nucleophilic attack at the C-2 carbon from the back side of the epoxide ring to form Compound I as shown in figure 3. Because (−)-epicatechin is not able to take the conformation to make an intramolecular nucleophilic substitution, a phenol attacks directly the C-2 carbon from the back side against the ether bond to form Compound I. Thus, it should be considered that the formation of Compound I from (+)-catechin takes place mainly through two S_N2 reactions. Furthermore, an excess phenol attack to the diphenylmethane carbon charging on the cation to eliminate catechol from the B-ring of compound I. On the other hand, the interflavanoid bond is much more subject to cleavage by a phenol than the

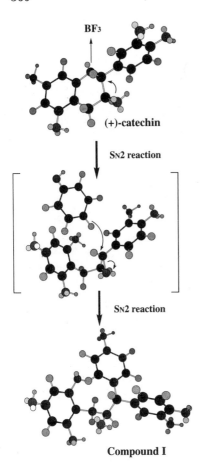

BF₃

(+)-catechin

S_N2 reaction

S_N2 reaction

Compound I

Figure 3. Phenolation of (+)-catechin in the presence of boron trifluoride.

pyran ring of proanthocyanidins as shown in figure 4. The resulting diphenylmethane compound eliminates phloroglucinol from the A-ring by the similar reaction described above.

The amounts of nuclei liberated by NER of the bark PAC are shown in table 2. On ABE, phloroglucinol and resorcinol nuclei were liberated in a molar ratio of about 1:3 from the A-ring, and catechol and pyrogallol nuclei in a molar ratio of about 1:2 from the B-ring. Judging from the chemical structures of proanthocyanidin dimers and trimers in *Acacia mearnsii* bark,[27] these ratios would indicate that the extender units consist mainly of profisetinidin and prorobinetinidin types, and the terminal unit is a procyanidin type. On the other hand, KBE consists of the procyanidin type as can be seen in most coniferous bark,[28] and Quebracho extracts (QE) consist of the profisetinidin type.

The molecular weight of each eluate fraction separated by LH-20 gel column chromatography and their IC_{50} values for GTase are shown in table 3. From the IC_{50} value of each ME fraction, which shows almost the same average molecular weights, the inhibitory effect of KBE-ME is the greatest, and those of ABE-ME and QE-ME are similar. Therefore, the hydroxylation patterns of A- and B-rings

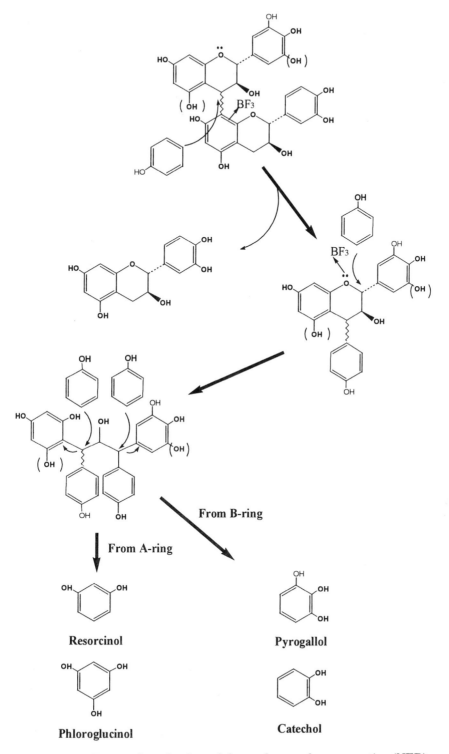

Figure 4. Proposed mechanism of the nucleus exchange reaction (NER).

Table 2. Phenyl nucleus composition of several proanthocyanidins analyzed by the NER method

Sample	A-ring		B-ring		Phl./Res.	Pyr./Cat.
	Phl.	Res.	Cat.	Pyr.		
ABE-ME	15.1	49.5	22.1	38.1	0.31	1.7
-AE	11.3	33.1	17.3	35.6	0.34	2.1
QE-ME	5.3	40.6	64.4	2.0	0.13	0.03
-AE	3.3	43.3	63.6	2.3	0.08	0.04
LBE-EE	9.4	0.6	43.8	2.2	15.6	0.05
-ME	20.5	0.5	65.0	2.8	41.0	0.04
-AE	23.4	0.6	70.3	3.1	39.0	0.04

* These values are represented by mol % to a unit of flavan-3-ol.

Phl.: phloroglucinol   Res.: resorcinol
Cat.: catechol   Pyr.: pyrogallol

Sample name: refer to Table 1

Table 3. The influence of molecular weight of fractionated bark extracts on GTase inhibition activity

Sample	$\overline{Mn}$	$\overline{Mw}$	Dp	IC50 (μg/ml)
ABE-EE	355	720	2.4	250
-ME	1045	1495	3.9	15.0
-AE	2474	4235	11.4	1.7
KBE-EE	325	750	2.2	13.0
-ME	1150	1585	4.0	5.0
-AE	2852	4825	13.5	1.0

$\overline{Mn}$: Number avarage molecular weight $\overline{Mw}$: Weight average molecular weight
Dp: Degree of polymerization
IC50 : The concentration (μ g/ml) giving 50% inhibition of GTase.
Sample name: refer to Table 1

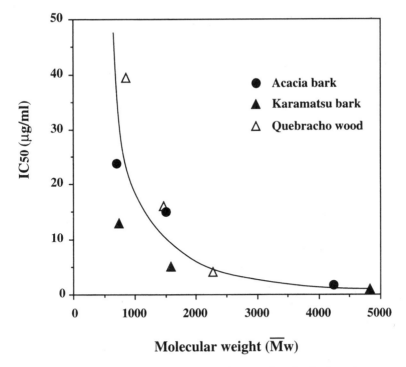

Figure 5. The relationship between molecular weight of polyphenols separated by LH-20 column chromatography and their GTase inhibitory activity.

presumably influence the extent of GTase inhibition. The molecular weight of these fractions increased in the following order, EE, ME, and AE fraction. Judging from the molecular weight of the top peak of GPC, these fractions consist of dimer to trimer, trimer to pentamer, and greater than hexamer, respectively. As shown in figure 5, their IC_{50} values showed a tendency to decrease with an increase of the molecular weight of the polyphenols, and the rapid increase of an inhibitory effect was shown in the neighborhood of average molecular weight 1,500 corresponding to oligomers from pentamer to hexamer of proanthocyanidins.

5. THE RELATIONSHIP BETWEEN HYDROXYLATION PATTERN AND THEIR GTASE INHIBITORY EFFECTS OF SYNTHESIZED PAC

Four PAC were synthesized non-enzymatically according to the method described in our previous paper.[29] The procedure of the synthesis is shown in figure 6. Taxifolin, fustin, and ampeloptin were obtained from the heartwood of *Larix leptolepis*, *Rhus succedanea* L, and *Salix sachalinensis* Fr. Schm, respectively. Naringenin was purchased from Wako Pure Chemical Industries LTD. These flavanoids were reduced with sodium borohydride to obtain 4-hydroxyflavan derivatives.

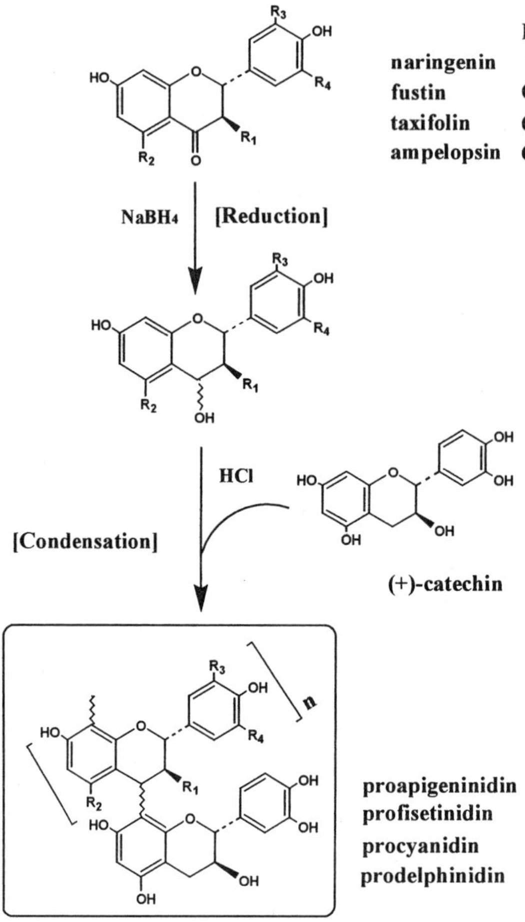

	R1	R2	R3	R4
naringenin	H	OH	H	H
fustin	OH	H	OH	H
taxifolin	OH	OH	OH	H
ampelopsin	OH	OH	OH	OH

NaBH₄ [Reduction]

(+)-catechin

[Condensation]

	R1	R2	R3	R4
proapigeninidin	H	OH	H	H
profisetinidin	OH	H	OH	H
procyanidin	OH	OH	OH	H
prodelphinidin	OH	OH	OH	OH

Figure 6. Synthesis scheme of proanthocyanidins.

Procyanidin (PC), profisetinidin (PF), prodelfinidin (PD), and proapigeninidin (PA) oligomers were synthesized by condensation of the corresponding 4-hydroxyflavan derivatives with (+)-catechin under acidic conditions. These oligomers were separated by Sephadex LH-20 gel column chromatography in a manner similar to that performed on the bark extracts. The hydroxylation patterns of the A- and B-rings of these synthesized PAC and their IC_{50} values are shown in table 4. The AE fraction has a higher inhibitory activity than the ME fraction in all oligomers indicating that the activity depends on their molecular weight similarly to the bark PAC. PC-ME and PF-ME show the same IC_{50} value of 4.2 (μg/mL), which indicates that the hydroxyl group in the A-ring does not influence the inhibitory activity. On the other hand, B-ring hydroxylation patterns resulted in different inhibition levels when comparing PA-ME, PC-ME, and PD-ME. The oligomers containing 3′,4′-dihydroxyl groups in the B-ring showed especially high inhibition, whereas

Table 4. The relationship between hydroxylation patterns of synthesized proanthocyanidins and GTase inhibition

Hydroxylation patterns		Synthesized polyphenols	IC 50 (µg/ml)
A-ring	B-ring		
HO — structure with OH, CH₃, OH	structure with OH	Proapigeninidin-ME	12.7
		-AE	6.2
HO — structure with O, CH₃	OH, OH structure	Profisetinidin-ME	4.2
		-AE	3.3
HO — structure with O, CH₃, OH	OH, OH structure	Procyanidin-ME	4.2
		-AE	2.0
HO — structure with O, CH₃, OH	OH, OH, OH structure	Prodelphinidin-ME	300<
		-AE	75.3

oligomers containing 3′,4′,5′-trihydroxyl groups showed little or no inhibition. These results would explain the reason that *Acacia* proanthocyanidins having a high proportion of pyrogallol B-ring units showed lower inhibition activity than *Karamatsu* proanthocyanidins as described above. Haslam suggested that the o-dihydroxyl functionality of procyanidins are the site of a combination with proteins and also showed that polymeric proanthocyanidins containing many sites for such a combination have a greater ability to precipitate proteins than low molecular weight proanthocyanidins.[30] If this specific association is caused by the hydrogen bond between polyphenols and proteins, stronger hydrogen bonds must be formed by pyrogallolic hydroxyl groups than by catecholic hydroxyl groups. However, the experimental results were contrary to this expectation. In the formation of polyphenol-protein complexes, hydrophobic interaction in binding sites is regarded as important as hydrogen bonding. Therefore, the balance of these two types of interactions in the polyphenol molecules would participate in the extent of inhibition of GTase.

The inhibition mode of the non-enzymatically synthesized procyanidin (PC) oligomer showing high inhibitory effects for GTase was examined from the Lineweaver-Burk plots shown in figure 7. The Km value, 5.9×10^{-5} M, obtained by addition of the PC oligomer was similar to the control, which indicated that the mode of inhibition was most closely correlated with the classical pattern of non-competitive inhibition in which polyphenol and substrate were assumed to bind

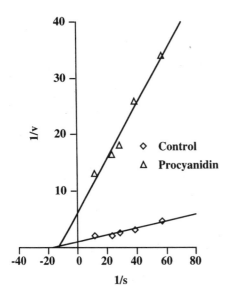

Figure 7. Lineweaver-Burk plots of GTase with and without synthesized procyanidin.

simultaneously to the GTase. From the facts presented above, polyphenols would interact with GTase through hydrogen-bonding and hydrophobic interaction at all sites except the reactive center, causing steric changes in the GTase and inhibition of glucan synthesis.

6. GTASE INHIBITORY EFFECTS OF CONVERTED COMPOUNDS FROM CATECHINS BY EXTRACELLULAR ENZYMES OF FUNGI

Basidiomycetes are available as foods and medicines for everyday living, and at the same time, have a significant function of decomposing plant materials in natural surroundings. This function suggests that basidiomycetes are capable of bioconversion of plant biomass into useful materials. We have examined chemical structures and inhibitory activities on GTase, relating to the generation of dental caries, of bioconverted compounds from (+)-catechins and (−)-epicatechin obtained from bark, by *Coriolus versicolor* and *Pycnoporus coccineus* known as wood-rotting fungi.

As a result, dimer Y-2 **(5)**, Y-3 **(2)** from (+)-catechin and trimer Y-4 **(6)**, dimer Y-5 **(7)** from (−)-epicatechin were produced by their extracellular enzymes (fig. 8). These compounds were isolated as amorphous yellowish compounds after Sephadex LH-20 column chromatography and preparative HPLC. From the data of ^{1}H-, ^{13}C-, ^{1}H-^{1}H COSY, NOESY, and HMBC NMR experiments, these compounds were identified as having condensed rings formed by oxidative coupling between the A and B ring of catechins and to have conjugated carbonyl groups in the molecules. Dehydrodicatechin-A, was also isolated from the reaction mixtures of (+)-catechin treated with peroxidase-H_2O_2 system and, as described in our previous paper,[10] so that the conversion in this system is also due to the phenol oxidase,

Table 5. BSA adsorption, IC_{50} on GTase activity, growth inhibition of *S. sobrinus*, and inhibition of plaque formation by polyphenols

Sample	BSA adsorption (%) [a]	GTase inhibition $IC_{50}(\mu g/ml)$ [b]	growth inhibition of *S.sobrinus* (%) [c]	inhibition of plaque formation(%) [d]
Y-2	88	64	49	48
Y-3	87.5	14	54	72
Y-4	0	189	0	10
Y-5	93	14	55	70
KBE	91.8	5	9	87
ABE	89.9	11	0	60
PC-oligomer	87.0	3	0	87
PC-dimer	0	>300	4	0
(+)-catechin	0	>300	0	0
(-)-epicatechin	0	>300	0	0

a) Test tube contained 200 μl of BSA solution (1mg/ml) and 200 μl of polyphenol solution (3mg/ml) was placed at room temperature for 1 hour, then it was centrifuged. Another test tube contained the supernatant (200 μl) and Ba(OH)2 100 mg was heated at 120°C for 10min, then 3ml of ninhydrin solution was added. After having been heated at 108°C for 5min, absorbance of the solution at 570 nm was measured.

b), c), d)   Refer to analytical method

probably laccase, in the extracellular enzymes of fungi. BSA adsorption, IC_{50} of GTase, growth inhibition of *S. sobrinus*, and inhibition of plaque formation by the enzyme-converted compounds and bark PAC are shown in table 5. As mentioned elsewhere, inhibitory activity of bark PAC on GTase increases with increasing of PAC's molecular weight though monomeric and dimeric classes of PAC do not show the activities. However, converted compounds from catechins showed high activities in spite of the dimeric classes; Y-3 and Y-5 were especially remarkable. On the other hand, trimeric compound Y-4 did not show the inhibitory effect. It is obvious that the inhibitory effects on GTase of a series of these converted compounds do not necessarily correlate to the molecular weight differently from that correlation of bark PAC. To understand these conflicting results, the protein-adsorbing ability of these compounds was examined by using bovine serum albumin (BSA). Table 5 shows IC_{50} values and the BSA adsorption of converted compounds and other polyphenols. The higher the inhibitory effect of GTase, the more the compound obviously can make precipitates with BSA. Therefore, manifestation of inhibitory action of polyphenols on GTase is intimately related to the phenomenon of protein precipitation. Some studies related to polyphenol-protein interactions have been examined in haze-forming of beer.[31,32,33] In those studies, it has been reported that polyphenols make hydrogen bonding and hydrophobic bonding with proline in proteins to produce precipitates. The inhibition of GTase by PAC may also occur in order to make these bondings between PAC and proline in GTase. However, the differences of procyanidin dimer and enzyme-converted dimers are particularly interesting for precipitating ability with BSA and inhibitory activity on GTase. Figure 8 shows molecular models of a PC dimer and Y-3 built with Chem 3D

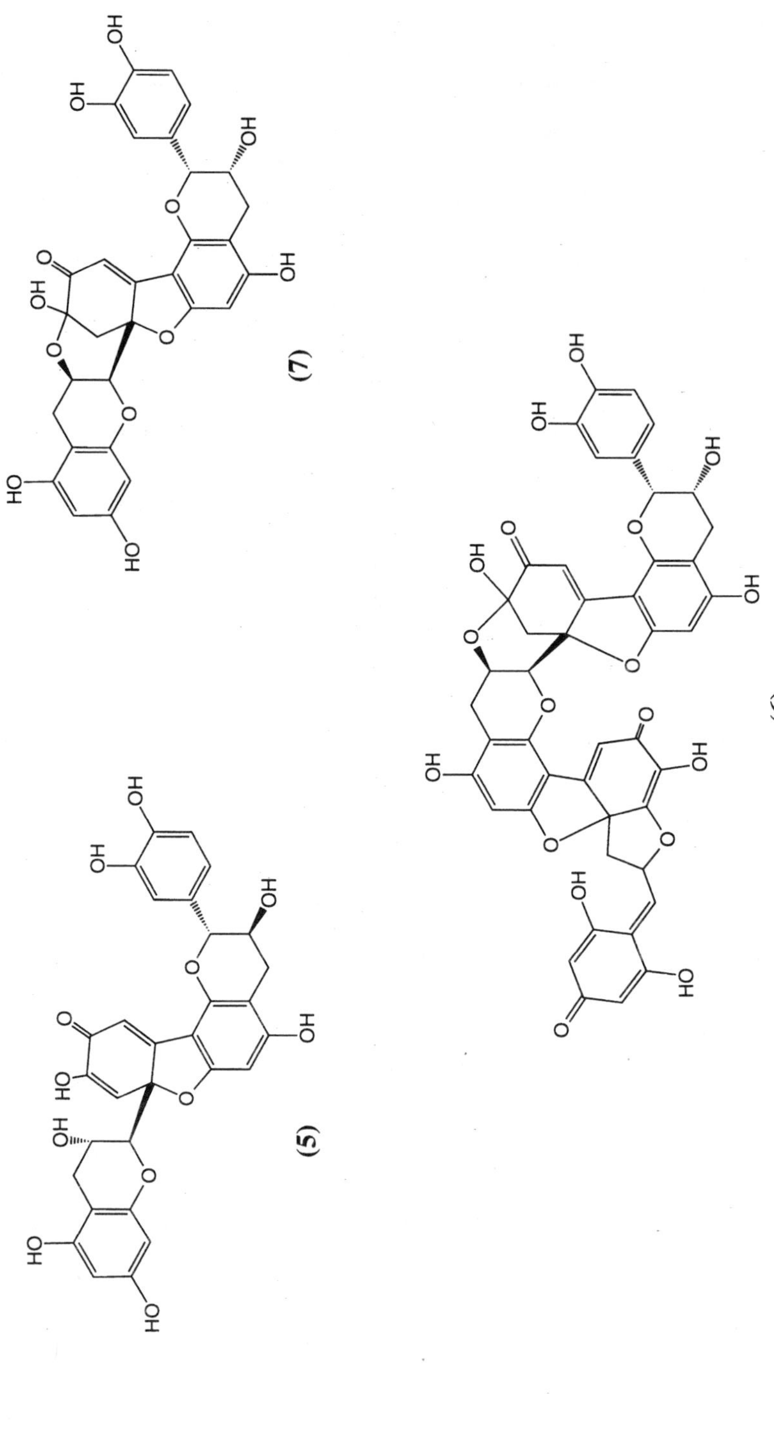

Figure 8. Structures of fused A- B-ring oligomers from oxidative coupling.

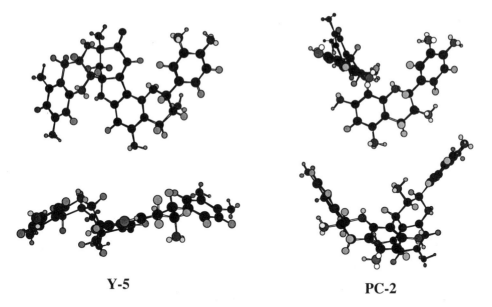

Y-5   PC-2

Figure 9. Three-dimensional structures of Y-5 and PC-2.

software. Because of the condensed ring, the molecule of Y-3 obviously has a more planar structure than that of the PC dimer. Furthermore, the MM2 energy calculation of enzyme-converted compounds resulted in higher stretch-bend energy than found for the PC dimer and monomers, indicating that these compounds have strained structures. Therefore, it was presumed that the planar and strained structure of polyphenols are important factors on GTase inhibitory action.

Antimicrobial activity against *streptococci* and adherent inhibition of plaque on tooth surfaces by polyphenols are thought to be necessary for dental caries prevention as well as GTase inhibition. Results of our studies are summarized in table 5. The compounds Y-2, Y-3, and Y-5 show especially high growth inhibition of *S. sobrinus* and high inhibition of plaque formation compared to flavan-3-ols and procyanidin and a dimer. On the other hand, bark PAC does not inhibit growth of *S. sobrinus*, but inhibits plaque formation. As a consequence, such a small molecular size as enzyme-converted compounds may pass through the cell membrane to inhibit GTase and the growth of *S. sobrinus*, whereas bark PAC may not show antimicrobial activity but GTase inhibition on the surface of the cell membrane of *S. sobrinus*.

7. ANALYTICAL METHODS FOR BARK PAC

Gel permeation chromatography (GPC) was recorded with a JASCO TRIRO-TAR system with Shodex GPC columns KF-802 and KF-804 (4.6 mmϕ × 250 mm) using tetrahydrofuran as an eluent. The chromatogram was calibrated against standard polystyrenes (Molecular weights: 2,000 and 9,000), (+)-catechin, and synthesized PF dimer. Averages of molecular weights (Mn, Mw) were calculated by an integrator, Jasco 807-IT, from the molecular weight distributions obtained by the GPC measurement.

The nuclear exchange reactions (NER) were carried out in a 2 mL glass ampule with 10 mg of polyphenol sample and 500 μL of the degradation reagent at a temperature of 80 °C or 150 °C. The constitution of the degradation reagent was benzene : phenol : BF^3-phenol complex = 10 : 19 : 3 (v/v/v). After the reaction, the reaction vessel was cooled to room temperature, and the reaction mixture was transferred quantitatively into a beaker with ethyl ether. A known amount of internal standard (dibenzyl in benzene) was added. Ether insoluble materials were filtered and washed with ethyl ether. The filtrate and washings were combined in a separatory funnel to which an excess of saturated brine was added. After the extraction (30 mL × 3), the ether layer was dried over sodium sulfate and evaporated up to a small volume (4–5 mL. One hundred μL of the ether solution was transferred into a small vial to which a few drops of pyridine and 100 μL of N,O-bis(trimethylsilyl) acetamide were added. After 1 hour derivatization reaction at room temperature, the TMS derivatives were analyzed quantitatively using gas liquid chromatography (GLC). GLC was performed using a Yanagimoto G-180 using a methyl silicone capillary column (Guadrex S2006, 0.25 mm I.D. × 25 m length × 0.25 mm film thickness). Column temperature of the of GLC was maintained at 80 °C for 2.5 min. and then increased to the final temperature of 250 °C at a rate of 2 °C/min.

Streptococcus sobrinus 6715 was grown for 16 hr at 37 °C in 5 L of Todd Hewitt (TH) broth. After centrifugation of the liquid medium at 5,000 rpm for 15 min, the cells were collected and then extracted with 75 mL of 8 M urea at 20 °C for 1 hr with stirring. The crude enzyme solution containing urea was dialyzed against 10 mM potassium phosphate buffer (pH 6) until the urea was removed entirely. One ml of the crude enzyme solution was pipetted into a microtube and stored in a freezer at −80 °C.

Insoluble glucan synthesized by GTase was measured turbidimetrically with a spectrophotometer (Jasco V-520) by determining the increase in A550. GTase was incubated in 3 mL of 0.1 M phosphate buffer (pH 6.0) containing 1 percent sucrose, 0.1 percent sodium azide, 0.5 percent dextran T-10, and in the presence or absence of polyphenols at 37 °C for 3 hr. The volume of the crude GTase solution used in the assay was determined by that giving an absorbance of 1.0 at 550 nm. Inhibition rate is expressed by the following equation:

$$\text{Inhibition rate } (\%) = 100 \times (AC - AP)/AC$$

Ac and Ap represent absorbance obtained in the control and in the polyphenol dose, respectively. IC$_{50}$ means the polyphenol concentration (μg/mL) giving 50 percent inhibition of GTase. The EE, ME, and AE fractions of Acacia, Karamatsu, and synthesized proanthocyanidins were tested for the IC$_{50}$ activity.

To examine the effect of polyphenols on cell growth, the filter-sterilized preparation was added to 2 ml of Tryptose broth (Difco Laboratories), which was then inculated with 100μL of a seed culture of *S. sobrinus*. After incubation at 37 °C for 18 hr, the growth of cells was determined tubidimetrically by measuring the optical density of the culture at 550 nm.

Sucrose-dependent cellular adherence was determined according to the method described by Hamada et al.[5] Mutans streptococci were grown at 37 °C at a 30° angle to the horizon for 18 h in 3 mL of brain heart infusion broth (Difco Laboratories. Detroit, Michigan) containing 1 percent sucrose. The quantity of adherent cells after gentle washing of each test tube was determined turbidimetrically and expressed as a percentage of the total cell mass (percentage of total cell adherence).

8. CONCLUSIONS

Seventy percent acetone aqueous bark extracts of several Japanese trees, *Acacia mearnsii* (Acacia), *Larix leptolepis* (Karamatsu), *Cryptomeria japonica* (Sugi), *Chamaecyparis obtusa* (Hinoki), and *Pinus thunbergii* (Kuromatsu) inhibited the activity of glucosyltransferase (GTase) derived from *Streptococcus sobrinus*, which causes dental caries. In particular, the bark extracts of Karamatsu showed a stronger activity (IC_{50}—much lower) than green tea or oolong tea extracts. The bark extracts inhibited not only GTase activity but also the adherence of the growing cells to a glass surface, whereas it did not show the antimicrobial activity of *S. sobrinus*. After the fractionation of the extracts by Sephadex LH-20 column chromatography, the GTase inhibitory activity increased as the molecular weight of proanthocyanidins contained in the extracts increased. The inhibition was very closely related to the hydroxylation patterns in the B-rings of synthesized proanthocyanidins, where the compound with a catechol ring showed higher activity than with a pyrogallol B-ring. Judging from the Lineweaver-Burk plots, its inhibition mode was a non-competitive type.

The converted yellowish compounds from (+)-catechin and (−)-epicatechin by the crude extracellular enzymes of white rot fungi were also examined by GTase inhibition. These compounds obtained by preparative HPLC were identified as dimers and trimer with condensed aromatic rings by radical oxidative coupling of catechins and showed high GTase inhibitory activities in spite of low molecular weights.

In summary:

70 percent acetone aqueous extracts of the barks of several Japanese trees showed high inhibitory effect of GTase. Especially *Larix leptolepis* (Karamatsu) extracts having procyanidin type polyphenols were remarkable for their high activity.

GTase inhibitory effect increased as the molecular weight of the proanthocyanidins was increased, and the extent of the inhibition was affected by the B-ring hydroxylation patterns of synthesized proanthocyanidins.

Extracellular enzymes converted compounds from catechin to compounds with demonstrated high inhibitory activity in spite of their low molecular weights. Their structures were estimated to be planar and strained as judged by MM2 molecular-energy calculations.

ACKNOWLEDGMENT

The author is deeply grateful to Emeritus Professor Dr. Isao Abe, Faculty of Bioresources, Mie University, for his valuable discussions. He also deeply grateful to Dr. H. Ono, Dr. K. Nakahara and Dr. T. Tanaka, Suntory Research Center, for advice on the GTase preparation and inhibitory assay. Furthermore, he would like to thank Dr. S. Ohara, Forestry and Forest Products Research Institute, for guiding NMR techniques and giving a generous gift of *Acacia mearnsii* bark extracts.

REFERENCES

1. Scalbert, A.; Haslam, E. Polyphenols and chemical defense of the leaves of *Quercus robur*. *Phytochemistry* :3191 (1987).

2. Sakanaka, S. Inhibitory effects of green tea polyphenols on glucan synthesis and cellular adherence of cariogenic streptococci. *Agric. Biol. Chem.* :2925 (1990).

3. Wu-yuan, C.D.; Chen, C.Y.; Wu, R.T. Gallotannins inhibit growth, water-insoluble glucan synthesis, and aggregation of mutans streptococci. *J. Dent. Res.* :51 (1988).

4. Ooshima, T.; Minami, T.; Hamada, S. Oolong tea polyphenols inhibit experimental dental caries in SPF rats infected with mutans streptococci. *Caries Res.* :124 (1993).

5. Koga, T.; Okahashi, N.; Asakawa, H.; Hamada, S. Adherence of *Streptococcus mutans* to tooth surfaces; *In:* Hamada, S.; Michalek, S.M.; Kiyono, H.; Menaker, L.; McGhee, J.R. (eds.) Molecular microbiology and immunobiology of *Streptococcus mutans*. Elsevier, Amsterdam. pp. 111–120 (1986).

6. Koga, K.; Nakahara, K.; Ono, H. Anti-caries activity of catechin oligomer synthesized by peroxidase. *Foods and Food Ingredients Journal* :62 (1993).

7. Hemingway, R.W.; Karchesy, J.J.; McGraw, G.W.; Wielesek, R.A. Heterogeneity of interflavanoid bond location in loblolly pine bark procyanidins. *Phytochemistry* :275 (1983).

8. Sakanaka, S.; Kim, M.; Taniguchi, M.; Yamamoto, T. Antibacterial substances in Japanese tea extract against *Streptococcus mutans*, a cariogenic bacterium. *Agric. Biol. Chem.* :2307 (1989).

9. Nakahara, K.; Kawabata, S.; Ono, H.; Ogura, K.; Tanaka, T.; Ooshima, T.; Hamada, S. Inhibitory effect of oolong tea polyphenols on glucosyltransferases of mutans streptococci. *Applied and Environmental Microbiology* :968 (1993).

10. Hamada, S.; Kontani, M.; Hosono, H.; Ono, H.; Tanaka, T.; Ooshima, T.; Mitsunaga, T.; Abe, I. Peroxidase-catalyzed generation of catechin oligomers that inhibit glucosyltransferase from *Streptococcus sobrinus*. *FEMS Microbiology Letters* :35 (1996).

11. Kakiuchi, N.; Hattori, M.; Nishizawa, M. Inhibitory effect of various tannins on glucan synthesis by glucosyltransferase from *Streptococcus mutans*. *Chem. Pharm. Bull.* :720 (1986).

12. Swain, T.; Hillis, W.E. The phenolic constituents of *Prunus domestica*. *J. Sci. Food Agric.* :63 (1959).

13. Broadhust, R.B.; Jones, W.T. Analysis of condensed tannins using acidified vanillin. *J. Sci. Food Agric.* :788 (1978).

14. Mitsunaga, T.; Doi, T.; Kondo, Y.; Abe, I. Color development of proanthocyanidins in vanillinhydrochloric acid reaction. *J. Wood Sci.* :125 (1998).

15. Samejima, M.; Yoshimoto, T. General aspects of phenolic extractives from coniferous barks. *Mokuzai Gakkaishi* :491 (1981).

16. Funaoka, M., Abe, I. Degradation of protolignin by the nuclear-exchange method. *Mokuzai Gakkaishi* :671 (1985).

17. Abe, I.; Funaoka, M.; Kodama, M. Approaches by the phenyl nucleus-exchange method. *Mokuzai Gakkaishi* :582 (1987).

18. Bate-Smith, E.C. Phytochemistry of proanthocyanidins. *Phytochemistry* :1107 (1975).

19. Brown, B.R.; Shaw, M.R. Reactions of flavanoids and condensed tannins with sulphur nucleophiles. *J. Chem. Soc., Perkin Trans. I* :2036 (1974).

20. Karchesy, J.J.; Hemingway, R.W. Loblolly pine bark polyflavanoids. *J. Agric. Food. Chem.* :222 (1980).

21. Nonaka, G.; Hsu, F.; Nishioka, I. Structures of dimeric, trimeric, and tetrameric procyanidins from *Areca catechu* L. *J. Chem. Soc., Chem. Commun.* :781 (1981).

22. Betts, M.J.; Brown, B.R.; Shaw, M.R. Reaction of flavonoids with mercaptoacetic acid. *J. Chem. Soc. Chem. Commun.* :1178 (1969).

23. Fletcher, A.C.; Porter, L.J.; Haslam, E.; Gupta, R.K. Plant procyanidins. Part 3. Conformational and configurational studies of natural procyanidins. *J. Chem. Soc., Perkin Trans. I* :1628 (1977).

24. Mitsunaga, T.; Abe, I. The phenolation of monomeric compound of condensed tannin by Lewis-Acid catalyst. *Mokuzai Gakkaishi* :585 (1992).

25. Mitsunaga, T.; Abe, I.; Nogami, T. The formation of ring-opened compound from (+)-catechin by BF3 catalyst. *Mokuzai Gakkaishi* :328 (1993).

26. Mitsunaga, T.; Abe, I.; Ohara, S. The chemical structure and stereochemistry of phenol adduct compound derived from (+)-catechin upon phenolation. *Mokuzai Gakkaishi* :100 (1994).

27. Ohara, S.; Suzuki, K.; Ohira, T. Condensed tannins from *Acacia mearnsii* and their biological activities. *Mokuzai Gakkaishi* :1363 (1994).

28. Samejima, M.; Yoshimoto, T. Systematic studies on the stereochemical composition of proanthocyanidins from coniferous bark. *Mokuzai Gakkaishi* :67 (1982).

29. Mitsunaga, T.; Yoshida, A.; Abe, I. Analysis on the pyran-ring of the condensed tannins by the nucleus exchange reaction. *Mokuzai Gakkaishi* :193 (1995).

30. Haslam, E. Polyphenol-protein interactions. *Biochem. J.* :285 (1974).

31. Asano, K.; Ohtsu, K.; Shinagawa, K. Affinity of proanthocyanidins and their oxidation products for haze-forming proteins of beer and the formation of chill haze. *Agric. Biol. Chem.* :1139 (1984).

32. Asano, K.; Hashimoto, N. Characterization of haze-forming proteins of beer and their roles in chill haze formation. *J. Am. Soc. Brewing Chemists* :147 (1982).

33. Siebert, K.J.; Troukhanova, N.V.; Lynn, P.Y. Nature of polyphenol-protein interaction. *J. Agric. Food Chem.* :80 (1996).

ENHANCEMENT OF ANTIMICROBIAL ACTIVITY OF TANNINS AND RELATED COMPOUNDS BY IMMUNE MODULATORY EFFECTS

Herbert Kolodziej,[a] Oliver Kayser,[a] Klaus Peter Latté,[a] and Albrecht F. Kiderlen[b]

[a] Institute of Pharmacy II
Pharmaceutical Biology
Free University Berlin
D-14195 Berlin
GERMANY

[b] Robert Koch-Institute
Nordufer 20
D-13353 Berlin
GERMANY

1. INTRODUCTION

Tannins presumably represent the most ubiquitous group of all plant phenols, being commonly found in a large array of woody and some herbaceous higher plant species.[1] Their presence in significant amounts in the wood, bark, leaves, and fruits of many plants of predominantly woody habit has suggested a crucial role of polyphenols at the ecological and evolutionary level.[2–4] One of their proposed and now well-documented functions includes protecting plants against herbivores such as phytophagous insects and vertebrate herbivores. It should be noted that the impact of plant polyphenols on herbivores ranges from negative to positive interactions, indicating that the effects may be strongly dependent on the individual herbivore, but structure-function relationships have also been assumed to explain the observed contrasting impacts on plant-consuming animals.[5] Recently, the action of polyphenols on herbivore-associated microbes has been suggested to mediate plant-herbivore interactions.[6] The mode of action of tannins in protecting plants against herbivores is generally attributed to polyphenol-protein interactions.[7] Polyphenols are also established as remarkable antimicrobials and potent

Plant Polyphenols 2: Chemistry, Biology, Pharmacology, Ecology, Edited by Gross et al. Kluwer Academic / Plenum Publishers, New York, 1999

inhibitors of viral infections in many ecological systems and in a number of *in vitro* experiments.[8-10] Although there is good evidence for the adaptation of some herbivores and microbes to this constitutive chemical defense,[11] plants may have benefited from accumulating large quantities of polyphenols in their tissues.

The recorded uses of traditional herbal medicines rich in polyphenols as effective antiseptic drugs appear to depend, at least in part, on their astringent action. Thus, antimicrobial activity of polyphenols may be postulated to have potential therapeutic benefits in condition, in which such associated complexation processes can take place. For example, the medical treatment of dental caries or infectious skin diseases can be seen in this light. In the latter, the production of an impervious layer under which the natural healing processes can occur takes place also. Although there is no firm experimental evidence, it is generally assumed that similar direct interactions of these plant metabolites with pathogens or their secreted products come off internally *in vivo*, as concluded from established polyphenol toxicity toward microorganisms in a number of biological experiments.[12] It is reasoned that such interactions may readily occur in the digestive tract, thus providing a rational but simplified explanation for beneficial effects in gastrointestinal upsets such as diarrhoea. Owing to the as yet unanswered key questions regarding absorption and metabolism of polyphenols, similar direct interactions at other biological targets are unlikely. Even if tannins are degraded under the gut conditions as some investigators suggest,[13] the possibly absorbed fragments may only benefit organisms when they represent biologically active metabolites functioning at appropriate sites *in vivo*. Although absorption of polyphenols is still a subject of debate, small phenolic molecules such as flavan-3-ol gallates have been shown to be readily absorbed into rat portal vein.[14,15]

In the course of our study on medicinal plants, we have been investigating medicinally used *Pelargonium* species of therapeutic potential for treatment of infections of the respiratory tract.[16,17] Antimicrobial and immune modulatory activities of phenolic and other constituents have been identified as the underlying active principles. Encouraged by our recent results, we extended our studies to a series of tannins and related compounds. Stimulation of the non-specific immune system by ingested polyphenolic metabolites present in herbal preparations may well provide a scientific basis for claimed remedial effects in infectious conditions.

The work described in this paper evaluates the antimicrobial potential of polyphenols, their possible role as NO-inducer in host defense, and their capabilities to stimulate cytokine production such as tumour necrosis factor (TNF) and interferon (IFN).

2. UMCKALOABO—PHYTOCHEMICAL AND PHARMACOLOGICAL FACETS

The information about the therapeutic use of Umckaloabo in traditional medicine for the cure of infectious respiratory diseases and its present utilization in modern phytotherapy in Europe prompted our studies on the effects of polyphenols and related compounds regarding anti-infectious properties, and may justify a brief illustration of the initiative project.

Botanically, Umckaloabo originates from the South African species *Pelargonium sidoides* DC. and *Pelargonium reniforme* Curt. (Geraniaceae).[18] Initially, only the latter species has been suggested to form the plant source of this traditional medicine,[19] but, according to present evidence, the origin of umckaloabo is strongly associated with the morphologically closely related species *P. sidoides*.[17] Umckaloabo is highly valued by traditional practitioners and the Southern African native population for its curative properties. The roots of *P. sidoides* and *P. reniforme* are used in traditional medicine as an antidiarrheic and as general remedy for treatment of colds and lung infection including tuberculosis. In addition, Umckaloabo is claimed to provide a cure for gastrointestinal disorders, hepatic complaints, dysmenorrhea, and polymenorrhea.[20] The aerial parts of these *Pelargoniums* are employed as a wound-healing agent.

For the purpose of this paper, we will limit our background information to phenolic metabolites. Other compounds are briefly mentioned, where appropriate. In our hands, compositional studies of distinct *Pelargonium* extracts have hitherto led to the characterization of a total of 26 various metabolites from the roots and another 45 compounds from the aerial parts including phenolic acids, hydroxycinnamic acid derivatives, flavan-3-ols, coumarins, flavonoids, phytosterols and lignans (table 1).[20,21] Most of the identified phenolic acids and hydroxycinnamic acid derivatives, representing metabolites that occur widely distributed in angiosperms, have been encountered in relatively low yields. Noteworthy, however, is the presence of gallic acid (**1**) and its methyl ester (**2**) in fairly high amounts, with somewhat higher concentrations in *P. reniforme*.

Known compounds among the flavan-3-ols included the 2,3-*trans* configured monomers (+)-afzelechin, (+)-catechin and (+)-gallocatechin, representing the putative precursors of associated oligomeric and polymeric proanthocyanidins. Amazingly, we have found only traces of condensation products of low molecular weights.

Table 1. Isolated and characterized constituents from *Pelargonium sidoides* and *Pelargonium reniforme*

Class of compounds	P. sidoides		P. reniforme	
	root	Herb[a]	root	herb
Coumarins	8	1	6	2
Phenolic acids	4	4	5	9
Ellagitannins	—	2	—	10
Flavonoids	1	3	—	21
Flavan-3-ols	3	2 (traces)	2	
Hydroxycinnamic acids	2		2	4
Phytosterols	1	—	1	—
Lignans				2

[a] Phytochemical studies of the aerial parts still in progress

A substantial proportion of the proanthocyanidins present in the roots is represented by oligomeric and polymeric forms. These consist of chains of at least seven chain extender units as concluded from gel permeation chromatography, with (+)-catechin and (+)-gallocatechin entities as dominating units. This finding suggested that these polyphenols are less likely candidates of the underlying biological principle related to respiratory tract infections.

On the other hand, the fairly high concentrations of proanthocyanidins (ca 9 percent) occurring in this indigenous medicine may explain the traditional use against gastrointestinal upsets such as diarrhea. Owing to the high degree of polymerization, absorption of these tannins from the digestive tract is highly unlikely, thereby acting as an effective astringent useful in these conditions. For example, the hitherto assumed precipitation of proteins in the epithelial surface of the gut should form a protective layer along the intestinal lumen. Also, some of the beneficial antisecretory effects that tannins exert in these conditions may well follow from their interaction with toxins produced by pathogenic bacteria in the intestine.[22]

As can be seen from table 1, the two *Pelargonium* species express conspicuously distinct tannin variations, with proanthocyanidins and hydrolyzable tannins as representatives for roots and aerial parts, respectively. The presence of the latter in the leaves may similarly explain their employment as wound healing agents in traditional medicine.

The dominating constituents of the traditional medicine Umckaloabo are represented by a wealth of coumarins.[16,21] Noteworthy is the high degree of aromatic functionalization including hydroxyl and methoxyl groups, an oxygenation pattern that is very uncommon in plants, but apparently typical for the genus *Pelargonium*. Apart from the widely distributed disubstituted scopoletin, all the coumarins possess tri- and tetrasubstituted oxygenation patterns on the aromatic nucleus.

The therapeutic efficacy and tolerability of *Pelargonium*-containing phytopharmaceuticals (Umckaloabo®, trademark, ISO-Arzneimittel, Ettlingen, Germany), advanced from the traditional medicine, have been confirmed in several clinical studies.[23–25] For example, a prospective open multicenter study was carried out in the period October 1993–February 1995 in 742 children suffering from acute bronchitis. Patients with acute exacerbations of chronic bronchitis were included in the study. The symptoms of coughing, expectoration, difficulty in breathing, wheezing, and chest pain were assessed on a five-point verbal rating scale running from 0 = absent to 4 = very severe. The total score thus obtained was used to assess the course of the treatment. Presence or absence of general concomitant symptoms such as lack of appetite, vomiting, diarrhea, headache, and fever were also asked about. For all individual symptoms, the response rate (remission/improvement) was over 80 percent (fig. 1). Adverse events were reported in only 13 patients, but reversed without any sequels.

This modern herbal drug has also been found to be extremely effective for the treatment of acute tonsillitis, one of the commonest conditions of children and young adults. In a randomized controlled trial involving 60 children aged between 6 and 10 years with acute tonsillitis, the response rate after 4 days of treatment

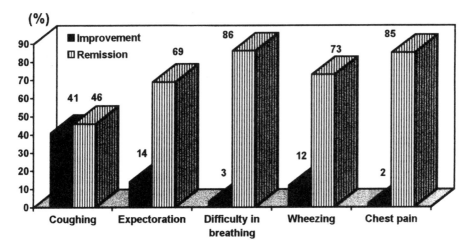

Figure 1. Alteration in respiratory symptoms in 742 children with acute bronchitis after a maximum of 14 days treatment with Umckaloabo®.

with Umckaloabo® was 76 percent compared to that of 30 percent in symptomatic treatment. A similar picture emerged from other recent clinical trials.

With a detailed phytochemical work on the constituents of umckaloabo as a basis, we next turned our attention to pharmacological studies for the identification of the underlying active principle. Following the well-documented therapeutic use of this herbal drug to cure infectious diseases, we first evaluated the antibacterial potential of *Pelargonium* extracts and isolated constituents against a panel of microorganisms including three Gram-positive (*Staphylococcus aureus*, *Steptococcus pneumoniae*, and beta-hemolytic *Streptococcus* 1451) and five Gram-negative bacteria (*Escherichia coli*, *Klebsiella pneumoniae*, *Proteus mirabilis*, *Pseudomona aeruginosa*, *Haemophilus influenzae*), pathogens that are primarily responsible for numerous respiratory tract infections.[26] In general, the tested compounds showed potent to moderate antibacterial activities, with minimum inhibitory concentrations (MICs) ranging from 200–2,000 μg/mL against the panel of microorganisms. As regards the phenolic constituents, gallic acid methyl ester (**2**) represented the most potent candidate with MICs of 250–500 μg/mL against all of the tested bacteria but *Streptococcus pneumoniae* (1,000 μg/mL), followed by gallic acid (**1**) with MICs ranging from 500–2,000 μg/mL, whereas (+)-catechin (**10**) unexpectedly proved to be inactive (>8,000 μg/mL) (table 2).

Clearly, their antibacterial activity is considerably inferior to commercial antibiotics like penicillin G, and the results cannot satisfactorily explain the documented efficacy in the claimed conditions. Respiratory infections are frequently primarily due to viruses including coxsackie-, parainfluenza-, influenza-, echo-, adeno- and rhinoviruses. This has led to the assumption of associated immune stimulatory activities of this herbal medicine. In fact, strong evidence for interferon-like activities and TNF inducing capabilities of *Pelargonium* constituents has subsequently been provided (fig. 2). Prominent cytoprotective effects were

Table 2. Evaluation of antibacterial activity of phenols isolated from Umckaloabo (agar dilution method; MIC values in μg/mL)

Compound	E. coli	Klebs. Pneum.	Staph. aureus	Pseud. aerug.	Prot. mirab.	β-hem. Strept.	Strep. pneum.	Haem. influenz
(1)	2000	1000	500	500	500	2000	2000	2000
(2)	500	250	250	250	250	500	1000	500
(10)	>8000	>8000	>8000	>8000	>8000	>8000	>8000	>8000
Penicillin G	25	5	5	5	6	166	16	16

observed for gallic acid (**1**). At 100 μg/mL, not only complete inhibition of cytopathic effects was noticed, but also proliferation of fibroblasts. In contrast, (+)-catechin (**10**) did not show any significant interferon-like activities. Noteworthy also are the considerable TNF-inducing capabilities of both (**1**) and (**10**) at 5 μg/mL. Higher concentrations resulted in a dramatic decrease of TNF production, as shown for (+)-catechin, or toxic effects on macrophages as observed for gallic acid.

The antimicrobial and immunomodulatory properties of this herbal drug thus provide a sound scientific basis for its outstanding efficacy. Similar combined pharmacological effects may be postulated for other polyphenolic medicinal agents. As gallic acid (**1**) is widely distributed in the plant kingdom and most frequently encountered in plants in ester form such as hydrolyzable tannins, this finding prompted us to extend our studies on antibacterial activities and immune responses of plant constituents to a series of tannins and related compounds.

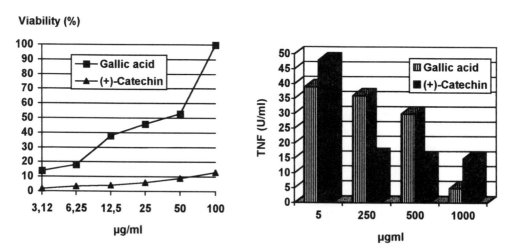

Figure 2. Cytoprotective effects (relative to IFN) of gallic acid and (+)-catechin on macrophages infected with *Leishmania* parasites, and its TNF α induction capability as assessed in the fibroblast/EMC virus test system.

3. ANTIMICROBIAL ACTIVITIES OF TANNINS AND RELATED COMPOUNDS

Tannins and related polyphenols have long been recognized to possess quite potent antibiotic activities.[7] For example, significant antimicrobial activities have been noted for the bark of *Okoubaka aubrevillei*,[27] the roots of *Krameria lappacea*,[28] *Potentilla erecta*,[29] and *Machaerium floribundum*.[30] Noteworthy are also the reported antimicrobial effects of tea extracts and their isolated constituents against a number of pathogens including *Vibrio cholerae*.[31,32] In a recent authoritative review, tannin toxicity for fungi, bacteria, and yeasts is summarized, and some reasonable mechanisms are presented to explain tannin antimicrobial activity.[12] These include inhibition of microbial enzymes, deprivation of substrates, and metal ions required for microbial growth or direct action on bacterial membranes. Tea catechins [(–)-epigallocatechin gallate and (–)-epicatechin] have been reported to damage the lipid bilayer of liposomes and bacterial cell membranes causing leakage of small molecules and aggregation of liposomes and bacterial cells.[33] It has thus been suggested that these polyphenols perturb the lipid bilayers and disrupt the barrier function. Also, the antibacterial activity appears higher against Gram-positive bacteria compared with that against Gram-negative bacteria. The low susceptibility of the latter group of microorganisms can be explained to some extent by the presence of negatively charged lipopolysaccharides, although the tight diffusion barrier of the outer membrane might as well contribute to higher resistance.

Most of the previous antimicrobial studies have only been carried out with crude tannin fractions. The MIC values found in the literature for tannin mixtures may be taken as indication of the antimicrobial potential of distinct polyphenolic plant sources. As regards structure-activity relationships, the information is limited to the oxygenation pattern on the B ring of proanthocyanidins and the presence of galloyl groups in the series of flavan-3-ols.[12] According to present evidence, antimicrobial activity apparently increases with the number of hydroxyl functions on the B ring and galloylation of parent molecules. Comparison between hydrolyzable and condensed tannins suggested similar antimicrobial activities.

Although the series of tannins and related compounds tested for their antimicrobial potencies in this study does not reflect the amazing structural diversity of this group of natural products, each sample represents a chemically defined compound of high purity. Because of the general character of this investigation, the panel of bacteria was limited to those that can be cultivated on standard Mueller-Hinton agar and do not require special growth factors as is the case for *Streptococcus pneumoniae*, β-hemolytic *Streptococcus*, and *Haemophilus influenzae*. The choice of test organisms, however, included two yeasts, *Candida albicans* and *Cryptococcus neoformans*, so that the range of the microorganisms employed is not only limited to bacteria. Minimum inhibitory concentrations (MICs) of the samples were determined for susceptible microorganisms by a standard twofold dilution technique[34] using Mueller-Hinton Broth, with penicillin G (bacteria) and nystatin (fungi) as reference agents. Inocula were prepared in the same medium by diluting microbial suspensions in broth after overnight incubations. Inhibition of growth was judged by comparison with a control culture prepared without any test sample. MIC values were determined as the lowest concentration of the samples completely inhibiting macroscopic growth of microorganisms. In this context, it should be pointed out that extreme care has been paid to the determination of MIC values and that the indicated data reflect the lower antimicrobial activity of respective sequential dilution steps.

As can be seen from table 3, gallic acid methyl ester (2) shows higher antibacterial activity relative to the parent phenol, gallic acid (1), an observation that is consistent with previous reports.[35] This finding may be attributed to the difference in polarity, facilitating the permeation through the cell wall. Comparison of the activity of 3-O-galloyl shikimic acid (4) with its 3,5-di-O-galloyl analog (5) indicated some dependency on the degree of galloylation (fig. 3). However, perusal of the MIC values of all tested galloyl esters clearly showed that neither the number of galloyl groups nor the molecular size, two structural features that have been repeatedly found to be major contributing factors toward pharmacological activities, are apparently significant parameters regarding antimicrobial potency. In addition, the present data suggest greater susceptibility for yeasts when compared to bacteria.

The picture that emerged from examining the antimicrobial potencies of the proanthocyanidins (figs. 4, 5, and 6) clearly indicated only weak to moderate activities (table 4), irrespective of the configuration of both relative 2,3-*cis* and 2,3-*trans* stereochemistry of constituent units and orientation of the interflavanyl linkage. It also appeared that an increase in molecular weights, i.e., dimers (12)–(17) and (21)–(27) vs. the trimer (16) and hexamer (17), at least at these levels, did not

Table 3. Antimicrobial activity of simple galloyl esters and hydrolyzable tannins (broth microdilution method; MIC values in μg/mL)

Compound	*E. coli*	*Klebs. Pneum.*	*Bac. subtilis*	*Staph. aureus*	*Pseud. aerug.*	*Prot. mirab.*	*Cand. albic.*	*Crypt. neofor.*
(1)	2000	1000	500	500	500	500	1000	250
(2)	500	250	500	250	250	250	1000	500
(4)	>2000	>2000	>2000	>2000	>2000	>2000	>2000	>2000
(5)	500	500	500	500	1000	1000	500	500
(6)	500	1000	500	250	1000	1000	500	500
(7)	>2000	>2000	>2000	>2000	>2000	>2000	>2000	>2000
(8)	1000	1000	2000	250	1000	1000	500	250
(9)	1000	2000	1000	1000	1000	1000	500	125
Penicillin G	125	125	125	125	50	16	—	—
Nystatin	—	—	—	—	—	—	16	16

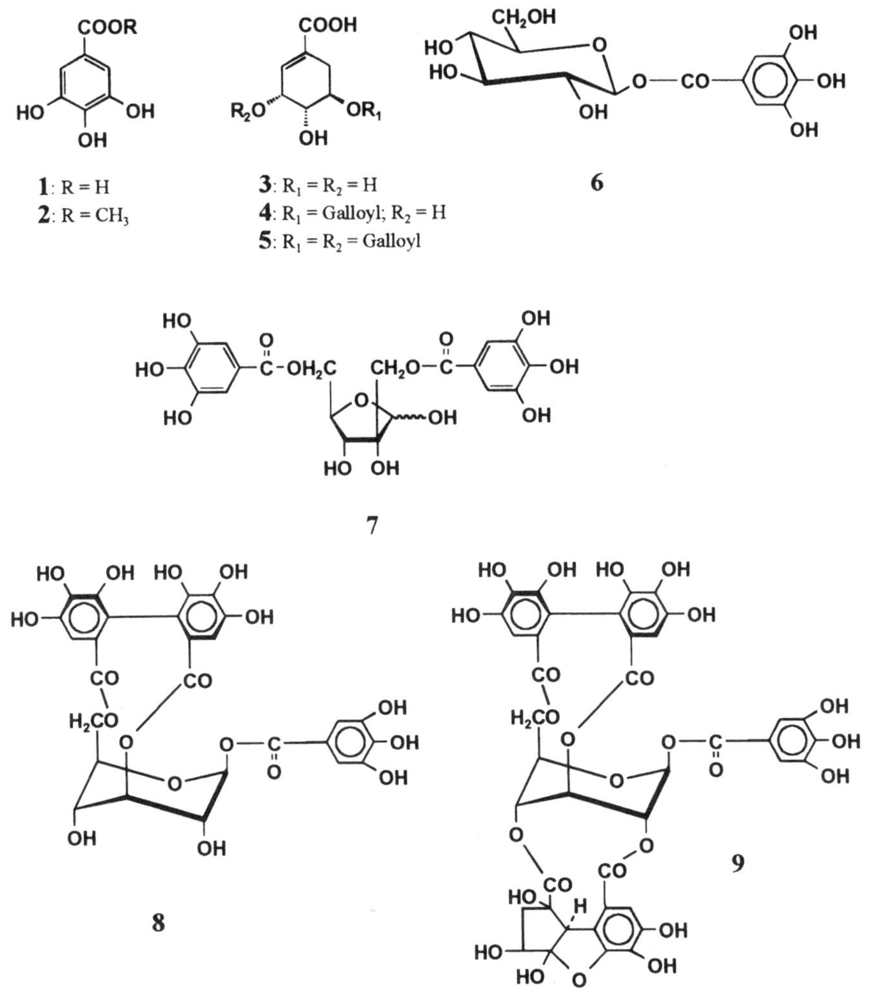

1: R = H
2: R = CH₃

3: R₁ = R₂ = H
4: R₁ = Galloyl; R₂ = H
5: R₁ = R₂ = Galloyl

6

7

8

9

Figure 3. Simple galloyl esters and hydrolyzable tannins.

enhance antimicrobial potency. For the oligomer (**17**), also the presence of galloyl groups at C-3 of flavan-3-ol entities did not confer some tannin toxicity. The similar negligible antimicrobial effects, observed for the doubly-linked dimer A-2 (**18**), indicated that molecular rigidity imposed by the introduction of an additional ether linkage between the two flavanol units does not enhance the antimicrobial properties. With the exception of the dimer epicatechin-4β,8-catechin (**12**) and oligomer (**17**), which proved to be potentially active against both *Candida albicans* and *Cryptococcus neoformans*, no significant differences were observed for the tested polyphenols regarding their antibacterial and antifungal activities.

It appears from this study that the antimicrobial potencies of this group of widely distributed secondary products in the plant kingdom are less pronounced

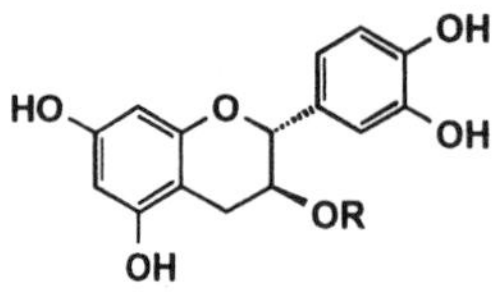

10: R = H
11: R = Galloyl

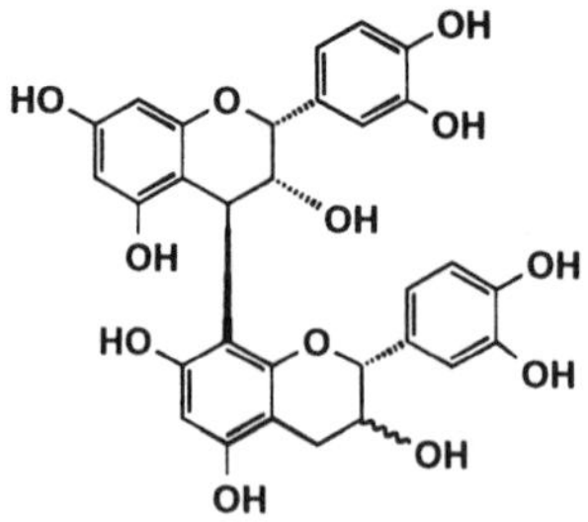

12: ⌇ = |
13: ⌇ = ⋮

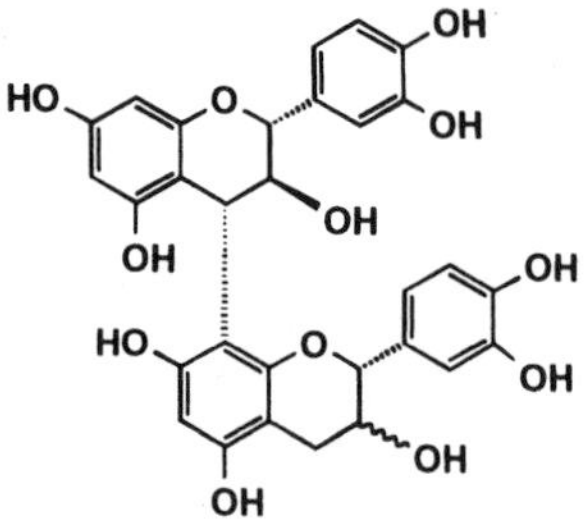

14: ⌇ = |
15: ⌇ = ⋮

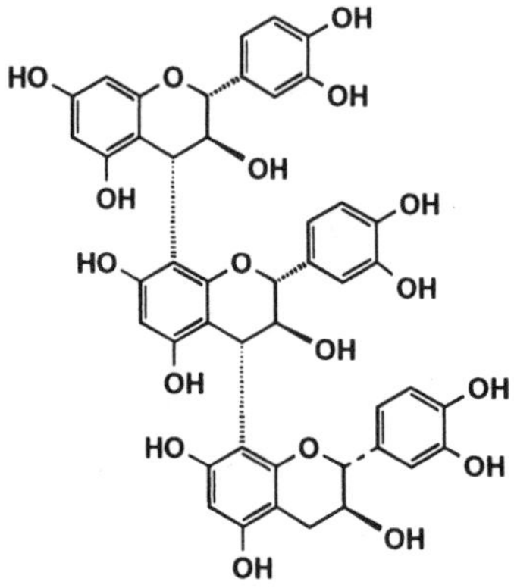

16

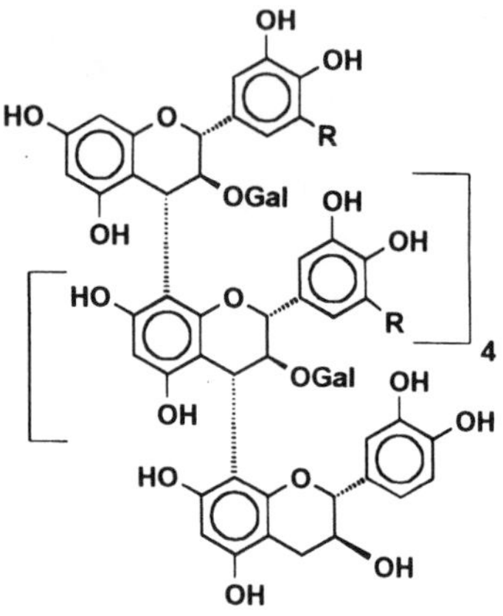

17

Figure 4. Flavan-3-ols and proanthocyanidins.

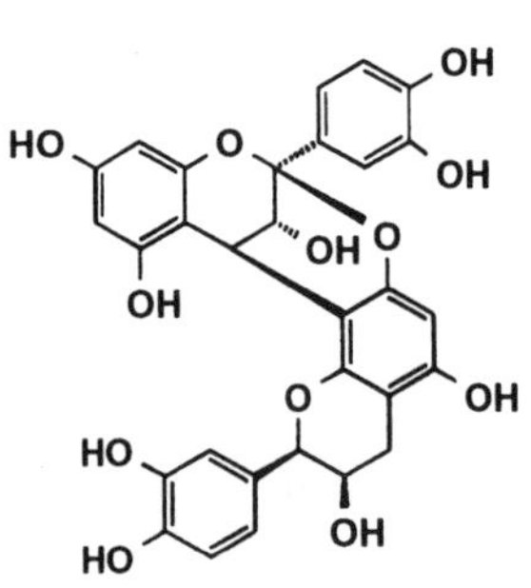

18

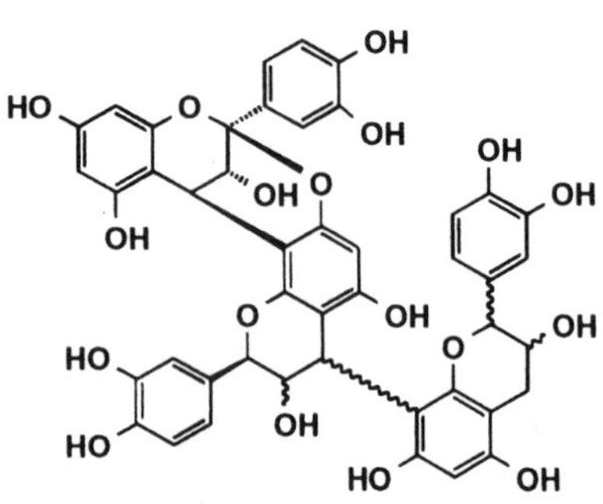

19: ⌇ = |
20: ⌇ = ⋮

Figure 5. A-Type proanthocyanidins.

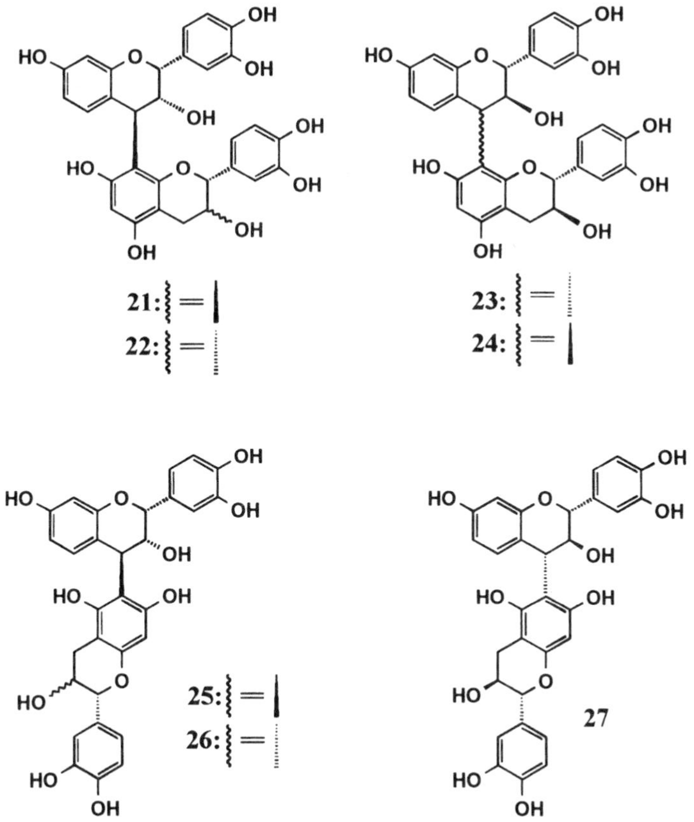

Figure 6. 5-Deoxy proanthocyanidins.

than commonly anticipated, at least within the range of compounds tested and for these microbial species. However, significant differences in antimicrobial activities may well result from a number of parameters including sample purity, use of crude fractions, oxidation state as reflected by quinone intermediates, susceptibility of test microorganisms, and test methods. Nevertheless, polyphenols do have the potential for affecting microorganisms. A plausible explanation of moderate antimicrobial activities of distinct polyphenols, but significantly stronger inhibitory potencies, in ecological systems and in fields of traditional medicine may be concluded from their presence in plants and herbal preparations in large concentrations and from chemical heterogeneity, which could well counterbalance their low toxicity for microorganisms.

4. IMMUNE MODULATORY ACTIVITIES

As regards herbal medicines, the claimed medicinal efficacy in infectious conditions could also be explained by stimulation of the nonspecific immune system

Table 4. Antimicrobial activity of flavan-3-ols and proanthocyanidins (broth microdilution method; MIC values in μg/mL)

Compound	E. coli	Klebs. Pneum	Bac. Subtilis	Staph. aureus	Pseud. aerug.	Prot. mirab.	Cand. albic.	Crypt. neofor.
Flavan-3-ols								
(10)	>8000	>8000	>8000	>8000	>8000	>8000	n.d.	n.d.
(11)	1000	1000	1000	2000	2000	>2000	1000	1000
B-Types								
(12)	1000	1000	1000	1000	1000	2000	500	250
(13)	2000	2000	1000	1000	1000	>2000	1000	500
(14)	2000	2000	1000	1000	500	1000	1000	500
(15)	1000	1000	1000	1000	1000	1000	1000	500
(16)	1000	2000	1000	1000	1000	1000	1000	500
(17)	2000	1000	1000	1000	2000	1000	500	500
A-Types								
(18)	1000	2000	2000	500	1000	2000	2000	2000
5-Deoxy Analogs								
(21)	1000	1000	1000	1000	1000	1000	1000	1000
(22)	1000	1000	1000	1000	1000	2000	1000	1000
(23)	1000	1000	2000	1000	1000	2000	1000	1000
(24)	1000	1000	1000	1000	1000	2000	1000	1000
(25)	1000	1000	1000	1000	1000	1000	1000	1000
(26)	1000	2000	1000	2000	500	2000	1000	500
(27)	1000	2000	1000	2000	1000	2000	1000	1000
Reference Agents								
Penicillin G	125	62	62	62	125	62	—	—
Nystatin	—	—	—	—	—	—	16	16

n.d. = not determined

induced by polyphenolic constituents present in the plant material. Accordingly, herbal drugs possessing the capability to also stimulate phagocytes would be more effective *in vivo* than predicted from results of simple antibiotic *in vitro* tests.

To assess the immunostimulating potential of tannins and related compounds, several bioassays were employed including an *in vitro* infection model with macrophages infected with the obligate intracellular parasite, *Leishmania dono-vani* (NO release),[36,37] a fibroblast/EMC virus test system (IFN production),[20] and a classical cytotoxic TNF assay.[38] Any cytotoxic effects of the samples on macrophages, fibroblasts, and *Leishmania* parasites were determined using the MTT assay.[39]

First experiments have focused on the evaluation of polyphenols on nitric oxide (NO) release in the *Leishmania* infection model, using the Griess assay.[36] Phagocytes, representing an integral part of the immune system, are known to produce reactive oxygen species that have potent antimicrobial activity. For example, the role of NO formed from L-arginine as an effector molecule in macrophage cytotoxicity of a number of microorganisms has been demonstrated.[40] Of the series of polyphenols tested (figs. 3, 4, 5, and 6), only gallic acid (**1**) (EC_{50} 0.71 µg/mL) and its methyl ester (**2**) (EC_{50} 3.71 µg/mL) proved to be potent NO-inducers in this biological system. Compared to the stimulus rIFN-γ/LPS, the inducing effect of these compounds accounted for 36 and 14 percent, respectively. NO production increased in a dose-dependent manner and reached a plateau in the range of 12.5 up to 50 µg/mL. Higher concentrations had an opposite effect by slightly decreasing the intracellular NO production. It should also be noted that toxic effects on macrophages were observed for considerably higher sample concentrations, i.e., >500 µg/mL.

To get a clue regarding the role of the inducible NO-synthase in cytotoxic mechanisms against *Leishmania* parasites in stimulated macrophages, the activity of the enzyme was blocked by addition of the well-known inhibitor $N^{(G)}$-monomethyl-L-arginine (L-NMMA) to the incubations, and the relative number of surviving organisms of *L. donovani* was determined using the MTT assay. Leishmanicidal effects attributable to sample-induced NO-concentrations were only moderate for gallic acid (**1**) and its methyl ester (**2**), as evidenced by similar survival rates of the parasites in experiments with and without NO-inhibitor, whereas all remaining samples were inactive. Thus, this finding clearly indicated that intracellular killing of *Leishmania* amastigotes is primarily due to oxygen-independent cytotoxicity mechanisms. It also supports the previous view that both oxygen-dependent and -independent cytotoxic mechanisms, although to various extents, are involved in this immune response.[41]

The direct toxic effects of the samples on the obligate intracellular parasite, *L. donovani*, were after 72 h tested in culture in the absence of macrophages (table 5). The promastigote forms were incubated with a range of concentrations up to 25 µg/mL of the samples, and their leishmanicidal activity was determined using the MTT assay. As reference agent, sodium stibogluconat (Pentostam®) was assayed under identical experimental conditions. Under the experimental conditions, the tested polyphenols did not show any antiprotozoal effects.

TNF α/β plays an important role in the network of cytokines including the activation of immune cells and is required for host defense against certain pathogens.[42,43] This factor was assayed by a protocol of spectrophotometrically determining the survival rate of actinomycin D pretreated L 929 (TNF α) fibroblasts in a cytotoxic TNF α/β assay.[38] Following an incubation period, the cytopathic effect of the TNF-sensitive L929 cells is reflected by cell lysis. At the host cell subtoxic concentration of 5 µg/mL, the TNF-inducing potential of the polyphenols examined proved to be quite remarkable (tables 5 and 6). Although both hydrolyzable (fig. 3) and condensed tannins (figs. 4, 5, and 6) showed activities ranging from moderate to fairly high potent actions, the former group appeared to be significantly more powerful regarding this particular biological activity.

588   *Kolodziej*

Starting with simple phenols, comparison of the effects of gallic acid (**1**) and its methyl ester (**2**) on the release of TNF from macrophages indicated only moderate stimulatory activities. A similar moderate TNF inducer proved to be shikimic acid (**3**). However, introduction of galloyl groups as reflected in (**4**) and (**5**) dramatically enhanced the amount of the induced cytokine released, indicating a strong dependency of the stimulatory activity on the degree of galloylation. This assumption is further substantiated by similar significant TNF production induced by the digalloylated compound hamamelitannin (**7**). Enhancement in potency similarly applied to structurally related molecules possessing a HHDP moiety derived from two adjacent galloyl groups, as reflected by corilagin (**8**) and phyllantusiin C (**9**). From a closer look on the molecular structures, it appeared that all the potentially active polyphenols have a distinct structural feature in common: the presence of two hydroxy-functionalized benzene elements (galloyl group, HHDP moiety) linked by a spacer assumed to place them in a position suitable for binding either to the TNF receptor or specific kinases involved in this signalling pathway.[44] This hypothesis is presently examined by molecular modelling experiments.

In the series of flavanol-3-ols and proanthocyanidins, the monomer (+)-catechin (**10**) showed similar moderate stimulatory potential as indicated above for simple phenols. Again, monogalloylation resulted in enhanced TNF production (**10** vs. **11**). This finding is consistent with recent reports on IL-1 stimulating activity of epicatechin- and epigallocatechin-3-O-gallate.[45,46]

Table 5. Leishmanicidal activity, cytotoxicity toward bone marrow macrophages (BMM), and TNF induction capability of simple galloyl esters and hydrolyzable tannins [EC_{50} values (μg/mL); sample concentration 5 μg/mL]

Compound	Leishmanicidal Activity		BMM- Toxicity	TNF[a]
	extracellular	Intracellular		
(1)	>25.0	4.4	15.6	0.0256 (39)
(2)	>25.0	12.5	17.8	0.028 (35)
(3)	>25.0	6.7	>25.0	0.0282 (34)
(4)	>25.0	3.7	10.7	0.006 (146)
(5)	>25.0	0.9	>25.0	0.0032 (306)
(6)	>25.0	2.7	>25.0	0.0281 (37)
(7)	>25.0	2.9	>25.0	0.0028 (350)
(8)	>25.0	8.8	8.83	0.00348 (287)
(9)	>25.0	7.53	>25.0	0.0028 (350)
Reference Agents				
Pentostam	2.58	7.84	—	—
IFN/LPS	—	—	—	(184)

[a] U/mL in parentheses

Table 6. Leishmanicidal activity, cytotoxicity toward bone marrow
macrophages (BMM), and TNF induction capability of proanthocyanidins
[EC$_{50}$ values (μg/mL); sample concentration 5 μg/mL]

Compound	Leishmanicidal Activity		BMM- Toxicity	TNF[a]
	Extracellular	**intracellular**		
Flavan-3-ols				
(10)	>25.0	14.6	19.7	0.0208 (48)
(11)	>25.0	4.74	>25.0	0.0039 (256)
B-Types				
(12)	>25.0	4.12	>25.0	0.0057(175)
(13)	>25.0	2.9	>25.0	0.0084 (118)
(14)	>25.0	1.08	>25.0	0.0119 (84)
(15)	>25.0	0.51	>25.0	0.0065 (153)
(16)	>25.0	8.47	>25.0	0.0054 (183)
(17)	>25.0	0.897	>25.0	0.0054 (185)
A-Types				
(18)	>25.0	1.33	>25.0	0.0125 (80)
(19)	>25.0	1.18	>25.0	0.0185 (53)
(20)	>25.0	1.46	>25.0	0.0135 (74)
5-Deoxy Analogs				
(21)	>25.0	1.18	>25.0	0.0074 (134)
(22)	>25.0	3.12	>25.0	0.0156 (64)
(23)	>25.0	1.10	>25.0	0.0156 (64)
(24)	>25.0	0.55	>25.0	0.0080 (125)
(25)	>25.0	3.34	>25.0	0.0156 (64)
(26)	>25.0	3.76	>25.0	0.0050 (200)
(27)	>25.0	0.83	>25.0	0.0059 (167)
Reference Agents				
Pentostam	2.58	7.84	—	—
IFN/LPS	—	—	—	(184)

[a] U/ml in parentheses

Within the group of proanthocyanidins, dimers (**12**)–(**15**), trimer (**16**), and the galloylated oligomer (**17**) showed similar TNF-inducing capability, indicating that an increase in the flavanyl chain length did not enhance TNF production. Noteworthy is the fairly high potency of (4,8)-linked dimeric proanthocyanidins with flavan-3-ol units of 'mixed' stereochemistry (**12, 15** and **21**) relative to those containing either 2,3-*trans* (**14** and **23**) or 2,3-*cis* configurated moieties (**13** and **22**) exclusively. However, comparison between the sets of proanthocyanidins with 'upper' catechin and epicatechin entities suggested that the TNF release showed a tendency to increase with the presence of 2,3-*cis* configurated chain extender units (**12** and **21**), associated with 4β interflavanyl linkages in the molecules. The significance of the orientation of the interflavanyl linkage is reflected by the higher stimulatory activity of fisetinidol-4β,8-catechin (**24**) relative to that of its 4α,8-analog (**23**). Unlike the (4,8)-linked proanthocyanidins, prominent TNF production was observed for (4,6)-coupled 5-deoxy compounds possessing upper and lower flavanyl moieties of the same relative 2,3-configuration (**26, 27**). A further group of compounds examined, possessing doubly linked units, showed much less stimulatory activity. This observation indicated that molecular rigidity imposed by the presence of two interflavanyl linkages is not favorable in this respect.

For the IFN assay,[47] reaction mixtures containing sample-treated L 929 (IFN) fibroblasts were incubated with mouse pathogenic encephalomyocarditis virus (EMCV). Inhibition of the cytopathic effect (CPE) was determined spectrophotometrically, using crystal violet as staining reagent for viable cells and an recombinant murine IFN-γ standard (100 U/mL) as positive control. Following an incubation period, nonprotected cells appeared totally lysed due to viral effects. However, it should be noted that this functional assay does not discriminate between IFN-α, -β, or -γ.

Of the series of polyphenols tested, significant cytoprotective effects were only observed for gallic acid (**1**) and the hydrolyzable tannins corilagin (**8**) and phyllantusiin C (**9**), whereas the remaining samples exhibited only negligible effects at all subtoxic concentrations up to 25 μg/mL (fig. 7). Clearly, this experimental design was less suited to claim interferon-like activities. Thus, in a modified procedure, supernatants of similar, but uninfected, incubations were transferred to sensitive cells infected with mouse pathogenic encephalomyocarditisvirus (EMCV). Addition of albumine prior to the transfer step precluded possible virus-polyphenol interactions by precipitating traces of samples present in the supernatants. Again, gallic acid (**1**), corilagin (**8**), and phyllantusiin C (**9**) provided for effective inhibitory agent of viral effects on fibroblasts, with EC$_{50}$ values of 17.9, 14.4, and 22.4 μg/mL, respectively. These results provide strong evidence for interferon-like activities of distinct hydrolyzable tannins, whereas proanthocyanidins

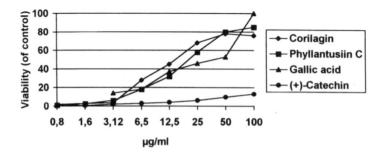

Figure 7. Cytoprotective effects (IFN) of polyphenols as assessed in the fibroblast/EMC virus test system.

appeared to be inactive in this respect. From these limited data, conclusions regarding structure-activity relationships are difficult to make. It may be postulated that the presence of galloyl moieties significantly contributes to inhibition of cytopatic effects. However, there are a few compounds that exhibited only low cytoprotective properties, despite the presence of galloyl groups. On the other hand, gallic acid itself showed significant inhibitory activities of cytopathic effects. The possibility that gallic acid was released from parent derivatives was ruled out by careful examination of incubations for co-occurring traces of this phenol. IFN induction then seems to be correlated with the whole structure of each polyphenolic compound, but distinct structural features may be identified by extending the range of hydrolyzable tannin samples.

5. CONCLUSIONS

For *Pelargonium* species of therapeutic potential for treatment of respiratory tract infections, antibacterial and immune modulatory effects of phenolic and other constituents have been identified as the underlying active principles. This finding encouraged evaluation of a series of tannins and related compounds for their antimicrobial activity, their effects on NO production and cytokine (TNF, IFN) release, using a number of cellular test systems.

Although tannins are generally recognized as quite potent antimicrobial agents, this study indicated that the potential of chemically defined polyphenols to inhibit the growth of microorganisms is apparently only moderate. Effective antimicrobial protection to plants may result from both their accumulation in large amounts in distinct plant tissues and the chemical heterogeneity of the produced polyphenols.

The low tannin toxicity can similarly not satisfactorily explain the claimed efficacy of traditional herbal medicines rich in polyphenols as anti-infectious agents. The indicated parameters, polyphenol concentration and composition, define, at least in part, the quality of such herbal medicines and provide for potential therapeutic benefits in the non-specific medical treatment of infectious conditions such as diarrhea and some skin diseases based upon inhibition of pathogens and their extracellular harmful products. Owing to the lack of precise knowledge of the absorption and metabolism of ingested polyphenols, direct action on pathogens at various internal biological targets *in vivo* appears highly unlikely. On the other hand, the demonstrated stimulation of the nonspecific immune system provides for a scientific basis of claimed remedial effects of herbal medicines rich in polyphenols in infectious conditions. The general picture that is now beginning to emerge is that anti-infectious protection in a human body may be attributed to both direct bactericidal effects, whenever possible, and stimulation of the nonspecific immune system. In this context, some previous reports on the activation of phagocytes, immune cells, and production of interleukine-1 by polyphenols deserve explicit mention.[27,45,46,48–51] However, since several immunological factors are associated with the immune response, further relevant experiments are needed to understand the roles of polyphenols in the stimulation of immune reactions.

ACKNOWLEDGMENTS

The authors are grateful to Prof. Daneel Ferreira, University of Orange Free State, Bloemfontein, South Africa, for the generous gift of 5-deoxy proanthocyanidins.

REFERENCES

1. Porter, L.J. Flavans and proanthocyanidins. *In*: Harborne, J.B. (ed.). The flavonoids: advances in research since 1986. Chapman and Hall, London, p. 23 (1994).

2. Feeny, P. Plant apparency and chemical defense. *Recent Adv. Phytochemistry* 10:1 (1976).

3. Rhoades, D.F.; Cates, R.G. Toward a general theory of plant antiherbivore chemistry. *Recent Adv. Phytochemistry* 10:168 (1976).

4. Zucker, W.V. Tannins: does structure determine function? An ecological perspective. *Am. Nat.* 121:331 (1983).

5. Schultz, J.C.; Hunter, M.D.; Appel, H.M. Antimicrobial activity of polyphenols mediates plant-herbivore interactions. *In*: Hemingway, R.W.; Laks, P.E. (eds.). Plant polyphenols: synthesis, properties, significance. Plenum Press, New York, p. 621 (1992).

6. Haslam, E. Plant polyphenols—vegetable tannins revisited. Cambridge University Press, Cambridge (1989).

7. Okuda, T.; Yoshida, T.; Hatano, T. Hydrolysable tannins and related polyphenols. *Fortschritte Chem. Org. Naturst.* 66:1 (1995).

8. Scholz, E. Pflanzliche gerbstoffe—pharmakologie und toxikologie. *Dtsch. Apotheker Ztg.* 134:3167 (1994).

9. Büechi, S. Antivirale gerbstoffe—pharmakologische und klinische untersuchungen. *Dtsch. Apotheker Ztg.* 138:1265 (1998).

10. Hagerman, A.E.; Butler, L.G. Tannins and lignins. *In*: Rosenthal, G.A.; Berenbaum, M.R. (eds.) Herbivores: their interaction with secondary plant metabolites. Academic Press, New York, p. 355 (1991).

11. Scalbert, A. Antimicrobial properties of tannins. *Phytochemistry* 30:3875 (1991).

12. Clausen, Th.P.; Reichardt, P.B.; Bryant, J.P.; Provenza, F. Condensed tannins in plant defense: a perspective on classical theories. *In*: Hemingway, R.W.; Laks, P.E. (eds.) Plant polyphenols: synthesis, properties, significance. Plenum Press, New York, p. 639 (1992).

13. Okushio, K.; Matsumoto, N.; Kohri, T.; Suzuki, M.; Nanjo, F.; Hara, Y. Absorption of (-)-epigallocatechin gallate into rat portal vein. *Biol. Pharm. Bull.* 18:190 (1995).

14. Okushio, K.; Matsumoto, N.; Kohri, T.; Suzuki, M.; Nanjo, F.; Hara, Y. Absorption of tea catechins into rat portal vein. *Biol. Pharm. Bull.* 19:326 (1996).

15. Kayser, O.; Kolodziej, H. Highly oxygenated coumarins from *Pelargonium sidoides*. *Phytochemistry* 39:1181 (1995).

16. Kolodziej, H.; Kayser, O. *Pelargonium sidoides* DC.—Neueste erkenntnisse zum verständnis des phytotherapeutikums umckaloabo. *Z. Phytother.* 19:141(1998).

17. Kolodziej, H.; Kayser, O.; Gutmann, M. Arzneilich verwendete Pelargonien aus Südafrika. *Dtsch. Apotheker Ztg.* 135:853 (1995).

18. Bladt, S. Umckaloabo-Droge der afrikanischen volksmedizin. *Dtsch. Apotheker Ztg.* 117:1655 (1977).

19. Kayser, O. Phenolische inhaltsstoffe von *Pelargonium sidoides* DC. und untersuchungen zur wirksamkeit der Umcka-Droge (*P. sidoides* DC. und *P. reniforme* CURT.). Shaker Verlag, Aachen (1997).

20. Latté, K.P. Ph. D. Thesis, Berlin (in preparation).

21. Hör, M.; Rimpler, H.; Heinrich, M. Inhibition of intestinal chloride secretion by proantho-cyanidins from *Guazuma ulmifolia*. *Planta Med.* 61:208 (1995).

22. Heil, Ch.; Reitermann, U. Atemwegs-und HNO-infektionen. *Therapiew. Pädiatrie*; 7:523 (1994).

23. Dome, L.; Schuster, R. Umckaloabo—eine phytotherapeutische alternative bei akuter bron-chitis im kindesalter? *Ärztezeitschr. Naturheilv.* 37:216 (1996).

24. Haidvogl, M.; Schuster, R.; Heger, M. Akute bronchitis im kindesalter. *Z. Phytother.* 17:300 (1996).

25. Kayser, O.; Kolodziej, H. Antibacterial activity of extracts and constituents of *Pelargonium sidoides* and *Pelargonium reniforme*. *Planta Med.* 63:508 (1997).

26. Wagner, H.; Kreutzkamp, B.; Jurcic, K. Inhaltsstoffe und pharmakologie der *Okoubaka aubrevillei* rinde. *Planta Med.* 51:404 (1985).

27. Scholz, E.; Rimpler, H. Proanthocyanidins from *Krameria triandra* root. *Planta Med.* 55:379 (1989).

28. Pourrat, A.; Coulet, M.; Pourrat, H. Activities bacteriostatique et agglutinante de complexes tanniques extraits de la tormentille, du Fraisier et de l'Eglantier. *Ann. Pharm.Franc.* 21:55 (1963).

29. Waage, S.K.; Hedin, P.A.; Grimley, E. A biologically active procyanidin from *Machaerium floribundum*. *Phytochemistry* 23:2785 (1984).

30. Toda, M.; Okubo, S.; Ikigai, H.; Shimamura, T. The protective activity of tea against infec-tion by *Vibrio cholerae* O 1. *J. Appl. Bacteriol.* 70:109 (1991).

31. Toda, M.; Okubo, S.; Hiyoshi, R.; Shimamura, T. The bactericidal activity of tea and coffee. *Lett. Appl. Microbiol.* 8:123 (1989).

32. Ikigai, H.; Nakae, T.; Hara, Y.; Shimamura, T. Bactericidal catechins damage the lipid bilayer. *Biochim. Biophys. Acta* 1147:132 (1993).

33. Vanden Berghe, D.A.; Vlietinck, A.J. Screening methods for antibacterial and antiviral agents from higher plants. *In*: Harborne, J.B.; Dey, P.M. (eds.) Methods in plant biochem-istry, Vol. 6. Academic Press, London, p. 47 (1991).

34. Saxena, G.; McCutcheon, A.R.; Farmer, S.; Towers, G.H.N.; Hancock, R.E.W. Antimicrobial constituents of *Rhus glabra*. *J. Ethnopharmacol.* 42:95 (1994).

35. Kiderlen, A.F.; Kaye, P.M. A modified colorimetric assay of macrophage activation for intracellular cytotoxicity against Leishmania parasites. *J. Immunol. Methods* 127:11 (1990).

36. Kayser, O.; Kiderlen, A.F.; Kolodziej, H. Inhibition of luminol-dependent chemiluminescence and NO release by a series of oxygenated coumarins in murine macrophages infected with *Leishmania donovani*. *Pharm. Pharmacol. Lett.* 7:71 (1997).

37. Wagner, H.; Jurcic, K. Assays for immunomodulation and effects on mediators of inflam-mation. *In*: Harborne, J.B.; Dey, P.M. (eds.) Methods in plant biochemistry, Vol. 6. Acade-mic Press, London, p. 195 (1991).

38. Carmichael, J.; DeGraff, W.G.; Gazdar, A.F.; Minna, J.D.; Mitchell, J.B. Evaluation of tetrazolium-based colorimetric assay: assessment of chemosensitivity testing. *Cancer Res.* 47:936 (1987).

39. Nussler, A.K.; Biliar, T.R. Inflammation, immunoregulation, and inducible nitric oxide syn-thase. *J. Leuk. Biol.* 54:171 (1993).

40. Pesanti E.L. Interaction of cytokines and alveolar cells with *Pneumocystis carinii in vitro*. *J. Infect. Dis.* 163:611 (1991).

41. Stuehr, D.J.; Nathan, C.F. A macrophage product responsible for cytostasis and respiratory inhibition in tumor target cells. *J. Exp. Med.* 169:1543 (1989).

42. Rege, A.A.; Huang, K.; Aggarwal, B.B. Tumour necrosis factor. *In:* Galvani, D.W.; Cawley, J.C. (eds.) Cytokine therapy. Cambridge University Press, Cambridge, p. 152 (1992).

43. Mander, Th.; Hill, S.; Highes, A.; Rawlins, Ph.; Clark, Ch.; Gammon, G.; Foxwell, B.; Moore, M. Differential effects on TNF α production by pharmacological agents with varying molecular sites of action. *Int. J. Immunpharmac.* 19:451 (1997).

44. Sakagami, H.; Asano, K.; Hara, Y.; Shimamura, T. Stimulation of human monocyte and polymorphonuclear cell iodination and interleukin-1 production by epigallocatechin gallate. *J. Leukocyte Biol.* 51:478 (1992).

45. Sakagami, H.; Asano, K.; Tanuma, S.-I.; Hatano, T.; Yoshida, T.; Okuda, T. Stimulation of monocyte iodination and IL-1 production by tannins and related compounds. *Anticancer Res.* 12:377 (1992).

46. Marcucci, F.; Klein, B.; Kirchner, M.; Zawatzky, R. Production of high titers of interferon gamma by prestimulated spleen cells. *Eur. J. Immunol.* 12:787 (1982).

47. Daniel, P.T.; Falcioni, F.; Berg, A.U.J.; Berg, P.A. Influence of cianidanol on specific and nonspecific immune mechanisms. *Meth. Find. Exptl. Clin. Pharmacol.* 8:139 (1986).

48. Rauch, G. The immunoenhancing effect of cianidanol on macrophages and on the T-cell system. *Meth. Find. Exptl. Clin. Pharmacol.* 8:147 (1986).

49. Murayama, T.; Kishi, N.; Koshiura, R.; Takagi, K.; Furukawa, T.; Miyamoto, K.-I. Agrimoniin, an antitumor tannin of *Agrimonia pilosa* Ledeb., induces interleukin-1. *Anticancer Res.* 12:1471 (1992).

50. Rohrbach, M.S.; Kreofsky, T.J.; Vuk-Pavlovic, Z.; Lauque, D. Cotton condensed tannin: a potent modulator of alveolar macrophage host-defense function. *In*: Hemingway, R.W.; Laks, P.E. (eds). Plant polyphenols—synthesis, properties, significance. Plenum Press, New York, p. 803 (1992).

INDUCTION OF APOPTOSIS AND ANTI-HIV ACTIVITY BY TANNIN- AND LIGNIN-RELATED SUBSTANCES

Hiroshi Sakagami[a], Kazue Satoh[b], Yoshiteru Ida[b],
Noriko Koyama[a,c], Mariappan Premanathan[d],
Rieko Arakaki[d], Hideki Nakashima[d], Tsutomu
Hatano[e], Takuo Okuda[e], and Takashi Yoshida[e]

[a] Department of Dental Pharmacology
Meikai University School of Dentistry
Sakado, Saitama 350-0283
JAPAN

[b] Analysis Center
School of Pharmaceutical Sciences
Showa University, Hatanodai
Shinagawa-ku, Tokyo 142-8555
JAPAN

[c] Second Department of Oral Surgery
Meikai University School of Dentistry
Sakado, Saitama 350-0283
JAPAN

[d] Department of Microbiology and Immunology
Kagoshima University Dental School
Sakuragaoka, Kagoshima 890-8544
JAPAN

[e] Faculty of Pharmaceutical Sciences
Okayama University
Tsushima, Okayama 700-8530
JAPAN

Plant Polyphenols 2: Chemistry, Biology, Pharmacology, Ecology, Edited by
Gross et al. Kluwer Academic / Plenum Publishers, New York, 1999

1. INTRODUCTION

Tannins and lignins are two major classes of polyphenolic compounds widely distributed in the plant kingdom. Tannins are largely classified into hydrolyzable and condensed tannins.[1] Hydrolyzable tannins have structures in which a polyalcohol (mainly glucose) is esterified with polyphenolic carboxylic acids, such as galloyl, hexahydroxydiphenoyl (HHDP) (a dimer of the galloyl group), valoneoyl (a trimer of the galloyl group), or dehydrohexahydroxydiphenoyl groups (an oxidized metabolite of the HHDP group). Condensed tannins are composed of flavan units, mostly (+)-catechin, (−)-epicatechin, or their analogues, condensed with each other via carbon-carbon bonds.

Lignins are phenylpropenoids formed by oxidative polymerization of p-coumaryl, coniferyl, and sinapyl alcohols, and are amorphous cell wall polymers covalently bound to cellulose and hemicelluloses.[2] We have previously reported diverse biological activities of these polyphenols (table 1).[3,4] We summarize here our comparative studies on the ability of tannins and lignins to induce apoptosis and anti-HIV (human immunodeficiency virus) activity.

Table 1. Comparison of the biological activity of tannins and lignins

	Tannin	Lignin	
		Natural	Synthetic
In vivo:			
Antitumor activity		o	
Antimicrobial activity		o	
Antiparasite activity		o	
In vitro:			
Antiviral activity			
HIV	o	o	o
Influenza virus		o	o
HSV	o	o	
Rotavirus		o	
Enterovirus		o	
Macrophage activation		o	o
Cytokine production (TNF, IL-1)	o	o	o
Stimulation of iodination	o	o	o
Radical generation	o	o	o
Methionine oxidation	o		
H_2O_2 production	o		
O_2^- scavenging activity	o	o	o
OH radical scavenging activity		o	
Stimulation of radical intensity of ascorbate		o	
Stimulation of cytotoxic activity of ascorbate		o	

2. APOPTOSIS INDUCTION

Hydrolyzable tannins induced apoptotic cell death characterized by internucleosomal DNA cleavage and apoptotic body in human promyelocytic leukemic HL-60 cells, as judged by agarose gel electrophoresis and fluorescence activated cell sorter (fig. 1).[5] The apoptosis-inducing activity between monomeric, dimeric, trimeric, and tetrameric hydrolyzable tannins was comparable with each other and was generally higher than that of condensed tannins.[5] Natural lignified materials and synthetic lignins (dehydrogenation polymers of phenylpropenoids) showed much weaker activity (table 2).

The highest activity was detected in gallic acid, a component unit of tannins (table 2).[5,6] Gallic acid increased the oxidation potential (as judged by NO monitor)[7] and stimulated the oxidation of methionine to methionine sulfoxide (as judged by

Table 2. Induction of apoptosis and anti-HIV activity by tannins and lignins

	Apoptosis -inducing activity (EC_{50}: µg/ml)	Anti-HIV activity (SI^a)	ref.
\<Tannins\>			5, 28
Gallic acid (MW 170)	5	<1	
Curcumin (MW 368)	0.5	<1	
Procyanidin B-1 (MW 579)	151	<1	
B-2 digallate (MW 883)	318	<1	
B-3 (MW 579)	579	<1	
B-3 monogallate (MW 731)	307	<1	
C-1 (MW 866)	866	<1	
Hydrolyzable tannins			
Monomers (MW 484-1255)	61 ± 19 (n=3)	1.8 ± 2.8 (n=21)	
Dimers (MW 1571-2282)	65 ± 21 (n=3)	2.3 ± 3.2 (n=39)	
Trimers (MW 2354-2658)	69 ± 45 (n=3)	3.4 ± 3.7 (n=4)	
Tetramers (MW 3140-3745)	69 ± 27 (n=2)	7.3 ± 6.5 (n=3)	
Condensed tannins (MW 290-1764)	129 ± 63 (n=3)	1.1 ± 0.4 (n=8)	
\<Lignins\>			5, 30, 31, 32
Natural lignified materials			
Fr. VI from pine cone extract	516	62	
Fr. VII from pine cone extract	492	4	
Alkali-lignin	488	115	
Lignin sulfonate	1077	124	
Synthetic lignins			
DHP-pCA	243	105	
DHP-FA	364	178	
AZT	161 (µM)	4095	

[a]Selectivity index (SI)= CC_{50}/EC_{50}
Each value represents mean ± S.D. n; number of samples

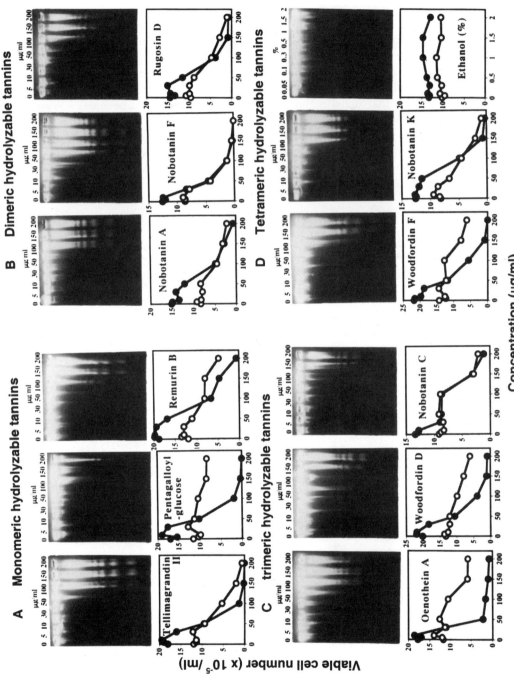

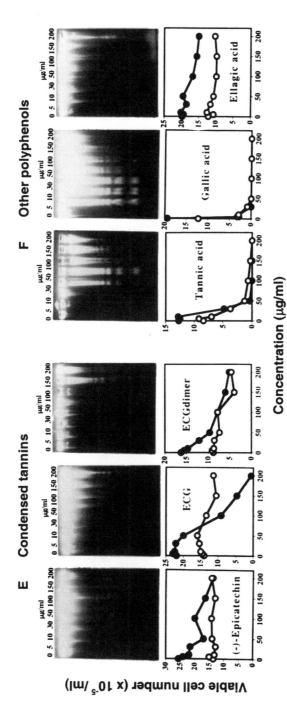

Figure 1. Effect of tannins on induction of cytotoxicity and DNA fragmentation in HL-60 cells. HL-60 cells were treated with each compound for 6 hours (for DNA fragmentation assay) or 8 (■) or 24 hours (●) (for cytotoxicity assay).

amino acid analyzer).[8] Cysteine, N-acetyl cysteine, and glutathione significantly reduced the radical intensity and cytotoxic activity of gallic acid,[9] suggesting the prooxidant action of gallic acid. The apoptosis-inducing activity of gallic acid was also considerably reduced by catalase,[10] suggesting the involvement of H_2O_2 in gallic acid-induced apoptosis. This was supported by our recent finding that gallic acid produced H_2O_2 in an amount necessary for the apoptosis induction (Arakawa and others, unpublished data). More lipophilic curcumin, widely used as a spice and coloring agent in food, induced apoptotic cell death in HL-60 more potently than gallic acid, but the apoptosis-inducing activity of curcumin was not affected by catalase.[10] Furthermore, the radical generation and superoxide anion (O_2^-) scavenging activities of curcumin were much weaker than those of gallic acid.[10] Curcumin might thus induce apoptosis not by the oxygen-mediated mechanism, but rather by modulating the nuclear transcription factor[11] or protein kinases[12,13] after incorporation into the cells.

Gallic acid induced the rapid increase in intracellular Ca^{2+} concentration, and the effect of gallic acid was significantly reduced by removal of Ca^{2+} from the culture medium.[5] This suggests that changes in Ca^{2+} metabolism are involved in gallic acid-induced apoptosis. Gallic acid induced internucleosomal DNA fragmentation in four human myelogenous leukemic cell lines (HL-60, ML-1, U-937, THP-1) (Group I cells), but not in human T-cell leukemia (MOLT-4), erythroleukemia (K-562) and eosinophilic (EOL-1) cell lines (Group II).[5,14] Other apoptosis-inducing agents, such as sodium ascorbate, H_2O_2, etoposide, TNF, hyperthermia, and UV irradiation, also induced internucleosomal DNA fragmentation in Group I cells.[15] This selective DNA fragmentation in Group I cells might be explained by the fact that chromatin DNA in Group I cells was generally more sensitive to digestion by endonucleases, such as micrococcal nuclease and DNase I, as compared with that of Group II cells.[16]

3. ROLE OF RADICALS FOR APOPTOSIS INDUCTION

Both tannins (procyanidins, structures are shown in figure 2) and lignins (either natural, commercial, or synthetic) produced ESR radical signals, and the radical intensity of tannins (fig. 3A) generally exceeded that of lignins (fig. 3C). The radical intensity of these polyphenols was increased exponentially with increasing pH (fig. 3B, D[17,18]).

Gallic acid derivatives that produce radicals (methyl gallate, ethyl gallate, *n*-propyl gallate, isoamyl gallate) induced apoptotic cell death in HL-60 cells, whereas those that do not produce radicals (methyl tri-*O*-methylgallate, tri-*O*-methyl gallic acid, methyl tri-*O*-acetylgallate) were inactive.[7] Similarly, ascorbic acid derivatives that produce radicals (g = 2.0064, hfc = 0.17 mT) (sodium L-ascorbate, L-ascorbic acid, D-isoascorbic acid, sodium 6-β-*O*-galactosyl-L-ascorbate, sodium 5,6-benzylidene-L-ascorbate) induced apoptosis, whereas those that do not produce radicals (L-ascorbic acid-2-phosphate magnesium, L-ascorbic acid 2-sulfate) were inactive.[19] This suggests the possible role of radicals for apoptosis induction.

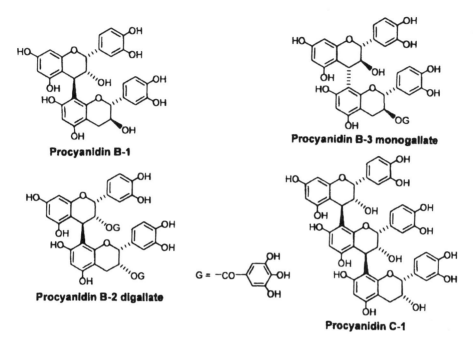

Figure 2. Structure of procyanidins.

4. INTERACTION WITH ASCORBATE

Figure 4A shows that the radical intensity of sodium ascorbate was inhibited or enhanced by lower and higher concentrations of procyanidins. Gallic acid showed a similar dose effect.[17,20] Lower concentrations of sodium ascorbate inhibited both the prooxidant action (Koyama and others, unpublished data) and apoptosis-inducing activity of gallic acid.[17] These data suggest interaction between sodium ascorbate and gallic acid, which might modify their individual biological activity. These findings were unexpected, since both ascorbate and gallate acted as prooxidants, and their cytotoxic activities were reduced by antioxidants (cysteine, N-acetyl cysteine, glutathione).[9] Similar interaction was found between sodium ascorbate and dopamine[21] and between sodium 5,6-benzylidene-L-ascorbate and dopamine.[22] These compounds interfered with each other, reducing their individual radical intensity, prooxidant action (methionine oxidation, H_2O_2 generation), and cytotoxic activity. Studies on the interaction between other redox compounds and polyphenols are underway to generalize these findings.

Natural lignified materials [*Pinus parviflora* Sieb. et Zucc. extract [PC-Fr. VI, PC-Fr. VII),[3,5] *Crataegus cuneata* Sieb et. Zucc. extracts,[23] *Ceriops decandra* (Griff.) Ding Hou extracts,[24] *Acer nikoense* Maxim. extracts[25]], commercial lignins (alkalilignin, lignin sulfonate), and synthetic lignin [dehydrogenation polymers of *p*-coumaric acid (DHP-*p*CA) and that of ferulic acid (DHP-FA)] only slightly reduced the ascorbate radical intensity at lower concentrations, but more potently

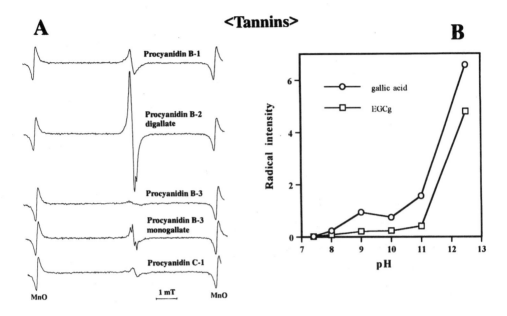

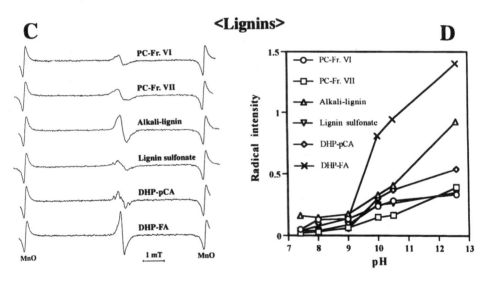

Figure 3. (A, C) ESR spectra of tannins (procyanidins) and lignins (3 mg/mL) in 0.1 M KOH. (B, D) Effect of pH on the radical intensity of procyanidins and lignins. Samples (6 mg/mL) were mixed with equal volume of 0.1 M Tris-HCl buffer (pH 7.4–8.0), 0.1 M NaHCO$_3$/Na$_2$CO$_3$ buffer (pH 9.0–10.5) or in 0.1 M KOH (pH 12.5). After 1 min, the radical intensity was measured by ESR spectroscopy.

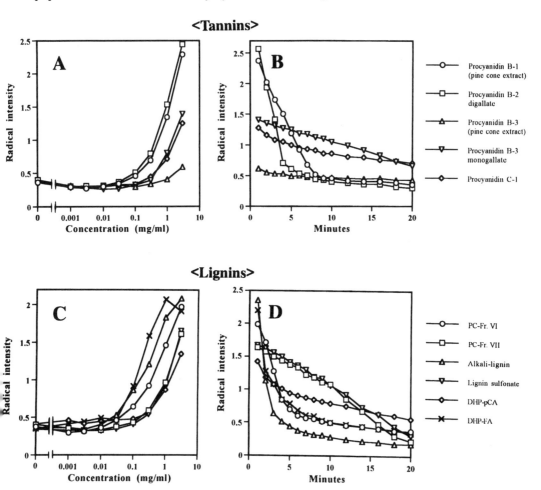

Figure 4. (A, C) Dose effect of procyanidins and lignins on the radical intensity of sodium ascorbate (3 mM) in 0.08 M Tris-HCl, pH 8.0. (B, D) Stability of the ascorbate radical in the presence of 3 mg/mL of each proanthocyanidin lignin.

enhanced the radical intensity at higher concentrations (fig. 4C). It is notable that ascorbate was more unstable in the presence of lignins (fig. 4D) than in the presence of tannins (fig. 4B).

The cytotoxic activity of sodium ascorbate was synergistically enhanced by higher concentrations of natural lignins (PC-Fr. VI, PC-Fr. VII) (figs. 5A, B). Commercial lignins and PSK showed similar and more pronounced synergistic effects with ascorbate,[26] possibly due to the stimulation of hypoxia.[27] On the other hand, synthetic lignins (DHP-pCA, DHP-FA) additively enhanced the cytotoxic activity of sodium ascorbate (figs. 5C, D). This suggests the involvement of the polyphenolic structure integrated with polysaccharides for the synergistic stimulation of ascorbate action.

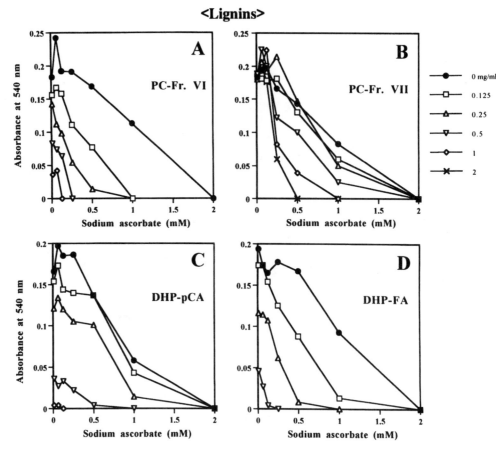

Figure 5. Effect of lignins on the apoptosis-inducing activity of sodium ascorbate. Near confluent human oral squamous carcinoma HSC-2 cells were incubated for 24 h with the indicated concentrations of sodium ascorbate in the presence of 0–2 mg/mL lignin, and the relative viable cell number (absorbance at 540 nm) was determined by crystal violet staining method.

5. O_2^- SCAVENGING ACTIVITY

The O_2^- scavenging activity of tannins was generally higher than that of lignins. Two procyanidins (procyanidin B-1, B-3) isolated from pine cone extract of *Pinus parviflora* Sieb. et Zucc. and their related compounds (procyanidin B-2 digallate, B-3 monogallate and C-1) effectively scavenged the O_2^- generated in the hypoxanthine-xanthine oxidase reaction (fig. 6A). The SOD activity of these substances ranged from 1131–1429 U/mg. Gallic acid showed similar, but slightly lower, SOD activity (571 U/mg) (table 3). On the other hand, the SOD activity of lignins from *Pinus parviflora* Sieb. et Zucc. (PC-Fr. VI, PC-Fr. VII), *Crataegus cuneata* Sieb et. Zucc.,[23] *Ceriops decandra* (Griff.) Ding Hou.,[24] pine seed shell of

Table 3. O_2^- scavenging activity of polyphenols

Substances		SOD activity (unit/mg)	Ref
\<Tannins\>			
Procyanidin B-1		1238	
B-2 digallate		1238	
B-3		1131	
B-3 monogallate		1429	
C-1		1280	
Gallic acid		571	10
\<Lignins\>			
Pinus parviflora Sieb. et Zucc.	Fr. VI	21	
	VII	8	
Crataegus Cuneata Sieb et. Zucc. extracts	Fr. I	44	23
	II	8	
	III	13	
	IV	6	
Ceriops decandra (Griff.) Ding Hou. extracts	Fr. I	53	24
	II	15	
	III	5	
	IV	12	
	V	8	
Pine seed shell extracts of *Pinus parvilora* Sieb. et. Zucc. (SPN)		88	24
Acer nikoense Maxim. extracts	Fr. I	3	25
	II	3	
	III	0.3	
Alkali-lignin (commercial lignin)		7	
Lignin sulfonate (commercial lignin)		6	
DHP-pCA (synthetic lignin)		5	
DHP-FA (synthetic lignin)		24	

Pinus parvilora Sieb. et. Zucc. (SPN),[24] and *Acer nikoense* Maxim.,[25] commercial lignins (alkali-lignin, lignin sulfonate) and synthetic lignin (DHP-pCA, DHP-FA), was one or two orders less (0.3–88 U/mg) (fig. 6B, table 3). Lignins also scavenged the hydroxyl radical, generated by Fenton reaction.[23,25] Lignins and tannins might thus alleviate the pathogenesis induced either by O_2^- or OH radicals.

6. ANTI-HIV ACTIVITY

Although several hydrolyzable tannins inhibited both the cytophatic effect of HIV and the expression of HIV antigen *in vitro*, most of the tannins failed to display significant anti-HIV activity. Lower molecular weight compounds, such as gallic acid, curcumin, and procyanidins (B-1, B-2 digallate, B-3, B-3 monogallate, C-1), showed essentially no anti-HIV activity (table 2). The anti-HIV activity was slightly

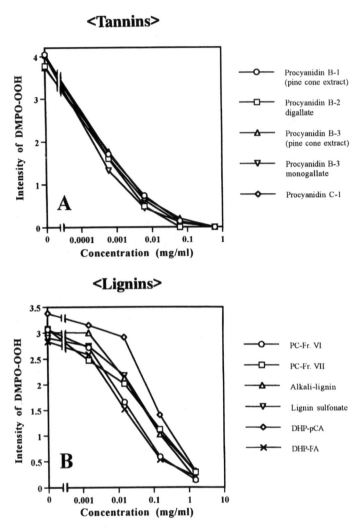

Figure 6. O_2^- scavenging activity of tannins and lignins. Reaction mixtures (200 μL) contain: (A) 2 mM HX in PBS 50 μL; 0.5 mM DETAPAC 20 μL; 5% DMPO (in PBS) 50 μL; sample in DMSO, 30 μL; H₂O, 20 μL; 0.5 U/mL XO (in PBS) 30 μL (for tannins); (B) 2 mM HX 50 μL; 0.5 mM DETAPAC 20 μL; 5% DMPO (in PBS) 50 μL; sample (in PBS) 50 μL; 0.5 U/mL XO (in PBS) 30 μL (for lignins).

increased with the extent of polymerization of the molecule (monomer < dimer < trimer < tetramer). This suggests that anti-HIV activity might at least in part depend on the molecular weight of each compound. The number and the type of functional groups present in the molecule seem to be another determinant of anti-HIV activity. Potent compounds generally contain HHDP and/or the valoneoyl groups, but none has dehydrohexahydroxydiphenoyl, chebuloyl, dehydrodigalloyl, isodehydrodigalloyl, lactonized valoneoyl, hellinoyl, euphorbinoyl,

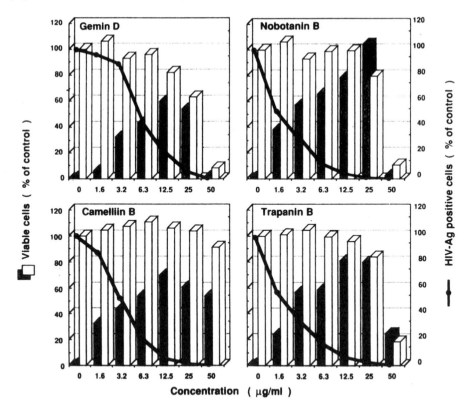

Figure 7. Inhibition by four hydrolyzable tannins of HIV-induced cytopathic effects and HIV-specific antigen expression. Viable HIV-1-infected (black bars) and mock-infected (open bars) MT-4 cells in the presence of test compounds were expressed as percentage of mock-infected controls with no test compound. The number of HIV-1 antigen-positive cells (solid circles), determined by indirect immunofluorescence and laser flow cytometry, was expressed as percentage of virus-infected and compound-free positive control cells (data from ref. 28).

dehydroeuphorbinoyl, or woodfordinoyl groups.[28] The specific stereostructure of HHDP and valoneoyl groups might also affect the anti-HIV activity. Figure 7 shows that several potent hydrolyzable tannins (gemin D, nobotanin B, camelliin B, trapanin B) significantly reduced the HIV-induced cytopathogenic effect and HIV-specific antigen expression. In contrast, condensed tannins did not show any detectable anti-HIV activity (table 2).[28] We have previously shown that oligomeric tannins inhibited the binding of HIV to the target cells.[28] Oligomeric hydrolyzable tannins (nobotanin B, E, K), which have HHDP and valoneic groups, also inhibited glucocorticoid-induced depoly(ADP-ribosyl)ation of HMG14 and 17 and histone H1 and reduced mouse mammary tumor virus (MMTV) mRNA synthesis,[29] whereas condensed tannins were much less effective inhibitors of the poly(ADP-ribose)glycohydrolase activity and suppressors of MMTV transcription. The

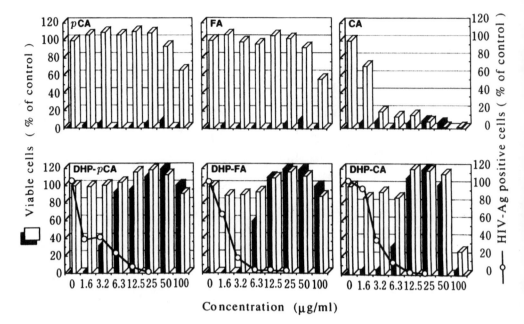

Figure 8. Inhibition of HIV-1-induced cytopathogenic effect and HIV-specific antigen expression by DHPs of phenylpropenoic acid. Symbols are the same as described in figure 7 (data from ref. 32).

possibility should thus be tested that hydrolyzable tannins might inhibit poly(ADP-ribose)glycohydrolase activity and thereby suppress the HIV transcription.

Anti-HIV activity of natural lignified materials (SI = 4–50) was generally higher than that of tannins (SI = 1–15) (table 2).[30] The anti-HIV activity of lignin almost completely disappeared after destruction of the polyphenolic portion with sodium chlorite, but was not affected by breaking the linkages in polysaccharide contaminants with sulfuric or trifluoroacetic acid.[31] Dehydrogenation polymers of *p*-coumaric, ferulic, and caffeic acids showed anti-HIV activity slightly higher than that of natural lignins (SI (CC_{50}/EC_{50}) = 10–100), whereas their respective monomers were inactive (fig. 8).[32] These data suggest the importance of a highly polymerized structure of phenylpropenoids for anti-HIV activity. Lignins also inhibited the binding of HIV to the target cells and poly(ADP-ribose)glycohydrolase activity.[33]

We have previously reported that both tannins and lignins potently stimulated the iodination (incorporation of radioactive iodine into the acid-insoluble fraction) of myeloperoxidase (MPO)-positive cells.[34,35,36,37] MPO-halide-hydrogen peroxidase has been reported to play a significant role in the bactericidal induction.[38,39] We thought this system might also work in the anti-HIV activity induction by polyphenols. However, natural and synthetic lignins effectively inhibited the cytopathic effect of HIV infection in both MPO-positive and MPO-negative cells.[40] Furthermore, HIV infection caused significant reduction of MPO activity. These data suggest that MPO might not be involved in the anti-HIV activity induction by lignins.

7. CONCLUSIONS

We found that tannins produce radicals, scavenge O_2^-, oxidize methionine, increase the oxidation potential, and induce apoptosis more potently than lignins. This indicates the involvement of oxygen-mediated mechanisms for the action of tannins. Water-solubility might also affect the biological activity of each compound: the water-soluble compounds such as gallic acid could be more susceptible to extracellular oxidation, whereas fat-soluble compounds, such as curcumin, might be more easily incorporated into the cells. Further studies are necessary for the identification of second messengers induced after treatment with tannins.

On the other hand, lignins induce anti-HIV activity more potently than tannins. The anti-HIV activity of lignins might not be mediated by oxygen radicals (such as O_2^-, MPO-HOCl), but rather be induced by the direct inactivation of HIV or the inhibition of HIV expression. Furthermore, lignins synergistically enhanced the apoptosis-inducing activity of ascorbate by enhancing the radical generation and decomposition of ascorbate. This suggests the possible application of lignins for cancer therapy.

We have recently found that fractionation of n-BuOH extracts of *Pinus parviflora* Sieb. et Zucc. enriched the anti-HIV activity (SI = 48). Further purification of these substances is underway. The present study may further emphasize the medicinal efficacy of polyphenols, which are abundant in the natural kingdom. Since biological activity of these natural products might be modified by various physiological factors (temperature, pH, metal ions, chelators, redox enzymes, serum proteins, polysaccharides, saliva components, etc.).[41] Interaction between these substances should thus be pursued.

REFERENCES

1. Okuda, T.; Yoshida, T.; Hatano, T. Chemistry and biological activities of tannins in medicinal plants. *In:* Wagner, H.; Farnsworth, N.R. (eds.). Econ. med. plant res. Vol. 5, Academic Press, p. 130 (1991).

2. Kirk, T.M.; Obst, J.R. Lignin determination. *In:* Wood, W.A.; Kelog, S.T. (eds.). Biomass. Methods Enzymol. 160:87 (1988).

3. Sakagami, H.; Kawazoe, Y.; Komatsu, N.; Simpson, A.; Nonoyama, M.; Konno, K.; Yoshida, T.; Kuroiwa, Y.; Tanuma, S. Antitumor, antiviral and immunopotentiating activities of pine cone extracts: Potential medicinal efficacy of natural and synthetic lignin-related materials (Review). *Anticancer Res.* 11:881 (1991).

4. Sakagami, H.; Sakagami, T.; Takeda, M. Antiviral properties of polyphenols. *Polyphenol Actualites* 12:30 (1995).

5. Sakagami, H.; Kuribayashi, N.; Iida, M.; Sakagami, T.; Takeda, M.; Fukuchi, K.; Gomi, K.; Ohata, H.; Momose, K.; Kawazoe, Y.; Hatano, T.; Yoshida, T.; Okuda, T. Induction of DNA fragmentation by tannin- and lignin-related substances. *Anticancer Res.* 15:2121 (1995).

6. Inoue, M.; Suzuki, R.; Koide, T.; Sakaguchi, N.; Ogihara, Y.; Yabu, Y. Antioxidant, gallic acid, induces apoptosis in HL-60RG cells. *Biochem. Biophys. Res. Commun.* 204:898 (1994).

7. Sakagami, H.; Satoh, K.; Hatano, T.; Yoshida, T.; Okuda, T. Possible role of radical intensity and oxidation potential for gallic acid-induced apoptosis. *Anticancer Res.* 17:377 (1997).

8. Sakagami, H.; Satoh, K.; Kadofuku, T.; Takeda, M. Methionine oxidation and apoptosis induction by ascorbate, gallate and hydrogen peroxide. *Anticancer Res.* 17:2565 (1997).

9. Satoh, K.; Sakagami, H. Effect of cysteine, N-acetyl-L-cysteine and glutathione on cytotoxic activity of antioxidants. *Anticancer Res.* 17:2175 (1997).

10. Nogaki, A.; Satoh, K.; Iwasaka, K.; Takano, H.; Takahama, M.; Ida, Y.; Sakagami, H. Radical intensity and cytotoxic activity of curcumin and gallic acid. *Anticancer Res.* 18:3487 (1998).

11. Huang, T.S.; Lee, S.C.; Lin, J.K. Suppression of c-Jun/AP-1 activation by an inhibitor of tumor promotion in mouse fibroblast cells. *Proc. Natl. Acad. Sci., USA* 88:5292 (1991).

12. Liu, J.Y.; Lin, S.J.; Lin, J.K. Inhibitory effects of curcumin on protein kinase C activity induced by 12-*O*-tetradecanoyl-phorbol-13-acetate in NIH 3T3 cells. *Carcinogenesis* 14:857 (1993).

13. Korutla, L.; Kumar, R. Inhibitory effect of curcumin on epidermal growth factor receptor kinase activity in A431 cells. *Biochim. Biophys. Acta.* 1224:597 (1994).

14. Noda, H.; Sakagami, H.; Kokubu, F.; Kurokawa, M.; Tokunaga, H.; Takeda, M.; Adachi, M. Induction of apoptosis in human eosinophilic leukemic cell line (EOL-1). *Int. Arch. Allergy Immunol.* 114 (suppl 1):84 (1997).

15. Yanagisawa-Shiota, F.; Sakagami, H.; Kuribayashi, N.; Iida, M.; Sakagami, T.; Takeda, M. Endonuclease activity and induction of DNA fragmentation in human myelogenous leukemic cell lines. *Anticancer Res.* 15:259 (1995).

16. Kuribayashi, N.; Sakagami, H.; Iida, M.; Takeda, M. Chromatin structure and endonuclease sensitivity in human leukemic cell lines. *Anticancer Res.* 16:1225 (1996).

17. Sakagami, H.; Satoh, K. The interaction between two antioxidants, sodium ascorbate and gallic acid: radical intensity and apoptosis induction. *Anticancer Res.* 16:1231 (1996).

18. Satoh, K.; Ao, Y.; Asano, K.; Sakagami, H. Rapid determination of ascorbate by ESR spectroscopy. *Anticancer Res.* 17:3479 (1997).

19. Sakagami, H.; Satoh, K.; Ohata, H.; Takahashi, H.; Yoshida, H.; Iida, M.; Kuribayashi, N.; Sakagami, T.; Momose, K.; Takeda, M. Relationship between ascorbyl radical intensity and apoptosis-inducing activity. *Anticancer Res.* 16:2635 (1996).

20. Satoh, K.; Sakagami, H. Ascorbyl radical scavenging activity of polyphenols. *Anticancer Res.* 16:2885 (1996).

21. Sakagami, H.; Satoh, K.; Ida, Y.; Hosaka, M.; Arakawa, H.; Maeda, M. Interaction between sodium ascorbate and dopamine. *Free Radic. Biol. Med.* 25:1013 (1998).

22. Satoh, K.; Ida, Y.; Kochi, M.; Sakagami, H. Interaction between sodium 5,6-benzylidene-L-ascorbate and dopamine. *Anticancer Res.* 18:3565–3570 (1999).

23. Satoh, K.; Anzai, S.; Sakagami, H. Enhancement of radical intensity and cytotoxic activity of ascorbate by *Crataegus Cuneata* Sieb et. Zucc. extracts. *Anticancer Res.* 18:2749 (1998).

24. Sakagami, H.; Kashimata, M.; Toguchi, M.; Satoh, K.; Odanaka, Y.; Ida, Y.; Premanathan, M.; Arakaki, R.; Kathiresan, K.; Nakashima, H.; Komatsu, N.; Fujimaki, M.; Yoshihara, M. Radical modulation activity of lignins from a Mangrove plant, *Ceriops decandra* (Griff.) Ding Hou. *In Vivo* 12:327 (1998).

25. Satoh, K.; Anzai, S.; Sakagami, H. Radical scavenging activity of *Acer nikoense* Maxim. extract. *Anticancer Res.* 18:833 (1998).

26. Satoh, K.; Sakagami, H.; Nakamura, K. Enhancement of radical intensity and cytotoxic activity of ascorbate by PSK and lignins. *Anticancer Res.* 16:2981 (1996).

27. Sakagami, H.; Satoh, K.; Aiuchi, T.; Nakaya, K.; Takeda, M. Stimulation of ascorbate-induced hypoxia by lignin. *Anticancer Res.* 17:1213 (1997).

28. Nakashima, H.; Murakami, T.; Yamamoto, N.; Sakagami, H.; Tanuma, S.; Hatano, T.; Yoshida, T.; Okuda, T. Inhibition of human immunodeficiency viral replication by tannins and related compounds. *Antiviral Res.* 18:91 (1992).

29. Tsai, Y.-J.; Aoki, T.; Maruta, H.; Abe, H.; Sakagami, H.; Hatano, T.; Okuda, T.; Tanuma, S. Mouse mammary tumor virus gene expression is suppressed by oligomeric ellagitannins, novel inhibitors of poly(ADP-ribose)glycohydrolase. *J. Biol. Chem.* 267:14436 (1992).

30. Lai, P.K.; Donovan, J.; Takayama, H.; Sakagami, H.; Tanaka, A.; Konno, K.; Nonoyama, M. Modification of human immunodeficiency viral replication by pine cone extracts. *AIDS Res. Human Retroviruses* 6:205 (1990).

31. Lai, P.K.; Oh-hara, T.; Tamura, Y.; Kawazoe, Y.; Konno, K.; Sakagami, H.; Tanaka, A.; Nonoyama, M. Polymeric phenylpropenoids are the active components in the pine cone extract that inhibit the replication of type-1 human HIV in vitro. *J. Gen. Appl. Microbiol.* 38:303 (1992).

32. Nakashima, H.; Murakami, T.; Yamamoto, N.; Naoe, T.; Kawazoe, Y.; Konno, K., Sakagami, H. Lignified materials as medicinal resources. V. Anti-HIV (human immunodeficiency virus) activity of some synthetic lignins. *Chem. Pharm. Bull.* 40:2102 (1992).

33. Tanuma, S.; Tsai, Y.-J.; Sakagami, H.; Konno, K.; Endo, H. Lignin inhibits (ADP-ribose)$_N$ glycohydrolase activity. *Biochem. Int.* 19:1395 (1989).

34. Sakagami, H.; Hatano, T.; Yoshida, T.; Tanuma, S.; Hata, N.; Misawa, Y.; Ishii, N.; Tsutsumi, T.; Okuda, T. Stimulation of granulocytic cell iodination by tannins and related compounds. *Anticancer Res.* 10:1523 (1990).

35. Sakagami, H.; Asano, K.; Tanuma, S.; Hatano, T.; Yoshida, T.; Okuda, T. Stimulation of monocyte iodination and IL-1 production by tannins and related compounds. *Anticancer Res.* 12:377 (1992).

36. Sakagami, H.; Kawazoe, Y.; Oh-hara, T.; Kitajima, K.; Inoue, Y.; Tanuma, S.; Ichikawa, S.; Konno, K. Stimulation of human peripheral blood polymorphonuclear cell iodination by lignin-related substances. *J. Leuk. Biol.* 49:277 (1991).

37. Sakagami, H.; Oh-hara, T.; Kohda, K.; Kawazoe, Y. Lignified materials as a potential medicinal resources. IV. Dehydrogenation polymers of some phenylpropenoids and their capacity to stimulate polymorphonuclear cell iodination. *Chem. Pharm. Bull.* 39:950 (1991).

38. Klebanoff, S.J. Iodination of bacteria, a bactericidal mechanism. *J. Exp. Med.* 126:1063–1077 (1967).

39. Klebanoff, S.J. Myeloperoxidase-halide-hydrogen peroxidase antibacterial system. *J. Bacteriol.* 95:2131–2138 (1968).

40. Kunisada, T.; Sakagami, H.; Takeda, M.; Naoe, T.; Kawazoe, Y.; Ushijima, H.; Muller, W.E.G.; Kitamura, T. Effect of lignins on HIV-induced cytopathogenicity and myeloperoxidase activity in human myelogenous leukemic cell lines. *Anticancer Res.* 12:2225 (1992).

41. Sakagami, H.; Satoh, K. Modulation factors of radical intensity and cytotoxic activity of ascorbate (Review). *Anticancer Res.* 17:3513–3520 (1997).

POLYPHENOLS AND CANCER

ANTITUMOR ACTIVITIES OF ELLAGITANNINS ON TUMOR CELL LINES

Ling-Ling Yang,[a] Ching-Chiung Wang,[a] Kun-Ying Yen,[a] Takashi Yoshida,[b] Tsutomo Hatano,[b] and Takuo Okuda[b]

[a] Graduate Institute of Pharmacognosy Science
Taipei Medical College
Taipei, Taiwan,
R.O.C.

[b] Faculty of Pharmaceutical Sciences
Okayama University
Tsushima, Okayama 700-8530
JAPAN

1. INTRODUCTION

Recent phytochemical studies on tannins and related polyphenols in medicinal plants traditionally used in China, Japan, and other countries have led to the characterization of numerous hydrolyzable tannins with diverse structures and have enabled studies on biological activities of individual tannins of defined structure.[1] The pharmacological properties found include inhibitory effects on various enzymes including reverse transcriptase,[2] histidine decarboxylase,[3] protein kinase C,[4] and lipoxygenase,[5] as well as inhibition of autoxidation of ascorbic acid[6,7] and lipid peroxidation via the free radical scavenging effects.[8-11] Antitumor,[12-18] antiviral,[19-22] and anti-inflammatory activities[23,24] have also been reported for some tannins. The biological activity of these compounds depends on their structure. For example, inhibition of replication of human immunodeficiency virus (HIV) was exhibited only by some oligomeric hydrolyzable tannins.[25] Similarly, strong host-mediated antitumor activity against Sarcoma-180 tumors in mice was exhibited by some ellagitannin oligomers including agrimoniin, oenothein B, and woodfordin C, whereas monomeric ellagitannins and condensed tannins showed negligible activity in the same assay system.[13-15] On the other hand, some tannins exhibited moderate selective cytotoxicity in *in vitro* assays using the PRMI-1951 melanoma cells,[26] although most of the tannins that have been tested were inactive in assay

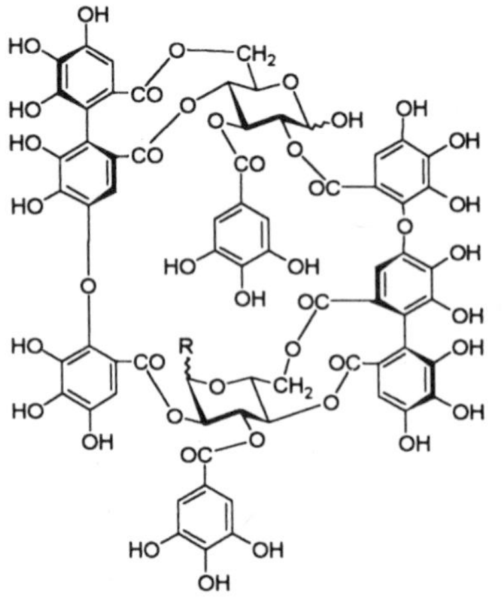

Oenothein B (**1**): R = OH
Woodfordin C (**2**): R = (α)-O-Galloyl

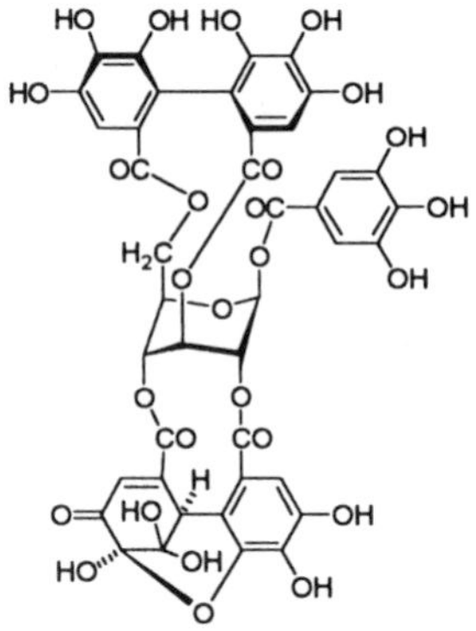

Geraniin (**3**)

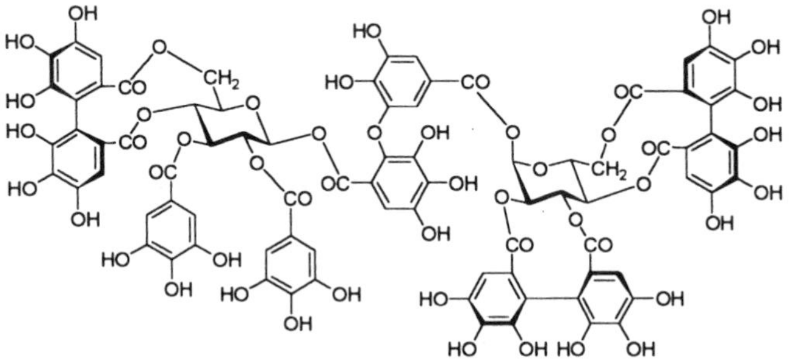

Gemin A (**4**)

Figure 1. Structures of the hydrolyzable tannins oenothein B, woodfordin C, geraniin, and gemin A.

systems using lung carcinoma (A0549), ileocecal adenocarcinoma (HCT-80), epidermoid carcinoma nasopharynx (KB), and medulloblastoma (TE-671) tumor cells.

In our recent study on polyphenols in plants grown in Taiwan,[27-31] we isolated two new dimeric hydrolyzable tannins, cuphiins D_1 and D_2 in addition to the major tannins, oenothein B (fig. 1, **1**), and woodfordin C (fig. 1, **2**), from *Cuphea hyssopifolia* leaves.[32] In this study, we evaluated the cytotoxicity of woodfordin C and related dimeric and monomeric hydrolyzable tannins against six human tumor cell lines (KB, oral epidermoid carcinoma, prostate carcinoma, cervical carcinoma, promyelocytic leukemia, and hepatoma carcinoma) and a human normal cell line (amnion tissue WISH cells). The mechanism for the cytotoxicity of woodfordin C (**2**), which was selectively toxic for tumor cells, was also investigated.

2. MATERIALS AND METHODS

The 34 ellagitannins tested (table 1) were isolated from natural sources.[1,32] Adriamycin, 5-fluorouracil, and MTT (3-[4,5-dimethylthiazol-2-yl]-2,5-iphenyl-tetrazolium bromide) were purchased from Sigma Industries (St. Louis, Missouri, USA).

The human oral epidermoid carcinoma (KB), human normal amnion tissue (WISH), and human promyelocytic leukemia (HL-60) cell lines were obtained from American Type Cell Culture (ATCC) (Rockville, Maryland, USA). The human hepatocellular carcinoma (Hep 3B and HCC 36), human prostate carcinoma (DU-145), and human cervical carcinoma HeLa cell lines were gifts from the Cancer Cell Line Center of Taiwan University.

Seven-week-old male ICR mice (weighing 28–32 g each) were purchased from the National Laboratory Animal and Research Center (Taipei, Taiwan). These animals were pathogen-free and were kept in environmentally controlled quarters (temperature maintained at 24 °C with 12 h light-dark cycle) for at least 1 week before use. All laboratory food pellets were obtained from the National Laboratory Animal and Research Center (Taipei, Taiwan).

The KB, Hep 3B, HCC 36, DU-145, HeLa and WISH cells were maintained in Dulbecco's Modified Eagle Medium (DMEM) supplemented with 10 percent fetal bovine serum, 100 mg/L streptomycin, and 100 IU/mL penicillin. HL-60 and Sarcoma-180 cell lines were maintained in supplemented RPMI-1640. All cell cultures were incubated at 37 °C in a humidified atmosphere of 5 percent CO_2. When the cells became confluent, they were washed with phosphate-buffered saline (PBS), trypsinized with 0.25 percent trypsin in PBS, washed with fresh culture medium, and transferred to 96-well plates (3×10^4 cells/mL) for the following *in vitro* assay.

Stock solutions of tannins ($4 \times 10^4 \, \mu g/mL$) were prepared by dissolving tannins in DMSO. Adriamycin was dissolved in DMSO at a concentration of 4 mg/ml and then stored at 4 °C until use. Serial dilutions of the test compounds were prepared in the culture medium.

For *in vivo* assays, test compounds and 5-fluorouracil were diluted with distilled water to concentrations of 10 mg/mL and 100 mg/mL, respectively. For *in vitro* cytotoxicity assays, cell suspensions were prepared at a concentration of 3×10^4 cells/mL, and 0.2 mL of the suspension was incubated in a 96 well plate for 24 h. After the culture medium was discarded and the cells washed with PBS, 0.2 mL of medium containing tannins at appropriate concentrations were added, and the cultures were then incubated for 48 h. For control experiments, cells were treated with medium containing equivalent concentrations of DMSO, the culture medium was discarded, and the remaining compounds were removed by thorough washing with fresh medium. The number of surviving cells was evaluated by tetrazolium (MTT) assay using a MRX microplate reader at 600 nm. The activity of the mitochondria, reflecting cellular growth and viability, was evaluated by the measurement of the optical density for each well at 600 nm, using a MRX multiwell spectrophotomer (USA).

The Cytotoxicity Index percentage (CI) was calculated according to the following equation:

$$CI = [1 - (T / C)] \times 100,$$

Table 1. Cytotoxicity index (%) of hydrolyzable tannins[a]

Compound[b]	KB	Hep 3B	HCC 36	HeLa	DU-145	WISH
Monomers						
Camelliatannin A	0	20	26	0	37	13
Casuarinin	39	0	15	0	15	15
Chebulinic acid	0	9	3	12	6	0
Corilagin	1	3	50	15	42	10
Gemin D	3	4	14	NT	0	16
Geraniin (3)	10	29	18	17	94	9
Pedunculagin	37	0	12	12	0	10
Remurin A	45	36	10	0	18	21
Strictinin	2	0	29	2	49	0
Tellimagrandin I	30	4	6	0	0	8
Tellimagrandin II	30	22	17	0	0	12
Dimers						
Agrimoniin	38	13	4	19	0	42
Coriariin A	33	40	25	27	0	35
Cornusiin A	33	45	14	18	4	45
Euphorbin A	8	0	6	2	0	13
Euphorbin B	14	0	1	0	0	14
Euphorbin C	35	24	8	0	47	19
Euphorbin D (5)	3	22	23	15	82	15
Euphorbin E	0	3	22	0	16	25
Euphorbin F	26	10	7	0	0	23
Euphorbin G	3	24	27	22	59	14
Gemin A (4)	42	15	9	0	83	20
Hirtellin A	46	14	0	12	0	42
Hirtellin B	26	35	1	54	NT	29
Nobotanin B	32	12	14	17	14	21
Rugosin D	26	34	23	NT	0	34
Rugosin E	26	29	17	14	27	63

[a] n=3.

[b]Concentration of tannin=20 µg/ml.

Table 1. *Continued*

Compound[b]	KB	Hep 3B	HCC 36	HeLa	DU 145	WISH
Macrocyclic dimers						
Hirtellin C	37	42	1	0	3	16
Woodfordin C (**2**)	48	22	25	85	80	34
Oenothein B (**1**)	59	82	51	36	54	79
Oligomers						
Cornusiin C (**7**)	41	0	0	13	70	35
Trapanin A (**6**)	35	4	25	NT	80	22
Woodfordin D	46	21	5	4	0	27
Trapanin B	24	32	22	9	0	23

[a] n=3.

[b] Concentration of tannin=20 µg/ml.

where T and C represent the mean optical densities of the treated group and the vehicle control group, respectively. Based on the dose-response curve for the CI, concentrations of the test compounds giving 50 percent of growth inhibition (IC_{50} value) were determined.

The mouse ascites tumor cell line S-180 was maintained by intraperitoneal passage at weekly intervals in male ICR mice. In experiments with the ascites-type S-180 cells, 1×10^6 cells in 0.2 mL RPMI 1640 were injected subcutaneously with a 26-gauge needle into the abdominal cavity of ICR mice (5 mice per group) on day 0. The injection volumes of the test compounds or 5-fluorouracil were 0.1 mL per 10 g body weight, and the final concentration was 10 mg per kg body weight. Intraperitoneal treatment with the test compounds or 5-fluorouracil was done 3 days before (day –3; pretreatment group) or 1 day after (day 1; post-treatment group) inoculation with tumor cells. For control experiments, distilled water was intraperitoneally injected into mice on day –3 or day 1.

The antitumor effect was defined as the percent increase in life span (ILS) calculated according to the following equation: ILS = [(T – C) / C] × 100% where T and C represent the mean number of survival days of the treated group and the vehicle control group, respectively. Data are presented as the mean ± standard deviation (SD). Student's *t*-test was used for comparison of the number of survival days between the test and control groups.

DNA, RNA, and protein syntheses were measured by the cellular incorporation of ^{3}H-thymidine (NEN, NET-027), ^{3}H-uridine (NEN, NET-174), and ^{3}H-leucine (NEN, NET-460), respectively. HL-60 cells (1×10^5 cells/well) and DU-145 cells (3×10^4 cells/well) were seeded into microtiter plates and then incubated with medium containing various concentrations of woodfordin C for 36 h. The cells were exposed to 2 µCi/mL radioactive precursors during the last 3 h of the incubation.

The incorporation of labeled precursors was stopped by adding 10 percent trichloroacetic acid (Sigma) and refrigerating the plates at 4 °C. After 20 min, the plates were washed with ethanol and dried. The residues were dissolved in 1 percent sodium dodecyl sulfate (SDS, Merck, USA) in 0.3M NaOH at 60 °C for 30 min. The solutions were then counted in 2 mL of Biofluor cocktail (Ecoscint H, National Diagnostics) in a liquid scintillation counter (Beckman, LS-6500). All experiments were performed in triplicate.

After the appropriate treatment, HL-60 cells (2×10^6 cells/well) were harvested by centrifugation and washed with phosphate buffered saline (PBS). The cells were fixed with chilled 80 percent ethanol for 30 min, washed with PBS, and then treated with 0.25 mL of 0.5 percent Triton X-100 solution containing 1 mg/mL RNase A at 37 °C for 1 h, and 0.25 mL of 50 μg/mL propidium iodide was added to the sample for 30 min in the dark. Samples were run through a FACScan (Becton Dickinson). Results are presented as the number of cells versus the amount of DNA indicated by the intensity of fluorescence.

HL-60 cells (2×10^6 cells/well) exposed to woodfordin C for 36 h were collected in tubes and then washed with PBS. The cells were incubated for 10 min in 200 μL of lysis buffer (50 mM Tris-HCl, pH 8.0, 10 mM EDTA, 0.5 percent Sarkosyl) at room temperature and centrifuged at 10,000 × G for 10 min at 4 °C. The supernatant was incubated 3 h at 56 °C with 250 μg/mL proteinase K. Cell lysates were then treated with 2 mg/mL RNase A and incubated at 56 °C for 1.5 h. DNA was extracted with chloroform/phenol/isoamyl alcohol (25:24:1), precipitated from the aqueous phase by centrifugation at 14,000 × G for 30 min at 0 °C. An aliquot (10–20 μL) of this solution was transferred to a 2-percent agarose gel containing 0.5 μg/mL of ethidium bromide, and electrophoresis was carried out at 80 V for 2 h with TBE (×0.5) as running buffer. The DNA in gel was visualized under UV light.

3. RESULTS AND DISCUSSION

The cytotoxicity indices of the tested compounds at 20 μg/mL are shown in table 1. Although the indices for KB cells were below 50 percent for all of the tannins tested except for oenothein B (1), the following tannins exhibited cytotoxic effects at indices greater or equal to 50 percent on several cell lines: corilagin (50 percent) to HCC 36 cells; oenothein B (54 percent) (1), woodfordin C (2) (80 percent), geraniin (fig. 1, 3) (94 percent), gemin A (fig. 1, 4) (83 percent), euphorbin D (fig. 2, 5) (82 percent), trapanin A (fig. 2, 6) (80 percent), cornusiin C (fig. 2, 7) (70 percent), and euphorbin G (59 percent) to DU-145 cells; oenothein B (1) to Hep3B cells (82 percent) and HCC36 cells (51 percent). Woodfordin C (2) with a macrocyclic structure, which exhibited strong cytotoxicity against HeLa cells (85 percent) as well as DU-145 cells, was less toxic to normal human cells (WISH; 34 percent). Similar selective cytotoxicity to tumor cells was observed in **3–6**.

The IC_{50} values of tannins with potent cytotoxic effects were then determined, and the results are shown in tables 2 and 3. Among the tested compounds, woodfordin C (2) was the most selective cytotoxic agent for tumor cells. *In vivo*

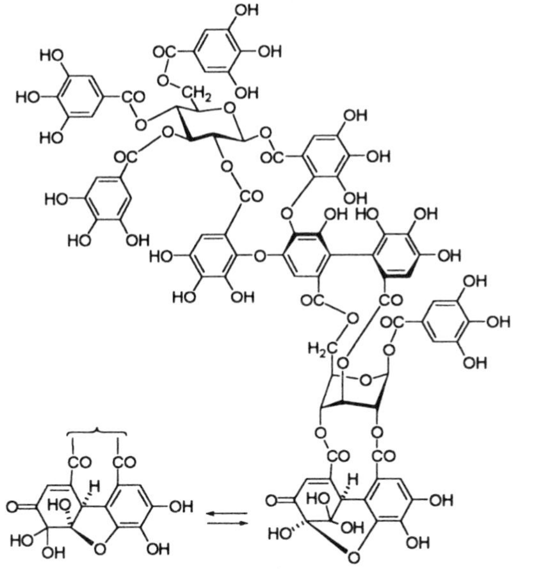

Euphorbin D (5)

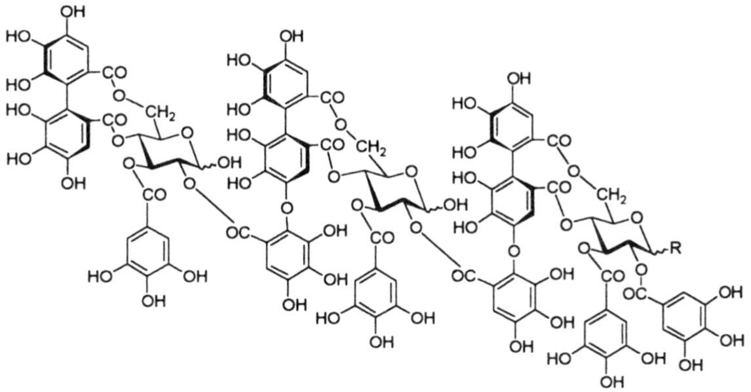

Trapanin A (6): R = (β)-O-Galloyl
Cornusiin C (7): R = OH

Figure 2. Structures of the hydrolyzable tannins euphorbin D, trapanin A, and cornusiin C.

antitumor activity of woodfordin C (**2**) on S-180 tumor-bearing mice is shown in table 4. Upon intraperitoneal injection of **2**, the pretreatment and post-treatment groups survived longer than the control group injected with 5-fluorouracil.

The antitumor mechanism of woodfordin C (**2**) was then analyzed using HL-60 and DU-145 cell lines by comparisons of cell morphology, cell cycle progression, cell metabolism, and DNA fragmentation.

Table 2. The IC_{50} of ellagitannins on DU-145 cell line[a]

Compound	IC_{50} (μg/ml)
Geraniin (**3**)	69.3
Gemin A (**4**)	59.6
Euphorbin D (**5**)	61.9
Euphorbin G	63.9
Oenothein B (**1**)	69.3
Woodfordin C (**2**)	65.0
Cornusiin C (**7**)	51.4

[a] n=3.

Table 3. The IC_{50} of oenothein B and woodfordin C on Hep 3B, HeLa and WISH cell lines[a]

Compound	IC_{50} (μg/ml)		
	Hep 3B	HeLa	WISH
Oenothein B (**1**)	19.0	79.3	67.2
Woodfordin C (**2**)	34.2	34.2	102.5

[a] n=3.

When adherent epithelial cells, DU-145, were treated with woodfordin C (**2**), the cells showed a remarkable number of vacuoles (fig. 3-A2), whereas untreated HL-60 cells exhibited typical non-adherent, fairly round morphology (fig. 3-B1). After incubation with $30\,\mu$g/mL of **2** for 36 h, some of the cells still appeared normal, whereas others exhibited dramatic morphological alteration characteristic of apoptosis (fig. 3-B2). Numerous apoptotic bodies, which were membrane-enclosed vesicles that had bud off cytoplasmic extensions, were also observed. The presence of

Table 4. Antitumor effects of Woodfordin C on S-180 bearing mice[a]

Compound	ILS[b]	
	Pre-treatment	Post-treatment
Woodfordin C (**2**)	34	21
5-Fluorouracil	20	17

[a] n=5.

[b] ILS: The percentage increase in the life span; ILS = $[(T-C)/C] \times 100\%$

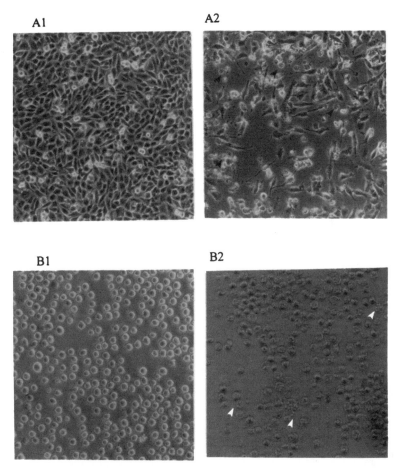

Figure 3. Morphological features of DU-145 (A) and HL-60 (B) cells before and after treatment with woodfordin C (**2**): Cells before treatment (A1, B1); (2) woodfordin C 150 μg/mL for 24 h on DU-145 cells (A2) and 30 μg/mL for 36 h on HL-60 cells (B2). Small arrows indicate the cells in apoptosis and white arrows indicate vacuoles (100×).

apoptotic cells, as well as intact cells that excluded trypan blue dye, suggests that the cells were not necrotic.

The effects of woodfordin C (**2**) on the cell cycle were also investigated. Woodfordin C showed nonspecific inhibitory effects on cellular metabolism and blocked progression of all phases of the cell cycle in DU-145 cells. When the HL-60 cell line was treated with **2** for 12 h, the DNA content frequency histograms showed a Sub-G_1 peak on HL-60 cells. However, this was not observed for DU-145 cells (fig. 4).

In order to analyze the cytotoxic action of woodfordin C on cell metabolism, the HL-60 cell line was cultured in the presence of woodfordin C for 36 h, and uptakes of [^{3}H]thymidine, [^{3}H]uridine and [^{3}H]leucine by the cells were measured (figs. 4

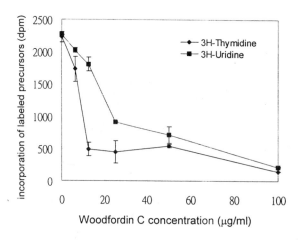

Figure 4. The uptake of [³H]thymidine and [³H]uridine in HL-60 cells treated with woodfordin C (**2**).

and 5). As shown in fig. 5, woodfordin C (**2**) inhibited DNA, RNA biosyntheses dose dependently, but gave no effect on protein biosynthesis.

When HL-60 cells were treated with 15 to 60 μg/mL of **2** for 36 h, increased DNA fragmentation was observed, and a typical DNA fragment ladder on the agarose gel electrophoresis is shown in figure 6.

4. CONCLUSIONS

Cytotoxic effects of 34 ellagitannins including 23 oligomeric tannins on six human tumor cell lines (KB, HeLa, HCC 36, Hep 3B, HL-60 and DU-145) and human amnion normal tissue cell line (WISH) were evaluated by the MTT stain method. Although cytotoxicity of ellagitannins at a concentration of 20 μg/mL depended largely on the nature of cell lines and on structure of each tannin, considerable amounts of dimers including woodfordin C (**2**), geraniin (**3**), gemin A (**4**), euphorbin D (**5**), trapanin A (**6**), and euphorbin G exhibited over 50 percent cytotoxicity to DU-145 cells, whereas they were less toxic to normal human cells.

Woodfordin C (**2**), a dimeric ellagitannin with a macrocyclic structure, exhibited significant cytotoxic effects on DU-145 and HL-60 cells, and this tannin also exhibited antitumor effects on S-180 tumor-bearing mice (ILS 34 percent). Investigations of the cytotoxic mechanism of woodfordin C (**2**) revealed a strong inhibitory effect on DNA synthesis rather than on RNA and protein syntheses in DU-145 and HL-60 cell lines. The cell cycle inhibition of **2** was analyzed using fluorescence-activated cell-sorting analysis, and treatment with woodfordin C for 12 h caused accumulation of HL-60 cells in the sub-G_1 stage as well as DNA fragmentation, although DNA fragmentation was not observed in the DU-145 cell line. These findings suggest that woodfordin C (**2**) induces apoptosis in HL-60 cells and necrosis in DU-145 cells. As woodfordin C was reported to exhibit antitumorgenic properties including inhibitory effect on DNA-dependent topoisomerase-II[33] and

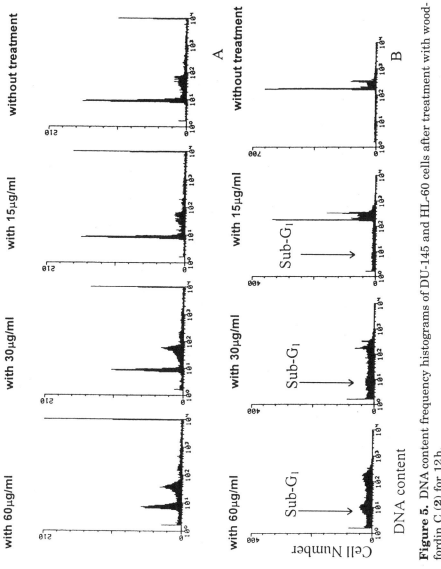

Figure 5. DNA content frequency histograms of DU-145 and HL-60 cells after treatment with woodfordin C (**2**) for 12h.

Figure 6. Effect of woodfordin C (**2**) on induction of DNA fragmentation in HL-60 cells after treatment for 36 h. Lane 1, control: Lane 2, 60 μg/mL: Lane 3, 30 μg/mL: Lane 4, 15 μg/mL.

host-mediated antitumor activity,[13] this polyphenol might be a promising candidate for further *in vivo* study as an antitumor agent.

REFERENCES

1. Okuda, T.; Yoshida, T.; Hatano, T. Hydrolyzable tannins and related polyphenols. *Prog. Chem. Org. Nat. Prod.* 66:1 (1995).

2. Kakiuchi, N.; Hattori, M.; Namba, T.; Nishizawa, M.; Yamagishi, T.; Okuda, T. Inhibitory effect of tannins on reverse transcriptase from RNA tumor virus. *J. Nat. Prod.* 48:614 (1985).

3. Kakiuchi, N.; Hattori, M.; Nishizawa, M.; Yamagishi, T.; Okuda, T.; Namba, T. Studies on dental caries prevention by traditional medicines. VIII. Inhibitory effect of various tannins on glucan synthesis by glucosyltransferase from *Streptococcus mutans. Chem. Pharm. Bull.* 34:720 (1986).

4. Ginsburg, I.; Mitra, R.S.; Gibbs, D.F.; Varani, J.; Kohen, R. Killing of endothelial cells and release of arachidonic acid. Synergistic effects among hydrogen peroxide, membrane-damaging agents, cationic substances, and proteinases and their modulation by inhibitors. *Inflammation.* 17:295 (1993).

5. Goswami, S.K.; Kinsella, J.E. Inhibitory effects of tannic acid and benzophenone on soybean lipoxygenase and ram seminal vesicle cyclooxygenase. *Prostaglandins Leukotrienes and Medicine.* 17:223 (1985).

6. Yoshida, T.; Koyama, S.; Okuda, T. Effects of the interaction of tannins with Co-existing substances. I. Inhibitory effects of tannins on cupric ion-catalyzed autoxidation of ascorbic acid. *Yakugaku Zasshi.* 101:695 (1981).

7. Fujita, Y.; Komagoe, K.; Sasaki, Y.; Uehara, I.; Okuda, T.; Yoshida, T. Studies on inhibition mechanism of autoxidation by tannins and flavonoids. I. Inhibition mechanism of tannins on Cu(II)-catalyzed autoxidation of ascorbic acid. *Yakugaku Zasshi.* 107:17 (1987).

8. Fujita, Y.; Komagoe, K.; Uehara, I.; Morimoto, Y.; Hara, R. Studies on inhibition mechanism of autoxidation by tannins and flavonoids. IV. Acceleration mechanism of methyl linoleate peroxidation induced by the presence of tannin and Cu(II) ion. *Yakugaku Zasshi.* 108:625 (1988).

9. Okuda, T.; Kimura, Y.; Yoshida, T.; Hatano, T.; Okuda, H.; Arichi, S. Studies on the activities of tannins and related compounds from medicinal plants and drugs. I. Inhibitory effects on lipid peroxidation in mitochondria and microsomes of liver. *Chem. Pharm. Bull.* 31:1625 (1983).

10. Kimura, Y.; Okuda, H.; Okuda, T.; Yoshida, T.; Hatano, T.; Arichi, S. Studies on the activities of tannins and related compounds of medicinal plants and drugs. II. Effects of various tannins and related compounds on adrenaline-induced lipolysis in fat cells. (1). *Chem. Pharm. Bull.* 31:2497 (1983).

11. Kimura, Y.; Okuda, H.; Mori, K.; Okuda, T.; Arichi, S. Studies on the activities of tannins and related compounds from medicinal plants and drugs. IV. Effect of various extracts of *Geranii herba* and geraniin on liver injury and lipid metabolism in rats fed peroxidized oil. *Chem. Pharm. Bull.* 32:1866 (1984).

12. Miyamoto, K.; Kishi, K.; Koshiura, R.; Yoshida, T.; Hatano, T.; Okuda, T. Relationship between the structures and the antitumor activities of tannins. *Chem. Pharm. Bull.* 35:814 (1987).

13. Miyamoto, K.; Nomura, M.; Murayama, T.; Furukawa, T.; Hatano, T.; Yoshida, T.; Koshiura, R.; Okuda, T. Antitumor activities of ellagitannins against sarcoma-180 in mice. *Biol. Pharm. Bull.* 16:379 (1993).

14. Miyamoto, K.; Nomura, M.; Sasakura, M.; Matsui, E.; Koshiura, R.; Murayama, T.; Furukawa, T.; Hatano, T.; Yoshida, T.; Okuda, T. Antitumor activity of oenothein B, a unique macrocyclic ellagitannin. *Japanese Journal of Cancer Research.* 84:99 (1993).

15. Murayama, T.; Kishi, N.; Koshiura, R.; Takagi, K.; Furukawa, T.; Miyamoto, K. N Agrimoniin, an antitumor tannin of *Agrimonia pilosa* Ledeb., induces interleukin-1. *Anticancer Research.* 12:1471 (1992).

16. Gali, H.U.; Perchellet, E.M.; Klish, D.S.; Johnson, J.M.; Perchellet, J.P. Antitumor-promoting activities of hydrolyzable tannins in mouse skin. *Carcinogenesis.* 13:715 (1992).

17. Yoshida, T.; Chou, T.; Nitta, A.; Miyamoto, K.; Koshiura, R.; Okuda, T. Woodfordin C, a macro-ring hydrolyzable tannin dimer with antitumor activity, and accompanying dimers from *Woodfordia fruticosa* flowers. *Chem. Pharm. Bull.* 38:1211 (1990).

18. Miyamoto, K.; Kishi, N.; Koshiura, R. Antitumor effect of agrimoniin, a tannin of *Agrimonia pilosa* Ledeb., on transplantable rodent tumors. *Jpn. J. Pharmacol.* 43:187 (1987).

19. Kakiuchi, N.; Wang, X.; Hattori, N.; Okuda, T.; Namba, T. Circular dichroism studies on the ellagitannins-nucleic acids interaction. *Chem. Pharm. Bull.* 35:2875 (1987).

20. Takechi, M.; Tanaka, Y.; Takehara, M.; Nonaka, G.; Nishioka, I. Structure and antiherpetic activity among the tannins. *Phytochemistry.* 24:2245 (1985).

21. Fukuchi, K.; Sakagami, H.; Okuda, T.; Hatano, T.; Tamura, S.; Kitajima, K.; Inoue, Y.; Inoue, S.; Ichikawa, S.; Nonoyama, M.; Konno, K. Inhibition of herpes simplex virus infection by tannins and related compounds. *Antiviral Research.* 11:285 (1989).

22. Hatano, T.; Yasuhara, T.; Miyamoto, K.; Okuda, T. Anti-human immunodeficiency virus phenolics from licorice. *Chem. Pharm. Bull.* 36:2286 (1988).

23. Wagner, H. Search for new plant constituents with potential antiphlogistic and antiallergic activity. *Planta Medica.* 55:235 (1989).

24. Kimura, Y.; Okuda, H.; Okuda, T.; Hatano, T.; Agata, I.; Arichi, S. Studies on the activities of tannins and related compounds from medicinal plants and drugs. VI. Inhibitory effects of caffeoylquinic acids on histamine release from rat peritoneal mast cells. *Chem. Pharm. Bull.* 33:690 (1985).

25. Nakashima, H.; Murakami, T.; Yamamoto, N.; Sakagami, H.; Tanuma, S.; Hatano, T.; Okuda T. Inhibition of human immunodeficiency viral replication by tannins and related compounds. *Antiviral Research.* 18:91 (1992).

26. Kashiwada, Y.; Nonaka, G.I.; Nishioka, I.; Chang, J.J.; Lee, K.H. Antitumor agents 129. Tannins and related compounds as selective cytotoxic agents. *J. Nat. Prod.* 55:1033 (1992).

27. Yoshida, T.; Namba, O.; Lu, C.F.; Yang, L.L.; Yen, K.Y.; Okuda, T. Tannins of euphorbiaceous plants. X. Antidesmin A, a new dimeric hydrolyzable tannin from *Antidesma pentandrum* var. *barbatum. Chem. Pharm. Bull.* 40:338 (1992).

28. Chang, C.W.; Yang, L.L.; Yen, K.Y.; Hatano, T.; Yoshida, T.; Okuda, T. New γ-pyrone gluco-side, and dimeric ellagitannins from *Gordonia axillaris*. *Chem. Pharm. Bull.* 43:1922 (1994).

29. Yoshida, T.; Nakazawa, T.; Hatano, T.; Yang, R.C.; Yang, L.L.; Yen, K.Y.; Okuda, T. Dimeric hydrolysable tannin from *Camellia oleifera*. *Phytochemistry*. 37:241 (1994).

30. Chen, L.G.; Yang, L.L.; Yen, K.Y.; Hatano, T.; Yoshida, T.; Okuda, T. Tannins of euphorbia-ceous plants. XIII. New hydrolyzable tannins having phloroglucinol residue from *Glochidion rubrum* Blume. *Chem. Pharm. Bull.* 43:2088 (1995).

31. Lee, M.H.; Yang, L.L.; Yen, K.Y.; Hatano, T.; Yoshida, T.; Okuda, T. Two macrocyclic hydrolysable tannin dimers from *Eugenia uniflora*. *Phytochemistry*. 44:1343 (1997).

32. Chen, L.G.; Yang, L.L.; Yen, K.Y.; Hatano, T.; Okuda, T.; Yoshida, T. Macrocyclic ellagitan-nin dimers, cuphiin D_1 and D_2, and accompanying tannins from *Cuphea hyssopifolia*. *Phytochemistry*. 50:307 (1999).

33. Motegi, A. K.; Kuramochi, H.; Kobayashi, F.; Ekimoto, H.; Takahashi, T.; Kadota, S.; Takamori, Y.; Kikuchi, T. Woodforticosin (Woodfordin C), a new inhibitor of DNA topoiso-merase II. *Biochem. Pharmacol.* 44:1961 (1992).

CHEMICAL CONSTITUENTS OF MAINLY ACTIVE COMPONENT FRACTIONATED FROM THE AQUEOUS TEA NON-DIALYSATES, AN ANTITUMOR PROMOTER

Yoshiyuki Nakamura,[a] Michiaki Matsuda,[a] Takeshi Honma,[a] Isao Tomita,[a] Naomi Shibata,[b] Tsutomu Warashina,[b] Tadataka Noro,[b] and Yukihiko Hara[c]

[a] Laboratory of Health Science
School of Pharmaceutical Sciences
University of Shizuoka
52-1 Yada, Shizuoka-shi 422-8526
JAPAN

[b] Graduate School of Nutritional and Environmental Sciences
University of Shizuoka
52-1 Yada, Shizuoka-shi 422-8526
JAPAN

[c] Food Research Laboratories
Mitsui Norin Co., Ltd.
223-1 Miyahara, Fujieda-shi 426-0133
JAPAN

1. INTRODUCTION

Much attention has been focused on the primary prevention of cancer using functional ingredients of edible plants.[1] The components of tea leaves (*Camellia sinensis*), particularly green tea catechins, are known to be potent cancer chemopreventive agents. Phase I study for their clinical application is now going on in the United States. Tea catechins represented by (–)-epigallocatechin gallate (EGCG) have potential to inhibit/suppress not only the initiation and promotion of cancer but also to cause regression of tumorigenic cells and to inhibit the metastasis of cancer cells.[2–6] Here, we introduce a new antitumor promoting component, tea aqueous non-dialysates (TNDs), as a candidate for the primary prevention of

Plant Polyphenols 2: Chemistry, Biology, Pharmacology, Ecology, Edited by
Gross et al. Kluwer Academic / Plenum Publishers, New York, 1999

cancer,[7-9] although only tea catechins have been used in many studies on the anti-cancer effects of tea leaves. In this paper, we describe the properties of antitumor promoting effects and the identification of principally active component of TNDs and its chemical constituents.

2. PREPARATION OF TNDs

Green tea was kindly provided by the Tea Council of the Shizuoka Prefecture. Black tea was obtained from Mitsui Norin Co., Japan. Oolong tea was a gift from Ito-en Co., Japan. Pu-erh tea, a mold-fermented tea, came from Yunnan Province, China. TNDs were available from various kinds of tea leaves, such as green, black, oolong, and Pu-erh teas. Crude tea extracts were prepared by extraction with 500 mL of boiling water per 50 g of tea leaves for 10 min and were lyophilized.[7] TNDs were obtained by dialysis of the aqueous fraction remaining after sequential frac-tionation of the crude tea extract with chloroform ($CHCl_3$), ethyl acetate (EtOAc), and n-butanol (BuOH) as shown in figure 1. They are light brown or brown colored powders with yields of 0.42, 2.01, 1.78, and 4.52 percent based on the dry weight of green, black, oolong, and Pu-erh teas. As shown in table 1, they are named as green tea non-dialysate (GTND), black tea non-dialysate (BTND), oolong tea non-dialysate (OTND), and Pu-erh tea (PTND), respectively. The large-scale produc-tion of TNDs was carried out at a pilot plant to obtain sufficient amounts for the assay of their anticarcinogenic effects on chemical carcinogenesis in mice and for the identification of active principle(s). In the pilot plant production, TNDs were prepared from hot water infusions of green and black tea leaves, followed by extraction with ethyl acetate, n-butanol, and finally purification by ultrafiltration (molecular size: 10,000). We found that the quality of TNDs produced in the lab-oratory and in the pilot plant were almost the same when they were checked by

Table 1. Yields of crude extracts and the six fractions from green, black, oolong, and Pu-erh teas

Extracts / fractions	% yields (w/w of dry tea leaves)			
	Green tea	Black tea	Oolong tea	Pu-erh tea
Crude extract	23.6	22.2	15.2	15.4
$CHCl_3$ sol. fraction	0.68	1.56	0.74	1.04
EtOAc sol. fraction	2.10	2.22	3.35	0.56
BuOH sol. fraction	2.08	4.14	2.02	1.06
Aqueous fraction	16.2	11.0	8.74	9.92
Dialysate	14.8	8.83	7.52	4.12
Non-dialysate	0.42	2.01	1.78	4.52

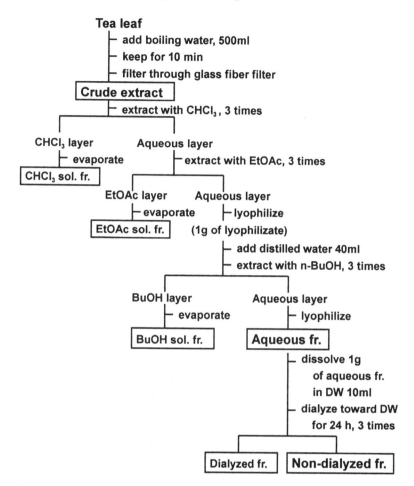

Figure 1. Fractionation of tea extracts. Aqueous fractions were dialyzed with a seamless cellulose tube (molecular size: 12,000; Wako Pure Chemical Ind., Ltd., Tokyo, Japan), and then dialysate and non-dialysate were lyophilized.

thin-layer chromatography as well as in their antitumor promoting activity assayed using mouse epidermal JB6 cells described in the next paragraph.

3. ANTITUMOR PROMOTING AND ANTICARCINOGENIC PROPERTIES OF TNDS

TNDs suppressed the neoplastic transformation induced by 12-*O*-tetradecanoylphorbol-13-acetate (TPA) in mouse epidermal JB6 cells, as did EGCG.[7] Although the tumor suppressing effects were about the same in BTND, OTND, and PTND in a dose-dependent manner, the activity of GTND was a little weaker (fig. 2). EGCG decreased not only TPA-induced soft agar colony induction

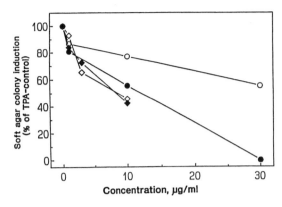

Figure 2. Effects of TNDs on the promotion of neoplastic transformation induced by TPA in mouse epidermal JB6 cells. Anchorage-independent soft agar colony induction in the presence of TNDs is expressed as a percent of the TPA-control. Each point is the mean of two or three independent experiments and $\bigcirc$, $\bullet$, $\diamond$, and $\blacklozenge$ represent the curve with green, black, oolong, and Pu-erh tea nondialysates.

in JB6 promotion-sensitive (p^+) clones,[10] but also soft agar colony formation in JB6 tumorigenic variants. This suggests that the effect of EGCG is "chemotherapeutic" because EGCG is rather more toxic toward the transformants (JB6 tumorigrenic variants) than the p^+ clones at the higher concentration of 10–30 $\mu g/mL$ or more. The critical point is that the suppressing effects of TNDs differ largely from those of EGCG. For example, BTND shows only about one-tenth the cytotoxicity of EGCG in p^+ cells, and BTND shows no selective effect on the growth of the transformants. Therefore, these results suggest that BTND and TNDs could suppress the process of tumor promotion.

In contrast to EGCG, TNDs normalized the morphological changes caused by the tumor promoter TPA in the JB6 promotion-sensitive clone (p^+).[8] The JB6 p^+ cells were growing flatly adhered on the culture vessel wall surface in normal culture. Addition of a promoter such as TPA to the cells of a monolayer sheet changes the cells morphologically. The cells become spherical (rounding), and caudate structures such as nerve processes (tailing) began to develop among the cells. TNDs used simultaneously with TPA inhibited these morphological changes (both the rounding and tailing characteristics). The number of flat cells increased (increases in flatness and extension), cell adhesion resumed, and a monolayer sheet again formed. BTND was most effective among TNDs. EGCG was ineffective. These morphological changes in cells were inhibited either by treatment with TNDs simultaneously or within 6 hours after TPA treatment. When TNDs were added to the transformants of JB6 family, the cells remained morphologically unchanged. These findings suggest that progression of TPA-induced morphological changes in cells is interrupted by the addition of TNDs during the early stage of promotion and that the cells are morphologically restored to a normal appearance.

TNDs strengthened the cytoskeletal actin microfilaments in JB6 cells.[9] The morphology of cultured cells is determined by the cytoskeleton and structures such as the extracellular matrix. The effects of TNDs on these factors were investigated. F-actin, the main structure of the cytoskeleton, a contractile protein, is bound to extracellular matrices such as fibronectin via desmosomes. F-actin is believed to be related to cell division or a cellular signal transduction system for cell proliferation, as well as cell morphology and motility.[11] Concerning the relation to carcinogenesis, Pollack et al.[12] reported that the distribution of actin changes along with transformation. Long bundles of microfilaments, so-called stress fibers, were observed by fluorescent staining of F-actin in monolayer-cultured JB6 p$^+$ cells. When the monolayer-cultured cells were incubated with TPA, cellular stress fibers quickly disappeared, and actin microfilaments became localized around the cellular and nuclear membranes. However, when the cells were incubated with TPA and TNDs, the TND$_s$ blocked the reduction of cellular stress fibers and cell shape became morphologically normal. BTND was most effective among TNDs. As noticed before, EGCG was not active. TNDs also showed effects on the cellular fibronectin (FN), a component of extracellular matrix, as shown in cytoskeletal stress fibers in JB6 cells. According to Zerlauth and Wolf,[13] when the neoplastic transformation of JB6 p$^+$ cells was induced by TPA, cellular FN was released from the cell surface into the medium. The loss of FN from the cell surface was closely associated with the tumor promotion processes. From these findings, we examined the hypothesis that TNDs act on FN of the extracellular matrix to induce the expression, send a signal to F-actin via desmosomes, and inhibit tumor promotion.

GTND inhibits mouse duodenal carcinogenesis. Mice with duodenal tumors induced with *N*-ethyI-*N*'-nitro-*N*-nitrosoguanidine (ENNG) were given drinking water containing 0.005 or 0.05 percent of GTND and examined for anticarcinogenic effects. Figure 1 summarizes the results. The number of tumor-bearing mice, total number of tumors, number of tumors per mouse, and tumor size were all lower in the 0.05 percent group. The total number of tumors was inhibited (46 percent of that in the ENNG control group). In the 0.005-percent group, the tumor incidence was not decreased, but the ratio of early stage tumorigenic lesions was found to be high on histopathological investigation. The malignancy in the 0.005-percent group was lower than that in the 0.05-percent group. TNDs were superior to Polyphenon 100 (total catechin level is 91.2 percent, containing 53.9 percent of EGCG, Mitsui Norin Co.) used as the control in inhibiting tumorigenesis and in histopathological findings. These findings suggest the usefulness of TNDs in the primary prevention of cancer.

4. IDENTIFICATION OF THE MAINLY ACTIVE FRACTION(S) OF GTND AND BTND

The chemical or enzymatic modifications of TNDs by hydrolysis with hydrochloric acid, β-glucosidase, and tannase or oxidation with potassium permanganate and polyphenoloxidase diminished the antitumor promoting activity shown in table 2. This suggests that the sugars and polyphenols in the chemical structure of TNDs are essential for their activity.

The antitumor-promoting effects of TNDs were characterized by the inhibitory effects on the soft agar colony induction (neoplastic transformation) and morphological changes induced by 12-*O*-tetradecanoyfphorbol-13-acetate (TPA) in mouse epidermal JB6 cells. Therefore, we conducted experiments on the identification of the active components/fractions from the TNDs by screening these two activities.

Table 2. Effects of chemical or enzymatic modifications on the antitumor promoting activities of TNDs

| Treatment | TLC analysis (color ; Rf value) | | Inhibition of TPA-induced tumor promotion | |
	Saccharides (phenol -sulfuric acid)	Phenols (1% FeCl₃)	Inhibition of soft agar colony induction (30 µ g/ml)	Inhibition of morphological changes (30 µ g/ml)
Green tea				
Control	brown ; -	blue ; -	+ +	+ +
Oxidation by				
KMnO₄[c]	brown ; UC	NC	+	+
Polyphenol-oxidase[d]	brown ; UC	NC	±	+
Hydrolysis by				
HCl[e]	brown ; C	blue ; C	+	−
Glucosidase[f]	brown ; C	blue ; C	−	−
Tannase[g]	NC	NC	−	+
Green tea				
Control	brown ; -	blue ; -	+ + +	+ + +
Oxidation by				
KMnO₄[c]	brown ; UC	NC	+	+ +
Polyphenol-oxidase[d]	brown ; UC	NC	+	+ +
Hydrolysis by				
HCl[e]	brown ; C	blue ; C	+ +	±
Glucosidase[f]	brown ; C	blue ; C	+	±
Tannase[g]	brown ; C	blue ; C	+	+ +

TNDs were undertaken by the chemical or enzymatic modification indicated in the above column of treatment, and the reaction mixtures were dialyzed through the Visking cellulose tube toward distilled water. The non-dialysates were examined by TLC-analysis and JB6 promotion assay.

a NC: no color detected, C and UC indicate changed and unchanged in Rf values.

b −, ±, +, + +, and + + + indicate no, marginal, weak, moderate, and strong inhibitions on the soft agar colony induction or the morphological chances induced by TPA in JB6 cell lines.

c treated with 1% KMnO₄ at room temperature for 1 hour.

d treated with tea polyphenol oxidase at 37°C for 2.5 hours.

e treated with 1 mol/L hydrochloric acid at 85°C for 1 hour.

f treated with β -glucosidase at 37°C for 6 hours.

g treated with tannase at 37°C for 1 hour.

As TNDs are expected to contain polysaccharides and protein, we examined their separation from the aqueous TND solution by adding increasing concentrations of ethanol (EtOH). It was found that the active principles were in the 60 percent EtOH-supernatant of GTND (GTND-60EtOHsup) and in the 50 percent EtOH-supernatant of BTND (BTND-50EtOHsup), shown in table 3. Both of the supernatants showed more inhibitory effects of soft agar colony induction and cell shape changes induced by TPA in JB6 Cl 41 and Cl 22 cells than crude TNDs did.

Table 3. Preliminary separation of active components of TNDs

Fractions	Yield (g)	SA-assay (% of TPA-control at 10 μ g/ml)	Inhibition of cell shape changes
GTND	(10.0)	81	++
EtOH precipitation			
60%EtOH ppt.	2.51	83	−
60%EtOH sup.	7.81	73	++
BTND	(10.0)	64.9	+++
EtOH precipitation			
60%EtOH ppt.	4.69	80.1	−
60%EtOH sup.	4.99	52.0	+++

The GTND-60EtOHsup and BTND-50EtOHsup were loaded on Toyopearl HW65F columns (32 mm i.d. × 240 mm) and eluted by 1,000 mL of water and 700 mL of 50 percent EtOH, step by step, elution of sugars was monitored with phenol-sulfuric acid and polyphenols were detected with FeCl$_3$. The patterns of fractionation are summarized in figure 3. The apparently pure polysaccharide component eluted first did not show any antitumor-promoting effect. The following C and E fractions were effective against the soft agar colony induction and cell shape alteration by TPA in tumor promotion-sensitive JB6 clones shown in tables 4 and 5. The E-fraction of BTND was separated into two parts by re-chromatography. We found that the highest and most characteristic activities in the peak E1 were eluted at the front. Subsequent fractions isolated as elute E2 exhibited strongest inhibition of soft agar colony induction by TPA but did not show any inhibitory effect on the cell shape changes of JB6 clones.

The chemical or enzymatic modification of C, E, E1, and E2 fractions diminished their inhibitory effects on the TPA-induced alternation of JB6 cells shown in table 6. But the extent of the reduction of the effects of C, E1, and E2 fractions from BTND on the TPA-induced alteration were stronger in the C and E1 fractions than in the E2 fraction. Therefore, we decided to estimate the comparative chemical constitution of C and E1.

The chemical constitutions of the principally active C and E1 fractions were investigated in more detail. We determined the sugar contents in the TND-fractions using the phenol-sulfuric acid method, expressed as D-glucose equivalents. Total polyphenolics were estimated by colorimetric method with FeCl$_3$, expressed as (−)-epicatechin. Gallic acid was determined by HPLC after the hydrolysis with tannase and the extraction with EtOAc. Identification of sugar components in TND-fractions was undertaken by GC-MS (Hewlett Packard 5890 and JEOL JMS-AX505W, SPB-1 capillary column) after the hydrolysis with HCl. TMS-

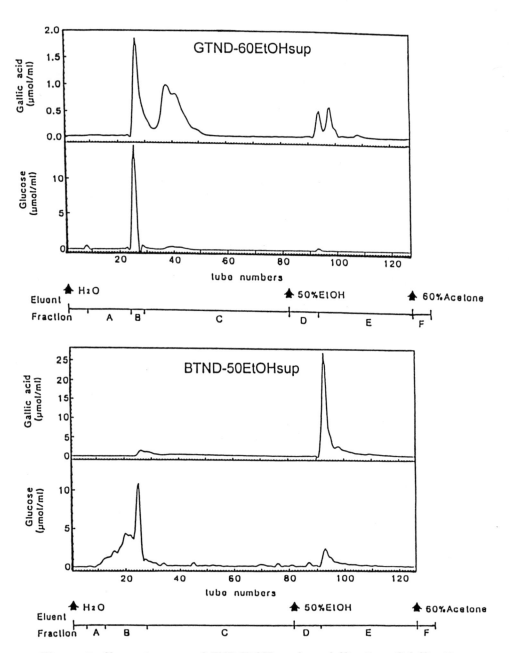

Figure 3. Chromatograms of TND-EtOHsup by gel-filtration. Gel-filtration performed on Toyopearl HW65F column (32 × 240 mm) by stepwise elution of indicated solvents. Charging 500 mg of GTND-60EtOHsup/BTND-50EtOHsup gave six fractions indicated, respectively.

Table 4. Fractionation of GTND-60EtOHsup by gel-filtration

Fractions	A	B	C	D	E	F
Yield (%)	0.7	79.5	9.3	0.6	4.6	1.2
Color	white	light brown	light brown	brown	brown	brown
TLC-detection (Rf) phenol-sulfuric acid (sugars)	0.0	0.0 0.51 0.64	0.64 0.83	0.51 0.64 0.83	0.64 0.83	0.83
$FeCl_3$ (polyphenols)		0.64	0.51 0.64 0.83	0.51 0.64 0.83	0.51 0.64 0.83	0.51 0.64 0.83
SA-assay (% of TPA-control at 10 μg/ml)	ND	61.5	52.5	ND	47.9	ND
Inhibition of cell shape changes	—	+	++	+	++	—

ND: not determined.

Underlined figures represent major peaks.

Table 5. Fractionation of BTND-50EtOHsup by gel-filtration

Fractions	A	B	C	D	E	F	E1	E2
Yield (%)	0.8	28.2	25.5	4.2	45.9	7.8	13.6	66.8
Color	white	light brown	light brown	brown	brown	brown	brown	brown
TLC-detection (Rf) phenol-sulfuric acid (sugars)	0.0	0.49 0.68	0.49 0.68 0.84	0.49 0.68 0.84	0.49 0.68 0.84	0.49 0.68 0.84	0.49 0.64	0.64 0.85
$FeCl_3$ (polyphenols)		0.68 0.84	0.49 0.68 0.84	0.49 0.68 0.84	0.49 0.68 0.84	0.49 0.68 0.84	0.43 0.64 0.85	0.46 0.64 0.85
SA-assay (% of TPA-control at 10 μg/ml)	ND	97.0	42.7	25.0	37.1	35.6	20.6	12.8
Inhibition of cell shape changes	—	+	++	+	+++	—	+++	—

ND: not determined.

Underlined figures represent major spots.

Table 6. Enzymatic modifications of TND-fractions reduce the antitumor
promoting activities

Sample modification	% Activities (concentration, μ g/ml)		Sample modification	% Activities (concentration, μ g/ml)	
GTND-60EtOHsup			BTND-50EtOHsup		
C			C		
Control	100	(30)	Control	100	(10)
+Glucosidase	29.2		+Glucosidase	18.5	
+Tannase	33.7		+Tannase	3.6	
E			E1		
Control	100	(10)	Control	100	(10)
+Glucosidase	23.4		+Glucosidase	27.7	
+Tannase	54.9		+Tannase	32.5	
			E2		
			Control	100	(10)
			+Glucosidase	53.6	
			+Tannase	37.3	

derivatives were determined by a Shimadzu GC 4CM (FID) with 5-percent silicone GS-101 on Uniport HP60/80 column (3 mm i.d. × 200 mm).

UV spectra of TND-EtOH supernatants and C- and E-TND fractions from green and black teas showed maximum absorption at 205 and 270 nm, and weaker, broad absorption at 300 to 360 nm (data not shown). This suggested that flavonol exists in TND fractions, associated with NMR spectra of BTND-E1 and E2. D-Glucose, D-galactose, D-mannose, arabinose, gallic acid, and quinic acid were identified in the TND fractions of C, E, E1, and E2 by GC-MS. Contents of sugars and polyphenolics in TND fractions determined by GLC and HPLC were approximately 13–30 percent (as D-glucose), and maximally 74 percent (as gallic acid), respectively, shown in table 7. It is noteworthy that the ratio of sugars and polyphenolics in active fractions of C and E1 is 1:2, approximately, by colorimetric determination.

5. CONCLUSIONS

We have described the properties of antitumor-promoting effects of aqueous tea non-dialysates (TNDs) that are found in tea leaves. To identify the principally active component in TNDs and its chemical constituent, we conducted the following investigations.

The chemical or enzymatic modifications of TNDs, such as oxidation with polyphenoloxidase, $KMnO_4$ and hydrolysis with β-glucosidase, tannase, and HCl of TNDs diminished their activities. Those results indicate that polyphenols and sugars are required to express the antitumor-promoting activities of TNDs. Main

Table 7. Contents of sugars and polyphenols in the active TND fractions

Determined by / Samples	Phenol-sulfuric acid	GC-MS	5% FeCl₃		HPLC
	As glucose mg/mg (μ mol/mg)	Sugars (Hydrolysis & TMS-ders.)	As (-)-epicatechin mg/mg (μ mol/mg)	As gallic acid* mg/mg (μ mol/mg)	Gallic acid mg/mg (μ mol/mg)
GTND					
60EtOHsup	0.31 (1.7)	ND**	0.24 (0.83)	0.16 (0.88)	ND
C	0.30 (1.7)	Glc, Gal, Rha	0.89 (3.0)	0.56 (3.0)	ND
E	0.15 (0.84)	Glc	0.87 (3.0)	0.55 (2.9)	ND
BTND					
50EtOHsup	0.29 (1.6)	ND	0.88 (3.0)	0.56 (3.0)	ND
C	0.19 (1.1)	Glc, Gal, Man, Ara	0.78 (2.7)	0.48 (2.5)	21.0 (0.11)
E	0.13 (0.74)	ND	1.1 (3.7)	0.72 (3.8)	ND
E1	0.14 (0.77)	Glc, Gal	0.37 (1.3)	0.22 (1.2)	20.5 (0.11)
E2	0.14 (0.78)	Glc, Gal	1.0 (3.6)	0.68 (3.6)	73.8 (0.39)

* including (-)-epicatechin phenolic residues, 3,4,5-trihydroxyphenyl and 3,4,5-trihydroxylbenzoyl residues.
** not determined.
Glc: *D*-glucose, Gal: *D*-galactose, Man: *D*-mannose, Rha: *L*-Rhamnose, Ara: Arabinose

activities are found in the 60-percent ethanol supernatant of GTND (GTND-60EtOHsup) and in 50-percent ethanol supernatant of BTND (BTND-50EtOHsup) by ethanol precipitation. Active fractions of C and E (E1) are obtained from GTND-60EtOHsup and BTND-50EtOHsup by gel-filtration with a Toyopearl HW65F column. The activities of C and E1 are also diminished by both chemical/enzymatic oxidation and hydrolysis. Active TND-fractions were shown to contain polyphenolic compounds of flavonols and gallates, quinic acid, and carbohydrates of D-glucose and D-galactose unit. Colorimetric determinations of sugars and polyphenolics in C and E1 indicate that the existence of them is at the ratio of approximately 1:2. As of now, we conclude that TNDs are chemically assumed to be a complex mixture of tannins with these polyphenols and sugars.

Our studies have shown that tea leaves have various other components that suppressed carcinogenesis. We hope that these components will be investigated in the future. Meanwhile, we should consider using all tea leaf components together for better human health.

ACKNOWLEDGMENTS

This study was a collaborative effort with Drs. H. Sugimoto, K. Nakano, T. Arika, and K. Saitou of the Research Institute of Toxicology and Safety Evaluation, Kaken Pharmaceutical Manufacturer Co., Fujieda-shi, Japan. This work was supported in part by grants-in-aid from the Program for Promotion of Basic Research Activities for Innovative Biosciences (PROBRAIN) in Japan, the Shizuoka Prefectural Foundation for the Promotion of Science in Japan, and the Chiyoda Mutual Life Foundation in Japan.

REFERENCES

1. Begley, S. Beyond vitamins. *Newsweek*, April 25, p. 42 (1994).

2. Yoshizawa, S.; Horiuchi, T.; Suganuma, M.; Nishiwaki, N.; Yatsunami, J.; Okabe, S.; Okuda, T.; Muto, Y.; Frenkel, K.; Troll, W.; Fujiki, H. Penta-O-galloyl-D-β-glucose and (−)-epigallocatechin gallate, a cancer preventive agent. *In*: Huang, M.T.; Ho, C.T.; Lee, C.Y. (eds.) Phenolic compounds in foods and health. II. Antioxidants and cancer prevention. American Chemical Society, Washington, DC, p. 316 (1992).

3. Yang, C.S.; Wang, Z.Y. Tea and cancer. *J. Natl. Cancer Inst.* 85:1038 (1993).

4. Wang, Z.Y.; Huang, M.T.; Ho-C.T.; Chang, R.; Ma, W.; Ferraro, T.; Reuhl, K.R.; Yang, C.S.; Conney, A.H. Inhibitory effects of green tea on the growth of established skin papillomas in mice. *Cancer Res.* 52:6657 (1994).

5. Taniguchi, S.; Fujiki, H.; Kobayashi, H.; Go, H.; Miyado, K.; Sadano, H.; Shimokawa, R. Effect of (−)-epigallocatechin gallate the main constituent of green tea, on lung metastasis with mouse B16 melanoma cell lines. *Cancer Lett.* 65:51 (1992).

6. Isemura, M.; Suzuki, Y.; Satoh, K.; Narumi, K.; Motomiya, M. Effects of catechins on the mouse lung carcinoma cell adhesion to the endothelial cells. *Cell Biol. Int.* 17:559 (1993).

7. Nakamura, Y.; Harada, S.; Kawase, I.; Matsuda, M.; Tomita, I. Inhibitory effect of tea ingredients on the *in vitro* tumor promotion of mouse epidermal JB6 cells. *In*: Yamanishi, T. et al. (eds.) Proceedings of international symposium on tea science, Shizuoka, Japan, p. 205 (1991).

8. Nakamura, Y.; Tomita, I. Antimutagens/antiprompters in edible plants and the modes of actions: disulfides and tea components. *Environ. Mutat. Res. Commun.* 17:107 (1995).

9. Nakamura, Y.; Kawase, I.; Harada, S.; Matsuda, M.; Honma, T.; Tomita, I. Antitumor promoting effects of tea aqueous non-dialysates in mouse epidermal JB6 cells. *In*: Ohigashi, H. et al. (eds.) Food factors for cancer prevention. Springer-Verlag, Tokyo, Japan, p. 138 (1997).

10. Colburn, N.H.; Lerman, M.; Srinivas, L.; Nakamura, Y.; Gindhart, T.G. Membrane and genetic events in tumor promotion: studies with promoter resistant variants of JB6 cells. *In*: Fujiki, H.; Sugimura T. (eds.), Cellular interaction by environmental tumor promoters. Scientific Societies Press, Tokyo, p. 155 (1984).

11. Takahashi, K.; Heine, U.I.; Junker, I.L.; Colburn, N.H.; Rice, I.M. Role of cytoskeleton changes and expression of H-ras oncogene during promotion of neoplastic transformation in mouse epidermal JB6 cells. *Cancer Res.* 46:5923 (1986).

12. Pollack, R.; Osborn, M.; Weber, K. Patterns of organization of actin and myosin in normal and transformed cultured cells. *Proc. Natl. Acad. Sci. USA*, 72:994 (1985).

13. Zerlauth, G.; Wolf, G. Release of fibronectin is linked to tumor promotion: response of promotable and non-promotable clones of a mouse epidermal cell line. *Carcinogenesis* 6:73 (1985).

HOST-MEDIATED ANTICANCER ACTIVITIES OF TANNINS

Ken-ichi Miyamoto,[a] Tsugiya Murayama,[b] Tsutomu Hatano,[c]
Takashi Yoshida,[c] and Takuo Okuda[c]

[a] Department of Hospital Pharmacy School of Medicine
Kanazawa University
Kanazawa 920
JAPAN

[b] Department of Microbiology
Kanazawa Medical University
Ishikawa 920-02
JAPAN

[c] Faculty of Pharmaceutical Sciences
Okayama University
Okayama 700-8530
JAPAN

1. INTRODUCTION

A number of medicinal plants have been traditionally used for treatment of various ailments, and it has been suggested that tannin may be the active agent of many of them. Among the major activities of tannins found are antioxidant and radical scavenging activities.[1,2] These are the basic activities underlying the action of tannin-rich medicinal plants, which are effective in preventing and treating many diseases such as arteriosclerosis, heart dysfunction, and liver injury by inhibiting lipid-peroxidation.[3,4] The inhibition of hepatotoxins[5] and mutagens[6] and the antitumor-promoter action of polyphenols[7,8] are also correlated with their antioxidant activity. Most of these actions have been shown using comparatively low-molecular-weight tannins, including epigallocatechin gallate (EGCG). On the other hand, hydrolyzable tannin monomers, oligomers, and galloylated condensed tannins inhibit the replication of herpes simplex virus[9] and human immunodeficiency virus[10,11] by blocking virus adsorption to the target cells and inhibition of reverse transcriptase activity of the virus. It is also

Plant Polyphenols 2: Chemistry, Biology, Pharmacology, Ecology, Edited by
Gross et al. Kluwer Academic / Plenum Publishers, New York, 1999

known that many plants containing tannins are effective against cancer and tumors.[12–15]

In 1985, we first found that a dimeric ellagitannin agrimoniin is a main substance for anticancer activity of *Agrimonia pilosa*.[16,17] Thereafter, we evaluated anticancer activity of a hundred tannins isolated from various plants and studied their action mechanisms.[18–23] As presented here, tannins exhibit chemical-structure specific anticancer activities through enhancement of the host-defense potential.

2. ANTICANCER ACTIVITIES

We developed a simple screening system to examine the host-mediated anti-cancer activity of a large number of tannins and related polyphenols, using a small amount of compounds isolated and purified from many medicinal plants.[18] Each compound was intraperitoneally (i.p.) injected into six mice in a group at 5 or 10 mg/kg once 4 days before the i.p. inoculation of sarcoma-180 (S-180) cancer cells (1×10^5 cells/head), and the survival time of cancer-bearing mice was observed. The anticancer activity of tannins and related polyphenols is shown in table 1. Condensed tannins, caffeic acid derivatives, bergenin derivatives, and gallotannins exhibited negligible or no anticancer activity. Monomeric ellagitannins, except for rugosin A and tellimagrandin II, also showed very weak activities. Rugosin A and tellimagrandin II cured one and three out of six mice, respectively. Moreover, nearly half of the oligomeric ellagitannins (20 out of 36) and all macrocyclic ellagitannin oligomers exhibited potent anticancer activities. It is clear that the presence of free phenolic hydroxyl groups is essential for the anticancer activity; nonacosa-*O*-methylcoriariin was inactive, whereas the parent compound coriariin A showed a strong anticancer activity. In addition, the activity of dehydroellagitannins, either monomers or dimers, was generally weak. The data for OK-432 (Picibanil), a immunomodulator,[24–26] were obtained under similar experimental conditions. There are many tannins that show a similar action or exceed OK-432 in anticancer activity. Table 2 summarizes the evaluation of the anticancer activitiy of 108 compounds tested.

As mentioned above, active compounds were found among ellagitannins. Figure 1 shows the chemical structures of oligomeric ellagitannins, which exhibited extremely potent anticancer activities. Agrimoniin is a dimer of potentillin. Coriariin A, hirtellin B, and tamarixinin A consist of rugosin A and tellimagrandins I and/or II as their monomer units. All macrocyclic ellagitannins such as oenotheins A and B have tellimagrandin I or II in their structure, and they also exhibited a potent anticancer activity. Laevigatin B and euphorbin C-Hy (fig. 2) were negative, although they have a potentillin or tellimagrandin unit together with other monomer units. Consequently, oligomeric ellagitannins with a potent anticancer activity consist of only potentillin, rugosin A, and tellimagrandins I and/or II as the monomer units (structures are shown in figure 3). It was also found that oligomeric tannins, having a casuarictin (β-anomer of potentillin) unit, had generally a low anticancer activity, except nobotanin H, which is a conjugate of casuarictin and tellimagrandin II. Among these monomeric ellagitannins,

Table 1. Anticancer activity of tannins and related polyphenols against S-180 in mice[1-6]

Compound	Dose (mg/kg)	%ILS[a]	60-day survivors
Condensed tannins and related compounds			
1 (-) Epicatechin	5	35.5	0
2 (-) Epigallocatechin	5	-19.4	0
3 (-) Epicatechin gallate (ECG)	10	38.7	0
4 (-) Epigallocatechin gallate (EGCG)	10	-1.9	0
	5	18.1	0
5 ECG-(4β→8)-ECG	10	-1.9	0
	5	48.7	0
6 ECG-(4β→8)-ECG-(4β→8)-ECG	10	48.4	0
	5	16.8	0
7 ECG-(4β→6)-ECG	5	57.4	0
8 ECG-(4β→6)-ECG-(4β→6)-ECG	5	13.3	0
Caffeic acid derivatives			
9 Chlorogenic acid	5	-3.2	0
10 3,5-Di-*O*-Caffeoylquinic acid	5	23.0	0
11 Rosmarinic acid	5	21.8	0
Bergenin and its derivatives			
12 Bergenin	5	-5.2	0
13 Dimethylbergenin	5	19.4	0
14 11-*O*-Galloylbergenin	5	31.0	0
Gallotannins and related polyphenols			
15 3-*O*-Digalloylquinic acid	5	6.8	0
16 3-*O*-Trigalloylquinic acid	5	52.6	0
17 3-*O*-Tetragalloylquinic acid	5	134.2	1
18 3-*O*-Pentagalloylquinic acid	5	15.0	0
19 3-*O*-Hexagalloylquinic acid	5	38.9	0
20 3-*O*-Heptagalloylquinic acid	5	65.5	0
21 Quercitrin-2″-*O*-gallate	5	-19.4	0
22 1,2,6-Tri-*O*-galloyl-β-D-glucose	5	15.7	0
23 1,2,3,6-Tetra-*O*-galloyl-β-D-glucose	10	27.2	0
24 Penta-*O*-galloyl-β-D-glucose	10	81.9	0

tellimagrandin II, and rugosin A showed potent anticancer activities, but potentillin, casuarictin, and tellimagrandin I had a very low or no anticancer activity. Thus, although it is not always necessary that the active oligomeric tannins consist of active monomer units, the structures of the monomer units play an important role in anticancer activity of the oligomers. The increment in the activity is not proportional to the degree of polymerization. For example, oenothein B, a macrocyclic dimer of tellimagrandin I, exhibited a potent anticancer activity, but addition of the tellimagrandin I or 2,3-digalloylglucose unit to the dimer molecule to form a trimer (oenothein A and woodfordin E) decreased the activity of the molecule. Woodfordin F, a tetramer, did not cause regression of tumor growth (table 1).

Table 1. *Continued*

Compound	Dose (mg/kg)	%ILS	60-day survivors
Monomeric ellagitannins			
25 Alnusiin	5	74.3	0
26 Castalagin	5	41.8	0
27 Casuarictin	10	47.1	0
28 Casuarinin	5	8.2	0
29 Corilagin	10	36.1	0
30 1-*O*-Galloyl-2,2:3,6-bis-*O*-HHDP-β-D-glucose	10	25.0	0
31 Gemin D	5	1.8	0
32 Guavin A	5	29.5	0
33 Isovalolaginic acid	5	-3.3	0
34 Pedunculagin	10	27.3	0
35 Potentillin	10	-1.9	0
36 Praecoxin A	5	70.8	0
37 Praecoxin B	10	-17.2	0
38 Pterocarinin C	5	9.2	0
39 Punicalagin	5	41.6	0
40 Punicalin	5	43.4	0
41 Roxbin B	5	20.2	0
42 Rugosin A	10	25.2	1
	5	110.3	1
43 Rugosin C	5	91.7	0
44 Strictinin	5	44.2	0
45 Tellimagrandin I	10	35.2	0
46 Tellimagrandin II	10	18.1	3
	5	73.1	0
47 Teroblongin	5	16.8	0
48 Vescalagin	5	58.2	0

HHDP: hexahydroxydiphenoyl

These results support the hypothesis that both the composition of the monomer units and their appropriate molecular size are important for the anticancer activity of oligomeric ellagitannins.

In 1985, we reported that the methanol extract from root of *Agrimonia pilosa* Ledeb. exhibited activity against several rodent cancers[12] and that its active constituent was agrimoniin,[13,18] which was already isolated from Rosaceous medicinal plants by Okuda et al.[27] Oenothein B was first isolated from *Oenothera erythrosepala*[28] and then from the flower of *Woodfordia fruticosa*.[29,30] Table 3 indicates that the acute toxicities of agrimoniin and oenothein B in female ddY mice were low; particularly when orally administered, these tannins did not show any toxicity. In order to see the anticancer activity against murine mammary carcinoma MM2 in C3H/He mice, each tannin was administered at a dosage of about one-tenth of the LD50 for each administration route on different days before the cancer inoculation. The intraperitoneal (i.p.), intravenous (i.v.), and oral (p.o.)

Table 1. *Continued*

Compound	Dose (mg/kg)	%ILS	60-day survivors
Dimeric ellagitannins			
49 Agrimoniin	10	136.2	3
50 Camelliin A	10	-16.4	2
51 Campthotin A	5	48.7	0
52 Coriariin A	5	238.0	3
* Nonacosa-*O*-methyl-coriariin	5	21.3	0
53 Coriariin C	10	156.2	2
54 Coriariin E	10	53.9	0
55 Cornusiin A	10	181.8	2
56 Eumaclin A	10	8.4	0
57 Euphorbin C-Hy	10	-6.5	0
58 Gemin A	10	138.3	1
59 Hirtellin A	10	150.0	1
60 Hirtellin B	10	114.2	3
61 Isorugosin D	5	146.5	2
62 Laevigatin B	10	-9.7	0
63 Laevigatin C	10	-1.9	0
64 Medinillin B	5	108.8	1
65 Medinillin C	10	-5.3	0
66 Nobotanin A	10	126.6	0
67 Nobotanin B	5	33.6	0
68 Nobotanin F	5	76.4	0
69 Nobotanin G	10	36.5	0
70 Nobotanin H	10	-30.2	3
71 Nobotanin I	10	19.6	0
72 Roxibin A	5	53.5	0
73 Rugosin D	10	171.5	1
74 Rugosin E	10	234.7	2
75 Rugosin F	10	35.2	0
76 Tamarixinin A	10	75.0	4
77 Woodfordin A	10	-3.8	0
78 Woodfordin B	10	56.3	1
79 Woodfordin H	10	64.0	1
Other oligomeric ellagitannins			
Trimer			
80 Cornusiin C	10	108.2	0
81 Nobotanin C	10	125.0	1
82 Nobotanin E	5	41.0	0
83 Nobotanin J	10	21.6	0
Tetramer			
84 Nobotanin K	10	31.6	2

Table 1. *Continued*

Compound	Dose (mg/kg)	%ILS	60-day survivors
Macrocyclic ellagitannins			
Dimer			
85 Camelliin B	10	36.5	2
86 Hirtellin C	10	154.0	0
87 Oenothein B	10	196.0	4
88 Woodfordin C	10	59.5	1
Trimer			
89 Oenothein A	10	102.7	1
90 Woodfordin D	10	123.0	1
91 Woodfordin E	10	77.4	2
Tetramer			
92 Woodfordin F	10	179.0	0
Dehydro ellagitannins			
Monomer			
93 Chebulagic acid	10	-18.4	1
94 Chebulinic acid	5	3.4	0
95 Dehydrogeraniin	5	35.4	0
96 Furosinin	5	23.8	0
97 Geraniin	10	23.8	0
98 Granatin A	5	3.5	0
99 Granatin B	5	66.4	0
100 Isoterchebin	5	57.4	0
101 Mallotusinic acid	5	77.0	0
Dimer			
102 Antidesmin A	10	35.5	0
103 Euphorbin A	10	45.2	0
104 Euphorbin B	10	18.9	1
105 Euphorbin C	10	-27.1	0
106 Euphorbin D	10	0.0	0
107 Euphorbin E	10	27.1	0
108 Euphorbin G	10	34.0	0
Cf. OK-432	100 KE/kg (10 mg/kg)	79.2	4/12

Each compound (5 or 10 mg/kg) was i.p. injected once into six mice in a group 4 days before the i.p. inoculation of S-180 (10^5 cells/mouse).

[a][(T-C)/C] x 100, T: survival days of treated mice, C: survival days of untreated control mice, excluding 60-day survivors. The control mice died at 12.9 ± 0.8 days after the cancer cell inoculation.

Table 2. Summary for anticancer activity of polyphenols against S-180 in mice

	Number of tannins tested	Anticancer activity[a]				
		−	±	+	++	+++
Condensed tannins and related polyphenols	8	3	5	0	0	0
Hydrolyzable tannins and related polyphenols	100	37	25	9	22	7
Caffeic acid derivatives	3	3	0	0	0	0
Bergenin derivatives	3	2	1	0	0	0
Gallotannins	10	5	3	1	1	0
Ellagitannins						
Monomers	24	11	8	3	1	1
Oligomers	36	9	7	2	13	5
Macrocyclic oligomers	8	0	0	2	5	1
Dehydroellagitannins	16	7	6	1	2	0

[a]Evaluation: -, < 30 %ILS; ±, 30 ≤ %ILS < 70; +, ≥ 70 %ILS with no survivor; ++, one or two survivors; +++, > three survivors out of six mice at a dosage of 5 or 10 mg/kg.

premedications of agrimoniin caused cancer regression even when the treatment began 14 days before the cancer inoculation (table 4). Agrimoniin was also effective in the postmedication by each administration route (table 5). Oenothein B also exhibited an anticancer activity against MM2 carcinoma by i.p. administration in both medication schedules. The effects of these tannins by i.p. administration were stronger than that of OK-432, a streptococcal preparation with a potent immune-stimulatory activity.[24–26] However, oral administration of oenothein B did not show any anticancer activity (tables 4 and 5). This may be attributed to the difficulty of absorption of the macrocyclic tannin through the intestine and relatively easy absorption of agrimoniin. Agrimoniin and oenothein B were effective on the solid type Meth-A fibrosarcoma (table 6), as well as MH134 hepatoma, while OK-432 showed negligible effect on the cancer.[18,31]

3. IMMUNOMO DURATORY EFFECTS OF AGRIMONIIN

Tannins were classically known to have capability to bind to protein,[32] metals,[33] alkaloids,[34] polymers, etc. through their polyhydroxy groups in the molecule. Therefore, when the tannins are administered to the animal, they should bind to some components in the host. Agrimoniin showed only a weak *in vitro* cytotoxicity against MM2 cells in the presence of calf serum (IC_{50}: 63 µg/mL), but was cytotoxic in the absence of the serum (IC_{50}: 2.6 µg/mL) (fig. 4).[18] These data indicate that the tannin easily binds to serum components such as protein, and its unbound fraction is very small in the animal. It is therefore difficult to attribute the anticancer activity of the tannins only to their direct cytotoxicity on cancer cells.

When agrimoniin was administered by injection (i.p.), the total number of peripheral white blood cells significantly increased 4 to 10 days after the injection.

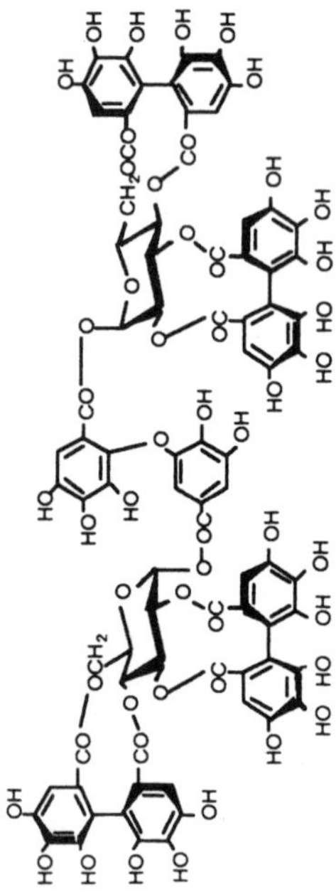

Coriariin A

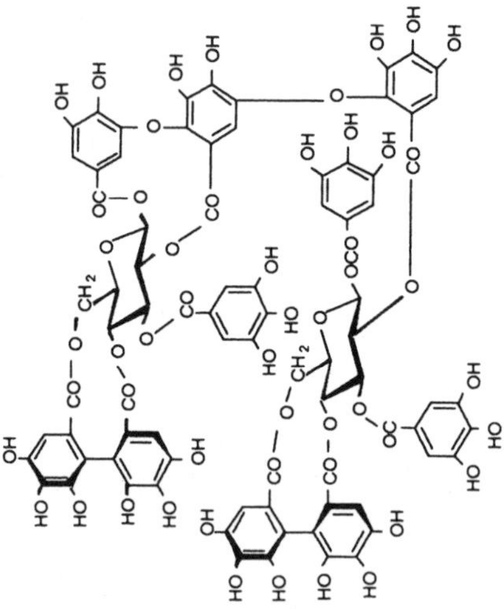

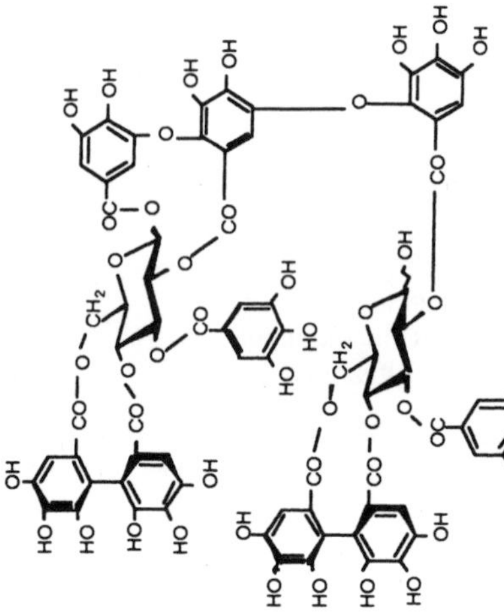

Tamarixinin A

Agrimoniin

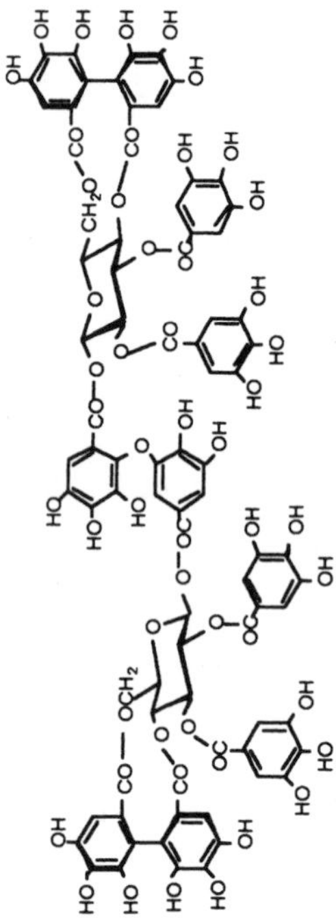

Hirtellin B

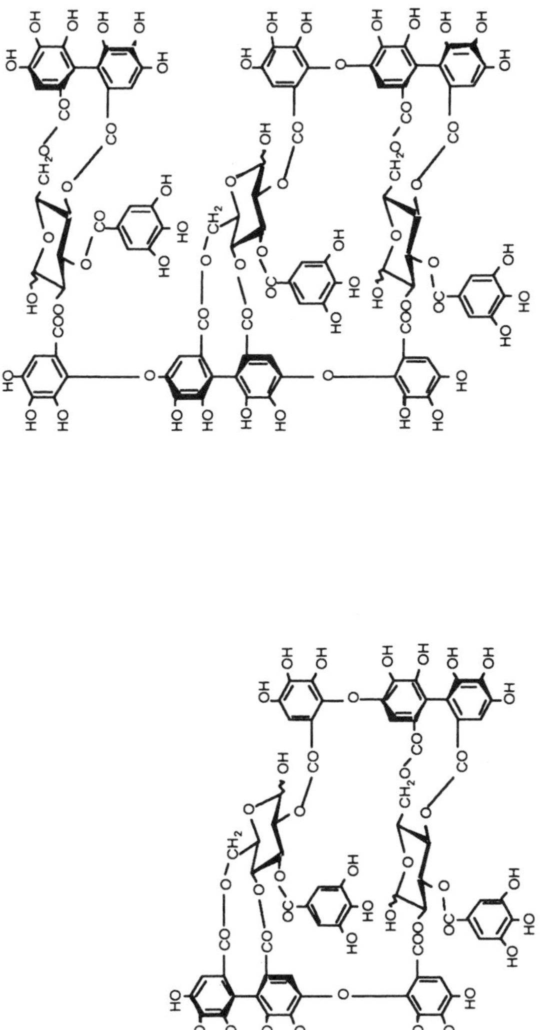

Oenothein A

Oenothein B

Figure 1. Chemical structures of oligomeric ellagitannins with a potent anticancer activity.

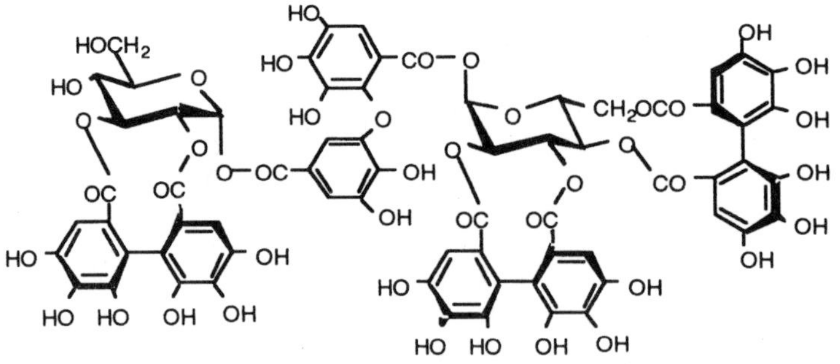

Laevigatin B

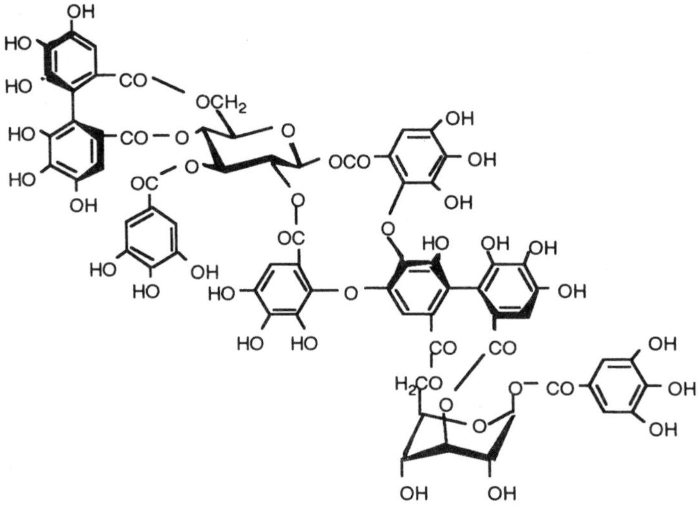

Euphorbin C-Hy

Figure 2. Chemical structures of dimeric ellagitannins with a weak anticancer activity.

In the cell population, the percentage of monocytes and lymphocytes was markedly higher than that of the non-treated control.[18] After the administration of agrimoniin, the number of peritoneal exudate cells (PEC) (fig. 5) and also spleen weight (fig. 6) also increased with a peak at 4 days in a dose-dependent manner. The non-adherent spleen cells (lymphocytes) had the ability to take up ^{3}H-TdR, indicating

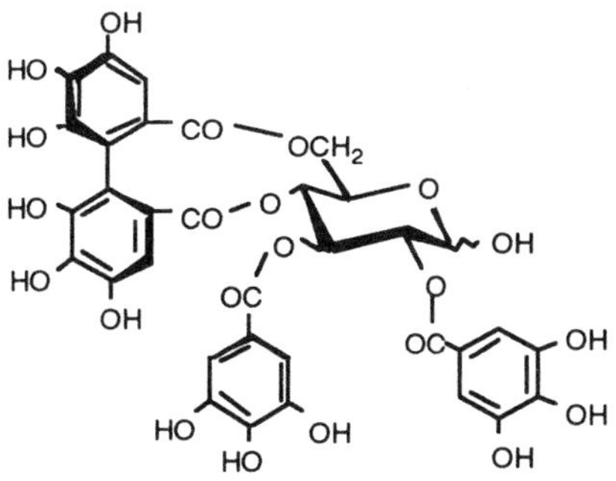

Tellimagrandin I

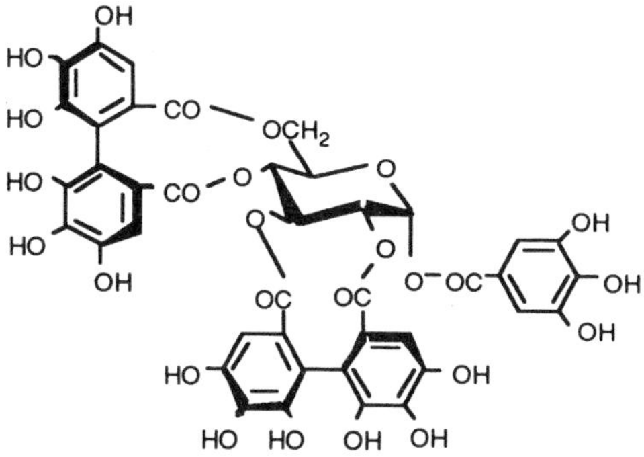

Potentillin

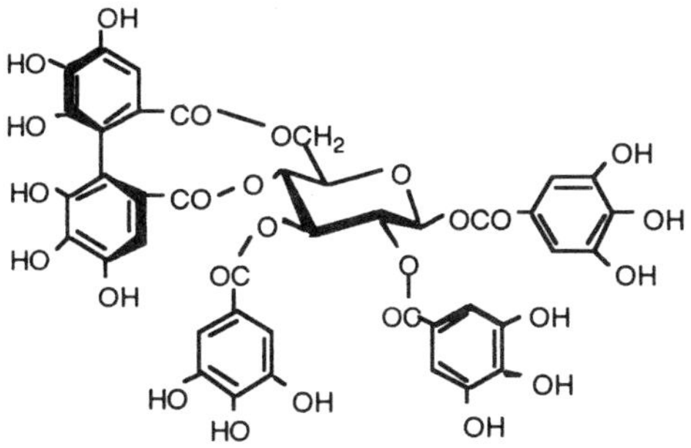

Tellimagrandin II

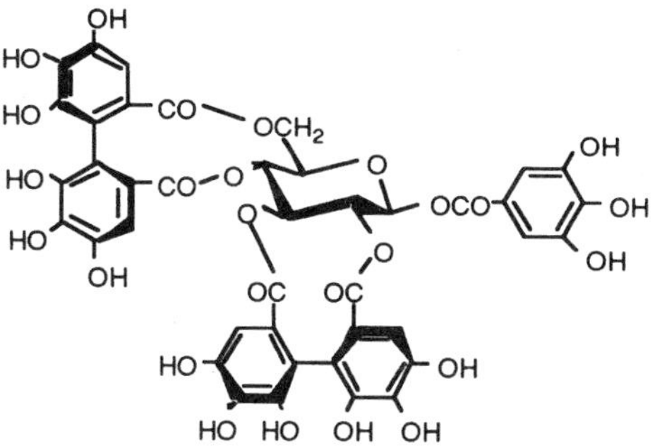

Casuarictin

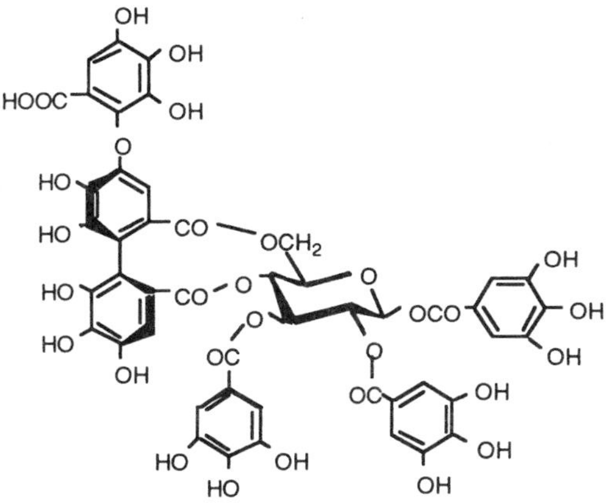

Rugosin A

Figure 3. Chemical structures of some monomeric ellagitannins.

Table 3. Acute toxicity of agrimoniin and oenothein B in mice

Route	LD$_{50}$(mg/kg)		
	i.p.	i.v.	p.o.
Agrimoniin	102.7 (88.2-119.1)	35.4 (30.7-41.5)	>1000
Oenothein B	79.2 (67.2-92.8)	——	>1000

( ): 95 percent confidence interval.

Table 4. Anticancer effects of premedication with agrimoniin and oenothein B on mammary carcinoma MM2 in mice

Compound	Route	Treatment schedule	Dose (mg/kg)	%ILS	60-day survivors
Agrimoniin	i.p.	on day -14	10	-	6
		on day -7	10	153	5
		on day -4	10	-	6
			3	6	3
			1	2	2
		on day -1	10	-	6
	i.v.	on day -14	3	14	4
		on day -7	3	17	4
		on day -4	3	14	4
		on day -1	3	0	2
	p.o.	on day -1 4	300	-1	4
		on day -7	300	34	5
		on day -4	300	7	2
		on day -1	300	9	1
Oenothein B	i.p.	on day -4	10	192	5
			3	74	1
			1	5	0
	p.o.	on days -7, -4	300	3	0
OK-432	i.p.	on day -4	100 (KE/kg)	64	3

MM2 cells (5x10^5) were i.p. inoculated into six female C3H/He mice in a group on day 0, and the treatment with a tannin or OK-432 was done in the indicated treatment schedules. The anticancer activity was evaluated on day 60.

Table 5. Anticancer effects of postmedication with agrimoniin and oenothein B on MM2 cancer-bearing mice

Compound	Route	Treatment schedule	Dose (mg/kg)	%ILS	60-day survivors
Agrimoniin	i.p.	on days 1,4,7	10	-	6
			3	16	0
			1	10	0
	i.v.	on days 1,4,7	10	16	2
			3	16	3
			1	56	3
	p.o.	on days 1-7	300	45	1
			100	7	2
			30	47	2
			10	19	0
Oenothein B	i.p.	on days 1,4,7	10	127	4
			3	40	1
			1	43	0
	p.o.	on days 1-7	300	2	0
OK-432	i.p.	on days 1,4,7	100 (KE/kg)	26	4

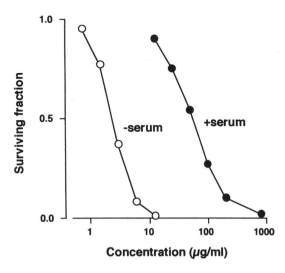

Figure 4. Direct cytotoxicity of agrimoniin against MM2 cells. MM2 cells (2×10^5 cells/mL) were treated with varying concentrations of agrimoniin in the absence (−serum) or presence (+serum) of 10 percent fetal calf serum for 2 hr. The treated cells were incubated for 48 hr, and vaible cell number was counted.

Table 6. Anticancer effects of agrimoniin and oenothein B on Meth-A solid type cancer-bearing mice

	Single dose (mg/kg)	Cancer weight (mean ± S.D., g)	% inhibition [a]	Regressors[b]
Agrimoniin	10	2.38 ± 0.36	30.2	0
	3	1.72 ± 0.54	49.8	2
	1	3.23 ± 0.48	5.4	0
Control	-	3.42 ± 0.45	-	0
Oenothein B	10	1.36 ± 0.57	64.9	0
	3	1.52 ± 0.53	60.7	0
	1	3.19 ± 0.42	17.6	0
OK-432	100 (KE/kg)	3.29 ± 0.63	15.0	0
Control	-	3.87 ± 0.57	-	0

Meth A cells (1×10^6) were subcutaneously inoculated at the left inguinal region of ten female BALB/c mice in a group on day 0, and the i.p. treatment with a tannin or OK-432 was done on day 8 to 14 once a day. The anticancer activity was evaluated on day 25 for agrimoniin and day 21 for oenothein B and OK-432.

[a][(C - T)/C] x 100, T: cancer weight of the test group, C: cancer weight of the vehichle control group.

[b]Number of primary cancer regressors out of ten mice in a group at the terminal of the experiments.

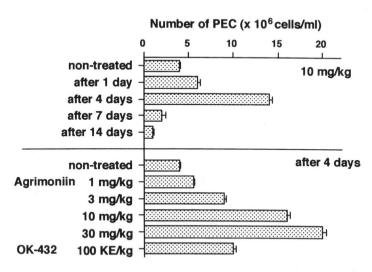

Figure 5. Changes of number of peritoneal exudate cells in mice treated with agrimoniin. Upper panel: experiments were done at the indicated days after the i.p. injection with agrimoniin (10 mg/kg). Lower panel: experiments were done at 4 days after the i.p. injection with the indicated doses of agrimoniin or OK-432. Data are the mean ± S.D. of triplicate determinations, of five mice.

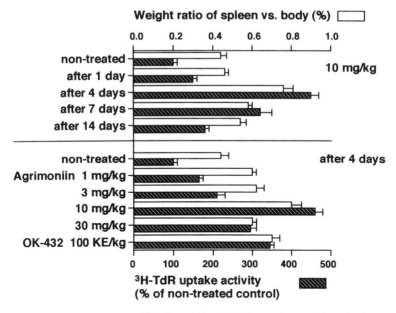

Figure 6. Spleen weight and ^{3}H-thymidine (TdR) uptake activity of spleen cells from mice treated with agrimoniin. The experiment protocol was the same mentioned in figure 5.

the self-growing activity of the cells (fig. 6). In addition, agrimoniin enhanced the cytotoxic potential of several effector cells with different induction kinetics[20]: the natural killer (NK) cell activity as an earlier response, the cytostatic activity of macrophage, and the antibody-dependent macrophage-mediated cytotoxic activity as a later response (fig. 7). These results indicate that the tannin activates macrophage and lymphocytes and exhibits the anticancer activity. However, these *in vivo* effects of agrimoniin were not observed in the *in vitro* experiment, except for the effect on macrophage, which may be described in detail in the following section. For example, table 7 shows that agrimoniin exhibited very weak migration activity against spleen cells from non-treated mice, compared with *E. coli* LPS and concanavarin A. This suggests that the action of the tannin is quite different from those of known mitogens and immunopotentiators.

Agrimoniin potentiated the cell-mediated immunity such as NK cell activation and induction of cytotoxic macrophage *in vivo* and exhibited strong anticancer activity against rodent cancers. We found that agrimoniin stimulated macrophage to produce and secret interleukin-1 (IL-1).[21,23] Figure 8 shows that agrimoniin induced IL-1 secretion from mice macrophage by not only *in vivo* treatment but *in vitro* treatment, in a concentration-dependent manner, with a similar or higher potency as that by *E. coli* LPS. Agrimoniin also stimulated human peripheral mononuclear cells to secret IL-1β, and this activation was detected as early as 4 hr after stimulation and was maintained at very high levels, the same as those after LPS-stimulation (fig. 9). These findings indicate that the target of the tannin

Table 7. *In vitro* stimulation of spleen cells from
normal mice

Agent	Concentration (μg/ml)	^{3}H-TdR uptake x 10^3 cpm (mean ± SE)
Agrimoniin	30	2.33 ± 0.61
	10	4.04 ± 0.41
	3.0	2.17 ± 0.25
	1.0	1.46 ± 0.04
E. coli LPS	1.0	35.2 ± 1.0
	0.1	18.8 ± 0.8
Concanavarin A	3.0	14.8 ± 0.2
	0.3	2.15 ± 0.22
Control		1.39 ± 0.10

Spleen cells (1 x 10^6) from normal C3H/He mice were incubated
with each agent for 24 hr and were further incubated with
0.1 μCi ^{3}H-TdR for 24 hr.

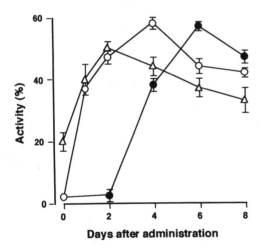

Figure 7. Cytotoxic activities of peritoneal exudate cells (PEC) from mice treated with agrimoniin. PEC were harvested from mice on the day indicated after an i.p. injection of agrimoniin (10 mg/kg). NK cell activity (△) was defined as cytolytic activity of nonadherent PEC against YAC-1 cells. Cytostatic activity (○) was determined as the growth inhibitory activity of adherent PEC agaist MM2 cells. Antibody-dependent cell-mediated cytotoxicity of adherent PEC (●) was assayed in the presence of anti-MM2 serum.

to exhibit anticancer activity is monocyte-macrophages but not lymphocytes and cancer cells.

Figure 10 shows the IL-1β-inducer activities of several tannins in human macrophage. Monocytes-macrophages were treated with each tannin (10 μM) for 4 hr and cultured for 24 hr. Following this, IL-1β in the culture supernatant was measured.[23] Structural units of condensed tannins, ECG and EGCG, having no host-mediated anticancer activity, induced IL-1β a little. Monomeric ellagitannins increased IL-1β production by about twofold over the non-stimulated basal

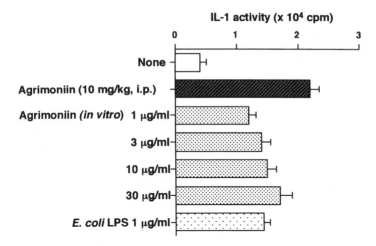

Figure 8. Secretion of IL-1 activity of macrophages treated with agrimoniin *in vivo* and *in vitro*. Adherent PEC from normal mice or agrimoniin (10 mg/kg, i.p.)-treated mice were cultured with or without the indicated concentrations of agrimoniin or LPS, and IL-1 activity in the culture supernatant was determined as the ^{3}H-TdR uptake activity of thymocytes. Data are the mean ± S.E.

Figure 9. Kinetics of IL-1β secretion from human peripheral monocytes treated by agrimoniin or LPS *in vitro*. Cells were incubated with agrimoniin (30 μg/mL, ●), LPS (10 μg/mL, ◇) or media alone (○) for the indicated times. IL-1 β in the culture supernatant was measured using an ELISA technique. Data are the mean ± S.E.

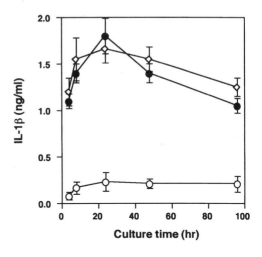

production. Among them, tellimagrandin II showed anticancer activity and produced IL-1β with a similar extent to other monomeric tannins, which showed less activity. The oligomeric ellagitannins, agrimoniin, hirtellin B, and oenotheins A and B, which exhibited strong anticancer activity, more potently stimulated the IL-1β production, but inactive oligomers, laevigatin B, and euphorbin C-Hy, effected less IL-1β production than the monomers. These results indicate that the correlation between their anticancer activity and IL-1β induction remains unclear. The tannins may be unstable, but it is unlikely that oligomeric tannins act on the target cells

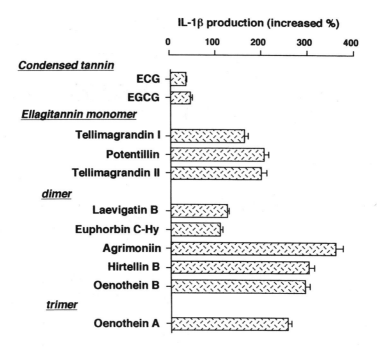

Figure 10. IL-1 β inducer activities of tannins. Human peripheral monocytes were treated with each tannin (10 μM) for 4 hr, then cultured for 24 hr. IL-1 β in the culture supernatant was measured and represented the percent of increased amount against control (305 ± 45 pg/mL). Data are the mean ± S.E.

after degradation to the monomers in the culture and in the host animals, for the following reasons: 1) the IL-1-inducer activities of agrimoniin, a dimer of potentillin unit, hirtellin B, a dimer of tellimagrandin II unit, oenothein B, a dimer of telimagrandin I and II unit, and oenothein A, a trimer of tellimagrandin I unit, were not additive of those of the corresponding monomers, 2) tellimagrandin II and potentillin produced more IL-1β than laevigatin B and euphorbin C-Hy, which have one of them in their molecule, 3) similarly, tellimagrandin II showed anticancer activity, but euphorbin C-Hy did not, 4) in contrast, potentillin was ineffective, while its dimer agrimoniin was a strong anticancer compound. This indicates that the tannin molecule may need tellimagrandins, potentillin, and their related structure units to show the host-mediated anticancer activity, but the whole structure and conformation of the oligomers are more important.

It is well documented that IL-1 causes macrophage and lymphocytes to induce IL-2, IL-2 receptors, and other lymphokines and potentiates many functions of host defense against cancers.[35,36] Moreover, there is evidence that IL-1 increases the binding of NK cells to tumor targets.[37] Since IL-1 induces interferon, which synergizes with IL-1 with respect to its action on NK cells,[38] one could view both mechanisms as an efficient aspect of host defense against cancers. Additionally, agrimoniin has a potency to induce IL-8 production by macrophage, but not IL-2 secretion by lymphocytes (unpublished data), resulting in an increase in the number of white blood cells.[18] Consequently, tannins initially stimulate monocytes-macrophage and act in host defense against cancer cells through complex action of activated immunocytes.

4. CONCLUSIONS

One hundred eight tannins and related polyphenols were intraperitoneally injected into mice at 4 days before intraperitoneal inoculation of sarcoma-180 cells and their anticancer activities were evaluated. The condensed tannins and related compounds all showed negligible activity. As regards the hydrolyzable tannins, active compounds were found among ellagitannins. In particular, dimeric ellagitannins, consisting of only potentillin, tellimagrandins I and II, and rugosin A as the monomer unit, exhibited a potent anticancer activity. Macrocyclic ellagitannins were all active. The host-mediated anticancer activity of agrimoniin, a dimeric ellagitannin isolated from *Agrimonia pilosa*, was presented in detail. Agrimoniin increased the number of cytotoxic effector cells and their activities in mice. This tannin induced interleukin-1 secretion from monocyte-macrophage in both *in vivo* and *in vitro* treatments. Consequently, the anticancer activity of tannins is attributable to potential of the host-defense via activation of macrophage.

Our research indicates that anticancer activity is one of the important biological activities of tannins. Oligomeric ellagitannins, which consist of monomer units, i.e., potentillin, tellimagrandin I and II, and related structures, showed a potent anticancer activity. It is notable that these tannins potentiate the immune-defense of the host through stimulation of monocyte-macrophage and IL-1 production. A part of the inhibitory activities by tannins on cancer initiation and promotion may be based on the potentiation of the host-immunity. The effects of tannin-rich plant foods on diseases due to decreased immunity may also be documented. This agrees with the concept underlying Oriental medicine. It is hoped that tannin preparations are developed for clinical application to cancer prevention and treatment.

REFERENCES

1. Hatano, T.; Edamatsu, R.; Hiramatsu, M.; Mori, A.; Fujita, Y.; Yasuhara, T.; Yoshida, T.; Okuda, T. Effects of tannins and related polyphenols on superoxide anion radical, and on 1,1-diphenyl-2-picrylhydrazyl radical. *Chem. Pharm. Bull.* 37:2016 (1989).

2. Okuda, T.; Yoshida, T.; Hatano, T. Antioxidant effects of tannins and related polyphenols. *In*: ACS Symposium Series No. 507. American Chemical Society, Washington, DC, p. 87 (1992).

3. Kimura, Y.; Okuda, H.; Okuda, T.; Yoshida, T.; Hatano, T.; Arichi, S. Studies on the activities of tannins and related compounds of medicinal plants and drugs. II. Effects of various tannins and related compounds on adrenaline-induced lipolysis in fat cells. *Chem. Pharm. Bull.* 31:2497 (1983).

4. Hong, C.-Y.; Wang, C.-P.; Huang, S.-S.; Hsu, F.-L. The inhibitory effect of tannins of lipid peroxidation of rat heart mitochondria. *J. Pharm. Pharmacol.* 47:138 (1995).

5. Hikino, H.; Kiso, Y.; Hatano, T.; Yoshida, T.; Okuda, T. Antihepatoxic actions of tannins. *J. Ethnopharmacol.* 14:19 (1985).

6. Okuda, T.; Mori, K.; Hayatsu, H. Inhibitory effect of tannins on direct-acting mutagens. *Chem. Pharm. Bull.* 32:3755 (1984).

7. Yoshizawa, S.; Horiuchi, T.; Fujiki, H.; Yoshida, T.; Okuda, T. Antitumor promoting activity of (–)epigallocatechin gallate, the main constituent of 'tannin' in green tea. *Phytotherapy Res.* 1:44 (1987).

8. Nishida, H.; Omori, M.; Fukutomi, Y.; Ninomiya, M.; Nishiwaki, S.; Suganuma, M.; Moriwaki, H.; Muto, Y. Inhibitory effects of (–)-epigallocatechin gallate on spontaneous hepatoma in C3H/HeNCrj mice and human hepatoma-derived PLC/PRF/5 cells. *Jpn. J. Cancer Res.* 85:221 (1994).

9. Fukuchi, K.; Sakagami, H.; Okuda, T.; Hatano, T.; Tanuma, S.; Kitajima, K.; Inoue, Y.; Inoue, S.; Ichikawa, S.; Nonoyama, M.; Konno, K. Inhibition of herpes simplex virus infection by tannins and related compounds. *Antiviral Res.* 11:285 (1989).

10. Asanaka, M.; Kurimura, T.; Koshiura, R.; Okuda, T.; Mori, M.; Yokoi, H. Tannins as candidate for anti-HIV drug. 4th International Conference on Immunopharmacology. Osaka, p. 35 (abstr.) (1988).

11. Nakashima, H.; Murakami, T.; Yamamoto, N.; Sakagami, H.; Tanuma, S.; Hatano, T.; Yoshida, T.; Okuda, T. Inhibition of human immunodeficiency viral replication by tannins and related compounds. *Antiviral Res.* 18:91 (1992).

12. Akamatsu, K. Shintei wakanyaku. Ishiyakushuppan, Tokyo (1970).

13. Kimura, K. Shinchukotei kokuyaku-honzokomoku. Shunyodo-Shoten, Tokyo (1974).

14. Chiang, S. New medical college dictionary of Chinese crude drugs. Shanghai Scientific Technologic Publisher, Shanghai (1997).

15. Kondo, K. Cancer therapy in China today. Shizensha, Tokyo (1997).

16. Koshiura, R.; Miyamoto, K.; Ikeya, Y.; Taguchi, H. Antitumor activity of methanol extract from roots of *Agrimonia pilosa* Ledeb. *Jpn. J. Pharmacol.* 38:9 (1985).

17. Miyamoto, K.; Koshiura, R.; Ikeya, Y.; Taguchi, H. Isolation of agrimoniin, an antitumor constituent, from the roots of *Agrimonia pilosa* Ledeb. *Chem. Pharm. Bull.* 33:3977 (1985).

18. Miyamoto, K.; Kishi, N.; Koshiura, R. Antitumor effect of agrimoniin, a tannin of *Agrimonia pilosa* Ledeb., on transplantable rodent tumors. *Jpn. J. Pharmacol.* 43:187 (1987).

19. Miyamoto, K.; Kishi, N.; Koshiura, R.; Yoshida, T.; Hatano, T.; Okuda, T. Relationship between the structures and antitumor activities of tannins. *Chem. Pharm. Bull.* 35:814 (1987).

20. Miyamoto, K.; Kishi, N.; Murayama, T.; Furukawa, T.; Koshiura, R. Induction of cytotoxicity of peritoneal exudate cells by agrimoniin, an immunomodulatory tannin of *Agrimonia pilosa* Ledeb. *Cancer Immunol. Immunother.* 27:59 (1988).

21. Murayama, T.; Kishi, N.; Koshiura, R.; Takagi, K.; Furukawa, T.; Miyamoto, K. Agrimoniin, an antitumor tannin of *Agrimonia pilosa* Ledeb., induces interleukin-1. *Anticancer Res.* 12:1417 (1992).

22. Miyamoto, K.; Nomura, M.; Murayama, T.; Furukawa, T.; Hatano, T.; Yoshida, T.; Koshiura, R.; Okuda, T. Antitumor activities of ellagitannins against sarcoma-180 in mice. *Biol. Pharm. Bull.* 16:379 (1993).

23. Miyamoto, K.; Murayama, T.; Nomura, M.; Hatano, T.; Yoshida, T.; Furukawa, T.; Koshiura, R.; Okuda, T. Antitumor activity and interleukin-1 induction by tannins. *Anticancer Res.* 13:37 (1993).

24. Okamoto, H.; Shoin, S.; Koshimura, S. Streptolysin S-forming and antitumor activities of group A streptococci. *In*: Jeljaszewicz, J.; Wadstrom, T. (eds.). Bacterial toxins and cell membrane. Academic Press, New York, p. 259 (1978).

25. Kai, S.; Tanaka, J.; Nomoto, K.; Torisu, M. Studies on the immunopotentiating effects of a streptococcal preparation, OK-432. I. Enhancement of T cell-mediated immune response of mice. *Clin. Exp. Immunol.* 37:98 (1979).

26. Murayama, T.; Natsuume-Sakai, S.; Ryoyama, K.; Koshimura, S. Studies on the properties of a streptococcal preparation, OK-432 (NSC-B116209), as an immunopotentiator. II. Mechanism of macrophage activation by OK-432. *Cancer Immunol. Immunother.* 12:141 (1982).

27. Okuda, T.; Yoshida, T.; Kuwahara, M.; Memon, M.U.; Shingu, T. Tannins of Rosaceous medicinal plants. I. Structures of potentillin, agrimonic acids A and B, and agrimoniin, a dimeric ellagitannin. *Chem. Pharm. Bull.* 32:2165 (1984).

28. Hatano, T.; Yasuhara, T.; Matsuda, M.; Yazaki, K.; Yoshida, T.; Okuda, T. Oenothein B, a dimeric hydrolyzable tannin with macrocyclic structure, and accompanying tannins from *Oenothera erythrosepala*. *J. Chem. Soc., Perkin Trans. 1.* :2735 (1990).

29. Yoshida, T.; Chou, T.; Nitta, A.; Miyamoto, K.; Koshiura, R.; Okuda, T. Woodfordin C, a macro-ring hydrolyzable tannin dimer with antitumor activity, and accompanying dimers from *Woodfordia fruticosa* flowers. *Chem. Pharm. Bull.* 38:1211 (1990).

30. Yoshida, T.; Chou, T.; Matsuda, M.; Yasuhara, T.; Yazaki, K.; Hatano, T.; Okuda, T. Antitumor hydrolyzable tannins of macro-ring structure with anti-tumor activity. *Chem. Pharm. Bull.* 39:1157 (1991).

31. Miyamoto, K.; Nomura, M.; Sasakura, M.; Matsui, E.; Koshiura, R.; Murayama, T.; Furukawa, T.; Hatano, T.; Yoshida, T.; Okuda, T. Antitumor activity of oenothein B, a unique macrocyclic ellagitannin. *Jpn. J. Cancer Res.* 84:99 (1993).

32. Haslam, E. Plant polyphenols—vegetable tannins revisited. Cambridge University Press, Cambridge (1989).

33. Okuda, T.; Mori, K.; Shiota, M.; Ida, K. Effects of the interaction of tannins with co-existing substances. II. Reduction of heavy metal ions and solubilization of precipitate. *Yakugaku Zasshi* 102:734 (1982).

34. Okuda, T.; Mori, K.; Shiota, M. Effects of the interaction of tannins with co-existing substances. III. Formation and solubilization of precipitates. *Yakugaku Zasshi* 102:854 (1982).

35. Durum, S.K.; Schmidt, J.A.; Oppenheim, J.J. IL-1, an immunological perspective. *Annu. Rev. Immunol.* 3:263 (1985).

36. Dinarello, C.A. Biology of interleukin 1. *Fed. Am. Soc. Exp. Biol. J.* 2:108 (1988).

37. Herman, J.; Dinarello, C.A.; Kew, M.C.; Rabson, A.R. The role of interleukin 1 (IL-1) in tumor-NK cell interactions: Correction of defective NK cell activity in cancer patients by treating target cells with IL-1. *J. Immunol.* 135:2882 (1985).

38. Dempsey, R.A.; Dinarello, C.A.; Mier, J.W.; Rosenwasser, L.J.; Allegretta, M.; Brown, T.E.; Parkinson, D.R. The differential effects of human leukocytic pyrogen/lymphocyte-activating factor, T cell growth factor, and interferon on human natural killer activity. *J. Immunol.* 129:2504 (1982).

INHIBITORY EFFECTS OF HYDROLYZABLE TANNINS ON TUMOR PROMOTING ACTIVITIES INDUCED BY 12-O-TETRADECANOYL-PHORBOL-13-ACETATE (TPA) IN JB6 MOUSE EPIDERMAL CELLS

Tadataka Noro,[a] Takeshi Ohki,[a] Yuko Noda,[b]
Tsutomu Warashina,[a] Kazuko Noro,[a]
Isao Tomita,[b] and Yoshiyuki Nakamura[b]

[a] Graduate School of Nutritional and Environmental Sciences
University of Shizuoka
52-1 Yada
Shizuoka, 422-8526
JAPAN

[b] School of Pharmaceutical Sciences
University of Shizuoka
52-1 Yada
Shizuoka, 422-8526
JAPAN

1. INTRODUCTION

Our studies are currently focused on the constituents of tea. Tea is a flavorful healthy food whose anticancer properties are the subject of many ongoing investigations. The main components of tea are phenolic compounds such as catechins, as well as caffeine. Theaflavins are the components of black tea. Theaflavins are phenolic compounds; also, strictinin is a hydrolyzable tannin of tea (fig. 1). It has been reported that tea extracts show anticancer effects, and these extracts contain many kinds of phenolic components. On the other hand, Professors Fujiki and Yoshida et al.[1,2] reported that hydrolyzable tannins, geraniin, penta-O-galloyl-β-D-glucose, etc., showed inhibitory effects on tumor promotion induced by teleocidin on mouse epidermis. In this case, the initiator of carcinogenesis was 7,12-dimethyl-benz[a]anthracene (DMBA). This suggested that hydrolyzable tannins have

Plant Polyphenols 2: Chemistry, Biology, Pharmacology, Ecology, Edited by
Gross et al. Kluwer Academic / Plenum Publishers, New York, 1999

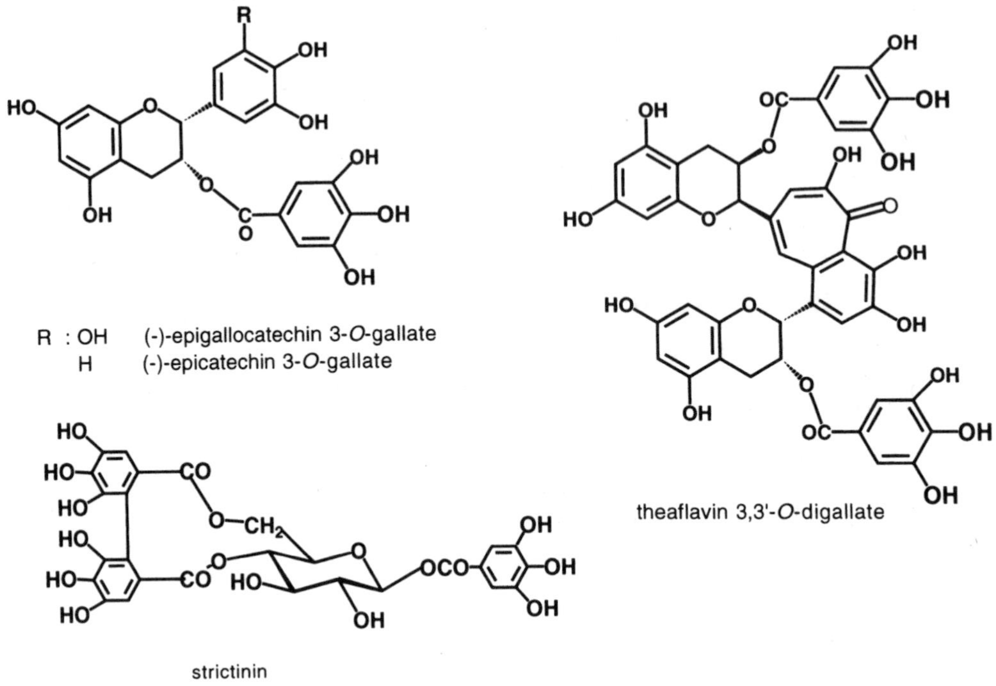

R : OH   (-)-epigallocatechin 3-O-gallate
    H   (-)-epicatechin 3-O-gallate

theaflavin 3,3'-O-digallate

strictinin

Figure 1. Chemical structure of hydrolyzable tannins.

antitumor promotion effects. Therefore, we focused on the inhibitory effects of hydrolyzable tannins on tumor promoting activities in other assay systems.

This paper reports on the inhibitory effects of hydolyzable tannins, such as 1,2,3,6-tetra-O-galloyl-β-D-glucose (**1**), 1,2,3,4,6-penta-O-galloyl-β-D-glucose (**2**), corilagin (**3**), and nupharin C (**4**) on the neoplastic transformation to be induced by TPA in JB6 mouse epidermal cells (fig. 2).

2. MOUSE EPIDERMAL JB6 CELLS

JB6 cells were established from the BALB/c mouse primary epidermal cell culture system by N.H. Colburn et al.[3,4] in 1978. JB6 cells are sensitive to promotion of neoplastic transformation by 12-O-tetradecanoylphorbol-13-acetate (TPA) without an initiator. We used this cell line for our experiments to assay the inhibitory effects of the hydrolyzable tannins on tumor promotion. The assay procedure for testing the effect of the test samples on promotion of neoplastic transformation by TPA is as follows. Anchorage-dependent JB6 cells change to anchorage-independent cells by the promoter TPA. Many colonies of neoplastically transformed JB6 cells, anchorage-independent colonies, can be observed on the 0.33 percent soft agar test plates. On the other hand, if the test samples are inhibitors of cancer promotion to be induced by TPA, when anchorage-dependent JB6 cells are cultured together with TPA and test samples,

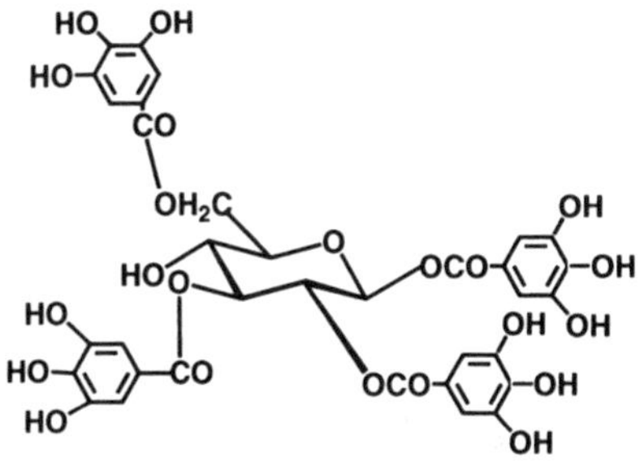

1,2,3,6-tetra-*O*-galloyl-β-D-glucose (**1**)

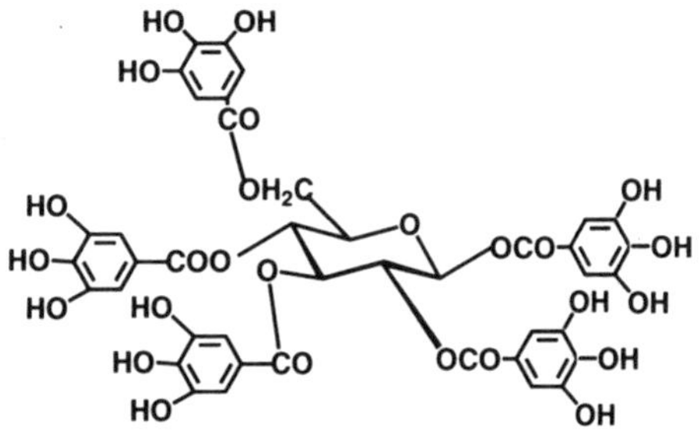

1,2,3,4,6-penta-*O*-galloyl-β-D-glucose (**2**)

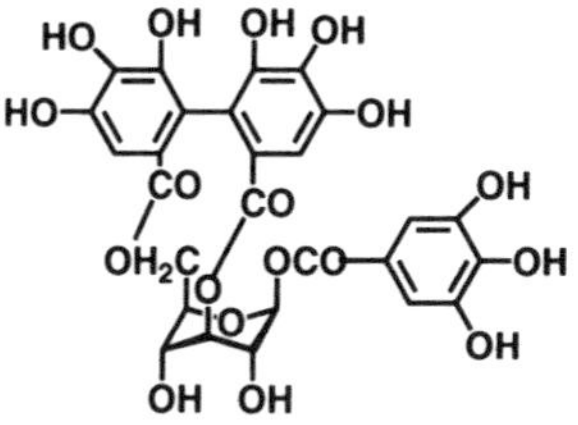

corilagin (**3**)

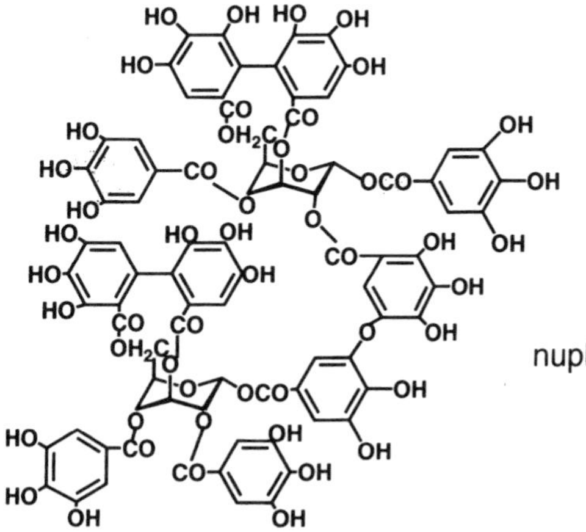

nupharin C (**4**)

Figure 2. Chemical structure of **1,2,3** and **4**.

anchorage-independent colonies on the soft agar test plates will grow very little. JB6 cell lines of Cl22-2a and Cl41 (clones 22-2a and 41) were used to evaluate the ability of test samples as to inhibitory effects on the promotion of neoplastic transformation by TPA.

3. PREPARATION OF HYDROLYZABLE TANNINS

We prepared four hydrolyzable tannins. One of them, 1,2,3,6-tetra-*O*-galloyl-β-D-glucose (TGG) (**1**), was extracted from the dried leaves of *Arctostaphylos uva-ursi* (Ericaceae) as shown in figure 3. Corilagin (**3**) was extracted from the same plants. Compounds **1**[5] and **3**[6,7] were identified by comparison with data in the literature (fig. 3).

1,2,3,4,6-Penta-*O*-galloyl-β-D-glucose (PGG) (**2**) was prepared by methanolysis of tannic acid, Chinese gallotannin, of *Rhus javanica* (Anacardiaceae) as shown in figure 4. For methanolysis, 0.5N acetate buffer (pH 6.0) was used. Compound **2**[8,9] was identified by comparison with data in the literature (fig. 4).

Another hydrolyzable tannin, nupharin C (**4**), was extracted from the rhizomes of *Nuphar japonicum* (Nymphaeaceae) shown in figure 5. The rhizomes were

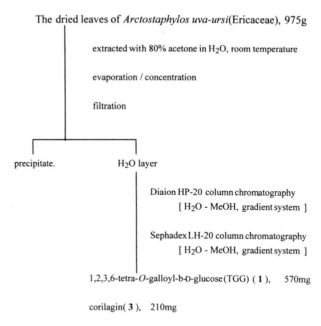

Figure 3. Extraction of **1** and **3**.

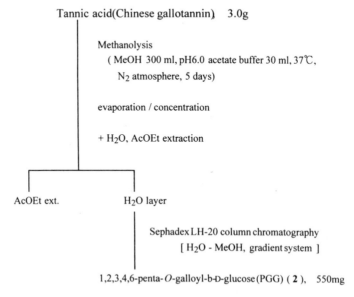

Figure 4. Preparation of **2**.

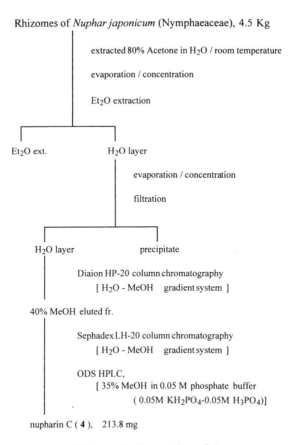

Rhizomes of *Nuphar japonicum* (Nymphaeaceae), 4.5 Kg

extracted 80% Acetone in H_2O / room temperature

evaporation / concentration

Et_2O extraction

Et_2O ext. H_2O layer

evaporation / concentration

filtration

H_2O layer   precipitate

Diaion HP-20 column chromatography
[H_2O - MeOH   gradient system]

40% MeOH eluted fr.

Sephadex LH-20 column chromatography
[H_2O - MeOH   gradient system]

ODS HPLC,
[35% MeOH in 0.05 M phosphate buffer
(0.05M KH_2PO_4-0.05M H_3PO_4)]

nupharin C (**4**),   213.8 mg

Figure 5. Extraction of **4**.

extracted with 80 percent acetone in water at room temperature. The extract was concentrated and was separated into components on Dia-ion HP-20 and Sephadex LH-20 column chromatography. Finally, nupharin C was purified by ODS HPLC. The elution solvent was 35 percent methanol in 0.05 M phosphate buffer (0.05 M KH_2PO_4-0.05 M H_3PO_4). Compound **4**[10] was identified by comparison with data in the literature (fig. 5).

4. ASSAY METHODS

For determination of the maximal dose cytotoxicity test[3,4] of samples, tester JB6 cells were seeded in Eagle's minimum essential medium supplemented with 8 percent heat-inactivated fetal bovine serum at a density of 1×10^4 cells per well in 24 well plates. These were exposed to several doses of each test sample with or without TPA (1 ng/mL, 1.6×10^{-9} M) after 20–24 hrs from the seeding.[3,4] The cells, while growing in a logarithmical phase, were trypsinized and counted after 3 or 4 days of treatment. Maximal dose was determined to a concentration of 90 percent survival against the control. JB6 cells growing logarithmically in monolayer culture were trypsinized and suspended in 0.33 percent agar medium containing 10 percent serum and 1 ng/mL TPA (1.6×10^{-9} M) with or without test samples. In duplicated 60 mm petri dishes, 1.5 mL of the suspension of containing 1×10^4 cells was plated over a bottom 0.5 percent agar layer containing the same concentration of TPA with or without test samples. Soft agar grouped colonies of eight or more cells were scored after 14 days.

5. EFFECTS OF HYDROLYZABLE TANNINS ON SOFT AGAR COLONY INDUCTION IN JB6 CELLS

The effects (percent of TPA control) of hydrolyzable tannins on the soft agar colony induction in JB6 cells are shown in table 1 and figures 6 through 9. In table 1, the upper lines are results by Cl22-2a, and the lower lines are results by Cl41. The "Ave." rows in table 1 are the average of the results for both Cl22-2a and Cl41. All four hydrolyzable tannins showed strong inhibitory effects against TPA-induced cancer promotion in JB6 cells. Nupharin C was particularly active as it inhibited cancer promotion by about 60 percent at the concentration of 3 μg/mL (table 1 and figures 6 through 9).

6. CONCLUSIONS

The antitumor promoting activities (percent inhibition of TPA control) of four hydrolyzable tannin test samples at concentrations of 3 μg/ml were as follows:

Table 1. Effects of hydrolyzable tannins on the soft agar induction in JB6 cells (percent of TPA control)

Concentration(mg/ml)		0.0	0.1	0.3	1.0	3.0	10.0
TGG(1)	Cl22-2a	100		81.4	75.7	54.2	
	Cl41	100		88.1	55.7	44.3	
	Ave.(Cl22,Cl41)	100		84.8	65.7	49.3	
PGG(2)	Cl22-2a	100			70.8	46.1	13.2
	Cl41	100			74.7	47.0	18.5
	Ave.(Cl22,Cl41)	100			72.8	46.6	15.8
corilagin(3)	Cl22-2a	100	77.3		64.7	51.0	
	Cl41	100	81.8		69.5	53.6	
	Ave.(Cl22,Cl41)	100	79.6		67.1	52.3	
nupharin C(4)	Cl22-2a	100		80.6	72.1	43.9	
	Cl41	100		82.6	48.7	35.2	
	Ave.(Cl22,Cl41)	100		81.6	60.4	39.6	

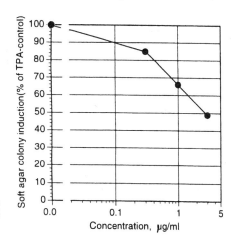

Figure 6. Effect of hydrolyzable tannin, TGG (**1**) on the soft agar induction in JB6 cells (percent of TPA control).

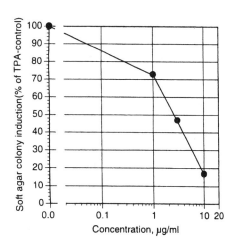

Figure 7. Effect of hydrolyzable tannin, PGG (**2**) on the soft agar induction in JB6 cells (percent of TPA control).

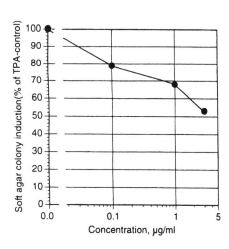

Figure 8. Effect of hydrolyzable tannin, corilagin (**3**) on the soft agar induction in JB6 cells (percent of TPA control).

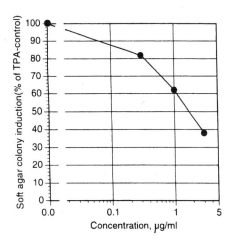

Figure 9. Effect of hydrolyzable tannin, nupharin C (**4**) on the soft agar induction in JB6 cells (percent of TPA control).

1,2,3,6-tetra-*O*-galloyl-β-D-glucose (**1**), TGG = 50.7 percent inhibition; 1,2,3,4,6-penta-*O*-galloyl-β-D-glucose (**2**), PGG = 53.4 percent inhibition; corilagin (**3**) = 47.7 percent inhibition: and nupharin C (**4**) = 60.4 percent inhibition. These compounds showed almost no effects on the cell growth and the soft agar colony formation of JB6 tumorigenic cells at this concentration. These results indicate that some hydrolyzable tannins have suppressing effects on tumor promotion induced by TPA in JB6 mouse epidermal cells. The number of hydroxyl groups in the molecule is closely related to these activities.

ACKNOWLEDGMENTS

This work was supported by a grant from the Program for Promotion of Basic Research Activities for Innovative Biosciences (PROBRAIN) in Japan.

REFERENCES

1. Yoshizawa, S.; Fujiki, H.; Horiuchi, T.; Sugiura, M.; Suganuma, M.; Nishiwaki, S.; Okabe, S.; Okuda, T. *In:* Huang, M.T.; Ho, C.T., Lee, C.Y. (eds.). Phenolic compounds in food and health II. Antioxidant and cancer prevention Am. Chem. Soc., Washington DC, pp. 316–325 (1992).

2. Horiuchi, T.; Fujiki, H.; Yamashita, K.; Suganuma, M.; Sugimura, T.; Yoshida, T.; Okuda, T. Inhibition of tumor promotion by tannins. Abstracts of papers, 32nd Annual Meeting of the Japanese Society of Pharmacognosy, Okayama, October, p. 11 (1985).

3. Colburn, N.H.; Bruegge, W.F.V.; Bates, J.R.; Gray, R.H.; Rossen, J.D.; Kelsey, W.H.; Shimada, T. Correlation of anchorage-independent growth with tumorigenicity of chemically transformed mouse epidermal cells. *Cancer Res.* 35:624–634 (1978).

4. Nakamura, Y.; Colburn, N.H.; Gindhart, T.D. Role of reactive oxygen in tumor promotion: implication of superoxide anion in promotion of neoplastic transformation in JB-6 cells by TPA. *Carcinogenesis*, 6:229–235 (1985).

5. Nishizawa, M.; Yamagishi, T.; Nonaka, G.; Nishioka, I. Tannins and related compounds. Part 9. Isolation and characterization of polygalloylglucoses from turkish gall (*Quercus infectoria*). *J. Chem. Soc., Perkin Trans.* I:961–965 (1983).

6. Tanaka, T.; Nonaka, G.; Nishioka, I. Punicafolin, an ellagitannin from the leaves of *Punica granatum. Phytochemistry* 24:2075–2078 (1985).

7. Yoshida, T.; Okuda, T. ^{13}C nuclear magnetic resonance spectra of corilagin and geraniin. *Heterocycles* 14:1743–1749 (1980).

8. Nishizawa, M.; Yamagishi, T.; Nonaka, G.; Nishioka, I.; Nagasawa, T.; Oura, H. Tannins and related compounds. XII. Isolation and characterization of galloylglucoses from *Paeoniae radex* and their effect on urea-nitrogen concentration in rat serum. *Chem. Pharm. Bull.* 31:2593–2600 (1983).

9. Armitage, R.; Bayliss, G.S.; Gramshaw, J.W.; Haslam, E.; Haworth, R.D.; Jones, K.; Rogers, H.J.; Searle, T. Gallotannins. Part III. The constitution of chinese, turkish, sumach, and tara tannins. *J. Chem. Soc.* 1842 (1961).

10. Ishimatsu, M.; Tanaka, T.; Nonaka, G.; Nishioka, I.; Nishizawa, M.; Yamagishi, T., Tannins and related compoumds. LXXIX. Isolation and characterization of novel dimeric and trimeric hydrolyzable tannins, nupharin C, D, E and F, from *Nuphar japonicum* DC. *Chem. Pharm. Bull.* 37:1735–1743 (1989).

PLANT LIGNANS AND HEALTH: CANCER CHEMOPREVENTION AND BIOTECHNOLOGICAL OPPORTUNITIES

Joshua D. Ford, Laurence B. Davin, and Norman G. Lewis

Institute of Biological Chemistry
Washington State University
Pullman, Washington 99164-6340
USA

1. INTRODUCTION

A considerable body of evidence has accumulated in the past two decades suggesting that the consumption of certain lignan-containing foodstuffs significantly reduces the risk of breast, colon, and prostate cancers, i.e., via a so-called chemoprevention mechanism.[1] Accordingly, the lignans of primary interest in this chapter are the diphenolic compounds from higher plants, which are converted into the "mammalian" lignans enterodiol (**1**) and enterolactone (**2**) (fig. 1) by the action of colonic bacteria. This chapter thus focuses on the major mammalian lignans so formed and their roles in cancer chemoprevention. A summary of what is currently known about plant lignan biosynthesis in *Forsythia intermedia*, *Linum flavum*, western red cedar (*Thuja plicata*), and flax (*Linum usitatissimum*) is also presented, together with how this knowledge can be exploited for the benefit of humanity.

2. MAMMALIAN LIGNAN CHEMOPREVENTION OF VARIOUS HUMAN CANCERS

The diets of people living in Western countries, which are typically low in dietary fiber and high in animal fat/protein, have been viewed as contributing significantly to the increased incidences of breast, prostate, and colon cancers.[2-5] On the other hand, individuals living in the Western Hemisphere, but who are sustained on either vegetarian or semi-vegetarian diets, have significantly lower incidence rates of such diseases.[6-14] In addition to these (semi)vegetarians, it would appear that some measure of protection is conferred to the populations of various Asian countries since they also have low incidence rates of such cancers. However, immigrants from Asian countries, after consuming typical Western

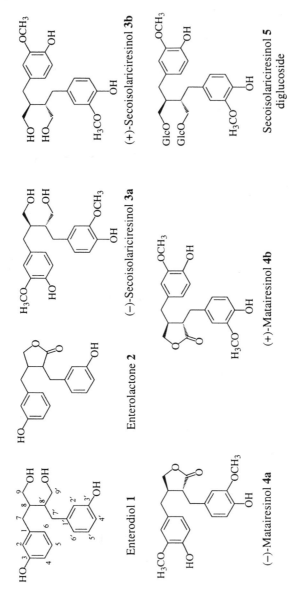

Figure 1. The so-called "mammalian" lignans enterodiol (**1**) and enterolactone (**2**) and their plant lignan precursors, (+) and (−)-secoisolariciresinols (**3a/3b**), (+)- and (−)-matairesinols (**4a/4b**) and secoisolariciresinol diglucoside (**5**).

diets over prolonged periods, subsequently develop these diseases at the same levels of the general population, suggesting that they had developed no special genetic protection.[15] This was shown, for example, in the increased incidence rates of such cancers in Japanese people who had immigrated to the United States (Hawaii) and who became "assimilated" in terms of typical Western dietary preferences.

Initially, the reductions in cancer frequencies were attributed simply to the differences (increases) in the amount of dietary fiber ingested.[6-11] However, this could only be partially correct since comparisons showed that total fiber intake for Japanese and Western diets was roughly equivalent.[12] Recognition was eventually paid to two classes of plant-derived phenolics, the so-called "mammalian" lignans and isoflavonoids, detected first in urine and ultimately attributed with significant roles as cancer-preventive substances.[12-37] Although several thousand lignans have thus far been discovered in the plant kingdom,[38] only a very few are positively correlated with "chemoprotection or chemoprevention" against cancers and other disorders.

The two most important "mammalian" lignans, which are present in urinary excretions, bile, and blood plasma, are enterodiol (**1**) and enterolactone (**2**) (see fig. 1). They are considered to be mainly, if not exclusively, formed from the plant lignans, secoisolariciresinol (**3**) and matairesinol (**4**), respectively, following dietary ingestion and subsequent metabolism by gastrointestinal flora such as *Clostridia* sp.[22,39-44]

Flax seed (*L. usitatissimum*) is currently the richest source of the plant-derived "mammalian" lignan precursors. Its primary mammalian lignan precursor is (+)-secoisolariciresinol (**3b**), which is stored as the conjugate, (+)-secoisolariciresinol diglucoside (**5**) (SDG). Indeed, flaxseed contains levels 75–800 times greater than any other plant food[43] and has been widely examined for its cancer preventive effects. Moreover, *in vitro* studies using cultured human fecal microflora confirmed a metabolic pathway for the formation of the mammalian lignans from SDG (**5**). That is, intestinal bacteria hydrolyze the sugar moiety to release secoisolariciresinol (**3**), followed by loss of both hydroxyl groups at C-4/C-4′ and demethylation at C-3/C-3′ to give rise to the mammalian lignan enterodiol (**1**). The latter is then presumed to be oxidized by the gastrointestinal flora, at least in part, to give enterolactone (**2**), although it may also arise from catabolism of the plant lignan, matairesinol (**4**), the amounts of which can vary depending upon the plant species under consideration. Once formed, both enterodiol (**1**) and enterolactone (**2**) can undergo enterohepatic circulation[40,41,45] as shown in figure 2, and a striking positive correlation has been established as regards their presence and the reduced incidence rates of breast, colon, and prostate cancers.[46-51]

In order to provisionally account for how the plant lignans secoisolariciresinol (**3**) and matairesinol (**4**) are converted into the "mammalian" lignans enterodiol (**1**) and enterolactone (**2**), the following hypothesis is proposed in figure 3. As shown, the conversions contemplated to be engendered by the gastrointestinal flora involve demethylation, dehydration, and perhaps even racemization, with the actual sequence of events involving any or all of the permutations illustrated. Note also that the proposed dehydration of a 1,4-dihydrophenyl intermediate (**6**) and (**9**)

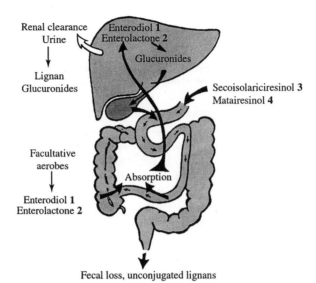

Figure 2. The physiological enterohepatic circulation of lignans in man. Lignans are biosynthesized most probably from ingested precursors by intestinal bacteria and absorbed from the intestine. After transfer to the liver, they are efficiently conjugated to glucuronic acid and excreted both in urine and bile as conjugates. Conjugated lignans are reabsorbed from the intestine and that which escapes reabsorption becomes deconjugated during passage through the intestine and excreted in the faeces in unconjugated form. The arrows in the diagram are not intended to indicate that the exact sites of absorption of lignans are known but merely to illustrate the enterohepatic circulation occurring. (Redrawn from Setchell et al.[40])

has precedence in the plant literature, e.g., in the conversion of arogenic acid into phenylalanine.[52]

Several distinct mechanisms have also been proposed to account for the anti-cancer, or cancer-preventing, actions of the "mammalian" lignans. These include: modulating enzyme activities related to sex hormone metabolism; increasing sex hormone binding globulin levels in the body; inhibiting cell proliferation; competing for estrogen binding sites; and displaying antioxidant activities.[1,29,39,53–60] Selected examples of these properties are discussed further below to illustrate their roles in health maintenance. For example, enterodiol (**1**) and enterolactone (**2**) have been suggested as modulating the onset of breast cancer because of their structural similarities to the estrogens estradiol (**10**) and diethylstilbestrol (**11**) (see fig. 4). In support of this contention, epidemiological studies have revealed a lower urinary excretion (a marker of production in the colon) of enterodiol (**1**) and enterolactone (**2**) in breast cancer patients and populations with a higher risk of breast cancer (non-vegetarians) than those at low risk (vegetarians).[1,29,39,53,54]

Additionally, oral administration of SDG (**5**) has also been documented as having antitumor effects at the early promotion stage of tumorigenesis in Sprague-Dawley female rats. The ingestion of purified SDG (**5**) at 1.5 mg/day for 20 weeks,

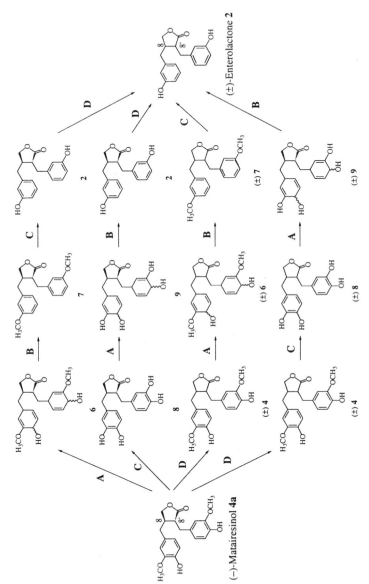

Figure 3. Possible catabolic pathways to enterolactone (**2**) from matairesinol (**4**) via action of gastrointestinal flora. (A) Reductase-catalyzed formation of 1,4-dihydrophenyl intermediate (**6**) and (**9**), (B) dehydratase-catalyzed generation of phenyl intermediates (**7**) and (**2**) (or via spontaneous dehydration), (C) demethylase-catalyzed conversions affording (**2**) and (**8**), and (D) racemization at C-8/C-8'. N.B. comparable conversions are considered to result from the conversion of secoisolariciresinol (**3**) into enterodiol (**1**).

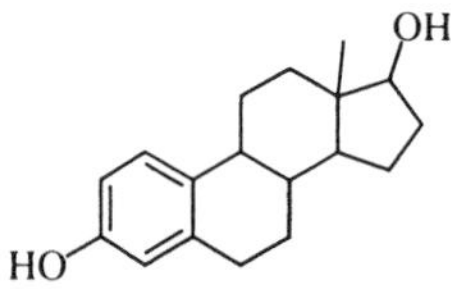

Estradiol **10** 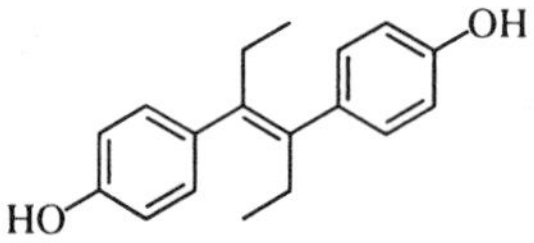   Diethylstilbestrol **11** 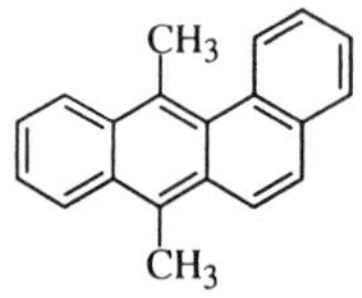   Dimethylbenzanthracene **12**

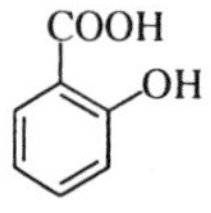

Salicylic acid **13** 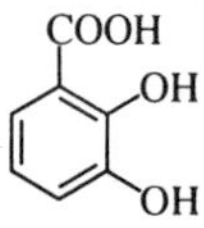   2,3-Dihydroxybenzoic acid **14** 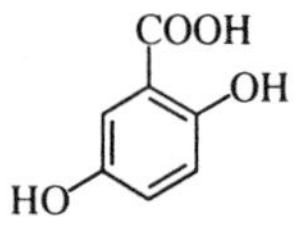   2,5-Dihydroxybenzoic acid **15**

Figure 4. The estrogens (**10**) and (**11**), the carcinogen, dimethylbenzanthracene (**12**), and salicylic acid (**13**) and its derivatives (**14**) and (**15**).

beginning 1 week after treatment with the carcinogen dimethylbenzanthracene (**12**) resulted in a 37 percent reduction ($p < 0.05$) in the number of tumors per tumor-bearing rat and a 46 percent reduction ($p < 0.05$) in the number of tumors per number of rats in each group. Additionally, this study showed that urinary "mammalian" lignan (**1**) and (**2**) excretion significantly increased ($p < 0.0001$) with SDG (**5**) treatment, revealing that the conversion of SDG (**5**) into the "mammalian" lignans (**1**) and (**2**) had also occurred.[54]

Concerning possible antiestrogenic effects of mammalian lignans, enterodiol (**1**) has been shown to stimulate the *in vitro* synthesis of sex hormone binding globulin (SHBG) by HepG2 liver carcinoma cells. Indeed, for comparative purposes, 10 times the concentration of estradiol (**10**) is required for the same effect. This is considered as antiestrogenic since increased SHBG concentrations are associated with decreased free estradiol (**10**) concentrations and/or the removal of free estradiol (**10**).[55,56] Enterolactone (**1**) also inhibits the estradiol-stimulated cell growth of estrogen dependent MCF-7 breast cancer cells at the same concentrations required for direct stimulation of cell growth.[57] Additionally, "mammalian" lignans (**1**) and (**2**) competitively inhibit aromatase, the rate-limiting enzyme in estrogen biosynthesis.[58,59]

Aside from this, SDG **5** has antioxidant properties, e.g., to scavenge oxygen radicals.[60] For example, the production of $HO^\bullet$ via photolysis of H_2O_2 with ultraviolet light in the presence of salicylic acid (**13**) results in the formation of the $OH^\bullet$ adduct products 2,3-dihydroxybenzoic acid (DHBA) (**14**) and 2,5-DHBA (**15**) (see fig. 4). When SDG (**5**) was introduced, however, a concentration dependent decrease in the formation of 2,3-DHBA (**14**) and 2,5-DHBA (**15**) was noted. The decrease in $HO^\bullet$ adduct products was due to scavenging of $HO^\bullet$ by SDG (**5**) and not by scavenging to form 2,3-DHBA (**14**) and 2,5-DHBA (**15**).[60]

In general, the lignans are widely distributed through vascular plants with structures of several thousand having now been determined.[61] They are produced in various tissues ranging from fruits, flowers, and seeds to roots, stems, and leaves.[62] They are normally found in dimeric form (two phenylpropanoids linked

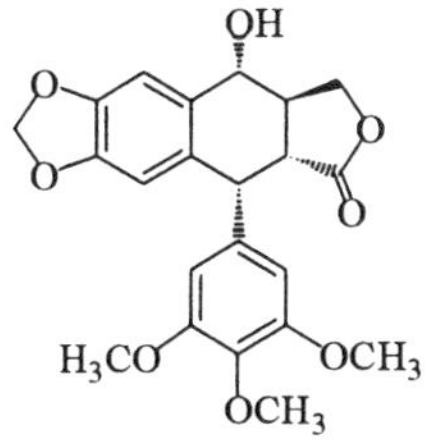

Figure 5. Podophyllotoxin (**16**), etoposide (**17**) and teniposide (**18**).

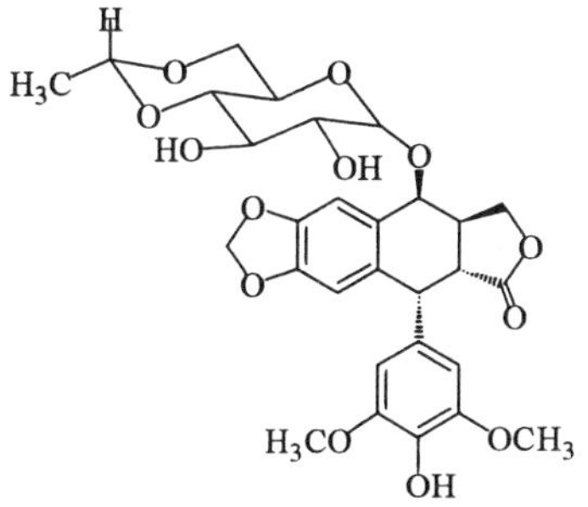

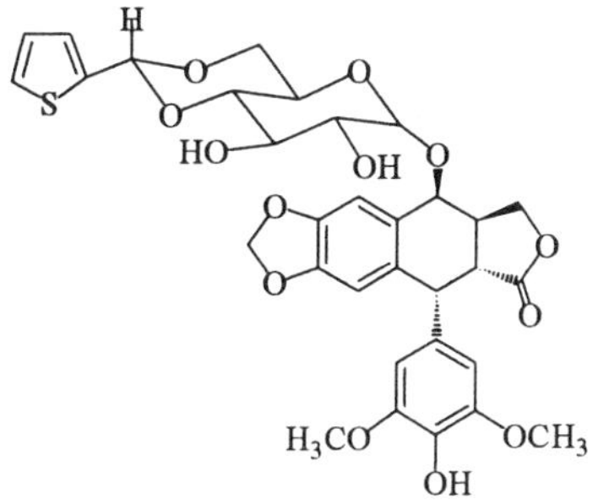

together) and are optically active. Many have been shown to demonstrate important physiological functions in defense (via their potential fungicidal,[63,64] antiviral,[65,66] bactericidal,[63,64,67–69] antifeedant,[70,71] and antioxidant[72–75] properties). From a health perspective, and for the purpose of this chapter, two of the most important lignans are secoisolariciresinol (**3**) and matairesinol (**4**). They are not only the presumed precursors of enterodiol (**1**) and enterolactone (**2**), but also of the antiviral lignan podophyllotoxin (**16**), which is either used directly in the treatment of venereal warts or is derivatized to give etoposide (**17**) and teniposide (**18**) (see fig. 5), with the latter two substances being used extensively today in cancer treatment.[62]

One focus of our research has thus been to delineate the biochemical pathways to the "mammalian" lignan precursors secoisolariciresinol (**3**), secoisolariciresinol diglucoside (**5**), and matairesinol (**4**). This was undertaken as a necessary first step prior to biotechnologically increasing the levels of these lignans in staple dietary foodstuffs, e.g., cereals, which typically produce them in low amount. As indicated below, the studies to date have employed *Forsythia*, *L. flavum*, western red cedar (*T. plicata*), and flax (*L. usitatissimum*) as model systems, in order to facilitate both biochemical and genetic analyses.

3. LIGNAN BIOSYNTHESIS

The biosynthetic pathway leading to secoisolariciresinol (**3**) and matairesinol (**4**) was elucidated using *F. intermedia* as a model system[38,61,76–85] (fig. 6). It initially involves the first known example of regio- and stereoselective intermolecular phenoxy radical coupling of two molecules of *E*-coniferyl alcohol (**19**) to yield (+)-pinoresinol (**20a**).[79] Sequential enantiospecific reduction of this intermediate then occurs to consecutively generate (+)-lariciresinol (**21a**) and then (−)-secoisolariciresinol (**3a**),[79,80,82,83] whose dehydrogenation affords (−)-matairesinol (**4a**).[86–88] The latter is presumed to be the precursor of podophyllotoxin (**16**)[89] in *Podophyllum hexandrum* (fig. 5) and plicatic acid (**22**) (fig. 7) in western red cedar (*T. plicata*) heartwood.[90]

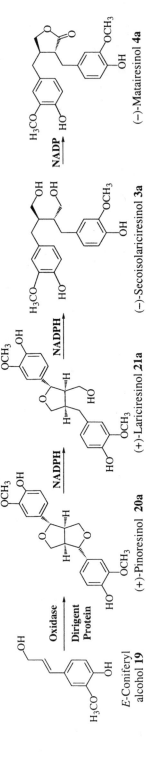

Figure 6. Biosynthetic pathway to lignans (–)-secoisolariciresinol (**3a**) and (–)-matairesinol (**4a**) in *Forsythia intermedia.*

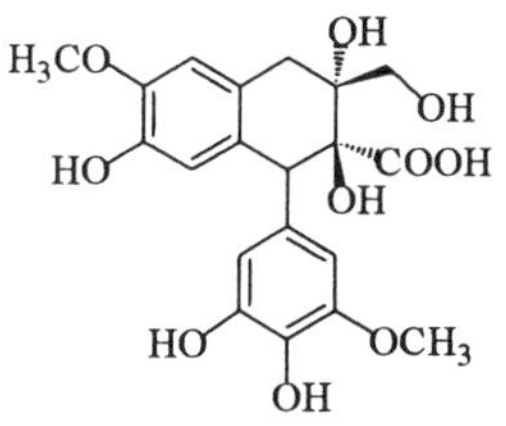

Plicatic acid **22** 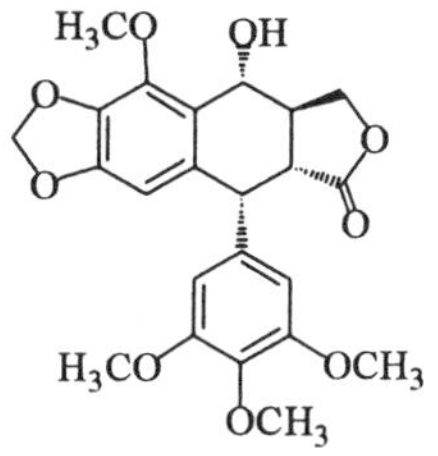   (−)-5-Methoxypodophyllotoxin **23**

Figure 7. Plicatic acid (**22**) from western red cedar heartwood and 5-methoxy-podophyllotoxin (**23**) from *Linum flavum*.

Interestingly, in *Forsythia* species, secoisolariciresinol (**3**) and matairesinol (**4**) are found as their (−)-enantiomers (**3a** and **4a**), whereas in flaxseed, the secoiso-lariciresinol (**3**) is present essentially only as the (+)-form (**3b**).[61] Furthermore, as described below, both secoisolariciresinol (**3**) enantiomers have been obtained by expressing the corresponding recombinant proteins in *E. coli*. This was done as a first step toward establishing in the future the relative efficacies of each enan-tiomeric form as a cancer "chemoprevention" agent. Each biochemical step to the "mammalian lignan" precursors (**3**), (**5**), and (**4**) is described below.

3.1 Stereoselective Phenolic Coupling and Dirigent Proteins

The stereoselectivity of bimolecular phenoxy radical coupling was first discov-ered using a *Forsythia* species.[77] Initial work established that an insoluble *For-sythia* stem residue preparation, remaining following removal of soluble proteins was able to catalyze the 8–8′ stereoselective coupling of two molecules of *E*-coniferyl alcohol (**19**) to give optically pure (+)-pinoresinol (**20a**).[79] In that partic-ular case, an ~78 kDa dirigent protein engendering the formation of (+)-pinoresinol (**20a**) was subsequently solubilized and purified to apparent homogeneity. The term, dirigent protein (Latin: *dirigere*, to guide), was introduced to define this new class of proteins, since they confer specificity to intermolecular phenoxy radical coupling reactions.[79] However, bimolecular coupling stereoselectivity was only observed when oxidative capacity (e.g., with laccase) was provided, since by itself the dirigent protein had no detectable catalytically active (oxidative) center. Thus, the role of the dirigent protein apparently involves the binding and orientation of the free-radical species, generated by one-electron oxidants such as laccase, following which stereoselective coupling then occurs.

SDS-PAGE analysis of the denatured 78 kDa dirigent protein gave a subunit M_r ~27 kDa, suggesting that the native protein existed as a trimer. It was also established to be a glycoprotein, with pI's ranging from 8–8.5, due to several forms having similar N-terminal amino acid sequences. The gene encoding the (+)-pinoresinol (**20a**) forming dirigent protein was then cloned,[91] with its func-tionally competent recombinant protein subsequently expressed in a baculovirus/

Spodoptera cell culture system. Interestingly, analysis of the dirigent protein gene revealed that it had encoded a secretory pathway protein of 18 kDa; the remainder (~9 kDa) of the subunit was derived as a glycosylation modification. Significantly, the dirigent protein gene displayed no homology to any other protein or enzyme of known function, an observation in harmony with its unique biochemical role. Moreover, SDS-PAGE analysis indicated that the recombinant dirigent protein from the baculovirus/ *Spodoptera* cell culture system was secreted into the medium and had a molecular mass that was very similar to the native protein originally isolated from *F. intermedia*. Importantly, when a suitable oxidant such as laccase was, the stereoselective coupling of coniferyl alcohol (**19**) to afford (+)-pinoresinol (**20a**) was again demonstrated, thereby confirming the authenticity of the gene. Since then, dirigent protein homologues have been obtained from various species, including western red cedar (*T. plicata*),[91] western hemlock (*Tsuga heterophylla*),[91] and *Eucommia ulmoides* (M.K. Kim, Washington State University, unpublished results).

In flaxseed, however, the same stereoselectivity of coniferyl alcohol (**19**) derived bimolecular phenoxy radical coupling to give (+)-pinoresinol (**20a**) was not observed. Instead, preliminary experiments revealed that incubation of cell-free extracts of flaxseed with *E*-[g-³H]-coniferyl alcohol (**19**) gave the opposite enantiomeric form, (–)-pinoresinol (**20b**). Currently, a cDNA library from flaxseed is being screened to obtain this dirigent protein gene; it will be of substantial interest to establish how both the *Forsythia* and *Linum* dirigent protein genes and the corresponding dirigent proteins differ in engendering the formation of both the (+)- and (–)- forms of pinoresinol (**20a**) and (**20b**), respectively.

3.2 Pinoresinol/Lariciresinol Reductase

In all plant species examined to date, namely *F. intermedia*, *L. flavum* (which biosynthesizes 5-methoxypodophyllotoxin (**23**), see fig. 7), western red cedar (*T. plicata*), and flax (*L. usitatissimum*), the formation of secoisolariciresinol (**3**) from pinoresinol (**20**) is catalyzed by a NADPH-dependent bifunctional reductase, pinoresinol/lariciresinol reductase. For example, in *F. intermedia*, it catalyzes the sequential benzylic ether reduction of the furano rings of (+)-pinoresinol (**20a**) to afford first (+)-lariciresinol (**21a**) and then (–)-secoisolariciresinol (**3a**). With *F. intermedia*, the corresponding protein was purified ~3,200-fold to apparent electrophoretic homogeneity from a soluble crude protein extract; this was achieved by employing a series of affinity, hydrophobic interaction, hydroxyapatite, gel filtration, and ion exchange chromatographic steps.[80] The purified protein was demonstrated to be a type A NADPH-dependent reductase.

The respective gene (called plr-Fi1) was cloned from a *Forsythia* cDNA library,[80] with its fully functional recombinant protein then over-expressed in *E. coli* using a pET-based expression system (pSBETa vector)[92] as summarized in figures 8A (upper portion) and 8B (panel I). That is, it was found that the only products formed following incubation of the recombinant pinoresinol/lariciresinol reductase with (±)-pinoresinols (**20a/b**) in the presence of NADPH were (+)-lariciresinol (**21a**)

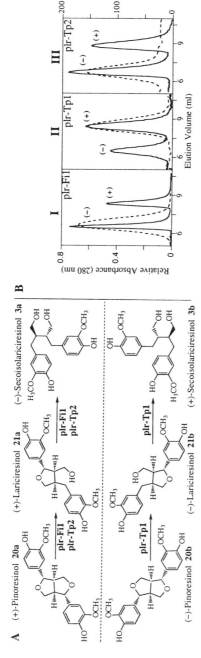

Figure 8. A. Reactions catalyzed by 3 distinct recombinant pinoresinol reductases encoded by genes from *Forsythia* (plr-Fi1) and western red cedar (plr-Tp1 and plr-Tp2). Gene abbreviation over the arrow indicates the reaction catalyzed by that particular protein. **B.** Chiral HPLC enantiomeric analysis of [7,7'-³H]secoisolariciresinols (**3**) following incubation of (±)-pinoresinols (**20a/b**), in the presence of [4A-³H]NADPH with recombinant plr-Fi1 (Panel I), plr-Tp1 (Panel II) and plr-Tp2 (Panel III), respectively. Unlabeled (±)-(**3a/b**) were added as radiochemical carriers [UV absorbance: —; ³H: -----]. (Redrawn from Gang et al.[81])

and (−)-secoisolariciresinol **3a**, i.e., only (+)-pinoresinol (**20a**) and (+)-lariciresinol (**21a**), and not (−)-pinoresinol (**20b**) nor (−)-lariciresinol (**21b**), served as substrates. Thus, the recombinant enzyme catalyzed exactly the same enantiospecific conversion as for the native plant protein from *Forsythia*.[80]

In a complementary manner, in a study of 5-methoxypodophyllotoxin (**23**) biosynthesis in *L. flavum*, it was also demonstrated that the same pathway occurred, i.e., (+)-pinoresinol (**20a**) was converted into (−)-secoisolariciresinol (**3a**) via (+)-lariciresinol (**21a**); the corresponding antipodes (**20b**) and (**21b**) did not serve as substrates (Z.-Q. Xia, Washington State University, unpublished results).

Genes encoding pinoresinol/lariciresinol reductase homologues were also obtained from western red cedar (*T. plicata*) and flax (*L. usitatissimum*). With western red cedar, a reverse transcriptase PCR screening strategy resulted in four cDNAs being identified, two of whose genes were inserted into *E. coli*, with the corresponding recombinant proteins being subsequently obtained. Each was found to display *different* pinoresinol/lariciresinol reductase activities in terms of their relative enantiospecificities.[90] The first (plr-Tp1) catalyzed the consecutive reduction of (−)-pinoresinol (**20b**) into (−)-lariciresinol (**21b**), which was then further reduced to give (+)-secoisolariciresinol (**3b**) (see figure 8A, lower portion); figure 8B, Panel II shows a typical chiral HPLC chromatogram of the (+)-secoisolariciresinol (**3b**) so formed. On the other hand, the second reductase (plr-Tp2) converted (+)-pinoresinol (**20a**) into (+)-lariciresinol (**21a**) and then into (−)-secoisolariciresinol (**3a**) (figures 8A upper portion and 8B panel III). These radiochemical data were further substantiated using deuterated substrates (data not shown).[90] Thus, in western red cedar, parallel pathways to lignan formation were observed, but with enzymes of *different* enantiospecificities; the full biological significance of which now needs to be established.

In developing flaxseed, it was also found that (−)-pinoresinol (**20b**) was effectively converted into (+)-secoisolariciresinol (**3b**), this in turn being the precursor of the cancer-preventing secoisolariciresinol diglucoside (SDG) (**5**). The gene encoding this reductase has since been obtained from a cDNA library constructed from gene transcripts of two week-old developing flaxseeds. The 936 base pair gene displays 73.8 percent similarity and 60.5 percent identity at the amino acid level to that of the *Forsythia* (+)-pinoresinol/(+)-lariciresinol reductase, plr-Fi1. Preliminary assays of the flax recombinant protein, expressed in Invitrogen's® pBad-Topo TA expression system using Top 10 *E. coli* strains under the induction of arabinose, have confirmed the NADPH-dependent conversion of (−)-pinoresinol (**20b**) into (+)-secoisolariciresinol (**3b**) (J.D. Ford, Washington State University, unpublished results). Thus, the biochemical pathway in flaxseed from coniferyl alcohol (**19**) to the cancer preventing agent (+)-secoisolariciresinol (**3b**) has been fully delineated (fig. 9).

3.3 Matairesinol Dehydrogenase

In plant species such as *Forsythia*, *L. flavum*, and western red cedar, the dehydrogenation of secoisolariciresinol (**3**) to matairesinol (**4**), is catalyzed by an NADP⁺ dependent secoisolariciresinol dehydrogenase.[87,88,93] With *F. intermedia*, an ~37 kDa

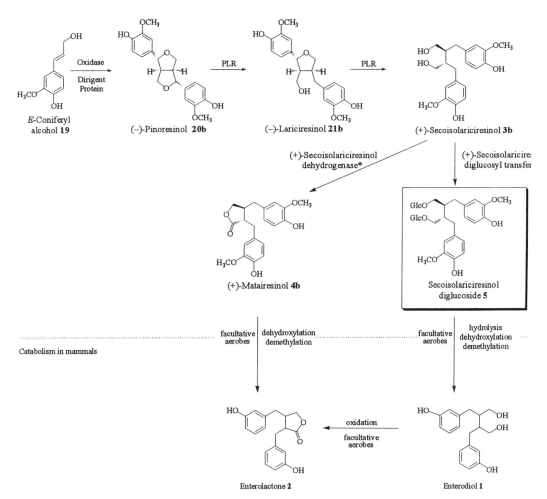

Figure 9. Proposed lignan biosynthetic pathway in flaxseed and catabolism of SDG (**5**) in the gastrointestinal tract of mammals. PLR: pinoresinol/lariciresinol reductase, *: proposed pathway in flaxseed.

protein was purified and its encoding gene cloned.[88] Interestingly, this enzyme is specific for only (–)-secoisolariciresinol (**3a**) as a substrate, with (–)-matairesinol (**4a**) being formed as product. The gene encoding (–)-secoisolariciresinol dehydrogenase was also obtained using a PCR-guided strategy, coupled with the screening of the *F. intermedia* cDNA library with ^{32}P labeled probes. When expressed in *E. coli* under the induction of isopropyl-β-D-thiogalactopyranoside (IPTG), the resulting recombinant matairesinol dehydrogenase was capable of enantiospecifically converting (–)-secoisolariciresinol (**3a**) into (–)-matairesinol (**4a**),[88] i.e., thus completing the elucidation of the biochemical pathway to the presumed precursors of the "mammalian" lignans. In *L. flavum*, an analogous conversion of (–)-secoisolariciresinol (**3a**) into (–)-matairesinol (**4a**) was also demonstrated, although both the formation and presence of matairesinol (**4**) have yet to be documented in flaxseed.

3.4 Secoisolariciresinol Diglucosyl Transferase

Using flaxseed as an enzyme source, the purification and characterization of the UDPglucose-glucosyltransferase capable of the conversion of secoisolariciresinol (**3**) into SDG (**5**) are underway (fig. 9). Thus far, assays conducted on a partially purified putative secoisolariciresinol diglucosyl transferase (SDGT) in the presence of UDP[1-³H]glucose and (±)-secoisolariciresinols (**3a/b**) were found to yield [³H]-SDG (**5**). Moreover, ¹H NMR spectroscopic analysis of the acetylated derivative of the enzymatic product further confirmed the enzymatic formation of SDG (**5**). Indeed, maximum SDGT activity was observed at week two of flaxseed development and appears to be localized in the seedcoat (J.D. Ford, Washington State University, unpublished results). The SDG **5** itself can be solubilized from flaxseed with a chloroform extraction to remove oils followed by treatment with dioxane:ethanol (1:1). Treatment of the resulting precipitate with strong base (2 percent sodium hydroxide) yields approximately 2 percent SDG (**5**) (H.B. Wang, Washington State University, unpublished results) based on the original seed weight. Hydrolysis (10 percent HCl, 100 °C) of the purified diglucoside **5** gives enantiomerically pure (+)-secoisolariciresinol (**3b**). Figure 9 also shows the catabolism of the flax lignans in the gastrointestinal tract of mammals to afford the "mammalian" lignans (**1**) and (**2**).

4. CONCLUSIONS

The biochemical pathways to the plant lignans secoisolariciresinol (**3**), matairesinol (**4**), and secoisolariciresinol diglucoside (**5**) have now been delineated. Additionally, the genes encoding all of the biochemical steps to (**3**) and (**4**) are in hand. Accordingly, future work can now be directed to developing strategies for enhancing the levels of these lignans in staple dietary foodstuffs and hence increased chemoprotection against breast, prostate, and colon cancers. The strategies for increasing their levels may include: over-production of mammalian lignan precursors; increasing the concentration of rate-limiting enzymes thereby facilitating enhanced carbon flow into the pathway;[94–96] creating a new branch pathway from a preexisting phenylpropanoid pathway, manipulating regulatory genes, and/or selecting for regulatory mutants.[97,98]

ACKNOWLEDGMENTS

The authors thank the United States Department of Energy (DE-FG03-97ER20259), the United States Department of Agriculture (9603622), the National Science Foundation (MCB09631980), and the National Aeronautics and Space Administration (NAG100164) for generous support of these studies.

REFERENCES

1. Adlercreutz, H. Western diet and western diseases: some hormonal and biochemical mechanisms and associations. *Scand. J. Clin. Lab. Invest. Suppl.* 201:3–23 (1990).

2. Drasar, B.S.; Irving, D. Environmental factors and cancer of the colon and breast. *Br. J. Cancer* 27:167–172 (1973).

3. Armstrong, B.; Doll, R. Environmental factors and cancer incidence and mortality in different countries, with special reference to dietary practices. *Int. J. Cancer* 15:617–631 (1975).

4. Howell, M.A. The association between colorectal cancer and breast cancer. *J. Chron. Dis.* 29:243–261 (1976).

5. Reddy, B.S.; Cohen, L.A.; McCoy, G.D.; Hill, P.; Weisburger, J.H.; Wynder, E.L. Nutrition and its relationship to cancer. *Adv. Cancer Res.* 32:237–345 (1980).

6. Burkitt, D.P. Colonic-rectal cancer: fiber and other dietary factors. *Am. J. Clin. Nutr.* 31:S58-S64 (1978).

7. Trowell, H. Western diseases and dietary changes with special reference to fibre. *Trop. Doct.* 9:133–142 (1979).

8. Smith, R.L. Recorded and expected mortality among Japanese of the United States and Hawaii, with reference to cancer. *J. Natl. Cancer Inst.* 17:459–473 (1956).

9. Nomura, A.; Henderson, B.E.; Lee, J. Breast cancer and diet among the Japanese in Hawaii. *Am. J. Clin. Nutr.* 31:2020–2025 (1978).

10. Dunn, J.E., Jr. Cancer epidemiology in populations of the United States—with emphasis on Hawaii and California—and Japan. *Cancer Res.* 35:3240–3245 (1975).

11. Muir, C.; Waterhouse, J.; Powell, M.T.; Whelan, S. Cancer incidence in five continents. Vol. 5. International Agency for Research on Cancer, Lyon, France (1987).

12. Adlercreutz, H. Does fiber-rich food containing animal lignan precursors protect against both colon and breast cancer? An extension of the "fiber hypothesis". *Gastroenterology* 86:761–764 (1984).

13. Adlercreutz, H.; Honjo, H.; Higashi, A.; Fotsis, T.; Hämäläinen, E.; Hasegawa, T.; Okada, H. Urinary excretion of lignans and isoflavonoid phytoestrogens in Japanese men and women consuming a traditional Japanese diet. *Am. J. Clin. Nutr.* 54:1093–1100 (1991).

14. Adlercreutz, H. Diet and sex hormone metabolism. *In*: Rowland, I.R. (ed.). Nutrition, toxicity, and cancer. CRC Press, Boca Raton, FL, pp. 137–195 (1991).

15. Barnes, S.; Grubbs, C.; Setchell, K.D.R.; Carlson, J. Soybeans inhibit mammary tumors in models of breast cancer. *Prog. Clin. Biol. Res.* 347:239–253 (1990).

16. Setchell, K.D.R.; Lawson, A.M.; Axelson, M.; Adlercreutz, H. The excretion of two new phenolic compounds during the human menstrual cycle and in pregnancy. *Res. Steroids* 9:207–215 (1981).

17. Setchell, K.D.R.; Lawson, A.M.; Mitchell, F.L.; Adlercreutz, H.; Kirk, D.N.; Axelson, M. Lignans in man and in animal species. *Nature* 287:740–742 (1980).

18. Stitch, S.R.; Toumba, J.K.; Groen, M.B.; Funke, C.W.; Leemhuis, J.; Vink, J.; Woods, G.F. Excretion, isolation and structure of a new phenolic constituent of female urine. *Nature* 287:738–740 (1980).

19. Setchell, K.D.R.; Bull, R.; Adlercreutz, H. Steroid excretion during the reproductive cycle and in pregnancy of the vervet monkey (*Cercopithecus aethiops pygerythrus*). *J. Steroid Biochem.* 12:375–384 (1980).

20. Setchell, K.D.R.; Lawson, A.M.; Conway, E.; Taylor, N.F.; Kirk, D.N.; Cooley, G.; Farrant, R.D.; Wynn, S.; Axelson, M. The definitive identification of the lignans trans-2,3-bis(3-hydroxybenzyl)-γ-butyrolactone and 2,3-bis(3-hydroxybenzyl)butane-1,4-diol in human and animal urine. *Biochem. J.* 197:447–458 (1981).

21. Setchell, K.D.R.; Lawson, A.M.; Borriello, S.P.; Harkness, R.; Gordon, H.; Morgan, D.M.L.; Kirk, D.N.; Adlercreutz, H.; Anderson, L.C.; Axelson, M. Lignan formation in man. Microbial involvement and possible roles in relation to cancer. *Lancet* July 4, 4–7 (1981).

22. Axelson, M.; Setchell, K.D.R. The excretion of lignans in rats. Evidence for an intestinal bacterial source for this new group of compounds. *FEBS Lett.* 123:337–342 (1981).

23. Axelson, M.; Sjövall, J.; Gustafsson, B.E.; Setchell, K.D.R. Origin of lignans in mammals and identification of a precursor from plants. *Nature* 298:659–660 (1982).

24. Axelson, M.; Kirk, D.N.; Farrant, R.D.; Cooley, G.; Lawson, A.M.; Setchell, K.D.R. The identification of the weak oestrogen equol [7-hydroxy-3-(4'-hydroxyphenyl)chroman] in human urine. *Biochem. J.* 201:353–357 (1982).

25. Adlercreutz, H.; Fotsis, T.; Heikkinen, R.; Dwyer, J.T.; Woods, M.; Goldin, B.R.; Gorbach, S.L. Excretion of the lignans enterolactone and enterodiol and of equol in omnivorous and vegetarian postmenopausal women and in women with breast cancer. *Lancet* 1295–1299 (1982).

26. Axelson, M.; Sjövall, J.; Gustafsson, B.E.; Setchell, K.D.R. Soya—a dietary source of the non-steroidal oestrogen equol in man and animals. *J. Endocr.* 102:49–56 (1984).

27. Bannwart, C.; Fotsis, T.; Heikkinen, R.; Adlercreutz, H. Identification of the isoflavonic phytoestrogen daidzein in human urine. *Clin. Chim. Acta* 136:165–172 (1984).

28. Bannwart, C.; Adlercreutz, H.; Fotsis, T.; Wähälä, K.; Häse, T.; Brunow, G. Identification of O-desmethylangolensin, a metabolite of daidzein, and of matairesinol, one likely precursor of the animal lignan, enterolactone, in human urine. *Finn. Chem. Lett.* 45:120–125 (1984).

29. Adlercreutz, H.; Höckerstedt, K.; Bannwart, C.; Hämäläinen, E.; Fotsis, T.; Bloigu, S. Association between dietary fiber, urinary excretion of lignans and isoflavonic phytoestrogens, and plasma non-protein bound sex hormones in relation to breast cancer. *In*: King, R.J.B.; Lippman, M.E.; Raynaud, J.-P. (eds.). Progress in cancer research and therapy. Vol. 35. Raven Press, Ltd, New York, pp. 409–412 (1988).

30. Adlercreutz, H. Lignans and phytoestrogens: possible preventive role in cancer. *Front. Gastrointest. Res.* 14:165–176 (1988).

31. Barnes, S. The chemopreventive properties of soy isoflavonoids in animal models of breast cancer. *Breast Cancer Res. Treat.* 46:169–179 (1997).

32. Lee, H.P.; Gourley, L.; Duffy, S.W.; Estève, J.; Lee, J.; Day, N.E. Dietary effects on breast-cancer risk in Singapore. *Lancet* 337:1197–1200 (1991).

33. Kelly, G.E.; Nelson, C.; Waring, M.A.; Joannou, G.E.; Reeder, A.Y. Metabolites of dietary (soya) isoflavones in human urine. *Clin. Chim. Acta* 223:9–22 (1993).

34. Hutchins, A.M.; Lampe, J.W.; Martini, M.C.; Campbell, D.R.; Slavin, J.L. Vegetables, fruits and legumes: effect on urinary isoflavonoid phytoestrogen and lignan excretion. *J. Am. Diet. Assoc.* 95:769–774 (1995).

35. Kirkman, L.M.; Lampe, J.W.; Campbell, D.R.; Martini, M.C.; Slavin, J.L. Urinary lignan and isoflavonoid excretion in men and women using vegetable and soy diets. *Nutr. Cancer* 24:1–12 (1995).

36. Hutchins, A.M.; Slavin, J.L.; Lampe, J.W. Urinary isoflavonoid phytoestrogen and lignan excretion after consumption of fermented and unfermented soy products. *J. Am. Diet. Assoc.* 95:545–551 (1995).

37. Herman, C.; Adlercreutz, H.; Goldin, B.R.; Gorbach, S.L.; Höckerstedt, K.A.V.; Watanabe, S.; Hämäläinen, E.K.; Markkanen, M.H.; Mäkelä, T.H.; Wähälä, K.T.; Hase, T.A.; Fotsis, T. Soybean phytoestrogen intake and cancer risk. *J. Nutr.* 125:757–770 (1995).

38. Gang, D.R.; Dinkova-Kostova, A.T.; Davin, L.B.; Lewis, N.G. Phylogenetic links in plant defense systems: lignans, isoflavonoids and their reductases. *In*: Hedin, P.A.; Hollingworth, R.M.; Masler, E.P.; Miyamoto, J.; Thompson, D.G. (eds.). Phytochemicals for pest control. Vol. 658. ACS symposium series, Washington, DC, pp. 58–89 (1997).

39. Setchell, K.D.R.; Borriello, S.P.; Hulme, P.; Kirk, D.N.; Axelson, M. Nonsteroidal estrogens of dietary origin: possible roles in hormone-dependent disease. *Am. J. Clin. Nutr.* 40:569–578 (1984).

40. Setchell, K.D.R.; Lawson, A.M.; Borriello, S.P.; Adlercreutz, H.; Axelson, M. Formation of lignans by intestinal microflora. *Falk Symp.* 31 (Colonic Carcinog.):93–97 (1982).

41. Borriello, S.P.; Setchell, K.D.R.; Axelson, M.; Lawson, A.M. Production and metabolism of lignans by the human faecal flora. *J. Appl. Bacteriol.* 58:37–43 (1985).

42. Setchell, K.D.R.; Adlercreutz, H. Mammalian lignans and phyto-oestrogens. Recent studies on their formation, metabolism and biological role in health and disease. *In:* Rowland, I.R. (ed.). Role of the gut flora in toxicity and cancer. Academic Press, London, pp. 315–345 (1988).

43. Thompson, L.U.; Robb, P.; Serraino, M.; Cheung, F. Mammalian lignan production from various foods. *Nutr. Cancer* 16:43–52 (1991).

44. Kurzer, M.S.; Lampe, J.W.; Martini, M.C.; Adlercreutz, H. Fecal lignan and isoflavonoid excretion in premenopausal women consuming flaxseed powder. *Cancer Epidemiology, Biomarkers and Prevention* 4:353–358 (1995).

45. Setchell, K.D.R. Discovery and potential clinical importance of mammalian lignans. *In:* Cunnane, S.C.; Thompson, L.U. (eds.). Flaxseed in human nutrition. AOCS Press, Champaign, IL, pp. 82–98 (1995).

46. Thompson, L.U.; Rickard, S.E.; Orcheson, L.J.; Seidl, M.M. Flaxseed and its lignan and oil components reduce mammary tumor growth at a late stage of carcinogenesis. *Carcinogenesis* 17:1373–1376 (1996).

47. Adlercreutz, H.; Mazur, W. Phyto-oestrogens and western diseases. *Ann. Med.* 29:95–120 (1997).

48. Rickard, S.E.; Thompson, L.U. Chronic exposure to secoisolariciresinol diglycoside alters lignan disposition in rats. *J. Nutr.* 128:615–623 (1998).

49. Landstrom, M.; Zhang, J.X.; Hallmans, G.; Aman, P.; Bergh, A.; Damber, J.E.; Mazur, W.; Wähälä, K.; Adlercreutz, H. Inhibitory effects of soy and rye diets on the development of Dunning R3,327 prostate adenocarcinoma in rats. *Prostate* 36:151–161 (1998).

50. Sung, M.K.; Lautens, M.; Thompson, L.U. Mammalian lignans inhibit the growth of estrogen-independent human colon tumor cells. *Anticancer Res.* 18:1405–1408 (1998).

51. Evans, B.A.J.; Griffiths, K.; Morton, M.S. Inhibition of 5α-reductase in genital skin fibroblasts and prostate tissue by dietary lignans and isoflavonoids. *J. Endocrinol.* 147:295–302 (1995).

52. Jensen, R.A. Tyrosine and phenylalanine biosynthesis: relationship between alternative pathways, regulation and subcellular location. *In:* Conn, E.E. (ed.). Recent advances in phytochemistry. Vol. 20. Plenum Press, New York, NY, pp. 57–81 (1986).

53. Cunnane, S.C.; Thompson, L.U. (eds.). Flaxseed and human nutrition. AOCS Press, Champaign, IL, 384 p. (1995).

54. Thompson, L.U.; Seidl, M.M.; Rickard, S.E.; Orcheson, L.J.; Fong, H.H.S. Antitumorigenic effect of a mammalian lignan precursor from flaxseed. *Nutr. Cancer* 26:159–165 (1996).

55. Adlercreutz, H.; Mousavi, Y.; Clark, J.; Höckerstedt, K.; Hämäläinen, E.; Wähälä, K.; Mäkelä, T.; Hase, T. Dietary phytoestrogens and cancer: *in vivo* and *in vitro* studies. *J. Steroid Biochem. Molec. Biol.* 41:331–337 (1992).

56. Martin, M.E.; Haourigui, M.; Pelissero, C.; Benassayag, C.; Numez, E.A. Interactions between phytoestrogens and human sex steroid binding protein. *Life Sci.* 58:429–436 (1996).

57. Mousavi, Y.; Adlercreutz, H. Enterolactone and estradiol inhibit each other's proliferative effect on MCF-7 breast cancer cells in culture. *J. Steroid Biochem. Molec. Biol.* 41:615–619 (1992).

58. Wang, C.; Mäkelä, T.; Hase, T.; Adlercreutz, H.; Kurzer, M.S. Lignans and flavonoids inhibit aromatase enzyme in human preadipocytes. *J. Steroid Biochem. Molec. Biol.* 50:205–212 (1994).

59. Adlercreutz, H.; Bannwart, C.; Wähälä, K.; Mäkelä, T.; Brunow, G.; Hase, T.; Arosemena, P.J.; Kellis Jr, J.T.; Vickery, L.E. Inhibition of human aromatase by mammalian lignans and isoflavonoid phytoestrogens. *J. Steroid Biochem. Molec. Biol.* 44:147–153 (1993).

60. Prasad, K. Hydroxyl radical-scavenging property of secoisolariciresinol diglucoside (SDG) isolated from flax-seed. *Mol. Cell. Biochem.* 168:117–123 (1997).

61. Lewis, N.G.; Davin, L.B. Lignans: biosynthesis and function. *In*: Barton, Sir D.H.R.; Nakanishi, K.; Meth-Cohn, O. (eds.). Comprehensive natural products chemistry. Vol. 1. Elsevier, London, pp. 639–712 (1999).

62. Ayres, D.C.; Loike, J.D. Chemistry and pharmacology of natural products. Lignans. Chemical, biological and clinical properties. Cambridge University Press, Cambridge, England 402 p. (1990).

63. El-Feraly, F.S.; Cheatham, S.F.; Breedlove, R.L. Antimicrobial neolignans of *Sassafras randaiense* roots. *J. Nat. Prod.* 46:493–498 (1983).

64. Nitao, J.K.; Nair, M.G.; Thorogood, D.L.; Johnson, K.S.; Scriber, J.M. Bioactive neolignans from the leaves of *Magnolia virginiana*. *Phytochemistry* 30:2193–2195 (1991).

65. Schröder, H.C.; Merz, H.; Steffen, R.; Müller, W.E.G.; Sarin, P.S.; Trumm, S.; Schulz, J.; Eich, E. Differential *in vitro* anti-HIV activity of natural lignans. *Z. Naturforsch.* 45c:1215–1221 (1990).

66. Fujihashi, T.; Hara, H.; Sakata, T.; Mori, K.; Higuchi, H.; Tanaka, A.; Kaji, H.; Kaji, A. Anti-human immunodeficiency virus (HIV) activities of halogenated gomisin J derivatives, new nonnucleoside inhibitors of HIV type 1 reverse transcriptase. *Antimicrob. Agents Chemother.* 39:2000–2007 (1995).

67. Ito, K.; Iida, T.; Ichino, K.; Tsunezuka, M.; Hattori, M.; Namba, T. Obovatol and obovatal, novel biphenyl ether lignans from the leaves of *Magnolia obovata* Thunb. *Chem. Pharm. Bull.* 30:3347–3353 (1982).

68. Hattori, M.; Hada, S.; Watahiki, A.; Ihara, H.; Shu, Y.-Z.; Kakiuchi, N.; Mizuno, T.; Namba, T. Studies on dental caries prevention by traditional medicines. X. Antibacterial action of phenolic components from mace against *Streptococcus mutans*. *Chem. Pharm. Bull.* 34:3885–3893 (1986).

69. Whiting, D.A. Lignans, neolignans, and related compounds. *Nat. Prod. Rep.* 4:499–525 (1987).

70. Harmatha, J.; Nawrot, J. Comparison of the feeding deterrent activity of some sesquiterpene lactones and a lignan lactone towards selected insect storage pests. *Biochem. Syst. Ecol.* 12:95–98 (1984).

71. Munakata, K.; Marumo, S.; Ohta, K.; Chen, Y.-L. Justicidin A and B, the fish-killing components of *Justicia hayatai* var. *decumbens*. *Tetrahedron Lett.* 47:4167–4170 (1965).

72. Belmares, H.; Barrera, A.; Castillo, E.; Ramos, L.F.; Hernandez, F.; Hernandez, V. New rubber antioxidants and fungicides derived from *Larrea tridentata* (creosote bush). *Ind. Eng. Chem. Prod. Res. Dev.* 18:220–226 (1979).

73. Fauré, M.; Lissi, E.; Torres, R.; Videla, L.A. Antioxidant activities of lignans and flavonoids. *Phytochemistry* 29:3773–3775 (1990).

74. Oliveto, E.P. Nordihydroguaiaretic acid: a naturally occurring antioxidant. *Chem. Ind.* 677–679 (1972).

75. Osawa, T.; Nagata, M.; Namiki, M.; Fukuda, Y. Sesamolinol, a novel antioxidant isolated from sesame seeds. *Agric. Biol. Chem.* 49:3351–3352 (1985).

76. Chu, A.; Dinkova, A.; Davin, L.B.; Bedgar, D.L.; Lewis, N.G. Stereospecificity of (+)-pinoresinol and (+)-lariciresinol reductases from *Forsythia intermedia*. *J. Biol. Chem.* 268:27026–27033 (1993).

77. Davin, L.B.; Bedgar, D.L.; Katayama, T.; Lewis, N.G. On the stereoselective synthesis of (+)-pinoresinol in *Forsythia suspensa* from its achiral precursor, coniferyl alcohol. *Phytochemistry* 31:3869–3874 (1992).

78. Davin, L.B.; Lewis, N.G. Lignin and lignan biochemical pathways in plants: an unprecedented discovery in phenolic coupling. *An. Acad. bras. Ci.* 67 (Supl. 3):363–378 (1995).

79. Davin, L.B.; Wang, H.-B.; Crowell, A.L.; Bedgar, D.L.; Martin, D.M.; Sarkanen, S.; Lewis, N.G. Stereoselective bimolecular phenoxy radical coupling by an auxiliary (dirigent) protein without an active center. *Science* 275:362–366 (1997).

80. Dinkova-Kostova, A.T.; Gang, D.R.; Davin, L.B.; Bedgar, D.L.; Chu, A.; Lewis, N.G. (+)-Pinoresinol/(+)-lariciresinol reductase from *Forsythia intermedia*: protein purification, cDNA cloning, heterologous expression and comparison to isoflavone reductase. *J. Biol. Chem.* 271:29473–29482 (1996).

81. Gang, D.R.; Fujita, M.; Davin, L.B.; Lewis, N.G. The "abnormal lignins": mapping heartwood formation through the lignan biosynthetic pathway. *In*: Lewis, N.G.; Sarkanen, S. (eds.). Lignin and lignan biosynthesis. Vol. 697. ACS symposium series, Washington, DC, pp. 389–421 (1998).

82. Katayama, T.; Davin, L.B.; Lewis, N.G. An extraordinary accumulation of (−)-pinoresinol in cell-free extracts of *Forsythia intermedia*: evidence for enantiospecific reduction of (+)-pinoresinol. *Phytochemistry* 31:3875–3881 (1992).

83. Katayama, T.; Davin, L.B.; Chu, A.; Lewis, N.G. Novel benzylic ether reductions in lignan biogenesis in *Forsythia intermedia*. *Phytochemistry* 33:581–591 (1993).

84. Lewis, N.G.; Davin, L.B. Evolution of lignan and neolignan biochemical pathways. *In*: Nes, W.D. (ed.). Isopentenoids and other natural products: evolution and function. Vol. 562. ACS symposium series, Washington, DC, pp. 202–246 (1994).

85. Lewis, N.G.; Davin, L.B.; Dinkova-Kostova, A.T.; Fujita, M.; Gang, D.R.; Sarkanen, S. Patent: "Recombinant pinoresinol/lariciresinol reductase, recombinant dirigent protein, and methods of use". PCT/US97/20391, p. 146 (1997).

86. Umezawa, T.; Davin, L.B.; Yamamoto, E.; Kingston, D.G.I.; Lewis, N.G. Lignan biosynthesis in *Forsythia* species. *J. Chem. Soc., Chem. Commun.* 1405–1408 (1990).

87. Umezawa, T.; Davin, L.B.; Lewis, N.G. Formation of lignans (−)-secoisolariciresinol and (−)-matairesinol with *Forsythia intermedia* cell-free extracts. *J. Biol. Chem.* 266:10210–10217 (1991).

88. Xia, Z.-Q.; Costa, M.A.; Davin, L.B.; Lewis, N.G. Purification and characterization of (−)-secoisolariciresinol dehydrogenase; cloning and recombinant expression. (submitted) (1999).

89. Kamil, W.M.; Dewick, P.M. Biosynthetic relationship of aryltetralin lactone lignans to dibenzylbutyrolactone lignans. *Phytochemistry* 25:2093–2102 (1986).

90. Fujita, M.; Gang, D.R.; Davin, L.B.; Lewis, N.G. Recombinant pinoresinol/lariciresinol reductases from western red cedar (*Thuja plicata*) catalyze opposite enantiospecific conversions. *J. Biol. Chem.* 274:618–627 (1999).

91. Gang, D.R.; Costa, M.A.; Fujita, M.; Dinkova-Kostova, A.T.; Wang, H.-B.; Burlat, V.; Martin, W.; Sarkanen, S.; Davin, L.B.; Lewis, N.G. Regiochemical control of monolignol radical coupling: a new paradigm for lignin and lignan biosynthesis. *Chemistry and Biology* 6:143–151 (1999).

92. Schenk, P.M.; Baumann, S.; Mattes, R.; Steinbiß, H.-H. Improved high-level expression system for eukaryotic genes in *Escherichia coli* using T7 RNA polymerase and rare ^{Arg}tRNAs. *BioTechniques* 19:196–200 (1995).

93. Umezawa, T.; Davin, L.B.; Lewis, N.G. Formation of the lignan, (−)-secoisolariciresinol, by cell-free extracts of *Forsythia intermedia*. *Biochem. Biophys. Res. Commun.* 171:1008–1014 (1990).

94. Hain, R.; Bieseler, B.; Kindl, H.; Schroder, G.; Stöcker, R. Expression of stilbene synthase gene in *Nicotiana tabacum* results in synthesis of the phytoalexin resveratrol. *Plant Mol. Biol.* 15:325–335 (1990).

95. Hain, R.; Reif, H.-J.; Krause, E.; Langebartels, R.; Kindl, H.; Vornam, B.; Wiese, W.; Schmelzer, E.; Schreier, P.H.; Stöcker, R.H.; Stenzel, K. Disease resistance results from foreign phytoalexin expression in a novel plant. *Nature* 361:153–156 (1993).

96. Hamill, J.D.; Robins, R.J.; Parr, A.J.; Evans, D.M.; Furze, J.M.; Rhodes, M.J. Over-expressing a yeast ornithine decarboxylase gene in transgenic roots of *Nicotiana rustica* can lead to enhanced nicotine accumulation. *Plant Mol. Biol.* 15:27–38 (1990).

97. Ellis, B.E.; Kuroki, G.W.; Stafford, H.A. (eds.). Genetic engineering of plant secondary metabolism. Plenum Press, New York, NY, 368 p. (1994).

98. Verma, D.P.S. Control of plant gene expression. CRC Press, Boca Raton 579 p. (1993).

POLYPHENOLS IN COMMERCE

BLACK TEA DIMERIC AND OLIGOMERIC PIGMENTS— STRUCTURES AND FORMATION

Alan P. Davies,[a] Chris Goodsall,[a] Ya Cai,[a]
Adrienne L. Davis,[a] John R. Lewis,[a] John Wilkins,[a]
Xiaochun Wan,[b] Mike N. Clifford,[c] Chris Powell,[a]
Andrew Parry,[a] Ambalavanar Thiru,[a] Robert Safford,[a]
and Harry E. Nursten[d]

[a] Unilever Research
Colworth Laboratory, Sharnbrook
Bedford MK44 1LQ
ENGLAND

[b] Faculty of Light Industry
Anhui Agricultural University
Hefei, Anhui 230036
CHINA

[c] Food Safety Research Group
University of Surrey
Guildford, Surrey GU2 5XH
ENGLAND

[d] Department of Food Science and Technology
University of Reading
Whiteknights, PO Box 226
Reading RG6 2AP
ENGLAND

1. INTRODUCTION

Many of us consume tea daily, but it is surprizing how few know where it comes from or how it is made. Tea comes from the tea plant the botanical name for which is *Camellia sinensis* (L.) O. Kuntze.[1] This is a white-flowered evergreen indigenous to the rainforests of Assam, Northwest Burma and South West China, but now

Plant Polyphenols 2: Chemistry, Biology, Pharmacology, Ecology, Edited by
Gross et al. Kluwer Academic / Plenum Publishers, New York, 1999

cultivated in over 30 countries around the world in places as far apart as Argentina, China, Papua New Guinea, and Turkey. However, since most tea leaves are plucked by hand, the largest plantations are most often found in countries with low manual labor costs. Two main varieties of tea are cultivated; *Camellia sinensis* var *sinensis* and *C. sinensis* var *assamica*. There is some controversy as to whether these are varieties or two distinct species.[1] *C. sinensis* var. *sinensis* L, is grown mainly in China and Japan; it is a slow-growing bush, has small narrow leaves and is used for green tea manufacture. The second variety *Camellia sinensis* var. *assamica* is grown primarily in India and Africa, is fast growing, has a treelike habit, large broad leaves, and is the source material for black tea. In its cultivated state, it is pruned to approximately 3 feet for ease of harvesting the young shoots (flush), from which the tea is manufactured. This process (plucking) is carried out every 7–14 days during the growing season.

The tea bush and in particular the young shoots (flush) contain a high concentration of flavonoids and oxidative enzymes, and it is because of this feature that the flush is used for tea manufacture. The flavonoids are principally the flavan-3-ols and their corresponding 3-*O*-gallate esters, and the composition in the flush will vary depending on plant variety and growing environment. The major flavonoids present are epicatechin (**1**), epigallocatechin (**2**), epicatechin-3-*O*-gallate (**3**), epigallocatechin-3-*O*-gallate (**4**), catechin (**5**), and gallocatechin (**6**) (table 1 and fig. 1). In the variety *sinensis* the flavonoid level (percent dry wt) is in the range 10–20 percent, whereas in *assamica* it is 10–25 percent (fig. 1). Additionally the ratio of flavan-3-*O*-gallates to flavan-3-ols is higher in *assamica* compared with the variety *sinensis*. Hence, it is not normal practise to process green tea from *assamica*, nor black tea from *sinensis*, since the characteristics of the teas thus produced are not acceptable to the consumer.

The sensory characteristics of black tea are very distinct from those of green tea and are, in part, the result of oxidative transformation of the flavonoids present in the fresh leaves (flush). During the manufacturing process, the flavonoids, which are located within the vacuoles of the intact leaf cells, are released and come into contact with plant oxidases, which facilitates phenol oxidation. The oxidized phenols produced are mainly theaflavins and thearubigins, the former being dimers bearing a benzotropolone ring system, while the latter are an ill-defined mixture of flavonoid polymers. These species generate the unique color and taste properties of teas, and their composition and mode of formation have attracted much attention[2] over the years.

In this chapter, two aspects of black tea composition will be discussed: a) structures of the oxidized flavonoid derivatives (polyphenols) and a mechanistic interpretation of their formation and b) biochemical/chemical pathways for the conversion of flavonoids to their oxidized derivatives during tea manufacture.

2. STRUCTURAL CHEMISTRY OF FLAVONOID DIMERS AND OLIGOMERS

Flavan-3-ols such as epicatechin (**1**) possess a 1,3-dihydroxy phenyl group (Ring A) and an *o*-dihydroxy phenyl group (Ring B). Both groups can be oxidized, with the

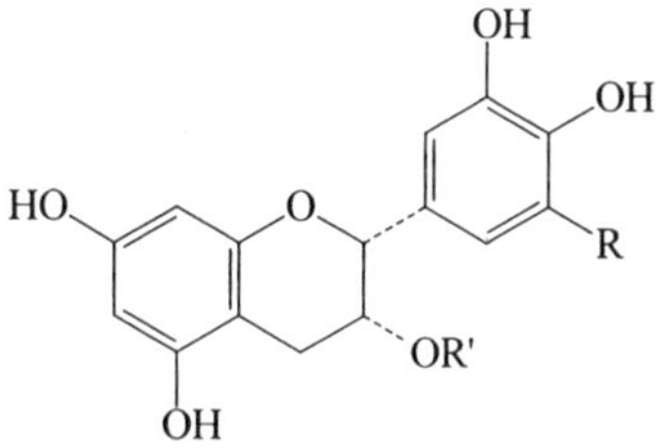

1	R = H, R' = H;	Epicatechin
2	R = OH, R' = H;	Epigallocatechin
3	R= H, R'= gallate;	Epicatechin-3-*O*-gallate
4	R = OH, R'= gallate;	Epigallocatechin-3-*O*-gallate

| 5 | R = H, R' = H; | Catechin |
| 6 | R=OH, R' = OH; | Gallocatechin |

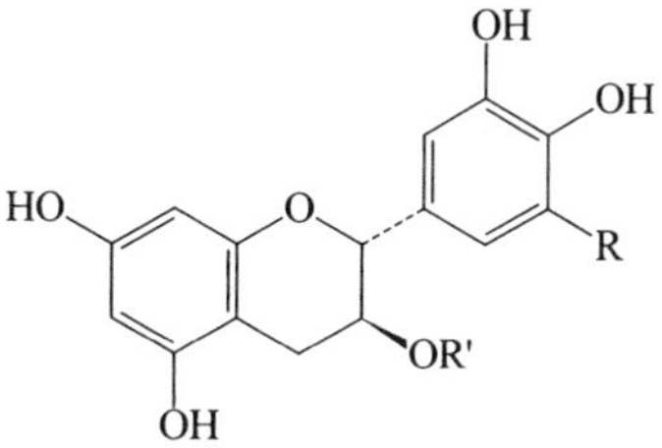

Gallate =

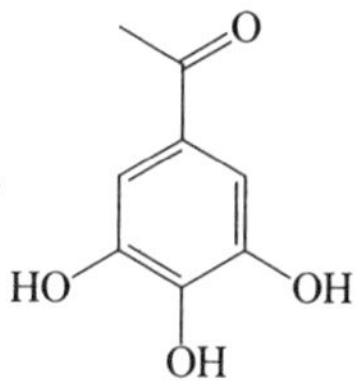

Figure 1. Structures of the flavonoids in *Camellia sinensis*.

o-dihydroxy phenyl group being much, more susceptible. Oxidation can be mediated with enzymes (oxidases), metal ions, and occurs autocatalytically at pH approximately 6.0 and above. One-electron oxidation generates a free radical that can couple with other radicals to form C—C or C—O bonds from which theasinensins[3] (C—C bond formation) are probably derived. *O*-Dihydroxy phenyl rings and *o*-trihydroxy phenyl rings can, through two one-electron oxidations, lead to the highly reactive *o*-quinones. These species can react further via intermolecular cycloaddition reactions to form products which contain a benzotropolone ring system and have intense yellow/orange colors. These are the basic structural requirements and mechanisms for the formation of dimeric flavonoid derived pigments.

The theaflavins are a group of polyphenol pigments found in black tea, which make important contributions to the properties such as color[4] and taste.[5,6] Studies

Table 1. Flavonoid compositions of *C. sinensis* var. *sinensis* and *C.*
sinensis var. *assamica*

Component	Var. sinensis (percent)	Var. assamica (percent)
Flavonoids	10 – 20	10 - 25
Epicatechin	0 - 2	1 - 3
Epigallocatechin	0 - 4	0 - 6
Epicatechin gallate	0 - 4	1 - 6
Epigallocatechin gallate	7 - 11	9 - 13
Catechin	< 1.0	0 - 2
Gallocatechin	< 1.0	0 - 2
Theasinensin A	Trace	Trace
Theasinensin B	Trace	Trace
Proanthocyanidin gallates (dimers)	Trace	Trace

on the structures of these pigments have been of interest since the early work of
Roberts.[2]

Theaflavin (TF) (**7**) was first isolated four decades ago[7] and its structure determined 10 years later.[8,9] These, and other studies[10–14] have demonstrated that
theaflavin and its gallate esters possess a hydroxy-substituted benzotropolone ring
system, formed from oxidative coupling of the B-rings of a pair of flavan-3-ols (or
flavan-3-*O*-gallates). The major theaflavins present in black tea are theaflavin (**7**),
theaflavin-3-*O*-gallate (**8**), theaflavin-3'-*O*-gallate (**9**) and theaflavin-3,3'-*O*-
digallate (**10**). Their precursors are identified in table 2.

Stereoisomers of theaflavin, namely isotheaflavin (**11**) and neotheaflavin (**12**)
have also been identified in black tea[15,16] (fig. 2) (see table 2 for precursors), as
have a number of closely related dimers, including theaflavic acids[17] (**13 a,b,c**) (fig.
3) and theaflagallins[18] (**14 a,b,c**) (fig. 4).

Examination of HPLC chromatograms of black tea extracts in the region where
theaflavin occurs revealed a number of small peaks that could not be accounted
for by known compounds. Solvent extraction of black tea leaf and separation of the

Table 2. Precursors to benzotropolone dimers found in tea

Precursors	Benzotropolone Dimer
EC + EGC	Theaflavin
EC + EGCG	Theaflavin-3-*O*-gallate
ECG + EGC	Theaflavin-3'-*O*-gallate
ECG + EGCG	Theaflavin-3,3'-*O*-digallate
EC + GC	Isotheaflavin
C + EGC	Neotheaflavin
C + EGCG	Neotheaflavin-3-*O*-gallate
GC + ECG	Isotheaflavin-3'-*O*-gallate
C + GC	"B" Theaflavin
EC + Ga	Epitheaflavic acid
C + Ga	Theaflavic acid
ECG + Ga	Epitheaflavic acid-3-gallate
ECG + ECG	Theaflavate A
EC + ECG	Theaflavate B

components using a combination of Sephadex LH-20 column chromatography and preparative reverse phase HPLC (Novapak C18 column) yielded a pure sample of isotheaflavin-3'-*O*-gallate (**15**) and neotheaflavin-3-*O*-gallate (**16**) as the main components in a mixture with epitheaflagallin-3-*O*-gallate (**14b**). A combination of UV-visible spectroscopy (spectrum very similar to other theaflavins), electrospray mass spectrometry (MW = 716 Daltons) and ^{1}H and ^{13}C NMR (benzotropolone ring system, gallate ester, *cis*-2,3 and *trans*-2,3 relative configurations identified)[19-21] were used to identify the structures (fig. 5).

Further supportive evidence was provided by chemical synthesis from a mixture of (+)-gallocatechin and (−)-epicatechin-3-*O*-gallate, which produced a reaction mixture in which one component had HPLC retention time identical with that of isotheaflavin-3'-*O*-gallate. A mixture of (+)-catechin and (−)-epigallocatechin-3-*O*-gallate provided a reaction mixture in which a component had an HPLC retention time equivalent to that of neotheaflavin-3-*O*-gallate.

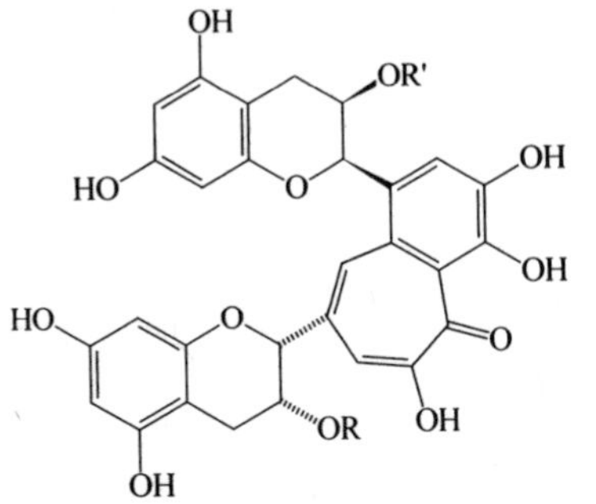

7	R = R' = H;	Theaflavin
8	R = gallate, R' = H;	Theaflavin-3-*O*-monogallate
9	R = H, R' = gallate;	Theaflavin-3'-*O*-monogallate
10	R = R' = gallate;	Theaflavin-3,3'-*O*-digallate

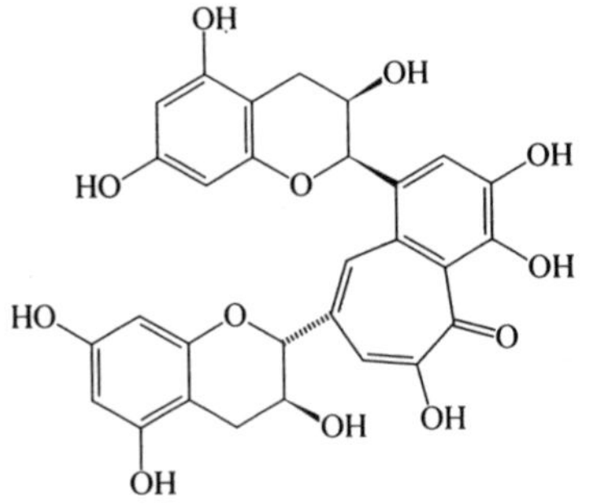

11   Isotheaflavin

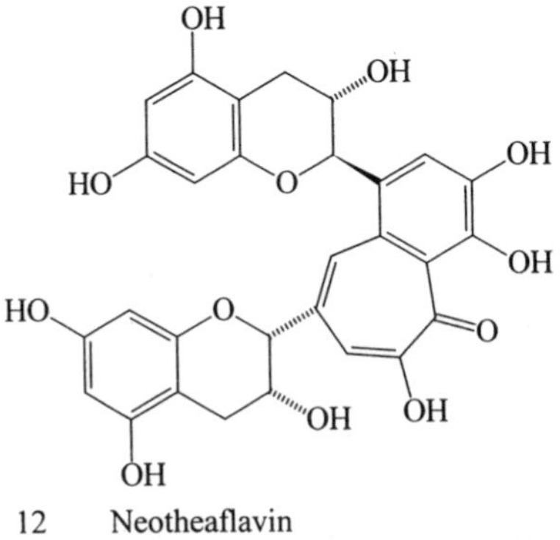

12   Neotheaflavin

Figure 2. Theaflavin, its isomers and gallate esters.

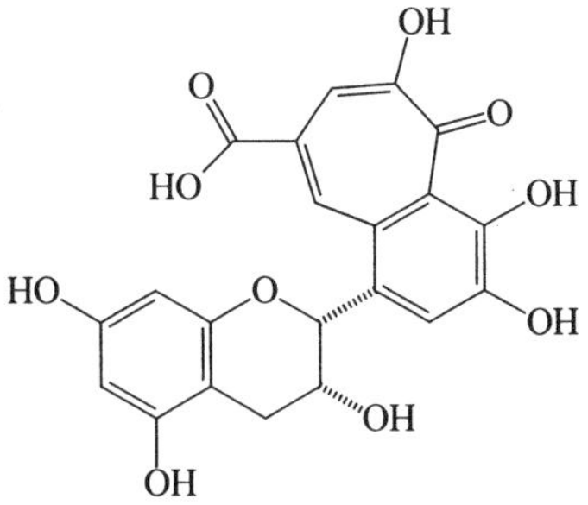

13 a   Epitheaflavic Acid

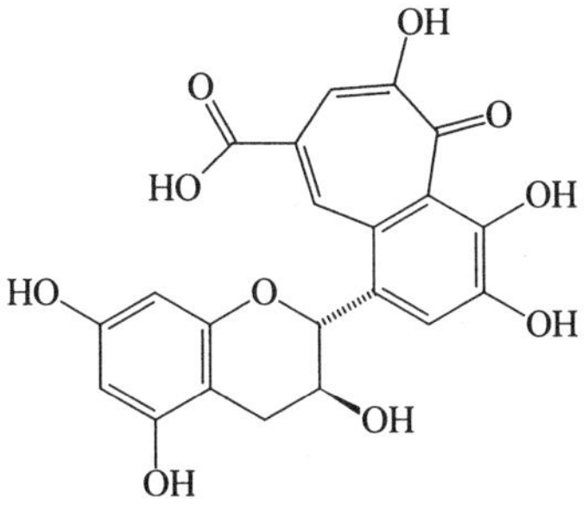

13 b   Theaflavic Acid

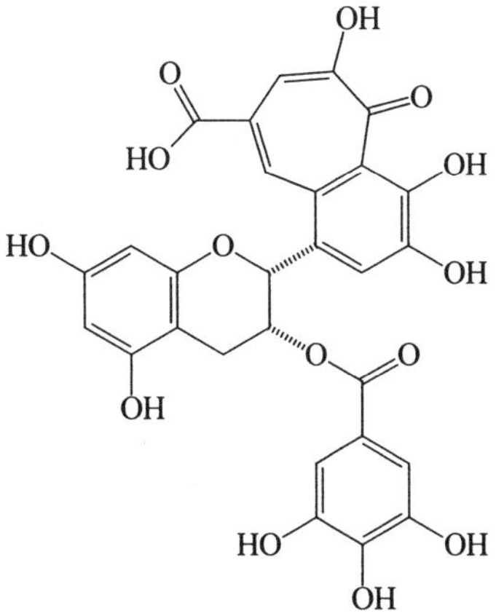

13 c   Epitheaflavic Acid-3-*O*-gallate

Figure 3. Epitheaflavic acid, its isomer and gallate ester.

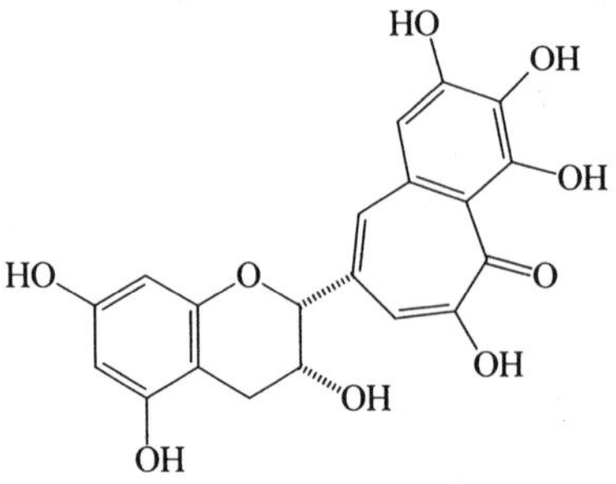

14 a   Epitheaflagallin

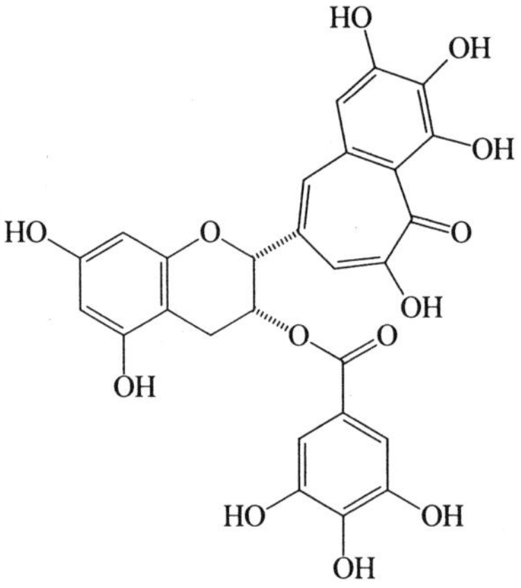

14 b   Epitheaflagallin-3-*O*-gallate

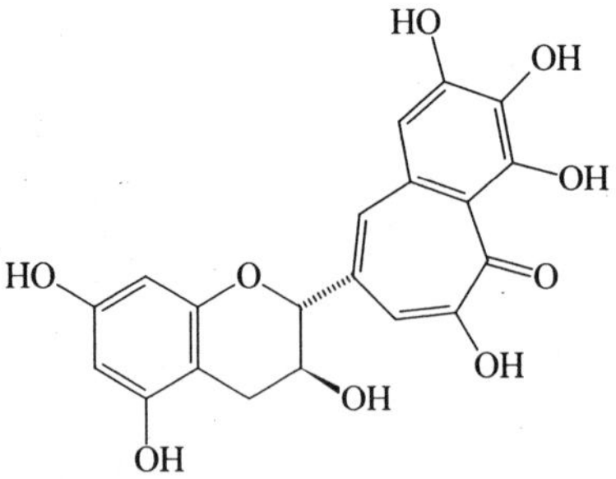

14 c   Theaflagallin

Figure 4. Epitheaflagallin, its isomer and gallate ester.

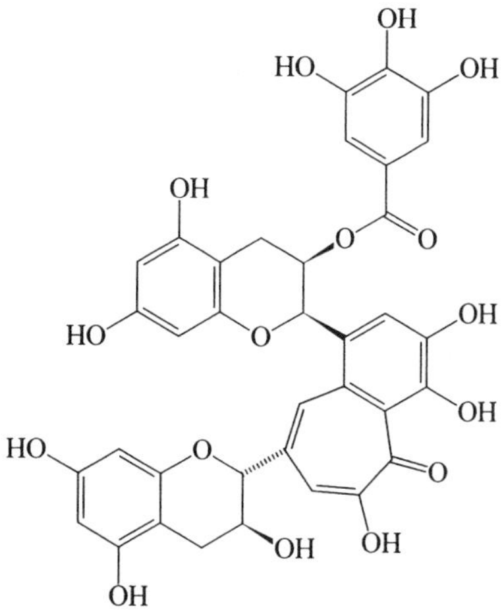

15   Isotheaflavin-3'-*O*-gallate

Figure 5. Structures of isotheaflavin-3'-*O*-gallate and neotheaflavin-3-*O*-gallate.

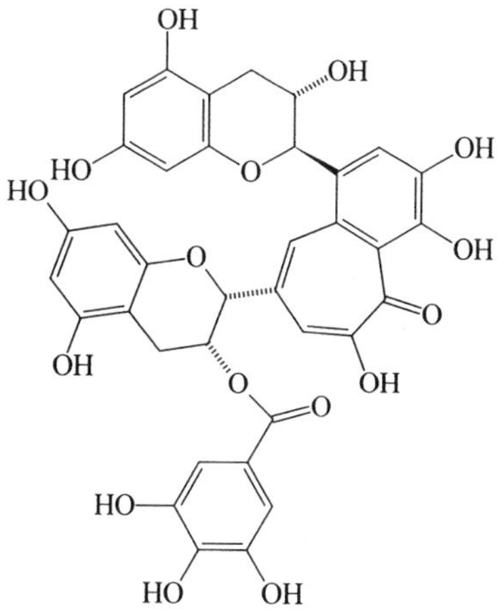

16   Neotheaflavin-3-*O*-gallate

The formation of a benzotropolone ring system is determined by the availability and structure of an *o*-dihydroxy phenyl and *o*-trihydroxy phenyl. In the case of theaflavins, these are provided by the catechin and gallocatechin B rings, respectively. It has been proposed[22] that the gallate ester ring could also take part in benzotropolone ring formation by its reaction with the B ring of epicatechin. In effect, epicatechin-3-*O*-gallate alone would be expected, under oxidative conditions, to form a dimer bearing a benzotropolone ring (**17**).

Chemical oxidation, using potassium ferricyanide, of epicatechin-3-*O*-gallate produced a single HPLC resolvable product, which was purified by column chromatography (Sephadex LH-20) and semi-preparative HPLC (Hypersil ODS). The single component product was named theaflavate A.[23] Electrospray mass spectrometry indicated a MW = 852 implying that two ECG molecules have been coupled via a benzotropolone ring system (loss of CH_4O). The structure (**17**) proposed for theaflavate A was confirmed by ^1H and ^{13}C NMR (fig. 6). Attempts to establish the presence of theaflavate A in black tea by spiking an infusion of black tea with the material and analysing by HPLC were inconclusive. The complexity of the HPLC chromatogram in the theaflavate A region (particularly

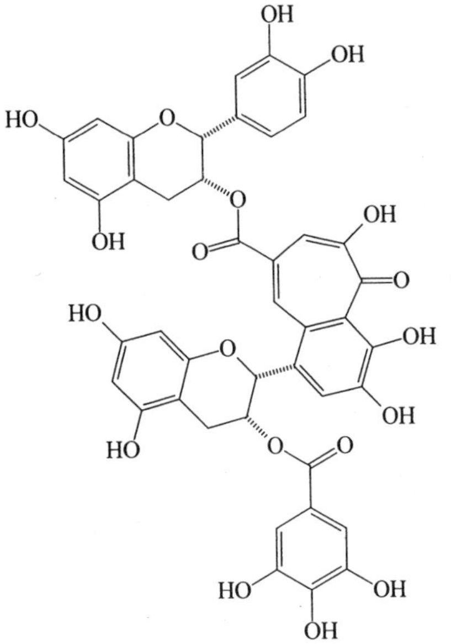

17   Theaflavate A

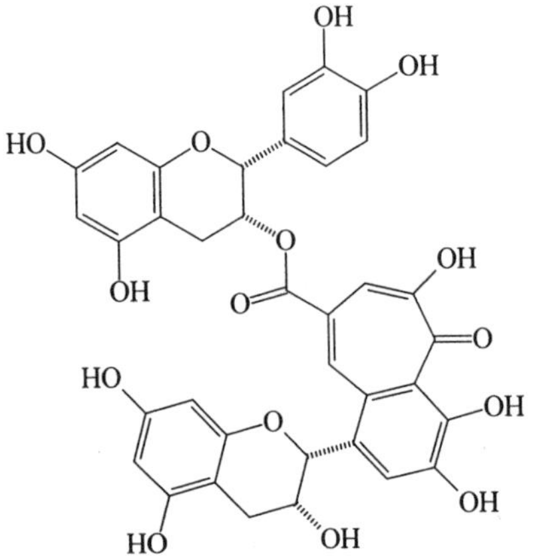

18   Theaflavate B

Figure 6. Structures of theaflavates A and B.

theaflavin-3'-O-gallate) made it difficult to identify any peak enhancement resulting from spiking. However, a compound containing the same benzotropolone skeleton has subsequently been isolated from black tea.[19] It is the non-gallated version of theaflavate A and is likely to have been formed via oxidative coupling of epicatechin-3-O-gallate and epicatechin. The product was named theaflavate B (**18**). The observation that gallate esters (o-trihydroxy benzoate) can be oxidized to form benzotropolone units suggests that they could also be precursors for the oxidation process leading to thearubigins.

Theaflavins, theaflavates, theaflavic acid, and other dimeric pigments containing a benzotropolone ring are relatively apolar. They partition into the ethyl acetate phase of a water/ethyl acetate solvent mixture. Studies on the separation and characterization of the thearubigins (species mostly present in the aqueous phase of the ethyl acetate/water solvent mixture) have led to the isolation of a minor quantity of a yellow pigment named theacitrin (**19**).[24] Extensive application of preparative HPLC (Hypersil ODS) has achieved purification of one product (theacitrin A) and identified the presence of two others (theacitrins B and C).[25] Rigorous analysis of ^{1}H and ^{13}C NMR spectral data from theacitrin A, together with molecular weight determination by electrospray mass spectrometry (MW = 760 Daltons) and IR analysis to confirm carbonyl location provided a structure consistent with all data (**19**). Mass spectral and UV-visible spectral analysis of the theacitrins B and C indicate that they are the corresponding 3'-O-gallate and 3,3'-O-di-gallate esters, respectively (**20, 21**) (fig. 7).

The identification of theacitrin A is of great interest, since it represents a new class of colored, dimeric flavan-3-ol derived polyphenolic compounds present in black tea. It has some structural features in common with oolongtheanin,[26] which also has a flavan-3-ol and a flavan-3-O-gallate joined together by a three fused ring system. However, oolongtheanin has a MW 28 Daltons less than theacitrin.

It is also of interest to note that oxidation of catechins during tea fermentation takes place principally on their B-ring and to a lesser extent the gallate ester ring, leading to the formation of a variety of dimeric compounds such as theaflavins,

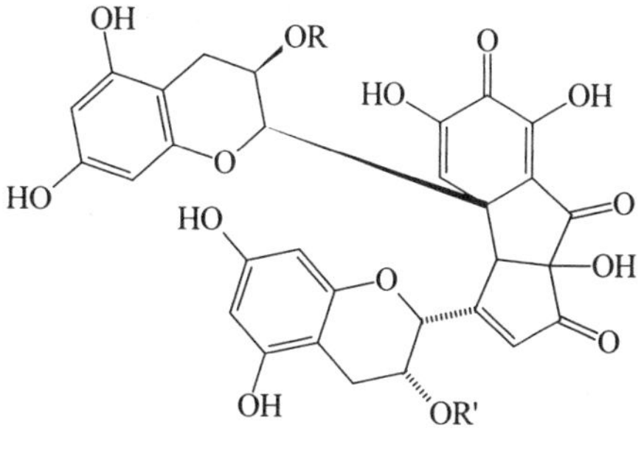

19	R = gallate, R' = H	Theacitrin A
20	R = H, R' = gallate	Theacitrin B
21	R = gallate, R' = gallate	Theacitrin C

Figure 7. Structures of theacitrin A, B and C.

theasinensins,[3] theaflavates,[19,23] oolongtheanin[26] and theacitrins.[25] Most of these contribute to the color of black tea and additionally they are likely to be the building blocks from which thearubigins are formed via further oxidation. The reasons for the diversity of molecular structures and complexity of thearubigins are therefore self evident and explain the lack of progress achieved in elucidating their structures.

The requirement of an *o*-dihydroxyphenyl B ring of epicatechin and an *o*-trihydroxyphenyl B ring of epigallocatechin for oxidative condensation to theaflavin was proposed[2] in 1962. The structure of theaflavin and mechanism for its formation were proposed independently by Takino et al.[8] and Ollis et al.[9] Their mechanisms were based on that proposed earlier[27,28] for the formation of purpurogallin, derived from catechol and pyrogallol. Most mechanisms invoke the oxidation of a combination of the *o*-dihydroxy and *o*-trihydroxy B rings to the corresponding reactive orthoquinones (fig. 8); this step is catalyzed by oxidase enzymes or facilitated by certain metal ions. These products subsequently react chemically to form dimers, oligomers, or polymers.

An alternative mechanism for theaflavin formation is currently being proposed.[29] This involves a form of molecular organization of the two B rings prior to the ring coupling, in order to direct the cyclization to the product formed experimentally.

For theaflavate formation, the mechanism proposed for theaflavins can be considered, but in which the *o*-trihydroxyphenyl ring of the B-ring is replaced by the gallate ester. The first step—oxidation of *o*-phenols to *o*-quinones—may need further consideration since gallic acid is not readily oxidized by polyphenol oxidase (PPO), one of the key enzymes in leaf. An alternative route could proceed via PPO catalyzed coupled oxidations, in which a simple catechin when oxidized to the *o*-quinone would subsequently oxidize the gallate ester to the *o*-quinone and itself be reduced back to catechin[30] (fig. 9). Oxidation by peroxidase is an unlikely route, since hydrogen peroxide, the required electron acceptor, is not available.

Theacitrin, with its fused three-ring system has a more complex structure than the benzotropolone unit. It is proposed[25] that the two substrates, EGC and EGCG are oxidized to two free radical quinones that couple through a C—C bond. This intermediate could then undergo intermolecular cyclization, rearrangement and hydration, finally leading to the formation of theacitrin B (fig. 10).

The major group of polyphenols present in black tea are structurally ill defined and are referred to as thearubigins. They constitute between 10 and 20 percent of the dry weight of black tea,[2] and due to their hot water solubility represent approximately 30–60 percent of the solids in liquors after infusion.[31]

Three groups of thearubigins have been isolated:[32] a) SI, soluble in ethyl acetate; b) SIa, soluble in water and diethyl ether and c) SII, which are soluble in water. Recent NMR studies on fraction SI type thearubigins, which were obtained in the ethyl acetate fraction of a water/ethyl acetate solvent mixture and subsequently separated using Sephadex LH-20 chromatography have provided information on the structural elements from which these compounds are constructed.[33] The ¹H and ¹³C NMR spectral data revealed little information, since the signals were very broad, which in itself is indicative of high molecular weight species.

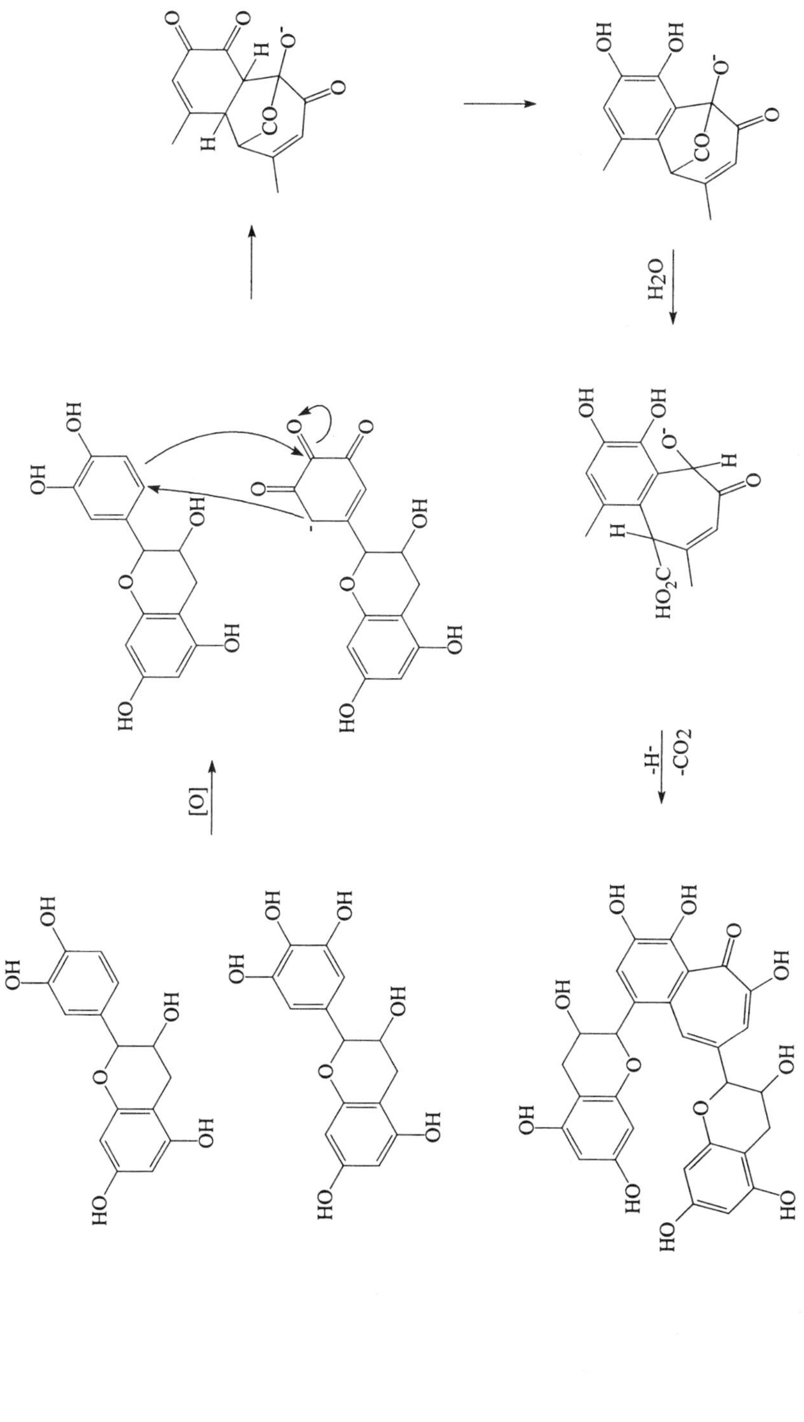

Figure 8. Proposed mechanism for theaflavin formation (D. Sant, Ph.D Thesis, University of Sheffield, 1973).

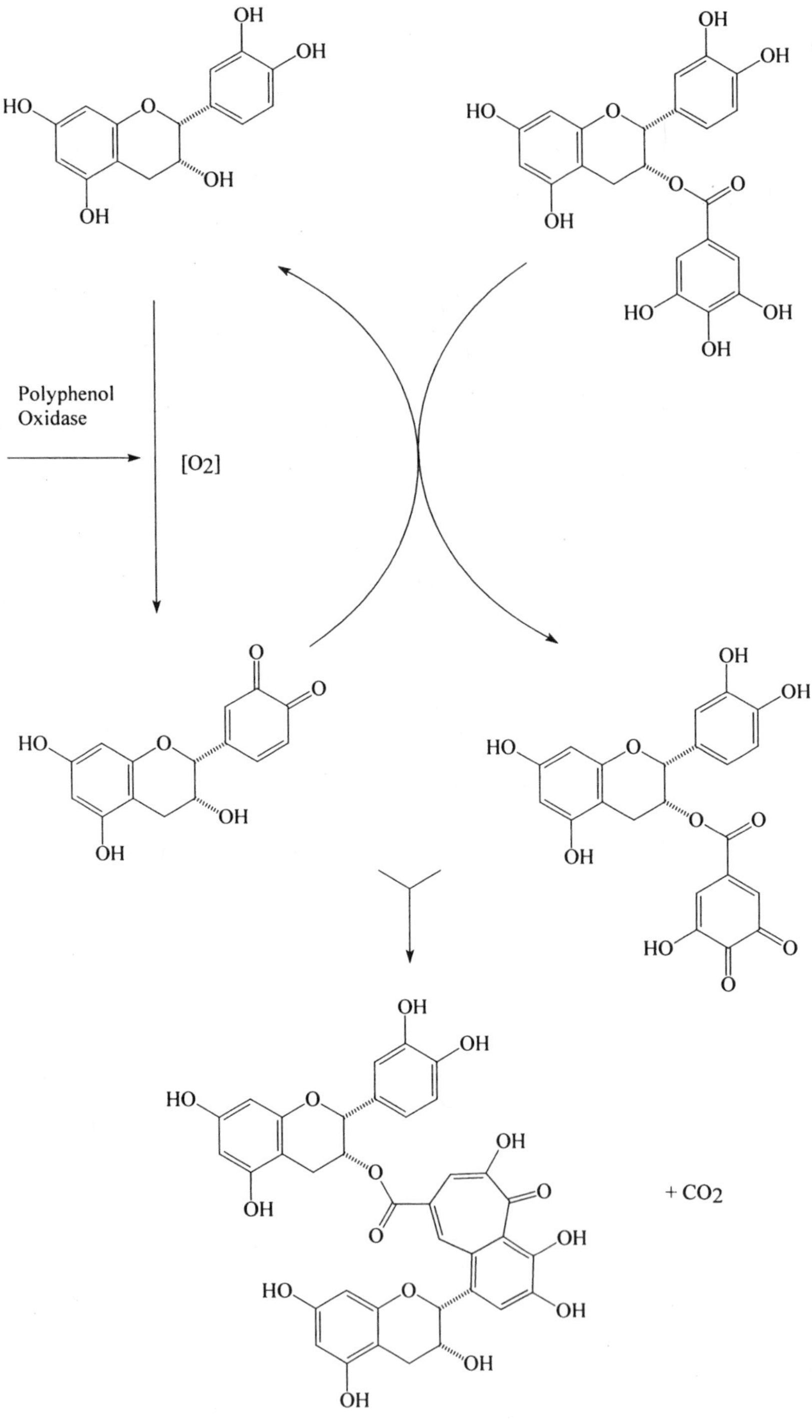

Figure 9. Proposed mechanism of formation of theaflavate B.

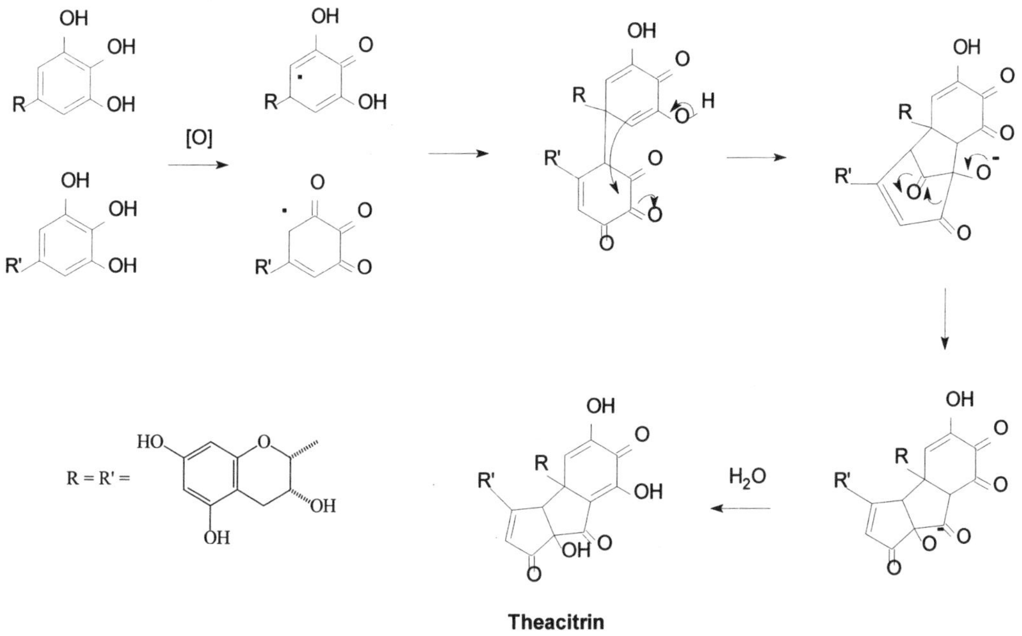

Theacitrin

Figure 10. Proposed mechanism of formation of theacitrins.

However, in-depth analysis of the 2D HMQC and HMBC NMR data revealed resonances characteristic of the following structures: a) benzotropolone; b) C2—C2′ bonds; c) gallate esters; d) A-ring, C-ring, dihydroxy- and trihydroxy-phenyl B-rings and e) C4—C8′ and C4—C6′ bonds. In conjunction with these NMR studies, a series of mass spectrometry experiments were carried out to establish an accurate molecular weight for these fractions. A reliable molecular weight was, however, not obtained.

These findings support the proposals of Nursten et al.[34] and Ozawa et al.[35] both of whom suggested that the SI type thearubigins are composed of flavan-3-ols, flavan-3-ol gallates, and benzotropolone moieties linked together by hydrolyzable and non-hydrolyzable interflavonoid bonds (fig. 11).

These data have provided a constructive step forward in our understanding of the SI type thearubigins. However, without accurate measurements of the molecular weights of the compounds, it is not possible to propose complete structures containing the chemical features identified.

3. BIOCHEMISTRY OF TEA FERMENTATION

During the fermentation stage of tea manufacture, the flavan-3-ols present in the tea leaf are oxidized to dimers (including theaflavins, theacitrins) and polymers (e.g., thearubigins), which are responsible for many of the properties of black tea infusion. The key enzymatic reaction during this process is the oxidation of

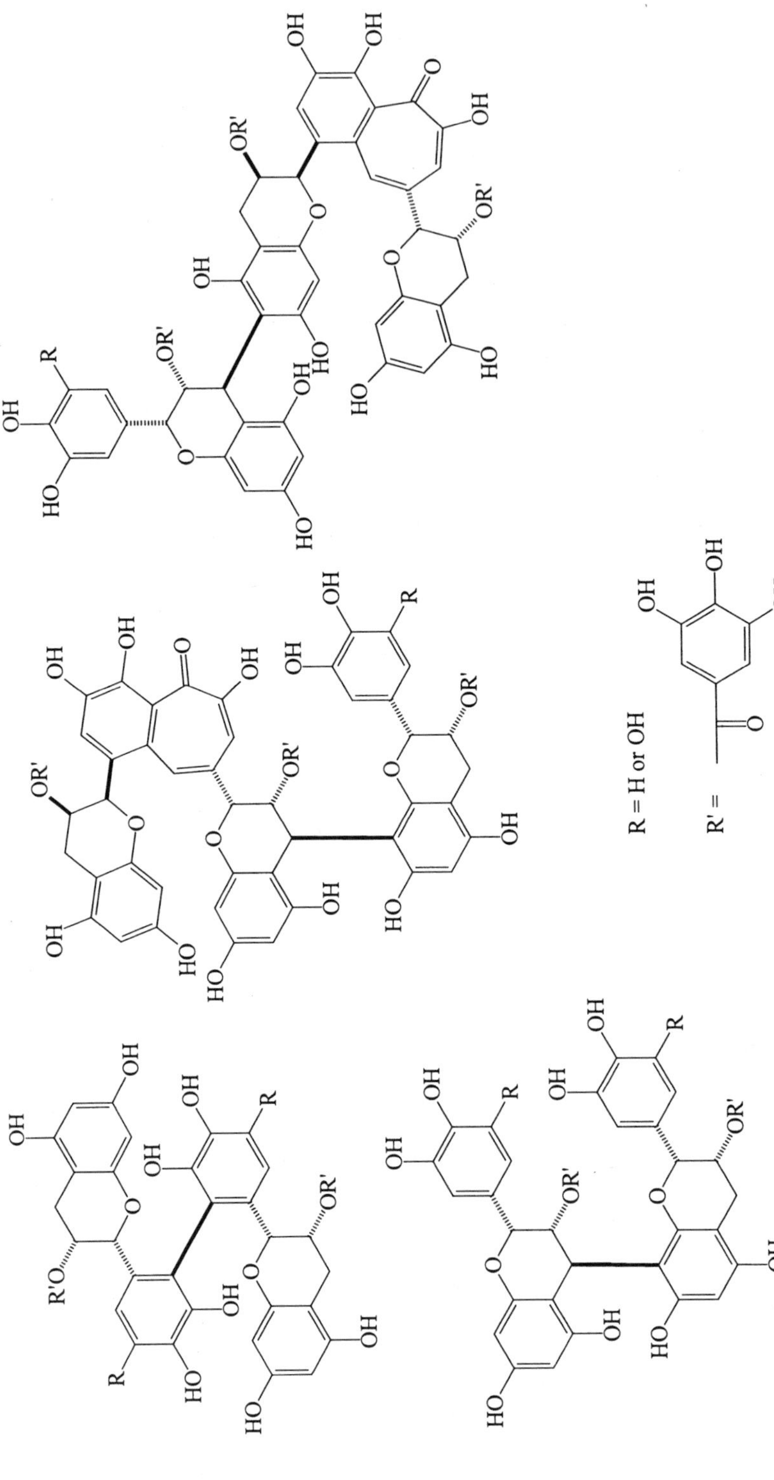

Figure 11. Thearubigin fragments.

the *o*-diphenol catechins to reactive *o*-di-quinones, which then oxidatively couple to generate the polyphenolic polymers that are characteristic of black tea. Although a range of other biochemical reactions accompany the oxidation of polyphenols during tea fermentation and contribute to tea aroma/flavor, this chapter will focus only on the oxidative coupling of flavan-3-ols to give dimeric and polyphenolic species.

Tea leaves are known to contain two classes of enzyme capable of oxidizing *o*-diphenol catechins to *o*-diquinones; polyphenol oxidase (PPO) and peroxidase (POD) (fig. 12). PPO utilises molecular oxygen, whereas POD requires the presence of hydrogen peroxide as an electron acceptor for activity. Catalase, which is universally present in plants,[36] catalyzes the dismutation of H_2O_2 to water and O_2, and therefore has the capacity to regulate POD activity *in vivo*. Although most data in the literature support the role of PPO as the key oxidative enzyme,[37–40] POD has long been thought to have a role in tea fermentation.[37,41] POD enzymes from a variety of sources have been shown to be capable of oxidizing catechins *in vitro*, and tea POD has been implicated in the oxidation of theaflavins to thearubigins.[40,42–45] Purified theaflavins are stable in the presence of tea PPO, but are readily oxidized by horseradish POD.[41] It was suggested that the nature of the pigments formed during the fermentation of tea leaves in aqueous suspension was dependent on the relative activity of PPO and POD,[41] which was in turn governed by the availability of oxygen and hydrogen peroxide. POD activity in fresh tea leaves has been reported to be higher than that of PPO, and it was postulated that control of POD activity could enhance the quality of the product.[44] PPO and POD from tea have different properties (pH optimum, temperature optimum, stability, etc.) and the activities of both are affected by the tea manufacturing process (plucking, withering, maceration, fermentation and firing).

The significance of much of the previous work is unclear, as the presence of an enzyme in an extract does not prove it is active in the tissue or during manufacture, but only that it could potentially be active. Addition of hydrogen peroxide to so-called "slurry fermentations", where macerated leaf is allowed to oxidize in aqueous suspension, under defined conditions, can lead to an alteration in the liquor characteristics,[40,46] but it was not proven that POD played a role under normal conditions. A wide variety of endogenous enzymes may liberate hydrogen peroxide, but in the intact plant elaborate systems have evolved to prevent the

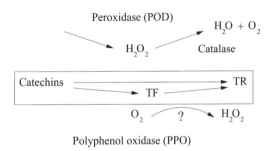

Figure 12. Enzymes potentially involved in catechin oxidation during black tea manfacture.

accumulation of such reactive oxygen species. However, under certain circumstances, such as in response to pest or pathogen attack, hydrogen peroxide can be found at elevated levels.[47] The involvement of POD will crucially depend on the availability of hydrogen peroxide during fermentation. Whether hydrogen peroxide is present or can be synthesized during tea manufacture is not known, although there is evidence that during the oxidation of phenolics by PPO, hydrogen peroxide is generated, probably via superoxide.[48,49]

In order to determine the relative importance of these enzymes in phenol oxidation, reliable and appropriate methods for their extraction and assay have been developed. Individual enzyme assays, designed to measure total extractable activities of PPO, POD, and catalase against endogenous tea substrates, were optimized and validated.

The maximum extractable activities of PPO, POD and catalase from a range of tissues from a series of Kenyan tea clones were determined. Enzyme activities varied significantly between different tissues, with PPO activities covering an eightfold and POD about a threefold range. Catalase showed less variation, but was present at higher levels in those tissues containing highest POD activities. In equivalent tissues of a series of tea clones, the PPO levels varied by over a sevenfold, POD about a twofold and catalase by an eightfold range. It was therefore shown that PPO, POD, and catalase with activities relevant to catechin oxidation are present in fresh tea tissues, and that the ratios of the three enzymes can vary significantly.[50]

The fermentation behavior of some of these clones was investigated (fig. 13). The clones were fermented in aqueous suspension[40,46] (slurry fermentation) allowing control of pH, temperature and gas exchange during fermentation, easy sampling of the system and facilitating the introduction of exogenous compounds. The data were not conclusive, but suggested that the rate of accumulation of theaflavins correlates with PPO activity (fig. 13), whereas no such relationship existed for POD. Such data support a critical (rate limiting) role for PPO in the formation of theaflavins but cannot alone exclude a role for POD. The activities of all three enzymes decreased during manufacture, with between 35–75 percent of the activity remaining after fermentation.

More direct evidence was sought for the relative contributions of PPO and POD in tea fermentation. If the fermentation was made anoxic by passing nitrogen gas through the system, no catechin oxidation or theaflavin formation occurred. This demonstrated that PPO was of prime importance in mediating catechin oxidation, and that POD alone could not catalyze catechin oxidation under these conditions. However, these results do not prove that POD makes no contribution during normal aerobic fermentation when PPO is active, and when H_2O_2 is more likely to be generated.

As extractable POD activity persists during fermentation, one must consider why the enzyme does not function when PPO is inhibited. Possible reasons for its lack of activity are: a) insufficient levels of co-substrate hydrogen peroxide; b) presence of an endogenous inhibitor; or c) a lack of accessibility to substrates.

To assess the availability of H_2O_2, the level in fresh leaves, macerated fresh leaf (dhool), and fermenting dhool were measured and shown to be 70–90 nmol/g

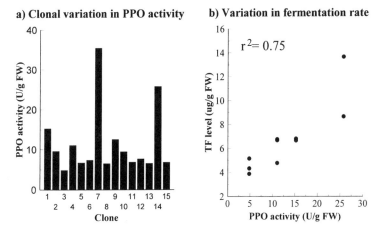

Figure 13. a) Activity of PPO in different African tea clones: b) variation in fermentation rate with PPO activity.

fresh weight. Thus the steady state level of peroxide was low, and sufficient to oxidize only 0.3 percent of the leaf catechins. This finding could be misleading as peroxide consuming enzymes (e.g., catalase) will presumably keep the steady state level low. To further the understanding, evidence was required to establish whether H_2O_2 was being produced during fermentation.

Tropolone, a copper chelator, is known to inhibit PPO and has been used to discriminate between PPO and POD activity.[51–53] It was observed that inclusion of 50 mM tropolone in the fermentation would inhibit both PPO and catalase (table 3), but not POD. Therefore, inclusion of tropolone in a slurry fermentation should inactivate both PPO and catalase and any H_2O_2 produced should be available to POD. There was no evidence of catechin oxidation in the presence of tropolone, but 91 percent oxidation occurred if H_2O_2 was added to the system (fig. 14). This implied that POD remains potentially active during the fermentation but cannot function due to a lack of H_2O_2. This also demonstrated that the lack of POD activity was not due to separation from substrates nor presence of inhibitors. It is still possible that H_2O_2 is only generated during aerobic fermentations as a byproduct of PPO activity or that tropolone was inhibiting other H_2O_2 generating systems.

To test for H_2O_2 generation during PPO mediated oxidation, the dhool was heat treated prior to fermentation, a treatment that specifically reduces catalase activity but not PPO or POD (table 3). This treatment did not influence fermentation indicating that H_2O_2 was not available to POD in the absence of catalase. A final piece of evidence that POD does not function is that gallic acid levels do not change during aerobic fermentation of thermally treated dhool (fig. 15). It is well known that gallic acid is not a substrate for tea PPO, but is oxidized by POD.[54,55] Addition of H_2O_2 led to rapid gallic acid oxidation. However, if 10 mM sodium azide was

Table 3. Extent of inhibition of tea PPO, POD and catalase with
different treatments

Treatment	Percent inhibition		
Addition to fermentation	PPO	POD	Catalase
$N_{2(g)}$	100	0	0
Tropolone (50 mM)	100	0	95
Sodium azide (10 mM)	99.8	99.5	100
Heat treated dhool	44	37	95
(60 °C, N_2, 15 min)			

also added to inhibit all oxidative enzyme activity (table 3), the oxidation of gallic
acid on peroxide addition was substantially reduced, which demonstrated that the
oxidation was an enzymic reaction and not a chemical oxidation. This demonstrated that whereas POD activity persists throughout fermentation, H_2O_2 generation is insufficient to allow any POD mediated catechin oxidation to occur.

4. FACTORS INFLUENCING THEAFLAVIN FORMATION

The studies on the chemistry of theaflavin formation have been complemented
by investigations into the factors influencing enzyme-mediated theaflavin

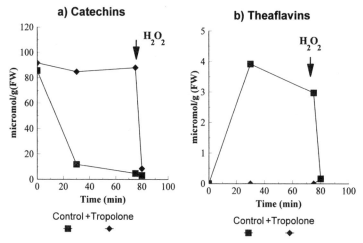

Figure 14. Effect of PPO and catalase inhibition on fermentation, with addition of peroxide after 75 minutes.

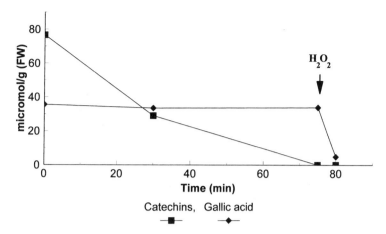

Figure 15. Gallic acid level during aerobic fermentation with peroxide addition after 75 min.

formation. The approach has been to use an *in vitro* model system to study the formation of theaflavins from pairs of catechins in the presence of isolated tea PPO. The formation of theaflavin (TF) from epigallocatechin and epicatechin was a particularly high yielding reaction (fig. 16), and TF formation accounted for a large proportion of oxidized catechins. Although a 1:1 ratio of epigallocatechin and epicatechin might be thought optimal for TF formation, the two catechins were oxidized at different rates. The effect of altering this ratio was found to have little affect on the extent of TF formation, but did significantly effect the rate at which maximal TF formation occurred (fig. 17).

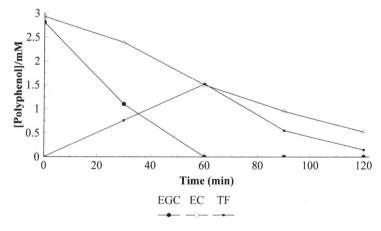

Figure 16. Formation of TF from an equimolar mixture of EGC and EC in the presence of tea PPO.

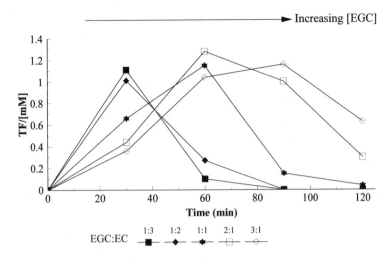

Figure 17. Effect of altering EGC:EC ratio on TF formation.

A range of theaflavins was shown to form from equimolar mixtures of the appropriate catechin pairs in the presence of tea PPO (fig. 18). There was considerable variation in the yield of the different theaflavin species; TF was the most readily formed species and neotheaflavin-3-*O*-gallate the least readily formed. The differences in yield were due to the ease of formation of the different species and not to differences in the extent of catechin oxidation (data not shown). Stereochemistry

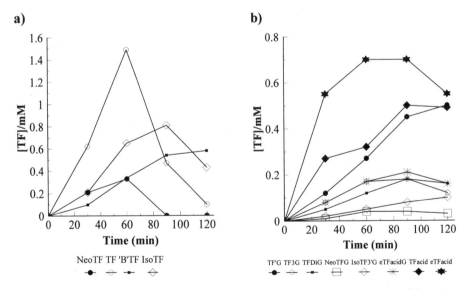

Figure 18. Formation of a range of theaflavins from appropriate catechin pairs; precursors are detailed in table 2.

has some influence on theaflavin formation since TF isomers (neotheaflavin and isotheaflavin) were not formed as readily as TF. Gallated catechins form theaflavins less readily than non-gallated catechins, a result likely to be due to steric hindrance or possibly stereoelectronic effects. Since neotheaflavin-3-*O*-gallate is a gallated theaflavin species formed from a mixture of C and EGCG, it is perhaps not surprizing that this is the least readily formed of the theaflavins investigated.

Theaflavins are clearly not stable end products of fermentation and can be oxidized further to thearubigin species. In the case of epigallocatechin and epicatechin, the oxidation of TF by epicatechin occurred as soon as the epigallocatechin was exhausted. TF oxidation could be prevented by the addition of epigallocatechin to the incubation, when further synthesis of TF occurred through the reaction of epicatechin with epigallocatechin. These incubations also demonstrated that the accumulation of TF is not a balance of synthesis and degradation, rather TF is formed and later oxidized.

The mechanism of theaflavin oxidation was studied further using purified TF. Theaflavin was not a direct substrate for tea PPO, but was readily oxidized if epicatechin was also present (fig. 19), confirming previous results.[56,57] The most likely mechanism is that TF is oxidized chemically by EC quinone generated enzymically from epicatechin by PPO. Varying the initial epicatechin concentration demonstrated that there was no stoichiometric relationship between TF and epicatechin oxidation. This indicated that epicatechin was not combining directly with TF to give benzotropolone-containing catechin trimers, rather epicatechin quinone was acting as an intermediate in enzymatic TF oxidation, and epicatechin was regenerated.

In fact, all dihydroxy B-ring catechin quinones (C, EC, ECG) could facilitate TF oxidation (fig. 20), whereas the trihydroxy B-ring gallocatechins (EGC, EGCG,

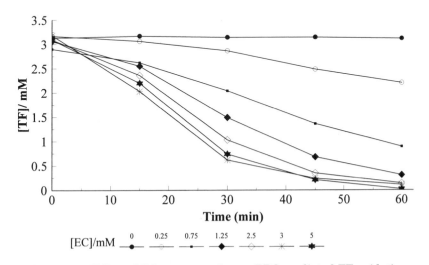

Figure 19. Effect of EC concentration on PPO mediated TF oxidation.

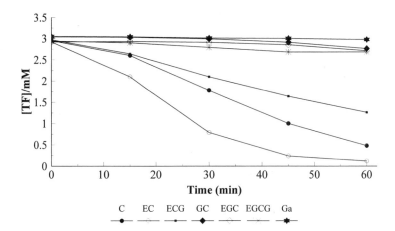

Figure 20. Ability of a range of catechins to facilitate TF oxidation.

GC) could not. This is probably due to the relative redox potentials of the different species, as the gallocatechins have lower redox potential compared with catechins.

As epicatechin quinone can react with epigallocatechin quinone to give TF and can also facilitate the further oxidation of TF, it is surprising that TF accumulation is not a balance of synthesis and oxidation. After 30 minutes oxidation of a mixture of epigallocatechin and epicatechin (fig. 16), epigallocatechin and TF are present at approximately equal concentration in the incubation, and consequently epicatechin quinone is equally likely to interact with either, yet no TF oxidation occurs. We suggest that epicatechin quinone may oxidize TF to a reactive intermediate, but that this intermediate is, in turn, able to oxidize epigallocatechin and thus regenerate TF. This scheme is illustrated in figure 21. In this way the rate of epigallocatechin oxidation is increased and, as epicatechin is cycling, apparent epicatechin oxidation is reduced. This may account for the different rates of epigallocatechin and epicatechin disappearance shown in figure 16.

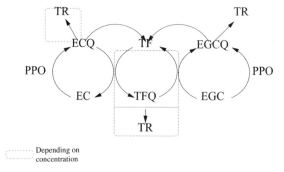

Figure 21. Proposed reaction pathways of EGC + EC model oxidation.

5. CONCLUSIONS

Black tea is produced from the young leaves of *Camellia sinensis* var. *assamica* by a process involving five stages—plucking, withering, maceration, fermentation, and drying. During this process, particularly the fermentation stage, the flavonoids (mainly the flavan-3-ols; catechins) present in the leaves undergo oxidative transformation to dimers and oligomers/polymers. The major dimers are theaflavins, which are well characterized, while the polymers are uncharacterized and referred to as thearubigins.

In this chapter we have described recent advances in characterizing the structures of dimeric polyphenol species responsible for some of the color properties of black tea, namely: isotheaflavin-3′-*O*-gallate, neotheaflavin-3-*O*-gallate, theaflavate B and theacitrin A. Investigations on the oligomers have identified a number of partial structures for the thearubigins, but no complete structure has been identified.

The formation of the dimers/oligomers/polymers has been shown to be the result of oxidation of catechins by polyphenol oxidase using oxygen as the electron acceptor. Peroxidase although present in leaf and capable of oxidizing catechins cannot function since hydrogen peroxide, the necessary electron acceptor, is not available. Details on the influence of catechin types, stereochemistry at C2—C3, and gallate ester substituents on the formation of theaflavins have also been elucidated. Theaflavins have also been shown to be unstable and are oxidized further by flavan-3-ols (epicatechin).

REFERENCES

1. Banerjee, B. Botanical classification of tea. *In*: Willson, K.C.; Clifford, M.N. (eds.) Tea: cultivation to consumption. Chapman and Hall, London (1992).

2. Roberts, E.A.H. Economic importance of flavonoid substances: tea fermentation. *In*: Geissman, T.A. (ed.) The chemistry of flavonoid compounds, Pergaman Press, Oxford, London and New York, p. 468 (1962).

3. Nonaka, G.I.; Kawahara, O.; Nishioka, I. Tannins and related compounds. XV. A new class of dimeric flavan-3-ol gallates, theasinensins A and B, and proanthocyanidin gallates from green tea leaf. *Chem. Pharm. Bull.* 31:3906 (1983).

4. Roberts, E.A.H. The chemistry of tea manufacture. *J. Sci. Food Agric.* 9:381 (1958).

5. Millin, D.J.; Crispin, D.J.; Swaine, D. Non-volatile components of black tea and their contribution to the character of the beverage. *J. Agr. Food Chem.* 17:717 (1969).

6. Powell, C.; Clifford, M.N.; Opie, S.; Ford, M.A.; Robertson, A.; Gibson, C. Tea cream formation: the contribution of black tea phenolic pigments determined by HPLC. *J. Sci. Food Agric.* 63:77 (1992).

7. Roberts, E.A.H.; Myers, M. The phenolic substances of manufactured tea. VI. The preparation of theaflavin and theaflavin gallate. *J. Sci. Food Agric.* 10:175 (1959).

8. Takino, Y.; Ferretti, A.; Flanagan, V.; Gianturco, M.; Vogel, M. Structure of theaflavin, a polyphenol of black tea. *Tetrahedron Lett.* :4019 (1965).

9. Ollis, W.D.; Brown, A.G.; Haslam, E.; Falshaw, C.P.; Holmes, A. The constituents of theaflavin. *Tetrahedron Lett.* :1193 (1966).

10. Nakagawa, M.; Torii, H. Studies on the flavan-3ols in tea. Part IV. Enzymic oxidation of flavan-3-ols. *Agr. Biol. Chem.* 29:278 (1965).

11. Takino, Y.; Imagawa, H. Studies on the mechanism of the oxidation of tea leaf catechins (IV). *Agr. Biol. Chem.* 28:125 (1964).

12. Takino, Y.; Imagawa, H. Crystalline Reddish orange pigment of manufactured black tea. *Agr. Biol. Chem.* 28:255 (1964).

13. Takino, Y.; Imagawa, H.; Horikawa, H.; Tanaka, A. Studies on the mechanism of the oxidation of tea leaf catechins—formation of a reddish/orange pigment and its spectral relationship to some benzotropolone derivatives. *Agr. Biol. Chem.* 28:64 (1964).

14. Collier, P.D.; Bryce, T.; Mallows, R.; Thomas, P.E.; Frost, D.J.; Korver, O.; Wilkins, C.K. The theaflavins of black tea. *Tetrahedron Letts.* 29:125 (1973).

15. Collier, P.D.; Bryce, T.; Mallows, R.; Thomas, P.E.; Frost, D.J.; Wilkins, C.K. Three new theaflavins from black tea. *Tetrahedron Letts.* 6:463 (1972).

16. Coxon, D.T.; Holmes, A.; Ollis, W.D. Isotheaflavin. A new black tea pigment. *Tetrahedron Letts.* :5241 (1970a).

17. Coxon, D.T.; Holmes, A.; Ollis, W.D. Theaflavic acid and epitheaflavic acids. *Tetrahedron Letts.* :5247 (1970b).

18. Nonaka, G.I.; Hashimoto, F.; Nishioka, I. Tannins and related compounds XXXVL. Isolation and structure of theaflagallins, new red pigments from black tea. *Chem. Pharm. Bull.* 34:61 (1986).

19. Lewis, J.R.; Davis, A.L.; Cai, Y.; Davies, A.P.; Wilkins, J.P.G. Theaflavate B, isotheaflavin-3'-O-gallate and neotheaflavin-3-O-gallate: Three new polyphenolic pigments from black tea. *Phytochemistry* 49:2511 (1998).

20. Davis, A.L.; Cai, Y.; Davies, A.P. ¹H and ¹³C NMR assignment of theaflavin, theaflavin monogallate and theaflavin digallate. *Magnetic Resonance in Chemistry* 33:549 (1995).

21. Davis, A.L.; Cai, Y.; Davies, A.P.; Lewis, J.R. ¹H and ¹³C NMR assignments of some green tea polyphenols. *Magnetic Resonance in Chemistry* 34:887 (1996).

22. Bailey, B.G.; Nursten, H.E.; McDowell, I. The chemical oxidation of catechins and other phenolics: A study of the formation of black tea pigments. *J. Sci. Food Agric.* 63:455 (1993)

23. Wan, X.; Nursten, H.E.; Cai, Y.; Davis, A.L.; Wilkins, J.P.G.; Davies, A.P. A new type of tea pigment—from the chemical oxidation of epicatechin gallate and isolated from tea. *J. Sci. Food. Agric.* 74:401 (1997).

24. Powell, C.; Clifford, M.N.; Opie, S.C.; Gibson, C.L. Fractionation of thearubigins and isolation of theacitrins *In*: Brouillard, R.; Jay, M.; Scalbert, A. (eds.) Polyphenols 94. XVIIe Journees Internationales Groupe Polyphenols. INRA Editions, Paris, France p. 279 (1995).

25. Davis, A.L.; Lewis, J.R.; Cai, Y.; Powell, C.; Davies, A.P.; Wilkins, J.P.G.; Pudney, P.; Clifford, M.N. A polyphenolic pigment from black tea. *Phytochemistry* 46:1397 (1997).

26. Hashimoto, F.; Nonaka, G.-I.; Nishioka, I. Tannins and related compounds. LXIX) Isolation and structure elucidation of B-B'-linked bisflavanoids, theasinensins D—G and oolongtheanin from oolong tea. *Chem. Pharm. Bull.* 36:1684 (1988).

27. Horner, L.; Durckheimer, W.; Weber, K.-H. Zur kenntnis der o-chinone XIX. Hydrolyse studien an 2-substituierten 1,3 dicarbonyl verbindungen als beitrag zum mechanismus der purpurogallinbildung. *Chem. Ber.* 94:2881 (1961).

28. Horner, L.; Durckheimer, W.; Weber, K.-H.; Dolling, K. Zur kenntnis der o-chinone. XXIV. Synthese, struktur und eigenschaften von 1'2'-dihydroxy-6,7-benzotropolonen. *Chem. Ber.* 97:312 (1964).

29. Haslam, E. Quinone tannin and oxidative polymerisation. *In:* Haslam, E. Practical polyphenolics. From structure to molecular recognition and physiological action. Cambridge University Press, p. 335 (1998).

30. Haslam, E.; Lilley, T.H.; Warminski, E.; Liao, H.; Cai, Y.; Martin, R.; Gaffney, S.H.; Goulding, P.N.; Luck, G. Polyphenol complexation—a study in molecular recognition. *In:* Ho, C.-T.; Lee, C.Y.; Huang M.-T. (eds.) Phenolic compounds in food and their effects on

health. I. Analysis, occurrence and chemistry. A.C.S Symposium Series 506, Washington DC. p. 8 (1992).

31. Wickremasinghe, R.L. *In:* Chichester C.O.; Mrak E.M.; Stuart G.F. (eds.) Advances in food Research. Academic Press, New York, p. 229 (1978).

32. Roberts, E.A.H.; Cartright, R.A.; Oldschool, M. Fractionation and paper chromatography of water soluble substances from manufactured tea. *J. Sci. Food Agric.* 22:72 (1957).

33. Davis, A.L.; Cai, Y.; Lewis, J.R.; Davies, A.P. (unpublished results).

34. Nursten H.E.; Cattell, D.J. Fractionation and chemistry of ethyl acetate soluble thearubigins from black tea. *Phytochemistry* 15:1967 (1976).

35. Ozawa, T.; Kataoka, M.; Morikawa, K.; Negishi, O. Elucidation of the partial structure of polymeric thearubigins from black tea by chemical degradation. *Biosci. Biotech. Biochem.* 60:2023 (1996).

36. Willekens, H.; Inze, D.; Van Montagu, M.; van Camp, W. Catalases in plants. *Molecular Breeding* 1:207 (1995).

37. Hara, Y.; Luo, S.J.; Wickremasinghe, R.L.; Yamanishi, T. Biochemistry of processing black tea. *Food Rev. Int.* 11:457 (1995).

38. Jain, J.C.; Takeo, T. A review: The enzymes of tea and their role in tea making. *J. Food Biochem.* 8:243 (1984).

39. Sanderson, G.W.; Coggon, P. Use of enzymes in the manufacture of black tea and instant tea. *In:* Ory, R.L.; St. Angelo, A.J. (eds.) Enzymes in food and beverage processing. American Chemical Society, Washington DC, p. 12 (1977).

40. Robertson, A. The chemistry and biochemistry of black tea production-the non-volatiles. *In:* Willson, K.C.; Clifford, M.N. (eds.) Tea: cultivation to consumption. Chapman and Hall, London (1992).

41. Dix, M.A.; Fairley, C.J.; Millin, D.J.; Swaine, D. Fermentation of tea in aqueous suspension. Influence of tea peroxidase. *J. Sci. Food Agric.* 32:920 (1981).

42. Cloughley, J.B. The effect of temperature on enzyme activity during the fermentation phase of black tea manufacture. *J. Sci Food Agric.* 31:924 (1980).

43. Finger, A. *In-vitro* studies on the effect of polyphenol oxidase and peroxidase on the formation of polyphenolic black tea constituents. *J. Sci. Food Agric.* 66:293 (1994).

44. Mahanta, P.K.; Baruah, S.K.; Baruah, H.K.; Kalita, J.N. Changes of polyphenol oxidase and peroxidase activities and pigment composition of some manufactured black teas (*Camellia sinensis* L.). *J. Agric. Food Chem.* 41:272 (1993).

45. Stagg, G.V. Beverages: tea. *Rep. Prog. Appl. Chem.* 59:451 (1974).

46. Millin, D.J.; Swaine, D. Fermentation of tea in aqueous suspension. *J. Sci. Food Agric.* 32:905 (1981).

47. Bolwell, G.P.; Butt, V.S.; Davies, D.R.; Zimmerlin, A. The origin of the oxidative burst in plants. *Free Raical Res.* 23:517 (1995).

48. Jiang, Y.; Miles, P.W. Generation of H_2O_2 during enzymic oxidation of catechin. *Phytochemistry* 33:29 (1993).

49. Ricard-Forget, F.C; Gauillard, F. Oxidation of chlorogenic acid, catechins, and 4-methylcatechol in model solutions by combinations of pear (*Pyrus communis* Cv. Williams) polyphenol oxidase and peroxidase: a possible involvement of peroxidase in enzymatic browning. *J. Agric. Food Chem.* 45:2472 (1997).

50. Parry, A.D.; Goodsall, C.W.; Safford, D.S. The involvement of polyphenol oxidase and peroxidase in the oxidation of polyphenols during the manufacture of black tea. *In:* Vercauteren, J.; Cheze, C.; Dumon, M.C.; Weber, J.F. (eds.) Polyphenol Communications 96. Bordeaux University, France. p. 499 (1996).

51. Kahn, V. Tropolone—a compound that can aid in differentiating between tyrosinase and peroxidase. *Phytochemistry* 24:915 (1985).

52. Kahn, V.; Andrawis, A. Inhibition of mushroom tyrosinase by tropolone. *Phytochemistry* 24:905 (1985).

53. Valero, E.; Garcia-Moreno, M.; Varon, R.; Garcia-Carmona, F. Time-dependent inhibition of grape polyphenol oxidase by tropolone. *J. Agric. Food Chem.* 39:1043 (1991).

54. Roberts, E.A.H.; Meyers, M. The phenolic substances of manufactured tea. VIII. Enzymic oxidation of polyphenolic mixtures. *J. Sci. Food Agric.* 11:158 (1960).

55. Coggon, P.; Moss, G.A.; Sanderson, G.W. The biochemistry of tea fermentation: oxidative degallation and epimerization of the tea flavanol gallates. *J. Agr. Food Chem.* 21:727 (1973).

56. Bajaj, K.L.; Anan, T.; Tsushida, T.; Ikegaya, K. Effects of (–)-epicatechin on oxidation of theaflavins by polyphenol oxidase from tea leaves. *Agric. Biol. Chem.* 51:1767 (1987).

57. Opie S.C.; Clifford M.N.; Robertson A. The role of (–)-epicatechin and polyphenol oxidase in the coupled oxidative breakdown of theaflavins. *J. Sci. Food Agric.* 63:435 (1993).

CATECHIN AND PROCYANIDIN LEVELS IN FRENCH WINES: CONTRIBUTION TO DIETARY INTAKE

Stéphane Carando and Pierre-Louis Teissedre

Centre de Formation et de Recherche en Oenologie
Université de Montpellier 1
Faculté de Pharmacie
34060 Montpellier Cedex 2
FRANCE

1. INTRODUCTION

A great many epidemiological studies indicate that a diet rich in flavonoids can reduce the incidence of coronary artery disease. Regular moderate consumption of wine can contribute to this phenomenon. Flavonoids in wine and food have been shown to be antioxidant and anti-aggregant *in vitro* and could indeed help protect against coronary disease. However, the epidemiological studies in this field are based on data concerning the flavonoid composition of foods, and the contribution of a regular, moderate consumption of wine remains difficult to quantify. In this study, we have tried to obtain a first estimation of catechin and procyanidin contents. We also discuss the metabolism of these molecules and the appearance of (+)-catechin in the plasma.

2. MODERATE CONSUMPTION OF WINE AND CORONARY ARTERY DISEASE

Numerous epidemiological studies have observed a negative correlation between moderate consumption of alcoholic beverages and coronary artery disease. Often, a "U"- or "J"-shaped graph is described,[1] which indicates lower mortality for moderate consumers than for abstainers or heavy drinkers. The ethanol content of such consumption would inhibit platelet aggregation and increase plasma high-density lipoprotein (HDL) levels. However, among the group of moderate consumers, it is possible to further distinguish different subgroups. The large variability in the composition of alcoholic beverages and in their modes of consumption has led epidemiologists to consider three different groups: consumers of

Plant Polyphenols 2: Chemistry, Biology, Pharmacology, Ecology, Edited by
Gross et al. Kluwer Academic / Plenum Publishers, New York, 1999

wine, beer and spirits respectively. The results of the various studies carried out suggest that all these groups of moderate consumers show a lower risk of coronary mortality than either abstainers or heavy drinkers but that the difference is even more pronounced in wine drinkers (fig. 1).

In one of the best known studies, Renaud and De Lorgeril[7] suggest an explanation of the phenomenon particularly favorable to the French population with regard to cardiovascular disease, known as the "French Paradox," which was first described in 1987 by Richard.[8] The results of the Monica program,[9] a worldwide coronary artery disease (CAD) surveillance system organized by the World Health Organization (WHO), confirm that mortality levels provoked by coronary artery disease are much lower in France than in other industrialized countries, even though the consumption of saturated fats in France is much the same and blood cholesterol levels are generally higher. Furthermore, other factors associated with risk of coronary artery disease, such as arterial blood pressure, body weight and smoking, are no lower in France than in the other countries. This is the French Paradox.

Renaud and De Lorgeril[7] used WHO and the Organization for the Cooperation and Economical Development (OCED) data to study which food parameters could be correlated with cardiovascular disease (CVD) mortality levels. Among the numerous different foods, only the consumption of dairy fats (between 1980 and 1985) showed positive correlation with CVD mortality (1987 levels), but this correlation was highly significant ($r = 0.73$, $p < 0.001$). However, although the consumption of dairy fats is the same in France as in the UK, Australia and Germany, mortality from coronary artery disease is lower in France. Using multifactor analysis, Renaud & De Lorgeril[7] revealed that in 17 wine-drinking countries, the only dietary factor correlating significantly with coronary artery disease mortality, apart from dairy fats, was wine ($r = 0.87$, $p < 0.001$). Moreover, wine consumption had a negative coefficient indicating a protective effect. It was thus realized that the French case was not paradoxical when wine consumption was taken into account. In this study, the protection afforded by wine consumption was also detected in Switzerland and other industrialized countries. From this, attention turned to the non-alcoholic fractions of wine.

Wine, particularly red wine, is an important source of polyphenols, which are capable of inhibiting the processes behind coronary artery disease. This hypothesis is supported by the results of recent epidemiological studies concerning foodstuff polyphenols, particularly flavonoids. A correlation was also noticed between increasing levels of flavonoid ingestion from fruit and vegetables and reduction in coronary artery disease. The studies carried out by Hertog et al.,[10] Knekt et al.[11] and Rimm et al.[12] reveal the benefits of a diet rich in flavonoids. In parallel to these epidemiological studies, biochemical data on wine flavonoids, particularly flavanols should be considered. Differences in the ability of wine, beer and spirits to protect against cardiovascular disease could be explained by the specific action of flavonoids and their metabolites in wine, which are practically absent from beer and spirits.

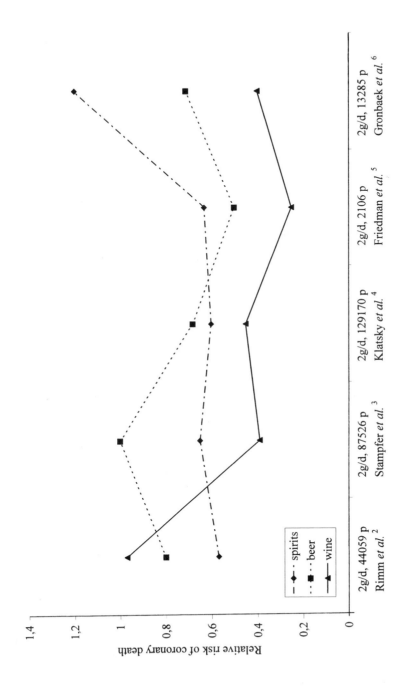

Figure 1. Risk of mortality from coronary artery disease from epidemiological studies separating the consumption of wine, beer and spirits.[2-6]

3. BIOCHEMICAL PROPERTIES OF FLAVANOLS

From this family of compounds, we are particularly interested in catechins and procyanidins. (+)-Catechin and (–)-epicatechin are the basic units of this group. The procyanidins are formed from the association of several of these monomeric units: 2–5 units for catechin oligomers, over 5 units for catechin polymers. The procyanidins differ in the position and configuration of their monomeric linkages. The structures of procyanidin dimers B1, B2, B3 and B4, and trimers C1 and C2 are the best known (fig. 2). These molecules possess a structure that confers them with an antioxidant property which can inhibit the processes leading in the long term to arteriosclerosis and arterial thrombosis.

The flavonoids are the most lipophilic of the natural antioxidants; but less so than α-tocopherol. The α-tocopherol seems to be located in the lipid membrane within the phospholipid bilayer, whereas the flavonoids are probably mainly located at the polar surface of the bilayer. The aqueous, i.e., transported in the plasma, free radicals would therefore be captured more easily by the flavonoids than by the less accessible α-tocopherol. Thus, the flavonoids could be concentrated near to the membranous surface of the low-density lipoprotein (LDL) particles, ready to capture the oxygenated aqueous free radicals. They would in this way prevent the consumption of lipophilic α-tocopherol and thus delay oxidation of the lipids contained in the LDL. Moreover, if, as research suggests, the initiation phase and the propagation phase of lipid peroxidation take place, respectively, at the surface and the interior of the membranes, then the flavonoids could well hinder the correct course of the reaction by limiting the initiation phase.

Catechins and procyanidins have been shown *in vitro* to be powerful inhibitors of LDL oxidation, more so than α-tocopherol,[13] and of platelet aggregation.[14] Moreover, it has recently been shown that the consumption of wine by humans leads to an increase in the antioxidant capacity of plasma.[15]

Other studies have been carried out on the antioxidant activity—through inhibition of copper-catalyzed oxidation of human LDL—of a selection of California wines, made from cabernet-sauvignon, merlot, zinfandel, petite syrah, pinot-noir, sauvignon blanc and chardonnay. The relative inhibition of LDL oxidation (calculated with respect to the total phenol concentration of each sample) varied from 46 to 100 percent for red wines and from 3 to 6 percent for white wines.[16] The antioxidant activity of wines made with long maceration times exclusively from Rhône valley syrah and grenache varieties was 60 percent relative inhibition of LDL oxidation.[17]

Comparison of these results with the different California red wines showed that the Rhodanian wines made from syrah and grenache gave values of relative inhibition of human LDL oxidation identical to those of cabernet-sauvignon but less than those of petite syrah (80 percent) or merlot (74 percent) (fig. 3). In contrast, the syrah wines produced with short extractions gave a low mean value of relative inhibition (16 percent) which is nevertheless four times higher than that obtained for California white wines (4.4 percent). It was deduced from this that each phenolic compound of wine could play a role in the protection against LDL oxidation. All the properties and studies reported support the present hypotheses

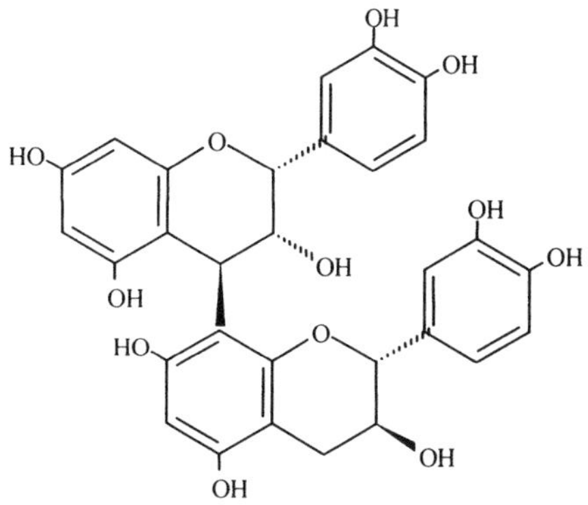

Procyanidin B1

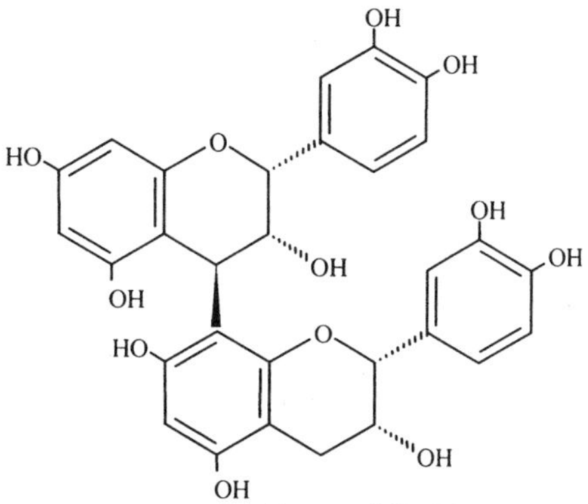

Procyanidin B2

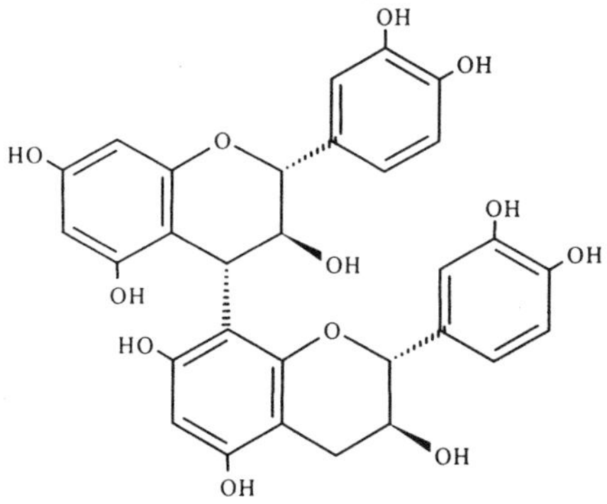

Procyanidin B3

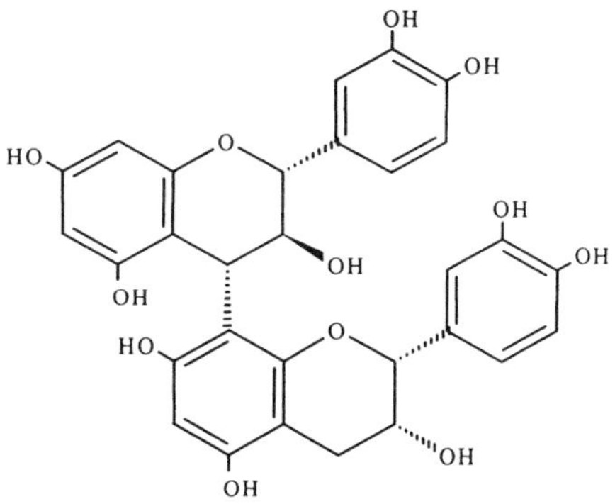

Procyanidin B4

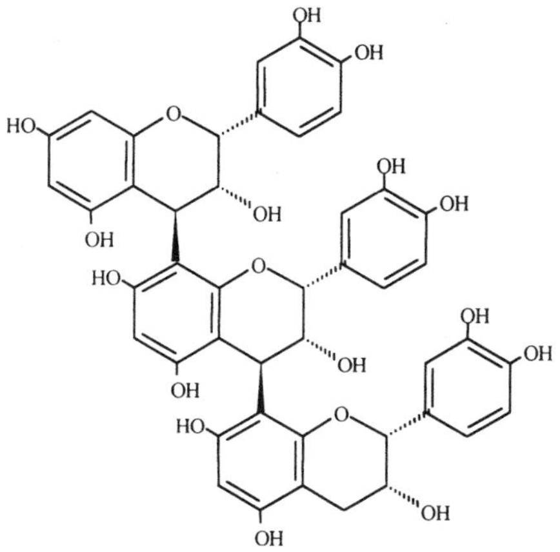

Procyanidin C1

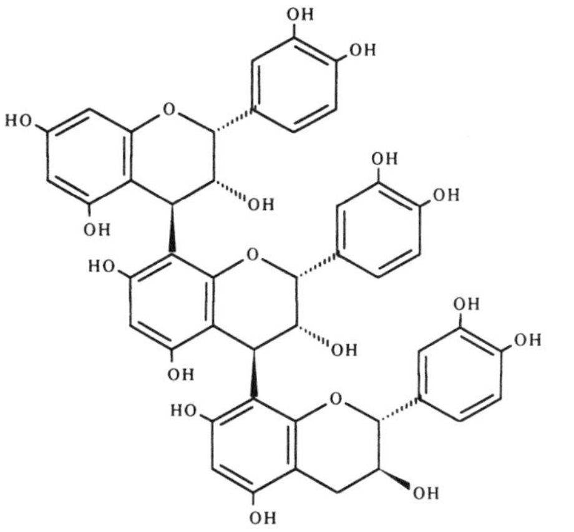

Procyanidin trimer 2

Figure 2. Dimeric and trimeric procyanidins.

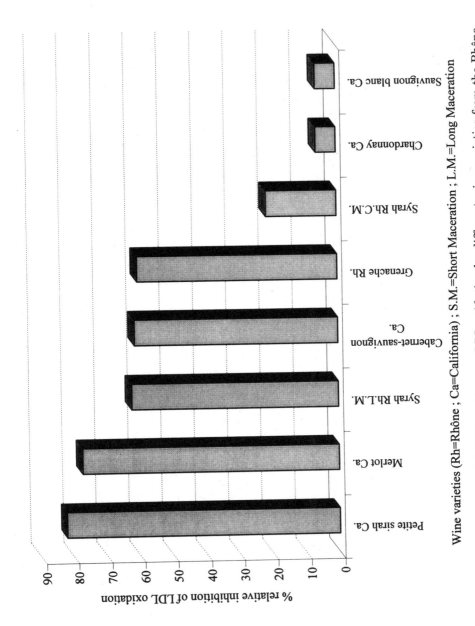

Figure 3. Percentage inhibition of human LDL oxidation by different wine varieties from the Rhône Valley and California.

Wine varieties (Rh=Rhône ; Ca=California) ; S.M.=Short Maceration ; L.M.=Long Maceration

for explaining the reduced risk of mortality from coronary artery disease in moderate and regular consumers of wine (in particular red wine).

4. CATECHIN AND PROCYANIDIN LEVELS IN FRENCH WINES

In order to refine the epidemiological studies on wine consumption, it is of interest to improve understanding of the flavonoid composition of wine so as to be able to evaluate the relative contribution of this consumption. Analytical studies on flavonoids in wine revealed that this beverage contains high levels and that the different winemaking techniques are important factors. However, the results cited here concern wines made either from a unique variety,[18] from California[19] or from micro-vinification[20] and not intended for consumption. The aim of this study was to analyze the catechin and procyanidin content of wines from all French regions made from diverse varieties and using varied vinification techniques, i.e., an analysis reflecting the great diversity of French wines. The results obtained were used to evaluate the catechin and procyanidin intake from regular, moderate wine consumption. A sampling of commercial French wines composed of 95 red wines, 57 white wines and 8 rosé wines was analyzed. These wines came from the different French wine-growing regions and are commonly consumed.

The procyanidin dimers B1, B2, B3, B4 and trimers C1, C2 are not commercially available and must be isolated, purified and certified before they can be used as standards. These compounds were obtained during work by Teissedre et al.[13] at the University of California, Davis. For HPLC analysis, a Hewlett Packard 1090 HPLC system fitted with 3 low-pressure pumps and a UV diode-array detector was used. Programming of analytical conditions, storage of data and treatment of results were effected using a linked computer. The analytical conditions used were those described by Lamuela-Raventos et al.[21]

Red wines had generally much higher levels of phenolics than white wines. This difference is the result of different winemaking techniques. The flavanoids in grapes, present mainly in the pips, diffuse into the must during the course of maceration. The maceration phase constitutes a fundamental difference in the vinification of red and white wine. It is essential for the production of red wine and is deliberately avoided for the production of white, where rapid pressing separates the liquid phase from the solid matter. Thus, the red wines studied had high overall levels of flavanoids. The sum of monomers, procyanidin dimers and trimers analyzed was 557.9 mg/L, but there were considerable differences between the red wines (fig. 4). Levels in the white and rosé wines were lower: 15.2 mg/L and 17.1 mg/L, respectively.

All the compounds studied were present in each of the 95 red wines analyzed. For the monomers, the mean concentration (sum of (+)-catechin and (−)-epicatechin) was 190.2 mg/L. The quantity of procyanidin dimers (sum of B1, B2, B3, B4) was 274.3 mg/L and that of procyanidin trimers (sum of C1, T2) was 93.4 mg/L. The white wines (n = 57) and rosé wines (n = 8) showed low concentrations of monomers (15.1 mg/L and 17.1 mg/L, respectively). The mean contents of catechin monomers and procyanidin dimers and trimers are given in figure 5. The presence of dimers and trimers in the white and rosé wines could not be determined through

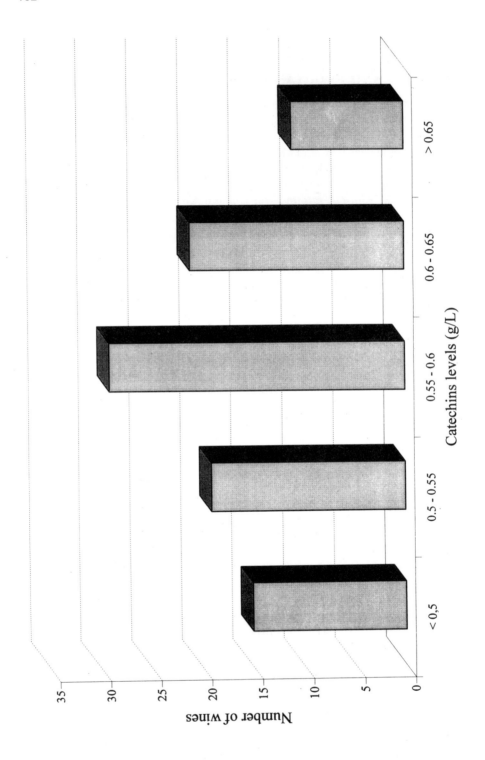

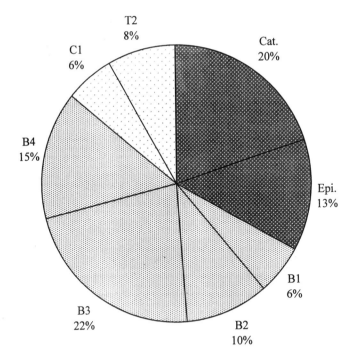

Cat. : (+)-Catechin
Epi. : (-)-Epicatechin

Figure 5. Proportions of monomers, dimers, and trimers in French red wines.

direct injection in HPLC since the levels of these compounds were at the limit of detection by UV. Further analysis would require pre-concentration of the samples.

5. ESTIMATION OF CATECHIN AND PROCYANIDIN INTAKE AND BLOOD PLASMA LEVELS IN MODERATE WINE CONSUMERS IN FRANCE

Over the last few years, the consumption of wine in France has fallen considerably. In 1986, the mean consumption was 305 mL/person/day.[22] This level fell sharply to 180 mL/person/day in 1995.[23] An estimation of the intake of catechins and procyanidins was calculated from these latest consumption figures and our own analytical results on the catechin (monomers, dimers, trimers) content of 160 wines. Although the levels of these compounds vary considerably from one wine to another, the regular consumption of the different products available to the consumer tends toward a mean value that we considered to be close to that calculated for the 160 wines.

For this reason, this estimation can only be considered for regular consumption of different wines over a sufficiently long period of time (which remains to be determined statistically). The consumption of 180 mL of red wine for which the mean catechin and procyanidin concentration is 557.9 mg/L gives a mean daily intake of 100.4 mg of these compounds. This reasoning applied to each type of wine (red, white, rosé) for regular (daily), moderate (180 mL) consumption gives estimations of catechin and procyanidin intake of 100.4 mg for red wines, 2.7 mg for white wines, and 3.1 mg for rosé wines.

Although epidemiological studies note a correlation between the reduction of risk of coronary artery disease and the regular, moderate consumption of wine, no causal relationship has been demonstrated. The flavonoids in wine constitute a considerable LDL antioxidant and platelet anti-aggregant potential, but it is difficult to know whether this potential is slightly, partially or totally realized during metabolism. It would appear necessary, therefore, to supply irrefutable biochemical proof of the envisaged protective mechanisms. Although compounds such as (+)-catechin, (–)-epicatechin, procyanidin dimers B1, B2, B3, B4 and trimers C1, C2 inhibit LDL oxidation *in vitro*,[13] no account can be taken of the large variability in the composition of blood plasma in order to verify whether these compounds or their metabolites can be sufficiently active *in vivo* to afford protection, even though epidemiological studies suggest that they do. Current research concerns the fate of these compounds in human plasma after wine consumption. The intestinal flora is likely to metabolize some of these compounds. The first step consists of identifying the intact molecules and/or their metabolites in the plasma. This analysis is very difficult to carry out because these compounds have different stabilities depending on the physical and biological conditions. A method developed recently associates a simple, rapid phase of plasma sample preparation with a highly sensitive detection method: fluorescence.[24] This method has enabled us to identify (+)-catechin in plasma and to quantify it for several persons having consumed wine with their meal. The very large variability in plasma (+)-catechin concentrations (260–810 ng/mL) is illustrated in figure 6.

At present, this method has only been applied to (+)-catechin, but it would be of interest in the future to apply it to the analysis of other compounds. It would also be interesting to determine whether the plasma (+)-catechin comes directly from the wine (+)-catechin or whether it is also likely to arise from procyanidin metabolism.

6. CONCLUSIONS

This estimation of catechin and procyanidin intake (60 mg/person/day) concerns the 8 compounds analyzed. It would, of course, be important in the future to investigate other compounds, such as epigallocatechin, B1-3-*O*-gallate or B2-3-*O*-gallate, in order to refine this estimation. Furthermore, it would be desirable to extend this estimation to the non-flavonoid components of wine. However, it will also be very important in the future to obtain data on the fate and bioavailability of these compounds in the plasma. It has been established that (+)-catechin is present in the plasma after a meal during which wine had been consumed, but these data must now be correlated with an LDL antioxidant activity *in vivo* so that the pharmacological action of this tracer molecule can be definitively established *in vivo*. This is exactly what we are trying to establish in a research program

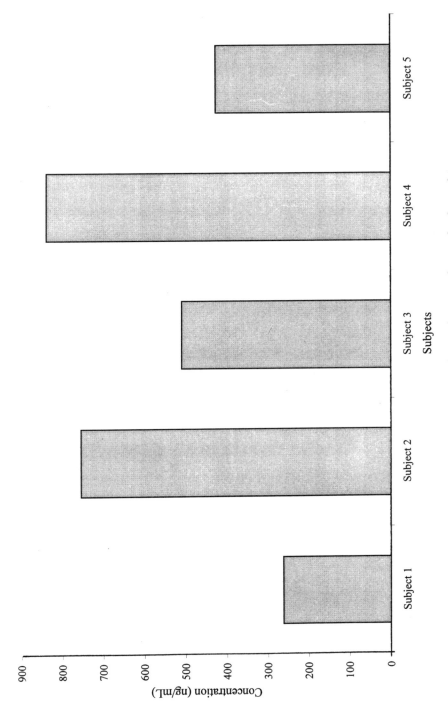

Figure 6. Concentrations of (+)-catechin in human plasma for five subjects.

entitled "Wine and Health: Biology and Vascular Pathology" within our multidisciplinary team in Montpellier from the University, INRA, and INSERM.

ACKNOWLEDGMENTS

This work was supported by the European Institute for Research and Communication on Wine, Health and Society, France. Thanks to F. Gasc for glassware technical assistance.

REFERENCES

1. Klatsky, A.L.; Armstrong, M.A. Alcoholic beverage choice and risk of coronary artery disease mortality: do red wine drinkers fare best? *Am. J. Epidemiol.*, 71:467–469 (1993).

2. Rimm, E.B.; Giovannucci, E.L.; Willett, W.C.; Colditz, G.A.; Aschiero, A.; Rosner, B.; Stampfer, M.J. Prospective study of alcohol consumption and risk of coronary disease in men. *Lancet* 338:464–468 (1991).

3. Stampfer, M.J.; Colditz, G.A.; Willett, W.C.; Speizer, F.E.; Hennekens, C.H. A prospective study of moderate alcohol consumption and the risk of coronary disease and stroke in women. *New Engl. J. Med.* 319:267–273 (1988).

4. Klatsky, A.L.; Friedman, G.D.; Siegelab A.B. Alcohol consumption before myocardial infarction: results from the Kaiser-Permanente epidemiologic study of myocardial infarction. *Ann. Intern. Med.* 81:294–301 (1974).

5. Friedman, L.A.; Kimball, A.W. Coronary heart disease mortality and alcohol consumption in Framingham. *Am. J. Epidemiol.* 124:481–489 (1986).

6. Gronbaek, M.; Deis, A.; Sorensen, T.I.A.; Becker, U.; Schnohr, P.; Jensen, G. Mortality associated with moderate intakes of wine, beer or spirits. *B.M.J.* 310:1165–1169 (1995).

7. Renaud, S.; de Lorgeril, M. Wine, alcohol, platelets, and the French paradox for coronary heart disease. *Lancet* 339:1523–1526 (1992).

8. Richard, J.L. Les facteurs de risque coronarien. Le paradoxe français. *Arch. Mal. Coeur.* numéro spécial avril, 17–21 (1987).

9. World health statistics annual. World Health Organization, Geneva, (1989).

10. Hertog, M.G.L.; Feskens, E.J.M.; Hollman, P.C.H.; Katan, M.B.; Kromhout, D. Dietary antioxidant flavonoids and risk of coronary heart disease. The Zutphen Elderly Study. *Lancet* 342:1007–1011 (1993).

11. Knekt, P.; Jarvinen, R.; Reunanen, A.J.; Maatela, J. Flavonoid intake and coronary mortality in Finland: a cohort study. *B.M.J.* 312:478–481 (1996).

12. Rimm, E.B.; Katan, M.B.; Ascherio, A.; Stampfer, M.J.; Willett, W.C. Relation between intake of flavonoids and risk for coronary heart disease in male health professionals. *Ann. Inter. Med.* 155:391–396 (1996).

13. Teissedre, P.L.; Frankel, E.N.; Waterhouse, A.L.; Peleg, H.; German, J.B. Inhibition of *in vitro* human LDL oxidation by phenolic antioxidants from grapes and wines. *J. Sci. Food Agric.* 122:157–168 (1996).

14. Ruf, J.C.; Berger, J.L.; Renaud, S. Platelet rebound effect of alcohol withdrawal and wine drinking in rats, atheriosclerosis. *Thrombosis and Vascular Biology* 15:140–144 (1995).

15. Fuhrman, B.; Lavy, A.; Aviram, M. Consumption of red wine with meals reduces the susceptibility to human plasma and low density lipoprotein to lipid peroxidation. *Am. J. Clin. Nutr.* 61:549–554 (1995).

16. Frankel, E.N.; Waterhouse, A.L.; Teissedre, P.L. Principal phenolic phytochemicals in selected California wines and their antioxidant activity in inhibiting oxidation of human low-density lipoprotein. *J. Agric. Food Chem.* 43:890–894 (1995).

17. Teissedre, P.L.; Waterhouse, A.L.; Frankel, E.N. Principal phytochemicals in french syrah and grenache rhône wines and their antioxidant activity in inhibiting oxidation of human low density lipoproteins. *J. International Sciences Vigne et Vin* 29(4):205–212 (1995).

18. Archier, P.; Cohen, S.; Roggero, J.P. Composition phénolique de vins issus de monocépages. *Sci. Alim.* 12:453–466 (1992).

19. Waterhouse, A.L.; Teissedre, P.L. Levels of phenolics in California varietal wines. *In*: Watkins, T.C. (ed.) Wine nutritional and therapeutic benefits. American Chemical Society Symposium Series 661, Washington DC. pp. 12–23 (1997).

20. Bourzeix, M.; Weyland, D.; Heredia, N. A study of catechins and procyanidins of grape custers, the wine and other by-products of the wine. *Bull O.I.V.* 59:171–254 (1986).

21. Lamuela-Raventos, R.M.; Waterhouse, A.L. A direct hplc separation of wine Phenolics. *Am. J. Enol. Vitic.* 45:1–5 (1994).

22. Darret, G.; Couzy, F.; Antoine, J.M.; Magliola, C.; Mareschi, J.P. Estimation of minerals and trace elements provided by beverages for the adult in France. *Ann. Nutr. Metab.* 30:335–344 (1986).

23. Boulet, D.; Laporte, J.P.; Aigrin, P.; Lalanne, J.B. The development of behaviour of wine consumption in France. *ONIVINS infos*, 26:72–112 (1995).

24. Carando, S.; Teissedre, P.L.; Cabanis, J.C. Comparison of (+)-catechin determination in human plasma by high-performance liquid chromatography with two types of detection: fluorescence and ultraviolet. *J. Chromatogr. B.* 707:195–201 (1998).

FUNCTIONAL PROPERTIES OF HOP POLYPHENOLS

Denis De Keukeleire,[a] Luc De Cooman,[a] Haojing Rong,[a]
Arne Heyerick,[a] Jogen Kalita,[b] and Stuart R. Milligan[b]

[a] Faculty of Pharmaceutical Sciences
University of Gent
B-9000 Gent
BELGIUM

[b] School of Biomedical Sciences
King's College London
London, SE1 1UL
UNITED KINGDOM

1. INTRODUCTION

The cultivated hop (*Humulus lupulus* L.) is a dioecious plant of the Cannabaceae family, which is grown in most of the temperate climate zones of the world. The day-length requirements, related to the amount of light hops needed during the growing season for flowering, restrict commercial cultivation of hops to a latitude between 35° and 55° in both hemispheres. The main commercial cultivars of interest to the worldwide market are grown in Europe and the Northwest region of the United States of America in the Northern Hemisphere, and in Australia, New Zealand, and South Africa in the Southern Hemisphere. The most important commercial hops (ca. 40 varieties) are all cultivars of the *Humulus lupulus* L. species (table 1). The only other species, *Humulus japonicus* L., is an ornamental plant. Wild hops are found ubiquitously.

Hops are currently almost exclusively used for brewing. Traditional uses of hops for health and medicinal purposes refer, among others, to antibacterial, sedative, and estrogenic properties, and these have been highlighted throughout history in pharmacopoeia and authoritative reference works. Vegetative material has found application as organic fertilizer, whereas tender hop shoots are harvested for use as a delicate vegetable in gastronomy.

During winter, the aboveground parts of the plant die, but the rootstock is perennial. The useful commercial life is between 12 and 20 years. As the root system of adult plants can extend to more than 1.5 m in depth and 2 m laterally,

Table 1. Botanical description of hops

Division	Spermatophyta
Subdivision	Angiospermae
Class	Dicotyledones
Subclass	Dilleniidae
Order	Urticales
Family	Cannabaceae (hemp family)
Genus	*Humulus*
Species	*lupulus* L.

rich and deep soils are required for efficient growing. In early spring, numerous shoots are produced from the buds on the upper part of the rootstock. Most of these shoots are removed by pruning, whereas the remaining shoots (three to six) are trained up strings when they have reached a length of about 1 m in mid-spring. Twining is done in a clockwise direction around the supports that each year are tied to the fixed structures of poles and wires up to a height of 8 m. Newly bred dwarf and low trellis (up to 2–3 m) hop varieties seem to be more cost-effective, since the comfort of low-wirework structures can be associated to highly automated harvesting using specially designed hop-picking machines.

Growth occurring from April to July in the Northern Hemisphere is vigorous and fast, thereby requiring large amounts of fertilizer and a regular supply of water either by natural rainfall or via irrigation systems. During July and August, the flowers of the female plants develop, forming the hop cones. In some hop-growing areas, particularly England, male plants are cultivated together with the females in order to produce seeds in the hop cones following pollination, and, accordingly, to improve crop yields. In other parts of the world, however, hop growers are obliged by law to remove wild males that may grow near the hop farms for production of seedless hops, i.e., hops that contain less than 2 percent seeds by weight. Although the issue is controversial, pertinent observations indicate that the relatively high amount of seed lipids can give rise to rancid flavors and to less stable foam heads on beer.

Throughout the growing season, a number of pests such as aphids (*Phorodon humuli*) and red spider mites (*Tetranychus urticae*) and diseases including downy mildew (*Pseudoperonospora humuli*), powdery mildew (*Sphaerotheca humuli*), and verticillium wilt (*Verticillium albo-atrum* and *V. dahliae*) may damage the hops. Wilt can lead to severe deterioration, but the other pests and diseases are particularly obnoxious when they attack after the hop cones have been formed. Protection programs are necessary in all hop-growing regions, and a number of effective and approved pesticides are available for use in hop cultivation.

The significance of crop protection became clearly apparent in the summer of 1997 in the hop-growing region in the State of Washington. Following a severe winter and an unusually cool spring, the summer temperatures were moderate to hot with the humidity being uncharacteristically high. As a result, the first ever appearance of powdery mildew was noted in early June. The rapid spread of the mold had a big impact on both the quantity and quality of the 1997 Washington crop. Eventually, partial control was achieved, but almost 10 percent of the total USA acreage needed to be cut down and burned. The origin of the outbreak of the powdery mildew has not been clearly established. It is likely to be an ever-present threat and, although the normally dry weather conditions could mitigate against its uncontrolled spread, much thought is going into ensuring early identification and containment of any similar outbreak. Although the resistance against the pests and diseases depends very much on the particular hop variety, not a single hop cultivar has been developed until now that is completely pest- and disease-free.

Once the hop cones have reached ripening, the crop is harvested using picking machines. Hop cones at harvest have a moisture content of 75–80 percent (w/w) and, in order to prevent deterioration before storage or processing, reduction to less than 12 percent (w/w) is necessary. Drying is achieved by hot air dryers (kilns), which are usually installed on the hop farms. This process is very critical with respect to hop quality, as the high temperature, particularly above 65 °C, may accelerate oxidative decomposition of major constituents. Even for dried hops, low-temperature storage is recommended prior to processing or further use.

2. SECONDARY METABOLITES IN HOPS

Hops contain hundreds of secondary metabolites comprising many different groups of organic compounds.[1] The major components are shown in table 2. Of particular interest are: a) hop resins, which mainly contain hop acids; b) hop oil; c) hop polyphenols.

These three classes are important as biochemical markers to differentiate hop varieties.[2,3] For this purpose, variances in the content of individual

Table 2. Composition of dried hop cones (percent w/w)

Components	Percent w/w
α-Acids	2–19
β-Acids	2–10
Hop oil	0.5–3.0[a]
Polyphenols and tannins	3–6
Monosaccharides	2
Amino acids	0.1
Proteins	15
Lipids and fatty acids	1–5
Pectins	2
Ash—salts	10
Cellulose—lignins	40–50
Water	8–12

[a]Percent v/w.

components and the absence or presence of particular substances can be exploited.

Hop oils and hop polyphenols represent intricate mixtures, but individual constituents occur also in many other plants. Conversely, secondary metabolites, identical to the hop acids, have hitherto not been discovered in any other plant, although similar structures have been reported, such as ceratiolin from *Ceratiola ericoides*[4] and hyperforin from *Hypericum perforatum*.[5] A very intriguing feature is the exceptionally high content of hop acids, up to 25 percent or even more, of the dry weight of the hop cones in cultivated varieties, resulting from excretion of the hop acids in exterior organs, the so-called lupulin glands. Attempts to produce hop acids in artificial media, e.g., cell suspension cultures, have failed until now.

The hop acids consist of two related series, the α-acids (α because they were discovered first) and β-acids, respectively. These weakly acidic compounds, occurring as pale-yellowish solids in the pure state, are poorly soluble in water and have almost no bitter taste. Most important are the α-acids or humulones. To date, five analogs of the α-acids (isomers and homologs) have been characterized, i.e., humulone (**1a**), cohumulone (**1b**), adhumulone (**1c**), prehumulone (**1d**), and posthumulone (**1e**) (fig. 1). The major component of the α-acids mixture, humulone, is usually referred to as n-humulone or "normal" humulone. The relative amount of the adhumulones within the α-acids is fairly constant between varieties (ca. 15 percent), whereas the relative amounts of humulone and cohumulone are variety-dependent (20–50 percent).[6] Cohumulone has been associated with a poor hop quality, although this issue is not proven unambiguously. Pre- and posthumulone are minor constituents. Detailed analysis by HPLC-MS reveals the presence of yet other related compounds in small concentrations.[7]

The β-acids or lupulones also comprise five analogs, corresponding to those of the α-acids, namely, lupulone (**2a**), colupulone (**2b**), adlupulone (**2c**), and the less

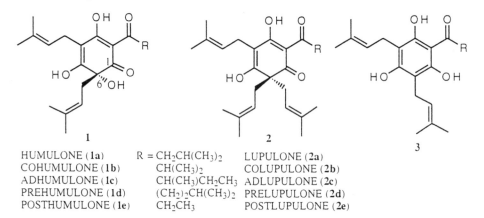

HUMULONE (**1a**)	R = CH$_2$CH(CH$_3$)$_2$	LUPULONE (**2a**)
COHUMULONE (**1b**)	CH(CH$_3$)$_2$	COLUPULONE (**2b**)
ADHUMULONE (**1c**)	CH(CH$_3$)CH$_2$CH$_3$	ADLUPULONE (**2c**)
PREHUMULONE (**1d**)	(CH$_2$)$_2$CH(CH$_3$)$_2$	PRELUPULONE (**2d**)
POSTHUMULONE (**1e**)	CH$_2$CH$_3$	POSTLUPULONE (**2e**)

Figure 1. Chemical structures of the α-acids, the β-acids, and the 6-deoxy-α-acids.

important prelupulone (**2d**) and postlupulone (**2e**) (fig. 1). Both series of hop acids appear to have common biochemical precursors, the 6-deoxy-α-acids (**3**). However, the relative proportions of α-acids and β-acids, as well as the content of co-homologs, depend strongly on the hop variety and, for a given variety, on the growing conditions.

Separation of the individual hop acids is achieved by high-efficient separation techniques, such as HPLC[1] or MEEKC (microemulsion electrokinetic chromatography) with diode array detection (DAD).[6] This technique can be applied to whole hops in a fully automated sequence involving supercritical fluid extraction (SFE at a high density of carbon dioxide)-MEEKC-DAD.[8] By using two-step SFE, the hop acid analysis can be combined with analysis of the hop oil, thereby providing complementary information for identification of hop cultivars and hop quality control.

The hop oil represents a small, volatile fraction of hops (0.5–3.0 V/W), in which over 300 components have been positively or tentatively identified. Our current understanding is that the hop aroma is most likely the result of the interaction of many different constituents. The complex mixture of volatiles, occurring together with the hop acids in the lupulin glands, can be separated by CGC (capillary gas chromatography). Identification of individual components is achieved either by the use of reference compounds or, in a more comprehensive fashion, by CGC-MS.[9]

The hop oil components comprise different classes. Hop oils are usually divided into an apolar (hydrocarbon) fraction (40 to 80 percent) and a polar (oxygenated or sulfur-containing) fraction, e.g., by solid-phase extraction (SPE) on silica, followed by CGC-MS. A fully automated analysis of hop oils by SFE (at a low density of carbon dioxide)-SPE-CGC-MS has been developed.[8]

The hydrocarbon fraction consists mainly of the monoterpene myrcene (**4**) and the sesquiterpenes caryophyllene (**5**), humulene (**6**), and farnesene (**7**) (fig. 2). Many high-valued hops have significant percentages of humulene and low myrcene and caryophyllene contents. Humulene is considered to be partly responsible for the pleasant hop aroma, although this may have no impact on beer flavor as this compound must be readily oxidized during wort boiling. The presence or absence of farnesene has been used as a quality criterion. Myrcene and the sesquiterpene hydrocarbons (at least 40 members have been detected) are notoriously reactive. Oxidative processes followed by rearrangements give rise to a number of derivatives, including tricyclic sesquiterpenes, which may be of interest for the differentiation of hop varieties.[10]

Linalool (**8**) (fig. 2), the major monoterpene alcohol in hops, and geraniol (**9**) are oxygenated terpenes reputed for their floral scents. Esters such as 2-methylpropyl isobutyrate (**10**) and 2-methylbutyl isobutyrate (**11**) convey fruity notes to hops. Fatty acids, e.g., 2-methylbutyric acid (**12**), characterize the cheesy aroma of old hops. Hop oils often contain a range of organosulfur compounds, which may have an adverse effect on beer flavor. These include thiols, sulfides, polysulfides, thioesters, thiophenes, and episulfides such as 4,5-epithiocaryophyllene (**13**) and 1,2-epithiohumulene (**14**). Although sulfur compounds are present in very low quantities in hops, some have flavor thresholds of a few microgram/kg,

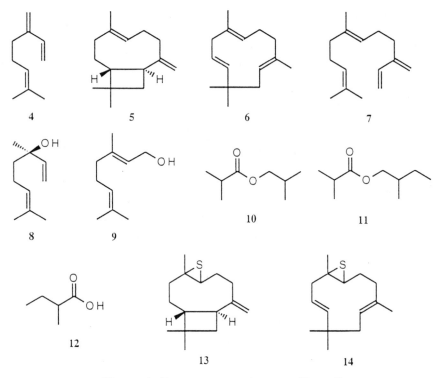

Figure 2. Important constituents of hop oil.

or even lower. At present, the relative influence of individual sulfur compounds on hop aroma is still speculative.

Early chromatographic explorations disclosed that hop polyphenols were complex mixtures consisting predominantly of flavanoids and that the constituents of malt differed appreciably from those of hops.[11–13] About 70–80 percent of the polyphenols in wort are derived from malt and only 20–30 percent from hops. At least 100 components may be separated by HPLC-DAD.[14] Groups of substances that can be determined are given in table 3, together with the respective concentration ranges in hops. Research on hop polyphenols has been fragmentary, and conclusive results are scarce. As it was found that the properties of the polyphenols, in general, depend significantly on the molecular weight, attention has been focused on particular molecular weight ranges.

Higher molecular weight polyphenols include proanthocyanidins, such as procyanidin B3 (**15**) and prodelphinidin B3 (**16**), as well as oligomers composed of (+)-catechin (**17**) and (–)-epicatechin (**18**) units (fig. 3). In the relevant literature, these polyphenols are also known as tannoids, which are defined as polyphenols with a molecular weight between 500 and 3,000.[15] Lower molecular weight polyphenols include flavonols, mainly kaempferol (**19**) and quercetin (**20**), which occur to a large extent as glycosides and are partly transferred from hops to wort.

Table 3. Groups of polyphenols in hops

Group	Concentration/mg·kg⁻¹
Hydroxybenzoic acids	<100
Hydroxycinnamic acids	100–300
Proanthocyanidins	600–1,500
Flavanols	300–1,100
Quercetin glycosides	500–2,000
Kaempferol glycosides	500–1,700
Flavonols	<100–200

The presence of at least 12 different flavonol glycosides in different relative amounts among nine hop varieties was ascertained by HPLC-Ion Spray-Tandem Mass Spectrometry.[16] Exact identification of individual constituents was possible only with reference standards, but information on both the nature of the aglycone and the carbohydrate units could be obtained by MS/MS parent ion scanning. Well-known phytoestrogenic isoflavones such as genistein (**21**) and daidzein (**22**) may also be present, whereas prenylated flavonoids are represented by the chalcone,

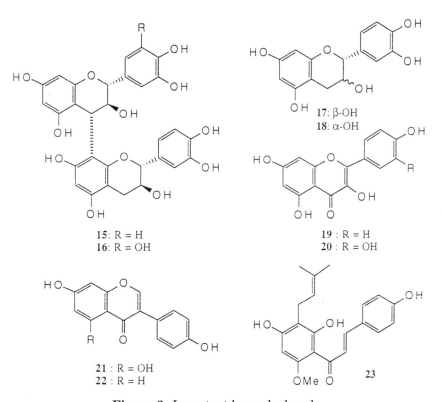

Figure 3. Important hop polyphenols.

xanthohumol (**23**). The composition of the polyphenols depends on the hop variety, cultivation area, harvesting technique, and degree of aging. Drying hops in air at 60 °C or 80 °C results in a drastic decrease of flavanols and proanthocyanidins.

3. FUNCTIONS OF HOP POLYPHENOLIC DERIVATIVES

Although the α-acids (**1**, fig. 1) and β-acids (**2**) themselves are vinylogous acids rather than polyphenols, many researchers include these important hop constituents in the group of hop polyphenols in view of the direct connotation with phloroglucinol derivatives, namely, the 6-deoxy-α-acids (**3**).

The α-acids as such do not have a direct use, although some medicinal properties have been claimed. Conversely, they are precursors of a wide variety of derivatives, which are formed, either by oxidative processes, or, more importantly, by isomerization to iso-α-acids.[1] Particularly, the iso-α-acids are key components in beer brewing and, from all hop-derived compounds known, the iso-α-acids receive by far the most attention. The transformations of the hop α-acids during wort boiling have been studied in great detail, and our laboratory has significantly contributed to the current knowledge in the field.[1,17] In fact, current developments in modern hop technology are based to a great extent on original findings by the Gent hop research school.

The most important chemical conversion occuring during wort boiling is the thermal isomerization of the α-acids (**1**) to the iso-α-acids (**24** + **25**) via an acyloin-type ring contraction (fig. 4). Each α-acid gives rise to two epimeric iso-α-acids, which are distinguished as *trans*-iso-α-acids (**24**) and *cis*-iso-α-acids (**25**), respectively, depending on the spatial arrangement of the tertiary alcohol function at C(4) and the prenyl side chain at C(5). The term *trans* indicates that these groups point to opposite faces of the five-membered ring, while *cis* refers to these groups being located on the same face.

The main α-acid, humulone (**1a**), is isomerized in boiling wort to a mixture of *trans*-isohumulone (**24a**) and *cis*-isohumulone (**25a**) in a ratio *trans*:*cis* of 32 : 68. Consequently, 6 iso-α-acids (*trans*-isohumulone and *cis*-isohumulone, *trans*-

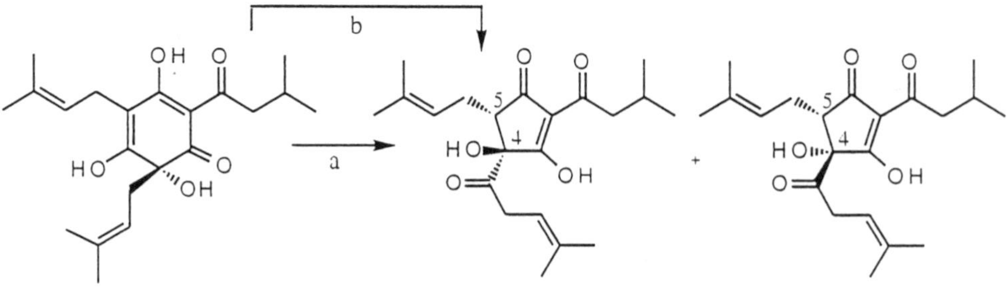

a: boiling wort; b: hv (UV-A light).

Figure 4. Conversion of humulone to the isohumulones.

isocohumulone and *cis*-isocohumulone, *trans*-isoadhumulone and *cis*-isoadhumulone) are present in beer resulting from isomerization of the three major α-acids, humulone, cohumulone, and adhumulone, respectively. The concentrations of the individual iso-α-acids depend on the isomerization conditions and the solubilities in the aqueous matrix. The *trans*-iso-α-acids can be readily isolated from the mixture of *trans*- and *cis*-iso-α-acids via selective precipitation of the dicyclohexylammonium salts.[18]

Iso-α-acids may also be formed by heating solid salts of α-acids, e.g., magnesium(II) humulates, or by irradiation of the α-acids in the wavelength region of 350–366 nm (UV-A light). When excited by light, humulone (**1a**) undergoes a stereoselective oxa-di-π-methane photorearrangement, which affords exclusively *trans*-isohumulone (**24a**) (fig. 4).[19] Likewise, irradiation of the mixture of α-acids leads to the 3 *trans*-iso-α-acids (*trans*-isohumulone, *trans*-isocohumulone, and *trans*-isoadhumulone).

In traditional brewing practice, wort is boiled with whole hops for 1.5–2 h, but the inconsistency both of the hop quality and the isomerization yield has led to drastic changes during the last two decades. In modern hop technology, whole hops are extracted using liquid or supercritical carbon dioxide. Due to the selectivity of this medium, the extracts consist mainly of hop acids and hop oil. In appropriate conditions of temperature and pressure, extracts can be obtained, which are almost free of hop oil. A further processing step involves separation of α-acids and β-acids based on distinct differences in acidity. Next, the α-acids are thermally converted to the iso-α-acids in almost quantitative yield using alkaline conditions.[1,17]

Such pre-isomerized hop extracts can be used to modify bitterness levels in beer at any stage in the course of the brewing process, e.g., during wort boiling, prior to fermentation, post-fermentation, or even just before end-filtration. It has proven possible to brew excellent beers, which are bittered exclusively with pre-isomerized hop extracts. By exploiting the inherent advantages of such modern hop products, brewers are able to control judiciously the consistency of their beers in an economically feasible way.

The iso-α-acids exhibit an intensely bitter taste (threshold value estimated at 6 mg/L). Although iso-α-acids are rather labile, they survive the boiling process with surprisingly minor decomposition. The iso-α-acids constitute the quantitatively most important hop-derived fraction present in beer (from about 10 up to 100 mg/L, or even higher), and they impart to a great extent the bitter taste to beer. This organoleptic feature is particularly noticeable in lager beers and, in general, in beers that are brewed using pale malts.

Additionally, the iso-α-acids exhibit interesting features that relate to typical beer characteristics. The tensioactive properties are essential to stabilize beer foam. According to current knowledge, the iso-α-acids exhibit their stabilizing effect, on the one hand by complexation with foam-active barley polypeptides via hydrophobic associations, on the other hand with metal ions via formation of stable chelates.[20] This amphiphilic behavior creates favorable conditions for crosslinking surface-adsorbed proteins, thus fortifying the film around foam bubbles. As a consequence, a well-hopped beer is characterized by a strong foam head. Furthermore, iso-α-acids inhibit the growth of gram-positive bacteria, thereby protecting beer against spoilage by these microorganisms. In fact, the use of hops in the Middle Ages was mainly intended to prolong the shelf-life of beer rather than to add flavor. Iso-α-acids act, in combination with other hop constituents, as natural preservatives in beer.

Degradation of the iso-α-acids occurs, even at ambient temperature, and this gradual decomposition may be partly responsible for the changes in the beer flavor on storage. Aging off-flavors include stale and cardboard flavors, which are due to oxidized volatiles, such as 3-hydroxybutanal. Another unfavorable feature pertains to the light-sensitivity of the iso-α-acids, leading to the so-called 'lightstruck flavor'. The light-sensitive chromophore in iso-α-acids is the acyloin group, which is composed of the tertiary alcohol function at C(4) and the carbonyl group of the side chain at C(4). Activation of, for example, the isohumulones **24a** and **25a** with UV light causes bond cleavage by a Norrish Type I photoreaction, giving rise to a ketyl-acyl radical pair. Subsequent loss of carbon monoxide from the acyl radical and recombination of the resulting allyl radical with a thiol radical gives rise to formation of 3-methylbut-2-ene-1-thiol (**26**, MBT, fig. 5), next to dehydrohumulinic acid (**27a**).[21,22] Due to its offending smell and taste, MBT has become known as 'skunky thiol'. The flavor threshold of the thiol is so low that even concentrations of less than 10 ng/L are deleterious to the beer quality. The thiol is formed also by exposure of beer to visible light or sunlight. Since the iso-α-acids do not absorb in the visible region, the reaction is sensitized by light-absorbing substances present in the matrix, including riboflavin (vitamin B_2). It is obvious that beer must be protected against the influence of light by storage in the dark or in brown or green glass bottles. Alternatively, reduced iso-α-acids may be used to inhibit or even prevent formation of the lightstruck flavor.

The first applications of reduced iso-α-acids were aimed at exploiting the enhanced resistance of these compounds to light for the production of light-stable

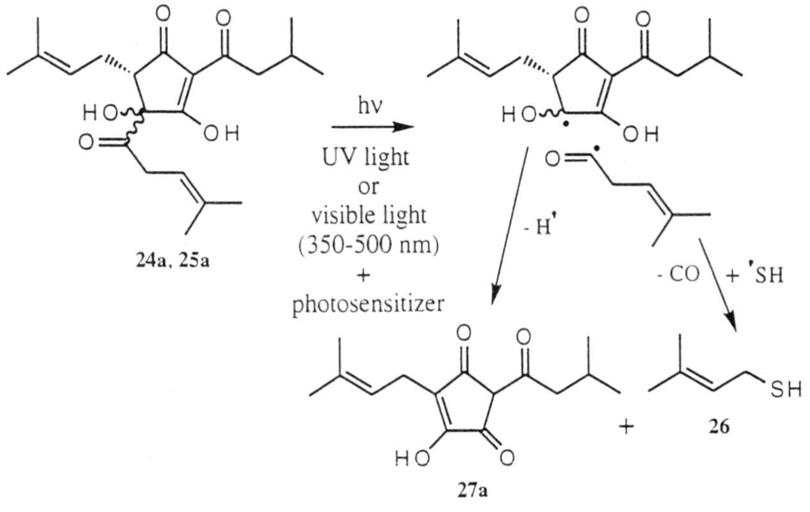

Figure 5. Formation of the lightstruck flavor in beer.

beers. It was observed, however, that these reduced substances showed additional desirable features, including increased bitterness intensity and enhanced ability to stabilize beer foam. As a consequence, hop products based on reduced iso-α-acids are gaining widespread acceptance.[23] Three major types of reduced iso-α-acids should be considered, depending on the number of hydrogen atoms (dihydro, tetrahydro, hexahydro) incorporated during reduction. The bitter impact of the dihydroiso-α-acids is only 60–80 percent compared to the iso-α-acids, the hexahydroiso-α-acids are similar to iso-α-acids, and the tetrahydroiso-α-acids have a high impact on bitter flavor, approximately 1.7–1.9 times that of iso-α-acids. Reduced hop products have a foam-enhancing effect. This is particularly pronounced for the tetrahydroiso-α-acids, which highlights the significance of hydrophobic interactions in stabilizing beer foam.

Dihydroiso-α-acids, also known as rho-iso-α-acids, are formed by reduction of the carbonyl group in the side chain at C(4) using sodium borohydride. Since formation of the secondary alcohol function is accompanied by formation of a new chiral center, two epimeric dihydroiso-α-acids arise from each iso-α-acid, e.g., compounds **28a** and **29a** (fig. 6) from *trans*-isohumulone (**24a**). Consequently, a mixture of 12 dihydroiso-α-acids may result from reduction of the six major iso-α-acids. However, hop products based on dihydroiso-α-acids are usually less complex due to particular reaction conditions and selectivities. The dihydroiso-α-acids are

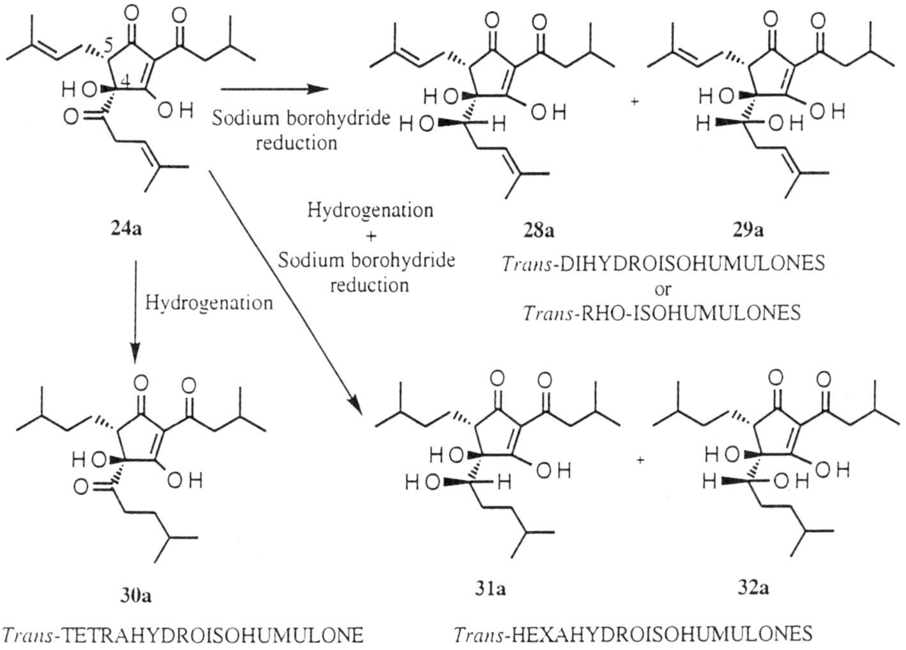

Figure 6. Chemical structures of reduced isohumulones.

light-stable as the light-sensitive acyloin group has been converted to a light-stable diol.

Tetrahydroiso-α-acids are obtained by hydrogenation of the double bonds in the side chains of the iso-α-acids, e.g., *trans*-tetrahydroisohumulone (**30a**) from *trans*-isohumulone (**24a**). Commercially available tetrahydroiso-α-acids are, in general, produced directly from α-acids as a mixture of six tetrahydroiso-α-acids as a result of consecutive hydrogenation and isomerization. Since the acyloin group is still present, photochemical reactions may occur on light exposure, although the specific allyl radical, intervening in the formation of 3-methyl-but-2-ene-1-thiol (**26**), can no longer be formed.

Hexahydroiso-α-acids are accessible by a combination of the aforementioned processes, i.e., simultaneous or successive reduction of the side-chain carbonyl group and hydrogenation of the side-chain double bonds in the iso-α-acids. *Trans*-isohumulone (**24a**) would afford *trans*-hexahydroisohumulones **31a** and **32a**, whereas the mixture of hexahydroiso-α-acids could in principle be composed of 12 isomers and homologs. Obviously, the absence of the light-sensitive acyloin group accounts for the resistance to light.

The hop β-acids are very labile, but their transformations during brewing are quite different from those of the α-acids. Since the tertiary alcohol function at C(6) is substituted by another prenyl group, the β-acids do not give the facile isomerization reaction of the α-acids affording the iso-α-acids. Instead, β-acids are very susceptible to a variety of oxidation reactions that mainly involve the double bonds of the prenyl side chains. Most oxidation products have unpleasant organoleptic properties, although some oxidized derivatives of the β-acids have a pronounced bitter taste. Such compounds may partly compensate for the losses of α-acids in aged hops, thereby maintaining to some extent the bittering potential.

An interesting recent application pertains to the addition of a so-called hop base-extract to reduce bacterial activity in sugar beet extraction.[24,25] This product originates from hops that have been extracted with supercritical carbon dioxide. After removal of the α-acids for further processing (e.g., isomerization to iso-α-acids), the residue contains mainly the β-acids, thus providing a refined hop extract, which is appropriate for a variety of uses based on the antibacterial properties of the β-acids.[26] Strong inhibition of the growth of gram-positive bacteria has been attributed to interference of the three prenyl groups with the function of the cell plasma membrane.[27,28]

4. FUNCTIONS OF HOP POLYPHENOLS

It can be expected that hop polyphenols influence beer quality. Although some indications point to a significant contribution of polyphenols to overall beer mouth-flavor, no conclusive proof has been provided yet. On the contrary, there is evidence that polyphenols impart harsh astringent flavors to beer. Also, the suggestion that low molecular weight polyphenolic substances, mainly flavonols, contribute reducing power to beer needs to be corroborated. Preliminary studies led to confusing results, and there is a definite need for investigations directed at defining factors governing the antioxidant potential with the aim to protect beer against oxidation and to improve the flavor stability of beer.

From a brewer's perspective, only a negative influence of hop (and malt) polyphenols on the colloidal stability of beer is evident.[29] Indeed, the first manifestation of loss of colloidal stability is observed as chill haze, which is a reversible association of higher molecular weight polyphenols and proteinaceous material. The chill haze redissolves when warmed up. The tendency to form chill haze increases progressively over time. Initially, simple monomeric and dimeric proanthocyanidins form hydrogen bonds with soluble proteins without causing turbidity. Oxidation and polymerization of simple flavanols during storage lead to formation of flavanol-type oligomers that are able to aggregate with soluble proteins, some of which can form chill haze on cooling. As the polyphenols continue to oxidize, larger complexes are formed, and supramolecular bonds are partly substituted for covalent bonds. Heating can no longer break these strong bonds, and the result is permanent (irreversible) haze. The validity of this model has been confirmed by a number of studies.[30–33] Furthermore, it may be noted that oxidized polyphenols contribute somehow to the beer color.

PVPP (polyvinyl-polypyrrolidone) has since long been identified as a highly effective agent for the longterm prevention of chill and permanent haze in beer.[34–35] PVPP strongly adsorbs flavanoid polyphenols and tannins to form insoluble complexes, thus removing active constituents from the tannin-protein complex. Treatment of beer allows brewers to routinely guarantee the haze-free shelf-life of beer for 12 months or longer. The treatment removes both tannoids and simple flavanoids, extends the lag phase, and decreases the rate of haze development. Alternative stabilization techniques tackle the problem from the protein side. Thus, haze-sensitive proteins can be adsorbed by silica gel and bentonite or precipitated by treatment with hydrolyzable tannic acids.[36] Combined treatment with PVPP and tannic acids should afford beers with very high colloidal stability.

Although the main use of hops is for brewing beer, hops have since ancient times been reputed for their bioactivity. This was expressed in the treatment of a variety of diseases and in the widespread use as a health-promoting plant. In particular, the sedative effect is prominent and, today, more than 100 hop-containing preparations are on the market. It can be taken for granted that hops exhibit a mild sedative action, and a great number of studies have been carried out to identify the sedative component(s), but, until now, no one has been able to provide conclusive evidence. Most attention has been given to volatile constituents of the hop oil, and 2-methylbut-3-en-2-ol has been suggested as the active compound.[37] The bacteriostatic action of hops, on the other hand, can be safely attributed to hop acids, particularly β-acids, and derivatives thereof. Much less attention was given to the polyphenols in hops, although one can expect that some of the known biological effects may be associated with particular polyphenolic classes. Therefore, we were very interested in examining hop polyphenols in more detail and embarked on a research program to particularly address the controversial issue pertaining to the estrogenic properties of hops.

The literature contains a number of reports on the possible estrogenic activity of hops ranging from suggestions that they may be one of the richest sources of plant estrogens to the failure to find any evidence of estrogenic activity. Circumstantial evidence over many years has linked hops with potential estrogenic activity in women. Menstrual disturbances were reportedly common among female hop pickers, an effect attributed to the estrogenic activity of hops in general and of xanthohumol (**23**, fig. 3) in particular.[37] Using an *in vivo* assay based on the ability of estrogens to stimulate the reproductive tract of immature mice, Koch and Heim[38] initially reported that crude extracts of hop cones contained estrogenic activity. Zenisek and Bednar[39] confirmed these findings and suggested that the estrogenic activity of hops was manyfold higher than that found in other plants; the extract rich in the β-acids reportedly contained the highest activity. Hesse et al.[40] used an *in vitro* receptor binding assay based on calf uterine estrogen receptors to investigate the potential estrogenic activity of a variety of plant extracts and concluded that hops must be considered as plants richest in estrogens. In contrast to these positive reports of estrogenic activity in hops, we were unable, using an *in vitro* assay, to confirm the suggestion that xanthohumol may be estrogenic. We could, furthermore, not detect any estrogenic activity in a supposedly enriched fraction of β-acids. Similarly, Fenselau and Talalay[41] found no evidence of any estrogenic activity in a wide range of hop fractions, including hop oils, α-acids, and β-acids, and extracts of various European and American hop varieties.

The current interest in human exposure to environmental and dietary estrogens led us to re-examine the proposal that hops may contain large amounts of compouns with estrogenic activity. It is our belief that previous discrepancies over the potential estrogenic activity of hops were likely to be due partly to insufficient identification of the hop materials, and partly to the nature and variety of assays used. *In vivo* assays (e.g., as used by Koch and Heim,[38] Zenisek and Bednar,[39] and Fenselau and Talalay[41]), were based on the stimulation of vaginal cornification or uterine growth in immature or ovariectomized mice. The advantage of such assays is that they measure overall biological activity, but they suffer from the disadvantage that their ability to detect the weak estrogenic activity characteristic of phytoestrogens is very dependent on the route and frequency of administration of the material. Negative results in these assays could easily be explained by an inappropriate treatment regimen. In addition, these *in vivo* assays are cumbersome and expensive. However, the *in vitro* receptor binding assay (as used by Hesse et al.) also presents problems. While the receptor assay detects binding to the estrogen receptor, binding activity does not necessarily bear any relation to biological activity.[42]

Our approach was based on *in vitro* assays, such as the one developed by Littlefield et al.[43] This assay exploits the ability of compounds with estrogenic activity to stimulate alkaline phosphatase in a human endometrial adenocarcinoma cell line (Ishikawa) established by Nishida et al.[44] A variant of this cell line (Ishikawa-Var I) is unresponsive to estrogens with respect to proliferation, but is sensitive to the stimulatory effect on alkaline phosphatase.[43] The estrogenic activity is assayed by monitoring the formation of *p*-nitrophenol at 405 nm following enzymatic hydrolysis of *p*-nitrophenylphosphate. The stimulation of alkaline phosphatase activity has been shown to be specific to estrogens; no other type of steroid, including androgens, progestins, mineralocorticoids, or glucocorticoids, produces the effect. The cells also respond to non-steroidal estrogens, including phytoestrogens.[45] The responses can be blocked by pure anti-estrogens, indicating that they are mediated by the estrogen receptor.[43]

Estrogenic activity detected using the Ishikawa cells was subsequently confirmed by a different *in vitro* assay using an estrogen-inducible yeast (*Saccharomyces cerevisiae*) screen, expressing the estrogen receptor.[46] Yeast cells do not normally contain an estrogen receptor. Therefore, the DNA sequence of the human estrogen receptor was stably integrated into the main chromosome of the yeast. The yeast cells also contain expression plasmids carrying the reporter gene *lac-Z* encoding the enzyme β-galactosidase, which is used to measure the receptors' activity. Thus, the chromogenic substrate, chlorophenol red β-D-galactopyranoside, is converted to a product that can be monitored at 540 nm.

Preliminary studies suggested that, while methanol/water extracts of hops produced toxic effects on the Ishikawa cells at high concentrations, they were able to stimulate alkaline phosphatase activity at lower dilutions. In our study involving a number of different hop varieties, all methanol/water extracts showed strong estrogenic activity (fig. 7) with dose-response curves that paralleled that of the 17β-estradiol standard. The stimulation of alkaline phosphatase activity could be blocked by both tamoxifen and a pure anti-estrogen. Extrapolating from the 17β-estradiol standard curve and allowing for the initial extraction procedure for the hops enabled a quantitative estimate of the relative estrogenic potency of the hop extracts to be made. A considerable spread of the estrogenic activity was noted across the different hop varieties, i.e., from 187 to more than 2,700 ng 17β-estradiol equivalents estrogenic activity per g dry weight hop cones. The highest activities were found in Hallertau Magnum (a main hop variety in Europe) and Galena (a major hop variety in the USA), the weakest in Cluster and Wye Target.

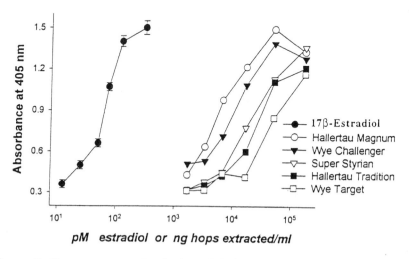

Figure 7. Dose response stimulation of akaline phosphatase activity by 17β-estradiol and extracts of selected hop varieties.

No estrogenic activity was detected in the hop acids fractions, or in iso-α-acids samples. Moreover, none of these latter samples could block the stimulation of alkaline phosphatase activity by 17β-estradiol.

The observation that very dilute hop extracts stimulated alkaline phosphatase activity in the estrogen-sensitive Ishikawa Var-I cell line, coupled with the ability of anti-estrogens to block this response, confirms that hops are endowed with pronounced estrogenic activity. Direct comparison of the estrogenic activity in hops with that reported for other plants is somewhat difficult. This issue is enhanced by the fact that phytoestrogens occur in plants largely as a mixture of glycosides and aglycones, and the relative activity of these compounds in the Ishikawa cell bioassay is unknown. However, some estimates can be made: Franke et al. (1995) have measured the amounts of various phytoestrogens, including isoflavones, in a variety of food items, using HPLC analysis of extracted, hydrolyzed material.[47] Since the relative activity of common phytoestrogens compared to 17β-estradiol in the Ishikawa assay is known, the estrogenic activity of these plants can be expressed in common terms. The data suggest that hops are one of the richest natural sources of phytoestrogens, at least equivalent to soybeans and second only to clover.

We then carried out experiments to study the chemical nature of the estrogenic activity in the polar hop extracts, which very likely contained polyphenolic constituents. This involved estrogenic activity-guided fractionation of the hop extracts and progressive refining of the fractions containing the activity. This approach has allowed us not only to refute the estrogenic activity of alleged 'hop hormones' such as xanthohumol (**23**, fig. 3),[48] but also to prove that the high activity of hops cannot be attributed to well-known phytoestrogenic isoflavones such as genistein (**21**) or daidzein (**22**).

The bioactivity of xanthohumol (**23**) deserves some attention. While we confirmed that the compound was devoid of any estrogenic activity, Tobe et al. patented a pharmaceutical composition for treating osteoporosis containing xanthohumol,[49] and, more recently, they isolated xanthohumol, as well as humulone (**1a**, fig. 1), as active ingredients that inhibit bone resorption in the pit formation assay.[50] If their finding could be corroborated, then our observations would indicate that the inhibition of bone resorption is not associated to any estrogenicity. Furthermore, Tabata et al. showed that xanthohumol inhibits rat liver diacylglycerol acyltransferase and also the formation of triacylglycerol in intact Raji cells.[51] Clearly, this chalcone is a very interesting bioactive molecule worthy of further investigation. Rosenblum et al.,[52] Sauerwein and Meyer,[53] and Lapcik et al.[54] detected isoflavones in beer in levels ranging from 1.26 to 29 nmol/L. Taking into account the concentration of genistein or daidzein in beer and their known estrogenic activity, we were unable to fit these data to the estrogenic activity measured in beers.[55] Our experiments indicate that the estrogenic fraction of hops, which is not of isoflavone nature, is to some extent transferred to beer and, indeed, we are able to profile the estrogenic fraction, as defined from our studies on polyphenolic hop extracts, in various beers. Significant differences are observed, not only with respect to the different beer types, but even for a selection of lager beers. The use of particular hop varieties, e.g., varieties rich in phytoestrogens, or hop products, e.g., pre-isomerized hop extracts or even reduced iso-α-acids, should determine whether or not hop-derived estrogens are present in beers.

We wondered whether the different estrogenic activities of the hop varieties studied reflect differences in the amount of a single phytoestrogen or in the profiles of various estrogenic components. Our studies have led to the identification of the hop estrogens via spectroscopic analysis of individual constituents of the estrogenic fraction following preparative HPLC-separation. Thus, the estrogenically most active hop compound is 8-prenylnaringenin (**33**, fig. 8), while isoxanthohumol (**34**) and, particularly, 6-prenylnaringenin (**35**) are much less active. In fact, 8-prenylnaringenin is more potent than any of the common phytoestrogens of the isoflavone type as found, e.g., in soy.[56]

The estrogenic activity of the novel phytoestrogen could be blocked using the anti-estrogen ICI 182,780, thereby indicating that the estrogenicity involves interaction with the estrogen receptor. The *in vitro* activity of 8-prenylnaringenin (**33**) was considerably greater than that of established phytoestrogens (fig. 9). The EC_{50}-values for 17β-estradiol (**36**, fig. 8), 8-prenylnaringenin (**33**), coumestrol (**37**), genistein (**21**, fig. 3), and daidzein (**22**) were 0.82, 4.2, 30, 194, and 1,452 nM, respectively, in the Ishikawa assay, and 0.29, 44, 72, 1,170, and 2,215 nM in the yeast screen. Interestingly, xanthohumol showed no estrogenic activity in either of the *in vitro* bioassays. The relatively high estrogenic potency of 8-prenylnaringenin was reflected in its ability to interact with the estrogen receptor in a competitive binding assay using rat uterine cytosol. The relative binding

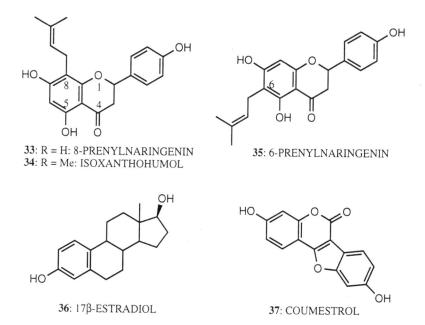

33: R = H: 8-PRENYLNARINGENIN
34: R = Me: ISOXANTHOHUMOL

35: 6-PRENYLNARINGENIN

36: 17β-ESTRADIOL

37: COUMESTROL

Figure 8. Chemical structures of hop phytoestrogens.

affinities of 17β-estradiol (**36**), 8-prenylnaringenin (**33**), coumestrol (**37**), and genistein (**21**) were 1, 0.023, 0.008, and 0.003, respectively.

The *in vivo* activity of 8-prenylnaringenin (**33**, fig. 8) was determined using an acute assay based on the ability of estrogens to induce a rapid rise in uterine vascular permeability in ovariectomized mice.[57] The estrogenic response can be rapidly quantified by monitoring the extravasation of intravenously administered [^{125}I]-labelled human serum albumin.[58,59] The dose-response relationship of 8-prenylnaringenin (**33**) was similar to that of coumestrol (**37**), with a large stimulatory effect being produced by 100 nmoles. The doses of genistein (**21**) required to produce a similar effect were at least tenfold higher. Recent findings indicate that 8-prenylnaringenin (**33**) is also orally active, since administration to the drinking water (100 μg/mL) stimulated epithelial mitoses in the uterus and vagina of ovariectomized mice.

The high estrogenic activity of hops may provide an explanation for reported menstrual disturbances in female hop workers in the past.[37] However, hop-picking is now largely done by machine, and the only natural exposure that humans are likely to have is via beer consumption. It was suggested that, indeed, the phytoestrogens in a variety of alcoholic beverages (bourbon, wine, and beer) may be of significance to human health.[60] Although phytoestrogens generally show a relatively weak activity, the presence of estrogen-active constituents in beer should be kept in perspective.

Some health-beneficial effects of moderate beer consumption have been claimed, and compounds such as xanthohumol (**23**, fig. 3) have been reported to

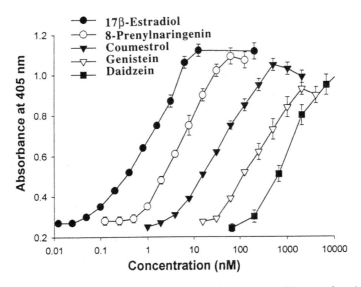

Figure 9. Comparison of the *in vitro* estrogenic activity of 8-prenylnaringenin with 17β-estradiol (E₂), coumestrol, genistein and daidzein. Estrogenic activity was determined by the stimulation of alkaline phosphatase activity in the human Ishikawa-Var I cell line, ref. 45; abscissa: molar log concentration (nM); ordinate: absorbance at 405 nm.

have anticancer properties. Reputed health-beneficial effects of other phytoestrogens include reduction in the incidence of breast cancer, cardiovascular diseases, prostate cancer, and menopausal symptoms.[61,62] Hops may provide an additional source of a very potent natural phytoestrogen, i.e., 8-prenylnaringenin (**33**), with such beneficial effects, and an important new economic use for hops—either as a food supplement, or as a phytopharmaceutical—is feasible. Extrapolating from our knowledge of other plant estrogens (e.g., genistein), 8-prenylnaringenin may also provide an interesting new probe for scientific and pharmaceutical investigations of basic cellular mechanisms (e.g., cell signalling pathways, significance of estrogen-responsive receptors, control of cell proliferation).

Consideration also needs to be given to the possibility that inappropriate or excessive exposure to hops may be deleterious to persons working in close association with hops. Such occupational exposure could cause hormonal disturbances both in men and women. It should be emphasized that, if any phytoestrogen in hops does affect human handlers, it is likely to be having its effect via inhalation or transdermally, rather than by the normal route for established phytoestrogens, which is by ingestion. The fragile, dusty nature of hops and the volatility of hop oils would facilitate transfer by the former routes. While exposure of humans to dietary phytoestrogens may normally be insufficient to induce any major reproductive effects, hop workers may suffer from a much higher exposure through these non-dietary routes.[63]

5. CONCLUSIONS

Hops are primarily used for flavoring beer. Hop polyphenolic derivatives are essential to beer bitterness, foam stability, and preservation of beer. On the other hand, they may give rise to off-flavors, such as the lightstruck flavor. The current state of hop technology allows optimized application of various hop products and stringent control of beer quality features associated with the use of hops.

On the other hand, the market for hops is under intense pressure, due to:

- a large excess of α-acids resulting from a rapid increase in the growing and application of so-called super-α-hops, which contain very high concentrations of α-acids (up to 19 percent),

- a much improved utilization of hops in the form of modern hop products,

- a worldwide tendency to reduce bitterness levels in beers, and

- a stagnating beer consumption on a worldwide basis.

Thus, alternative uses of hops would have considerable economic significance. Since hops have proven to be one of the richest natural sources of 8-prenylnaringenin, the most active plant estrogen known to date, they might be exploited in this respect. Identification of the full bioactivity spectrum, associated to hop polyphenols, may further extend the scope of hops as a most interesting agricultural crop.

ACKNOWLEDGMENTS

The authors are grateful to The British Council and the "Fund for Scientific Research-Flanders" (Belgium) for providing funds to establish the collaboration between DDK and SRM. DDK acknowledges support by the University of Gent (Special Research Fund, project number 01109394).

REFERENCES

1. Verzele, M.; De Keukeleire, D. Chemistry and analysis of hop and beer bitter acids. Elsevier, Amsterdam, The Netherlands (1991).

2. Green, C.P. Use of a chromatographic data system to identify varieties in binary mixtures of hops. *J. Inst. Brew.* 103:293 (1997).

3. De Cooman, L.; Everaert, E.; De Keukeleire, D. Quantitative analysis of hop acids, essential oils and flavonoids as a clue to the identification of hop varieties. *Phytochem. Anal.* 9:145 (1998).

4. Tanrisever, N.; Fronczek, F.R.; Fischer, N.H.; Williamson, G.B. Ceratiolin and other flavonoids from *Ceratiola ericoides. Phytochemistry* 26:175 (1987).

5. Bombardelli, E.; Morazonni, P. Hypericum perforatum. *Fitoterapia* 66:43 (1995).

6. Szücs, R.; Vindevogel, J.; Everaert, E.; De Cooman, L.; Sandra, P.; De Keukeleire, D. Separation and quantification of all main hop acids in different hop cultivars by microemulsion electrokinetic chromatography. *J. Inst. Brew.* 100:293 (1994).

7. Verzele, M.; Van De Velde, N. High-performance liquid chromatography with photodiode array detection of minor hop bitter acids in hops extracts and in beer. *J. Chromatogr.* 387:473 (1987).

8. Sandra, P.; Haghebaert, K.; Szücs, R.; David, F. Fully automated analysis of hops by characterization of the essential oil by SFE-SPE-CGC-MS and of the bitter principles by SFE-MEEKC-DAD. Sandra, P.; Devos, G. (eds.), Proc. Eighteenth international symposium on capillary chromatography, Riva del Garda, Italy, 1996, vol. III, pp. 1916–1924 (1996).

9. Sharpe, F.R.; Laws, D.R.J. The essential oil of hops: a review. *J. Inst. Brew.* 87:98 (1981).

10. Tressl, R.; Engel, K.-H.; Kossa, M.; Köppler, H. Characterization of tricyclic sesquiterpenes in hop (*Humulus lupulus*, var. Hersbrucker Spät). *J. Agric. Food Chem.* 31:892 (1983).

11. McMurrough, I.; McDowell, J. Chromatographic separation and automated analysis of flavanols. *Anal. Biochem.* 91:92 (1978).

12. McMurrough, I. High-performance liquid chromatography of flavonoids in barley and hops. *J. Chromatogr.* 218:683 (1981).

13. Jerumanis, J. Quantitative analysis of flavanoids in barley, hops and beer by high-performance liquid chromatography. *J. Inst. Brew.* 91:250 (1985).

14. Forster, A.; Beck, B.; Schmidt, R. Untersuchungen zu Hopfenpolyphenolen. Proc. 25th Congr. of the European Brewery Convention, Brussels, Belgium, 1995. Oxford University Press, Oxford, UK, pp. 143–150 (1995).

15. Chapon, L. Der begriff tannoide. *Monatsschr. f. Brauwiss.* (7/8):263 (1993).

16. Sägesser, M.; Deinzer, M. HPLC-ion apray-tandem mass spectrometry of flavonol glycosides in hops. *J. Am. Soc. Brew. Chem.* 54(3):129 (1996).

17. De Keukeleire, D. The effects of hops on flavour stability and beer properties. *Cerevisia Biotechnol.* 18(4):33 (1993).

18. Thornton, H.A.; Kulandai, J.; Bond, M.; Jontef, M.P.; Hawthorne, D.B.; Kavanagh, T.E. Preparation of *trans*-iso-alpha acids and use of their dicyclohexylamine salts as a standard for iso-alpha acids analysis. *J. Inst. Brew.* 99:473 (1993).

19. De Keukeleire, D.; Blondeel, G.M.A. The mechanism of the regio- and stereospecific photorearrangement of humulone to the beer bitter component *trans* isohumulone. *Tetrahedron Lett.* :1343 (1979).

20. Simpson, W.J.; Hughes, P.S. Stabilization of foams by hop-derived bitter acids. Chemical interactions in beer foam. *Cerevisia Biotechnol.* 19(3):39 (1994).

21. Kuroiwa, Y.; Hashimoto, N. Composition of sunstruck flavor substance and mechanism of its evolution. *Proc. Am. Soc. Brew. Chem.* :28 (1961).

22. Blondeel, G.M.A.; De Keukeleire, D.; Verzele, M. The photolysis of *trans*-isohumulone to dehydrohumulinic acid, a key route to the development of sunstruck flavour in beer. *J. Chem. Soc. Perkin Trans. 1* :2715 (1987).

23. Goldstein, H.; Ting, P. Post kettle bittering compounds: analysis, taste, foam and light stability. *In*: European Brewery Convention Monograph XXII. Verlag Hans Carl, Nürnberg, Germany, pp. 154–162 (1994).

24. Pollach, G.; Hein, W.; Hollaus, F. Einsatz von hopfenprodukte als bacteriostaticum in der zuckerindustrie. *Zuckerind.* 121:919 (1996).

25. Heim, W.; Pollach, G. Neue erkenntnisse beim einsatz von hopfenprodukte in der zuckerindustrie. *Zuckerind.* 122:940 (1997).

26. Cowles, J.M.; Goldstein, H.; Chicoye, E.; Ting, P.L. Process for separation of β-acids from extract containing α-acids and β-acids. *US Patent* 4 590 296 (1986).

27. Smith, N.A.; Smith, P. Antibacterial activity of hop bitter resins derived from recovered hopped wort. *J. Inst. Brew.* 99:43 (1993).

28. Haas, G.J.; Barsoumian, R. Antimicrobial activity of hop resins. *J. Food Protect.* 57(1):59 (1994).

29. McMurrough, I.; Madigan, D.; Kelly, R.J.; Smythe, M.R. The role of flavanoids in beer stability. *J. Am. Soc. Brew. Chem.* 54:141 (1996).

30. Gardner, R.J.; McGuinness, J.D. Complex polyphenols in brewing—a critical survey. *Techn. Quart. Master Brew. Assoc. Am.* 14:250 (1977).

31. Moll, M.; Fonknechten, G.; Carnielo, M.; Flayeux, R. Changes in polyphenols from raw materials to finished beer. *Techn. Quart. Master Brew. Assoc. Am.* 21:79 (1984).

32. Delcour, J.A.; Schoeters, M.M.; Meysman, E.W.; Dondeyne, P.; Moerman, E. The intrinsic influence of catechins and procyanidins on beer haze formation. *J. Inst. Brew.* 90:381 (1984).

33. Schneider, J.; Raske, W. The protein-polyphenol balance in the brewing process. *Brauwelt Int.* :228 (1997).

34. McMurrough, I.; Madigan, D.; Kelly, R.J. Evaluation of rapid colloidal stabilization with polyvinylpolypyrrolidone (PVPP). *J. Am. Soc. Brew. Chem.* 55(2):38 (1997).

35. McMurrough, I.; O'Rourke, T. New insight into the mechanism of achieving colloidal stability. *Techn. Quart. Master Brew. Assoc. Am.* 34(1):271 (1997).

36. Anger, H.-M. Assuring nonbiological stability of beer as an important factor for guaranteeing minimum shelf-life. *Brauwelt Int.* :142 (1996).

37. Verzele, M. 100 years of hop chemistry and its relevance to brewing. *J. Inst. Brew.* 92:32 (1986).

38. Koch, W.; Heim G. Hormone in hopfen und bier. *Brauwiss.* 8:132 (1953).

39. Zenisek, A.; Bednar, I.J. Contribution to the identification of estrogen activity in hops. *Am. Perfum. Arom.* 75(5):61 (1960).

40. Hesse, R.; Hoffman, B.; Karg, H.; Vogt, K. Untersuchungen über den nachweis von phytoöstrogenen in futterpflanzen und hopfen mit hilfe eines rezeptortests. *Zentralbl. Veterinarmed.* 28:422 (1981).

41. Fenselau, C.; Talalay, P. Is estrogenic activity present in hops? *Food Cosmetic. Toxicol.* 11:597 (1973).

42. Kupfer, D. Critical evaluation of methods for detection and assessment of estrogenic compounds in mammals: strengths and limitations for application to risk assessment. *Reprod. Toxicol.* 1:147 (1988).

43. Littlefield, B.A.; Gurpide, E.; Markiewicz, L.; McKinley, B.; Hochberg, R.B. A simple and sensitive microtiter plate estrogen bioassay based on stimulation of alkaline phosphatase in Ishikawa cells: estrogenic action of Δ^5 adrenal steroids. *Endocrinology* 127:2757 (1990).

44. Nishida, M.; Kasahara, K.; Kaneko, M.; Iwasaki, H. Establishment of a new human endometrial adenocarcinoma cell line, Ishikawa cells, containing estrogen and progesterone receptors. *Acta Obstet. Gynaec. Jap.* 37:1103 (1985).

45. Markiewicz, L.; Garey, J.; Adlercreutz, H.; Gurpide, E. *In vitro* bioassays of non-steroidal phytoestrogens. *J. Steroid Biochem. Molec. Biol.* 45:399 (1993).

46. Routledge, E.J.; Sumpter, J.P. Estrogenic activity of surfactants and some of their degradation products assessed using a recombinant yeast screen. *Environ. Toxicol. Chem.* 15:241 (1996).

47. Franke, A.A.; Custer, L.J.; Cerna, C.M.; Narala K. Rapid HPLC analysis of dietary phytoestrogens from legumes and from human urine. *Proc. Soc. Exp. Biol. Med.* 208:18 (1995).

48. De Keukeleire, D.; Milligan, S.R.; De Cooman, L.; Heyerick, A. The oestrogenic activity of hops (*Humulus lupulus* L.) revisited. *Pharm. Pharmacol. Lett.* 7(2/3):83 (1997).

49. Tobe, H.; Kitamura, K.; Komiyama, O. Pharmaceutical composition for treating osteoporosis containing xanthohumol. *Eur. Pat. Appl. nr. 95105766.0* (18 April 1995).

50. Tobe, H.; Muraki, Y.; Kitamura, K.; Komiyama, O.; Sato, Y.; Sugioka, T.; Maruyama, H.B.; Matsuda, E.; Nagai, M. Bone resorption inhibitors from hop extract. *Biosc. Biotech. Biochem.* 61:158 (1997).

51. Tabata, N.; Ito, M.; Tomoda, H.; Omura, S. Xanthohumols, diacylglycerol acyltransferase inhibitors, from *Humulus lupulus*. *Phytochemistry* 46:683 (1997).

52. Rosenblum, E.R.; Campbell, I.M.; Van Thiel, D.H.; Gavaler, J.S. Isolation and identification of phytoestrogens from beer. *Alcoholism: Clin. Exp. Res.* 1:843 (1992).

53. Sauerwein, H.; Meyer, H.D. Erfassung östrogenwirksamer substanzen in bier und in dessen rohstoffen. *Monatsschr. f. Brauwiss.* (7/8):143 (1997).

54. Lapcik, O.; Hill, M.; Hampl, R.; Wähälä, K.; Adlercreutz, H. Identification of isoflavonoids in beer. *Steroids* 63:14 (1998).

55. De Keukeleire, D.; De Cooman, L.; Heyerick, A.; Milligan, S.R. Hop-derived phytoestrogens in beer? Proc. 26th European Brewery Convention Congress, Maastricht, The Netherlands, 1997. Oxford University Press, Oxford, England, pp. 239–246 (1997).

56. De Keukeleire, D.; De Cooman, L.; Heyerick, A.; Rong, H.; University of Gent, Belgium, and Milligan, S.R.; Kalita, J.; King's College London, England, unpublished results (1998).

57. Milligan, S.R.; Balasubramanian, A.V.; Kalita, J.C. Relative potency of xenobiotic oestrogens in an acute *in vivo* mammalian assay. *Environ. Health Perspect.* 106:23 (1998).

58. Arvidson, N.G. Early oestrogen-induced changes in uterine albumin exchange in mice. *Acta Physiol. Scand.* 100:325 (1977).

59. Milligan, S.R. The sensitivity of the uterus of the mouse and rat to intraluminal instillation. *J. Reprod. Fert.* 79:251 (1987).

60. Gavaler, J.S.; Rosenblum, E.R.; Deal, S.R.; Bowie, B.T. The phytoestrogen congeners of alcoholic beverages: current status. *Proc. Soc. Exp. Biol. Med.* 208:98 (1995).

61. Knight, D.C.; Eden, J.A. A review of the clinical effects of phytoestrogens. *Obstet. Gynecol.* 87:897 (1996).

62. Cassidy, A.; Milligan, S.R. How significant are environmental oestrogens to women? *Climacteric* 1:229 (1998)

63. Verdeal, K.; Ryan, D.S. Naturally-occurring estrogens in plant foodstuffs—a review. *J. Food Protect.* 7:577 (1979).

MODIFICATION OF THE SOLUBILITY OF TANNINS: BIOLOGICAL SIGNIFICANCE AND SYNTHESIS OF LIPID-SOLUBLE POLYPHENOLS

Takashi Tanaka,[a] Zhi-Hong Jiang,[a] Gen-ichiro Nonaka,[b] and Isao Kouno[a]

[a] School of Pharmaceutical Sciences
Nagasaki University
Nagasaki 852-8521
JAPAN

[b] Saga Prefectural Institute for Pharmaceutical Research
Shuku-machi, Tosu, Saga 841-0052
JAPAN

1. INTRODUCTION

Tannins are believed to play a significant role in plant defense systems because they possess a severe astringent taste and decrease nutritional value of the plant by precipitating proteins and inhibiting digestive enzymes.[1-3] Usually, tannins exist as a complex mixture of slightly varying but basically very similar compounds, which are derived from several key compounds such as catechin and pentagalloylglucose, by oxidation, substitution, hydrolysis, and so forth.[3] The structural variation leads to differences in their physico-chemical properties including water solubility,[4] chemical reactivity,[5-7] and the ability of complex formation with proteins.[8-10] Among these properties, the water solubility is an important factor in showing their biological activities.

2. INSOLUBILIZATION OF PROANTHOCYANIDINS IN PERSIMMON FRUITS

Japanese persimmon (*Diospyros kaki*) protects its seeds from animal feeding by accumulating astringent proanthocyanidins in the fruits. However, the fruits become sweet and soft in early winter, and crows and monkeys disperse them. The astringent fruits can also be edible after artificial removal of astringency by warm water, alcohol vapor, or CO_2 gas treatment. During these anaerobic treatments, the water-soluble proanthocyanidins were gradually transformed into insoluble, less astringent forms. Matsuo and Itoo showed that the tannins of persimmon fruit are polymeric proanthocyanidins with a large molecular weight (ca. 1.38×10^4 daltons on average) and suggested that the insolubilization of the tannin is caused by condensation with acetaldehyde, based on the finding that the water-soluble proanthocyanidins form gels on treatment with acetaldehyde *in vitro*.[11] In order to obtain unambiguous chemical evidence for the mechanism of de-astringency *in vivo*, we extended the thiol degradation reaction using 2-mercaptoethanol directly to the insoluble tannins remaining in the fleshy debris.[12]

Thiol degradation reaction of the aqueous acetone extract of astringent persimmon fruit yielded four flavan-3-ol thioethers **1–4** as major products along with minor dimeric products **5–8** (fig. 1 A and fig. 2). However, after complete removal of the astringency in persimmon fruit by treatment with 35 percent ethanol, the aqueous acetone extract no longer yielded any thioethers on thiol degradation, indicating the absence of proanthocyanidins in the extract. On the other hand, direct thiol degradation of the remaining fleshy debris afforded the thioethers **1–8** in reasonable yields that are comparable to those obtained from the extract of the astringent fruit (fig. 1B). This fact indicated that the proanthocyanidins were insolubilized in the fruit flesh without appreciable chemical changes. The difference of the HPLC profiles was the appearance of some small peaks designated as **9a, b–14a, b** in the chromatogram of the products from the insolubilized tannin. The structures of these products were established as 4β-(2-hydroxyethylthio)flavan-3-ols possessing a 1-(2-hydroxyethylthio)ethyl group at C-8 or C-6 position of A-ring [each product was a mixture of two epimers (a and b) with respect to the C_2 unit, which could not be separated]. The C_2 unit attached to the flavan A-ring of these products originates from the acetaldehyde formed *in vivo* by enzymic oxidation of ethanol. This was confirmed by the following experiments: (1) The extract of astringent fruit was treated with acetaldehyde to form a gel. Subsequent thiol degradation yielded products **9a, b** and **11a, b**. (2) Removal of astringency by treatment with 30 percent hexadeuterioethanol and subsequent thiol degradation of the flesh afforded a product (**9c, d**) corresponding to **9a, b**, 50 percent of the molecules of which possess a C_2 unit having deuterium analogues. These results evidently showed that the C_2 unit is derived from endogenous and exogenous ethanol via acetaldehyde and that the proanthocyanidin molecules are linked through the C_2 unit in the insolubilized tannin (fig. 3). In addition to oxidation of ethanol, acetaldehyde is also supplied by decarboxylation of pyruvate by pyruvate decarboxylase. When astringency is removed by treatment with CO_2 gas, the production of acetaldehyde is explained not only by anaerobic respiration but also through fixation of atmospheric CO_2 into malate by phosphoenolpyruvate carboxylase and malate dehydrogenase (dark CO_2 fixation process), subsequently leading to acetaldehyde production.[13] Since acetaldehyde is a ubiquitous compound in nature, the reaction between flavan-3-ols and aldehyde seems to have significance in plant physiology and food science. Actually, some recent reports presumed the presence of catechin-acetaldehyde or anthocyanin-acetaldehyde condensation products in red wine.[14,15]

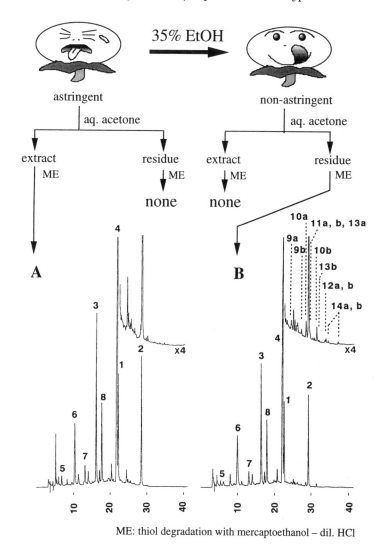

ME: thiol degradation with mercaptoethanol – dil. HCl

Figure 1. Reversed-phase HPLC profiles of thiol degradation products of persimmon proanthocyanidins.

3. REACTION OF FLAVAN-3-OLS WITH AMINO ACIDS IN THE PRESENCE OF ALDEHYDE

A model reaction of acetaldehyde with (–)-epigallocatechin 3-O-gallate (**15**)—a major extension unit of the persimmon proanthocyanidin—afforded a complex mixture of somewhat unstable compounds.[12] Among them, dimeric compounds having a C-6—C-6 or a C-6—C-8 linkage through C_2 unit could be isolated (**16** and **17**, respectively); however, the isomer having a C-8—C-8 linkage (**18**) was too

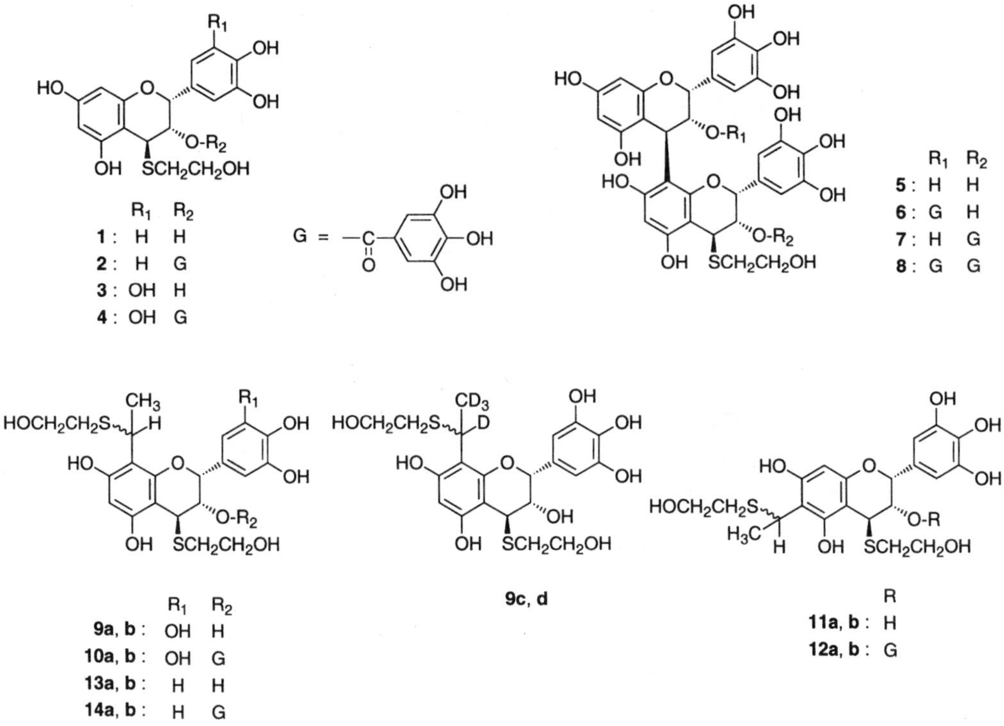

Figure 2. Structures of the thiol degradation products.

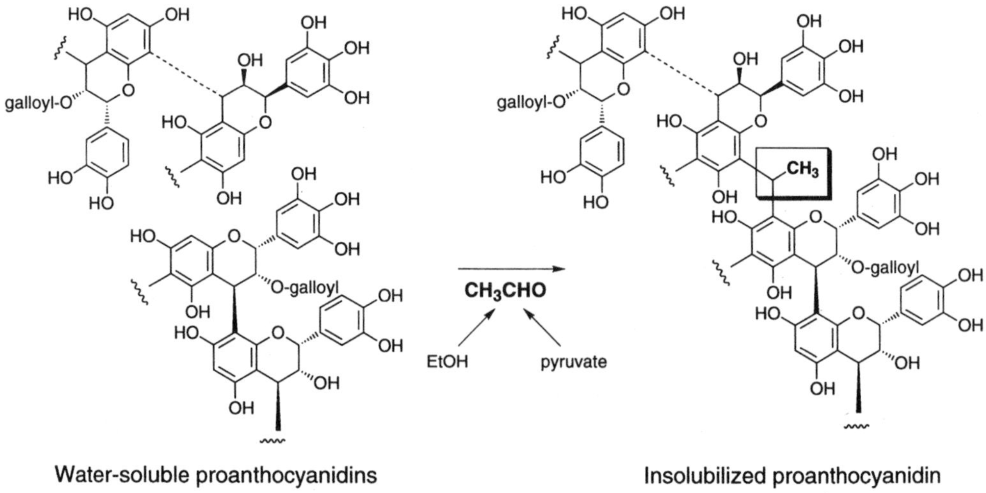

Water-soluble proanthocyanidins   Insolubilized proanthocyanidin

Figure 3. Insolubilization of persimmon proanthocyanidins in the presence of acetaldehyde.

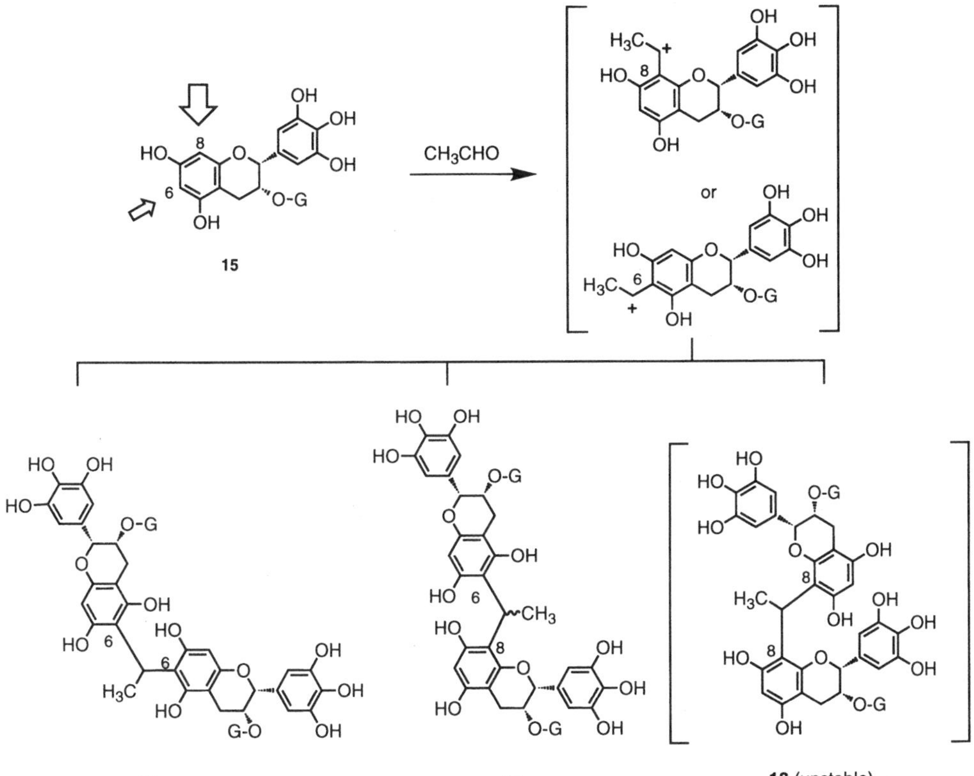

Figure 4. Reaction of epigallocatechin 3-O-gallate with acetaldehyde.

unstable to be isolated, even though the C-8 position is more reactive than C-6 position (fig. 4). The former two products also gradually decomposed to give a complex mixture containing epigallocatechin 3-O-gallate. These results suggested that the linkage between the C_2 unit and proanthocyanidin A-ring is somewhat unstable and easily cleaved. Actually, it was known that after artificial removal of astringency under anaerobic conditions, treatment of the persimmon fruits with diluted acid makes the taste astringent again. The instability of the flavan-3-ol-aldehyde condensation products (**16–18**) reflects high reactivity of the carbon atom between two phloroglucinol rings. Our research has now focused on the potential importance of the reactivity of the flavan-3-ol-aldehyde condensation products with various nucleophiles such as thiols and amino acids.

Treatment of glycine with a solution containing an excess of catechin (**19**) and formaldehyde under weakly acidic conditions afforded a complex mixture of products containing some polymeric substances. Thin-layer chromatography of the reaction mixture showed no ninhydrin-positive spots in the mixture. The disappearance of amino acid was only observed in the presence of both catechin and formaldehyde. An attempt to separate the products failed because of their complexity and instability. Among 22 common amino acids examined, only L-proline gave a product (**20**) that could be isolated by chromatography on high porous polystyrene gel. The yield based on catechin was 80 percent when a large excess of proline and formaldehyde was used. The ^{1}H and ^{13}C NMR spectra showed signals due to two proline moieties and two benzylic methylenes (δ_C 54.8, 2C) together with a catechin moiety. The disappearance of the H-6 and H-8 signals of catechin A-ring and the ^{13}C-chemical shifts of the benzylic methylene indicated that nitrogen atoms of the proline moieties were linked to the C-6 and C-8 positions through the methylene carbons. This was corroborated by negative ion FAB MS, which showed the [M-H]$^-$ peak at m/z 543. From the spectroscopic observation, the structure of the product was concluded to be formulated as **20**. Reaction using acetaldehyde instead of formaldehyde under similar conditions gave a mixture of some products. Although purification of the products failed, the ^{1}H-NMR spectral analysis suggested the presence of stereoisomers with respect to the C_2 unit of 6, 8-disubstituted products, which is analogous to **20**, together with mono-substituted products.

Formation of the polymeric substance by reaction with other amino acids is deducible from the structure of **20** as shown in figure 5: the primary amino group attacked the methylene carbon between the two phloroglucinol rings and the resulting secondary amine reacted with another benzylic methylene. With multifunctional amino acids such as lysine and cysteine, the mechanism should be more complex.

Occurrence of similar covalent bond formation between flavan-3-ol and protein in the presence of aldehyde was also presumed; however, sufficient chemical evidence has not been shown yet, because flavan-3-ols and proanthocyanidins associate with proteins by non-covalent interaction, i.e., hydrophobic and hydrogen bonding. Although insufficient to prove, the following simple experiment suggested a possibility of the covalent interaction: chicken meat (its light color was suitable for a coloration test) was treated with a wine model solution (10 percent ethanol, 5 percent tartaric acid, pH 4) containing catechin (0.02 percent) and acetaldehyde (1 percent) for 1 hour. After successive washing with water, aqueous alcohol and aqueous acetone, the meat was treated with 9 percent ferric chloride solution to give a dark green coloration. Without acetaldehyde, the meat was only darkened slightly. This result indicated a possibility that the insolubilization of proanthocyanidins with acetaldehyde in plant tissue is not only self-polymerization but also cross-reaction with other maclomolecules.

4. SYNTHESIS AND ANTIOXIDANT ACTIVITY OF LIPID-SOLUBLE DERIVATIVES OF TEA POLYPHENOLS

Polyphenols are potent protective agents against coronary heart disease and atherosclerosis, because they possess a radical scavenging activity that inhibits oxidation of plasma low-density lipoproteins.[16–18] Epigallocatechin 3-O-gallate (**15**), the most active radical scavenger in tea polyphenols, inhibited lipid peroxidation in the liposome lipid bilayer caused by a water-soluble radical initiator [2,2'-azobis (2-amidinopropane) dihydrochloride (AAPH)] (table 1). However, **15** was not effective against the peroxidation caused by the lipophilic radical initiator [2,2'-azobis

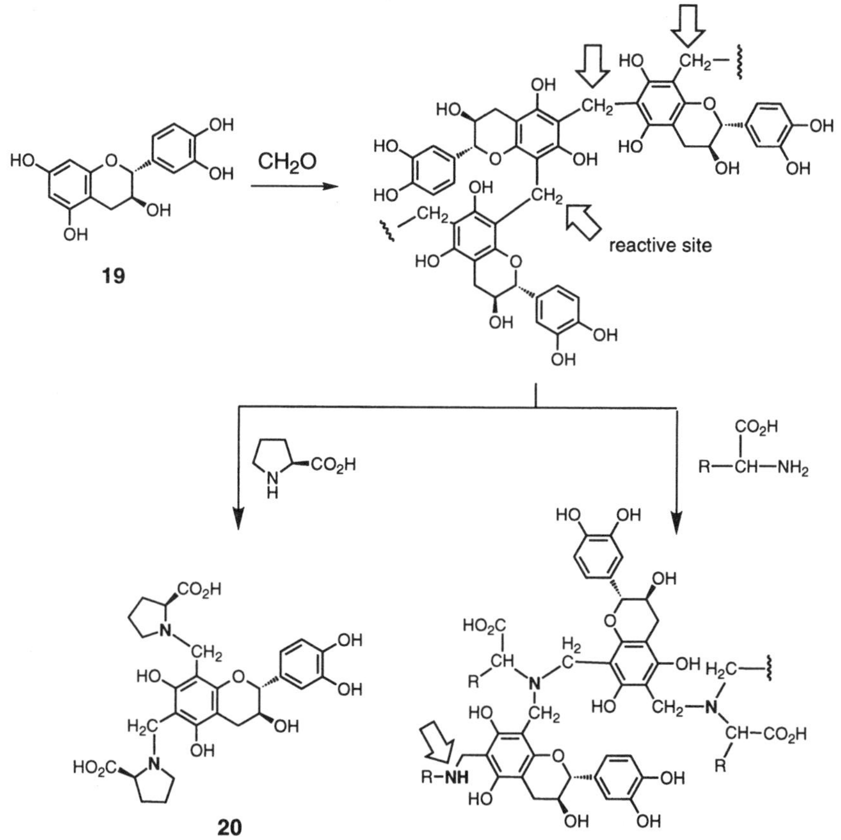

Figure 5. Reaction of catechin with amino acids in the presence of formaldehyde.

(2,4-dimethylvaleronitrile) (AMVN)], because **15** does not penetrate into the hydrophobic region of the lipid bilayer. We have applied the reaction of alkylthiols with flavan-3-ol-aldehyde condensation products to synthesis amphipathic antioxidants having radical scavenging ability in both a hydrophilic and hydrophobic environment.[19] The derivatives **21–26** having two substituents on the A ring were prepared from **15** by simple and one-pot reaction with various thiol compounds in the presence of formaldehyde. Derivatives having three substituents on the A and C rings (**27** and **28**) were synthesized in a similar manner from 4β-(2-hydroxyethylthio)-epigallocatechin 3-O-gallate (**4**). Furthermore, a derivative **29** with a long-chain alkyl group on the galloyl group was obtained by reaction of **15** with *n*-octadecylisocyanate. Distribution between n-octanol and water [partition coefficient value (P) = concentration in n-octanol phase / concentration in water phase (table 1)] indicated that hydrophobicity of these derivatives was much higher than that of **15** and **4**.

Table 1. Inhibition activity of tea polyphenol derivatives against lipid peroxidation of egg-PC liposome caused by water-soluble (AAPH) and lipid-soluble (AMVN) radical initiators and their partition coefficient (P).[a]

	IC_{50} (µM)		P
	AAPH	AMVN	(27°C)
15	20.8	>500	8
4	20.5	>500	12
21	7.9	60.0	10000
22	9.6	432.0	4000
23	11.6	>500	480
24	12.0	>500	9
27	135.1	19.4	89000
28	8.8	60.8	1600
29	24.6	22.1	1300
pentagalloylglucose (**30**)	10.0	>500	92
castalagin (**33**)	0.0	>500	0.01
n-octyl gallate	8.6	188.3	25000
α-tocopherol	40.3	5.9	–

[a] Partition coefficient values between n-octanol and water were evaluated by comparison of HPLC peak area.

Inhibition activities (IC_{50} value) of the derivatives against lipid peroxidation of egg-phosphatidylcholine (PC) liposome were evaluated by measuring the concentration of thiobarbitulic acid reactive substance (table 1).[19] Compared to **15**, pentagalloylglucose (**30**) and *n*-octyl gallate, compounds **21**, **27**, **28**, and **29** strongly inhibited lipid peroxidation caused by AMVN. In particular, **27** and **29**, which have three C_8-alkyl chains and a C_{18}-alkyl chain, respectively, showed the strongest inhibition activity. On the other hand, the activities of **22**, **23**, and **24** were similar to that of **15**. These results apparently suggested that the derivatives having long-alkyl chains penetrated into the hydrophobic region of the lipid bilayer and scavenged radicals.

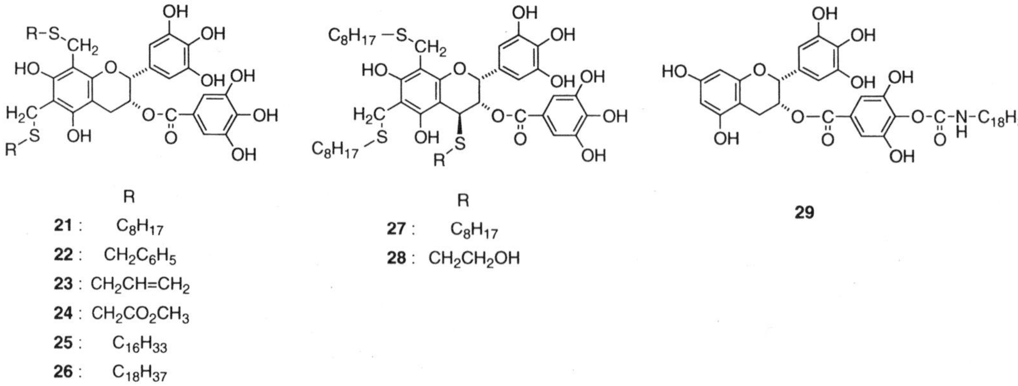

Figure 6. Derivatives of epigallocatechin 3-O-gallate.

Compounds **25** and **26** having C_{16} and C_{18} alkyl chains, respectively, strongly aggregated and precipitated the lipids; hence, the IC_{50} values could not be evaluated. The reason for the aggregation was presumed to be penetration of two long-alkyl chains into two different bilayers and adsorption of the epigallocatechin-3-O-gallate moiety on the surface of phosphatidylcholine bilayers,[20] which resulted in formation of an interbilayer bridge between two opposing bilayers (fig. 7). The following observations also supported the adsorption of **15** on lipid bilayer: i) on ultrafiltration of an aqueous solution, the permeability of **15** through membrane filter was reduced in the presence of liposome; ii) the ^{1}H-NMR signals of **15** in deuterium oxide were significantly broadened by addition of liposome and, in a similar NMR experiment using paeoniflorine (**36**), a water-soluble glycoside having a hydrophobic benzoyl group, such signal broadening was not observed by addition of liposome; iii) electron microscopic observation showed morphological change (adhesion and fusion) of liposome in the presence of high concentration of **15**.

Compound **22** having π-electron rich benzyl groups showed strong inhibition activity against AAPH and relatively low activity against AMVN despite its high hydrophobicity. This fact suggested that this compound was located on the surface of the bilayer, and electrostatic interaction between π-electron of the aromatic rings and $-N(CH_3)_3$ groups on the phosphatidylcholine head groups is of importance for the adsorption.[20] Since polyphenols are a multiple hydrogen-bond donor, hydrogen bonding with hydrogen-bond-accepting groups on the bilayer probably increases the overall strength of the adsorption. From these results, it was presumed that the epigallocatechin 3-O-gallate moiety of the derivatives such as **21** and **28** was electrostatically adsorbed on the ionic surface of bilayer, and the alkyl chains were anchored into the hydrophobic region (fig. 7). The most hydrophobic derivative **27** having three C_8 alkyl chains showed the highest inhibition activity against AMVN and the lowest activity against AAPH. This derivative was supposed to be present

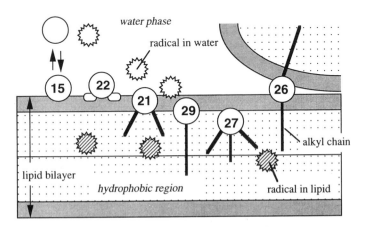

Figure 7. Plausible interaction of derivatives of tea polyphenol with lipid bilayer.

in the hydrophobic region. Compound **29** having a C_{18}-long chain alkyl group on its galloyl group did not aggregate lipids and showed high inhibition activity against lipid peroxidation caused by both water-soluble and lipid-soluble radical generators. For this derivative, masking of a phenolic hydroxy group of the galloyl residue by a carbamoyl ester may weaken the interaction with the membrane surface, and the molecule may penetrate deeper into the hydrophobic region.

As the summary of this section, the compounds **21**, **28**, and **29** are the first tea polyphenol derivatives with strong inhibition activity against lipid peroxidation in the liposome lipid bilayer caused by both lipid-soluble and water-soluble radical generators. The relation between the structure and the activity suggested that, in addition to hydrophobicity, the interaction of the polyphenol moiety with surface of the lipid bilayer affected penetration of the derivatives into the hydrophobic region of the bilayer.

5. SOLUBILIZATION OF TANNINS

Although purified polymeric proanthocyanidins are hardly soluble in water, the aqueous extract of the astringent persimmon fruits contains them abundantly. There should be solubilization mechanisms. Some biologically active tannins contained in crude drugs such as pentagalloylglucose (**30**) of peony root (*Paeonia lactiflora*) and oligomeric proanthocyanidins of rhubarb (*Rheum palmatum*) also show low water solubility when they are purified; hence, the solubilization of tannins may be important from a pharmacological viewpoint.

Biosynthesis of hydrolyzable tannins starts from galloylation of glucose,[21] and the galloylation decreases water solubility of tannins.[4] Comparison of distribution between *n*-octanol and water (15 °C) of 1-mono- (P = 0.01), 1,6-di- (0.5), 1,2,6-tri- (4), 1,2,3,6-tetra- (36), and 1,2,3,4,6-pentagalloylglucoses (160) clearly showed that addition of galloyl groups increases hydrophobicity of the molecule, and pentagalloylglucose is the most hydrophobic of simple gallotannins.

Although many plants have been found to accumulate hydrolyzable tannins, plants accumulating pentagalloylglucose in normal tissue are rare. Paeonia species and gall of Rhus species accumulate polygalloylglucoses, which possess extra galloyl groups on pentagalloylglucose through depside linkages. Although pentagalloylglucose (**30**) forms gel in water at room temperature due to the low water solubility, a mixture of the polygalloylglucoses does not form a gel in the similar conditions. Many other plants containing hydrolyzable tannins tend to metabolize pentagalloylglucose into ellagitannins by oxidative couplings between two adjacent galloyl groups (fig. 8). The oxidative C—C coupling between galloyl groups in ellagitannin metabolism decreases the distribution coefficients, that is, formation of biphenyl linkage increases the water solubility. The increase of water solubility is possibly related to the compartmentation of tannins in cell vacuoles. In this respect, the effect of the oxidative coupling in gallotannin may be similar to that of glycosidation of flavonoids.

Rhoipteleanin A (**35**)[22] is the first dimeric ellagitannin biogenetically derived by stereospecific intermolecular C-C oxidative coupling between two molecules of 1(β)-O-galloylpedunculagin (**32**). The stereospecificity of the intermolecular radical

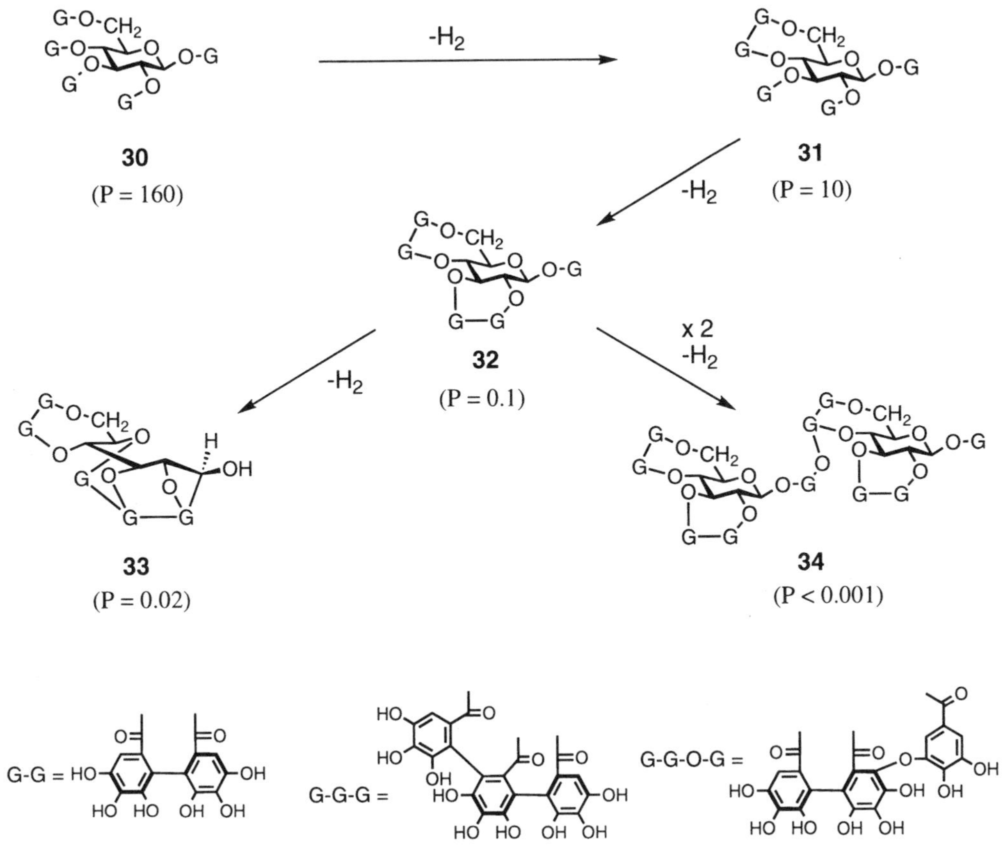

(P = partition coefficent between *n*-octanol-water at 15°C)

Figure 8. Biogenesis of ellagitannins and their partition coefficients.

coupling indicated that the relative orientation of the two molecules of the precursor **32** is stereochemically regulated prior to the coupling. Enzymes probably regulate the coupling, because the types of dimeric ellagitannins are usually specific to each plant, even if they are derived from the same biogenetic precursors. However, it is possible that the regulation occurs spontaneously by hydrophobic self-association of **32** prior to the enzymatic oxidation.

The ¹H-NMR spectrum of **32** in deuterium oxide at various concentrations showed a large upfield shift of the glucose H-5 signal and, reversely, a lowfield shift of the aromatic signal of the pyrogallol ring attached to the glucose C-2 with higher concentration.[23] The remaining proton signals were all shifted upfield, and the changes in chemical shift were much smaller than that of glucose H-5. Since it is known that the hydrophobic association of hydrolyzable tannins occurs preferentially at the galloyl group attached to the anomeric position of the glucose core,[4,24] the shifts of the specific proton signals of **32** (upfield shift of glucose H-5 and downfield shift of one of the HHDP protons) were considered to be caused by the anisotropic effect of the galloyl group of the other molecule of **32** in the complex. Therefore, the two molecules of **32** seem to be arranged so as to mutually cover the most hydrophobic site of the molecule, i.e., around H-5 of the glucopyranose ring, with the galloyl group in the most effective manner (fig. 9). In this stereochemically regulated association, the galloyl group of **32** is probably situated above glucose H-5 and at the side of the aromatic proton of the pyrogallol ring linked to the glucose C-2 of the other molecule of **32**. If the enzymatic intermolecular C-C coupling between the galloyl group and the pyrogallol ring attached to H-2 occurs in this complex, the resulting biphenyl bond is restricted to S configuration, which is in agreement with that of **35**.

As mentioned above, aqueous solution of pentagalloylglucose forms a gel at room temperature. The gel formation reflects how firmly the molecules are associated by themselves in aqueous solution. Addition of sucrose did not dissolve the gel; however, very interestingly, addition of paeoniflorin (**36**), a monoterpenoid glycoside coexisting with polygalloylglucoses in the hot-water extracts of peony root, completely destroyed the gel.[4] The increase of water solubility of pentagalloylglucose in the presence of paeoniflorin was shown in figure 10. Presence of 50 molls of paeoniflorin decreased the n-octanol-water distribution coefficient 100 times. Actually, the water solubility of pentagalloylglucose in the aqueous extract of peony root is much higher than that of pentagalloylglucose. The increase of water solubility was explained by regiospecific association with paeoniflorin: the ^{1}H NMR spectrum of the mixture in D_2O showed large upfield shifts of the signals due to the benzoyl group of **36** and the anomeric proton and the galloyl group at C-1 position of **30**, which can be attributed to intermolecular anisotropic effect (fig. 11). In addition, intermolecular NOEs were observed between the benzoyl group of **36** and the anomeric proton of **30**. It was presumed that the benzoyl group of **36** covered the most hydrophobic site of the pentagalloylglucose molecule and interfered with the hydrophobic self-association.[25] This association resulted in an increase of total entropy of the system. In addition to hydrophobic interaction, π-π interaction[26] is probably important in this association.

Similarly, the presence of water-soluble glycosides increased the water solubility of polymeric proanthocyanidins:[4] rhubarb polymeric proanthocyanidins,[27] substantially insoluble to water in a pure state, could be dissolved by addition of rhein 8-O-glucoside potassium salt (**37**) under neutral conditions (fig. 12). Because the addition of sucrose did not change the solubility, it was suggested that association between polymeric proanthocyanidins and the anthraquinone glycoside caused the increase of solubility. The following experiment supported occurrence of the association: the ^{1}H-NMR spectrum of **37** with various concentrations of procyanidin B-2 3,3'-di-O-gallate (**38**), which is also a major constituent of rhubarb, showed an upfield shift of H-6 and H-5 [$\Delta\delta_H$: H-6, 0.41; H-5, 0.37 (**37**, 0.016M; **38**, 0.032M] and a downfield shift of H-2 and H-4 ($\Delta\delta_H$: H-4, −0.1; H-2, −0.155).[4] On the other hand, chemical shift change in **38** was mainly observed at galloyl group of the lower unit [$\Delta\delta_H$: 3'-O-galloyl, 0.79; 3-O-galloyl, 0.25 (**38**, 0.008M; **37**, 0.16M)]. From these Dd_H values, the interaction was deduced to occur mainly between the 3'-O-galloyl group of the **38** and around C-5 and C-6 of the

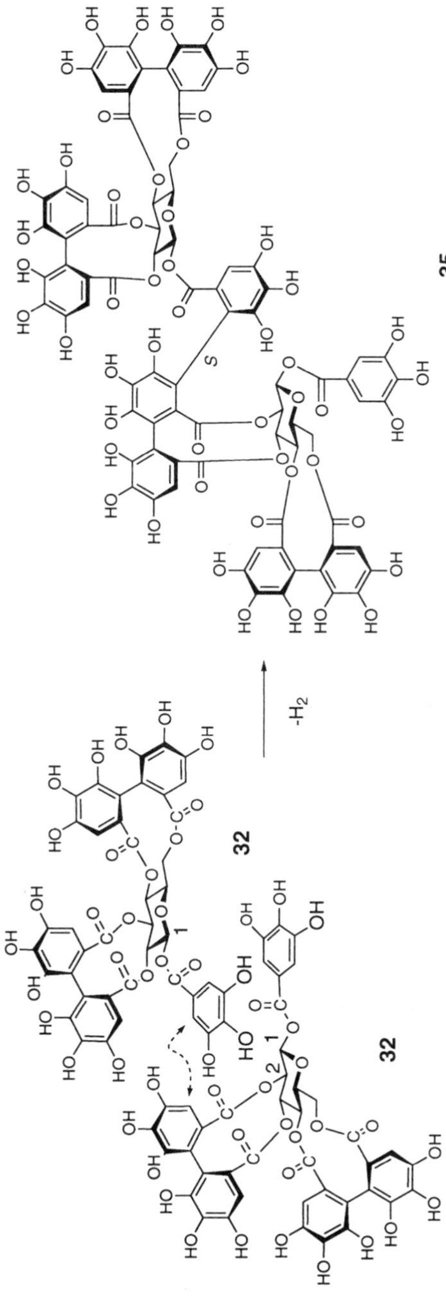

Figure 9. Biogenesis of rhoipteleanin A (**35**).

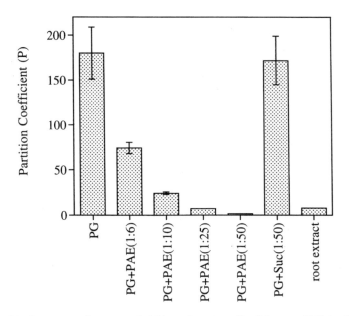

Figure 10. Increase of water solubility of pentagalloylglucose (**30**) in the presence of paeoniflorin (**36**). Pentagalloylglucose: PG; paeoniflorin: PA; sucrose: Suc.

anthraquinone nucleus of **37**. When the ^{1}H-NMR spectrum of **37** was measured in the presence of the polymeric proanthocyanidins (37.2 mg, polymer 2 mg in 0.7 mL of D$_2$O), H-5 of the **37** showed the largest upfield shift ($\Delta\delta_H$ 0.13). Since C-5 is the position where the association with **38** occurs preferentially, this result suggests that hydrophobic interaction is an important factor in increasing the solubility of polymeric proanthocyanidins.

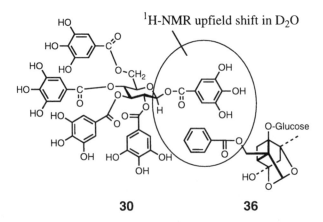

Figure 11. Association between pentagalloylglucose (**30**) and paeoniflorin (**36**).

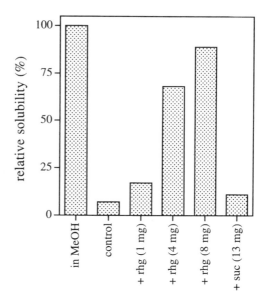

Figure 12. Solubility of polymeric proanthocyanidins of rhubarb in the presence of rhein 8-O-glucoside potassium salt (rhg) [polymeric proanthocyanidin (2 mg) in water (1 mL), pH 6, 25 °C].

6. CONCLUSIONS

Absorption of tannins by the digestive tract has not been clarified, except for some flavan-3-ols.[28] However, some biological activities of tannins, such as antihypertensive activity by polymeric proanthocyanidins of *Areca catechu*[29] and effect on uremic toxins of rhubarb tannins,[30,31] gallotannins,[32] and condensed tannins of *Ephedra* sp.[33] were shown by oral administration. We need to direct our attention

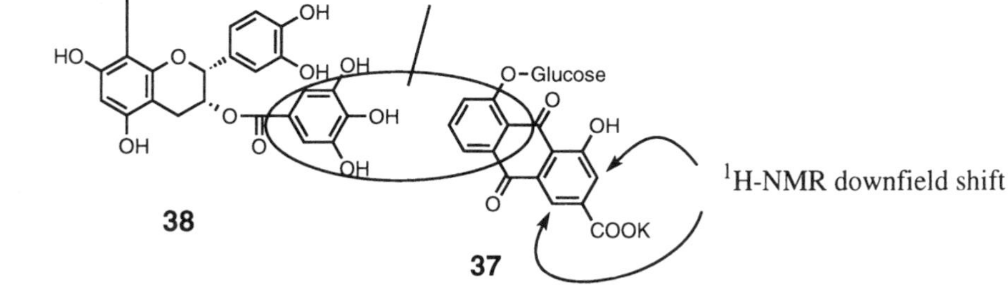

Figure 13. Association between anthraquinone glycoside (**37**) and procyanidin dimer gallate (**38**) of rhubarb.

to the pharmacological mechanisms of these biological activities of tannins in the future. Recently, preventive effects of the gallotannin fraction of peony root on neuron damage after oral administration were shown to be enhanced by combination with paeoniflorin (**36**), although use of **36** alone had no effect.[34] As for flavonoids, it was reported that the bioavailability of an isoflavonoid in the extract of an oriental crude drug was much higher than that in a pure state.[35] These results indicated that purification does not always increase pharmacological activity and suggested the importance of covalent and non-covalent interaction with coexisting compounds. The oriental medical system uses combinations of crude drugs in which tannins are sometimes solubilized by association with water-soluble glycosides including saponins and occasionally precipitated with some alkaloids. In the traditional medical system, solubility of the chemical constituents of the crude drugs may hold significance for biological activity. In addition, tannins possess various useful properties besides antioxidant activity; therefore, modifications of the tannins such as preparation of the hydrophobic derivatives of tea polyphenols may lead to useful new materials.

REFERENCES

1. Clausen, T.P.; Reichardt, P.B.; Bryant, J.P.; Provenza F. Condensed tannins in plant defense: a perspective on classical theories. *In*: Hemingway, R.W.; Laks, P.E.(eds.) Plant polyphenols-synthesis, properties, significance, Plenum Press, New York, p. 639 (1992).

2. Beart, J.E.; Lilley, T.H.; Haslam, E. Plant polyphenols-secondary metabolism and chemical defence: some observations. *Phytochemistry* 24:33 (1985).

3. Zucker W.V. Tannins: does structure determine function? An ecological perspective. *The American Naturalist* 121:335 (1983).

4. Tanaka, T.; Zhang, H.; Jiang, Z.; Kouno, I. Relationship between hydrophobicity and structure of hydrolysable tannins, and association of tannins with crude drug constituents in aqueous solution. *Chem. Pharm. Bull.* 45:1891(1997).

5. Tanaka, T.; Fujisaki, H.; Nonaka, G.; Nishioka, I. Reaction of dehydroellagitannins with L-cysteine methyl ester. *Heterocycles*, 33:375 (1992).

6. Tanaka, T.; Kouno, I.; Nonaka, G. Glutathione-mediated conversion of the ellagitannin geraniin into chebulagic acid. *Chem. Pharm. Bull.*, 44:34 (1996).

7. Tanaka, T.; Fujisaki, H.; Nonaka, G.; Nishioka, I. Tannins and related compounds. CXVIII. Structures, preparation, high-performance liquid chromatography and some reactions of dehydroellagitannin-acetone condensates. *Chem. Pharm. Bull.*, 40:2937 (1992).

8. McManus, J.P.; Davis, K.G.; Beart, J.E.; Gaffney, S.H.; Lilley, T.H.; Haslam, E. Polyphenol interactions, part 1. Introduction; some observations on the reversible complexation of polyphenols with proteins and polysaccharides. *J. Chem. Soc. Perkin Trans.* 1:1429 (1985).

9. Charlton, A.J.; Baxter, N.J.; Lilley, T.H.; Haslam, E.; McDonald, C.J.; Williamson, M.P. Tannin interactions with a full-length human salivary proline-rich protein display a stronger affinity than with single proline-rich repeats. *FEBS Letters* 382:289 (1996).

10. Baxter, N.J.; Lilley, T.H.; Haslam, E.; Williamson, M.P. Multiple interactions between polyphenols and a salivary proline-rich protein repeat result in complexation and precipitation. *Biochemistry* 36:5566 (1997).

11. Matsuo, T.; Itoo, S. A model experiment for de-astringency of persimmon fruit with high carbon dioxide treatment: *in vitro* gelation of kaki-tannin by reacting with acetaldehyde. *Agric. Biol. Chem.* 46:683 (1982).

12. Tanaka, T.; Takahashi, R.; Kouno, I.; Nonaka, G. Chemical evidence for the de-astringency (insolubilization of tannins) of persimmon fruit. *J. Chem. Soc. Perkin Trans.* 1:3013 (1994).

13. Pesis, E.; Ben-Arie, R. Carbon dioxide assimilation during postharvest removal of astringency from persimmon fruit. *Physiol. Plant.* 67:644 (1986).

14. Saucier, C.; Little, D.; Glories, Y. First evidence of acetaldehyde-flavanol condensation products in red wine. *Am. J. Enol. Vitic.*, 48:370 (1997).

15. Francia-Aricha, E.M.; Guerra, M.T.; Rivas-Gonzalo, J.C.; Santos-Buelga, C. New anthocyanin pigments formed after condensation with flavanols. *J. Agric. Food Chem.* 45:2262 (1997).

16. Pearson, D.A.; Frankel, E.N.; Aeschbach, R.; German, J.B. Inhibition of endothelial cell mediated low-density lipoprotein oxidation by green tea extracts. *J. Agric. Food Chem.* 46:1445 (1998).

17. Rice-Evans, C.A.; Miller, N.J. Antioxidant activities of flavonoids as bioactive components of food. *Biochem. Soc. Trans.* 24:790 (1996).

18. Frankel, E.N.; Kanner, J.; German, J.B.; Parks, E. Kinsella, J.E. Inhibition of oxidation of human low-density lipoprotein by phenolic substances in red wine. *Lancet.* 341:454 (1993).

19. Tanaka, T.; Kusano, R.; Kouno, I. Synthesis and antioxidant activity of novel amphipathic derivatives of tea polyphenol. *Bioorg. Med. Chem. Lett.* in press (1998).

20. Huh, N.W.; Porter, N.A.; McIntosh, T.J.; Simon, S.A. The interaction of polyphenols with bilayers: conditions for increasing bilayer adhesion. *Biophys. J.* 71:3261 (1996).

21. Gross, G.G. Enzymes in the biosynthesis of hydrolyzable tannins. *In:* Hemingway, R.W.; Laks, P.E. (eds.) Plant polyphenols-synthesis, properties, significance, Plenum Press, New York, p. 43 (1992).

22. Jiang, Z.; Tanaka, T.; Kouno, I. Rhoipteleanins A and E, aimeric ellagitannins formed by intermolecular C-C oxidative coupling from *Rhoiptelea chiliantha*. *J. Chem. Soc., Chem. Commun.* :1467 (1995).

23. Tanaka, T.; Jiang, Z.; Kouno, I. Structures and biogenesis of rhoipteleanins, ellagitannins formed by stereospecific intermolecular C-C oxidative coupling, isolated from *Rhoiptelea chiliantha*. *Chem. Pharm. Bull.* 45:1915 (1997).

24. Spencer, C.M.; Cai, Y.; Martin, R.; Gaffney, S.H.; Goulding, P.N.; Magnolato, D.; Lilley, T.H.; Haslam, E. Polyphenol complexation-some thoughts and observations. *Phytochemistry* 27:2397 (1988).

25. Spencer, C.M.; Cai, Y.; Martin, R.; Lilley, T.H.; Haslam, E. Metabolism of gallic acid and hexahydroxydiphenic acid in higher plants, Part 4; polyphenol interactions part 3. spectroscopic and physical properties of esters of gallic acid and (S)-hexahydroxydiphenic acid with D-glucopyranose ($^4C_1$). *J. Chem. Soc. Perkin Trans 1.* :651 (1990).

26. Hunter, C.A.; Sanders, K.M. The nature of π-π interactions. *J. Am. Chem. Soc.* 112:5525 (1990).

27. Nonaka, G.; Nishioka, I.; Nagasawa, T.; Oura, H. Tannins and related compounds. I. Rhubarb(1). *Chem. Pharm. Bull.* 29:2862 (1981).

28. Nakagawa, K.; Okuda, S.; Miyazawa, T. Dose-dependent incorporation of tea catechins, (−) epigallocatechin-3-gallate and (−)-epigallocatechin, into human plasma, *Biosci. Biotech. Biochem.* 61:1981 (1997).

29. Inokuchi, J.; Okabe, H.; Yamauchi, T.; Nagamatsu, A.; Nonaka, G.; Nishioka, I. Antihypertensive substance in seeds of *Areca catechu* L. *Life Sci.* 38:1375 (1986).

30. Yokozawa, T.; Suzuki, N.; Oura, H.; Nonaka, G.; Nishioka, I. Effect of extracts obtained from rhubarb in rats with chronic renal failure. *Chem. Pharm. Bull.* 34:4718 (1986).

31. Yokozawa, T.; Fujioka, K.; Oura, H.; Nonaka. G.; Nishioka. I. Effects of rhubarb tannins on uremic toxins. *Nephron* 58:155 (1991).

32. Yokozawa, T.; Fujioka, K.; Oura, H.; Tanaka, T.; Nonaka, G.; Nishioka, I. Uraemic toxin reduction: a newly found effect of hydrolyzable-type tannin-containing crude drug and gallotannin. *Phytotherapy Research* 9:327 (1995).

33. Yokozawa, T.; Fujioka, K.; Oura, H.; Tanaka, T.; Nonaka, G.; Nishioka, I. Decrease in uraemic toxins, a newly found beneficial effect of Ephedrae Herba. *Phytotherapy Research* 9:382 (1995).

34. Tsuda, T.; Sugaya, A.; Ohguchi, H.; Kishida, N.; Sugaya, E. Protective effects of peony root extract and its components on neuron damage in the hippocampus induced by the cobalt focus epilepsy model. *Exp. Neurol.* 146:518 (1997).

35. Keung W. M.; Lazo, O.; Kunze, L.; Vallee, B. L. Potentiation of the bioavailability of daidzin by an extract of Radix puerariae. *Proc. Natl. Acad. Sci. USA* 93:4284 (1996).

SULFUR DIOXIDE DECOLORIZATION OR RESISTANCE OF ANTHOCYANINS: NMR STRUCTURAL ELUCIDATION OF BISULFITE-ADDUCTS

Bénédicte Berké, Catherine Chèze, Gérard Deffieux, and Joseph Vercauteren

Laboratory of Pharmacognosy—GESNIT, EA 491
University Victor Segalen Bordeaux 2
146, rue Léo Saignat
33076 Bordeaux
FRANCE

1. INTRODUCTION

Use of anthocyanins as coloring agents in food is limited, since it is well known that they lose their original color in the presence of nucleophiles. This phenomenon occurs in aqueous alkaline media (OH^- addition), but most of all in the presence of bisulfite, commonly used as a preservative. Even though this has been observed for many years, the mechanism is not thoroughly understood. The chalcone form (3), in equilibrium[1,2] with the carbinol (2) resulting from hydroxyl addition on the flavylium (1), was postulated. Addition of bisulfite readily occurs, as it does with carbonyls leading to a chalcone-bisulfite adduct.[3,4] More recent works, by means of UV-visible,[5] kinetic,[6,7] and thermodynamic measurements,[8] have shown that adducts could result from nucleophilic addition of SO_2 either on the C-2 or C-4 carbon of the flavylium, thus refuting the necessity of opening into a chalcone (fig. 1).

However, these investigators could not go further in the structure knowledge of adducts, since only UV-visible spectroscopy was used, whereas flavylium transformation in aqueous solution has been studied by NMR.[9–11] The use of NMR allowed us to investigate sulfur dioxide reactivity toward anthocyanins. We could go further in defining the structure of adducts, using generally one and two-dimensional ^1H and ^{13}C NMR techniques, together with new experiments on ^{33}S at natural abundance. It was thus possible to determine that the decolorization was due to SO_2 addition rather than to the formation of hemiketal forms [fig. 1, (2)].

Plant Polyphenols 2: Chemistry, Biology, Pharmacology, Ecology, Edited by Gross et al. Kluwer Academic / Plenum Publishers, New York, 1999

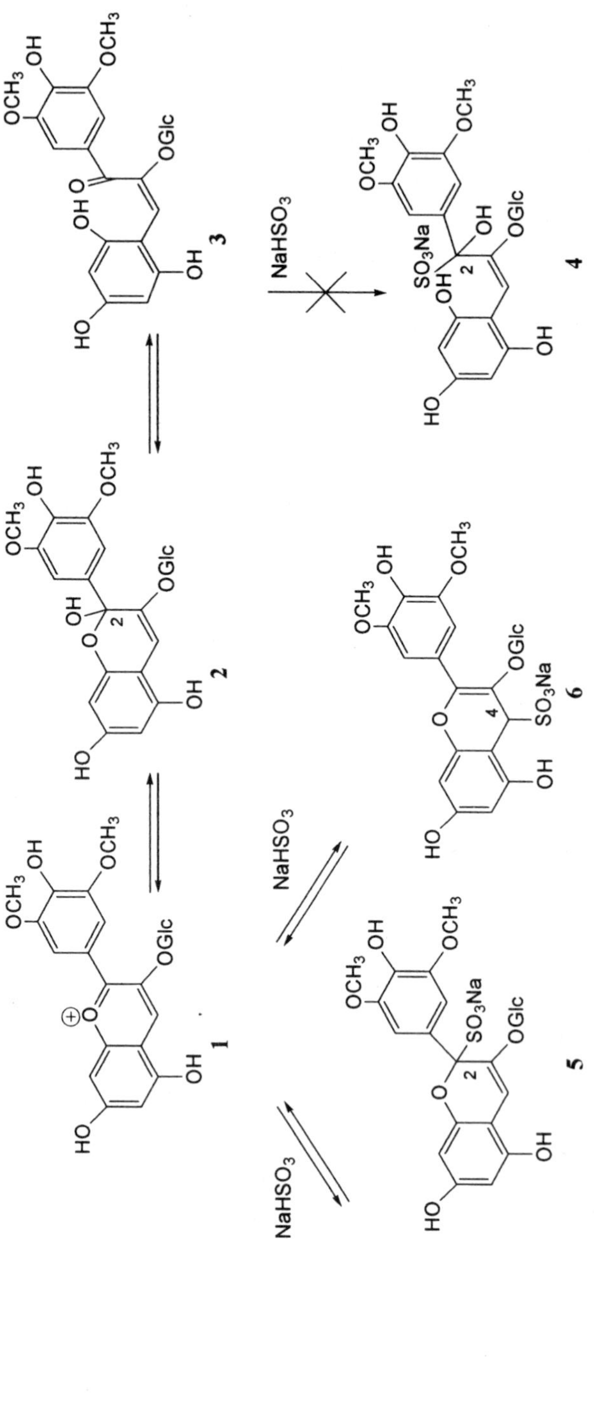

Figure 1. Proposed mechanism for bisulfite addition to anthocyanin.

2. MAJOR ANTHOCYANINS OF *VITIS VINIFERA*: FULL NMR ASSIGNMENTS

Flavylium salts have already been used to study the SO_2 decolorization.[5-7] We chose to analyze this phenomenon using 3-*O*-β-D-glucosyl malvidin (**1**), the major anthocyanin[12,13] of *Vitis vinifera*, since our work is part of a study on sulfur dioxide reactivity toward wine constituents. We purified **1** from *Vitis vinifera* extracts by preparative reverse-phase HPLC. Along with this main flavylium, we isolated enough material of four other anthocyanins to obtain good NMR data on each. They differ only by the pattern of substitution of B-ring (fig. 2) and are the 3-*O*-β-D-glucosyl derivatives of peonidin (**7**), delphinidin (**8**), petunidin (**9**), and cyanidin (**10**).

Among them, only 3-*O*-glucosides of cyanidin (**10**),[14-16] delphinidin (**8**),[14] and peonidin (**7**)[16] have their ^1H and ^{13}C NMR spectra fully assigned, and our data are in good agreement with those published in the literature. Total ^1H and ^{13}C NMR assignments of 3-*O*-β-D-glucosyl malvidin (**1**) had to be done to facilitate the study of its reaction with sulfur dioxide. These assignments could be made using 2D homo and heteronuclear ^1H and ^{13}C NMR experiments.

The ^1H NMR spectrum of **1** showed two equivalent (δ_H 7.80, H-2', H-6') and two non-equivalent aromatic protons (δ_H 6.58 and δ_H 6.83, H-6 and H-8), in addition to those of the glucosyl residue (δ_H 3.43 to 3.91, 5.3) and the two methoxy groups (δ_H 3.94, 6H). The most deshielded resonance at δ_H 8.9 was assigned to the H-4 proton. The COSY[17] spectrum was used to assign the spin system of the glucose residue, starting from the unambiguous signal of the anomeric proton H-1". The doublet at δ_H 5.3 (J = 7 Hz) is characteristic of a β-glucopyranoside. To complete the assignments and proof of structure of anthocyanins, there are two main difficulties to overcome: a) define the site of glycosylation; and b) discriminate between the H-6 and H-8 resonances. Two common strategies[18] are used. In the HMBC[19] spectrum,

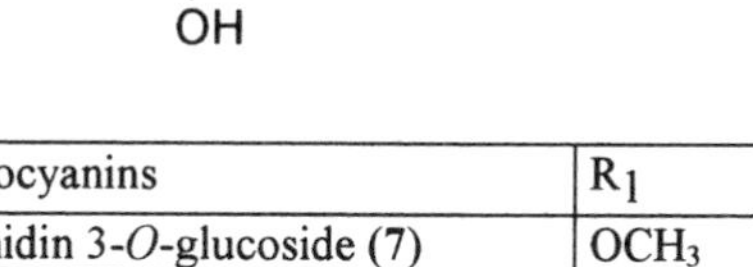

anthocyanins	R_1	R_2
peonidin 3-*O*-glucoside (**7**)	OCH₃	H
delphinidin 3-*O*-glucoside (**8**)	OH	OH
petunidin 3-*O*-glucoside (**9**)	OCH₃	OH
cyanidin 3-*O*-glucoside (**10**)	OH	H

Figure 2. Major anthocyanins of *Vitis vinifera*.

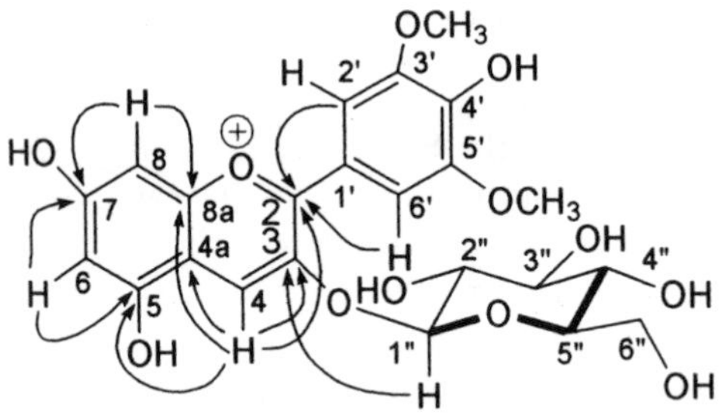

Figure 3. HMBC correlations of 3-*O*-β-D-glucosyl malvidin.

a quaternary oxygenated olefinic carbon atom at $\delta_C 148$ displays long-range connectivities with the anomeric proton (H-1″) as well as with the H-4 proton. These correlations allow assignment of the signal at $\delta_C 148$ to C-3 thus bearing the glycosyl residue. The ROESY[20] spectrum was used to discriminate between the two residual aromatic A-ring resonances. Connectivity between the sharp intense signal (δ_H 7.8, H-2′/H-6′) and the resonance at δ_H 6.8 can only be due to the H-8 proton. Thus, the C-8 and C-6 chemical shifts can be deduced from the HMQC[21] and the C-5, C-7 and C-8a aromatic oxygenated carbon atoms from the HMBC experiments (fig. 3). The complete NMR data for these five anthocyanins are summarized in the following tables.

3. NMR STUDY OF ADDUCTS BETWEEN MALVIDIN GLUCOSIDE AND SULFUR DIOXIDE

Adducts **6′** and **6″** were prepared[22] from a solution of **1** (50 mg, 0.095 mmole in 0.5 mL D$_2$O/DCl 5%) in a NMR tube. Sodium bisulfite was added until absorption in the visible spectrum ($\lambda = 537$ nm) became weak (416 mg, 4 mmoles). ^{1}H and ^{13}C

Table 1. ^{1}H NMR of anthocyanidin monoglucosides **1, 7, 8, 9** and **10**

	1	7	8	9	10
H-4	8.90 *s*	8.96 *s*	8.95 *s*	8.92 *s*	9.00 *s*
H-6	6.58 *s*	6.62 *s*	6.62 *s*	6.62 *d* (1.9)	6.65 *s*
H-8	6.83 *s*	6.84 *s*	6.84 *s*	6.82 *d* (1.9)	6.87 *s*
H-2'	7.80 *s*	8.13 *s*	7.74 *s*	7.86 *d* (2.2)	8.05 *s*
H-3' or R1	3.94 *s*	3.94 *s*	-	-	-
H-5' or R2	3.94 *s*	7.00 *d* (8.7)	-	3.95	7.00 *d* (8.6)
H-6'	7.80 *s*	8.16 *d* (8.7)	7.74 *s*	7.70 *d* (2.2)	8.22 *d* (8.6)
H1″	5.30 *d* (7.6)	5.28 *s* (7.7)	5.30 *d* (7.8)	5.30 *d* (7.7)	5.27 *d* (7.7)
H2″	3.62 *m*	3.65 *m*	3.70 *m*	3.67 *m*	3.69 *m*
H3″	3.55 *m*	3.57 *m*	3.55 *m*	3.57 *m*	3.54 *m*
H4″	3.43 *m*	3.45 *m*	3.47 *m*	3.46 *m*	3.44 *m*
H5″	3.55 *m*	3.55 *m*	3.55 *m*	3.56 *m*	3.54 *m*
H6a″	3.91 *m*	3.95 *m*	3.90 *m*	3.92 *m*	3.91 *m*
H6b″	3.71 *dd* (12; 6)	3.72 *dd* (12; 6)	3.74 *m*	3.74 *dd* (12; 5.9)	3.71 *m*

Table 2. ^{13}C NMR of anthocyanidin monoglucosides **1, 7, 8,
9** and **10**

	1	7	8	9	10
C-2	164.1	164.2	164.3	163.6	164.4
C-3	146.2	147.8	148.7	145.7	147.4
C-4	137.5	137.6	137.0	136.5	137.1
C-4a	114.2	113.9	113.4	113.4	113.4
C-5	160.3	159.5	157.7	159.7	157.8
C-6	104.0	103.9	103.5	103.5	103.8
C-7	172.0	171.2	170.4	170.6	170.5
C-8	95.9	95.6	99.8	95.2	95.0
C-8a	158.6	158.1	153.9	157.6	155.8
C-1'	120.3	121.4	120.1	119.9	121.3
C-2'	111.2	115.6	114.0	109.3	117.5
C-3'	150.6	149.9	147.6	149.8	145.8
R1	57.6	57.3	-	57.2	-
C-4'	147.3	156.9	149.0	145.2	145.6
C-5'	150.6	118.0	147.6	147.4	118.6
R2	57.6	-	-	-	-
C-6'	111.2	129.2	114.0	113.7	128.2
C-1"	104.5	104.3	103.7	103.7	103.8
C-2"	75.5	75.3	75.0	74.9	74.8
C-3"	79.3	79.2	79.2	78.8	78.8
C-4"	71.7	71.5	71.5	71.1	71.1
C-5"	78.8	78.6	78.5	78.2	78.1
C-6"	62.7	62.7	62.4	62.4	62.4

NMR spectra on the crude mixture revealed two sets of resonances in a 1:1 ratio for which the formation of diastereomers is postulated. The HPLC analysis revealed a facile reaction of the adducts back to the original flavylium (**1**). This prevented further isolation attempts. The structures of the two diastereomers **6'** and **6"** result from the spectra recorded on the crude mixture. Complete proton and carbon assignments are reported in table 3 for both isomers, but structural elucidation of only one diastereomer is discussed. The two major modifications of ^{1}H NMR spectra of **6'** and **6"**, when compared to the spectrum of **1**, are related to the loss of conjugation of the flavylium ion indicated by a large upfield shift ($\Delta\delta$ppm = 3.3) of the H-4 resonance (δ_H 5.6, **6'**) and to the newly formed stereocenter at C-4, now bearing the sulfonate or sulfinate group, either in the β- or α-position. The large chemical shift difference of the closest proton to this center (the anomeric H-1" proton of the glucosyl residue) appeared at δ_H 5.3 in **6'** and δ_H 4.8 in **6"**.

The J-modulated[23] ^{13}C NMR spectrum revealed a resonance at δ_C 56.9 characteristic of a methine carbon in a benzylic carbon position and substituted by an heteroatom that is assigned to the C-4 carbon. HMQC experiment showed correlations between δ_H 5.6 and δ_C 56.9 in isomer **6'** and, thus confirmed this assignment. The C-2 at δ_C 146.4 is deduced unambiguously from its HMBC correlations with the H-4 proton and the H-2'/H-6' protons as well (fig. 4). The H-4 proton shows other correlations with two oxygenated aromatic carbon atoms at δ_C 156.6 and δ_C 157.1 (C-8a, C-5) and with an aromatic carbon at δ_C 102.0 (C-4a).

Table 3. ^{1}H and ^{13}C NMR of sulfur dioxide adducts of an anthocyanidin monoglucoside **6′** and **6″**

	6′		6″	
	^{1}H	^{13}C	^{1}H	^{13}C
2	/	146.4	/	147.5
3	/	131.4	/	133.0
4	5.6*s*	56.9	5.55 *s*	61.25
4a	/	102.0	/	103.0
5	/	157.1	/	157.4
6	6.52*s*	102.4	6.55 *s*	102.4
7	/	159.6	/	159.8
8	6.55 *s*	98.7	6.62 *s*	98.7
8a	/	156.6	/	156.8
1′	/	126.1	/	125.4
2'	7.45 *s*	108.9	7.35 *s*	109.4
3'	/	149.9	/	150.0
4'	/	137.6	/	138.0
5'	/	149.9	/	150.0
6'	7.45 *s*	108.9	7.22 *s*	109.4
OCH3	4.05 *s*	59.7	4.1 *s*	59.7
1"	5.3 *d* (8 Hz)	102.3	4.8	102.3
2"	3.62	75.9	3.65	76.2
3"	3.54	78.2	3.56	78.5
4"	3.68	72.1	3.6	72.2
5"	3.77	78.5	3.6	78.6
6"	3.93, 3.85	63.3	3.93, 3.78	63.4

There are two possible sites for addition of bisulfite on the flavylium on C-4 (fig. 1, **6**) or on C-2 (fig. 1, **5**). The expected NMR chemical shifts of the concerned carbon atoms would be as indicated in figure 5. Without doubt, the resonance of C-2 at δ_C 146.4 is only compatible with bisulfite addition at C-4.

We still need to discriminate between the two possible types of adducts resulting from bisulfite addition through its oxygen (leading to a sulfinate) or its sulfur (giving a sulfonate) (fig. 6).

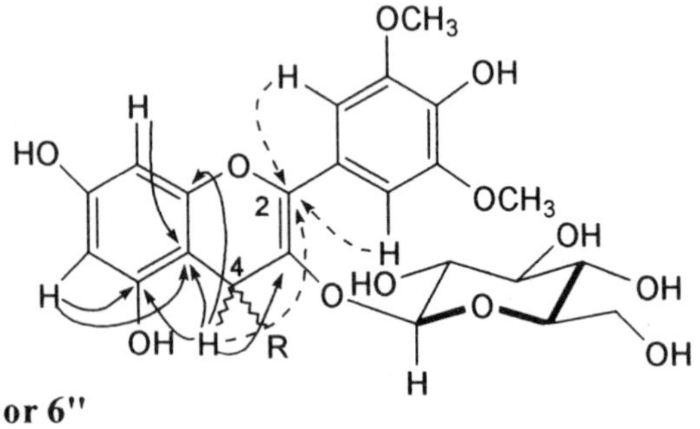

Figure 4. HMBC correlations used to assign C-2 (---) and C-4 (—).

6′ or 6″

R = SO$_3$Na ou OSO$_2$Na

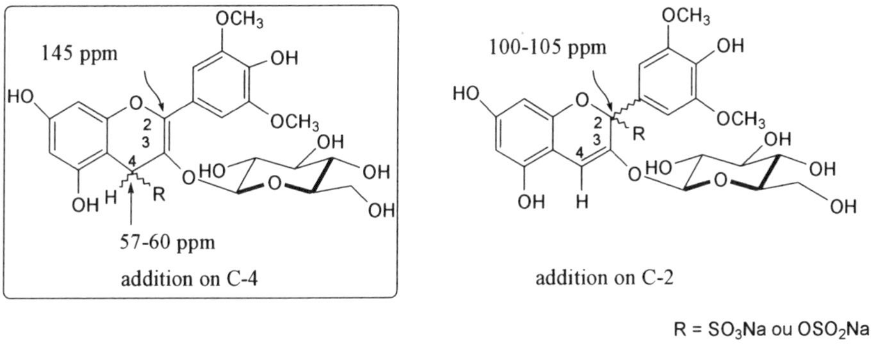

Figure 5. Bisulfite adducts on C-4 carbon and on C-2 carbon.

Due to the lack of examples of such derivatives in the literature, we could not deduce the nature of adducts from the measured ^{13}C NMR data of C-4 δ_C 56.9. Taking advantage of a newly developed analytical tool based on ^{33}S NMR,[24] we are able to discriminate between the two and thus to establish the complete structure of adducts. To measure NMR spectra of ^{33}S, we met many difficulties, because of its low magnetogyric ratio (2.0517×10^7 rad $T^{-1}s^{-1}$) and of the short relaxation times (due to its quadrupole moment, I = 3/2). However, the most important problem was related to the huge acoustic ringing effect,[25] leading to a severe baseline distortion after Fourier transformation. Special pulse sequences, based on increasing the preacquisition delay, are used to overcome such a phenomenon, but that can completely mask the sulfur signal. The best results were obtained with an adaptation of the "aring sequence"[a] that even proved to be successful at recording the S(IV) NMR signal. The line width of ^{33}S signal is a powerful analytical tool since it is linked to the sulfur atom environment. In symmetrical compounds

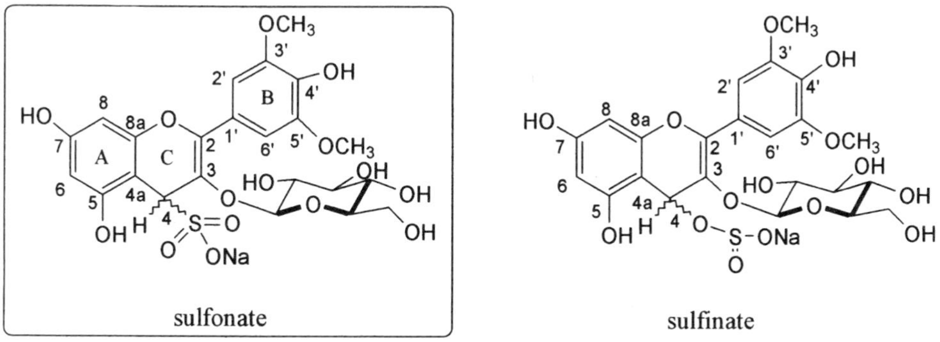

Figure 6. Bisulfite addition on **1** leading to sulfonate or sulfinate.

[a] From Bruker pulse program library: delay − 90°$_x$ − τ − 90°$_{-x}$ − τ − 90°$_x$ − (acquire). SFO$_1$ 38.387 MHz, 90° pulse length 18 μs, delay d1 50 ms, τ 13 μs, AQ 16 ms, SW 31250 Hz, NS 5315280, TD 1K, RG 4K, sulfolane as external reference

(sulfonates), it is narrow (as small as a tenth of a Hz), but in unsymmetrical compounds such as sulfinates lines broadening typically is as large as 1,000 Hz. In our case, the measured values of δ_S (−40, relative to sulfolane) and $W_{1/2} = 40$ Hz, clearly prove the sulfonate nature of the adducts **6′** and **6″** (fig. 6).

The nucleophilic addition of bisulfite to the C-4 carbon could occur on either its *si* or *re* face, leading to the observed diastereomers (fig. 7). Even though, the chiral centers of the glucose residue could have been a source of enantioselectivity, by a glucosyl-bisulfite intermediate formation prior to addition, the isomers are formed in equal amounts (from NMR data). No C-2 adduct previously suggested by Glories[26] could be correct. However, the C-4 sulfonate adducts correlate well with the former suggestions made by Jurd[6] and Brouillard.[8] In a similar case, nucleophilic addition of methoxide ion on diphenylpyrylium has been reported[27] to take place on the more accessible C-4 position because of its lower steric hindrance with respect to the C-2 position. This argument could also apply in our case. When the competition between water and bisulfite as nucleophile is possible, it seems very much to favor the latter, since only bisulfite-adducts are detected on NMR spectra. This observation is in good agreement with Brouillard's study.[8]

It is generally accepted that pigment properties of anthocyanins strongly resist bisulfite decolorization when the C-4 carbon bears a bulky substituent, as is the case in a 4-methyl or a 4-phenyl flavylium[28] (**11**). This idea should be reconsidered since the recent isolation of two new anthocyanin pigments[29] have very different coloring properties even though they both are 4-substituted derivatives. Vitisin A (**12**) is described[30] as highly resistant to decolorization, whereas vitisin B (**13**) is partly bleached by bisulfite (fig. 8).

In the course of this study on bisulfite reactivity toward wine constituents,[24] especially among the tested carbonyls, we observed that it is deeply dependent on

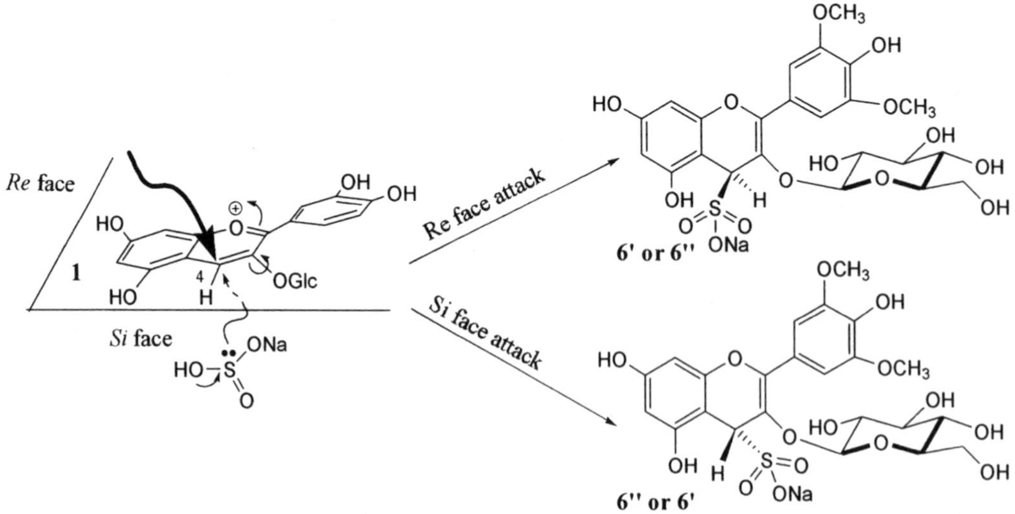

Figure 7. Mechanism of the diastereomer formation **6′** or **6″**.

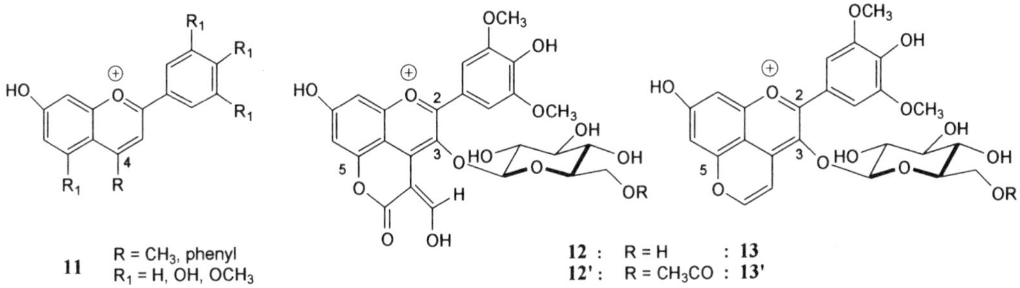

Figure 8. C-4 substituted anthocyanins **11**, **12** and **13**.

the size and hindrance of the substituents borne by the carbonyl. In the case of butanedione, for example, bisulfite could react only once, whereas glyoxal readily reacted in a 2 molar ratio. Bearing this in mind, the resistance of vitisine A (**12**) to bisulfite decolorization could be better explained by a first addition of bisulfite to the acrylic functionalities (fig. 9), under their non-enolized forms (brackets), either at C-1′ forming **12a**, or at C-3′ forming **12b**. This addition increases the steric hindrance at C-4, preventing any further addition at this center but, most importantly, leaving the flavylium chromophore. The partial decolorization of vitisine B (**13**) could result, even if it is substituted at C-4, from an addition of bisulfite to either of the accessible electrophilic carbons of the benzopyrilium system, i.e., C-4 or C-2′ (fig. 10). Whatever the site of this addition, the result is the destruction the flavylium chromophore.

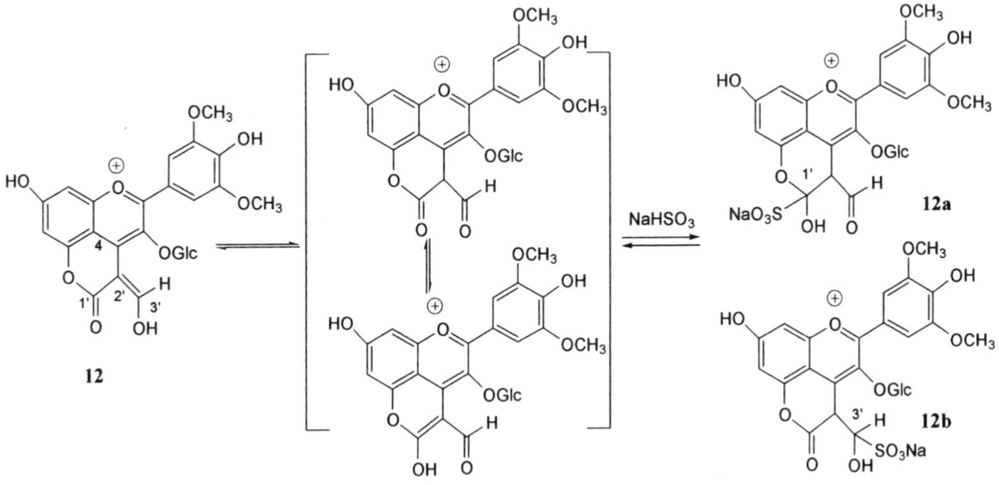

Figure 9. Proposed mechanism for vitisine A (**12**) complete resistance to decolorization.

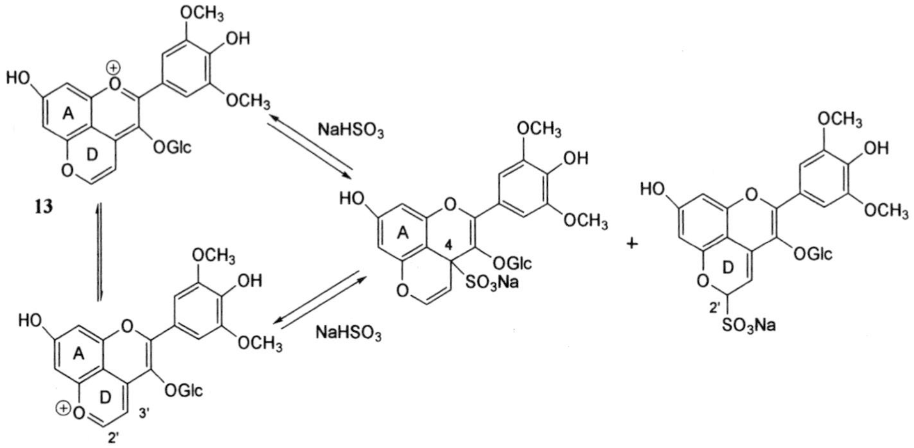

Figure 10. Proposed mechanism for vitisine B (**13**) partial resistance to decolorization.

4. CONCLUSIONS

We have shown that ^{33}S NMR is a valuable technique to prove the formation of sulfonate derivatives, rather than sulfinates. The complete framework of 4α- and 4β-sulfonate malvidin 3-*O*-glucoside adducts (**6′**) and (**6″**) was securely obtained via the ^{1}H and ^{13}C 1D and 2D NMR measurements. The increasing knowledge on the reactivity of bisulfite with electrophiles afforded by these studies allowed us to give a tentative explanation for the big differences in behavior between Vitisin B (**13**) and Vitisin A (**12**) with regard to their decolorization by bisulfite. It is thus possible to reach simple anthocyanin structure modifications allowing their use as food coloring additives, all the more as they show favorable health properties.[31]

ACKNOWLEDGMENTS

Financial support for this study was provided by the C.I.V.B. (Conseil Interprofessionnel du Vin de Bordeaux), which is gratefully acknowledged. The authors are very indebted to Pr. R. Brouillard for his great interest to this work.

REFERENCES

1. Brouillard, R.; Dubois, J-E. Mechanism of the structural transformations of anthocyanins in acidic media. *J. Am. Chem. Soc.* 99:1359 (1977).

2. Brouillard, R.; Delaporte, B. Chemistry of anthocyanin pigments. 2. Kinetic and thermodynamic study of proton transfer, hydration and tautomeric reactions of malvidin 3-glucoside. *J. Am. Chem. Soc.* 99:8461 (1977).

3. Genevois, M.L. The isomeric keto-enol forms of anthocyanidins and anthocyanins. *Bull. Soc. Chim. Biol.* 38:7 (1956).

4. Ribereau-Gayon, P. Recherche sur les anthocyanines des végétaux. Application au genre *Vitis*. Librairie générale de l'enseignement, Paris (1959).

5. Jurd, L. Reactions involved in sulfite bleaching of anthocyanins. *J. Food Sci.* 29:16 (1964).

6. Timberlake, C.F.; Bridle, P. Effect of substituents on the ionisation of flavylium salts and anthocyanins and their reactions with sulphur dioxide. *Chem. Ind.* 1965 (1966).

7. Timberlake, C.F.; Bridle, P. Flavylium salts, anthocyanidins and anthocyanins. II. Reactions with sulphur dioxide. *J. Sci. Food Agric.* 18:479 (1967).

8. Brouillard, R.; El Hage Chahine, J-M. Chemistry of anthocyanin pigments. 6. Kinetic and thermodynamic study of hydrogen sulfite addition to cyanin. Formation of a highly stable Meisenheimer-type adduct derived from a 2-phenylbenzopyrylium salt. *J. Am. Chem. Soc.* 102:5375 (1980).

9. Cheminat, A.; Brouillard, R. NMR investigation of 3-O-(β-D-glucosyl)malvidin. Structural transformations in aqueous solutions. *Tetrahedron Lett.* 27:4457 (1986).

10. Santos, H.; Turner, D.L.; Lima, J.C.; Figueiredo, P.; Pina, F.S.; Maçanita, A.L. Elucidation of the multiple equilibria of malvin in aqueous solution by one- and two-dimensional NMR. *Phytochemistry* 33:1227 (1993).

11. Pina, F.; Benedito, L.; Melo, M.J.; Parola, A.J.; Bernardo, A. Photochemistry of 3,4'-dimethoxy-7-hydroxyflavylium chloride. Photochromism and excited-state proton transfer. *J. Chem. Soc., Faraday Trans.* 92:1693 (1996).

12. Wulf, L.W.; Nagel, C.W. High-pressure liquid chromatographic separation of anthocyanins of *Vitis vinifera*. *Amer. J. Enol. Viticult.* 29:42 (1978).

13. Bakker, J.; Timberlake, C.F. The distribution of anthocyanins in grape skin extracts of port wine cultivars as determined by high performance liquid chromatography. *J. Sci. Food Agric.* 36:1315 (1985).

14. Andersen, Ø.M. Semipreparative isolation and structure determination of pelargonidin 3-O-α-L-rhamnosyl-(1-2-β-D-glucopyranoside and other anthocyanins from the tree *Dacrycarpus dacrydioides*. *Acta Chem. Scand., Ser B* 42:462 (1988).

15. Johansen, O.-P.; Andersen, Ø.M.; Nerdal, W.; Aksnes, D.W. Cyanidin 3-[6-(*p*-coumaroyl)-2-(xylosyl)-glucoside]-5-glucoside and other anthocyanins from fruits of *Sambucus canadensis*. *Phytochemistry* 30:4137 (1991).

16. Van Calsteren, M.-R.; Cormier, F.; Bao Do, C.; Laing, R.R. ^{1}H and ^{13}C NMR assignments of the major anthocyanins from *Vitis vinifera* cell suspension culture. *Spectroscopy* 9:1 (1991).

17. Aue, W.P.; Bartholdi, E.; Ernst, R.R. Two-dimensional spectroscopy. Application to nuclear magnetic resonance. *J. Chem. Phys.* 64:2229 (1976).

18. Pedersen, A.T.; Andersen, Ø.M. Difficult" ^{1}H and ^{13}C NMR signals in structure elucidation of anthocyanins. *In*: Brouillard, R., Jay, M., Scalbert A. (eds.). Polyphenols 94, les colloques n° 69. INRA edition, Paris, p. 429 (1995).

19. Bax, A.; Summers, M.F. ^{1}H and ^{13}C assignments from sensitivy-enhanced detection of heteronuclear multiple-bond connectivity by 2D multiple quantum NMR. *J. Am. Chem. Soc.* 108:2093 (1986).

20. Bax, A.; Davis, D.G. Practical aspects of two-dimensional transverse NOE spectroscopy. *J. Magn. Reson.* 63:207 (1985).

21. Bax, A.; Subramanian, S. Sensitivity-Enhanced two-dimensional heteronuclear shift correlation NMR spectroscopy. *J. Magn. Reson.* 67:565 (1986).

22. Berké, B.; Chèze, C.; Vercauteren, J.; Deffieux, G. Bisulfite addition to anthocyanins: revisited structures of colourless adducts. *Tetrahedron Lett.* 39: 5771 (1998).

23. Le Coq, C.; Lallemand, J.Y. Precise carbon-13 NMR multiplicity determination. *J. Chem. Soc., Chem. Commun.* 150 (1981).

24. Berké, B. Le devenir du dioxyde de soufre dans les vins de Bordeaux. Ph.D. Thesis, University of Bordeaux 2 (1998).

25. Gerothanassis, I.P. Methods of avoiding the effects of acoustic ringing in pulsed Fourier transform nuclear magnetic resonance spectroscopy. *Progr. Nucl. Magn. Reson. Spectrosc.* 19:267 (1987).

26. Glories, Y. La couleur des vins rouges. 1re partie. Les équilibres des anthocyanes et des tanins. Connaiss. *Vigne Vin* 18:195 (1984).

27. Bersani, S.; Doddi, G.; Fornarini, S.; Stegel, F. Reactions of heteroaromatic cations with nucleophilic reagents. Addition of methoxide ion to 2,6-diphenyl- and 4-methoxy-2,6-diphenylpyrylium cations. *J. Org. Chem.* 43:4112 (1978).

28. Timberlake, C.F.; Bridle, P. Flavylium salts resistant to sulphur dioxide. *Chem. Ind.* 1489 (1968).

29. Bakker, J.; Bridle, P.; Honda, T.; Kuwano H., Saito, N.; Terahara, N.; Timberlake, C.F. Identification of an anthocyanin occurring in some red wines. *Phytochemistry* 44:1375 (1997)

30. Bakker, J.; Timberlake, C.F. Isolation, identification, and characterization of new color-stable anthocyanins occurring in some red wines. *J. Agric. Food Chem.* 45:35 (1997).

31. Rechkemmer, G.; Pool-Zobel, B.L. Health aspects of anthocyanins and anthocyanidins. *Pflanz. Nahrungsm.* 31:219 (1996).

AN ENVIRONMENTAL ASSESSMENT OF VEGETABLE TANNERY BIOSOLIDS AS AN ORGANIC FERTILIZER AND SOIL CONDITIONER

Thomas E. Herlihy

Gannett Fleming
800 Leonard Street
Clearfield, Pennsylvania 16833
USA

1. INTRODUCTION

Driven by global competition, environmental regulations, and a strong sense of stewardship, three of the five remaining vegetable tanneries in the United States sponsored a longterm unique solid waste cooperative research and development project. Vegetable leather tanneries cure raw hides with plant tannins (predominantly wattle, quebracho, and chestnut bark extracts) rather than with the more common and potentially environmentally harsher chrome-based chemicals. This ongoing project has an end goal of a "near zero" tannery waste stream.

Vegetable tanning is an intensive water-use industry; raw cattle hides must be cleaned, rehydrated, dehaired, soaked in tanning baths, rinsed, and rewetted. On average, 240 gallons of water are used for each complete hide tanned. Treating this large volume of wastewater to meet strict regulatory limits has forced the participating tanneries to construct in-house wastewater treatment plants (WWTP) to remove the biological loading prior to direct discharge. A byproduct of the WWTP remediation is the generation of approximately 20 pounds per hide of a biological sludge known as biosolids. Biosolids are traditionally thought of as waste products, requiring disposal in whatever method is the most efficient and economical.

The three Pennsylvania-based vegetable tanneries, however, joined together on a 4-year research and development program to find a new innovative solution to their common biosolids problem. Economic and environmental regulatory pressure was mounting on the tanneries to commit to a longterm, regulatory-oriented, biosolids management program. As an example, at the start of the project, $82/ton was the cheapest waste disposal option for the region near the largest remaining tannery (landfill tipping fee + trucking + regulatory compliance), which, coupled

Plant Polyphenols 2: Chemistry, Biology, Pharmacology, Ecology, Edited by
Gross et al. Kluwer Academic / Plenum Publishers, New York, 1999

with their biosolids' production of 12,000 tons/year, resulted in a nearly $1,000,000 per year disposal choice.

Given the organic, nutrient-rich composition of the material, it was quickly determined that the most resource-efficient and economical option was to have the biosolids "delisted" as a regulated solid waste and sold as an organic fertilizer/soil conditioner. To become delisted by the appropriate agencies, the biosolids had to be proven to "pose no greater threat of harm to human health or the environment than commonly recognized materials used for similar purpose". Using this as criteria, data were gathered for cautious state regulators (mindful of their responsibilities) on the environmental effect of land-applied vegetable tannery biosolids. Once delisted, the material would be changed from a regulated waste product to a registered fertilizer and coproduct of the vegetable tanning process.

The objective of the research portion of this project was, therefore, to determine whether the wastewater biosolids from the three vegetable leather tanneries were as environmentally safe as traditional fertilizers and soil conditioners. To meet this objective, extensive data were gathered on the effects of biosolids on plant, soil, surface, and groundwater resources. Empirical data were collected during a $3^1/_2$-year series of laboratory and field, plant growth, water quality, and soil chemistry experiments. In a greenhouse, over 2,000 samples of 13 plant species were cultivated in tannery biosolids with measurements taken of plant growth, soil chemistry, and plant tissue analysis. Additional research apparatus was constructed to determine the effects of biosolids on surface water runoff and groundwater recharge quality. All experiments compared the biosolids to control groups of commercial fertilizers and an untreated soil.

The initial assumption was that land application of biosolids would increase the organic matter content of the receiving soil. Research has shown that as the organic content of a soil is raised, it will have increased water-holding capacity, improved tilth (porosity), greater nutrient-holding potential, and a larger buffering capacity.[1] Increased availability of soil micronutrients/minerals, increased activity of soil microorganisms, and improved longterm fertility of the soil are also reported benefits from the addition of organic matter such as the biosolids.[2]

2. MATERIALS AND METHODS

Depending on the tannery, the biosolids are either the product of activated sludge, aerobic wastewater treatment plants, or simply the settleable solids removed prior to a Sequencing Batch Reactor (SBR). The activated sludges are dewatered in a plate and frame filter press, and the settleable solids are removed by centrifuging. Biosolids that come from the filter presses emerge as solid, non-greasy, reddish-brown or black cakes, bearing the impression of the filter plate, whereas the centrifuged biosolids have the consistency of moist putty and a light brown appearance. The biosolids are composed primarily of carbon, calcium, phosphorous, and nitrogen, which originated from waste animal proteins, lime from the dehairing process, and the tree bark extracts (tannin) used in the curing of raw hides. The nitrogen in the biosolids is in part from the proteins in the raw hide, hair, and leather solids that have been dissolved and precipitated out by the tannin during the tanning process. Biosolids are an organic-rich material with an exceptionally low heavy metal content; see table 1 for an elemental analysis.

All data referring to plant biomass are mean values calculated from samples receiving the same treatment and are expressed in dry mass, grams. See figure 1, for a representative schematic of a sample pot in which all plants were cultivated. All treatments were replicated five times, and plant biomass samples were taken flush with the soil surface (no root mass) and dried 24 hours at 103 °C ± 6°. After drying, the samples were removed from the oven and weighed immediately in accordance with accepted standards.[3] All empirical data were statistically analyzed using standard analysis of variance (ANOVA), and Fishers' protected least significant difference (lsd) at the 95 percent confidence level.

Surface runoff water-quality samples were produced using a rainfall simulator, which generated a 17 cm/hr rain event to start overland flow on 33 runoff trays;

Table 1. Elemental analysis of vegetable tannery biosolids

Parameter	Units	Facility 1	Facility 2	Facility 3
Nitrogen	mg/kg	12,800	36,500	18,000
Phosphorous	mg/kg	8,500	28,400	14,100
Potassium	mg/kg	300	700	300
pH	std.	8.8	7.9	12.3
Carbon	mg/kg	345,900	231,300	265,500
Magnesium	mg/kg	9,300	1,700	26,200
Calcium	mg/kg	196,100	128,700	208,900
Sodium	mg/kg	4,600	4,600	7,700
Iron	mg/kg	2,400	1,500	1,400
Aluminum	mg/kg	17,000	8,700	1,200
Manganese	mg/kg	334.8	282.4	47.1
Cadmium	mg/kg	0.62	0.5	0.63
Chromium	mg/kg	6.2	44.1	9.4
Copper	mg/kg	37.0	27.7	26.7
Lead	mg/kg	6.2	2.52	3.1
Mercury	mg/kg	0.09	0.05	0.07
Nickel	mg/kg	4.6	2.5	1.6
Zinc	mg/kg	91.0	133.6	36.1

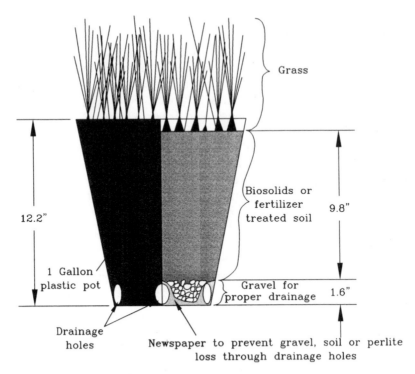

Figure 1. Diagram of typical turfgrass sample pot.

see figure 2, for schematic of the collection apparatus. The runoff from untreated soil, soil amended with biosolids, and soil amended with fertilizer/lime was analyzed for seven parameters: total nitrogen, total dissolved nitrogen, total phosphorous, total dissolved phosphorous, chromium, cadmium, and nickel. Samples were collected twice during the project, on day 25 of the growing season, and at the end of 13 weeks (day 90). The runoff trays were (83 × 33 × 15 cm) and cultivated with perennial ryegrass on a slope of 13 percent throughout this experiment. Each company's biosolids were in nine trays in three replications of three treatments at approximately 25, 50, and 100 tons/acre, based on the material's nitrogen and solids content. The control groups in this study were three replications of a fertilizer/lime treatment (150 lb/ac nitrogen, 350 lb/ac phosphate, 260 lb acre potash [potassium], and 7,500 lb/acre of lime[CaCO$_3$ equivalent]) and three replications of an untreated soil.

Data on the effect of land applied biosolids on the groundwater recharge were developed using soil columns, which simulated infiltration underneath soil amended with biosolids. The same treatments and number of samples (33) used in the runoff portion of this project were repeated in this groundwater study. Figure 3 is a schematic of the soil columns used in this study, which were 36″ long and made from 4″ diameter PVC pipe. A perforated end cap held the soil in place, and the 24″ under the treatment layer was packed in 2-inch lifts, to approximately the same

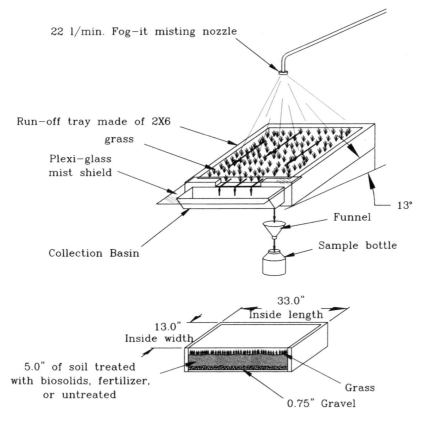

Figure 2. Schematic of runoff collection apparatus.

moist bulk density as the parent soil. ($\sim$1.25 gm/cm^3). Infiltrated water was not collected from the perforations near the edge of the end cap so that water that might have "short circuited" the infiltration process (soil) and seeped only down the soil-pipe interface would be ignored. The infiltrated water was collected and analyzed for six parameters: total dissolved nitrogen, nitrate nitrogen, total dissolved phosphorous, chromium, cadmium, and nickel. Infiltrate samples were collected the same 2 days as the runoff samples (day 25 and 90). The University of Maine's Environmental Chemistry laboratory analyzed all water-quality samples.

The biosolids effect on soil chemistry was determined from the analysis of 48 samples, taken from four different treatments (12 replications): untreated soil, soil amended with fertilizer/lime, and soil amended with biosolids at 30 and 80 tons/acre. Both application rates exceed the normal 20 t/ac used in practice but were tested here as worst case scenarios. Soil samples were analyzed for pH, organic matter, phosphorous, potassium, magnesium, calcium, manganese, iron, copper, zinc, sodium, lead, chromium, cadmium, and nickel, by the Agricultural Analytical Services Laboratory (AASL) of Pennsylvania State University, State College, Pennsylvania.

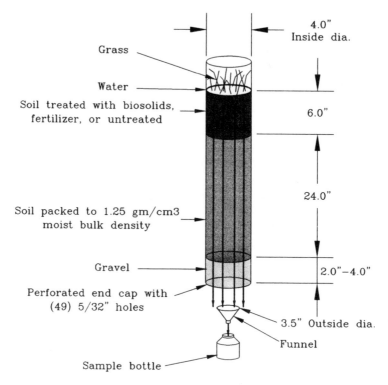

Figure 3. Schematic of infiltration apparatus.

The elemental response of plant tissue to cultivation in biosolids-amended soil was measured using 39 samples of bean and corn. The analysis was designed to detect nutrient imbalances or accumulation of heavy metals. Tissue samples were collected from plants cultivated in 100 percent biosolids, soil amended with biosolids at 80 tons/acre, untreated soil, soil amended with fertilizer/lime, and potting soil. Plant tissue specimens were analyzed by the AASL for: nitrogen, phosphorous, potassium, calcium, magnesium, manganese, iron, copper, boron, aluminum, zinc, cadmium, nickel, lead, and chromium. The accepted ranges for plant tissue parameters were taken from The Pennsylvania State University 1993–1994 Agronomy Guide.[4]

3. RESULTS AND DISCUSSION

Plant growth data for the 13 species cultivated will not be presented here, but results were similar in nature to those for three species of turfgrass mentioned in the discussion that follows. There was no significant difference in bluegrass biomass produced between samples that received biosolids, treatments 1–3, and samples that received traditional fertilizer/lime, treatment 4, table 2. Compared to the untreated soil (control), however, the biosolids and fertilizer treatments did significantly increase the amount of biomass produced. The biosolids and fertilizer treatments increased plant biomass over untreated soil by 300 percent to 400 percent. In these first trials, fertilizers were added to the biosolids to compensate for the material's potassium (K) deficiency and a possible initial nitrogen deficit from the bacteria's metabolic demand to start decomposition. Surprisingly, considering that potassium is a macronutrient, the fertilizer additions did not significantly affect the amount of biomass produced. Fertilizer additions to biosolids were discontinued in later trials. Eighty tons of biosolids contains ~800 pounds of total nitrogen, and through testing, it is believed to be equivalent to 120 lbs of mineral nitrogen (tests indicated the organic nitrogen is between 10–15 percent available the first year).

The fine fescue samples cultivated in the traditional fertilizer treatment (150, 350, 270, and 7,500 lbs/acre of N, P, K and lime, respectively) produced significantly more biomass than samples cultivated in the three biosolids treatments and the untreated soil treatment (table 2). In comparison to untreated soil, the three biosolids treatments' significantly increased the resulting biomass (100 percent yield increase).

There was no significant difference in the mass of ryegrass produced between samples that received biosolids, treatments 1–3, and samples that received traditional fertilizer/lime, treatment 4 (table 2). In comparison to the untreated soil (control group), the biosolids and fertilizer treatments significantly increased plant biomass (75–100 percent yield increase). In this trial, as with the other turfgrass species, there was again no significant difference between samples cultivated in biosolids alone and biosolids with a starter 30 lb/ac of nitrogen (urea 46-0-0).

Table 2. Turfgrass trials of plant biomass production analysis (shoot only)

Treatment #	Description	Bluegrass, lsd = 2.7		Fine Fescue, lsd = 1.4		Ryegrass, lsd = 1.3	
		Biomass (gm)	Rank	Biomass (gm)	Rank	Biomass (gm)	Rank
1	80 tons/acre of biosolids w/K fertilizer at 260 lbs/acre .	6.4	1	8.7	2	9.1	1
2	80 tons/acre of biosolids w/30 lbs/acre of starter urea nitrogen	5.4	3	8.4	3	8.7	2
3	80 tons/acre biosolids	4.4	4	8.0	4	8.0	4
4	Fertilizer N-P-K and lime at 150, 350, 270 and 7,500 lbs/acre respectively.	6.3	2	10.3	1	8.6	3
5	Untreated soil	1.4	5	4.2	5	4.6	5

Two reasons are proposed to explain the lack of significant differences: (1) There may not be an initial nitrogen demand for tannery biosolids as there is for other sludges, or compost; (2) the physical characteristics of the biosolids compensate for any chemical characteristic deficiency—the water-holding capacity and tilth of the biosolids-amended soil (physical characteristics) provide such a good medium for plant germination and growth that the initial nitrogen deficiency is overshadowed.

Nonpoint-source pollution from the land application of biosolids or fertilizers usually takes the form of excess nitrogen, potassium, and phosphorous being carried in the runoff from fields and stimulating the eutrophication of the nearby bodies of water. The biosolids do not contain any appreciable amount of potassium, so only nitrogen and phosphorous were examined in this study. Runoff from biosolids-amended soil was consistently lower in nitrogen and phosphorous than the runoff from fertilizer-amended soil (see tables 3 and 4). Less nitrogen was released into the runoff from biosolids-amended soil containing 1,000 lbs/acre of total nitrogen (over 95 percent organic) than from fertilizer-amended soil treated with only 150 lbs/acre of total nitrogen. The same was true for the phosphorous where concentrations were lower in the runoff from biosolids-amended soil containing 1,800 lbs/acre of total phosphorous than from fertilizer-amended soil treated with 350 lbs/acre of total phosphorous.

The runoff water quality data for the biosolids also exhibited the expected trend of having lower concentrations of nitrogen and phosphorous with the lower application rates. The consistent runoff nitrogen levels from day 25 to 90 strongly indicate a slow and steady nutrient release, whereas the runoff from the fertilizer-amended soil declined over 70 percent from a large initial slug. These data support the hypothesis that the physical and chemical characteristics of the biosolids make it a longterm, slow-release form of nitrogen (compared to traditional fertilizers). This pattern of release was repeated even more strongly with the dissolved phosphorous, but the trend did not have as strong a correlation in total phosphorous (table 4). It is hypothesized that the nitrogen in the biosolids is almost exclusively released during the warm growing season because it is during this time that the

Table 3. Analysis of nitrogen content in runoff samples

Treatment	Initial nitrogen applied (lbs/acre)	Total nitrogen day 25 (mg/l)	Total nitrogen day 90 (mg/l)	Dissolved nitrogen day 25 (mg/l)	Dissolved nitrogen day 90 (mg/l)
Fertilizer and lime	150	6.26	2.42	2.95	0.91
Biosolids ~ 100 t/ac	1000	1.48	1.57	0.91	0.54
Biosolids ~ 50 t/ac	500	1.38	1.71	0.82	0.61
Biosolids ~ 25 t/ac	250	0.95	1.28	0.59	0.62
Untreated soil	0	1.63	1.12	0.52	0.56

Table 4. Analysis of phosphorous content in runoff samples

Treatment	Initial P applied (lbs/acre)	Total P day 25 (mg/l)	Total P day 90 (mg/l)	Dissolved P day 25 (mg/l)	Dissolved P day 90 (mg/l)
Fertilizer and lime	350	4.84	0.90	2.84	0.05
Biosolids ~ 100 t/ac	1800	0.39	0.14	0.09	0.08
Biosolids ~ 50 t/ac	900	0.36	0.13	0.04	0.03
Biosolids ~ 25 t/ac	450	0.15	0.04	0.01	0.02
Untreated soil	0	1.19	0.04	0.02	0.02

organic nitrogen in the biosolids is converted to plant usable nitrate and ammonia by microbial decomposition. This temperature correlation would limit nutrient runoff and leachate during the critical early spring period.

The analysis of runoff water quality from untreated soil, soil amended with fertilizer/lime, and biosolids-amended soil revealed very low levels of cadmium, chromium, and nickel. Cadmium levels were below the detection limit in all samples, and nickel was found in only trace amounts (<0.02 mg/L) from the traditionally fertilized trays at day 25. Chromium was detected in small amounts (0.012 mg/L) in the runoff of traditionally fertilized trays sampled on day 25 and day 90. Chromium was also detected in a lesser amount (0.009 mg/L) in the runoff from the soil amended with biosolids at 100 tons/acre tray in the day 90 sample.

The quality of groundwater recharge also revealed that the biosolids are an environmentally friendly, low-impacting, source of plant nutrients. Again, less nitrogen leached from the 1,000 lbs/acre of total nitrogen biosolids application than from the fertilizer/lime application of only 150 lbs/acre (table 5). The levels of

Table 5. Analysis of nitrogen content in the infiltrate samples

Treatment	Initial nitrogen applied (lbs/acre)	Dissolved nitrogen day 25 (mg/l)	Dissolved nitrogen day 90 (mg/l)	Nitrate nitrogen day 25 (mg/l)	Nitrate nitrogen day 90 (mg/l)
Fertilizer and lime	150	13.40	7.63	9.87	6.83
Biosolids ~ 100 t/ac	1000	3.90	5.68	3.89	6.05
Biosolids ~ 50 t/ac	500	3.54	2.31	2.98	1.85
Biosolids ~ 25 t/ac	250	5.18	1.08	4.93	0.65
Untreated soil	0	8.18	5.56	6.76	1.46

dissolved phosphorous entering the groundwater were equivalent to those from biosolids-amended soil containing 1,800 lbs/acre, 900 lbs/acre, and 450 lbs/acre of total phosphorous and from fertilizer-amended soil containing only 350 lbs/acre of total phosphorous. An application of biosolids at 100 tons/acre was taken as an extreme worst case scenario, and it contains 850 lbs/acre more nitrogen and 1,450 lbs/acre more phosphorous than an application of traditional fertilizer, yet no greater nutrient loading is experienced by the surrounding water environment. The data coupled with the runoff information support the hypothesis that the nutrients contained in the biosolids are less mobile and must therefore be available for building longterm soil fertility. This strong conclusion was well received and accepted by the environmental and agricultural regulatory agencies.

The initial high concentration of nitrogen from the untreated soil samples was judged to be due to the new exposure of the existing soil minerals. The soil used to pack the columns was extensively handled and sieved. Therefore, new "faces" of the soil peds were exposed to weathering for the first time inside the soil columns. It was also judged that the biosolids-amended soils had lower concentrations of nitrogen than untreated soil due to the organic matter in the biosolids absorbing nitrogen from the passing infiltrate water.

The cadmium, chromium, and nickel content of the biosolids infiltrate samples were as low or lower than the fertilizer infiltrate samples with one exception. The cadmium contents were equivalent in the 100 tons/acre biosolids sample (0.0022 mg/L) and in the fertilizer/lime infiltrate sample (0.0020 mg/L) collected 90 days after the treatments were applied. Infiltrate from the fertilizer/lime treated soil also had an initial peak of 0.04 mg/l of nickel before declining to 0.018 mg/L by day 90.

Normal ranges for the elemental composition of bean and corn tissue were taken from the Pennsylvania State University 1993–1994 Agronomy Guide. The guide states that the normal ranges are a "sufficiency range," and "observed ranges for some elements can be up to 10 times greater with little impact on plant yield." Table 6 reports the tissue test for bean samples cultivated in untreated soil, soil treated with 80 tons/acre of biosolids, soil treated with lime and fertilizer, and potting soil. All the test parameters fall within or nearly within the accepted normal range. There was little difference in the elemental composition between the samples grown in biosolids and fertilizer, except for the concentration of potassium. Potassium is the one macronutrient that the biosolids do not have in appreciable amounts, so more research is needed to learn why there is a larger potassium content in the plants cultivated in biosolids-amended soil.

Visual observations and qualitative analysis of over 2,000 greenhouse samples and over 1,000 acres of biosolids-amended crops have revealed no discernible macro-, or micro-nutrient imbalances.

Soil analyses measured changes in: pH, soil nutrients, organic matter, and metal accumulations. The lime component of the biosolids acted as a buffer significantly raising the soil pH, and the biosolids applied at 80 dry tons/acre significantly increased the amount of organic matter in the soil at the end of the growing season (table 7). The biosolids significantly raised the soil phosphorous and potash levels, and no significant accumulations of metals were found in the soil (table 8).

Table 6. Elemental tissue analysis of bean samples

Parameter	Unit	Normal range	Biosolids	Fertilizer and lime	Untreated soil	Potting soil	Fisher's lsd $P \leq 0.05$
Nitrogen	(%)	4.25-6.00	3.74	3.65	2.13	2.08	0.88
Phosphorous	(%)	0.25-0.75	0.27	0.12	0.14	0.45	0.10
Potassium	(%)	2.25-4.00	1.58	0.66	1.25	2.45	0.44
Calcium	(%)	1.50-2.50	3.05	1.97	0.92	1.77	0.28
Magnesium	(%)	0.30-1.00	0.52	1.55	0.19	0.69	0.13
Manganese	ppm	50-300	242	134	598	310	123
Iron	ppm	50-300	77	80	71	210	74
Copper	ppm	7-31	10	6	9	10	2.9
Boron	ppm	20-775	36	9	25	20	6.2
Aluminum	ppm	10-200	35	41	142	20	30.3
Zinc	ppm	20-200	26	13	65	32	9.2
Cadmium	ppm	0.1-2.4	0.05	0.05	0.27	0.04	0.05
Nickel	ppm	0.02-5.0	0.59	0.31	7.13	2.05	1.3
Lead	ppm	0.20-20	0.25	0.25	0.25	0.25	0.0
Chromium	ppm	0.03-5.0	0.17	0.13	0.13	0.26	0.08

% = Percent of Dry Mass   ppm = Parts Per Million

Table 7. Soil conditioning parameters, at the end of 11 weeks

Treatment #	Description	pH (lsd = 0.60) pH (std)	Rank	Organic matter (lsd = 0.3) OM (%)	Rank
1	Untreated soil.	5.0	4	2.9	3
2	Soil With 30 t/ac biosolids	6.6	3	3.0	2
3	Soil With 80 t/ac biosolids	7.1	1	3.6	1
4	Fertilizer and lime	6.8	2	2.4	4

Table 8. Soil chemistry extractable analysis

Parameter	Units	Biosolids 30 t/ac	Biosolids 80 t/ac	Fertilizer and lime	Untreated soil	lsd
Phosphorous	lbs/ac	27.0	35.0	19.0	22.3	8.1
Potassium	lbs/ac	223.6	222.3	111.8	75.4	71.0
Magnesium	lbs/ac	176.0	252.0	896.0	64.0	404.4
Calcium	lbs/ac	2371	4142	2033	361	934
Manganese	lbs/ac	47.7	40.1	33.8	114.3	25.0
Iron	lbs/ac	44.3	35.4	39.6	74.7	13.0
Copper	lbs/ac	0.8	1.0	0.8	0.8	0.23
Zinc	lbs/ac	2.3	3.2	1.8	3.1	1.4
Sodium	lbs/ac	160.4	371.0	19.3	50.0	15.1
Lead	lbs/ac	2.5	2.3	2.6	2.1	0.7
Chromium	lbs/ac	28.5	28.5	29.3	35.7	7.0
Nickel	lbs/ac	0.4	0.2	0.5	1.5	0.4
Cadmium	lbs/ac	0.08	0.08	0.08	0.16	0.0

4. CONCLUSIONS

A 4-year research project sponsored by three cooperating vegetable tanneries assessed the environmental impact of land applied tannery wastewater biosolids on soil chemistry, surface water runoff, groundwater recharge, and plant tissue composition. Research has shown the biosolids to be an environmentally beneficial, low-impacting source of plant nutrients and a very effective soil conditioner/plant growth supplement. The organic macronutrients (nitrogen and phosphorous) in the biosolids were found to be less mobile and more slowly released than traditional mineral fertilizers, giving the material a lower nonpoint-source pollution potential. Results of this project were used to convince appropriate environmental and agricultural regulatory agencies that the biosolids were not a waste product but rather a valuable and useful "coproduct" of the manufacturing (tanning) process. Each tanneries' biosolids have been "delisted" as a regulated waste, and the materials are now licensed organic fertilizers.

The yields of plant biomass and the health ratings of samples treated with biosolids were always significantly higher than the untreated soil control samples. In most cases, the biosolids also proved to be as effective as or more effective than commercial fertilizers/soil conditioners in increasing soil fertility and supporting healthy plant growth.

The elemental analysis of the plant tissue samples indicated no increased uptake of metals between the plants cultivated in biosolids-amended soil (including 100 percent biosolids) and plants cultivated in untreated soil and soil amended with traditional fertilizer/lime. Measured plant tissue parameters were considered within an acceptable "sufficiency" range with no phytotoxic levels observed or recorded for any of the elements tested. This quantitative data coupled with the numerous qualitative plant health observations led to the regulatory conclusion that the biosolids "pose no greater threat to plant health than other similarly used products."

Soil chemistry analyses of tannery biosolids-amended soil revealed raised nutrient levels, pH, and organic matter content. Soil analyses also confirmed no increased metal accumulations between the biosolids-amended soil and untreated soil/soil amended with traditional fertilizer/lime. Biosolids-amended soils by definition have a larger organic matter content, and therefore, increased water-holding capacity, tilth (porosity), nutrient-holding capacity, buffering capacity, availability of soil minerals, activity of soil microorganisms, and both short- and longterm improvement in soil fertility.

The water quality assessment showed that vegetable tannery biosolids were a clean, low-impacting source of plant nutrients and had a lower nonpoint-source pollution potential than commercial fertilizers. Nitrogen and phosphorous contents of the surface runoff and groundwater recharge from biosolids-amended soils were less than that of the fertilizer-amended soil, even in the extreme case where the total nitrogen application rate was nearly seven times greater for the biosolids than for the fertilizer (1,000 lbs/acre biosolids-N vs. 150 lbs/acre fertilizer-N).

The results of this research project and subsequent delisting of the biosolids as a waste product have:

1. Provided a longterm environmental solution for utilization of 15,000 tons/year of biosolids that will not have to enter limited landfill space.

2. Reduced operating costs. The tanneries' save over $1.1 million dollars each and every year in disposal costs that are now used for capital improvement projects.

3. Placed the tanneries, who are large regional employers, in a much firmer financial position. As an example, the largest tannery employs over 200 in a town of 1,300 representing over $4,000,000 in direct salaries alone to the local economy and tax structure.

4. Provided area farmers and mine reclamation companies a valuable source of plant nutrients and soil pH adjustment at a greatly reduced price. In a typical sale, a farmer/reclaimer receives material worth approximately $170.00 per acre in fertilizer and lime equivalent for $20.00 per acre.

5. Promoted the agricultural and mine reclamation use of tannery biosolids. This is, in essence, recycling vegetable extracts and animal proteins back into the soil from which they came, closing the loop.

REFERENCES

1. Kuhlman, Lester R. Value of composting cattle feedlot manure. Australian Lot Feeders Association Conference (BEEFEX 92) Coffs Harbour, New South Wales, Australia (1992).

2. Stevens, F.J. Organic matter-micronutrient reactions in soil. *In*: Luxmoore, R.J. (ed.) Micronutrients in agriculture. Soil Science Society of America (1991).

3. American Society of Agricultural Engineers Standards 1993. Standard Engineering Practices Data, ASAE St. Joseph MI (1993).

4. Pennsylvania State University Agronomy Guide, 1993–1994, University Park, PA (1994).

TANNIN AUTOCONDENSATION AND POLYCONDENSATION FOR ZERO EMISSION TANNIN WOOD ADHESIVES

Antonio Pizzi

ENSTIB
University of Nancy 1
Epinal
FRANCE

1. INTRODUCTION

Recently, the ionic and radical mechanisms of the reaction of autocondensation to hardened resins of polyflavonoid tannins induced by weak Lewis acids, in particular silicic acid, but also others have been described.[1-7] The relative balance of their contribution to hardening of the two types of mechanism differs according to the type of tannin used—with the difference in structure of the tannin repeating unit, with the degree of polymerization of the tannin, and particularly with the level of colloidal state of the tannin solution.[5-7] Autocondensation of flavonoid tannins by cellulose-induced catalysis has also been identified and described for both ionic and radical mechanisms.[3,7]

The application of such a reaction of autocondensation for wood adhesives based on tannins hardened without the use of an aldehyde has also been examined.[8] The results of this latter investigation revealed that, notwithstanding the differences in structure and behavior of different polyflavonoid tannins, effective wood particleboard bonding could be obtained based just on the autocondensation reaction of tannins. The results showed, however, that tannins hardened in this manner only yielded bonds of interior-grade quality, whereas the reaction of tannins as phenolic materials with formaldehyde has always traditionally yielded weather- and boil-proof networks, hence exterior-grade bonds.[8,9] The reasons for such results have mainly been ascribed to the low crosslinking density of tannin networks obtained exclusively through their reaction of autocondensation.[1,8]

The autocondensation reaction of tannins to yield resins, however, still holds great interest as bonded wood products that do not emit formaldehyde (as none has been added) can be produced.[8,10] To examine whether exterior-grade hardened tannin networks with no (or very low) formaldehyde emission properties can be obtained, it is necessary to determine how reduced amounts of traditional coreac-

Plant Polyphenols 2: Chemistry, Biology, Pharmacology, Ecology, Edited by
Gross et al. Kluwer Academic / Plenum Publishers, New York, 1999

tants and hardeners for polyflavonoid tannins, which react by two electrons, ionic mechanisms to form polycondensates, could be used together with the mainly radical autocondensation reaction.

The influence and effects caused on polyflavonoid tannins radical autocondensation by the presence of the most common tannin resins coreactants and hardeners such as paraformaldehyde, formalin, urea, paraformaldehyde + urea, hexamethylenetetramine (hexamine), furfuryl alcohol, and polymeric 4,4′-diphenylmethane diisocyanate (MDI) were investigated both by ESR and TMA studies of the networks formed.[1-9]

2. RESULTS AND DISCUSSION

The work described here follows from the chance finding in 1993 that SiO_2 and silicates dissolved in sodium hydroxide solution when added in small percentages to a concentrated tannin solution cause rapid hardening in alkaline and also in neutral and very acid environment if silica is predissolved. This effect can be described by a schematic gel time graph as a function of pH of the tannin solution when 3 percent silica is added (fig. 1).[1]

Solid state ^{13}C and ^{29}Si NMR spectra and X-ray diffraction studies have shown that part of the Si is transitorily in coordination No. 5, and that in the final hardened mass, some of the Si conserves only some reflections of its crystalline structure.[1] The ^{29}Si NMR spectrum in particular indicated that only Si bound to an

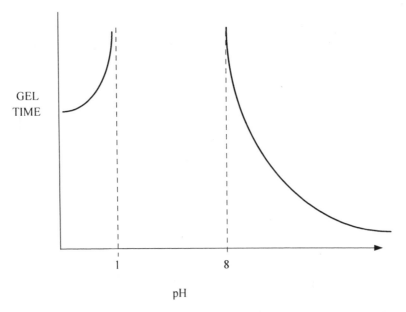

Figure 1. Schematic 100 °C gel time trend as a function of pH of a concentrated (40 percent—50 percent) tannin solution to which silica predissolved in alkali has been added.

aromatic oxygen, hence to a flavonoid phenolic oxygen atom would be able to give the type of shift observed (fig. 2), whereas ^{13}C NMR showed that the flavonoid heterocycle had opened during the reaction.[1] Potentiometric titrations finally indicated that the B-ring of the flavonoid was not directly involved in the complex formed by the flavonoid with silicic acid.[1] The mechanism proposed initially, an ionic one, is shown in figure 3.[1,10]

Electron spin resonance (ESR) investigation showed that a radical mechanism indeed existed and varied in importance for different types of flavonoid tannins and at different pHs. At different pH's, a considerable surge in intensity of the phenoxide radical signal appeared on addition of silicic acid, sometimes up to 10 times greater than for the tannin alone.[4] The peak intensity decreased as a function of time without any other type of radical appearing indicating that some reaction of autocondensation due to radical quenching existed. More interesting, the phenoxide radical signal showed a well-defined fine structure, which, on addition of silicic acid, changed in a well-defined pattern (fig. 4).[4-7] The variation shown in figure 4 indicated the existence of an equilibrium between different radical forms of the flavonoid, which was proposed to be as shown in figure 5 and where the radical species 4 never appeared to form.[5] Thus, the peaks further away from the center of symmetry of the ESR spectrum were found to represent phenoxide radicals on the A-rings of the flavonoid tannins, whereas the peaks closer to the center

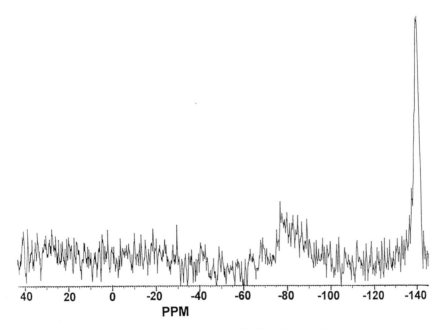

^{29}Si-NMR (CP-MAS) of pecan nut tannin extract hardened with 4% SiO$_2$

Figure 2. Solid phase CP-MAS ^{29}Si NMR spectrum of pH 12 pecan nut tannin extract water solution of 43 percent concentration hardened with 4-percent silica smoke (Aerosyl).

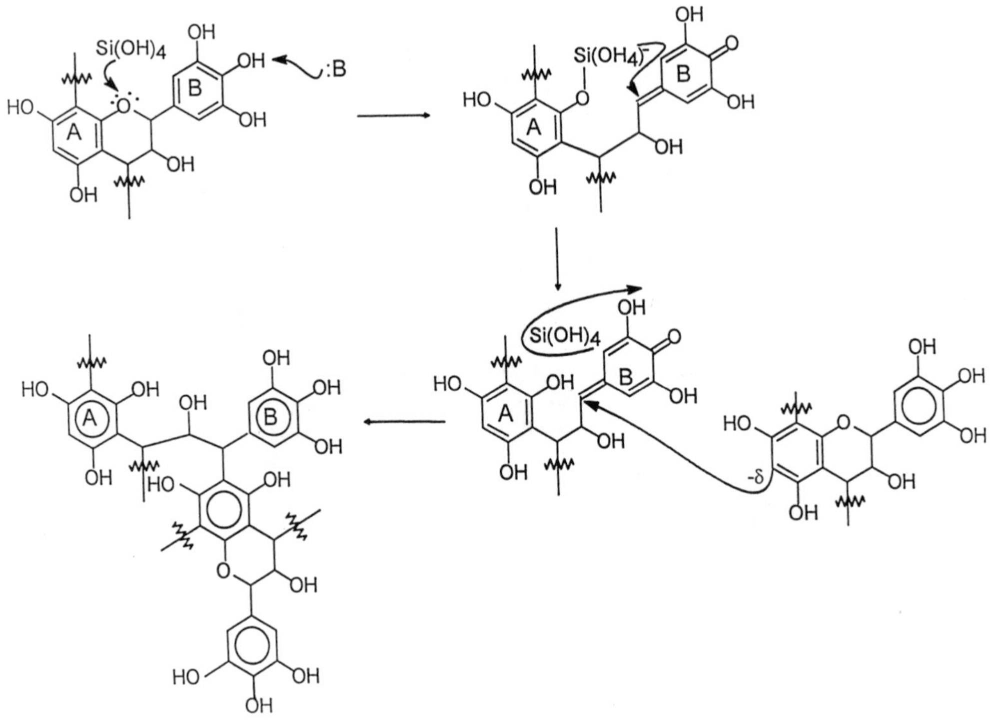

Figure 3. Proposed mechanism of the interaction of silica and silicates with tannin solutions leading to partial tannin autocondensation.

of symmetry were found to represent phenoxide radicals on the flavonoid tannin B-rings.[6] A clear example of this is shown in figure 4. Kinetics of radical formation and decay for all the different peaks, and thus for all the different phenoxide radicals observed in the ESR spectra, were done, and relevant kinetic equations and coefficients for several tannins have now been established and already reported.[5-7] Assignment of different ESR peaks to different flavonoid radical anion species was obtained, although not all peaks could be exactly assigned.[6]

The very extensive kinetic study of different natural tannin extracts, and even of tannin extract that had been modified by sulfitation, by complete elimination of carbohydrates or even by known treatments to prepare adhesive intermediates, indicated that four important factors determine the rate of the reaction. These are in order of relative importance:

1. The extent of the colloidal state of the tannin solution. The higher is the level of colloidality the more extensive is the radical reaction. This is true both in the case when the high colloidal state is induced by the polymeric carbohydrates present as well as in the case in which this high level of colloidality is inherent to the tannin itself.

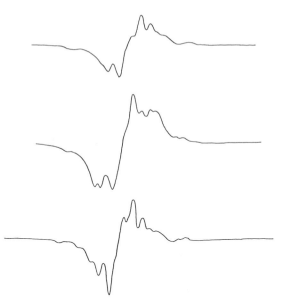

Figure 4. Fine structure of the ESR phenoxide radical signal of the radical-anions of a prodelphinidin tannin (pecan nut tannin extract solution) at pH 12: (a) by itself; (b) 5 minutes after addition of 10-percent fine silica powder; (c) 15 minutes after the same 10-percent silica addition. Note the variation in intensity of the different signals.

2. The degree of polymerization of the tannin (hence, the number average DP), with higher DP rendering more noticeable the radical reaction.

3. The presence or not of a lignocellulosic surface: the reaction being induced by a lignocellulosic surface almost as strongly as by silicic acid. Furthermore, the radical reaction was observed to continue once it had been started by contact with the lignocellulosic material even when it was separated from the inducing substrate.

4. The relative structure of the tannin flavonoid repeating unit, thus the type of A- and B-rings present.

Of these, the first three (and especially the first one) are the most important factors influencing the radical reaction. The fourth factor has some less defined influence only.

The system of autocondensation was then used to make laboratory wood particleboard, at first without any silicic acid addition, just relying on the autocondensation induced by the lignocellulosic substrate used.[8,10] The results, shown in figure 6, indicate that while lignocellulosic induction of the reaction is more than sufficient for procyanidin and prodelphinidin tannins such as pine tannin and pecan tannin to prepare interior grade particleboard, this is not the case for prorobinetinidin/profisetinidin tannins such as quebracho and mimosa tannin extracts. Figure 6 also shows the much lower pH onset of the radical reaction for pecan tannin in relation to that of pine tannin where more drastic conditions need to be used to achieve panels of sufficiently high internal bond strength. Figure 7 instead shows that a considerable improvement in internal bond strength can be

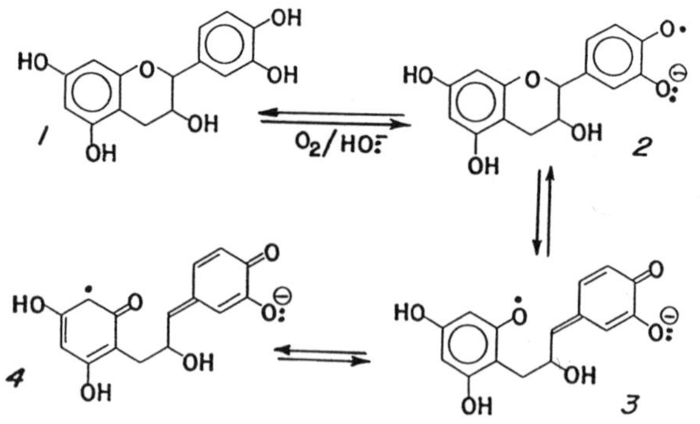

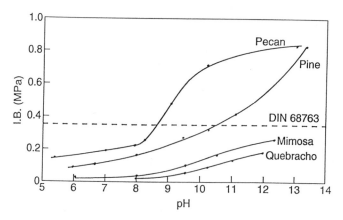

Figure 5. Equilibrium between different phenoxide radical-anions species proposed to explain the relative variation in ESR peak intensity as a function of the advancement of the reaction.

obtained for mimosa (and also for quebracho—not reported here) if a small amount of silicic acid is added. From the same figure, it is equally evident that addition of silicic acid to pine and pecan tannin causes a decrease in particleboard strength. This has been ascribed to such an acceleration of the formation of the network to cause early immobilization and thus lesser density of crosslinking (an effect well known in PF and other wood adhesives resins) as well as precuring in the hot particleboard press. Finally, procyanidin and prodelphinidin tannins are better used without silicic acid acceleration due to the lignocellulosic surface being more than sufficient under the correct conditions to give excellent interior-grade dry internal bond strength of the panels, whereas prorobinetinidin/profisetinidin tannins such as quebracho and mimosa tannin extracts need some silicic acid addition to achieve a dry internal bond strength of the panel satisfying relevant standards.

Figure 6. Dry internal bond (IB) strengths as a function of tannin solution pH of laboratory particleboard prepared with four different tannin extracts without any aldehyde hardener, using only the lignocellulosic substrate-induced tannin autocondensation.

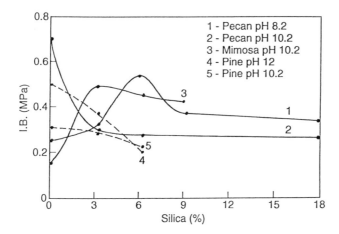

Figure 7. Dry internal bond (IB) strengths of laboratory wood particleboard bonded with different tannins without any aldehyde hardener, using both Lewis acid-induced (silica; boric acid) and substrate-induced tannin autocondensation, as a function of weight percentage of silica addition.

These results are good but only for interior use panels.[8] While a no-aldehyde, natural, interior panel adhesive has a definite "niche" market in completely environmentally friendly applications, tannins, due to their cost structure and their characteristics, have traditionally been used as exterior wood adhesives. However, exterior grade tannin-bonded wood panels of much lower hardener content could be obtained by combining autocondensation and polycondensation with traditional ionic hardeners. As a consequence, tannin autocondensation was studied in the presence of a number of traditional ionic hardeners, such as formalin, paraformaldehyde, paraformaldehyde + urea, hexamethylenetetramine (hexamine), furfuryl alcohol, polymeric diisocyanates (pMDI) and pMDI + paraformaldehyde with and without silicic acid to determine if synergic or interference effects occurred between the different types of reactions. At first, the reactions were followed by ESR, and the variations of relative peak intensity for all the peaks as a function of time yielded variations on the theme of two types of curve (fig. 8).[17]

The peculiar shape of one of these basic curves indicates that several complex reactions are at play, especially when ionic reactants are present. Three initial hypotheses[17,18] could be advanced:

1. The presence by transfer or other reactions of radical species different from the phenoxyls. This possibility is unlikely because no other types of radical species are shown in the ESR spectra.

2. The presence of ionic equilibria with the added coreactants on top of the known equilibria between the flavonoid radical anions.

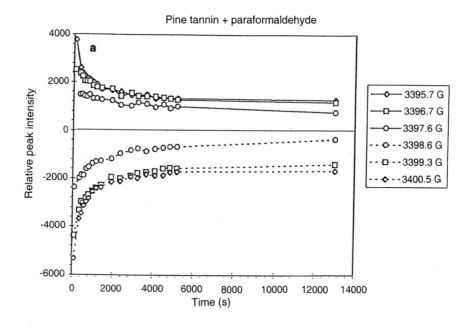

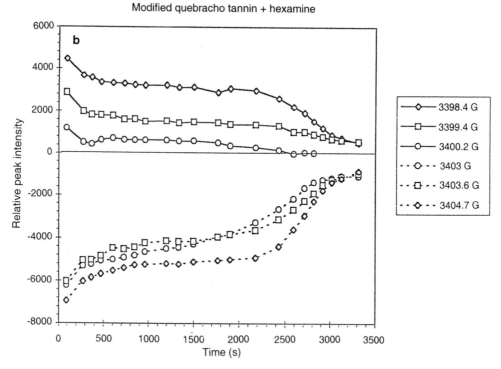

Figure 8. Examples of the two different shapes of the curves of variation of relative ESR peak intensities as a function of time: (a) pine tannin extract + paraformaldehyde; (b) modified quebracho tannin extract + hexamethylenetetramine. The curves of all the peaks of the ESR spectra are shown.

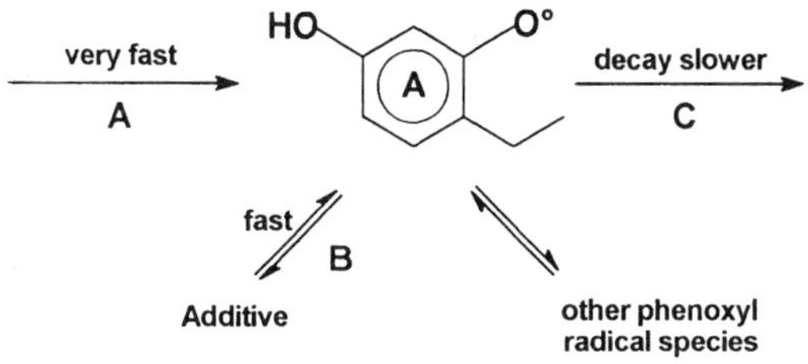

Figure 9. A generalized mechanism explaining the relative variation in intensity of ESR phenoxyl radicals peaks when in presence of ionic hardeners. This mechanism has severe limitations (see text).

3. The presence of ionic reactions with the added coreactants, which would subtract radicals to the main equilibrium between the flavonoid radical-anion species.

The existence of either 2) and 3) above is rather likely. An acceptable generalized mechanism is shown in figure 9.

Thus, after at first a fast radical surge followed by radical decay at a slower rate, transfer or quenching by the coreactants of the phenoxyl radicals will give a decreasing radical concentration. As the phenoxyl radical concentration becomes lower due to the slower decay rate, the coreactant ↔ phenoxyl equilibrium shifts and tends to reestablish the concentration of phenoxyl radicals; if reaction B is fast and reaction C is slower, then at first an increase followed by a plateau in phenoxyl radicals concentration is obtained. If the radical decay reaction and the reaction of the equilibrium have similar rates, then a plateau is again reached. As the store of radicals entrapped in whatever manner in the additive becomes exhausted, only the radical decay reaction will be visible, and the curve of radical concentration as a function of time will assume the descending slope appearance noticeable in the end parts of the graphs. Such a mechanism would explain well the effect of the coreactants, but as no coreactant radicals are observed, the equilibrium can only be based on ionic reactions. The problem of such an approach is that, on reversing the equilibrium, phenoxyl radicals must again be produced, a difficult occurrence to explain. The concept that needs to be introduced in the mechanism proposed is then that of an unknown species not detectable by ESR. This chemical species might not be detectable by ESR because either it is not a radical, or because it is a diradical (hence the two unpaired electron spins would cancel each other, and the species would also not be paramagnetic). This is an acceptable occurrence for phenoxyl radicals.[17,18]

The situation is simpler as regards the initial interference between tannins radical autocondensation and formaldehyde both in the form of paraformaldehyde and of formalin solution where the mechanism is that shown in figure 10.

In these cases, the radical decay curves should not present the plateau effect, and this is what is always observed.[17,18] Many more interesting conclusions have been derived from this series of experiments, which are too long and too involved to report here.[17,18] Equally, tables presenting the existence or not of the plateau effect for prorobinetinidin/profisetinidin type tannins and for procyanidins (pine tannin) have already been reported.[17,18]

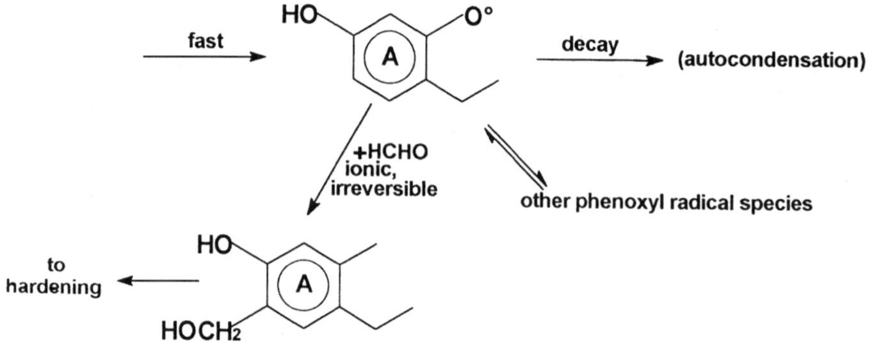

Figure 10. A mechanism explaining well the simpler case of the relative variation in intensity of ESR phenoxyl radicals peaks when in presence of formaldehyde used as an ionic hardener.

Extensive thermomechanical analyses of glue mixes based on two types of tannin, namely, a prorobinetinidin/profisetinidin tannin and a procyanidin tannin with the same types of ionic reactants already described, have also been carried out. This work allowed both the characterization of the final, hardened networks formed as well as an approximate determination of the relative proportions of auto-condensation and polycondensation in the formation of the final hardened network (table 1).[19–21]

This was a difficult exercise because the usual interpretation of the single variation in modulus as a function of curing time and temperature lost any significance in complex systems such as tannins in which two to four different phenolic nuclei coexist. The thermograms obtained presented "first derivative" peaks and successive network tightening corresponding to both the faster reacting phenolic nuclei, namely, resorcinol and phloroglucinol. However, they also presented later and at higher temperatures less marked first derivative peaks corresponding to further reaction and network tightening due to the less reactive flavonoid B-ring phenolic nuclei. It was possible by model compound reactions to also identify these weaker contributions, some of them such as catechol type B-rings almost being to the limit of detection of the instrument. In figures 11a,b as well as in figures 12a,b, thermograms of the reaction of two types of tannin and of their respective A-ring model compounds with formaldehyde are shown.

The work was further complicated by internal successive reactions that occurred for both tannins and simple phenolic model compounds due to the presence of the methylene ether to methylene bridges rearrangement ($-CH_2-O-CH_2- \rightarrow -CH_2-$) and its consequences.[19–21] This can be summarized as shown in figure 13.[19–21]

If one considers that the above can (and sometimes does) occur for all the phenolic nuclei present in a tannin extract, then it is easy to understand that the examples presented in figures 11 a,b and 12 a,b are particularly understandable ones for the sake of this presentation. This work allowed us also to define which

Table 1. Relative percentages of the contribution of polycondensation and autocondensation to the total modulus for pine and quebracho tannins in the presence of different hardeners

	Polycondensation (%)	Autocondensation (%)	Total modulus(GPa)
Pine tannin			
Tannin alone,pH 4.5	-	100	1.50
Tannin alone,pH 9.5	-	100	1.74
+paraformaldehyde, pH 4.5	63	37	1.56
+paraformaldehyde, pH 7.3	89	11	2.28
+hexamine, pH 4.5	71	29	1.55
+hexamine, pH 9.5	91	9	1.88
+paraform./urea, pH 4.5	47	53	1.70
+paraform./urea, pH 7.3	80	20	2.45
+MDI/paraform.,pH 7.3	80	20	2.46
+SiO$_2$, pH 4.5	-	100	1.60
+SiO$_2$, pH 9.5	-	100	1.23
Quebracho tannin			
Tannin alone, pH 7.3	-	100	1.37
Tannin alone, pH 10.3	-	100	1.58
+paraformaldehyde,pH 7.3	84	16	1.94
+hexamine, pH 7.3	55	45	0.81
+hexamine, pH 10.3	92	8	1.37
+furfuryl alcohol, pH 7.3	43	57	1.56
+furfuryl alcohol, pH 10.3			
+paraform./urea, pH 7.3	75	25	2.13
+MDI, pH 7.3	84	16	1.94
+MDI, pH 10.3	88	12	1.88
+MDI/paraform., pH 7.3	91	9	2.22
+SiO$_2$, pH 7.3	-	100	1.86
+SiO$_2$, pH 10.3	-	100	2.08

were the best ultimate strength hardeners, and under which conditions, for the two main types of flavonoid tannins usable for adhesives.[19,21,22]

The results in table 1 show what is the approximate relation between tannin autocondensation and tannin polycondensation with ionic hardeners as formaldehyde. Autocondensation is always shown to participate to the formation of the final, hardened tannin network, more or much less markedly according to the

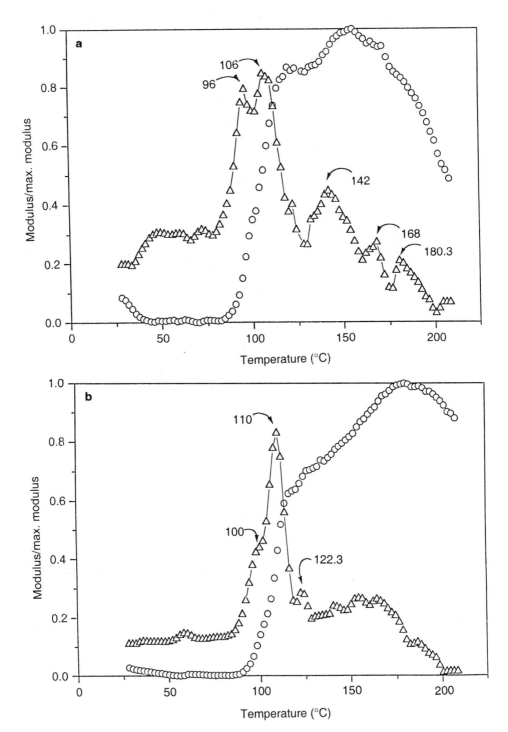

Figure 11. Curves of the variation of the ratio modulus/max. modulus (O) as a function of temperature at a constant heating rate of 10 °C/minute, and curve of its first derivate (Δ) of beech joints bonded with (a) phloroglucinol monomer/paraformaldehyde at pH 4.5 and (b) resorcinol monomer/paraformaldehyde at pH 7.3.

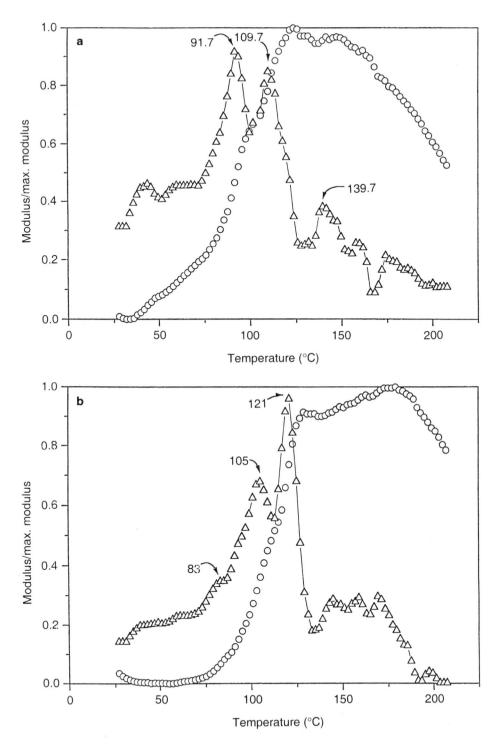

Figure 12. Curves of the variation of the ratio modulus/max. modulus (O) as a function of temperature at a constant heating rate of 10 °C/minute, and curve of its first derivate (Δ) of beech joints bonded with (a) pine tannin (procyanidin) extract/paraformaldehyde at pH 4.5 and (b) modified quebracho tannin extract/paraformaldehyde at pH 7.3.

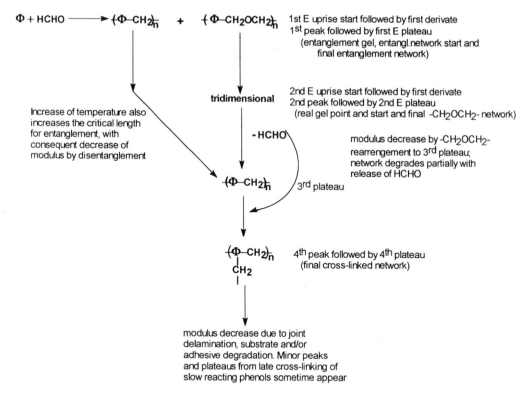

Figure 13. Schematic diagram of the series of reactions and cured tannin adhesives internal network rearrangements occurring during thermomechanical analysis of the reaction of tannins with formaldehyde. Correspondence of main reactions and rearrangements with features on the modulus increase curve are also reported.

hardener used. The proportion of the network due to polycondensation appears to be related to the extent of water resistance of the final network, whereas the contribution of tannin autocondensation appears to be only limited to the dry strength of the network. The results obtained provided a theoretical justification to allow the marked decrease of formaldehyde hardener in tannin adhesives without any loss of exterior performance and with even lower formaldehyde emission than the already low emissions obtained with standard tannin adhesives formulations.[19–21] In the case of procyanidin type tannins such as pine tannin, it was not necessary to even add the silica catalyst, since under the correct conditions, whatever the hardener used, tannin autocondensation is always present and fast enough to contribute to the strength of the final hardened network.

Tannin-bonded particleboard panels of exterior grade quality were then made by coupling autocondensation and polycondensation reactions and using much lower amounts of aldehyde-type hardeners. The most promising systems tried that gave acceptable exterior grade results[23] were:

(a) Quebracho tannin extract +3% paraformaldehyde + 3%SiO₂ + urea

(b) Quebracho tannin extract + UF 1:2 resin + SiO₂

(c) Quebracho tannin extract + PF 1:2.5 resin + SiO₂

(d) Quebracho tannin extract + UF 1:2 resin + PF 1:2.5 resin + SiO₂

The same systems (a) to (d) give good results for exterior particleboard when using pine tannin but without any SiO_2.

3. CONCLUSIONS

Comparative kinetics of the radical contribution to autocondensation induced by silicic acid and by activation via cellulose surface effects on different polyflavonoid tannins indicate the parameters of relative importance in promoting tannin autocondensation reactions. These parameters were shown to be the colloidal state of the tannin solution, the catalytic activation induced by a cellulosic surface, the degree of polymerization of the polyflavonoid, and the structure of the tannin involved. It is the combination of these parameters that determine the extent of the autocondensation reaction. Furthermore, Electron Spin Resonance (ESR) and thermomechanical analysis (TMA) studies were carried out to detect any interference by ionic hardening mechanisms and ionic coreactants on polyflavonoid tannins autocondensation. These studies indicated that in certain cases hardening by ionic coreactants can be coupled with the simultaneous hardening of the tannins by induced autocondensation. Some coreactants tend to depress the tannin radical autocondensation while still leaving a small contribution of this reaction to the formation of the final crosslinked network. Other coreactants instead appear to enhance formation of the final network by synergy between ionic and radical mechanisms, whereas still others do not show any interference between the two types of reaction. Under the conditions prevalent in the application of tannins to wood adhesives, the reaction of autocondensation has been shown to always contribute, to a greater or lesser extent, to the formation of the final hardened network. The extent of its contribution is quantified for several different wood adhesives hardeners. On the basis of the mechanisms describing the interaction between polycondensation and autocondensation that are proposed and discussed, interior grade wood particleboard panels with no aldehyde hardeners were prepared. Exterior grade wood particleboard panels needing a much lower level of aldehyde or other hardeners were also produced. Both types of products satisfied relevant industrial standards. The approach presented allows a different route to the preparation of tannin-bonded wood panels of very low or zero formaldehyde emission.

The work presented here has advanced preparation of tannin-based adhesives for exterior-grade particleboard and other wood panels of much reduced aldehyde content and emission. Tannin-based adhesives have also been developed for interior-grade particleboard and other wood panels that contain no aldehyde at all, hence are of E0 formaldehyde emission class and are totally environment friendly. A well-defined market niche for these adhesives already exists. The fundamentals of these studies have also allowed knowledge to advance on the application of flavonoid tannins, especially in their use as antioxidants as well as in the preparation of tannin-based cement superplasticizers.[24,25]

REFERENCES

1. Meikleham, N.; Pizzi, A.; Stephanou, A. Induced accelerated autocondensation of polyflavonoid tannins for phenolic polycondensates—Part 1: ^{13}C NMR, ^{29}Si NMR, X-ray and polarimetry studies and mechanism. *J. Appl. Polymer Sci.* 54:1827 (1994).

2. Pizzi, A.; Meikleham, N.; Stephanou, A. Induced accelerated autocondensation of polyflavonoid tannins for phenolic polycondensates—Part 2: cellulose effect and application. *J. Appl. Polymer Sci.* 55:929 (1995).

3. Pizzi, A.; Meikleham, N. Induced accelerated autocondensation of polyflavonoid tannins for phenolic polycondensates—Part 3: CP-MAS ^{13}C NMR of different tannins and models. *J. Appl. Polymer Sci.* 55:1265 (1995).

4. Merlin, A.; Pizzi, A. An ESR study of the silica-induced autocondensation of polyflavonoid tannins. *J. Appl. Polymer Sci.* 59:945 (1996).

5. Masson, E.; Merlin, A.; Pizzi, A. Comparative kinetics of the induced radical autocondensation of polyflavonoid tannins—Part 1: modified and non-modified tannins. *J. Appl. Polymer Sci.* 60:263 (1996).

6. Masson, E.; Pizzi, A.; Merlin, A. Comparative kinetics of the induced radical autocondensation of polyflavonoid tannins—Part 2: flavonoid units effects. *J. Appl. Polymer Sci.* 64:243 (1997).

7. Masson, E.; Pizzi, A.; Merlin, A. Comparative kinetics of the induced radical autocondensation of polyflavonoid tannins-Part 3: micellar reactions vs. cellulose surface catalysis. *J. Appl. Polymer Sci.* 60:1655 (1996).

8. Pizzi, A.; Meikleham, N.; Dombo, B.; Roll, W. Autocondensation-based, zero emission, tannin adhesives for particleboard. *Holz Roh Werkstoff* 53:201 (1995).

9. Pizzi, A. Wood adhesives chemistry and technology, Vol. 1. Marcel Dekker Inc., New York (1983).

10. Pizzi, A. Advanced wood adhesives technology, Marcel Dekker Inc., New York (1994).

11. Sealy-Fisher, V.J.; Pizzi, A. Increased pine tannin extraction and wood adhesives development by phlobaphenes minimization. *Holz Roh Werkstoff* 50:212 (1992).

12. Pizzi, A.; Stephanou, A. Comparative and differential behaviour of pine vs. pecan nut tannin adhesives for particleboard. *Holzforschung und Holzverwertung* 45(2):30 (1993).

13. Pizzi, A.; Valenzuela, J.; Westermeyer, C. Low-formaldehyde emission, fast- pressing, pine and pecan tannin adhesives for exterior particleboard. *Holz Roh Werkstoff* 52:311 (1994).

14. Trosa, A.; Pizzi, A. Industrial hardboard and other panels binder from tannin/furfuryl alcohol in absence of formaldehyde. *Holz Roh Werkstoff* 56:213 (1998).

15. Pizzi, A.; Walton, T. Non-emulsifiable, water-based diisocyanate adhesives for exterior plywood. *Holzforschung* 46:541 (1992).

16. Pizzi, A.; von Leyser, E.P.; Valenzuela, J.; Clark, J.G. The chemistry and development of pine tannin adhesives for exterior particleboard. *Holzforschung* 47:164 (1993).

17. Garcia, R.; Pizzi, A.; Merlin, A. Ionic polycondensation effects on the radical autocondensation of polyflavonoid tannins. *J. Appl. Polymer Sci.* 65:2623 (1997).

18. Garcia, R. Developpement de resines thermodurcissables et de haute performance a base de tannins. Doctoral thesis, University Henri Poincaré Nancy 1, Nancy, France (1998).

19. Garcia, R.; Pizzi, A. Polycondensation and autocondensation networks in polyflavonoid tannins-Part 1: final networks. *J. Appl. Polymer Sci.* 70:1083 (1998).

20. Garcia, R.; Pizzi, A. Polycondensation and autocondensation networks in polyflavonoid tannins—Part 2: polycondensation vs. autocondensation. *J. Appl. Polymer Sci.* 70:1093 (1998).

21. Garcia, R.; Pizzi, A. Cross-linked and entanglement networks in thermomechanical analysis of polycondensation resins. *J. Appl. Polymer Sci.* 70:1111 (1998).

22. Pizzi, A. On the correlation of some theoretical and experimental parameters in polycondensation cross-linked networks. *J. Appl. Polymer Sci.* 63:603 (1997).

23. Trosa, A. Developpement de resines phenoliques synthetiques thermodurcissables diluées avec resines naturelles et de dechet en utilisant des durcisseurs nouveaux pour basse emission de formaldehyde. Doctoral thesis, University Henri Poincaré Nancy 1, Nancy, France (1999).

24. Noferi, M.; Masson, E.; Merlin, A.; Pizzi, A.; Deglise, X. Antioxydant characteristics of hydrolysable and flavonoid tannins-an ESR kinetic study. *J. Appl. Polymer Sci.* 63:475 (1997).

25. Kaspar, H.R.E.; Pizzi, A. Industrial plasticizing/dispersion aids for cement based on polyflavonoid tannins. *J. Appl. Polymer Sci.* 59:1181 (1996).

POLYPHENOLS AND ECOLOGY

CONDENSED TANNINS OF SALAL (*GAULTHERIA SHALLON* PURSH): A CONTRIBUTING FACTOR TO SEEDLING "GROWTH CHECK" ON NORTHERN VANCOUVER ISLAND?

Caroline M. Preston

Pacific Forestry Centre
Natural Resources Canada
Victoria, BC V8Z 1M5
CANADA

1. INTRODUCTION

The west coast of British Columbia, a region of high rainfall and mild climate, is noted for large areas of old-growth temperate rainforest. Logging operations starting on northern Vancouver Island during the 1960s generated large cutover areas with poor regeneration and slow growth of seedlings, concern for which stimulated research into forest ecology, harvesting practices, and silviculture. For over a decade, research has been coordinated and carried out by the umbrella group SCHIRP (Salal-Cedar-Hemlock-Integrated-Research-Program).[1]

On northern Vancouver Island, the native forests form a mosaic of two types of stands. Old-growth "CH" stands, dominated by western red cedar (*Thuja plicata* Donn.) and western hemlock (*Tsuga heterophylla* (Raf.) Sarge), have mixed ages, a fairly open canopy, and a dense understory of the ericaceous shrub salal (*Gaultheria shallon* Pursh). They have been largely undisturbed for centuries, and cedars are up to 1,000 years old. By contrast, the second-growth "HA" stands are dominated by western hemlock and amabalis fir (*Abies amabilis* Dougl.). Many of these forest stands originated following a windstorm in 1906, and there is little understory under the dense canopy. Soils in the whole region are characterized by deep forest floors, often with a large woody component.[1-2]

After clearcutting and slashburning of CH sites, seedlings initially grow well, but within 5–8 years, growth rates decline, coincident with vigorous resprouting of the ericaceous shrub salal (*Gaultheria shallon* Pursh) and severe decline in N and P availability.[3] The problem is much less severe on sites previously occupied by HA stands. Several factors may contribute to the poor nutrient status and regrowth problems on the CH cutovers, but this discussion focuses on the effects of salal, a plant that can spread rapidly and resprout from underground rhizomes. In recently clearcut and burned sites, belowground salal biomass increased from 1,908 kg ha^{-1} at 2 years to 11,415 kg ha^{-1} at 8 years,[4] whereas nutrient availability declined.[3] A study with ^{15}N-labelled fertilizer showed reduced N uptake by tree seedlings in the presence of dense salal in CH cutovers.[5]

The effects of salal in these sites are similar to those observed for other ericaceous shrubs that inhibit tree growth, and in extreme cases, cause conversion from forest to permanent heath.[6–8] In addition to direct competition for mineral nutrients and moisture, these shrubs may exert allelopathic effects, reducing germination of tree seedlings and inhibiting mycorrhizal associations of the tree species, as well as contributing to the development of an acidic, poorly decomposed mor humus. Many ericaceous shrubs, including salal[1,9] and some conifers,[10] can use their mycorrhizal associations to access organic forms of N and P, thus enhancing their competitive ability. The accumulation of mor humus of low nutrient availability has long been associated with plants with high tannin or polyphenolic content.[11–14] This has recently been interpreted as a beneficial strategy by which plants in nutrient-limited environments conserve nutrients by building up a thick humus from which they can access organic forms of N and P.[10]

A previous study[2] using ^{13}C nuclear magnetic resonance (NMR) spectroscopy showed high condensed tannin content in salal litter and flower stalks. It also indicated that humus from undisturbed CH stands had a higher tannin content than from HA. However, the NMR results were far from quantitative, and reported here is some basic work toward understanding the role of salal tannins in these ecosystems.

2. METHODS

Soil and plant samples were obtained from two research installations in CH cutovers near Port McNeill on northern Vancouver Island. The SCHIRP site is a major installation with different tree species, stocking density, scarification, and fertilization treatments.[1] The ^{15}N site was used to study the fate of ^{15}N-labeled fertilizer (one application of 200 kg N ha^{-1}) in three tree species (western hemlock, western red cedar, and Sitka spruce (*Picea sitchensis* Bong. Carr.) with and without salal competition.[5] At the final 6-year sampling in September 1996, there were 12 single-tree plots (1 m radius), of which 6 had undergone salal control for 9 years by repeated clipping, and 6 had uncontrolled salal growth. Humus was sampled by depth in 10 cm increments from two pits per plot.

Bulk samples of salal flower-heads and foliage for extraction were collected from the ^{15}N and SCHIRP sites. The survey of tannin levels in salal components and humus was done on samples from the ^{15}N plots as was the extraction of humus. From the ^{15}N plots with uncontrolled dense salal growth, we prepared composite samples of salal foliage, litter, young stems, old woody stems, and roots (mostly rhizomes). The plots with salal control had only a few small shoots, but a composite sample was prepared of foliage and stems.

Salal and humus samples from the ^{15}N plots were ovendried at 70 °C. As far as possible, all live roots were removed from humus samples by sieving and handpicking. Total C was determined by automatic combustion using a Leco model CR12 carbon analyzer. Total N was determined either by a LECO FP-228 N analyzer or by Kjeldahl digestion.

Purified condensed tannins were prepared from salal flower-heads and salal foliage, and tannin-rich fractions from humus using standard methods:[15] pre-extraction with hexane, extraction with 70 percent v/v aqueous acetone, and cleanup of the aqueous phase by several washings with $CHCl_3$ and ethyl acetate. After solvents were removed by rotary evaporation, the crude extract was freeze-dried and then purified by chromatography on Sephadex LH-20 (Pharmacia, Uppsala, Sweden) using 50 percent aqueous methanol to remove sugars and low-molecular weight phenolics, followed by 70 percent v/v aqueous acetone to elute the condensed tannin. Acetone was removed by rotary evaporation, and the aqueous phase was freeze-dried to give the purified tannin.

Salal flower-heads were collected from two CH cutover areas (SCHIRP and ^{15}N sites) in June 1994 and stored at 4 °C for a few days before extraction. These were 5–6 cm long segments of stalk with attached flowers. In open-grown salal, the white flowers have a pinkish cast due to tiny red droplets, and the flower-bearing stalks are red and sticky. In order for us to focus on exterior tannins, the flower-heads (approximately 100 g field-moist weight) were extracted without drying, grinding, or pre-extraction with hexane, and then processed and purified in the standard manner. Some bright pink and purple fractions were collected during elution with methanol/water but were not characterized.

Salal foliage was collected in July 1996 from the same sites as the flower-heads and combined into one bulk sample. Damaged or diseased leaves were discarded, the remainder ground through a Wiley mill in the field-moist state, and a sample of 382 g (ovendry basis) was immediately extracted. After standard processing and LH-20 purification, the yield of purified tannin was 5.84 g, or 1.53 percent of the dry weight.

The humus extraction was carried out on samples collected in September 1996 from the 0–10 and 10–20 cm depths of the ^{15}N plots. From a total of 48 samples (previously ovendried at 70 °C), 10 samples giving the highest tannin levels in the PA assay (0.25 to 0.4 percent) were combined into a composite sample, which was extracted and processed following the conventional procedure for plant material. Washing the extract with ethyl acetate produced a high proportion of stable emulsion. This was processed separately including purification on Sephadex LH-20. The H_2SO_4 hydrolysis of humus fraction 2B and a brown-rot lignin (highly decomposed coarse woody debris)[16] was carried out as reported previously.[17]

The procedures were similar to those reported elsewhere.[17-18] The freshly prepared reagent was 5 percent concentrated HCl in n-butanol (v/v), with a total water content of 5 percent v/v and 200 mg/L of Fe^{2+} added as solid $FeSO_4.7H_2O$. This composition was chosen after preliminary experiments to optimize color development. To prepare calibration curves, tannin from salal foliage was dissolved in methanol and aliquots of 0.05–3.0 mg dried in screw-cap test tubes at 70 °C. For the assay, 5 mL reagent was added to the test tubes, which were stirred with a vortex mixer and heated in a water bath at 95 °C for 1 hour. After cooling, the solutions were transferred to disposable cuvettes and scanned from 430 to 630 nm. Absorbances were determined at the peak maximum, which was within ±2 nm of 555 nm, with allowance for sloping baselines.

Samples were hydrolyzed directly as dry powders or after extraction to determine soluble and residual tannins. For extraction, samples of 50 mg (dry weight) were weighed into plastic screw-cap centrifuge tubes. Twenty mL of acetone/water was added, and the tubes shaken for 2 h. They were centrifuged and the solution decanted into a 50 mL volumetric flask. Another 20 mL was added to the tube, the extraction was repeated, and the combined extracts brought to 50 mL. For the PA assay, aliquots (typically 0.5 to 1.5 mL) were added to screw-cap test tubes and dried at room temperature with a stream of air. Preliminary experiments showed that ovendrying of the aqueous acetone extracts at 70 °C seriously degraded the response. (This was not a problem with tannin standards in methanol.) The extraction residues were transferred into screw-cap test tubes and hydrolyzed. Samples to be analyzed without extraction were weighed into the screw-cap test tubes using approximately 50 mg for humus. Reagent was added to the dried aliquots or powders, and the analysis carried out as described above, except that the tubes were centrifuged after hydyrolysis.

Carbon-13 NMR spectra of solid-state samples were obtained at 75.47 MHz with cross-polarization and magic-angle spinning (CPMAS NMR) on a Bruker MSL 300 spectrometer (Bruker Instruments Inc., Karlsruhe, Germany). Dry, powdered samples of up to 200 μg were packed into a zirconium oxide rotor of 7 mm OD. Acquisition conditions were: spinning rate 4.7 KHz, 1 ms contact time, 2 s recycle time, and approximately 5,000 scans. Free induction decays were processed with 30–40 Hz linebroadening and baseline correction. Chemical shifts are reported relative to tetramethylsilane (TMS) at 0 ppm, with the reference frequency set using adamantane. Dipolar dephased (DD) spectra were generated by inserting a delay period of 45–50 μs without ^{1}H decoupling between the cross-polarization and acquisition portions of the CPMAS sequence. Most DD spectra and some normal spectra were acquired using the TOSS pulse sequence for Total Suppression of Spinning Sidebands since they have a serious impact on the aromatic and carboxyl regions at 300 MHz.

Salal spectra were divided into chemical shift regions as follows: 0–50 ppm alkyl C; 50–58 ppm methoxyl C; 58–93 ppm O-alkyl C; 93–112 ppm di-O-alkyl C and some aromatic; 112–140 ppm aromatic C; 140–165 ppm phenolic C, and 165–190 ppm carboxyl C. Spectra of humus fractions and lignin were divided into fewer regions: 0–50, 50–58, 58–93, 93–140, 140–165, 165–190, and 190–220 ppm. Areas of the chemical shift regions were determined from the integral curves and were expressed as percentages of total area ("relative intensity"). As several factors influence the intensity distributions in CPMAS and CPMAS-TOSS NMR spectra, the relative intensities were only used for comparative purposes.

The solution spectrum of the salal foliage tannin was obtained on the same instrument in 50:50 D_2O:acetone using a 10 mm sample tube with inverse-gated decoupling, 45° pulse, 0.2 s acquisition time, and 2 s relaxation delay. Spectra were also processed with line broadening and baseline correction, with chemical shifts reported relative to TMS. Spectra of the flower-head tannins were obtained on a Bruker WM 250 spectrometer at 63 MHz for ^{13}C using similar conditions.

3. PROPERTIES OF SALAL TANNINS AND THE N-15 SITE

Salal foliage yielded 1.53 percent by weight purified tannin. The ^{13}C solution NMR spectrum (fig. 1b) showed approximately equal proportions of prodelphinidin (PD) and procyanidin (PC), about 80 percent *cis* relative stereochemistry, and a chain length of approximately 5 units.[19–20] The salal flower-heads yielded 2.48 and 3.78 percent purified tannin, for the ^{15}N and SCHIRP sites, respectively. This is a high yield, considering that only surface material was extracted. Figure 1a shows the spectrum of the tannin from the ^{15}N site. Both flower-head tannins had approximately one-third PD and two-thirds PC content and almost entirely *cis* stereochemistry. For the ^{15}N site, the terminal C3 peak at 67 pm was barely detectable, whereas the chain length was approximately 10 for the SCHIRP site.

Salal leaves, roots, and young stems contained 11 to 21 percent by weight condensed tannin (table 1), with the highest level found in the leaves of the small residual shoots in the salal-removal plots. Old woody stems contained only 5.5 percent tannin, much lower than younger stems, whereas litter showed a dramatic loss with only 0.8 percent tannin. The bulk of the tannin (89 percent or more) was found in the extractable fraction, except for litter for which only 27 percent was extractable.

The ^{13}C CPMAS NMR spectra are shown in figure 2 and the intensity distributions in table 2a. Based on previous NMR studies of plant biopolymers,[16–17,19–24] the foliage (fig. 2a) is high in cutin and surface lipids, as shown by the strong signals in the alkyl region (0–50 ppm) with maximum intensity at 30 ppm. Cellulose and other polysaccharides account for most of the intensity in the O-alkyl and di-O-alkyl regions (50–93 and 93–112 ppm). However, there is also a weak signal

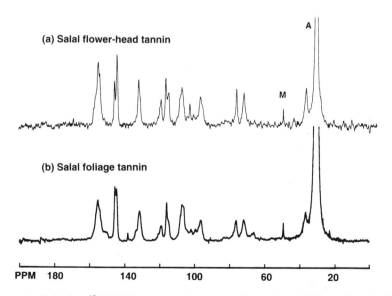

Figure 1. Solution ^{13}C NMR spectra of tannins from (a) salal flower-heads from ^{15}N site and (b) salal foliage in acetone/D$_2$O; M, methanol; A, acetone.

Table 1. Condensed tannins (mg/g) in salal components (composite samples from ^{15}N lots)

Sample	Extractable	Residue	Total
	% of Total		mg/g
No salal control (dense growth)			
Roots/rhizomes	95.9	4.1	120.9
Leaves	89.0	11.0	175.3
Stems (young)	94.2	5.8	130.4
Stems (old, woody)	93.9	6.1	55.3
Litter	26.5	73.5	7.5
Salal removal plots (residual shoots)			
Leaves	92.7	7.3	214.5
Stems	95.7	4.3	114.9

for the methoxyl C of lignin (a mixture of guaiacyl and syringyl lignin) at 57 ppm, whereas the signal for anomeric C1 of cellulose is underlain by broader peaks from condensed tannins and hemicelluloses giving rise to the shoulder at 98–100 ppm. The aromatic and phenolic regions indicate a higher proportion of condensed tannin than lignin. This is based on the fairly sharp feature at 131 ppm and the well-resolved split phenolic signal with approximately equal intensities at 145 and 154 ppm. Carboxyl, amide, and ester carbons (including those in acetate, cutin, and proteins) give rise to the peak in the carboxyl region.

During the delay in the DD spectra, carbons with attached hydrogens lose intensity rapidly, except for those with some molecular mobility in the solid state. This includes methyl and methoxyl groups that can rotate and CH$_2$ in long-chains. The DD spectra are also consistent with high tannin and cutin content, since they show the diagnostic peak for condensed tannins at 106 ppm but a very weak signal for the methoxyl of lignin. The alkyl signal also retains some intensity due to the molecular mobility of the waxy materials.

The two foliage samples were similar, with only slight differences in the intensity distributions (table 2a). Foliage of residual shoots from the salal-removal plots (fig. 2b) had a slightly lower proportion of O-alkyl C and was higher in alkyl and aromatic C. This indicates lower foliage quality in the smaller, stressed leaves (lower carbohydrates and higher cutin/waxes and tannins (21 percent vs 18

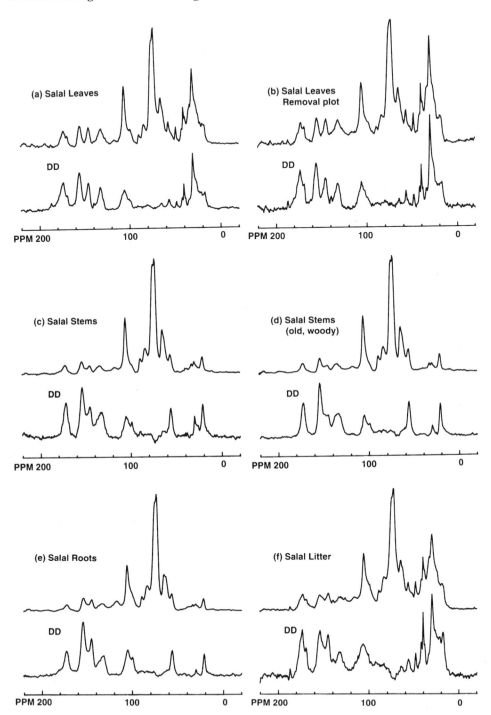

Figure 2. Carbon-13 CPMAS NMR spectra of salal components as indicated. Normal CPMAS (top) and DD-TOSS (bottom).

Table 2. Relative areas of ^{13}C-CPMAS NMR spectra of salal components
and humus fractions

(a) Salal Components

	Alkyl	O-CH$_3$	O-alkyl	Di-O-alkyl	Aromatic	Phenolic	Carbonyl
	0-50	50-58	58-93	93-112	112-140	140-165	165-190
No salal control (dense growth)							
Leaves	29.2	3.0	38.3	11.8	6.0	6.7	5.0
Litter	30.4	4.1	38.4	12.0	6.6	5.1	3.4
Stems	10.5	3.9	54.5	14.2	7.0	5.8	4.1
Stems-old	7.2	4.3	56.4	13.5	6.5	6.3	5.8
Roots	5.7	3.9	61.6	14.2	5.7	6.2	2.7
Salal removal plots (residual shoots)							
Leaves	31.1	4.5	34.1	11.4	7.8	6.6	4.5
Stems	9.1	4.7	54.4	13.6	7.1	6.0	5.1

(b) Humus Fractions and Lignin

	Alkyl	OCH$_3$	O-alkyl	Di-O-alkyl + Aromatic	Phenolic	Carboxyl
	0-50	50-58	58-93	93-140	140-165	165-210
Humus	19.0	7.1	35.5	24.1	6.5	7.8
Residue	16.4	6.3	37.9	24.5	6.6	8.3
Extract	18.8	8.4	31.5	27.0	7.8	6.5
Fr. 2[a]	21.8	6.1	12.8	33.7	17.6	8.0
Fr. 2B[a]	22.9	6.6	13.0	32.2	16.9	8.4
Fr. 2B-H[a]	23.2	11.8	9.9	33.4	16.9	4.8
Fr. 4B[a]	20.7	7.9	18.0	31.5	14.1	7.8
Lignin[a]	7.7	11.9	28.9	31.8	15.9	3.8
Lignin-H[a]	13.5	10.3	17.3	32.7	18.7	7.5

[a] TOSS spectra

percent, table 1). The litter sample was made up of leaves that had been on the ground for an indeterminate time and were brown, thin and dry, but still intact. The transition from fresh foliage (fig. 2a, 2b) to old litter (fig. 2f) is accompanied by a general broadening of the peaks, especially for the phenolic and aromatic (note the loss of resolution between 145 and 154 pm). On the other hand, the intensity distribution hardly changes from fresh leaves, and the DD spectrum still has a prominent feature at 106 ppm. These minor changes in the NMR spectrum are in contrast to the PA results, which show a dramatic drop in tannin content from leaves to litter (table 1). Tannins constitute up to 21 percent of salal weight; after senescence, they are rapidly lost or rendered invisible to the PA assay but may still give a characteristic signature in the NMR spectra.

The spectra of young stems from both plots, of older woody stems, and of roots (mostly rhizomes) are similar. Compared to foliage and litter, they have a much higher proportion of O-alkyl C and less cutin and aromatic C. The main alkyl peak is at 21 ppm, most likely from acetate, and there is a stronger peak from methoxyl at 57 ppm. The phenolic region still has two peaks at 145 and 154 ppm, but the latter has higher intensity, and the peaks are broader with less splitting. This is associated with a higher lignin content, as shown by the stronger methoxyl signals in both the normal and DD spectra. Although it is premature to use the NMR spectra to quantify proportions of salal lignin and tannin, the spectra do give consistent indications. With increasing proportions of tannin/lignin (old stems to fresh stems to roots), the ratio of the 144/154 ppm peaks increases in both the normal

Table 3. Tannin levels (mg/g) in humus from ^{15}N Plots. Data are means from two plots and two pits per plot

	Uncontrolled Salal		Salal Control	
	0-10 cm	10-20 cm	0-10 cm	10-20 cm
Cedar	2.36	1.45	0.97	0.46
Hemlock	3.49	1.82	1.43	0.64
Spruce	2.28	1.34	2.17	1.50

and DD spectra. For old stems, the peak at 144 ppm is reduced to a shoulder in the DD spectrum. With increasing proportions of lignin, the methoxyl peak becomes stronger; in addition, the aromatic peak becomes broader, and its maximum shifts from 131 to 135 ppm (although the latter cannot be seen at the scale of figure 2).

The PA analysis of 48 individual humus samples from the N-15 plots showed tannin contents up to 0.44 percent. For plots with western red cedar and western hemlock seedlings, mean values for tannin were lower at 10–20 cm than at 0–10 cm, and lower in salal removal plots compared to those with uncontrolled salal growth (table 3). For the Sitka spruce plots, however, there was much less decline of tannin with depth, while at 10–20 cm, tannin content was higher for plots with salal removal. The spruce trees also had the least growth of the three species. However, results for all analyses were very variable, and a more extensive field study is in progress.

One interesting field observation was that the plots with salal removal appeared to have subsided in height compared to their surroundings. Further, during analysis of the samples from 10 cm increments, we found that the salal-removal plots had more occurrence of mineral soil (i.e., less than 15 percent C) at 10–20 cm. With aboveground clipping for several years, salal belowground biomass declines, and the rhizomes die and decay. Thus, salal control may have resulted in loss of humus, whether by general reduction of input or the specific effects of salal polyphenolics in humus building.[10]

NMR spectra were obtained for eight of the humus samples, four of which are shown in figure 3. Our purpose was to see whether the concentration extremes found in the tannin assay were reflected in the NMR spectra. The spectra have features similar to those previously observed for humus[2,25–26] and also to those of the salal components described above, and therefore will not be described in detail. All of the alkyl C signals show a slight splitting, with maxima at 30 and 33 ppm. Both of these come from CH_2 in long chains, but the former is more persistent in the DD spectra, indicating that it comes from components with greater molecular motion in the solid state.[26] Coarse woody debris is a prominent feature of these

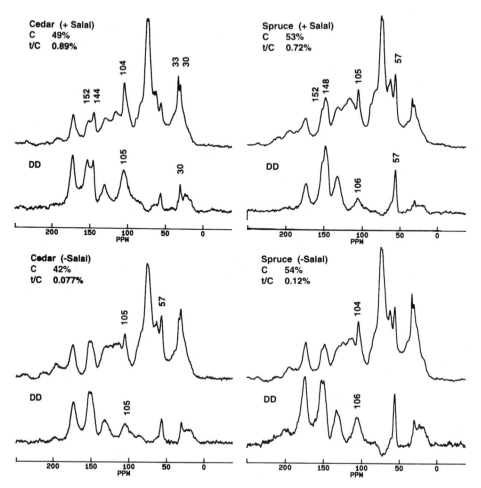

Figure 3. Carbon-13 CPMAS NMR spectra of humus from plots with uncontrolled salal growth (+Salal) and with continuous salal removal (–Salal). Normal CPMAS (top) and DD-TOSS (bottom).

old-growth forests, and three of the four samples have features indicating a high content of guaiacyl lignin (methoxyl C at 57 ppm, and phenolic peak with maximum at 147 ppm and shoulder at 152 ppm).[16,21]

Tannin contents (expressed as a percentage of C content) range from 0.077 percent to 0.89 percent for the samples in figure 3. All have a broad peak at 105–106 ppm in the DD spectra, but only the cedar (+ salal) sample has a distinct peak at 144 ppm, which comes from two carbons in the B ring of both PC and PD structures.[19,20] This feature can also be seen in figure 4a, the composite humus used for extraction. The spruce (+ salal) humus has a similar percent t/C content (0.72 percent), but the phenolic and aromatic region is dominated by the lignin

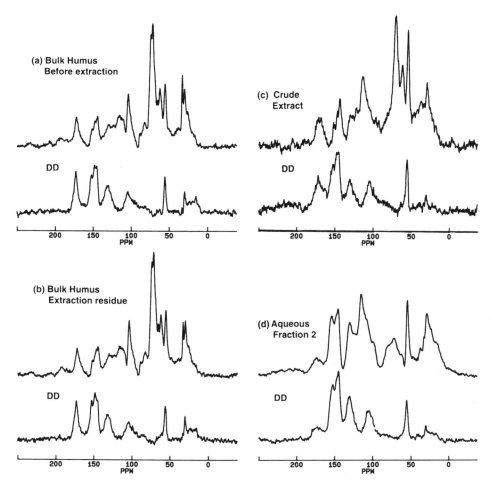

Figure 4. Carbon-13 CPMAS NMR spectra of composite humus before (a) and after (b) extraction, crude extract (c) and Fraction 2 from aqueous phase (d). Normal CPMAS (top) and DD-TOSS (bottom) except Fraction 2 (both TOSS).

component. Based on the eight spectra obtained, humus samples with higher tannin assays are more likely to have a distinct peak or shoulder at 144 ppm, although this feature may be masked by high lignin content. On the other hand, none of the samples with low tannin content had a distinct peak at 144 ppm. The 154 ppm signal from three carbons of the A ring was never clearly detected in the phenolic region of the humus spectra, but all samples had the diagnostic feature for tannins, a broad peak at 106 ppm in the DD spectrum.[24]

Tannins must have a significant input to humus in many forest ecosystems, but it is not known how they may be decomposed and/or incorporated into humic structures. Clearly, they leave traces in the humus NMR, in some cases, far beyond what may be indicated by low PA assays. Where decomposition is restricted by factors such as acidity, low temperature, high moisture, lack of earthworm activity, or plant inputs high in "recalcitrant" components, it tends to proceed with an increase in the ratio of alkyl/O-alkyl C in the NMR spectrum.[25] The retention of spectral features typical of condensed tannins may be another indicator of restricted decomposition, although one easily masked by the lignin peaks or lost in poor-quality spectra. A previous NMR study of humus from CH and HA stands found that these tannin "fingerprints" were a little stronger in CH than in HA stands where decomposition and nutrient cycling are more rapid.[2] The marked reduction of tannin content from fresh foliage to litter to humus is consistent with other studies,[18,27-31] whereas persistence of tannin features in the spectra may be associated with more complex, insoluble, or modified tannins ("brown polyphenolic pigments").[11,14] In contrast to the low tannin levels usually found in humus, we detected 3 to 4 percent tannins in the forest floor of black spruce stands in northern Ontario (abstracts, this meeting).

4. TANNIN-RICH FRACTIONS FROM HUMUS

Extracting and characterizing condensed tannins from humus have met with little success.[28,31-32] We used the procedure for plant material to extract a composite humus assaying at 0.25 percent tannin, the only problem being a large amount of stable emulsion that was processed separately. Table 4 shows that the sum of the two crude fractions accounted for 2.169 percent of mass and 25.40 percent of tannin in the humus (N was not determined). For the sum of the fractions from the Sephadex LH-20 column, recoveries were 1.86 percent of mass, 26.90 percent of tannin, and 1.07 percent of N. Fractions 2 and 2B with tannin assays of 64.4 and 50.7 mg/g accounted for 10.55 percent and 8.73 percent, respectively, of humus tannin. Fractions 3 and 4 had even higher tannin assays (281 and 178 mg/g), but the yields were very small. It is noteworthy that all of the fractions were depleted in N relative to the starting humus.

Spectra from the humus fractionation are shown in figures 4 and 5, and the relative areas of the NMR spectra in table 2b. Extraction removed less than 10 percent of the humus mass, so that the residue spectrum (fig. 4b) was similar to that of the starting humus (fig. 4a). The residue has less intensity at 33 ppm, the peak corresponding to long-chain alkyl C of a more rigid nature. The intensity distribution of the crude extract differed only slightly from the starting humus, being slightly higher in alkyl C and lower in O-alkyl C. However, there were differences within the 93–140 ppm region (di-O-alkyl + aromatic C): the original humus had a sharp peak for di-O-alkyl C at 105 ppm, whereas the crude extract had a stronger aromatic signal at 115 ppm, and very little intensity for di-O-alkyl C. This indicates that most of the extract O-alkyl C does not originate from easily recognizable carbohydrate moieties. Purification with Sephadex LH-20 removed much of the O-alkyl intensity.

Figures 4d and 5a show spectra of fractions 2 and 2B with 64.4 and 50.7 mg/g tannin. Spectra of fractions 2, 2B, and 4B (not shown) were similar, with minor variations of the relative peak areas (table 2b). They contain high proportions of alkyl, aromatic, and phenolic C and a sharp methoxyl peak. The features indicate

Table 4. Analysis and percentage recovery of mass, N, and tannins in the humus extraction

	Mass		Tannin				Nitrogen			
	g	%	mg/g	mg	%	E^d	mg/g	mg	%	E
Initial	391.60	100.00	2.5	979	100.00	1	7.66	2999.7	100.00	1
Residue	353.80	90.35	1.2	425	43.41	0.48	8.32	2943.6	98.13	1.08
AQ[a], Crude	4.500	1.149	29.3	131.85	13.47	11.7				
1 MeOH	2.053	0.524	9.4	19.29	1.97	3.8	6.2	12.73	0.42	0.81
2 Acetone	1.614	0.412	64.0	103.30	10.55	25.6	2.4	3.87	0.13	0.31
3 Acetone	0.027	0.007	281.0	7.58	0.77	112.4				
4 Acetone	0.080	0.020	178.2	14.31	1.46	71.3				
Sum, Fr AQ	3.774	0.963		144.48	14.75			16.60	0.55	
EM[b], Crude	4.000	1.020	29.2	116.80	11.93	11.7				
1 MeOH	1.024	0.261	4.9	5.02	0.51	1.96	5.2	5.32	0.18	0.68
2A Acetone	0.525	0.134	19.8	10.40	1.06	7.92	3.1	1.63	0.05	0.40
2B Acetone	1.686	0.430	50.7	85.48	8.73	20.28	5.2	8.77	0.29	0.68
4A Acetone	0.280	0.072	64.5	18.06	1.85	25.8				
Sum, Fr. EM	3.515	0.897		118.96	12.15			15.72	0.52	
Sum, Fractions	7.289	1.860		263.44	26.90	14.46		32.32	1.07	
Recovery[c]	361.09	92.21		688.44	70.31			2975.92	99.20	

[a] aqueous fraction; [b] emulsified fraction; [c] Sum of residue plus Sephadex LH-20 fractions; [d] enrichment factor

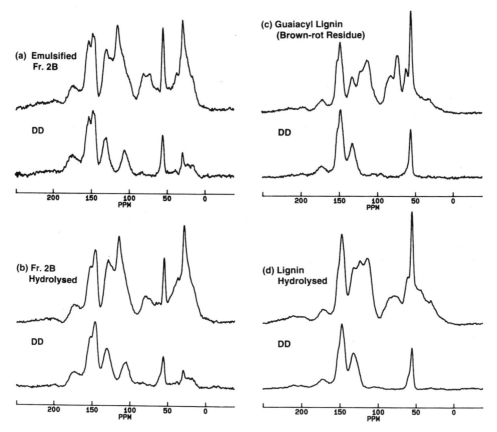

Figure 5. Carbon-13 CPMAS NMR spectra of fraction 2B (a) and brown-rot lignin (c) and after acid hydrolysis (b, d). CPMAS-TOSS (top) and DD-TOSS (bottom).

a high proportion of lignin-derived structural components. For fraction 2 (fig. 4d), the phenolic peak is more characteristic of tannin, with maxima at 145 and 154 ppm. The tannin characteristics of the phenolic region become more attenuated from fraction 2 to 2B and 4B, but all have a broad signal at 106 ppm in the DD spectra. The methoxyl peak of lignin is also present in both normal and DD spectra. There is no distinct peak in the di-O-alkyl region, and the broad O-alkyl signal most likely originates from the side-chain of lignin and the C2 and C3 of tannins rather than carbohydrate. There is also a high proportion of alkyl C with little internal mobility as it is largely lost in the DD spectra. This may have originated from cutin or from microbial biomass produced during decomposition. Overall, these fractions appear like leaves that have undergone loss of most extractives, carbohydrates, and protein. It is interesting to speculate that they may be fragments of a foliar biopolymer composite of lignin, tannin, and cutin. This is consistent with a previous study showing that the so-called "Klason lignin" of foliage incorporated these three insoluble biopolymers,[17] and it is inconsistent with a tannin-protein complex.

Acid hydrolysis of fraction 2B caused remarkably little change in the NMR spectrum (figs. 5a, b, table 2b). By contrast, brown-rot lignin (figs. 5c, d) produced the classic spectrum of Klason lignin, with the changes caused by loss of side chains and increased condensation through C—C coupling and CH_2 bridges.[21] This indicates a highly stable, rigid, cross-linked structure for these tannin-rich humus fractions.

5. CONCLUSIONS

Although this study originated with concerns about the effects of salal on forest regeneration on northern Vancouver Island, the results should have more general applicability. Salal is one of several ericaceous shrubs associated with soil degradation and reversion to heathland. Salal components are high in condensed tannins (of which 89–95 percent is extactable by acetone/water) from a low of 5 percent in old woody stems to a high of 21 percent in foliage. Old salal litter had less than 1 percent tannin, of which only 26.5 percent was extractable, but the NMR spectra were not consistent with such a dramatic drop in tannin content.

The PA assay showed up to 0.4 percent tannin in humus, and the characteristic tannin phenolic fingerprint could be detected in some humus with higher tannin contents. This was sometimes masked by high lignin content in the humus, but a distinctive tannin peak could still be seen at 106 ppm in DD spectra. Longterm salal control by clipping resulted in a decrease in humus tannin content, as well as some humus subsidence. Tannin-rich fractions (5–6 percent CT assay) prepared from the humus accounted for about 25 percent of the humus tannins. Their spectra suggested a structural complex of lignin, tannin, and cutin, and they were depleted in both carbohydrate C and in N.

Results from this study, previous SCHIRP studies, and other heath/forest ecosystems indicate that a variety of factors operate to push a site toward acid mor humus, low nutrient availability, and domination by ericaceous shrubs. High plant tannin content may be one such factor, whose influence increases when other circumstances are inimical to decomposition, including acid parent material, low temperature, excessive or insufficient moisture, and lack of earthworm activity. Retention of chemically recognizable condensed tannin in humus may be both a symptom and cause of restricted decomposition and nutrient cycling. A better understanding of the transport and transformation (chemical or biotic) of tannins in these ecosystems is essential in developing management strategies to enhance the desired direction of ecosystem development.

ACKNOWLEDGMENTS

We thank the members of the SCHIRP group over many years, co-op students Andrea Ellis, Carol Cutforth, and Derek Wong, graduate student L. de Montigny, and PFC technicians K. McCullough and A. Harris for invaluable field and laboratory assistance.

REFERENCES

1. Prescott, C.E.; Weetman, G.F. Salal Cedar Hemlock Integrated Research Program: a synthesis, Faculty of Forestry, University of British Columbia, Vancouver (1994).

2. de Montigny, L.E.; Preston, C.M.; Hatcher, P.G.; Kögel-Knabner, I. Comparison of humus horizons from two ecosystem phases on Northern Vancouver Island using ^{13}C CPMAS NMR spectroscopy and CuO oxidation. *Can. J. Soil Sci.* 73:9 (1993).

3. Messier, C. Factors limiting early growth of western redcedar, western hemlock and Sitka spruce seedlings on ericaceous-dominated clearcut sites in coastal British Columbia. *For. Ecol. Manage.* 60:181 (1993).

4. Messier, C.; Kimmins, J.P. Above- and below-ground vegetation recovery in recently clearcut and burned sites dominated by *Gaultheria shallon* in coastal British Columbia. *For. Ecol. Manage.* 46:275 (1991).

5. Chang, S.X.; Preston, C.M.; McCullough, K.; Weetman, G.F.; Barker, J. Effect of understory competition on distribution and recovery of ^{15}N applied to a western red cedar-western hemlock clearcut site. *Can. J. For. Res.* 26:313 (1996).

6. Bernier, N. Altitudinal changes in humus form dynamics in a spruce forest at the montane level. *Plant Soil* 178:1 (1996).

7. de Montigny, L.E.; Weetman, G.F. The effects of ericaceous plants on forest productivity. *In*: Titus, B.D.; Lavigne, M.B.; Newton, P.F.; Meades, W.J. (eds.). The silvics and ecology of boreal spruces. *For. Can. Inf. Rep.* N-X-271, p. 83 (1990).

8. Titus, B.D.; Sidhu, S.S.; Mallik, A.U. A summary of some studies on *Kalmia angustifolia* L.: a problem species in Newfoundland forestry. *Can. For. Ser. Inf. Rep.* N-X-296 (1995).

9. Leake, J.R.; Read, D.J. The effects of phenolic compounds on nitrogen mobilization by ericoid mycorrhizal systems. *Agric. Ecosys. Environ.* 29:225 (1989).

10. Northup, R.R.; Dahlgren, R.A.; Yu, Z. Intraspecific variation of conifer phenolic concentration on a marine terrace soil acidity gradient; a new interpretation. *Plant Soil* 171:255 (1995).

11. Handley, W.R.C. Further evidence for the importance of residual leaf protein complexes in litter decomposition and the supply of nitrogen for plant growth. *Plant Soil* 15:37 (1961).

12. Howard, P.J.A.; Howard, D.M. Ammonification of complexes prepared from gelatin and aqueous extracts of leaves and freshly-fallen litter of trees on different soil types. *Soil Biol. Biochem.* 9:1249 (1993).

13. Kuiters, A.T. Role of phenolic substances from decomposing forest litter in plant-soil interactions. *Acta. Bot. Neerl.* 39:329(1990).

14. Toutain, F. Activité biologique des sols, modalités et lithodépendance. *Biol. Fert. Soils* 3:31 (1987).

15. Jones, W.T.; Broadhurst, R.B.; Lyttleton, J.W. The condensed tannins of pasture legume species. *Phytochemistry* 15:1407 (1976).

16. Preston, C.M.; Trofymow, J.A.; Niu, J.; Fyfe, C.A. ^{13}C CPMAS NMR spectroscopy and chemical analysis of coarse woody debris in coastal forests of Vancouver Island. *For. Ecol. Manage.* 111:51 (1998).

17. Preston, C.M.; Trofymow, J.A.; Sayer, B.G.; Niu, J. ^{13}C nuclear magnetic resonance spectroscopy with cross-polarization and magic-angle spinning investigation of the proximate analysis fractions used to assess litter quality in decomposition studies, *Can. J. Bot.* 75:1601 (1997).

18. Tiarks, A.E.; Meier, C.E.; Flagler, R.B.; Steynberg, E.C. Sequential extraction of condensed tannins from pine litter at different stages of decomposition. *In*: Hemingway, R.W.; Laks, P.E (eds.) Plant polyphenols—synthesis, properties, significance. Plenum Press, New York. p. 597 (1992).

19. Czochanska, Z.; Foo, L.Y.; Newman, R.H.; Porter, L.J. Polymeric proanthocyanidins. Stereochemistry, structural units, and molecular weight. *J. Chem. Soc. Perkin Trans. I*:2278 (1980).

20. Newman, R.H.; Porter, L.J. Solid-state ^{13}C-NMR studies on condensed tannins. *In*: Hemingway, R.W.; Laks, P.E (eds.) Plant polyphenols—synthesis, properties, significance. Plenum Press, New York. p. 339 (1992).

21. Preston, C.M.; Sollins, P.; Sayer, B.G. Changes in organic components for fallen logs in old-growth Douglas-fir forests monitored by ^{13}C nuclear magnetic resonance spectroscopy. *Can. J. For. Res.* 20:1382 (1990).

22. Preston, C.M.; Sayer, B.G. What's in a nutshell: an investigation of structure by carbon-13 cross-polarization magic-angle spinning nuclear magnetic resonance spectroscopy. *J. Agric. Food Chem.* 40: 206 (1992).

23. Hatcher, P.G. Chemical structural studies of natural lignin by dipolar dephasing solid-state ^{13}C nuclear magnetic resonance. *Org. Geochem.* 11:31 (1987).

24. Wilson, M.A.; Hatcher, P.G. Detection of tannins in modern and fossil barks and in plant residues by high-resolution solid-state ^{13}C nuclear magnetic resonance. *Org. Geochem.* 12:539 (1988).

25. Baldock, J.A.; Preston, C.M. Chemistry of carbon decomposition processes in forests as revealed by solid-state ^{13}C NMR. *In*: Kelly, J.M.; McFee, W.W. (eds.) Carbon forms and functions in forest soils, *Soil Sci. Soc. Am.*, Madison, WI, p. 89 (1995).

26. Kögel-Knabner, I.; Hatcher, P.G.; Tegelaar, E.W.; de Leeuw, J.W. Aliphatic components of forest soil organic matter as determined by solid-state ^{13}C NMR and analytical pyrolysis. *Sci. Tot. Environ.* 113:89 (1992).

27. Beck, G.; Dommergues, Y.; Van den Driessche, R. L'effet litière. II. Étude expérimentale du pouvoir inhibiteur des composés hydrosolubles des feuilles et des litières forestières vis-a-vis de la microflore tellurique. *Oecol. Plant.* 4:237 (1969).

28. Coulson, C.B.; Davies, R.I.; Lewis, D.A. Polyphenols in plant, humus, and soil I. Polyphenols of leaves, litter, and superficial humus from mull and mor sites. *J. Soil Sci.* 11:20 (1960).

29. Nikolai, V. Phenolic and mineral content of leaves influences decomposition in European forest ecosystems. *Oecologia* 75:575 (1988).

30. Racon, L.; Sadaka, N.; Gil, G.; le Petit, J.; Matheron, R.; Poinsot-Balageur, N.; Sigoillot, J.C.; Woltz, P. Histological and chemical changes in tannin compounds of evergreen oak litter. *Can. J. Bot.* 66:663 (1988).

31. Schofield, J.A.; Hagerman, A.E.; Harold, A. Loss of tannins and other phenolics from willow leaf litter. *J. Chem. Ecol.* 24:1409 (1998).

32. Rice, E.L.; Pancholy, S.K. Inhibition of nitrification by climax ecosystems. II. Additional evidence and possible role of tannins. *Am. J. Bot.* 60:691 (1973).

CONSTITUENT AND INDUCED TANNIN ACCUMULATIONS IN ROOTS OF LOBLOLLY PINES

Charles H. Walkinshaw

Southern Research Station
USDA Forest Service
Pineville, Louisiana 71360
USA

1. INTRODUCTION

Loblolly pine (*Pinus taeda* L) has become the most important source of wood fiber in the southern United States. This tree is an excellent competitor and recovers well from a variety of adverse conditions.[1,2] The value of loblolly pine as a plantation tree has encouraged many researchers to devote their careers to the study of this species. It is surprising then that the root system of loblolly pine has received so little attention! I felt that an anatomical study of loblolly pine roots would be valuable and chose to measure tannin abundance as a primary trait to evaluate root health at the microscopic level.

This histological study of tannin in pine roots follows extensive microscopy of pine cells in culture and in stem tissue begun in 1963.[3] Of particular interest throughout my studies is the manner in which tannins are contained in cellular organelles until tissue death or cell membrane rupture. These tannins are nonspecifically absorbed by cell walls and organelles. These processes are easily observed in loblolly pine roots that have a high turnover rate, especially cortical cells of mycorrhizal roots.

Nutritional tests with pine callus cells indicate that cells accumulate tannin within cell membranes of the endoplasmic reticulum.[4-6] When nutrients such as nitrogen are altered, many callus cells fill with tannin. This seems also to occur in pine roots on certain sites. The hypothesis of this study was that two processes of tannin accumulation occur in root tissue. Constituent tannins accumulate in all root tissues, whereas the process controlling induced tannin is related to stress. Also, it appears that tannin has a close relationship to the deposition of cellulose and lignin in cell walls. After observing many thin sections, I focused on anatomical differences between constituent tannin accumulation in periderm cells and induced tannin in cells responding to stress. This paper describes the types of

Plant Polyphenols 2: Chemistry, Biology, Pharmacology, Ecology, Edited by
Gross et al. Kluwer Academic / Plenum Publishers, New York, 1999

tannins in roots and quantifies their occurrence in plantation-grown loblolly pines.

2. MATERIALS AND METHODS

Observations of tannin anatomy were made of roots from 5- to 40-year-old loblolly pines in plantations. Ten trees were selected at random from each plantation. Each plantation was sampled from one to six times. A total of 2,619 roots were collected in Louisiana, 623 in Mississippi, and 1,221 in North Carolina.

After the observational study, tannin in roots from different levels of soil compaction and organic matter was quantified. The soil treatments have been described in detail.[7] Fourteen roots from 5 to 10 trees in 9 soil treatments were collected for a total of 1,017 roots.

An experiment to measure effects of severe stress on tannin accumulation used 1,447 roots from 68 trees that were progeny of control-pollinated parents. Treatments were felling, girdling, and severe pruning of crowns. Experimental trees were in their twelfth growing season and were located near the plantation's center. In all cases, roots were collected from a 25- by 25-cm area located 1 m from the stem of the pine. Roots were taken from a depth of 2 to 20 cm and the roots were gently shaken to remove excess soil. The center 2 to 4 cm of each root was excised and placed into formalin-acetic acid-alcohol (FAA) fixative for 14 days.[8] Fixed-root specimens were recut to 1 to 3 mm, dehydrated in alcohol series, embedded in paraffin, and cut into 7 to 10 micron transverse sections. Sections from 12 to 18 roots were mounted on a slide. Nine slides were stained with a variety of schedules, including Papanicolaou's schedule, an acid-Schiff procedure, toluidine blue, ferric salt, safranin-aniline blue, acid fuchsin, Giemsa, Congo red, and Groett's methenamine.[9–11] Polarized light was used to differentiate tannins and primary cell walls from secondary cell walls, sclereid cells, and starch granules.[12]

Root traits were defined as follows: *Cortex shedding*: Cortical cells are dead and either remain attached to the root or are released into the soil. The shed is invaded by many soil microbes. *Periderm forming*: This is the first stage of coating residual cortical cells with a protective layer of tannin and cellulose:lignin complex. This is constituent tannin. *Bark formation*: Bark cells encompass the root. *Degradation of starch*: Starch grains are in 50 percent of the cells. *Induced tannin*: Tannin-containing cells accumulate in cortex, rays, and inner xylem. These cells number from 10 to over 100. *Dead mycorrhizae*: Short roots that shed in cortical cell death. *New lateral roots*: Roots that originate in the xylem and phloem. *Infection*: Cambial or fiber cells infected. Invasion usually kills the tissues and the root. *Starch grains*: Number of plastids per cell with starch as viewed at a single focal length at 100 to 500 diameters.

For statistical analysis in soil compaction studies, trees within plots were used as replicates. This does not fit the rigid requirement of randomization. In the stress studies, analysis of variance was used to separate the means of the treatments. Regression analysis was used to approximate relationships between scored traits.

3. CONSTITUENT TANNINS

The constituent tannin component of the periderm appears immediately after the shed of the root cortex. The proportion of roots with this tannin was statistically different when soil bulk density was altered (table 1). As soil compaction decreased and organic matter increased, the synthesis of bark increased. The anatomy of tannin in periderm cells was similar for all soil treatments. This similarity was seen for both the tannin and cellulose and lignin containing cells in roots and stems of loblolly pines.

A typical loblolly pine root in secondary growth (fig. 1) uses starch (fig. 2) and produces periderm (bark) cells (figs. 3,4). Many layers accumulate to reduce the initial vulnerability that is created by the shedding of cortical cells. Roots that are 1 mm or shorter may not synthesize sufficient cellulose and lignin elements. Excess tannin cells temporarily accumulate (fig. 5). The incidence of periderm without tannin (fig. 6) occurred in less than one root per thousand.

Tannin in bark cells persisted in all stress treatments and well beyond cortical and cambial death (figs. 7,8). Five or more layers of cells with apparently normal tannin were seen in dead roots.

4. INDUCED TANNIN

Induced tannin accumulated in all parts of the root tissue. It was significantly higher ($p = 0.001$) in roots of loblolly pines that were 1 mm in diameter. For undisturbed trees, the proportion of roots with non-periderm tannin was 0.80, 0.37, and 0.05 for roots 0–1 mm, 1.1–10 mm, and 11 mm or greater in diameter, respectively. The mean for proportion of tannin in 2,000 roots was 0.12. The proportion of roots with induced tannin was negatively correlated with numbers and size of starch grains in parenchyma cells ($R = -0.56$ and -0.50, respectively). All sectioned roots from undisturbed trees were normal using the criteria of stain affinity for nuclei and cambial elements.

To test the effects of mild stress on tannin accumulation, roots were collected from 6-year-old lobllolly pines. Of the variety of soil treatments that were installed

Table 1. Effect of soil treatments on incidence of root processes

Trait	Severe compaction, O.M. removed	No compaction O.M. left	Probability significance
	-----------------Percentage of root sections-----------------		
Periderm/tannin present	52	28	0.002
Bark cells with tannin	43	70	0.008
Depletion of starch	29	6	0.008

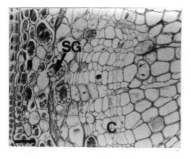

Figure 1. Typical transverse view of cells in a loblolly pine root. Note cambium (C) and starch grains (SG). The scale bar is 50 microns in all figures.

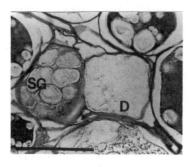

Figure 2. Stages of starch grain (SG) utilization. Depleted of starch (D).

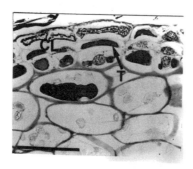

Figure 3. Cellulose: lignin (CL) and tannin (T) components of the developing periderm of a root in secondary growth.

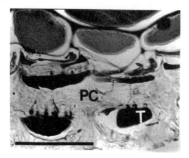

Figure 4. Completed periderm cells (PC) with constituent tannin (T).

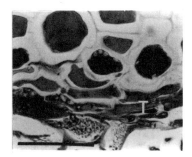

Figure 5. Excess tannin (T) accumulating at the site of periderm formation.

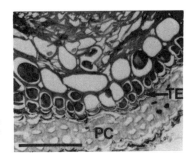

Figure 6. Periderm cells (PC) that lack tannin. Note developing tannin elements (TE) in new periderm.

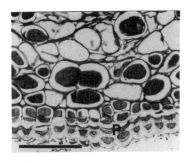

Figure 7. Intact periderm (P) in a root starved for 5 months.

Figure 8. Multilayered bark cells enclosing a dead root. Cortical cells (CC) lack structure.

in standard plots, only the severely compacted soil and soil with total aboveground biomass removed before planting showed differences. These rather extreme treatments did not affect tannin accumulation directly. However, periderm formation slowed. Induced tannin was low in all the root samples. Roots were healthy, and wounds were not common.

Roots were starved by felling or girdling stems. Starch was reduced in roots by removing 50 to 67 percent of the crown. Other roots were killed by prescribed burning.[13] After 4 or 5 months, roots from all felled trees began to accumulate more tannin than uncut controls. Roots from felling in the winter were statistically different after 5 months. Those from spring felling were statistically different after 4 months. The proportion of large roots with induced tannin for winter felling was 0.17 as compared to 0.54 for spring felling and 0.05 for the control. After 6 months, the mean for the large roots of trees felled in the winter was 0.83 (table 2). Tree values for proportion of roots within a sampling period ranged from 0.20 to 1.00, but they were not significantly different. However, ANOVA with month as the class showed month after felling to be significant (P = 0.001). Proportion of roots with tannin 0–6 months after winter felling was not correlated to the number or condition of starch grains in roots (table 3). Observations of nuclei and cambial cells showed 61 percent of the roots were dead 6 months after winter felling. Since tannin was highest in roots 5 months after felling, this treatment period was used in sampling girdled or crown-pruned trees. The proportion of roots with tannin was 0.31 for girdled and 0.50 for severely pruned trees. Nuclei and cambial cells were normal in both treatments and not significantly different from the controls. Number and size of starch grains were significantly lower than controls and approximately twice as high as winter felling tree roots in the 5-month collection.

The number of cells with induced tannin in roots was highly variable. Roots from control trees and trees in soil treatments had low induced tannin. However, roots that were starved for at least 4 to 5 months or subjected to high heat contained high amounts of induced tannin. Examples of induced tannin

Table 2. Incidence of observed variables following winter felling of loblolly pines[a]

Month after felling	Starch use	Induced tannin	Cambium abnormal
	---------------------------Proportion of roots with trait-------------------		
0	0.94	0.26	0.0092
1	0.95	0.53	0.0
3	0.96	0.53	0.024
4	0.78	0.38	0.027
5	0.30	0.63	0.384
6	0.0	0.83	0.611

[a] Traits were tabulated for 612 roots that were sectioned and stained.

Table 3. Relationship between starch and tannin after winter felling[a]

Month after fell	Number of roots	Starch grains vs tannin	Starch grain use vs tannin
		------------Pearson correlation coefficients------------	
0	109	-0.34	-0.02
1	130	-0.22	-0.21
3	126	-0.27	-0.19
4	112	-0.38	-0.34
5	117	-0.38	-0.35

[a] Roots from non-felled trees averaged -0.50 and -0.24 for the two comparisons.

accumulations are illustrated in figures 9–12. The tannin in figures 11 and 12 is accumulated in cortical-derived and modified thin-walled cells. As starvation progressed in stress treatments, cell structure disintegrated, and tannin bound to cell wall elements (fig. 13).

Fire was used to cause rapid death of roots. A thinned 40-year-old plantation was prescription-burned, and roots were collected at 2 to 20 cm depth after 30 days. Roots without lesions were pooled from 10 trees, fixed for light microscopy. Examination of 46 roots revealed 22 with dead nuclei and cambial elements. Their structure was normal, but induced tannin coated the cell walls of the cortex (fig. 14).

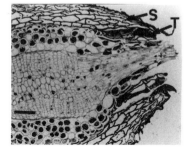

Figure 9. Emergence of a lateral root that is protected by tannin (T). Note shed (S) composed of dead cortex.

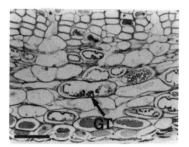

Figure 10. Granular tannin (GT) that is accumulating in the cortex of a root in nutritional stress.

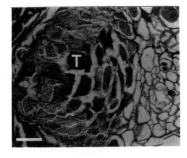

Figure 11. Tannin (T) accumulating in a wound of a root. Note the absence of a secondary wall matrix.

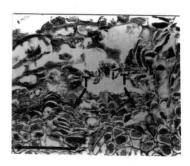

Figure 12. Tannin (T) being deposited to seal a severe wound affecting 40 percent of a pine root. Elements of tannin appear similar to those in figure 3.

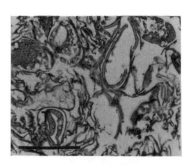

Figure 13. Appearance of cortex:cambial region of a root that died from starvation. Note tannin (T) has bound to dead cell organelles.

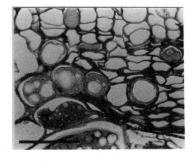

Figure 14. Pine root that has died from the heat of a burn. Much of the tissue morphology has been preserved.

Plastids were embedded in tannin, but their starch appeared to be intact. Walls of cambial cells were coated with tannin in some roots. Induced tannin that was contained within membrane-bound bodies in cells was released in most of the roots killed by high temperature. This pattern was quite different from the effects of drought on tannins reported by Pizzi and Cameron.[14]

5. CONCLUSIONS

The histological observations and analyses of deposition of tannin in loblolly pine roots confirm that two anatomical systems govern the accumulation. The first system is highly reliable since less than 0.1 percent of root periderms lacked tannin. Tannin in the periderm of roots often was deposited without the cellulose and lignin complex. Constituent tannin is apparently extremely durable in the periderm. Neither contact with soil microbes nor death of the underlying cortex generally changed the appearance of this tannin to a significant extent. Oxidations from orange to black deposits were seen in only a few roots. The formation of the tannin, cellulose, and lignin periderm appeared sensitive to soil conditions. The unfavorable silviculture treatments caused a significant delay in bark formation that led to an increased time of exposure for roots with shed bark. Roots less than 1 mm in diameter appeared to compensate for this via increased tannin excretion. Tissues from roots of trees that were felled, girdled, and severely pruned were ideal for determining the relationships between starch utilization and induced tannin accumulation. However, I did not find a statistically significant relationship. I repeatedly observed many starch grains in cells with tannin. Likewise, there was an absence of both starch and induced tannin in many roots. Cells in the process of tannin accumulation seemed to be using a source of carbon other than starch grains. Nuclei in such cells stained as though they were actively metabolizing.

Staining of tannin was not a necessary emphasis in this study.[15] Emphasis was placed on using stains to classify as functional the nucleus and other cell components. These staining schedules often enhanced the natural color of the tannins. The most interesting finding was the discovery that nuclei were active in cells that appeared filled with tannin (figs. 3,4). This verified my earlier report about stem tissues and further indicated that tannin is normally isolated from other organelles in the pine cell.[16] This feature appears of major importance in containing tannin in wounds without inhibiting adjacent cells.

Continuous generation and repair of the periderm and cortical tissues of roots are needed to overcome adverse effects by soil and microbial actions.[17,18] Sealing the many wounds from dying root hairs, mycorrhizae, and short roots is an important role for induced tannin. Roots are especially vulnerable to microbes during shedding of cortical cells. The induction of tannins causes the conversion of the root cortical cell into an elongated tannin body (figs. 3,4,5). The cellulose and lignin unit in the periderm (fig. 3) did not form in conjunction with induced tannin synthesis (see figs. 11 and 12 for mass of tannin cells). The induced tannin was in thin primary cell walls. The mass of tannin cells was generally disordered (fig. 11). The stress appeared to be reflected in lignin synthesis. Stress-induced phenylpropanoid metabolism as a precursor of tannin has been recently reviewed by Dixon and

Paiva.[19] Also, Stafford discussed relationships between tannins and lignins[20] in general. However, the relationship between lignin and tannin formation appears to be different for constituent and induced tannin. This is an exciting area for further histochemical research.

REFERENCES

1. Clark, A.; Saucier, J.R. Tables for estimating total-tree weights, stem weights, and volumes of planted and natural southern pines in the southeast. Georgian Forest Research Paper 79:1–23 (1990).

2. Wright, J.W. Introduction to forest genetics. Academic Press, New York, 463 pp (1976).

3. Walkinshaw, C.H.; Jewell, F.F.; Walker, N.M. Callus culture of fusiform rust-infected slash pine. *Pl. Dis. Reptr.* 49:616 (1965).

4. Walkinshaw, C.H. Cellular changes that promote tannin formation in slash pine. *In:* Hemingway, R.W.; Laks, P.E. (eds.). Plant polyphenols: synthesis, properties, significance. Plenum Press, New York, pp. 97–110 (1992).

5. Baur, P.S.; Walkinshaw, C.H. Fine structure of tannin accumulations in callus cultures of *Pinus elliottii* (slash pine). *Can. J. Bot.* 52:615 (1974).

6. Hall, R.H.; Baur, P.S.; Walkinshaw, C.H. Variability in oxygen consumption and cell morphology in slash pine tissue cultures. *Forest Sci.* 18:298 (1972).

7. Tiarks, A.E.; Kimble, M.S.; Elliott-Smith, M.L. The first location of a national, long-term forest soil productivity study: Methods of compaction and residue removal. *In:* Proc. 6th Biennial Southern Silvicultural Research Conference VI:431 (1991).

8. Sass, J.E. Botanical microtechnique. Iowa State Press, Ames, Iowa, 228 p. (1951).

9. Chayen, J.; Bitensky, L.; Butcher, R.G. Practical histochemistry. John Wiley and Sons, New York, 271 pp. (1973).

10. Hass, E. 50 diagnostic special stains for surgical pathology. All-Type Editorial, Los Angeles, CA, 86 pp. (1980).

11. Preece, A. A manual for histologic technicians (3rd ed.). Little, Brown and Co., Boston, 263 pp. (1972).

12. Walkinshaw, C.H. Cell necrosis and fungus content in fusiform rust-infected loblolly, longleaf, and slash pine seedlings. *Phytopathology* 68:1705 (1978).

13. Fuller, M. Forest fires. John Wiley and Sons, New York, 238 pp. (1991).

14. Pizzi, A.; Cameron, F.A. Flavonoid tannins-structural wood components for drought-resistant mechanisms of plants. *Wood Sci. Technol.* 20:119 (1986).

15. Lees, G.L.; Suttill, N.H.; Gruber, M.Y. Condensed tannins in sainfoin.1. A histological and cytological survey of plant tissues. *Can. J. Bot.* 71:1147 (1993).

16. Walkinshaw, C.H. Are tannins resistance factors against rust fungi? *In:* Hemingway, R.W.; Karchesy, J.J. (eds). Chemistry and significance of condensed tannins. Plenum Press, New York (1989).

17. Atkinson, D.; Last, F. Growth, form and function of roots and root systems. *Scottish For.* 48:153 (1994).

18. Coutts, M.P. Developmental processes in tree root systems. *Can. J. For. Res.* 17:761 (1987).

19. Dixon, R.A.; Paiva, N.L. Stress-induced phenylpropanoid metabolism. *Pl. Cell* 7:1085 (1995).

20. Stafford, H.A. Proanthocyanidins and the lignin connection. *Phytochemistry* 27:1 (1988).

GEOCHEMISTRY OF TANNIN: METHODS AND APPLICATIONS

Peter J. Hernes and John I. Hedges

School of Oceanography
University of Washington
Seattle, Washington 98195-7940
USA

1. INTRODUCTION

Although a considerable body of knowledge now exists about tannin in general, the geochemistry of tannin is in its infancy. The geochemical perspective of tannin is quite different from previous work, taking as its prime motivation the role of tannin in the carbon cycle both quantitatively and in terms of biomarker potential. Net terrestrial and marine primary production is approximately 100×10^{17} gCy^{-1}.[1] Of this global primary production, only 0.2 percent passes through the ecosystem and is ultimately preserved in marine sediments. Carbon preservation is of interest because of its link to the atmosphere. For every mole of organic carbon buried, one mole of carbon dioxide is removed, and one mole of oxygen released. Ultimately, all of the oxygen in the atmosphere is due to organic carbon preservation and pyrite formation, and considerable effort has been made to understand all the factors and feedback mechanisms in these relationships.[2]

The role of tannin in the carbon cycle and carbon preservation is largely unknown. Riverine input of terrestrial organic carbon to the marine environment is approximately twice that of preservation in the sediments, suggesting that terrestrial organic carbon could potentially comprise a large component of preserved organic carbon. Stable carbon isotopes have shown that this is not the case, however, which presents a paradox: Riverine organic carbon represents the most refractory terrestrial organic carbon and yet more than 80 percent is remineralized upon entering the ocean.[3,4] Because tannins are quantitatively important (the fourth most abundant component of vascular plant tissue), their study may be an important piece in the solution of this paradox.

In addition to quantitative importance, molecular level tannins represent a unique class of biomarkers both in structure and reactivity that may lead to new understanding of carbon diagenesis. Hydrolyzable tannin, with its carbohydrate component and gallic acid derivatives, is very different from the polyflavanoid

Plant Polyphenols 2: Chemistry, Biology, Pharmacology, Ecology, Edited by
Gross et al. Kluwer Academic / Plenum Publishers, New York, 1999

nature of condensed tannin (fig. 1). As the name implies, the former is more soluble than the latter, and therefore the latter would be more likely to persist. Monomers of condensed tannin are also much more varied and stereochemically active than the basic building blocks of hydrolyzable tannin, and this has led some to attribute much different functionality within plants of the two types.[5] The combination of lower solubility and greater monomeric diversity highlights the biomarker potential of condensed tannins.

Few geochemical studies are aimed specifically at tannins. Historically, interest in tannins has stemmed from industrial uses. In recent decades, tannin research has expanded to natural products (where much progress has been made in terms of tannin structure and taxonomic surveys) and in a much more limited fashion to forestry. Research in forestry has tended to focus on the delivery of tannins to the soil (via leaves, needles, bark, and wood), the degradation of tannin in the soil, and the impact of tannin on nutrient recycling and organic matter degradation within the soil.[6] The methods employed are primarily bulk measurements of tannins, including ^{13}C NMR, various spectrophotometric methods, and tannin binding assays. Findings include the loss of 84 percent of extractable tannins from pine litter after 1 year and 92 percent loss after 2 years,[7] little difference between tannin content of trees vs. logs after 1 year on the ground,[8] and the control of nitrogen release from pine litter by tannin and less effective organic matter degradation when tannins are present.[9,10] The latter is of particular concern to forestry because of its impact on young trees in recent clearcuts and the overall decline in site quality (see Preston in this volume).[11] Despite the impacts of tannin

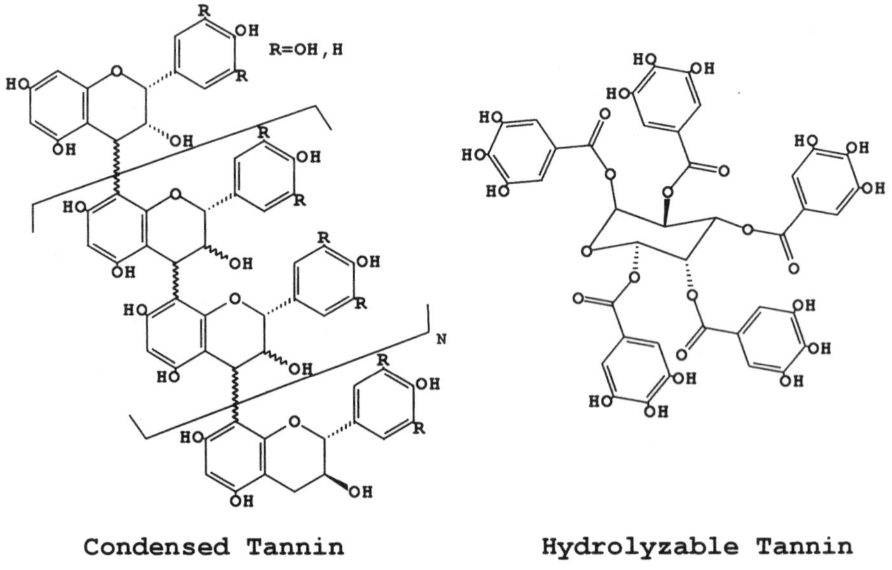

Condensed Tannin   **Hydrolyzable Tannin**

Figure 1. Structures of typical condensed and hydrolyzable tannins.

on soils, little or no research has been done at the molecular level, and even the bulk measurements have been questioned as to their accuracy or meaningfulness.[12,13]

Tannin geochemistry research in riverine and marine systems is virtually nonexistent, limited primarily to the monitoring of tannin leaching or degradation from leaves via bulk measurements.[14,15] We are unaware of any molecular-level studies, in spite of the fact that tannins constitute up to 20 percent of the leaf tissue[14] that is a major form of terrigenous organic matter cycling in these systems. For comparison with other organic biomarkers, lignin accounts for approximately 5 percent of leaf tissue,[14,16] carbohydrates 40–50 percent,[14,17] cutin 5–10 percent,[18] amino acids 5–10 percent,[19] lipids 5–10 percent,[15] and pigments ~1 percent.[20] Among other sources of organic matter to natural systems, tannin accounted for up to 17 percent of outer bark (western hemlock), 4 percent of inner bark (Douglas-fir), 1 percent of sapwood (Douglas-fir), and 0.6 percent of heartwood (western red cedar).[8] Tannins are a major component of leaf tissue and bark, and therefore can have a significant impact on the bulk properties of organic mixtures, such as aromaticity, organic carbon:nitrogen ratios, phenolic OH, color, and reactivity. Because of this importance, Benner et al.[14] concluded "Clearly, molecular-level methods for the characterization and quantification of tannins . . . in nonwoody vascular plant tissues need to be developed. Parallel application of such methods to soil and sediment samples may explain, in part, why a comparably large fraction of these organic materials remain uncharacterized."

Unlike carbohydrates, lipids, amino acids, and pigments—which are ubiquitous in organic matter and have both marine and terrestrial sources—tannins (along with lignin and cutin) are uniquely terrestrial. Thus, in addition to bulk importance, tannins have the potential to provide source information that is complementary to lignin and cutin. For instance, monocots are not distinguishable based on lignin composition. Although monocots do have a unique cutin signature, this signature can be overwhelmed by nonwoody gymnosperm or dicotyledon tissues.[18] Monocots, however, uniquely produce *ent*-epicatechin (2S and (+) optical rotation), a monomer found in condensed tannins.[21] *Ent*-epicatechin should be clearly resolvable from (–)-epicatechin with a chiral gas chromatography column. In addition, propelargonidin-containing polymers are much more common in monocots than dicots.[21] Hydrolyzable tannins, on the other hand, are notably absent from monocots. Though flavones can be found in angiosperms, they do not appear to be present in gymnosperms. In addition, condensed tannin dimers appear to contain more species-dependent taxonomic information among angiosperms that may be useful in certain environments in which the potential sources are more constrained.[22]

A challenge in measuring tannins at the molecular level in natural samples is that these polyphenols are very reactive and may undergo significant changes once organic matter senesces and becomes part of the litter. For instance, vicinal diols (such as those on the B-ring in many condensed tannins) can be easily oxidized to quinones in basic conditions.[23] Quinones, in turn, can participate in many condensation reactions, primarily due to the electron withdrawing nature of the carbonyl functional group. The non-carbonyl carbons become electron-poor and can form bonds very readily with electron-rich carbons (such as those on the A-ring of condensed tannin or other phenols) or any other strong nucleophile. Alternatively, the carbonyl groups on quinones and flavones can react directly with amines via a Schiff base reaction to form nitrogenous condensation products.

The hydroxy groups in tannin are also responsible for its complex-forming nature. A tannin polymer (either condensed or hydrolyzable) has a plethora of hydroxy groups to hydrogen-bond with proteins and amino acids and to complex with metals. Interestingly, a condensed tannin must be sufficiently large before significant protein complexation takes place—monomers, dimers, and other small oligomers apparently are not able to form enough cross-bridges to strongly complex with proteins.[24]

The ability to complex with proteins and amino acids leads to another geochemically significant trait of tannin mentioned previously—inhibition of organic matter degradation. At the right concentration, tannins are toxic. At $15\,\mathrm{mg\,L^{-1}}$, tannins have been known to cause fish kills.[25] Concentrations ranging from 325 to $3,000\,\mathrm{mg\,L^{-1}}$ have been reported to be inhibitory to methanogenic bacteria.[25] Concentrations of 1–2 percent tannin have been shown to reduce the overall decomposition of organic materials applied to soil.[25] In addition to complexing bacterial exoenzymes and directly slowing degradation, tannin may also bind up nitrogen sources used by heterotrophs for growth.

Despite the challenges of measuring tannins, there may be a wealth of information to be obtained from them specifically because of their highly reactive nature. Because of their redox and photochemical sensitivity, it may be possible to use tannins as an indicator of the environmental history of associated organic matter. For instance, tannins present in anoxic sediments may be able to tell us whether the sediments have been under constant or intermittent anoxia. In the Rio Negro, there is evidence for significant photochemical degradation in the upper few centimeters of the water column,[26] as well as evidence for a significant fraction of dissolved tannin in the Rio Negro that could account for this degradation.

Finally, tannins have long been suspected as precursors to humic materials via "autoxidation" when neutral to alkaline pH conditions prevail.[25] Again, this is due to the ease of formation of quinones and subsequent condensation reactions. Because so little is known about molecular-level tannins in natural samples, the role of tannins in humification is still largely theoretical. Perhaps molecular-level tannin analyses of natural samples coupled with isolation of humic substances and bulk characterization by NMR will provide a first look at the empirical relationship of tannins to humification.

2. METHODS

Several available techniques for measuring tannins can be used in geochemical studies, including (but not limited to) ^{13}C-NMR estimates,[14] protein-binding assays,[27] Folin-Denis spectrophotometric analyses,[28] anthocyanidin assays,[29] and an acid hydrolysis technique that we have developed by combining and modifying several hydrolysis conditions used by others.[30] The strengths and weaknesses of each as they relate to geochemistry will be briefly discussed. More in-depth discussion of analytical methods can be found in Hagerman and Butler.[31]

^{13}C-NMR is a bulk method for estimating total tannins, either directly by integrating peaks attributable to tannin,[32] or indirectly by subtracting independently measured lignin content from total phenolics.[14] When carbon content of natural samples permits, solid-state ^{13}C-NMR can provide this information nondestructively. A primary disadvantage of ^{13}C-NMR is that while it is good for identifying functional groups, structural information can be sparse. Preston et al.[33] showed that exposing a sample to harsh acid treatments can significantly alter the chemistry of the sample without changing the ^{13}C-NMR. ^{13}C-NMR is also subject to interference from paramagnetic metals such as iron and manganese present in soils and sediments.[34-36]

The Folin-Denis colorimetric technique is a bulk method for estimating total tannin that utilizes the oxidizing property of tannin to produce color.[28] An advantage to Folin-Denis analyses is the relative ease with which it can be carried out. Unlike ^{13}C-NMR, this technique does not necessarily measure all of the tannin, but only that readily available as an oxidizing agent toward the Folin-Denis reagent. In this sense, then, the Folin-Denis technique is useful as an estimate of the potential reactivity of the tannin in natural environments, but may be limited in detecting reactions that have already occurred. A primary disadvantage of this method is that it is most often calibrated using tannic acid, when in fact, tannic acid (or other hydrolyzable tannin) is not necessarily a major component of the sample. Another disadvantage is that other compounds capable of oxidizing the Folin-Denis reagent such as ascorbic acid or peptides will lead to overestimates of the tannin content. On the other hand, it is usually carried out on an extract, as opposed to the whole sample, which can lead to underestimates.

Protein-binding assays (e.g., precipitation of tannin with protein, or radial diffusion of tannin through protein gels) are a third method for measuring total tannin.[27] The advantages and disadvantages are similar to Folin-Denis. Protein-binding assays are easy and give an indication of the potential reactivity of the tannin in natural environments, in this case, their ability to complex with proteins. As with Folin-Denis, protein-binding assays may not provide much information about prior tannin reactions. The use of appropriate calibration standards is an important consideration, since different tannin compounds will have different complexing ability. Again, there is potential interference from other compounds present in the samples leading to overestimates, while the use of extracts may lead to underestimates of tannin in a particulate source.

The anthocyanidin assay of Porter et al.[29] is another colorimetric technique specific for the extender units of condensed tannin oligomers and polymers. This technique relies on acid hydrolysis to release the extender units as carbocations, which then convert to cyanidins, pelargonidins, and delphinidins (fig. 2). The latter can then be detected spectrophotometrically. Because condensed tannins are more likely to persist in natural samples than more labile and soluble hydrolyzable tannins, this technique is applicable to soils and sediments when molecular information is desired.[11] In principle, cyanidins, pelargonidins, and delphinidins are distinguishable from each other due to respective differences in absorption

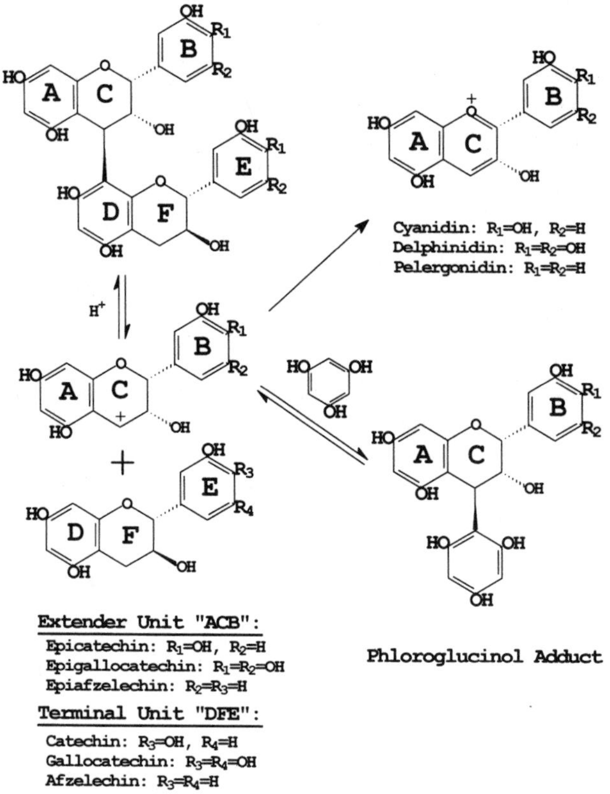

Figure 2. Acid hydrolysis of a condensed tannin dimer with and without phloroglucinol.

maxima, but in practice they are not well resolved spectrophotometrically. The primary advantage of this technique is that it provides specific molecular information. Disadvantages are, again, that appropriate standards for calibration must be obtained (preferably a well-characterized oligomer), stereochemical information is lost upon conversion of the carbocations, and the terminal unit of the oligomers or polymers is not detected. The latter is crucial toward calculating degree of polymerization—a potentially important geochemical indicator of environmental history of a sample.

Acid-hydrolysis utilizing nucleophilic capture of the extender unit carbocations (fig. 2) provides the least ambiguous measure of tannins and potentially the most diagnostic information. Unfortunately, no standard hydrolysis conditions are outlined in the literature. Hydrolysis usually has been carried out on extractions, with many different acids, solvents, and nucleophiles at varying concentrations, temperatures, and times, and utilizing many different chromatographic techniques. Advantages to these techniques are that stereochemistry of the extender units is preserved, all monomers can be unambiguously resolved, and since both extender and terminal units can be measured, one can calculate an average degree of polymerization by dividing total molecular tannin by the total terminal units. Degree of polymerization may be an important diagenetic parameter because of its relation to leachability, complexation with proteins or metals, or lability. The primary disadvantage is incomplete hydrolysis leading to underestimates of tannin.

With this in mind, our goals for molecular-level analysis of tannins in natural samples included high sensitivity (low background), applicability to various forms and matrices, no analytical artifacts, ease and safety of analysis, quantification, and reproducibility.[30] For ease and safety considerations, we opted to use ~0.25 M phloroglucinol as the nucleophile instead of the more common sulfur nucleophiles. Because of its relative safety and the potential for evaporation if so desired, hydrochloric acid (1.0 M) was chosen for hydrolysis. Instead of dioxane:water solvents, we opted to used acetone:water (70:30, v/v), which we also deemed most appropriate for analyses to be done on whole samples since acetone:water is the typical solvent system used for extracting tannin from whole samples. The hydrolysis is carried out at 30 °C, the reaction mixture extracted with ethyl acetate and run through sodium sulfate drying columns, and then the ethyl acetate evaporated to dryness before redissolving the sample in pyridine. Because gas chromatographs (GCs) are ubiquitous and commonly coupled to mass spectrometers, we felt it was desirable to carry out the chromatography on a GC if feasible, with trimethylsilyl (TMS) derivatization. We used splitless injection on an HP 5890 GC equipped with electronic pressure control and a DB35ms capillary column. Utilizing a temperature ramp of 4 °C per minute from 200 °C to 320 °C and a column head pressure ramp of 1 psi per minute from 13 psi to 30 psi, most compounds have eluted by 40 minutes. Hematoxylin is used as the internal standard. More detailed information can be found in Hernes and Hedges.[30]

3. DEGRADATION OF MANGROVE LEAVES IN A TROPICAL ESTUARY

None of the analytical techniques mentioned above are mutually exclusive of each other, and in geochemical studies especially, a combination of two or more techniques will provide the most insight. A good example of this approach involves tannin analyses of mangrove leaves degrading in a tropical estuary.[14,37] Benner et al.[14] collected a series of mangrove leaves, including green and yellow senescent leaves, attached to trees and and yellow, orange, brown, and black leaves that had been submerged in water for different time periods. Leaves were sorted by color, which also proved to be the best predictor of relative degradation. Mass per unit area of the leaves was used for mass balance comparisons between leaf colors. ^{13}C-NMR and the Folin-Denis method (results of the latter expressed in tannic acid equivalents [TAE]) were used to estimate tannin content (fig. 3). Using only ^{13}C-NMR, one would conclude that the tannin content of the leaves remained unchanged at approximately 20 percent. Folin-Denis analyses of the same samples indicate that tannin content varies greatly, dropping quite dramatically between orange leaves (approximately 1 week in the water) and brown leaves (3–4 weeks in the water). Used together, Benner et al.[14] hypothesized that there were two

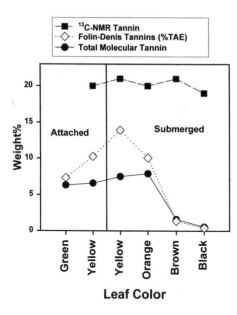

Figure 3. Percent tannin in mangrove leaves, including percent of carbon attributable to tannin as determined by ^{13}C-NMR,[14] weight percent tannin in Tannic Acid Equivalents [TAE] as determined by the Folin-Denis method on leaf extracts,[14] and weight percent total molecular tannin as determined by acid hydrolysis.[37]

types of tannin present: hydrolyzable tannin as measured by the Folin-Denis analysis, and another form as represented by the difference between the two techniques. Benner et al.[14] further suggested that the large drop shown in the Folin-Denis measurements could come from leaching. A subsequent laboratory experiment showed that leaching alone could not account for all of the decrease in tannin as measured by the Folin-Denis method. Thus, Benner et al.[14] hypothesized that in the latter brown and black stages of degradation, microbial degradation was the dominant factor in the decrease of tannin as measured by the Folin-Denis method. Unfortunately, neither method reveals much about the molecular makeup of the tannin, which Benner et al.[14] concluded was necessary to gain further insight into the nature of tannin diagenesis in these samples. Subsequently, Hernes et al.[37] utilized the hydrolysis technique described above to add a molecular perspective to the study.

In contrast to the supposition by Benner et al.[14] that the Folin-Denis method was measuring hydrolyzable tannin, Hernes et al.[37] showed that condensed tannin could account for much or all of the Folin-Denis tannin (fig. 3). This is important information since the structures and functions of condensed and hydrolyzable tannin are quite different.[5] Analytically, these contrasting results highlight the problem of using a hydrolyzable tannin such as tannic acid for calibrating the Folin-Denis method. Even so, the molecular method confirmed the drop in tannin content shown by the Folin-Denis method between the orange and brown leaves. Further, both the molecular method and the Folin-Denis method show a tannin increase in the early stages of diagenesis that cannot be accounted for by conservative behavior (figs. 3 and 4).

While the early increase in tannin as measured by the Folin-Denis method could be explained by interference from a non-tannin component that is formed or

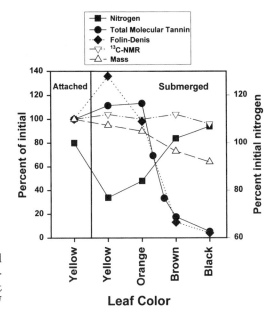

Figure 4. Percent of tannin and nitrogen along the diagenetic pathway relative to the initial amount present in yellow senescent leaves.[14,37]

concentrated during senescence and early degradation, the molecular data are considerably less ambiguous. If both techniques are measuring increased amounts of tannin, it is most likely due to an analytical artifact or matrix effect as opposed to an actual increase in the amount of tannin present. Since the two techniques involve very different chemistries—reduction of hydroxyl groups with the Folin-Denis method vs. cleavage of the interflavan bond between extender units with acid hydrolysis—both chemistries must be accounted for when explaining such an artifact or effect. Because total nitrogen was inversely proportional to tannin content (fig. 4), Hernes et al.[37] hypothesized that the initial "increase" was due to tannin-masking by complexations with water-soluble nitrogenous compounds present in the leaves. Complexations would tie up hydroxyl groups and may "lock" tannin in place or prevent proton access to the interflavan bond, thus inhibiting hydrolysis. As the leaves were leached of nitrogenous compounds, the tannin was unmasked, and the tannin measurements showed an increase.

In the latter stages of diagenesis, as nitrogen content begins to increase, measurable tannin decreases. While complexation and masking may again play some role, Benner et al.[14] concluded from the leaching experiment that microbial degradation is more important than leaching in the brown and black stages, which suggests that any nitrogen-tannin interactions in the latter stages may involve covalent bonding as opposed to complexation. Once again, molecular information sheds some light on a possible mechanism. The mean degree of polymerization in condensed tannin is likely to be affected by leaching and diagenesis. One would predict that the degree of polymerization would increase with both leaching and degradation, the former because of the higher solubility of lower molecular weight tannins and the latter because larger oligomers and polymers have been found

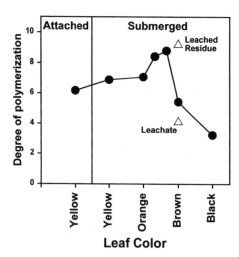

Figure 5. Average degree of polymerization in mangrove leaves.[37]

generally to be more refractory than smaller ones.[38,39] Hernes et al.[37] showed that this is, indeed, the case for leaching, as the degree of polymerization increased from 6.2 to 9.2 when yellow senescent leaves were leached (fig. 5). However, between the orange and brown leaves, and again between the brown and black leaves, the degree of polymerization actually decreased from 8.8 to 5.4, and then from 5.4 to 3.3. Why this discrepancy?

When polymers are degraded, it is generally assumed that the degraded pieces are consumed and either respired or converted into other forms (i.e., incorporated into biomass) by biological organisms. However, in this case, ^{13}C-NMR indicates that polyphenolic carbon content of the brown and black leaves is similar to that of yellow leaves both quantitatively from a mass balance perspective and qualitatively,[14] suggesting that tannin is not being consumed and not being incorporated into biomass. At the same time, the nitrogen content of the brown and black leaves is actually higher than that of the yellow leaves before submersion (fig. 4). What could account for nitrogen immobilization? Both the conversion but retention of tannin phenolic carbon and the immobilization of nitrogen could be explained by tannin/nitrogen complexation and covalent bonding.[37] Complexation of tannin with nitrogen could explain nitrogen immobilization as well as the decrease in degree of polymerization. Jones et al.[24] found that for complexation of tannin to proteins to occur, the oligomeric length had to be at least four. Thus, the larger oligomers become complexed, whereas the smaller oligomers do not. Once complexed, covalent bonds could form between the tannin and nitrogen, perhaps in reactions analogous to melanoidon reactions. An alternative explanation to the decrease in degree of polymerization simply involves hydrolysis of interflavan bonds, perhaps due to acidic microenvironments within the leaves as degradation progresses.[37] Reactive carbocations would perhaps be more likely to bond with nitrogenous compounds, once again offering an explanation for nitrogen immobilization. Since the tannin phenolic carbon would remain in the sample in each case, the change would not be detected by ^{13}C-NMR.

4. CONCLUSIONS

The study of tannin in natural samples shows great promise for addressing geochemical questions. While many techniques have been used for measuring bulk tannin in natural samples, the need for a molecular detailed perspective prompted the development of an acid hydrolysis technique based on modification of literature hydrolysis conditions for natural products. In addition to complementing ^{13}C-NMR and Folin-Denis data on mangrove leaves from a tropical estuary, molecular tannin measurements highlight the potential importance of understanding the role of nitrogen in the diagenetic pathway of these leaves.

REFERENCES

1. Hedges, J.I. Global biogeochemical cycles: progress and problems. *Marine Chemistry* 39:67 (1992).

2. Berner, R.A. Biogeochemical cycles of carbon and sulfur and their effect on atmospheric oxygen over Phanerozoic time. *Palaeogeography, Palaeoclimatology, Palaeoecology* 73:97 (1989).

3. Aller, R.C.; Blair, N.E.; Xia, Q.; Rude, P.D. Remineralization rates, recycling, and storage of carbon in Amazon shelf sediments. *Continental Shelf Research* 16:753 (1996).

4. Emerson, S.; Hedges, J.I. Processes controlling the organic carbon content of open ocean sediments. *Paleoceanography* 3:621 (1988).

5. Zucker, W.V. Tannins: does structure determine function? An ecological perspective. *The American Naturalist* 121:335 (1983).

6. Horner, J.D.; Gosz, J.R.; Cates, R.G. The role of carbon-based plant secondary metabolites in decomposition in terrestrial ecosystems. *The American Naturalist* 132:869 (1988).

7. Tiarks, A.E.; Meier, C.E.; Flagler, R.B.; Steynberg, E.C. Sequential extraction of condensed tannins from pine litter at different stages of decomposition. *In*: Hemingway, R.W.; Laks, P.E. (eds.) Plant polyphenols—synthesis, properties, significance. Plenum Press, New York, p. 597 (1992).

8. Kelsey, R.G.; Harmon, M.E. Distribution and variation of extractable total phenols and tannins in the logs of four conifers after 1 year on the ground. *Canadian Journal of Forest Resources* 19:1030 (1989).

9. DeMontigny, L.E.; Preston, C.M.; Hatcher, P.G.; Kögel-Knabner, I. Comparison of humus horizons from two ecosystem phases on northern Vancouver Island using ^{13}C CPMAS NMR spectroscopy and CuO oxidation. *Canadian Journal of Soil Science* 73:9 (1993).

10. Northup, R.R.; Yu, Z.; Dahlgren, R.A.; Vogt, K.A. Polyphenol control of nitrogen release from pine litter. *Nature* 377:227 (1995).

11. Preston, C.M. Condensed tannins of salal (*Gaultheria shallon* Pursh): a contributing factor to seedling "growth-check" on northern Vancouver Island? *In*: Gross, G.G.; Hemingway, R.W.; Yoshida, T. (eds.) Plant polyphenols 2: chemistry, biology, pharmacology, ecology. Plenum Press, New York (in press).

12. Mole, S.; Waterman, P.G. A critical analysis of techniques for measuring tannins in ecological studies. *Oecologia* 72:137 (1987).

13. Preston, C.M., personal communication.

14. Benner, R.; Hatcher, P.G.; Hedges, J.I. Early diagenesis of mangrove leaves in a tropical estuary: Bulk chemical characterization using solid-state ^{13}C NMR and elemental analyses. *Geochimica et Cosmochimica Acta* 54:1991 (1990).

15. Suberkropp, K.; Godshalk, G.L.; Klug, M.J. Changes in the chemical composition of leaves during processing in a woodland stream. *Ecology* 57:720 (1976).

16. Hedges, J.I.; Weliky, K. Diagenesis of conifer needles in a coastal marine environment. *Geochimica et Cosmochimica Acta* 53:2659 (1989).

17. Cowie, G.L.; Hedges, J.I. Carbohydrate sources in a coastal marine environment. *Geochimica et Cosmochimica Acta* 48:2075 (1984).

18. Goñi, M.A.; Hedges, J.I. Potential applications of cutin-derived CuO reaction products for discriminating vascular plant sources in natural environments. *Geochimica et Cosmochimica Acta* 54:3073 (1990).

19. Cowie, G.L.; Hedges, J.I. Sources and reactivities of amino acids in a coastal marine environment. *Limnology and Oceanography* 37:703 (1992).

20. Tissot, B.P.; Welte, D.H. Petroleum formation and occurrence. Springer-Verlag, New York (1978).

21. Ellis, C.J.; Foo, L.Y.; Porter, L.J. Enantiomerism: A characteristic of the proanthocyanidin chemistry of the monocotyledonae. *Phytochemistry* 22:483 (1983).

22. Haslam, E. Plant polyphenols—vegetable tannins revisited. Cambridge University Press, Cambridge (1989).

23. Laks, P.E. Chemistry of the condensed tannin B-ring. *In:* Hemingway, R.W.; Karchesy, J.J. (eds.) Chemistry and significance of condensed tannins, Plenum Press, New York, p. 249 (1989).

24. Jones, W.T.; Broadhurst, R.B.; Lyttleton, J.W. The condensed tannins of pasture legume species. *Phytochemistry* 15:1407 (1976).

25. Field, J.A., Lettinga, G. Biodegradation of tannins. *In*: Sigel, H.; Sigel, A. (eds.) Metal ions in biological systems; degradation of environmental pollutants by microorganisms and their metalloenzymes. Marcel Dekker, Inc., New York, 28:61 (1992).

26. Amon, R.M.W.; Benner, R. Photochemical and microbial consumption of dissolved organic carbon and dissolved oxygen in the Amazon River system. *Geochimica et Cosmochimica Acta* 60:1783 (1996).

27. Hagerman, A.E. Radial diffusion method for determining tannin in plant extracts. *Journal of Chemical Ecology* 13:437 (1987).

28. Folin, O.; Denis, W. A colorimetric method for the determination of phenols (and phenol derivatives) in urine. *Journal of Biological Chemistry* 22:305 (1915).

29. Porter, L.J.; Hrstich, L.N.; Chan, B.C. The conversion of procyanidins and prodelphinidins to cyanidin and delphinidin. *Phytochemistry* 25:223 (1986).

30. Hernes, P.J.; Hedges, J.I. Determination of condensed tannin monomers in plant tissues, soils, and sediments by capillary gas chromatography of acid hydrolysis extracts. [unpublished results]

31. Hagerman, A.E.; Butler, L.G. Choosing appropriate methods and standards for assaying tannin. *Journal of Chemical Ecology* 15:1795 (1989).

32. Newman, R.H.; Porter, L.J. Solid-state ^{13}C-NMR studies on condensed tannins. *In*: Hemingway, R.W.; Laks, P.E. (eds.) Plant polyphenols—synthesis, properties, significance. Plenum Press, New York. p. 339 (1992).

33. Preston, C.M.; Trofymow, J.A.; Sayer, B.G.; Niu, J. ^{13}C nuclear magnetic resonance spectroscopy with cross-polarization and magic-angle spinning investigation of the proximate analysis fractions used to assess litter quality in decomposition studies, *Canadian Journal of Botany* 75:1601 (1997).

34. Skjemstad, J.O.; Clarke, P.; Taylor, J.A.; Oades, J.M.; Newman, R.H. The removal of magnetic materials from surface soils. A solid-state ^{13}C CP/MAS n.m.r. study. *Australian Journal of Soil Research* 32:1215 (1994).

35. Preston, C.M. The application of NMR to organic matter inputs and processes in forest ecosystems of the Pacific Northwest. *The Science of the Total Environment* 113:107 (1992).

36. Baldock, J.A.; Oades, J.M.; Waters, A.G.; Peng, X.; Vassallo, A.M.; Wilson, M.A. Aspects of the chemical structure of soil organic materials as revealed by solid-state ^{13}C NMR spectroscopy. *Biogeochemistry* 16:1 (1992).

37. Hernes, P.J.; Benner, R.; Hedges, J.I. Tannin diagenesis in mangrove leaves from a tropical estuary: A novel molecular approach. [unpublished results]

38. Field, J.A.; Lettinga, G. Treatment and detoxification of aqueous spruce bark extracts by *Aspergillus niger*. *Water Science and Technology* 24(3/4):127 (1991).

39. Grant, W.D. Microbial degradation of condensed tannins. *Science* 17:1137 (1976).

EFFECT OF CONDENSED TANNINS IN THE DIETS OF MAJOR CROP INSECTS

Alister D. Muir, Margaret Y. Gruber, Christopher F. Hinks,
Garry L. Lees, Joseph Onyilagha, Julie Soroka,
and Martin Erlandson

Agriculture and Agri-Food Canada
Saskatoon Research Centre
Saskatoon, Saskatchewan S7N 0X2
CANADA

1. INTRODUCTION

The role of leaf hydrolyzable and condensed tannins in plant defenses against herbivorous insects has been the subject of substantial investigation.[1-8] Most of these early studies focused on the role of hydrolyzable tannins rather than condensed tannins in insect/plant interactions. Of the insects considered in these studies, acridids, particularly ambivorous species, appear to be distinguished by their relative tolerance to tannic acid.[9,10] In contrast, the graminivorous locust, *Locusta migratoria*, suffered damage to the tissues of the digestive system when tannin was included in its diet.[9]

In these studies, hydrolyzable tannins were associated with plant apparency, i.e., the continuing presence of a species in space and time that defines the perennial habit.[11] Thus hydrolyzable tannins are widely found in trees and shrubs in which, typically, there is a progressive accumulation through the growing season. Less attention has been given to the presence of condensed tannins in herbaceous perennials, although their presence has been documented in a number of legumes, including some of economic importance. These include sainfoin[12] and several trefoil species.[13-16] In some of these studies, it was demonstrated that the nutritional status of the plant influenced tannin concentration, and nutrient deficiency was associated with tannin accumulation. Temperature and light intensity also influenced tannin accumulation.[17,18]

The capacity of tannins to chemically combine with proteins has been given a particular prominence in studies of the differential benefits of tannins to plants and livestock. For example, by binding with the relatively abundant protein found in legume foliage, condensed tannins slow protein digestion and help prevent

Plant Polyphenols 2: Chemistry, Biology, Pharmacology, Ecology, Edited by
Gross et al. Kluwer Academic / Plenum Publishers, New York, 1999

gastric bloating that can have fatal results in cattle.[14,19] However, negative effects of tannins on monogastric mammals have been associated with the reduction of protein digestibility[20] as well as with non-competitive inhibition, particularly of tryptic enzymes.[21,22] Consequently, the capacity of tannins to bind proteins was extended to herbivorous insects in explanation of their functional role in plant defense.[23,24] The generalization that tannin expression was a quantitative defensive secondary product with a universal effect has since been qualified with the discovery that insects have evolved a number of ways of countering its inimical effects. These adaptations include alkaline conditions in the midgut,[23] the presence of surfactants in the gut fluid,[25,26] high cation concentrations,[27] reducing conditions in the midgut,[28] or increased production of the peritrophic membrane.[29,30]

Herbivorous insects and livestock may be in direct competition for forage. Because condensed tannins (CT) can have a negative effect on herbivorous insects but have a beneficial effect in the prevention of bloating in cattle, it is important to understand the quantitative relationships of these two phenomena. Hence this study was initiated as part of a broad study of forage legume tannins to provide information on the general effects of condensed tannin from *Lotus* spp. on the polyphagous grasshopper *Melanoplus sanguinipes* (Fab.), an insect that causes significant damage to many legume and cereal crops on the Canadian prairies. Subsequently, the scope of the study was expanded to include the impact of dietary condensed tannin on growth and viability of major canola insect pests. Feeding trials have suggested that grasshoppers will be a significant pest for canola as the crop expands into the wheat belt (Olfert, personal communication). Other insect pests of canola included in this study are flea beetles (*Phyllotreta cruciferae*), which are an endemic specialist feeder responsible for a major portion of the annual Canadian canola crop loss, bertha armyworm (*Mamestra configurata*), a polyphagous insect responsible for cyclical severe losses to canola and flax crops, and diamondback moth (*Putella xylostella*), a specialist feeder that can also cause economic losses to canola when blown into Canada on winds from the United States and Mexico.

2. MATERIALS AND METHODS

Lotus corniculatus [cv. Empire (SL-644)] and a low tannin selection (SL-694), *L. uliginosus* var. *glabrusculus* (SL-3502), *L. uliginosus* (SL-7569), wheat [*Triticum aestivum* (cv. Katepwa)], and oats [*Avena sativa* (cv. Harmon)] were grown under field conditions at the Saskatoon Research Centre at Saskatoon.

Mature field plants of *L. uliginosus* were selected for differences in foliar condensed tannin concentration and moved to the greenhouse for further propagation and testing.[18] These were grouped into high- and low condensed tannin-containing plants, and two representatives from each group were cloned by rooted cuttings to provide 10 plants. The plants were grown in a soilless mix.[31] Once established, they were divided between growth chambers at L/D 16: 8 and 20/15 °C and L/D 16: 8 and 30/25 °C, both illuminated by a mixture of incandescent and fluorescent lights emitting $300\,\mu E.\,m^{-2}\,sec^{-1}$.

Leaf samples were taken from individual plants and assayed using the vanillin-HCl microassay using catechin as a standard.[32] During tests for flavan-3-ol monomers (catechin) and polymers, each clone was treated as a sample and was sub-sampled three times. To obtain a representative sample, individual leaves from the upper half of stems were picked randomly and 200 mg weighed into each of three 50-mL centrifuge tubes to which 20 mL of 1 percent HCl-methanol was added immediately. The fresh samples were ground for 30 sec at setting 5 using a Brinkmann Polytron109 PT 10–35 homogenizer equipped with a PT 20 generator probe. Homogenates were extracted in 1 percent HCl/methanol at 4 °C for 2 h, then centrifuged for 10 min at 6,000 × G retaining the supernatant for assay. Condensed tannin levels were recalculated or remeasured by the butanol:HCl assay using a condensed tannin polymer fraction purified from *L. uliginosus* as a standard[33] and only the butanol-HCl presented.[18] For dry weight determination, 1.5 g of leaves was harvested and dried for 72 h at 70 °C.

Frass was pooled from individual grasshoppers for each concentration of dietary condensed tannin, and then pooled frass samples were assayed for condensed tannin in the same manner as the plant material. Frass tannin was calculated as a percentage of the condensed tannin content of each respective diet. The guts of grasshoppers fed diets of powdered wheat seedlings augmented with purified condensed tannins were gently washed free of gut contents and sectioned longitudinally. Squashes of the sections were assayed qualitatively using butanol:HCl hydrolysis to localize tannin retained within specific gut regions.

Condensed tannin was extracted from fresh *L. uliginosus* foliage with 70 percent acetone in water containing 0.1 percent ascorbic acid. The acetone fraction was reduced to the aqueous phase, which was partitioned against petroleum ether and ethyl acetate until the organic phase was clear. Traces of ethyl acetate were removed from the resulting aqueous phase by evaporation, and the ethyl acetate-free aqueous phase was diluted to 50 percent with methanol and applied to a Sephadex LH-20 column equilibrated in 50-percent methanol. The column was washed with 50-percent methanol until the eluate was clear, and then the condensed tannin fraction eluted with 70-percent acetone. The condensed tannin fraction was concentrated to the aqueous phase, freeze dried, and stored in a desiccator in the dark until required. The condensed tannin content of the polymer fraction was assessed by three different assays (butanol-HCl, vanillin-HCl, and Prussian Blue) using catechin and a preparation of sainfoin leaf polymer as standards.[34] Commercial quebracho tannin (Tannin Corporation, Peabody, Massachusetts) was purified by chromatography over Sephadex LH-20 as described for extraction of condensed tannins.

Grasshopper hatchlings of a non-diapause strain of the lesser migratory grasshopper, *M. sanguinipes*, were supplied from a laboratory colony.[35] The hatchlings were allowed to feed briefly on lettuce until enough of them had hatched (within 2 hr) to supply the experiment; thus, the new grasshoppers were naive to any of the plants used in the study. Grasshoppers at this stage weighed 4.5 ± 0.5 mg.

In the initial experiment, hatchling grasshoppers (n = 25) or adult female grasshoppers (n = 10) were fed for 5 days on either fresh cut foliage from the *Lotus* field plot, on a standard performance diet of wheat (Katepwa) seedlings grown to a height of 16 ± 1 cm, or on low protein oat seedlings. Fresh foliage was provided to the grasshoppers on each of five consecutive days. In the second experiment, hatchling grasshoppers were fed leaves from *L. uliginosus* plants grown at either the 20/15 or 30/25 °C temperature regimes, and the diet was changed daily. In the third experiment, hatchling grasshoppers were fed 1.5 g samples of lyophilized, standard wheat seedlings that were milled and then formulated with extracted *L. uliginosus* condensed tannin to provide tannin concentrations of 1-, 5-, 10-, and 20 percent w/w. In formulating each diet, the condensed tannin (dissolved in methanol) was thoroughly mixed with the powder. The mixture was then dried under reduced pressure at 40 °C, remilled, and the powder from each diet was divided into 80 mg aliquots and compressed into 8 mm diameter pellets. The control diet for this study was formulated by adding the same volume of methanol to lyophilized wheat powder and processing as previously described.

Groups of 25 hatchling grasshoppers were fed each of the three experimental diets in quart mason jars.[36] Adult grasshoppers were fed in larger cages. Each experiment was replicated six times. At the end of the fifth day of each experiment, grasshoppers were removed from their cages, killed by freezing, counted, and weighed.

Flea beetle (*Phyllotreta cruciferae*) insect feeding assays were conducted using late summer field-collected adults. Each test consisted of 10 beetles placed in a cell containing *Brassica napus* leaf disks painted with solutions of the purified tannin or a control solvent. Each test diet was replicated 15 times. No-choice feeding trials in which control and treated leaf disks were in separate cells were used to determine physiological effects of tannins. Choice tests in which control and treated leaf disks were in the same cell were used to assess a combination of feeding preference and deterrence. The feeding deterrence index (FDI) was calculated in the choice bioassays using the following formula $(C\text{-}T)/(C + T) \times 100$, where C = area consumed of leaf disk spread with control solvent and T = area consumed of disk spread with tannin solution.

Bertha armyworm larvae were raised from eggs and reared on an artificial diet until transferred to test diets. Diamondback moth adults (male and female) were allowed to mate in insect cages containing 3–4-week-old *B. napus* plants. The resulting eggs were allowed to hatch on the *B. napus* plants, and the emerging larvae were collected for use in the feeding bioassays. For both species of larvae, the experimental design was the same as for flea beetles except only five 3rd instar larvae per diet replicate were used.

Groups of 10 adult fledgling grasshoppers were fed for 5 days on either a high-condensed tannin diet (*Lotus uliginosus*), a low-condensed tannin diet (*L. corniculatus*), wheat, or oat seedlings (high- and low-protein performance controls, respectively), and then killed by freezing. The whole gut was removed, ground, and processed individually for the analysis of the endopeptidases, trypsin, and chymotrypsin as previously described. [37]

Percentages of grasshopper survival were compared without transformation. Weights of grasshoppers from the feeding experiments were analyzed using ANOVA[38] and the means separated using the Waller-Duncan K-ratio *t*-test. Similarly, enzyme activities were analyzed using ANOVA and the means separated by a Duncan's multiple range test. The growth response of grasshoppers to a pelleted wheat powder diet supplemented with condensed tannin was subjected to regression analysis using a general linear model. Canola insect feeding deterrence measurements were analyzed using standard errors of the means.

3. RESULTS AND DISCUSSION

The feeding deterrence index of condensed tannin polymer purified from either *L. uliginosis* leaves or from quebracho was examined using canola pests in choice bioassays. Feeding deterrence (32 and 51 percent) was observed when 3rd instar diamondback moth larvae were fed *B. napus* disks spread with 0.3 percent tannin (w/FW disk) from LH-20 fractionated *L. uliginosis* leaf extracts or quebracho tannin, respectively (fig. 1). This tannin concentration was too low to effectively deter bertha armyworm feeding, but was very effective in deterring late summer field-collected flea beetles. The flea beetles had a feeding deterrence of greater than 90 percent on *B. napus* leaf disks to which quebracho condensed tannins were topically applied at 0.3 percent w/FW disk (fig. 1). *L. uliginosis* tannins strongly

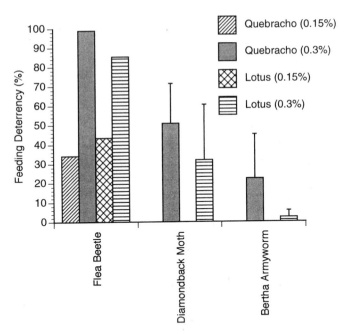

Figure 1. Feeding deterrence of purified condensed tannins to canola insect pests. The condensed tannins were applied topically to leaf disks of *B. napus*. Feeding deterrence was calculated as percent difference in consumption of the leaf disks relative to zero tannin controls. Each bar represents the mean feeding difference of three experiments, each replicated 15 times with five larvae per replicate (bertha armyworm and diamondback moth). Flea beetle experiments were replicated 15 times with 10 beetles per replicate.

inhibited the beetles at greater than 80 percent deterrence for the same tannin concentration. Both types of tannin were also effective at deterring flea beetles at 0.15 percent (w/FW disk) (fig. 1). Although not statistically significant, the quebracho tannin appeared more effective at deterring the feeding of the generalist bertha armyworm. There was no significant difference in deterrence between either tannin source to the specialist insects (flea beetle and diamondback moth).

When disks with 0.3 percent applied tannin were fed to 3rd instar larvae of diamondback moth or bertha armyworm in preliminary no-choice feeding bioassays, diamondback larvae avoided the tannin-smeared disks. As a result, the average weight gain of these diamondback larvae was measurably lower than control larvae after 4 days feeding, and larval weight continued to drop until pupation (data not shown). The tannin diet appeared to have no effect on bertha armyworm larval weight gain over a 7-day period. There was no change in the onset of pupation as a function of diet with either insect species. These experiments are being repeated using *Lotus japonicus* mutants which express a range of tannin content as a more realistic model for assessing the deterrent effect of condensed tannins.

The condensed tannin content among four species/accessions of *Lotus* sampled from field-grown plants in August ranged from 0.3 to 56.2 mg/g DW, whereas the wheat seedlings contained no detectable levels of tannins (table 1). The weights of hatchling grasshoppers fed foliage of *Lotus* (*L. corniculatus* or *L. uliginosus* var. *glabrusculus*) harvested in August or wheat seedlings were not significantly different, while those fed *L. uliginosus* were significantly lighter (table 1).

The condensed tannin content of leaves selected from cloned field-grown plants of *L. uliginosus* grown in a growth chamber, ranged from 7.2–20.7 to 123.3–195.3

Table 1. Survival and weight of hatchling grasshoppers fed on diets of cut foliage of field grown *Lotus* spp and wheat (*Triticum*) seedlings

Dietary Plant (Accession number)	Tannin content (mg/gDW)	Grasshoppers percent Survival	Mean Dry Weight (mg)*
L. corniculatus (SL-694)	0.5	92	9.9[a]
L. corniculatus (SL-644)	0.3	84	9.4[a]
L. uliginosus var. *glabrusculus* (SL-3502)	39.8	90	9.0[a]
L. uliginosus (SL-7569)	56.2	92	7.9[b]
T. aestivum Katepwa	0	100	9.7[a]

* Significant at the 0.05 level, Duncan's Multiple Range Test

mg/g DW at 14 and 81 days after establishment, respectively. The condensed tannin content of clones grown under the 30/20 °C regimen were slightly higher than those grown under 20/15 °C regimen. These results confirm the hypothesis that temperature influenced condensed tannins expression in *L. uliginosus* insofar as condensed tannins accumulated more rapidly and attained greater concentrations at the higher temperature regime.[18]

Survival of grasshoppers fed the clones of *L. uliginosus* was variable (fig. 2). A higher survival of grasshoppers could be correlated (R = 0.92) with increases in condensed tannin concentration among those fed the clones grown at 20 °C (fig. 2a). In contrast, there was a trend to lower survival among grasshoppers fed clones at grown at 30 °C as the condensed tannin content increased (fig. 2b). Higher condensed tannin concentration in these clones could also be correlated with reduced

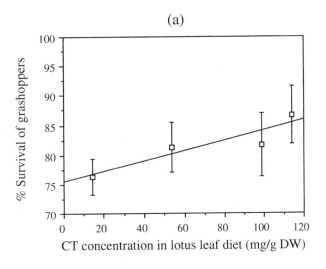

Figure 2. Survival of hatchling grasshoppers and the concentration of condensed tannin in diets of *Lotus uliginosus* leaves: (a) plants grown in a growth chamber under L/D 16: 8 and 20/15 °C; (b) plants grown in a chamber under L/D 16: 8 and 30/25 °C. Each datum point represents the mean value with standard error bars from six replicates comprising 25 hatchling grasshoppers.

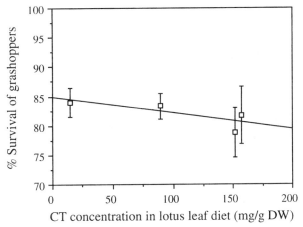

rates of weight gain of grasshoppers (R = 0.88 and 1.0 for plants grown at 20 °C and 30 °C, respectively).

There was little difference in survival among grasshoppers fed pellets comprising lyophilized, milled leaves of wheat seedlings to which *L. uliginosus* condensed tannin was added at concentrations of 10, 50, or 100 mg/g DW. However, survival fell sharply in those grasshoppers fed pellets that contained 200 mg/g DW (fig. 3). Mean weight of grasshoppers was not significantly different (P < 0.05) between those fed wheat pellets without the addition of tannin and those fed pellets to which condensed tannin comprised 10 mg/g. All higher concentrations of tannin negatively and proportionately reduced the growth rate of grasshoppers, with a high correlation (R = 0.99) between concentration and decline in weight (fig. 3).

Examination of survival among grasshoppers in the experiment in which they were fed fresh foliage of *L. uliginosus* revealed two seemingly contradictory trends. However, the positive trend in survival among grasshoppers fed on plants grown at the lower temperature, in contrast to the negative trend among those fed plants grown at the higher temperature, may merely reflect the different concentrations of condensed tannins in the two groups of plants and the threshold of toxicity of condensed tannins. Alternatively, beneficial effects of dietary tannins mediated through a negative effect on insect pathogens, could occur in grasshoppers fed diets with a low condensed tannin content, although this has yet to be investigated. Such interactions have been documented with other insects.[39,40] Furthermore, low CT diets were demonstrated to stimulate increased consumption of food,[41] a response that may have had the added benefit of reducing facultative pathogens in the grasshopper guts (Hinks and Erlandson, unpublished data).

Briggs[42] reported that protein concentration of *L. corniculatus* diets manipulated by fertilizer application, rather than dietary condensed tannin concentration, determined feeding behavior of *Spodoptera eridania*. In other studies, the quantity and quality of protein have been shown to modify physiological responses of insects to secondary chemicals.[43–45] In this context, the continued decline in the

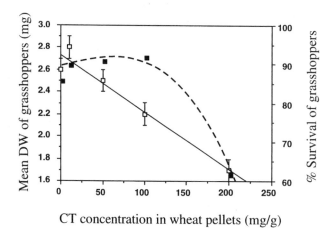

Figure 3. Survival ■ (B) and mean weight □ (G) of hatchling grasshoppers as a function of diet supplemented with *Lotus uliginosus* condensed tannin. (n = 6, ±SE, each replicate contained 25 grasshoppers).

ability of grasshoppers to gain weight on field-grown *L. uliginosus* at a late sampling date, despite condensed tannin concentrations slightly lower than those recorded earlier in the season, may have been due to deterioration in the quality or quantity of protein of the plants. Similarly, changes in protein content or composition may underpin the positive effect found on grasshopper survival with the plants grown at 20/15 °C despite significant concentrations of condensed tannins. In addition, survival is more variable and a less reliable indicator of grasshopper performance than is growth rate.[46]

The effect of condensed tannin on protein nutrition was examined in newly moulted adult female grasshoppers. Females fed *L. uliginosis* field-grown foliage (3 percent tannin) for 5 days utilized 26.5 percent of the leaf nitrogen (i.e., protein), compared with 30.9 percent of N from field-grown *L. corniculatus* foliage (1 percent tannin). The lower N utilization by adult females fed the high tannin *L. uliginosis* diet confirmed an earlier study[46] and was accompanied by a 1.2-fold increase in the consumption ratio, as well as a twofold increase in total midgut trypsin activity (fig. 4). Similarly, a 1.8-fold increase in adult gut tryptic activity was observed with a diet of low-protein oat seedlings (cv. Harmon) compared with high-protein wheat (c.v Katepwa). Chymotrypsin activity was also elevated in low-protein oat diets and in high-tannin *L. uliginosus* diets, although not to the same extent as trypsin activity. Our results confirm an earlier report comparing grasshopper digestive enzyme activity for several grass species.[37]

Condensed tannins have been reported to function as non-competitive inhibitors of digestive proteinases in monogastric mammals.[20–22] Hypermodulation of digestive proteinases in insects, a compensatory response, has been associated with the presence in the diet of proteinase inhibitors.[47] Hypermodulation of

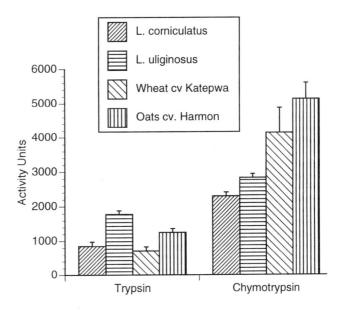

Figure 4. Activity of the principle gut endopeptidases in adult female grasshoppers (*M. sanguinipes*) fed diets varied in tannin and protein content. Wheat seedlings (high protein, zero tannin), oat seedings (low protein, zero tannin), *L. corniculatus* (high protein, low tannin), or *L. uliginosus* (high protein, high tannin) (n = 12).

digestive proteinases has been observed also in grasshoppers fed a diet of low-protein oats seedlings.[37]

One of the effects of prolonged exposure to proteinase inhibitors is growth retardation. Thus, the negative effects of condensed tannins in *Lotus* diets on the growth of grasshoppers may be due to interference with digestion, either directly or indirectly. Increased proteinase production and the concomitant loss of amino acids comprising these enzymes will become attritional when there is insufficient dietary protein available as a substrate for the enzymes to balance the amino acids required for their synthesis. A similar effect can be expected if the amino acid composition of the available protein is imbalanced. Herbivores responding to dietary constituents by hypermodulating digestive enzymes suffer a protein-attritional effect related to the composition of these enzymes, which have a high proportion of sulphur amino acids. When this attritional effect is severe in *M. sanguinipes*, protein withdrawal from the tissues occurs, accompanied by a higher rate of both excretion of uric acid and its accumulation in the fat body (Hinks, unpublished observations). Increased uric acid accumulation results from the metabolism of amino acids, which have become surplus through the creation of the imbalance. In the adult female grasshopper fed high concentrations of dietary condensed tannins, protein withdrawal is also manifested by atrophied ovaries and infertility (Hinks, unpublished observations).

The model for the interaction between plant proteinase inhibitors and digestive proteinases in animals was derived from studies of monogastric mammals in which a characteristic syndrome was hypertrophy and hyperplasia of the pancreas. This syndrome was considered to be a direct result of the hypermodulation of the proteinases secreted by the pancreas.[48] For example, in a recent study of the effects of dietary tannins on poultry, growth retardation, hypermodulation of trypsin, and hypertrophy of the pancreas were reported.[49] The latter response was proportional to the tannin content of the diet.

The higher protein content of *Lotus* foliage compared with oats could actually mitigate against the attritional effects of prolonged hypermodulation of grasshopper proteases in response to dietary tannins. Attenuation due to high protein content may explain why condensed tannins do not always affect digestion of protein.[9,30] Indeed, the differences in activities of both trypsin and chymotrypsin appear to be proportional to condensed tannin concentration in our experiments; however, this requires confirmation using diets covering a wider range of condensed tannin concentrations. Nevertheless, our results imply that the grasshopper may compensate for low available protein caused by sublethal concentrations of dietary tannin by increasing digestive activity and food consumption activity. These results may explain why sublethal concentrations of tannins result in increased damage to the plant rather than providing a protective effect.

Little is known of the extent to which condensed tannins are hydrolyzed and absorbed in the guts of herbivores, but it has been suggested that in mammals it is the products of tannin hydrolysis that are toxic.[50] Grasshoppers appear to have a mechanism to complex or possibly metabolize condensed tannins. However, the precise fate of the tannin has yet to be determined. For example, proportionately more condensed tannin was recovered from the frass of grasshoppers fed pelleted wheat diets formulated with low concentrations of condensed tannin compared to diets formulated with higher concentrations of tannin. The percentages of total dietary condensed tannin recovered from the frass using a butanol:HCl assay were 167, 97, 84, and 77 percent from 1st instar grasshoppers fed diets to which *Lotus* condensed tannin was added to give final concentrations of 1, 5, 10, and 20 percent w/w, respectively. However, condensed tannin measured similarly by butanol:HCl hydrolysis could not be detected in guts dissected from 1st, 3rd, or 5th instar nymphs fed equivalent diets and washed free of digesta.

Reduction in the percentage of condensed tannins recovered in the frass at dietary concentrations of 5 percent or more paralleled the negative effect on growth, and suggests a loss of condensed tannins by hydrolysis or by covalent binding to protein. Changes also occurred at sublethal condensed tannin concentrations, including a seemingly contradictory increase in frass tannin content above the 1 percent dietary level. Difficulties arise in collecting and interpreting these frass data because of color inhancement, which can occur under special circumstances with these chemical assays and because of complex gut interactions involving condensed tannins such as peritrophic membrane production, increased enzyme synthesis, and tannin binding to soluble plant proteins. The use of radiolabeled tannins in insect diets may shed light on whether degradation products of condensed tannins are responsible for the toxic effects.

Butler and Rogler[44] presented evidence that the target of dietary tannin in monogastric animals is within the body rather than the digestive system, and that absorbable, low molecular weight polyphenols are primarily responsible. For example, in rats, it appears that low molecular weight materials absorbed from tannin degradation have a greater toxic effect than the direct interaction of tannins with proteinases.[44] Protein digestion in rats was also much less affected by tannins than was the efficiency of conversion of new biomass from the food. However, findings are also consistent with the creation of an imbalance in available amino acids such as that reported to occur when proteinase inhibitors are present in the diet. Accordingly, an alternative explanation may be postulated in which either accompanying low-molecular weight phenolics or the products of condensed tannin hydrolysis interfere in some manner with the secretogogue mechanism for digestive proteinases. Unfortunately, little is known of how the secretogogue mechanism functions in insects, and until it is defined by the appropriate experimentation, interference with it by plant secondary chemicals will remain an area for speculation.

While considering the controversy over the mechanims of toxicity of tannins on insects, Berenbaum[51] made several important observations. First, he observed a significant dichotomy in the interpretation of several studies that related to methodology. Reduced protein digestibility was usually demonstrated in studies where the insects were fed artificial diets supplemented with tannin, whereas no effect on protein digestibility was apparent in studies of insects fed intact tannin-bearing leaves. However, in the present study, we observed a negative effect of condensed tannins on grasshopper growth and survival over a similar range of condensed tannin concentration in both types of feeding experiments. Although the negative effects of tannin on grasshopper growth were more pronounced with the powdered diets compared to the leaf diets, this is likely the result of the lower nutritional value of wheat leaves compared to *Lotus* leaves. This low nutritional value is typical of the leaves of cereals compared to those of legumes.

The second of Berenbaum's observations[51] was the importance of distinguishing between toxicity and deterrence. Distinguishing between deterrence, which is often manifested as a reduced rate of growth or a reduced rate of consumption, and toxicity, which can be manifest as either a reduced rate of growth or increased mortality at higher concentrations, presents a complex problem. For example, the observed deterrence may be a response to sublethal levels of a toxin, or simply avoidance of morphological features. Plants with a high nutritional quality may ameliorate or conceal the effects of toxins by providing sufficient nutrients to allow the insect to overcome the effects of compounds like condensed tannins. Ideally, for any putative toxin, a dose-response relationship should be demonstrated and the toxic mechanism identified.

Collectively, our results imply that when condensed tannins are present at moderate concentrations in crop plants, these compounds appear to act as feeding deterrents for the more specialist insects such as diamondback moth and flea beetles. However, significantly higher concentrations of condensed tannins are required to deter the feeding of the more generalist feeding insects, grasshopper and bertha armyworm.

4. CONCLUSIONS

The impact of dietary condensed tannin on insect feeding preferences and deterrence and on insect growth and viability was examined for four major Canadian prairie crop insects. In choice feeding preference bioassays using grasshoppers (*Melanoplus sanguinipes* Fab.), 2nd instar nymphs tended to eat more low-tannin lotus (*Lotus uliginosis*) foliage. However, when the tannin concentrations exceeded typical field plant levels, the nymphs tended to eat less. Grasshopper hatchling weight gain was negatively correlated with tannin concentration in no-choice assays with a *L. uliginosis* diet and with a tannin-enriched wheat diet. Wheat diets with a concentration of >10 percent tannin were lethal. Female grasshoppers fed *L. uliginosis* (3 percent tannin) for 5 days in the no-choice assays utilized 26.5 percent of the leaf nitrogen, compared with 30.9 percent of N from *L. corniculatus* foliage (1 percent tannin). The lower N utilization was accompanied by a 1.2-fold increase in the consumption ratio, as well as a twofold increase in total midgut trypsin and chymotrypsin activity, suggesting that the grasshoppers compensated for low available protein caused by tannin in their diet by increasing their feeding activity. A feeding deterrence index (FDI) of greater than 90 percent was observed when adult flea beetles (*Phyllotreta cruciferae*) fed in choice bioassays on *Brassica napus* leaf disks to which quebracho tannins were topically applied at 0.3 percent w/FW disk. Isolated *L. uliginosis* tannins strongly inhibited (80 percent FDI) the beetles at a similar concentration. Similarly, feeding inhibition (32 and 51 percent FDI) was also observed when 3rd instar diamondback moth (*Putella xylostella*) larvae were fed *B. napus* disks spread with 0.3 percent *L. uliginosis* or quebracho tannins (w/FW disk). This latter concentration was too low to effectively deter bertha armyworm (*Mamestra configurata*) feeding. Our results imply that crop plants that contain condensed tannin will deter populations of the more generalist feeding insects, grasshopper and bertha armyworm, when present at relatively high concentrations in plants. Lower concentrations of tannin appear to deter the more specialist insects, diamondback moth and flea beetles.

ACKNOWLEDGMENTS

We thank L. Braun, D. Hupka, W.B. Martin, A. Aubin, N. Suttill, and F. Xia for technical assistance, and the Western Grains Research Foundation for financial support. This is AAFC Contribution No. 1289.

REFERENCES

1. Bernays, E.A.; Cooper-Driver, G.; Bilgener, M. Herbivores and plant tannins. *Adv. Ecol. Res.* 19:263 (1989).

2. Mole, S. Polyphenolics and the nutritional ecology of herbivores. *In*: Cheeke, P.R. (ed.). Toxicants of plant origin. Vol IV. Phenolics, CRC Press Inc., Boca Raton, FL. p. 191 (1989).

3. Schultz, J.C. Tannin-insect interactions. *In*: Hemingway, R.W.; Karchesy, J.J. (eds.). Chemistry and significance of condensed tannins, Plenum Press, New York. p. 417 (1989).

4. Dini, J.; Owen-Smith, N. Condensed tannin in *Eragrostis chloromelas* leaves deters feeding by a generalist grasshopper. *Afr. J. Range For. Sci.* 12:49 (1995).

5. Mallampalli, N.; Barbosa, P.; Weinges, K. Effects of condensed tannins and catalpol on growth and development of *Compsilura concinnata* (Diptera: Trachinidae) reared in gypsy moth (Lepidoptera: Lymantriidae). *J. Entomol. Sci.* 31:289 (1996).

6. Mansour, M.H.; Zohdy, N.M.; El-Gengaihi, S.E.; Amr, A.E. The relationship between tannin concentration in some cotton varieties and susceptibility to piercing sucking insects. *J. App. Entomol.* 121:321 (1997).

7. Klocke, J.A.; Chan, B.G. Effects of condensed tannin on feeding and digestion in the cotton pest *Heliothis zea*. *J. Insect Physiol.* 28:911 (1982).

8. Ayres, M.P.; Clausen, T.P.; MacLean, S.F.; Redman, A.M.; Reichardt, P.B. Diversity of structure and antiherbivore activity in condensed tannins. *Ecology.* 78:1696 (1997).

9. Bernays, E.A. Tannins, an alternative viewpoint. *Entomol. Exp. Appl.* 24:44 (1978).

10. Bernays, E.A.; Chamberlain, D.; McCarthy, P. The differential effects of ingested tannic acid on different species of Acridoidea. *Ent. Exp. Appl.* 7:158 (1980).

11. Rhoades, D.F.; Cates, R.G. Toward a general theory of plant antiherbivore chemistry. *In*: Wallace, J.W.; Mansell, R.L. (eds.) Biochemical interaction between plants and insects, Plenum Press, New York. p. 168 (1975).

12. Jones, W.T.; Broadhurst, R.B.; Lyttleton, J.W. The condensed tannins of pasture legume species. *Phytochemistry.* 15:1407 (1976).

13. Kelman, W.M.; Tanner, G.J. Foliar condensed tannin levels in lotus species growing on limed and unlimed soils in South-Eastern Australia. *Proc. N.Z. Grassland Assn.* 52:51 (1990).

14. Barry, T.N.; Duncan, S.J. The role of condensed tannins in the nutritional value of *Lotus pedunculatus* with sheep. 1. Voluntary intake. *Br. J. Nutrition.* 51:485 (1984).

15. Jones, W.T.; Anderson, L.B.; Ross, M.D. Bloat in cattle. XXXIX. Detection of protein precipitants (flavolans) in legumes. *N.Z. J. Agric. Res.* 16:441 (1973).

16. Ross, M.D.; Jones, W.T. Bloat in cattle. XL. Variation in flavanol content in lotus. *N.Z. J. Agric. Res.* 17:191 (1974).

17. Dudt, J.F.; Shure, D.J. The influence of light and nutrients on foliar phenolics and insect herbivory. *Ecology.* 75:86 (1994).

18. Lees, G.L.; Hinks, C.F.; Suttill, N.H. Effect of high temperature on condensed tannin accumulation in leaf tissues of big trefoil (*Lotus uliginosus* Schkuhr.). *J. Agric. Food Chem.* 65:415 (1994).

19. Barry, T.N.; Forss, D.A. The condensed tannin content of vegetative *Lotus pedunculatus*, its regulation by fertilizer application, and its effect upon protein solubility. *J. Sci. Food Agric.* 34:1047 (1984).

20. Hagerman, A.E.; Butler, L.G. The specificity of proanthocyanidin-protein interactions. *J. Biol. Chem.* 256:4494 (1981).

21. Griffiths, D.W. The inhibition of digestive enzymes by polyphenolic compounds. *In*: Friedman, M. (ed.). Nutritional and toxicological significance of enzyme inhibitors in foods, Plenum Press, New York and London. p. 509 (1986).

22. Tamir, M.; Alumot, E. Inhibition of digestive enzymes by condensed tannins from green and ripe carobs. *J. Sci. Food Agric.* 20:199 (1969).

23. Feeny, P.P. Effect of oak leaf tannins on larval growth of the winter moth *Operophtera brumata*. *J. Insect Physiol.* 14:805 (1968).

24. Goldstein, J.L.; Swain, T. The inhibition of enzymes by tannins. *Phytochemistry.* 4:185 (1965).

25. De Veau, E.J.I.; Schultz, J.C. Reassessment of interaction between gut detergents and tannins in Lepidoptera and significance for gypsy moth larvae. *J. Chem. Ecol.* 18:1437 (1992).

26. Martin, M.M.; Martin, J.S. Surfactants: Their role in preventing the precipitation of proteins by tannins in insect guts. *Oecologia.* 61:342 (1984).

27. Martin, M.M.; Rockholm, D.C.; Martin, J.S. Effects of surfactants, pH, and certain cations on precipitation of proteins by tannins. *J. Chem. Ecol.* 11:485 (1985).

28. Appel, H.M.; Martin, M.M. Gut redox conditions in herbivorous lepidopteran larvae. *J. Chem. Ecol.* 16:3277 (1990).

29. Barnehenn, R.V.; Martin, M.M. The protective role of the peritrophic membrane in the tannin-tolerant larvae of *Orgyia leucostigma* (Lepidoptera). *J. Insect Physiol.* 38:973 (1992).

30. Bernays, E.A.; Chamberlain, D.J.; Leather, E.M. Tolerance of acridids to ingested condensed tannin. *J. Chem. Ecol.* 7:247 (1981).

31. Stringham, G.R. Genetics of four hypocotyl mutants in *Brassica campestris* L. *J. Hered.* 62:248 (1971).

32. Dalrymple, E.J.; Goplen, B.P.; Howarth, R.E. Inheritance of tannins in birdsfoot trefoil. *Crop Sci.* 24:921 (1984).

33. Porter, L.J.; Hrstich, L.N.; Chan, B.G. The conversion of procyanidins and prodelphinidins to cyanidin and delphinidin. *Phytochemistry.* 25:223 (1986).

34. Bae, H.D.; McAllister, T.A.; Muir, A.D.; Yanke, J.; Bassendowski, K.A.; Cheng, K.-J. Selection of a method of condensed tannin analysis for studies with rumen bacteria. *J. Agric. Food Chem.* 41:1256 (1993).

35. Hinks, C.F.; Erlandson, M.A. Rearing grasshoppers and locusts, review, rationale and update. *J. Orthoptera Res.* 3:1 (1994).

36. Westcott, N.D.; Hinks, C.F.; Olfert, O. Dietary effects of secondary plant compounds on nymphs of *Melanoplus sanguinipes* (Orthoptera, Acrididae). *Ann. Entomol. Soc. Amer.* 85:304 (1992).

37. Hinks, C.F.; Cheeseman, M.T.; Erlandson, M.A.; Olfert, O.; Westcott, N.D. The effects of kochia, wheat and oats on digestive proteinases and the protein economy of adult grasshoppers, *Melanoplus sanguinipes*. *J. Insect Physiol.* 37:417 (1991).

38. SAS-Institute, SAS/STAT user's guide, version 6, 4th Ed. 1989: SAS Institute, Cary, N.C.

39. Keating, S.T.; Yendol, W.G.; Schultz, J.C. Relationship between susceptibility of gypsy moth larvae (*Lepidoptera, Lymantriidae*) to a baculovirus and host plant foliage constituents. *Environ. Entomol.* 17:952 (1988).

40. Schultz, J.C.; Hunter, M.D.; Appel, H.M. Antimicrobial activity of polyphenols mediates plant-herbivore interactions. *In*: Hemingway, R.W.; Laks, P.E. (eds.). Plant polyphenols: synthesis, properties, significance, Plenum Press, New York. p. 621 (1992).

41. Hinks, C.F.; Hupka, D.; Olfert, O. Nutrition and the protein economy in grasshoppers and locusts. *Comp. Biochem. Physiol.* 104A:133 (1993).

42. Briggs, M.A. Relation of *Spodoptera eridania* choice to tannins and proteins of *Lotus corniculatus*. *J. Chem. Ecol.* 16:1557 (1990).

43. Broadway, R.M.; Duffey, S.S. The effect of plant protein quality on insect digestive physiology and the toxicity of plant proteinase inhibitors. *J. Insect Physiol.* 34:1111 (1988).

44. Butler, L.G.; Rogler, J.C. Biochemical mechanisms of the antinutritional effects of tannin. *In*: Ho, C.-T.; Lee, C.Y.; Huang, M.-T. (eds.) Phenolic compounds in food and their effects on health I. Analysis, occurrence and chemistry. American Chemical Society, Washington, DC. p. 298 (1992).

45. Reese, J.C. Interactions of allelochemicals with nutrients in herbivore food. *In*: Rosenthal, G.A.; Janzen, D.H. (eds.) Herbivores, their interactions with secondary plant metabolites, Academic Press, New York. p. 309 (1979).

46. Hinks, C.F.; Olfert, O. Cultivar resistance to grasshoppers in temperate cereal crops and grasses, a review. *J. Orthopt. Res.* 1:1 (1993).

47. Broadway, R.M.; Duffey, S.S. Plant proteinase inhibitors, mechanism of action and effect on the growth and digestive physiology of larval *Heliothis zea* and *Spodoptera exigua*. *J. Insect Physiol.* 32:827 (1986).

48. Birk, Y. Protein protease inhibitors of plant origin and their significance in nutrition. *In*: Huisman, J.; van der Poel, T.F.B.; Liener, I.E. (eds.) Recent advances of research in antinutritional factors in legume seeds, Pudoc, Wageningen. p. 83 (1989).

49. Ahmed, A.E.; Smithard, R.; Ellis, M. Activities of enzymes of the pancreas, and the lumen and mucosa of the small intestine in growing broiler cockerels fed on tannin-containing diets. *Br. J. Nutr.* 65:189 (1991).

50. Freeland, W.J.; Calcott, P.H.; Geiss, D.P. Allelochemicals, minerals and herbivore population size. *Biochem. Syst. Ecol.* 13:195 (1985).

51. Berenbaum, M. Post-ingestive effects of phytochemicals on insects: on paraclesus and plant products. *In*: Miller, J.R.; Miller, T.A. (eds.) Insect-plant interactions, Springer-Verlag, New York. p. 122 (1986).

PHENOLIC CONTENT AND ANTIOXIDANT ACTIVITY: A STUDY ON PLANTS EATEN BY A GROUP OF HOWLER MONKEYS (*ALOUATTA FUSCA*)

Gilda Guimarães Leitão,[a] Luciana Lopes Mensor,[b]
Luciene Ferreira G. Amaral,[a] Núbia Floriano,[a]
Vania L. Garcia Limeira,[c] Fábio de Sousa Menezes,[b]
and Suzana Guimarães Leitão[b]

[a] Núcleo de Pesquisas de Produtos Naturais
Universidade Federal do Rio de Janeiro
Rio de Janeiro, RJ
BRAZIL

[b] Departamento de Produtos Naturais e Alimentos, Faculdade
de Farmácia
Universidade Federal do Rio de Janeiro
Rio de Janeiro, RJ
BRAZIL

[c] Museu Nacional
Universidade Federal do Rio de Janeiro
Rio de Janeiro, RJ
BRAZIL

1. INTRODUCTION

Tannins comprise a large group of complex substances that are widely distributed in the plant kingdom. They are usually subdivided into two classes: the hydrolyzable tannins and the non-hydrolyzable or condensed tannins. Such division is based on the identity of the phenolic moieties involved and on the way they are linked. The hydrolyzable tannins consist of gallic acid or hexahydroxydiphenic acid and their derivatives esterified with glucose. The condensed tannins are formed basically of flavonoid polymers linked or not to carbohydrates or proteins. Both classes of tannins are widely distributed in nature. Reports on the utility and

Plant Polyphenols 2: Chemistry, Biology, Pharmacology, Ecology, Edited by
Gross et al. Kluwer Academic / Plenum Publishers, New York, 1999

safety of the longterm ingestion of quantities of plant material rich in condensed tannins are contradictory and somewhat difficult to reconcile.[1]

A free radical has an unpaired electron in its outer orbit that makes it unstable and causes it to react rapidly with any substance in the vicinity. Its half-life is measured in fractions of a second, and such reactions often result in a cascade of free radical formation in a multiplying effect, rather like an atomic explosion. These highly reactive molecules thus produce new compounds rapidly. Ultraviolet light, X-rays, or any other form of high-energy ionizing radiation knocks electrons out of orbiting pairs, thus causing formation of free radicals.[2] Because free radical reactions normally occur in tissue and are an essential factor in health, it is their control that is all important.[3]

Briefly, free radicals are characterized by enhanced biological activity with potential to damage cellular structures and to affect the composition of biological membranes together with changes in enzyme activity resulting in cell function and substance metabolism impairment. Several diseases such as cancer, cardiovascular diseases, arthritis, inflammatory diseases, and others have their etiology based on production and accumulation of free radicals.[4-6] This knowledge has focused many studies on a search for modern approaches to treat and prevent diseases using compounds with free radical scavenging activity. Numerous compounds useful as antioxidants have been studied as food additives and/or medicines. The therapeutic uses of the free radical scavengers attempt to reduce free radical injury either by intervening in their generating processes or by scavenging already formed free radicals. However, the application of some free radical scavengers is limited by their toxicity and side effects.[7]

Recent research conducted with condensed tannins has shown that these compounds are very effective antioxidants and that they serve as active free radical scavengers.[8] Such compounds, extracted from seeds of *Vitis vinifera* L. or from barks and needles of *Pinus pinaster* Soland, are now commonly marketed as dietary supplements. They are thought to be effective in improving such conditions as peripheral arterio-venous circulation, capillary fragility, retinopathies, inflammatory collagen disease, and others. *Camellia sinensis*, which contains up to 30 percent of these polyphenolic compounds, is also being recommended for its antioxidative properties. Condensed tannins from the above-named sources have been shown to be apparently devoid of toxicity in both short- and longterm tests conducted with small animals.[1]

The antioxidant and the antitumor-promotion activities of several hydrolyzable tannins, including a commercial tannic acid mixture, were examined in mouse skin treated with 12-*O*-tetradecanoylphorbol-13-acetate *in vivo*. The results showed that the antioxidant effects of hydrolyzable tannins are essential but not sufficient for their antitumor-promotion activity.[9] Two aspects of the mammalian herbivory plant interaction can be pointed out: plant defense against herbivory (by its secondary metabolites) and the herbivore's mechanisms to avoid or minimize secondary metabolites' toxic effects. Chemical defense is one of the numerous strategies used by plants to deter mammalian herbivory.[10] The great majority of deterrent substances are products of secondary metabolism. These substances may be bitter-tasting, poisonous, offensively odored, or they may have antinutritional effects.[11] The three main classes of secondary compounds are phenolics, including the well-known tannins, nitrogen-containing metabolites, and terpenoids.[10]

It is possible to distinguish two categories of plant chemical defenses: the quantitative ones and the qualitative ones. Quantitative defenses tend to reduce the herbivore's ability to digest or assimilate nutrients (e.g., tannins may reduce protein assimilation). Essential oils, resins, and condensed tannins are considered as quantitative defense against herbivory. Qualitative secondary compounds are toxic after a certain quantity is consumed (e.g., alkaloids, cyanogenic glycosides).[11,12]

Tannins may reach concentrations up to 60 percent of the dry weight of a plant.[11] They are polyphenols capable of forming strong indigestible complexes with proteins although, according to Butler,[13] inhibition of digestive enzymes in the intestine, where many other proteins compete for binding tannins, may be insignificant. Tannins are also considered effective feeding deterrents, but the mechanism of these effects are not well established. Their effects on animal diet include loss of weight and low efficiency, characterizing thus an antinutritional effect.

It is well known that mammalian herbivores do not simply consume any plant they encounter, rather, they select plants higher in certain nutritional components and lower in others.[12] Also, they consume more than a single plant during a single feeding. Consequently, the proportions of different plants in the diet differ from the proportions of plants available in the environment. Explanations for this behavior include the need to diversify diet in order to overcome the problem of plant secondary metabolites and need to balance nutrient intake. This behavior was well observed by Limeira[14] in her study of a group of *Alouatta fusca*. Other factors such as the abundance and distribution of plants, their nutrient content, the digestive capability and capacity of herbivory, the presence of predators, and learned behavior can significantly influence feeding strategy.

Howler monkeys are the second heaviest of the New World monkeys with adult males averaging 7 Kg and adult females 6 Kg. Also, they are the most widely distributed genus, ranging from about 18°N in the Mexican State of Veracruz to about 27°S in the Argentinean State of Corrientes and perhaps 28°S in the State of Rio Grande do Sul, Brazil.[15] The brown howler, *Alouatta fusca*, is endemic to the Atlantic Forest, occurring from the State of Bahia (Brazil) to the Northeast of Argentina.[16] Species of this genus are particularly known as the most folivorous of the neotropical primates, although they are not exclusively so. The howler monkeys are not highly specialized in the digestion of leaves.[15,17] According to Bauchop,[18] *Alouatta* species do not have morphological adaptations like sacculed stomachs in their digestive apparatus as do the folivorous primates of the Old World (families Cercopithecidae and Indriidae). Thus, they are less efficient in digesting mature leaves.

Howlers consume items such as young leaves (more digestible and rich in protein), fruits (rich in carbohydrates), and flowers (can be rich either in protein or carbohydrates) so they can have a nutritional balance in their diet. Based on these findings, Nagy and Milton[19] concluded that howlers have a great need to diversify their diet in terms of number of plant species consumed. In the last 10 years, many populations of *A. fusca* have been studied in remainder fragments of the Atlantic Forest with different degrees of conservation or different habitats. Limeira[14] studied a group of four brown howlers in the state of Rio de Janeiro (district of Comendador Levy Gasparian) and has observed that, in contrast to other studies carried out in São Paulo[20] and Rio Grande do Sul,[21] the diet of that group of *A. fusca* was less diversified in terms of plant species, in all the four seasons, probably due to the degrees of fragmentation and degradation of their habitat. The results of that study have demonstrated that the diet of that groups of brown howlers was very selective. Two plant species, *Apuleia leiocarpa* (Leguminosae-Caesalpinosoideae) and *Brosimum guianense* (Moraceae), accounted for 55 percent of the time spent on feeding, followed by *Platypodium elegans* (Leguminosae-Papilionoideae), with 11 percent. The remaining 34 percent was distributed among 34 other different species. This preference may be due to the abundance of those plants in their home range allied to their chemical composition, since two other plant species that are very widespread too, *Anadendnthera peregrina* (Leguminosae-Mimosoideae) and *Pseudopiptadenia contorta* (Leguminosae-Mimosoideae), were less consumed.

2. OBJECTIVES

The study on the diet of a group of *A. fusca* carried out by Limeira[22] showed that three plant species, *Apuleia leiocarpa* (Leguminosae-Caesalpinioideae), *Brosimum guianense* (Moraceae), and *Platypodium elegans* (Leguminosae-Papilionoideae) are the major components of the diet of those animals (66 percent). The relative percentage varies seasonally. In a 3-year study (1995–1997), we examined the possible relationship between seasonal variation of leaf condensed tannins (CT) and total phenolics (TP) content and foliage consumption by *A. fusca* (data obtained from April 1993 to March 1994). During these 3 years, we observed that the maximum content of TP for the five plants culminated with the driest and wettest months for each period. In general, highest values for CT occurred always in the same period as for TP showing, as expected, a good correlation between the two parameters. Minimum CT and TP values generally occur in the transition months between wet and dry seasons. In general, the highest values for CT were observed for *B. guianense*, *P. contorta*, and *A. peregrina*.[23] According to the data obtained for leaf preference, it was not possible to correlate foliage consumption by *A. fusca* with CT variation.

In order to verify whether there is any relationship between levels of CT, consumed plants, and antioxidant activity, the antioxidant activity was measured for the three most consumed plants as well as for two other plants, *Anadendnthera peregrina* (Leguminosae-Mimosoideae) and *Pseudopiptadenia contorta* (Leguminosae-Mimosoideae), which are practically not consumed by the howlers but are abundant in the region.

3. MATERIALS AND METHODS

Leaves of *Apuleia leiocarpa* (Leguminosae-Caesalpinioideae), *Brosimum guianense* (Moraceae), *Platypodium elegans* (Leguminosae-Papilonoideae), *Anadenanthera peregrina* (Leguminosae-Mimosoideae), and *Pseudopiptadenia contorta*

(Leguminosae-Mimosoideae) were collected in August, 1997, at Mata Boa Vista by the margins of the Paraibuna River, district of Levy Gasparian, Rio de Janeiro State, Brazil.

Plants were dried at the site of collection and reduced to small fragments in a food processor. Five hundred mg of each plant were extracted separately with 5×5 mL of two different solvent solutions: MeOH/H_2O 1:1 (M/W) and Me_2CO/H_2O 7:3 (A/W).

Samples from each of the five plants (10 g) obtained as described previously were extracted with the two solvent solutions mentioned above and additionally, crude ethanolic extracts were prepared from 10 g of each plant.

The methanol/water and acetone/water extracts were analyzed by the Folin Denis method[24] and by the proanthocyanidin method[25] in order to determine total phenolics and condensed tannins. All measurements were made in triplicate. The concentrations of these compounds (mg/mL) were calculated from standard curves developed, respectively, using tannic acid and quebracho tannin. A mixture of 0.5 mL of the extracts and 4 mL of *n*-butanol-HCl (5 percent) was heated for 2 hours at 95 °C. After 30 minutes, the absorbance was measured at 550 nm on a Shimadzu UV-2.200 recording spectrophotometer.

An aliquot of 0.5 mL of the extracts prepared from methanol/water and acetone/water was added to 0.5 mL of the Folin Denis reagent[26] and 1 mL of Na_2CO_3 saturated solution, under agitation. The mixture was diluted to 10 mL and, after 1 hour, the absorbance measured at 725 nm on a Shimadzu UV-2.200 recording spectrophotometer.

The crude ethanolic, methanol/water and acetone/water extracts, from each of the five plants were diluted to final concentrations of 250, 125, 50, 25, 10, and 5 μg/mL in ethanol, beginning from a 1.0 mg/mL stock solution. One mL of the DPPH (2,2-diphenyl-1-picryl-hydrazyl-hydrate, Sigma®) 0.3 mM ethanolic solution was added to 2.5 mL of sample solution and allowed to react for 30 minutes at room temperature. The absorbance was then measured at 518 nm on a Shimadzu UV-2.200 recording spectrophotometer. Ethanol (1.0 mL) plus plant extract solution (2.5 mL) was used for blank. A DPPH solution (1.0 mL; 0.3 mM) plus ethanol (2.5 mL) was used as the control. *Ginkgo biloba* extract—EGb 761 (from Tebonin® 40 mg Oral Solution) was used as standard antioxidant. The absorbance values were converted into percentage of antioxidant activity (AA percent) using the following formula:

$$AA\% = 100 - \{[(ABS_{SAMPLE} - ABS_{BLANK}) \times 100]/ABS_{CONTROL}\}$$

Statistical analyses were made in order to determine: 1) differences in antioxidant activity between the two extraction methods, M/W and A/W (Student's t test); 2) any correlation between antioxidant activity and tannin content (Spearman r test).

4. RESULTS AND DISCUSSION

Antioxidant activity (AA), condensed tannins (CT), and total phenolics (TP) content, measured for all plants extracted with different methodologies, are reported in table 1.

Table 1. Relationships between tannin contents and antioxidant activity

PLANT/ FAMILY	*ANTIOXIDANT ACTIVITY (%)			CONDENSED TANNINS (mg/mL)		TOTAL PHENOLICS (mg/mL)	
	EtOH	M/W	A/W	M/W	A/W	M/W	A/W
Anadenanthera peregrina (Leguminoseae-Mimosoideae)	91.66	87.80	90.80	0.009	0.337	0.182	0.702
Apuleia leiocarpa (Leguminoseae-Caesalpinioideae)	38.85	9.88	18.88	0.018	0.068	0.170	0.219
Brosimum guianense (Moraceae)	41.02	69.21	45.55	0.078	0.113	0.465	0.688
Platypodium elegans (Leguminoseae-Papilionoideae)	22.82	15.96	30.01	0.047	0.017	0.269	1.052
Pseudopiptadenia contorta (Leguminoseae-Mimosoideae)	91.92	88.72	87.05	0.166	0.875	0.731	0.679

* Measured for a 125 µg/mL solution

Figures 1–5 show a comparison between AA vs. extract concentration (µg/mL) for each of the five plants using the three different extraction methodologies. AA for *Ginkgo biloba* extract, used as standard, is also shown. Figure 6 summarizes the AA for all plants extracted with M/W and A/W. Statistical treatment of data from figures 1–6 (Student's t test) shows significant differences between AA values obtained in both solvent solutions—M/W and A/W.

The AA values obtained for *B. guianense* and *P. contorta* extracted with M/W were significantly higher than A/W ones (p < 0.05 and p < 0.10, respectively). For *P. elegans*, *A. peregrina*, and *A. leiocarpa*, M/W values were smaller than A/W ones (p < 0.05; p < 0.05 and p < 0.10, respectively). No significant correlation was found between AA and the two phenolic parameters, CT and TP, using the Spearman r test (p > 0.10). The same procedure was used for CT and TP values in order to verify differences between the two extraction methodologies. In fact, there were statistical differences between M/W and A/W (p > 0.10) for all plants. As a general rule, the mixture A/W was more efficient in extraction of CT than M/W, except for *P. elegans*. This is also true for TP, with *P. contorta* the exception in this case.

Based on results obtained from data shown in table 1 and figures 1–6, as well as the statistical analyses, we can say that the different extractants (M/W; A/W) highly influence AA and CT/TP contents. Several methodologies are described in the literature for extraction of CT that take into account different extractants. According to Waterman,[25] "different plants with different phenolics of different polarities may be best extracted with different solvents optimized to meet their needs". He suggests, as a general recommendation, the use of a mixture of water and one of the following: methanol, ethanol, or acetone. Aqueous methanol

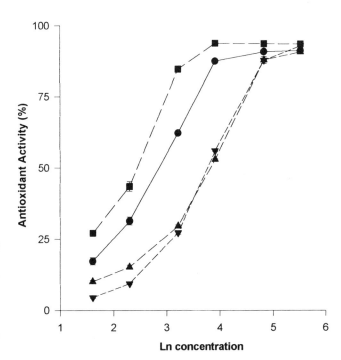

Figure 1. Antioxidant activity (percent) of *Anadenanthera peregrina* extracts (■ Ethanolic; ▲ Methanol/Water; ● Acetone/Water) measured for 6 different concentrations (Ln) using DPPH Photometric Assay. *Ginkgo biloba* extract (▼ EGb 761) was used as standard.

(50 percent) is a popular choice. Although some investigators do not recommend acetone containing extracts for the assay of CT, when a survey of a large number of different species is to be made, it may not be possible to optimize for each of them, and therefore an arbitrary choice may have to be made. The statistical difference observed for CT and TP values with the different extractants (M/W and A/W) could be explained in terms of solubility, which is linked to the structure of the tannin. Comparing the two solvents (A/W and M/W) used for the extraction of CT, it is possible to observe their marked differences in extracting capability for *A. peregrina* and *P. contorta*. The other three plants have no such marked difference. Coincidentally, they are the three most eaten plants in the diet of *A. fusca*, especially *B. guianense*. This may suggest that the tannins of *B. guianense* are structurally different from those of *A. peregrina* and *P. contorta*. Such structural differences could be influencing the palatability of these plants in relation to *A. fusca*.[23] Differences in tannin structure can lead to different depolymerized products and rates of depolymerization, both of which may affect herbivore preferences.[27]

The use of DPPH provides an easy and rapid way to evaluate antioxidants, but when this method is used, the interpretation of data must be done carefully. Recent studies demonstrated that the interaction of a potential antioxidant with DPPH depends on its structural conformation. The number of DPPH molecules that are

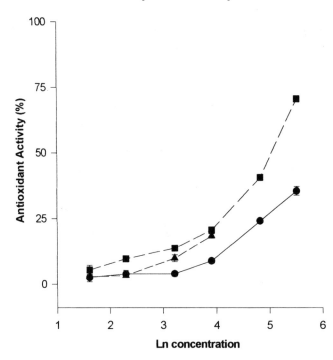

Figure 2. Antioxidant activity (percent) of *Apuleia leiocarpa* extracts (■ Ethanolic; ▲ Methanol/Water; ● Acetone/Water) measured for 6 different concentrations (Ln) using DPPH Photometric Assay. *Ginkgo biloba* extract (▼ EGb 761) was used as standard.

reduced seems to be correlated with the number of available hydroxyl groups.[28] It is strongly suggested that DPPH free radical abstracts the phenolic hydrogen of the electron-donating molecule, and this could be the general mechanism of the scavenging action of antiperoxidative flavonols, for example.[29] Therefore, as we did not find any correlation between AA and CT/TP (Spearman r test) in these plants, we can infer that these phenolics are not reducing DPPH molecules to the same extent, due to differences in their structures.

Differences observed for AA of plants with the same amount of total phenolics in *A. leiocarpa* and *A. peregrina* (M/W) may be explained by differences in the type of phenolics in each extract. Substances other than phenolic compounds could be influencing these results. It is interesting to note in figures 1–6 that AA measured for *A. peregrina* and *P. contorta* are higher than the *Ginkgo biloba* extract used as a standard. Amazingly, they are the two plants that are practically never eaten by the brown howler studied by Limeira,[14] and coincidentally they contain the highest levels of CT (not to the same extent) extracted in A/W. Among the three most eaten plants—*A. leiocarpa*, *P. elegans* and *B. guianense*—this latter showed the highest concentration of CT, which is a medium value when compared with those of *A. peregrina* and *P. contorta*. These results reinforce the theory that a certain degree of astringency might indicate absence of harmful chemicals and even stimulate ingestion, as observed by Gottlieb and coworkers.[30]

Brosimum guianense

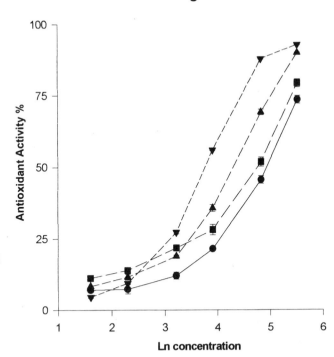

Figure 3. Antioxidant activity (percent) of *Brosimum guianense* extracts (■ Ethanolic; ▲ Methanol/Water; ● Acetone/Water) measured for 6 different concentrations (Ln) using DPPH Photometric Assay. *Ginkgo biloba* extract (▼ EGb 761) was used as standard.

Platypodium elegans

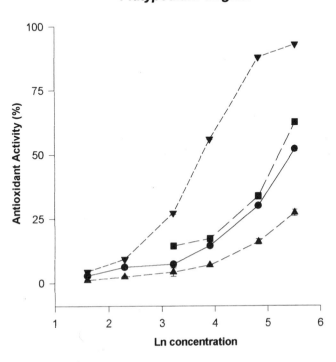

Figure 4. Antioxidant activity (percent) of *Platypodium elegans* extracts (■ Ethanolic; ▲ Methanol/Water; ● Acetone/Water) measured for 6 different concentrations (Ln) using DPPH Photometric Assay. *Ginkgo biloba* extract (▼ EGb 761) was used as standard.

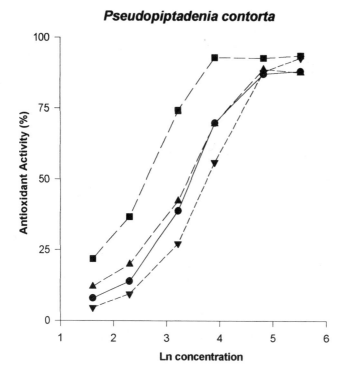

Figure 5. Antioxidant activity (percent) of *Pseudopiptadenia contorta* extracts (■ Ethanolic; ▲ Methanol/Water; ● Acetone/Water) measured for 6 different concentrations (Ln) using DPPH Photometric Assay. *Ginkgo biloba* extract (▼ EGb 761) was used as standard.

5. CONCLUSIONS

Antioxidant activity and tannin content were compared to investigate a possible correlation between the two parameters. Three different extracts (ethanolic, methanol/water 1:1, and acetone/water 7:3) from five plants consumed by a group of howler monkeys, *Alouatta fusca*, were studied: *Apuleia leiocarpa* (Leguminosae-Caesalpinioideae), *Brosimum guianense* (Moraceae), *Platypodium elegans* (Leguminosae-Papilonoideae), *Anadenanthera peregrina* (Leguminosae-Mimosoideae), and *Pseudopiptadenia contorta* (Leguminosae-Mimosoideae). 2,2-Diphenyl-1-picryl-hydrazyl-hydrate radical scavenging test, Folin Denis method, and proanthocyanidin method were used to determine antioxidant activity, total phenolics, and condensed tannins, respectively. No correlations between antioxidant activity as measured by the 2,2-diphenyl-1-picryl-hydrazyl-hydrate radical scavenging test and phenolic contents in the extracts were statistically significant.

Based on findings discussed in this work, we can suggest that the observation of the brown howler monkey's feeding behavior can lead us to plants that are potentially biologically active, as *A. peregrina* and *P. contorta*. Briefly, the target in the discovery of new drugs from plant origin, based on primate feeding behavior, should be the investigation of those plants that are less consumed by these animals.

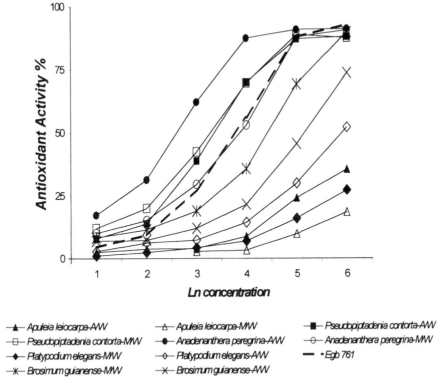

Figure 6. Antioxidant activity in acetone/water and methanol/water extracts. EGb 761 was plotted as standard curve.

"*L'omo e la scimmia*

L'Omo disse a la Scimmia:
-Sei brutta, dispettosa:
ma come sei ridicola!
Ma quanto sei curiosa!

Quann'io te vedo, rido:
Rido nun se sa quanto! . . .
La Scimmia disse: -Sfido!
T'arissomijo tanto!"

Trilussa

REFERENCES

1. Robbers, J.E.; Speedie, M.K.; Tyler, V.E. Pharmacognosy and pharmacobiotechnology. Williams and Wilkins—A Waverly Company, Baltimore, pp. 139–143 (1996).

2. Chedekel, M.R.; Zeise, L. Sunlight, melanogenesis and radicals in the skin. *Lipids* 23(6):587–591 (1988).

3. Demopoulos, H.B. Control of free radicals in the biologic systems. *Fed. Proc.* 32:1903–1908 (1973).

4. Halliwell, B. Current status review. Free radicals, reactive oxygen species and human disease: a critical evaluation with special reference to atherosclerose *Br. J. Exp. Path.* 70:737–757 (1989).

5. Xiong, Q.; Kadota, S.; Tani, T.; Namba, T. Antioxidative effects of phenylethanoids from *Cistanche deserticola. Biol. Pharm. Bull.* 19:1580–1585 (1996).

6. Terada, L.S.; Rubistein, J.D.; Lesnefsky, E.J.; Horwitz, L.D.; Leff, J.A.; Ropine, J.E. Existence and participation of the xanthine oxidase in reperfusion injury of ischemic rabbit myocardium. *Am. J. Physiol.* 260:805–810 (1991).

7. Haramaki, N.; Aggarwal, S.; Kawasada, T.; Droy-Lefaix, M.T.; Packer, L. Effects of natural antioxidant *Ginkgo biloba* extract on myocardial ischemia-reperfusion injury. *Free Rad. Biol. Med.* 16:789–794 (1994).

8. Spilková, J.; Dusek, J. Natural substances with antioxidant activity. *Ceska. Slav. Farm.* 45(6):296–301 (1996).

9. Gali, H.U.; Perchellet, E.M.; Klish, D.S.; Johnson, J.M.; Perchellet, J.P. Hydrolyzable tannins: potent inhibitors of hydroperoxide production and tumor promotion in mouse skin treated with 12-O-tetradecanoylphorbol-13-acetate *in vivo. Int. J. Cancer* 51:425–432 (1992).

10. Harborne, J.B. The chemical basis of plant defense. *In*: Pablo, R.T.; Robbins, C.T. (eds.) Plant defenses against mammalian herbivory. CRC Press, Inc., Boca Raton, p. 45 (1991).

11. Luckner, M. Secondary metabolites in microorganisms, plants and animals. Springer-Verlag, Germany (1990).

12. Belovsky, G.E.; Schmitz, O.J. Mammalian herbivory optimal foraging and the role of plant defenses. *In*: Pablo, R.T.; Robbins, C.T. (eds.) Plant defenses against mammalian herbivory. CRC Press, Inc., Boca Raton, p. 1 (1991).

13. Butler, L.G. Antinutritional effects of condensed and hydrolyzable tannins. *In*: Hemingway, R.W.; Laks, P.E. (eds.) Plant polyphenols: synthesis, properties, significance. Plenum Press, New York, p. 693 (1992)

14. Limeira, V.L. A.G. Comportamento alimentar, padrão de atividades e uso do espaço por *Alouatta fusca* (Primates, Platyrrhini) em fragmento degradado de Floresta Atlântica noestado do Rio de Janeiro. MSC Thesis, Museu Nacional/UFRJ, Rio de Janeiro (1996).

15. Neville, M.K.; Glander, K.E.; Braza, F.; Rylands, A.B. The howling monkeys, genus *Alouatta. In*: Mittermeier, R.A.; Rylands, A.B.; Coimbra-Filho, A.; Fonseca, G.A.B. (eds.) Ecology and behavior of neotropical primates. World Wildlife Fund, Washington, DC, vol. 2, p. 349 (1988).

16. Hirsch, A.; Landau, E.C.; Tedeschi, A.C.; Menegheti, J.O. Estudo comparativo das espécies do Gênero *Alouatta* Lacèpéde, 1799 (Platyrrhini, Atelidae) e sua distribuição geográfica na América do Sul. *In*: Rylands, A.B.; Bernardes, A.T. (eds.), A primatologia do Brasil, vol. 3. Fundação Biodiversitas, Belo Horizonte, pp. 239–262 (1991).

17. Milton, K. The foraging strategy of howler monkeys. Columbia University Press, New York (1980)

18. Bauchop, T. Digestion of leaves in vertebrate arboreal folivores. *In*: Montgomery, G.G. (ed.), The ecology of arboreal folivores. Smithsonian Inst., Washington, DC pp. 193–204 (1978).

19. Nagy, K.A.; Milton, K. Aspects of dietary quality, nutrient assimilation and water balance in wild howler monkeys (*Alouatta palliata*). *Oecologia (Berl.)* 39:249–258 (1979).

20. Chiarello, A.G. Diet of the brown howler monkey *Alouatta fusca* in a semi-deciduous forest fragment of Southeastern Brazil. *Primates* 35(1):25–34 (1994).

21. Cunha. Aspectos sócio-ecológicos de um grupo de bugios (*Alouatta fusca clamitans*) do Parque Estadual de Itapuã, RS. MSC Thesis, Universidade Federal do Rio Grande do Sul. (1994).

22. Limeira, V.L.A.G. Behavioral ecology of *Alouatta fusca clamitans* in a degraded Atlantic forest fragment in Rio de Janeiro. *Neotropical Primates* 5(4):116–117 (1997).

23. Amaral, L.F.G. Estudo químico ecológico das plantas da dieta de um grupo de Barbados (Primatas: Cebidae). PhD Thesis, Universidade Federal do Rio de Janeiro, Núcleo de Pesquisas de Produtos Naturais (1997).

24. Oates, J.F.; Waterman, P.G.; Choo, G.M. Food selection by the South Indian leaf monkey, *Presbytis johnii*, in relation to leaf chemistry. *Oecologia*, 45:45–56 (1980).

25. Waterman, P.G.; Mole, S. Analysis of phenolic plant metabolites—methods in ecology series. Blackwell Scientific Publications, Oxford (1994).

26. Mole, S.; Waterman, P.G. A critical analysis of techniques for measuring tannins in ecological studies I. Techniques for chemically defining tannins. *Oecologia* 72:137–147 (1987).

27. Clausen, T.P.; Provenza, F.D.; Burritt, E.A.; Reichardt, P.B.; Bryant, J.P. Ecological implications of condensed tannin structure: a case study. *Journal of Chemical Ecology* 16(8):2381–2392 (1990).

28. Brand-Williams, W.; Cuvelier, M.E.; Berset, C. Use of a free radical method to evaluate antioxidant activity. *Lebensm.-Wiss. A-Technol.* 28:25–30 (1995).

29. Ratty, A.K.; Sunamoto, J.; Das, N.P. Interaction of flavonols with 1,1-diphenyl-2-picrylhydrazyl free radical, lipossomal membranes and soybean lipoxygenase-1. *Biochemical Pharmacology* 37(6):989–995 (1988).

30. Gottlieb, O.R.; Borin, M.R. de M.B.; Bosisio, B.M. Trends of plant use by humans and non-human primates in Amazonia. *American Journal of Primatology* 40:189–195 (1996).

TANNINS AS NUTRITIONAL CONSTRAINTS FOR ELK AND DEER OF THE COASTAL PACIFIC NORTHWEST

Edward E. Starkey,[a] Patricia J. Happe,[b]
Maria P. Gonzalez-Hernandez,[c] Karen M. Lange,[d]
and Joseph J. Karchesy[e]

[a] USGS Forest and Rangeland Ecosystem Science Center
3200 SW Jefferson Way
Corvallis, Oregon 97331
USA

[b] Olympic National Park
600 E. Park Ave.
Port Angeles, Washington 98362
USA

[c] Santiago de Compostela University
Department of Crop Production
EPS, campus Lugo
27002-Lugo,
SPAIN

[d] Department of Forest Science
Oregon State University
Corvallis, Oregon 97331
USA

[e] Department of Forest Products
Oregon State University
Corvallis, Oregon 97331
USA

1. INTRODUCTION

In the coastal Pacific Northwest, diets of elk (*Cervus elaphus roosevelti*) and deer (*Odocoileus hemionus columbianus*) are dominated by shrubs, ferns, and forbs

Plant Polyphenols 2: Chemistry, Biology, Pharmacology, Ecology, Edited by
Gross et al. Kluwer Academic / Plenum Publishers, New York, 1999

897

Table 1. Common and scientific names of
plants

Common	Scientific
Deer fern	*Blechnum spicant*
Oceanspray	*Holodiscus discolor*
Red huckleberry	*Vaccinium parvifolium*
Salal	*Gaultheria shallon*
Salmonberry	*Rubus spectabilis*
Sword fern	*Polystichum munitum*
Vine maple	*Acer circinatum*
Youth-on-age	*Tolmiea menziesii*
Wood sorrel	*Oxalis oregana*

during most seasons. Many species of these plants contain relatively high concentrations of tannins, which might be expected to reduce dietary quality by reducing protein and dry matter digestibility (DMD).[1,2,3] We attempt to assess the significance of tannins as nutritional constraints on elk and deer populations based on our previous and ongoing studies of food habits and nutritional ecology of cervids (elk and deer) in the region.[4-11] Throughout the paper, we use common names, with scientific names given in table 1.

2. EFFECTS OF TANNINS ON DEER AND ELK

Both hydrolyzable and condensed tannins are present in plants consumed by elk and deer[7,10,12] (fig. 1). In western Washington, vine maple and salmonberry contain significant levels of hydrolyzable tannins, whereas concentrations of condensed tannins are greatest in ferns, forbs, and conifers. In many species, tannin content of leaves decreases from spring through summer, fall, and winter. Stems of shrubs, which are important forage sources during winter, contain very low concentrations of tannins.[1]

Some tannins have direct physiological effects on herbivores. Hydrolyzable tannins, or products of their digestive decomposition, may be absorbed through the intestinal wall and cause physiological damage.[13,14] Condensed tannins may also be absorbed and be toxic to some species of mammals.[15]

Most studies of tannins and cervids have focused on nontoxic effects on availability of protein and to a lesser extent energy. Hydrolyzable and condensed tannins combine with protein to form tannin-protein complexes during mastication and digestion of forage.[16] Formation of these stable complexes may reduce the amount of digestible protein available for herbivores. Digestibility of forage may also be reduced if rumen microflora are negatively influenced by tannins, or if microbial protein binds with tannin. An index of protein-precipitating capacity is commonly determined by the capacity of known amounts of plant extracts or tannins to precipitate bovine serum albumin (BSA) from a

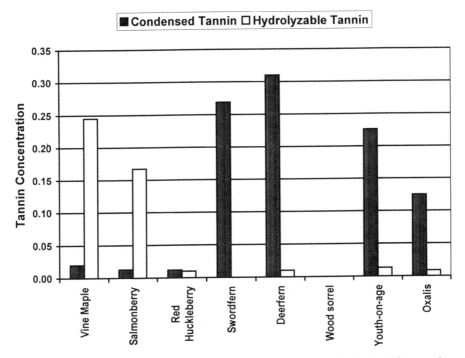

Figure 1. Condensed tannin (catechin equivalent) and hydrolyzable tannin (tannic acid equivalent) concentrations (mg tannin/mg plant dry matter) during summer in understory species of an old-growth forest, western Washington.

solution[17,18] or agarose gel.[19] In this paper, we refer to the capacity to precipitate proteins as astringency.[20]

Many plants consumed by cervids of the region are astringent (figs. 2 and 3). Leaves of salmonberry and vine maple (shrubs) are quite astringent in most seasons, whereas wood sorrel (a forb) and red huckleberry (a shrub) precipitate very little BSA in any season. There is considerable variation among seasons, but astringency is generally greatest in spring and intermediate in summer. Protein-precipitating capacity of ferns increases greatly in spring, with sword fern attaining the greatest astringency we've measured for any species and season. Stems of shrubs have very low astringency in all seasons.[7,10]

Protein-precipitating capacity also varies with forest stand type. Understory plants in clearcuts have greater astringency than those growing in closed canopy forests in Alaska,[21] western Oregon[10] (fig. 4) and Washington.[7] Availability of light is an obvious difference between clearcuts and old-growth forests. Because of differences in canopy coverage, plants growing in clearcuts are commonly exposed to more sunlight than those in old-growth forests. Increased light availability results in increased production of phenolic compounds[22–24] and increased astringency.[25] Thus, it seems likely that differences in astringency of plants growing in clearcuts

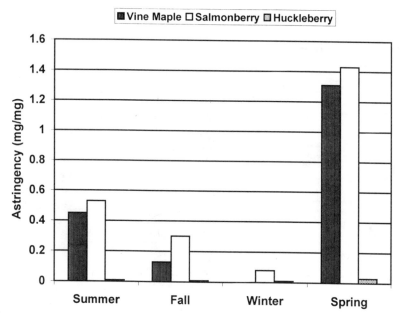

Figure 2. Seasonal variation in astringency (mg BSA precipitated/mg plant tissue) of current annual growth of shrubs growing in the understory of an old-growth forest of western Washington.

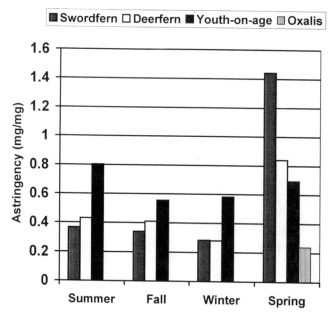

Figure 3. Seasonal variation in astringency (mg BSA precipitated/mg plant tissue) of ferns and forbs growing in the understory of an old-growth forest, western Washington.

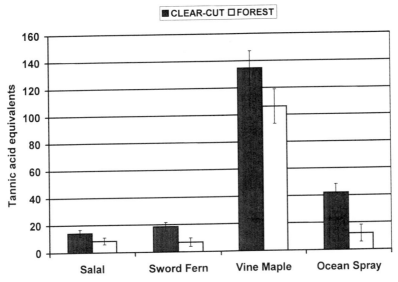

Figure 4. Astringency of leaves of shrubs (mg/g of tannic acid equivalent) growing in clearcuts and forests of western Oregon during summer (determined by radial diffusion technique of Hagerman[19]).

and old-growth forests are associated with differences in canopy cover and availability of sunlight.

Elk and deer of the coastal Northwest have adapted to an environment in which many forage species contain tannins with significant capacity to precipitate dietary proteins. An important adaptive strategy is reflected in food habits that shift seasonally to take advantage of forages such as grasses or forbs with lower tannin content and/or capacity to precipitate proteins (fig. 5). For example, as astringency of many forage plants increases in spring and summer, consumption of wood sorrel by deer and elk increases greatly.[5] Wood sorrel has relatively low concentrations of tannins and very little astringency.[10] Thus, cervids take advantage of a forage species that contains high levels of digestible protein, at a time when many other species contain seasonally maximum levels of tannin and astringency.

Deer and elk of the coastal Northwest cannot restrict their diets to only those plants with low tannin content; they must consume a variety of forest understory species to meet their nutritional needs. In more open environments, elk consume relatively great quantities of grass[26] that generally contains very little tannin. In coastal forest habitats, grasses often are not abundant, and cervids must rely on other species, many of which contain significant concentrations of tannins.

Elk and deer produce salivary proteins that form relatively stable complexes with tannins during mastication and digestion. These proteins minimize nitrogen loss by efficiently binding with tannins, minimize the absorption of tannins by forming stable tannin-protein complexes that are excreted in the feces, and reduce the effect of tannins on fermentation in the rumen.[27]

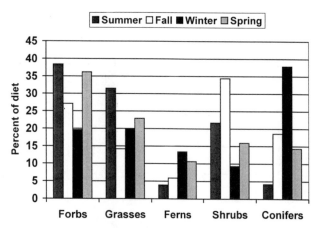

Figure 5. Seasonal diets of elk in the Hoh Valley of western Washington.

To quantify the impact of tannins on nutrient availability for elk and deer, Robbins et al.[1,2] conducted feeding trials with known quantities of tannins and developed the following predictive equations that allow the prediction of protein and dry matter digestibility:

$$(1) \quad DP = -3.87 + 0.9283X - 11.82Y$$

where Z is digestible protein in grams per 100 grams of feed, X is crude protein content as percent dry matter, and Y is amount of bovine serum albumin precipitated, in milligrams per milligram of forage dry matter, as determined by the assay of Martin and Martin.[16]

$$(2) \quad DMD = [(0.923\, e^{-0.0451\,A} - 0.03\, B)(NDF)] + [(-16.03 + NDS) - 2.8(11.82Y)]$$

where DMD = dry matter digestibility (grams/100 g of feed), A = (lignin + cutin) content as percentage of NDF (neutral detergent fiber), B = biogenic silica content, NDS = neutral detergent solubles (100 − NDF), Y = BSA precipitation (mg/mg of forage dry matter). NDF and lignin are determined with sequential detergent analysis.

3. TANNINS AS NUTRITIONAL CONSTRAINTS FOR ELK AND DEER

We applied the above equations to forage plants of western Washington to predict the influence of astringency on forage quality during summer. We first solved Equations (1) and (2) using astringency (Y) and other nutritional variables measured by Happe et al.[7] and Happe.[10] We then set Y = 0 and solved the equations to predict DMD and DP without the influence of tannin astringency.

Astringency reduced the DMD of vine maple, salmonberry, and youth-on-age from >80 percent to about 60 percent, representing a reduction in available energy of about 25 percent. Dry matter digestibility of sword fern and deer fern was

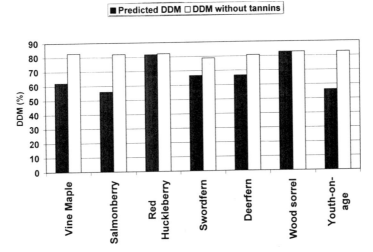

Figure 6. Effect of tannin astringency on dry matter digestibility (DMD) of common summer forages for elk and deer in western Washington.

reduced from 80 percent to about 65 percent, and DMD of red huckleberry and wood sorrel was virtually unaffected (fig. 6).

Astringency greatly reduced DP of vine maple, salmonberry, sword fern, deer fern, and youth-on-age. Digestible protein of salmonberry was reduced from 12 percent to about 2 percent, and nearly all DP of vine maple became unavailable. As for DMD, DP of red huckleberry and wood sorrel was unaffected (fig. 7). In general, astringency of tannin was predicted to have a relatively greater effect on DP than on DMD. Similar findings in Alaska[3] and the Cascade mountains of western Oregon[12] suggest that astringency does not greatly limit the availability of energy for cervids.

Forest stand type also influences nutritional quality of forage species. Happe et al.[7] compared seasonal DP values for vine maple, salmonberry, red huckleberry, and sword fern growing in both clearcuts and old-growth forests of western Washington (figs. 8 and 9). During all seasons except spring, vine maple, salmonberry, and sword fern growing in clearcuts had astringency in excess of that predicted to completely precipitate all their protein, resulting in negative values for digestible protein. In spring, vine maple and red huckleberry contained about 2.5 percent and 15 percent DP, respectively. In old-growth forests, DP of these species was generally greater than in clearcuts. Except for vine maple in fall, salmonberry in winter, and sword fern in spring, the predicted DP for all species in old-growth understories was >0 in all seasons. Red huckleberry was consistently the best source of DP. Elk and deer consuming these species in old-growth forests would be more likely to obtain adequate levels of digestible protein than those feeding in clearcuts. However, grass and forbs are often relatively abundant in young clearcuts,[9] and elk and deer may reduce the influence of tannins by increasing the proportion of these species in their diets.

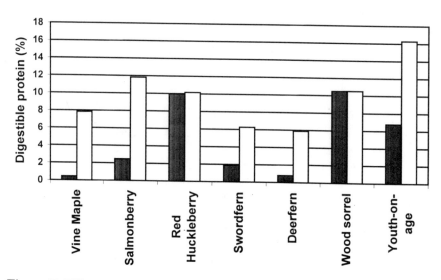

Figure 7. Effect of tannin astringency on digestible protein of common summer forages for elk and deer in western Washington.

Although energy requirements vary with age, sex, reproductive status, weather, and cover availability, ruminants in general cannot maintain body weight on diets with DMD coefficients less than 50 percent.[29] In the Rocky Mountains, elk require a dietary DMD of 47 percent to meet energy requirements.[30]

We found tannin astringency reduced DMD moderately, but current annual growth of all shrub species analyzed contained >60 percent DMD in the summer. Although we did not calculate DMD percentages for winter forages, astringency of many forage plants is at an annual minimum during the winter (figs. 2 and 3). Also, shrub stems in winter contain few tannins and digestibility is generally not reduced by astringency.[3] Thus, the DMD of these important forage resources would not be significantly reduced by tannins. We conclude that most forage plants in western Oregon and Washington provide adequate levels of digestible energy to meet the seasonal needs of deer and elk.

Elk cows require a diet containing 5–6 percent crude protein for maintenance and early pregnancy, increasing to nearly 12 percent during lactation.[23] Solving Equation (1) for these crude protein values, assuming no astringency (Y = 0), yields dietary requirements of approximately 1 percent DP for maintenance and early pregnancy, and about 7 percent DP for lactation. Hobbs et al.[27] concluded that elk diets containing less than 5 percent crude protein (<1 percent DP) would not meet metabolic requirements for protein and would likely result in reduced rates of carbohydrate digestion.

During winter, protein availability appears to be a significant nutritional limitation for cervids in the region. For some shrub species growing in clearcuts, there

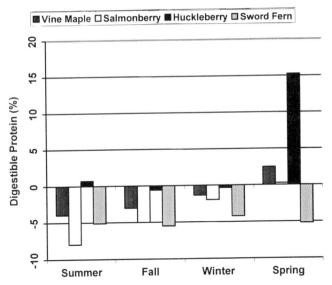

Figure 8. Seasonal variation in digestible protein of understory plants growing in clearcuts of western Washington.

is more than enough tannin in these plants to bind completely with all available protein (fig. 8). In old-growth forests, the DP of these shrubs is greater, but only marginally meets the predicted minimum requirement of 1 percent (fig. 9).

During summer, DP of shrubs growing in clearcuts is even lower than in winter. Red huckleberry in spring provides adequate DP to support lactation, but DP decreases to <1 percent in summer when demands of lactation are greatest. In old-growth forests, DP of red huckleberry, wood sorrel, and youth-on-age in spring and summer are adequate to meet the protein needs of lactation.

Although tannins negatively influence DP of many forage plants, cervid populations are able to optimize the nutrient availability in seasonal diets by includ-

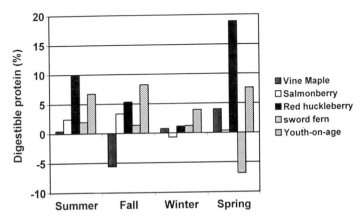

Figure 9. Seasonal variation in digestible protein of understory plants growing in old-growth forests of western Washington.

ing a mix of species, some with low tannin content. Grasses and selected species of forbs that contain essentially no tannins are important dietary components during spring and summer. For example, wood sorrel represented nearly 30 percent of spring and summer diets of elk and deer on the Olympic Peninsula.[5] Because of wood sorrel's very low tannin astringency, it is an important source of dietary protein that is required to support lactation and recovery from nutritional deficits incurred during winter.

4. CONCLUSIONS

In the coastal Pacific Northwest, many forage species contain significant levels of astringent tannins with the potential to greatly reduce the availability of both energy and protein for elk and deer. The influence of these tannins is partially mitigated by salivary proteins that neutralize tannins by forming relatively stable complexes with them. These salivary proteins are relatively effective in reducing the influence of tannins on energy availability (DMD) but less effective in mitigating the influence on protein digestibility. Thus, deer and elk must optimize the nutrient availability in seasonal diets by including a mix of species, some with low tannin content. They must be able to select optimum seasonal diets that balance nutritional requirements with tannin content and digestibility of energy and protein.

Digestible protein is likely to be a significant limiting factor for some deer and elk populations in the region, especially if grasses or forbs with low tannin content are unavailable. Because of increases in astringency associated with reductions in canopy coverage, digestible protein may be limiting for populations inhabiting areas dominated by clearcuts. Thinning and selective harvest may provide management tools to provide for timber harvest, increased growth of trees, and an optimum forage resource for elk and deer. Further research is required to describe the response of astringency over a wide range of canopy coverages and exposures to solar radiation.

Although deer and elk are widely distributed in much of the region, nutritional quality may constrain their productivity. Elk populations in coastal Oregon and Washington are generally less productive than those of the Rocky Mountains.[31] Diets of many Rocky Mountain elk populations include significant proportions of grasses and forbs that contain few tannins. We hypothesize that the relatively low productivity of coastal Pacific Northwest elk is at least partly a result of nutrient limitations caused by the high tannin content of their diets.

REFERENCES

1. Robbins, C.T.; Hanley, T.A.; Hagerman, A.E.; Hjeljord, O.; Baker, D.L.; Schwartz, C.C.; Mautz, W.W. Role of tannins in defending plants against ruminants: reduction in protein availability. *Ecology* 68:98–107 (1987).

2. Robbins, C.T.; Mole, S.; Hagerman, A.E.; Hanley, T.A. Role of tannins in defending plants against ruminants, reduction in dry matter digestion. *Ecology* 68:1606–1615 (1987).

3. Hanley, T.A.; Robbins, C.T.; Hagerman, A.E.; McArthur, C. Predicting digestible protein and digestible dry matter in tannin-containing forages consumed by ruminants. *Ecology* 73:537–541 (1992).

4. Leslie, D.M., Jr. Nutritional ecology of cervids in old-growth forests in Olympic National Park. Washington. Ph.D. Thesis, Oregon State University, Corvallis, 141 pp. (1983).

5. Leslie, D.M., Jr.; Starkey, E.E.; Vavra, M. Elk and deer diets in old-growth forests of western Washington. *Journal of Wildlife Management* 48:762–775 (1984).

6. Leslie, D.M., Jr.; Starkey, E.E.; Smith, B.G. Forage acquisition by sympatric cervids along an old-growth sere. *Journal of Mammalogy* 68:430–434 (1987).

7. Happe, P.J.; Jenkins, K.J.; Starkey, E.E.; Sharrow, S.H. Nutritional quality and tannin astringency of browse in clearcuts and old-growth forests. *Journal of Wildlife Management* 54:557–566 (1990).

8. Jenkins, K.J.; Starkey, E.E. Food habits of Roosevelt elk. *Rangelands* 13:261–265 (1991).

9. Jenkins, K.J.; Starkey, E.E. Winter forages and diets of elk in old-growth and regenerating coniferous forests in western Washington. *American Midland Naturalist* 130:299–313 (1993).

10. Happe, P.J. Ecological relationships between cervid herbivory and understory vegetation in old-growth Sitka spruce-western hemlock forests in western Washington. Ph.D. Thesis. Oregon State University, Corvallis 147 p. (1993).

11. Lange, K.M. Nutrient and tannin concentrations of shrub leaves in managed and unmanaged forests of the Oregon Coast Range: implications for herbivores. M.S. Thesis, Oregon State University, Corvallis (1998).

12. Friesen, C.A. The effect of broadcast burning on the quality of winter forage for elk, western Oregon. M.S. Thesis, Oregon State University, Corvallis 89 p. (1991).

13. Mould, E.D.; Robbins, C.T. Nitrogen metabolism in elk. *Journal of Wildlife Management* 45:323–334 (1981).

14. Robbins, C.T. Wildlife feeding and nutrition. Academic Press, New York, 343 p. (1983).

15. Bryant, J.P.; Reichardt, P.B.; Clausen, T.P. Chemically mediated interactions between woody plants and browsing mammals. *Journal of Range Management* 45:18–24 (1992).

16. Hagerman, A.E.; Robbins, C.T.; Weerasuriya, Y.; Wilson, T.C.; McArthur, C. Tannin chemistry in relation to digestion. *Journal of Range Management* 45:57–62 (1992).

17. Asquith, T.N.; Butler, L.G. Use of a dye-labelled protein as a spectrophotometric assay for protein precipitants such as tannin. *Journal of Chemical Ecology* 11:2535–2544 (1985).

18. Martin, J.S.; Martin, M.M. Tannin assays in ecological studies: Lack of correlation between phenolics, proanthocyanidins and protein-precipitating constituents in mature foliage of six oak species. *Oecologia* 54:205–211 (1982).

19. Hagerman, A.E. Radial diffusion method for determining tannin in plant extracts. *Journal of Chemical Ecology* 13:437–449 (1987).

20. Haslam, E. Plant polyphenols: Vegetable tannins revisited. *In*: Chemistry and pharmacology of natural products. Phillipson, J.D.; Ayres, D.C.; Baxter, H. (eds.) Cambridge University Press, Cambridge (1989).

21. Van Horne, B.; Hanley, T.A.; Cates, R.G.; McKendrick, J.D.; Horner, J.D. Influence of seral stage and season on leaf chemistry of southeastern Alaska deer forage. *Canadian Journal of Forest Research* 18:90–99 (1988).

22. Shure, D.J.; Wilson, L.A. Patch-size effects on plant phenolics in successional openings of the southern Appalachians. *Ecology* 74:55–67 (1993).

23. Dudt, J.F.; Shure, D.J. The influence of light and nutrients on foliar phenolics and insect herbivory. *Ecology* 75:86–98 (1994).

24. Mole, S.; Ross, J.A.M.; Waterman, P.G. Light-induced variation in phenolic levels in foliage of rain-forest plants. I. Chemical changes. *Journal of Chemical Ecology* 13:1–21 (1988).

25. Mole, S.; Waterman, P.G. Light-induced variation in phenolic levels in foliage of rain-forest plants. II. Potential significance to herbivores. *Journal of Chemical Ecology* 14:23–34 (1988).

26. Nelson, J.R.; Leege, T.A. Nutritional requirements, food habits, and diet quality of elk. *In:* Thomas, J.W. (ed.) The ecology of North American elk. Wildlife Management Institute, Washington, D.C. (1980).

27. Robbins, C.T.; Hagerman, A.E.; Austin, P.J.; McArthur, C.; Hanley, T.A. Variation in mammalian physiological responses to a condensed tannin and its ecological implications. *Journal of Mammalogy* 72:480–486 (1991).

28. Austin, P.J.; Suchar, L.A.; Robbins, C.T.; Hagerman, A.E. Tannin-binding proteins in saliva of deer and their absence in saliva of sheep and cattle. *Journal of Chemical Ecology* 15:1335–1347 (1989).

29. Amman, A.P.; Cowan, R.L.; Mothershead, C.L.; Baumgard, B.R. Dry matter and energy intake in relation to digestibility in a white-tailed deer. *Journal of Wildlife Management* 37:195–201 (1973).

30. Hobbs, N.T.; Baker, D.L.; Ellis, J.E.; Swift, D.M. Composition and quality of elk winter diets in Colorado. *Journal of Wildlife Management* 45:156–171 (1981).

31. Starkey, E.E.; DeCalesta, D.S.; Witmer, G.W. Management of Roosevelt elk habitat and harvest. *Transactions of the North American Wildlife and Natural Resource Conference* 47:353–362 (1982).

CONCLUDING REMARKS

THE 3ʳᵈ TANNIN CONFERENCE—RETROSPECT AND PROSPECT

Georg G. Gross,[a] Richard W. Hemingway,[b]
and Takashi Yoshida[c]

[a] Department of General Botany
University of Ulm
D-89069 Ulm
GERMANY

[b] Southern Research Station
USDA Forest Service
Pineville, Louisiana 71360
USA

[c] Faculty of Pharmaceutical Sciences
Okayama University
Tsushima, Okayama 700-8530
JAPAN

It is rather trivial to conclude that a symposium entitled "The 3rd Tannin Conference" might have been nothing else than the continuation of two earlier events that dealt with more or less the same topics. The first of these, the "North American Tannin Conference", was held in Port Angeles, Washington, in 1988 with the objective, as stated in the foreword to the book developed from the contributions to this meeting,[1] "to bring together people with a common interest in condensed tannins and to promote interdisciplinary interactions that will lead to a better understanding of these important substances". This event apparently filled a gap in communication of scientists in this field, stimulating a "2nd North American Tannin Conference" that took place in Houghton, Michigan, in 1991. After a significant pause, the "3rd Tannin Conference" was held in Bend, Oregon, in 1998 and it provided the basis for this book.

At first glance, a strong continuity appears to be characteristic of these meetings according to their common title "Tannin Conference". However, by analogy to the permanent phylogeny of living biological systems, also these symposia were subject to significant evolution. It is apparent that the previous term "North

Plant Polyphenols 2: Chemistry, Biology, Pharmacology, Ecology, Edited by
Gross et al. Kluwer Academic / Plenum Publishers, New York, 1999

911

American" has been deleted from the title of the third meeting after recognizing the growing international significance and acceptance of this series. Less apparent, but more significant, are pronounced changes with respect to the number of symposium attendants and particularly to the emphasized topics, as documented by the size and contents of the books developed from these meetings.

Sixty-nine participants were recorded for the initial Tannin Conference, and their contributions were published in a book comprising 34 chapters on 553 pages under the title "*Chemistry and Significance of Condensed Tannins*".[1] Accordingly, the overwhelming majority of these papers dealt with structure determinations, chemical reactivities and technical applications of a class of flavonoid derivatives that is preferentially called proanthocyanidins at present. The 2nd Tannin Conference could already welcome more than 100 attendants. The proceedings of this meeting, published as "*Plant Polyphenols. Synthesis, Properties, Significance*",[2] not only grew considerably in size to as much as 64 chapters on 1,053 pages, but also now covered the previously widely neglected classes of hydrolyzable tannins and other polyphenolic compounds. Notably, over-proportional increase of the (bio)synthesis section from three to seven chapters and particularly of the section "Biological Significance" from seven to as much as 19 contributions has to be recognized.

Right from the beginning, it was planned that this most recent book of the series, developed from the "3rd Tannin Conference" and entitled "*Plant Polyphenols 2. Chemistry, Biology, Pharmacology and Ecology*",[3] should not exceed the size of the preceding volume. As a consequence of the increasing popularity of the Tannin Conferences, documented by more than 150 participants from all continents attending the third meeting of this series, it became inevitable to separate submitted contributions into about 50 oral lectures of presumed general significance, thought to provide the basis of this book, and into about 60 posters relevant to more specialized questions. There were many harsh decisions to be made in this selection process, and the editors apologize to the authors for real or imaginary injustices. It should be mentioned in this connection that extended two-page abstracts of all contributions to the "Third Tannin Conference", i.e., both oral and poster presentations, have been published in a separate booklet[4] (229 pages) that is available through the conference editors for all those wishing to receive additional information about subjects and authors beyond the topics described in this book.

Inspection of the Table of Contents of this 3rd Tannin Conference proceedings book, "*Plant Polyphenols 2. Chemistry, Biology, Pharmacology and Ecology*", reveals that its main emphasis has again changed as compared to the preceding volume. The chapters dealing with biosynthesis, chemical and biomimetic syntheses, biotechnological applications, and molecular biology of tannins have considerably gained in extent. Particularly noteworthy are the newly introduced sections related to the potential medical implications of plant polyphenols, as exerted by their manifest antioxidant, antimicrobial, and anti-tumor activities, which clearly demonstrate the scientific trends that have emerged over the past years and which certainly will dominate this field in years to come. It will be interesting to learn what topics emerge in an eventual 4th Tannin Conference. Not much fantasy is required to predict that tannins, or plant polyphenols in general, will gain considerable impact on many facets of human life and welfare, and these developments will certainly be reflected by the topics of a future meeting of this style.

In contrast to all these apparent changes, we appreciate the existence of a true fix-point in this endeavour, namely the constant support of Plenum Press (now Kluwer Academic / Plenum Publishers) over the past decade, which has made it possible to publish the proceedings of the three Tannin Conferences, making the results and ideas discussed at these meetings available for the international tannin family. It is hoped that this fruitful relationship will persist also in the future, allowing more projects like this to become reality.

REFERENCES

1. Hemingway, R.W.; Karchesy, J.J. (eds.). Chemistry and significance of condensed tannins. Plenum Press, New York (1989).

2. Hemingway, R.W.; Laks, P.E. (eds.). Plant polyphenols. Synthesis, properties, significance. Plenum Press, New York (1992).

3. Gross, G.G.; Hemingway, R.W.; Yoshida, T. (eds.). Plant polyphenols 2. Chemistry, biology, pharmacology and ecology. Kluwer Academic / Plenum Publishers, New York (1999).

4. Gross, G.G.; Hemingway, R.W.; Yoshida, T. (eds.). Abstracts of the 3rd tannin conference. Phytochemical Connections Inc., Alexandria, LA (1998).

AUTHOR INDEX

915

SUBJECT INDEX